Polymer Chemistry

Polymer Chemistry
THE BASIC CONCEPTS

Paul C. Hiemenz

California State Polytechnic University
Pomona, California

MARCEL DEKKER,INC.   New York and Basel

Library of Congress Cataloging in Publication Data

Hiemenz, Paul C., [date]

Polymer chemistry.

Includes index.
1. Polymers and polymerization. I. Title.
QD381.H52   1984   547.7   83-25207
ISBN 0-8247-7082-X

MARCEL DEKKER, INC.
270 Madison Avenue, New York, New York 10016

Current printing (last digit):
10 9 8 7 6 5

PRINTED IN THE UNITED STATES OF AMERICA

Preface

Physical chemistry has been defined as that branch of science that is fundamental, molecular, and interesting. I have tried to write a polymer textbook that could be described this way also. To the extent that one subscribes to the former definition and that I have succeeded in the latter objective, then the approach of this book is physical chemical. As a textbook, it is intended for students who have completed courses in physical and organic chemistry. These are the prerequisites which define the level of the book; no special background in physics or mathematics beyond what is required for physical chemistry is assumed. Since chemistry majors generally study physical chemistry in the third year of the undergraduate curriculum, this book can serve as the text for a senior-level undergraduate or a beginning graduate-level course. Although I use chemistry courses and chemistry curricula to describe the level of this book, students majoring in engineering, materials science, physics, and various specialties in the biological sciences will also find numerous topics of interest contained herein.

Terms like "fundamental," "molecular," and "interesting" have different meanings for different people. Let me explain how they apply to the presentation of polymer chemistry in this text.

The words "basic concepts" in the title define what I mean by "fundamental." This is the primary emphasis in this presentation. Practical applications of polymers are cited frequently—after all, it is these applications that make polymers such an important class of chemicals—but in overall content, the stress is on fundamental principles. "Foundational" might be another way to describe this. I have not attempted to cover all aspects of polymer science, but the topics that have been discussed lay the foundation—built on the bedrock of organic and physical chemistry—from which virtually all aspects of the subject are developed. There is an enormous literature in polymer science; this book is intended to bridge the gap between the typical undergraduate background in polymers—which frequently amounts to little more than occasional "relevant" examples in other courses—and the professional literature on the subject.

Accordingly, the book assumes essentially no prior knowledge of polymers, and extends far enough to provide a usable level of understanding.

"Molecular" describes the perspective of the chemist, and it is this aspect of polymeric materials that I try to keep in view throughout the book. An engineering text might emphasize processing behavior; a physics text, continuum mechanics; a biochemistry text, physiological function. All of these are perfectly valid points of view, but they are not the approach of this book. It is polymer molecules—their structure, energetics, dynamics, and reactions—that are the primary emphasis throughout most of the book. Statistics is the type of mathematics that is natural to a discussion of molecules. Students are familiar with the statistical nature of, say, the kinetic molecular theory of gases. Similar methods are applied to other assemblies of molecules, or in the case of polymers, to the assembly of repeat units that comprise a single polymer molecule. Although we frequently use statistical arguments, these are developed quite thoroughly and do not assume any more background in this subject than is ordinarily found among students in a physical chemistry course.

The most subjective of the words which (I hope) describe this book is "interesting." The fascinating behavior of polymers themselves, the clever experiments of laboratory researchers, and the elegant work of the theoreticians add up to an interesting total. I have tried to tell about these topics with clarity and enthusiasm, and in such a way as to make them intelligible to students. I can only hope that the reader agrees with my assessment of what is interesting.

This book was written with the student in mind. Even though "student" encompasses persons with a wide range of backgrounds, interests, and objectives; these are different than the corresponding experiences and needs of researchers. The following features have been included to assist the student:

1. Over 50 solved example problems are sprinkled throughout the book.
2. Exercises are included at the end of each chapter which are based on data from the original literature.
3. Concise reviews of pertinent aspects of thermodynamics, kinetics, spectrophotometry, etc. are presented prior to developing applications of these topics to polymers.
4. Theoretical models and mathematical derivations are developed in enough detail to be comprehensible to the student reader. Only rarely do I "pull results out of a hat," and I scrupulously avoid saying "it is obvious that . . ."
5. Generous cross-referencing and a judicious amount of repetition have been included to help unify a book which spans quite a wide range of topics.
6. SI units have been used fairly consistently throughout, and attention is paid to the matter of units whenever these become more than routine in complexity.

The book is divided into three parts of three chapters each, after an introductory chapter which contains information that is used throughout the book.

In principle, the three parts can be taken up in any order without too much interruption in continuity. Within each of the parts there is more carryover from chapter to chapter, so rearranging the sequence of topics within a given part is less convenient. The book contains more material than can be covered in an ordinary course. Chapter 1 plus two of the three parts contain about the right amount of material for one term. In classroom testing the material, I allowed the class to decide—while we worked on Chapter 1—which two of the other parts they wished to cover; this worked very well.

Material from Chapter 1 is cited throughout the book, particularly the discussion of statistics. In this connection, it might be noted that statistical arguments are developed in less detail further along in the book as written. This is one of the drawbacks of rearranging the order in which the topics are covered. Chapters 2 through 4 are concerned with the mechanical properties of bulk polymers, properties which are primarily responsible for the great practical importance of polymers. Engineering students are likely to have both a larger interest and a greater familiarity with these topics. Chapers 5 through 7 are concerned with the preparation and properties of several broad classes of polymers. These topics are closer to the interests of chemistry majors. Chapters 8 through 10 deal with the solution properties of polymers. Since many of the techniques described have been applied to biopolymers, these chapters will have more appeal to students of biochemistry and molecular biology.

Let me conclude by acknowledging the contributions of those who helped me with the preparation of this book. I wish to thank Marilyn Steinle for expertly typing the manuscript. My appreciation also goes to Carol Truett who skillfully transformed my (very) rough sketches into effective illustrations. Lastly, my thanks to Ron Manwill for preparing the index and helping me with the proofreading. Finally, let me acknowledge that some errors and/or obscurities will surely elude my efforts to eliminate them. I would appreciate reports about these from readers so that these mistakes can eventually be eliminated.

Paul C. Hiemenz

Contents

SOME IMPORTANT CLASSES OF POLYMERS
AND POLYMERIZATION REACTIONS

SOME PROPERTIES OF POLYMER SOLUTIONS AND THEIR RELATION TO POLYMER CHARACTERIZATION

1

The Chains and the Averages of Polymers
An Overview

> The *polymers*, those giant molecules,
> Like starch and polyoxymethylene,
> Flesh out, as protein serfs and plastic fools,
> The kingdom with life's stuff
>
> *The Dance of the Solids*, by John Updike

1.1 Introduction

Science tends to be plagued by clichés which make invidious comparisons of its efforts: "they can cure such and such a dreaded disease, but they cannot do anything about the common cold" or "we know more about the surface of the moon than the bottom of the sea." If such comparisons were popular in the 1920s, the saying might have been, "we know more about the structure of the atom than about those messy, sticky substances called polymers." Indeed, Millikan's determination of the electron's charge, Rutherford's idea of the nuclear atom, and Bohr's model of the hydrogen atom were all well-known concepts before the notion of truly covalent macromolecules was accepted. This was the case in spite of the great importance of polymers to human life and activities. Our bodies, like all forms of life, depend on polymer molecules: carbohydrates, proteins, nucleic acids, and so on. From the earliest times, polymeric materials have been employed to satisfy human needs: wood; hide; natural resins and gums; and fibers like cotton, wool, silk, and so forth.

Attempts to characterize polymeric substances had been made, of course, and high molecular weights were indicated, even if they were not too accurate. Early workers tended to be more suspicious of the interpretation of the colligative properties of polymeric solutions than to accept the possibility of high molecular weight compounds. Faraday had already arrived at C_5H_8 as the empirical formula of rubber in 1826, and isoprene was identified as the product

1

resulting from the destructive distillation of rubber in 1860. The idea that a natural polymer such as rubber somehow "contained" isoprene emerged, but the nature of its involvement was more elusive.

During the early years of this century, organic chemists were enjoying success in determining the structures of ordinary-sized organic molecules, and this probably contributed to their reluctance to look beyond structures of convenient size. Physical chemists were interested in intermolecular forces at this period, and the idea that polymers were the result of some sort of association between low molecular weight constituent molecules prevailed for a long while.

Staudinger is generally credited as being the father of modern polymer chemistry, although a foreshadowing of his ideas can be traced through older literature. In 1920 Staudinger proposed the chain formulas we accept today, maintaining that structures are held together by covalent bonds which are equivalent in every way to those in low molecular weight compounds. There was a decade of controversy before these ideas began to experience widespread acceptance. Staudinger was awarded the Nobel Prize in 1953 for his work with polymers.

By the 1930s, Carothers began synthesizing polymers using well-established reactions of organic chemistry such as esterification and amidation. His products were not limited to single ester or amide linkages, however, but contained many such groups: They were *poly*esters and *poly*amides. Physical chemists also got in on the act. Kuhn and Guth and Mark† were soon applying statistics and crystallography to describe the multitude of forms a long-chain molecule could assume.

Our purpose in this introduction is not to trace the history of polymer chemistry beyond the sketchy version above, instead, the objective is to introduce the concept of polymer chains which is the cornerstone of all polymer chemistry. In the next few sections we shall introduce some of the categories of chains, some of the reactions that produce them, and some aspects of isomerism which multiply their possibilities. A common feature of all of the synthetic polymerization reactions is the random nature of the polymerization steps. Likewise, the twists and turns the molecule can undergo along the backbone of the chain produce shapes which are only describable as averages. As a consequence of these considerations, another important part of this chapter is an introduction to some of the statistical concepts which also play a central role in polymer chemistry.

---

† No enterprise as rich as polymer science has only one "father." Herman Mark is one of those to whom the title could readily be applied. An interesting interview with Professor Mark appears in the *Journal of Chemical Education*, 56:83 (1979).

1.2 How Big Is Big?

The term *polymer* is derived from the Greek words *poly* and *meros*, meaning many parts. We noted in the last section that the existence of these parts was acknowledged before the nature of the interaction which held them together was known. Today we realize that ordinary covalent bonds are the intramolecular forces which keep the polymer molecule intact. In addition, the usual type of intermolecular forces—hydrogen bonds, dipole-dipole interactions, and London forces—hold assemblies of these molecules together in the bulk state. The only thing that is remarkable about these molecules is their size, but that feature is remarkable indeed.

One of the first things we must consider is what we mean when we talk about the size of a polymer molecule. Ordinarily, the molecular weight is meant, although a closely related concept, called the degree of polymerization, may also be used in this context. A variety of experimental techniques are available for determining the molecular weight of a polymer. We shall discuss one such method in Sec. 1.7 and defer most of the rest to Part III of this book, which is concerned with the properties of polymer solutions. The expression *molecular weight* should almost always be modified by the word *average*. This, too, is something we shall take up presently. For now, we assume that a polymer molecule has a molecular weight M, which can be anywhere in the range 10^3–10^7 or more in magnitude. We shall omit units when we write molecular weights in this book, but the student is advised to attach the units g mol^{-1} or amu molecule^{-1} to these quantities when they appear in problem calculations.

Since polymer molecules are made up of chains of repeating units, after the chain itself comes the repeat unit as a structural element of importance. Many polymer molecules are produced by covalently bonding together only one or two types of repeating units. These units are the parts from which chains are generated; as a class of compounds they are called monomers. Throughout this book, we shall designate the molecular weight of a repeat unit M_0.

The degree of polymerization of a polymer is simply the number of repeat units in a molecule. The degree of polymerization n is given by the ratio of the molecular weight of the polymer to the molecular weight of the repeat unit:

$$n = \frac{M}{M_0} \tag{1.1}$$

One type of polymerization reaction is the addition reaction in which successive repeat units add on to the chain. No other product molecules are formed, so the weight of the monomer and that of the repeat unit are identical in this case. A second category of polymerization reaction is the condensation reaction, in which one or two small molecules like water or HCl are eliminated for each chain linkage formed. In this case the molecular weight of the monomer and the

repeat unit are somewhat different. For example, suppose an acid (subscript A) reacts with an alcohol (subscript B) to produce an ester linkage and a water molecule. The molecular weight of the ester—the repeat unit if an entire chain is built up this way—differs from the combined weight of the reactants by twice the molecular weight of the water; therefore

$$n = \frac{M}{M_0} = \frac{M}{M_A + M_B - 2M_{H_2O}} \tag{1.2}$$

The end units in a polymer chain are different from the units that are attached on both sides to other repeat units. The end units are attached to the chain on one side only; the valence which would continue the chain if this were not the end must be capped off in some other way. We see this situation in the n-alkanes: Each end of the chain is a methyl group and the middle parts are methylene groups. Of course, the terminal group does not have to be a hydrogen as in the alkanes; indeed, it is generally something different. Our interest in end groups is concerned with the question of what effect they introduce into the evaluation of n through Eq. (1.2). The following example examines this through some numerical calculations.

Example 1.1

As a polymer prototype consider an n-alkane molecule consisting of n methylenes and 2 methyl groups. How serious an error is made in M for different n's if the difference in molecular weight between methyl and methylene groups is ignored?

Solution

The effect of different end groups on M can be seen by comparing the true molecular weight with an approximate molecular weight, calculated on the basis of a formula $(CH_2)_{n+2}$. These M's and the percentage difference between them are listed here for several values of n:

n	M_{true}	$M_{approx.}$	Percent difference
1	44	42	4.5
5	100	98	2.0
10	170	168	1.2
50	730	728	0.3
100	1,430	1,428	0.14
500	7,030	7,028	0.028
1000	14,030	14,028	0.014

Although the difference is almost 5% for propane, it is closer to 0.1% for the case of n = 100, which is about the threshold for polymers. The precise values of these numbers will be different, depending on the specific repeat units and end groups present. For example, if M_0 = 100 and M_{end} = 80, the difference is 0.39% in a calculation like those above for n = 100.

•

The example shows that the contribution of the ends becomes progressively less important as the number of repeat units in a structure increases. By the time polymeric molecular sizes are reached, the error associated with failure to distinguish between segments at the end and those within the chain is generally less than experimental error. In Sec. 1.7 we shall consider a method for polymer molecular weight determination which is based on chemical analysis for the end groups in a polymer. A corollary of the present discussion is that the method of end group analysis is applicable only in the case of relatively low molecular weight polymers.

Not all polymers are built up from bonding together a single kind of repeating unit. At the other extreme, protein molecules are polyamides in which n amino acid repeat units are bonded together. Although we might still call n the degree of polymerization in this case, it is less useful, since an amino acid unit might be any one of some 20-odd molecules that are found in proteins. In this case the molecular weight itself, rather than the degree of polymerization, is generally used to describe the molecule. When the actual content of individual amino acids is known, it is their sequence that is of special interest to biochemists and molecular biologists.

We began this section with an inquiry into how to define the size of a polymer molecule. In addition to the molecular weight or the degree of polymerization, some linear dimension which characterizes the molecule could also be used for this purpose. For purposes of orientation, let us again consider a hydrocarbon molecule stretched out to its full length but without any bond distortion. There are several features to note about this situation:

1. The tetrahedral geometry of the bonding at the carbon atoms has bond angles of 109.5°.
2. The equilibrium bond length between adjacent singly bonded carbon atoms is 0.154 nm or 1.54 Å.
3. Because of the possibility of rotation around carbon–carbon bonds, a molecule possessing many such bonds will undergo many twists and turns along the chain.
4. The fully extended molecular length is not representative of the spatial extension that a molecule actually displays. The latter is sensitive to environmental factors, however, so the extended length is convenient for our present purposes to provide an idea of the spatial size of polymer molecules.

A fully extended hydrocarbon molecule will have the familiar zigzag profile

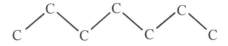

in which the hydrogens extend in front of and in back of the plane containing the carbons, also making an angle of 109.5° between them at the carbon atoms. The chain may be pictured as a row of triangles resting corner to corner. The length of the row equals the product of the number of triangles and the length of the base of each. Although it takes three carbons to define one of these triangles, one of these atoms is common to two triangles; therefore the number of triangles is the same as the number of *pairs* of carbon atoms, except where this breaks down at the ends of the molecule. If the chain is sufficiently long, this end effect is inconsequential. The law of cosines can be used to calculate the length of the base of each of these triangles: $[2(0.154)^2(1 - \cos 109.5°)]^{1/2} =$ 0.252 nm. If the repeat unit of the molecule contributes two carbon atoms to the backbone of the polymer—this is the case for vinyl polymers—the extended chain length is given by n(0.252) nm. For a polymer with $n = 10^4$, this corresponds to 2.52 μm. Objects which actually display linear dimensions of this magnitude can be seen in an ordinary microscope, provided that they have suitable optical properties to contrast with their surroundings. Note that the distance between every other carbon atom we have used here is also the distance between the substituents on these carbons for the fully extended chains.

We shall see in Sec. 1.10 that, because of all the twists a molecule undergoes, the actual end-to-end distance of the jumbled molecule increases as $n^{1/2}$. With the same repeat distance calculated above, but the square root dependence on n, the actual end-to-end distance of the coiled chain with $n = 10^4$ is closer to $(10^4)^{1/2}$ 0.252 nm = 25 nm. If we picture one end of this jumbled chain at the origin of a coordinate system, the other end might be anywhere on the surface of a sphere whose radius is given by this end-to-end distance. This spherical geometry comes about because the random bends occurring along the chain length can take the end of the chain anywhere in a spherical domain whose radius depends on the chain length (as $n^{1/2}$).

The above discussion points out the difficulty associated with using the linear dimensions of a molecule as a measure of its size: It is not the molecule alone that determines its dimensions, but also the shape in which it exists. Linear arrangements of the sort described above exist in polymer crystals, at least for some distance, although not over the full length of the chain. We shall take up the structure of polymer crystals in Chap. 4. In the solution and bulk states, many polymers exist in the coiled form we have also described. Still other structures are important, notably the helix, which we shall discuss in Sec. 1.11. The overall shape assumed by a polymer molecule is greatly affected

by the molecule's environment. The shape of a molecule in solution plays a role in determining many properties of polymer solutions. From a study of these solutions, some conclusions can be drawn regarding the shape of the molecule in that environment. The discussion of polymer solutions is taken up in Part III.

Figure 1.1 is a rather remarkable photograph which shows individual polystyrene molecules as spherical blobs having average diameters of about 20 nm. The picture is an electron micrograph in which a 10^{-4}% solution of polystyrene was deposited on a suitable substrate, the solvent evaporated, and the contrast enhanced by shadow casting. There is a brief discussion of both electron microscopy and shadowing in Sec. 4.7. Several points should be noted in connection with Fig. 1.1:

1. The circular cross section of the polymer blobs does not prove that the polymer existed in solution as a tangled coil (although this is the case). The shape displayed by the particles in the photograph is probably due in part to surface tension occurring during the drying of the sample.
2. Since the preparation of the specimen began with such a dilute solution, there seems to be little doubt that the particles are individual polymer molecules rather than clusters thereof. The diameters of the blobs are of the right order of magnitude for random structures, although this comparison must be used cautiously in view of item (1).
3. The fact that the dimensions of the various blobs is not identical should not surprise us. Again, we acknowledge that this may be an artifact arising from sample preparation. At the same time, however, we recall that the nature of polymerization reactions is such that a range of molecular weights is obtained. Accordingly, we expect a range of linear dimensions as well.

We conclude this section by questioning whether there is a minimum molecular weight or linear dimension that must be met for a molecule to qualify as a polymer. Although a dimer is a molecule for which n = 2, no one would consider it a polymer. The term *oligomer* has been coined to designate molecules for which n < 10. If they require a special name, apparently the latter are not full-fledged polymers either. At least as a first approximation, we shall take the attitude that there is ordinarily no discontinuity in behavior with respect to some observed property as we progress through a homologous series of compounds with different n values. At one end of the series we may be dealing with a simple low molecular weight compound, and at the other end with a material that is unquestionably polymeric. The molecular weight and chain length increase monotonically through this series, and a variety of other properties smoothly increase also. This point of view emphasizes a continuity with familiar facts concerning the properties of low molecular weight compounds. There are some properties, on the other hand, which follow so closely from the chain structure of polymers that the property is simply not observed until a certain

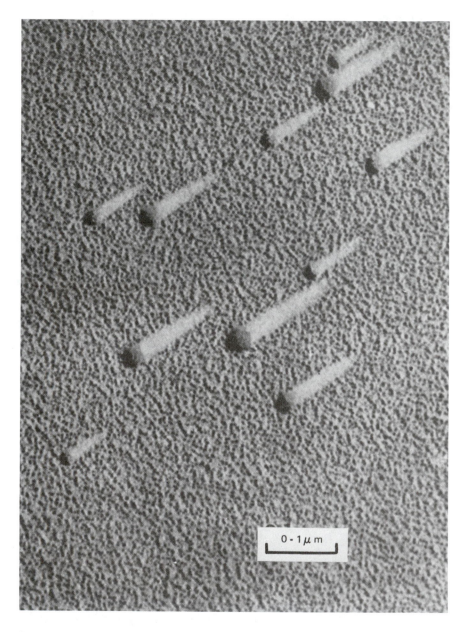

Figure 1.1 Individual polystyrene molecules as seen in an electron microscope. "Tails" are the result of shadow casting, which is used to enhance the visibility of the particles. [From M. J. Richardson, *Proc. Royal Soc.* 279A:50 (1964).]

critical molecular size has been reached. This critical size is often designated by a threshold molecular weight. The elastic behavior of rubber and several other so-called mechanical properties fall into this latter category. In theoretical developments, large values of n are often assumed to justify neglecting end effects, using certain statistical methods and other mathematical approximations. With these ideas in mind, M = 1000 is a convenient round number for designating a compound to be a polymer, although it should be clear that this cutoff is arbitrary.

1.3   Linear and Branched Polymers, Homopolymers, and Copolymers

The preceding section was based on the implicit assumption that polymer chains are linear. In evaluating both the degree of polymerization and the extended chain length, we assumed that the chain had only two ends. While linear polymers are important, they are not the only type of molecules possible. Branched and cross-linked molecules are also important. When we speak of a branched polymer, we refer to the presence of additional polymeric chains issuing from the backbone of a linear molecule. Substituent groups such as methyl or phenyl groups on the repeat units are not considered branches. Branching is generally introduced into a molecule by intentionally adding some monomer with the capability of serving as a branch. Consider the formation of a polyester. The presence of difunctional acids and difunctional alcohols allows the polymer chain to grow. These difunctional molecules are incorporated into the chain with ester linkages at both ends of each. Trifunctional acids or alcohols, on the other hand, produce a linear molecule by reacting two of their functional groups. If the third reacts and the resulting chain continues to grow, a branch has been introduced into the original chain. Adventitious branching sometimes occurs as a result of an atom being abstracted from the original linear molecule, with chain growth occurring from the resulting active site. Molecules with this kind of accidental branching are generally still called linear, although the presence of significant branching has profound effects on some properties of the polymer, most notably the tendency to undergo crystallization.

The amount of branching introduced into a polymer is an additional variable that must be specified for the molecule to be fully characterized. When only a slight degree of branching is present, the concentration of junction points is sufficiently low that these may be simply related to the number of chain ends. For example, two separate linear molecules have a total of four ends. If the end of one of these linear molecules attaches itself to the middle of the other to form a T, the resulting molecule has three ends. It is easy to generalize this result. If a molecule has ν branches, it has $\nu + 2$ chain ends if the branching is relatively low. Branched molecules are sometimes described as either combs or

stars. In the former, branch chains emanate from along the length of a common backbone; in the latter, all branches radiate from a central junction.

If the concentration of junction points is high enough, even branches will contain branches. Eventually a point is reached at which the amount of branching is so extensive that the polymer molecule becomes a giant three-dimensional network. When this condition is achieved, the molecule is said to be cross-linked. In this case, an entire macroscopic object may be considered to consist of essentially one molecule. The forces which give cohesiveness to such a body are covalent bonds, not intermolecular forces. Accordingly, the mechanical behavior of cross-linked bodies is much different from those without cross-linking.

Just as it is not necessary for polymer chains to be linear, it is also not necessary for all repeat units to be the same. We have already mentioned molecules like proteins where a wide variety of different repeat units are present. Among synthetic polymers, those in which a single kind of repeat unit are involved are called homopolymers, and those containing more than one kind of repeat unit are copolymers. Note that these definitions are based on the repeat unit, not the monomer. An ordinary polyester is not a copolymer, even though two different monomers, acids and alcohols, are its monomers. By contrast, copolymers result when different monomers bond together in the same way to produce a chain in which each kind of monomer retains its respective substituents in the polymer molecule. The unmodified term *copolymer* is generally used to designate the case where two different repeat units are involved. Where three kinds of repeat units are present, the system is called a terpolymer; where there are more than three, the system is called a multicomponent copolymer. The copolymers we discuss in this book will be primarily two-component molecules. We shall discuss copolymers in Chap. 7, so the present remarks are simply for purposes of orientation.

The moment we admit the possibility of having more than one kind of repeat unit, we require additional variables to describe the polymer. First, we must know how many kinds of repeat units are present and what they are. This is analogous to knowing what components are present in a solution, although the similarity ends there, since the repeat units in a polymer are bonded together and not merely mixed. To describe the copolymer quantitatively, the relative amounts of the different kinds of repeat units must be specified. Thus the empirical formula of a copolymer may be written $A_x B_y$, where A and B signify the individual repeat units and x and y indicate the relative number of each. From a knowledge of the molecular weight of the polymer, the molecular weights of A and B, and the values of x and y, it is possible to calculate the number of each kind of monomer unit in the copolymer. The sum of these values gives the degree of polymerization of the copolymer. Note that we generally do not call n_A and n_B the degrees of polymerization of the individual units. The inadvisability of the latter will become evident presently. The following example illustrates some of the ways of describing a copolymer.

Example 1.2

A terpolymer is prepared from vinyl monomers A, B, and C; the molecular weights of the repeat units are 104, 184, and 128, respectively. A particular polymerization procedure yields a product with the empirical formula $A_{3.55}B_{2.20}C_{1.00}$. The authors[†] of this research state that the terpolymer has "an average unit weight of 134" and "the average molecular weight per angstrom of 53.5." Verify these values.

Solution

The empirical formula gives the relative amounts of A, B, and C in the terpolymer. The total molecular weight of this empirical formula unit is given by adding the molecular weight contributions of A, B, and C: $3.55(104) + 2.20(184) + 1.00(128) = 902$ amu per empirical formula unit. The total number of chain repeat units possessing this total weight is $3.55 + 2.20 + 1.00 = 6.75$ repeat units per empirical formula unit. The ratio of the total molecular weight to the total number of repeat units gives the average molecular weight per repeat unit:

$$\frac{902}{6.75} = 134 \text{ amu per repeat unit}$$

Since the monomers are specified to be vinyl monomers, each contributes two carbon atoms to the polymer backbone, with the associated extended length of 0.252 nm per repeat unit. Therefore the total extended length of the empirical formula unit is

$$6.75(0.252 \text{ nm}) = 1.70 \text{ nm} = 17.0 \text{ Å}$$

The ratio of the total weight to the total extended length of the empirical formula unit gives the average molecular weight per length of chain:

$$\frac{902}{17} = 53 \text{ amu per Å}$$

Note that the average weight per repeat unit could be used to evaluate the overall degree of polymerization of this terpolymer. For example, if the molecular weight were 43,000, the corresponding degree of polymerization would be

$$\frac{43,000 \text{ amu molecule}^{-1}}{134 \text{ amu (repeat unit)}^{-1}} = 321 \text{ repeat units per molecule}$$

•

[†] A. Ravve and J. T. Khamis, *Addition and Condensation Polymerization Processes*, Advances in Chemistry Series, Vol. 91, American Chemical Society Publications, Washington, D.C., 1969.

With copolymers, it is not sufficient merely to describe the empirical formula to characterize the molecule. Another question that can be asked concerns the distribution of the different kinds of repeat units in the molecule. Starting from monomers A and B, the following distribution patterns are obtained in linear polymers:

1. Random. The A—B sequence is governed strictly by chance, subject only to the relative abundances of repeat units. For equal proportions of A and B we have:

    −AAABABAABBABBB−

2. Alternation. A regular pattern of alternating repeat units:

    −ABABABABABAB−

3. Block. A long, uninterrupted sequence of each monomer is the pattern:

    −AAAAAAAAAAAAAABBBBBBBBBBBBBBBAAAAAAAAAA−

If a copolymer is branched with different repeat units occurring in the branches and the backbone, we have the following:

4. Graft. This segregation is accomplished by first homopolymerizing the backbone. This is dissolved in the second monomer, with sites along the original chain becoming the origin of the comonomer side-chain growth:

In a cross-linked polymer, the junction units are different kinds of monomers than the chain repeat units, so these molecules might be considered to be still another comonomer. While the chemical reactions which yield such cross-linked substances are copolymerizations, the products are described as cross-linked rather than as copolymers. In this instance, the behavior due to cross-linking takes precedence over the presence of an additional type of monomer in the structure.

It is apparent from items (1)–(3) above that linear copolymers—even those with the same proportions of different kinds of repeat units—can be very different in structure and properties. In classifying a copolymer as random, alternating, or block, it should be realized that we are describing the average character of the molecule; accidental variations from the basic patterns may be present. In Chap. 7 we shall see how an experimental investigation of the sequence of repeat units in a copolymer is a valuable tool for understanding copolymerization reactions. This type of information along with other details of structure are collectively known as the microstructure of a polymer.

1.4 Addition, Condensation, and Natural Polymers

In the last section we examined some of the categories into which polymers can be classified. Various aspects of molecular structure were used as the basis for classification in that section. Next we shall consider the chemical reactions that produce the molecules as a basis for classification. The objective of this discussion is simply to provide some orientation and to introduce some typical polymers. For this purpose a number of polymers may be classified as either addition or condensation polymers. Each of these classes of polymers are discussed in detail in Part II of this book, specifically Chaps. 5 and 6 for condensation and addition, respectively. Even though these categories are based on the reactions which produce the polymers, it should not be inferred that only two types of polymerization reactions exist. We have to start somewhere, and these two important categories are the usual place to begin.

These two categories of polymer can be developed along several lines. For example, in addition-type polymers the following applies:

1. The repeat unit in the polymer and the monomer have the same composition, although, of course, the bonding is different in each.
2. The mechanism of these reactions places addition polymerizations in the kinetic category of chain reactions, with either free radicals or ionic groups responsible for propagating the chain reaction.
3. The product molecules usually have a carbon chain backbone, with pendant substituent groups

$$-C-C-C-C-$$
$$\quad | \quad\quad |$$
$$\quad X \quad\quad X$$

in condensation-type polymers:

4. The polymer repeat unit arises from reacting together two different functional groups which usually originate on different monomers. In this case the repeat unit is different from either of the monomers. In addition, small molecules are often eliminated during the condensation reaction. Note the words *usual* and *often* in the previous statements; exceptions to both statements are easily found.
5. The mechanistic aspect of these reactions is summarized by saying that the reactions occur in steps. Thus the formation of an ester linkage is not essentially different between two small molecules or in a polyester.
6. The product molecules have the functional groups formed by the condensation reactions interspersed regularly along the backbone of the polymer molecule:

$$-C-C-Y-C-C-Y-$$

Next let us consider a few specific examples of these classes of polymers.

The addition polymerization of a vinyl monomer $CH_2=CHX$ involves three distinctly different steps. First, the reactive center must be initiated by a suitable reaction to produce a free radical or an anion or cation reaction site. Next, this reactive entity adds consecutive monomer units to propagate the polymer chain. Finally, the active site is capped off, terminating the polymer formation. If one assumes that the polymer produced is truly a high molecular weight substance, the lack of uniformity at the two ends of the chain—arising in one case from the initiation, and in the other from the termination—can be neglected. Accordingly, the overall reaction can be written

$$nCH_2=CHX \longrightarrow \{CH_2-CHX\}_n \tag{1.A}$$

Again we emphasize that end effects are ignored in writing reaction (1.A). These effects as well as the conditions of the reaction and other pertinent information will be discussed when these reactions are considered in Chap. 6. Table 1.1 is a list of several important addition polymers, showing the polymerization reaction in the manner of reaction (1.A). Also included in Table 1.1 are the molecular weights of the repeat units and the common names of the polymers. The former will prove helpful in many of the problems in this book; the latter will be discussed in the next section. Poly(ethylene oxide) and poly(ε-caprolactam) have been included in this list as examples of the hazards associated with classification schemes. They resemble addition polymers because the molecular weight of the repeat unit and that of the monomer are the same; they resemble condensation polymers because of the heteroatom chain backbone. The reaction mechanism, which might serve as arbiter in this case, can be either chain or step, depending on the reaction conditions! These last reactions are examples of ring-opening polymerizations, still another category of classification.

The requirements for formation of condensation polymers are twofold: The monomers must possess functional groups capable of reacting to form the linkage and they ordinarily require more than one reactive group to generate a chain structure. The functional groups can be distributed such that two difunctional monomers with different functional groups react or a single monomer reacts which is difunctional with one group of each kind. In the latter case especially, but also with condensation polymerization in general, the tendency to form cyclic products from intramolecular reactions may compete with the formation of polymer. This possibility is discussed in Sec. 5.10. Condensation polymerizations are especially sensitive to impurities. The presence of monofunctional reagents introduces the possibility of a reaction product forming which would not be capable of further growth. If the functionality is greater than 2, on the other hand, branching becomes possible. Both of these modifications dramatically alter the product compared to a high molecular weight linear product. When reagents of functionality less than or greater than 2 are added in carefully measured and controlled amounts, the size and geometry of product

Table 1.1 Reactions by Which Several Important Addition Polymers are Produced[a]

Monomer and M_0 (g mol^{-1})	Polymer repeat unit and name(s)
1. $nCH_2{=}CH_2$ $\longrightarrow$ ($M_0 = 28.0$)	$\{CH_2{-}CH_2\}_n$ polyethylene

2. $nCH_2{=}CH$ $\longrightarrow$
       |
      C_6H_5
 ($M_0 = 104$)
	$\{CH_2{-}CH\}_n$
	      |
	     C_6H_5
	 polystyrene

3. $nCH_2{=}CHCl$ $\longrightarrow$ ($M_0 = 62.5$)	$\{CH_2{-}CHCl\}_n$ poly(vinyl chloride), "vinyl"

4. $nCH_2{=}CH$ $\longrightarrow$
       |
      CN
 ($M_0 = 53.9$)
	$\{CH_2{-}CH\}_n$
	     |
	    CN
	 polyacrylonitrile, "acrylic"

5. $nCH_2{=}CCl_2$ $\longrightarrow$ ($M_0 = 97.0$)	$\{CH_2{-}CCl_2\}_n$ poly(vinylidene chloride)

6. CH_3
        |
 $nCH_2{=}C$ $\longrightarrow$
      |
   $O{=}C{-}O{-}CH_3$
 ($M_0 = 100$)
	CH_3
	      |
	 $\{CH_2{-}C\}_n$
	     |
	 $O{=}C{-}O{-}CH_3$
	 poly(methyl methacrylate), Plexiglas, lucite

7. CH_3
        |
 $nCH_2{=}C$ $\longrightarrow$
      |
      CH_3
 ($M_0 = 56.0$)
	CH_3
	      |
	 $\{CH_2{-}C\}_n$
	     |
	     CH_3
	 polyisobutylene

8. $nCF_2{=}CF_2$ $\longrightarrow$ ($M_0 = 100$)	$\{CF_2{-}CF_2\}_n$ polytetrafluoroethylene, Teflon
9. $nCH_2{-}CH_2$ $\longrightarrow$ $\backslash$ O $/$ ($M_0 = 44.0$)	$\{CH_2{-}CH_2{-}O\}_n$ poly(ethylene oxide), carbowax

10. O H
      || |
 $nC{-}(CH_2)_5{-}N$ $\longrightarrow$
  $\lfloor$_____$\rfloor$
 ($M_0 = 113$)
	O H
	      || |
	 $\{(CH_2)_5{-}C{-}N\}_n$
	
	 poly(ϵ-caprolactam), nylon-6

[a]Molecular weights of repeat units and common names of the products are included for future reference. Reactions 9 and 10 are ring-opening polymerizations and are discussed in Sec. 5.10.

molecules can be manipulated. When such reactants enter as impurities, the undesired results can be disasterous! Marvel[†] has remarked that more money has been wasted in polymer research by the use of impure monomers than in any other manner.

Table 1.2 lists several examples of condensation reactions and products. Since the reacting monomers can contain different numbers of carbon atoms between functional groups, there are quite a lot of variations possible among these basic reaction types.

The inclusion of poly(dimethyl siloxane) in Table 1.2 serves as a reminder that polymers need not be organic compounds. The physical properties of inorganic polymers follow from the chain structure of these molecules, and the concepts developed in this volume apply to them and to organic polymers equally well. For example, poly(dimethyl siloxane) shows a very low viscosity compared to other polymers of comparable degree of polymerization. We shall see in Chap. 2 that this is traceable to its high chain flexibility, which, in turn, is due to the high concentration of chain backbone atoms with no substituents. We shall not examine the classes and preparations of the various types of inorganic polymers in this text. References in inorganic chemistry should be consulted for this information.

We conclude this section with a short discussion of naturally occurring polymers. Since these are of biological origin, they are also called biopolymers. Although our attention in this volume is primarily directed toward synthetic polymers, it should be recognized that biopolymers, like inorganic polymers, have physical properties which follow directly from the chain structure of their molecules. The synthesis by and contribution to living organisms by these biopolymers is the subject matter of other disciplines, such as biochemistry and molecular biology. The reader who desires information on this aspect of these naturally occurring polymers should consult references in these other disciplines.

As examples of natural polymers, we consider polysaccharides, proteins, and nucleic acids. Another important natural polymer, polyisoprene, will be considered in Sec. 1.6.

Polysaccharides are macromolecules which make up a large part of the bulk of the vegetable kingdom. Cellulose and starch are, respectively, the first and second most abundant organic compounds in plants. The former is present in leaves and grasses; the latter in fruits, stems, and roots. Because of their abundance in nature and because of contemporary interest in renewable resources, there is a great deal of interest in these compounds. Both cellulose and starch are hydrolyzed by acids to D-glucose, the repeat unit in both polymer chains.

[†] Carl S. Marvel, another pioneer in polymer chemistry, reminisces about the early days of polymer chemistry in the United States in the *Journal of Chemical Education*, 58:535 (1981).

Table 1.2   Reactions by Which Several Important Condenstation Polymers are Produced [a]

1. Polyester

$$n \ HO(CH_2)_2OH + n \ HOOCC_6H_4COOH \rightarrow$$

$$\left[O{-}(CH_2)_2{-}O{-}\overset{\overset{\displaystyle O}{\|}}{C}{-}\!\!\left\langle \bigcirc \right\rangle\!\!{-}\overset{\overset{\displaystyle O}{\|}}{C} \right]_n + 2n \ H_2O$$

$M_0 = 192$

poly(ethylene terephthalate), Terylene, Dacron, Mylar

$$n \ HO\underset{\underset{\displaystyle C_6H_{13}}{|}}{C}H(CH_2)_{10}COOH \rightarrow \left[O\underset{\underset{\displaystyle C_6H_{13}}{|}}{C}H(CH_2)_{10}\overset{\overset{\displaystyle O}{\|}}{C} \right]_n + n \ H_2O$$

$M_0 = 282$

poly(12-hydroxystearic acid)

2. Polyamide

$$n \ H_2N(CH_2)_6NH_2 + n \ ClCO(CH_2)_4COCl \rightarrow \left[NH{-}(CH_2)_6{-}\underset{\underset{\displaystyle H}{|}}{N}{-}\overset{\overset{\displaystyle O}{\|}}{C}{-}(CH_2)_4{-}CO \right]_n + 2nHCl$$

$M_0 = 226$

poly(hexamethylene adipamide), nylon 6,6

3. Polyurethane

$$n \ HO(CH_2)_4OH + n \ O{=}C{=}N(CH_2)_6N{=}C{=}O \rightarrow$$

$$\left[\overset{\overset{\displaystyle O}{\|}}{C}{-}\underset{\underset{\displaystyle H}{|}}{N}(CH_2)_6\underset{\underset{\displaystyle H}{|}}{N}{-}\overset{\overset{\displaystyle O}{\|}}{C}{-}O(CH_2)_4O \right]_n$$

$M_0 = 258$

poly(tetramethylene hexamethylene urethane), spandex, Perlon U

4. Polycarbonate

$$n \ ClCOCl + n \ HO{-}\!\!\left\langle \bigcirc \right\rangle\!\!{-}\underset{\underset{\displaystyle CH_3}{|}}{\overset{\overset{\displaystyle CH_3}{|}}{C}}{-}\!\!\left\langle \bigcirc \right\rangle\!\!{-}OH \rightarrow$$

$$\left[O{-}\!\!\left\langle \bigcirc \right\rangle\!\!{-}\underset{\underset{\displaystyle CH_3}{|}}{\overset{\overset{\displaystyle CH_3}{|}}{C}}{-}\!\!\left\langle \bigcirc \right\rangle\!\!{-}O{-}\overset{\overset{\displaystyle O}{\|}}{C} \right]_n + 2nHCl$$

$M_0 = 244$

poly(4,4-isopropylidenediphyenylene carbonate), bisphenol A polycarbonate, lexan

5. Inorganic

$$n \ (CH_3)_2SiCl_2 + n \ H_2O \rightarrow \left[\underset{\underset{\displaystyle CH_3}{|}}{\overset{\overset{\displaystyle CH_3}{|}}{Si}}{-}O \right]_n + 2n \ HCl$$

$M_0 = 74.0$

poly(dimethyl siloxane)

---

[a] Molecular weights of repeat units and common names of the products are included for future reference.

The configuration of the glucoside linkage is different in the two, however. Structures [I] and [II], respectively, illustrate that the linkage is a β-acetal–hydrolyzable to an equitorial hydroxide–in cellulose, and an α-acetal–hydrolyzable to an axial hydroxide–in amylose, a starch:

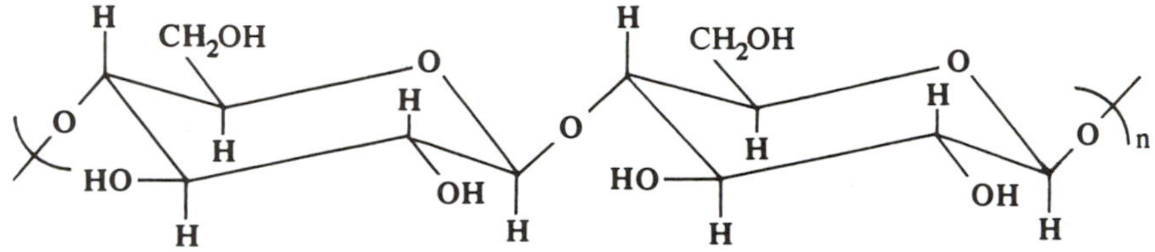

[I]

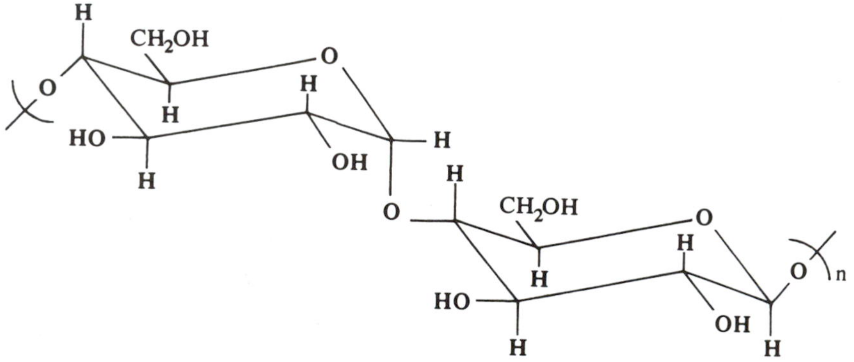

[II]

Amylopectin and glycogen are saccharides similar to amylose, except with branched chains.

The cellulose molecule contains three hydroxyl groups which can react and leave the chain backbone intact. These alcohol groups can be esterified with acetic anhydride to form cellulose acetate. This polymer is spun into the fiber acetate rayon. Similarly, the alcohol groups in cellulose react with CS_2 in the presence of strong base to produce cellulose xanthates. When extruded into fibers, this material is called viscose rayon, and when extruded into sheets, cellophane. In both the acetate and xanthate formation, some chain degradation also occurs, so the resulting polymer chains are shorter than those in the starting cellulose.

As noted above, proteins are polyamides in which α-amino acids make up the repeat units, as shown by structure [III]:

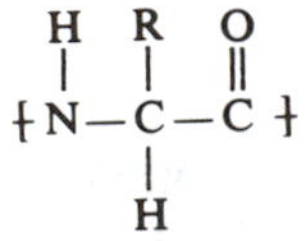

[III]

These molecules are also called polypeptides, especially in cases where $M \stackrel{\sim}{<}$ 10,000. The various amino acids differ in R groups. The nature of R, the name, and the abbreviation used to represent some of the more common amino acids are listed in Table 1.3. In proline (Pro) and hydroxyproline (Hyp), the nitrogen and the α-carbon are part of the five-atom pyrrolidine ring. Since some of the amino acids carry substituent carboxyl or amino groups, protein molecules are charged in aqueous solutions, and hence migrate in electric fields. This is the basis of electrophoresis as a means of separating and identifying proteins.

It is conventional to speak of three levels of structure in protein molecules:

1. Primary structure refers to the sequence of amino acids in the polyamide chain.
2. Secondary structure refers to the shape of the molecule as a whole, particularly to those aspects of structure which are stabilized by intramolecular hydrogen bonds.
3. Tertiary structure also refers to the overall shape of a molecule, especially to structures stabilized by disulfide bridges (cystine) formed by the oxidation of cysteine mercapto groups.

Hydrogen bonding stabilizes some protein molecules in helical forms, and disulfide cross-links stabilize some protein molecules in globular forms. We shall consider helical structures in Sec. 1.11 and shall learn more about ellipsoidal globular proteins in the chapters concerned with the solution properties of polymers, especially Chap. 9. Both secondary and tertiary levels of structure are also influenced by the distribution of polar and nonpolar amino acid molecules relative to the aqueous environment of the protein molecules. Nonpolar amino acids are designated in Table 1.3.

The three levels of structure listed above are also useful categories for describing nonprotein polymers. Thus details of the microstructure of a chain is a description of the primary structure. The overall shape assumed by an individual molecule as a result of the rotation around individual bonds is the secondary structure. Structures that are locked in by chemical cross-links are tertiary structures.

Examples of the effects and modification of the higher-order levels of structure in proteins are found in the following systems:

1. Collagen is the protein of connective tissues and skin. In living organisms, the molecules are wound around one another to form a three-strand helix stabilized by hydrogen bonding. When boiled in water, the collagen dissolves and forms gelatin, apparently establishing a new hydrogen bond equilibrium with the solvent. This last solution sets up to form the familiar gel when cooled, a result of shifting the hydrogen bond equilibrium.
2. Keratin is the protein of hair and wool. These proteins are insoluble because of the disulfide cross-linking between cystine units. Permanent waving of

Table 1.3 Name, Abbreviation, and R Group for Some Common Amino Acids

Name	Abbreviation	R group
Alanine[a]	Ala	$-CH_3$
Arginine	Arg	$-CH_2CH_2CH_2NHCNH_2$, with $\overset{NH}{\overset{\|}{}}$
Aspartic acid	Asp	$-CH_2COOH$
Cysteine	Cys	$-CH_2SH$
Glutamic acid	Glu	$-CH_2CH_2COOH$
Glycine	Gly	$-H$
Histidine	His	
Isoleucine[a]	Ile	$-CH(CH_3)CH_2CH_3$
Leucine[a]	Leu	$-CH_2CH(CH_3)_2$
Lysine	Lys	$-CH_2CH_2CH_2CH_2NH_2$
Methionine[a]	Met	$-CH_2CH_2SCH_3$
Phenylalanine[a]	Phe	
Serine	Ser	$-CH_2OH$
Threonine	Thr	$-CHOHCH_3$
Tryptophan[a]	Trp	
Tyrosine	Tyr	
Valine[a]	Val	$-CH(CH_3)_2$

[a] Nonpolar R groups.

hair involves the rupture of these bonds, reshaping of the hair fibers, and the reformation of cross-links which hold the chains in the new positions relative to each other. We shall see in Chap. 3 how such cross-linked networks are restored to their original shape when subjected to distorting forces.

3. The globular proteins albumin in eggs and fibrinogen in blood are converted to insoluble forms by modification of their higher-order structure. The process is called denaturation and occurs, in the systems mentioned, with the cooking of eggs and the clotting of blood.

Ribonucleic acid (RNA) and deoxyribonucleic acid (DNA) are polymers in which the repeat units are substituted polyesters. The esters are formed between the hydrogens of phosphoric acid and the hydroxyl groups of a sugar, D-ribose in the case of RNA and D-2-deoxyribose in the case of DNA. The sugar rings in DNA carry four different kinds of substituents: adenine and guanine, which are purines, and thymine and cytosine, which are pyramidines. The familiar double-helix structure of the DNA molecule is stabilized by hydrogen bonding between pairs of substituent base groups. The replication of these molecules, the template model of their functioning, and their role in protein synthesis and the genetic code make the study of these polymers among the most exciting and actively researched areas in all science. A single paragraph scarcely does justice to the fundamental importance, brilliant research, and extensive body of knowledge centered around these materials. As for inorganic polymers, however, we refer the reader who seeks more information about these substances to texts in biochemistry or molecular biology.

1.5 Polymer Nomenclature

Considering that a simple compound like C_2H_5OH is variously known as ethanol, ethyl alcohol, grain alcohol, or simply alcohol, it is not too surprising that the vastly more complicated polymer molecules are also often known by a variety of different names. The International Union of Pure and Applied Chemistry (IUPAC) has recommended a system of nomenclature based on the structure of the monomer or repeat unit. A semisystematic set of trivial names is also in widespread usage; these latter names seem even more resistant to replacement than is the case with low molecular weight compounds. Synthetic polymers of commercial importance are often widely known by trade names which seem to have more to do with marketing considerations than with scientific communication. Polymers of biological origin are often described in terms of some aspect of their function, preparation, or characterization.

If a polymer is formed from a single monomer, as in addition and ring-opening polymerizations, it is named by attaching the prefix *poly* to the name

of the monomer. In the IUPAC system, the monomer is named according to the IUPAC recommendations for organic chemistry, and the name of the monomer is set off from the prefix by enclosing the former in parentheses. Variations of this basic system often substitute a common name for the IUPAC name in designating the monomer. Whether or not parentheses are used in the latter case is influenced by the complexity of the monomer name; they become more important as the number of words in the monomer name increases. The polymer $\{CH_2-CHCl\}_n$ is called poly(1-chloroethylene) according to the IUPAC system; it is more commonly called poly(vinyl chloride) or polyvinyl chloride. Acronyms are not particularly helpful but are an almost irresistible aspect of polymer terminology, as evidenced by the initials PVC, which are so widely used to describe the polymer just named. The trio of names—poly(1-hydroxyethylene), poly(vinyl alcohol), and polyvinyl alcohol—emphasizes that the polymer need not actually be formed from the reaction of the monomer named; this polymer is formed by the hydrolysis of poly(1-acetoxyethylene), otherwise known as poly(vinyl acetate). These same alternatives are used in naming polymers formed by ring-opening reactions; for example, poly(6-aminohexanoic acid), poly(6-aminocaproic acid) and poly(ϵ-caprolactam) are all more or less acceptable names for the same polymer.

Those polymers which are the condensation product of two different monomers are named by applying the preceding rules to the repeat unit. For example, the polyester formed by the condensation of ethylene glycol and terephthalic acid is called poly(oxyethylene oxyterphthaloyl) according to the IUPAC system, as well as poly(ethylene terephthalate) or polyethylene terephthalate.

The polyamides poly(hexamethylene sebacamide) and poly(hexamethylene adipamide) are also widely known as nylon-6,10 and nylon-6,6, respectively. The numbers following the word *nylon* indicate the number of carbon atoms in the diamine and dicarboxylic acid, in that order. On the basis of this same system, poly(ϵ-caprolactam) is also known as nylon-6.

Most of the polymers in Tables 1.1 and 1.2 are listed with more than one name. Also listed are some of the patented trade names by which these substances—or materials which are mostly of the indicated structure—are sold commercially.

Some commercially important cross-linked polymers go virtually without names. These are heavily and randomly cross-linked polymers which are insoluble and infusible and therefore widely used in the manufacture of such molded items as automobile and household appliance parts. These materials are called resins and, at best, are named by specifying the monomers which go into their production. Often even this information is sketchy. Examples of this situation are provided by phenol–formaldehyde and urea–formaldehyde resins, for which typical structures are given by structures [IV] and [V], respectively:

[IV]   [V]

1.6 Positional, Stereo, and Geometrical Isomerism

In this section we shall consider three types of isomerism which are encountered in polymers. These are positional isomerism, stereo isomerism, and geometrical isomerism. We shall focus attention on synthetic polymers and shall, for the most part, be concerned with these types of isomerism occurring singly, rather than in combination. The synthetic and analytical aspects of stereo isomerism will be considered in Chap. 7. Our present concern is merely to introduce the possibilities of these isomers and some of the vocabulary associated with them.

Positional isomerism is conveniently illustrated by considering the polymerization of a vinyl monomer. In such a reaction, the adding monomer may become attached to the growing chain in either of two orientations:

$$\sim CH_2-CHX + CH_2=CHX \longrightarrow \begin{cases} \sim CH_2-CHX-CH_2-CHX & [VI] \\ \\ \sim CH_2-CHX-CHX-CH_2 & [VII] \end{cases} \qquad (1.B)$$

Structures [VI] and [VII], respectively, are said to arise from head-to-tail or head-to-head orientations. In this terminology, the substituted carbon is defined to be the head of the molecule, and the methylene is the tail. Tail-to-tail linking is also possible. The term *orienticity* is also used to describe positional isomerism.

For most vinyl polymers, head-to-tail addition is the dominant mode of addition. Variations from this generalization become more common for polymerizations which are carried out at higher temperatures. Head-to-head addition is also somewhat more abundant in the case of halogenated monomers such as vinyl chloride. The preponderance of head-to-tail additions is understood to arise from a combination of resonance and steric effects. In many cases the ionic or free-radical reaction center occurs at the substituted carbon due to the possibility of resonance stabilization or electron delocalization through the substituent group. Head-to-tail attachment is also sterically favored, since the substituent groups on successive repeat units are separated by a methylene

carbon. At higher temperatures of polymerization, larger amounts of available thermal energy make the less-favored states more accessible. In vinyl fluoride, no resonance stabilization is possible and steric effects are minimal. This monomer adds primarily in the head-to-tail orientation at low temperatures and tends toward a random combination of both at higher temperatures. The styrene radical, by contrast, enjoys a large amount of resonance stabilization in the bulky phenyl group and polymerizes almost exclusively in the head-to-tail mode. The following example illustrates how chemical methods can be used to measure the relative amounts of the two positional isomers in a polymer sample.

Example 1.3

1,2-Glycol bonds are cleaved by reaction with periodate; hence poly(vinyl alcohol) chains are broken at the site of head-to-head links in the polymer. The fraction of head-to-head linkages in poly(vinyl alcohol) may be determined by measuring the molecular weight before (subscript b) and after (subscript a) cleavage with periodate according to the following formula: Fraction = $44(1/M_a - 1/M_b)$. Derive this expression and calculate the value for the fraction in the case of $M_b = 10^5$ and $M_a = 10^3$.

Solution

Begin by recognizing that a molecule containing x of the head-to-head links will be cleaved into x + 1 molecules upon reaction. Hence if N is the number of polymer molecules in a sample of mass w, the following relations apply before and after cleavage: $N_a = (x + 1)N_b$ or $w/M_a = (x + 1)(w/M_b)$. Solving for x and dividing the latter by the total number of linkages in the original polymer gives the desired ratio. The total number of links in the mass w of polymer is $w/n_b M_0$. Therefore the ratio is $x/n_b = M_0 (1/M_a - 1/M_b)$. For polyvinyl alcohol M_0 is 44, so the desired formula has been obtained. For the specific data given, $x/n_b = 44(10^{-3} - 10^{-5}) = 0.044$, or about 4% of the additions is in the less favorable orientation. We shall see presently that the molecular weight of a polymer is an average which is different, depending on the method used for its determination. The present example used molecular weights as a means for counting the number of molecules present. Hence the sort of average molecular weight used should also be one which is based on counting.[†]

●

[†]A physical chemistry laboratory experiment based on this principle is found in D. P. Shoemaker, C. W. Garland, J. I. Steinfels, and J. W. Nibler, *Experiments in Physical Chemistry*, 4th ed., McGraw-Hill, New York, 1981.

The second type of isomerism we discuss in this section is stereo isomerism. Again we consider the number of ways a singly substituted vinyl monomer can add to a growing polymer chain:

$$\text{(1.C)}$$

Structures [VIII] and [IX] are not equivalent; they would not superimpose if the extended chains were overlaid. The difference has to do with the stereochemical configuration at the asymmetric carbon atom. Note that the asymmetry is more accurately described as pseudoasymmetry, since two sections of chain are bonded to these centers. Except near chain ends, which we ignore for high polymers, these chains provide local symmetry in the neighborhood of the carbon under consideration. The designations D and L or R and S are used to distinguish these structures, even though true asymmetry is absent.

We use the word *configuration* to describe the way the two isomers produced by reaction (1.C) differ. It is only by breaking bonds, moving substituents, and reforming new bonds that the two structures can be interconverted. This state of affairs is most readily seen when the molecules are drawn as fully extended chains and then examining the side of the chain on which substituents lie. The configurations are not altered if rotation is allowed to occur around the various bonds of the backbone to change the shape of the molecule to a jumbled coil. We shall use the term *conformations* to describe the latter possibilities for different molecular shapes. The configuration is not influenced by conformational changes, but the stability of different conformations may be affected by differences in configuration. We shall return to these effects in Sec. 1.11.

In the absence of any external influence such as a catalyst which is biased in favor of one configuration over the other, we might expect structures [VIII] and [IX] to occur at random with equal probability as if the configuration at each successive addition were determined by the toss of a coin. Such, indeed, is the ordinary case. However, in the early 1950s, stereospecific catalysts were discovered; Ziegler and Natta received the Nobel Prize for this discovery in 1963.

Following the advent of these catalysts, polymers with a remarkable degree of stereoregularity have been formed. These have had such a striking impact on polymer science that a large part of Chap. 7 is devoted to a discussion of their preparation and characterization. For now, only the terminology involved in their description concerns us. Three different situations can be distinguished along a chain containing pseudoasymmetric carbons:

1. Isotactic. All substituents lie on the same side of the extended chain. Alternatively, the stereoconfiguration at the asymmetric centers is the same, say, −DDDDDDDDD−.
2. Syndiotactic. Substituents on the fully extended chain lie on alternating sides of the backbone. This alternation of configuration can be represented as −DLDLDLDLDLDL−.
3. Atactic. Substituents are distributed at random along the chain, for example, −DDLDLLLDLDLL−.

Figure 1.2 shows sections of polymer chains of these three types; the substituent R equals phenyl for polystyrene and methyl for polypropylene. The general term for this stereoregularity is *tacticity*, a term derived from the Greek word meaning "to put in order."

Polymers of different tacticity have quite different properties, especially in the solid state. One of the requirements for polymer crystallinity is a high degree of microstructural regularity to enable the chains to pack in an orderly manner. Thus atactic polypropylene is a soft, tacky substance, whereas both isotactic and syndiotactic polypropylenes are highly crystalline.

No polymer is ever 100% crystalline; at best, patches of crystallinity are present in an otherwise amorphous matrix. In some ways, the presence of these domains of crystallinity is equivalent to cross-links, since different chains loop in and out of the same crystal. Although there are similarities in the mechanical behavior of chemically cross-linked and partially crystalline polymers, a significant difference is that the former are irreversibly bonded while the latter are reversible through changes of temperature. Materials in which chemical cross-linking is responsible for the mechanical properties are called thermosetting; those in which this kind of physical cross-linking operates, thermoplastic.

The final type of isomerism we take up in this section involves various possible structures which result from the polymerization of 1,3-dienes. Three important monomers of this type are 1,3-butadiene, 1,3-isoprene, and 1,3-chloroprene, structures [X]–[XII], respectively:

$$CH_2{=}CH{-}CH{=}CH_2 \qquad CH_2{=}\overset{\overset{\textstyle CH_3}{|}}{C}{-}CH{=}CH_2 \qquad CH_2{=}\overset{\overset{\textstyle Cl}{|}}{C}{-}CH{=}CH_2$$

[X]   [XI]   [XII]

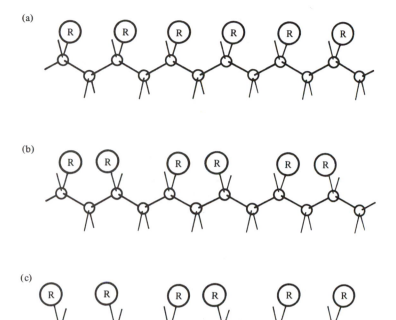

Figure 1.2 Sections of polymer chains of differing tacticity: (a) isotactic (b) syndiotactic (c) atactic.

To illustrate the possible modes of polymerization of these compounds, consider the following reactions of isoprene:

1. 1,2- and 3,4-Polymerizations. As far as the polymer chain backbone is concerned, these compounds could just as well be mono-olefins, since the second double bond is relegated to the status of a substituent group. Because of the reactivity of the latter, however, it might become involved in cross-linking reactions. For isoprene, 1,2- and 3,4-polymerizations yield different products:

$$CH_2=\overset{\overset{\textstyle CH_3}{|}}{C}-CH=CH_2$$

(1.D)

$$\begin{array}{c} \overset{\overset{\textstyle CH_3}{|}}{-CH_2-\underset{\underset{\textstyle CH_2}{\underset{\|}{CH}}}{C}-}_n \end{array}$$

$$\begin{array}{c} \overset{}{-CH_2-\underset{\underset{\textstyle CH_2}{\underset{\|}{C-CH_3}}}{CH}-}_n \end{array}$$

[XIII]   [XIV]

These differences do not arise from 1,2- or 3,4-polymerization of butadiene. Structures [XIII] and [XIV] can each exhibit the three different types of tacticity, so a total of six structures can result from this monomer when only one of the olefin groups is involved in the backbone formation.

2. 1,4-Polymerization. This mode of polymerization gives a molecule with double bonds along the backbone of the chain. Again using isoprene as the example,

$$n\ CH_2=\overset{\overset{\textstyle CH_3}{|}}{C}-CH=CH_2 \longrightarrow -CH_2-\overset{\overset{\textstyle CH_3}{|}}{C}=CH-CH_2-_n$$

(1.E)

As in all double-bond situations, the adjacent chain sections can be either cis or trans—structures [XV] and [XVI], respectively—with respect to the double bond, producing the following geometrical isomers:

$$\left[\begin{array}{c} \overset{\textstyle CH_3}{\diagdown}\ \ \ \ \overset{\textstyle H}{\diagup} \\ C = C \\ \overset{\diagup}{-CH_2}\ \ \ \ \overset{\diagdown}{CH_2-} \end{array}\right]_n \qquad \left[\begin{array}{c} \overset{\textstyle CH_3}{\diagdown}\ \ \ \ \overset{\textstyle CH_2}{\diagup} \\ C = C \\ \overset{\diagup}{-CH_2}\ \ \ \ \overset{\diagdown}{H} \end{array}\right]_n$$

[XV]   [XVI]

Figure 1.3 shows several repeat units of cis-1,4-polyisoprene and trans-1,4-polyisoprene. Natural rubber is the cis isomer of 1,4-polyisoprene, and gutta-percha is the trans isomer.

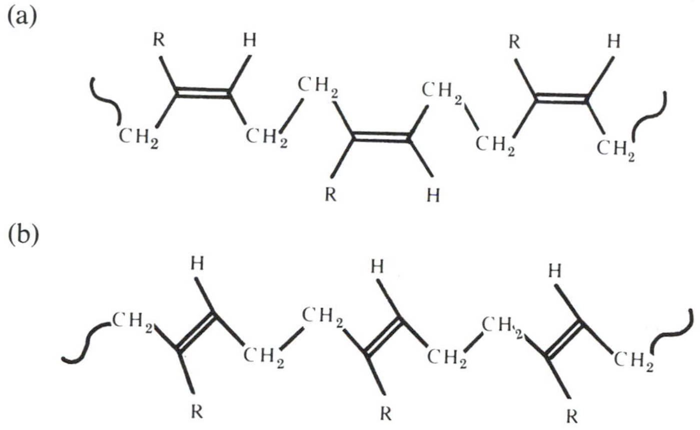

(a)

(b)

Figure 1.3 1,4-polyisoprene with R=CH$_3$ (a) cis isomer: natural rubber (b) trans isomer: gutta-percha.

3. Polymers of chloroprene (structure [XII]) are called neoprene and copolymers of butadiene and styrene are called SBR, an acronym for styrene–butadiene rubber. Both are used for many of the same applications as natural rubber. Chloroprene displays the same assortment of possible isomers as isoprene; the extra combinations afforded by copolymer composition and structure in SBR offsets the fact that structures [XIII] and [XIV] are identical for butadiene.

4. Although the conditions of the polymerization reaction may be chosen to optimize the formation of one specific isomer, it is typical in these systems to have at least some contribution of all possible isomers in the polymeric product, except in the case of polymers of biological origin, like natural rubber and gutta-percha.

   Throughout the past few sections we have ignored the uniqueness of chain ends and focused attention on the repeating portion of the chain. Generally speaking, this will be our attitude throughout this book. An exception is the material of the next section, where we consider the quantitative analysis for end groups as a method for polymer molecular weight determination.

1.7 Molecular Weight by End Group Analysis

The high molecular weight of a polymer is one of the most immediate consequences of the chain structure of these molecules. As indicated in Sec. 1.2, it is also the basis for describing the size of the polymer molecule, either directly or through the degree of polymerization. Most methods for the determination

of the molecular weight of a polymer depend on the properties of polymer solutions. These methods will be taken up in Part III. In this section, however, we shall at least introduce the topic of molecular weight determination by considering those methods based on end group analysis.

The terminal groups of a polymer chain are different in some way from the repeat units that characterize the rest of the molecule. If some technique of analytical chemistry can be applied to determine the number of these end groups in a polymer sample, then the average molecular weight of the polymer is readily evaluated. In essence, the concept is no different than the equivalent procedure applied to low molecular weight compounds. The latter is often included as an experiment in general chemistry laboratory classes. The following steps outline the experimental and computational essence of this procedure:

1. The mass of the sample is determined. Only an analytical balance is required for this.
2. A suitable functional group is assayed in the same sample. In general chemistry and many polymer applications, this is merely the titration of acid groups with a base, or vice versa. Note that only volumetric glassware and a method for end point determination are required to do this.
3. From the volume and concentration of the base, the number of equivalents of the neutralized acid group is readily calculated.
4. The number of grams in a sample divided by the number of equivalents in the same sample gives the gram-equivalent weight of the material.
5. If the number of equivalents per mole is known, the molecular weight is calculated from the equivalent weight by multiplying the latter by the number of equivalents per mole.
6. The method of end group analysis for molecular weight determination is not only simple to understand, but can also be done with ordinary laboratory equipment in many instances.

One limitation of this method that should immediately come to mind is the restriction to relatively low molecular weight polymers. This is a corollary of the fact that chain ends are inconsequential for very long chains. Hence, the sensitivity of the method decreases as the molecular weight of the polymer increases. As a general rule, molecular weights in the neighborhood of 25,000 represent the upper limit for applicability of this method.

Condensation polymers such as polyesters and polyamides are especially well suited to this method of molecular weight determination. For one thing, the molecular weight of these polymers is usually less than for addition polymers. Even more pertinent to the method is the fact that the chain ends in these molecules consist of unreacted functional groups. Using polyamides as an example, we can readily account for the following possibilities:

1. A linear molecule has a carboxyl group at one end and an amino group at the other, such as poly(ϵ-caprolactam):

$$HOOC-(CH_2)_5 \{ NHCO-(CH_2)_5 \}_n NH_2$$

In this case there is one functional group of each kind per molecule, and each could titrated as a double check of the method.

2. If a polyamide is prepared in the presence of a large excess of diamine, the average chain will be capped by an amine group at each end:

$$H_2N-R\{NHCO-R\}_n NH_2$$

In this case only the amine can be titrated, and two ends are counted per molecule.

3. If a polyamide is prepared in the presence of a large excess of dicarboxylic acid, the average chain will have a carboxyl group at each end:

$$HOOC-R\{CONH-R\}_n COOH$$

Only acid groups are titrated and two ends are counted per molecule.

The foregoing examples point out certain critical aspects of the end group method for molecular weight determination. At least for certain functional groups and down to a certain limit of concentration, end groups can be analyzed. Proceeding from this information to the molecular weight of the polymer requires either knowing or assuming something about the nature of the chain ends in the specific sample under consideration. The alternatives enumerated above emphasize that this must be considered explicitly for each sample. In addition to the variations listed above, the following points should also be considered:

1. Have any amino groups been acylated by reaction with an acid catalyst?
2. Has any decarboxylation occurred as a result of elevated temperatures?
3. Have any ring structures formed from the reaction of two ends of the same molecule?
4. Can branching be ruled out, since it would obviously make the number of chain ends per molecule an unknown quantity?

Comparable but equally specific considerations must be applied to other condensation polymer systems. The following example is an illustration of the application of these ideas to the molecular weight of polyamides.

Example 1.4

Aqueous caprolactam is polymerized alone and in the presence of sebacic acid (S) or hexamethylenediamine (H).[†] After a 24-hr reaction time, the polymer is isolated and the end groups are analyzed by titrating the carboxyl groups with KOH in benzyl alcohol and the amino groups with p-toluenesulfonic acid in trifluoroethanol. The number of milliequivalents of carboxyl group per mole caprolactam converted to polymer, [A*], and the number of milliequivalents of amino groups per mole caprolactam converted to polymer, [B*], are given below for three different runs:

Additive	[A*]	[B*]
None	12.0	11.6
S	39.9	2.4
H	1.87	35.1

Calculate the molecular weights in each case and comment briefly on the results.

Solution

Note the units of the end group concentrations [A*] and [B*]: milliequivalents per mole polymerized monomer. The reciprocal is therefore moles of monomer per milliequivalent of end groups. Accordingly, the molecular weight is given by

$$M = \frac{\text{moles caprolactam}}{\text{mEq specific end group}} \times \frac{113 \text{ g caprolactam}}{\text{moles caprolactam}} \times \frac{1000 \text{ mEq}}{1 \text{ Eq}}$$

$$\times \frac{\text{number of equivalents of specific end groups}}{\text{polymer molecule}}$$

Only the last factor is a little tricky; it is also different with and without additives. With no additive, polycaprolactam can be represented A*BABAB ... ABAB*, where the A and B are acid and base groups, respectively, and those marked with the asterisk are those analyzed. Thus every molecule has one of each. In this case, then, we use the average of 12.0 and 11.6 as the end group concentration, and unity as the number of ends of each kind to obtain

---

[†]H. K. Reimschussel and G. J. Dege, *J. Polym. Sci.* 9:2343 (1971).

$$M = \frac{1}{11.8} \times 113 \times 1000 \times 1 = 9580 \text{ g mol}^{-1}.$$

With either S or H as additives, two different kinds of chain are present: A*BAB ... ABAB* and A*BABAB ... ABABAA* with S or B*BABAB ... ABAB* with H. In these cases the total number of acid and base groups counts the total number of chain ends. There are two equivalents of total chain ends per mole of chains. Therefore, for S, [total ends] = 39.9 + 2.4 = 42.3 and

$$M = \frac{1}{42.3} \times 113 \times 1000 \times 2 = 5340 \text{ g mol}^{-1}$$

and for H, [total ends] = 1.87 + 34.1 = 37.0 and

$$M = \frac{1}{37.0} \times 113 \times 1000 \times 2 = 6110 \text{ g mol}^{-1}$$

Several comments come to mind:

1. As polymers go, none of these is a particularly high molecular weight.
2. No consideration is made of the fact that S and H contribute to M a molecular weight different from M_0.
3. The results of end group analyses must be examined on a system-by-system basis for correct interpretation.
4. The presence of monomers with two functional groups of the same kind limits chain growth and decreases the molecular weight.
5. The molecular weights obtained by this method are averages. This is particularly evident from the situations where additives are present. In these cases, two different kinds of chains result, with those terminated by the same end group being stunted in growth compared to the normal polycaprolactam. Yet it is the total weight of polymer and the total number of ends that are used to evaluate M. This must result in an average molecular weight.

•

In principle, the same concept could be used to determine the molecular weights of materials from any class of polymers, and not just condensation polymers. In practice, the method is less suitable for polymers of the addition type, since these generally have considerably higher molecular weights and less-certain end groups. Modern instrumental techniques, especially spectroscopic techniques, sometimes have sufficient sensitivity to offset the lower concentration of chain ends in addition polymers. The use of radioisotopes in this connection also extends the range of molecular weights which can be determined by this approach. No general principles beyond those already discussed are involved in the more sophisticated methods. As in the cases discussed here, specific consideration must be given to the number of measured chain ends per molecule as well as to the specific analytical chemistry associated with the experimental method employed.

Note that the method of end group analysis is inapplicable to copolymers, since the presence of more than one repeat unit adds extra uncertainty as to the nature of chain ends. The above example included the remark that the molecular weights calculated in the example were average values. In the next section we shall examine this point in greater detail.

1.8 Molecular Weight Averages

The example in the last section argued that polymers of more than one molecular weight must be present when a polymer is prepared in the presence of an additive with two functional groups of the same kind. While this is true, it presents a far too limited explanation for the presence of chains with a distribution of molecular weights in all polymer preparations. We describe this state of affairs by saying that the polymer shows polydispersity with respect to molecular weight or degree of polymerization. Some degree of polydispersity is the normal state for all polymer systems. To see how this comes about, we only need to think of the reactions between monomers that lead to the formation of polymers in the first place. Random encounters between reactive species are responsible for chain growth, so statistical descriptions are appropriate for the resulting product. The situation is reminiscent of the distribution of molecular velocities in a sample of gas. In that case, also, random collisions impart extra energy to some molecules while reducing the energy of others. Only the average energy makes sense in this latter situation and only the average molecular weight makes sense for polymers.

In Chaps. 5 and 6 we shall examine the distribution of molecular weights for condensation and addition polymerizations in some detail. For the present, our only concern is how such a distribution of molecular weights is described. The standard parameters used for this purpose are the mean and standard deviation of the distribution. Although these are well-known quantities, many students are familiar with them only as results provided by a calculator. Since statistical considerations play an important role in several aspects of polymer chemistry, it is appropriate to digress into a brief examination of the statistical way of describing a distribution.

Suppose that a variable of interest, the molecular weight in the present context, displays an assortment of values. Then it is convenient to divide the observed range of molecular weights into 10–20 categories, called classes, and to hypothetically sort the molecules by molecular weight into these different classes. We choose the molecular weight at the center of each class as typifying that class and call that value the class mark. Thus for category 1, the class mark has the value M_1; for category 2, M_2, and so on. After all of the molecules have been sorted into these molecular weight classes, we imagine counting the number that have accumulated in each category. Accordingly, we might find

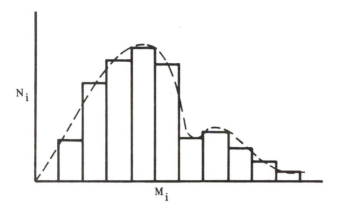

Figure 1.4 Histogram showing the number of molecules N_i having the molecular weight M_i for classes indexed i. The broken line shows how the distribution would be described by a continuous function.

N_1 in class 1, N_2 in class 2, and so forth. In class i there will be N_i molecules of molecular weight M_i. One way of representing the results of this type of inventory procedure is to construct a bar graph called a histogram, like that shown in Fig. 1.4. If we were doing an actual classification instead of a hypothetical one—say, classifying beans instead of molecules—we would lose detail if we used less than about 10 categories, and the effort of classification would be prohibitive if more than about 20 categories were used. As Fig. 1.4 illustrates, however, a number of classes in the range 10–20 does give a reasonable picture of the distribution of M values. Furthermore, as the number of classes increases with each spanning a narrower range, the resulting histogram comes closer and closer to outlining a smooth curve. In the limiting case each class becomes one of the infinitesimally thin slices of calculus. At one extreme, then, the histogram enables us to classify actual data, and at the other it leads to the mathematics of continuous distribution functions. The broken line in Fig. 1.4 shows how a continuous function might look which describes the same system as the histogram approximates. In Chap. 8 we shall see that polymer preparations may be fractionated into molecular weight classes so that a histogram representation of the original distribution is a possibility and not merely a hypothetical exercise.

On the basis of the concepts and notation introduced here, we see that there are several ways of describing a polydisperse system:

1. A continuous distribution function is a mathematical function which gives N as a function of M. This is the most general way of describing the

distribution, since, in principle, all other aspects of the polydispersity can be derived from the continuous distribution function.

2. The histogram is a graphical device which is both attainable in practice and also an approximation to a theoretical distribution function.

3. The mean can be evaluated from the classified data of the histogram; it measures the center of the distribution. The mean (whose symbol is an overbar) is defined as

$$\overline{M}_n = \frac{\Sigma_i N_i M_i}{\Sigma_i N_i} \tag{1.3}$$

This quantity is also called the number (subscript n) average molecular weight.

4. The standard deviation can also be evaluated from the same classified data; it measures the width of the distribution. The standard deviation σ is defined as

$$\sigma = \left(\frac{\Sigma_i N_i (M_i - \overline{M})^2}{\Sigma_i N_i} \right)^{\frac{1}{2}} \tag{1.4}$$

Note that σ^2 has the significance of being the mean value of the square of the deviations of individual M_i values from the mean $\overline{M}$. Accordingly, σ is sometimes called the root mean square (rms) deviation.

In both Eqs. (1.3) and (1.4), the summations are carried out over all classes of data. From a computational point of view, standard deviation may be written in a more convenient form by carrying out the following operations. First both sides of Eq. (1.4) are squared; then the difference $M_i - \overline{M}$ is squared to give

$$\sigma^2 = \frac{\Sigma_i N_i M_i^2}{\Sigma_i N_i} - 2\overline{M} \frac{\Sigma_i N_i M_i}{\Sigma_i N_i} + \overline{M}^2 \tag{1.5}$$

Recalling the definition of the mean, we recognize the first term on the right-hand side of Eq. (1.5) to be the mean value of M^2 and write

$$\sigma^2 = \overline{M^2} - 2\overline{M}^2 + \overline{M}^2 \tag{1.6}$$

It is important to realize that $\overline{M^2} \neq \overline{M}^2$. An alternative to Eq. (1.4) as a definition of standard deviation is, therefore,

$$\sigma = (\overline{M^2} - \overline{M}^2)^{\frac{1}{2}} \tag{1.7}$$

We shall make use of this relationship presently.

Since $\Sigma_i N_i$ represents the total number of molecules N_t in the population we are describing, each of the coefficients in Eqs. (1.3) and (1.4) is the fraction f_i of the total number of molecules in category i:

$$f_i = \frac{N_i}{N_t} \tag{1.8}$$

Introducing this notation means that Eqs. (1.3) and (1.4) may be written as

$$\bar{M} = \Sigma_i f_i M_i \tag{1.9}$$

and

$$\sigma = [\Sigma_i f_i(M_i - \bar{M})^2]^{\frac{1}{2}} \tag{1.10}$$

where the fractions f_i are the weighting factors used in the definition of the average. In the mean and standard deviation, the number fraction is the weighting factor involved.

Finally we define a quantity known as the kth moment of the distribution. In terms of molecular weight,

$$\text{kth moment} = \Sigma_i f_i (M - M_s)^k \tag{1.11}$$

The numerical value of the exponent k determines which moment we are defining, and we speak of these as moments about the value chosen for M_s. Thus the mean is the first moment of the distribution about the origin ($M_s = 0$) and σ^2 is the second moment about the mean ($M_s = \bar{M}$). The statistical definition of moment is analogous to the definition of this quantity in physics. When $M_s = 0$, Eq. (1.11) defines the average value of M^k; this result was already used in writing Eq. (1.6) with k = 2.

Throughout this discussion we have used the numerical fraction of molecules in a class as the weighting factor for that portion of the population. This restriction is not necessary; some other weighting factor could be used equally well. As a matter of fact, one important type of average encountered in polymer chemistry is the case where the mass fraction of the ith component is used as the weighting factor. Defining the mass of material in the ith class as m_i, we write

$$\bar{M}_w = \frac{\Sigma_i m_i M_i}{\Sigma_i m_i} \tag{1.12}$$

This quantity is called the weight average molecular weight, reflecting the chemist's customary carelessness about distinguishing between mass and weight, and is given the symbol $\bar{M}_w$. By contrast, the mean, where number fractions are used, is called the number average molecular weight and is given the symbol $\bar{M}_n$.

The mass of material in a particular molecular weight class is given by the product of the class mark molecular weight and the number of molecules in the class:

$$m_i = N_i M_i \qquad (1.13)$$

Substituting this result into Eq. (1.12) gives

$$\bar{M}_w = \frac{\Sigma_i N_i M_i^2}{\Sigma_i N_i M_i} = \frac{\Sigma_i f_i M_i^2}{\Sigma_i f_i M_i} \qquad (1.14)$$

which shows that the weight average molecular weight may also be regarded as the ratio of the second moment of the distribution to the first moment, where each of these moments are taken with respect to the origin of the distribution.

The weight average molecular weight of a distribution will always be greater than the number average. This is true because the latter merely counts the contribution of molecules in each class, whereas the former weights their contribution in terms of mass. Thus those molecules with higher molecular weights contribute relatively more to the average when mass fraction rather than number fraction is used as the weighting factor. For all polydisperse systems

$$\frac{\bar{M}_w}{\bar{M}_n} > 1 \qquad (1.15)$$

and the amount by which this ratio deviates from unity is a measure of the polydispersity of a sample. In the event that all of the molecules in a sample have the same molecular weight, the summations in Eqs. (1.3) and (1.4) would each consist of a single term and their ratio would equal unity. Such a sample is said to be monodisperse.

In connection with Eq. (1.4), we noted that the standard deviation measures the spread of a distribution; now we see that the ratio $\bar{M}_w/\bar{M}_n$ also measures this polydispersity. The relationship between these two different measures of polydispersity is easily shown. Equation (1.14) may be written as

$$\Sigma_i f_i M_i^2 = \bar{M}_w \Sigma_i f_i M_i = \bar{M}_w \bar{M}_n \qquad (1.16)$$

The left-hand side of this expression equals $\overline{M^2}$, so Eq. (1.16) may also be written as

$$\overline{M^2} = \bar{M}_w \bar{M}_n \qquad (1.17)$$

Substituting this result into Eq. (1.7) gives

$$\sigma = (\bar{M}_w \bar{M}_n - \bar{M}^2)^{1/2} = \bar{M}_n \left(\frac{\bar{M}_w}{\bar{M}_n} - 1 \right)^{1/2} \tag{1.18}$$

This result shows that the square root of the amount by which the ratio $\bar{M}_w/\bar{M}_n$ exceeds unity equals the standard deviation of the distribution relative to the number average molecular weight. Thus if a distribution is characterized by $\bar{M}_n = 10,000$ and $\sigma = 3000$, then $\bar{M}_w/\bar{M}_n = 1.09$. Alternatively, if $\bar{M}_w/\bar{M}_n = 1.50$, then the standard deviation is 71% of the value of $\bar{M}_n$. This shows that reporting the mean and standard deviation of a distribution or the values of M_n and $\bar{M}_w/\bar{M}_n$ gives equivalent information about the distribution. We shall see in a moment that the second alternative is more easily accomplished for samples of polymers. First, however, consider the following example in which we apply some of the equations of this section to some numerical data.

Example 1.5

The first and second columns of Table 1.4 give the number of moles of polymer in six different molecular weight fractions. Calculate $\bar{M}_n$ and $\bar{M}_w$ for this polymer and evaluate σ using both Eqs. (1.7) and (1.18).

Solution

Evaluate the product $n_i M_i$ for each class; this is required for the calculation of both $\bar{M}_n$ and $\bar{M}_w$. Values of this quantity are listed in the third column of Table 1.4. From $\Sigma_i n_i M_i$ and $\Sigma_i n_i$, $\bar{M}_n = 734/0.049 = 15,000$. The matter of significant figures will not be strictly adhered to in this example.

The products $m_i M_i$ are mass-weighted contributions and are listed in the fourth column of Table 1.4. From $\Sigma_i m_i$ and $\Sigma_i m_i M_i$, $\bar{M}_w = 113 \times 10^5/734 = 15,400$.

The ratio $\bar{M}_w/\bar{M}_n$ is found to be $15,400/15,000 = 1.026$ for these data. Using Eq. (1.18), we have $\sigma/\bar{M}_n = (1.026 - 1)^{1/2} = 0.162$ or $\sigma = 0.162(15,000) = 2430$.

To evaluate σ via Eq. (1.7), differences between M_i and $\bar{M}$ must be considered. The fifth and sixth columns in Table 1.4 list $(M_i - \bar{M}_n)^2$ and $N_i (M_i - \bar{M}_n)^2$ for each class of data. From $\Sigma_i N_i$ and $\Sigma_i N_i (M_i - \bar{M}_n)^2$, $\sigma^2 = 28.1 \times 10^4/0.049 = 5.73 \times 10^6$, and $\sigma = 2390$.

The discrepancy between the two values of σ is not meaningful in terms of significant figures. The standard deviation is 2400.

We shall see that, as polymers go, this is a relatively narrow molecular weight distribution.

•

Table 1.4 Some Classified Molecular Weight Data for a Hypothetical Polymer Used in Example 1.5

N_i (mol)	M_i ($g\,mol^{-1}$)	m_i (g)	$m_i M_i \times 10^{-5}$ ($g^2\,mol^{-1}$)	$(M_i - \bar{M})^2 \times 10^{-6}$ ($g^2\,mol^{-2}$)	$N_i(M_i - \bar{M})^2 \times 10^{-4}$ ($g^2\,mol^{-1}$)
0.003	10,000	30	3.0	25	7.50
0.008	12,000	96	11.5	9	7.20
0.011	14,000	154	21.6	1	1.10
0.017	16,000	272	43.5	1	1.70
0.009	18,000	162	29.2	9	8.10
0.001	20,000	20	4.0	25	2.50
$\Sigma = 0.049$		$\Sigma = 734$	$\Sigma = 113$	$\Sigma = 70$	$\Sigma = 28.10$

The preceding discussion and example are based on the premise that classified molecular weight data are available. While this is sometimes the case, the average molecular weight of a polydisperse system is usually the information available to characterize the sample. The significant thing about this, however, is the fact that different experimental techniques yield different averages. We shall see in Chap. 8, for example, that osmotic pressure experiments can be interpreted to give the number average molecular weight; in Chap. 10 we shall see that light scattering produces a weight average. Hence these different experimental methods applied to the same sample will provide $\bar{M}_n$ and $\bar{M}_w$, thus yielding some statistical information about the molecular weight distribution.

It is important to realize that this difference in averaging is an intrinsic part of the way different experimental methods "see" a polydisperse system. Accordingly, osmometry effectively counts while light scattering effectively weighs particles of different molecular weight. The overall value of the measured property is the average effect produced by the polydisperse system, with the weighting factors which are appropriate to the specific method. When the measured properties are interpreted in terms of the average polymer responsible for the effect, the difference in averaging procedure shows up in the result. A light-scattering experiment and an osmotic pressure experiment conducted on the same sample will produce two different molecular weights, with light scattering producing the higher value. If we were not aware of this situation, we might very well blame the difference on experimental error instead of recognizing that it is the natural consequence of polydispersity.

Table 1.5 lists the different molecular weight averages most commonly encountered in polymer chemistry. Table 1.5 also includes the definition of these averages for easy reference, some experimental methods that produce them, and cross-references to sections of this volume where the specific techniques are discussed. Note that end group analysis produces a number average molecular weight, since it is a technique based on counting. This is especially evident when we compare end group analysis with the procedure for evaluating $\bar{M}_n$ in Example 1.5. In the latter, $\bar{M}_n$ is given by dividing the total mass of the sample by the total number of moles of polymer it contains. This is exactly what is done in end group analysis.

Table 1.5 also includes two additional types of molecular weight average besides those already discussed. The following remarks describe some features of these two, the z-average molecular weight, and the viscosity average:

1.   The z-average molecular weight is defined by the equation

$$\bar{M}_z = \frac{\Sigma_i N_i M_i^3}{\Sigma_i N_i M_i^2} \qquad (1.19)$$

Table 1.5   Summary of the Molecular Weight Averages Most Widely Encountered in Polymer Chemistry

Average	Definition	Methods	Section
$\bar{M}_n$	$\dfrac{\Sigma_i N_i M_i}{\Sigma_i N_i}$	Osmotic pressure and other colligative properties End group analysis	8.8 1.7
$\bar{M}_w$	$\dfrac{\Sigma_i N_i M_i^2}{\Sigma_i N_i M_i}$	Light scattering Sedimentation velocity	10.7 9.10
$\bar{M}_z$	$\dfrac{\Sigma_i N_i M_i^3}{\Sigma_i N_i M_i^2}$	Sedimentation equilibrium	9.10
$\bar{M}_v$	$\left(\dfrac{\Sigma_i N_i M_i^{1+a}}{\Sigma_i N_i M_i}\right)^{1/a}$	Intrinsic viscosity	9.5

From this definition, we see it is the ratio of the third moment of the distribution about the molecular weight origin to the second moment about the origin.

2. The weighting factors used in the determination of the z-average molecular weight are $N_i M_i^2$, which means that higher molecular weight molecules are weighted even more heavily in this average than is the case in the weight average. For the same distribution, the order of the averages is $\bar{M}_z > \bar{M}_w > \bar{M}_n$.

3. The viscosity average is defined by the equation

$$\bar{M}_v = \left(\frac{\Sigma_i N_i M_i^{a+1}}{\Sigma_i N_i M_i}\right)^{1/a} \tag{1.20}$$

where the exponent a is characteristic of the system under investigation and generally lies in the range $0.5 < a < 1.0$. Note that $\bar{M}_v = \bar{M}_w$ when $a = 1$.

4. The viscosity average molecular weight is not an absolute value, but a relative molecular weight based on prior calibration with known molecular weights for the same polymer–solvent–temperature conditions. The parameter a depends on all three of these; it is called the Mark-Houwink exponent, and tables of experimental values are available for different systems.

5. For the viscosity average molecular weight, $\bar{M}_v^a$ may be considered the ratio of the $(a + 1)$th moment to the first moment of the distribution. Alternatively, $\bar{M}_v^a$ may be viewed as the weight average value of M^a in view of Eq. (1.13).

Note that a statistical study could be done on an electron micrograph like that shown in Fig. 1.1. The dimensions of the blobs could be converted to volumes and then to masses with a knowledge of the density of the deposited polymer. This approach could be organized into a table of classified data from which any of these averages could be calculated.

1.9   The Binomial Distribution and the Random Walk

In this section, we shall examine another application of statistics to polymer chemistry. This time we shall consider the average spatial dimensions of an isolated polymer molecule, especially with regard to its dependence on molecular weight. In this context, we disregard the fact that polymer samples show polydispersity: The distribution of dimensions we are concerned with would be present even if all polymer chains were identical in length. This is because successive chain units are able to rotate along the bonds of the backbone to acquire a jumbled conformation. No two molecules have identical shapes or dimensions, even if the sample is monodisperse. Likewise, the conformation of any particular molecule continuously changes with time owing to thermal fluctuations. Only a statistical description is adequate to characterize this situation.

As a consequence of these various possible conformations, the polymer chains exist as coils with spherical symmetry. Our eventual goal is to describe these three-dimensional structures, although some preliminary considerations must be taken up first. Accordingly, we begin by discussing a statistical exercise called a one-dimensional random walk.

We start this exercise by considering the placement of n successive repeat units in a polymer chain along a straight line, say, the x axis. We assume that the chain is perfectly flexible and that it excludes a negligible volume so that more than one repeat unit can be placed on the same site. We anchor one end of the chain at the origin of the axis and propose to use the toss of a coin to decide on the placement of successive units; that is, if the coin turns up heads (subscript H), we place the next unit one step ahead—an increment of +1 in the x direction, where 1 is the length of the repeat unit. If the coin shows tails (subscript T), the next unit is placed back a step for a change in x value of −1. How far from the origin will the other end of the chain lie after n units are placed on the basis of this random walk? A possible sequence for 10 tosses might be HTTHHTHHTH in a particular exercise; this amounts to 6 steps forward and 4 backward for a net displacement of +2. In this example, the two ends of the chain are separated by only 2 1, even though the fully extended chain would have a length of 10 1. Since the outcome of 10 tosses could be different in another trial of the same exercise, it is clear that we must turn to statistics to describe the average placement of the chain.

For a one-dimensional random walk, the probability of n_H heads after n moves is supplied by application of the bionomial distribution formula:

$$P(n_H, n) = \frac{n!}{n_H! \, n_T!} \; p_H{}^{n_H} p_T{}^{n_T} \qquad (1.21)$$

in which the p's are the probabilities of either a head or tail in a single toss and the subscripted n's are the number of heads and tails in the specific exercise. For a fair coin $p_H = p_T = \frac{1}{2}$, but for the time being we shall continue using the more general formula.

The binomial distribution function is one of the most fundamental equations in statistics and finds several applications in this volume. To be sure that we appreciate its significance, we make the following observations about the plausibility of Eq. (1.21):

1. To evaluate the probability that one event *and* another will occur, we multiply the probabilities of the individual events. Thus the probability of tossing two heads is $p_H{}^2$ and that of tossing n_H heads is $p_H{}^{n_H}$.

2. Likewise, the probability of tossing n_T tails is $p_T{}^{n_T}$. The probability of tossing n_H heads and n_T tails is $p_H{}^{n_H} p_T{}^{n_T}$ by the same principle.

3. The probability calculated so far is too low because it describes one specific sequence of heads and tails. From the point of view of net displacement, the sequence does not matter. Hence the above results must be multiplied by the number of different ways this outcome can arise. Instead of tossing one coin n times, we could toss n coins drawn at random from a piggy bank. For the first, we have a choice of n to draw from; for the second, n − 1; for the third, n − 2, and so on. The total possible ways the toss could be carried out is given by the product of these different choices, that is by n!

4. This suggests that $n! p_H{}^{n_H} p_T{}^{n_T}$ gives the desired probability, but now we have gone too far in the opposite direction and overcounted the probability. To appreciate this fact, we recognize that among the n! ways the coins could be tossed, we have included a number of ways that yield the same net outcome achieved through different sequences of tosses. For example, the n! count would include HHHTTT, HTHTHT, and HTTHHT as different, although they each consist of the same number of heads and tails.

5. To correct for this overcounting, we cancel out the number of ways n_H heads can be permuted and the number of ways n_T tails can be permuted. Using the same logic as in item (3), these redundant possibilities are given by $n_H!$ and $n_T!$, respectively. Dividing the result in item (4) by these factorials gives Eq. (1.21).

If we apply Eq. (1.21) to the problem of determining the number of heads occurring in 10 tosses, we find that the probability of, say, 6 heads is given by $(10!/6!4!) \frac{1}{2}^{10} = 0.205$. That is, heads would occur in 6 of every 10 tosses about 20% of the time. If a large number of 10-mer molecules were being positioned along an axis according to this hypothetical procedure, in about 20% of the cases the end-to-end distance would be 2 l. For an outcome of 8 heads and 2 tails, corresponding to a net forward displacement of 6 steps or an end-to-end distance of 6 l, the probability is $(10!/8!2!) \frac{1}{2}^{10} = 9.77 \times 10^{-4} \cong 0.1\%$.

It is an easy matter to rewrite Eq. (1.21) in terms of the probability of a displacement x occurring after n tosses. We recognize the following:

1. Each toss is either a head or a tail:

$$n = n_H + n_T \tag{1.22}$$

2. The displacement x after n steps of length l is

$$x = (n_H - n_T)l \tag{1.23}$$

3. Solving Eqs. (1.22) and (1.23) for n_H and n_T gives

$$n_H = \frac{1}{2}(n + x/l) \tag{1.24}$$

and

$$n_T = \frac{1}{2}(n - x/l) \tag{1.25}$$

4. For a fair coin $p_H = p_T = \frac{1}{2}$; there Eq. (1.21) becomes

$$P(n_H, n) = \frac{n!}{[(n + x/l)/2]! \, [(n - x/l)/2]!} \, \frac{1}{2}^n \tag{1.26}$$

This result enables us to calculate the probability of any specified outcome for the one-dimensional random walk. We shall continue to develop this one-dimensional relationship somewhat further, since doing so will produce some useful results.

5. For high molecular weight polymers, n is large and the logarithm of large factorials is accurately given by Sterling's approximation,

$$\ln y! \cong y \ln y - y \tag{1.27}$$

for large y.

6. Taking logarithms of both sides in Eq. (1.26) and applying Sterling's approximation gives

$$-\ln P(x, n) = \frac{nl + x}{2l} \ln \left(1 + \frac{x}{nl}\right) + \frac{nl - x}{2l} \ln \left(1 - \frac{x}{nl}\right) \qquad (1.28)$$

7. The number of steps is always much larger than the displacement x, since there is a good deal of back-and-forth cancellation. Hence the ratio x/nl is less than unity and the logarithms may be approximated by the leading terms of a series expansion

$$\ln \left(1 + \frac{x}{nl}\right) \cong \frac{x}{nl} - \frac{1}{2} \left(\frac{x}{nl}\right)^2 + \cdots \qquad (1.29)$$

in which x can be either positive or negative.

8. Applying Eq. (1.29) to Eq. (1.28) gives

$$\ln P(x, n) \cong - \frac{x^2}{2nl^2} + \cdots \qquad (1.30)$$

or

$$P(x, n) = k \exp \left(\frac{-x^2}{2nl^2}\right) \qquad (1.31)$$

where the factor k is a constant called a normalization factor.

9. Well-behaved probability functions total unity when they are summed over all possible outcomes. Since Eq. (1.31) is a continuous function—this has been accomplished by getting rid of the factorials—this sum may be written as an integral over all possible values of x:

$$k \int_{-\infty}^{\infty} \exp \left(\frac{-x^2}{2nl^2}\right) dx = 1 \qquad (1.32)$$

A probability function which satisfies this criterion is said to be normalized. This will be accomplished when a value of k which satisfies Eq. (1.32) is found.

10. The integral in Eq. (1.32) is known as a gamma function and may be found in tables of integrals. The result of the integration is that

$$k = (2\pi nl^2)^{-\frac{1}{2}} \qquad (1.33)$$

Therefore an expression which is equivalent to Eq. (1.21) for the case of large n's is

$$P(x, n) = (2\pi n l^2)^{-\frac{1}{2}} \exp\left(\frac{-x^2}{2n l^2}\right) \tag{1.34}$$

Since we have ended up with a continuous distribution function, it is more appropriate to multiply both sides of Eq. (1.34) by dx and to say that the equation gives the probability of x values between x and x + dx for n steps of length l.

In the next section we shall adapt this probability function to the description of a three-dimensional coil. We conclude this section by noting that Eq. (1.21) may be approximated by two other functions which are used elsewhere in this book. For these general relationships we define ν to be the number of successes— that is, some specified outcome such as tossing a head—out of n tries and define p as the probability of success in a single try. In this amended notation, Eq. (1.21) becomes

$$P(\nu, n) = \frac{n!}{\nu!(n - \nu)!} \, p^\nu (1 - p)^{n-\nu} \tag{1.35}$$

In terms of this notation we have the following:

1. The average number of successes is given by

$$\bar{\nu} = np \tag{1.36}$$

and the standard deviation of the distribution of successes is

$$\sigma = [np(1 - p)]^{\frac{1}{2}} \tag{1.37}$$

2. If n is large and p is very small, Eq. (1.35) is approximated by the Poisson distribution

$$P_{Pois} = \frac{e^{-\bar{\nu}} \, \bar{\nu}^\nu}{\nu!} \tag{1.38}$$

3. As an example of the Poisson distribution, consider the case of n = 1000, p = 0.01, and ν = 5. According to the binomial distribution, P_{bin} = (1000!/995!5!)(0.01)^5(0.99)^{995} = 0.0375. By Eq. (1.36), $\bar{\nu}$ = (0.01)(1000) = 10 and P_{Pois} = e^{-10} 10^5/5! = 0.0378.

4. If n is large, but without any particular restriction on p, Eq. (1.35) is approximated by the Gaussian or normal distribution

$$P_{norm} = \frac{1}{(2\pi)^{\frac{1}{2}}\sigma} \exp\left[-\frac{1}{2}\left(\frac{\nu - \bar{\nu}}{\sigma}\right)^2\right] \tag{1.39}$$

5. As an example of the normal distribution, consider the case of $n = 60$, $p = 0.20$, and $\nu = 10$. According to the binomial distribution, $P_{bin} = (60!/\ 50!10!)(0.2)^{10}(0.8)^{50} = 0.110$. By Eq. (1.36), $\bar{\nu} = (0.2)(60) = 12$, and by (1.37), $\sigma = [(60)(0.2)(0.8)]^{\frac{1}{2}} = 3.10$, therefore

$$P_{norm} = \frac{1}{3.10\,(2\pi)^{\frac{1}{2}}} \exp\left[-\frac{1}{2}\left(\frac{10 - 12}{3.10}\right)^2\right] = 0.105$$

6. The proof that these expressions are equivalent to Eq. (1.35) under suitable conditions is found in statistics textbooks. We shall have occasion to use the Poisson approximation to the binomial in discussing crystallization of polymers in Chap. 4, and the distribution of molecular weights of certain polymers in Chap. 6. The normal distribution is the familiar bell-shaped distribution that is known in academic circles as "the curve." We shall use it in discussing diffusion in Chap. 9.

1.10 Size Parameters for Random Coils

Next let us apply random walk statistics to three-dimensional chains. We begin by assuming isolated polymer molecules which consist of perfectly flexible chains.

To isolate polymer chains from one another, we consider a solution which is sufficiently dilute that the domains of the individual polymer molecules are well separated from each other. For the present, we assume the solvent has no influence on the polymer but merely supports the molecule. In fact, this is not generally the case, although it can be achieved by proper choice of solvent or temperature.

The chain is imagined to consist of n units connected to one another by perfectly flexible joints. Figure 1.5a shows the ith and $(i + 1)$th unit along the backbone of a vinyl polymer according to this model. The three carbon atoms involved in these bonds are numbered for ease of discussing the model. We focus attention at carbon 2, which we picture as lying at the center of a sphere. By hypothesis, the chain is perfectly flexible at C_2 so that the angle θ between the two bonds can have any value between $0°$ and $360°$. Likewise, regardless of the value of θ, there is complete freedom of rotation around bond i connecting C_1 and C_2; that is, the angle ϕ may take on all values between $0°$ and $360°$.

This kind of perfect flexibility means that C_3 may lie anywhere on the surface of the sphere. According to the model, it is not even excluded from C_1. This model of a perfectly flexible chain is not a realistic representation of an actual polymer molecule. The latter is subject to fixed bond angles and experiences some degree of hindrance to rotation around bonds. We shall consider the effect of these constraints, as well as the effect of solvent–polymer interactions, after we explore the properties of the perfectly flexible chain. Even in this revised model, we shall not correct for the volume excluded by the polymer chain itself.

If a chain consists of m bonds of this sort, then there are m + 1 atoms which can act as centers like C_2 in Fig. 1.5. The values of the angles at each successive center are assumed to be independent, so the chain can twist and bend in an enormous number of conformations along its length. This is how the random coil comes about. Figure 1.5b shows a somewhat longer segment of chain. Each carbon atom has the possibility of being anywhere on the surface of the sphere centered at the preceding carbon according to this model.

The one-dimensional random walk of the last section is readily adapted to this problem once we recognize the following connection. As before, we imagine that one end of the chain is anchored at the origin of a three-dimensional coordinate system. Our interest is in knowing, on the average, what will be the distance of the other end of the chain from this origin. A moment's reflection will convince us that the x, y, and z directions are all equally probable as far as the perfectly flexible chain is concerned. Therefore one-third of the repeat units will be associated with each of the three perpendicular directions

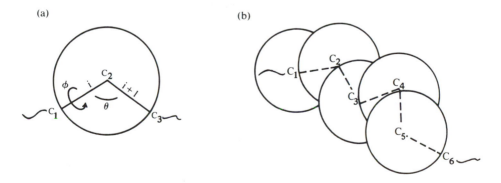

Figure 1.5 Placement of successive polymer segments connected by perfectly flexible joints. In (a), the ith and (i + 1)th bond can be moved through angles ϕ and θ so that carbon 3 can lie anywhere on the surface of a sphere. In (b), the pattern is illustrated for a longer portion of chain.

of the coordinate system. The probability of a displacement x in that direction is given by Eq. (1.34), with n replaced by n/3. Since the x, y, and z directions are equivalent, the same expression holds for $P(y, n/3)$ and $P(z, n/3)$. Thus the probability that the loose end of the chain will be found in a volume element dx dy dz located at a specific set of x, y, z values is the probability that is has *all* of the following:

1. A value of x between x and x + dx.
2. A value of y between y and y + dy.
3. A value of z between z and z + dz.

According to the rules for compounding probabilities, this is given by

$$P(x, y, z, n) \, dx \, dy \, dz = P(x, n/3) \, P(y, n/3) \, P(z, n/3) \, dx \, dy \, dz \qquad (1.40)$$

Substituting Eq. (1.34) with n replaced by n/3 for each of the probabilities in (1.40) gives

$$P(x, y, z, n) \, dx \, dy \, dz = \left(2\pi \frac{n}{3} l^2\right)^{-3/2} \exp\left(-\frac{3(x^2 + y^2 + z^2)}{2nl^2}\right) dx \, dy \, dz$$
$$(1.41)$$

The x, y, and z coordinates of the loose end of the chain can be related to the radial distance r from the origin by

$$x^2 + y^2 + z^2 = r^2 \qquad (1.42)$$

In addition, the volume element of interest is not the box dx dy dz shown in Fig. 1.6a but, rather, a spherical shell of radius r and thickness dr as shown in Fig. 1.6b. The result of expressing the volume element in spherical coordinates and integrating over all angles is the replacement

$$dx \, dy \, dz \longrightarrow 4\pi r^2 \, dr \qquad (1.43)$$

Making these substitutions gives the probability of finding one end of a perfectly flexible chain of n units a distance r from the other end by

$$P(r, n) \, dr = \left(2\pi \frac{n}{3} l^2\right)^{-3/2} 4\pi r^2 \exp\left(\frac{-3r^2}{2nl^2}\right) dr \qquad (1.44)$$

Several features of this expression should be noted:

1. The probability is normalized, since Eq. (1.44) was assembled from separately normalized components.

(a)   (b)

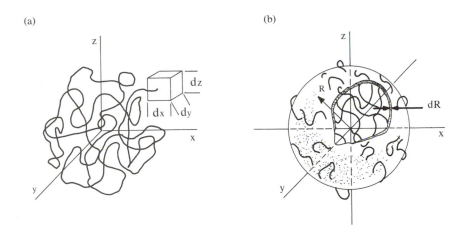

Figure 1.6   A flexible coil attached at the origin at one end and (a) in a volume element dx dy dz at the other end and (b) in a spherical shell of volume $4\pi r^2$ dr. (Reprinted from Ref. 4, p. 116.)

2. The factor containing r^2 increases with increasing r and reflects the fact that there are more locations to place the loose chain end within larger spherical shells, but . . . .
3. The exponential factor decreases with increasing r and reflects the fact that large displacements become decreasingly probable.
4. The overall probability function for the end-to-end distance is the product of these two considerations. Starting at r = 0, the probability increases owing to the r^2 term, passes through a maximum, then decreases as the exponential factor takes over at large r values.
5. This factor is reminiscent of the radial distribution function for electron probability in an atom and the Maxwell distribution of molecular velocities in a gas, both of which pass through a maximum for similar reasons.

With this probability expression, it is an easy matter to calculate the average dimensions of a coil. Because of the back-and-forth character of the x, y, and z components of the random walk, the average end-to-end distance is less meaningful than the average of r^2. The latter squares positive and negative components before averaging and gives a more realistic parameter to characterize the coil. To calculate $\overline{r^2}$, we remember Eq. (1.11) and write

$$\overline{r^2} = \sum_i f_i r_i^2 \tag{1.45}$$

Now we recognize that the weighting factor f_i is precisely what the probability function gives, so we write Eq. (1.45) as

$$\overline{r^2} = \int_0^\infty P(r, n) r^2 \, dr \tag{1.46}$$

where the sum has been replaced by an integral, since we are using a continuous probability function. Substituting Eq. (1.44) for $P(n, r)$ gives

$$\overline{r^2} = 4\pi \left(2\pi \frac{n}{3} l^2\right)^{-3/2} \int_0^\infty r^4 \exp\left(\frac{-3r^2}{2nl^2}\right) dr \tag{1.47}$$

This is also a gamma function and may be solved with the help of a table of integrals. Evaluation of the integral gives the simple result

$$\overline{r^2} = nl^2 \tag{1.48}$$

which is the relationship we have sought.

The coil dimensions are characterized by the root-mean-square (rms) end-to-end distance which Eq. (1.48) shows to increase with the square root of the degree of polymerization. According to Eq. (1.1), the latter is directly proportional to M. Hence two chains of the same polymer compared under the same conditions would have to differ by a factor of 4 in M to show a factor of 2 difference in r_{rms}. We shall return to an examination of the length of the repeat unit l in the next section. Before doing that, however, there is another parameter of considerable importance which is used to characterize polymer coil dimensions, namely, the radius of gyration. It turns out that the radius of gyration r_g is the way chain dimensions enter the theories of viscosity and light scattering, as we shall see in Chaps. 9 and 10, respectively.

For a body that consists of n masses m_i, each separated by a distance r_i from the axis of rotation of the array, the radius of gyration is defined

$$r_g = \left(\frac{\Sigma_i m_i r_i^2}{\Sigma_i m_i}\right)^{\frac{1}{2}} \tag{1.49}$$

We may therefore think of r_g^2 as the weight average value of r^2, by analogy with Eq. (1.12). As a reminder of how the radius of gyration comes to be defined this way, recall that the moment of inertia I of this same body is given by

$$I = \sum_i m_i r_i^2 \tag{1.50}$$

There exists some radial distance from the axis of rotation at which all of the mass could be concentrated to produce the same moment of inertia that the actual distribution of mass possesses. This distance is defined to be the radius of gyration. According to this definition,

$$I = r_g^2 \sum_i m_i = \sum_i m_i r_i^2 \tag{1.51}$$

which leads directly to Eq. (1.49). The relationship between the radius of gyration and the actual dimensions of a body depends on the geometry of that body. Examples of these relationships are derived in many elementary physics textbooks for bodies of simple geometry; Table 10.1 lists several of these results that are pertinent to polymer chemistry. Let us briefly examine what the relationship is for a randomly coiled polymer molecule.

We desire to use the probability function derived above, so we recognize that the mass contribution of the volume element located a distance r from an axis through the center of mass is the product of the mass of a chain unit m_0 times the probability of a chain unit at that location as given by Eq. (1.44). For this purpose, however, it is not the distance from the chain end that matters but, rather, the distance from the center of mass. Therefore we temporarily identify the jth repeat unit as the center of mass and use the index k to count outward toward the chain ends from j. On this basis, Eq. (1.49) may be written as

$$(r_g^2)_j = \frac{\displaystyle\sum_{k=1}^{j} m_0 P(k,r) r^2 + \sum_{k=1}^{n-j} m_0 P(k,r) r^2}{\displaystyle\sum_{i=1}^{n} m_0} \tag{1.52}$$

where the terms in the numerator correspond to the two portions of chain on either side of j, and the denominator equals nm_0. We may cancel the mass of the individual chain units to give

$$(r_g^2)_j = \frac{1}{n}\left(\sum_{k=1}^{j} P(k,r) + \sum_{k=1}^{n-j} P(k,r) \right) r^2 \tag{1.53}$$

Next we recognize that any one of the repeat units can lie at the center of mass of the coil. To incorporate this last consideration, we sum Eq. (1.53) for all values of j and then divided by $\Sigma_j = n$:

$$\overline{r_g^2} = \sum_{j=1}^{n} (r_g^2)_j = \frac{1}{2n^2} \sum_{j=1}^{n}\left(\sum_{k=1}^{j} P(k,r) + \sum_{k=1}^{n-j} P(k,r) \right) r^2 \tag{1.54}$$

This last result includes the factor ½, since all segments are counted twice by the combination of Eqs. (1.53) and (1.54). The double summation also makes the two terms in the brackets equal to two times either one of them. This procedure *averages* over all segments fulfilling the role of center of mass; therefore it is appropriate to identify the quantity evaluated by Eq. (1.59) by the symbol $\overline{r_g^2}$.

Next we may substitute Eq. (1.44) for $P(k, r)$ in Eq. (1.54):

$$\overline{r_g^2} = \frac{1}{n^2} \sum_{j=1}^{n} \sum_{k=1}^{j} \int_{0}^{\infty} 4\pi r^4 \, (2\pi k l)^{-3/2} \exp\left(\frac{-3r^2}{2kl^2}\right) \, dr \tag{1.55}$$

This integral is the same as Eq. (1.47) and its value is given by (1.48). Substituting the latter, with k replacing n, yields

$$\overline{r_g^2} = \frac{1}{n^2} \sum_{j=1}^{n} \sum_{k=1}^{j} k l^2 \tag{1.56}$$

For polymers in which the degree of polymerization is large—a condition already assumed by the use of Eq. (1.48)—the remaining summations may be replaced by integrals

$$\overline{r_g^2} = \int_{0}^{n} \left(\int_{0}^{j} \frac{l^2}{n^2} \, k \, dk \right) dj = \int_{0}^{n} \frac{l^2}{n^2} \, \frac{j^2}{2} \, dj \tag{1.57}$$

or

$$\overline{r_g^2} = \frac{1}{6} n l^2 \tag{1.58}$$

Comparing this last result with Eq. (1.48) shows that

$$\overline{r_g^2} = \frac{1}{6} \overline{r^2} \tag{1.59}$$

As should be expected, both $(\overline{r_g^2})^{1/2}$ and r_{rms} show the same dependence on the degree of polymerization or molecular weight. Since the radius of gyration can be determined experimentally through the measurement of viscosity or light scattering, it is through this quantity that we shall approach the evaluation of l.

1.11   Application to Polymer Chains

In this section we compare actual polymer chains with the perfectly flexible model discussed in the last section. There are four respects in which an actual molecule differs from the idealized model:

1.   The angle formed between successive bonds along the chain backbone—θ in Fig. 1.5a—is not free to assume all values, but is fixed at a definite angle depending on the nature of the bond. For the tetrahedral angle associated with carbon-carbon single bonds, $\theta = 109.5°$.
2.   The rotation of one carbon-carbon bond around another—say, the $(i + 1)$th around the ith in Fig. 1.5a—is subject to steric hindrance, so that not all values of ϕ are equally probable.
3.   Actual polymer repeat units occupy finite volumes and therefore exclude other segments from occupying the same space.
4.   To obtain isolated polymer chains, a solvent must be present. The solvent might be selectively excluded or imbibed by the coil, depending on the free energy of interaction, and thereby perturb the coil dimensions.

At first glance it seems problematic whether anything is salvageable from the random walk model with so many areas of difference.

The strategy for rescuing the model based on a highly idealized picture of the chain depends on the following insight. Fixed bond angles and steric hindrance may limit the flexibility of the chain at any specific bond, but these restrictions can be overcome if we consider a somewhat longer segment of the polymer chain; that is, Fig. 1.5a with a full range of values for θ and ϕ is a highly unrealistic model for the backbone of a polymer molecule, even though it is an attractive basis for a theoretical derivation. On the other hand, the six-carbon segment shown in Fig. 1.5b—even if restricted in the allowed possibilities for θ and ϕ—does possess very nearly perfect flexibility. If five restricted bonds are still insufficiently flexible, we may subdivide the actual polymer chain into still longer subsections until they are long enough to qualify as "perfectly flexible." If the subsections so defined have a degree of polymerization ν and an effective length 1^*, then the actual polymer chain contains $n/\nu = n^*$ such units. All of the relationships of the last section may now be applied to this chain—which *is* perfectly flexible—consisting of n^* segments of length 1^*. Of course, n^* is still directly proportional to n, so we continue to expect the average coil dimensions to increase as $M^{1/2}$; the effective length of an individual step can be regarded as a proportionality constant. Since the radius of gyration can be measured under certain circumstances, the effective step length can be evaluated experimentally through the proportionality between $\overline{r_g^2}$ and n.

With the results of the preceding section thus salvaged, let us look in more detail at the specific areas of discrepancy between the perfectly flexible model

and an actual molecule. At least those deviations which arise from the actual restrictions on θ and ϕ can be dealt with quantitatively.

We begin by considering the fact that θ in Fig. 1.5a is limited to the value 109.5° for the carbon–carbon bonds along the backbone of a vinyl polymer chain. This means that carbon 3 in Fig. 1.5 may not, in fact, lie anywhere on the surface of a sphere, but is limited to positions on the rim of a cone, as shown in Fig. 1.7a. As before, the cone is generated by rotating the (i + 1)th bond between C_2, and C_3 around the ith bond. We shall return to the question of hindrance to this rotation presently. Figure 1.7b shows a longer segment of chain with this restricted bond angle and shows that a high degree of flexibility is still attainable over a somewhat longer portion of chain.

The approach to dealing with this quantitatively is to consider adding together the projections of individual bonds onto the direction of the first and averaging. If the length of the individual bonds is l, then the projection of the second bond onto the direction of the first is $l \cos \theta$, and the projection of the ith onto the direction of the first is $l(\cos \theta)^i$. For large values of n, the average value of r^2 obtained by adding and averaging these projections is

$$\overline{r^2} = nl^2 \ \frac{1 - \cos \theta}{1 + \cos \theta} \ = nb^2 \tag{1.60}$$

(a)   (b)

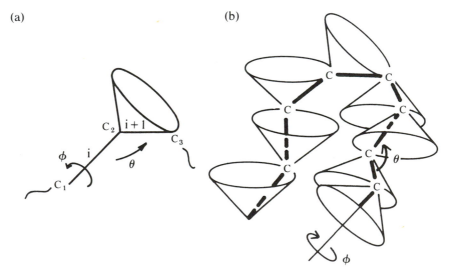

Figure 1.7   Placement of successive polymer segments connected at fixed bond angles. (a) Carbon 3 can lie anywhere on the rim of the cone. (b) This effect is illustrated for a longer portion of chain. [Panel (a) reprinted from Ref. 4, p. 118.]

where b corrects the step length for the effect of fixed bond angles. In the case of tetrahedral bond angles, $\cos 109.5° = -0.333$ so $(1 - \cos\theta)/(1 + \cos\theta) = 2.00$ and $b = \sqrt{2}$. This result shows that the restriction on bond angles has the effect of increasing the rms end-to-end distance by about 40% over the perfectly flexible case.

The formalism that we have set up to describe chain flexibility readily lends itself to the problem of hindered rotation. Figure 1.8a shows a sawhorse representation of an ethane molecule in which the angle of rotation around the bond is designated by ϕ. Because of electron repulsion between the atoms bonded to

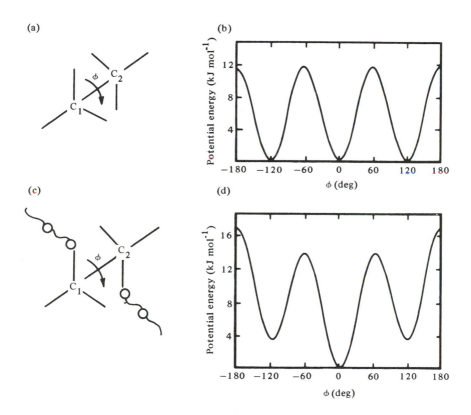

Figure 1.8 Hindered rotation around a carbon–carbon bond. (a) The definition of ϕ (from $\phi = 0$) in terms of the ethane molecule. (b) The potential energy as a function of ϕ. (c) Here ϕ is shown (from $\phi = 0$) for a carbon–carbon bond along a polyethylene backbone. (d) The potential energy for case (c) shown as a function of ϕ. [Panels (b) and (d) reprinted with permission from W. J. Taylor, *J. Chem. Phys.* 16:257 (1948).]

carbons 1 and 2, the potential energy of this molecule varies with the value of ϕ as shown in Fig. 1.8b. If we define $\phi = 0$ to be the conformation shown in Fig. 1.8a, then the potential energy shows the following:

1. A maximum after a rotation through $\pm 60°$ when the hydrogens on adjacent carbon atoms are eclipsed.
2. A minimum after rotation through $\pm 120°$ when the hydrogens are staggered.
3. A maximum after rotation through $\pm 180°$ when they are eclipsed again.
4. Potential energy barriers which are equal in height, on the order of 12 kJ mol^{-1}, for ethane.

For a carbon–carbon bond located along a polymer backbone, the preceding molecular representation must be modified to Fig. 1.8c. The chain segments on either side of the bond of interest are substituents for which the amount of steric hindrance follows a slightly different pattern than for the unsubstituted ethane. Using the same convention for ϕ, we see the following in Fig. 1.8d:

5. The potential energy shows a minimum at $\phi = 0°$, which is defined as the trans conformation.
6. The potential energy shows secondary minima after rotation through $\pm 120°$. These are called gauche conformations, and for polyethylene they are on the order of 4 kJ mol^{-1} higher in energy than the trans conformation.
7. The potential energy barrier associated with the chain segment passing a hydrogen is about 12 kJ mol^{-1} relative to the trans energy, and that of the two chain segments passing each other is about 16 kJ mol^{-1} relative to trans.

A still more intricate pattern of potential energy may be expected if the repeat units of the polymer chain carry other substituents, such as the phenyl groups in polystyrene, but these examples establish the general method for quantitatively describing the effects of steric hindrance on rotation.

Because of the variation in potential energy with the angle of rotation, not all locations on the rim of the cones in Fig. 1.7b are equally favored. The probability of a particular angular position depends on the potential energy at that location, V_ϕ, and an averaging procedure which considers this angular variation must be used to modify Eq. (1.60). The result of this procedure is

$$\overline{r^2} = nl^2 \left(\frac{1 - \cos \theta}{1 + \cos \theta} \right) \left(\frac{1 + \overline{\cos \phi}}{1 - \overline{\cos \phi}} \right) = nb^2 \left(\frac{1 + \overline{\cos \phi}}{1 - \overline{\cos \phi}} \right) \tag{1.61}$$

In this expression, $\overline{\cos \phi}$ is the average value of $\cos \phi$; the weighting factor used to evaluate the average is given by the Boltzmann factor $\exp(-V_\phi/RT)$, where R is the gas constant in the units of V_ϕ and T is in degrees Kelvin. Note that the correction factor introduced by these considerations reduces to unity if

$\overline{\cos \phi} = 0$, the case for free rotation, and increase as $\overline{\cos \phi}$ increases. Qualitatively, then, the effect of hindered rotation is to introduce still further expansion of coil dimensions.

In Chap. 9 we shall apply Eq. (1.44) to describe the probability that a particle has diffused a distance r after n thermal collisions with other molecules. In that context the particle may return to the sites of previous occupancy with no difficulty, but an error is clearly made by the same assumption regarding placement of chain segments. In the random walk derivation, we assumed that no excluded volume effect was operative regarding chain placement. There has been a good deal of research directed toward circumventing this clearly erroneous assumption. We shall not go into any detail in this matter, except for the following assessment of the magnitude of the effect. In computer simulation studies it has been found that the fraction of nonintersecting chains decreases by about 4% per segment added to the chain so that only 1 chain out of about 10^7 could attain a length of 400 units without intersection. The result of this sort of interference is to rule out still more conformations, just as restrictions on θ and ϕ did. Thus we have uncovered still one more factor which tends to increase coil dimensions above the value predicted by Eq. (1.48). Instead of examining the attempts that have been made to deal quantitatively with this source of error, let us look instead at conditions under which it might be offset.

At the beginning of this section we enumerated four ways in which actual polymer molecules deviate from the model for perfectly flexible chains. The three sources of deviation which we have discussed so far all lead to the prediction of larger coil dimensions than would be the case for perfect flexibility. The fourth source of discrepancy, solvent interaction, can have either an expansion or a contraction effect on the coil dimensions. To see how this comes about, we consider enclosing the spherical domain occupied by the polymer molecule by a hypothetical boundary as indicated by the broken line in Fig. 1.9. Only a portion of this domain is actually occupied by chain segments, and the remaining sites are occupied by solvent molecules which we have assumed to be totally indifferent as far as coil dimensions are concerned. The region enclosed by this hypothetical boundary may be viewed as a solution, an we next consider the tendency of solvent molecules to cross in or out of the domain of the polymer molecule.

The thermodynamic criterion for spontaneity is that the change in Gibbs free energy is negative for those processes which occur spontaneously. Thus if the free energy in the domain of the coil is lowered by increasing the number of polymer--solvent contacts, then solvent will be imbibed and the dimensions of the domain will increase as shown in Fig. 1.9. On the other hand, if the free energy is lowered by decreasing the number of polymer–solvent contacts, then solvent is squeezed out and the coil dimensions decrease. It is in this way that polymer--solvent interactions affect the rms values of the end-to-end distance

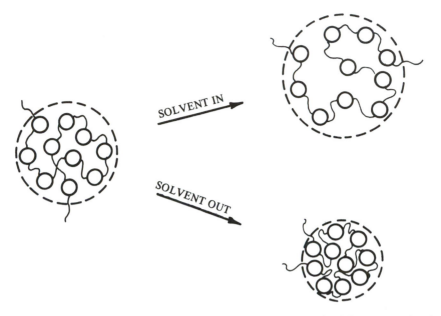

Figure 1.9 The spherical domain of a polymer molecule either expanding by imbibing solvent or contracting by excluding solvent.

and r_g. Quantitative discussion of this point must be deferred until the properties of polymer solutions are considered in Chap. 8. In the meanwhile, however, we can at least introduce some of the vocabulary that is used to describe these situations:

1. A "good solvent" is the technical as well as descriptive term used to identify a solvent which tends to increase coil dimensions. Since this is a consequence of thermodynamically favorable polymer–solvent interactions, good solvents also dissolve polymers more readily in the first place.
2. By contrast, a "poor solvent" is the technical description of a solvent which tends to decrease coil dimensions.
3. The relative goodness of a solvent depends on the temperature as well as the nature of the polymer–solvent system. Solvent goodness can be decreased by changing (generally lowering) the temperature or by adding a poorer solvent, which can result in the precipitation of polymer. Since higher molecular weight polymers are less soluble, controlled regulation of solvent goodness can be employed to fractionate a polydisperse polymer sample.

That state of affairs in which the poorness of the solvent exactly compensates for the excluded volume effect is called a Θ condition or Flory condition, after

P. J. Flory[†] whose outstanding overall contributions in polymer chemistry won him the Nobel Prize in 1974. Flory's book *Principles of Polymer Chemistry* contains an admirable discussion of these topics.

In summary, we see that the first two sources of deviation can be dealt with quantitatively, while the last two are dispatched by joining them in compensation for one another. (If you can't beat 'em, join 'em!) By convention, the coil dimensions under Θ conditions are given the subscript 0, so we write

$$\overline{r_0^2} = 6\overline{r_{g,0}^2} = nl_0^2 \tag{1.62}$$

where l_0 is the step length under these so-called unperturbed conditions. We shall see in Chap. 8 that Θ conditions can be achieved experimentally, so $\overline{r_{g,0}^2}$ and l_0 are measurable quantities. Table 1.6 lists some values of l_0 calculated from experimental data via Eq. (1.62). Table 1.6 also includes the ratio of these experimental l_0 values to b, the length of a carbon–carbon bond, corrected for fixed bond angles (or whatever distance is appropriate as the length of the repeat unit for non-vinyl polymers). The value of this ratio may be taken as a measure of the factor $(1 + \overline{\cos\phi})/(1 - \overline{\cos\phi})$ in Eq. (1.61). The following example will help clarify the meaning of these quantities.

Example 1.6

A polystyrene sample of molecular weight 10^6 shows an rms end-to-end distance under unperturbed conditions equal to 735 Å. In polystyrene $M_0 = 104$ and the length of the carbon–carbon bond along the backbone is 0.154 nm. Use these data to verify the numbers given for this polymer in Table 1.6.

Solution

For polystyrene of molecular weight 10^6, the degree of polymerization is $10^6/104 = 9620$. Since there are two carbon–carbon bonds associated with each repeat unit, the total number of bonds comprising the random walk is $2(9620) = 1.92 \times 10^4$. According to Eq. (1.48), for a random walk of 1.92×10^4 steps of length 0.154 nm, $\overline{r^2} = (1.92 \times 10^4)(0.154)^2 = 455$ nm^2. This is based on the assumption of unrestricted values for θ and ϕ. Limiting θ to the tetrahedral angle leads to the corrected value [Eq. (1.60)], which is double this, or 910 nm^2. Therefore, for fixed bond angles but free rotation around bonds, $r_{rms} = 30.2$ nm = 302 Å. The actual value for this quantity is subject to hindred rotation, so

---

[†]Paul J. Flory is another of the "fathers" of polymer chemistry. An interview with Flory appears in the *Journal of Chemical Education*, 54:341 (1977).

Table 1.6 Values of l_0 and b for Some Common Polymers and l_0/b Ratios Which Measure Steric Hindrance via Eq. (1.61)

Polymer	l_0 (nm)	Length of bond, corrected for fixed angles, b (nm)	l_0/b
Polyisobutylene	0.421	$0.154\sqrt{2} = 0.218$	1.93
Polystyrene	0.530	0.218	2.44
Poly(methyl methacrylate)	0.481	0.218	2.20
Poly(acrylic acid)	0.426	0.218	1.96
Poly(dimethyl siloxane)	0.444	0.278	1.60
Natural rubber	0.684	0.402^a	1.71
Gutta-percha	0.849	0.580^a	1.46

[a]These values are the total length of the repeat unit, based on appropriate bond lengths and bond angles.
Source: Data from Ref. 3.

the ratio $735/302 = 2.44$ equals the correction factor for this effect, $[(1 + \overline{\cos\phi})/(1 - \overline{\cos\phi})]^{1/2}$.

•

The values of the ratio l_0/b are thus seen to be quantitative measures of the hindrance to rotation in these polymers. The following observations are pertinent to the trends in this behavior:

1. Poly(dimethyl siloxane) offers the least steric hindrance of the polymers listed; every other atom along the backbone of the chain is devoid of substituents in this case.
2. Polystyrene with its bulky phenyl substituents shows the largest amount of hindrance of the polymers listed here.
3. In the two polyisoprene isomers, the length of the repeat unit and the steric hindrance factor vary oppositely for the two isomers. The greater end-to-end distance in the trans isomer is the dominant influence on the order of l_0 values.

In a good solvent, the end-to-end distance is greater than the l_0 value owing to the coil expansion resulting from solvent imbibed into the domain of the polymer. The effect is quantitatively expressed in terms of an expansion factor α defined by the relationship

$$\alpha = \frac{l}{l_0} = \frac{r_g}{r_{g,0}} = \frac{r_{rms}}{r_{rms,0}} \tag{1.63}$$

We shall defer a quantitative discussion of this expansion factor until the discussion of solutions in Chaps. 8 and 9.

We end this section with a disclaimer. In spite of the importance of the random coil in polymer chemistry, it must be conceded that this is not the only structure that polymer molecules assume. We have already noted that globular proteins may be cross-linked into tertiary structures best described as ellipsoids of revolution. In crystals, polyethylene folds back and forth along the backbone of the chain. Something on the order of 40 segments exist in the fully extended form in these crystals before the chain folds back on itself and repeats the process. In vinyl polymers the substituents are too bulky to fit into a crystal with only a 0.252-nm repeat distance as provided with the chain in the fully extended zigzag form. In these cases the chains accommodate substituents by twisting subsequent repeat units along the length of the otherwise extended backbone.

The helix is a geometrical structure to which certain polymer molecules are especially well suited. As an example, we consider an isotactic vinyl polymer in which successive repeat units assume alternating trans and gauche conformations. The tacticity is an aspect of the primary structure and the alternating gauche-trans conformations describe a specific situation that we may regard as secondary structure. Our present purpose is to show that this results in a still higher order of structure—tertiary structure—which has helical geometry. Figure 1.10a is an illustration of this conformation for an R-substituted vinyl polymer. This is the observed conformation for such molecules in the crystalline state when R is $-CH_3$, $-C_2H_5$, $-C_3H_7$, $-C_6H_5$, and so on. In Fig. 1.10 the ribbon connects successive R groups, tracing a helix containing three repeat units per turn. To see how such a structure is generated, examine Fig. 1.10a through the following steps (or, better yet, construct a model of the structure):

1. Number successive bonds along the backbone $1, 2, 3$, and so on.
2. Bonds 1 and 3 are rotated to trans positions with respect to bond 2. All odd-numbered bonds are in similar situations with respect to the even-numbered bonds.
3. Bonds 1 and 3 are in axial positions in the helix, with bond 2 oriented equatorially.
4. Bonds 2 and 4 are rotated to gauche positions with respect to bond 3. All even-numbered bonds are in similar conformations with respect to the odd-numbered bonds.
5. A consistent direction of twist is maintained; clockwise or counterclockwise rotations are equally probable.

This particular sequence of conformations—trans bonds that advance the helix along the axis, alternating with gauche bonds which provide the twist—takes the chain through a series of relatively low potential energy states and generates a structure with minimum steric hindrance between substituents. If the polymer series is extended to include bulkier substituents, for example,

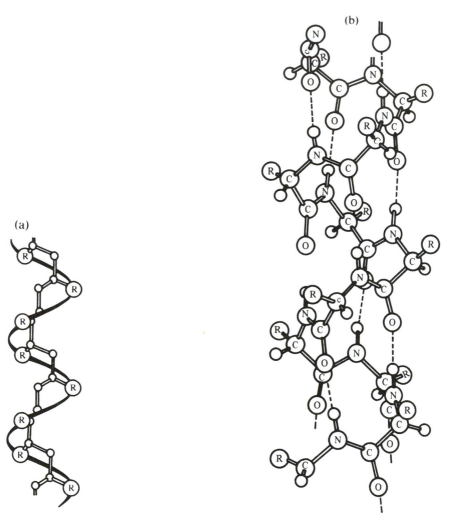

Figure 1.10 Helical conformations in polymer molecules. (a) A vinyl polymer with R substituents has three repeat units per turn. (b) The α helix of the protein molecule is stabilized by hydrogen bonding. [From R. B. Corey and L. Pauling, *Rend. Inst. Lombardo Sci.* 89:10 (1955).]

$R = -CH_2 CH(CH_3)_2$, there occurs a modest deviation from a strict $0°$-$120°$ alternation which characterizes the trans–gauche sequence. This produces a helical structure with seven repeat units occurring in two turns. Even bulkier substituents, for example, *o*-methyl phenyl, produce still more open helices

with four repeat units per turn as a result of even greater deviation from the rotational positions of lowest potential energy.

In addition to existing as helices in crystals, there is evidence that certain vinyl polymers also show some degree of regular alternation between trans and gauche conformations in solution. In solution, the chain is free from the sort of environmental constraints that operate in a crystal, so the length of the helical sequence in a dissolved isotactic vinyl polymer may be relatively short.

Many polypeptides also assume helical structures in the crystalline state and—in equilibrium with unwound random conformations—in solution. In this case the helix is stabilized by hydrogen bonding between the N–H and O=C groups on successive turns of the helix. Fixed bond angles and bond lengths restrict stable helices to those containing either 3.7 or 5.1 repeat units per turn. The former, commonly called the α helix, is shown in Fig. 1.10b; it possesses about 18 amino acid residues in every five turns. Many globular proteins, for example, myoglobin, contain helical sections in which the helix is interrupted occasionally, the chain bends through a kink, then the helix resumes. Finally, there is the most famous helical structure of all: the DNA double helix. Random coils are not without competition as the structure of polymer molecules!

1.12 Preview of Things to Come

In this chapter we have focused attention on various aspects of individual polymer molecules. In the next three chapters we shall examine some properties of assemblies of polymer molecules. Our interest in these chapters will be mostly directed toward samples of pure polymer; assemblies of high and low molecular weight molecules—polymer solutions—will be discussed in Part III of this book.

We shall divide our discussion of bulk polymers on the basis of the mechanical states of these materials; that is, instead of classification into states based on thermodynamic criteria alone, we favor a system based on the way samples respond to deforming forces. Accordingly, we shall discuss the viscous state in Chap. 2, the elastic and viscoelastic states in Chap. 3, and the glassy and crystalline states in Chap. 4. There are good reasons for choosing these mechanical states as the basis for classification and, incidentally, for presenting this material right at the start of this text. If polymers did not possess the mechanical properties they have, they would, in all likelihood, still be in the category of laboratory curiosities—or disasters. Instead, their sluggish flow, elastic snap, plastic flexibility, and toughness in wear are all marketable properties. Each of these properties is ultimately traceable to the chain structure of polymer molecules. It is not the *amide* in its IUPAC name that makes nylon the important material that it is; it is the *poly* which is primarily responsible!

In Chap. 2, we consider polymers whose molecular weight and/or temperature gives them liquid flow behavior. We shall begin by quantitatively examining what we mean by viscosity and then consider how this property depends on, say, how fast the material is stirred and on the characteristics of the polymer itself. It should not be too surprising that the length of the chain is the most important polymer property in this regard. All we need to do is think of our experience with a plate of spaghetti to get an idea of what happens when we try to induce some motion in a mass of tangled flexible "chains."

Polymers are not alone in displaying elastic properties, but their elasticity is unique in both origin and magnitude. For one thing, the elastic restoring force which acts in polymers increases with increasing temperature, whereas the opposite is true in, say, metals. Again, this is a consequence of the chain structure of the polymer. Stretching a live snake to a "fully extended conformation" may be problematic under the best of conditions, but it will resist all the more if we warm it up besides. We shall see in Chap. 3 that polymer chains behave in pretty much the same way.

Chapter 4 is concerned with the solid forms of polymers. Crystals of any material consist of highly ordered arrays of molecules. In view of the assortment of variations that can occur within polymer chains, it is not surprising that only those polymers with a high degree of chain regularity crystallize. Even crystallizable polymers fall short of complete crystallinity. Those polymers which do not crystallize, whether partly or totally, still solidify in the mechanical sense at a temperature called the glass transition temperature. Below this temperature the thermal energy available for chain motion is inadequate to allow much relative motion between chains. Low molecular weight materials also form glasses, but the chain structure of polymers makes this behavior one to which polymers are particularly susceptible.

In the next group of chapters—indeed, throughout the book—we shall be talking about a variety of properties of individual polymer molecules and assemblies of these molecules. One concept that should be fairly clear from the discussions of this chapter is that the word *average* should be used as a modifier in almost every statement we make about polymers. This is true in many areas of chemistry: Even an assembly of relatively simple water molecules possesses an assortment of isotopic possibilities; translational, rotational, vibrational, and electronic energy states; and different extents of hydrogen bonding and ionization. Still we speak of *the* water molecule as if it were an invariant species. What we are really talking about is an *average* molecule. Unless we specifically wish to underscore the statistical aspect of a statement, however, the word *average* is generally implied rather than stated. So it will be in this volume. One of the themes of this chapter is the diversity of structures—even down to something as fundamental as molecular weight—that polymers display. Accordingly, the word *average* is even more appropriate as a modifier in polymer

chemistry than in chemistry as a whole. More often than not, however, the actual word is implied rather than stated. Remembering that this is the case is one of the major lessons of this chapter.

Problems

Whenever feasible, the problems in this book are based on data from the original literature. In many instances the values given have been estimated from graphs, transformed from another functional representation, or changed in units. Therefore these quantities do not necessarily reflect the accuracy of the original work, nor is the given number of significant figures always justified. Finally, the data may be used for purposes other than were intended in the original measurement.

1. Cohen and Ramos[†] describe some phase equilibrium studies of block copolymers of butadiene (B) and isoprene (I). One such polymer is described as having a 2:1 molar ratio of B to I with the following microstructure:

    B — 45% cis-1,4; 45% trans-1,4; 10% vinyl.
    I — over 92% cis-1,4.

    Draw the structure of a portion of this polymer consisting of about 15 segments and having approximately the composition of this polymer.
2. Hydrogenation of polybutadiene converts both cis and trans isomers to the same linear structure and vinyl groups to ethyl branches. A polybutadiene sample of molecular weight 168,000 was found[‡] by infrared spectroscopy to contain double bonds consisting of 47.2% cis, 44.9% trans, and 7.9% vinyl. After hydrogenation, what is the average number of backbone carbon atoms between ethyl side chains?
3. Landel[§] used a commercial material called Vulcollan 18/40 to study the rubber-to-glass transition of a polyurethane. This material is described as being "prepared from a low molecular weight polyester which is extended and crosslinked ... by reacting it with naphthalene-1,4-diisocyanate and 1,4-butanediol. The polyester ... is prepared from adipic acid and a mixture of ethylene and propylene glycols." Draw the structural formula of a portion of the cross-linked polymer which includes the various possible linkages that this description includes. Remember that isocyanates react with active hydrogens; use this fact to account for the cross-linking.

[†]R. E. Cohen and A. R. Ramos, *Macromolecules* 12:131 (1979).
[‡]W. E. Rochefort, G. G. Smith, H. Rachapudy, V. R. Raju, and W. W. Graessley, *J. Polym. Sci. Polym. Phys.* 17:1197 (1979).
[§]R. F. Landel, *J. Colloid Sci.* 12:308 (1957).

4. Some polymers are listed below by either IUPAC[†] (I) names or acceptable trivial (T) names. Draw structural formulas for the repeat units in these polymers and propose an alternative name in the system other than the one given.

   (a) Polymethylene (I).
   (b) Polyformaldehyde (T).
   (c) Poly(phenylene oxide) (T).
   (d) Poly[(2-propyl-1,3-dioxane-4,6-diyl)methylene] (I).
   (e) Poly(1-acetoxyethylene) (I).
   (f) Poly(methyl acrylate) (T).

5. Star polymers are branched molecules with a controlled number of linear arms anchored by one central molecular unit acting as a branch point. Schaefgen and Flory[‡] prepared poly(ϵ-caprolactam) four- and eight-pointed stars using cyclohexanone tetrapropionic acid and dicyclohexanone octapropionic acid as branch points. These authors present the following stoichiometric definitions/relations to relate the molecular weight of the polymer to the concentration of unreacted acid groups in the product. Provide the information required for each of the following steps:

   (a) The product has the formula $R\{-CO[-NH(CH_2)_5CO-]_y-OH\}_b$. What is the significance of R, y, and b?
   (b) If Q is the number of equivalents of multifunctional reactant which reacts per mole of monomer and L represents the number of equivalents of unreacted (end) groups per mole monomer, then $\bar{y} = (1 - L)/(Q + L)$. Justify this relationship, assuming all functional groups are equal in reactivity.
   (c) If M_0 is the molecular weight of the repeat unit and M_b is the molecular weight of the original branch molecule divided by b, then the number average molecular weight of the star polymer is $\bar{M}_n = b\{M_0[(1 - L)/(Q + L)] + M_b\}$. Justify this result and evaluate M_0 and M_b for the b = 4 and b = 8 stars.
   (d) Evaluate $\bar{M}_n$ for the following molecules:

	Q	L
b = 4	0.2169	0.0018
b = 8	0.134	0.00093

6. Batzer[§] has reported the following data for a fractionated polyester made from sebacic acid and 1,6-hexanediol:

[†] IUPAC Macromolecular Nomenclature Commission, *Macromolecules* 6:149 (1973).
[‡] J. R. Schaefgen and P. J. Flory, *J. Am. Chem. Soc.* 70:2709 (1948).
[§] H. Batzer, *Makromol. Chem.* 5:5 (1950).

	Fraction no.								
	1	2	3	4	5	6	7	8	9
Mass (g)	1.15	0.73	0.415	0.35	0.51	0.34	1.78	0.1	0.94
$M \times 10^{-4}$ (g mol^{-1})	1.25	2.05	2.40	3.20	3.90	4.50	6.35	4.1	9.40

Evaluate $\overline{M}_n$, $\overline{M}_w$, and $\overline{M}_z$ from these data.

7. At 25°C, the Mark–Houwink exponent for poly(methyl methacrylate) has the value 0.69 in acetone and 0.83 in chloroform. Calculate (retaining more significant figures than strictly warranted) the value of $\overline{M}_v$ that would be obtained for a sample with the following molecular weight distribution if the sample were studied[†] by viscometry in each of these solvents:

$N_i \times 10^3$ (mol)	1.2	2.7	4.9	3.1	0.9
$M_i \times 10^{-5}$ (g mol^{-1})	2.0	4.0	6.0	8.0	10.0

8. In addition to r_g and r_{rms}, another way of characterizing coil dimensions is to consider which end-to-end distance has the greatest probability of occurring for specified n and l values. Derive an expression for this most probable value of r, r_m, from Eq. (1.44). Compare the ratio r_{rms}/r_m to the ratio u_{rms}/u_m from the kinetic molecular theory of gases (consult, say, a physical chemistry textbook), where u is molecular velocity. Comment briefly on the significance of the comparison.

9. Random walk statistics also describe the path followed by a diffusing particle. In this case, the number of steps which a particle takes is proportional to the time allowed for diffusion: n = Kt, with K the proportionality constant. A totally independent approach to diffusion gives $\overline{r^2} = 2Dt$ as the relationship between the mean-square displacement, time, and the diffusion coefficient of the particle. Use this result to evaluate the constant K. The diffusion coefficients of low molecular weight solutes through low molecular weight solvents, polymers through low molecular weight solvents, and polymers through polymers are on the order of 10^{-9}, 10^{-11}, and 10^{-17} m^2 sec^{-1}, respectively. Calculate the time required for r_{rms} to be 10 nm in each case.

10. For two different esters of cellulose, l_0 values are listed[‡] here in two different temperature regions, 30 and 130–140°C:

Ester	T(°C)	l_0 (Å)
Cellulose tributyrate	30	38.6
Cellulose tributyrate	130	14.1
Cellulose tricaprylate	140	17.3

[†] S. N. Chinai, J. D. Matlock, A. L. Resnick, and R. J. Samuels, *J. Polym. Sci.* 17:391 (1955).
[‡] Ref. 3.

Taking the length per repeat unit (i.e., bond angles already considered) as 0.78 nm in each instance, evaluate the factors $(1 + \cos \phi)/(1 - \cos \phi)$ and $\cos \phi$ for each polymer. Ignoring the difference between 130 and 140°C, do you find the difference in steric hindrance between the tributyrate and tricaprylate to be what you expected? Is the effect of temperature on the l_0 value of cellulose tributyrate what you expected? Briefly explain each answer. For each polymer, calculate r_{rms} if $n = 10^4$; also do this for the hypothetical chain with no restrictions to rotation and having the same repeat length.

11. Strauss and Williams[†] have studied coil dimensions of derivatives of poly(4-vinylpyridine) by light-scattering and viscosity measurements. The derivatives studied were poly(pyridinium) ions quaternized y% with n-dodecyl groups and $(1 - y)\%$ with ethyl groups. Experimental coil dimensions extrapolated to Θ conditions and expressed relative to the length of a freely rotating repeat unit are presented here for the molecules in two different environments:

	$r_{rms,0}/n^{1/2}b$	
	In isopropanol	In water
y	with LiBr	with KBr
0	–	2.54
4.8	2.60	2.88
10.3	2.77	2.26
16.3	2.80	1.86
34.1	2.98	1.23

(a) Write three unit structural formulas for poly(4-vinyl pyridine) and for the molecule in which y = 34.1%.

(b) In both solvents the limiting value of $r_{rms,0}/n^{1/2}b$ as $y \to 0$ is similar in magnitude to the value of this quantity for polystyrene as given in Table 1.6. Briefly explain why this might be expected.

(c) The trends in $r_{rms,0}/n^{1/2}b$ with increasing y are different in the two solvents. Propose brief explanations for each of these trends. (Hint: The "micelle" concept is helpful for the water case; look this up in Chap. 6 or surface chemistry texts if unfamiliar.)

12. The special stability of the helix with three repeat units per turn for isotactic vinyl polymers in which the substituent is not too large was discussed in connection with Fig. 1.10a. Rodriguez[‡] has suggested a simple paper model which is also helpful in illustrating this structure. The following steps summarize this method; additional details are available in this readily accessible reference.

[†] U. P. Strauss and B. L. Williams, *J. Phys. Chem.* 65:1390 (1961).
[‡] F. Rodriguez, *J. Chem. Educ.* 45:507 (1968).

(a) Construct a triangular prism mandrel by folding construction paper or, even better, transparent plastic sheeting. Axial bonds of the helix lie along the edges—labeled A, B, and C—of this prism.

(b) Equatorial bonds lie on the faces of the prism—labels AB, BC, and CA. The bonds on the faces make an angle of 109.5° with the previous bond in the plane of that face.

(c) To maintain proper bond lengths, the width of each face should be 94.3% (sin 70.5° = 0.943) of the bond length.

(d) Use a felt-tip pen to mark bond 1 along edge A. Bond 2 is then drawn on face AB (measure angle, 109.5°). Bond 3 goes along edge B; bond 4 on face BC, and so on.

(e) Slots are cut into the edges of the mandrel, bisecting the bond angles along the chain. Cards are inserted into the slots onto which chain substituents have been drawn to scale.

Construct such a model for isotactic polypropylene. Estimate the volume of an isobutyl group on the scale of your model and examine whether interference between successive substituents would occur if this were the R group present.

13. The intrinsic viscosity of a solution of particles shaped like ellipsoids of revolution is given by the expression

$$\frac{14}{15} + \frac{(a/b)^2}{15[\ln(2a/b) - 1.5]} + \frac{(a/b)^2}{5[\ln(2a/b) - 0.5]}$$

(Simha equation), where a/b is the length/diameter ratio of these cigar-shaped particles. Doty et al.[†] measure the intrinsic viscosity of poly(γ-benzyl glutamate) in a chloroform–formamide solution and obtained (approximately) the following results:

Degree of polymerization	98	317	1240	1650
Intrinsic viscosity (dimensionless as in expression above)	13	66	740	1060

Use the Simha equation and these data to criticize or defend the following proposition: These polymer molecules behave like rods whose diameter is 16 Å and whose length is 1.5 Å per repeat unit. The molecule apparently exists in fully extended form in this solvent rather than as random coils.

[†]P. Doty, A. M. Holtzer, J. H. Bradbury, and E. R. Blout, *J. Am. Chem. Soc.* 76:4493 (1954).

Bibliography

1. Allcock, H. R., and Lampe, F. W., *Contemporary Polymer Chemistry*, Prentice-Hall, Englewood Cliffs, N.J., 1981.
2. Billmeyer, F. W., Jr., *Textbook of Polymer Science*, 2nd ed., Wiley-Interscience, New York, 1971.
3. Flory, P. J., *Principles of Polymer Chemistry*, Cornell University Press, Ithaca, N.Y., 1953.
4. Hiemenz, P. C., *Principles of Colloid and Surface Chemistry*, Marcel Dekker, New York, 1977.
5. Seymour, R. B., and Carraher, C. E., Jr., *Polymer Chemistry, an Introduction*, Marcel Dekker, New York, 1981.

BULK POLYMERS AND THEIR MECHANICAL BEHAVIOR

2

The Viscous State

The fond voyeur
And narcissist alike devoutly peer
Into disorder, the disorder
Being covalent bondings that prefer
Prolonged viscosity and spread loose nets
Photons slip through

The Dance of the Solids, by John Updike

2.1 Introduction

In this chapter we examine the flow behavior of bulk polymers in the liquid state. Such substances are characterized by very high viscosities, a property which is directly traceable to the chain structure of the molecules. All substances are viscous, even low molecular weight gases. The enhancement of this property due to the molecular structure of polymers is one of the most striking features of these materials.

Before we are in a position to discuss the viscosity of polymer melts, we must first give a quantitative definition of what is meant by viscosity and then say something about how this property is measured. This will not be our only exposure to experimental viscosity in this volume—other methods for determining bulk viscosity will be taken up in the next chapter and the viscosity of solutions will be discussed in Chap. 9—so the discussion of viscometry will only be introductory. Throughout we shall be concerned with constant temperature experiments conducted under nonturbulent flow conditions.

There are three major aspects of polymer viscosity discussed in this chapter. First, we shall consider the fact that most bulk polymers display shear-dependent viscosity; that is, this property does not have a single value but varies with the shearing forces responsible for the flow. Second, the molecular weight dependence of polymer viscosity is examined. We may correctly expect a

considerable sensitivity to this parameter, since it is a direct measure of chain length. Third, we shall investigate the effect of chain branching on viscosity.

Our approach in this chapter is to alternate between experimental results and theoretical models to acquire familiarity with both the phenomena and the theories proposed to explain them. We shall consider a model for viscous flow due to Eyring which is based on the migration of vacancies or holes in the liquid. A theory developed by Debye will give a first view of the molecular weight dependence of viscosity; an equation derived by Bueche will extend that view. Finally, a model for the snakelike wiggling of a polymer chain through an array of other molecules, due to deGennes, Doi, and Edwards, will be taken up.

There are a number of important concepts which emerge in our discussion of viscosity. Most of these will come up again in subsequent chapters as we discuss other mechanical states of polymers. The important concepts include free volume, relaxation time, spectrum of relaxation times, entanglement, the friction factor, and reptation. Special attention should be paid to these terms as they are introduced.

2.2 The Coefficient of Viscosity

Before we attempt to discover the molecular origin of the viscosity of a polymer melt, we must first define more quantitatively what we mean by viscosity. As a place to begin, we visualize the fluid as a set of infinitesimally thin layers moving parallel to each other, each with a characteristic velocity. In addition, we stipulate that those fluid layers which are adjacent to nonflowing surfaces have the same velocity as the rigid surfaces. This is another way of stating that there is no slipping at the interface between the stationary and flowing phases and is an experimental fact of fluid behavior at such boundaries. Now suppose we consider a sample of fluid which is maintained at constant temperature, sandwiched between two rigid parallel plates of area A as shown in Fig. 2.1. If a force F is applied to the top plate as shown in Fig. 2.1, that plate and the layer of fluid adjacent to it will accelerate until a steady velocity is reached. As long as the deforming force continues to be applied, this velocity is unchanged. This time-independent behavior is called stationary-state flow and will be our primary concern. During the acceleration that precedes the stationary state, the velocity is a function of time. For our purposes, we shall simply wait until the stationary state is reached and not even question how long it will take. Force per unit area is called stress and is given the symbol σ; we often attach the subscript s to F and σ as a reminder that these operate in a shear mode in this discussion.

In the experiment described in Fig. 2.1, the bottom plate remains in place and the nonslip condition stipulated above requires that the layer of fluid adjacent to the bottom plate also have zero velocity. This situation clearly

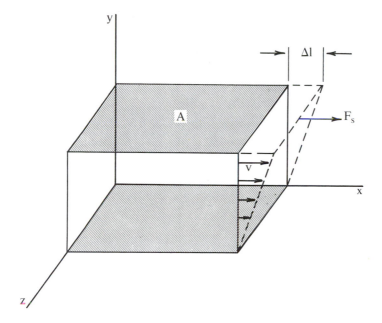

Figure 2.1 The relationship between the applied force per unit area and the velocity used in the definition of viscosity.

requires that the velocity of the fluid vary from layer to layer across the gap between the two rigid plates.

To formulate this mathematically, we write that the top and bottom of these imaginary fluid layers are separated by a distance Δy and differ from each other in velocity by Δv. The ratio $\Delta v/\Delta y$ has units of reciprocal time and is called either the velocity gradient or the rate of shear. The former name is self-explanatory, and the latter may be understood by considering the actual deformation the sample undergoes under the shearing force. During a short time interval Δt, the top layer moves a distance Δl relative to the bottom layer. Accordingly, Δv may be written $\Delta v = \Delta l/\Delta t$ and the velocity gradient may be expressed as $\Delta v/\Delta y = (\Delta l/\Delta t)/\Delta y = (\Delta l/\Delta y)/\Delta t$. The shear displacement Δl divided by the distance over which it vanishes to zero, in this case Δy, is called the shear strain, which we represent by γ. These relationships show that the ratio $\Delta v/\Delta y$ also describes the rate at which the shear strain develops, or, more simply, the rate of shear $\Delta\gamma/\Delta t$, or $\dot{\gamma}$. For the present, we shall represent the velocity gradient by either $\Delta v/\Delta y$ or $\dot{\gamma}$, but the more compact symbol for this quantity is used exclusively in later sections. In summary,

$$\frac{\Delta v}{\Delta y} = \dot{\gamma} = \frac{\Delta l/\Delta y}{\Delta t} \tag{2.1}$$

Now let us invoke our experience with liquids of different viscosities, say, water and molasses, and imagine the magnitude of the shearing force that would be required to induce the same velocity gradient in separate experiments involving these two liquids. Our experience suggests that more force is required for the more viscous fluid. Since the area of the solid plates in this liquid sandwich is also involved, we can summarize this argument by writing a proportionality relation between the shear force per unit area σ_s and the velocity gradient

$$\sigma_s = \eta \, \frac{\Delta v}{\Delta y} = \eta \dot{\gamma} \tag{2.2}$$

where η is called the coefficient of viscosity of the fluid or, more simply, its viscosity. Equation (2.2) implies that the velocity gradient is exactly the same throughout the liquid. Since this may not be the case over macroscopic distances, our best assurance of generality is to consider the limiting case in which Δy and therefore Δv approach zero. In the limit of these infinitesimal increments Δv and Δy become dv and dy, respectively, so Eq. (2.2) becomes

$$\sigma_s = \eta \, \frac{dv}{dy} \tag{2.3}$$

Equation (2.3) is called Newton's law of viscosity and those systems which obey it are called Newtonian.

Equation (2.3) describes a straight line of zero intercept if σ_s is plotted versus the velocity gradient. Such a plot is shown in Fig. 2.2. Since the coefficient of viscosity is the slope of this line, this quantity has a single value for Newtonian liquids. Liquids of low molecular weight compounds and their solutions are generally Newtonian, but quite a few different variations from this behavior are also observed. We shall not attempt to catalog all of these variations, but shall only consider the other pattern of behavior shown in Fig. 2.2. This example of non-Newtonian behavior is described as pseudoplastic and is often observed when the material under study is a fluid polymer. Since Eq. (2.3) defines the coefficient of viscosity as the slope of a plot of σ_s versus velocity gradient, it is clear from Fig. 2.2 that pseudoplastic substances are not characterized by a single viscosity. The apparent viscosity at a particular velocity gradient is given by the ratio $\sigma_s/(dv/dy)$. Inspection of Fig. 2.2 reveals that pseudoplastic materials appear less viscous at high rates of shear than at low rates.

The amount of curvature in plots like Fig. 2.2 is a measure of the deviation from Newtonian behavior. We can use the logic of calculus to argue that such curvature becomes less apparent as we examine progressively smaller segments of the line. This statement leads us to two important conclusions:

1. If our objective is to examine non-Newtonian behavior, we must design experiments which permit the relationship between σ_s and dv/dy to be studied over as wide a range as possible. This topic is taken up in the next section.

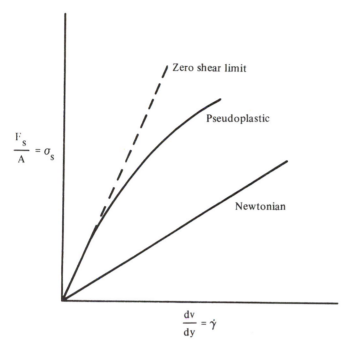

Figure 2.2 Comparison of Newtonian and pseudoplastic behavior.

2. If we wish to avoid the complication of non-Newtonian behavior, we must focus attention on a relatively narrow range of values for dv/dy.

Which range should be considered? The answer is the region near the origin of a plot like Fig. 2.2 for pseudoplastic materials. The slope of the tangent to a pseudoplastic curve at the origin is called the viscosity at zero rate of shear. Note that this is an extrapolation to a limit rather than an observation at zero shear (which corresponds to no flow). We shall use the symbol η_N to indicate the viscosity of a polymer in the limit of zero shear, since the behavior is Newtonian (subscript N) in this region.

To see another interpretation of viscosity, we multiply both sides of Eq. (2.2) by $\Delta v/\Delta y$:

$$\sigma_s \frac{\Delta v}{\Delta y} = \eta \left(\frac{\Delta v}{\Delta y} \right)^2 \tag{2.4}$$

To make sense of this result, we remember that v as introduced in Fig. 2.1 is actually $\Delta x/\Delta t$. Thus the product $F(\Delta v/\Delta y)$ can be written $F(\Delta x/\Delta t)/\Delta y$, and $F(dv/dy)/A$ becomes $F \Delta x/A \Delta y \Delta t$. The product of a force and the distance through which it operates equals an energy ΔE, and the product of A and Δy

equals the volume element ΔV upon which the shearing force described in Fig. 2.1 operates. Therefore $F \Delta x/A \Delta y \Delta t$ is the same as $\Delta E/\Delta V \Delta t$. Defining the increment in shear energy dissipated per unit volume by the symbol ΔW, we obtain

$$\frac{\Delta W}{\Delta t} = \eta \left(\frac{\Delta v}{\Delta y}\right)^2 \tag{2.5}$$

As in the parallel case of going from Eq. (2.2) to Eq. (2.3), we take the limit of infinitesimal increments and write

$$\frac{dW}{dt} = \eta \left(\frac{dv}{dy}\right)^2 \tag{2.6}$$

The deforming forces which induce flow in fluids are not recovered when these forces are removed. These forces impart kinetic energy to the fluid, an energy which is dissipated within the fluid. This is the origin of the idea that viscosity represents an internal friction which resists flow. This friction originates from the way molecules of the sample interact during flow.

We conclude this section with a consideration of the units required for η by Eqs. (2.3) and (2.6). To do this, we rewrite these equation in terms of the units of all quantities except η. The units of η must make the expressions dimensionally correct. Force has units of mass times acceleration, or mass length time^{-2}, and area is length^2. Since the velocity gradient has units time^{-1}, the dimensional statement of Eq. (2.3) is

$$\frac{\text{mass length time}^{-2}}{\text{length}^2} = (\eta)\,\text{time}^{-1}$$

Similarly, the rate of energy dissipation in Eq. (2.6) has units energy volume^{-1} time^{-1}, so the dimensions of that equation are

$$\frac{\text{mass length}^2 \text{ time}^{-2}}{\text{length}^3 \text{ time}} = (\eta)\,\text{time}^{-2}$$

In order to satisfy these two defining equations, η must have units mass length^{-1} time^{-1}. In SI units this is kg m^{-1} sec^{-1}. The cgs unit 1 g cm^{-1} sec^{-1} is defined to be 1 poise (1 P). Note that 1 kg m^{-1} s^{-1} = 10 P. At room temperature, water has a viscosity of about 10^{-3} kg m^{-1} sec^{-1}, and other low molecular weight liquids also have viscosities of this magnitude. The viscosity of a polymer depends on the architecture of the polymer molecule and, especially, on the molecular weight of the polymer, as we shall see in Sec. 2.9. To get a feel for the magnitude of polymer viscosities, note that $\eta_N = 3.3 \times 10^4$ kg m^{-1} sec^{-1} at 200°C for a polystyrene sample with $\overline{M}_w = 3.71 \times 10^5$ and that $\eta_N = 7.5 \times 10^5$ kg m^{-1} sec^{-1} at 25°C for poly(dimethyl siloxane) with $\overline{M}_w = 1.34 \times 10^6$.

2.3 Measuring Viscosity: Concentric Cylinder Viscometers

As the numbers cited in the preceding section suggest, the range of viscosities spanned between low and high molecular weight substances is very broad and no single experimental procedure is equally suitable over the entire range. In Chap. 9 we shall discuss the capillary viscometer as a means of measuring the viscosity of dilute polymer solutions. For undiluted fluid polymer samples, our concern in this chapter, this type of viscometer is not particularly useful. For the present we shall focus our attention on a device called the concentric cylinder viscometer. Some of our reasons for interest in this type of viscometer are the following:

1. The basic design is a direct extension of the discussion of the preceding section.
2. The range of applicability is exceptionally wide, approximately 10^{-1}-10^{11} kg m^{-1} sec^{-1}.
3. The design permits different velocity gradients to be considered, so that pseudoplasticity can be investigated if desired.
4. A number of technically important viscosity-measuring devices may be thought of as variants of this basic apparatus.

The illustration that enabled us to define the coefficient of viscosity also suggests a modification which would be experimentally useful. Suppose the two rigid parallel plates of Fig. 2.1 and the intervening layers of fluid were wrapped around the z axis to form two concentric cylinders, with the fluid under consideration in the gap between them. The required velocity gradient is then established by causing one of these cylinders to rotate while the other remains stationary. The velocity is now in the direction described by the angle θ in Fig. 2.3a, and its gradient is in the radial direction r. Thus the velocity gradient in this arrangement may be written dv_θ/dr. As a practical matter, the outer cylinder is part of a cup which holds the fluid, while the inner cylinder is a coaxial bob suspended within the outer cup. Suppose the cup is centered on a turntable which rotates with an angular velocity ω, measured in radians per second. The viscous fluid now transmits a force to the suspended bob which can be measured in terms of the torque on a torsion wire. This arrangement is sketched in Fig. 2.3b. In this representation, the outer cylinder has a radius R and the inner cylinder has a radius fR, where f is some fraction. The closer to unity this fraction is, the narrower will be the gap between the cylinders and the more closely the apparatus will approximate the parallel plate model in terms of which η was defined.

A formal mathematical analysis of the flow in the concentric cylinder viscometer yields the following relationship between the experimental variables and the viscosity:

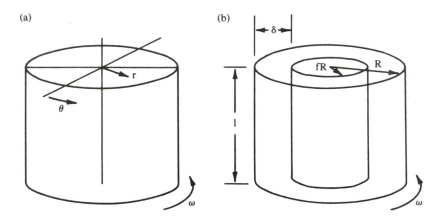

Figure 2.3 Definition of variables for concentric cylinder viscometers; (a) the rotating cylinder and (b) the coaxial cylinders.

$$\text{Torque} = 4\pi\eta l R^2 \omega \frac{f^2}{1 - f^2} \tag{2.7}$$

This equation can be cast into a more recognizable from by assuming that f is very close to unity. In that case we have the following:

1. The radius R also applies to the entire fluid sample. Since torque equals the product of force and R, canceling out one power of R leaves the shearing force acting on the fluid on the left-hand side of Eq. (2.7).
2. The remaining factor R times ω on the right-hand side of Eq. (2.7) can be replaced by the linear velocity v_θ.
3. The factor $1 - f^2$ can be replaced by $2(1 - f)$, since $1 - f^2 = (1 + f)(1 - f)$ and $1 + f \to 2$ as $f \to 1$.
4. The area of contact A between the cylinders and the fluid is $2\pi Rl$; therefore $4\pi l R/R(1 - f^2) = A/(1 - f)R$.
5. The product $(1 - f)R$ is the width of the gap, δ.

Introducing these substitutions in Eq. (2.7) gives

$$\frac{F_s}{A} = \sigma_s = \eta f^2 \frac{v_\theta}{\delta} \tag{2.8}$$

Since v_θ is the difference in velocity between the inner and outer cylinders and δ is the difference in the radial location of the two rigid surfaces, Eq. (2.8) becomes

$$\sigma_s = \eta \; \frac{dv_\theta}{dr} \qquad\qquad (2.9)$$

in the limit as $f \to 1$. This is identical to Eq. (2.3) and is the result we anticipated in rolling Fig. 2.1 into a cylinder. Equation (2.7) is more flexible than Eq. (2.8), since its applicability is not limited to vanishingly small gaps.

While conceptually easy to understand, there are several difficulties associated with the concentric cylinder viscometer. Particularly at high rates of shear, it is difficult to maintain constant temperature conditions. Although experimental data can be corrected for some temperature variation, this is an undesirable complication. In the analysis presented above, it appears irrelevant whether the outer cylinder rotates while the force is measured on the inner cylinder or vice versa. It seems that only the direction of the gradient—a matter of sign— would be affected by the reversal. It turns out, however, that there is a greater tendency toward turbulent flow when the inner cylinder rotates. The derivation of Eq. (2.7) is based on the assumption of laminar flow. Instrument design is simplified by using a rotating center spindle so variations of the latter are widely encountered.

A final objection to concentric cylinder viscometers is that the velocity gradient is not uniform throughout the gap. Since a very wide range of shear rates is accessible, it may be quite acceptable to use the average value in the gap to describe a particular experiment. The nonuniformity of the velocity gradient can result in some interesting phenomena for pseudoplastic materials. If the velocity gradient is different in different parts of the gap, then it follows that a pseudoplastic material will show different viscosities in the sheared sample. This manifests itself in a tendency for the sample to flow out of the region of high shear, perhaps climbing the wall where the shear is less. A whole assortment of phenomena of this sort have been studied. Such observations are sometimes called Weissenberg effects.

In the light of the difficulties described above, it is not surprising that in many areas of technology viscosity is measured with rotating devices for which the geometry deviates some from precisely coaxial cylinders. Instruments with modified geometry are simpler to construct and therefore open to such conveniences as interchangeable spindles as well as variable speeds. Although they may lack the precise geometry which permits the absolute determination of viscosity from experimentally measurable parameters, they are sturdy and versatile and permit reproducible observations to be made. These observations may serve as the basis for comparison of relative viscosities of samples measured with the same apparatus, or they may be calibrated with substances of known viscosity so that unknown viscosities may be evaluated. Figure 2.4 is an example of one such commercial apparatus. The instrument shown is the Brookfield digital viscometer, and—with up to eight different speeds and seven different spindles—can measure 56 different average ranges for dv/dr.

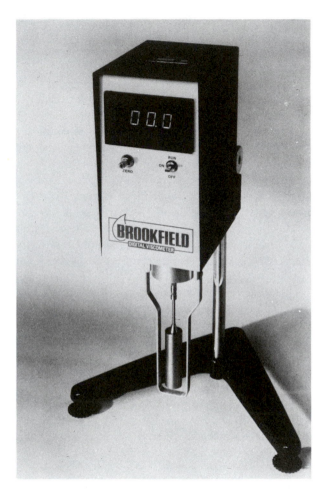

Figure 2.4   A commercial instrument, the Brookfield Digital Viscometer, based on the geometry of the concentric cylinder viscometer. (Photo courtesy of Brookfield Engineering Laboratories, Inc., Stoughton, Mass. 02072.)

2.4   The Power Law for Pseudoplastic Liquids

Figure 2.5 shows some actual experimental data for σ_s versus $\dot{\gamma}$, measured on a sample of polyethylene at $126°C$. Note that the data are plotted on log–log coordinates. In spite of the different coordinates, Fig. 2.5 is clearly an example of pseudoplastic behavior as defined in Fig. 2.2. In this and the next several sections, we discuss shear-dependent viscosity. In this section the approach is strictly empirical, and its main application is in correcting viscosities measured

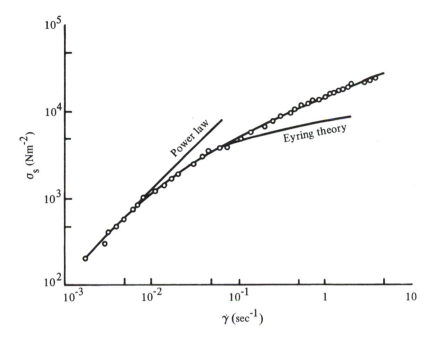

Figure 2.5 Shearing force per unit area versus shear rate. The experimental points are measured for polyethylene, and the labeled lines are drawn according to the relationship indicated. (Data from J. M. McKelvey, *Polymer Processing*, Wiley, New York, 1962.)

at one rate of shear for use under somewhat different shear conditions. In subsequent sections we shall turn to a model for the origin of pseudoplasticity at the molecular level. Although the actual data points in Fig. 2.5 do not describe a straight line, it is apparent that portions of the data over at least an order of magnitude variation in $\dot{\gamma}$ do display linearity. The line labeled "power law" in Fig. 2.5 is an example of such a linear domain. The straight line shown has been drawn with a slope of unity. It can be seen from Fig. 2.5 that other straight lines representing portions of the data at higher rates of shear would be described by lines with slopes less than unity. This is characteristic of pseudoplasticity.

Since any straight line obeys the equation $y = mx + b$, the equation for any tangent drawn in Fig. 2.5 will be

$$\log \sigma_s = m \log \dot{\gamma} + \text{const.} \tag{2.10}$$

where m is the local slope. Taking the antilogs in Eq. (2.10) converts that result to

$$\sigma_s = K\dot{\gamma}^m \tag{2.11}$$

For the straight line in Fig. 2.5 where m = 1.0, this equation expresses direct proportionality between σ_s and $\dot{\gamma}$, the condition of Newtonian behavior. In the non-Newtonian region where m < 1, Eq. (2.11) may describe the data over an order of magnitude or so. Next we consider the relationship between the constant K and viscosity. If Eq. (2.11) is solved for K and the resulting expression multiplied and divided by $\dot{\gamma}^{1-m}$, we obtain

$$K = \frac{\sigma_s}{\dot{\gamma}^m} \frac{\dot{\gamma}^{1-m}}{\dot{\gamma}^{1-m}} = \frac{\sigma_s}{\dot{\gamma}} \dot{\gamma}^{1-m} = \eta\dot{\gamma}^{1-m} \tag{2.12}$$

Since K is a constant, two different $\eta-\dot{\gamma}$ combinations measured in the same regime can be related to each other in the form given by Eq. (2.12) because both factors must equal the same constant; that is,

$$\eta\dot{\gamma}^{1-m} = \eta_r \dot{\gamma}_r^{1-m} \tag{2.13}$$

where one of the $\eta-\dot{\gamma}$ combinations is distinguished by a subscript, the r signifying a reference or calibration point. Rearranging, this result becomes

$$\eta = \eta_r \left(\frac{\dot{\gamma}}{\dot{\gamma}_r}\right)^{m-1} \tag{2.14}$$

When m = 1.0, as in Fig. 2.5, the exponent becomes zero and the viscosity is independent of $\dot{\gamma}$; when m = 0.7, a factor of 10 change in $\dot{\gamma}$ results in a decrease of viscosity by a factor of 2. This is approximately the case for the data in Fig. 2.5 for $\dot{\gamma}$ values between 10^{-2} and 10^{-1} sec^{-1}. Equation (2.14) and its variations are called power laws. Relationships of this sort are valuable empirical tools for extrapolating either F/A or η over modest ranges of $\dot{\gamma}$. In such an application, the exponent m − 1 and the proportionality constant $\eta_r \dot{\gamma}_r^{1-m}$ are evaluated in a calibration experiment, and Eq. (2.14) is nothing more than a two-parameter empirical equation.

Equation (2.14) has the advantage of simplicity; its drawback is that we learn nothing about either the nature of viscosity or the nature of the sample from the result. In the next few sections we shall propose and develop a molecular model for the flow process. The goals of that development will be not only to describe the data, but also to do so in terms of parameters which have some significance at the molecular level. Before turning to this, it will be helpful if we consider a bit further the form of Eq. (2.14).

The power law developed above uses the ratio of the two different shear rates as the variable in terms of which changes in η are expressed. Suppose that instead of some reference shear rate, values of $\dot{\gamma}$ were expressed relative to some other rate, something characteristic of the flow process itself. In that case Eq. (2.14) or its equivalent would take on a more fundamental significance. In the model we shall examine, the rate of flow is compared to the rate of a chemical reaction. The latter is characterized by a specific rate constant; we shall see that such a constant can also be visualized for the flow process. Accordingly, we anticipate that the molecular theory we develop will replace the variable $\dot{\gamma}/\dot{\gamma}_r$ by a similar variable $\dot{\gamma}/k_R$, where k_R is the rate constant for the flow process.

Figure 2.5 reveals that polymer viscosity approaches Newtonian behavior for sufficiently low rates of shear. From an empirical point of view, this simply means that $m \to 1$ as $\dot{\gamma} \to 0$. From a molecular point of view, in the region of Newtonian behavior the rate of shear is small compared to the rate constant for the flow process. When molecular displacements occur very much faster than the rate of shear ($\dot{\gamma} \ll k_R$), the molecules show maximum efficiency in dissipating the applied forces. When the molecules cannot move fast enough to keep pace with the external forces, they couple with and dissipate those forces to a lesser extent. Thus there is a decrease in viscosity from its upper, Newtonian limit with increasing $\dot{\gamma}/k_R$. The rate constant for the flow process is therefore seen to define a standard against which the rate of shear is to be judged large or small. In the next section we shall consider a molecular model in terms of which this rate constant can be analyzed.

We shall consider a number of different models in this volume. Before proceeding, it might be helpful to remember the status of the model in the methodology of science. Several things come to mind:

1. Models are the handiwork of theoreticians and may be mechanical or electrical analogs, pictorial representations, or purely mathematical constructs.
2. There must be ongoing comparison of experimental results and modelistic predictions. Experiment and theory feed each other.
3. Models necessarily involve approximations: They focus on one facet of behavior to the exclusion of others.
4. Models may succeed in some areas and not in others. This is common and must be remembered in evaluating the success or failure of a model.
5. Models are open to revision. Ideally, the good ones get better and the poor ones are abandoned.
6. There is a tendency among both students and researchers to place more faith in the model than it deserves. Successful models may unify many observations for us and hence be highly valued. The observation always has priority, however.

7. Parameters of interest evaluated via a model may reflect the model as well as the system under consideration.

The statistical model of the random coil discussed in Chap. 1 illustrates many of these items.

2.5   A Model for Flow in Liquids

We begin developing a model for viscous flow for a liquid of small molecules which we can visualize as filling space like beads might fill a beaker. Describing such a beaker as "full" means that it can hold no more beads, and not that it contains no empty space. There are empty spaces in such an assembly due both to the loose, imperfect packing that would ordinarily result from tossing beads into a beaker and to the interstitial spaces between the beads, even in dense packing. The analogous space in a liquid is called the free volume and is responsible for the fact that liquids are compressible. Figure 2.6 represents such an array of spherical particles. We wish to describe the flow of this liquid in terms of steps occurring at the molecular level. We can picture one such elementary step in the flow of the liquid in Fig. 2.6 from left to right by considering the displacement of particle P to hole H. Note that it is equivalent to focus on the hole and consider it being displaced to the left. Since there are fewer holes to account for, the latter process may be easier.

If we were required to pack beads in a beaker, we know from experience that by jostling the container we could achieve some compaction or decrease in free volume. In fact, we can picture the flow of a huge array of beads through a pipe by considering the beaker as a volume element in that pipe. By vibration, the beads are jostled downward; that is, the holes work their way to the top.

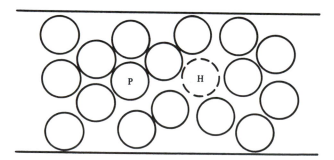

Figure 2.6   Model of a particle P, a hole H, and the bottleneck between them in a liquid of spherical molecules.

Now we imagine the bottom of the beaker being lowered by a small amount such that the lower layers of beads can loosen up a bit. Additional vibration will result in additional compaction as the newly formed holes are worked out. The bottom is then lowered again and the process repeated until the whole array has been moved along an appreciable distance. The entire process can be described in terms of opening holes and moving them through the array. In terms of Fig. 2.6 the exchange of positions between particle P and hole H is the elementary step of this mechanism.

The beads and the holes are necessary to the previous discussion, but by themselves are insufficient to describe the process. The vibration also plays a crucial role. Here, too, our analogy extends to molecular systems, with temperature being the measure of kinetic energy. Because of the free volume present in a liquid, we can imagine each molecule as enclosed by a cage of its neighbors. Because we are concerned with temperatures above absolute zero, these molecules possess thermal energy which is distributed through a variety of different storage modes. When we discuss molecules in the gaseous state where the interactions between neighbors can be neglected, we think of the molecules as moving from place to place, as tumbling in all directions as they move, and as having some parts which oscillate against other parts. We speak of these as translational, rotational, and vibrational modes of energy storage, respectively. An important result of classical physics is the equipartition principle, which says that the energy stored in each of these modes is some multiple of kT, where k is the Boltzmann constant, 1.38×10^{-23} J K^{-1} molecule^{-1}, and T is the absolute temperature. Note that this result may be scaled up to apply to a mole of particles simply by multiplying by Avogadro's number N_A. The product of N_A and k is the gas constant R; in SI units R is 8.314 J K^{-1} mol^{-1}. In discussions of thermal energy, kT and RT are often used interchangeably, the understanding being that the former applies "per molecule" while the latter is "per mole."

The equipartition principle is a classic result which implies continuous energy states. Internal vibrations and to a lesser extent molecular rotations can only be understood in terms of quantized energy states. For the present discussion, this complication can be overlooked, since the sort of vibration a molecule experiences in a cage of other molecules is a sufficiently loose one (compared to internal vibrations) to be adequately approximated by the classic result.

Next let us consider the source of resistance to flow. Examination of the path between particle P and hole H in Fig. 2.6 reveals the source of this resistance. The moving particle must pass through the bottleneck formed by neighboring particles along the path between P and H. All of the particles in the assembly are equivalent inasmuch as each vibrates with a certain energy around an equilibrium position separated from neighbors by an average distance. Neither the energy nor the separation are identical for all molecules, but average values for

these quantities are characteristic of the sample. This means that adjacent molecules must be dislodged from their equilibrium positions by the migration of P to H. An expenditure of energy is needed to do this.

Figure 2.7 shows a representation of this situation. The ordinate is an energy axis and the abscissa is called the reaction coordinate and represents the progress of the elementary step. In moving from P to H, the particle simply moves from one equilibrium position to another. In the absence of any external forces, the energy of both the initial and final locations should be the same as shown by the solid line in Fig. 2.7. Between the two minima corresponding to the initial and final positions is the energy barrier arising from the dislodging of the particles neighboring the reaction path from *their* positions of minimum energy.

We conclude this section by noting that the picture we have developed can be extended, with some modification, to polymers relatively easily. From a pictorial point of view, we have compared our system to a beaker of beads. To obtain a representation of polymers we need only to string the beads together. The number of beads on a string gives the degree of polymerization of the polymer. If we string together only a few of the beads and leave most free, the resulting system can be compared to a solution of polymers in a solvent of low

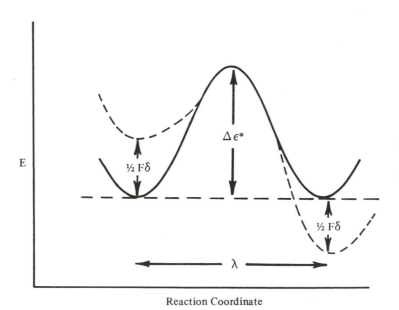

Reaction Coordinate

Figure 2.7 Potential energy as a function of location along the reaction coordinate. The solid line describes an undisturbed liquid; the broken line applies to liquids subjected to shearing force.

molecular weight molecules. On the other hand, tying all the beads into strings of more or less uniform length represents a polymer melt. All this can be accomplished in our model without any modification of the basic picture, except that the bead now represents a flow segment rather than an individual molecule. Now, however, the movement of one flow segment necessarily involves adjustment of the location of other segments in the chain in addition to dislodging those in the flow path. This is another consideration which makes it easier to focus attention on the holes, since these are not entrained with other units. We shall see in subsequent sections that a good deal of effort has been directed toward understanding the movement of chains through polymer melts. For our present purposes, however, it is sufficient to continue thinking of the elementary flow step as a flow segment and a hole exchanging positions. Note that the flow segment and the repeat unit of the polymer molecule are not necessarily the same.

2.6 Flow As a Rate Process

In this section we shall examine the analogy between the flow of a liquid and the rate of a chemical reaction. This approach has been developed extensively by Eyring and co-workers and has been applied to a wide variety of deformation processes and systems.

At first glance it may seem strange to compare the rate of fluid flow to the rate of a chemical reaction, but the following analogy may help. The array of particles shown in Fig. 2.6 may be thought of as the "atoms" in a large molecule. Particle P and hole H are clearly different kinds of "atoms." The exchange of positions between P and H is thus comparable to an atomic rearrangement: an isomerization! The rate of such a chemical reaction is generally directly proportional to the concentration of the reacting species C and the proportionality constant is called the specific rate constant:

$$\text{Rate} = k_R C \tag{2.15}$$

This is a phenomenological equation, relating observables; what we seek is the molecular basis for this expression. Eyring's approach focuses attention on the energy barrier along the reaction coordinate as the source of this insight. The molecular situation at the top of the energy barrier is described as an activated complex, so the corresponding theory is called the activated complex theory.

The activated complex theory has been developed extensively for chemical reactions as well as for deformation processes. The full details of the theory are not necessary for us. Instead, it is sufficient to note that k_R can be written as

$$k_R = Ae^{-\Delta\epsilon^*/kT} \tag{2.16}$$

where A gives the frequency with which molecules cross the energy barrier at infinite temperature. The exponential multiplier indicates what fraction of that frequency is achieved at temperature T. As indicated in Fig. 2.7, $\Delta\epsilon^*$ is the height of the energy barrier above the equilibrium energy. It is significant that $\Delta\epsilon^*$ occurs in ratio with kT in Eq. (2.16), since it is thermal energy that allows molecules to pass such barriers. Thus an energy barrier that may be prohibitively high at low temperatures is less of an obstacle at higher temperatures. Note that the exponent must be dimensionless, so that the units of $\Delta\epsilon^*$ and k must be self-consistent; both can be expressed on either a per-molecule or per-mole basis. In the latter case kT becomes RT, as previously noted.

It is generally true and especially clear in the case of the barrier represented by the solid line in Fig. 2.7 that movement in either the forward or the backward direction across the barrier is possible. Hence the *net* rate of the process is written as the difference between the rate forward (subscript f) and the rate backward (subscript b):

$$\text{Net rate} = A\left[(e^{-\Delta\epsilon^*/kT})_f - (e^{-\Delta\epsilon^*/kT})_b\right]C \tag{2.17}$$

where the high-temperature frequency A is taken to be the same in both directions. If we apply Eq. (2.17) to the flow process described by the solid line in Fig. 2.7, we see that the height of the energy barrier is the same in both directions; hence the two exponentials are identical and their difference is zero. There is no net flow, since forward and reverse processes cancel, although at the molecular level fluctuations in density occur. When we examine light scattering in Chap. 10, we shall see that these fluctuations play a role in the discussion of that subject. In the next section we consider how this situation is modified by the application of an external force.

2.7 The Effect of Shearing Forces

In this section we continue our discussion of flow viewed as a rate process, applying our model explicitly to polymer melts. We begin by considering how the energy barrier to flow is modified by the application of an external force. Figure 2.8 will enable us to do this in the case of a shearing force F. Figure 2.8 represents a tangle of polymer molecules. A portion of one chain is shown; its neighbors occupy the shaded region of the diagram. Critical to the development of this picture are the holes labeled H. On the average such vacancies are assumed to be separated by the distance λ, as shown in Fig. 2.8. We are interested in the displacement of the molecule from left to right by suitable segments of the chain exchanging positions with hole H. By appropriate adjustments of position, the effect of this displacement is damped out over the distance λ.

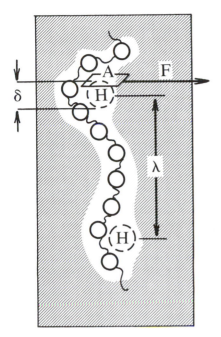

Figure 2.8   Model for the displacement of a polymer chain in relation to a hole H.

The affected portions of the chain define the flow segment according to this picture.

We can imagine that the barrier to this displacement occurs at the midpoint of the forward step $\delta/2$, where δ characterizes the dimensions of the symmetrical holes. Since the product of a force and a distance is energy, we see that the forward motion is facilitated by an amount of energy given by $F\delta/2$. Another way of thinking of this is to imagine that the height of the barrier to the forward motion is lowered by this amount. This is equivalent to starting from an initial state which is higher in energy by $F\delta/2$ than the unstressed fluid. This situation is indicated by the broken line in Fig. 2.7. A corollary of this is that the backward motion is impeded by this extra energy in addition to $\Delta\epsilon^*$. Hence the height of the backward barrier is increased by $F\delta/2$. This is equivalent to initiating the backward migration from a lower-energy state; it is also shown by the broken line in Fig. 2.7.

Under these circumstances the two exponentials in Eq. (2.17) are no longer identical but become

$$\text{Net rate} = A\left[\exp\left(\frac{-(\Delta\epsilon^* - F\delta/2)}{kT}\right) - \exp\left(\frac{-(\Delta\epsilon^* + F\delta/2)}{kT}\right)\right]C = k_R' C$$

$$(2.18)$$

where the prime on the k_R reminds us that this expression applies when an external force is applied in the direction of the flow. Writing the sum of exponents as the product of two exponentials enables us to rearrange Eq. (2.18) to

$$k'_R = Ae^{-\Delta\epsilon^*/kT}(e^{F\delta/2kT} - e^{-F\delta/2kT}) = k_R(e^{F\delta/2kT} - e^{-F\delta/2kT})$$

(2.19)

This expression gives us the rate constant for the net rate of forward flow according to the activated complex theory.

Now let us consider the significance of the factor $\delta k_R/\lambda$. According to Fig. 2.8, δ gives the forward distance a molecule moves in a single flow step and k_R gives the frequency of such steps. The product of the two therefore gives the velocity of forward displacement. Equation (2.1) shows that the rate of shear is given by the velocity at which a shear displacement is propagated divided by the distance over which the displacement develops. In the notation of this model, the latter is λ. Therefore, $\delta k_R/\lambda$ is identical to the rate of shear:

$$\dot{\gamma} = \frac{\delta k_R}{\lambda}$$

(2.20)

Combining Eqs. (2.19) and (2.20) gives

$$\dot{\gamma} = \frac{\delta k_R}{\lambda}(e^{F\delta/2kT} - e^{-F\delta/2kT})$$

(2.21)

The difference in exponentials which occurs in Eq. (2.21) is directly related to the hyperbolic sine function

$$\sinh x = \tfrac{1}{2}(e^x - e^{-x}) = y$$

(2.22)

Accordingly, Eq. (2.21) can be rewritten as

$$\dot{\gamma} = \frac{2\delta k_R}{\lambda} \sinh\left(\frac{F\delta}{2kT}\right)$$

(2.23)

Even though they look very different, Eqs. (2.21) and (2.23) are identical.

If we multiply and divide the argument of the sinh by A, we achieve the desired result: a relationship between F/A and $\dot{\gamma}$. Remembering that F/A is the same as σ_s, we write

$$\dot{\gamma} = \frac{2\delta k_R}{\lambda} \sinh \frac{\delta A\sigma_s}{2kT}$$

(2.24)

Since the sinh function is less familiar than simple exponentials, an example illustrating Eq. (2.24) may be helpful.

Example 2.1

The σ_s (in N m^{-2}) versus $\dot{\gamma}$ (in sec^{-1}) behavior of poly(dimethyl siloxane) [$\bar{M}_w = 3.2 \times 10^6$] at 25°C has been found to follow the expression

$$\dot{\gamma} = 5.75 \times 10^{-9} \sinh\left(3.60 \times 10^{-3}\frac{F}{A}\right)$$

(a) Find $\dot{\gamma}$ when $\sigma_s = 681$ N m^{-2} and (b) Find σ_s when $\dot{\gamma} = 1.70 \times 10^{-6}$ sec^{-1}.

Solution

(a) Evaluate $\dot{\gamma}$ directly from the expression given, using tables of the sinh function: $3.60 \times 10^{-3}\sigma_s = (3.60 \times 10^{-3})681 = 2.45$; sinh $2.45 = 5.75$ from tables of sinh x versus x; or by direct calculation, $5.75 \times 10^{-9}(5.75) = 3.31 \times 10^{-8}$ sec^{-1}. The experimental[†] value was 3.34×10^{-8} sec^{-1}. (b) Evaluate σ_s from Eq. (2.22), using the inverse sinh: $\dot{\gamma}/5.75 \times 10^{-9} = 1.70 \times 10^{-6}/5.75 \times 10^{-9} = 296$; sin x = 296. We may use tables of sinh x versus x or the identity given in Table 2.1 to find the inverse sinh. The procedure is analogous to finding an antilog: $\sinh^{-1} 296 = x = 6.38$; $6.38 = 3.60 \times 10^{-3}\sigma_s$; $\sigma_s = 1.77 \times 10^3$ N m^{-2}. The experimental[‡] value was 1.78×10^3 N m^{-2}.

•

Table 2.1 Some Useful Relationships Involving the Hyperbolic Sine and Inverse Hyperbolic Sine Function

Hyperbolic sine	Inverse hyperbolic sine
Notation y = sinh x	x = sinh^{-1} y
Definition–identity $\sinh x = \frac{1}{2}(e^x - e^{-x})$	$\sinh^{-1} y = \ln(y + \sqrt{y^2 + 1})$
Series expansion $\sinh x = x + \dfrac{x^3}{3!} + \dfrac{x^5}{5!} + \cdots$	$\sinh^{-1} y = y - \dfrac{1}{2}\dfrac{y}{3} + \dfrac{1}{2}\dfrac{3}{4}\dfrac{y^5}{5} - \cdots$
Limiting values $x \to 0, \ \sinh x \to 0;$	$y \to 0, \ \dfrac{1}{y}\sinh^{-1} y = 1;$
$x \to \infty, \ \sinh x \cong \dfrac{1}{2}e^x \to \infty$	$y \to \infty, \ \dfrac{1}{y}\sinh^{-1} y = 0$
Miscellaneous $\sinh(-x) = -\sinh x$; not periodic	

[†] D. J. Plazek, W. Dannhauser, and J. D. Ferry, *J. Colloid Sci.* 16:101 (1961).
[‡] Ibid.

The example shows that Eq. (2.24) successfully describes the behavior of this polymer under shear over more than two orders of magnitude.

Introduction of the inverse hyperbolic sine function encourages us to take Eq. (2.24) a bit further and derive an expression for η itself. Before continuing, let us remember the following:

1. $\sinh^{-1} y$ is the inverse or arc function, *not* the reciprocal.
2. While less familiar, finding $\sinh^{-1} y$ is no more complicated than finding an antilog.
3. Numerical values of $\sinh x$ versus x are tabulated in standard sources (e.g., *The Handbook of Tables for Mathematics*, Chemical Rubber Company) and the tables can be used to evaluate either $\sinh x$ or $\sinh^{-1} y$.
4. The identities for both $y = \sinh x$ and $x = \sinh^{-1} y$ in Table 2.1 may be used to calculate these quantities directly, eliminating the need to interpolate.

With this change in notation, Eq. (2.24) becomes

$$\sigma_s = \frac{2kT}{\delta A} \sinh^{-1} \frac{\lambda \dot{\gamma}}{2\delta k_R} = \frac{2kT}{\delta A} \sinh^{-1} \beta \dot{\gamma} \tag{2.25}$$

where we have defined β to replace the cluster of constants:

$$\beta = \frac{\lambda}{2\delta k_R} \tag{2.26}$$

Finally, if we multiply and divide the right-hand side of Eq. (2.25) by $\beta \dot{\gamma}$, we obtain

$$\sigma_s = \frac{2kT\beta}{\delta A} \left(\frac{\sinh^{-1} \beta \dot{\gamma}}{\beta \dot{\gamma}} \right) \dot{\gamma} \tag{2.27}$$

Comparing this result with Eq. (2.2) gives

$$\eta = \frac{2kT\beta}{\delta A} \frac{\sinh^{-1} \beta \dot{\gamma}}{\beta \dot{\gamma}} \tag{2.28}$$

We shall refer to Eq. (2.28) as the Eyring viscosity equation.

In the next section we shall examine the testing and applications of this result; for now we justify the efforts of this section by observing that Eq. (2.28) reduces to two simple cases when the limits presented in Table 2.1 are considered:

$$\eta \rightarrow \eta_N = \frac{2kT\beta}{\delta A} \qquad \text{as } \beta \dot{\gamma} \rightarrow 0 \tag{2.29}$$

and

$$\eta \rightarrow 0 \qquad \text{as } \beta\dot{\gamma} \rightarrow \infty \tag{2.30}$$

Regardless of the value of β, these limits are obtained at sufficiently small or large values of shear rate; hence we have successfully shown the following:

1. Polymers display Newtonian behavior (η_N = constant) at sufficiently low rates of shear [Eq. (2.29)].
2. Polymers display pseudoplasticity, with apparent viscosity decreasing as $\dot{\gamma}$ increases [Eq. (2.30) and Fig. 2.2].
3. The details of how η varies between these two limits are given by Eq. (2.28).
4. $\beta\dot{\gamma}$, which is proportional to $\dot{\gamma}/k_R$, is a natural variable for describing the dependence of σ_s or η on $\dot{\gamma}$. This result was anticipated in Sec. 2.4.

2.8 Testing the Eyring Equation

While the Eyring equation gives a relationship between σ_s and $\dot{\gamma}$ which allows for both Newtonian and pseudoplastic behavior, the result does not appear too promising. To begin with, the inverse hyperbolic sine function is discouraging and, worse yet, the relationship involves four adjustable parameters: k_R, δ, λ, and A. Having this many parameters imparts a flexibility to the theory which enhances the chance of fitting it to experimental data. On the other hand, we might be hard pressed to find suitable values for these parameters to evaluate η via Eq. (2.25). Furthermore, the possibility of extracting useful molecular information from an equation which contains so many parameters seems slim. To assess this situation more thoroughly, let us reexamine the parameters in Eq. (2.25) and the grouping in which they occur.

Probably the easiest place to begin is with a consideration of the product δA. As defined by Fig. 2.8, δ is the average diameter of a hole and A is its average cross section.

Accordingly, the product δA is about the same as V_h, the volume of the hole. For liquids of low molecular weight, V_h is on the order of 0.5% the volume of the liquid at room temperature, increasing to 2–3% at the boiling point, and larger yet at still higher temperatures. This identification makes δA easier to visualize, but it still leaves us with four parameters: k_R, δ, λ, and V_h.

Next we recognize that δ and λ always appear as a ratio in our theory. If we argue that the hole and the polymer chain have comparable cross-sectional areas, we can multiply both the numerator and denominator of the λ/δ ratio by this cross section and convert it into the ratio V_s/V_h, where V_s is the volume of the flow segment of length λ. While we know neither of these volumes directly, there are indications that V_s/V_h may be on the order of 10–20 for many linear

polymers. Since only the ratio concerns us, we may think of λ/δ or the equivalent V_s/V_h as a single parameter which describes how much larger than the hole the flow segment is. This leaves us with three parameters: k_R, V_h, and V_s/V_h.

Of the adjustable parameters in the Eyring viscosity equation, k_R is the most important. In Sec. 2.4 we discussed the desirability of having some sort of natural rate compared to which rates of shear could be described as large or small. This natural standard is provided by k_R. The parameter k_R entered our theory as the factor which described the frequency with which molecules passed from one equilibrium position to another in a flowing liquid. At this point we will find it more convenient to talk in terms of the period of this vibration rather than its frequency. We shall use τ to symbolize this period and define it as the reciprocal of k_R. In addition, we shall refer to this characteristic period as the relaxation time for the polymer. As its name implies, τ measures the time over which the system relieves the applied stress by the relative slippage of the molecules past one another. In summary,

$$\text{Relaxation time} = \tau = \frac{1}{k_R} \propto \beta \qquad (2.31)$$

From now on, we shall use the product $\tau\dot{\gamma}$ instead of $\dot{\gamma}/k_R$ as the natural variable to describe η. Small values of $\tau\dot{\gamma}$ mean that the molecule responds rapidly compared with the duration of the shearing force. Regardless of the value of τ, this condition is met at sufficiently low rates of shear. In terms of Eq. (2.29), polymers are expected to show Newtonian behavior as long as the rate of shear is low enough to allow the molecules to respond; when the velocity gradient is too large, the molecules are unable to keep up, and non-Newtonian behavior results. Thus it is not surprising that low molecular weight liquids are generally Newtonian: They can respond to shear much faster than the more entangled polymer molecules.

Of the various parameters introduced in the Eyring theory, only τ—or β, which is directly proportional to it—will be further considered. We shall see that the concept of relaxation time plays a central role in discussing all the deformation properties of bulk polymers and thus warrants further examination, even though we have introduced this quantity through a specific model.

The concept of free volume is taken up again in Chap. 4.

In the following example we illustrate how values for β and τ can be extracted from experimental data.

Example 2.2

Test the ability of Eq. (2.28) to represent the data of Fig. 2.5, and evaluate β. Estimate τ, recalling that $\beta = \frac{1}{2}(V_s/V_h)\tau \cong 10\tau$.

Table 2.2 Calculated Values for the Evaluation of Eyring Parameters for the Data and the Theoretical Curve of Fig. 2.5

Part I: Evaluation of parameters

$\dot{\gamma}$ (sec^{-1})	F/A (N m^{-2})	η (kg m^{-1} sec^{-1})	η/η_N
0.002	245	123,000	1.00
0.004	491	123,000	1.00
0.008	981	123,000	1.00
0.02	1,960	98,000	0.80
0.04	2,940	73,600	0.60
0.08	4,420	55,200	0.45
0.2	7,650	38,300	0.31
0.4	9,520	23,800	0.19
0.8	14,200	17,800	0.14

Part II: Calculation of theoretical curve

$\dot{\gamma}$ (sec^{-1})	$\beta\dot{\gamma}$ (β = 75 sec)	sinh^{-1} ($\beta\dot{\gamma}$)	$(1/\beta\dot{\gamma})$ sinh^{-1} ($\beta\dot{\gamma}$)	η (kg m^{-1} sec^{-1})	F/A (N m^{-2})
0.002	0.150	0.149	0.993	122,000	245
0.004	0.300	0.296	0.987	121,000	481
0.008	0.600	0.569	0.946	116,000	932
0.02	1.50	1.195	0.797	97,700	1,950
0.04	3.00	1.818	0.606	74,300	2,970
0.08	6.00	2.492	0.415	50,900	4,070
0.2	15.00	3.402	0.227	27,800	5,570
0.4	30.0	4.095	0.137	16,800	6,720
0.8	60.0	4.788	0.080	9,800	7,850

Solution

Apply Eq. (2.27) to some of the data points to evaluate the apparent viscosity at different $\dot{\gamma}$'s. The first section of Table 2.2 shows the results of such calculations. Note that the calculated η's are constant at low $\dot{\gamma}$ values, indicating Newtonian behavior. Table 2.2 also expresses all η values relative to the Newtonian limiting value η_N. Comparison of Eqs. (2.28) and (2.29) shows that η/η_N values decrease from the Newtonian limit by the fraction sinh^{-1} ($\beta\dot{\gamma}$)/$\beta\dot{\gamma}$.

Next, by trial and error, we try to find a value for y such that sinh^{-1} y matches one of the η/η_N fractions in Table 2.2, say η/η_N = 0.80. This is easily done using either tables of sinh functions or the equation given in Table 2.1. The following results show that it is possible to place y within a range and then narrow that range without much difficulty. Remember, it is the *inverse* sinh values we are examining:

		Approximation			
	1		2		3
Trial y	1	2	1.40	1.60	1.50
$\sinh^{-1} y$	0.88	1.44	1.14	1.25	1.20
$(1/y) \sinh^{-1} y$	0.88	0.72	0.81	0.78	0.80

Note that the ratio η/η_N equals 0.80 at $\dot{\gamma} = 0.02$ sec^{-1}, so we identify y = 1.50 with $\beta\dot{\gamma}$ when $\dot{\gamma} = 0.02$ sec^{-1}. Therefore $\beta = 1.50/0.02 = 75$ sec and $\tau \cong 7.5$ sec.

Once a value for β has been obtained, we can calculate the value of $(1/\beta\dot{\gamma})$ $\sinh^{-1}$ $(\beta\dot{\gamma})$ for a range of $\beta\dot{\gamma}$ values. These fractions are tabulated in the second part of Table 2.2. We multiply η_N by these fractions to obtain η values and multiply these by $\dot{\gamma}$ to obtain σ_s. Calculated values of $\dot{\gamma}$ and σ_s are also tabulated and the latter are plotted in Fig. 2.5.

Examination of Table 2.2 and Fig. 2.5 enables us to judge the success of the Eyring equation. In the region of first deviations from Newtonian behavior the agreement between theory and experiment is excellent. Since the parameters are fitted with these data, no great significance should be attached to the fact that the agreement occurs here rather than elsewhere. Perhaps an equally good fit could be obtained with other parameters for a different range of experimental values. For the parameters used, theory and experiment agree to within 0.5% over an order of magnitude variation of $\dot{\gamma}$ and within 10% for one and a half orders of magnitude. On the log-log scale of Fig. 2.5 the agreement is excellent up to a $\dot{\gamma}$ value of about 1.0 sec^{-1}, at which point theoretical viscosities are less than experimental. One useful application of the Eyring equation is to extrapolate shearing forces, viscosities, and so on, over modest ranges from the conditions where they are measured.

This guarded assessment does not mean that the effort involved in deriving the Eyring equation was wasted—far from it. Several points might be noted:

1. Additional trial and error manipulation of the data might yield agreement over a somewhat wider range of conditions with slightly modified parameters.
2. The system considered in this example was chosen for convenience, not as either a best or worst case example. Eyring and co-workers have published numerous examples in which the theory (slightly modified) fits experimental data quite well.
3. The phenomenon under consideration *is* complicated and the theory developed in the last section is fairly simple—involved, but not really difficult.
4. We have successfully discovered that the transition from Newtonian to pseudoplastic behavior is governed by the product $\tau\dot{\gamma}$, or the relative values of the shear rate and the rate of molecular response.

5.· We have found an alternative to the power law, Eq. (2.14), which describes experimental data as well as the latter. In the Eyring approach, however, the curve-fitting parameters have a fundamental significance in terms of a model for the flow process at the molecular level.

Next let us briefly consider how the Eyring equation might be modified to improve its agreement with experiment. As noted in item (4) above, the relaxation time plays a very important role in the theoretical development and the practical application of the Eyring model. A little reflection readily convinces us that it is also the greatest weakness of the theory. Specifically, it seems highly improbable that the slithering of a polymer chain through a tangle of similar chains would be characterized by a single relaxation time. For some of the flow segments, the barrier to displacement may be the consequence of weak London forces which attract all molecules to each other. For other segments the barrier may be much higher owing to a virtual knotting of chains together through entanglements. The response time of a heavily entangled chain is expected to be significantly longer than that for a segment which moves the same way as a nontangled small molecule. This leads us to the idea that a whole spectrum of relaxation times rather than a single value might be a more suitable way to describe the flow process.

Suppose we divide the flow segments into classes according to relaxation times and index the various states by the subscript i. Thus the relaxation time and the component of shear stress borne by the segments in class i are τ_i and F_i, respectively. The applied shear force is related to the F_i's through

$$F = \sum_i f_i F_i \tag{2.32}$$

where f_i represents the fraction of the surface occupied by segments in the ith category. We might question why the summation of effects is conducted over F rather than by assigning to different classes different values of shear rate. The explanation is that all types of segments must experience the same rate of shear; otherwise different classes of segments would move at different rates and the flow process would segregate the different regions. Instead, all experience the same shear rate, with different amounts of stress.

Equations (2.27) and (2.32) may now be combined to give

$$\sigma_s = \left(\frac{2kT}{V_h} \sum_i f_i \beta_i \frac{\sinh^{-1}(\beta_i \dot{\gamma})}{\beta_i \dot{\gamma}} \right) \dot{\gamma} \tag{2.33}$$

In writing this result, we have assumed that the same value of V_h characterizes all classes and have indicated the variation in τ by attaching the subscript i to values of β. The summation is carried out over the entire spectrum of relaxation

times. According to this modification, the viscosity of the system is given by the bracketed terms in Eq. (2.33) rather than by Eq. (2.28).

One of our previous complaints was that we had more parameters than we knew what to do with; Eq. (2.33) makes this problem even worse. It turns out, however, that using only two or three terms of Eq. (2.33) results in a usable equation with improved curve-fitting ability. Techniques have been developed for extracting acceptable parameters from experimental data in these cases (see Problem 4). Figure 2.9, for example, shows data collected from a sample of natural rubber, analyzed according to a two-term version of Eq. (2.33). The line in Fig. 2.9 is drawn according to the equation

$$\sigma_s = (1.50 \times 10^4)(6.01 \times 10^{-5}) \frac{\sinh^{-1}(6.01 \times 10^{-5}\ \dot{\gamma})}{6.01 \times 10^{-5}\ \dot{\gamma}}$$

$$+ (1.80 \times 10^4)(6.80 \times 10^{-3}) \frac{\sinh^{-1}(6.80 \times 10^{-3}\ \dot{\gamma})}{6.80 \times 10^{-3}\ \dot{\gamma}}$$

Note that the second term involves a β value which is about 100 times larger than the β value of the first term. Since the proportionality between β and τ is expected to be about the same for both classes of flow elements, this ratio also

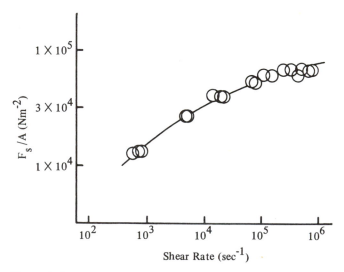

Figure 2.9 F_s/A versus shear rate for natural rubber. The line is drawn according to a two-term version of the Eyring theory. (Redrawn from Ref. 5.)

describes the relative values for the two relaxation times. Furthermore, the weighting factors for the two terms, $2kTf/V_h$, are about the same magnitude, suggesting equal surface concentration of two types of flow segments according to this analysis. This two-term, four-parameter expression accounts for the observed pseudoplasticity over three orders of magnitude of shear rate. The complication of additional terms not only results in a better fit of experimental results, but also makes sense at the molecular level. Especially when subjected to a wide range of impulse times for experimental force, an array of polymer molecules will respond with more than one characteristic relaxation time.

One of the striking omissions from our discussion has been an explicit consideration of polymer molecular weight on the viscous behavior of the sample. This omission will be corrected in the next section.

2.9 The Molecular Weight Dependence of Viscosity: Experimental Aspects

A basic theme throughout this book is that the long-chain character of polymers is what makes them different from their low molecular weight counterparts. Although this notion was implied in several aspects of the discussion of the shear dependence of viscosity, it never emerged explicitly as a variable to be investigated. It makes sense to us intuitively that longer chains should experience higher resistance to flow. Our next task is to examine this expectation quantitatively, first from an empirical viewpoint and then in terms of a model for molecular motion.

Figure 2.10 shows a plot of viscosity versus degree of polymerization for several polymers. Several things should be observed about this graph:

1. Figure 2.10 is drawn on log-log coordinates; more than two orders of magnitude in degree of polymerization are covered.
2. The viscosity values have been displaced vertically from each other by arbitrary amounts for display purposes.
3. The family of curves consist of two straight-line portions, with a change of slope occurring at a degree of polymerization in the range 10^2–10^3.
4. Numerous additional systems could be added to this display; the behavior is quite general.

One of the most striking features of Fig. 2.10 is the change of slope. Let us begin our discussion of this feature by examining the slopes of the linear portions of these lines on either side of the break. Representing the equation of a straight line by $y = ax + b$, we can write an empirical equation for the phenomena shown in Fig. 2.10 as

$$\log \eta = a \log n + b \qquad (2.34)$$

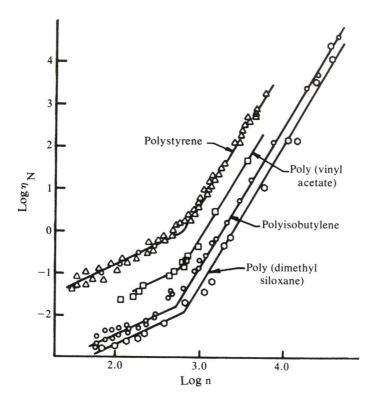

Figure 2.10 Log viscosity versus log molecular weight for several polymers. The lines are displaced vertically for display purposes. [From T. G. Fox and V. R. Allen, *J. Chem. Phys.* 41:344 (1964) with permission.]

where n is the degree of polymerization and a and b are the slope and intercept, respectively, of the linear regions. It is convenient to think of b as a logarithmic quantity also and write b = log B, so Eq. (2.34) becomes

$$\log \eta = a \log n + \log B \qquad (2.35)$$

Taking the antilog of both sides yields

$$\eta = Bn^a = B'M^a \qquad (2.36)$$

where the second equation follows from the direct proportion between n and the molecular weight M.

The two straight-line portions of the individual graphs are seen to be remarkably parallel, with slopes close to unity below the break and about 3.4 at higher

molecular weights. Since the slope gives the power to which the viscosity depends on molecular weight, we see that the dependence is one of direct proportion for low molecular weights—not too striking a result. For high molecular weights, on the other hand, viscosity shows a spectacular 3.4-power dependence on M. To appreciate the magnitude of this effect, we note that two samples of the same polymer differing by a factor of 2 in molecular weight will differ by a factor of more than 10 in viscosity if they fall on the high side of the break in Fig. 2.10. This extreme sensitivity of polymer viscosity to molecular weight is one of the things that makes a discussion of viscosity so important in polymer chemisty.

The empirical results contained in Fig. 2.10 present us with three challenges:

1. To account for a first-power dependence of viscosity on molecular weight for lower molecular weights.
2. To account for a 3.4-power dependence at higher molecular weights.
3. To account for the transition between the two.

It should be noted that a log–log plot condenses the data considerably and that the transition between a first-power and a 3.4-power dependence occurs over a modest range rather than at a precise cutoff. Nevertheless, the transition is read from the intersection of two lines and is identified as occurring at a degree of polymerization or molecular weight designated n_c or M_c, respectively.

In the next few sections we shall examine some of the theoretical effort which has been directed toward an understanding of these exponents. The principal difference in the theoretical approach to the two regimes of Fig. 2.10 involves the absence or presence of chain entanglements. In discussing the Eyring theory, we saw that different relaxation times might be associated with the relatively loose associations arising from London forces, while the knotting together of chains would result in more sluggish motion. A little reflection will convince us that polymers are not essentially different from any other class of compounds with respect to the first of these interaction modes: All molecules attract one another. The knotting mechanism, by comparison, is unique to long-chain structures, implying an ability of molecules to become entangled. Thus we expect that some sort of critical chain length must be exceeded before the mechanism of entanglement becomes effective. Although precise incorporation of this notion into a theoretical treatment of the phenomena is still incomplete, there is no doubt that this is the source of the breaks in the curves in Fig. 2.10. The subscript c which we have used to designate the degree of polymerization and molecular weight at the break point indicates a critical molecular weight above which entanglement effects make significant contributions to the observed viscosity.

Throughout this section we have spoken of molecular weight as if a single value characterized all samples. This contrasts sharply with the position taken in

Sec. 1.8, where polydispersity in ordinary samples was emphasized. Polydispersity clearly complicates things, especially in the neighborhood of n_c, where a significant number of molecules are too short to show entanglement effects while an equally significant fraction are entangled. We simply note that any study conducted with the intention of a molecular interpretation should be conducted on a sample with as sharp a distribution as possible.

The effect of polydispersity on the fitting of data by Eq. (2.36) may be treated as follows. We assume that, in the absence of entanglements, the observed viscosity of a polydisperse system is simply the average of viscosity contributions of each of the molecular weight fractions present, each weighted by the mass of that particular fraction in a unit volume of sample. The justification for using this weight average instead of some other rests on the theory to be presented in the next section [Eq. (2.56)]. Thus we write

$$\eta_i = K \rho_i M_i \tag{2.37}$$

where η_i, ρ_i, and M_i are, respectively, the viscosity, density, and molecular weight of the ith category of molecules; K is a constant. The experimental density and experimental viscosity are also related through the same relationship, except that some average of the molecular weight—we call it the viscosity average $\bar{M}_v$—is involved:

$$\eta_{ex} = K \rho_{ex} \bar{M}_v \tag{2.38}$$

We further note that $\rho_{ex} = \Sigma_i \rho_i$ and below the threshold for entanglements $\eta_{ex} = \Sigma_i \eta_i$. Substituting these into a combination of Eqs. (2.37) and (2.38) gives

$$\bar{M}_v = \frac{\Sigma_i \rho_i M_i}{\Sigma_i \rho_i} = \frac{\Sigma_i m_i M_i}{\Sigma_i m_i} \tag{2.39}$$

Equation (2.39) is the weight average molecular weight as defined in Sec. 1.8. It is important to note that this result, $\bar{M}_v = \bar{M}_w$, applies only in the case of nonentangled chains where η is directly proportional to M. A more general definition of $\bar{M}_v$ for the case where $\eta \propto M^a$ is

$$\bar{M}_v = \left(\frac{\Sigma_i m_i M_i^a}{\Sigma_i m_i} \right)^{1/a} = \left(\frac{\Sigma_i N_i M_i^{1+a}}{\Sigma_i N_i M_i} \right)^{1/a} \tag{2.40}$$

but this latter result does not apply in the event of chain entanglements (i.e., a > 1), since the simple additivity of viscosity contributions breaks down in

that case. As we shall see in Chap. 9, however, values of less than unity are often encountered in the study of the viscosity of polymer solutions. Entanglement is not a problem in dilute systems, so Eq. (2.40) is widely used in connection with solution viscosities.

2.10 A Theory for Viscosity in the Absence of Entanglements

Figure 2.11a shows a velocity gradient in a flowing liquid. Our interest is in the effect of this gradient on a (nonentangled) polymer molecule, regardless of where in the shear apparatus the molecule is located. We wish to set up a coordinate system centered on the molecule of interest. Suppose the center of mass of a polymer molecule is located in the layer labeled v_p in Fig. 2.11a. This is to be our reference point, and we wish to examine the polymer's motion relative to this point. Accordingly, we subtract v_p from each of the velocity vectors of Fig. 2.11a and shift the origin of our coordinate system to the center of mass. Figure 2.11b shows the result of this shift. This transformation enables us to see that the velocity gradient will cause the polymer molecule to rotate in a

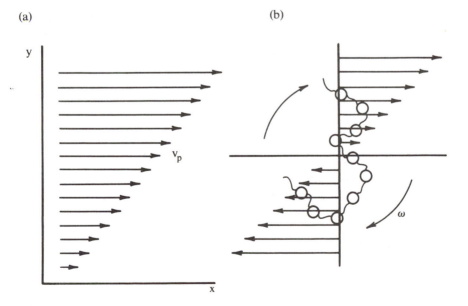

(a)   (b)

Figure 2.11 (a) The velocity gradient in a flowing liquid. (b) Velocities relative to the center of mass of a polymer molecule.

clockwise direction as shown in Fig. 2.11. Since the velocity gradients in Fig. 2.11 are the result of an applied shearing force, it is apparent that the induced rotation uses energy that was intended to produce translational motion. This is effectively a dissipation of that energy; hence Eq. (2.6) will permit us to evaluate η if the rate of energy dissipation can be related to the rate of shear.

To convert this strategy into an actual result, we must examine the mechanics of two-dimensional rotation. When the shear force is first applied, the molecule experiences an acceleration. Within a short time, however, the shear force and the force of viscous resistance to the particle movement equalize and no further net acceleration occurs. This means that the particle rotates with a constant average angular velocity. This final situation is called a stationary-state condition and is the subject of our attention. Our first problem, then, is to evaluate this average angular velocity. Even though the particle is in a stationary state, its velocity is not absolutely constant, but only constant on the average. To see that this is the case, we return to an inspection of Fig. 2.11b. Those polymer segments along the y axis bear the full brunt of the velocity gradient as an inducement to rotation. It can be shown (e.g., see Schultz's *Polymer Materials Science*) by an analysis of the torque balance under stationary-state conditions that for segments along the y axis (subscript y), $\omega_y = dv/dy$. At the same time those segments which lie along the x axis (subscript x), experience no difference in velocity relative to the center of mass of the molecule: $\omega_x = 0$. The angular velocity of a segment depends on its location relative to the center of mass in the molecule. Because of the symmetry of the rotation, the average angular velocity of a segment is the mean of these two extremes:

$$\omega = \frac{\omega_x + \omega_y}{2} = \frac{1}{2}\frac{dv}{dy} \tag{2.41}$$

Next, suppose we consider the tangential velocity v of segment i in a polymer molecule. The segment is located a distance r from the center of mass of the molecule and possesses an average angular velocity ω. The situation is sketched in Fig. 2.12a. Since $v = r\omega$, it follows that the x and y components of the velocity are given by

$$v_{x,i} = r\omega \sin\theta \tag{2.42}$$

and

$$v_{y,i} = r\omega \cos\theta \tag{2.43}$$

as shown in Fig. 2.12b.

(a)

(b)

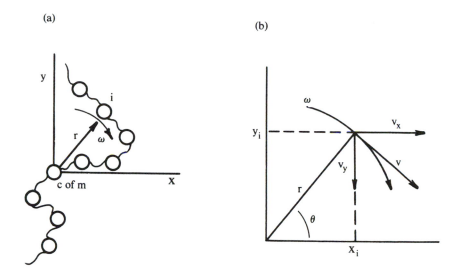

Figure 2.12 (a) Description of segment i in a polymer molecule relative to the center of mass. (b) Velocity components of segment i.

Now we must consider how the frictional force experienced by this segment is related to the tangential velocity whose components are given by these expressions.

In Chap. 9 we shall discuss in considerable detail a parameter called the molecular friction factor f. For velocities that are not too great, the friction factor expresses the proportionality between the frictional force a particle experiences and its velocity:

$$F_{vis} = fv \tag{2.44}$$

This is precisely the kind of thing we are looking for. Before proceeding, let us summarize some important properties of the friction factor:

1. It has the units required by Eq. (2.44), namely, kg sec^{-1} in the SI system.
2. For spherical particles of radius R moving through a medium of viscosity η, Stokes showed that the friction factor is given by

$$f = 6\pi\eta R \tag{2.45}$$

3. For particles of any shape at an absolute temperature T, Einstein showed that f is related to the experimental diffusion coefficient D by the expression

$$f = \frac{kT}{D} \tag{2.46}$$

where k is the Boltzmann constant.

4. The illustrious names associated with these expressions for f show that it is a factor of considerable importance and not merely a concept conjured up for the present discussion.

Strictly speaking, the friction factor as described above applies to a freely moving particle. We shall apply the same defining expression, Eq. (2.44), to the friction factor *per segment*, but shall use the symbol ζ to signify the latter. Note that neither Eq. (2.45) nor Eq. (2.46) can be exactly applied to ζ, since the segments do not move independently. Applying the concept of the segmental friction factor to the ith segment gives

$$F_{vis,i} = \zeta v_i \tag{2.47}$$

Next Eq. (2.47) can be substituted into Eqs. (2.42) and (2.43) to give the x and y components of the viscous force on segment i:

$$F_{x,i} = \zeta r \omega \sin \theta \tag{2.48}$$

and

$$F_{y,i} = \zeta r \omega \cos \theta \tag{2.49}$$

In connection with Eq. (2.6), we used the fact that the product of a viscous force and a velocity gives a rate of energy dissipation, so $F_{x,i} v_{x,i} + F_{y,i} v_{y,i}$ equals the rate of energy dissipation by segment i. Thus the energy loss per second for the ith segment $(\Delta W / \Delta t)_i$ is

$$\left(\frac{\Delta W}{\Delta t}\right)_i = \zeta r^2 \omega^2 (\sin^2 \theta + \cos^2 \theta) \tag{2.50}$$

Substituting Eq. (2.41) and recalling that $\sin^2 \theta + \cos^2 \theta = 1$, Eq. (2.50) becomes

$$\left(\frac{\Delta W}{\Delta t}\right)_i = \frac{1}{4} \zeta r_i^2 \left(\frac{dv}{dy}\right)^2 \tag{2.51}$$

For a polymer molecule consisting of n segments, this result must be summed over all the segments in the molecule to give the energy dissipated per second per polymer molecule $(\Delta W/\Delta t)_p$:

$$\left(\frac{\Delta W}{\Delta t}\right)_p = \sum_{i=1}^{n}\left(\frac{\Delta W}{\Delta t}\right)_i = \frac{1}{4}\,\zeta\sum_{i=1}^{n} r_i^2\left(\frac{dv}{dy}\right)^2 \tag{2.52}$$

The summation in Eq. (2.52) corresponds to n times the square of a two-dimensional radius of gyration $r_{g,2D}$ since there is no rotation in the z direction. Our next step, therefore, is to eliminate the summation in Eq. (2.52) and replace it with an expression for $r_{g,2D}$ and then $r_{g,3D}$.

The three-dimensional radius of gyration of a random coil was discussed in Sec. 1.10 and found to equal one-sixth the mean-square end-to-end distance of the polymer [Eq. (1.59)]. What we need now is a connection between two- and three-dimensional radii of gyration. Since the molecule has spherical symmetry $r_{g,3D}^2 = r_{g,x}^2 + r_{g,y}^2 + r_{g,z}^2 = 3r_{g,x}^2$. If only two of these contributions are present, we obtain $(2/3)r_{g,3D}^2 = r_{g,2D}^2$. We use this result and Eq. (1.59) to rewrite Eq. (2.52) in terms of the mean-square end-to-end distance $\overline{r^2}$ in the polymer coil:

$$\left(\frac{\Delta W}{\Delta t}\right)_p = \left(\frac{1}{4}\right)\left(\frac{2}{3}\right)\left(\frac{1}{6}\right)\zeta n\overline{r^2}\left(\frac{dv}{dy}\right)^2 \tag{2.53}$$

This expression gives the rate of energy dissipation per molecule while we seek an expression for the rate per unit volume. In the absence of molecular coupling through entanglement, we can multiply Eq. (2.53) with the number of molecules per unit volume to achieve the desired result. The latter is given by $\rho N_A/M$, where ρ is the density, N_A is Avogadro's number, and M is the molecular weight of the polymer. Equation (2.6) shows that the coefficient of viscosity equals the proportionality factor between the rate of energy dissipation per unit volume and the square of the velocity gradient; therefore we have shown that

$$\eta = \frac{n\zeta\overline{r^2}}{36}\,\frac{\rho N_A}{M} \tag{2.54}$$

Next we consider a substitution for the mean-square end-to-end distance $\overline{r^2}$. At first glance this seems easy, since Eq. (1.62) gives $\overline{r^2} = nl_0^2$. There are two problems associated with the use of this substitution:

1. The derivation of Eq. (1.62) explicitly requires that n be large, while the notion of nonentanglement which underlies (2.54) precludes large values of n. Since we have not confronted the question of how large is large in either of these derivations, we proceed with this substitution, hoping for a compatibility of n values.

2. The subscript 0 on l implies Θ conditions, a state of affairs characterized in Chap. 1 by the compensation of chain-excluded volume and solvent effects on coil dimensions. In the present context we are applying this result to bulk polymer with no solvent present. We shall see in Chap. 9, however, that coil dimensions in bulk polymers and in solutions under Θ conditions are the same.

With this substitution, Eq. (2.54) becomes

$$\eta = \frac{n^2 \zeta l_0^2}{36} \frac{\rho N_A}{M} \qquad (2.55)$$

Using the molecular weight of the repeat unit M_0 as the proportionality factor between n and M, we write

$$\eta = \frac{\zeta l_0^2 \rho N_A}{36 M_0^2} M = \frac{\zeta l_0^2 \rho N_A}{36 M_0} n \qquad (2.56)$$

This result was published by Debye in 1946. Since we shall also encounter a light-scattering equation associated with his name, we shall refer to Eq. (2.56) and its variations as the Debye viscosity equation.

It is the presence of ρ in the proportionality factor between η and M in the Debye viscosity equation that justifies the use of this quantity as a weighting factor in the definition of the viscosity average molecular weight [e.g., Eq. (2.37)]. Of the various parameters which appear in Eq. (2.56), only ζ is unfamiliar. While many polymers differ relatively little in l_0, ρ, or even M_0, it turns out that variation in ζ span a wide range. Defining a quantity such as ζ is easy, but if we attempt to assign a numerical value to it, we find ourselves at a loss. Accordingly, the following example gives us an idea of the magnitude of ζ on the basis of some experimental viscosity data.

Example 2.3

At 217°C (490 K) a polyisobutylene sample of $\bar{M}_w = 25,000$ has a viscosity of about 3 P. Estimate the segmental friction factor for polyisobutylene at this temperature, taking $l_0 = 5.9 \times 10^{-10}$ m and estimating $\rho = 1.0$ g cm^{-3}.

Solution

First verify that Eq. (2.56) is applicable; this is the case, since $M < M_c$:

$$\text{Number of molecules per unit volume} = \frac{\rho N_A}{\overline{M}_n}$$

$$= \frac{(1 \text{ g cm}^{-3})(6.02 \times 10^{23} \text{ mol}^{-1})}{(2.5 \times 10^4 \text{ g mol}^{-1})} \cong 2.4 \times 10^{19} \text{ cm}^{-3}$$

and the mean-square end-to-end distance

$$\overline{r^2} = \frac{M}{M_0} l_0^2 = \frac{25{,}000}{56} (5.9 \times 10^{-10} \text{ m})^2 = (450)(5.9 \times 10^{-10})^2$$

$$= 1.6 \times 10^{-16} \text{ m}^2$$

Substituting into Eq. (2.56) and solving for ζ yields

$$\zeta = \frac{36\eta}{(M/M_0) l_0^2 (\rho N_A/M)(M/M_0)}$$

$$= \frac{36 (0.3 \text{ kg m}^{-1} \text{ sec}^{-1})}{(1.6 \times 10^{-16} \text{ m}^2)(2.4 \times 10^{-25} \text{ m}^{-3})(450)} = 6.3 \times 10^{-12} \text{ kg sec}^{-1}$$

For the same polymer this parameter has values of 4.47×10^{-8} and 5.01×10^{-11} kg sec^{-1} at 298 and 398 K, respectively. Since density is far less sensitive to temperature, these results show that the primary temperature dependence of viscosity is described by the temperature dependence of ζ.

•

2.11 The Segmental Friction Factor

The segmental friction factor introduced in the derivation of the Debye viscosity equation is an important quantity. It will continue to play a role in the discussion of entanglement effects in the theory of viscoelasticity in the next chapter, and again in Chap. 9 in connection with solution viscosity. Now that we have an idea of the magnitude of this parameter, let us examine the range of values it takes on.

To the extent that the segmental friction factor ζ is independent of M, then Eq. (2.56) predicts a first-power dependence of viscosity on the molecular weight of the polymer in agreement with experiment. A more detailed analysis of ζ shows that segmental motion is easier in the neighborhood of a chain end because the wagging chain end tends to open up the structure of the melt and

thus facilitate flow. The number of chain ends per unit volume increases as the molecular weight decreases, suggesting that ζ decreases with decreasing molecular weight. Conversely, the condition that ζ is independent of molecular weight is best realized toward higher molecular weights, where the effect of chain ends on the free volume is less. Since Eq. (2.56) does not apply in the presence of entanglements, we expect the assumption that ζ is independent of M to work best in the vicinity of M_c. At $M \ll M_c$, ζ is sensitive to M, decreasing as M decreases. For example, in studies of poly(vinyl acetate) at $40°C$, ζ drops to two-thirds its value at $M = 8 \times 10^5$ for a sample with $M = 2 \times 10^5$.

A much simpler experimental situation for examining ζ arises from a study of the diffusion of small molecules through polymer samples. The small probe molecule is chosen to be similar in size to the repeat unit of the polymer. In diffusing through the polymer, the probe molecule must push its way past individual polymer segments, and presumably experiences comparable resistance to displacement. By suitable analytical techniques—for example, monitoring the progress of an isotopically labeled probe—the diffusion coefficient of the small foreign molecule can be measured. Since this molecule does move freely, its friction factor can be evaluated from the experimental diffusion coefficient via Eq. (2.46). The friction factor so obtained is then identified with the segmental friction factor. Although sensitive to the size and shape of the probe, quite acceptable agreement may be achieved between these "apparent" friction factors and those actually determined from mechanical measurements. For example, n-butane, isobutane, and n-pentane show "ζ" values of 3.47×10^{-8}, 7.76×10^{-8}, and 3.80×10^{-8} kg sec^{-1} in diffusion experiments into poly-isobutelene at $25°C$, while the latter has a ζ value of 4.47×10^{-8} kg sec^{-1} when measured directly.

Equation (2.56) not only enables us to understand the basis for the first-power dependence of η on M, but also presents us with a new and important theoretical parameter, the segmental friction factor. We shall see in the next chapter that it is a quantity which can also be extracted from measurements of the viscoelasticity of polymers.

Table 2.3 lists an assortment of polymers in order of decreasing ζ values. The numbers listed were obtained from viscoelastic measurements. Extensive calculations show that viscoelastic and viscosity experiments agree on the magnitude of ζ values for a given sample, although there is a consistent tendency for the values determined from viscosity to be somewhat higher than those determined from viscoelastic measurements. We shall not pursue the cause of this discrepancy, but merely note that it is not sufficient to significantly modify the order of ranking presented in Table 2.3.

The polymers compared in Table 2.3 were not all studied at the same temperature; instead, each was measured at a temperature $100°C$ above its respective glass transition temperature T_g. We shall discuss the latter in considerable detail

Table 2.3 Segmental Friction Factors Ranked in Order of Decreasing Values for Polymers Compared 100°C Above Their Respective Glass Transition Temperatures

Polymer	T_g (K)	$l_0 \times 10^{10}$ (m)	ζ (kg sec^{-1})
Methyl methacrylate	379	6.9	1.58×10^{-7}
Methoxyethyl methacrylate	293	7.0	9.12×10^{-8}
2-Ethyl butyl methacrylate	284	6.5	5.37×10^{-8}
Ethylene glycol monomethacrylate	366	7.0	4.27×10^{-8}
Ethyl methacrylate	335	5.9	3.98×10^{-8}
Butyl rubber	205	5.9	3.47×10^{-8}
Propoxyethyl methacrylate	253	7.0	2.51×10^{-8}
Polyisobutylene	205	5.9	2.14×10^{-8}
n-Butyl methacrylate	300	6.4	1.70×10^{-8}
n-Hexyl methacrylate	268	7.5	6.61×10^{-9}
n-Octyl methacrylate	253	7.0	4.27×10^{-9}
1-Hexene	218	7.0	3.31×10^{-9}
Urethane rubber	238	~12	9.55×10^{-10}
1,4-Butadiene	172	6.0	6.92×10^{-10}
Methyl acrylate	276	6.8	5.75×10^{-10}
Vinyl acetate	305	6.9	4.79×10^{-10}
cis-Isoprene	200	6.8	3.24×10^{-10}
Styrene	373	7.4	1.12×10^{-10}
1,2-Butadiene	261	7.6	9.77×10^{-11}
Vinyl chloride	347	6.0	3.47×10^{-11}
Dimethyl siloxane	150	6.2	3.16×10^{-11}

Source: Values from Ref. 3.

in Chap. 4. For now it is sufficient to note that this temperature is characteristic of the polymer, just as the critical temperature is a useful reference point in a discussion of liquids and gases. Thus comparing polymers a fixed temperature difference from their individual T_g value is analogous to the comparison of gases at comparable differences from their respective critical temperatures. The law of corresponding states informs us that gases behave similarly when thus compared; a similar principle is being used here.

Inspection of the results presented in Table 2.3 reveals several interesting facts:

1. The samples surveyed span a range of more than three orders of magnitude in ζ values.
2. No significant variation in l_0 value is responsible for this ranking, and no overall pattern in T_g is detectable.

3. All the entries above the broken line in Table 2.3 are disubstituted, and those below, with the exception of poly(dimethyl siloxane), are monosubstituted. Those in the former category consistently have larger values of ζ than the latter.

4. Within a homologous series such as the *n*-butyl-, *n*-hexyl-, and *n*-octyl-substituted methacrylates there is a trend for those polymers with the bulkier substituents to show smaller values of ζ. Note that there is also a correlation between substituent size and T_g within this series.

5. Some data on copolymers of ethylene and propylene (not shown) is also consistent inasmuch as the values of ζ decrease as the ethylene (unsubstituted) content increases.

It will be recalled from the definition of the friction factor, Eq. (2.47), that this parameter gives the force required to impart a unit velocity to a polymer chain segment. This force increases with increasing number of substituents, apparently because the latter interfere with free rotation along the chain backbone and thus impede chain motion. Another effect that arises from the presence of substituents and increases with their bulkiness is a decrease in the efficiency of chain packing in the sample or an increase in free volume. As noted in item (4) above, increasing free volume by bulky substituents decreases ζ. It is gratifying that experiment agrees with our intuitive expectations in this matter, that chain mobility is enhanced by open structure in the liquid and maximum flexibility for the chain.

2.12 Dealing with Entanglements: The Bueche Theory

In our discussion of Fig. 2.10 we saw that the presence of entanglements leads to a change in the molecular weight dependence of viscosity. Note that the onset of entanglements results in a change of slope for the plot of η versus M, not a discontinuity in the plot. With entanglements, η becomes considerably more sensitive to chain length, but it does not jump to infinity. We point this out, since in some ways an entanglement is like a crosslink inasmuch as the fates of two or more chains become interdependent. If entanglement coupling were identical to chemical crosslinking, the entire sample would become essentially one giant molecule and the resulting network would be a nonflowing gel. This is not the case and implies that the chains, though entangled, are still able to move, somehow slipping and sliding past one another.

We begin by assuming that entanglements occur at uniform intervals along the polymer chain, with M_e and n_e representing the molecular weight and degree of polymerization, respectively, between entanglement points. Thus each molecule in the liquid has M/M_e entanglements, or M/M_e molecules somehow attached to it. Our goal is to examine how this encumbrance will increase the force necessary

to move some reference molecule through the sample. We might expect M_c to be at least two times M_e if we picture the first entanglement occurring above M_c as dividing the chain into two portions of average length M_e. This is a very primitive picture and more detailed theories have been developed on the relationship between M_c and M_e.

As with previously developed models, the applied shearing force causes a certain molecule—we shall call it the original molecule (subscript 0)—to move with a velocity v_0. Through entanglements, some fraction s of that velocity is transmitted to another group of molecules entangled with the original. These are called primary couplings and are indicated by the subscript 1. Therefore the velocity of the chains involved through primary coupling with the original chain is

$$v_1 = sv_0 \tag{2.57}$$

In a similar fashion, a fraction of the velocity of the molecules with first-order coupling is transmitted to other molecules entangled with the latter. This is called second-order coupling (subscript 2). Still higher orders of effect radiate from the original molecule in the manner suggested by Fig. 2.13. Because of the

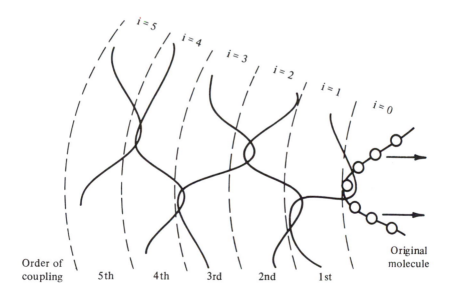

Figure 2.13   Model of several orders of coupling through entanglements according to Bueche theory.

slippage that occurs at each higher-order coupling, successively smaller fractions of the original velocity are transmitted to the higher couplings. For example, the velocity of the molecules entrained through second-order coupling is

$$v_2 = sv_1 = s^2 v_0 \tag{2.58}$$

and for order i the velocity is

$$v_i = s^i v_0 \tag{2.59}$$

Since the slippage factor is a fraction, Eq. (2.59) states in mathematical terms something we realize must be the case, namely, that the effects of entanglements on the neighbors of the original molecule must diminish as we move away from that molecule to prevent the coupling from producing an infinite viscosity.

By analogy with Eq. (2.47), we write the force of viscous resistance experienced by a molecule in an array of uniform molecules of degree of polymerization n as

$$F_{vis} = n\zeta v_0 + C_1 n\zeta s v_0 + C_2 n\zeta s^2 v_0 + \cdots = n\zeta v_0 \left(1 + C_1 s + C_2 s^2 \cdots\right) \tag{2.60}$$

where ζ is the segmental friction factor and the C_i values are constants which count the number of effective couplings of order i. The velocity of the original molecule will disappear from our equations just as it did in the Debye theory; hence our next consideration must be the constants C_i.

If this approach is to have any success, the weighting factors C_i must also decrease with increasing i to avoid a catastrophic increase in viscosity due to the proposed web of entanglements. We shall not detail the entire derivation of these C_i values as developed by Bueche but shall only note the following points:

1. Since each molecule has M/M_e entanglements, and each could entrain a different molecule, an upper limit for the number of couplings of order i is given by $(M/M_e)^i$.
2. This last factor overcounts the number of couplings, since the random placement of chain segments makes it improbable that each entanglement will involve a new molecule. Thus an entanglement may be redundant; the chain might already be coupled to the original molecule.
3. Redundancy of another type may arise if a molecule is coupled to the tangle of moving molecules more than once and in couplings of different order. Not only is one coupling sufficient to induce movement, but also the coupling of higher order will have precedence; that is, if a chain is coupled to the original molecule with both a second-order and a fifth-order coupling, it will move with the second-order fraction of v_0. This mode of

attenuation increases with i, since the likelihood of prior involvement increases with i.

Bueche was able to incorporate these ideas into a quantitative theory, the mathematical details of which need not concern us. The result is complex, but simplifies when applied to polymers of very large molecular weight. In this limit the Bueche theory predicts

$$\eta \cong \frac{(\rho N_A)^2}{1728} \frac{l_0^5 \zeta}{M_e^2} \left(\sum_i s^i (2i - 1)^{3/2} \right) n^{3.5} \tag{2.61}$$

in which the complicated summation is a numerical constant for a specific value of the slippage factor s.

Equation (2.61) predicts a 3.5-power dependence of viscosity on molecular weight, amazingly close to the observed 3.4-power dependence. In this respect the model is a success. Unfortunately, there are other mechanical properties of highly entangled molecules in which the agreement between the Bueche theory and experiment are less satisfactory. Since we have not established the basis for these other criteria, we shall not go into specific details. It is informative to recognize that Eq. (2.61) contains many of the same factors as Eq. (2.56), the Debye expression for viscosity, which we symbolize η_D. If we factor the Bueche expression so as to separate the Debye terms, we obtain

$$\eta = \eta_D \frac{\rho N_A l_0^3 M_0 S}{48 M_e^2} n^{2.5} = \eta_D \frac{\rho N_A l_0^3 S}{48 M_0} \left(\frac{M}{M_e} \right)^2 n^{1/2} \tag{2.62}$$

where we have used the symbol S to signify the summation involving the slippage factors. In the second form of Eq. (2.62), two of the additional 2.5 powers of n associated with entanglements are expressed in terms of M/M_e. Since the latter gives the number of entanglements per molecule, Eq. (2.62) shows that the effect of entanglements on viscosity increases with the square of the number of entanglements per chain. It must be remembered that Eqs. (2.61) and (2.62) are derived on the premise of entanglements. Note that Eq. (2.62) does not reduce to the Debye limit for $M < M_c$.

Our discussion of Fig. 2.10 and the Debye theory show that polymer chains below a certain critical length move independently and produce a viscosity which depends on the first power of the molecular weight. Longer chains result in entanglements and a greatly increased sensitivity of viscosity to molecular weight. Figure 2.13 on which the Bueche theory is based proposes a model for the effects of entanglement, but we almost feel that it goes overboard. On one side of M_c, chains move independently; on the other, they are meshed in a web which would show infinite viscosity if it were not for the attenuation of the

coupling that Bueche carefully introduced. The problem is that independent chain mobility vanishes at the onset of entanglement according to this model. Since the latter is totally responsible for viscosity below M_c, we might find a model for entanglements more plausible if it continued to allow for the independent movement, however encumbered, of a polymer molecule for $M > M_c$. The concept of reptation introduced in the next section provides just such a model.

2.13 The Reptation Model

Figure 2.14a shows a portion of a polymer chain in which the molecule displays entanglement coupling to two other molecules. The molecule of interest is pictured as passing through a tube or sleeve, which, in turn, weaves between the obstacles presented by entangled neighbors. In terms of this picture, two types of motion can immediately be distinguished: conformational changes occurring within the confines of the tube and escape from the tube by a slalomlike motion through the sleeve. The latter is called reptation. The tube is an artifact which makes it easier to describe and distinguish between these two modes of motion.

The tube is a construct which we might continue to sketch around an emerging chain as it diffuses out of the original sleeve. Instead, it is convenient to start with the tube initially in place and consider how long it takes for the molecule to escape. The initial entanglements which determine the contours of the tube comprise a set of constraints from which the molecule is relaxing, even if only to diffuse into another similar set. Accordingly, we identify this reptation time as a relaxation time τ for the molecule.

In order to draw some conclusions about viscosity from the reptation model, it is again necessary to anticipate some results from Chap. 9 on diffusion. The

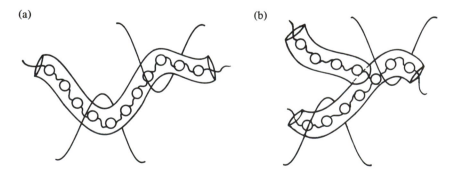

Figure 2.14   Reptation model for entanglements for (a) a linear molecule and (b) a branched molecule.

result we need relates the distance x diffused by a particle, the migration time t, and the diffusion coefficient D:

$$\overline{x^2} = 2Dt \qquad (2.63)$$

If we were able to directly observe the meanderings of an individual molecule for a time t as implied by Eq. (2.63), we would find no agreement as to the displacement from one observation to another. Only the average displacement has any significance, and Eq. (2.63) states that it is specifically the squares of the displacements in a large number of experiments that are to be averaged. The situation is reminiscent of the velocities of molecules in a gas sample. Individual molecules may vary widely in velocity, yet the average of the square of individual velocities is a unique property of the gas at the temperature under consideration. Equation (2.63) was derived by Einstein and was applied to the displacements of microscopic colloidal particles by J. Perrin as the basis for an early determination of Avogadro's number. Before we apply Eq. (2.63) to reptation, there are several additional aspects of this relationship that are worth noting:

1. Since the diffusion coefficient is constant for a given material, Eq. (2.63) shows that the time required for a displacement increases with the square of the distance traveled. This can be understood by thinking that the displacement criterion would be met by finding the diffused particle anywhere on the surface of a sphere of radius x after time t if it started at the origin. The surface area of a sphere is proportional to the square of its radius.

2. An alternative way to look at this result is to ignore the averaging procedure required by Eq. (2.63) and regroup the factors $D = (1/2)(x/t)x$. Since x/t is the diffusion velocity, this version shows that the latter decreases as the displacement increases. This apparently strange result arises from the fact that we are dealing with *net* displacement and a diffusing particle follows a very irregular path. Accordingly, longer and longer times are required for larger net displacements.

3. A final interpretation of the regrouped expression given in item (2) is that 2D equals the diffusion velocity for a particle undergoing unit displacement, $x = 1$ m in the SI system.

With these ideas in mind, let us consider how long it would take for a polymer chain to escape from the tube shown in Fig. 2.14 by reptation.

Without implying anything about the shape of the tube, it is clear that it has the same length as the polymer chain itself, that is, nl_0. This is the displacement of interest in calculating the escape time from the tube. Therefore Eq. (2.63) becomes

$$(nl_0)^2 = 2D_{tube}\,\tau \qquad (2.64)$$

where t has been replaced by τ in recognition of the fact that this characteristic time is in fact the relaxation time. The diffusion coefficient within the tube has been designated on the assumption that sliding along its own contour is an easier mode of motion for a polymer than net translational movement would be. Next we use Eq. (2.46) to replace D_{tube} with the friction factor for the chain in the tube:

$$(nl_0)^2 = 2 \frac{kT}{f_{tube}} \tau \tag{2.65}$$

Since the tube friction factor measures the force needed to impart a unit velocity to the chain along the tube direction, we can think of applying this force, one segment at a time, to the diffusing chain. Since the friction factor per segment is ζ, Eq. (2.65) becomes

$$(nl_0)^2 = \frac{2kT}{n\zeta} \tau \tag{2.66}$$

This simple derivation gives us the desired result, a relationship between the relaxation time and the degree of polymerization:

$$\tau = \frac{l_0^2 \zeta}{2kT} n^3 = \tau_0 n^3 \tag{2.67}$$

where the segmental relaxation time τ_0 is seen to be $\zeta l_0^2/2kT$ by letting n = 1.

The following example will give us a feel for the magnitude of the quantities involved, as well as for the reptition process itself.

Example 2.4

Assuming that Eq. (2.67) applies to small molecules in the limit as n → 1, calculate τ_0, using $D \cong 3 \times 10^{-9}$ m^2 sec^{-1} for a typical low molecular weight molecule. Use this value of τ_0 to estimate τ for a polymer with n = 10^4. Based on Eq. (2.63), evaluate D_{tube} and D_p, the net diffusion coefficient for bulk polymer, from these results.

Solution

Assume that kT/ζ can be replaced by D when Eq (2.67) is applied to small molecules. Use a representative value from Table 2.3, say, 6×10^{-10} m, as a value for l_0 and the D value given to estimate τ_0:

$$\tau_0 = \frac{l_0^2}{2D} = \frac{(6 \times 10^{-10} \text{ m})^2}{2(3 \times 10^{-9} \text{ m}^2 \text{ sec}^{-1})} = 6 \times 10^{-11} \text{ sec}$$

The corresponding quantity for a polymer is larger by n^3 according to Eq. (2.67). For $n = 10^4$,

$$\tau = \tau_0 n^3 = (6 \times 10^{-11} \text{ sec})(10^4)^3 = 60 \text{ sec}$$

Diffusion coefficients are obtained by dividing the square of the length of distance covered by twice this time [Eq. (2.64)]. The length of the reptation tube is $n l_0$; therefore

$$D_{\text{tube}} = \frac{(n l_0)^2}{2\tau} = \frac{[10^4 (6 \times 10^{-10} \text{ m})]^2}{2(60 \text{ sec})} = 3 \times 10^{-13} \text{ m}^2 \text{ sec}^{-1}$$

The net displacement of the molecule is very much less than the length of the tube because of the jumbled configuration of the chain. In time τ the net displacement is on the order of the root mean square of the end-to-end distance [Eq. (1.62)]; that is, in Eq. (2.63), $\overline{x^2} = \overline{r_0^2} = n l_0^2$, so for net diffusion

$$D_{\text{net}} = \frac{n l_0^2}{2\tau} = \frac{10^4 (6 \times 10^{-10} \text{ m})^2}{2(60 \text{ sec})} = 3 \times 10^{-17} \text{ m}^2 \text{ sec}^{-1}$$

Note that the diffusion coefficient for a polymer through an environment of low molecular weight molecules is typically on the order of magnitude of 10^{-11} m^2 sec^{-1}. If the first subscript indicates the diffusing species, and the second the surrounding molecules, and P stands for polymer and S for small molecules, we see that the order of diffusion coefficients is $D_{\text{S,S}} > D_{\text{P,S}} > D_{\text{P,tube}} > D_{\text{P,P}}$, a sequence which makes sense in terms of relative frictional resistance.

•

Before continuing with an application of the reptation model to the viscosity of polymer melts, it is interesting to note that the third-power dependence of relaxation time on degree of polymerization [Eq. (2.67)] has been verified in experiments in which polymer chains are trapped in a network. In one study, for example, styrene –butadiene (S–B) copolymers consisting of blocks of S and of B were mixed with homopolymers of B, the latter with variable molecular weights. The temperature is then lowered such that the S blocks experience phase separation, effectively crosslinking the copolymer network and trapping the chains of pure B. The relaxation behavior of B was then monitored and found to depend on the 3.1 power of n_B—quite acceptable agreement with the predictions of Eq. (2.67). Note the assortment of polymer concepts and phenomena involved in this experiment:

1. Homopolymers versus copolymers.
2. Copolymer microstructure (block structure).

3. Controlled chain length (either by preparation or fractionation).
4. Effective crosslinking by phase separation (could also be accomplished chemically or by crystallization).
5. Direct determination of relaxation time through viscoelastic studies (all mechanical properties involve this important parameter).
6. Tailoring polymeric systems to suit an experimental model.

The behavior of molecules trapped in a network satisfies us that we are on the right track; we continue by applying this relaxation time to viscous behavior.

In connection with a discussion of the Eyring theory, we remarked that Newtonian viscosity is proportional to the relaxation time [Eqs. (2.29) and (2.31)]. What is needed, therefore, is an examination of the nature of the proportionality between the two. At least the molecular weight dependence of that proportionality must be examined to reach a conclusion as to the prediction of the reptation model of the molecular weight dependence of viscosity.

Apart from some numerical coefficients, the details of which we shall forgo, the required proportionality factor involves kT and the concentration of entanglement points. Multiplying Eq. (2.67) by these factors, we obtain

$$\eta \propto kT \left(\frac{M}{M_e}\right)\left(\frac{\rho N_A}{M}\right)\left(\frac{l_0^2 \zeta}{2kT}\right) n^3 \tag{2.68}$$

where M/M_e gives the number of entanglements per chain and $\rho N_A/M$ gives the number of chains per unit volume. Since these additional involvements of molecular weight cancel, the relaxation time alone determines the power dependence of viscosity on molecular weight according to the reptation model.

The reptation model, developed largely by de Gennes, Doi, and Edwards, is thus seen to predict a very strong dependence of viscosity on molecular weight, although not quite the 3.4-power dependence observed experimentally. Although not quite as successful in this regard as the Bueche theory, the reptation model is vastly simpler and is consistent with experiments in other areas. Finally, reptation provides us with an excellent model for the discussion of the role of chain branching on viscosity. This is the topic of the next section.

2.14 The Effects of Branching

In the foregoing discussions of theoretical models and experimental results, we have focused on linear polymers. We have seen the effect of chain substituents on viscosity. All other things being equal, bulky substituents tend to decrease ζ and thereby lower η. The effect is primarily due to the opening up of the liquid because of the steric interference with efficient packing arising from the substituents. With side chains of truly polymeric character, the picture is quite different.

Figure 2.14b suggests how the reptation picture for linear chains must be modified by the presence of side chains. While we can visualize the unbranched molecule sliding along its reptation tube relatively easily, the polymeric side chain interferes considerably. Since the side chain is enclosed in its own tube, sliding the main chain along entails withdrawing the side chain from its sleeve and inserting it into a new, equivalent one located a step farther along the contour of the main tube. How long does it take to move this branch chain from one sleeve to another?

If the side chain were an independent molecule, its escape time from its sleeve would be given by Eq. (2.67), with n replaced by the degree of polymerization n_s of the side chain. This is not the case, however, since the loose end of the side chain must retrace its steps out of the sleeve before the other end, the branch point along the backbone, can move. Our interest is in the probability of such a withdrawal path occurring among all the possible modifications of conformation the side chain can undergo. It is clear that the probability of such a path being followed will decrease with increasing n_s, and detailed theoretical analysis of the problem suggests that the probability P depends on the length of the side chain as follows:

$$P = Ae^{-an_s} \tag{2.69}$$

where A and a are constants.

Aside from the side chains, the movement of the backbone along the main reptation tube is still given by Eq. (2.67). With the side chains taken into account, the diffusion velocity must be decreased by multiplying by the probability of the side-chain relocation. Since the diffusion velocity is inversely proportional to τ, Eq. (2.67) must be divided by Eq. (2.69) to give the relaxation time for a chain of degree of polymerization n carrying side chains of degree of polymerization n_s:

$$\tau = A\tau_0 n^3 e^{an_s} \tag{2.70}$$

Comparing Eqs. (2.68) and (2.70) suggests that the effect of branching on η might be an exponential increase with the length of the side chain, provided that the entanglement mechanism is operative for both the branched and unbranched samples being compared:

$$\eta_{br} = \eta_{unbr} e^{an_s} \tag{2.71}$$

or

$$\ln \frac{\eta_{br}}{\eta_{unbr}} = an_s + \text{const.} \tag{2.72}$$

We must be careful in assessing the experimental results on the viscosity of branched polymers. If we compare two polymers of identical molecular weight, one branched and the other unbranched, it is possible that the branched one would show lower viscosity. Two considerations enter the picture here. First, since the side chains contribute to the molecular weight, the backbone chain

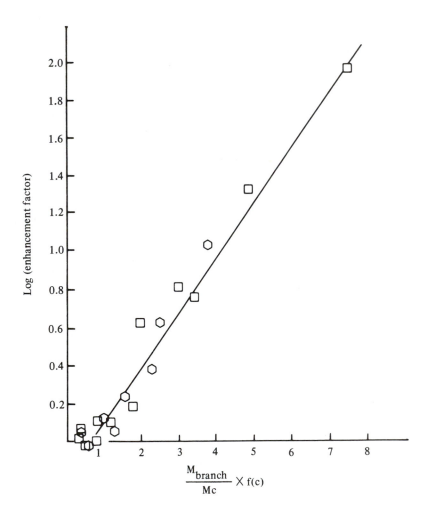

Figure 2.15 Log of viscosity enhancement factor versus parameter measuring branch length for polyisoprene. [Data from W. W. Graessley, T. Masuda, J. E. I. Roovers, and N. Hadjichristidis, *Macromolecules* 9:127 (1976).]

is shorter for the branched polymer. Second, the branched molecule is more compact and may actually be below the entanglement threshold in terms of chain length if not molecular weight. It would then be totally inappropriate to compare its behavior with a linear polymer having entanglements.

Figure 2.15 is a plot of experimental results which is consistent with the relationship given by Eq. (2.72). The systems investigated as the basis for Fig. 2.15 were branched polyisoprene samples. The polymers are called star molecules, consisting of four or six arms of equal length. Some were measured in bulk, and others in solution. The viscosity enhancement factor plotted in Fig. 2.15 is different than a simple ratio of experimental viscosities, but measures essentially the same thing. Likewise, the abscissa values have been multiplied by a function of the concentration of the polymer to reduce data from different experimental conditions to a single line. Since this axis is proportional to n_s, however, the data in Fig. 2.15 are consistent with the predictions of Eq. (2.72). The same research obtained similar results for three-arm polybutadiene stars. Note that the molecular weights of the branches in Fig. 2.15 are listed as multiples of M_c, indicating that the molecules are above the entanglement threshold in all cases.

The molecules used in the study described in Fig. 2.15 were model compounds characterized by a high degree of uniformity. When branching is encountered, it is generally in a far less uniform way. As a matter of fact, traces of impurities or random chain transfer during polymer preparation may result in a small amount of unsuspected branching in samples of ostensibly linear molecules. Such adventitious branched molecules can have an effect on viscosity which far exceeds their numerical abundance. It is quite possible that anomalous experimental results may be due to such effects.

Problems

1. The following are approximate σ_s (in dyne cm^{-2}) versus $\dot{\gamma}$ data[†] for three different samples of polyisoprene in tetradecane solutions of approximately the same concentration:

[†]W. W. Graessley, T. Masuda, J. E. L. Rovers, and H. Hadjichristidis, *Macromolecules* 9:127 (1976).

sample:	P-4	S-6	H-3
$\bar{M}_w$ (g mol^{-1})	1.61×10^6	1.95×10^6	1.45×10^6
Description	Linear	Four-armed star	Six-armed star
C (g cm^{-3})	0.0742	0.0773	0.0778
$\dot{\gamma}$ (sec^{-1})	$\sigma_s \times 10^{-3}$	$\sigma_s \times 10^{-3}$	$\sigma_s \times 10^{-2}$
0.6	0.7	–	–
0.8	0.9	–	–
1	1	0.3	–
2	2	0.6	–
4	4	1	0.15
8	5	2	0.3
10	6	3	0.4
20	7	4	0.8
60	–	7	2
100	–	8	4

From plots of these data, estimate the Newtonian viscosity of each of the solutions and the approximate rate of shear at which non-Newtonian behavior sets in. Are these two quantities better correlated with the molecular weight of the polymer or the molecular weight of the "arms"?

2. Wagner and Dillon[†] have described a low-shear viscometer in which the inside diameter of the outer, stationary cylinder is 30 mm and the outside diameter of the inner, rotating cylinder is 28 mm; the rotor is driven by an electromagnet. The device operates at 135°C and was found to be free of wobble and turbulence for shear rates between 3 and 8 sec^{-1}. The conversion of Eq. (2.7) to Eq. (2.9) shows that $F/A = (\eta)(dv/dr)$ (instrument constant) for these instruments Evaluate the instrument constant for this viscometer.

3. A fluid of viscosity η is confined within the gap between two concentric cylinders as shown in Fig. 2.3b. Consider a cylindrical shell of radius r, length l, and thickness dr located within that gap.
   (a) What is the torque acting on the shell if torque is the product of force and the distance from the axis and $F/A = \eta r \, d\omega/dr$?
   (b) Under stationary-state conditions, the torques at r and at r + dr must be equal, otherwise the shell would accelerate. This means that the torque must be independent of r. Show that this implies the following variation of ω with r: $\omega = -B/2r^2 + C$, where B and C are constants.
   (c) Evaluate the constant B by noting that $\omega = \omega_{ex}$, the experimental velocity, at r = R and $\omega = 0$ at r = fR.
   (d) Combine the results of (a), (b), and (c) to obtain Eq. (2.7).

[†]H. L. Wagner and J. G. Dillon, *Polym. Prepr.* 22:260 (1981).

4. The two-term version of equaton (2.33) contains four parameters. It may be written as

$$\frac{F}{A} = \left(\alpha_1 \beta_1 \frac{\sinh^{-1}(\beta_1 \dot{\gamma})}{\beta_1 \dot{\gamma}} + \alpha_2 \beta_2 \frac{\sinh^{-1}(\beta_2 \dot{\gamma})}{\beta_2 \dot{\gamma}} \right) \dot{\gamma} = \eta \dot{\gamma}$$

It is sometimes observed that β_1 is sufficiently small that $\sinh^{-1}(\beta_1 \dot{\gamma})/\beta_1 \dot{\gamma} = 1$ for all $\dot{\gamma}$ values, in other words, an exclusively Newtonian contribution. In such a case $\eta = \eta_N + \alpha_2 \beta_2 \sinh^{-1}(\beta_2 \dot{\gamma})/\beta_2 \dot{\gamma}$. Consider both limiting values of $(1/\beta\dot{\gamma}) \sinh^{-1}(\beta\dot{\gamma})$ from Table 2.1 to suggest a procedure whereby η_N, α_2, and β_2 could be evaluated from experimental data describing F/A versus $\dot{\gamma}$, assuming that this model applies and that data are available over a sufficiently wide range of $\dot{\gamma}$ values.

5. The bulk viscosity of polystyrene ($\overline{M}_w = 371{,}000$) at $200°C$ was measured by Graessley and Segal[†] at different rates of shear. At low rates of shear $\eta_N = 330{,}000$ P and drops off with $\dot{\gamma}$ approximately as follows:

$\dot{\gamma}$ (sec^{-1})	0.03	0.1	0.3	1	3
η/η_N	0.89	0.76	0.56	0.35	0.22

Does a β value of 1, 10, or 100 sec work best in Eq. (2.28) to describe the variation of η with $\dot{\gamma}$ for this sample? Note that no single-term version of the Eyring theory gives a totally acceptable fit.

6. A slightly different but useful way of defining the viscosity average molecular weight is the following:

$$\overline{M}_v{}^a = \frac{\Sigma_i f_i M_i M_i{}^a}{\Sigma_i f_i M_i}$$

where $f_i M_i$ is the weighting factor used to average $M_i{}^a$. A satisfactory way of treating many polymer distributions is to define

$$f_i = \frac{1}{\overline{M}_n} e^{-M_i/\overline{M}_n}$$

Then

$$\overline{M}_v{}^a = \frac{\displaystyle\int_0^{\infty} f_i M_i{}^{a+1} \, dM_i}{\displaystyle\int_0^{\infty} f_i M_i \, dM_i}$$

Combine the last two expressions and integrate to express $\overline{M}_v$ in terms of $\overline{M}_n$ and a. The integrals are standard forms and are listed in integral tables as gamma functions.

[†]W. W. Graessley and L. Segal, *A.I.Ch.E. J.* 16:261 (1970).

7. Pearson et al.[†] determined the viscosity average molecular weight of polyisoprene samples working with solutions of the polymer in toluene. For this system, the a value in Eq. (2.40) is known to be 0.74. Use the results of the last problem and tables of gamma functions (e.g., *CRC Handbook of Tables for Mathematics*) to convert the following viscosity average molecular weights to number average molecular weights: 13,300, 31,300, 80,000, 205,000, 220,000, 270,000, 490,000, 560,000, and 1,150,000. If $(M_c)_n$ for polyisoprene is 10^4, estimate the number of entanglements per chain each of these samples would display in the bulk.

8. At $200°C$ the Newtonian viscosities of polystyrene samples of different molecular weights were studied by Spencer and Dillon[‡] and the following results were reported:

$\bar{M}_w \times 10^{-3}$ (g mol^{-1})	η_N (P)
86	3.50×10^3
162	4.00×10^4
196	6.25×10^4
360	4.81×10^5
490	1.89×10^6
508	1.00×10^6
510	1.64×10^6
560	3.33×10^6
710	6.58×10^6

Use these data to determine the exponent of M in the relationship between η_N and M.

9. The Newtonian viscosities of polystyrene samples were measured at $183°C$ and the following results were reported by Graessley and Segal[§]:

$\bar{M}_w$ (g mol^{-1})	117,000	179,000	217,000	242,000
η_N (P)	25,700	109,000	190,000	295,000

What power dependence on M does η display according to these results? Comment on the significance of the 3.4-power law according to these data and the results of the last problem.

10. A semiempirical way of extending the Debye equation to molecular weights above M_c is given by multiplying the right-hand side of Eq. (2.56) by $(M/M_c)^{2.4}$. For a poly(dimethyl siloxane) sample of $\bar{M}_w = 1.34 \times 10^6$,

---

[†] D. S. Pearson, A. Mera, and W. E. Rochefort, *Polym. Prepr.* 22:102 (1981).
[‡] R. S. Spencer and R. E. Dillon, *J. Colloid Sci.* 4:241 (1949).
[§] W. W. Graessley and L. Segal, *A.I.Ch.E. J.* 16:261 (1970).

Plazek et al.[†] measured η_N to be 7.50×10^6 P at $25°C$ when $\rho = 0.974$ g cm^{-3}. Using 24,500 as the value for M_c and taking the value of l_0 from Table 2.3, estimate the value of ζ which is consistent with these data. How does this value compare with the ζ value listed in Table 2.3?

11. Once the value of the constant and the a value in Eq. (2.36) have been evaluated for a particular system, viscosity measurements constitute a relatively easy method for determining the molecular weight of a polymer. Criticize or defend the following proposition: Since viscosity is so highly dependent on molecular weight for $M > M_c$, a 10% error in η will result in a 34% error in M above M_c, but only a 10% error in M below M_c.

12. A polystyrene sample of relatively narrow molecular weight distribution ($\bar{M}_w/\bar{M}_n = 1.07$) has a molecular weight of 275,000. Its viscosity was found[‡] to be 10^7 kg m^{-1} sec^{-1}. Evaluate the S factor in Eq. (2.62) which is compatible with the result, taking $M_e = 16,000$ and using $\zeta = 10^{-8}$ kg sec^{-1} and $\rho = 1.0$ g cm^{-3} at $160°C$, the temperature of the experiment. Evaluate the first 10 terms of the summation in Eq. (2.61) for s = 0.4 and 0.5 and compare with the experimental S value.

13. A sphere of density ρ_2 and radius R falling through a medium of density ρ_1 and viscosity η experiences three kinds of forces: gravitational, buoyant, and frictional. The latter is given by Eqs. (2.44) and (2.45). During most of the fall (excluding the very beginning and the very end of the path), these forces balance. Use this condition to derive an equation showing how this stationary-state velocity of fall is related to R, ρ_1, ρ_2, and η. This is the basis for the so-called falling-ball viscometer.

14. Plazek et al.[†] measured the viscosities of a poly(dimethyl siloxane) sample of $\bar{M}_w = 4.1 \times 10^5$ over a range of temperatures using the falling-ball method. Stainless steel ($\rho_2 = 7.81$ g cm^{-3}) balls of two different diameters, 0.1590 and 0.0966 cm, were used at $25°C$, where $\rho_1 = 0.974$ g cm^{-3} and $\eta = 8.64 \times 10^4$ P. Use the result derived in the last problem to calculate the ratio of the stationary-state settling velocities for the two different balls. How long would it take the smaller ball to fall a distance of 15 cm under these conditions?

15. At concentrations roughly five times higher than those described in Problem 1, the same polymer samples show[§] the following Newtonian viscosities:

	P-4	S-6	H-3
C (g cm^{-3})	0.331	0.326	0.329
η_N (P)	1.35×10^6	3.8×10^7	4.7×10^5

[†] D. J. Plazek, W. Dannhauser, and J. D. Ferry, *J. Colloid Sci.* 16:101 (1961).
[‡] S. Onogi, T. Masuda, and K. Kitagawa, *Macromolecules* 3:109 (1970).
[§] W. W. Graessley, T. Masuda, J. E. L. Roovers, and N. Hadjichristidis, *Macromolecules* 9:127 (1976).

Criticize or defend the following proposition: In dilute solutions, branching affects viscosity only inasmuch as the branched molecule has a more compact shape. At higher concentrations, the effect of branching is closer to a bulk effect.

Bibliography

1. Bueche, F., *Physical Properties of Polymers*, Interscience, New York, 1962.
2. deGennes, P. G., *Scaling Concepts in Polymer Physics*, Cornell University Press, Ithaca, N.Y., 1979.
3. Ferry, J. D., *Viscoelastic Properties of Polymers*, Wiley, New York, 1980.
4. Krausz, A. S., and Eyring, H., *Deformation Kinetics*, Wiley, New York, 1975.
5. Ree, T., and Eyring, H., in *Rheology*, Vol. 2 (F. R. Eirich, Ed.), Academic, New York, 1958.
6. Schultz, J., *Polymer Materials Science*, Prentice-Hall, Englewood Cliffs, N.J., 1974.

3

The Elastic and Viscoelastic States

Our time has seen
The synthesis of polyisoprene
And many cross-linked helixes unknown
To *Robert Hooke*; but each primoridal bean
Knew cellulose by heart . . .

The Dance of the Solids, by John Updike

3.1 Introduction

In this chapter we examine the elastic behavior of polymers. We shall see that this behavior is quite different from the elasticity displayed by metals and substances composed of small molecules. This is a direct consequence of the chain structure of the polymer molecules. In many polymers elasticity does not occur alone, but coupled with viscous phenomena. The combination of these effects is called viscoelasticity. We shall examine this behavior as well.

The elastic and viscoelastic properties of materials are less familiar in chemistry than many other physical properties; hence it is necessary to spend a fair amount of time describing the experiments and the observed response of the polymer. There are a large number of possible modes of deformation that *might* be considered: We shall consider only elongation and shear. For each of these we consider the stress associated with a unit strain and the strain associated with a unit stress; the former is called the modulus, the latter the compliance. Experiments can be time independent (equilibrium), time dependent (transient), or periodic (dynamic). Just to define and describe these basic combinations takes us into a fair amount of detail and affords some possibilities for confusion. Pay close attention to the definitions of terms and symbols.

We shall rely heavily on models again in this chapter; this time they are of two different types. We shall consider elasticity in terms of a molecular model in which the chains are described by random flight statistics. The phenomena of

133

viscoelastic behavior are also discussed in terms of models, but here they are mechanical equivalents to the whole sample and do not involve molecular parameters. Finally, the Rouse theory for viscoelasticity involves a model that describes the vibrations of molecules in mechanical terms. Like the composite behavior it describes, the beads and springs of the Rouse theory comprise a combination model.

Temperature is an important variable in the discussion of viscoelasticity. For reasons of pedagogy, most consideration of this variable is deferred to the next chapter. This does not mean that temperature is unimportant to the present discussion, only that the agenda is full enough without it.

3.2 Elastic Deformation

The various elastic and viscoelastic phenomena we discuss in this chapter will be developed in stages. We begin with the simplest: the case of a sample that displays a purely elastic response when deformed by simple elongation. On the basis of Hooke's law, we expect that the force of deformation—the stress—and the distortion that results—the strain—will be directly proportional, at least for small deformations. In addition, the energy spent to produce the deformation is recoverable: The material snaps back when the force is released. We are interested in the molecular origin of this property for polymeric materials but, before we can get to that, we need to define the variables more quantitatively.

A quantitative formulation of Hooke's law is facilitated by considering the rectangular sample shown in Fig. 3.1a. If a force F is applied to the face of area A, the original length of the block L_0 will be increased by ΔL. Now consider the following variations:

1. Imagine subdividing the block into two portions perpendicular to the direction of the force, as shown in Fig. 3.1b. Each slice experiences the same force as before, and the same net deformation results. A deformation $\frac{1}{2} \Delta L$ is associated with a slice of length $L_0/2$. The same argument could be applied for any number of slices; hence it is the quantity $\Delta L/L_0$ which is proportional to the force.

2. Imagine subdividing the face of the block into two portions of area $A/2$. A force only half as large would be required for each face to produce the same net distortion. The same argument could be applied for any degree of subdivision; hence it is the quantity F/A which is proportional to $\Delta L/L_0$.

3. Force per unit area along the axis of the deformation is called the uniaxial tension or stress. We shall use the symbol σ as a shorthand replacement for F/A and attach the subscript t to signify tension. The elongation, expressed as a fraction of the original length $\Delta L/L_0$ is called the strain. We shall use γ_t as the symbol for the resulting strain (subscript t for tension). Both σ

(a)   (b)   (c)

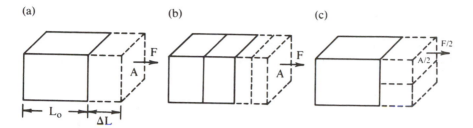

Figure 3.1 (a) The force F applied to area A extends the length of the block from L_0 by an amount ΔL. Parts (b) and (c) illustrate the argument that $F/A \propto \Delta L/L$.

and γ were encountered in the last chapter under conditions of shear. We shall take up shear deformation presently.

With these considerations in mind, we write

$$\sigma_t = E\gamma_t = E\left(\frac{\Delta L}{L_0}\right)_t \qquad (3.1)$$

where the proportionality constant E is called the tensile modulus or Young's modulus. Remember, it will be different for different substances and for a given substance at different temperatures. For the present we assume that E is independent of time, but we shall remove this constraint eventually. Since γ is dimensionless, E has the same units as F/A, namely, force length^{-2}, or Nm^{-2} in the SI system. Note that Eq. (3.1) also applies to the case of compression.

There is another aspect of tensile deformation to be considered. The application of a distorting force not only stretches a sample, but it also causes the sample to contract at right angles to the stretch. If w and h represent the width and height of area A in Fig. 3.1, both contract by the same fraction, a fraction which is related in the following way to the strain:

$$\frac{\Delta w}{w} = \frac{\Delta h}{h} = \mu\frac{\Delta L}{L_0} \qquad (3.2)$$

where the constant μ is called Poisson's ratio. Poisson's ratio is also a property of the material; it may be written as

$$\mu = \frac{1}{2}\left(1 - \frac{1}{V}\frac{dV}{d\gamma}\right) \qquad (3.3)$$

where V is the volume of the sample. Thus if the volume does not change on elongation, the fractional contraction in each of the perpendicular directions is half the fractional increase in length. It takes two parameters to describe the response of a sample to tensile force, for example, E and μ. Other constants can be defined, but two are needed for the general case, regardless of how they are defined.

Although we are not specifically interested in time-dependent behavior in this section, it will be helpful to examine the temporal development of the elongation experiment described above. When the force is first applied, the sample begins to stretch, starting from $\Delta L = 0$. Eventually an elongation consistent with the applied force is reached. It is this point in the history of our experiment that we are describing. The time development of the strain is not our concern for the present. Except for the difference between tensile and shear forces, the situation described in Fig. 3.1 resembles Fig. 2.1, the defining figure for viscosity. The sample behaves differently, however. In the case of viscosity, the deformation is more accurately called flow: It is not recovered when the force is removed. For purely elastic behavior, the sample does not flow, and the deformation is fully recoverable. For substances composed of small molecules, nearly perfect Newtonian behavior is associated with the liquid state. Likewise, the solid state is associated with elastic behavior, although the origin of the elasticity is quite different in substances composed of small molecules than for polymers. Many polymers show the superpositioned effects of elastic deformation and viscous flow and in this sense are intermediate between simple liquids and solids. We shall avoid words like *solid* and *liquid* altogether, since the superficial appearance of a sample can be misleading. A specimen can look solid, for example, yet achieve flow deformation over a long period of time. Alternatively, it may appear to flow, but eventually reach an equilibrium deformation. Observation over a long period of time is required for accurate characterization. For the present, we shall assume that our tensile experiment has had all the time it needs to reach its equilibrium deformation. We shall postpone time-dependent viscoelastic behavior until we have dealt with elasticity alone.

Throughout this chapter we shall take a very liberal attitude toward time. When time-dependent properties are examined, we shall examine a time range in seconds spanning 18 powers of 10. On this scale, if the shortest time were 1 sec, the longest would be 3×10^{10} years! Obviously the shortest time must be very much less than 1 sec, but measuring extremely short times is as much of a challenge as measuring extremely long times. The way out of this predicament is to conduct measurements over a range of temperatures. A variation of 100°C in temperature may have the same effect on the viscoelastic properties of a substance as does a time span of 10^{10} sec. Since the former is readily achieved, it is clearly the method of choice. Results determined at different temperatures are subsequently reduced to a common temperature. The extraordinary time scales cited above are the result. The procedure for this reduction

to a common temperature is taken up in the next chapter. In this chapter we shall mostly describe isothermal data—in the reduced form if not experimental— and discuss very long and very short times with equal ease. To maintain contact with reality in the face of this strategy, the time–temperature equivalency should be borne in mind. Although not our primary concern in this chapter, we shall have more than one occasion to point out this equivalency.

Next let us consider the differences in molecular architecture between polymers which exclusively display viscous flow and those which display a purely elastic response. To attribute the entire effect to molecular structure we assume the polymers are compared at the same temperature. Crosslinking between different chains is the structural feature responsible for elastic response in polymer samples. If the crosslinking is totally effective, we can regard the entire sample as one giant molecule, since the entire volume is permeated by a continuous network of chains. This result was anticipated in the discussion of the Bueche theory for chain entanglements in the last chapter, when we observed that viscosity would be infinite with entanglements if there were no slippage between chains.

The details of chemical crosslinking need not concern us, but some examples will illustrate materials with the potential to display elasticity:

1. Natural rubber, cis-1,4-polyisoprene, cross-linked with sulfur. This reaction was discovered by Goodyear in 1839, making it both historically and commercially the most important process of this type. This reaction in particular and crosslinking in general are also called vulcanization.
2. Vinyl polymers cross-linked with divinyl monomers, for example, polystyrene polymerized in the presence of divinyl benzene.
3. Condensation polymers prepared with some monomer of functionality greater than 2, for example, a polyester formed with some glycerol or tricarboxylic acid.
4. Polyethylene cross-linked by irradiation with high-energy electrons.
5. Polysilicones cross-linked by reaction with benzoyl peroxide.

The concentration of crosslink junctions in the network is also important: if too low, flow will be possible; if too high, the maximum attainable elongation will be decreased. From the point of view of theoretical analysis, the length of chain between crosslink points must be long enough to be described by random flight statistics.

Stretching a polymer sample tends to orient chain segments and thereby facilitate crystallization. The incorporation of different polymer chains into small patches of crystallinity is equivalent to additional crosslinking and changes the modulus accordingly. Likewise, the presence of finely subdivided solid particles, such as carbon black in rubber, reinforces the polymer in a way that imitates the effect of crystallites. Spontaneous crystal formation and reinforcement

by fillers are both important in technology, but since they complicate the picture of the polymer network, we shall not consider them further at this time.

Merely having a cross-linked polymer does not guarantee good elastic properties. For the latter, the polymer chains should also have relatively unhindered rotation along the chain backbone so that distortions occur rapidly in response to the applied force. Thus a high local mobility is desirable, while the overall mobility of the chains must be blocked so that they will snap back when the force is removed. High chain mobility also works in favor of promoting crystallinity, and the latter is undesirable if high elasticity is sought. So optimization in the context of mechanical properties consists in finding a polymer with good chain mobility, but not so good as to produce crystallization. From a practical point of view, the glass transition temperature should be well below the temperature at which the elasticity is desired. The glass transition temperatures of a number of polymers are listed in Table 2.3. Inspection of Table 2.3 shows that a number of hydrocarbon polymers—such as 1,4-butadiene, cis-isoprene, butyl rubber, and polyisobutylene—have low values of T_g and thus the potential for a wide working range. We shall have more to say about the glass transition temperature and related phenomena in the next chapter.

The tightrope situation that arises from balancing high mobility, low crystallinity, and optimum crosslinking is often dealt with by using copolymers rather than homopolymers. With chain composition as an additional variable, molecules can be tailored better for specific application situations.

Polymers with the mechanical and chemical properties we have discussed in this section are called elastomers. In the next couple of sections we shall examine the thermodynamic basis for elasticity and then apply these ideas to cross-linked polymer networks.

3.3 Thermodynamics and Elasticity

It is not particularly difficult to introduce thermodynamic concepts into a discussion of elasticity. We shall not explore all of the implications of this development, but shall proceed only to the point of establishing the connection between elasticity and entropy. Then we shall go from phenomenological thermodynamics to statistical thermodynamics in pursuit of a molecular model to describe the elastic response of cross-linked networks.

We begin by remembering the mechanical definition of work and apply that definition to the stretching process of Fig. 3.1. Using the notation of Fig. 3.1, we can write the increment of elastic work we associated with an increment in elongation dL as

$$dw_e = F \, dL \tag{3.4}$$

It is necessary to establish some conventions concerning signs before proceeding further. When the applied force is a tensile force and the distortion is one of stretching, F, dL, and dw_e are all defined to be positive quantities. Thus dw_e is positive when elastic work is done on the system. The work done by the sample when the elastomer snaps back to its original size is a negative quantity.

The classical formulation of the first law of thermodynamics defines the change dU in the internal energy of a system as the sum of heat dq absorbed by the system plus the work dw done on the system:

$$dU = dq + dw \tag{3.5}$$

The element of work is generally written $-p\ dV$, where p is the external pressure, but with the possibility of an elastic contribution, it is $-p\ dV + F\ dL$. With this substitution Eq. (3.5) becomes

$$dU = dq - p\ dV + F\ dL \tag{3.6}$$

A consistent sign convention has been applied to the pressure–volume work term: A positive dV corresponds to an expanded system, and work is done by the system to push back the surrounding atmosphere.

The phenomenological definition for the change in entropy associated with the isothermal, reversible absorption of an element of heat dq is

$$dS = \frac{dq}{T} \tag{3.7}$$

This relationship can be used to replace dq by T dS in Eq. (3.6), since the infinitesimal increments implied by the differentials mean that the system is only slightly disturbed from equilibrium and the process is therefore reversible:

$$dU = T\ dS - p\ dV + F\ dL \tag{3.8}$$

Remember that in this relationship, as in all thermodynamic equations, temperature must be expressed in degrees Kelvin.

A quantity of great importance in chemical thermodynamics is the Gibbs free energy G. The latter is defined in terms of enthalpy H as

$$G = H - TS \tag{3.9}$$

where

$$H = U + pV \tag{3.10}$$

Combining the last two results and taking the derivative gives

$$dG = dU + p\,dV + V\,dp - T\,dS - S\,dT \tag{3.11}$$

Comparing Eq. (3.11) with Eq. (3.8) enables us to replace several of these terms by F dL,

$$dG = V\,dp - S\,dT + F\,dL \tag{3.12}$$

thus establishing the desired connection between the stretching experiment and thermodynamics.

Since G is a state variable and forms exact differentials, an alternative expression for dG is

$$dG = \left(\frac{\partial G}{\partial p}\right)_{T,L} dp + \left(\frac{\partial G}{\partial T}\right)_{p,L} dT + \left(\frac{\partial G}{\partial L}\right)_{p,T} dL \tag{3.13}$$

Comparing Eqs. (3.12) and (3.13) enables us to write

$$F = \left(\frac{\partial G}{\partial L}\right)_{p,T} \tag{3.14}$$

Note this is the same derivation that yields the important results $V = (\partial G/\partial p)_T$ and $S = -(\partial G/\partial T)_p$ when no elastic work is considered. It would be inappropriate for a book like this to digress into thermodynamics any further than this. The two relationships cited above are derived in almost every thermodynamics text; the student is advised to consult a suitable reference and review this material if the treatment above is too abbreviated.

Now we return to the definition of F provided by Eq. (3.14) and differentiate Eq. (3.9) with respect to L, keeping p and T constant:

$$\left(\frac{\partial G}{\partial L}\right)_{p,T} = \left(\frac{\partial H}{\partial L}\right)_{p,T} - T\left(\frac{\partial S}{\partial L}\right)_{p,T} \tag{3.15}$$

The left-hand side of this equation gives F according to Eq. (3.14); therefore

$$F = \left(\frac{\partial H}{\partial L}\right)_{p,T} - T\left(\frac{\partial S}{\partial L}\right)_{p,T} \tag{3.16}$$

This expression is sometimes called the equation of state for an elastomer in analogy to

$$p = \left(\frac{\partial U}{\partial V}\right)_T - T \left(\frac{\partial S}{\partial V}\right)_T \tag{3.17}$$

the thermodynamic equation of state. Note the parallel roles played by length and volume in these two expressions.

Equation (3.16) shows that the force required to stretch a sample can be broken into two contributions: one that measures how the enthalpy of the sample changes with elongation and one which measures the same effect on entropy. The pressure of a system also reflects two parallel contributions, except that the coefficients are associated with volume changes. It will help to pursue the analogy with a gas a bit further. The internal energy of an ideal gas is independent of volume: The molecules are noninteracting so it makes no difference how far apart they are. Therefore, for an ideal gas $(\partial U/\partial V)_T = 0$ and the thermodynamic equation of state becomes

$$p = -T \left(\frac{\partial S}{\partial V}\right)_T \tag{3.18}$$

By analogy, an ideal elastomer is defined as one for which $(\partial H/\partial L)_{p,T} = 0$; in this case Eq. (3.16) becomes

$$F = -T \left(\frac{\partial S}{\partial L}\right)_{p,T} \tag{3.19}$$

Although defined by analogy to an ideal gas, the justification for setting $(\partial H/\partial L)_{p,T} = 0$ cannot be the same for an elastomer as for such a gas. All molecules attact one another, and this attraction is not negligible in condensed phases such as our sample. What the ideality condition requires in an elastomer is that there be no change in the enthalpy of the sample as a result of the stretching process. As long as deformations are not too large, the polymer chain responds to deforming forces by a combination of rotations along the chain backbone that straighten out some of the kinks in the chain and increase the overall end-to-end distance. One way of thinking of this is to imagine that some bond near the middle of a randomly jumbled coil experiences a rotation of 180°, all other angles remaining fixed. Since the nature of the random flight process tends to bring the ends of the chain together, this rotation would have the opposite effect and increase the end-to-end distance. A chain of degree of polymerization n can achieve exactly the same effect by rotations of 180°/n at each repeat unit. If the chain is long enough, this amounts to a slight disturbance of each bond from its equilibrium position. The foregoing implies a cooperative effect along the backbone, while it is actually the average of a

variety of rotations that produces the net effect. Nevertheless, considerable modification of the end-to-end distance in a chain can be accomplished with the expenditure of relatively little energy. To get an idea of the numerical magnitude of this energy, let us consider the following example.

Example 3.1

A cross-linked polymer has a density of 0.94 g cm^{-3} at 25°C and a molecular weight between crosslinks of 28,000. The conformation of one bond in the middle of the molecule changes from trans to gauche, and the molecule opens up by 120°. In *n*-butane, the trans to gauche transformation requires about 3.3 kJ mol^{-1}. Estimate a value for ΔH of stretching based on this model, and use the law of cosines to estimate the magnitude of the "opening up" that results.

Solution

We need to know the number of chains per unit volume and must calculate the result on this basis:

$$\frac{0.94 \text{ g}}{\text{cm}^3} \times \frac{\text{moles polymer}}{28,000 \text{ g}} = 3.4 \times 10^{-5} \text{ moles polymer cm}^{-3}$$

$$\Delta H_{t \to g} = \frac{3300 \text{ J}}{\text{moles monomer}} \times \frac{1 \text{ repeat unit}}{\text{polymer chain}} \times \frac{3.4 \times 10^{-5} \text{ mol}}{\text{cm}^3}$$

$$= 0.12 \text{ J cm}^{-3}$$

In the present discussion it is the relative magnitudes of ΔH and ΔS which are significant. Several considerations lead us to conclude that this value of ΔH represents an upper limit for this quantity:

1. The same 120° opening can be achieved by a gauche → gauche transition with no difference in energy.
2. Only those bonds in the direction of stretch are elongated; in perpendicular directions they contract. The latter outnumber the former by two to one.
3. Not only does this calculation overcount the number of chains involved, but it makes no attempt to consider any compensation from the perpendicular direction.

We can use the law of cosines to estimate the effect on the end-to-end distance as follows. Picture two vectors of equal length b extending from the center of the chain to each of the ends. In the trans conformation of the center bond the

angle between these vectors is $180°$ (see Fig. 1.8c), and the end-to-end distance is 2b. In the gauche position at the center bond the angle between these vectors is $60°$, and the law of cosines gives

$$r^2 = 2b^2 - 2b^2 \cos 60° = b^2 \quad \text{or} \quad r = b$$

as the end-to-end distance. In terms of this oversimplified model, a single trans → gauche rearrangment changes the end-to-end separation by a factor of 2. We shall estimate ΔS for this process in Example 3.2.

•

For large deformations or for networks with strong interactions—say, hydrogen bonds instead of London forces—the condition for an ideal elastomer may not be satisfied. There is certainly a heat effect associated with crystallization, so $(\partial H/\partial L)_{p,T}$ would not apply if stretching induced crystal formation. The compounds and conditions we described in the last section correspond to the kind of system for which ideality is a reasonable approximation.

Since entropy plays the determining role in the elasticity of an ideal elastomer, let us review a couple of ideas about this important thermodynamic variable:

1. Equation (3.7) gives a simple procedure for evaluating the entropy change accompanying a change of state. At the normal boiling point of a liquid, for example, the heat is absorbed reversibly and equals the heat of vaporization ΔH_v. Since T is constant, the entropy of vaporization is $\Delta H_v/T$. For benzene, for example, $\Delta S_v = (30.8 \text{ k J mol}^{-1})/353 = 87 \text{ J K}^{-1} \text{ mol}^{-1}$.

2. While this is an easy calculation to make, Eq. (3.7) does little to clarify exactly what ΔS means. Phenomenological proofs that ΔS as defined by Eq. (3.7) is a state variable often leave us with little more than a lament for the inefficiency of spontaneous processes.

3. Most chemistry students acquire a much better "feel" for entropy by considering it a measure of the amount of disorder in a system. For example, the fact that ΔS_v is positive immediately makes sense, since the vapor phase is more disordered than the liquid.

4. A quantitative way of dealing with the degree of disorder in a system is to define something called the thermodynamic probability Ω which counts the number of ways in which a particular state can come about. Thus situations we characterize as relatively disordered can come about in more ways than a relatively ordered state, just as an unordered deck of cards compared to a deck arranged by suits.

Chemists learn to use the thermodynamic probability almost instinctively in a qualitative manner; it is quantitatively related to entropy through an equation due to Boltzmann:

$$S = k \ln \Omega \tag{3.20}$$

where k is Boltzmann's constant. As usual, k is replaced by R for calculations on a per mole basis. The difference in entropy between two states of different thermodynamic probability is

$$\Delta S = S_2 - S_1 = k \ln \left(\frac{\Omega_2}{\Omega_1} \right) \tag{3.21}$$

and this is positive when $\Omega_2 > \Omega_1$ and negative when $\Omega_2 < \Omega_1$.

For the evaporation process we mentioned above, the thermodynamic probability of the gas phase is given by the number of places a molecule can occupy in the vapor. This, in turn, is proportional to the volume of the gas (subscript g): $\Omega_g \propto V_g$. In the last chapter we discussed the free volume in a liquid. The total free volume in a liquid is a measure of places for molecules to occupy in the liquid. The thermodynamic probability of a liquid (subscript l) is thus $V_l \propto V_{l,\,free}$. Based on these ideas, the entropy of the evaporation process can be written as

$$\Delta S_v = k \ln \left(\frac{V_g}{V_{l,\,free}} \right) \tag{3.22}$$

for one molecule, or

$$\Delta S_v = k \ln \left(\frac{V_g}{V_{l,\,free}} \right)^N = Nk \ln \left(\frac{V_g}{V_{l,\,free}} \right) \tag{3.23}$$

for N molecules. If N equals Avogadro's number, Eq. (3.23) becomes

$$\Delta S_v = R \ln \left(\frac{V_g}{V_{l,\,free}} \right) \tag{3.24}$$

Neither the volume occupied by a mole of gas at the boiling point nor the free volume of a liquid vary too widely from substance to substance. Taking the former to be about 30,000 ml and the latter to be about 3 ml gives

$$\Delta S_v = R \ln \left(\frac{30,000}{3} \right) = (8.314) \ln 10^4 = 77 \text{ J K}^{-1} \text{ mol}^{-1} \tag{3.25}$$

A great many liquids have entropies of vaporization at the normal boiling point in the vicinity of this value (see benzene above), a generalization known as Trouton's rule. Our interest is clearly not in evaporation, but in the elongation of elastomers. In the next section we shall apply Eq. (3.21) to the stretching process for a statistical—and therefore molecular—picture of elasticity.

3.4 Entropy Elasticity

By combining random flight statistics from Chap. 1 with the statistical definition of entropy from the last section, we shall be able to develop a molecular model for the stress--strain relationship in a cross-linked network. It turns out to be more convenient to work with the ratio of stretched to unstretched lengths L/L_0 than with γ itself. Note the relationship between these variables:

$$L = L_0 + \Delta L \tag{3.26}$$

or

$$\alpha - 1 = \gamma_t \tag{3.27}$$

where the elongation α is defined to be L/L_0 while $\gamma_t = \Delta L/L_0$.

The next step in the development of a model is to postulate a perfect network. By definition, a perfect network has no free chain ends. An actual network will contain dangling ends, but it is easier to begin with the perfect case and subsequently correct it to a more realistic picture. We define ν as the number of subchains contained in this perfect network, a subchain being the portion of chain between the crosslink points. The molecular weight and degree of polymerization of the chain between crosslinks are defined to be M_c and n_c, respectively. Note that these same symbols were used in the last chapter with different definitions.

Finally, we assume that the network undergoes an affine deformation. This means that each volume element within the sample deforms in exact proportion to the overall specimen. Since a volume element can be inscribed around a subchain of the network, affine deformation requires that the coordinates describing the ends of subchains also be changed in proportion to changes in the macroscopic dimensions of the sample. Such a situation is represented in Fig. 3.2. In Fig. 3.2 the solid lines represent the volume element and the broken lines suggest the overall sample. Figure 3.2a indicates an unstretched sample; Fig. 3.2b, the same sample after elongation in the z direction ($\alpha = z/z_0$). Assuming that no volume change accompanies the deformation, the x and y dimensions of the volume element as well as the macroscopic specimen are decreased by $1/\alpha^{1/2}$ compared to the original dimensions (subscript 0):

$$\text{Volume} = x_0 y_0 z_0 = \left(\frac{1}{\alpha^{1/2}} x_0\right) \left(\frac{1}{\alpha^{1/2}} y_0\right) (\alpha z_0) \tag{3.28}$$

Next let us apply random walk statistics to the subchain before and after stretching.

(a)   (b)

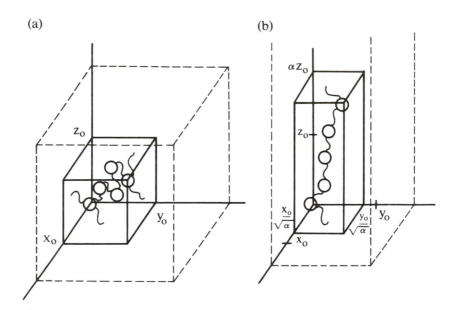

Figure 3.2 The affine deformation of a polymer subchain. (a) The original coordinates of the end of the subchain are x_0, y_0, and z_0. (b) The same coordinates as in (a) are $x_0/\sqrt{\alpha}$, $y_0/\sqrt{\alpha}$, and αz_0.

Equation (1.41) gives the probability of finding one end of a chain with degree of polymerization n in a volume element dx dy dz located at x, y, and z if the other end of the chain is located at the origin. We can use this relationship to describe the unstretched chain shown in Fig. 3.2a; all that is required is to replace n by n_c, the degree of polymerization of the subchain. Therefore for the unstretched chain (subscript u) we write

$$P_u \, dx \, dy \, dz = \left(\frac{2}{3} \pi n_c l_0^2\right)^{-3/2} \exp\left(\frac{-3(x^2 + y^2 + z^2)}{2n_c l_0^2}\right) dx \, dy \, dz \quad (3.29)$$

After stretching, the z coordinate is increased by α, while x and y are decreased by $\alpha^{-1/2}$, as described by Eq (3.28) and shown in Fig. 3.2b. For this situation (subscript s) the probability becomes

$$P_s \, dx \, dy \, dz = \left(\frac{2}{3} \pi n_c l_0^2\right)^{-3/2} \exp\left(\frac{-3(x^2/\alpha + y^2/\alpha + \alpha^2 z^2)}{2n_c l_0^2}\right) dx \, dy \, dz$$

$$(3.30)$$

A moment's reflection will convince us that these probabilities can be used as thermodynamic probabilities in Eq. (3.21) to calculate the entropy change on stretching:

$$S_s - S_u = \Delta S = k(\ln P_s - \ln P_u) \tag{3.31}$$

The only limitations imposed by using these expressions are that the subchains are long enough to be described by random flight statistics and that the elongation is not so large that the average end-to-end distance in the subchain ceases to be large compared to $n_c l_0$. This last requirement is imposed by the approximation introduced in the derivation of Eq. (1.31). Combining the last three results and rearranging gives

$$\Delta S_i = \frac{-3k\,[(1/\alpha - 1)\,x^2 + (1/\alpha - 1)\,y^2 + (\alpha^2 - 1)\,z^2]}{2n_c\,l_0{}^2} \tag{3.32}$$

for the entropy change of a typical subchain, say, the ith one.

In the volume elements describing individual subchains, the x, y, and z dimensions will be different, so Eq. (3.32) must be averaged over all possible values to obtain the average entropy change per subchain. This process is also easily accomplished by using a result from Chap. 1. Equation (1.62) gives the mean-square end-to-end distance of a subchain as $n_c l_0{}^2$, and this quantity can also be written as $\overline{x^2} + \overline{y^2} + \overline{z^2}$; therefore

$$\overline{x^2} + \overline{y^2} + \overline{z^2} = n_c\,l_0{}^2 \tag{3.33}$$

Because of the random nature of the distribution of repeat units, each contribution to the left-hand side of this expression is equal and

$$\overline{x^2} = \overline{y^2} = \overline{z^2} = \frac{1}{3}\,n_c\,l_0{}^2 \tag{3.34}$$

Using this value for the average value of x^2, y^2, and z^2 in Eq. (3.32) and rearranging gives

$$\overline{\Delta S} = -\frac{3k}{2}\left[\frac{2}{3}\left(\frac{1}{\alpha} - 1\right) + \frac{1}{3}(\alpha^2 - 1)\right] \tag{3.35}$$

or

$$\overline{\Delta S} = -\frac{k}{2}\left(\alpha^2 + \frac{2}{\alpha} - 3\right) \tag{3.36}$$

This expression gives the average entropy change per chain; to get the average for the sample, we multiply by the number ν of subchains in the sample. The total entropy change is

$$\Delta S = -\frac{k\nu}{2}\left(\alpha^2 + \frac{2}{\alpha} - 3\right) \tag{3.37}$$

The quantity in parentheses is always positive for $\alpha > 1$, the case of elongation, making $\Delta S < 0$ for stretching. Therefore ΔS is positive for the opposite process, showing that entropy alone is sufficient to explain the elastomer's snap. To get an idea of the magnitude of this entropy effect, consider the following example.

Example 3.2

Calculate ΔS for a 100% elongation of a polymer sample at 25°C using Eq. (3.37) and compare ΔS with the value of ΔH calculated for the same process in Example 3.1.

Solution

From Example 3.1, $\nu = 3.4 \times 10^{-5}$ mol (expressed per cubic centimeter of sample), and for a 100% elongation $\alpha = 2$. Substitution into Eq. (3.37) gives

$$\Delta S = -\frac{1}{2}\left(\frac{8.314\ \text{J}}{\text{K mol}}\right)\left(\frac{3.4 \times 10^{-5}\ \text{mol}}{\text{cm}^3}\right)\left(2^2 + \frac{2}{2} - 3\right)$$

$$= -2.8 \times 10^{-4}\ \text{J K}^{-1}\ \text{cm}^{-3}$$

To compare with ΔH at 25°C, ΔS must be multiplied by the temperature:

$$T\ \Delta S = 298(-2.8 \times 10^{-4}) = -0.084\ \text{J cm}^{-3}$$

Comparison of this result with Example 3.1 shows that $T\ \Delta S$ and ΔH are of the same order of magnitude. Example 3.1 stressed that the value of ΔH estimated there was an upper limit. Experimental results show that the assumption of ideality—while a slight oversimplification—generally introduces an error of less than 10%. Comparison of the two examples shows how much more difficult it is to deal with ΔH than with ΔS in these systems.

Recalling that this is an ideal elastomer, we apply Eq. (3.19) to Eq. (3.37) to obtain

$$F = -T\left(\frac{\partial S}{\partial L}\right)_T = -\frac{T}{L_0}\left(\frac{\partial \Delta S}{\partial \alpha}\right) = \frac{kT\nu}{L_0}\left(\alpha - \frac{1}{\alpha^2}\right) \tag{3.38}$$

Dividing both sides of the equation by the cross-sectional area of the sample gives

$$\sigma_t = \frac{F}{A} = kT \frac{\nu}{V} \left(\alpha - \frac{1}{\alpha^2} \right) \tag{3.39}$$

where the product of the cross section and L_0 has been set equal to the volume V of the sample. Note that σ_t changes sign at $\alpha = 1$, as physically required.

Although it is in this form that we compare theoretical predictions with experiment in the next section, it is instructive to express Eq. (3.39) in terms of L/L_0 and then differentiate the result with respect to L:

$$d\sigma_t = kT \frac{\nu}{V} \left(\frac{1}{L_0} + 2L_0^2 L^{-3} \right) dL \tag{3.40}$$

Next we multiply and divide the right-hand side by L to get

$$d\sigma_t = \left[kT \frac{\nu}{V} \left(\alpha + \frac{2}{\alpha^2} \right) \right] \frac{dL}{L} \tag{3.41}$$

Comparing this result with Eq. (3.1) shows that the quantity in brackets equals Young's modulus for an ideal elastomer in a perfect network. Since the number of subchains per unit volume, ν/V, is also equal to $\rho N_A/M_c$, where M_c is the molecular weight of the subchain, the modulus may be written as

$$E = \frac{RT\rho}{M_c} \left(\alpha + \frac{2}{\alpha^2} \right) \tag{3.42}$$

There are several things to notice about this result:

1. The modulus increases with temperature. This behavior is verified by experiment. By contrast, the modulus of metals decreases with increasing T. The difference arises from the fact that entropy is the origin of elasticity in polymers but not in metals.
2. The modulus increases as M_c decreases. This effect on M_c is brought about by increased crosslinking and is consistent with intuitive expectations in this regard.
3. The modulus is not independent of α. For $\alpha \gg 1$, the first term in parentheses predominates and the modulus is directly proportional to α. For $\alpha \ll 1$, the first term is insignificant and E varies as α^{-2}.
4. An interesting limit for small deformation (i.e., $\alpha \cong 1$) is

$$E_{\alpha \cong 1} = \frac{3RT\rho}{M_c} \tag{3.43}$$

showing the condition for Young's modulus to be constant and its value when that condition is met.

5. Carrying out a parallel differentiation on Eq. (3.38) and examining the case of $\alpha \cong 1$ yields

$$dF = \frac{3kT\nu}{L_0{}^2} dL \tag{3.44}$$

Applied to a single chain, $\nu = 1$ and $L_0{}^2$ is replaced by Eq. (1.62) to give

$$dF = \frac{3kT}{nl_0{}^2} dL = k_H \, dL \tag{3.45}$$

This equation shows that at small deformations individual chains obey Hooke's law with the force constant $k_H = 3kT/nl_0{}^2$. This result may be derived directly from random flight statistics without considering a network.

A typical cross-linked polymer at room temperature has $\rho \cong 1$ g cm^{-3} and $M_c \cong 10^4$. According to Eq. (3.43), Young's modulus for such a polymer is $3(8.314$ J K^{-1} mol$^{-1})(300$ K$)(10^3$ kg m$^{-3})/10$ kg mol^{-1} = 7.5×10^5 N m^{-2} which is on the order of the observed magnitude of this quantity for polymers. By contrast, for metals E is on the order of 10^{11} N m^{-2}.

3.5 Experimental Behavior of Elastomers

We have already observed that the entropy theory of elasticity predicts a modulus of the right magnitude and possessing the proper temperature coefficient. Now let us examine the suitability of Eq. (3.39) to describe experimental results in detail.

Figure 3.3 shows a plot of σ_t versus α at 20°C for a sample of natural rubber cross-linked with sulfur. The points are experimental and the solid line is drawn according to Eq. (3.39). The agreement is seen to be very good for α values between 0.4 and 1.3. For greater elongations the theoretical force is initially too large, although at $\alpha \gtrsim 5$ the theoretical line falls below the experimental. Remember, the sample is already increased in length by 30% at $\alpha = 1.3$; however, since elongations approaching 1000% are possible for actual elastomers, the range of success is modest. There is no doubt that the dependence of the modulus on α is correctly given by Eq. (3.42) for small elongations and compressions.

Over the range of fit, matching the dependence of σ_t on α produces only a numerical coefficient and does not constitute a test of the predicted dependence of the modulus on either T or M_c. We have already confirmed that Young's modulus actually does increase with temperature for elastomers. Next let us

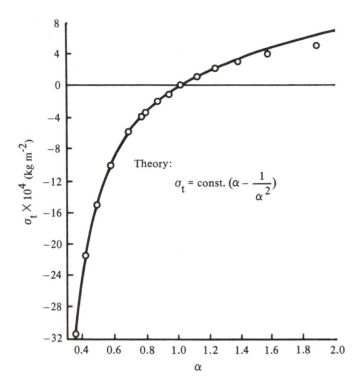

Figure 3.3 Comparison of experiment (points) and theory [Eq. (3.39)] for the entropy elasticity of a sample of cross-linked natural rubber. [From L. R. G. Treloar, *Trans. Faraday Soc.* 40:59 (1944).]

consider experiments which test the predicted effect of M_c on this quantity. The best way to proceed is by considering at the same time the fact that real elastomers do not form perfect networks. By definition, the perfect network has no loose ends. An actual network is cross-linked at random, and a certain fraction of subchains are only dangling appendages to the network. The number of subchains in this category should be subtracted from the total number of subchains, since they do not contribute to the elasticity of the network. Loose ends relax from deformation over a period of time—like the branch chains discussed in connection with viscosity—and do not oppose the deformation with a restoring force. Our problem, then, is to subtract the number of ineffective subchains from the total number ν, since the latter overcounts.

Suppose the un-cross-linked polymer chain has a molecular weight M which, upon crosslinking, is divided into subchains of molecular weight M_c. This means that each subchain is a fraction M_c/M of the original chain. Since the crosslink

points occur at random along the original chain, two of the subchains will be in terminal positions; that is, they will be held to the network by only one cross-link and thus not contribute to elasticity. Therefore $2M_c/M$ is the fraction of subchains which are ineffective and should be subtracted from the total number to give an improved representation of elasticity. With this modification, Eq. (3.39) becomes

$$\sigma_t = \frac{RT\rho}{M_c} \left(1 - \frac{2M_c}{M}\right)\left(\alpha - \frac{1}{\alpha^2}\right) \tag{3.46}$$

Note that as $M \to \infty$, the absolute number of chain ends per unit volume decreases, as does the chain-end correction.

Figure 3.4 is a display of data which confirms the predictions of Eq. (3.46). For T and α constant—the case in these experiments—Eq. (3.46) predicts that a plot of σ_t versus $1/M$ should yield a straight line with an intercept proportional to $1/M_c$. The elastomers on which the experiments were conducted were co-polymers of isobutylene containing a small amount of isoprene. The polymers

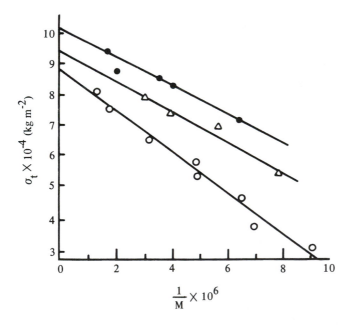

Figure 3.4 σ_t versus M^{-1} for polyisobutylene samples with three different degrees of crosslinking (T and α constant). [Reprinted with permission from P. J. Flory, *Ind. Eng. Chem.* 38:417 (1946). Copyright 1946, American Chemical Society.]

were fractionated into samples of different molecular weight prior to crosslinking. The crosslink sites, the isoprene units, are distributed at random through the polymer; after vulcanization, samples are available with the same values of M_c but different values of M. Elasticity measurements on the latter are plotted in Fig. 3.4. Figure 3.4 shows that copolymers with three different degrees of crosslinking agree with theoretical predictions very well. The sample represented by the open circles in Fig. 3.4 yields a value of M_c = 37,000 when analyzed according to Eq. (3.46). Independent measurement of this quantity through solubility studies gives a value of 35,000. In view of the complexity of the phenomena involved, this must be regarded as excellent agreement.

The entropy theory for elasticity is thus seen to correctly describe the dependence of the modulus on T, ν, and α, at least for small elongations. For distortion of the sample to many times its original length, crystallization due to chain alignment may cause the breakdown of the theory. At intermediate elongations, however, the model should still apply, provided that we have not limited its validity by mathematical approximations. In writing Eq. (3.32) we observed that large deformations might invalidate the assumption that the end-to-end distance in the subchain is small compared to $n_c l_0$. Without going into full mathematical details, let us examine how this constraint can be lifted.

The first step is to write an expression for the energy stored in a network U as a result of the elongation α. This may be written as

$$U = \int_{x_0}^{x_0/\sqrt{\alpha}} F_x \, dx + \int_{y_0}^{y_0/\sqrt{\alpha}} F_y \, dy + \int_{z_0}^{\alpha z_0} F_z \, dz \tag{3.47}$$

where the F's are the x, y, and z components of the distorting force and x_0, y_0, and z_0 are the initial dimensions of the sample. Assuming that the applied force is in the z direction, as in Fig. 3.1, we note that the contraction in the x and y directions is much less than the elongation in z. The x and y components of the force are therefore assumed to obey Hooke's law; that is, $F_x = k_H x$, with k_H given by Eq. (3.45). A similar expression describes F_y, but F_z is described differently because of the greater elongation it experiences. For the latter the the force component may be shown to be

$$F_z = \frac{kT}{l_0} \left[3 \frac{r_z}{n_c l_0} + \frac{9}{5} \left(\frac{r_z}{n_c l_0} \right)^3 + \frac{297}{175} \left(\frac{r_z}{n_c l_0} \right)^5 + \cdots \right] \tag{3.48}$$

in which r_z is the z component of the average end-to-end distance in the subchain. Note that Eq. (3.48) gives a force which is identical to the x and y components if $r_z \ll n_c l_0$. Equation (3.48) gives the first three terms in the series expansion of the inverse Langevin function of $r_z/n_c l_0$; this describes

the chain-end separation without assuming stringent limitations on $r/n_c l_0$. Equations (3.48) and (3.45) with $n = n_c$ can be substituted into Eq. (3.47) and the result integrated and then multiplied by ν to give the elastic energy stored per unit volume of elastomer. The derivative of this last quantity with respect to α equals σ_t. When the indicated operations are carried out, the following result is obtained:

$$\sigma_t = \frac{\nu kT}{V} \left\{ -\frac{1}{\alpha^2} + \frac{1}{3} \sqrt{n_c} \left[3 \frac{\alpha}{\sqrt{n_c}} + \frac{9}{5} \left(\frac{\alpha}{\sqrt{n_c}} \right)^3 + \frac{297}{175} \frac{5}{3} \left(\frac{\alpha}{\sqrt{n_c}} \right)^5 + \cdots \right] \right\}$$

$$(3.49)$$

For the case of $\alpha \ll n_c^{1/2}$, terms higher than first order in this expression are negligible and the result becomes identical to Eq. (3.39). However, the derivation of Eq. (3.49) does not limit its applicability to these small values of α. Figure 3.5 shows that experimental results are still not exactly described by Eq. (3.49), although the general shape of the theoretical and experimental lines is similar.

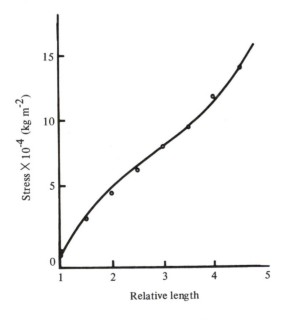

Figure 3.5   Comparison of experiment (points) and theory [Eq. (3.49)] for the entropy elasticity of the same sample shown in Fig. 3.3. [Reprinted with permission from H. M. James and E. Guth, *J. Chem. Phys.* 11:455 (1943).]

In connection with Eq. (3.45) we noted that the deformation of individual chains can be studied directly from random flight statistics. Using equivalent expressions for the x, y, *and* z components of force and following the procedure outlined above gives a more rigorous derivation of Eq. (3.39) than that presented in the last section.

Even better agreement between theory and experiment has been obtained in other theories by abandoning the notion of affine deformation and recognizing that shorter subchains experience a greater strain than do longer subchains for a given stress. We shall not pursue this development any further, however, and shall turn next to a consideration of other types of deformation.

3.6 The Shear Modulus and the Compliances

Until now we have restricted ourselves to consideration of simple tensile deformation of the elastomer sample. This deformation is easy to visualize and leads to a manageable mathematical description. This is by no means the only deformation of interest, however. We shall consider only one additional mode of deformation, namely, shear deformation. Figure 3.6 represents an elastomer sample subject to shearing forces. Deformation in the shear mode is the basis

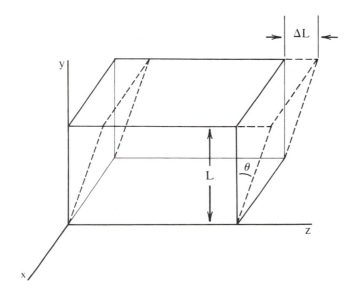

Figure 3.6   Definition of variables to define the shear deformation of an elastic body.

for the definition of viscosity. An examination of elasticity according to the same mode of deformation will make it easy to combine the two for a description of viscoelasticity. We shall begin, however, by describing the purely elastic response to a body under conditions of shear.

By analogy with Eq. (3.1), we seek a description for the relationship between stress and strain. The former is the shearing force per unit area, which we symbolize as σ_s, as in Chap. 2. For shear strain we use the symbol γ_s; it is the rate of change of γ_s that is involved in the definition of viscosity in Eq. (2.2). As in the analysis of tensile deformation, we write the strain $\Delta L/L$, but this time ΔL is in the direction of the force, while L is at right angles to it. These quantities are shown in Fig. 3.6. It is convenient to describe the sample deformation in terms of the angle θ, also shown in Fig. 3.6. For distortion which is independent of time—we continue to consider only the equilibrium behavior—stress and strain are proportional with proportionality constant G:

$$\sigma_s = G\gamma_s = G\,\frac{\Delta L}{L} = G\tan\theta \cong G\theta \qquad (3.50)$$

where the last relationship is based on the fact that $\tan\theta \cong \theta$ for small θ. The factor G is called the shear modulus or the modulus of rigidity. In SI units, G also has units $N\,m^{-2}$.

The shear modulus for an ideal elastomer in a perfect network is not difficult to derive:

1. Equation (3.47) is used with each of the force components given by (3.44). The limits of integration are different for affine deformation under shear at constant volume. In terms of the coordinates shown in Fig. 3.6, there is no change in x, z increases by α, and y decreases by $1/\alpha$.

2. Integration, multiplication by ν, and differentiation with respect to γ yields $\sigma_s = (kT\nu/V)\gamma_s = (RT\rho/M_c)\gamma_s$ and, by comparison with Eq. (3.50), makes the shear modulus

$$G = \frac{RT\rho}{M_c} \qquad (3.51)$$

3. This result is the shear equivalent to Eq. (3.42) for tensile deformation. Note the modulus is a constant independent of strain for shear, while this is only true for $\alpha \cong 1$ in the case of tension as shown by Eq. (3.43).

A general relationship between Young's modulus and the shear modulus is

$$G = \frac{E}{2(1+\mu)} \qquad (3.52)$$

where μ is Poisson's ratio defined by Eq. (3.3). For constant volume deformations $\mu = \frac{1}{2}$ and $G = E/3$. This same result is obtained when Eq. (3.51) is

compared with Young's modulus under conditions where the latter is a constant, namely, Eq. (3.43).

As long as the moduli are constants, it makes no difference in either a tensile or shear experiment which variable, stress or strain, is independent and which is dependent; that is, we could apply a constant force and measure the strain or induce a constant strain and measure the force responsible. The modulus is the ratio of the stress to the strain. If the ratio were calculated as the ratio of the strain to the stress, the reciprocal of the modulus would result. The latter is called the compliance and is given the symbols D and J for tensile and shear conditions, respectively. When they are independent of time, the moduli and compliances for a particular deformation are simply reciprocals.

The situation is not so simple when these various parameters are time dependent. In the latter case, the moduli, designated by $E(t)$ and $G(t)$, are evaluated by examining the (time dependent) value of σ needed to maintain a constant strain γ_0. By constrast, the time-dependent compliances $D(t)$ and $J(t)$ are determined by measuring the time-dependent strain associated with a constant stress σ_0. Thus whether the deformation mode is tension or shear, the modulus is a measure of the stress required to produce a unit strain. Likewise, the compliance is a measure of the strain associated with a unit stress. As required by these definitions, the units of compliance are the reciprocals of the units of the moduli m^2 N^{-1} in the SI system.

These distinctions are summarized in Table 3.1 for handy reference. The nomenclature and notation are somewhat confusing, and the situation gets even worse when other sources are consulted. Not all authors use the same notation, so Table 3.1 is useful as a concordance.

As is often the case, models will assist us in understanding the time-dependent moduli and compliances. This time we turn to a purely mechanical model to characterize viscoelastic systems. In their simplest form, the models will consist of two units, a spring and a dashpot. The latter is a cup and bob arrangement where the piston moves through a viscous medium. These units embody the features of elastic and viscous response. We shall characterize the former by a Hookean modulus and the latter by a Newtonian viscosity. Although there are exceptions to these statements, the way the spring and dashpot units are arranged determines whether a modulus or compliance is most readily calculated from a model. For example, if the spring and dashpot are connected in series, we have the following:

1. The net deformation is the sum of the deformations of the individual units.
2. Dividing the stress by the net deformation gives the modulus.
3. A spring and dashpot in series is called a Maxwell model.

Alternatively, if the spring and dashpot are connected in parallel, the following holds:

Table 3.1 Summary of the Names and Notation for Moduli[a] and Compliances[b] Under Equilibrium, Transient, and Dynamic Conditions

Experimental condition	Tension	Shear	Comment
Equilibrium	1. $E = \sigma_t/\gamma_t$	$G = \sigma_s/\gamma_s$	Time independent; E = Young's modulus
	2. $D = 1/E = \gamma_t/\sigma_t$	$J = 1/G = \gamma_s/\sigma_s$	
Transient	1. $E(t) = \sigma_t(t)/\gamma_0$	$G(t) = \sigma_s(t)/\gamma_0$	Time dependent; Subscript 0, held constant in experiment
	2. $D(t) = \gamma_t(t)/\sigma_0$	$J(t) = \gamma_s(t)/\sigma_0$	
Dynamic	1. $E(\omega) = \sigma_t(\omega)/\gamma_0$	$G(\omega) = \sigma_s(\omega)/\gamma_0$	Frequency dependent; Subscript 0, maximum amplitude; all have storage (single prime) and loss (double prime) components
	2. $D(\omega) = \gamma_t(\omega)/\sigma_0$	$J(\omega) = \gamma_s(\omega)/\sigma_0$	

[a] Modulus: Stress per unit strain
[b] Compliance: Strain per unit stress

4. The deformations are the same for each unit and it is the stresses that are added.
5. Dividing the deformation by the stress gives the compliance.
6. A spring and dashpot in parallel is called a Voigt model.

Each of these models will be examined in subsequent sections as devices for describing specific transient experiments.

3.7 The Maxwell Model: Stress Relaxation

Our objectives in this section are twofold: to describe and analyze a mechanical model for a viscoelastic material, and to describe and interpret an experimental procedure used to study polymer samples. We shall begin with the model and then proceed to relate the two. Pay attention to the difference between the model and the actual observed behavior.

Suppose we consider a spring and dashpot connected in series as shown in Fig. 3.7a; such an arrangement is called a Maxwell element. The spring displays a Hookean elastic response and is characterized by a modulus G^*. The dashpot displays Newtonian behavior with a viscosity η^*. These parameters (superscript *) characterize the model; whether they have any relationship to the

(a) (b)

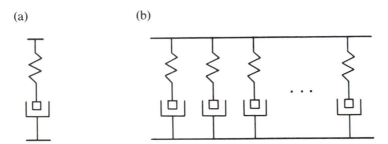

Figure 3.7 Maxwell models consisting of a spring and dashpot in series: (a) single unit and (b) set of units arranged in parallel.

properties of molecules remains to be seen. As a word of caution, it might be observed that other authors frequently use complex numbers for the mathematics of viscoelasticity and use the asterisk to label these complex numbers. We shall avoid the use of complex numbers; for us the asterisk simply labels the modulus or viscosity of a spring or dashpot.

Returning to the Maxwell element, suppose we rapidly deform the system to some state of strain γ_{tot} and secure it in such a way that it retains the initial deformation. Because the material possesses the capability to flow, some internal relaxation will occur such that less force will be required with the passage of time to sustain the deformation. Our goal with the Maxwell model is to calculate how the stress varies with time, or, expressing the stress relative to the constant strain, to describe the time-dependent modulus. Such an experiment can readily be performed on a polymer sample, the results yielding a time-dependent stress relaxation modulus. In principle, the experiment could be conducted in either a tensile or shear mode measuring $E(t)$ or $G(t)$, respectively. We shall discuss the Maxwell model in terms of shear.

In the Maxwell model, the two units are connected in series, so that each bears the full stress individually and the deformations of the elastic and viscous components are additive:

$$\gamma_{tot} = \gamma_{el} + \gamma_{vis} \tag{3.53}$$

In a shear experiment the first of these is given by Eq. (3.50). For the viscous component we do not have an expression for γ, only for the way γ varies with time. Hence it is not possible to develop this relationship any further as an explicit equation, but only as a differential equation. Differentiating Eq. (3.53) with respect to time, we obtain

$$\frac{d\gamma_{tot}}{dt} = \dot{\gamma}_{tot} = \dot{\gamma}_{el} + \dot{\gamma}_{vis} \tag{3.54}$$

and, differentiating (3.50) with G constant, we obtain

$$\dot{\gamma}_{el} = \frac{1}{G^*} \frac{d\sigma}{dt} \tag{3.55}$$

Substituting Eqs. (3.55) and (2.2) into Eq. (3.54) gives .

$$\dot{\gamma}_{tot} = \frac{1}{G^*} \frac{d\sigma}{dt} + \frac{1}{\eta^*} \sigma \tag{3.56}$$

Since the total strain in the experiment is held constant, $d\gamma_{tot}/dt = 0$ and Eq. (3.56) becomes

$$\frac{1}{G^*} \frac{d\sigma}{dt} + \frac{1}{\eta^*} \sigma = 0 \tag{3.57}$$

This is the fundamental differential equation for a shear stress relaxation experiment. The solution to this differential equation is an equation which gives σ as a function of time in accord with experiment.

Example 3.3

Show that the equation $\sigma = \sigma_0 \exp(-G^*t/\eta^*)$ is a solution to Eq. (3.57) where σ_0 is the value of σ at $t = 0$.

Solution

Demonstrating that a function solves a differential equation is easier than finding the solution. All that is required is to perform the indicated differentiation and verify the equality. Remember that σ_0, G^*, and η^* are constants:

$$\frac{d\sigma}{dt} = \sigma_0 e^{-G^*t/\eta^*} \left(- \frac{G^*}{\eta^*} \right)$$

Substitution of the proposed solution and its derivative into Eq. (3.57) gives

$$\frac{1}{G^*} \sigma_0 e^{-G^*t/\eta^*} \left(- \frac{G^*}{\eta^*} \right) + \frac{1}{\eta^*} \sigma_0 e^{-G^*t/\eta^*} = 0$$

as required. Note that G^*/η^* must have units time^{-1}. This is readily verified from the units of the individual quantities as follows: (kg m^{-1} sec^{-2})/(kg m^{-1} sec^{-1}) = sec^{-1}

•

The example verifies that the solution to this differential equation is

$$\sigma = \sigma_0 \, e^{-G^*t/\eta^*} = \sigma_0 \, e^{-t/\tau} \tag{3.58}$$

where η^*/G^* has the units of time and is identified as the experimental relaxation time and given the symbol τ.

Equation (3.58) can be written in terms of the shear modulus by dividing both sides of the equation by the constant strain:

$$G(t) = G_0 \, e^{-t/\tau} \tag{3.59}$$

where G_0 is the value of the time-dependent modulus at $t = 0$. Figure 3.8a shows a plot of this result in linear coordinates, and Fig. 3.8b does the same in log–log coordinates. Note that a semilog plot would be a straight line whose slope would measure τ. Similar results would be obtained for $E(t)$ in the case of tensile deformation. Several things about the relaxation time readily come to mind:

1. It is the time required for σ to drop to $1/e$ or 37% of its initial value.
2. As a ratio of η^*/G^*, τ is large when η^* is large and G^* is small and decreases with increasing modulus and decreasing viscosity.
3. The Newtonian viscosity is given by the product of the relaxation time and the Hookean modulus. This result was anticipated in the discussion of Eqs. (2.29) and (2.31).

In the last chapter we saw that the molecular relaxation time is the "yard-stick" against which times are measured. The experimental relaxation time plays a similar role. This is easily seen by an examination of Fig. 3.8b. For

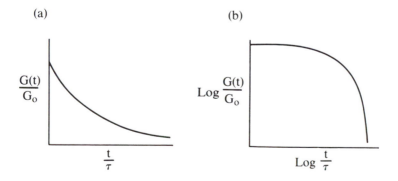

Figure 3.8 Time-dependent shear modulus [as $G(t)/G_0$] versus time (as t/τ): (a) linear coordinates and (b) log–log coordinates.

times very short compared to τ, the ratio $G(t)/G_0$ is essentially constant; that is, the elasticity totally predominates and the modulus is constant. For times which are long compared to τ, the modulus drops to zero, indicating totally viscous behavior. It is only for time scales within roughly one order of magnitude of τ that the combined viscoelastic behavior is observed. Two aspects of this will be developed in subsequent sections:

1. If the applied force varies sinusoidally with time, the period of the oscillation defines the time scale. Quite different mechanical responses are expected at different frequencies. This type of experiment will be described in Secs. 3.10 and 3.11.
2. Viscosity is considerably more sensitive to temperature than elasticity. By varying the temperature, the relaxation time of the polymer will be changed. Hence different mechanical response might be expected on a fixed laboratory time scale for samples examined at different temperatures.
3. Equivalent mechanical behavior can be achieved by either time (or frequency) or temperature manipulation. As noted in Sec. 3.2, results measured at different temperatures can be reduced to a common temperature to describe response over a wide range of times. We shall consider data reduced to a common temperature in this chapter and discuss the reduction process in Chap. 4.

Until now we have been describing the behavior of the Maxwell model in a relaxation experiment. How do these predictions compare with the observed behavior of polymers under the same experimental conditions? Figure 3.9 shows a plot of experimental results for two different un-cross-linked polymers of high molecular weight. One of the samples is polyisobutylene of $M = 1.56 \times 10^6$, the temperature to which the data has been reduced is $25°C$, and the time scale is in hours. The second polymer is polystyrene of $M = 2.0 \times 10^5$, the reduced temperature is $135°C$ and the units of time are seconds. The fact that the reduced temperatures are different should not concern us, nor is it significant that the moduli reported are Young's moduli while Eq. (3.59) was derived for shear. The issue is whether the model predicts the shape of the experimental curves. Since Fig. 3.9 is a double logarithmic plot, it should resemble Fig. 3.8b if the model applies. Our first reaction is that the agreement is not very good. Since these are the first examples of actual time-dependent data we have examined, it will be worthwhile to look at them in some detail. Moving from short to long times in Fig. 3.9, four regions of behavior can be distinguished:

1. At very short times the modulus is on the order of 10^{-10} N m^{-2}, comparable to ordinary window glass at room temperature. In fact, the mechanical behavior displayed in this region is called the glassy state, regardless of the chemical composition of the specimen. Inorganic and polymeric glasses

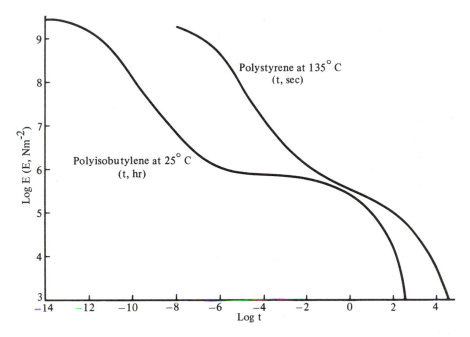

Figure 3.9 Log–log plots of modulus versus time for polyisobutylene at 25°C and polystyrene at 135°C. Note the different units of time for the two substances. (From data of A. V. Tobolsky and E. Catsiff and of H. Fujita and K. Ninomiya. From Ref. 4.)

are very similar in many properties and are often described as being hard, stiff, and brittle. These are the properties displayed by the polymers in Fig. 3.9 for very short periods of observation.

2. At somewhat longer times the modulus undergoes a gradual transition to a value lower by 3 or 4 orders of magnitude than its glassy value. Observed on this time scale, the material has a leathery consistency.

3. At still longer times a more or less pronounced plateau is encountered. The value of the plateau modulus is on the order of 10^6 N m^{-2}, comparable to the effect predicted for cross-linked elastomers in Sec. 3.4. This region is called the rubbery plateau and the sample appears elastic when observed in this time frame.

4. Eventually the modulus begins to drop off rapidly. This is an indication of the onset of flow rather than elastic resistance to deforming forces. The material would display viscous flow for very long periods of observation. Here again we can cite an analogy between inorganic glass and polymers.

The windows in some medieval cathedrals show greater thickness at the bottom than at the top, owing to the slow flow of the glass under the influence of gravity.

One of the things we immediately become aware of in examining a horizontal sweep across Fig 3.9 is that a similar range of behaviors would be encountered for an observation of some standardized duration, if temperature were the independent variable. As a matter of fact, the temperature at which the modulus drops from the glassy to rubbery values is the glass transition temperature. In an isothermal situation this same transition in behavior is characterized by the time required for certain relaxations. From Fig. 3.9 we can immediately conclude that polystyrene has a higher glass transition temperature than polyisobutylene. Table 2.3 lists the T_g values for these as 373 and 205 K, respectively. Next let us examine the meaning of these results.

3.8 Understanding the Modulus: Molecules and Models

The four regions of behavior shown by the polymers in Fig. 3.9 are fairly typical of high molecular weight un-cross-linked polymers. Let us examine the four regions of behavior, this time in terms of what is happening at the molecular level:

1. In the glassy state net movements of the chain backbone are impossible. We may consider this now from two points of view: The time required for movement of the chain is long, and the glassy state is observed when the time of observation is too short for such movements to occur. Alternatively, at low temperatures backbone motion is literally frozen out. The principal difference between a crystalline solid and a glassy solid is in the order (or lack thereof) of the molecular arrangement, not in the magnitude of the modulus. Both glassy and crystalline states are discussed in the next chapter.
2. At the transition between glassy and rubbery behavior, a distinct relaxation occurs. From one viewpoint, the molecules have enough time to jostle into more relaxed conformations; from another, they have enough thermal energy to do so.
3. In the rubbery plateau, a new impediment to movement must be overcome: entanglements along the polymer chain. In discussing the effects of entanglements in Chap. 2, we compared them to crosslinks. Is it any surprise, then, that rubbery behavior similar to that shown by cross-linked elastomers characterizes this region?
4. Chemical crosslinks and entanglements differ, however. The former is permanent, the latter transient. Given sufficient time, even the effects of entanglements can be overcome and stationary-state flow is achieved. An

increase of temperature facilitates the process and produces a similar effect. When the time-dependent modulus of chemically cross-linked polymers is investigated, similar results to Fig. 3.9 are obtained, except that the fourth region is not observed and the modulus of an elastomer is observed at equilibrium.

Now that we have reviewed the phenomena observed in a stress relaxation experiment and the molecular processes occurring therein, let us return to a discussion of the Maxwell model. When we first compared the predictions of the model with experimental observations in the last section, the verdict was rather pessimistic. The foregoing review of the molecular processes which occur reveals the source of the trouble: Two (at least) distinctly different relaxation processes are taking place. At the glass transition, localized displacements of chain segments become feasible. At termination of the rubbery zone, displacements which overcome the more long-range effects of entanglements are possible. Since these different modes of motion meet entirely different types of opposition, the relaxation times may be very different. We allowed for the possibility of more than one relaxation time in discussing the Eyring theory in the last chapter. This possibility is even more plausible in the present context, where significantly different relaxations are visualized.

To quantitatively examine the effect of having two relaxations, suppose we assemble two Maxwell elements in parallel. One of these is characterized by G_1^*, η_1^*, and τ_1; the other by G_2^*, η_2^*, and τ_2. For the sake of discussion we assume the two relaxation times to differ by many orders of magnitude, say, $\tau_1 = 10^{-8}$ sec and $\tau_2 = 10^3$ sec. The values of G^* are also different, say, $G_1^* = 10^{10}$ N m^{-2} and $G_2^* = 10^6$ N m^{-2}. Table 3.2 shows the result of calculating $G(t)$ according to Eq. (3.59) for each of these elements separately over a wide range of times. Table 3.2 also lists the sum of the two contributions. Because of the logarithmic time scale, the contribution of the individual Maxwell element is its respective G^* value for $t < \tau$. For $t > \tau$, the contribution to G rapidly approaches zero. Provided that the two relaxation times are sufficiently different, the total modulus for the system consisting of two Maxwell elements shows two distinct steps occurring in the vicinity of the individual τ's. Table 3.2 also shows the logarithms of this composite modulus for ease of comparison with Fig. 3.9. From this comparison, it is clear that the two-element mechanical model with widely spaced relaxation times is capable of approximating the behavior of actual polymers in terms of the general features observed. An improvement over Eq. (3.59), therefore, is

$$G(t) = G_{1,0}^* \, e^{-t/\tau_1} + G_{2,0}^* \, e^{-t/\tau_2} \tag{3.60}$$

Even with this modification, we note that the model predicts a drop off in modulus which is steeper than observed in the individual steps. This gradual

Table 3.2 Calculated Values of G(t) at Various Times Based on a Model Consisting of Two Maxwell Elements in Parallel ($\tau_1 = 10^{-8}$ sec, $G_1^* = 10^{10}$ N m^{-2}, and $\tau_2 = 10^3$ sec, $G_2^* = 10^6$ N m^{-2})

Log t (sec)	$10^{10}\, e^{-t/10^{-8}}$ (N m^{-2})	$10^6\, e^{t/10^3}$ (N m^{-2})	$\sum_{1+2} G\, e^{-t/\tau}$ (N m^{-2})	$\log \sum_{(1,2)}$ (N m^{-2})
-14.0	1×10^{10}	1×10^6	1×10^{10}	10.00
-12.0	1×10^{10}	1×10^6	1×10^{10}	10.00
-10.0	9.9×10^9	1×10^6	9.9×10^9	10.00
-8.0	3.68×10^9	1×10^6	3.68×10^9	9.57
-7.5	4.23×10^8	1×10^6	4.24×10^8	8.63
-7.0	4.54×10^5	1×10^6	1.45×10^6	6.16
-6.5	1.85×10^{-4}	1×10^6	1×10^6	6.00
-6.0	3.7×10^{-34}	1×10^6	1×10^6	6.00
-4.0	~ 0	1×10^6	1×10^6	6.00
-2.0	~ 0	1×10^6	1×10^6	6.00
0.0	~ 0	9.99×10^5	9.99×10^5	6.00
$+2.0$	~ 0	9.05×10^5	9.05×10^5	5.96
$+3.0$	~ 0	3.68×10^5	3.68×10^5	5.57
$+3.5$	~ 0	4.23×10^4	4.23×10^4	4.63
$+4.0$	~ 0	4.54×10^1	4.54×10^1	1.66

decline can be viewed as a whole series of steps of the general type described above, except now closely spaced in G^* and τ values. Such a set of steps will not be resolved as in Table 3.2, but slightly different elements will contribute to the modulus simultaneously.

The Maxwell model for a set of n elements is shown in Fig. 3.7b. Recall that each element is characterized by its own parameters, for the ith: G_i^*, η_i^*, and τ_i. For a set of n Maxwell elements, Eq. (3.59) becomes

$$G(t) = \sum_{i=1}^{n} G_i(t) = \sum_{i=1}^{n} G_{i,0}^* \, e^{-t/\tau_i} \tag{3.61}$$

Finally, if a very large number of relaxation times is present, the summation in Eq. (3.61) can be replaced by an integral:

$$G(t) = \int_0^\infty G(\tau) e^{-t/\tau} \, d\tau \tag{3.62}$$

where $G(\tau)$ is a continuous distribution function which gives the contribution of relaxation times between τ and $\tau + d\tau$. If $G(\tau)$ were known, the modulus could be predicted exactly. Since a logarithmic time scale is almost always used in this type of work, it is more common to multiply and divide the right-hand side of Eq. (3.62) by τ and regroup the terms:

$$G(t) = \int_0^\infty \tau G(\tau) e^{-t/\tau} \, \frac{d\tau}{\tau} = \int_0^\infty H(\tau) e^{-t/\tau} \, d \ln \tau \tag{3.63}$$

The factor $\tau G(\tau)$ is called the relaxation spectrum and is given the symbol $H(\tau)$.

In principle, the relaxation spectrum $H(\tau)$ describes the distribution of relaxation times which characterizes a sample. If such a distribution function can be determined from one type of deformation experiment, it can be used to evaluate the modulus or compliance in experiments involving other modes of deformation. In this sense it embodies the key features of the viscoelastic response of a spectrum. Methods for finding a function $H(\tau)$ which is compatible with experimental results are discussed in Ferry's *Viscoelastic Properties of Polymers*. In Sec. 3.12 we shall see how a molecular model for viscoelasticity can be used as a source of information concerning the relaxation spectrum.

3.9 The Voigt Model: Creep

In this section we consider a different experimental situation: the case of creep. In a creep experiment σ is maintained at a constant value and the time dependence of the strain is measured. Thus it is the exact inverse of the relaxation

experiment. As was the case with relaxation, it is convenient to proceed in terms of a model. Although a new model is not required—creep can be discussed in terms of the Maxwell model—it is convenient to approach the subject in terms of a different model. The components of the model are the same as those in the Maxwell model: the spring and the dashpot. This time, however, they are arranged in parallel rather than in series. This simple mechanical model is called a Voigt element and is sketched in Fig. 3.10a.

We shall follow the same approach as the last section, starting with an examination of the predicted behavior of a Voigt model in a creep experiment. We should not be surprised to discover that the model oversimplifies the behavior of actual polymeric materials. We shall continue to use a shear experiment as the basis for discussion, although a creep experiment could be carried out in either a tension or shear mode. Again we begin by assuming that the Hookean spring in the model is characterized by a modulus G^*, and the Newtonian dashpot by a viscosity η^*.

Since the strain is the same in both elements in the Voigt model, the applied stress (subscript 0) must equal the sum of the opposing forces arising from the elastic and viscous response of the model:

$$\sigma_0 = \sigma_{el} + \sigma_{vis} \tag{3.64}$$

Substituting Eqs. (2.2) and (3.50) into this relationship yields

$$\sigma_0 = G^*\gamma + \eta^* \frac{d\gamma}{dt} \tag{3.65}$$

In the following example we examine the solution to this differential equation.

(a)   (b)

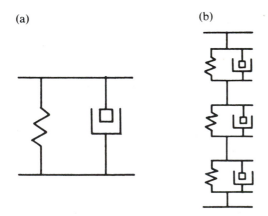

Figure 3.10   Voigt models consisting of a spring and dashpot in parallel: (a) simple Voigt unit and (b) set of units arranged in series.

Example 3.4

Verify that $\gamma(t) = (1/G^*)\sigma_0 + Be^{-G^*t/\eta^*}$ is a solution to Eq. (3.65) and evaluate the constant B for the boundary condition that $\gamma = 0$ at $t = 0$.

Solution

We verify the solution by carrying out the indicated differentiation and substituting back into the differential equation:

$$\frac{d\gamma}{dt} = -\frac{G^*B}{\eta^*} e^{-G^*t/\eta^*} \quad \text{for } \sigma_0, G^*, \text{ and } \eta^* \text{ constant}$$

If we substitute back into Eq. (3.65),

$$\sigma_0 = G^* \left(\frac{1}{B} \sigma_0 + Be^{-G^*t/\eta^*} \right) + \eta^* \left(-\frac{G^*B}{\eta^*} e^{-G^*t/\eta^*} \right)$$

If $\gamma = 0$ at $t = 0$, the solution becomes $0 = (1/G^*)\sigma_0 + B$ or $B = -\sigma_0/G^*$. Therefore we write $\gamma(t) = (\sigma_0/G^*)(1 - e^{-t/\tau})$ using the result of $\tau = \eta^*/G^*$. Dividing through by σ_0 converts $\gamma(t)$ to compliance as defined in Table 3.1.

•

The example shows that the shear compliance of the Voigt material is given by

$$J(t) = \frac{\gamma(t)}{\sigma_0} = \frac{1}{G^*} (1 - e^{-t/\tau}) = J(\infty)(1 - e^{-t/\tau}) \tag{3.66}$$

Figure 3.11a is a plot of this solution, showing how the strain develops exponentially with time. The maximum value of the compliance is attained as $t \to \infty$, hence the upper limits of compliance is designated $J(\infty)$. Since the deformation

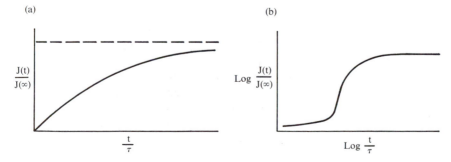

Figure 3.11 Time-dependent shear compliance [as $J(t)/J(\infty)$] versus time (as t/τ): (a) linear coordinates and (b) log–log coordinates.

only gradually approaches this limit, the ratio η^*/G^* is called the retardation time in this kind of experiment. The retardation time of the Voigt element continues to define the time scale of the experiment. A log–log representation of Eq. (3.66) exaggerates the shape of the solution and results in a rather sharp step in compliance at $t = \tau$. This is shown in Fig. 3.11b.

It is interesting to note that the Voigt model is useless to describe a relaxation experiment. In the latter a constant strain was introduced instantaneously. Only an infinite force could deform the viscous component of the Voigt model instantaneously. By constrast, the Maxwell model can be used to describe a creep experiment. Equation (3.56) is the fundamental differential equation of the Maxwell model. Applied to a creep experiment, $d\sigma/dt = 0$ and the equation becomes

$$\frac{d\gamma}{dt} = \frac{1}{\eta^*} \sigma_0 \tag{3.67}$$

Integrating Eq. (3.67) and using $\gamma = \gamma_0$ at $t = 0$ to evaluate the integration constant gives

$$\gamma(t) = \gamma_0 + \frac{\sigma_0}{\eta^*} t \tag{3.68}$$

or

$$J(t) = J(0) + \frac{1}{\eta^*} t \tag{3.69}$$

The Maxwell model thus predicts a compliance which increases indefinitely with time. On rectangular coordinates this would be a straight line of slope $1/\eta^*$, and on log–log coordinates a straight line of unit slope, since the exponent of t is 1 in Eq. (3.69).

As we did in the case of relaxation, we now compare the behavior predicted by the Voigt model—and, for that matter, the Maxwell model—with the behavior of actual polymer samples in a creep experiment. Figure 3.12 shows plots of such experiments for two polymers. The graph is on log–log coordinates and should therefore be compared with Fig. 3.11b. The polymers are polystyrene of molecular weight 6.0×10^5 at a reduced temperature of $100°C$ and cis-polyisoprene of molecular weight 6.2×10^5 at a reduced temperature of $-30°C$.

Since the compliance is essentially the inverse of the modulus, it is not surprising that the same four regions of mechanical behavior show up again here. The data for polystyrene is more fully developed, so we shall examine it:

1. At very short times the compliance is low and essentially constant. This is the glassy state where chain motion requires longer times to be observed.

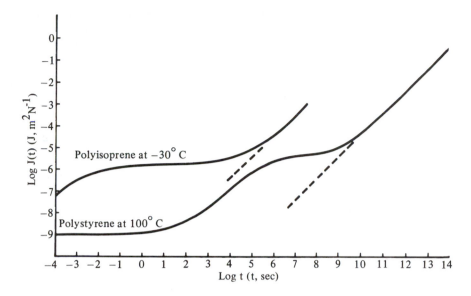

Figure 3.12   Log–log plots of compliance versus time for polystyrene at $100°C$ and cis-polyisoprene at $-30°C$. (Data of D. J. Plazek and V. M. O'Rourke and of N. Nemoto, M. Moriwaki, H. Odani, and M. Kurata from Ref. 4.)

The same mechanical behavior would be observed at longer times at lower temperatures.

2.   At longer times an increase in compliance marks the relaxation of the glassy state to the rubbery state. Again, an increase of temperature through T_g would produce the same effect.

3.   The plateau compliance is characteristic of rubbery behavior where chain entanglements play the role of effective crosslinks.

4.   The upswing in compliance from the rubbery plateau marks the onset of viscous flow. In this final stage the slope of the lines (the broken lines in Fig. 3.12) is unity, which means that the compliance increases linearly with time.

As was the case with the modulus, the transitions from one horizontal region of compliance to another is more gradual than that predicted by the model and shown in Fig. 3.11b.

The procedure we followed in adapting the relaxation model to experimental findings immediately suggests how to handle the discrepancies between the model and experiment in the case of creep:

1. The more gradual approach to equilibrium than the model predicts can be taken into account by imagining that the rise consists of a series of n smaller (and unresolved) steps. This is equivalent to expanding the model so that it consists of n Voigt elements as shown in Fig. 3.10b. Each of these Voigt elements is characterized by its own value for G^*, η^*, and τ.

2. With this modification of the model, Eq. (3.66) becomes

$$J(t) = \sum_{i=1}^{n} J_i(t) = \sum_{i=1}^{n} J_i(\infty) (1 - e^{-t/\tau_i}) \tag{3.70}$$

or

$$J(t) = \int_{0}^{\infty} J(\tau) (1 - e^{-t/\tau}) \, d\tau \tag{3.71}$$

if a continuous distribution of retardation times is considered [compare Eq. (3.62)].

3. In addition to the set of Voigt elements, a Maxwell element could also be included in the model. The effect is to include a contribution given by Eq. (3.69) to the calculated compliance. This long time flow contribution to the compliance is exactly what we observe for non-cross-linked polymers in Fig. 3.12.

4. Therefore, in the most general case, we write for an actual polymer

$$J(t) = J(0) + \frac{t}{\eta^*} + \int_{0}^{\infty} J(\tau) (1 - e^{-t/\tau}) \, d\tau \tag{3.72}$$

An advantage of having the relaxation spectrum defined by Eq. (3.63) is that it can be adapted to expressions like this to calculate mechanical behavior other than that initially measured.

The Maxwell and Voigt models of the last two sections have been investigated in all sorts of combinations. For our purposes, it is sufficient that they provide us with a way of thinking about relaxation and creep experiments. Probably one of the reasons that the various combinations of springs and dashpots have been so popular as a way of representing viscoelastic phenomena is the fact that simple and direct comparison is possible between mechanical and electrical networks, as shown in Table 3.3. In this parallel, the compliance of a spring is equivalent to the capacitance of a condenser and the viscosity of a dashpot is equivalent to the resistance of a resistor. The analogy is complete

Table 3.3   Comparison of Mechanical and Electrical Models Consisting of Different Arrangements of Springs and Dashpots or Their Equivalents, Capacitance and Resistance, Respectively

	Mechanical	Electrical
	F/A	Electromotive force
	γ	Charge
	$\dot{\gamma}$	Current
	J	Capacitance
	η	Resistance
Maxwell element	Series	Parallel
Voigt element	Parallel	Series

because electrical work done on a network is stored in capacitors and dissipated by resistors. Likewise, mechanical energy is stored and dissipated by the elastic and viscous units of a mechanical model system. The only word of caution about the use of this analogy is that the rules for combination are reversed—that is, series becomes parallel and vice versa—between the electrical and mechancial networks to produce the close correspondence between the storage and dissipative units. Electrical circuits can thus be designed and analyzed which embody the appropriate features of a model mechanical system. The usefulness of this analogy will be even more evident—although we shall not pursue it—in the next section, in which we take up periodic stress–strain relationships, the mechanical analog of alternating current.

3.10   Dynamic Viscoelasticity

The relaxation and creep experiments that were described in the preceding sections are known as transient experiments. They begin, run their course, and end. A different experimental approach, called a dynamic experiment, involves stresses and strains that vary periodically. Our concern will be with sinusoidal oscillations of frequency v in cycles per second (Hz) or ω in radians per second. Remember that there are 2π radians in a full cycle, so $\omega = 2\pi v$. The reciprocal of ω gives the period of the oscillation and defines the time scale of the experiment. In connection with the relaxation and creep experiments, we observed that the maximum viscoelastic effect was observed when the time scale of the experiment is close to τ. At a fixed temperature and for a specific sample, τ or the spectrum of τ values is fixed. If it does not correspond to the time scale of a transient experiment, we will lose a considerable amount of information about the viscoelastic response of the system. In a dynamic experiment it may

be possible to vary the frequency in such a way that the period and the range of τ values overlap optimally. This sort of dynamic mechanical test yields the maximum amount of information about a viscoelastic substance.

Suppose an oscillating strain of frequency ω is induced in a sample:

$$\gamma = \gamma_0 \sin(\omega t) \tag{3.73}$$

where γ_0 is the maximum amplitude of the strain. If the sample showed only elasticity and obeyed Hooke's law, the stress and strain would be exactly in phase:

$$\sigma = G\gamma = G\gamma_0 \sin(\omega t) = \sigma_0 \sin(\omega t) \tag{3.74}$$

where σ_0 is the maximum amplitude of the stress. The rate of change of the strain is given by

$$\frac{d\gamma}{dt} = \omega \gamma_0 \cos(\omega t) \tag{3.75}$$

If the sample behaves as a Newtonian liquid, the stress is

$$\sigma = \eta\dot{\gamma} = \eta\omega\gamma_0 \cos(\omega t) \tag{3.76}$$

Comparing this result with Eq. (3.74) shows that elastic and viscous forces— occurring separately—are 90° out of phase. In the more general case of the two occurring together, these two components are out of phase by some angle δ.

Next suppose we consider the effect of a periodically oscillating stress on a Voigt element of modulus G^* and viscosity η^*. Remember from the last section that for a Voigt element the applied stress equals the sum of the elastic and viscous responses of the model. Therefore, for a stress which varies periodically, Eq. (3.64) becomes

$$\sigma = \sigma_0 \cos(\omega t) = G^*\gamma + \eta^* \frac{d\gamma}{dt} \tag{3.77}$$

In this case, σ_0 is the maximum amplitude of the stress. The solution to this differential equation will give a functional description of the strain in this dynamic experiment. In the following example, we examine the general solution to this differential equation.

Example 3.5

Evaluate the constants B and C in the expression $\gamma = B \cos(\omega t) + C \sin(\omega t)$ so that this equation is a solution to Eq. (3.77).

Solution

Evaluate $d\gamma/dt$ and substitute this and γ into Eq. (3.77):

$$\frac{d\gamma}{dt} = -\omega B \sin(\omega t) + \omega C \cos(\omega t)$$

Therefore $\sigma_0 \cos(\omega t) = G^*[B \cos(\omega t) + C \sin(\omega t)] + \eta^*[-\omega B \sin(\omega t) + \omega C \cos(\omega t)]$. For an acceptable solution, coefficients of trigonometric functions must be the same on both sides of the equation. Therefore we can write $\sigma_0 = G^*B + \eta^*\omega C$ and $0 = -\eta^*\omega B + G^*C$. The second of these can be written $C = \eta^*\omega B/G^* = \tau\omega B$. Solving these two simultaneous equations for B and C yields

$$B = \frac{\sigma_0}{G^*(1 + \omega^2 \tau^2)} \quad \text{and} \quad C = \frac{\sigma_0 \tau \omega}{G^*(1 + \omega^2 \tau^2)}$$

With these values for the constants in the general equation, the latter satisfies the original differential equation.

•

The example shows that we have found an acceptable solution for $\gamma(t)$ for the case of a periodic applied stress. Dividing through by σ_0 gives the appropriate compliance:

$$J(t) = [G^*(1 + \omega^2 \tau^2)]^{-1} \cos(\omega t) + \omega \tau [G^*(1 + \omega^2 \tau^2)]^{-1} \sin(\omega t) \quad (3.78)$$

The frequency-dependent coefficients in this equation are given separate names and symbols to facilitate discussion. Remember it is these coefficients that determine the behavior of the system; the trigonometric functions merely describe the oscillations. The following can be said of the coefficient of the cosine term:

1. It is called the storage compliance and is given the symbol $J'(\omega)$.
2. It is directly related to the energy stored in the sample per cycle of oscillation.
3. It is defined by the expression

$$J'(\omega) = [G^*(1 + \omega^2 \tau^2)]^{-1} \quad (3.79)$$

The following can be said of coefficient of the sine term:

4.   It is called the loss compliance and is given the symbol $J''(\omega)$.
5.   It is directly related to the nonrecoverable or dissipated energy per cycle.
6.   It is defined by the expression

$$J''(\omega) = \omega\tau[G^*(1 + \omega^2\tau^2)]^{-1} \tag{3.80}$$

Identical relationships apply to the compliance in a dynamic tensile experiment. Equations (3.79) and (3.80) are plotted in Fig. 3.13.

It is not difficult to understand the behavior of $J'(\omega)$ and $J''(\omega)$ sketched in Fig. 3.13. At low frequencies the motion is slow and the energy loss due to viscous resistance is negligible. The displacement is due essentially to the spring, so the compliance is constant. In Eq. (3.79), $J'(\omega) \to 1/G^*$ as $\omega \to 0$. As the frequency increases, motion is faster and the applied stress becomes more evenly divided between the spring and the dashpot. The elastic portion of the

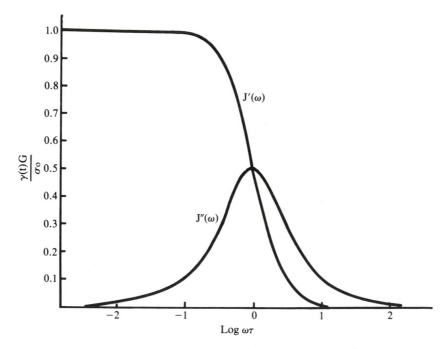

Figure 3.13   Storage compliance $J'(\omega)$ and loss compliance $J''(\omega)$ as functions of $\omega\tau$ as given by Eqs. (3.79) and (3.80).

compliance is therefore less, while the viscous part increases. At very large frequencies the resistance of the dashpot becomes large and effectively suppresses the motion of the spring completely. Also, because the viscosity is so high, the displacement of the dashpot is insignificant. Energy dissipation involves both force and displacement. At high frequencies the former is large but the latter small in the dashpot, so the loss compliance also decreases with increasing $\omega\tau$. In Eq. (3.80) $J''(\omega) \to (G^*\omega\tau)^{-1} \to 0$ as $\omega\tau \to \infty$.

A dynamic modulus can also be evaluated by following a procedure similar to that used for the dynamic compliance above. The modulus is most easily approached by considering a Maxwell element in which case the differential equation analog to Eq. (3.77) is

$$\omega\gamma_0 \cos(\omega t) = \frac{1}{G^*} \frac{d\sigma}{dt} + \frac{1}{\eta^*} \sigma \tag{3.81}$$

Equations (3.77) and (3.81) both have the same general form $dy/dt + Py = Q$, so the general solution—given in Example 3.5—is the same for both, although the values of the constants are different. When the constants are evaluated, the storage and loss components of the modulus are found to be

$$G'(\omega) = \omega^2\tau^2 G^*(1 + \omega^2\tau^2)^{-1} \tag{3.82}$$

and

$$G''(\omega) = \omega\tau G^*(1 + \omega^2\tau^2)^{-1} \tag{3.83}$$

respectively.

The dynamic viscosity is related to the loss component of the shear modulus through the result $\eta_{dyn} = G''/\omega$ As $\omega \to 0$, the dynamic viscosity approaches the zero shear viscosity of an ordinary liquid, η_N.

We commented above that the elastic and viscous effects are out of phase with each other by some angle δ in a viscoelastic material. Since both vary periodically with the same frequency, stress and strain oscillate with t, as shown in Fig. 3.14a. The phase angle δ measures the lag between the two waves. Another representation of this situation is shown in Fig. 3.14b, where stress and strain are represented by arrows of different lengths separated by an angle δ. Projections of either one onto the other can be expressed in terms of the sine and cosine of the phase angle. The bold arrows in Fig. 3.14b are the components of γ parallel and perpendicular to σ. Thus we can say that $\gamma \cos \delta$ is the strain component in phase with the stress and $\gamma \sin \delta$ is the component out of phase with the stress. We have previously observed that the elastic response is in phase with the stress and the viscous response is out of phase. Hence the ratio of

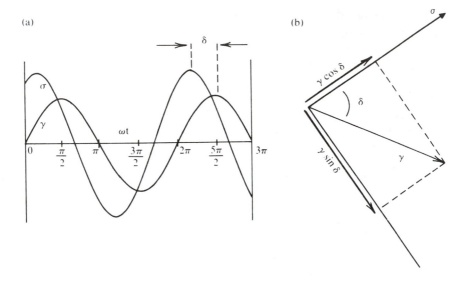

Figure 3.14 General representations of stress and strain out of phase by amount δ: (a) represented by oscillating functions and (b) represented by vectors.

the two measures the relative contributions of these effects. These two contributions are also measured by $J'(\omega)$ and $J''(\omega)$; therefore

$$\frac{\gamma \sin \delta}{\gamma \cos \delta} = \tan \delta = \frac{J''(\omega)}{J'(\omega)} \tag{3.84}$$

where the single parameter $\tan \delta$ is called the loss tangent. The loss tangent is also proportional to the energy loss per cycle, so both $J''(\omega)$ and $\tan \delta$ measure viscous dissipation of mechanical energy in a sample. Often the quantity $\pi \tan \delta$ is reported; it is called the logarithmic decrement.

3.11 The Dynamic Components: Measurement and Interpretation

The wide range of variables involved makes experimental viscoelasticity somewhat difficult to describe. In a particular study the range might be narrowed because of the specific character of the material under investigation. In more general terms, however, such wide ranges of variation are involved that no single experimental design is suitable for all purposes. Figure 3.12, for example, spans 9 orders of magnitude variation in $J(t)$ and 18 orders in t (no wonder log–log

coordinates are standard in this type of work!). In addition, samples can vary from viscous liquids to brittle solids, and there may be restrictions on the size or shape of available samples.

Both stress relaxation and creep experiments have their place in the study of viscoelasticity, but measurement and interpretation of the dynamic mechanical behavior of a polymer gives the most information on the most convenient time scale. We shall describe the experimental aspects of dynamic viscoelasticity only in broad conceptual terms. The actual implementation of these concepts embraces many different kinds of apparatus and experimental routines. In a conceptually simple apparatus for measuring the dynamic components, the sample is sandwiched between two surfaces, one of which is driven with a known periodic displacement. The force at the other side of the gap is measured by a strain gauge. The stress and strain at the two surfaces are proportional to the voltage outputs of electromechanical transducers monitoring each plate. These can be calibrated so that the maximum force F_0 and maximum displacement x_0 are read directly. The signals could be fed into a dual-beam oscilloscope so that the sinusoidal variation of both the stress and strain appear on the screen. The phase angle δ can be measured from this trace, as shown in Fig. 3.14a. Alternatively, the ratio of the two signals can be used to measure the loss tangent directly as indicated by Eq. (3.84). If A is the area of contact between the sample and the plates and h is the separation of the surfaces,

$$J''(t) = \frac{x_0/h}{F_0/A} \sin \delta$$

and

$$J'(t) = \frac{x_0/h}{F_0/A} \cos \delta$$

These equations reveal that sample dimensions must also be known with precision, and, of course, the specimens must be free from defects.

In another type of measurement, the parallel between mechanical and electrical networks can be exploited by using variable capacitors and resistors to balance the impedance of the transducer circuit. These electrical measurements readily lend themselves to computer interfacing for data acquisition and analysis.

A variety of commercial instruments are available for the determination of the viscoelastic behavior of samples. Figure 3.15 shows one such apparatus, the Rheovibron Viscoelastometer. This instrument also takes advantage of the complementarity that exists between time and temperature: It operates at four frequencies over a $175°C$ temperature range. With accessories, both the frequency range and the temperature range can be broadened still further.

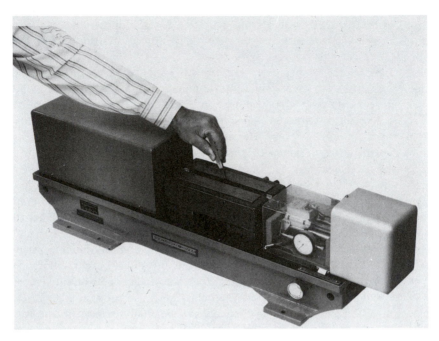

Figure 3.15 The Rheovibron Viscoelastometer, a commercially available instrument for the determination of the dynamic moduli and compliances. [Photo courtesy of Imass, Inc., Accord (Hingham), Mass. 02018.]

Figure 3.16a shows the storage and loss components of the compliance of crystalline polytetrafluoroethylene at 22.6°C. While not identical to the theoretical curve based on a single Voigt element, the general features are readily recognizable. Note that the range of frequencies over which the feature in Fig. 3.16a develops is much narrower than suggested by the scale in Fig. 3.13. This is because the sample under investigation is crystalline. For amorphous polymers, the observed loss peaks are actually broader than predicted by a

Figure 3.16 Some experimental dynamic components. (a) Storage and loss compliance of crystalline polytetrafluoroethylene measured at different frequencies. [Data from E. R. Fitzgerald, *J. Chem. Phys.* 27:1180 (1957).] (b) Storage modulus and loss tangent of poly(methyl acrylate) and poly(methyl methacrylate) measured at different temperatures. (Reprinted with permission from J. Heijboer in D. J. Meier (Ed.), *Molecular Basis of Transitions and Relaxations*, Gordon and Breach, New York, 1978.)

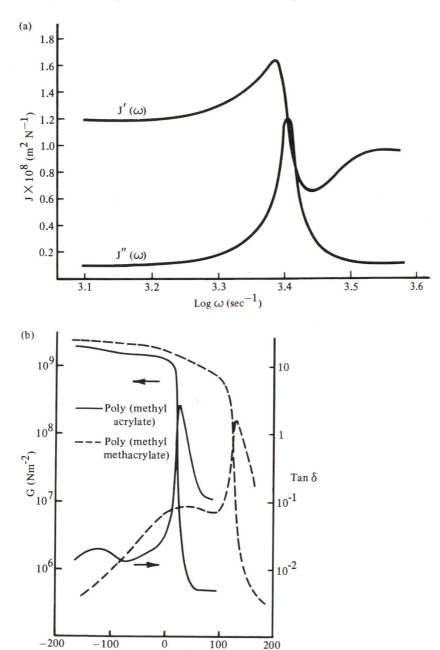

Figure 3.16

single element model. By now, the remedy to this latter situation is becoming commonplace: We simply replace the single element in the model by a set of n elements, each having parameters of G_i^*, η_i^*, and τ_i. For this general system, Eqs. (3.79) and (3.80) become

$$J'(\omega) = \sum_{i=1}^{n} [G_i^*(1 + \omega^2\tau_i^2)]^{-1} \tag{3.85}$$

and

$$J''(\omega) = \sum_{i=1}^{n} \omega\tau_i [G_i^*(1 + \omega^2\tau_i^2)]^{-1} \tag{3.86}$$

for storage and loss, respectively.

The abscissa in Fig. 3.16a is the experimental frequency on a log scale, while log $(\omega\tau)$ is plotted in Fig. 3.13. The maximum of the loss compliance occurs at $\omega\tau = 1$ (log 1 = 0) so $\omega = 1/\tau$ at the location of this feature. For the system in Fig. 3.16a, therefore, $\tau = 4 \times 10^{-4}$ sec.

Figure 3.16b is an equivalent diagram in which a different set of variables are employed:

1. The storage modulus rather than compliance is plotted. This is a trivial difference, but note that the modulus is measured at a single frequency, 1 Hz.
2. The loss tangent rather than the loss modulus is plotted, also at 1 Hz.
3. Temperature rather than frequency is the independent variable. We have previously noted the equivalency of these variables. Figure 3.16 may be taken as experimental evidence for that.
4. Temperature variation is not only easy to implement experimentally, but is also more familiar to chemists than the frequency of mechanical oscillations.
5. With T as the independent variable, the transition between glassy and rubbery behavior can be read directly at T_g. Note that T_g is about 100° lower for poly(methyl acrylate) than for poly(methyl methacrylate).
6. An additional interesting feature that emerges from Fig. 3.16b is the smaller loss tangent peak which occurs below T_g. Since backbone motion is essentially frozen out below T_g, this low-temperature peak must measure some energy-dissipating mode of motion that does not involve the main polymer chain. The fact that this feature is so much more prominent in poly(methyl methacrylate) than in poly(methyl acrylate) suggests that the motion involves the methyl group.

The parallel between the loss compliance spectrum—or those of the loss tangent or logarithmic decrement, which are equivalent—and an electromagnetic absorption spectrum should not be overlooked. Several areas of similarity and/or difference can be identified:

1. The electromagnetic spectrum measures the absorption of radiation energy as a function of the frequency of the radiation. The loss spectrum measures the absorption of mechanical energy as a function of the frequency of the stress–strain oscillation.
2. In the electromagnetic spectrum, the energy absorbed makes up the difference between two allowed energy states in the absorber. In the loss spectrum the frequency absorbed closely matches the frequency of dissipative modes of molecular motion in the sample.
3. The electromagnetic spectrum is a quantum effect and the width of a spectral feature is traceable to the Heisenberg uncertainty principle. The mechanical spectrum is a classical resonance effect and the width of a feature indicates a range of closely related τ values for the model elements.
4. The electromagnetic spectrum can be used to "fingerprint" molecules or functional groups. For example, hydroxyl and carbonyl groups absorb different frequencies of infrared radiation. A well-resolved mechanical spectrum can distinguish between the relaxations occurring in, say, crystalline and amorphous regions of the same sample.
5. The amount of a particular component in a sample can be monitored by examining the height of a spectral absorption peak: The reduction of an aldehyde to an alcohol would show up as a decrease in line intensity for the carbonyl and an increase for the hydroxyl peaks in the spectrum. Changes in the relative importance of different relaxation modes in a polymer can also be followed by the corresponding changes in a mechanical spectrum.
6. Theoretical analysis of certain features in the electromagnetic spectrum yields basic molecular parameters such as bond lengths and bond stiffness. We shall see presently that the mechanical spectra can be related to molecular parameters and not just modelistic characteristics as we have used until now.

The following example illustrates an application of the mechanical spectra.

Example 3.6

A diblock copolymer, 71% polyisoprene (I) by weight and 29% polybutadiene (B), was blended in different proportions into a 71%–29% mixture of the individual homopolymers. The loss tangent was measured as a function of temperature for various proportions of copolymer. Two peaks are observed:

Table 3.4 Temperature Coordinate and Relative Height (in Parenthesis) for the Two Loss Tangent Maxima Observed in Mixtures of Isoprene–Butadiene Block Copolymers with Homopolymers of These Two Repeat Units in the Same Proportion[a]

Weight percent copolymer	Temperature ($^\circ$C) and relative height (arbitrary units)	
	Peak 1	Peak 2
Pure homopolymers	Polyisoprene: −49	Polybutadiene: −82
0	−50 (2.5)	−84 (0.9)
20	−52 (2.6)	−84 (0.8)
40	−54 (2.5)	−85 (0.7)
60	−57 (2.8)	−86 (0.5)
80	−57 (2.6)	−86 (0.5)
100	−58 (2.6)	None

[a] Results interpreted in Example 3.6.

Source: Data from R. R. Cohen and A. R. Ramos, *Macromolecules* 12:131 (1979).

Their temperature coordinate and relative height are listed in Table 3.4. Briefly describe any trends or features of the data and propose an interpretation.

Solution

The system combines having three components and having constant composition, since the copolymer has the same composition as the homopolymer mixture. Listed below are some observations and possible interpretations:

1. The position of the loss tangent maximum is quite different for pure B and pure I, indicating distinctly different T_g values for the two.
2. The homopolymer mixture shows only a minor modification of the two T_g values. This suggests that the two homopolymers are concentrated in separate phases of almost pure polymer.
3. The butadiene peak is absent in the pure copolymer sample. Since this must be a single phase, this phase containing both B and I has a T_g of -58°C.
4. Increasing the copolymer content of the system lowers the T_g of both peaks, but the effect is far more pronounced for the peak associated with I. The latter is essentially constant in height, while the B peak diminishes in height with increasing copolymer. This suggests that the copolymer dissolves preferentially in the polyisoprene.

•

3.12   A Molecular Theory for Viscoelasticity: The Rouse Model

We have relied heavily on the use of models in discussing the viscoelastic behavior of polymers in the transient and dynamic experiments of the last few sections. The models were mechanical, however, and while they provide a way for understanding the phenomena involved, they do not explicitly relate these phenomena to molecular characteristics. To establish this connection is the objective of this section.

The molecular theory begins by subdividing the polymer molecules into subchains with the following properties:

1. The degree of polymerization of the subchain is n_s. If the degree of polymerization of the molecule as a whole is n, then there are n/n_s subchains per molecule. We symbolize the number of subchains per molecule as N_s. Other properties of the subchain—which, incidentally, should not be confused with the chains between crosslink points in elastomers—will also have the subscript s as they emerge.
2. The length of the subchain is sufficient to justify the use of random flight statistics in its description.
3. The mass of the subchain is pictured as concentrated in a bead, connected to adjacent beads by Hookean springs which, individually, obey Eq. (3.45).
4. The displacement of beads representing subchains is resisted by viscous forces which follow Eq. (2.47).
5. Whether the beads representing subchains are imbedded in an array of small molecules or one of other polymer chains changes the friction factor in Eq. (2.47), but otherwise makes no difference in the model. This excludes chain entanglement effects and limits applicability to $M < M_c$, the threshold molecular weight for entanglements.
6. The subchain of this model is an artifact about which we have no information. After developing expressions for the behavior of the subchains, we must describe the latter in terms of the actual polymer chains.

We refer to this model as the bead–spring model and to its theoretical development as the Rouse theory, although Rouse, Bueche, and Zimm have all been associated with its development.

We begin the mathematical analysis of the model, by considering the forces acting on one of the beads. If the sample is subject to stress in only one direction, it is sufficient to set up a one-dimensional problem and examine the components of force, velocity, and displacement in the direction of the stress. We assume this to be the z direction. The subchains and their associated beads and springs are indexed from 1 to N_s; we focus attention on the ith. The absolute coordinates of the beads do not concern us, only their displacements.

Accordingly, we designate z_i as the displacement of the ith bead in the direction of the stress. Since its immediate neighbors also experience some displacement, the springs connecting the ith bead to these neighbors are stretched by the amounts $z_{i+1} - z_i$ and $z_i - z_{i-1}$, respectively. As indicated above, these springs obey Hooke's law with $k_H = 3kT/n_s l_0^2$. The net elastic force on the ith bead is given by the difference between the forces exerted on it by the two connecting springs:

$$F_{el,i} = \frac{3kT}{n_s l_0^2} (z_{i+1} - z_i) - \frac{3kT}{n_s l_0^2} (z_i - z_{i-1}) = \frac{3kT}{n_s l_0^2} (z_{i+1} + z_{i-1} - 2z_i)$$

(3.87)

Note that the quantity $n_s l_0^2$ is the mean-square end-to-end distance in the subchain according to Eq. (1.62). We shall designate it $(\overline{r^2})_s$.

The phenomena under discussion are viscoelastic; we have only considered the elastic forces. Next we must incorporate viscous forces. As indicated above, we use Eq. (2.47) to express the proportionality between the viscous resistance to displacement and the velocity of the bead, dz_i/dt:

$$F_{vis,i} = \zeta_s \frac{dz_i}{dt} = n_s \zeta \frac{dz_i}{dt}$$

(3.88)

where ζ_s is the friction factor of the subchain. It equals $n_s \zeta$, where ζ is the segmental friction factor discussed in Sec. 2.11. Under stationary-state conditions (and assuming viscous forces to be much larger than inertial forces) the forces given by Eqs. (3.87) and (3.88) equal:

$$\zeta_s \frac{dz_i}{dt} = \frac{3kT}{(\overline{r^2})_s} (z_{i+1} + z_{i-1} - 2z_i)$$

(3.89)

for each of the N beads in the polymer, except the two end ones, which experience only one elastic force.

What makes this set of differential equations tricky to solve is the fact that the z values for the two beads adjacent to the ith bead must be taken into account.

The bead and spring model is clearly based on mechanical elements just as the Maxwell and Voigt models were. There is a difference, however. The latter merely describe a mechanical system which behaves the same as a polymer sample, while the former relates these elements to actual polymer chains. As a mechanical system, the differential equations represented by Eq. (3.89) have been thoroughly investigated. The results are somewhat complicated, so we shall not go into the method of solution, except for the following observations:

1. An important part of solving any differential equation is the specification of the boundary conditions. In the present case these can correspond to tension or shear and can be solved to give either a modulus or a compliance.

2. If the viscous contribution were absent, the solution to this problem would be characterized by a resonance frequency close to which the application of small driving forces results in displacements of large amplitude.

3. Including viscosity has the effect of damping the system of vibrators, broadening the range of resonating frequencies, and decreasing the amplitude of the resulting displacement.

4. The solutions describe the vibrational modes of the system. As waves, the solutions are characterized by integers p which essentially count the number of nodes along the chain in a particular mode of vibration. The upper limit of p corresponds to the number of subchains in the molecule N_s.

The shear relaxation modulus of the bead–spring system is given by an expression very much like Eq. (3.61), namely,

$$G(t) = \sum_{p=1}^{N_s} G_p \, e^{-t/\tau_p} \tag{3.90}$$

except that G_p and τ_p are the modulus and relaxation time, respectively, for the wave characterized by the integer p. The value of τ_p in the Rouse theory is given by

$$\tau_p = \left[\frac{24kT}{(\overline{r^2})_s \zeta_s} \sin^2 \left(\frac{p\pi}{2(N_s + 1)} \right) \right]^{-1} \tag{3.91}$$

Next we shall examine G_p and τ_p individually.

We return to Eq. (3.45) to describe the elasticity of a subchain. According to that equation, the force can be written

$$F = \frac{3kT}{\sqrt{n_s}\, l_0} \left(\frac{\Delta z}{\sqrt{n_s}\, l_0} \right) = \frac{3kT}{\sqrt{n_s}\, l_0} \gamma$$

where γ is the strain. Dividing both sides of this expression by the cross-sectional area of the subchain converts F to σ and $\sqrt{n_s}\, l_0$ to V_s, the volume of the subchain. This last quantity is given by the reciprocal of the number of subchains per unit volume, which, in turn, is the product of the number of polymer chains per unit volume and the number of subchains per molecule. Putting these results together gives $V_s = (N_s \, \rho N_A/M)^{-1}$. As a result of these considerations, we write the following expression for the tensile modulus:

$$E = \frac{F/A}{\gamma} = \frac{3kT}{V_s} = 3kT \frac{\rho N_A}{M} N_s \tag{3.92}$$

A value one-third this large applies to the shear modulus. We can eliminate N_s from this last result by observing that the limit of Eq. (3.90) as $t \rightarrow 0$ is $\sum_{p=1}^{N_s} G_p$, which becomes $N_s G_p$ if all values of G_p are assumed to be equal.

Expressing Eq. (3.92) in terms of shear and equating it to $N_s G_p$ gives

$$\frac{RT\rho}{M} = G_p \tag{3.93}$$

Although the notation is different, this result is identical to Eq. (3.51).

Next we turn our attention to τ_p. Equation (3.91) is messy enough for us to accept a certain degree of approximation if some simplification results. For small x, $\sin x \cong x$; therefore

$$\left[\sin^2\left(\frac{p\pi}{2(N_s + 1)}\right)\right]^{-1} \cong \frac{4(N_s + 1)^2}{\pi^2 p^2}$$

Substituting this approximation into Eq. (3.91) gives

$$\tau_p = \frac{(\overline{r^2})_s \zeta_s}{24kT} \frac{4(N_s + 1)^2}{\pi^2 p^2} \cong \frac{(\overline{r^2})_s \zeta_s}{6kT\pi^2 p^2} N_s^2 \tag{3.94}$$

where the second version assumes $N_s \gg 1$.

How are G_p and τ_p related to experimental quantities? We have repeatedly used the ratio of η to G as a definition of τ in this chapter. The experimental viscosity is related to the products of the individual G_p and τ_p values as follows:

$$\eta = \sum_{p=1}^{N_s} \tau_p G_p \tag{3.95}$$

Substituting Eqs. (3.93) and (3.94) into Eq. (3.95) gives

$$\eta = \frac{RT\rho}{M} \frac{(\overline{r^2})_s \zeta_s N_s^2}{6kT\pi^2} \sum_{p=1}^{N_s} \frac{1}{p^2} \tag{3.96}$$

The summation in this equation is known to be equal to $\pi^2/6$ for large values of N_s; therefore we write

$$\eta = \frac{RT\rho}{M} \frac{(\overline{r^2})_s \zeta_s N_s^2}{6kT\pi^2} \frac{\pi^2}{6} = \frac{\rho N_A}{M} \frac{(\overline{r^2})_s \zeta_s N_s^2}{36} \tag{3.97}$$

The remaining step is to eliminate those terms which refer to the subchain by expressing them in terms of the polymer molecule as a whole. Specifically, we recall that $(\overline{r^2})_s = n_s l_0{}^2$, $\zeta_s = n_s \zeta$, and $N_s = n/n_s$. Substituting into Eq. (3.97), we obtain

$$\eta = \frac{\rho N_A l_0{}^2 \zeta}{36 M_0} \, n \tag{3.98}$$

which is identical to the Debye viscosity equation, Eq. (2.56). Making this last group of substitutions in Eq. (3.94) gives

$$\tau_p = \frac{n^2 l_0{}^2 \zeta}{6 k T \pi^2 p^2} \tag{3.99}$$

Although the Rouse theory is the source of numerous additional relationships, Eq. (3.98) is a highpoint for us, because it demonstrates that the viscosity we are dealing with in the Rouse theory for viscoelasticity is the same quantity that we would obtain in a flow experiment. Several aspects of this statement deserve amplification:

1. Through the dashpot a viscous contribution was present in both the Maxwell and Voigt models and is essential to the entire picture of viscoelasticity. These have been the viscosities of mechanical units which produce equivalent behavior to that shown by polymers. While they help us understand and describe observed behavior, they do not give us the actual viscosity of the material itself.
2. The viscosity given by Eq. (3.98) not only follows from a different model than the Debye viscosity equation, but it also describes a totally different experimental situation. Viscoelastic studies are done on solid samples for which flow is not measurable. A viscous deformation is present, however, and this result shows that it is equivalent to what would be measured directly, if such a measurement were possible.
3. Although we still need to explain the use of this theory, Eq. (3.98) shows that segmental friction factors are accessible through viscoelastic studies. This fact was anticipated in the list of ζ values given in Table 2.3.

If we combine Eqs. (3.98) and (3.94), we can eliminate N_s from the latter to obtain an expression for the relaxation time of mode p which is free of any reference to the subchain:

$$\tau_p = \frac{6 M \eta}{\pi^2 \rho R T} \, \frac{1}{p^2} \tag{3.100}$$

Setting p = 1 in this expression enables us to identify the cluster of constants

$$\tau_1 = \frac{6M\eta}{\pi^2 \rho RT} \tag{3.101}$$

Now the relaxation times for all higher modes of vibration can be expressed relative to τ_1:

$$\tau_p = \tau_1 p^{-2} \tag{3.102}$$

This result shows that the highest modes of response have the shortest relaxation times and influence the initial response of the sample. Conversely, the longest relaxation time is τ_1, which we can identify with the terminal behavior of the sample. For example, in Fig. 3.9 the final collapse of the modulus at long times occurs at τ_1. An example will show how we can use this idea.

Example 3.7

Inspection of Fig. 3.9 suggests that for polyisobutylene at 25°C, τ_1 is about $10^{2.5}$ hr. Use Eq. (3.101) to estimate the viscosity of this polymer, remembering that M = 1.56 × 10^6. As a check on the value obtained, use the Debye viscosity equation, as modified here, to evaluate M_c, the threshold for entanglements, if it is known that ζ = 4.47 × 10^{-8} kg sec^{-1} at this temperature. Both the Debye theory and the Rouse theory assume the absence of entanglements. As a semi-empirical correction, multiply ζ by $(M/M_c)^{2.4}$ to account for entanglements. Since the Debye equation predicts a first-power dependence of η on M, inclusion of this factor brings the total dependence of η on M to the 3.4 power as observed.

Solution

Assume a density of 1.0 g cm^{-3} = 10^3 kg m^{-3}, since no density is given. This assumption will affect the viscosity but will cancel out of the estimation of M_c.

$$\tau_1 = 10^{2.5} \text{ hr} = 1.14 \times 10^6 \text{ sec}$$

$$\eta = \frac{\pi^2 \rho RT \tau_1}{6M}$$

$$= \frac{\pi^2 (10^3 \text{ kg m}^{-3})(8.314 \text{ J K}^{-1} \text{ mol}^{-1})(298 \text{ K})(1.14 \times 10^6 \text{ sec})}{6(1.56 \times 10^3 \text{ kg mol}^{-1})}$$

$$= 2.98 \times 10^9 \text{ kg m}^{-1} \text{ sec}^{-1}$$

Table 2.3 shows that $l_0 = 5.9 \times 10^{-10}$ m and $n = (1.56 \times 10^6)/56 = 2.79 \times 10^4$; therefore

$$\zeta_{app} = \frac{36 M_0 \eta}{\rho N_A l_0^2 n}$$

$$= \frac{36 \, (0.056 \text{ kg mol}^{-1})(2.98 \times 10^9 \text{ kg m}^{-1} \text{ sec}^{-1})}{(10^3 \text{ kg m}^{-3})(6.02 \times 10^{23})(5.9 \times 10^{-10} \text{ m})^2 \, (2.79 \times 10^4)}$$

$$= 1.09 \times 10^{-3} \text{ kg sec}^{-1}$$

$$\zeta_{app} = \zeta \left(\frac{M}{M_c} \right)^{2.4} \quad \text{hence} \quad \left(\frac{M}{M_c} \right)^{2.4} = \frac{1.09 \times 10^{-3}}{4.47 \times 10^{-8}} = 2.31 \times 10^4$$

$$\frac{M}{M_c} = 65.8 \quad \text{and} \quad M_c = \frac{M}{65.8} = \frac{1.56 \times 10^6}{65.8} = 23,700$$

and $n_c = 23,700/56 = 423$, which appears quite reasonable compared to the value observed in Fig. 2.10.

•

It should be observed that Eq. (3.102) may be viewed as a distribution function for relaxation times. In fact, if N_s is large enough, integer increments in p may be approximated as continuous p values. This makes τ_p continuous also. The significance of this is that Eq.(3.90) can be written as an integral in analogy with (3.62) if p is continuous:

$$G(t) = \int_0^\infty G_p \, e^{-t/\tau_p} \, dp \tag{3.103}$$

Comparison of Eqs. (3.63) and (3.103) reveals

$$H(\tau) \, d \ln \tau = G(\tau) \, d\tau = G_p \, dp \tag{3.104}$$

Taking the log of Eq. (3.102) and differentiating (remember that τ_1 is constant) gives

$$\frac{dp}{d \ln \tau} = -\frac{1}{2} \, p \tag{3.105}$$

Using Eq. (3.93) for G_p, substituting this and Eq. (3.105) into Eq. (3.104) gives $H(\tau) \, d \ln \tau = -(\frac{1}{2}) \, p \, (RT\rho/M) \, d \ln \tau$. Eliminating p via Eq. (3.100) yields

$$H(\tau) = -\left(\frac{3}{2} \frac{RT\rho\eta}{\pi^2 M} \right)^{1/2} \tau^{-1/2} \tag{3.106}$$

Table 3.5 Rouse Theory Expressions for the Modulus (entries labeled 1) and Compliances (entries labeled 2) for Tension and Shear Under Different Conditions[a]

Experimental condition	Tension	Shear
Transient	1. $E(t) = \Sigma_p E_p \, e^{-t/\tau_p}$ 2. $D(t) = \Sigma_p D_p \, (1 - e^{-t/\tau_p})$	1. $G(t) = \Sigma_p G_p \, e^{-t/\tau_p}$ 2. $J(t) = \Sigma_p J_p (1 - e^{-t/\tau_p})$
Dynamic components	1. $E'(\omega) = \Sigma_p E_p \omega^2 \tau_p^2 \, (1 + \omega^2 \tau_p^2)^{-1}$ $E''(\omega) = \Sigma_p E_p \omega \tau_p \, (1 + \omega^2 \tau_p^2)^{-1}$ 2. $D'(\omega) = \Sigma_p D_p \, (1 + \omega^2 \tau_p^2)^{-1}$ $D''(\omega) = \Sigma_p D_p \omega \tau_p \, (1 + \omega^2 \tau_p^2)^{-1}$	1. $G'(\omega) = \Sigma_p G_p \omega^2 \tau_p^2 (1 + \omega^2 \tau_p^2)^{-1}$ $G''(\omega) = \Sigma_p G_p \omega \tau_p \, (1 + \omega^2 \tau_p^2)^{-1}$ 2. $J'(\omega) = \Sigma_p J_p \, (1 + \omega^2 \tau_p^2)^{-1}$ $J''(\omega) = \Sigma_p J_p \omega \tau_p \, (1 + \omega^2 \tau_p^2)^{-1}$

[a] For the dynamic components the storage component is designated by a single prime, and the loss component by a double prime.

for the distribution of relaxation times. A theoretical result like this can be used in conjunction with Eq. (3.63) to derive an expression for the shear modulus or the moduli and, with minor differences, the compliances for other modes of deformation.

We observed above that the Rouse expression for the shear modulus is the same function as that written for a set of Maxwell elements, except that the summations are over all modes of vibration and the parameters are characteristic of the polymers and not springs and dashpots. Table 3.5 shows that this parallel extends throughout the moduli and compliances that we have discussed in this chapter. In Table 3.5 we observe the following:

1. $G(t)$ has the same form as Eq. (3.61) for a set of Maxwell elements.
2. $J(t)$ has the same form as Eq. (3.70) for a set of Voigt elements.
3. $J'(\omega)$ and $J''(\omega)$ have the same form as Eqs. (3.79) and (3.80) for a dynamic Voigt system.
4. $G'(\omega)$ and $G''(\omega)$ have the same form as Eqs. (3.82) and (3.83) for a dynamic Maxwell system.

The purpose of these comparisons is simply to point out how complete the parallel is between the Rouse molecular model and the mechanical models we discussed earlier. While the summations in the stress relaxation and creep expressions were included to give better agreement with experiment, the summations in the Rouse theory arise naturally from a consideration of different modes of vibration. It should be noted that all of these modes are overtones of the same fundamental and do not arise from considering different relaxation processes. As we have noted before, different types of encumbrance have different effects on the displacement of the molecules. The mechanical models correct for this in a way the simple Rouse model does not. Allowing for more than one value of ζ, along the lines of Example 3.7, is one of the ways the Rouse theory has been modified to generate two sets of τ_p values. The results of this development are comparable to summing multiple effects in the mechanical models. In all cases the more elaborate expressions describe experimental results better.

Problems

1. A constant force is applied to an ideal elastomer, assumed to be a perfect network. At an initial temperature T_i the length of the sample is l_i. The temperature is raised to T_f and the final length is l_f. Which is larger: l_i or l_f (remember F is a constant and $T_f > T_i$)? Suppose a wheel were constructed with spokes of this same elastomer. From the viewpoint of an observer, the spokes are heated near the 3 o'clock position—say, by exposure to sunlight—while other spokes are shaded. Assuming the torque produced can overcome any friction at the axle, would the observer see the wheel turn clockwise or counterclockwise? How would this experiment contrast, in magnitude and direction, with an experiment using metal spokes?

2. An important application of Eq. (3.39) is the evaluation of M_c. Flory et al.[†] measured the tensile force required for 100% elongation of synthetic rubber with variable crosslinking at $25°C$. The molecular weight of the un-cross-linked polymer was 225,000, its density was 0.92 g cm^{-3}, and the average molecular weight of a repeat unit was 68. Use Eq. (3.39) to estimate M_c for each of the following samples and compare the calculated value with that obtained from the known fraction of repeat units cross-linked:

Fraction cross-linked	0.005	0.010	0.015	0.020	0.025
F/A (lb-force in.$^{-2}$)	61.4	83.2	121.8	148.0	160.0

How important is the end group correction introduced in Eq. (3.46) for this system?

3. The purpose of this problem is to consider numerically the effect of including more than two Maxwell elements in the model for a relaxation experiment. Prepare a table analogous to Table 3.2 for a set of four Maxwell elements having the following properties:

Element number	1	2	3	4
G^* (N m^{-2})	10^9	10^8	10^7	10^6
τ (sec)	10^{-8}	10^{-7}	10^{-6}	10^{-5}

Evaluate G(t) for integral powers of 10 between 10^{-10} and 10^{-3} sec. Use the same table entries to evalute G(t) for a two-element Maxwell model consisting of elements 1 and 4 above. On the same graph plot both sets of results as log G(t) versus log t. Comment on the similarities and differences between the two curves.

4. Stress relaxation studies were conducted on samples of nylon yarn at a constant strain of 2% and the following results were obtained[‡]:

At $25°C$

log t (sec)	0.6	1.0	1.3	1.6	2.0	2.3	2.7	3.0	3.3	3.8
E x 10^{-10} (dyne cm^{-2})	4.00	3.85	3.73	3.60	3.47	3.37	3.25	3.17	3.08	2.98

At $45°C$

log t (sec)	0.6	1.0	1.5	1.75	2.15	2.65	3.15	3.40
E x 10^{-10} (dyne cm^{-2})	3.35	3.22	3.08	3.00	2.92	2.83	2.79	2.75

[†] P. J. Flory, N. Rabjohn, and M. C. Shaffer, *J. Polym. Sci.* 4:225 (1949).

[‡] T. Murayama, J. H. Dumbleton, and M. L. Williams, *J. Macromol. Sci. Phys.* B-1:1 (1967).

Plot E versus log t for both of these sets of data on the same graph. Now suppose that the units of the 45°C experiment are minutes instead of seconds (this is *not* the case, but we can pretend that it is). On the basis of this imaginary condition, each of the times at 45°C should be multiplied by the factor 60 sec/min to make the comparison with the 25°C data. Apply this "correction" to the 45°C data and plot on the original graph. Be sure to select a scale of the original graph so that "corrected" data can be accommodated; also, label various portions clearly. Briefly comment on the results of this manipulation.

5. Figure 3.12 shows that, at long times, the creep compliance is directly proportional to time for the polymers shown. For polystyrene ($M = 600,000$) at $100°C$, the following values describe the linear portion of the data[†]:

log $J(t)$ (m^2 N^{-1})	-1.84	-1.44	-1.05	-0.64	-0.24	$+0.16$
log t (sec)	12.6	13.0	13.4	13.8	14.2	14.6

Use Eq. (3.69) to evaluate the viscosity of the polymer at this temperature. For polystyrene $M_c \cong 30,000$. Use the viscosity value calculated above and the entanglement correction procedure introduced in Example 3.7 to estimate ζ for polystyrene. Take $\rho = 1.04$ g cm^{-3} and $l_0 = 7.4 \times 10^{-10}$ m for this calculation.

6. Equation (3.77) is the starting point for the derivation of the compliance in a dynamic experiment. This derivation is based on a Voigt model. Working from a Maxwell model with additive strains yields the analogous expressions for the modulus. Starting with Eqs. (3.73) and (3.54), derive Eq. (3.81); then use the function given for γ in Example 3.5 — this time written for σ — to verify that the storage and loss components of the dynamic modulus have the values given by Eqs. (3.82) and (3.83).

7. In a dynamic experiment $\gamma(t) = \gamma_0 \sin(\omega t)$. The power loss per cycle of oscillation is given by $\int_{\omega t=0}^{\omega t=2\pi} \sigma \, d\gamma$.
   (a) Evaluate the power loss per cycle if the material is a Hookean solid: $\sigma = G\gamma$.
   (b) Evaluate the power loss per cycle if the material is a Newtonian liquid: $\sigma = \eta(d\gamma/dt)$.
   (c) Briefly comment on the significance of these results.

8. Blends of poly(oxytetramethylene glycol) (PTMG) ($M = 1000$) with poly(tetramethylene terephthalate) (PTMT) were studied[†] for dynamic viscoelastic properties as a function of temperature. For the following percentages by weight PTMG, the temperature coordinate of the loss peak (T_g) and the median plateau modulus vary as follows:

[†] Data of D. J. Plazek and V. M. O'Rourke, quoted from Ref. 4.
[‡] R. W. Seymour, J. R. Overton, and L. S. Corley, *Macromolecules* 8:331 (1975).

Weight percent PTMG	20	25	35	50
T_{max} (°C)	19	10	-3	-35
Modulus (N m^{-2})	13	10	8	4

Criticize or defend the following propositions:

(a) The two polymers combined contribute different degrees of hardness to the blend in proportion to their relative abundance. PTMG is the "hard" component, and PTMT the "soft."

(b) Since a single loss peak is observed, phase separation into hard and soft components does not occur.

(c) Crystallization of PTMT increases the modulus. The crystals function as physical crosslinks in the sample.

9. For the tire cord–rubber composite system, the experimental loss tangent $(\tan \delta)_{ex}$ may be regarded as the sum of two contributions: the loss tangent associated with the individual constituents, assuming perfect adhesion between them, and the loss tangent associated with poor adhesion between the tire and the cord. The first of these contributions, $(\tan \delta)_{perfect}$, is calculated from the properties of the rubber and the cord measured separately. Using the following data[‡] to evaluate the adhesion contribution to the loss tangent:

	Nylon-6,6 cord		Poly(ethylene terephthalate) cord	
Adhesive?	Yes	No	Yes	No
$(\tan \delta)_{ex} \times 10^{-3}$:	86.3	108.7	74.4	84.2
$(\tan \delta)_{perfect} \times 10^{-3}$:	68.1	58.5	36.2	27.2

Briefly explain each of the following points: Does it make sense that the loss tangent is larger when a condition of poor adhesion exists between the cord and the rubber? Is the adhesive effective? Is it equally effective for both of the cord materials?

10. The following data were obtained on the same system described in Example 3.6.[§] This time the copolymer (C) concentration is fixed at 25% by weight and the proportions of polybutadiene (B) and polyisoprene (I) are varied:

[†] T. Murayama and E. J. Lawton, *J. Appl. Polym. Sci.* 17:669 (1973).

[‡] R. E. Cohen and A. R. Ramos, *Macromolecules* 12:131 (1979).

Weight percent			Temperature ($^\circ$C) and, in parentheses, relative height (arbitrary units)	
B	I	C	Peak 1	Peak 2
75	0	25	-78 (2.1)	-54 (1.3)
50	25	25	-76 (2.2)	-50 (1.8)
25	50	25	-85 (0.9)	-54 (2.7)
0	75	25		-51 (3.0)

Criticize or defend the following proposition: These data display the same trends described in the example and suggest the same interpretation.

11. Chains of polybutadiene were trapped in the network formed by cooling a butadiene–styrene copolymer until phase separation occurred for the styrene, effectively crosslinking the copolymer. At 25°C the loss modulus shows a maximum which is associated with the free chains. This maximum occurs† at the following frequencies for the indicated molecular weights of polybutadiene:

ω (Hz)	1.3×10^4	63	3.2	0.79	2.5×10^{-2}
$M_w \times 10^{-3}$ (g mol^{-1})	10	57	120	249	634

Evaluate the relaxation time associated with each of these molecular weights and verify that the molecular weight dependence of τ corresponds to the value given in Sec. 2.13.

Bibliography

1. Aklones, J. J., Mac Knight, W. J., and Shen, M., *Introduction to Polymer Viscoelasticity*, Wiley-Interscience, New York, 1972.
2. Alfrey, T., Jr., *Mechanical Behavior of High Polymers*, Interscience, New York, 1948.
3. Bueche, F., *Physical Properties of Polymers*, Interscience, New York, 1962.
4. Ferry, J. D., *Viscoelastic Properties of Polymers*, Wiley, New York, 1980.
5. Flory, P. J., *Principles of Polymer Chemistry*, Cornell University Press, Ithaca, N.Y., 1953.
6. Meares, P., *Polymers: Structure and Bulk Properties*, D. Van Nostrand, London, 1965.

†G. Kraus and K. W. Rollmann, *J. Polym. Sci. Phys.* 15:385 (1977).

4

The Glassy and Crystalline States

The average *polymer*
Enjoys a glassy state, but cools, forgets
To slump, and clouds in closely patterned minuets.

The Dance of the Solids, by John Updike

4.1 Introduction

In this chapter we examine some aspects of the solid states of polymers. There are two different forms in which a polymer can display the mechanical properties we associate with solids: as a crystal or as a glass. Not all polymers are able to crystallize; a high degree of microstructural regularity within the polymer chains is essential for crystallization to occur. Even in those polymers which do crystallize, the degree of crystallinity is less than 100%. Therefore in all cases there is at least some polymer, perhaps all, which is noncrystalline even though the temperature is well below the freezing point. If the temperature is lowered far enough, the glassy state is ultimately formed from the noncrystalline polymer. In the last chapter we saw that glasses are characterized by high values of the moduli and low values of the compliances.

In discussing the low-temperature (and this is only low compared to the region where the material is rubbery or viscous) behavior of polymers, two transitions are possible: the crystal–liquid transition and the glass–liquid transition. In the temperature range between these two transitions we may be dealing with a mixture of orderly crystals and chaotic amorphous polymer. It is possible to classify the literature on the subject as partly ordered and partly chaotic as well. When some unifying principle or perspective is operating, an orderly discussion is possible. When a catalog of diverse behaviors arising from diverse procedures is presented, the subject matter seems chaotic. From the viewpoint of the student trying to get oriented to the subject, the ordered approach is

definitely preferred. From the point of view of fidelity to the actual range of observable behaviors, some chaos is closer to the true picture.

To provide a rational framework in terms of which the student can become familiar with these concepts, we shall organize our discussion of the crystal–liquid transition in terms of thermodynamic, kinetic, and structural perspectives. Likewise, we shall discuss the glass–liquid transition in terms of thermodynamic and mechanistic principles. Every now and then, however, to impart a little flavor of the real world, we shall make reference to such complications as the prior history of the sample, which can also play a role in the solid behavior of a polymer.

A variety of experimental techniques have been employed to research the material of this chapter, many of which we shall not even mention. For example, pressure as well as temperature has been used as an experimental variable to study volume effects. Dielectric constants, indices of refraction, and nuclear magnetic resonance (NMR) spectra are used, as well as mechanical relaxations, to monitor the onset of the glassy state. X-ray, electron, and neutron diffraction are used to elucidate structure along with electron microscopy. It would take us too far afield to trace all these different techniques and the results obtained from each, so we restrict ourselves to discussing only a few types of experimental data. Our failure to mention all sources of data does not imply that these other techniques have not been employed to good advantage in the study of the topics contained herein.

The material of this chapter and, for that matter, of the two preceding chapters has wide applicability in the area of technology and manufacture. To do justice to this facet of the subject would require a book in itself, so we must settle for a few paragraphs concerned with industrial applications.

4.2 The Glass and Melting Transitions

In the last chapter time was the primary independent variable under consideration. We saw that at short times of observation, polymeric solids take on high values of the modulus, roughly three or four orders of magnitude higher than those shown in the rubbery state of these materials. The transition between the two values of the modulus occurs over a range of times. We also saw that temperature changes produce changes in mechanical properties which parallel those resulting from shifts of time scale. This change in mechanical behavior is called the glass transition, and, when monitored by temperature variation, it occurs as the glass transition temperature. We shall return to an examination of the equivalency of time and temperature with respect to effects on mechanical properties later in this chapter. To begin, however, it is desirable to consider some other properties of matter that change at T_g. Although quite a variety of observations are available to choose from, we shall consider the volume changes which occur in cooling a polymer.

Figure 4.1 illustrates schematically the range of possibilities for variation in specific volume with temperature. Remember that specific volume is the reciprocal of the density; it, rather than density, is chosen to describe these changes in anticipation of the molecular interpretations which follow in Sec. 4.9. Line ABDG in Fig. 4.1 shows how the specific volume changes upon freezing a low molecular weight compound. A few substances—water is the best known example—occupy a larger volume per unit mass in the solid state than in the liquid, and for these the transition at the melting point T_m would be a jump rather than a drop in volume. That is not the main point, however; instead, what is significant is that the transition occurs at a single temperature, the melting point T_m. The slopes of AB and DG measure the coefficients of thermal expansion of liquid and solid, respectively. The temperature variation of the latter is not our interest either; the point is that it is different for solids and liquids. Note that the coefficient of thermal expansion shows a discontinuity at the melting point.

An entirely different pattern of behavior is shown along line ABHI. In this case there is no discontinuity at T_m. The line AB which characterizes the liquid changes slope at T_g to become HI. Actually, the change in slope occurs over a range of temperatures (about $5°$), as suggested by Fig. 4.1, but extrapolation of the two linear portions permits T_g to be defined by this method. Region HI characterizes the glassy state, and the threshold for its appearance is the glass

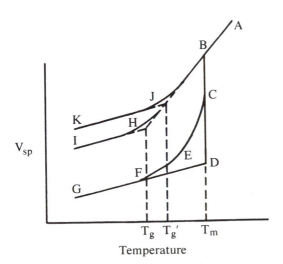

Figure 4.1 Schematic illustration of possible changes in the specific volume of a polymer with temperature. See text for a description of the various lettered phenomena.

transition temperature. The glassy state is also called the vitreous state, and its formation, vitrification. In the region BH, supercooled liquid exists.

In principle, each of the two lines we have discussed could describe the behavior of either high or low molecular weight compounds. This is not to say, however, that each is equally probable for the two classes of compounds. For low molecular weight compounds, special effort must be made to suppress crystallization and achieve glass formation. With polymers, on the other hand, the glassy state is always obtained whether a particular polymer is crystallizable or not. The mere fact that molecular structure allows the possibility of crystal formation does not mean that the latter occurs rapidly or completely: There is often a significant amount of amorphous polymer present, even in a partially crystalline sample. The line ABCEFG in Fig. 4.1 describes the situation of a partly crystalline, partly amorphous polymer. At T_m crystallization begins and the characteristic discontinuity occurs in specific volume (V_{sp}). The sharpness of T_m is not as pronounced for polymers as for low molecular weight compounds, as evidenced by the trailing off in the transition temperature between C and E. In the region EF the volume reflects the supercooling of the amorphous portion of the polymer. The change in slope between segments EF and FG occurs at T_g just as it would in the absence of crystallization. If partial crystallization occurs, the amount of amorphous material is decreased and the change in slope at T_g may be harder to detect in this case.

The line ABJK in Fig. 4.1 is a displaced variation of ABHI in which AB is liquid, BJ is supercooled liquid, and JK is glass. The experimental variable which causes region JK to be offset from HI is the cooling rate, ABJK being the course of the more quickly cooled polymer. Since T_g is identified from the change in slope, it is apparent that T_g is also displaced, appearing at higher temperatures for faster rates of cooling.

The low-temperature (remember that this is a relative term: $T_m = 317°C$ for polyacrylonitrile) behavior of linear polymers may conveniently be divided into three regimes:

1. Above T_m the material is liquid and its viscosity depends on the molecular weight of the polymer and the time scale of the observation, but it would be considered high by all standards.
2. Between T_m and T_g, depending on the regularity of the polymer and on the experimental conditions, this domain may be anything from almost 100% crystalline to 100% amorphous. The amorphous fraction, whatever its abundance, behaves like a supercooled liquid in this region. The presence of a certain degree of crystallinity mimics the effect of crosslinking with respect to the mechanical behavior of a sample.
3. Below T_g the material is hard and rigid with a coefficient of thermal expansion equal to roughly half that of the liquid. With respect to mechanical properties, the glass is closer in behavior to a crystalline solid than to a

liquid. In terms of molecular order, however, the glass more closely resembles the liquid. In this temperature region the noncrystalline fraction acquires the same glassy properties it would have if the crystallization had been suppressed completely. In general terms, there is no difference between linear and cross-linked polymers below T_g.

4. The location of T_g depends on the rate of cooling. The location of T_m is not subject to this variability, but the degree of crystallinity depends on the conditions of the experiment, as well as on the nature of the polymer. For example, if the rate of cooling exceeds the rate of crystallization, there may be no observable change at T_m, even for a crystallizable polymer.

The foregoing description of behaviors introduces us to the phenomena with which we shall be dealing in this chapter. As noted above, both high and low molecular weight compounds are capable of displaying these effects, but the chain structure of the polymer molecules is responsible for the reversed probability of the various possibilities. It is the specific identity of the polymer that anchors these transitions in some particular region of the temperature scale. The regularity of the microstructure of the polymer molecule, along with experimental conditions, determines the extent of crystallization. The glassy state is thus seen as a lowest common denominator shared by all polymers, since 100% crystallinity is virtually impossible. This promotes T_g to the position of importance assumed by T_m for low molecular weight compounds. The fact that the mechanical properties undergo such profound change at T_g also contributes to the significance of this parameter.

Some of the distinctions that we shall have to examine in more detail before proceeding much further are the considerations of order versus disorder, solid versus liquid, and thermodynamics versus kinetics. These dualities are taken up in the next section. With those distinctions as background, we shall examine both the glassy and crystalline states from both the experimental and modelistic viewpoint.

We conclude this section with a few remarks about the measurement of polymer density. This is about as unglamorous an experiment as one can imagine. As a property of matter, we take density very much for granted. The fact that it is conceptually simple, readily available for many materials, and relatively monotonous in its variations all contribute to this attitude. The types of phenomena represented schematically in Fig. 4.1 require careful experimentation on well-defined samples to yield reproducible results. The device that is used to follow volume changes upon cooling is called a dilatometer. The sample is placed in a bulb which is then filled with an inert liquid, generally mercury. The bulb is connected to a capillary so that changes in volume register as variations in the height of the mercury column, just as in a thermometer. For a constant temperature experiments, say, monitoring crystallization at T_m, the volume changes in the capillary correspond identically to changes occurring in

the sample. When temperature variation is involved, the expansion of the mercury due to the temperature change is superimposed on the expansion of the specimen and must be taken into account. To obtain meaningful results it is necessary to standardize the rate at which temperature changes are made and, of course, to have an accurately measured and uniform temperature in the bath surrounding the dilatometer. The sample must be immiscible with the displacement fluid and degassed to prevent entrapment of air. A gas bubble can really raise havoc in this kind of experiment.

4.3   The Thermodynamics of Crystallization: Round I

In dealing with experimental thermodynamics, one of the criteria that a true equilibrium has been established is to approach the state of interest from opposite directions. Accordingly, we are in the habit of thinking of the equilibrium melting point of a crystal or the equilibrium freezing point of the corresponding liquid as occurring at the same temperature. In dealing with polymer crystals, unfortunately, we are not so lucky as to observe this simple behavior. The transition liquid → crystal is so overshadowed by kinetic factors that some workers even question the value of any thermodynamic discussion of the transition in this direction. Furthermore, because of the kinetic complications occurring during the formation of the crystal, the transition crystal → liquid also becomes more involved.

   If polymers had infinite molecular weights and formed infinitely large crystals, the thermodynamics of the transition would be simpler, although the kinetics might very well be worse. Assuming, temporarily, that any kinetic complications can be overlooked, we will define the temperature of equilibrium (subscript e) between crystal and liquid for a polymer meeting the infinity criteria stated above as T_e^∞. Subsequently, we will use the superscript ∞ to indicate either infinite crystal dimension, infinite molecular weight, or both; it will be clear from the context which is meant. In such a case the melting point of the crystal would be T_e^∞ and the freezing point of the liquid would also be T_e^∞. The facts that actual molecular weights are less than infinite and that crystals have finite dimensions both tend to lower the equilibrium transition temperature below T_e^∞. The fact that kinetic complications also interfere means, in addition, that the temperature of crystallization T_c does not equal the temperature of melting T_m and that neither equals T_e^∞.

   Figure 4.2 illustrates some of these points for poly(1,4-cis-isoprene). The temperature at which the crystals are formed is shown along the abscissa, and the temperature at which they melt, along the ordinate. Note the following observations:

1.   The lower the crystallization temperature, the lower the melting point. These facts are connected through consideration of crystal dimensions.

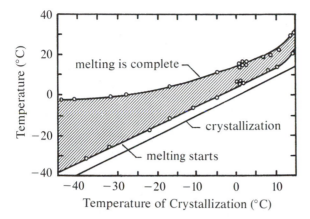

Figure 4.2 Melting temperature of crystals versus temperature of crystallization for poly(1,4-cis-isoprene). Note the temperature range over which melting occurs. [Reprinted with permission from L. A. Wood and N. Bekkedahl, *J. Appl. Phys.* 17:362 (1946).]

2. Melting occurs over a range of temperatures, as in Fig. 4.1. The range narrows as the crystallization temperature increases. This is probably due to a wider range of crystal dimensions and less perfect crystals under the lower temperatures of formation.

3. There is a suggestion of convergence of these lines in the upper right-hand portion of Fig. 4.2. For this polymer T_m^∞ is estimated to be 28°C—not an unreasonable point of convergence for the lines in Fig. 4.2.

We shall take up the kinetics of crystallization in detail in Secs. 4.5 and 4.6. For the present, our only interest is in examining what role kinetic factors play in complicating the crystal–liquid transition. In brief, the story goes like this. Polymers have a great propensity to supercool. If and when they do crystallize, it is an experimental fact that smaller crystal dimensions are obtained the lower the temperature at which the crystallization is carried out. The following considerations supply some additional details:

1. At and near T_m—which is a low temperature for the liquid state of a polymer, whatever its absolute value may be—the viscosity of a polymer melt will be quite high, which is a direct consequence of the encumbered movement of the polymer chains past one another.

2. Polymer crystals form by the chain folding back and forth on itself, with crystal growth occurring by the deposition of successive layers of these folded chains at the crystal edge. The resulting crystal, therefore, takes on a platelike structure, the thickness of which corresponds to the distance between folds.

3. This intricate mode of crystallization requires more time to accomplish than, say, the entry of small ions into growing salt crystals. This, coupled with low chain mobility due to viscous effects, makes the rate of crystallization slow and accounts in part for the fact that with rapid cooling—called quenching—the temperature drops below T_e^∞ without crystallization.

4. The thickness of the crystal plates or lamellae can be measured by x-ray diffraction studies and other methods. It is a well-established fact that thinner lamellae are formed at lower temperatures of crystallization.

5. The thickness depends on the supercooling, which, in turn, is the result of kinetic considerations. Accordingly, crystal thickness is related to T_c, but neither have much to do with T_e^∞.

6. The melting point T_m of the resulting crystal is less than it would be if the crystal had infinite dimensions (T_m^∞). This latter temperature approaches T_e^∞ as $M \to \infty$.

In summary, T_m gives a truer approximation to a valid equilibrium parameter, although it will be less than T_m^∞ owing to the finite dimensions of the crystal and the finite molecular weight of the polymer. We shall deal with these considerations in the next section. For now we assume that a value for T_e^∞ has been obtained and consider the simple thermodynamics of a phase transition.

We begin our application of thermodynamics to polymer phase transitions by considering the fusion (subscript f) process: crystal → liquid.

Figure 4.3a shows schematically how the Gibbs free energy of liquid (subscript l) and crystalline (subscript c) samples of the same material vary with temperature. For constant temperature–constant pressure processes the criterion for spontaneity is a negative value for ΔG, where the Δ signifies the difference final minus initial for the property under consideration. Applying this criterion to Fig. 4.3, we conclude immediately that above T_m^∞, $\Delta G_f = G_l - G_c$ is negative and the process is spontaneous, whereas below T_m^∞, $\Delta G_f > 0$ and it is the reverse process which is spontaneous. At T_m^∞ both phases have the same value of G; at this temperature $\Delta G = 0$ and a condition of equilibrium exists between the phases.

For a particular phase, an increment in Gibbs free energy dG can be expressed in terms of increments of pressure and temperature dp and dT; that is,

$$dG = V\,dp - S\,dT \tag{4.1}$$

where V and S are the volume and entropy of the phase, respectively. Since this is thermodynamics, T is always expressed in degrees Kelvin in these equations. Since G is a state variable and forms exact differentials, Eq. (4.1) can be interpreted to mean

$$\left(\frac{\partial G}{\partial T}\right)_p = -S \tag{4.2}$$

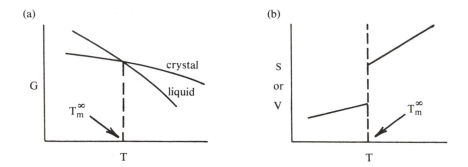

Figure 4.3 Behavior of thermodynamic variables at T_m^∞ for an idealized phase transition: (a) Gibbs free energy and (b) entropy and volume.

and

$$\left(\frac{\partial G}{\partial p}\right)_T = V \tag{4.3}$$

Figure 4.3b is a schematic representation of the behavior of S and V in the vicinity of T_m^∞. Although both the crystal and liquid phases have the same value of G at T_m^∞, this is not the case for S and V (or for the enthalpy H). Since these latter variables can be written as first derivatives of G and show discontinuities at the transition point, the fusion process is called a first-order transition. Vaporization and other familiar phase transitions are also first-order transitions. The behavior of V at T_g in Fig. 4.1 shows that the glass transition is not a first-order transition. One of the objectives of this chapter is to gain a better understanding of what else it might be. We shall return to this in Sec. 4.8.

At T_m^∞, $\Delta G_f = 0$, but ΔS_f, ΔV_f, and ΔH_f have nonzero values. For any constant-temperature process such as fusion,

$$\Delta G = \Delta H - T \Delta S \tag{4.4}$$

therefore at equilibrium

$$T_m^\infty = \frac{\Delta H_f}{\Delta S_f} \tag{4.5}$$

This fundamental relationship points out that the temperature at which crystal and liquid are in equilibrium is determined by the balancing of entropy and enthalpy effects. Remember, it is the difference between the crystal and

the liquid that is pertinent; sometimes these differences are not what we might first expect.

Table 4.1 lists values of T_m as well as ΔH_f and ΔS_f per mole of repeat units for several polymers. A variety of experiments and methods of analysis have been used to evaluate these data, and because of an assortment of experimental and theoretical approximations, the values should be regarded as approximate. We assume $T_m \cong T_m^\infty$. In general, both ΔH_f and ΔS_f may be broken into contributions H_0 and S_0 which are independent of molecular weight and increments $\Delta H_{f,1}$ and $\Delta S_{f,1}$ for each repeat unit in the chain. Therefore $\Delta H_f = H_0 + n\,\Delta H_{f,1}$, where n is the degree of polymerization. In the limit of $n \to \infty$, $\Delta H_f \cong n\,\Delta H_{f,1}$ and $\Delta S_f = n\,\Delta S_{f,1}$, so $T_m^\infty = \Delta H_{f,1}/\Delta S_{f,1}$. The values of $\Delta H_{f,1}$ and $\Delta S_{f,1}$ in Table 4.1 are expressed per mole of repeat units on this basis. Since no simple trends exist within these data, the entries in Table 4.1 appear in numbered sets, and some observations concerning these sets are listed here:

1. Polyethylene. The crystal structure of this polymer is essentially the same as those of linear alkanes containing 20-40 carbon atoms, and the values of T_m and $\Delta H_{f,1}$ are what would be expected on the basis of an extrapolation from data on the alkanes. Since there are no chain substituents or intermolecular forces other than London forces in polyethylene, we shall compare other polymers to it as a reference substance.

2. Poly(1,4-cis-isoprene). Although $\Delta H_{f,1}$ is slightly higher than that of polyethylene, it is still completely reasonable for a hydrocarbon. The

Table 4.1   Values of T_m, $\Delta H_{f,1}$, and $\Delta S_{f,1}$ for Several Polymers[a]

Polymer	T_m (°C)	Per mole repeat units	
		$\Delta H_{f,1}$ (J mol^{-1})	$\Delta S_{f,1}$ (J K^{-1} mol^{-1})
1. Polyethylene	137.5	4,020	9.8
2. Poly(1,4-cis-isoprene)	28	4,390	14.5
3. Poly(ethylene oxide)	66	8,280	22.4
4. Poly(decamethylene sebacate)	80	50,200	142.3
Poly(decamethylene azelate)	69	41,840	121.3
5. Poly(decamethylene sebacamide)	216	34,700	71.1
Poly(decamethylene azelamide)	214	36,800	75.3

[a] This notation indicates the contribution to the value for the fusion process (subscript f) per mole of repeat units (subscript 1). Entries are discussed by number in the text.
Source: Data from Ref. 5.

lower T_m is the result of a value of $\Delta S_{f,1}$, which is 50% higher than that of polyethylene. The low melting point of this polymer makes natural rubber a useful elastomer at ordinary temperatures.

3. Poly(ethylene oxide). Although $\Delta H_{f,1}$ is more than double that of poly- ethylene, the effect is offset by an even greater increase for $\Delta S_{f,1}$. The latter may be due to increased chain flexibility in the liquid caused by the regular insertion of ether oxygens along the chain backbone.

4. Polyesters. The next two polyesters have $\Delta H_{f,1}$ values an order of magni- tude higher than polyethylene. Our first thought might be to attribute this to a strong interaction between the polar ester groups. The repeat units of these compounds are considerably larger than in the reference compound, so the $\Delta H_{f,1}$ values should be compared on a per gram basis. When this is done, $\Delta H_{f,1}$ is actually less than for polyethylene. This suggests that the larger value for $\Delta H_{f,1}$ is the result of a greater number of methylene groups contributing London attraction for the polyesters, with the dipole–dipole interaction of the ester groups about the same in both liquid and crystal and therefore contributing little to ΔH_f. When compared on the basis of the number of bonds along the backbone, $\Delta S_{f,1}$ is not exceptional either. Accordingly, T_m is less than polyethylene for these esters.

5. Polyamides. The next two compounds are the amide counterparts of the esters listed under item (4). Although the values of $\Delta H_{f,1}$ are less for the amides than for the esters, the values of T_m are considerably higher. This is a consequence of the very much lower values of $\Delta S_{f,1}$ for the amides. These, in turn, are attributed to the low entropies of the amide in the liquid state owing to the effects of hydrogen bonding and chain stiffness arising from the contribution of the resonance form

$$
\begin{array}{cc}
\text{H} & \text{O} \\
| \oplus | \ominus \\
-\text{N}=\text{C}-
\end{array}
$$

These items show that it may be easier to rationalize an observation or trend than to predict it a priori. This state of affairs is not unique to polymers, how- ever. The following example gives another illustration of this type of reasoning.

Example 4.1

The melting points of a series of poly(α-olefin) crystals were studied. All of the polymers were isotactic and had chain substituents of different bulkinesses. Table 4.2 lists some results. Use Eq. (4.5) as the basis for interpreting the trends in these data.

Solution

The bulkiness of the substituent groups increases moving down Table 4.2. Also moving down the table, the melting points decrease, pass through a minimum, and then increase again. As is often the case with reversals of trends such as this, there are two different effects working in opposition in these data:

1. As the bulkiness of the substituents increases, the chains are prevented from coming into intimate contact in the crystal. The intermolecular forces which hold these crystals together are all London forces, and these become weaker as the crystals loosen up owing to substituent bulkiness. Accordingly, the value for the heat of fusion decreases moving down Table 4.2.
2. As the bulkiness of the chain substituents increases, the energy barriers to rotation along the chain backbone increase. As seen in Chap. 1, this decreases chain flexibility in the liquid state. It is flexibility which permits the molecules to experience a large number of conformations and therefore have high entropies. If the flexibility is reduced, the entropy change on melting is less than it would otherwise be. Accordingly, the entropy of fusion decreases moving down the table.

Table 4.2 Values of T_m for Poly(α-olefin) Crystals in Which the Polymer has the Indicated Substituent (Results are Discussed in Example 4.1)

Substituent	$T_m(^\circ C)$
$-CH_3$	165
$-CH_2 CH_3$	125
$-CH_2 CH_2 CH_3$	75
$-CH_2 CH_2 CH_2 CH_3$	-55
$-CH_2 \underset{\underset{CH_3}{\vert}}{CH} -CH_2 -CH_3$	196
$-CH_2 -\underset{\underset{CH_3}{\vert}}{\overset{\overset{CH_3}{\vert}}{C}} -CH_2 -CH_3$	350

Source: Data quoted from F. W. Billmeyer, *Textbook of Polymer Science*, 2nd ed., Wiley-Interscience, New York, 1971.

3. Since $T_m = \Delta H_f / \Delta S_f$, the observed behavior of this series of polymers may be understood as a competition between these effects. For the smaller substituents, the effect on ΔH_f dominates and T_m decreases with bulk. For larger substituents, the effect of ΔS_f dominates and T_m increases with bulk.

The polymers compared all have similar crystal structures but are different from polyethylene, which excludes the possibility for also including the latter in this series. Also note that the isotactic structure of these molecules permits crystallinity in the first place. With less regular microstructure, crystallization would not occur at all.

•

In the discussion of Table 4.1, we acknowledged that there might be some uncertainty in the values of the quantities tabulated, but we sidestepped the origin of the uncertainty. In the next section we shall consider one of these areas: the effect of crystal dimensions of the value of T_m.

4.4   The Thermodynamics of Melting: Round II

Whenever a phase is characterized by at least one linear dimension which is small, the properties of the surface begin to make significant contributions to the observed behavior. We shall examine the structure of polymer crystals in more detail in Sec. 4.7, but for now the following summary of generalizations about these crystals will be helpful:

1. Polymers crystallize in the form of thin plates or lamellae. The thickness of each of these lamellae is on the order of 10 nm.
2. The dimensions of the crystal plates perpendicular to the small dimensions depend on the conditions of the crystallization, but are many times larger than the thickness for a well-developed crystal.
3. The chain direction within the crystal is along the short dimension of the crystal, indicating that the molecule folds back and forth, fire hose fashion, with successive layers of folded molecules accounting for the lateral growth of the platelets.
4. A crystal does not consist of a single molecule, nor does a molecule reside exclusively in a single crystal.
5. The loop formed by the chain as it emerges from the crystal, turns around, and reenters the crystal may be regarded as amorphous polymer, but is insufficient to account for the total amorphous content of most crystalline polymers.
6. Polymer chain ends disrupt the orderly fold pattern of the crystal and tend to be excluded from the crystal and relegated to the amorphous portion of the sample.

7.   Variations of all the above points—all the way to the complete failure of a crystallizable polymer to actually crystallize—are possible, depending on experimental conditions.

Since the polymer crystal habit is characterized by plates whose thickness is small, surface phenomena are important. During the early development of the crystal, the lateral dimensions are also small and the effect is even more pronounced. The key to understanding this fact lies in the realization that all phase boundaries possess surface tension and that this surface tension measures the Gibbs free energy stored per unit area of the phase boundary. To get a preliminary feel for the importance of this, suppose we consider a spherical phase of radius r, density ρ, and surface tension γ. The total surface free energy associated with a particle such as this is given by the product of γ and the area of the sphere, or $\gamma(4\pi r^2)$. The total mass of material in the sphere is given by the product of the density and the volume of the sphere, or $\rho(4\pi r^3/3)$. The ratio of the former to the latter gives the Gibbs free energy arising from surface considerations, expressed per unit mass; that is, the surface Gibbs free energy per unit mass is $3\gamma/\rho r$. Since γ is small compared to most other chemical and physical contributions to free energy, surface effects are not generally considered when, say, the ΔG° of formation is quoted for a substance. The above argument shows that this becomes progressively harder to justify as the particle size of the material decreases. The emergence of a new phase implies starting from an r value of zero in the argument above, and the surface contribution to the energy becomes important indeed. Since two phases with their separating surface must already exist for γ to have any meaning, we are spared the embarrassment of the surface free energy becoming infinite at r = 0. Nevertheless, it is apparent from the foregoing that the effect of the surface free energy contribution is to increase G. Inspection of Fig. 4.3a shows that an increase in the G value for the crystalline phase arising from its small particle size has the effect of shifting T_m to lower temperatures. The smaller the particle size, the bigger the effect. This is the origin of all superheating, supercooling, and supersaturation phenomena: An equilibrium transition is sometimes overshot because of the difficulty associated with the initiation of a new phase. Likewise, all nucleation practices—cloud seeding, bubble chambers, and the use of boiling chips—are based on providing a site on which the emerging phase can grow.

To develop a more quantitative relationship between particle size and T_m, suppose we consider the melting behavior of the cylindrical crystal sketched in Fig. 4.4. Of particular interest in this model is the role played by surface effects. The illustration is used to define a model and should not be taken too literally, especially with respect to the following points:

1.   The geometry of the cylinder is a matter of convenience. Except for numerical coefficients, the results we shall obtain will apply to plates of any cross-sectional shape.

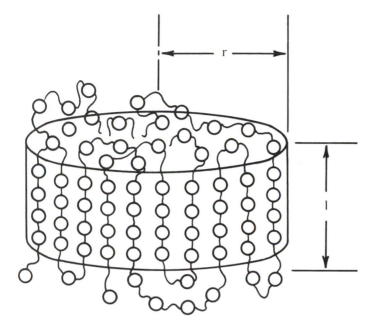

Figure 4.4 Idealized representation of a polymer crystal as a cylinder of radius r and thickness l. Note the folded nature of polymer chains in crystal.

2. The thickness of the plate, while small, is greater than the few repeat units shown.
3. The specific nature of the reentry loops is not the point of this illustration. The sketch shows both hairpin turns and longer loops. Problem 7 at the end of the chapter examines the actual nature of the reentry loop.

To develop this model into a quantitative relationship between T_m and the thickness of the crystal, we begin by realizing that for the transition crystal $\rightarrow$ liquid, ΔG is the sum of two contributions. One of these is ΔG^∞, which applies to the case of a crystal of infinite (superscript ∞) size; the other ΔG^s arises specifically from surface (superscript s) effects which reflect the finite size of the crystal:

$$\Delta G_f = \Delta G^\infty + \Delta G^s \qquad (4.6)$$

Now each of these can be developed independently.

As in the qualitative discussion above, let γ be the Gibbs free energy per unit area of the interface between the crystal and the surrounding liquid. This is undoubtedly different for the edges of the plate than for its faces, but we

shall not worry about this distinction. The area of each of the circular faces of the cylinder is πr^2, and the area of the edge is $2\pi rl$, where r is the radius of the face and l is the length of the side as shown Fig. 4.4. Since surface is destroyed by the melting process, the contribution of these considerations to ΔG_f is

$$\Delta G^s = -[2(\pi r^2) + 2\pi rl]\gamma = -2\pi r^2 \gamma \left(1 + \frac{1}{r}\right) \tag{4.7}$$

For the bulk effect we elect to proceed on the basis of a unit volume (superscript V) and immediately write

$$\Delta G^\infty = (\pi r^2 l)\Delta G_V^\infty \tag{4.8}$$

and

$$\Delta G_V^\infty = \Delta H_V^\infty - T^\infty \Delta S_V^\infty \tag{4.9}$$

In this last result we have retained the superscript ∞ on the T as a reminder that this refers to the infinitely extensive bulk phase; it describes the transition in the absence of surface complications.

When this infinitely extensive phase is in equilibrium with the melt, $\Delta G_V^\infty \to 0$ and $T^\infty \to T_m^\infty$. Accordingly, we can solve Eq. (4.9) for $\Delta S_V^\infty = \Delta H_V^\infty / T_m^\infty$ and substitute this back into Eq. (4.9):

$$\Delta G_V^\infty = \Delta H_V^\infty \left(1 - \frac{T^\infty}{T_m^\infty}\right) \tag{4.10}$$

This gives the values of ΔG_V at any temperature in terms of the two parameters ΔH_V^∞ and T_m^∞. Combining Eqs. (4.6)-(4.8) and (4.10) enables us to write

$$\Delta G_f = (\pi r^2 l)\Delta H_V^\infty \left(1 - \frac{T^\infty}{T_m^\infty}\right) - 2\pi r^2 \gamma \left(1 + \frac{1}{r}\right) \tag{4.11}$$

We can now drop the superscript ∞ on the T in the numerator, recognizing that it is merely the temperature at which we are evaluating ΔG for the process c → l for a crystal characterized by r and l and a polymer characterized by ΔH_V^∞, T_m^∞, and γ. When the value of this ΔG is zero, we have the actual melting point of the crystal of finite dimension T_m. That is,

$$(\pi r^2 l)\Delta H_V^\infty \left(\frac{T_m^\infty - T_m}{T_m^\infty}\right) = 2\pi r^2 \gamma \left(1 + \frac{1}{r}\right) \tag{4.12}$$

or

$$\Delta T = T_m^\infty - T_m = 2l^{-1} \frac{\gamma}{\Delta H_V^\infty} (1 + r^{-1}) T_m^\infty \qquad (4.13)$$

Note that this equation is dimensionally correct, since γ has units of energy area^{-1} and ΔH_V^∞ has units energy volume^{-1}. Therefore the units of $l\Delta H_V^\infty$ and γ cancel, as do the units of l and r, leaving only temperature units on both sides of the equation. All of the quantities on the right-hand side of the equation are positive (ΔH_V^∞ is the heat of fusion), which means that $T_m^\infty > T_m$, as anticipated. The difference ΔT derived in Eq. (4.13) is called the undercooling, since it represents the melting point lowering due to particle size. Several limiting cases of this equation are of note:

1. If $\gamma = 0$, $\Delta T = 0$, regardless of particle size. This is not expected, however, since chains emerging from a crystal face either make a highly constrained about-face and reenter the crystal or meander off into the liquid from a highly constrained attachment to the solid. In either case, a free-energy contribution is inescapable.
2. As $r \to 0$, $\Delta T \to \infty$, showing that the lateral dimensions of the plate are critical for very small crystals. This makes the crystal nucleation event especially crucial.
3. As $r \to \infty$, which describes well-developed crystals, Eq. (4.13) becomes

$$\Delta T = 2T_m^\infty \frac{\gamma}{\Delta H_V^\infty} l^{-1} \qquad (4.14)$$

which shows that an undercooling is still important because of the platelike crystal habit of polymers with limited crystal dimensions along the chain direction.

Equation 4.14 shows that a direct proportionality relationship should exist between crystal thickness l and the ratio $T_m^\infty/\Delta T$; a plot of l versus $T_m^\infty/\Delta T$ should result in a straight line of zero intercept with a slope proportional to $\gamma/\Delta H_V^\infty$. Figure 4.5 shows such a plot for polyethylene in which T_m^∞ was taken to be 137.5°C and the l values were determined by x-ray diffraction. While there is considerable and systematic divergence from the predicted form at large undercoolings, the data show a linear relationship for the higher-temperature region. In the following example we analyze the linear portion of Fig. 4.5 in terms of Eq. (4.14).

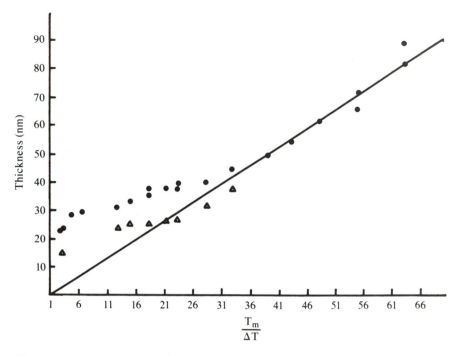

Figure 4.5 Crystal thickness versus $T_m/\Delta T$ for polyethylene. (Reprinted with permission from Ref. 5, copyright 1964, McGraw-Hill.)

Example 4.2

Use Eq. (4.14), the results in Fig. 4.5, and the data in Table 4.1 to estimate a value for γ for polyethylene. Figure 4.10 shows the unit cell of polyethylene; Fig. 4.10b shows the equivalent of two chains emerging from an area 0.740 by 0.493 nm^2. On the basis of the calculated value of γ and the characteristics of the unit cell, estimate the free energy of the fold surface per mole of repeat units.

Solution

Equation (4.14) predicts a straight line of zero intercept and slope $2\gamma/\Delta H_V^\infty$ when l is plotted versus $T_m/\Delta T$. The solid line in Fig. 4.5 has a slope of 650Å/51 = 12.75 Å. Therefore $2\gamma/\Delta H_V^\infty = 12.75 \times 10^{-10}$ m. The value of $\Delta H_{f,1}$ given in Table 4.1 is used for ΔH_V^∞ after the following change of units:

$$\Delta H_V^\infty = \frac{4020 \text{ J}}{\text{mole}} \times \frac{1 \text{ mole}}{28 \text{ g}} \times \frac{1 \text{ g}}{\text{cm}^3} \times \frac{10^6 \text{ cm}^3}{1 \text{ m}^3} = 1.44 \times 10^8 \text{ J m}^{-3}$$

Therefore $\gamma = \frac{1}{2}(12.75 \times 10^{-10} \text{ m})(1.44 \times 10^8 \text{ J m}^{-3}) = 0.091 \text{ J m}^{-2}$. From the data on the unit cell

$$\frac{0.091 \text{ J}}{\text{m}^2} \times \left(\frac{0.740 \times 0.493 \text{ nm}^2}{\text{unit cell}} \times \frac{1 \text{ m}^2}{10^{18} \text{ nm}^2}\right) \times \frac{1 \text{ unit cell}}{2 \text{ molecules}}$$

$$\times \frac{6.02 \times 10^{23} \text{ molecules}}{\text{mol}} = 9990 \text{ J mol}^{-1}$$

•

Although it applies to a totally different kind of interface, the value of γ calculated in the example is on the same order of magnitude as the γ value for the surface between air and liquids of low molecular weight.

Before concluding this section, there is one additional thermodynamic factor to be mentioned which also has the effect of lowering T_m. Since we shall not describe the thermodynamics of polymer solutions until Chap. 8, a quantitative treatment is inappropriate at this point. However, some relationships familiar from the behavior of low molecular weight compounds may be borrowed for qualitative discussion. The specific effect we consider is that of chain ends. The position we take is that they are "foreign species" from the viewpoint of crystallization.

In this context the repeat units in a polymer may be divided into two classes: those at the ends of the chain (subscript e) and the others which we view as being in the middle (subscript m) of the chain. The mole fraction of each category in a sample is x_e and x_m, respectively. Since all segments are of one type or the other,

$$x_m = 1 - x_e \tag{4.15}$$

the proportion of chain ends increases with decreasing molecular weight; hence x_e decreases with increasing M.

The attitude we adopt in this discussion is that only those chain segments in the middle of the chain possess sufficient regularity to crystallize. Hence we picture crystallization occurring from a mixture in which the concentration of crystallizable units is x_m and the concentration of solute or diluent is x_e. The effect of solute on the freezing (melting) point of a solvent is a well-known result: T_m is lowered. Standard thermodynamic analysis yields the relationship

$$\ln x_m = -\frac{\Delta H_f}{R} \left(\frac{1}{T_m} - \frac{1}{T_m^\infty}\right) \tag{4.16}$$

where R is the gas constant, ΔH_f is the heat of fusion, T_m^∞ is the melting point for a polymer of infinite molecular weight, and T_m is the melting point of the polymer under consideration. This result assumes that the solution is ideal, an assumption we shall not pursue in view of the purely analogous nature of this argument. Combining Eqs. (4.15) and (4.16), expanding the logarithms, and rearranging yields

$$\Delta T = T_m^\infty - T_m = \frac{R x_e\, T_m\, T_m^\infty}{\Delta H_f} \tag{4.17}$$

which shows that a freezing point depression is to be expected from an increased concentration of chain ends. At least qualitatively, the effect of other types of defects is expected to also lower T_m. Remember that in the present discussion T_m^∞ is the melting point for a polymer of infinite molecular weight without regard to the crystal size, while Eq. (4.14) is the melting point for a crystal of infinite dimension without regard to molecular weight. The two effects are therefore complementary and are both operative if both particle dimension and molecular weight are small enough to lower the freezing point appreciably.

Throughout this section we have focused attention on the thermodynamic considerations which help explain the lowering and broadening of polymer melting points. The same thermodynamic arguments can be applied to the raising and sharpening of this transition temperature through annealing. When a crystal is maintained at a temperature between the crystallization temperature and the equilibrium melting point, an increase in T_m is observed. This may be understood in terms of the melting of smaller, less-perfect crystals and the redeposition of the polymer into larger, more stable crystals. This is analogous to the procedure of digesting a precipitate prior to filtration. There is more to the story than this, however. The digestion analogy suggests that those crystals which are enlarged simply add more folded chains around their perimeter. In fact, x-ray diffraction studies reveal progressive thickening of lamellae as T_m increases. This means large-scale molecular reorganization within the crystal. Such rearrangements apparently require the molecule to snake along the chain axis, with segments being reeled in and out across the crystal surface. The process of annealing, therefore, not only involves crystal thickening, but also provides the opportunity to work out kinks and defects.

Direct electron-microscopic examination of single crystals subjected to annealing shows a tendency of the platelets to develop holes like Swiss cheese or to develop an irregular perimeter like an amoeba. Each of these transformations has the effect of decreasing the face area of the crystal while increasing the edge area. These modifications occur as a result of the system seeking a lower overall free energy and demonstrate that the assumption made in the derivation of Eq. (4.14) concerning the equality of the face and edge values of γ was an oversimplification.

The last two sections have examined various aspects of the transition at T_m from the thermodynamic viewpoint. In the next section we turn to a kinetic approach to the same transition.

4.5 The Kinetics of Crystallization: The Avrami Equation

The fundamental equilibrium relationships we have discussed in the last sections are undoubtedly satisfied to the extent possible in polymer crystallization, but this possibility is limited by kinetic considerations. To make sense of the latter, both the mechanisms for crystallization and experimental rates of crystallization need to be examined.

The rate of emergence of a new phase must be considered in terms of two other rates: the rate of nucleation and the rate of growth of the nucleated particle. The nucleus is the structural entity which constitutes the growth center and may appear as the result of random fluctuations within the liquid or from the presence of foreign bodies. While seeding by deliberately introduced particles is an important practical matter, it is not our primary concern. Instead we imagine the ongoing conformational changes that occur in liquid polymers. We recognize that among the myriad of shapes possible small regions with the order which is prerequisite for crystal growth will spontaneously and randomly form and redissolve. We may regard these flickering domains as potential embryos for crystal growth.

The rate law describing embryo growth can be written as the product of two factors:

1. A Boltzmann factor in which the energy of crystallization appears in a negative exponent. According to Eq. (4.11), this energy increases—hence the exponential decreases—with increasing r.
2. A preexponential factor in which the frequency of chain addition increases with the surface area of the embryo. This, in turn, increases with increasing r.
3. As the product of two factors which vary oppositely with increasing r, the rate of embryo growth passes through a maximum at some critical (subscript c) dimension.

From a kinetic point of view, therefore, an embryo of radius r_c is an essential intermediate and the value of ΔG_c is analogous to an energy barrier to be overcome along the path to crystallization. As the temperature is lowered, molecular motion slows, delaying the thermal disruption of these embryos. Crystal growth from a large number of nuclei randomly dispersed throughout the melt becomes possible. The migration of polymer molecules to sites of growth is also slowed by the lowering of temperature. While the foregoing considerations are general

for all nucleation phenomena, the high molecular weight of polymers aggravates the problem of crystal nucleation to the point that supercooling a liquid all the way to T_g without appreciable crystallization is commonplace.

In spite of these obstacles, crystallization does occur and the rate at which it develops can be measured. The following derivation will illustrate how the rates of nucleation and growth combine to give the net rate of crystallization. The theory we shall develop assumes a specific picture of the crystallization process. The assumptions of the model and some comments on their applicability follow:

1.   The crystals are assumed to be circular disks. This geometry is consistent with previous thermodynamic derivations. It has the advantage of easy mathematical description.

2.   The disks are assumed to lie in the same plane. While this picture is implausible for bulk crystallization, it makes sense for crystals grown in ultrathin films, adjacent to surfaces, and in stretched samples. A similar mathematical formalism can be developed for spherical growth and the disk can be regarded as a cross section of this.

3.   Nucleation is assumed to begin simultaneously from centers positioned at random throughout the liquid. This is more descriptive of heterogeneous nucleation by foreign bodies introduced at a given moment than of random nucleation. We shall dispense with the requirement of simultaneity below.

4.   Growth in the radial direction is assumed to occur at a constant velocity. There is ample experimental justification for this in the case of three-dimensional spherical growth.

Figure 4.6a represents the top view of an array of these disks after the crystals have been allowed to grow for a time t after nucleation. The two disks on the left are widely enough separated to still have room for further growth; the three on the right have impinged upon one another and can grow no more. We shall see in Sec. 4.7 that this latter situation can be observed microscopically.

Suppose we define the rate of radial growth of the crystalline disks as $\dot{r}$. Then disks originating from all nuclei within a distance $\dot{r}t$ of an arbitrary point, say, point x in Fig. 4.6a, will reach that point in an elapsed time t. If the average concentration of nuclei in the plane is N (per unit area), then the average number of fronts $\bar{F}$ which converge on x in this time interval is

$$\bar{F} = \pi(\dot{r}t)^2 N \tag{4.18}$$

If real growth fronts were to impinge on a point like this, their growth would terminate at x. Suppose we imagine point x to be charmed in some way such that any number of growth fronts can pass through it without interference.

If we were to monitor the number of (noninterfering) fronts which cross x in a series of observations, we would expect a distribution of values because of

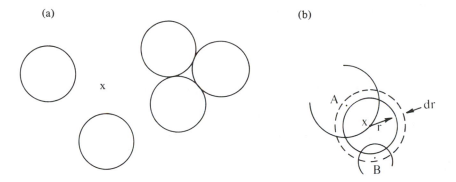

Figure 4.6 The growth of disk-shaped crystals. (a) All crystals have been nucleated simultaneously. All crystals have the same radius $\dot{r}t$ after an elapsed time t. (b) Nucleation is sporadic. Crystal A has had enough time to reach point x, while B has not, although both originate in the same ring a distance r from x.

the random placement of the nuclei. Furthermore, the distribution of F values is expected to pass through a maximum. Fronts arising from nuclei very close to x can easily cross x in the allotted time, but the area of melt under consideration in this case is small, so the number of fronts is small. As the area around the charmed point x is enlarged, a larger number of nuclei will be encompassed, so the number of fronts crossing x will increase. This increase is offset by the fact that fronts originating from more distant nuclei will require more time to reach x. Therefore the number of fronts which cross x (remember that these are free from interference by hypothesis) will increase, pass through a maximum, and decrease as we allow them to originate from all parts of the sample. This distribution of values for F is our next interest. If a distribution of values is possible, then it can always be characterized by an average value, $\bar{F}$ in this case.

We propose to describe the distribution of the number of fronts crossing x by the Poisson distribution function, discussed in Sec. 1.9. This probability distribution function describes the probability P(F) of a specific number of fronts F in terms of that number and the average number $\bar{F}$ as follows [Eq. (1.38)]:

$$P(F) = \frac{e^{-\bar{F}} \bar{F}^F}{F!} \tag{4.19}$$

Next we apply this distribution to the case where F = 0, that is, to the case where no fronts have crossed point x. There are several aspects to note about this situation:

1.   Since $\bar{F}^0 = 1$ and $0! = 1$, Eq. (4.19) becomes

$$P(0) = e^{-\bar{F}} \tag{4.20}$$

for $F = 0$.

2.   The condition of no fronts crossing x is automatically a condition of non-interference, so the special magic postulated for point x poses no problem.

3.   Since point x is nonspecific, Eq. (4.20) describes the fraction of observations in which no fronts cross any arbitrary point or the fraction of the area in any one experiment which is crossed by no fronts.

4.   This last interpretation makes $P(0)$ the same as the fraction of a sample in the amorphous state. It is conventional to focus on the fraction crystallized θ; therefore the fraction amorphous is $1 - \theta$ and

$$1 - \theta = P(0) = e^{-\bar{F}} \tag{4.21}$$

5.   Inverting and taking the logarithm of both sides of Eq. (4.21), we obtain

$$\ln \frac{1}{1 - \theta} = \bar{F} \tag{4.22}$$

Equations (4.18) and (4.22) both describe the same situation and can be equated to each other to give

$$\ln \left(\frac{1}{1 - \theta} \right) = \pi \dot{r}^2 N t^2 \tag{4.23}$$

or

$$\theta = 1 - \exp(-\pi \dot{r}^2 N t^2) \tag{4.24}$$

Remember the units involved here: For $\dot{r}$ they are length time^{-1}; for N, length^{-2}; and for t, time. Therefore the exponent is dimensionless, as required. The form of Eq. (4.24) is such that at small times the exponential equals unity and $\theta = 0$; at long times the exponential approaches zero and $\theta = 1$. In between, an S-shaped curve is predicted for the development of crystallinity with time. Experimentally, curves of this shape are indeed observed. We shall see presently, however, that this shape is also consistent with other mechanisms besides the one considered until now.

In terms of spontaneous crystallization, the assumption that N nuclei commence to grow simultaneously at $t = 0$ is the most unrealistic. We can modify the model to allow for sporadic, spontaneous nucleation by the following

argument. We draw a set of concentric rings in the plane of the disks around point x as shown in Fig. 4.6b. If the radii are r and r + dr for the rings, then the area enclosed between them is $2\pi r\,dr$. We postulate that spontaneous random nucleation occurs with a frequency $\dot{N}$, having units area^{-1} time^{-1}. The rate of formation of nuclei within the ring is therefore $\dot{N}2\pi r\,dr$.

We continue to assume that the crystals so nucleated display a constant rate of radial growth $\dot{r}$. This means that it takes a crystal originating in a ring of radius r around point x a time given by $r/\dot{r}$ to cross x. The crystal labeled A in Fig. 4.6b has had just enough growth time to reach x. On the other hand, a crystal nucleated in this ring after $t - r/\dot{r}$ will not have had time to grow to x. The crystal labeled B in Fig. 4.6b is an example of the latter. It is only nucleation events that occur up to $t - r/\dot{r}$ which have time to grow from the ring of radius r and cross point x by their growth front. The increment in this number of fronts for the ring of radial thickness dr is

$$dF = (\dot{N}2\pi r\,dr)\left(t - \frac{r}{\dot{r}}\right) \tag{4.25}$$

The average number of fronts crossing point x at a time of observation t is the sum of contributions from all rings which are within reach of x in time t. The most distant ring included by this criterion is a distance $\dot{r}t$ from x. The average number of fronts, therefore, is given by integrating Eq. (4.25) for all rings between r = 0 and r = $\dot{r}t$:

$$\bar{F} = 2\pi\dot{N}\int_{0}^{\dot{r}t} r\left(t - \frac{r}{\dot{r}}\right)dr \tag{4.26}$$

As far as this integration is concerned, $\dot{r}$ and t are constants, so Eq. (4.26) is readily evaluated to give

$$\bar{F} = \frac{1}{3}\pi\dot{N}\dot{r}^2 t^3 \tag{4.27}$$

As before, this quantity in relation to the degree of crystallinity is given by Eq. (4.22), so equating the latter to Eq. (4.27) gives

$$\ln\left(\frac{1}{1-\theta}\right) = \frac{\pi}{3}\dot{N}\dot{r}^2 t^3 \tag{4.28}$$

or

$$\theta = 1 - \exp\left(-\frac{\pi}{3}\dot{N}\dot{r}^2 t^3\right) \tag{4.29}$$

Equations (4.24) and (4.29) are equivalent, except that the former assumes instantaneous nucleation at N sites per unit area while the latter assumes a nucleation rate of $\dot{N}$ per unit area *per unit time*. It is the presence of this latter rate which requires the power of t to be increased from 2 to 3 in this case.

In deriving these results we have focused attention on growth fronts originating elsewhere and crossing point x. We would count the same number if the growth originated at x and we evaluated the number of nucleation sites swept over by the growing front. This change of perspective is immediately applicable to a three-dimensional situation as follows. Suppose we let N represent the number of sites per unit volume (note that this is a different definition than given above) and assume that a spherical growth front emanates from each. Then the average number of fronts which cross nucleation sites in time t is

$$\bar{F} = \frac{4}{3}\,\pi(\dot{r}t)^3 N \tag{4.30}$$

by analogy with Eq. (4.18). Following the same argument as produced Eq. (4.24), we obtain

$$\theta = 1 - \exp\left(-\frac{4}{3}\,\pi\dot{r}^3 Nt^3\right) \tag{4.31}$$

for the three-dimensional case of simultaneous nucleation (concentration N). Following the argument that produced (4.29), we obtain

$$\theta = 1 - \exp\left(-\frac{\pi}{3}\,\dot{r}^3\dot{N}t^4\right) \tag{4.32}$$

for the three-dimensional case of sporadic nucleation (rate $\dot{N}$).

Equations (4.24), (4.29), (4.31), and (4.32) all describe different mechanisms for crystallization, yet all have basically the same form:

$$\theta = 1 - \exp(-Kt^m) \tag{4.33}$$

where K is a cluster—different in detail for each case—of numerical and growth constants and m is the power dependence of time. Equation (4.33) is known as the Avrami equation, after one of the researchers who has made contributions in this area. Likewise, the exponent m in Eq. (4.33) is called the Avrami exponent. The latter takes on the values 2, 3, or 4 for the four mechanisms considered above. To acquire some numerical familiarity with the Avrami function, consider the following example.

Example 4.3

Three different crystallization systems show m values of 2, 3, and 4. Calculate the value required for K in each of these systems so that all will show $\theta = 0.5$ after 10^3 sec. Use these m and K values to compare the development of crystallinity with time for these three systems.

Solution

Solve Eq. (4.33) for K and evaluate at $t = 10^3$ sec for each of the m values: $K = [-\ln(1 - \theta)]/t^m$. For $m = 2$, $K = (\ln 0.5)/(10^3)^2 = 6.93 \times 10^{-7}$ sec^{-2}; for $m = 3$, $K = 6.93 \times 10^{-10}$ sec^{-3}; for $m = 4$, $K = 6.93 \times 10^{-13}$ sec^{-4}. Note that the units of K depend on the value of m. Solve Eq. (4.33) for t and evaluate at different θ's for the m and K values involved:

	t in seconds		
θ	$m = 2$ $(K = 6.93 \times 10^{-7})$	$m = 3$ $(K = 6.93 \times 10^{-10})$	$m = 4$ $(K = 6.93 \times 10^{-13})$
0.1	3.89×10^2	5.33×10^2	6.24×10^2
0.2	5.67×10^2	6.85×10^2	7.53×10^2
0.3	7.18×10^2	8.02×10^2	8.47×10^2
0.4	8.59×10^2	9.03×10^2	9.27×10^2
0.5	1×10^3	1×10^3	1×10^3
0.6	1.15×10^3	1.10×10^3	1.07×10^3
0.7	1.32×10^3	1.20×10^3	1.15×10^3
0.8	1.52×10^3	1.32×10^3	1.23×10^3
0.9	1.82×10^3	1.49×10^3	1.35×10^3

These three systems describe a set of crystallization curves that cross at $\theta = 0.5$ and $t = 10^3$ sec. For the case where $m = 2$, the time span over which the change occurs is widest (1430 sec from $\theta = 0.1$-0.9) and the maximum slope is flattest (7.8×10^{-4} sec^{-1} between $\theta = 0.4$ and 0.6). For $m = 4$, the range is narrowest (726 sec) and the maximum slope is steepest (1.4×10^{-3} sec).

•

Although two of the mechanisms presented above yield the same power dependence on t, it appears possible to eliminate certain mechanisms by experimentally testing the development of θ with time. A strategy for this is suggested by Eq. (4.28). Taking the logarithm of both sides of that equation gives

$$\ln\left[\ln\left(\frac{1}{1 - \theta}\right)\right] = 3 \ln t + \text{const.} \qquad (4.34)$$

or for the generalized Avrami equation

$$\ln\left[\ln\left(\frac{1}{1-\theta}\right)\right] = m \ln t + \text{const.} \tag{4.35}$$

Thus the slope of a plot of $\ln[\ln(1-\theta)^{-1}]$ versus $\ln t$ will have a slope equal to the Avrami exponent.

Before turning to an examination of this prediction, a few more complications must be mentioned. Until now we have considered the case of disk or spherical growth from simultaneous or sporadic nucleation. Are these the only possibilities? As might be anticipated, the answer is no. Other geometries for growth have been examined; we shall include only the case of the cylinder, which is limited in radius but not in length (i.e., one-dimensional growth) to produce fibrillar structures. Another modification which has been investigated concerns the rate-determining step of the crystallization process. Equation (4.18) and those following from it imply that a contact between the growing disk and the surrounding melt for time t is sufficient for crystallization. Another possibility is that allowance must be made for the diffusion of the molecules to (or from) the growth site. A way of dealing with this assumes that amorphous molecules must diffuse out of the crystal domain to allow space for the crystallizing molecules. For a crystal of radius r, the time required for molecules to diffuse out of this domain is, according to Eq. (2.63), $r = (2Dt)^{1/2}$. In Eq. (4.18) this radius is written $r = \dot{r}t$. Thus, if the growth rate is diffusion controlled, these two expressions for r can be equated and solved for $\dot{r}$:

$$\dot{r} = \left(\frac{2D}{t}\right)^{1/2} \tag{4.36}$$

If this result is substituted into the previous expressions containing $\dot{r}$, the effect is to replace $\dot{r}$ with $(2D)^{1/2}$ and to multiply those t's which accompany $\dot{r}$ by $t^{-1/2}$.

This rather complex array of possibilities is summarized in Table 4.3. Table 4.3 lists the predicted values for the Avrami exponent for the following cases:

1. Growth geometry: fibrillar rod, disk, and sphere.
2. Nucleation mode: simultaneous and sporadic.
3. Rate determination: contact and diffusion.

Those exponents which we have discussed explicitly are identified by equation number in Table 4.3. Other tabulated results are readily rationalized from these. For example, according to Eq. (4.24) for disk (two-dimensional) growth on contact from simultaneous nucleations, the Avrami exponent is 2. If the dimensionality of the growth is increased to spherical (three dimensional), the exponent becomes 3. If, on top of this, the mechanism is controlled by diffusion, the

Table 4.3 Summary of Exponents in the Avrami Equation for Different Crystallization Mechanisms

Avrami exponent	Crystal geometry	Nucleation mode	Rate determination	Equation[a]
0.5	Rod	Simultaneous	Diffusion	
1	Rod	Simultaneous	Contact	
1	Disk	Simultaneous	Diffusion	
1.5	Sphere	Simultaneous	Diffusion	
1.5	Rod	Sporadic	Diffusion	
2	Disk	Simultaneous	Contact	(4.24)
2	Disk	Sporadic	Diffusion	
2	Rod	Sporadic	Contact	
2.5	Sphere	Sporadic	Diffusion	
3	Sphere	Simultaneous	Contact	(4.31)
3	Disk	Sporadic	Contact	(4.29)
4	Sphere	Sporadic	Contact	(4.32)

[a] Equation numbers refer to those relationships discussed explicitly in text.

exponent becomes 1.5. If the nucleation is sporadic in addition, the exponent is increased to 2.5.

While there are several instances of redundancy among the Avrami exponents arising from different pictures of the crystallization process, there is also enough variety to make the experimental value of this exponent a valuable way of characterizing the crystallization process. In the next section we shall examine the experimental side of crystallization kinetics.

4.6 The Kinetics of Crystallization: Experimental Aspects

In order to carry out an experimental study of the kinetics of crystallization, it is first necessary to be able to measure the fraction θ of polymer crystallized. While this is necessary, it is not sufficient; we must also be able to follow changes in the fraction of crystallinity with time. So far in this chapter we have said nothing about the experimental aspects of determining θ. We shall now briefly rectify this situation by citing some of the methods for determining θ. It must be remembered that not all of these techniques will be suitable for kinetic studies.

Since the fractions of crystalline (subscript c) and amorphous (subscript a) polymer account for the entire sample, it follows that we may measure whichever of the two is easiest to determine and obtain the other by difference. Generally, it is some property P_c of the crystalline phase that we are able to

monitor. If this property can be measured for a sample which is 100% crystalline (superscript $°$) we can compare the value of P_c measured on an actual sample (no superscript) to evaluate θ:

$$\theta = \frac{P_c}{P_c{}^°} \tag{4.37}$$

This relationship is sketched in Fig. 4.7a, which emphasizes that P_c must vary linearly with θ and that $P_c{}^°$ must be available, at least by extrapolation. The heat of fusion is an example of a property of the crystalline phase that can be used this way. It could be difficult to show that the value of ΔH_f is constant per unit mass at all percentages of crystallinity and to obtain a value for $\Delta H_f{}^°$ for a crystal free from defects. Therefore, while conceptually simple, the actual utilization of Eq. (4.37) in precise work may not be easy.

Figure 4.7b shows a variation of this in which a property of the sample (no subscript) is found to vary linearly with θ, having a value P_a when $\theta = 0$ and a value P_c when $\theta = 1$. The slope of this line is simply $P_c - P_a$, since the difference of θ values is unity for this difference in P. The equation for the line in Fig. 4.7b is

$$P = P_a + \theta (P_c - P_a) \tag{4.38}$$

which can be solved for θ as a function of $P, P_a,$ and P_c:

$$\theta = \frac{P - P_a}{P_c - P_a} \tag{4.39}$$

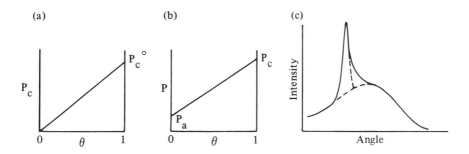

Figure 4.7 Various representations of the properties of a mixture of crystalline and amorphous polymer. (a) The monitored property is characteristic of the crystal and varies linearly with θ. (b) The monitored property is characteristic of the mixture and varies linearly with θ between P_a and P_c. (c) X-ray intensity is measured with the sharp and broad peaks being P_c and P_a, respectively.

If the amorphous component contributes nothing to the measured property (as with heat of fusion), then Eq. (4.39) reduces to Eq. (4.37). Specific volume is an example of a property which has been extensively used in this way to evaluate θ.

Figure 4.7c illustrates how x-ray diffraction techniques can be applied to the problem of evaluating θ. If the intensity of scattered x-rays is monitored as a function of the angle of diffraction, a result like that shown in Fig. 4.7c is obtained. The sharp peak is associated with the crystalline diffraction, and the broad peak, with the amorphous contribution. If the area A under each of the peaks is measured, then

$$\theta = \frac{A_c}{A_c + A_a} \tag{4.40}$$

An obvious difficulty here is deciding the location of the broken line portions of the peaks in the region of overlap. Some features of the infrared absorption spectrum may also be analyzed by the same procedure to yield values for θ.

As noted above, not all techniques which provide information regarding crystallinity are useful to follow the rate of crystallization. In addition to sufficient sensitivity to monitor small changes, the method must be rapid and suitable for isothermal regulation, quite possibly over a range of different temperatures. Specific volume measurements are especially convenient for this purpose. We shall continue our discussion using specific volume as the experimental method.

Although the extent of crystallinity is the variable under consideration, time is the experimental variable. Accordingly, what is done is to identify the specific volume of a sample at $t = 0$ (subscript 0) with V_a, the volume at $t = \infty$ (subscript ∞) with V_c, and the volume at any intermediate time (subscript t) with the composite volume. On this basis, Eq. (4.39) becomes

$$\theta = \frac{V_t - V_0}{V_\infty - V_0} \tag{4.41}$$

and the fraction amorphous becomes

$$1 - \theta = \frac{V_\infty - V_t}{V_\infty - V_0} \tag{4.42}$$

Figure 4.8a shows how this quantity varies with time for polyethylene crystallized at a series of different temperatures. Several aspects of these curves are typical of all polymer crystallizations and deserve comment:

(a)

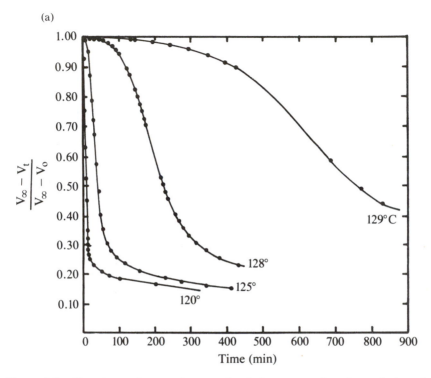

Figure 4.8 Fraction of amorphous polyethylene as a function of time for crystallizations conducted at indicated temperatures: (a) linear time scale and (b) logarithmic scale. Arrows in (b) indicate shifting curves measured at 126 and 130 to 128°C as described in Example 4.4. [Reprinted with permission from R. H. Doremus, B. W. Roberts, and D. Turnbull (Eds.) *Growth and Perfection of Crystals*, Wiley, New York, 1958.]

1. The decrease in amorphous content follows an S-shaped curve. The corresponding curve for the growth of crystallinity would show a complementary but increasing plot. This aspect of the Avrami equation was noted in connection with the discussion of Eq. (4.24).

2. The greater the undercooling, the more rapidly the polymer crystallizes. This is due to the increased probability of nucleation the more supercooled the liquid becomes. Although the data in Fig. 4.8 are not extensive enough to show it, this trend does not continue without limit. As the crystallization temperature is lowered still further, the rate passes through a maximum and then drops off as T_g is approached. This eventual decrease in rate is due to decreasing chain mobility which offsets the nucleation effect.

(b)

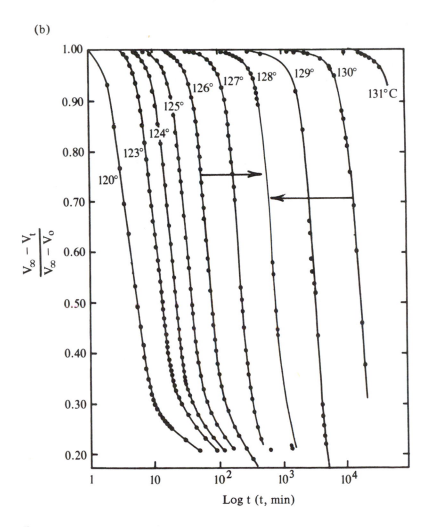

Figure 4.8   *(continued)*

3. Replotting the data on a lorgarithmic time scale as shown in Fig. 4.8b has an interesting effect. Figure 4.8b shows that this modification produces a far more uniform set of S curves. As a matter of fact, if the various curves are shifted along the horizontal axis, they may be superimposed over a wide portion of the transformation. The arrows in Fig. 4.8b show this displacement of the data at 126 and 130°C to correspond to the data at 128°C. This superpositioning is examined in the example below.

Example 4.4

From the lengths of the arrows drawn in Fig. 4.8b, estimate the change in time scale which will produce the same effect on the rate of crystallization as changing the temperature from 130 to 128°C. Do the same for a temperature change from 126 to 128°C.

Solution

At the values of $1 - \theta$ listed below, determine the difference between $(\log t)_{128}$ and $(\log t)_{130}$ and that between $(\log t)_{128}$ and $(\log t)_{126}$.

$1 - \theta$	0.9	0.8	0.7	0.6	0.5	0.4	0.3
$(\log t)_{128} - (\log t)_{126}$	1.00	1.00	1.03	1.00	0.97	0.98	0.97
$(\log t)_{128} - (\log t)_{130}$	-1.30	-1.33	-1.33	-1.35	-1.35	-1.35	—

The fact that these differences are as constant as they are for each of the shifts illustrates the validity of the superpositioning procedure. Now consider the average value of these differences:

	Average for $126 \to 128°C$	Average for $130 \to 128°C$
$\Delta \log t$	0.99	-1.34
antilog	10	$0.046 = 1/22$

The significance of these numbers is seen as follows. The average values of $\Delta \log t$ are to be added to the log t values at 126 or 130°C to superimpose the latter curves on the one measured at 128°C. Since these values are added to log t values, the effect is equivalent to multiplying the individual t values at 126 and 130°C by the appropriate antilogs to change the time scale in the individual runs to a common time scale. Using the case of $\theta = 0.5$ as an illustration, we see the following times are required to reach this level of crystallinity:

$T\ (°C)$	t observed (min)	t shifted (min)
126	80	$80 \times 10 = 800$
128	800	$800 \times 1 = 800$
130	17,500	$17,500 \times 0.046 = 800$

•

The preceding example of superpositioning is an illustration of the principle of time–temperature equivalency. We referred to this in the last chapter in connection with the mechanical behavior of polymer samples and shall take up the

latter quantitatively in Sec. 4.10. Although time–temperature equivalency with respect to mechanical properties is a different phenomenon, precisely the same formalism as illustrated above applies. In the above sample note that the absolute magnitude of the two separate shifts could be combined so that a change of temperature from 126 to 130°C would involve a difference in log t of about 2.33, the antilog of which is about 214. This shows that a 4°C increase in temperature is equivalent to an expansion in the time scale of the experiment by a factor of 214.

Next let us examine an experimental test of the Avrami equation and the assortment of predictions from its various forms as summarized in Table 4.3. Figure 4.9 is a plot of $\ln[\ln(1 - \theta)^{-1}]$ versus $\ln t$ for poly(ethylene terephthalate) at three different temperatures. According to Eq. (4.35), this type of

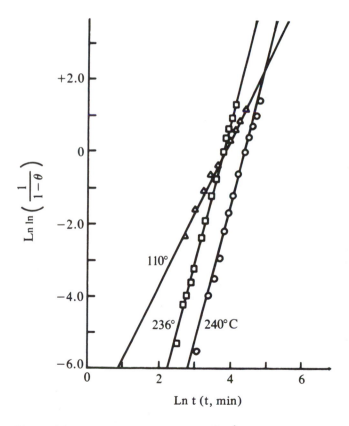

Figure 4.9 Log–log plot of $\ln(1 - \theta)^{-1}$ versus time for poly(ethylene terephthalate) at three different temperatures. [Reprinted from L. B. Morgan, *Philos. Trans. R. Soc. London* 247A:13 (1954).]

representation should yield a straight line, the slope of which corresponds to the Avrami exponent. The data in Fig. 4.9 show that linearity is indeed obtained and that the slope equals 2 when the crystallization is carried out at $110°C$ and changes to 4 at higher temperatures. The melting point of poly(ethylene terephthalate) is $267°C$, so when the undercooling is about $25°C$, three-dimensional growth with sporadic nucleation is indicated. With an undercooling of $150°C$, the mechanism of crystallization is clearly different, although it is not possible to identify the specific combination of factors responsible for the exponent 2. The values of K in Eq. (4.33) are best obtained analytically, once the exponent has been determined graphically. The two K values for the case where m = 4 in Fig. 4.9 are 2.94×10^{-7} min^{-4} at $236°C$ and 3.13×10^{-8} min^{-4} at $240°C$. The mechanism is the same in these two cases, but the rate is more than nine times faster when the temperature is lowered by $4°C$. Note that it is not possible to resolve K, which is a cluster of other nucleation and growth parameters, into its constituent factors, even when the value of the exponent identifies the mechanism unambiguously. At both 110 and $120°C$, m = 2 and the values of K are 7.93×10^{-4} and 7.45×10^{-3} min^{-2}, respectively. In this region the rate is about 10 times slower when the temperature is lowered by $10°C$. Thus both the value of m and the effect on K of changing temperature are different for these two regimes of behavior.

The testing of the Avrami equation reveals several additional considerations of note:

1. The multiple use of logarithms in the analysis presented by Fig. 4.9 obliterates much of the deviation between theory and experiment. More stringent tests can be performed by other numerical methods.
2. Deviations from the Avrami equation are frequently encountered in the long time limit of the data. This is generally attributed to secondary nucleation occurring at irregularities on the surface of crystals formed earlier.
3. Exponents other than integral multiples of one-half are observed. As a matter of fact, a method for determining the Avrami exponent which is based on graphical differentiation rather than logarithmic analysis yields instantaneous m values at particular values of θ rather than a single value averaged over the entire transition. When this method is used, it is found that m increases initially, eventually leveling off.
4. These unpredicted Avrami exponents may be indications that multiple mechanisms are operative and/or that $\dot{r}$ and/or $\dot{N}$ is itself a function of θ.

Experimental results are in general conformity with the Avrami equation, but the interpretation of various observations is still complicated in many instances. One intriguing observation is that the induction period for nucleation is inversely proportional to the length of time the liquid is held in the liquid state after previous melting. This dependence on prior history may be qualitatively understood

by assuming that the last traces of order from the first crystals persist for quite a while and affect the rate of nucleation. It is difficult to handle these complications quantitatively, however. Having looked at some aspects of both the thermodynamics and the kinetics of the crystallization process, it is time to take a look at the crystals themselves. We shall do this in the next section.

4.7 Levels of Structure: Crystal Morphology

In this section we shall examine the various levels of structure displayed by polymers. The term *morphology* is used to describe such features and may be applied to both molecular and macroscopic levels of organization. A general principle that underlies morphological inquiry is the idea of using as a "yardstick" some probe which matches within an order of magnitude or so the linear dimensions of the feature being explored. Accordingly, x-ray diffraction is the preferred technique for investigating the spacing of individual atoms or chains within a crystal, since the wavelength is on the order of 100 pm for this radiation. Both transmission and scanning electron microscopy are valuable techniques for examining individual polymer crystals. The thickness and lateral dimensions of the individual crystals as well as surface topography can be studied by these instruments. Three-dimensional aggregates of single-crystal lamellae have characteristic structures of their own. Since they are ordered on a larger scale of linear dimension, optical microscopy can be used as a probe of structure in those cases.

Although x-ray diffraction yields the most detailed information about crystal morphology at the molecular level, we shall restrict ourselves to a few limited remarks about observations derived by this method. The interpretation of x-ray diffraction patterns is a topic best left to experts, but we note that the characterization of the unit cell is one of the pieces of information which this type of measurement yields. Of the many polymers which have been investigated, none crystallizes so completely or has been studied as thoroughly as polyethylene. Figure 4.10 shows the arrangement of molecules in the polyethylene crystal and outlines the unit cell. By transposing the unit cell through three dimensions, the overall crystal structure is generated. X-ray studies show the dimensions of the unit cell for polyethylene as a = 736 pm, b = 492 pm, and c = 253.4 pm. In regard to this unit cell, we observe the following:

1. The c axis corresponds to both the short axis of the crystal and the axis along the molecular chain. The observed repeat distance in the c direction is what would be expected between successive substituents on a fully extended hydrocarbon chain with normal bond lengths and angles (see Sec. 1.2).
2. The distances between all hydrogen atoms is approximately the same in this structure, so there is no problem with overcrowding.

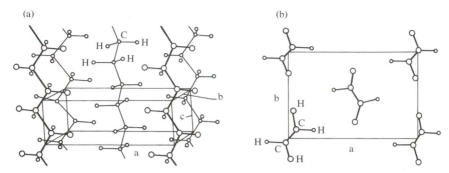

Figure 4.10 Crystal structure of polyethylene: (a) unit cell shown in relation to chains and (b) view of unit cell perpendicular to the chain axis. [Reprinted from C. W. Bunn, *Fibers from Synthetic Polymers*, R. Hill (Ed.), Elsevier, Amsterdam, 1953.]

3. While not overcrowded, the polyethylene structure uses space with admirable efficiency, the atoms filling the available space with 73% efficiency. For contrast, recall that close-packed spheres fill space with 74% efficiency, so polyethylene does about as well as is possible in its utilization of space.

One of the things that can be done with a knowledge of the unit cell dimensions is to calculate the crystal density. This is examined in the following example.

Example 4.5

Use the unit cell dimensions cited above to determine the crystal density of polyethylene. Examine Fig. 4.10 to decide the number of repeat units per unit cell.

Solution

Figure 4.10 shows that the equivalent of two ethylene units are present in each unit cell. Accordingly, the mass per unit cell is

$$\frac{2 \text{ repeat units}}{\text{unit cell}} \times \frac{1 \text{ mol repeat units}}{6.02 \times 10^{23} \text{ repeat units}} \times \frac{28.0 \text{ g}}{1 \text{ mol repeat units}}$$

$$= 9.30 \times 10^{-23} \text{ g (unit cell)}^{-1}$$

Since all angles in the cell are 90°, the volume of the unit cell is

$$\frac{736 \text{ pm} \times 492 \text{ pm} \times 253 \text{ pm}}{\text{unit cell}} \times \left(\frac{1 \text{ cm}}{10^{10} \text{ pm}}\right)^3$$

$$= 9.16 \times 10^{-23} \text{ cm}^3 \text{ (unit cell)}^{-1}$$

The density of the crystal is obtained from the ratio of these two quantities:

$$\rho = \frac{9.30 \times 10^{-23}}{9.16 \times 10^{-23}} = 1.015 \text{ g cm}^{-3}$$

A result such as this could be used in Eq. (4.39) as part of the information needed to determine θ.

•

Next, we consider the electron microscope as a tool for investigating crystallization in polymers. The electron microscope uses the de Broglie waves associated with accelerated electrons to produce an image just as visible light produces an image in an optical microscope. Electromagnets function as lenses for the electron beam and the image is formed on a phosphorescent screen or a photographic plate. Otherwise the electron microscope is completely analogous to the light microscope. The de Broglie wavelength of an electron under typical operating conditions in an electron microscope is on the order of a few picometers. In all types of microscopy it is the resolving power rather than the magnification per se which is the limiting factor. Light is diffracted from the edges of illuminated bodies and this diffraction blurs their boundaries. The resolving power measures the minimum separation between objects that will produce discernably different images in a microscope. As with many optical phenomena, this separation is on the order of the wavelength of the illuminating radiation. Therefore, the resolving power of an electron microscope is potentially smaller by some five orders of magnitude than that achieved by optical microscopes. Two adjunct procedures of electron microscopy find applicability in the study of polymer crystals: shadow casting and dark-field operation.

Shadow casting is used to improve the contrast between a sample and its background and between various details of the sample surface. Because polymer crystals are so thin and mostly consist of atoms of low atomic number (the number of electrons in an atom determines the extent of interaction with the electron beam), some sort of contrast enhancement is important. In the shadowing method the sample is placed in an evacuated chamber and a heavy metal is allowed to evaporate in the same chamber. The position of the metal source is such that the metal vapor strikes the sample at an oblique angle and condenses on the cool surface. The thin metal film thus formed literally casts shadows which enhance the image of the sample. If the angle of incidence of the heavy metal beam is known, the thickness of a crystal or the height of surface protubrances can be determined from the length of the shadow by simple trigonometry.

In dark-field electron microscopy it is not the transmitted beam which is used to construct an image but, rather, a beam diffracted from one facet of the object under investigation. One method for doing this is to shift the aperture of the microscope so that most of the beam is blocked and only those electrons

scattered into the chosen portion of the diffraction pattern contribute to the image. This decreases the intensity of the illumination used to produce the dark-field image and therefore requires longer exposure times, with the attendant modification or even degradation of the polymer. Nevertheless, dark-field operation distinguishes between portions of the sample with different orientations and therefore produces a more three-dimensional representation of the sample.

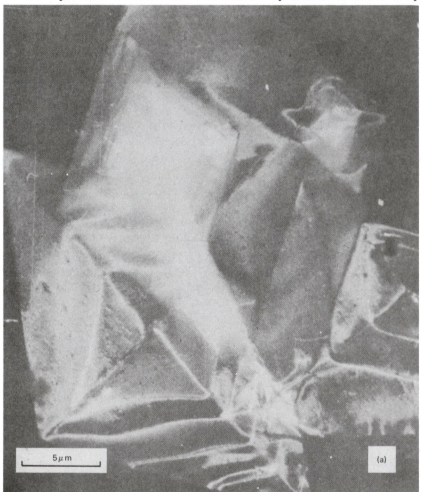

Figure 4.11 Electron micrographs of polyethylene crystals. (a) Dark-field illumination shows crystals to have a hollow pyramid structure. (Reprinted with permission from P. H. Geil, *Polymer Single Crystals*, Interscience, New York, 1963.) (b) Transmission micrograph in which contrast is enhanced by shadow casting [Reprinted with permission from D. H. Reneker and P. H. Geil, *J. Appl. Phys.* 31:1916 (1960).]

Figures 4.11a and b, respectively, are examples of dark-field and direct transmission electron micrographs of polyethylene crystals. The ability of dark-field imaging to distinguish between features of the object which differ in orientation is apparent in Fig. 4.11a. The effect of shadowing is evident in Fig. 4.11b, where those edges of the crystal which cast the shadows display sharper contrast.

The electron micrographs of Fig. 4.11 are more than mere examples of electron microscopy technique. They are the first occasion we have had to actually look at single crystals of polymers. Although there is a great deal to be learned from studies of single crystals by electron microscopy, we shall limit ourselves to just a few observations:

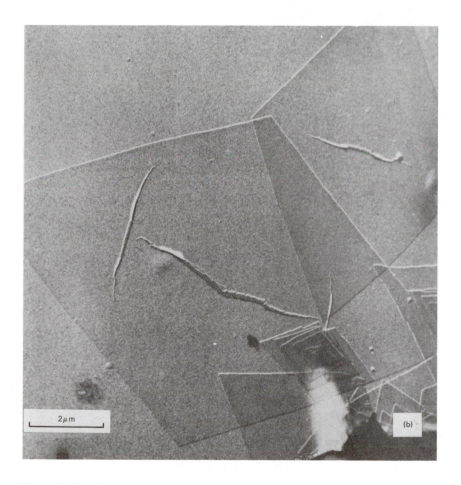

Figure 4.11 *(continued)*

1. Single crystals such as those shown in Fig. 4.11 are not observed in crystallization from the bulk. Crystallization from dilute solutions is required to produce single crystals with this kind of macroscopic perfection. Polymers are not intrinsically different from low molecular weight compounds in this regard.

2. Crystallization conditions such as temperature, solvent, and concentration can influence crystal form. One such modification is the truncation of the points at either end of the long diagonal of the diamond-shaped crystals seen in Fig. 4.11b. Twinning and dendritic growth are other examples of such changes of habit.

3. The polyethylene crystals shown in Fig. 4.11 exist as hollow pyramids made up of planar sections. Since the solvent must be evaporated away prior to electron microscopic observation, the pyramids become buckled, torn, and/ or pleated during the course of sample preparation. While the pyramidal morphology is clearly evident in Fig. 4.11a, there is also evidence of collapse and pleating. Likewise, the ridges on the apparently planar crystals in Fig. 4.11b are pleats of excess material that bunches up when the pyramids collapse.

4. Both hollow pyramids and corrugated pyramids are thoroughly documented and fairly well understood. Such structures are consistent with the notion that successive layers of folded chains do not fold at the same place, but offset this fold stepwise to generate the pyramid face. The polymer chains are perpendicular to the planar faces of the pyramid and are therefore tilted at an angle relative to the base of the pyramid.

The foregoing is by no means a comprehensive list of the remarkable structures formed by the crystallization of polymers from solution. The primary objective of this brief summary is the verification that single crystals can be formed and characterized, not only by x-ray diffraction, but by direct electron-microscopic observation. These single crystals are mostly formed from dilute solution, however, and our concern throughout previous parts of this chapter has been the crystallization of polymer from the melt. What is the relationship between these idealized single crystals and the morphological forms which result from crystallization of bulk polymer? We begin to answer this question by considering the microscopic examination of material resulting from the crystallization of polymer melt.

Suppose a bulk-crystallized polymer sample is observed in an optical microscope with the sample placed between Polaroid filters oriented at right angles to each other. In the absence of any sample, the light would be attenuated owing to the 90° angle between the vectors describing the light transmitted by the two filters. With a crystalline sample of polymer in place, however, a display like

Figure 4.12 Spherulites of poly(1-propylene oxide) observed through crossed Polaroid filters by optical microscopy. See text for significance of Maltese cross and banding in these images. [From J. H. MaGill, *Treatise on Materials Science and Technology*, Vol. 10A, J. M. Schultz (Ed.), Academic, New York, 1977, with permission.]

that shown in Fig. 4.12 is generally observed. The field of view becomes at least partially filled with domains called spherulites and which are described by the following general features:

1. They possess spherical symmetry around a center of nucleation. This symmetry projects a perfectly circular cross section if the development of the spherulite is not stopped by contact with another expanding spherulite.

2. A system of mutually impinging spherulites develop into an array of irregular polyhedra, the dimensions of which can be as large as a centimeter or so.
3. The individual spherulite shows up by the characteristic Maltese cross optical pattern under crossed Polaroids, although the Maltese cross is truncated in the event of impinging spherulites.
4. Superimposed on the Maltese cross may be such additional optical features as the banding seen in Fig. 4.12.
5. Spherulites have been observed in organic and inorganic systems of synthetic, biological, and geological origin, including moon rocks, and are therefore not unique to polymers.
6. A larger number of smaller spherulites are produced at larger undercoolings, a situation suggesting nucleation control. Various details of the Maltese cross pattern, such as the presence or absence of banding, may also depend on the temperature of crystallization.

On the basis of a variety of experimental observations, including an analysis of the ubiquitous Maltese cross, a number of aspects of the structure of spherulites have been elucidated. The spherulites are aggregates of lamellar crystals radiating from a nucleation site. The latter can be either a spontaneously formed single crystal or a foreign body. The spherical symmetry is presumably not present at the outset, but develops with time. Apparently, fibrous or lathlike crystals begin branching and fanning out as in dendritic growth. As the lamellae spread out radially and three dimensionally from the nucleus, branching of the crystallites continues to generate the spherical morphology. Figure 4.13 represents schematically the leading edge of some of these fibrils, one of which has just split.

The molecular alignment within these radiating fibers has been determined to be perpendicular to the radius of the individual spherulite. The individual lamellae are similar in organization to the planar faces of pyramidal single crystals; that is, they consist of ribbons on the order of 10–100 nm in thickness, built up from successive layers of folded chains. Growth is accomplished by the addition of successive layers of chains to the ends of the radiating laths. This feature is also indicated schematically in Fig. 4.13. The chain structure of polymer molecules suggests that a given molecule might become involved in more than one lamella and thus link radiating crystallites from the same or adjacent spherulites, as seen in Fig. 4.13. These interlamellar links are not possible in spherulites of low molecular weight compounds, which show poorer mechanical strength as a consequence.

The molecular chain folding is the origin of the Maltese cross which identifies the spherulite under crossed Polaroids. The Maltese cross is known to arise from a spherical array of birefringent particles through the following considerations:

1. The ordered polymer chains are consistently oriented perpendicularly to the radius of the spherulite.

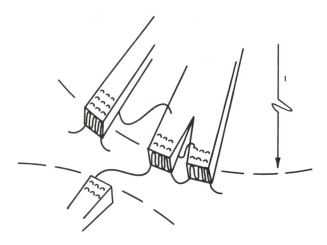

Figure 4.13 Schematic illustration of the leading edge of a lathlike crystal within a spherulite.

2. The index of refraction of most polymers is greater parallel to the chain than normal to the molecular axis. Substances showing this anisotropy of refractive index are said to be birefringent.
3. Items (1) and (2) mean that the refractive index in the tangential direction of the spherulite is generally greater than that along the radius.
4. This birefringence coupled with spherical geometry produces light extinction along the axis of each of the Polaroid filters, hence the 90° angle of the Maltese cross.
5. If the Polaroid filters are held fixed and the sample rotated between them, the Maltese cross remains fixed because of the symmetry of the spherulite.
6. Twisting of the lamellar ribbons along the radial direction is responsible for the banding superimposed on the Maltese cross in Fig. 4.12. From the spacing of the bands, the period of the twist can be calculated and is found to depend on crystallization conditions.

Spherulite morphology in a bulk-crystallized polymer is thus seen to involve ordering at several different levels of organization: individual molecules folded into crystallites which, in turn, are oriented into spherical aggregates. The overall kinetics of such crystallization are consistent with the model of spherical growth from either simultaneous or sporadic nucleations, as discussed in Sec. 4.5. Microscopic observations of growing spherulite fronts even show that the rate of radial growth is constant in isothermal experiments, as postulated in our discussion of crystallization kinetics. The complicated morphology of the spherulite makes it clear that the mechanistic details for bulk crystallization are

oversimplified if growth is visualized as a uniform deposition of polymer molecules on a spherical surface.

We conclude this brief presentation on the types of crystal morphology by briefly describing crystallization under applied stress. We noted this possibility in discussing deviations from the simple model for rubber elasticity under large deformations. Also, stress-induced crystallinity is important in film and fiber technology. When dilute solutions of polymers are stirred rapidly, unusual structures develop which are described as having a shish kebab morphology. These consist of chunks of folded chain crystals strung out along a fibrous central column. In both the "shish" and the "kebab" portions of the structure, the polymer chains are parallel to the overall axis of the structure. Polymers may have a number of obstacles to overcome to crystallize, but once these are surmounted, there seems to be no end to the surprising structures they can produce!

The objective of this chapter is to discuss both the crystalline and the glassy states of polymers. As noted above, only those polymers with the prerequisite order of molecular microstructure can crystallize, but all undergo the glass transition of whatever material remains amorphous at T_g. Thus we conclude this section on structure by observing that the glassy state lacks long-range order, although small domains of short-range order may be larger and last longer for glasses than for liquids because of the lower level of disruptive thermal energy below T_g. Within the glassy state there may be interspersed crystalline patches, but we shall take the position that the glassy state has essentially the same level of order—or lack of order—as the liquid state. With the morphology of the glassy state thus dismissed, we next turn to the question of how the glassy state does differ from the liquid state.

4.8 Thermodynamics and the Glass Transition

The kinetic nature of the glass transition should be clear from the last chapter, where we first identified this transition by a change in the mechanical properties of a sample in very rapid deformations. In that chapter we concluded that molecular motion could simply not keep up with these high-frequency deformations. The complementarity between time and temperature enters the picture in this way. At lower temperatures the motion of molecules becomes more sluggish and equivalent effects on mechanical properties are produced by cooling as by frequency variations. We shall return to an examination of this time–temperature equivalency in Sec. 4.10. First, however, it will be profitable to consider the possibility of a thermodynamic description of the transition which occurs at T_g.

In Secs. 4.3 and 4.4 we discussed the thermodynamics of the crystal → liquid transition. This and other familiar phase equilibria are examples of what are called first-order transitions. There are other less familiar but also well-known

transitions in nature that are not first-order thermodynamic processes. The disappearance of ferromagnetism at a temperature called the Curie point is an example of such a transition. Rather than a discontinuity in S, V, and H, as shown by first-order transitions, these variables merely change slope with increasing temperature. Since this is the sort of behavior which occurs at T_g, it is important to examine this second type of transition in some detail as well. It must be acknowledged from the outset that the applicability of these ideas to T_g is filled with controversy. Time is not a thermodynamic variable, yet we have already seen that the location of T_g depends on the rate at which the temperature variation is carried out. The full picture cannot be developed all at once, however, so for the present we shall ignore this aspect of the behavior and pursue the thermodynamics. We shall have more to say about the time-dependent aspects in later sections.

There is no discontinuity in volume, among other variables, at the Curie point, but there is a change in temperature coefficient of V, as evidenced by a change in slope. To understand why this is called a second-order transition, we begin by recalling the definitions of some basic physical properties of matter:

1. The coefficient of thermal expansion α:

$$\alpha = \frac{1}{V} \cdot \left(\frac{\partial V}{\partial T} \right)_p \tag{4.43}$$

2. The isothermal compressibility β:

$$\beta = - \frac{1}{V} \left(\frac{\partial V}{\partial p} \right)_T \tag{4.44}$$

Since V experiences a change of slope at the second-order transition, that is, $(\partial V/\partial T)_p$ and $(\partial V/\partial p)_T$ have different values on each side of the transition, it is α and β that show the discontinuities at the second-order transition rather than V itself. The term *second order* comes about since the quantities showing the discontinuity, α and β among others, may be written as second derivatives of G. That is, we apply Eq. (4.3) to Eqs. (4.43) and (4.44) to obtain

$$\alpha = \frac{1}{V} \left(\frac{\partial V}{\partial T} \right)_p = \frac{1}{V} \frac{\partial}{\partial T} \left[\left(\frac{\partial G}{\partial p} \right)_T \right]_p \tag{4.45}$$

and

$$\beta = - \frac{1}{V} \left(\frac{\partial V}{\partial p} \right)_T = - \frac{1}{V} \left(\frac{\partial^2 G}{\partial p^2} \right)_T \tag{4.46}$$

Figures 4.14a and b are the analogs of Figs. 4.3a and b; they schematically describe second- and first-order transitions, respectively. It is the discontinuity in these second-order properties that characterizes a second-order transition.

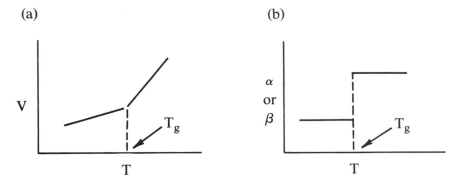

Figure 4.14   Behavior of thermodynamic variables at T_g for a second-order phase transition: (a) volume and (b) coefficient of thermal expansion α and isothermal compressibility β.

Another well-known thermodynamic result, the Clapeyron equation, applies to first-order transitions (subscript 1):

$$\frac{dp}{dT_1} = \frac{\Delta S_1}{\Delta V_1} \tag{4.47}$$

This expression describes the variation of the pressure–temperature coordinates of a first-order transition in terms of the changes in S and V which occur there. The Clapeyron equation cannot be applied to a second-order transition (subscript 2), because ΔS_2 and ΔV_2 are zero and their ratio is undefined for the second-order case. However, we may apply L'Hopital's rule to both the numerator and denominator of the right-hand side of Eq. (4.47) to establish the limiting value of dp/dT_2. In this procedure we may differentiate either with respect to p,

$$\frac{dp}{dT_2} = \frac{(\partial \Delta S_2/\partial p)_T}{(\partial \Delta V_2/\partial p)_T} = \frac{\Delta \alpha}{\Delta \beta} \tag{4.48}$$

or T,

$$\frac{dp}{dT_2} = \frac{(\partial \Delta S_2/\partial T)_p}{(\partial \Delta V_2/\partial T)_p} = \frac{\Delta C_p}{T_2} \frac{1}{V \Delta \alpha} \tag{4.49}$$

to generate some additional expressions. The derivatives of entropy involved in obtaining these equations will be found in any textbook treatment of the

second law. All of the Δ's in these equations refer to the difference in the value of the variable from one side (prime) of T_2 to the other (double prime):

$$\Delta\alpha = \alpha' - \alpha'' \tag{4.50}$$

$$\Delta\beta = \beta' - \beta'' \tag{4.51}$$

and

$$\Delta C_p = C_p' - C_p'' \tag{4.52}$$

The following example illustrates how results like these can be applied.

Example 4.6

On the assumption that the glass transition is a second-order thermodynamic transition, estimate the pressure dependence dT_g/dp of T_g using the following data for poly(vinyl chloride): $T_g = 347$ K, $V_{sp} = 0.75$ cm^3 g^{-1}, $\Delta\alpha = 3.1 \times 10^{-4}$ K^{-1} and $\Delta C_p = 0.068$ cal K^{-1} g^{-1}.[†]

Solution

Invert Eq. (4.49) and substitute. The ratio of gas constants is convenient for unit conversion:

$$\frac{dT_g}{dp} = \frac{T_g V \Delta\alpha}{\Delta C_p} = \frac{(347 \text{ K})(0.75 \text{ cm}^3 \text{ g}^{-1})(3.1 \times 10^{-4} \text{ K}^{-1})}{(0.068 \text{ cal g}^{-1} \text{ K}^{-1})}$$

$$\times \frac{1.99 \text{ cal}}{82 \text{ atm cm}^3} = 0.029 \text{ K atm}^{-1}$$

This quantity has been measured directly to be 0.016 K atm^{-1}. Note that a pressure change of 60 atm is needed to change T_g by 1 K.

•

Since Eqs. (4.48) and (4.49) both describe the same limit, they must be equal at the second-order transition point:

$$T_2 = \frac{\Delta C_p \Delta\beta}{V(\Delta\alpha)^2} \tag{4.53}$$

[†] J. M. O'Reilly, *J. Polym. Sci.* 57:429 (1962).

For a second-order transition Eq. (4.53) is the analog of Eq. (4.5), which is useful for first-order transitions. Equation (4.53) and the first- and second-order terminology are due to Ehrenfest.

Thermodynamics is an abundant source of relationships and does not let us down with respect to true second-order transitions. The reader will notice that the innocent-looking word *true* crept into the last sentence describing this second-order transition. The implication here is that the glass transition is not "truly" a second-order transition. A moment's reflection reveals that the source of this reservation is the doubt about equilibrium for the glass transition. Implicit throughout all of the thermodynamics in this chapter has been the notion that the phases on either side of a transition, whether first or second order, are in thermodynamic equilibrium. This enters the mathematical formalism right from the start: Eq. (4.1) assumes that the Gibbs free energy of a phase is described by only two variables, in that case p and T. While the glass transition is certainly affected by p and T, it is also dependent on the time of observation, as shown in Fig. 4.1. Because of this time dependence, the glass transition involves more than a simple second-order transition. Several generalizations do seem safe at this point, however:

1. The experimental value of T_g appears at progressively lower temperatures the more slowly the experiment is conducted.
2. Sufficient time must be allowed for thermodynamic equilibrium to be reached in other simpler reactions and transitions.
3. Changes in the conformation of polymer chain backbone occur much more slowly in the vicinity of T_g than most of the molecular processes that serve as examples of simpler equilibria.
4. If a T_g value corresponding to a true second-order transition exists, it is a value lower than those based on short-term observations. The latter should then be regarded as approximations to the true value.

4.9 Free Volume

The foregoing thermodynamic view of the glass transition assumes that equilibrium criteria are satisfied as a sample is cooled through T_g. In this section we shall set aside the issue of thermodynamic equilibrium and simply consider why the mechanical properties of polymers should undergo a dramatic change over a relatively narrow range of temperatures, an observation which was, in the last chapter, our introduction to T_g. In discussing Fig. 4.14, we saw that volume also shows a change in behavior at T_g—a change more typical of second-order than first-order transitions. Since volume is more readily visualized than the mechanical properties, we shall concentrate on the former in the following discussion. The emphasis is not merely a matter of convenience, however, since most attempts to understand T_g involve some concept of free volume.

There are two ways in which the volume occupied by a sample can influence the Gibbs free energy of the system. One of these involves the average distance of separation between the molecules and therefore influences G through the energetics of molecular interactions. The second volume effect on G arises from the contribution of free-volume considerations. In Chap. 2 we described the molecular texture of the liquid state in terms of a model which allowed for vacancies or holes. The number and size of the holes influence G through entropy considerations. Each of these volume effects varies differently with changing temperature and each behaves differently on opposite sides of T_g. We shall call free volume that volume which makes the second type of contribution to G.

On the basis of these ideas, the observed volume of a sample can be written as the sum of the volume occupied by the molecules (subscript 0) and the free volume (subscript f). Acknowledging that each of these is a function of temperature, we write

$$V(T) = V_0(T) + V_f(T) \tag{4.54}$$

The variation in the occupied volume with temperature arises from changes in the amplitude of molecular vibrations with changing T, a variation which affects the excluded volume of the molecules. The free volume, on the other hand, may be viewed as the "elbow room" that molecules require to undergo rotation and translational motion. These modes of motion also increase with increasing temperature, so the associated volume is also expected to increase with T. With decreasing temperature a point may be visualized at which V_f has decreased to a critical value below which the translational and rotational modes of motion are essentially crowded out. In the last chapter we attributed the dramatic difference in moduli between the glassy state and the rubbery state at T_g to an effective freezing out of translational modes of motion of chain backbone segments. According to this picture, then, we identify T_g as the temperature at which V_f reaches this critical size. It will simplify our discussion to assume that below T_g the quantity V_f is constant with further decreases in T. This assumption imposes no constraints on the behavior of V_0 below T_g. On the basis of these ideas, we can write the following expressions for the volume of the sample:

1. Below T_g

$$V(T < T_g) = V_0(T = 0) + V_f(T = 0) + \left(\frac{dV_0}{dT}\right)_g T \tag{4.55}$$

2. At T_g

$$V(T_g) = V_0(T = 0) + V_f(T = 0) + \left(\frac{dV_0}{dT}\right)_g T_g \tag{4.56}$$

3.   Above T_g

$$V(T > T_g) = V(T_g) + \left(\frac{d(V_0 + V_f)}{dT} \right)_1 (T - T_g) \tag{4.57}$$

The subscripts g and l on the coefficient of volume variation with T indicate that these are determined for the glass and liquid states, respectively. These relationships are indicated geometrically in Fig. 4.15. The broken line in Fig. 4.15 shows the temperature variation of V_0, and the solid line, the actual volume. Presumably V_0 does not undergo any significant change in its temperature variation at T_g, but continues at least some distance beyond T_g with the slope $(dV_0/dT)_g$. At any temperature the difference between the two lines gives the free volume at that temperature. Figure 4.15 illustrates by shading how the free volume funnels down with decreasing temperature to a value too small to allow those types of molecular motion with greater space requirements. Each of the differences used to describe the second-order phase transition as given by Eq. (4.53) can be qualitatively traced to the sort of change implied by Fig. 4.15. Around T_g, for example, the coefficient of expansion of the liquid and glassy states are,

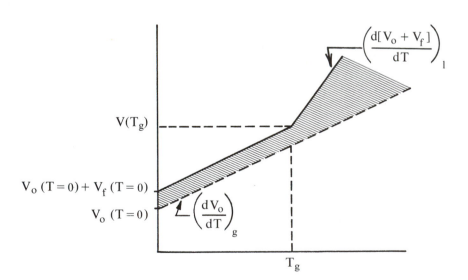

Figure 4.15   Geometrical representation of the temperature variation of the actual volume (solid line) and the "occupied" volume (broken line). The shaded difference indicates the free volume which decreases to a critical value at T_g.

respectively,

$$\alpha_1 = \frac{1}{V_g} \left(\frac{d(V_0 + V_f)}{dT} \right)_1 \quad \text{and} \quad \alpha_g = \frac{1}{V_g} \left(\frac{dV_0}{dT} \right)_g$$

therefore $\Delta\alpha$ is a measure of the opening up of V_f at T_g. The additional volume above T_g also accounts for the change in compressibility $\Delta\beta$, and the emergence of the associated modes of energy storage accounts for ΔC_p.

In discussing Fig. 4.1 we noted that the apparent location of T_g is dependent on the time allowed for the specific volume measurements. Volume contractions occur for a long time below T_g: The lower the temperature, the longer it takes to reach an equilibrium volume. It is the equilibrium volume which should be used in the representation summarized by Fig. 4.15. In actual practice, what is often done is to allow a convenient and standardized time between changing the temperature and reading the volume. Instead of directly tackling the rate of collapse of free volume, we shall approach this subject empirically, using a property which we have previously described in terms of free volume, namely, viscosity.

The following expression, known as the Dolittle equation, has been found to accurately describe the viscosity of low molecular weight compounds:

$$\eta = A \exp \left(\frac{B(V - V_f)}{V_f} \right) \tag{4.58}$$

where A and B are empirical constants. Dolittle showed that this expression described the viscosity of low molecular weight compounds better than the Arrhenius equation ($\eta = A' e^{-E^*/RT}$), which is generally used for this purpose. Although they relate η to different variables, both the Dolittle and Arrhenius equations have the same functional form. Just as the activation energy E^* measures the height of the energy barrier relative to thermal energy, so $V - V_f$ or V_0 is measured relative to V_f. If both V_0 and V_f are expressed per molecule, the exponent in the Dolittle equation measures the volume of a molecule (V_0 per molecule) relative to the volume of a hole (V_f per hole). When the molecule is large compared to the vacancy it must move into, the viscosity is high; when the molecule is small compared to the hole, the viscosity is low. This is the physical significance of the Dolittle equation, although we should remember that it is basically empirical in origin.

If we define f to be the fraction of volume contributed by the free volume V_f/V, then Eq. (4.58) can be written as

$$\eta = A \exp \left[B \left(\frac{1}{f} - 1 \right) \right] \tag{4.59}$$

Now the constant A can be eliminated from Eq. (4.59) by examining the ratio of two values of η with two different values of f (subscripts 1 and 2), remembering that B is a constant. In this case, we write

$$\frac{\eta_1}{\eta_2} = \exp\left\{ B\left[\left(\frac{1}{f_1} - 1 \right) - \left(\frac{1}{f_2} - 1 \right) \right] \right\} = \exp\left[B\left(\frac{1}{f_1} - \frac{1}{f_2} \right) \right] \quad (4.60)$$

or

$$\ln\left(\frac{\eta_1}{\eta_2} \right) = B\left(\frac{1}{f_1} - \frac{1}{f_2} \right) \quad (4.61)$$

which is analogous to the form into which the Arrhenius equation would be cast to evalute E*.

Next we assume that the state designated by the subscript 2 in Eq. (4.61) corresponds to T_g; we designate the fraction free volume at T_g by f_g. Likewise, state 1 is assumed to be above T_g, where f can be written

$$f = f_g + \alpha_f(T - T_g) \quad (4.62)$$

on the basis of Fig. 4.15. In this expression α_f is the coefficient of expansion of the free volume only. Substitution of f and f_g into Eq. (4.61) gives

$$\ln\left(\frac{\eta(T > T_g)}{\eta(T_g)} \right) = B\left(\frac{1}{f_g + \alpha_f(T - T_g)} - \frac{1}{f_g} \right) = \frac{-(B/f_g)(T - T_g)}{f_g/\alpha_f + (T - T_g)} \quad (4.63)$$

Things appear to have taken a strange turn: We started out discussing the free volume and have ended up with an equation which contains no volume at all! More specifically, we set out to examine the rate at which the free volume collapses at T_g. A final development of Eq. (4.63) will produce the desired result.

In Chap. 9 we shall examine the flow of a solution through a capillary tube. The rate of volume delivery in that case is given by Poiseuille's law [Eq. (9.29)], which states that the time required for a constant volume of liquid to drain out of the capillary is proportional to η/ρ. Accordingly, the viscosity is proportional to the product ρt, and when the delivery times for two liquids are compared in the same capillary,

$$\frac{\eta_1}{\eta_2} = \frac{\rho_1 t_1}{\rho_2 t_2} \quad (4.64)$$

This equation is the basis for viscosity determination by measuring flow times through a capillary. It can also be used to describe a single liquid at two different temperatures, as required for Eq. (4.63). Combining Eqs. (4.63) and (4.64) yields

$$\ln\left(\frac{t(T > T_g)}{t(T_g)}\right) = \frac{-(B/f_g)(T - T_g)}{f_g/\alpha_f + (T - T_g)} - \ln\left(\frac{\rho(T > T_g)}{\rho(T_g)}\right) \tag{4.65}$$

which is approximated by

$$\ln\left(\frac{t(T > T_g)}{t(T_g)}\right) \cong \frac{-(B/f_g)(T - T_g)}{f_g/\alpha_f + (T - T_g)} \tag{4.66}$$

since the variation of ρ with temperature is much less than the variation of η. To simplify the notation, we attach the subscript g to both T and t at the glass transition and use no subscript on either when the temperature exceeds T_g:

$$\ln t - \ln t_g = \frac{-C_1{'}(T - T_g)}{C_2 + (T - T_g)} \tag{4.67}$$

where the constants $C_1{'}$ and C_2 replace other clusters of constants. This is an important equation and we shall return to it in the next section. For the present time it is sufficient to note the following:

1. The left-hand side of the equation relates the times required for a specific displacement to occur at T_g and at a distance $T - T_g$ above T_g.
2. The right-hand side of the equation expresses this difference as a function of the temperature difference from T_g.
3. Since $T > T_g$, the negative sign assures that $\ln t_g > \ln t$; that is, the process takes longer at T_g.
4. As $T \to T_g$, the difference between the two times diminishes, dropping to zero when $T = T_g$.

These results make sense physically; what we must remember is that they follow from an analysis of the vanishing free volume.

Several additional observations are in order concerning the use of the Dolittle equation:

1. Although Eq. (4.58) describes the variation of viscosity over a wide range of conditions quite well, it tends to break down toward lower temperatures, precisely where we wish to apply it.
2. Although it is easy to discuss free volume, it is necessary to come up with a numerical value for this quantity in order to test these concepts. There is

some controversy as to which of several different methods of computation gives the best value for V_f and therefore V_0.

3. Although Eq. (4.67) is derived specifically for conditions of $T > T_g$, we shall apply it both above and below T_g in the next section.

4. It is possible to derive an expression equivalent to Eq. (4.67) starting from entropy rather than free volume concepts. We have emphasized the latter approach, since it is easier to visualize and hence to use for qualitative predictions about T_g.

5. Completely aside from the theories which attempt to explain it, the empirical usefulness of Eq. (4.67) is beyond doubt. We shall examine this in detail in the next section.

The collapse of the free volume below a critical size for molecular motion is the perspective on the glass transition that we have taken in this section. This concept is useful to understand the effects of polymer structure on T_g. Table 4.4 lists T_g values for many of the same polymers that were presented in Table 2.3, this time ranked in order of decreasing T_g values. Also listed are values of the clusters of constants shown in Eq. (4.66). Except for polystyrene and poly-(dimethyl siloxane), which fall out of place, the broken line divides the hydrocarbons from molecules containing more polar groups of atoms. The following

Table 4.4 Some Polymers Ranked in Order of Decreasing T_g Values, Along with Numerical Values for Quantities Appearing in the Dolittle Equation

Polymer	T_g (K)	B/f_g	f_g/α (K)	f_g
Molecules containing polar groups				
Methyl methacrylate	388	76.9	76.5	0.013
Styrene	373	31.3	50.8	0.032
Ethyl methacrylate	335	40.0	67.6	0.025
n-Butyl methacrylate	300	38.5	100	0.026
Methyl acrylate	276	41.7	45.3	0.024
n-Hexyl methacrylate	268	40.0	132	0.025
n-Octyl methacrylate	253	37.0	108	0.027
Urethane rubber	238	35.7	32.9	0.028
Hydrocarbons				
1-Hexene	218	50.0	20.6	0.020
Butyl rubber	205	38.5	108	0.026
Polyisobutylene	205	38.5	104	0.026
1,4-Butadiene	172	25.6	60.9	0.039
Dimethyl siloxane	150	14.1	68.9	0.071

Source: Data from J. D. Ferry, *Viscoelastic Properties of Polymers*, John Wiley and Sons, New York, 1980.

list of generalizations is based in part on Table 4.4 and on various additional observations:

1. Hydrocarbons without bulky side groups are held together by London forces, the weakest of intermolecular attractions. This means that the free volume tends to be large for these compounds, so a relatively large amount of cooling is necessary before the free volume collapses. Thus T_g is low for these compounds.

2. The effect of a bulky substituent like a phenyl group on the hydrocarbon chain apparently decreases chain flexibility sufficiently to allow more intimate alignment between molecules, less free volume, and therefore a high value for T_g.

3. In the methacrylate homologous series, the effect of side-chain bulkiness is just the opposite. In this case, however, the pendant groups are flexible and offer less of an obstacle to free rotation than the phenyl group in polystyrene. As chain bulk increases, molecules are wedged apart by these substituents, free volume increases, and T_g decreases.

4. The presence of two substituents rather than one (compare methyl methacrylate with methyl acrylate) increases chain stiffness, decreases V_f, and increases T_g.

5. An extra amount of free volume is associated with chain ends, which are capable of wagging in a way that is not possible in the middle of a chain. Accordingly, as molecular weight decreases, V_f increases, which, in turn, decreases T_g. The following expression has been found to describe this molecular weight dependence:

$$T_g = T_g^\infty - \frac{K}{M} \qquad (4.68)$$

where T_g^∞ is the glass transition temperature of a polymer of infinite molecular weight and K is a constant.

6. The effect of branching is to increase the number of chain ends and, therefore, free volume, which decreases T_g. Conversely, crosslinking ties together separate molecules, decreases the number of loose ends, and raises T_g.

7. Copolymers show different effects on T_g, depending on the microstructure of the molecule. Block copolymers show two values of T_g, each being the value appropriate to the respective homopolymer. Random copolymers show T_g values which are intermediate between the values of the homopolymers. It has been found that the following averaging procedure describes this intermediate T_g value:

$$\frac{1}{T_g} = w_1 \left(\frac{1}{T_g}\right)_1 + w_2 \left(\frac{1}{T_g}\right)_2 \qquad (4.69)$$

where the w's are weight fractions of components 1 and 2.

8.   The presence of low molecular weight molecules between polymer chains
     has the effect of pushing the chains apart, effectively increasing V_f. The
     resulting decrease in T_g can significantly widen the range of usefulness of a
     polymer, so the importance of such diluents—called plasticizers in view of
     this effect—in technology is very great.

4.10   Time–Temperature Equivalency

In this section we resume our examination of the equivalency of time and
temperature in the determination of the mechanical properties of polymers. In
the last chapter we had several occasions to mention this equivalency, but never
developed it in detail. In examining this, we shall not only acquire some practical
knowledge for the collection and representation of experimental data, but also
shall gain additional insight into the free-volume aspect of the glass transition.

In describing the various mechanical properties of polymers in the last
chapter, we took the attitude that we could make measurements on any time
scale we chose, however long or short, and that such measurements were made in
isothermal experiments. Most of the experimental results presented in Chap. 3
are representations of this sort. In that chapter we remarked several times that
these figures were actually the result of reductions of data collected at different
temperatures. Now let us discuss this technique; our perspective, however, will
be from the opposite direction: taking an isothermal plot apart.

Because so many orders of magnitude of time are spanned in comprehensive
studies of mechanical behavior, plots of experimental results like that shown
in, say, Fig. 3.9, require a long horizontal axis. As a graphical expediency, such
a figure could be cut in half at the midpoint. The second half of the data points
could then be shifted to the left and plotted beneath the larger, short-duration
moduli. Since a logarithmic time scale is used in Fig. 3.9, this compressed graph
is simply described by noting that for the second segment of the divided curve
the log t values should all be increased by log a, the amount by which the displaced section has been shifted. An even greater telescoping of the data is
accomplished by dividing the original full curve into three or more segments and
then shifting successive sections onto a single, shortened coordinate system.
Figure 4.16 shows schematically the results of such a procedure when the parent
curve is divided into three sections. Figure 4.16a shows the following decomposition of the full curve:

1.   The segment between t_2 and t_3 is shifted to the left by an amount log a
     so that log t_2 is displaced to overlap log t_1.
2.   The segment between t_3 and t_4 is shifted to the left by an amount log a$'$
     so that log t_3 is displaced to log t_1 also.

Figure 4.16b shows the following for the "stacked" segments:

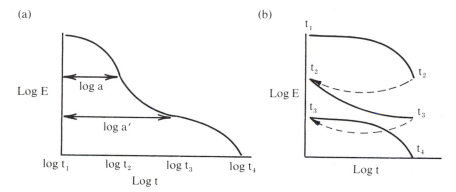

Figure 4.16 Schematic illustration showing how an experimental plot of modulus against log t (a) can be telescoped (b) by shifting successvie segments by an amount designated as log a.

3. t_1, t_2, and t_3 all overlap on the left edge of the log t scale, and t_2, t_3, and t_4 all overlap on the right.

4. In this case a narrower range of log t values is needed to represent the data and this can be narrowed still further by dividing the original curve into an even larger number of segments.

It is apparent that the curve in Fig. 4.16a could be reconstructed from the version in Fig. 4.16b with a knowledge of the displacements, log a and log a'.

The foregoing points are all fairly obvious to science and engineering students who have drawn and studied numerous graphs. The point of this discussion is not merely the telescoping of data by graphical manipulation; instead, it is an empirical fact that measurements of moduli or compliances over a modest range of times or frequencies at a variety of different temperatures produce a set of curves resembling Fig. 4.16b. If the full curve of Fig. 4.16a can be reconstructed from the pieces in Fig. 4.16b, then it must be possible to shift segments measured at different temperatures to produce a full curve by making the individual segments connect with one another. In such a procedure we have the following:

1. Measurements at one of the temperatures will be left unchanged. Data from other temperatures are shifted to bring them into line with the unmoved data.

2. The amount by which the individual isothermal results are shifted, log a_T for an experiment at temperature T, will be different for each T.

3. Shifts can be to either the right or the left; that is, log a_T can be either negative or positive.

4. The final composite curve is called the master curve at the temperature of the unmoved isotherm.

The isothermal curves of mechanical properties in Chap. 3 are actually master curves constructed on the basis of the principles described here. Note that the manipulations are formally similar to the superpositioning of isotherms for crystallization in Fig. 4.8b, except that the objective here is to connect rather than superimpose the segments. Figure 4.17 shows a set of stress relaxation moduli measured on polystyrene of molecular weight 1.83×10^5. These moduli were measured over a relatively narrow range of readily accessible times and over the range of temperatures shown in Fig. 4.17. We shall leave as an assignment the construction of a master curve from these data (Problem 10).

The procedure described above is an application of the time–temperature correspondence principle. By shifting a set of plots of modulus (or compliance) versus time (or frequency) at any temperature (subscript 1) along the log t axis, we obtain the value of that mechanical property at another time and temperature (subscript 2). Using the shear modulus as an example, the time–temperature correspondence principle states

$$G(t_1, T_1) = G(t_2, T_2) \tag{4.70}$$

at the point of overlap. We drop the subscript on time in the master curve; therefore Eq. (4.70) becomes

$$G\left(\frac{t}{a_{T_1}}, T_1\right) = G\left(\frac{t}{a_{T_2}}, T_2\right) \tag{4.71}$$

Note that subtracting an amount $\log a_T$ from the coordinate values along the abscissa is equivalent to dividing each of the t's by the appropriate a_T value. This means that times are represented by the reduced variable t/a_T in which t is expressed as a multiple or fraction of a_T which is called the shift factor. The temperature at which the master curve is constructed is an arbitrary choice, although the glass transition temperature is widely used. When some value other than T_g is used as a reference temperature, we shall designate it by the symbol T_0.

We shall presently examine the physical significance of the shift factors, since they quantitatively embody the time–temperature equivalence principle. For the present, however, we shall regard these as purely empirical parameters. The following list enumerates some pertinent properties of a_T:

1. At T_0, $a_T = 1$; no shift of time scale is required.
2. For $T > T_0$, $a_T < 1$, which corresponds to $\log a_T < 0$. These values are used when the curves are shifted to the right. The effect on the mechanical properties is equivalent to expanding the time scale, since we express time as t/a_T.
3. For $T < T_0$, $a_T > 1$, which corresponds to $\log a_T > 0$. These values are used

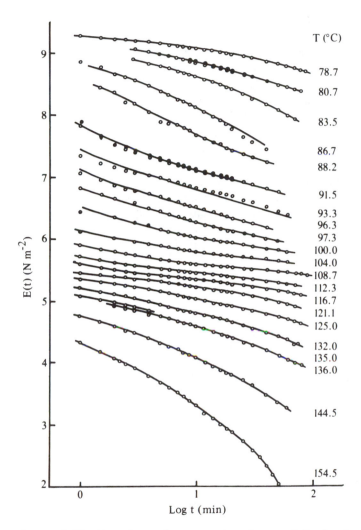

Figure 4.17 Experimental stress relaxation moduli of polystyrene measured over about two orders of magnitude in time at the temperatures indicated. [Reprinted with permission from H. Fujita and K. Ninomiya, *J. Polym. Sci.* 24:233 (1957).]

when curves are shifted to the left. The effect on the mechanical properties is equivalent to compressing the time scale.

All amorphous polymers show remarkably similar results when values of log a_T are plotted versus $T - T_0$. Manipulation of these data shows that the empirically determined shift factors can be fitted by the expression

$$\log a_T = \frac{-C_1(T - T_0)}{C_2 + (T - T_0)} \tag{4.72}$$

where C_1 and C_2 are constants. This equation is known as the WLF equation, for Williams, Landel, and Ferry, the researchers who discovered it. Note that when $T_0 = T_g$, the right-hand side of Eq. (4.72) is identical to Eq. (4.67). It is no wonder that this is the case; Eq. (4.67) is an expression for the difference in ln t values for a fixed amount of deformation at two different temperatures. This is precisely what $\log a_T$ does also. The only difference between Eqs. (4.67) and (4.72) is the use of natural logarithms in the former and logarithms to the base 10 in the latter. Accordingly, the constants in the numerator of the two equations are related as follows: $C_1 = C_1'/2.303$; C_2 is the same in both. Since it is easier to study mechanical properties and carry out time–temperature superpositioning than to study volume contraction, the constants listed in Table 4.4 were obtained from the former.

Rearrangement of Eq. (4.72) suggests a method for evaluating the constants C_1 and C_2 from experimental results. We write

$$\frac{T - T_g}{\log a_T} = -\frac{C_2}{C_1} - \frac{1}{C_1}(T - T_g) \tag{4.73}$$

which is an equation of the type y = mx + b if we identify y as $(T - T_g)/(\log a_T)$ and x as $T - T_g$. Then the slope m equals $-1/C_1$, and the intercept b equals $-C_2/C_1$. Thus if the empirical shift factors for different temperatures are analyzed according to Eq. (4.73), the slope and intercept can be used to evaluate C_1 and C_2:

$$C_1 = -\frac{1}{\text{slope}} \tag{4.74}$$

and

$$C_2 = -\frac{\text{intercept}}{\text{slope}} \tag{4.75}$$

The parameters f_g and α_f can be evaluated from C_1 and C_2 by virtue of the definition of the latter:

$$f_g = \frac{B}{2.303 C_1} \tag{4.76}$$

and

$$\alpha_f = \frac{B}{2.303 C_1 C_2} \tag{4.77}$$

The factor B in Eq. (4.58) is close to unity; hence the reciprocals of the

values B/f_g in Table 4.4 give the fraction of free volume in the various polymers at the glass transition temperature. According to this interpretation of the shift factors, the free volume of many polymers is remarkably constant and close to 2.5%. Despite the coherent picture presented here, there is controversy about the validity of this uniform value for V_f. It has been argued that the fraction of free volume at T_g should be up to five times larger than the WLF value.

We conclude this section with an example illustrating the application of the WLF equation using the data in Fig. 4.17.

Example 4.7

Use values of the constants for polystyrene from Table 4.4 to calculate the shift factors needed to connect those segments in Fig. 4.17 measured at 96.3 and 108.7°C, with the isotherm measured at $T_g = 100.0$°C. Are the values reasonable?

Solution

For polystyrene, $B/f_g = 31.3$ and $f_g/\alpha = 50.8$ from Table 4.4. Using Eq. (4.66), we obtain the following. At $t = 96.3$°C, or $T - T_g = -3.7$°C,

$$\ln a_T = \frac{-31.3(-3.7)}{50.8 - 3.7} = 2.46, \quad \log a_T = 1.07, \quad a_T = 11.8$$

At $T = 108.7$°C, or $T - T_g = 8.7$°C,

$$\ln a_T = \frac{-31.3(8.7)}{50.8 + 8.7} = -4.58, \quad \log a_T = -1.99, \quad a_T = 0.0103$$

At 96.3°C, $\log E \cong 6.5$ when $\log t = 1$. Subtracting the calculated value of $\log a_T$ from this gives a reduced value for $\log t \cong 0$. This is the time at which $\log E \cong 6.5$ at 100°C. Stated differently, $E \cong 10^{6.5}$ N m^{-2} after 10 min at 97.3°C and after 1 min at 100°C. Therefore $t/a_T = 10/11.8$ min = 0.85 min $\cong 1$ min, showing this to be a reasonable value.

At 108.7°C, $\log E \cong 5.9$ when $\log t = 0$. Subtracting the calculated value of $\log a_T$ from this gives a reduced value for $\log t \cong 2$. On the other hand, the time at which $\log E \cong 5.9$ at 100°C is about 20 min, or $\log t \cong 1.3$. This value is less satisfactory. In terms of reduced variables, $E \cong 10^{5.9}$ N m^{-2} after 1 min at 108.7°C and after 1/0.0103 min $\cong 100$ min at 100°C. This last value should be closer to 20 min.

•

The limited success of the previous example should not be a cause for alarm, since the constants listed in Table 4.4 are "best" values and will not necessarily

work equally well for all observations. The calculated values are certainly of the right magnitude, and we must bear in mind the difficulties involved in estimating values from a logarithmic scale. In addition, Eq. (4.66) was used in this example. This is an approximation based on the assumption that ρ does not change appreciably over the range of temperatures under examination. In precise work, the moduli being matched up should be corrected to the reference temperature by multiplying individual values by the ratio $T_0\rho_0/T\rho$ (subscript 0, reference state; no subscript, result being shifted) before attempting the super-positioning. This ratio is clearly very close to unity in the example.

4.11 Polymer Processing Technology

We conclude this chapter and wrap up the last three chapters with a few remarks about the application of the ideas contained herein to polymer technology. Chapters 2–4 have been concerned with various aspects of the mechanical states of polymers. The opinion was expressed in Chap. 1 that if polymers did not possess the mechanical properties they have, this whole class of compounds might be relegated to the category of laboratory curiosities. On the basis of any number of criteria—the number of scientists employed, the number of industries involved, the number of publications released, the number of patents issued—polymer science proves to be very viable indeed.

The history of individual polymers as well as the history of the discipline itself is one of symbiosis between pure science and applied technology. Haward (Ref. 3) makes this point clear in his discussion of polystyrene. When the latter emerged on the scene in the late 1930s to early 1940s, it was taken up by academic researchers but largely rejected by those in industry. It was a perfect polymer for fundamental, academic-type research. It could be prepared repro-ducibly and, because of its solubility, fractionated by molecular weight and characterized by solution properties. By contrast, its high T_g value made it brittle under many potential use conditions, and hence initially unattractive for applications. Eventually what Haward describes as an "irresistible combination of four favorable properties" led to industrial research into polystyrene applica-tions. The "favorable four" were low cost, good appearance, rigidity, and processability. Through the use of plasticizers, the development of dispersions of rubber in polystyrene (called high-impact polystyrene), and the development of copolymers, the initial objections to brittleness were overcome. In 1981 the styrene industry produced about 2.2×10^6 metric tons of monomer for use in homo- and copolymers.

From the point of view of technology, it is convenient to classify polymers as thermosetting and thermoplastic. The former "set" by chemical crosslinks introduced during fabrication and hence do not change appreciably in their deformability with changes in temperature. Thermoplastics, on the other hand, soften and/or melt on heating and can therefore be altered in shape by heating

or frozen in shape by cooling. The choice between these two classes is governed by the permanence or elasticity desired in the ultimate application of the material. We saw in Chap. 3 that the presence of crystalline regions and, to a lesser extent, entanglements can give thermoplastic materials similar mechanical properties to elastomers, at least under some conditions.

For pedagogical reasons we have attempted to concentrate on essentially pure polymers in these chapters. From the point of view of technology this creates a fairly narrow view of those systems where polymers find application. Dispersed particles are introduced into polymers as fillers in many applications. These are essentially two-phase systems and include such examples as colloidal carbon dispersed in rubber, pigments dispersed in coatings, and rubber dispersed in high-impact polystyrene. By contrast, plasticizers may be considered to form true, homogeneous solutions with polymers. The plasticizer is often present in relatively small amounts, so, from the point of view of the polymer as solute, these systems are often highly concentrated solutions. Both homogeneous and heterogeneous mixtures are probably more widely used in polymer technology than are pure bulk polymers. It is unfortunate that both the level and length of this book make it impossible to devote more space to these topics. In Chaps. 8–10 we shall return to the topic of polymer solutions, although the emphasis will be placed on systems which are dilute with respect to polymer.

One area of technology that has been the source and motivation for much polymer research, particularly with respect to mechanical properties, is the fiber industry. While many of the basic principles of the last three chapters find application in fiber technology, it must also be acknowledged at the outset that there is a great deal of empirical know-how in addition to fundamental science involved in the production of commercial fibers. The most widely used fiber-forming processes are called melt and wet (or solution) spinning. In both, the polymer is extruded from the small holes of a device called a spinneret to produce a crude fiber. In melt spinning molten polymer is used, while in wet spinning a solution is the source of the polymer. In melt spinning the polymer solidifies by air cooling after extrusion from the spinneret; in wet spinning the polymer solution is emitted into a nonsolvent which causes the precipitation of the polymer. In both cases the polymeric fibers are stretched before being wound onto a spool. Melt spinning is used with thermally stable polymers such as polyolefins, while wet spinning is useful for thermally unstable or sensitive polymers such as poly(acrylonitrile). High temperatures are required for melt spinning, and the cost of the required energy is a disadvantage of this method. The need for solvent recovery is the disadvantage of wet spinning.

When we speak of the solidification of the extruded polymer, we use the term in the broadest sense: It includes crystallization, vitrification, or both. The extent of the drawing of the fibers and the rate and temperature of the drawing affect the mechanical properties of the fiber produced. This conclusion should be evident from a variety of ideas presented in the last three chapters:

1.   At the molecular level, stretching extends polymer coils and facilitates crystallization.
2.   The nucleation, growth, and morphology of crystals are influenced by both temperature and stress.
3.   The presence of spherulites or smaller crystallites is comparable to cross-linking and affects not only the moduli and compliances, but also the ultimate properties such as yield strength and ultimate elongation.
4.   Both thermodynamic and kinetic aspects of mixed systems (e.g., the precipitation step in wet spinning) involve the properties of the other components (solvent and nonsolvent in wet spinning) as well as the polymer.

Two other processing techniques, namely, injection molding and calendering, should also be mentioned. In the first of these a heat-softened polymer is injected into a cavity or die, where it is allowed to cool and set. From the point of view of process efficiency, it is desirable to cool the product as rapidly as possible so that the mold can be refilled. From the point of view of the product, the rate of cooling determines the degree of crystallization. This influences the density, which, in turn, affects the linear dimensions of the finished product. In some instances, meeting tolerances specified for a product may be at cross-purposes with efficient equipment utilization. Calendering is the process of rolling a polymeric sample to produce films which can be used as such or layered on a substrate as a coating. As usual, time and temperature are important variables in this process. In addition, either uniaxial or biaxial stretching can be introduced to influence the extent of chain ordering and the attendant consequences of the latter.

4.12 Preview of Things to Come

In the last three chapters we have examined the mechanical properties of bulk polymers. Although the structure of individual molecules has not been our primary concern, we have sought to understand the influence of molecular properties on the mechanical behavior of polymeric materials. We have seen, for example, how the viscosity of a liquid polymer depends on the substituents along the chain backbone, how the elasticity depends on crosslinking, and how the crystallinity depends on the stereoregularity of the polymer. In the preceding chapters we took the existence of these polymers for granted and focused attention on their bulk behavior. In the next three chapters these priorities are reversed: Our main concern is some of the reactions which produce polymers and the structures of the products formed.

In the next group of chapters we shall discuss condensation or step-growth polymers and polymerizations in Chap. 5, addition or chain-growth polymers and polymerizations in Chap. 6, and copolymers and stereoregular polymers in Chap. 7. It should not be inferred from this that these are the only classes of polymers and polymerization reactions. Topics such as ring-opening polymeri-

zation, anionic polymerization, cationic polymerization, and diene polymerization could all be added to the list above and expanded into full chapters. As it is, the first three are relegated to sections within other chapters; the last is neglected entirely. The emphasis given to various topics was decided on the basis of the concepts that their inclusion would illustrate. It does not mean that the topics we have shortchanged are unimportant.

The mechanism of the reactions we shall discuss provides the basis for their classification. Thus the step-growth mechanism applies when chains form one link at a time, while the chain-growth mechanism applies when an avalanche of links form in succession. Even though the mechanisms are quite different in this regard, there are certain common features to their analysis. For example, regardless of the mechanism under consideration, the size of the molecule doing the reacting can be neglected. In both mechanisms, assembling a giant molecule from low molecular weight repeat units is a random process: Statistics play an important role in describing both.

In contrast with the last three chapters, descriptive reaction chemistry plays a large part in the next three chapters. The "steps" in step-growth polymerization are esterification, amidation, and other such reactions; the "chains" in the chain-growth mechanism involve free radicals, ions, and coordinated species as active centers. Reactive functional groups are required for step growth, and reactive double bonds for the chain mechanism. It is not just the chemistry of the monomers that distinguishes between the different polymerizations. Catalysts are also required for virtually all polymerizations, and these determine the nature of the reaction intermediate and thus dictate the mechanism as well as influence the rate of the reaction. Depending on the nature of the catalyst used, polystyrene can be polymerized via free-radical, anion, cation, or coordination species as intermediates. As expected, there are differences in the rates of these various possible reactions. In addition, the products also differ in molecular weight, molecular weight distribution, and, perhaps, tacticity. Generally these latter properties are as important or more important than the rate itself.

Specifically chemical considerations are especially evident in Chap. 7, where copolymers and stereoregular polymers are discussed. Since two monomers are required for the formation of a copolymers, the differences in their reactivity affects both the composition of the product and the distribution of components in it. Likewise, the catalysts that produce stereoregularity are highly specific, highly reactive, and poorly understood chemical reagents.

Problems

1. Illers and Hendus† measured the melting points of polyethylene crystals whose thickness was varied by controlling the conditions of crystallization

---

† K. H. Illers and H. Hendus, *Makromol. Chem.* 113:1 (1968).

and which was measured by x-ray diffraction. The following results were obtained:

T_m (°C)	139.4	137.5	136.0	134.9	131.9	127.9	117.0
l (Å)	1750	758	481	392	258	177	100

Prepare a plot of T_m versus l^{-1}, and from the intercept and slope respectively, evaluate T_m^∞ and γ from these data. Compare the values obtained with quantities given in Table 4.1 and Example 4.2 (use the latter as a source for ΔH_V^∞).

2. From the data in Fig. 4.8b, estimate the shift factors required to displace the data at $\theta = 0.5$ (consider only this point) so that all runs superimpose on the experiment conducted at 128°C at $\theta = 0.5$. Either a ruler or proportional dividers can be used to measure displacements. Criticize or defend the following proposition: Whether a buffered aqueous solution of H_2O_2 and I^-, containing small amounts of $S_2O_3^{2-}$ and starch, appears blue or colorless depends on both the time and the temperature. This standard general chemistry experiment could be used to demonstrate the equivalency of time and temperature. The pertinent reactions for the "iodine clock" are

$$H_2O_2 + 2I^- + 2H^+ \xrightarrow{\text{slow}} I_2 + 2H_2O$$

$$I_2 + 2S_2O_3^{2-} \xrightarrow{\text{fast}} 2I^- + S_4O_6^{2-}$$

$$I_2 + \text{starch} \xrightarrow{\text{fast}} \text{blue complex}$$

3. The crystallization of poly(ethylene terephthalate) at different temperatures after prior fusion at 294°C has been observed[†] to follow the Avrami equation with the following parameters applying at the indicated temperatures:

T (°C)	m	K (min)
110	2	3.49×10^{-4}
180	3	1.35
240	4	5.05×10^{-8}

Calculate the time required for θ to reach values $0.1, 0.2, \ldots, 0.9$ for each of these situations. Graph θ versus t using the results calculated at 110 and 240°C, plotting both in the same figure. Because of the much larger K at 180°C, the crystallization occurs much faster at this temperature than at

[†]F. D. Hartley, F. W. Lord, and L. B. Morgan, *Philos. Trans. R. Soc. London* 247A:23 (1954).

either 110 or 240°C. Multiply each of the times calculated at 180°C by the arbitrary constant 60 and plot the data thus shifted on the same coordinates as the other curves. What generalization appears concerning the relative slopes at $\theta = 0.5$?

4. Poly(ethylene terephthalate) was crystallized at 110°C and the densities were measured† after the indicated time of crystallization:

t (min)	ρ (g cm^{-3})	t (min)	ρ (g cm^{-3})
0	1.3395	35	1.3578
5	1.3400	40	1.3608
10	1.3428	45	1.3625
15	1.3438	50	1.3655
20	1.3443	60	1.3675
25	1.3489	70	1.3685
30	1.3548	80	1.3693

Using density as the property measured to determine crystallinity, evaluate θ as a function of time for these data. By an appropriate graphical analysis, determine the Avrami exponent (in doing this, ignore values of $\theta < 0.15$, since errors get out of hand in this region). Calculate (rather than graphically evaluate) the value of K consistent with your analysis.

5. The crystallization rate of isotactic polypropylene ($M_w = 181,000$, $T_m = 172°C$) was studied under various patterns of temperature change. Solids were melted at T_f, held at T_f for 1 hr, and then crystallized at T_c. The following Avrami exponents were observed.‡

	Avrami exponent		
T_f (°C)	$T_c = 150°C$	$T_c = 155°C$	$T_c = 160°C$
190	–	3.1	3.5
210	2.9	3.3	4.1
220	3.1	3.8	–
230	3.1	4.0	–

On the basis of these observations, criticize or defend the following propositions:

(a) When both T_f and T_c are low, the Avrami exponents are consistent with three-dimensional growth on contact with sporadic nucleation.

† A. Keller, G. R. Lester, and L. B. Morgan, *Philos. Trans. R. Soc. London* 247A:1 (1954).
‡ P. Parrini and G. Corrieri, *Makromol. Chem.* 62:83 (1963).

(b) The change in m can be interpreted as arising from a change in either the growth geometry or nucleation situation. That is, the change in m for $[T_f$ and T_c low$] \rightarrow [T_f$ and T_c high$]$ could arise from either the change spherical $\rightarrow$ disk geometry or the change sporadic $\rightarrow$ simultaneous nucleation.

(c) The changes in m are consistent with the idea that under some conditions nuclei from the original solid survive the period in the melt and nucleate the recrystallization.

6. The polymers listed below are all known to form unit cells in which all of the angles are $90°$. Use this fact plus the data† given to complete the following table:

Polymer	M_0	Unit cell dimensions (Å)			Number of repeat units per cell	Density (g cm^{-3})
		a	b	c		
Polystyrene	104.1	—— = ——		6.63	18	1.126
Polyisobutene	56.1	6.94	11.96	——	16	0.937
Poly(vinyl chloride)	62.5	10.11	5.27	5.12	4	——
Nylon 8	——	4.9	4.9	~22	2	1.038
Poly(methyl methacrylate)	100.1	21.08	12.17	10.55	——	1.23

7. Chemical evidence for chain folding in polyethylene crystals is obtained by etching polymer crystals with fuming nitric acid, which cleaves the chain at the fold surface. The resulting chain fragments are separated chromatographically and their molecular weights determined by osmometry. The folded chain is pictured as crossing through the crystal, emerging and folding back, then reentering and recrossing the crystal, and so on. According to this picture, the shortest chain showing up in the chromatograms should equal the crystal thickness in length. The second shortest chain exceeds twice this value by some amount which measures the length of the loop made by the chain outside the crystal. Molecular weights for the two shortest chains observed in an experiment‡ of this sort were 1260 and 2530. Since the cleaved chains end in nitro and carboxyl groups, 60 should be subtracted from each of these molecular weights to give the polyethylene chain weight. Calculate the degree of polymerization of each molecule and the chain length (use the length of the unit cell along the chain axis, 2.53 Å, as the distance per repeat unit). Compare the latter with the crystal thickness determined by x-ray diffraction, 105 Å. What does the ratio of chain lengths for the first and second peaks suggest about the tightness of folding?

---

† Data from Ref. 2.

‡ T. Williams, D. J. Blundell, A. Keller, and I. M. Ward, *J. Polym. Sci.* A-2,6:1613 (1968).

8. Hirai and Eyring[†] assembled the following data from diverse sources (scarcely any two pieces of data were measured in the same laboratory, much less on the same sample):

	V_{sp} $(cm^3 \, g^{-1})$	ΔC_p $(erg \, K^{-1} \, g^{-1})$	$\Delta \alpha$ (K^{-1})	$\Delta \beta$ $(cm^2 \, dyne^{-1})$
Rubber	1.1	5×10^6	4.0×10^{-4}	1×10^{-11}
Polystyrene	1.0	7.7×10^6	1.75×10^{-4}	3×10^{-12}
Polyisobutylene	1.1	4.0×10^6	4.5×10^{-4}	3×10^{-11}

Use these data to evaluate T_g, assuming that the latter is a true second-order transition. Compare your results with the values in Table 4.4 and comment on the agreement or lack thereof.

9. The time–temperature superpositioning principle was applied [‡] to the maximum in dielectric loss factors measured on poly(vinyl acetate). Data collected at different temperatures were shifted to match at $T_g = 28°C$. The shift factors for the frequency (in hertz) at the maximum were found to obey the WLF equation in the following form: $\log \omega + 6.9 = [19.6(T - 28)]/[42 + (T - 28)]$. Estimate the fractional free volume at T_g and α for the free volume from these data. Recalling from Chap. 3 that the loss factor for the mechanical properties occurs at $\omega \tau = 1$, estimate the relaxation time for poly(vinyl acetate) at 40 and 28.5°C.

10. Construct a master curve at 135°C from the data in Fig. 4.17 and determine the values of $\log a_T$ at each temperature. A relatively easy way to do this (especially without digital data) is to trace or photocopy Fig. 4.17 onto a transparent film. Place this transparency over a piece of paper and transfer the segment corresponding to T_0 to this paper. This is the beginning of the master curve. Slide the overlay to the right (or left) to connect segments measured at temperatures directly above (or below) T_0 with the segment at T_0. Now transfer (slipping a piece of carbon paper under the transparency is convenient) these segments to the paper on which the master curve is to be built up. On each segment (including T_0) place a mark at a convenient vertical reference, say, $\log t = 0$. Continue extending the master curve by this procedure, being careful to maintain a constant base line with perpendicular verticals. A ruler can be used to measure the displacement of the $\log t = 0$ mark at each temperature from $\log t = 0$ at T_0. These distances can then be converted to $\log a_T$ values by calibration of the horizontal displacements with the original figure.

11. Suppose you wanted to estimate the viscosity of a polystyrene sample at 125°C using the Debye viscosity equation, but the only available value

†N. Hirai and H. Eyring, *J. Polym. Sci.* 37:51 (1959).

‡ S. Matsuoka, G. E. Johnson, H. E. Bair, and E. W. Angerson, *Polym. Prepr.* 22:280 (1981).

of ζ is at $T_g + 100°C = 200°C$ (see Table 2.3). Comparing Eq. (4.71) with the Rouse theory [Eq. (3.90)] suggests that a_T and τ_p play comparable roles in each and that these two are proportional to each other: $a_T = Kn^2\zeta$ $l_0^2/6\pi^2 p^2 kT$ [Eq. (3.99)], with K the proportionality constant. According to this idea, the temperature dependence of ζ can be estimated from a knowledge of the shift factors: $a_{T_1}/a_{T_2} = (\zeta_{T_1}/T_1)/(\zeta_{T_2}/T_2)$. This neglects any temperature dependence of l_0. Use the constants for polystyrene from Table 4.4 to calculate log a_T at 125 and 200°C. From these values and ζ_{200}, estimate ζ_{125}. Use the Debye viscosity equation to estimate η_{125} for polystyrene, for which n = 100 (i.e., no entanglements). If the actual viscosity were 2.54 $\times$ 10^4 kg m^{-1} sec^{-1}, how much of an error would this imply for ζ_{125}? (Use $\rho = 1$ g cm^{-3} and $l_0 = 7.4 \times 10^{-10}$ m throughout.)

12. Williams and Ferry† measured the dynamic compliance of poly(methyl acrylate) at a number of temperatures. Curves measured at various temperatures were shifted to construct a master curve at 25°C, and the following shift factors were obtained:

T (°C)	log a_T	T (°C)	log a_T
25.00	0	54.90	-3.88
29.75	-0.98	59.95	-4.26
34.85	-1.80	64.70	-4.58
39.70	-2.42	69.50	-4.88
44.90	-3.00	80.35	-5.42
49.95	-3.47	89.15	-5.72

Test whether these data obey the WLF equation; if so, evaluate the constands C_1 and C_2. Note that $T_0 \neq T_g = 3°C$ in these data.

Bibliography

1. Aklonis, J. J., MacKnight, W. J., and Shen, M., *Introduction to Polymer Viscoelasticity*, Wiley-Interscience, New York, 1972.
2. Geil, P. H., *Polymer Single Crystals*, Interscience, New York, 1963.
3. Haward, R. N. (Ed.), *The Physics of Glassy Polymers*, Wiley, New York, 1973.
4. MaGill, J. H., *Treatise on Material Science and Technology*, Vol. 10, J. M. Schultz (Ed.), Academic, New York, 1977.
5. Mandelkern, L., *Crystallization of Polymers*, McGraw-Hill, New York, 1964.

---

† M. L. Williams and J. D. Ferry, *J. Colloid Sci.* 9:474 (1954).

SOME IMPORTANT CLASSES OF POLYMERS
AND POLYMERIZATION REACTIONS

5

Condensation or Step-Growth Polymerization

I'd get the urge to go splurging on hose, nylons, a dozen of those!
Now, rich or poor, we're enduring instead woolens which itch!
Rayons that spread!
I'll be happy When The Nylons Bloom Again!

When The Nylons Bloom Again, words by
George Marion, Jr. music by Thomas "Fats" Waller

5.1 Introduction

In Sec. 1.4 we discussed the classification of polymers into the categories of addition or condensation. At that time we noted that these classifications could be based on the following:

1. The stoichiometry of the polymerization reaction (small molecule eliminated?).
2. The composition of the backbone of the polymer (atoms other than carbon present?).
3. The mechanism of the polymerization (stepwise or chain reaction?).

It is the third of these criteria that offers the most powerful insight into the nature of the polymerization process for this important class of materials. We shall frequently use the terms *step-growth* and *condensation polymers* as synonyms, although by the end of the chapter it will be apparent that step-growth polymerization encompasses a wider range of reactions and products than either criteria (1) or (2) above would indicate.

The chapter is organized in a spiral fashion. First, we examine how the degree of polymerization and its distribution vary with the progress of the polymerization reaction, with the latter defined both in terms of stoichiometry and time. In the first round, we consider these topics for simple reaction

mixtures, those in which the proportions of reactants agree exactly with the stoichiometry of the reactions. After this we consider two important classes of condensation or step-growth polymers: polyesters and polyamides. In the second round of the spiral we consider nonstoichiometric proportions of reactants and monomers which introduce branching and crosslinking into the products. We then discuss some formaldehyde-based polymers which serve as examples of cross-linked systems. We conclude the chapter with a description of some more-or-less exotic step-growth polymers and a digression into a third class of polymers: those based on ring-opening reactions.

5.2 Condensation Polymers: One Step at a Time

As the name implies, step-growth polymers are formed through a series of steps and high molecular weight materials result from a large number of steps. Although our interest is in high molecular weight, long-chain molecules, a crucial premise of this chapter is that these molecules can be effectively discussed in terms of the individual steps which lead to the formation of the polymer. Thus polyesters and polyamides (among many others) are substances which result from the occurrence of many steps in which ester or amide linkages between reactants are formed. Central to our discussion is the idea that these steps may be treated in essentially the same way, whether they occur between small molecules or polymeric species. We shall return to a discussion of the implications and justification of this assumption throughout this chapter.

To see why the assumption of equal reactivity is so important to step-growth polymers, recall from Table 1.2 the kind of chemical reactions which produce typical condensation polymers:

1. Polyesters. Successive reactions between diols and dicarboxylic acids:

$$nHO-R-OH + nHOOC-R'-COOH \rightarrow \underset{}{\left(O-R-O-\overset{\overset{O}{\|}}{C}-R'-\overset{\overset{O}{\|}}{C} \right)_n} + 2nH_2O$$

$$(5.A)$$

2. Polyamides. Successive reactions between diamines and dicarboxylic acids:

$$nH_2N-R-NH_2 + nHOOC-R'-COOH \rightarrow \underset{}{\left(\overset{H}{\underset{|}{N}}-R-\overset{H}{\underset{|}{N}}-\overset{\overset{O}{\|}}{C}-R'-\overset{\overset{O}{\|}}{C} \right)_n} + 2nH_2O$$

$$(5.B)$$

3. General. Successive reactions between difunctional monomer A–A and difunctional monomer B–B:

$$nA-A + nB-B \rightarrow \left(A-A-B-B \right)_n + \cdots$$

$$(5.C)$$

Since the two reacting functional groups can be located in the same reactant molecule, we add the following:

4. Poly(amino acids).

$$nH_2N-\underset{\underset{\displaystyle R}{|}}{C}-COOH \rightarrow \left(N-\underset{\underset{\displaystyle R}{|}}{\overset{\overset{\displaystyle H}{|}}{C}}-\overset{\overset{\displaystyle O}{\parallel}}{C}\right)_n + nH_2O \qquad (5.D)$$

5. General.

$$nA-B \rightarrow (A-B)_n + \cdots \qquad (5.E)$$

Of course, in reactions (5.A) and (5.B) the hydrocarbon sequences R and R' can be the same or different, contain any number of carbon atoms, be linear or cyclic, and so on. Likewise, the general reactions (5.C) and (5.E) certainly involve hydrocarbon sequences between the reactive groups A and B. The notation involved in these latter reactions is particularly convenient, however, and we shall use it extensively in this chapter. It will become clear as we proceed that the stoichiometric proportions of reactive groups—A and B in the above notation—play an important role in determining the characteristics of the polymeric product. Accordingly, we shall confine our discussions for the present to reactions of the type given by (5.E), since equimolar proportions of A and B are assured by the structure of this monomer.

Table 5.1 presents a hypothetical picture of how reaction (5.E) might appear if we examined the distribution of product molecules in detail. Row 1 of Table 5.1 shows the initial pool of monomers, 10 molecules in this example. Row 2 shows a possible composition after a certain amount of reaction has occurred. We shall see in Sec. 5.4 that the particular condensations which

Table 5.1 Hypothetical Step-Growth Polymerization of 10 AB Molecules[a]

Line	Molecular species present
1	AB AB AB AB AB AB AB AB AB AB
2	AbaB AbaB AbaB AbaB AB AB
3	AbababaB AbaB AbabaB AB
4	AbababaB AbabababaB AB
5	AbababaB AbababababaB
6	AbababababababababaB

[a]A and B represent two different functional groups and ab is the product of their reaction with each other. Consult the text for a discussion of the line-by-line development of the reaction.

account for the difference between lines 1 and 2 are not highly probable. Our objective here is not to assess the probability of certain reactions—as a matter of fact, we would be at a loss to try this—but, rather, to consider some possibilities. Stoichiometrically, we can still account for the initial set of 10 A groups and 10 B groups; we indicate those that have reacted with each other as ab groups. The same conservation of atom groupings would be obtained if line 2 showed one trimer, two dimers, and three monomers instead of the four dimers and two monomers indicated. Other combinations could also be assembled. These possibilities indicate one of the questions that we shall answer in this chapter: How do the molecules distribute themselves among the different possible species as the reaction proceeds?

Line 3 of Table 5.1 shows the mixture after two more reaction steps have occurred. Again the components we have elected to show are an arbitrary possibility. Stoichiometric considerations rule out certain combinations and/or proportions of products, although these are less stringent for a more realistic picture involving a larger starting number of monomer molecules. For the monomer system we have chosen, the concentration of A and B groups in the initial monomer sample are equal to each other and equal to the concentration of monomer. In this case the assay of either A groups or B groups in the mixture may be used to monitor the progress of the reaction. Choosing the number of A groups for this purpose, we see that this quantity drops from 10 to 6 to 4, respectively, as we proceed through lines 1, 2, and 3 of Table 5.1. What we wish to point out here is the fact that the 10 initial monomers are now present in four molecules, so the number average degree of polymerization is only 2.5, even though only 40% of the initial reactive groups remain. Another question is thus raised: In general, how does the average molecular weight vary with the extent of the reaction?

The reaction mixture in line 4 of Table 5.1 is characterized by an average degree of polymerization of $10/3 = 3.3$, with only 30% of the functional groups remaining. This means that 70% of the possible reactions have already occurred, even though we are still dealing with a very low average degree of polymerization. Note that the average degree of polymerization would be the same if the 70% reaction of functional groups led to the mixture AbababababababaB and two AB's. This is because the initial 10 monomers are present in three molecules in both instances and is a consequence of the fact that we are using number averages to talk about these possibilities. The weight averages would be different in the two cases. This poses still another question: How does the molecular weight *distribution* vary with the extent of reaction?

By line 5, the reaction has reached 80% completion and the number average value of the degree of polymerization $\bar{n}_n$ is 5. Although we have considered this slowly evolving polymer in terms of the extent of reaction, another question starts to be worrisome: How long is this going to take?

Line 6 represents the end of the reaction as far as linear polymer is concerned. Of the 10 initial A groups, 1 is still unreacted, but this situation raises the possibility that the decamer shown in line 6—or for that matter, some other n-mer, including monomer—might form a ring compound, thereby eliminating functional groups without advancing the polymerization. Thus the question, What about rings?

It is an easy matter to generalize the procedure we have been following and express the number average degree of polymerization in terms of the extent of reaction, regardless of the initial sample size. We have been dividing the initial number of monomers present by the total number of molecules present after any extent of reaction. Each molecular species—whether monomer or polymer of any n—contains just one A group. The total number of monomer molecules is therefore equal to the initial (superscript 0) number of A groups ν_A^0; the total number of molecules at any extent of reaction (no superscript) is equal to the number of A groups ν_A present at that point. The number average degree of polymerization is therefore given by

$$\bar{n}_n = \frac{\nu_A^0}{\nu_A} \tag{5.1}$$

It is convenient to define the fraction of reacted functional groups in a reaction mixture by a parameter p, called the extent of reaction. Thus p is the fraction of A groups which have reacted at any stage of the process, and $1 - p$ is the fraction unreacted:

$$1 - p = \frac{\nu_A}{\nu_A^0} \tag{5.2}$$

or

$$p = 1 - \frac{\nu_A}{\nu_A^0} \tag{5.3}$$

Comparing Eqs. (5.1) and (5.2) enables us to write

$$\bar{n}_n = \frac{1}{1 - p} \tag{5.4}$$

This expression is consistent with the analysis of each of the lines in Table 5.1 as presented above and provides a general answer to one of the questions posed there. It is often a relatively easy matter to monitor the concentration of functional groups in a reaction mixture as we saw in discussing end group analysis as a method for molecular weight determination in Sec. 1.7. Equation (5.4) is

a quantitative summary of the end group method for determining n. Example 1.4 showed that the details of interpreting end group data depends on consideration of the specific polymer mixture. Accordingly, we reiterate that Eq. (5.4) assumes equal numbers of A and B groups, with none of either lost in nonpolymer reactions.

From line 6 in Table 5.1, we see that $\bar{n}_n$ = 10 when p = 0.9. The fact that this is also the maximum value for n is an artifact of the example. In a larger sample of monomers higher average degrees of polymerization are attainable. Equation (5.4) enables us to calculate that $\bar{n}_n$ becomes 20, 100, and 200, respectively, for extents of reaction of 0.950, 0.990, and 0.995. These considerations point out why condensation polymers are often of relatively modest molecular weight: Achieving the extents of reaction required for very high molecular weights may be difficult to accomplish.

Each of the questions raised in the last few paragraphs will be answered during the course of this chapter, some for systems considerably more involved than the one considered here. Before proceeding further, we should question one premise which underlies the entire discussion of Table 5.1: How does the chemical reactivity of A and B groups depend on the degree of polymerization of the reaction mixture? In Table 5.1 successive entries were generated by simply linking together at random those species present in the preceding line. We have thus assumed that, as far as reactivity is concerned, an A reacts as an A and a B as a B, regardless of the size of the molecule to which the group is attached. If this is valid, it results in a tremendous simplification; otherwise we shall have to characterize reactivity as a function of molecular weight, degree of polymerization, extent of reaction, and so on.

One of the most sensitive tests of the dependence of chemical reactivity on the size of the reacting molecules is the comparison of the rates of reaction for compounds which are members of a homologous series with different chain lengths. Studies by Flory and others on the rates of esterification and saponification of esters were the first investigations conducted to clarify the dependence of reactivity on molecular size. The rate constants for these reactions are observed to converge quite rapidly to a constant value which is independent of molecular size, after an initial dependence on molecular size for small molecules. The effect is reminiscent of the discussion on the uniqueness of end groups in connection with Example 1.1. In the esterification of carboxylic acids, for example, the rate constants are different for acetic, propionic, and butyric acids, but constant for carboxylic acids with 4–18 carbon atoms. This observation on nonpolymeric compounds has been generalized to apply to polymerization reactions as well. The latter are subject to several complications which are not involved in the study of simple model compounds, but when these complications are properly considered, the independence of reactivity on molecular size has been repeatedly verified.

The foregoing conclusion does not mean that the rate of the reaction proceeds through Table 5.1 at a constant value. The rate of reaction depends on the concentrations of reactive groups, as well as on the reactivities of the latter. Accordingly, the rate of the reaction decreases as the extent of reaction progresses. When the rate law for the reaction is extracted from proper kinetic experiments, specific reactions are found to be characterized by fixed rate constants over a range of $\bar{n}_n$ values.

Among the complications that can interfere with this conclusion is the possibility that the polymer becomes insoluble beyond a critical molecular weight or that the low molecular weight by-product molecules accumulate as the viscosity of the mixture increases and thereby shift some equilibrium to favor reactants. Note that we do not express reservations about the effect of increasing viscosity on the mobility of the polymer molecules themselves. Apparently it is not the migration of the center of mass of the molecule as a whole that determines the reactivity but, rather, the mobility of the chain ends which carry the reactive groups.

Figure 5.1 suggests that reactive end groups may be brought into contact by rotation around perhaps only a few bonds, an effect which is independent of chain length. In Chap. 2 we discussed the motion of chain segments in terms of entrapment within cages of neighboring groups. This same effect will promote the reaction of A and B once they find themselves in the same cage, indicated by the broken line in Fig. 5.1. It may take some time for the two reactive groups to diffuse together, but it will also take a while for them to diffuse apart; this provides the opportunity to react. The rate at which independent A and B groups convert to ab linkages is thus seen to depend on the relative rates of three processes: the rate at which the groups diffuse together, the rate at which they diffuse apart, and the rate at which the trapped A and B species react to form ab.

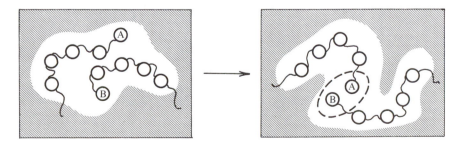

Figure 5.1   The reaction of A and B groups at the ends of two different chains. Note that rotations around only a few bonds will bring A and B into the same cage of neighboring groups, indicated by the broken line enclosure.

These considerations can be expressed quantitatively by writing the overall process in terms of the following mechanism

$$-A + -B \underset{k_o}{\overset{k_i}{\rightleftharpoons}} (-A + B-)$$

$$(5.F)$$

$$(-A + B-) \overset{k_r}{\rightarrow} -ab-$$

where the parentheses represent the state in which the two reactive groups are trapped together in the same cage, as seen in Fig. 5.1, and the k's are rate constants which typify the individual steps: k_i and k_o measure the rates of group diffusion in and out of the cage, respectively, and k_r measures the rate of reaction.

Since this is the first occasion we have had to examine the rates at which chemical reactions occur, a few remarks about mechanistic steps and rate laws seem appropriate. The reader who feels the need for additional information on this topic should consult the discussions which will be found in any physical chemistry text.

As a brief review we recall the following:

1. The rate of a process is expressed by the derivative of a concentration (square brackets) with respect to time, $d[\ ]/dt$. If the concentration of a reaction product is used, this quantity is positive; if a reactant is used, it is negative and a minus sign must be included. Also, each derivative $d[\ ]/dt$ should be divided by the coefficient of that component in the chemical equation which describes the reaction so that a single rate is described, whichever component in the reaction is used to monitor it.

2. A rate law describes the rate of a reaction as the product of a constant k, called the rate constant, and various concentrations, each raised to specific powers. The power of an individual concentration term in a rate law is called the order with respect to that component, and the sum of the exponents of all concentration terms gives the overall order of the reaction. Thus in the rate law Rate = $k[X]^1[Y]^2$, the reaction is first order in X, second order in Y, and third order overall.

3. A rate law is determined experimentally and the rate constant evaluated empirically. There is no necessary connection between the stoichiometry of a reaction and the form of the rate law.

4. A mechanism is a series of simple reaction steps which, when added together, account for the overall reaction. The rate law for the individual steps of the mechanism may be written by inspection of the mechanistic steps. The coefficients of the reactants in the chemical equation describing the step become the exponents of these concentrations in the rate law for

that step. The term *molecularity* is sometimes used instead of *order* to distinguish between the rates of mechanistic steps and experimental rate laws.

5. Frequently it is possible to write more than one mechanism which is compatible with an observed rate law. Thus ability to account for an experimental rate law is a necessary but not sufficient criterion for the correctness of the mechanism.

These ideas are readily applied to the mechanism described by reaction (5.F). To begin with, the rate at which ab links are formed is first order with respect to the concentration of entrapped pairs. In this sense the latter behaves as a reaction intermediate or transition state according to this mechanism. Therefore

$$\text{Rate of ab formation} = k_r[(-A + B-)] \tag{5.5}$$

These entrapped pairs, in turn, form at a rate given by the rate at which the two groups diffuse together minus the rate at which they either diffuse apart or are lost by reaction:

$$\frac{d[(-A + B-)]}{dt} = k_i[A][B] - k_o[(-A + B-)] - k_r[(-A + B-)] \tag{5.6}$$

The concentration of entrapped pairs is assumed to exist at some stationary-state (subscript s) level in which the rates of formation and loss are equal. In this stationary state $d[(-A + B-)]/dt = 0$ and Eq. (5.6) becomes

$$[(-A + B-)]_s = \frac{k_i}{k_o + k_r}[A][B] \tag{5.7}$$

where the subscript reminds us that this is the stationary-state value. Substituting Eq. (5.7) into Eq. (5.5) gives

$$\text{Rate of ab formation} = \frac{k_i k_r}{k_o + k_r}[A][B] \tag{5.8}$$

We shall have considerably more to say about this type of kinetic analysis when we discuss chain-growth polymerizations in Chap. 6.

According to the mechanism provided by reactions (5.F) and the analysis given by Eq. (5.8), the rate of polymerization is dependent upon the following:

1. The concentrations of both A and B, hence the reaction slows down as the conversion to polymer progresses, and . . . .
2. The three constants associated with the rates of the individual steps in reactions (5.F).

3. If the rate of chemical reaction is very slow compared to the rate of group diffusion ($k_r \ll k_i, k_o$), then Eq. (5.8) reduces to

$$\text{Rate of ab formation} = \frac{k_i}{k_o} k_r [A] [B] \tag{5.9}$$

4. The two constants k_i and k_o describe exactly the same kind of diffusional processes and differ only in direction. Hence they have the same dependence on molecular size, whatever that might be, and that dependence therefore cancels out.
5. The reaction step in mechanism (5.F) is entirely comparable to the same reaction in low molecular weight systems. Such reactions involve considerably larger activation energies than physical processes like diffusion and, hence, do proceed slowly.
6. If $k_r \gg k_i, k_o$, then Eq. (5.8) reduces to

$$\text{Rate of ab formation} = k_i [A] [B] \tag{5.10}$$

Note that the rate law in this case depends only on k_i and any size dependence for this constant would not cancel out.

Both Eqs. (5.9) and (5.10) predict rate laws which are first order with respect to the concentration of each of the reactive groups; the proportionality constant has a different significance in the two cases, however. The observed rate laws which suggest a reactivity that is independent of molecular size and the a priori expectation cited in item (5) regarding the magnitudes of different kinds of k values lend credibility to the version presented as Eq. (5.9).

Our objective in the preceding argument has been to justify the attitude that each ab linkage forms according to the same rate law, regardless of the extent of the reaction. While our attention is focused on the rate laws, we might as well consider the question, raised above, about the actual rates of these reactions. This is the topic of the next section.

5.3 Kinetics of Step-Growth Polymerization

In this section we turn to a consideration of the experimental side of condensation kinetics. The kind of ab links which have been most extensively studied are ester and amide groups, although numerous additional systems could also be cited. In many of these the carbonyl group is present and is believed to play an important role in stabilizing the actual chemical transition state involved in the reactions. The situation can be represented by the following schematic reaction:

$$R-\overset{\overset{\displaystyle O}{\|}}{C}-X + Y^- \longrightarrow \left[R-\overset{\overset{\displaystyle O}{|}}{\underset{\underset{\displaystyle X}{|}}{C}}-Y \right]^- \longrightarrow R-\overset{\overset{\displaystyle O}{\|}}{C}-Y + X^- \qquad (5.G)$$

in which the intermediate is stabilized by coordination with protons, metal ions, or other Lewis acids. The point of this is to emphasize that the kinds of reactions we are considering are often conducted in the presence of an acid catalyst, frequently something like a sulfonic acid or a metal oxide. The purpose of a catalyst is to modify the rates of a reaction, so we must be attentive to the situation with respect to catalysts. For the present we assume a constant concentration of catalyst and attach a subscript c to the rate constant to remind us of the assumption. Accordingly, we write

$$- \frac{d[A]}{dt} = k_c [A] [B] \qquad (5.11)$$

which is consistent with both Eqs. (5.9) and (5.10). We expect the constant k_c to be dependent on the concentration of the catalyst in some way which means that Eq. (5.11) may be called a pseudo-second-order rate law. We shall presently consider these reactions in the absence of external catalysts. For now it is easier to proceed with the catalyzed case.

Equation (5.11) is the differential form of the rate law which describes the rate at which A groups are used up. To test a proposed rate law and to evaluate the rate constant it is preferable to work with the integrated form of the rate law. The integration of Eq. (5.11) yields different results, depending on whether the concentrations of A and B are the same or different:

1. We define [A] and [B] as the instantaneous concentrations of these groups at any time t during the reaction, and $[A]_0$ and $[B]_0$ as the concentrations of these groups at $t = 0$.
2. If $[A]_0 = [B]_0$, the integration of Eq. (5.11) yields

$$\frac{1}{[A]} - \frac{1}{[A]_0} = k_c t \qquad (5.12)$$

3. If $[A]_0 \neq [B]_0$, the integration yields

$$\frac{1}{[A]_0 - [B]_0} \ln \left(\frac{[A] [B]_0}{[A]_0 [B]} \right) = k_c t \qquad (5.13)$$

Both of these results are readily obtained; we examine the less obvious relationship in item (3) in the following example.

Example 5.1

By differentiation, verify that Eq. (5.13) is a solution to Eq. (5.11) for the conditions given.

Solution

Neither $[A]_0$ nor $[B]_0$ are functions of t, although both $[A]$ and $[B]$ are. We write the latter two as $[A] = [A]_0 - x$ and $[B] = [B]_0 - x$. Substitute these results into Eq. (5.13) and differentiate:

$$d\left[\ln\left(\frac{[A]_0 - x}{[B]_0 - x}\right)\right] + d\left[\ln\left(\frac{[B]_0}{[A]_0}\right)\right] = ([A]_0 - [B]_0)\,k_c\,dt$$

$$\left(\frac{[B]_0 - x}{[A]_0 - x}\ \frac{([B]_0 - x)(-1) - ([A]_0 - x)(-1)}{([B]_0 - x)^2}\right)\,dx = \frac{([A]_0 - [B]_0)\,dx}{([A]_0 - x)([B]_0 - x)}$$

$$= ([A]_0 - [B]_0)\,k_c\,dt$$

$$\frac{dx}{dt} = k_c\,([A]_0 - x)([B]_0 - x) = k_c\,[A]\,[B]$$

Since $d[A]/dt = -\,dx/dt$ by the definition of x, this proves Eq. (5.13) to be a solution to Eq. (5.11). Equation (5.13) is undefined in the event $[A]_0 = [B]_0$, but in this case the expression is inapplicable anyhow. Since A and B react in a 1:1 proportion, their concentrations are identical at all stages of reaction if they are equal initially. In this case, Eq. (5.11) would reduce to a simple second-order rate law which integrates to Eq. (5.12).

•

We shall proceed on the assumption that $[A]_0$ and $[B]_0$ are equal. As noted above, the case of both reactive groups on the same molecule is a way of achieving this condition. Accordingly, we rearrange Eq. (5.12) to give the instantaneous concentration of unreacted A groups as a function of time:

$$[A] = \frac{[A]_0}{1 + k_c\,[A]_0\,t} \tag{5.14}$$

At this point it is convenient to recall the extent of reaction parameter defined by Eq. (5.3). If we combine Eqs. (5.2) and (5.14), we obtain

$$1 - p = \frac{1}{1 + k_c\,[A]_0\,t} \tag{5.15}$$

or

$$\frac{1}{1-p} = 1 + k_c \, [A]_0 \, t \tag{5.16}$$

Alternatively, we can incorporate Eq. (5.4) into the present discussion and write

$$\bar{n}_n = 1 + k_c \, [A]_0 \, t \tag{5.17}$$

These last expressions provide two very useful views of the progress of a condensation polymerization reaction with time. Equation (5.14) describes how the concentration of A groups asymptotically approaches zero at long times; Eq. (5.17) describes how the degree of polymerization increases linearly with time.

Equation (5.16) predicts a straight line when $1/(1 - p)$ is plotted against t. Figure 5.2 shows such a plot for adipic acid reacted with 1,10-decamethylene glycol and diethylene glycol. In both cases the esterifications were catalyzed by p-toluene sulfonic acid. Interpreting the slopes of these lines in terms of Eq. (5.16) and in the light of actual initial concentrations gives values of k_c equal to 0.013 and 0.097 kg eq^{-1} min^{-1}, respectively, for diethylene glycol at 109°C and decamethylene glycol at 161°C. Note that the units for these constants imply group concentrations expressed as equivalents per kilogram; mass rather than volume units are often used for concentration, since large volume changes may occur during polymerization.

Although the results presented in Fig. 5.2 appear to verify the predictions of Eq. (5.16), this verification is not free from controversy. This controversy arises because various workers in this field employ different criteria in evaluating the success of the relationships we have presented in fitting experimental polymerization data. One school of thought maintains that an adequate kinetic description of a process must apply to the data over a large part of the time of the experiment.

A second point of view maintains that a rate law correctly describes a process when it applies over a wide portion of the concentration change which occurs during a reaction. Each of these criteria seeks to maximize the region of fit, but the former emphasizes maximizing the range of t while the latter maximizes the range of p. Both standards tolerate deviations from their respective ideals at the beginning and/or the end of the experiment. Deviations at the beginning of a process are rationalized in terms of experimental uncertainties at the point of mixing or modelistic difficulties upon attainment of stationary-state conditions.

The existence of these two different standards for success would be of academic interest only if the analysis we have discussed applied to experimental

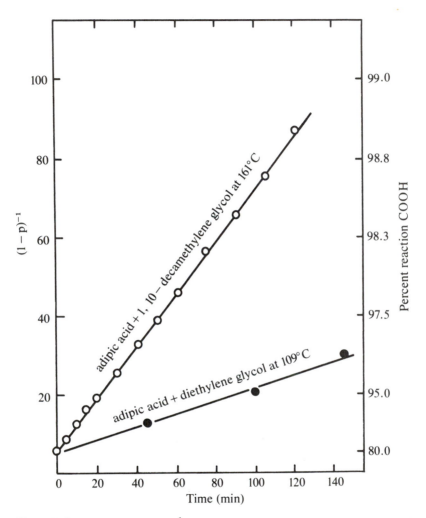

Figure 5.2 Plots of $(1 - p)^{-1}$ (left-hand ordinate) and p (right-hand ordinate) versus time for the catalyzed esterifications shown. [From Ref. 1, used with permission. Data cited from S. D. Hamann, D. H. Solomon, and J. D. Swift, *J. Macromol. Sci. Chem.* A2:153 (1968); P. J. Flory, *J. Am. Chem. Soc.* 61: 3334 (1939).]

results over most of the time range *and* over most extents of reaction as well. Unfortunately, this is not the case in all of the systems which have been investigated. Reference 5, for example, shows one particular set of data—adipic acid and diethylene glycol at 166°C, the same reactants at a different temperature as

the system in Fig. 5.2–analyzed according to two different rate laws. This system obeys one rate law between p = 0.50 and 0.85 that represents 15% of the duration of the experiment, and another rate law between p = 0.80 and 0.93 which spans 45% of the reaction time. These would be rated differently by the two different standards above. This sort of dilemma is not unique to the present problem, but arises in many situations where one variable undergoes a large percentage of its total change while the other variable undergoes only a small percent of its change. In the present context a way out of the dilemma is to accept the attitude that only the latter stages of the reaction are significant, since it is only beyond, say, p = 0.80 that it makes sense to consider the process as one of polymerization. According to this viewpoint, it is only at large extents of reaction that polymeric products are formed and, hence, the kinetics of *polymerization* should be based on a description of this part of the process. This viewpoint intentionally focuses attention on a relatively modest but definite range of p values. Since the reaction is necessarily slow as the number of unreacted functional groups decreases, this position tends to maximize the time over which the rate law fits the data.

Examination of the right-hand ordinate of Fig. 5.2 shows that the data presented there represent only about the last 20% of the range of p values. The zero of the time scale has thus been shifted to pick up the analysis of the reaction at this point.

We commented above that deviations at the beginning or the end of kinetics experiments can be rationalized, although the different schools of thought would disagree as to what constitutes "beginning" and "end." Now that we have settled upon the polymer range, let us consider specifically why deviations occur from a simple second-order kinetic analysis in the case of catalyzed polymerizations. At the beginning of the experiment, say, up to p = 0.5, the concentrations of A and B groups change dramatically, even though the number average degree of polymerization has only changed from monomer to dimer. By ordinary polymeric standards, we are still dealing with a low molecular weight system which might be regarded as the solvent medium for the formation of polymer. During this transformation, however, 50% of the very polar A groups and 50% of the very polar B groups have been converted to the less polar ab groups. Thus a significant change in the polarity of the polymerization medium occurs during, say, the first half of the change in p, even though an insignificant amount of true polymer has formed. In view of the role of ionic intermediates as suggested by reaction (5.G), the polarity of the reaction medium might very well influence the rate law during this stage of the reaction.

At the other end of the reaction, deviations from idealized rate laws are attributed to secondary reactions such as degradations of acids, alcohols, and amines through decarboxylation, dehydration, and deamination, respectively. The step-growth polymers which have been most widely studied are simple

condensation products such as polyesters and polyamides. Although we shall take up these classes of polymers specifically in Secs. 5.5 and 5.6, respectively, it is appropriate to mention here that these are typically equilibrium reactions as represented by the following equations:

$$-R_1-\overset{\overset{\text{O}}{\|}}{\text{C}}-OH + HO-R_2- \;\rightleftarrows\; -R_1-\overset{\overset{\text{O}}{\|}}{\text{C}}-O-R_2- + H_2O \tag{5.H}$$

and

$$-R_1-\overset{\overset{\text{O}}{\|}}{\text{C}}-OH + H_2N-R_2- \;\rightleftarrows\; -R_1-\overset{\overset{\text{O}}{\|}}{\text{C}}-\overset{\overset{\text{H}}{|}}{\text{N}}-R_2- + H_2O \tag{5.I}$$

In order to achieve large p's and high molecular weights, it is essential that these equilibria be shifted to the right by removing the by-product molecule, water in these reactions. This may be accomplished by heating, a partial vacuum, or purging with an inert gas, or some combination of the three. These treatments also open up the possibility of reactant loss due to volatility, which may accumulate to a significant source of error for reactions which are carried out to large values of p.

Until now we have been discussing the kinetics of catalyzed reactions. Losses due to volatility and side reactions also raise questions as to the validity of assuming a constant concentration of catalyst. Of course, one way of avoiding this issue is to omit an outside catalyst; reactions involving carboxylic acids can be catalyzed by these compounds themselves. Experiments conducted under these conditions are informative in their own right and not merely as means of eliminating errors in the catalyzed case. As noted in connection with the discussion of reaction (5.G), the intermediate is stabilized by coordination with a proton from the catalyst. In the case of autoprotolysis by the carboxylic acid reactant, the rate-determining step is probably the slow reaction of intermediate [I]:

$$\begin{bmatrix} \overset{\displaystyle H}{\underset{\displaystyle |}{\overset{\displaystyle O}{|}}} \\ R_1-\overset{|}{\underset{|}{C}}-OH \\ HO-R_2 \end{bmatrix}^{\oplus} {}^{\ominus}OOC-R_1$$

$$[I]$$

Since this intermediate involves an additional equivalent of acid functional groups, the rate law for the disappearance of A groups becomes

$$- \frac{d[A]}{dt} = k_u [A]^2 [B] \tag{5.18}$$

on the assumption that A represents carboxyl groups. In this case k_u is the rate constant for the uncatalyzed reaction. This differential rate law is the equivalent of Eq. (5.11) for the catalyzed reaction. Equation (5.18) is readily integrated for the case where $[A]_0 = [B]_0$, in which case it becomes

$$- [A]^{-3} d[A] = k_u dt \tag{5.19}$$

This integrates to

$$\frac{1}{[A]^2} - \frac{1}{[A]_0^2} = 2k_u t \tag{5.20}$$

where $[A]_0$ is the concentration of A at $t = 0$. Thus for the uncatalyzed case we have the following:

1. The rate law is third order.
2. Since $A/A_0 = 1 - p$, Eq. (5.20) may be written as

$$\frac{1}{(1 - p)^2} = 1 + 2k_u [A]_0^2 t \tag{5.21}$$

   which shows that a plot of $(1 - p)^{-2}$ increases linearly with t.
3. Since $A_0/A = \bar{n}_n$, Eq. (5.21) becomes

$$(\bar{n}_n)^2 = 1 + 2k_u [A]_0^2 t \tag{5.22}$$

   which shows that $\bar{n}_n$ increases more gradually with t than in the catalyzed case, all other things being equal.

Figure 5.3 shows the data for the uncatalyzed polymerization of adipic acid and 1,10-decamethylene glycol at 161°C plotted according to Eq. (5.21). The various provisos of the catalyzed case apply here also, so it continues to be appropriate to consider only the final stages of the conversion to polymer. From these results, k_u is about 4.3×10^{-3} kg^2 eq^{-2} min^{-1} at 161°C.

We conclude this section with a numerical example which serves to review and compare some of the important relationships we have considered.

Example 5.2

Assuming that $k_c = 10^{-1}$ kg eq^{-1} min^{-1}, $k_u = 10^{-3}$ kg^2 eq^{-2} min^{-1}, and $[A]_0 = 10$ eq kg^{-1}, calculate the times required for p to reach values 0.2, 0.4,

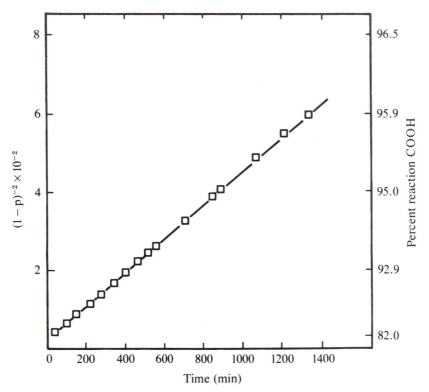

Figure 5.3 Plot of $(1 - p)^{-2}$ (left-hand ordinate) and p (right-hand ordinate) versus time for an uncatalyzed esterification. [From S. D. Hamann, D. H. Solomon, and J. D. Swift, *J. Macromol. Sci. Chem.* A2:153 (1968).]

0.6, and so on, for both catalyzed and uncatalyzed polymerizations, assuming that Eqs. (5.14) and (5.20), respectively, apply to the entire reaction. Compare the results obtained in terms of both the degree of polymerization and the fraction of unreacted A groups as a function of time.

Solution

Since we are asked to evaluate t, $\bar{n}_n$, and $[A]/[A]_0$ for specific values of p, it is convenient to summarize the following relationships:

1. Eq. (5.4): $\bar{n}_n = 1/(1 - p)$.
2. Eq. (5.2): $[A]/[A]_0 = 1 - p$.
3. Eq. (5.17): $t = (\bar{n}_n - 1)/k_c[A]_0 = \bar{n}_n - 1$ if catalyzed, since $10^{-1}(10) = 1$.

4.   Eq. (5.22): $t = (\bar{n}_n^2 - 1)/2k_u[A]_0^2 = (\bar{n}_n^2 - 1)(5)$ if uncatalyzed, since $2(10^{-3})(10)^2 = 0.2$.

Using these relationships the following table is developed:

			Time (min)	
p	$[A]/[A]_0$	$\bar{n}_n$	Catalyzed	Uncatalyzed
0.2	0.8	1.25	0.25	2.8
0.4	0.6	1.67	0.67	8.9
0.6	0.4	2.50	1.5	26
0.8	0.2	5.00	4.0	120
0.9	0.1	10.0	9.0	500
0.95	0.05	20.0	19	2.0×10^3
0.99	0.01	100	99	5.0×10^4
0.992	0.008	120	119	7.2×10^4
0.998	0.002	500	499	1.3×10^6

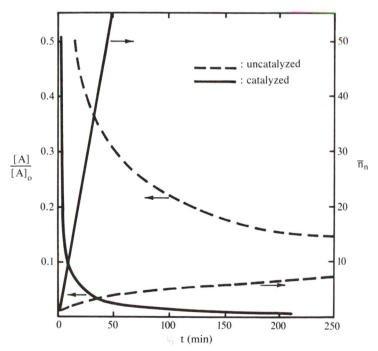

Figure 5.4 Comparison of catalyzed (solid lines) and uncatalyzed (broken lines) polymerizations using results calculated in Example 5.2. Here $1 - p$ (left-hand ordinate) and $\bar{n}_n$ (right-hand ordinate) are plotted versus time.

A graphical comparison of the trends appearing here is presented in Fig. 5.4. The importance of the catalyst is readily apparent in this hypothetical but not atypical system: To reach $\bar{n}_n$ = 5 requires 4 min in the catalyzed case and 120 min without any catalyst, assuming that the same rate law describes the entire reaction in each case.

•

The question posed in Sec. 5.2—how long will it take to reach a certain extent of reaction or degree of polymerization?—is now answered. As is often the case, the answer begins, "It all depends . . . ."

5.4 Distributions of Molecular Sizes

In this section we turn our attention to two other questions raised in Sec. 5.2, namely, how do the molecules distribute themselves among the different possible species and how does this distribution vary with the extent of reaction? Since a range of species is present at each stage of the polymerization, it is apparent that a statistical answer is required for these questions. This time, our answer begins, "On the average . . . ."

We shall continue basing our discussion on the step-growth polymerization of the hypothetical monomer AB. In Sec. 5.7 we shall take a second look at this problem for the case of unequal concentrations of A and B groups. For now, however, we assure this equality by considering a monomer which contains one group of each type. In a previous discussion of the polymer formed from this monomer, we noted that remnants of the original functional groups are still recognizable, although modified, along the backbone of the polymer chain. This state of affairs is emphasized by the notation Ababab . . . abaB in which the a's and b's of the ab linkages are groups of atoms carried over from the initial A and B reactive groups. In this type of polymer molecule, then, there are n - 1 a's and 1 A if the degree of polymerization of the polymer is n. The a's differ from the A's precisely in that the former have undergone reaction while the latter have not. At any point during the polymerization reaction the fraction of the initial number of A groups which have reacted to become a's is given by p, and the fraction which remains as A's is given by 1 - p. In these expressions p is the same extent of reaction defined by Eq. (5.3).

With these ideas in mind, we now turn to the question of evaluating the fraction of n-mers in a mixture as a function of p. The fraction of molecules of a particular type in a population is just another way of describing the probability of such a molecule. Hence our restated objective is to find the probability of an n-mer in terms of p. We symbolize this quantity P(n, p). Since the n-mer consists of n - 1 a's and 1 A, its probability is the same as the probability of finding n - 1 a's and 1 A in the same molecule. Recalling from Chap. 1 how such probabilities are compounded, we write

$$P(n, p) = p_a^{n-1} \, p_A^1 \tag{5.23}$$

where p_a and p_A are the probabilities of individual a and A groups, respectively. These probabilities, in turn, are given by their fractional abundance at any point in the reaction; that is, $p_a = p$ and $p_A = 1 - p$. Substituting these results into Eq. (5.23) gives

$$P(n, p) = p^{n-1} (1 - p) \tag{5.24}$$

The probability of an n-mer is converted to the number of n-mer molecules in the reaction mixture N_n by multiplying $P(n, p)$ and the total number of molecules N (no subscript) in the mixture after the reaction has occurred to the extent p:

$$N_n = N p^{n-1} (1 - p) \tag{5.25}$$

Note that N_n/N gives the mole fraction of n-mers in a mixture at an extent of reaction p. As we have seen before, $N = (1 - p) A_0$, since each molecule in the mixture contains one unreacted A group. Incorporating this result into Eq. (5.25) yields

$$N_n = p^{n-1} (1 - p)^2 \, N_0 \tag{5.26}$$

where N_0 is the total number of monomers present initially; $N_0 = A_0$ for AB monomers. This result may be used to evaluate the number of molecules of whatever degree of polymerization we elect to consider in terms of p and N_0. As such, it provides the answer to one of the questions posed earlier.

Figure 5.5 is a plot of the ratio N_n/N versus n for several values of p. Several features are apparent from Fig. 5.5 concerning the number distribution of molecules among the various species present:

1. On a number basis, the fraction of molecules decreases with increasing n, regardless of the value of p. The distributions in Table 5.1 are unrealistic in this regard.
2. As p increases, the proportion of molecules with smaller n values decreases and the proportion with larger n values increases.
3. The combination of effects described in item (2) tends to flatten the curves as p increases, but not to the extent that the effect of item (1) disappears.

The number average degree of polymerization for these mixtures is easily obtained by recalling the definition of the average from Sec. 1.8. It is given by the sum of all possible n values, with each multiplied by its appropriate weighting factor, provided by Eq. (5.24):

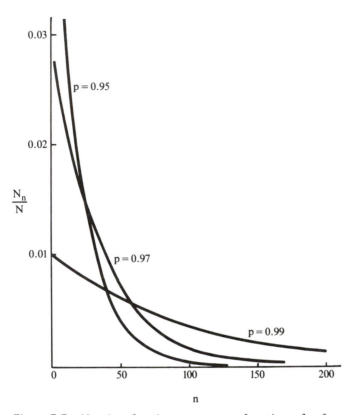

Figure 5.5 Number fraction n-mers as a function of n for several values of p.

$$\bar{n}_n = \sum_{n=1}^{N_0} nP(n, p) = \sum_{n=1}^{\infty} np^{n-1}(1-p) \tag{5.27}$$

Note that the upper limit of the second summation has been shifted from N_0 to ∞ for mathematical reasons. The change is of little practical significance, since Eq. (5.24) drops off for very large values of n. To simplify the summation in Eq. (5.27) consider the following steps:

1. Write out a few terms of the summation so that it appears less abstract:

$$\bar{n}_n = 1p^0(1-p) + 2p^1(1-p) + 3p^2(1-p) + \cdots$$

2. Factor the quantity $1 - p$ from each term:

$$= (1-p)(1p^0 + 2p^1 + 3p^2 + \cdots)$$

3. The remaining series of terms is given by $(1 + p + p^2 + \cdots)^2$, as is easily verified by squaring the latter:

$$= (1 - p)(1 + p + p^2 + p^3 + \cdots)^2$$

4. The series $1 + p + p^2 + \cdots$ is given by $1/(1 - p)$, as may be verified by long division:

$$= (1 - p)(1 - p)^{-2}$$

Simplification of the summation in Eq. (5.27) thus yields

$$\bar{n}_n = \frac{1}{1 - p} \tag{5.28}$$

Of course, this is the same result that was obtained more simply in Eq. (5.4). The earlier result, however, was based on purely stoichiometric considerations and not on the detailed distribution as is the present result.

Next we turn our attention to the distribution of the molecules by weight among the various species. A corollary of this is the determination of the weight average molecular weight and the ratio $\bar{M}_w/\bar{M}_n$.

We begin by recognizing that the weight fraction w_n of n-mers in the polymer mixture at any value of p equals the ratio of the mass of n-mer in the mixture divided by the mass of the total mixture. The former is given by the product nN_nM_0, where M_0 is the molecular weight of the repeat unit; the latter is given by N_0M_0. Therefore we write

$$w_n = \frac{nN_n}{N_0} \tag{5.29}$$

into which Eq. (5.26) may be substituted to give

$$w_n = np^{n-1}(1 - p)^2 \tag{5.30}$$

The weight fraction of n-mers is plotted as a function of n in Fig. 5.6 for several large value of p. Inspection of Fig. 5.6 and comparison with Fig. 5.5 reveals the following:

1. At any p, very small and very large values of n contribute a lower weight fraction to the mixture than do intermediate values of n. This arises because of the product nN_n in Eq. (5.29): N_n is large for monomers, in which case n is low, and N_n decreases as n increases. At intermediate values of n, w_n goes through a maximum.

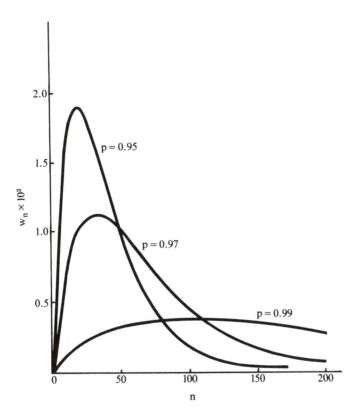

Figure 5.6   Weight fraction of n-mers as a function of n for several values of p.

2.   As p increases, the maximum in the curves shifts to larger n values and the
     tail of the curve extends to higher values of n.
3.   The effect in item (2) is not merely a matter of shifting curves toward
     higher n values as p increases, but reflects a distinct broadening of the
     distribution of n values as p increases.

The weight average degree of polymerization is obtained by averaging the
contributions of various n values using weight fractions as weighting factors in
the averaging procedure:

$$\bar{n}_w = \frac{\sum_{n=1}^{N_0} n w_n}{\sum_{n=1}^{N_0} w_n} = \frac{\sum_{n=1}^{\infty} n^2 p^{n-1}(1-p)^2}{\sum_{n=1}^{\infty} n p^{n-1}(1-p)^2} \tag{5.31}$$

where the upper limit on n has been extended to infinity as before. Again we
examine the simplification of the summations through the following steps:

1. Since each term in the numerator and the denominator contains the factor $(1 - p)^2$, this common factor cancels out:

$$\bar{n}_w = \frac{\sum_{n=1}^{\infty} n^2 p^{n-1}}{\sum_{n=1}^{\infty} n p^{n-1}}$$

2. Write out a few terms of the remaining series:

$$= \frac{1 + 4p + 9p^2 + 16p^3 + \cdots}{1 + 2p + 3p^2 + 4p^3 + \cdots}$$

3. This ratio can be reduced by long division:

$$= 1 + 2p + 2p^2 + 2p^3 + \cdots = (2 + 2p + 2p^2 + \cdots) - 1$$

$$= 2(1 + p + p^2 + \cdots) - 1$$

4. As noted above, $1 + p + p^2 + \cdots = 1/(1 - p)$; hence

$$= \frac{2}{1 - p} - 1$$

Performing these manipulations on Eq. (5.31) shows that

$$\bar{n}_w = \frac{1 + p}{1 - p} \tag{5.32}$$

which is the desired result.

We saw in Chap. 1 that the ratio $\bar{M}_w/\bar{M}_n$ is widely used in polymer chemistry as a measure of the width of a molecular weight distribution. If the effect of chain ends is disregarded, this ratio is the same as the corresponding ratio of n values:

$$\frac{\bar{M}_w}{\bar{M}_n} = \frac{\bar{n}_w}{\bar{n}_n} = 1 + p \tag{5.33}$$

where the ratio of Eq. (5.32) to Eq. (5.4) has been used. Table 5.2 lists values of $\bar{n}_w$, $\bar{n}_n$, and $\bar{n}_w/\bar{n}_n$ for a range of high p values. Note that $\bar{n}_w/\bar{n}_n \to 2$ as $p \to 1$. In light of Eq. (1.18), the standard deviation of the molecular distribution is equal to $\bar{M}_n$ for the polymer sample produced by this polymerization. In a manner of speaking, the molecular weight distribution is as wide as the average is high! The broadening of the distribution with increasing p is dramatically

Table 5.2 Values of $\bar{n}_n$, $\bar{n}_w$, and $\bar{M}_w/\bar{M}_n$
for Various Large Values of p

p	$\bar{n}_n$	$\bar{n}_w$	$\bar{n}_w/\bar{n}_n$
0.90	10.0	19.0	1.90
0.92	12.5	24.0	1.92
0.94	16.7	32.3	1.94
0.96	25.0	49.0	1.96
0.98	50.0	99.0	1.98
0.990	100	199	1.990
0.992	125	249	1.992
0.994	167	332	1.994
0.996	250	499	1.996
0.998	500	999	1.998

shown by comparing the values in Table 5.2 with the situation at a low p value, say, p = 0.5. At p = 0.5, $\bar{n}_n = 2$, $\bar{n}_w = 3$, and $\bar{n}_w/\bar{n}_n = 1.5$.

Since Eqs. (5.15) and (5.21), respectively, give p as a function of time for the catalyzed and uncatalyzed polymerizations, the distributions discussed in the last few paragraphs can also be expressed with time as the independent variable instead of p.

The results we have obtained on the basis of the hypothetical monomer AB are also applicable to polymerizations between monomers of the AA and BB type, as long as the condition [A] = [B] is maintained. We shall extend the arguments of this section to conditions in which [A] ≠ [B] in Sec. 5.7 and to monomers of the type A$\overset{A}{\underset{}{\downarrow}}$A in Sec. 5.8. In the meanwhile we interrupt this line of reasoning by considering a few actual condensation polymers as examples of step-growth systems. The actual systems we discuss will serve both to verify and reveal the limitations of the concepts we have been discussing. In addition, they point out some of the topics which still need clarification. We anticipate some of the latter points by noting the following:

1. When [A] ≠ [B], both ends of the growing chain tend to be terminated by the group which is present in excess. Subsequent reaction of such a molecule involves reaction with the limiting group. The effect is a decrease in the maximum attainable degree of polymerization.

2. When a monofunctional reactant is present—one containing a single A or B group—the effect is also clearly a decrease in the average degree of polymerization. It is precisely because this type of reactant can only react once that it is sometimes introduced into polymer formulations, thereby eliminating the possibility of long-term combination of chain ends.

                                          A
3. When monomers of the type A⊥A or even greater functionality are in-
    volved, the effect of their incorporation into the growing polymer chain is
    to introduce a branch point into the polymer.
4. In some cases of multifunctional monomers the possibility exists for
    branches on branches, which ultimately result in cross-linked products and
    effectively infinite molecular weights well before p reaches unity.

Polyesters and polyamides are two of the most studied step-growth polymers, as
well as being substances of great commercial importance. We shall consider
polyesters in the next section, and polyamides in Sec. 5.6.

5.5 Polyesters

The preceding discussions of the kinetics and molecular weight distributions in
the step-growth polymerization of AB monomers are clearly exemplified by the
esterification reactions of such monomers as glycolic acid or ω-hydroxydecanoic
acid. Therefore one method for polyester synthesis is the following:

1. Esterification of a hydroxycarboxylic acid.

Several other chemical reactions are also widely used for the synthesis of these
polymers. This list enumerates some of the possibilities and Table 5.3 illustrates
these reactions by schematic chemical equations:

2. Esterification of a diacid and a diol.
3. Ester interchange with alcohol.
4. Ester interchange with ester.
5. Esterification of acid chlorides.
6. Lactone polymerization.

We have not attempted to indicate the conditions of temperature, catalyst,
solvent, and so on, for these various reactions. For this type of information,
references that deal specifically with synthetic polymer chemistry should be
consulted. In the next few paragraphs we shall comment on the various routes
to polyester formation in the order summarized above and followed in Table 5.3.
    The studies summarized in Figs. 5.2 and 5.3 are examples of reaction 2 in
Table 5.3. While bifunctional reactants are the sort we have emphasized until
now, both monofunctional compounds and monomers with functionality greater
than 2 are present in some polymerization processes, either intentionally or
adventitiously. The effect of the monofunctional reactant is clearly to limit
chain growth. As noted above, a functionality greater than 2 results in branching. A type of polyester that includes mono-, di-, and trifunctional monomers
is the so-called alkyd resin. A typical example is based on the polymerization of
phthalic acid (or anhydride), glycerol, and an unsaturated monocarboxylic acid.
The following suggests the structure of a portion of such a polyester:

$$
\begin{array}{l}
CH_2-OH \\
| \\
CH-OH \\
| \\
CH_2-OH
\end{array}
\quad + \quad
\text{[phthalic acid]}
\quad + \quad
CH_3CH=(CH_2)_m\overset{\displaystyle O}{\overset{\|}{C}}-OH
\quad \rightarrow
$$

(5.J)

$$
\sim O-CH_2-CH-CH_2-O-\overset{O}{\overset{\|}{C}}-\underset{\text{[benzene ring]}}{C-C}-\overset{O}{\overset{\|}{C}}-O\sim
$$

$$
O\overset{\diagup C}{\diagdown}(CH_2)_mCH=CHCH_3
$$

The presence of the unsaturated substituent along this polyester backbone gives this polymer crosslinking possibilities through a secondary reaction of the double bond. These polymers are used in paints, varnishes, and lacquers, where the ultimate cross-linked product results from the oxidation of the double bond as the coating cures. A cross-linked polyester could also result from reaction (5.J) without the unsaturated carboxylic acid, but the latter would produce a gel in which the entire reaction mass solidified and is not as well suited to coatings applications as the polymer that crosslinks upon "drying."

Many of the reactions listed at the beginning of this section are acid catalyzed, although a number of basic catalysts are also employed. Esterifications are equilibrium reactions, and the reactions are often carried out at elevated temperatures for favorable rate and equilibrium constants and to shift the equilibrium in favor of the polymer by volatilization of the by-product molecules. An undesired feature of higher polymerization temperatures is the increased probability of side reactions such as the dehydration of the diol or the pyrolysis of the ester. Basic catalysts produce less of the undesirable side reactions.

Ester interchange reactions are valuable, since, say, methyl esters of dicarboxylic acids are often more soluble and easier to purify than the diacid itself. The methanol by-product is easily removed by evaporation. Poly(ethylene terephthalate) is an example of a polymer prepared by double application of reaction 4 in Table 5.3. The first stage of the reaction is conducted at temperatures below $200°C$ and involves the interchange of dimethyl terephthalate with ethylene glycol

$$
CH_3O\overset{O}{\overset{\|}{C}}-\text{[benzene]}-\overset{O}{\overset{\|}{C}}OCH_3 + 2HO-CH_2CH_2-OH \rightarrow HO-C_2H_4O\overset{O}{\overset{\|}{C}}-\text{[benzene]}-\overset{O}{\overset{\|}{C}}OC_2H_4-OH
$$

$$
+ 2 CH_3OH
$$

(5.K)

Table 5.3   Some Schematic Reactions for the Formation of Polyesters

1.   Esterification of a hydroxycarboxylic acid:

$$n \; HO\text{–}R\text{–}COOH \;\; \rightleftarrows \;\; \underset{\displaystyle \{O\text{–}R\text{–}\overset{\textstyle O}{\overset{\|}{C}}\}_n}{} + n \; H_2O$$

2.   Esterification of diacid and diol:

$$n \; HO\text{–}R_1\text{–}OH + n \; HOOC\text{–}R_2\text{–}COOH \;\; \rightleftarrows \;\; \{O\text{–}R_1\text{–}O\text{–}\overset{O}{\overset{\|}{C}}\text{–}R_2\text{–}\overset{O}{\overset{\|}{C}}\}_n + 2nH_2O$$

3.   Ester interchange with alcohol:

$$R_1\text{–}\overset{O}{\overset{\|}{C}}\text{–}OR_2 + R_3\text{–}OH \;\; \rightleftarrows \;\; R_1\text{–}\overset{O}{\overset{\|}{C}}\text{–}OR_3 + R_2\text{–}OH$$

4.   Ester interchange with ester:

$$R_1\text{–}\overset{O}{\overset{\|}{C}}\text{–}OR_2 + R_3\text{–}\overset{O}{\overset{\|}{C}}\text{–}OR_4 \;\; \rightleftarrows \;\; R_1\text{–}\overset{O}{\overset{\|}{C}}\text{–}OR_4 + R_3\text{–}\overset{O}{\overset{\|}{C}}\text{–}OR_2$$

5.   Esterification of acid chlorides (Schotten–Baumann reaction):

$$n \; HO\text{–}R_1\text{–}OH + ClCOR_2COCl \;\; \rightleftarrows \;\; \{O\text{–}R_1\text{–}O\text{–}\overset{O}{\overset{\|}{C}}\text{–}R_2\text{–}\overset{O}{\overset{\|}{C}}\} + 2nHCl$$

6.   Lactone polymerization:

$$n \; R\underset{\diagdown O}{\overset{\diagup C=O}{\big|}} \;\; \rightleftarrows \;\; \{O\text{–}R\text{–}\overset{O}{\overset{\|}{C}}\}_n$$

The rate of this reaction is increased by using excess ethylene glycol, and removal of the methanol is assured by the elevated temperature. Polymer is produced in the second stage after the temperature is raised above the melting point of the polymer, about 260°C.

$$n \ HO-C_2H_4OC-\underset{\|}{\overset{O}{}}\!\!\!\!\!\bigcirc\!\!\!\!\!-CO \ C_2H_4OH \ \rightarrow \ \{O-C_2H_4OC-\bigcirc-C\}_n + HO-C_2H_4-OH$$

(5.L)

The ethylene glycol liberated by reaction (5.L) is removed by lowering the pressure or purging with an inert gas. Because the ethylene glycol produced by reaction (5.L) is removed, proper stoichiometry is assured by proceeding via the intermediate, bis(2-hydroxyethyl) terephthalate; otherwise the excess glycol used initially would have a deleterious effect on the degree of polymerization. Poly(ethylene terephthalate) is more familiar by some of its trade names: Mylar as a film and Dacron, Kodel, or Terylene as fibers; it is also known by the acronym PET.

Ester interchange reactions like that shown in reaction 4 in Table 5.3 can be carried out on polyesters themselves to produce a scrambling between the two polymers. Studies of this sort between high and low molecular weight prepolymers result in a single polymer with the same molecular weight distribution as would have been obtained from a similarly constituted diol-diacid mixture by direct polymerization. This is true when the time-catalyst conditions allow the randomization to reach equilibrium. If the two prepolymers are polyesters formed from different monomers, the product of the ester interchange reaction will be a copolymer of some sort. If the reaction conditions (time-catalyst) favor esterification, the two chains will merely link together and a block copolymer results. If the conditions favor the ester interchange reaction, then a scrambled copolymer molecule results. These possibilities underscore the idea that the derivations of the preceding sections are based on complete equilibrium among all molecular species present during the condensation reaction.

Example 5.3

It has been hypothesized that cross-linked polymers would have better mechanical properties if interchain bridges were located at the ends rather than the center of chains. To test this, low molecular weight polyesters were synthesized†

---

†A. Szayna, *Ind. Eng. Chem. Prod. Res. Dev.* 2:105 (1963).

from a diol and two different diacids: one saturated, the other unsaturated. The synthetic procedure was such that the unsaturated acid units were located at either the center (centrene) or the ends (endene) of the chains. Some pertinent aspects of the overall experiment are listed here:

	Endene	Centrene
Step 1: 8 hr at about 150–200°C		
Maleic anhydride (mol)	0	2.0
Succinic anhydride (mol)	2.0	0
Diethylene glycol (mol)	3.0	3.0
Step 2: About ½ hr at about 120–130°C		
Maleic anhydride (mol)	2.0	0
Succinic anhydride (mol)	0	2.0
Catalyst	0	0
Step 3: 30% styrene + catalyst		
16 hr at 55°C + 1 hr at 110°C		
Elastic modulus	21,550	16,500

On the basis of these facts, do the following:

1. Comment on the likelihood that the comonomers are segregated as the names of these polymers suggest.
2. Sketch the structure of the average endene and centrene molecules, as well as the structure of one of the cross-linked products.
3. Comment on the results in terms of the initial hypothesis.

Solution

1. Since the reaction conditions are mild in step 2 (only 6% as much time allowed as in step 1 at a lower temperature) and no catalyst is present, it seems unlikely that any significant amount of ester scrambling occurs. Isomerization of maleate to fumarate is also known to be insignificant under these conditions.
2. The idealized structures of these molecules are

HOOC—$(CH_2)_2$—COO   OOC—CH=CH—COO   OOC—CH=CH—COO   OOC—$(CH_2)_2$—COOH

$(CH_2)_2$—O—$(CH_2)_2$ $(CH_2)_2$—O—$(CH_2)_2$ $(CH_2)_2$—O—$(CH_2)_2$

Centrene

HOOC—CH=CH—COO   OOC—$(CH_2)_2$—COO   OOC—$(CH_2)_2$—COO   OOC—CH=CH—COOH

$(CH_2)_2$—O—$(CH_2)_2$ $(CH_2)_2$—O—$(CH_2)_2$ $(CH_2)_2$—O—$(CH_2)_2$

Endene

$$\{CH_2—CH\}_n C—C \{CH_2—CH\}_{n'}$$

Cross–linked
Centrene

3. A cross-linked product with unsaturation at the chain ends does, indeed, have a higher modulus. This could be of commercial importance and indicates that industrial products might be formed by a nonequilibrium process precisely for this sort of reason.

•

Acid chlorides are generally more reactive than the parent acids, so polyester formation via reaction 5 in Table 5.3 can be carried out in solution and at lower temperatures, in contrast with the bulk reactions of the melt as described above. Again, the by-product molecules must be eliminated either by distillation or precipitation. The method of interfacial condensation, described in the next section, can be applied to this type of reaction.

The formation of polyesters from the polymerization of lactones (reaction 6 in Table 5.3) is a ring-opening reaction that may follow either a step-growth or chain mechanism, depending on conditions. For now our only concern is to note that the equilibrium representing this reaction in Table 5.3 describes polymerization by the forward reaction and ring formation by the back reaction. Rings clearly compete with polymers for monomer in all polymerizations. Throughout the chapter we have assumed that all competing side reactions including ring formation could be neglected. We shall examine the factors which influence the ring–polymer equilibrium in Sec. 5.10.

5.6 Polyamides

The discussion of polyamides parallels that of polyesters in many ways. To begin with, polyamides may be formed from an AB monomer, in this case amino acids:

1. Amidation of amino acids.

Additional synthetic routes which closely resemble the polyesters are also available. Several more of these are listed below and are illustrated by schematic reactions in Table 5.4:

2. Amidation of a diacid and a diamine.
3. Interchange reactions.
4. Amidation of acid chlorides.
5. Lactam polymerization.

We only need to recall the trade name of synthetic polyamides, nylon, to recognize the importance of these polymers and the reactions employed to prepare them. Remember from Sec. 1.5 the nylon system for naming these

Table 5.4 Some Schematic Reactions for the Formation of Polyamides

1. Amidation of amino acids:

$$n\ H_2N-R-COOH \ \rightleftarrows \ \left[N-R-\overset{\overset{O}{\parallel}}{\underset{\underset{H}{|}}{C}}\right]_n + nH_2O$$

2. Amidation of diacid and diamine:

$$n\ H_2N-R_1-NH_2 + n\ HOOC-R_2-COOH \ \rightleftarrows \ \left[N-R_1-N-\overset{O}{\overset{\parallel}{C}}-R_2-\overset{O}{\overset{\parallel}{C}}\right]_n + 2n\ H_2O$$

3. Interchange reactions:

$$R_1-\overset{O}{\overset{\parallel}{C}}-\overset{H}{\overset{|}{N}}-R_2 + R_3-\overset{O}{\overset{\parallel}{C}}-\overset{H}{\overset{|}{N}}-R_4 \ \rightleftarrows \ R_1-\overset{O}{\overset{\parallel}{C}}-\overset{H}{\overset{|}{N}}-R_4 + R_3-\overset{O}{\overset{\parallel}{C}}-\overset{H}{\overset{|}{N}}-R_2$$

4. Amidation of acid chlorides:

$$n\ H_2N-R_1-NH_2 + n\ ClCOR_2COCl \ \rightleftarrows \ \left[N-R_1-N-\overset{O}{\overset{\parallel}{C}}-R_2-\overset{O}{\overset{\parallel}{C}}\right]_n + 2n\ HCl$$

5. Lactam polymerization:

$$n\ R\begin{matrix} N-H \\ | \\ C=O \end{matrix} \ \rightleftarrows \ \left[N-R-\overset{O}{\overset{\parallel}{C}}\right]_n$$

compounds: the first number after the name gives the number of carbon atoms in the diamine, and the second, the number of carbons in the diacid.

The diacid–diamine amidation described in reaction 2 in Table 5.4 has been widely studied in the melt, in solution, and in the solid state. When equal amounts of two functional groups are present, both the rate laws and the molecular weight distributions are given by the treatment of the preceding sections. The stoichiometric balance between reactive groups is readily obtained by precipitating the 1:1 ammonium salt from ethanol:

$$H_2N-R_1-NH_2 \ + HOOC-R_2-COOH \longrightarrow \ \begin{array}{c} H_3N^{\oplus}-R_1-{^{\oplus}}NH_3 \\ {^{\ominus}}OOC-R_2-COO^{\ominus} \end{array} \qquad (5.M)$$

This compound is sometimes called a nylon salt. The salt $\rightleftarrows$ polymer equilibrium is more favorable to the production of polymer than in the case of polyesters, so this reaction is often carried out in a sealed tube or autoclave at about 200°C until a fairly high extent of reaction is reached; then the temperature is raised and the water driven off to attain the high molecular weight polymer.

The process represented by reaction 2 in Table 5.4 actually entails a number of additional equilibrium reactions. Some of the equilibria that have been considered include the following:

$$\sim NH_3^+ + {^-}OOC\sim \rightleftarrows \sim CONH_2\sim + H_2O \qquad (5.N)$$

$$\sim COOH + H_2O \rightleftarrows \sim COO^- + H_3O^+ \qquad (5.O)$$

$$\sim NH_2 + H_2O \rightleftarrows \sim NH_3^+ + OH^- \qquad (5.P)$$

$$2H_2O \rightleftarrows H_3O^+ + OH^- \qquad (5.Q)$$

$$\sim COOH + H^+ \rightleftarrows \sim C(OH)_2^+ \qquad (5.R)$$

$$\sim C(OH)_2^+ + H_2N\sim \rightleftarrows \sim CONH\sim + H^+ + H_2O \qquad (5.S)$$

Reaction (5.N) describes the nylon salt $\rightleftarrows$ nylon equilibrium. Reactions (5.O) and (5.P) show proton transfer with water between carboxyl and amine groups. Since proton transfer equilibria are involved, the self-ionization of water, reaction (5.Q), must also be included. Especially in the presence of acidic catalysts, reactions (5.R) and (5.S) are the equilibria of the acid–catalyzed intermediate described in general in reaction (5.G). The main point in including all of these equilibria is to indicate that the precise concentration of A and B

groups in a diacid–diamine reaction mixture is a complicated function of the moisture content and the pH, as well as the initial amounts of reactants introduced. Because of the high affinity for water of the various functional groups present, the complete removal of water is impossible: The equilibrium moisture content of molten nylon-6,6 at 290°C under steam at 1 atm is 0.15%. Likewise, the various ionic possibilities mean that at both high and low pH values the concentration of un-ionized carboxyl or amine groups may be considerably different from the total concentration—without regard to state of ionization—of these groups. As usual, upsetting the stoichiometric balance of the reactive groups has a lowering effect on the degree of polymerization attainable. The abundance of high-quality nylon products is evidence that these complications have been overcome in practice.

Amide interchange reactions of the type represented by reaction 3 in Table 5.4 are known to occur more slowly than direct amidation; nevertheless, reactions between high and low molecular weight polyamides result in a polymer of intermediate molecular weight. The polymer is initially a block copolymer of the two starting materials, but randomization is eventually produced.

As with polyesters, the amidation reaction of acid chlorides may be carried out in solution because of the enhanced reactivity of acid chlorides compared with carboxylic acids. A technique known as interfacial polymerization has been employed for the formation of polyamides and other step-growth polymers, including polyesters, polyurethanes, and polycarbonates. In this method the polymerization is carried out at the interface between two immiscible solutions, one of which contains one of the dissolved reactants, while the second monomer is dissolved in the other. Figure 5.7 shows a polyamide film forming at the interface between an aqueous solution of a diamine layered on a solution of a diacid chloride in an organic solvent. In this form interfacial polymerization is part of the standard repertoire of chemical demonstrations. It is sometimes called the nylon rope trick because of the filament of nylon produced by withdrawing the collapsed film.

The amidation of the reactive groups in interfacial polymerization is governed by the rates at which these groups can diffuse to the interface where the growing polymer is deposited. Accordingly, new reactants add to existing chains rather than interacting to form new chains. This is different than the bulk mechanism we have discussed elsewhere in this chapter, and it is evident that a higher molecular weight polymer results from this difference.

The HCl by-product of the amidation reaction is neutralized by also dissolving an inorganic base in the aqueous layer in interfacial polymerization. The choice of the organic solvent plays a role in determining the properties of the polymer produced, probably because of differences in solvent goodness for the resulting polymer. Since this reaction is carried out at low temperatures, the complications associated with side reactions can be kept to a minimum.

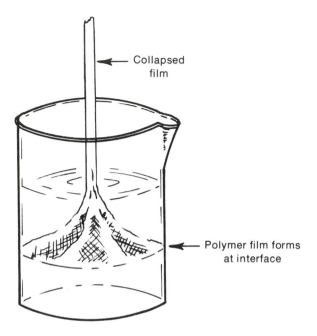

Figure 5.7 Sketch of an interfacial polymerization with the collapsed polymer film being withdrawn from the surface between the immiscible phases. [Redrawn with permission from P. W. Morgan and S. L. Kwolek, *J. Chem. Educ.* 36:182 (1959); copyright by the American Chemical Society.]

Polymer yield may be increased by increasing the area of the interface between the two solutions by stirring.

Lactam polymerization represented by reaction 5 in Table 5.4 is another example of a ring-opening reaction, the reverse of which is a possible competitor with polymer for reactants. We shall discuss this situation in Sec. 5.10.

The various mechanical properties of polyamides may be traced in many instances to the possibility of intermolecular hydrogen bonding between the polymer molecules and to the relatively stiff chains these substances possess. The latter, in turn, may be understood by considering still another equilibrium, this one among resonance structures along the chain backbone:

$$\sim\!\!\underset{\substack{\| \\ }}{C}\!\!-\!\!\underset{\substack{| \\ }}{N}\!\!\sim \; \rightleftarrows \; \sim\!\!\underset{\substack{| \\ }}{\overset{\ominus}{C}}\!\!=\!\!\underset{\substack{| \\ }}{\overset{\oplus}{N}}\!\!\sim \tag{5.T}$$

The combination of strong intermolecular forces and high chain stiffness accounts for the high melting points of polyamides through application of Eq. (4.5).

The remarks of this and the last section are only a small fraction of what might be said about these important materials. We have commented on some aspects of the polymerization processes and of the polymers themselves that have a direct bearing on the concepts discussed here and elsewhere in this volume. This material provides an excellent example of the symbiosis between theoretical and application-oriented points of view. Each stimulates and reinforces the other with new challenges, although it must be conceded that many industrial processes reach a fairly high degree of empirical refinement before the conceptual basis is quantitatively developed.

5.7 Stoichiometric Imbalance

We now turn to two of the problems we have sidestepped until now. In this section we consider the polymerization of reactants in which a stoichiometric imbalance exists in the numbers of reactive groups A and B. In the next section we shall consider the effect of monomers with a functionality greater than 2.

In prior sections dealing with the quantitative aspects of step-growth polymerization, we focused attention on monomers of the AB type to assure equality of reactive groups. The results obtained above also apply to AA and BB polymerizations, provided that the numbers of reactive groups are equal. There are obvious practical difficulties associated with the requirement of stoichiometric balance. Rigorous purification of monomers is difficult and adds to the cost of the final product. The effective loss of functional groups to side reactions imposes restrictions on the range of experimental conditions at best and is unavoidable at worst. These latter considerations apply even in the case of the AB monomer. We have already stated that the effect of the imbalance of A and B groups is to lower the eventual degree of polymerization of the product. A quantitative assessment of this limitation is what we now seek.

We define the problem by assuming the polymerization involves AA and BB monomers and that the B groups are present in excess. We define ν_A and ν_B to be the numbers of A and B functional groups, respectively. The number of either of these quantities in the initial reaction mixture is indicated by a superscript 0; the numbers at various stages of reaction have no superscript. The stoichiometric imbalance is defined by the ratio r, where

$$r = \frac{\nu_A^0}{\nu_B^0} \tag{5.34}$$

This ratio cannot exceed unity by definition of the problem.

As with other problems with stoichiometry, it is the less abundant reactant that limits the product. Accordingly, we define the extent of reaction p to be the fraction of A groups that have reacted at any point. Since A and B groups

react in a 1:1 proportion, the number of B groups that have reacted when the extent of reaction has reached p equals $p\nu_A^0$, which, in turn, equals $pr\nu_B^0$. The product pr gives the fraction of B groups that have reacted at any point. With these definitions in mind, the following relationships are readily seen:

1. The number of unreacted functional groups after the reaction reaches extent p is

$$\nu_A = (1 - p)\nu_A^0 \tag{5.35}$$

and

$$\nu_B = (1 - pr)\nu_B^0 = (1 - pr)\frac{\nu_A^0}{r} \tag{5.36}$$

2. The total number of chain ends is the sum of Eqs. (5.35) and (5.36):

$$\nu_{ends} = \left(1 - p + \frac{1 - pr}{r}\right)\nu_A^0 \tag{5.37}$$

3. The total number of chains is half the number of chain ends:

$$N_{chains} = \frac{1}{2}\left(1 + \frac{1}{r} - 2p\right)\nu_A^0 \tag{5.38}$$

4. The total number of repeat units distributed among these chains is the number of monomer molecules present initially:

$$N_{repeat\ units} = \frac{1}{2}\nu_A^0 + \frac{1}{2}\nu_B^0 = \frac{1}{2}\left(1 + \frac{1}{r}\right)\nu_A^0 \tag{5.39}$$

The number average degree of polymerization is given by dividing the number of repeat units by the number of chains, or

$$\bar{n}_n = \frac{1 + 1/r}{1 + 1/r - 2p} = \frac{1 + r}{1 + r - 2pr} \tag{5.40}$$

As a check that we have done this correctly, note that Eq. (5.40) reduces to the previously established Eq. (5.28) when r = 1.00.

One distinction that should be pointed out involves the comparison of Eqs. (5.1) and (5.40). In the former we considered explicitly the AB monomer, while the latter is based on the polymerization of AA and BB monomers. In both instances $\bar{n}_n$ is obtained by dividing the total number of monomer molecules initially present by the total number of chains after the reaction has occurred to extent p. Following the same procedure for different reaction

mixtures results in a different definition of the repeat unit. In the case of the AB monomer, the repeat unit is the ab entity, which differs from AB by the elimination of the by-product molecule. In the case of the AA and BB monomers, the repeat unit in the polymer is the aabb unit, which differs from AA + BB by two by-product molecules. Equation (5.1) counts the number of ab units in the polymer directly. Equation (5.40) counts the number of aa plus bb units. The number of aa plus bb units is twice the number of aabb units. Rather than attempting to formalize this distinction by introducing more complex notation, we simply point out that application of the formulas of this chapter to specific systems must be accompanied by a reflection on the precise meaning of the calculated quantity for the system under consideration. There is nothing particularly unique about this admonition: No equation should be used mindlessly by just plugging in numbers and submitting an answer.

The distinction pointed out in the last paragraph carries over to the evaluation of $\bar{M}_n$ from $\bar{n}_n$. We assume that the chain length of the polymer is great enough to make unnecessary any correction for the uniqueness of chain ends. In such a case the molecular weight of the polymer is obtained from the degree of polymerization by multiplying the latter by the molecular weight of the repeat unit. The following examples illustrate the distinction under consideration:

1. Polymerization of an AB monomer is illustrated by the polyester formed from glycolic acid. The repeat unit in this polymer has the structure $(-O-CH_2-\overset{\overset{\displaystyle O}{\|}}{C}-)$ and $M_0 = 58$. Neglecting end groups, we have $\bar{M}_n = 58\,\bar{n}_n$, with $\bar{n}_n$ given by Eq. (5.1).
2. Polymerization of AA and BB monomers is illustrated by butane-1,4-diol and adipic acid. The aabb repeat unit in the polymer has an M_0 value of 200. If Eq. (5.4) is used to evaluate $\bar{n}_n$, it gives the number of aa plus bb units; therefore $\bar{M}_n = 200(\bar{n}_n)/2$.
3. An equivalent way of looking at the conclusion of item (2) is to recall that Eq. (5.40) gives the (number average) number of monomers of both kinds in the polymer and multiply this quantity by the average molecular weight of the two kinds of units in the structure: $(88 + 112)/2 = 100$.

Equation (5.40) also applies to the case when some of the excess B groups present are in the form of monofunctional reagents. In this latter situation the definition of r is modified somewhat (and labeled with a prime) to allow for the fact that some of the B groups are in the BB-type monomers (unprimed) and some are in the monofunctional (primed) molecules:

$$r' = \frac{\nu_A}{\nu_B + 2\nu_B'} \tag{5.41}$$

The parameter r' continues to measure the ratio of the number of A and B groups; the factor 2 enters since the monofunctional reagent has the same effect on the degree of polymerization as a difunctional molecule with two B groups and, hence, is doubly effective compared to the latter. With this modification taken into account, Eq. (5.40) enables us to quantitatively evaluate the effect of stoichiometric imbalance or monofunctional reagents, whether these are intentionally introduced to regulate $\bar{n}_n$ or whether they arise from impurities or side reactions.

The parameter r varies between 0 and 1; as such it has the same range as p. Although the quantitative effect of r and p on n is different, the qualitative effect is similar for each: Higher degrees of polymerization are obtained the closer each of these fractions is to unity. Table 5.5 shows some values of n calculated from Eq. (5.40) for several combinations of (large values of) r and p. Inspection of Table 5.5 reveals the following:

1.   For any value of r, n is greater for larger values of p; this conclusion is the same whether the proportions of A and B are balanced or not.
2.   The final 0.05 increase in p has a bigger effect on n at r values that are closer to unity than for less-balanced mixtures.
3.   For any value of p, n is greater for larger values of r; stoichiometric imbalance lowers the average chain length of the preparation.
4.   A 0.05 increase in r produces a much bigger increase in n at p = 1.00 than in mixtures that have reacted to a lesser extent.

An interesting special case of Eq. (5.40) occurs when p = 1.00. In this case Eq. (5.40) becomes

$$\bar{n}_n = \frac{1 + r}{1 - r} \tag{5.42}$$

The following example illustrates some of the concepts developed in this section.

Table 5.5   Some Values of n Calculated by Eq. (5.40) for Several Values of r and p Close to Unity

			p	
r	0.95	0.97	0.99	1.00
0.95	13.5	18.2	28.3	39.0
0.97	15.5	22.3	39.9	65.7
0.99	18.3	28.7	66.8	199
1.00	20.0	33.3	100	∞

Example 5.4

It is desired to prepare a polyester with $\overline{M}_n$ = 5000 by reacting 1 mol of butane-1,4-diol with 1 mol of adipic acid.

1. Calculate the value of p at which the reaction should be stopped to obtain this polymer, assuming perfect stoichiometric balance and neglecting end group effects on $\overline{M}_n$.
2. Assuming that 0.5 mol % of the diol is lost to polymerization by dehydration to olefin, what would be the value of $\overline{M}_n$ if the reaction were carried out to the same extent as in (1)?
3. How could the loss in (2) be offset so that the desired polymer is still obtained?
4. Suppose the total number of carboxyl groups in the original mixture is 2 mol, of which 1.0% is present as acetic acid to render the resulting polymer inert to subsequent esterification. What value of p would be required to produce the desired polymer in this case, assuming no other stoichiometric imbalance?

Solution

The various expressions we have developed in this section relating p to the size of the polymer are all based on $\overline{n}_n$. Accordingly, we note that the average reactant molecule in this mixture has a molecular weight of 100 as calculated above. Therefore the desired polymer has a value of $\overline{n}_n$ = 50 based on this concept.

1. We use Eq. (5.28) for the case of equal numbers of A and B groups and find that $p = 1 - 1/n = 0.980$. Even though Eq. (5.28) is derived for an AB monomer, it applies to this case with the "average monomer" as the repeat unit.
2. Component AA is the diol in this case and ν_A = 0.995 mol; therefore $r = 0.995/1.00 = 0.995$. We use Eq. (5.40) and solve for $\overline{n}_n$ with p = 0.980 and r = 0.995:

$$\overline{n}_n = \frac{1.995}{1.995 - 2(0.995)(0.980)} = 44.5$$

   and $\overline{M}_n$ = 44.5(100) = 4450 g mol^{-1}
3. The effect of the lost hydroxyl groups can be offset by carrying out the polymerization to a higher extent of reaction. We use Eq. (5.40) and solve for p with $\overline{n}_n$ = 50 and r = 0.995:

$$p = \left(1 - \frac{1}{n}\right)\left(\frac{1+r}{2r}\right) = \left(1 - \frac{1}{50}\right)\frac{1.995}{1.990} = 0.9825$$

4.   The monofunctional reagent B′ is the acetic acid in this case and the number
     of monofunctional carboxyl groups is $2(0.010) = 0.020 = \nu'_B$. The number
     of B groups in BB monomers is $1.980 = \nu_B$. We use Eq. (5.41) to define r′
     for this situation, assuming the number of hydroxyl groups equals 2.00 mol:

$$r' = \frac{2.00}{1.980 + 2(0.020)} = 0.990$$

Equation (5.40) is now solved for p using n = 50 and r′ = 0.990:

$$p = \left(1 - \frac{1}{n}\right)\left(\frac{1 + r'}{2r'}\right) = \left(1 - \frac{1}{50}\right)\frac{1.990}{1.980} = 0.9849$$

It will be remembered from Sec. 5.3 that a progressively longer period of time
is required to shift the reaction to larger values of p. In practice, therefore, the
effects of side reactions and monofunctional reactants are often not compensated by longer polymerization times, but are accepted in the form of lower
molecular weight polymers.

   Next let us examine the effect of monomers with functionality greater than 2
on step-growth polymers.

5.8 Branching and Crosslinking

We noted above that the presence of monomer with a functionality greater
than 2 results in branched polymer chains. This in turn produces a three-
dimensional network of polymer under certain circumstances. The solubility
and mechanical behavior of such materials depend critically on whether the
extent of polymerization is above or below the threshold for the formation of
this network. The threshold is described as the gel point, since the reaction
mixture sets up or gels at this point. We have previously introduced the term
thermosetting to describe these cross-linked polymeric materials. Because their
mechanical properties are largely unaffected by temperature variations—in
contrast to thermoplastic materials which become more fluid on heating—step-
growth polymers that exceed the gel point are widely used as engineering
materials.

   For simplicity, we assume that the reaction mixture still contains only A
and B as reactive groups, but that either one (or both) of these is present (totally
or in part) in a molecule that contains more than two of the reactive groups.
We use f to represent the number of reactive groups in a molecule when this
quantity exceeds 2, and represent a multifunctional molecule A_f or B_f. For
example, the monomer $A \overset{A}{\underset{}{\downarrow}} A$ corresponds to f = 3. Several reaction possibilities
(all written for f = 3) come to mind in the presence of multifunctional reactants:

1. AA and BB plus either A_f or B_f:

$$AA + BB + A_3 \rightarrow Aabbaabba \left\langle \begin{array}{l} abbaabb \sim \\ abbaabba \end{array} \right. \left\langle \begin{array}{l} abb \sim \\ abbaa \sim \end{array} \right. \tag{5.U}$$

2. AA and B_f or BB and A_f:

$$AA + B_3 \rightarrow Aab \left\langle \begin{array}{l} baab \left\langle \begin{array}{l} baab \sim \\ baab \end{array} \right. \left\langle \begin{array}{l} baa \sim \\ baa \sim \end{array} \right. \\ baab \left\langle \begin{array}{l} baa \sim \\ baa \sim \end{array} \right. \end{array} \right. \tag{5.V}$$

3. AB with either AA and B_f or BB and A_f:

$$AB + BB + A_3 \rightarrow Abababa \left\langle \begin{array}{l} ababa \sim \\ ababa\underline{bb}ababa \end{array} \right. \left\langle \begin{array}{l} a \sim \\ a \sim \end{array} \right. \tag{5.W}$$

4. A_f and B_f:

$$A_3 + B_3 \rightarrow A \left\langle \begin{array}{l} a \sim \\ ab \end{array} \right. \left\langle \begin{array}{l} ba \\ b \sim \end{array} \right. \left\langle \begin{array}{l} ab \\ ab \sim \end{array} \right. \left\langle \begin{array}{l} b \sim \\ ba \end{array} \right. \left\langle \begin{array}{l} a \sim \\ ab \end{array} \right. \left\langle \begin{array}{l} b \sim \\ b \sim \end{array} \right. \tag{5.X}$$

5. Other possibilities include the involvement of still another functional group capable of reaction with A and/or B.

Reaction (5.W) is interesting inasmuch as either the AA or BB monomer must be present to produce crosslinking. Polymerization of AB with only A_f (or B_f) introduces a single branch point, but no more, since all chain ends are misoriented for further incorporation of branch points. Including the AA or BB molecule reverses this. The bb unit which accomplishes this in reaction (5.W) is underscored.

What we seek next is a quantitative relationship between the extent of the polymerization reaction, the composition of the monomer mixture, and the point of gelation. We shall base our discussion on the system described by reaction (5.U); other cases are derived by similar methods. To further specify the system we assume that A groups limit the reaction and that B groups are present in excess. Two parameters are necessary to characterize the reaction mixture:

1. The ratio of the initial number of A to B groups as defined by the factor r, given by Eq. (5.34). The total number of A groups from both AA and A_f are included in this application of r.

2. The fraction of A groups present in multifunctional molecules as defined by the ratio

$$\rho = \frac{\nu_{A \text{ from } A_f}}{\nu_{A, \text{ total}}} \tag{5.43}$$

There are two additional useful parameters which characterize the reaction itself:

3. The extent of reaction p is again based on the group present in limiting amount. For the system under consideration, p is the fraction of A groups that have reacted.

4. The probability that a chain segment is capped at both ends by a branch unit is described by the branching coefficient α. The branching coefficient is central to the discussion of gelation, since whether gelation occurs or not depends on what happens after capping a section of chain with a potential branch point.

The methods we consider have been developed by Stockmayer and Flory and have been applied to quite a variety of polymer systems and phenomena.

Our approach to the problem of gelation proceeds through two stages: First we consider the probability that AA and BB polymerize until all chain segments are capped by an A_f monomer; then we consider the probability that these are connected together to form a network. The actual molecular processes occur at random and not in this sequence, but mathematical analysis is feasible if we consider the process in stages. As long as the same sort of structure results from both the random and the subdivided processes, the analysis is valid.

The arguments we employ are statistical, so we recall that the probability of a functional group reacting is given by the fraction of groups that have reacted at any point, and the probability of a sequence of events is the product of their individual probabilities. We continue to use the concept that functional group reactivity is independent of the size of the molecule to which the group is attached. One additional assumption that enters when multifunctional monomers are considered is that all A groups in A_f are of equal reactivity.

Now let us consider the probability that a section of polymer chain is capped at both ends by potential branch points:

1. The first step is the condensation of a BB monomer with one of the A groups of an A_f molecule:

$$A_f + BB \longrightarrow A_{f-1} \, abB$$

Since all A groups have the same reactivity by hypothesis, the probability of this occurrence is simply p.

2. The terminal B group reacts with an A group from AA rather than A_f:

$$A_{f-1} \, abB + AA \longrightarrow A_{f-1} \, abbaaA$$

The fraction of unreacted B groups is rp, so this gives the probability of reaction for B. Since ρ is the fraction of A groups on multifunctional monomers, rp must be multiplied by $1 - \rho$ to give the probability of B reacting with an AA monomer. The total probability for the chain shown is the product of the probabilities considered until now:

$$p[rp(1 - \rho)]$$

3. The terminal A group reacts with another BB:

$$A_{f-1} \, abbaaA + BB \longrightarrow A_{f-1} \, abbaabB$$

The probability of this step is again p, and the total probability is

$$p[rp(1 - \rho)p]$$

4. Additional AA and BB molecules condense into the chain to give a sequence of n bbaa units:

$$A_{f-1} \, abbaabB + AA + BB \longrightarrow \longrightarrow \longrightarrow A_{f-1} \, a(bbaa)_n bB$$

We have just evaluated the probability of one such unit; the probability for a series of n units is just the product of the individual probabilities:

$$p[rp(1 - \rho)p]^n$$

5. The terminal B group reacts with an A group from a multifunctional monomer:

$$A_{f-1} \, a(bbaa)_n bB + A_f \longrightarrow A_{f-1} \, a(bbaa)_n bbaA_{f-1}$$

The probability of B reacting is rp and the fraction of these reactions with A_f molecules is rpρ. The probability of the entire sequence is

$$p[rp(1 - \rho)p]^n rp\rho$$

6. In the general expression above, n can have any value from 0 to ∞, so the probability for all possibilities is the sum of the individual probabilities.

Note that a different procedure is used for compounding probabilities here: the sum instead of the product. This time we are interested in *either* n = 0 *or* n = 1 *or* n = 2, and so forth. We have previously required the first A–B reaction *and* the second A–B reaction *and* the third A–B reaction and so on.

Since the branching coefficient gives the probability of a chain segment being capped by potential branch points, the above development describes this situation:

$$\alpha = \sum_{n=0}^{\infty} rp^2 \rho \, [rp^2(1-\rho)]^{\,n} \tag{5.44}$$

The summation applies only to the quantity in brackets, since it alone involves n. Representing the bracketed quantity by Q, we note that $\sum_{1}^{\infty} Q^n = 1 + Q + Q^2 + \cdots = 1/(1-Q)$; therefore

$$\alpha = \frac{rp^2 \rho}{1 - rp^2(1-\rho)} \tag{5.45}$$

We have now completed the first stage of the problem we set out to consider. We have arrived at the probability that chains are capped at both ends by potential branch points. The second stage of the derivation is concerned with the reaction between these chain ends via the remaining f − 1 reactive A groups. By hypothesis, the mixture contains an excess of B groups, so there are still unreacted BB monomers or other polymer chain segments with terminal B groups which can react with the A_{f-1} groups we have been considering. Table 5.6 shows the connecting of such groups—by BB molecules for simplicity—for several values of f. For each of the situations shown in Table 5.6, converting the boxed A BB A groups into a condensed abba sequence amounts to linking into a linear polymer the capped segments which had been separate until now. Since the capped ends have f − 1 remaining functional groups at this point, a linear condensation product results when any one of these groups reacts. Thus 1/(f − 1) is the probability of this particular eventuality.

Our interest from the outset has been in the possibility of crosslinking which accompanies inclusion of multifunctional monomers in a polymerizing system. Note that this does not occur when the groups enclosed in boxes in Table 5.6 react; however, any reaction *beyond* this for the terminal A groups will result in a cascade of branches being formed. Therefore a critical (subscript c) value for the branching coefficient occurs at

$$\alpha_c = \frac{1}{f-1} \tag{5.46}$$

For an extent of reaction corresponding to $\alpha > \alpha_c$, gelation is predicted to occur.

Table 5.6 Schematic Illustration Showing the Formation of a Linear Polymer by the Reaction of One of the f − 1 Reactive Groups at the End of a Portion of Polymer

Combining Eqs. (5.45) and (5.46) and rearranging gives the critical extent of reaction for gelation p_c as a function of the properties of the monomer mixture r, ρ, and f:

$$p_c = \frac{1}{[r + r\rho (f - 2)]^{1/2}} \tag{5.47}$$

This equation describes the extent of reaction at which the system is predicted to gel.

Equation (5.47) is of considerable practical utility in view of the commercial importance of three-dimensional polymer networks. Some reactions of the sort we have considered are carried out on a very large scale: Imagine the consequences of having a polymer preparation solidify in a large and expensive reaction vessel because the polymerization reaction went a little too far! Considering this kind of application, we might actually be relieved to know that Eq. (5.47) errs in the direction of underestimating the extent of reaction at

gelation. This comes about because some reactions of the multifunctional branch points result in intramolecular loops which are wasted as far as network formation is concerned. This possibility is readily apparent in Table 5.6.

As an example of the quantitative testing of Eq. (5.47), consider the polymerization of diethylene glycol (BB) with adipic acid (AA) in the presence of 1,2,3-propane tricarboxylic acid (A_3). The critical value of the branching coefficient is 0.50 for this system by Eq. (5.46). For an experiment in which r = 0.800 and ρ = 0.375, p_c = 0.953 by Eq. (5.47). The critical extent of reaction, determined by titration, in the polymerizing mixture at the point where bubbles fail to rise through it was found experimentally to be 0.9907. Calculating back from Eq. (5.45), the experimental value of p_c is consistent with the value α_c = 0.578.

Examination of several special cases of Eq. (5.47) gives us a better appreciation of the several factors which operate in that relationship:

1.   If $\nu_A{}^0 = \nu_B{}^0$ and r = 1,

$$p_c = \frac{1}{[1 + \rho (f - 2)]^{\frac{1}{2}}} \tag{5.48}$$

2.   Using the case of f = 3 and ρ = 0.3 as an example, p_c = 0.877 when r = 1 in contrast to, say, 0.886 when r = 0.98. This illustrates that failure to maintain stoichiometric balance continues to have a limiting effect on polymerization in this case also.

3.   If $\nu_A{}^0 < \nu_B{}^0$ (r < 1) and all A groups are present as A_f ($\rho = 1$),

$$p_c = \frac{1}{[r + r (f - 2)]^{\frac{1}{2}}} \tag{5.49}$$

4.   Using the case of f = 3 and r = 0.98 as an example, p_c = 0.714 when ρ = 1 in contrast to, say, 0.886 when ρ = 0.3. The higher the proportion of groups present as branch points, the lower the threshold for gelation.

It is apparent that numerous other special systems or effects could be considered to either broaden the range or improve the applicability of the derivation presented. Our interest, however, is in illustrating concepts rather than exhaustively exploring all possible cases, so we shall not pursue the matter of gelation further. Instead, we conclude this section with a brief examination of the molecular weight averages in the system generated from AA, BB, and A_f. For simplicity, we restrict our attention to the case of $\nu_A{}^0 = \nu_B{}^0$. As an approach to the number average degree of polymerization, it is useful to define the average functionality $\bar{f}$ of a monomer as

$$\bar{f} = \frac{\Sigma_i N_i f_i}{\Sigma_i N_i} \tag{5.50}$$

where N_i and f_i are the number of molecules and the functionality of the ith component in the reaction mixture, respectively. The summations are over all monomers. If N is the total number of molecules present in the reaction mixture at an extent of reaction p and N_0 is the number of molecules present initially, then $2(N_0 - N)$ is the number of functional groups that have reacted and $\bar{f}N_0$ is the total number of groups initially present. Two conclusions follow immediately from these concepts:

$$\bar{n}_n = \frac{N_0}{N} \tag{5.51}$$

and

$$p = \frac{2(N_0 - N)}{\bar{f}N_0} \tag{5.52}$$

Elimination of N from these expressions gives

$$\bar{n}_n = \frac{2}{2 - p\bar{f}} \tag{5.53}$$

This result is known as the Carothers equation. It is apparent that this expression reduces to Eq. (5.4) for the case of $\bar{f} = 2$. Furthermore, when $\bar{f}$ exceeds 2, as in the $AA/BB/A_f$ mixture under consideration, then $\bar{n}_n$ is increased over the value obtained at the same p for $\bar{f} = 2$. A numerical example will help clarify these relationships:

Example 5.5

An AA, BB, and A_3 polymerization mixture is prepared in which $v_A{}^0 = v_B{}^0 = 3.00$, with 10% of the A groups contributed by A_3. Use Eq. (5.53) to calculate $\bar{n}_n$ for p = 0.970 and p for $\bar{n}_n = 200$. In each case compare the results with what would be obtained if no multifunctional A were present.

Solution

Determine the average functionality of the mixture. The total number of functional groups is 6.00 mol, but the total number of molecules initially present must be determined. Using $3N_{A_3} + 2N_{AA} = 3.00$ and $3N_{A_3}/3 = 0.100$, we obtain that $N_{AA} = 1.350$ and $N_{A_3} = 0.100$. Since $N_B = 1.500$, the total number of moles initially present is $N_0 = 1.350 + 0.100 + 1.500 = 2.950$:

$$\overline{f} = \frac{0.100(3) + 1.350(2) + 1.500(2)}{2.950} = \frac{6.000}{2.950} = 2.034$$

Solve Eq. (5.53) with p = 0.970 and $\overline{f}$ = 2.034:

$$\overline{n}_n = \frac{2}{2 - 0.97(2.034)} = 73.8$$

For comparison, solve Eq. (5.4) with p = 0.970:

$$\overline{n}_n = \frac{1}{1 - 0.970} = 33.3$$

Solve Eq. (5.53) with n = 200 and $\overline{f}$ = 2.034:

$$p = \frac{(1 - 1/\overline{n}_n)(2)}{\overline{f}} = \frac{0.995(2)}{2.034} = 0.978$$

Solve Eq. (5.4) with n = 200:

$$p = \left(1 - \frac{1}{\overline{n}_n}\right) = \left(1 - \frac{1}{200}\right) = 0.995$$

For a fixed extent of reaction, the presence of multifunctional monomers in an equimolar mixture of reactive groups increases the degree of polymerization. Conversely, for the same mixture a lesser extent of reaction is needed to reach a specified $\overline{n}_n$ with multifunctional reactants than without them. Remember that this entire approach is developed for the case of stoichiometric balance. If the numbers of functional groups are unequal, this effect works in opposition to the multifunctional groups.

•

The Carothers approach, being an end group method, is limited to the number average degree of polymerization and gives no information concerning the breadth of the distribution. A statistical approach to the degree of polymerization yields expressions for both $\overline{n}_n$ and $\overline{n}_w$. Reference 2 contains a derivation of these quantities for the self-polymerization of A_f monomers. Although this specific system appears very different from the one we have considered, the essential aspects of the two different averaging procedures are applicable to the system we have considered as well. The results obtained for the A_f case are

$$\overline{n}_n = \frac{1}{1 - \alpha f/2} \tag{5.54}$$

and

$$\bar{n}_w = \frac{1 + \alpha}{1 - \alpha(f - 1)} \tag{5.55}$$

from which it follows that

$$\frac{\bar{n}_w}{\bar{n}_n} = \frac{(1 + \alpha)(1 - \alpha f/2)}{1 - \alpha(f - 1)} \tag{5.56}$$

The value of α to be used in these expressions is given by Eq. (5.45) for the specific mixture under consideration. At the point of gelation $\alpha_c = 1/(f - 1)$ according to Eq. (5.46). Equation (5.55) shows that $\bar{n}_w$ becomes infinite at this point while $\bar{n}_n$ remains finite. This merely means that there are still many molecules present at the gel point in addition to the network molecule of essentially infinite molecular weight. The ratio $\bar{n}_w/\bar{n}_n$ indicates an immense expansion of the degree of heterogeneity as $\alpha \to \alpha_c$. Expressions are also available to describe the distribution of molecules in that fraction of material that is not incorporated into the network beyond the point of gelation.

5.9 Some Cross-linked Step-Growth Polymers

In this section we examine some examples of cross-linked step-growth polymers. The systems we shall describe are thermosetting polymers of considerable industrial importance. The chemistry of these polymerization reactions is more complex than the hypothetical AB reactions of our models. We choose to describe these commercial polymers rather than model systems which might conform better to the theoretical developments of the last section both because of the importance of these materials and because the theoretical concepts provide a framework for understanding more complex systems, even if they are not quantitatively successful.

The materials we shall discuss are all polymers of formaldehyde which may be viewed as methylene glycol in the presence of water:

$$\underset{\substack{\diagup \diagdown \\ H \quad H}}{\overset{\overset{O}{\parallel}}{C}} + H_2O \;\rightleftharpoons\; HO{-}CH_2{-}OH \tag{5.Y}$$

This particular representation makes it easy to visualize formaldehyde as a step-growth monomer of functionality 2. Our principal interest is in the reactions of formaldehyde with the active hydrogens in phenol, urea, and melamine, compounds [II]−[IV], respectively:

[II]   [III]   [IV]

The hydrogen atoms shown in these monomers (only those underlined in phenol) are the active hydrogens, so these compounds have nominal functionalities of 3, 4, and 6, respectively. Note that a monosubstituted phenol would have a functionality of 2 and would be incapable of crosslinking.

At first glance it appears that these systems do conform fully to the discussion above; this is an oversimplification, however. The ortho and para hydrogens in phenol are not equal in reactivity, for example. In addition, the technology associated with these polymers involves changing the reaction conditions as the polymerization progresses to shift the proportions of several possible reactions. Accordingly, the product formed depends on the nature of the catalyst used, the proportions of the monomers, and the temperature. Sometimes other additives or fillers are added as well.

Industrial phenol-formaldehyde polymerization is a complex process, but the following reactions suggest the successive stages and the possible linkages involved:

1. Formation of methylol derivatives of phenol:

$$(5.Z)$$

these species may be regarded as the true monomers for the production of polymer. The reactions shown are catalyzed by either acid or base.

2. Formation of low molecular weight polymer with chains capped by phenol repeat units:

$$(5.AA)$$

Stopping the polymer at this point requires the ratio of formaldehyde to phenol to be less than unity. Both methylene and ether bridges are known to be present. The reaction is either acid or base catalyzed, and branching is uncommon at this stage. The products are variously known as A stage resins, novolacs, or resole prepolymers.

3. Formation of final cross-linked polymer:

$$(5.BB)$$

This reaction is carried out under base-catalyzed conditions and with a formaldehyde/phenol ratio greater than unity. The resulting product is called a C state resin or resite.

Not all of the hydrogens in phenol are equally reactive. Under acid conditions the quinoid structure

is stabilized as a reaction intermediate favoring the para additions of the protonated formaldehyde molecule. There are also indications that the reaction of one position alters the reactivity of others, so the reactivity depends on the extent of the reaction as well. Hence the assumption that we made in the last section concerning the equal reactivity of all A groups is not justified here. Nevertheless, the previously developed theory can be used to back-calculate from experimental results the effective parameters of the reaction mixtures. Table 5.7 lists some data for three different phenol–formadehyde resins formulated with various ratios of reactants. Weight and number average degrees of polymerization were measured experimentally on these products, and Eqs. (5.54) and (5.55) were used to evaluate α and f from these data. The variation in α reflects the differences in the polymerization mixture; the fact that f values turn out to be as nearly constant as they are suggests that phenol quite consistently displays this effective functionality. Strictly speaking, nonintegral f values are meaningless in terms of the theory presented in the last section, but differences in reactivity and the diversity of possible linkages are quantified to some extent by the concept of effective functionality defined in this way.

Table 5.7 Calculated Values of α and f for Phenol–formaldehyde Resins Formed from Different Proportions of Reactants and Based on Experimental Values of $\overline{n}_n$ and $\overline{n}_w$

Quantity	Resin I	Resin II	Resin III
Phenol–formaldehyde	1.667	1.227	1.176
$\overline{n}_{n,ex}$	2.12	3.51	4.28
$\overline{n}_{w,ex}$	3.64	8.89	11.68
α_{calc}	0.455	0.611	0.677
f_{calc}	2.32	2.35	2.27

Source: M. F. Drumm and J. R. LeBlanc, *Step-Growth Polymerizations*, D. H. Solomon (Ed.), Marcel Dekker, New York, 1972, p. 196.

In the polymerization scheme presented above, the A stage corresponds roughly to $\alpha < \alpha_c$, and the C stage to $\alpha > \alpha_c$; the products are soluble and fusible in the former, and insoluble and infusible in the latter. Presuming that the C stage is the product ultimately sought, it may be formed directly in one step or in two steps through subsequent reaction of the resole prepolymer. Phenol-formaldehyde resins find applications in the manufacture of molded products, as adhesives in products such as plywood, and as protective coatings. This type of polymer, first investigated by Baekeland in 1872, was one of the first synthetic polymers.

Polymers produced by the reactions of urea and melamine with formaldehyde are collectively known as amino resins. In most respects the preparation and properties of these polymers parallel the story of the phenolic resins. The nominal functionality (as opposed to the effective functionality) increases in the order phenol < urea < melamine. As might be anticipated from the effect of this order on crosslinking potential, the hardness and strength of C stage polymers follows the order melamine > urea > phenol. Other properties such as moisture resistance and color are traceable to chemical properties of the monomers other than explicitly polymeric considerations. The amino resins have less moisture resistance and lighter color than the corresponding phenolics. As with the phenolic resins, these polymers also proceed through the methylol intermediates with subsequent linkage through either methyl or ether bridges. They are also commercially available at various stages of polymerization. At the A stage they are used in the textile industry to impart "permanent press" characteristics to cotton and rayon. At the C stage they are used as molding and laminating resins, particulary in applications where their lack of color is an asset.

5.10 Rings 'n Things

From the beginning of this chapter we have acknowledged that ring formation is a possible reaction in the type of monomer mixtures we have considered. We have consistently ignored this complication. In this section we shall remedy this situation. At the same time we shall also consider the polymerizability of cyclic monomers. The formation of polyesters from lactones and of polyamides from lactams, reactions given in Tables 5.3 and 5.4, respectively, are examples of the latter. At first these two issues appear quite different, but the relative stability of rings versus linear structures is at the heart of each. Accordingly, we shall discuss ring formation as a competition to polymerization and ring opening as a mode of polymerization as two areas where the same general principles find application.

In the matter of ring stability it is the number of atoms in the ring that is of central importance. In the following equations l, m, and m′ represent the number of ring (l) and backbone (m, m′) *atoms* in the species shown. Note that for linear species m and m′ are proportional to the degree of polymerization but are more suitable for present purposes than the latter. The following general reactions are pertinent to the present discussion:

1. Monomers reacting to form a ring:

$$A\!-\!A + B\!-\!B \longrightarrow ab\frown ab_{[\,l\,]} \qquad\qquad\qquad (5.CC)$$

2. Monomers reacting to form a linear molecule:

$$A\!-\!A + B\!-\!B \longrightarrow A\!-\!ab\!-\!B_{[\,m\,]} \qquad\qquad\qquad (5.DD)$$

3. Cyclic molecules opening to a linear form:

$$ab\frown ab_{[\,l\,]} \longrightarrow A\!-\!ab\!-\!B_{[\,m\,]} \qquad\qquad\qquad (5.EE)$$

4. Rings combining with linear molecules to form longer linear species:

$$\overset{\frown}{ab}_{[\,l\,]} + A\!-\!B_{[\,m\,]} \longrightarrow A\!-\!ab\!-\!B_{[\,m'\,]} \qquad\qquad\qquad (5.FF)$$

Although the number of ring atoms is the structural feature upon which we focus attention, we shall use the criteria of thermodynamics and kinetics to assess the feasibility of the reactions listed above.

Reaction (5.EE) is particularly useful for the discussion of thermodynamic considerations because of the way differences in thermodynamic state variables are independent of path. Accordingly, if we know the value of ΔG for reaction (5.EE), we have characterized the following:

1. The spontaneity (or lack thereof) of reaction (5.EE) as written.
2. The spontaneity (or lack thereof) of its reverse: the cyclization of a linear molecule.
3. The difference between the values of ΔG for reactions (5.CC) and (5.DD), making it less important to characterize these latter reactions separately.

Reaction (5.FF) summarizes the way high molecular weight polymers form from ring-opening reactions. The reaction ignores the start of the whole process by not showing the origin of the initial linear species, but—as an end effect—that step represents a small part of the entire polymerization process. The reaction as shown is analogous to the neglect of chain initiation and termination in describing vinyl polymerization as was done in reaction (1.A).

Ring-opening polymerization is a distinct category of polymerization, sharing common features with both step-growth and addition polymerization. For example, step-growth polymer chains continue to grow throughout the entire reaction, as we saw in Sec. 5.3; chains continue to grow throughout the entire duration of the ring-opening reaction also. In step-growth polymerization reaction between groups on any molecular species is possible, as we saw in Sec. 5.2; only monomers react with growing chains in ring-opening polymerizations, as shown by reaction (5.FF). This latter feature also characterizes addition polymerization.

As a distinct class of polymerization, ring opening is entirely deserving of a chapter of its own in a polymer chemistry text. Ample material exists to justify such separate treatment. Unfortunately, no course or textbook is ever long enough to cover everything, and certain topics get shortchanged. Such is the fate of ring-opening polymerizations in this book.

We now turn specifically to the thermodynamics and kinetics of reactions (5.EE) and (5.FF). The criterion for spontaneity in thermodynamics is $\Delta G < 0$ with $\Delta G = \Delta H - T \Delta S$ for an isothermal process. Thus it is both the sign and magnitude of ΔH and ΔS and the magnitude of T that determine whether a reaction is thermodynamically favored or not. As usual in thermodynamics, the Δ's are taken as products minus reactants, so the conclusions apply to the reactions as written. If a reaction is reversed, products and reactants are interchanged and the sign of the ΔG is reversed also.

Several qualitative predictions can be made concerning the ring-opening reaction described by (5.EE):

1. The reaction involves breaking a bond, which requires the input of energy; hence we expect that $\Delta H > 0$.
2. Although there is no change in the number of molecules, the linear molecule can take on more conformations than the ring; hence we expect that $\Delta S > 0$.
3. Precise numerical values for either ΔH or ΔS will depend on the degree of strain in the ring structure, among other things.

4. The fact that both ΔH and ΔS are expected to be positive for reaction
    (5.EE) means that ΔG tends to be negative at higher temperatures and
    positive at lower temperatures. Thus linear molecules tend to be favored
    over rings at higher temperatures. We have seen elsewhere in this chapter
    that step-growth polymerizations are usually conducted at elevated
    temperatures.
5. The temperature at which ΔG changes sign (i.e., where $\Delta G = 0$) is given by
    the ratio $\Delta H/\Delta S$ if these quantities are independent of T. The temperature
    at which rings become favored depends on the specific values of ΔH and
    ΔS, and, again, this is a matter of ring size.

Even at the qualitative level of the discussion above, it is difficult to make
predictions regarding the spontaneity of the ring-opening polymerization
reaction (5.FF):

1. Regarding ΔH, one bond is broken and a similar bond is formed in a linear
    product. The linear molecule is unstrained, but the magnitude of the strain
    relief upon ring opening depends on the ring size.
2. Regarding ΔS, ring opening tends to increase entropy through the increase
    of possible conformations; the linkage of two separate molecules decreases
    the translational entropy. The latter effect is probably larger, but the net
    balance depends on ring size.

Thus consideration of both reactions (5.EE) and (5.FF) forces us to examine the
stability of cyclic structures as a function of ring size.

The rather special status of five- and six-membered rings is already familiar
from organic chemistry. Because of the general stability associated with rings
of this size, both reactions (5.EE) and (5.FF) are expected to be less favored
when the ring compounds have this size, that is, when $1 = 5$ or 6. To see the
basis for this conclusion, let us consider some of the evidence for stain in cyclic
compounds.

One of the classic methods for evaluating ring strain is to compare the heat
of combustion per methylene group of cyclic hydrocarbons as a function of ring
size. When this is done, $\Delta H_c/CH_2$ is found to decrease progressively and dramatically as the number of ring atoms increases from 3 to 6, increase through a slight
maximum for $7 \gtrsim 1 \gtrsim 11$, and then slowly drop back to the same minimum
observed for $1 = 6$. The following example also deals with cyclic alkanes and
relates the matter of ring size to the ring $\rightleftarrows$ polymer equilibrium.

Example 5.6

For the reaction

Cyclic alkane $(CH_2)_1$ (liq) $\rightleftarrows$ polymer (cryst)

the following values of $\Delta H°$ and $\Delta S°$ have been estimated at 25°C:

l	$\Delta H°$ (kJ mol^{-1})	$\Delta S°$ (J K^{-1} mol^{-1})
4	−105.0	−55.2
6	+2.9	+10.5
8	−34.7	+37.2

Evaluate $\Delta G°$ and K_{eq} at 25°C for each of the polymerization reactions and comment on the results.

Solution

We use the relationships $\Delta G° = \Delta H° - T \Delta S°$ and $\Delta G° = -RT \ln K_{eq}$, being careful about units: $T = 298$ K and $R = 8.314$ J K^{-1} mol^{-1}. Using the above values of $\Delta H°$ and $\Delta S°$, the following are readily calculated:

l	$\Delta G°$ (kJ mol^{-1})	$\ln K_{eq}$	K_{eq}
4	−88.5	35.7	3.3×10^{15}
6	−0.19	0.08	1.08
8	−45.8	18.5	1.1×10^{8}

Several observations come to mind:

1. For both $l = 4$ and $l = 8$, the polymer is favored over the ring to a far greater degree than for $l = 6$. In the latter case the difference between $\Delta H°$ and $T \Delta S°$ is so small that there is probably no justification for retaining as many figures as shown in the calculated results.
2. For $l = 4$, $\Delta H°$ makes a favorable contribution to $\Delta G°$, while the contribution of $\Delta S°$ is unfavorable. The effect of the former outweighs the latter.
3. For $l = 6$ the contribution of $\Delta H°$ is unfavorable and that of $\Delta S°$ is favorable. The specific values make $T = 298$ K very close to the equilibrium temperature. This implies that the reaction is shifted to favor polymer at higher temperatures and to favor the cyclic monomer at lower temperatures. Since the difference between $\Delta H°$ and $T \Delta S°$ is so small, the temperature dependence of $\Delta H°$ and $\Delta S°$ could alter this conclusion.
4. For $l = 8$ both $\Delta H°$ and $\Delta S°$ contribute favorably to $\Delta G°$.

•

Thus both combustion and polymerization data indicate a stabilization of rings in the neighborhood of $l = 6$.

Generally speaking, the replacement of carbon atoms in the rings by −O− or −NH− causes relatively minor alterations in the picture of ring stability. For

Table 5.8 Values for ΔH and ΔS for the Ring-Opening Polymerization, Reaction (5.FF), for Monomers with the Indicated Values of l

l	ΔH (kJ mol^{-1})	ΔS (J K^{-1} mol^{-1})
5	-5.4	-30.5
6	-4.6	-25.1
7	-12.6	-4.6
8	-23.8	-16.7
9	-40.2	$-$
13	-6.3	$-$

Source: G. Odian, *Principles of Polymerization*, 1st ed., McGraw-Hill, New York, 1970. Used with the permission of the McGraw-Hill Book Company.

evidence of this, we turn to the data in Table 5.8, which applies to the ring-opening polymerization reaction (5.FF). Table 5.8 lists values of ΔH and ΔS for the reaction of monomers with several different l values. The values of ΔH follow the same trend as discussed above for the combustion data. ΔS passes through a maximum as ring size increases. If we assume that ΔH and ΔS values in Table 5.8 apply at 400 K, $\Delta G = -12.3$ kJ mol^{-1} for l = 7, ϵ-caprolactam, while this quantity is $+3.7$ kJ mol^{-1} for l = 5 (2-pyrrolidone) and $+2.9$ kJ mol^{-1} for l = 6 (2-piperidone). The negative value of ΔG for the polymerization of ϵ-caprolactam results in the formation of the important product nylon-6. The positive values of ΔG for rings with l = 5 and l = 6 reflects the special stability of rings of this size.

Thermodynamics provides only one of the criteria for polymerizability. A favorable value for ΔG says nothing about the rate of a reaction. It is not particularly difficult to see that a reaction like the reverse of (5.EE) becomes kinetically improbable for large values of m, that is, for the cyclization of a large linear molecule. For the terminal A and B groups of a single chain to react with each other for ring closure, they must first diffuse together. We have seen in Chap. 1 that for long chains the end-to-end distance varies with the square root of the degree of polymerization, or with $m^{1/2}$ in the present context. The volume associated with a length of this magnitude varies with length^3, or $m^{3/2}$. Hence as m increases, the concentration of chain ends varies with $m^{-3/2}$. By contrast, the probability that an A group reacts with a B group from another molecule depends on the number of molecules present, and the number of chains in a reaction mixture varies with m^{-1}. Hence as m increases, reaction between different molecules becomes increasingly favored over ring closure simply by a dilution effect. Use of random walk statistics clearly limits the above

argument to chains which are considerably longer than others we have discussed in this section.

At the smaller end of the size range, kinetic studies show that the ease of formation of lactones, cyclic esters, and cyclic anhydrides is greatest in the neighborhood of five- or six-atom rings, drops to a minimum in the range of 8- to 12-membered rings, and then increases and levels off at a low plateau value for still larger rings. The broad minimum in the rate of formation of rings with $8 \lesssim 1 \lesssim 12$ is attributed to the crowding between chain hydrogens in the interior of the rings in this size range.

On the basis of both thermodynamic and kinetic evidence—both of which are interpretable in terms of the strain associated with rings of certain sizes or similar structural factors—we see that only rings with five or six atoms have any significant stability. Accordingly, we conclude the following:

1.   AA and BB monomers and also AB monomers invariably react to form predominantly linear structures in all but the rather special case where the ring structure in reaction (5.CC) has a value of $1 = 5$ or 6. This explains why so many of the monomers in step-growth polymerizations are tetra-, hexa-, and decamethylene compounds.
2.   The cyclization of a polymer that has already grown through the dimer or trimer stage is also insignificant.
3.   The formation of polymers by the ring-opening reaction (5.FF) requires an initiator to get the reaction started, and perhaps a catalyst to assure a suitable rate, but otherwise is quite feasible for $1 \lesssim 4$ and $1 \gtrsim 7$.

We conclude this section by citing some examples of ring-opening polymerizations. Table 5.9 lists several examples of ring-opening polymerizations. In addition to the reactions listed, we recall the polymerizations of lactones and lactams exemplified by equations in Table 5.3 and 5.4, respectively.

Ring-opening polymerizations are catalyzed by a wide variety of substances, including the bases OH^- and RO^- and the acids H^+ and BF_3; water is also used as a catalyst. The reactions proceed by the opening of the ring by the catalyst to form an active species,

$$\widehat{ab} + C \longrightarrow Ca-b* \qquad\qquad (5.GG)$$

followed by the subsequent attachment of additional monomers to the active site,

$$Cab* + \widehat{ab} \longrightarrow Cabab* \longrightarrow \longrightarrow polymer \qquad\qquad (5.HH)$$

Table 5.9 Some Typical Ring-Opening Polymerization Reactions

1. Epoxides:

$$n\ H_2C\overset{O}{\overset{\diagdown}{—}}CH_2 \longrightarrow \ —\!(CH_2\ CH_2\!-\!O)\!—_n$$

2. Cyclic formals (formaldehyde–cyclic ether dimer):

$$n\quad \underset{CH_2}{\overset{(CH_2)_5}{O\diagup\,\diagdown O}} \longrightarrow \ —\!(O\!-\!CH_2\!-\!O\!-\!(CH_2)_5)\!—_n$$

3. Cyclic sulfides (including S_8):

$$n\ \underset{S}{CH_2\!-\!CH\,CH_3} \longrightarrow \ —\!\left(CH_2\!-\!\underset{H}{\overset{CH_3}{C}}\!-\!S\right)\!—_n$$

4. Alkylenimines:

$$n\quad \underset{\underset{H}{N}}{CH_2\!-\!CH_2} \longrightarrow \ —\!\left(CH_2CH_2\!-\!\overset{H}{N}\right)\!—_n$$

5. Cyclic acetals (trioxane: cyclic trimer of formaldehyde):

$$\frac{n}{3}\quad \underset{O\!-\!CH_2}{\overset{O\!-\!CH_2}{CH_2\quad O}} \longrightarrow \ —\!(CH_2\!-\!O)\!—_n$$

6. Cyclic siloxanes:

$$\frac{n}{3}\quad \underset{O\!-\!Si(CH_3)_2}{\overset{O\!-\!Si\,(CH_3)_2}{Si\,(CH_3)_2\quad O}} \longrightarrow \ —\!\left(\underset{CH_3}{\overset{CH_3}{Si}}\!-\!O\right)\!—_n$$

5.11 More Rings

The search for substances which qualify for proposed applications has always been a driving force for the synthesis and characterization of new compounds. This is especially true in polymer chemistry, where it is the potential of polymers as engineering materials that often stimulates research. Polymeric materials frequently fail to be serviceable in engineering applications for one of the following reasons:

1. They lack resistance to solvents or other chemicals.
2. They lack thermal stability at high temperatures.
3. They are not sufficiently rigid.

To some extent each of these objections is met by the presence of either chemical or crystallite crosslinking in the polymer. Another approach which complements the former is to incorporate rings into the backbone of the chemical chain. As an example, contrast the polyesters formed between ethylene glycol and either suberic or terephthalic acid. Structures [V] and [VI], respectively, indicate the repeat units in these polymers:

$$\begin{matrix} & O & & O \\ & \parallel & & \parallel \\ \{O-(CH_2)_2-O-C-(CH_2)_6-C\} & & \end{matrix}$$

[V]

$$\{O-(CH_2)_2-O-C-\langle\bigcirc\rangle-C\}$$

[VI]

In both instances the two carboxyl groups are separated by six carbons, but these occur as six methylene groups in the suberate and as a benzene ring in the terephthalate. The suberate melts at about 62°C and the terephthalate at about 260°C. At least in part, this difference is traceable to the greater chain stiffness of the backbone containing the ring structure, which decreases the ΔS for the melting process and increases the melting point according to Eq. (4.5). This same stiffness enhances the crystallizability and increases the various deformation moduli both through the chain stiffness and the crystallinity. The improved interchain attraction in the stiffer chains decreases the solubility of these materials, and this in itself contributes to their diminished susceptibility to chemical attack. In addition, the backbone rings effectively provide multiple strands along the polymer so that there is a possibility for the polymer to remain intact even if one strand is cleaved.

Once the potential associated with this aspect of molecular architecture is recognized, the principles of the last section coupled with the richness of organic (and inorganic) chemistry suggest numerous synthetic possibilities. We shall not attempt to be comprehensive in discussing this facet of polymer chemistry; instead we cite only a few examples of step-growth polymers which incorporate

successively greater proportions of ring character into the chain backbone. These examples will also remind us that polyesters, polyamides, and formaldehyde polymers are not the only kinds of polymers possible from monomers with two or more functional groups.

Aromatic polyimides are the first example we shall consider of polymers with a rather high degree of backbone ring character. This polymer is exemplified by the condensation product of pyromellitic dianhydride [VII] and *p*-amino-aniline [VIII]:

[VII]   [VIII]

[IX]

[X]   (5.II)

This polymerization is carried out in the two stages indicated above precisely because of the insolubility and infusibility of the final product. The first-stage polyamide, structure [IX], is prepared in polar solvents and at relatively low temperatures, say, 70°C or less. The intermediate is then introduced to the intended application—for example, a coating or lamination—then the second-stage cyclization is carried out at temperatures in the range 150–300°C. Note the formation of five-membered rings in the formation of the polyimide, structure [X], and also that the proportion of acid to amine groups is 2:1 for reaction (5.II).

When the proportion of acid to amine groups is reversed—namely, 1:2—a process rather similar to reaction (5.II) yields a polymer which ultimately contains the five-membered imidazole ring. This reaction is also carried out in the stages listed below and illustrated by reaction (5.JJ):

1. The diphenyl ester of the diacid [XI] is used to prevent side reactions such as decarboxylation.
2. The condensation of the monomers with the elimination of water and the formation of a polyimine [XII] occurs at temperatures around 250°C.
3. This first-stage polymer is then introduced into the application environment, where the final cyclization reaction occurs.
4. The polyimidazole [XIII] is formed by heating to 350–400°C, with the elimination of phenol and ring closure:

[XI]   [XII]

[XIII]   (5.JJ)

Another interesting structure with a high degree of ring character along the backbone is the product obtained by the reaction of 1,4-cyclohexanedione [XIV] and pentaerythritol [XV]:

[XIV]   [XV]   (5.KK)

The backbone here is made up of six-membered rings in which the two rings share a common atom. These are called spiro polymers.

Lastly, we consider a class of compounds called ladder polymers, which are made up of a double-stranded backbone that is linked at regular intervals into rings so that the schematic structure is

Spiro polymers are also sometimes classified as ladder polymers, and molecules in which the ladder structure is interrupted by periodic single bonds are called semiladders. Consisting entirely of fused ring structures, ladder polymers possess very rigid chains with excellent thermal stability.

Discussion of ladder polymers also enables us to introduce a step-growth polymerization that deviates from the simple condensation reactions which we have described almost exclusively in this chapter. The Diels–Alder reaction is widely used in the synthesis of both ladder and semiladder polymers. In general, the Diels–Alder reaction occurs between a diene [XVI] and a dienophile [XVII] and yields an adduct with a ring structure [XVIII]:

$$\text{[XVI]} + \text{[XVII]} \longrightarrow \text{[XVIII]} \tag{5.LL}$$

[XVI]   [XVII]   [XVIII]

Since the six carbons shown above have 10 additional bonds, the variety of substituents they carry or the structures they can be a part of is quite varied, making the Diels–Alder reaction a powerful synthetic tool in organic chemistry. A moment's reflection will convince us that a molecule like structure [XVI] is monofunctional from the point of view of the Diels–Alder condensation. If the Diels–Alder reaction is to be used for the preparation of polymers, the reactants must be bis-dienes and bis-dienophiles. If the diene, the dienophile, or both are part of a ring system to begin with, a polycyclic product results. One of the first high molecular weight polymers prepared by this synthetic route was the product resulting from the reaction of 2-vinyl butadiene [XIX] and benzoquinone [XX]:

$$(5.MM)$$

The bifunctionality of the bis-diene and bis-dienophile monomers is apparent from the condensation product, structure [XXI], which still contains a diene and a dienophile in the same molecule. This polymer is crystalline, indicating a high degree of stereoregularity in the condensed rings. It decomposes to a graphitic material before melting.

It is also possible for a single monomer to behave as both diene and dienophile. Heating diacetylene [XXII] produces an infusible material which may be rationalized as follows:

[XXII] $$(5.NN)$$

Because of the versatility of the Diels–Alder reaction, an impressive assortment of complex ladder and semiladder polymers have been synthesized; this continues to be a fruitful area of research.

Problems

1. Howard[†] describes a model system used to test the molecular weight distribution of a condensation polymer: "The polymer sample was an acetic acid-stabilized equilibrium nylon-6,6. Analysis showed it to have the following end group composition (in equivalents per 10^6 g):acetyl = 28.9,

---

†G. J. Howard, *J. Polym. Sci.* 37:310 (1959).

amine = 35.3 and carboxyl = 96.5. The number average degree of polymerization is, therefore, 110 and the conversion degree (= extent of reaction) =
0.9909." Verify the self-consistency of these numbers.

2. Haward et al.[†] have reported some research in which a copolymer of styrene
   and hydroxyethylmethacrylate was cross-linked by hexamethylene di-
   isocyanate. Draw the structural formula for a portion of this cross-linked
   polymer and indicate what part of the molecule is the result of a condensation reaction and what part results from addition polymerization. These
   authors indicate that the crosslinking reaction is carried out in sufficiently
   dilute solutions of copolymer that the crosslinking is primarily intramolecular rather than intermolecular. Explain the distinction between these
   two terms and why concentration affects the relative amounts of each.

3. The polymerization of β-carboxymethyl caprolactam has been observed to
   consist of initial isomerization via a second-order kinetic process followed
   by condensation of the isomer to polymer:

The rate of polymerization is thus first order in ν_{NH_2} and first order in
$\nu_{(CO)_2O}$ or second order overall. Since $\nu_{NH_2} = \nu_{(CO)_2O}$, rate = kc^2, if
catalyzed; third order is expected under uncatalyzed conditions. The
indirect evaluation of c was accomplished by measuring the amount of
monomer reacted, and the average degree of polymerization of the mixture
was determined by viscosity at different times. The following data[‡] were
obtained at 270°C (the early part of the experiment gives nonlinear results).

t (min)	c (mole fraction)	t (min)	c (mole fraction)
20	0.042	90	0.015
30	0.039	110	0.013
40	0.028	120	0.012
50	0.024	150	0.0096
60	0.021	180	0.0082
80	0.018		

[†] R. N. Haward, B. M. Parker, and E. F. T. White, *Adv. Chem.* 91:498 (1969).
[‡] H. K. Reimschuessel, *Adv. Chem.* 91:717 (1969).

Graphically test whether these data indicate catalyzed or uncatalyzed conditions and evaluate the rate constant for polymerization at $270°C$. Propose a name for the polymer.

4. Examination of Fig. 5.5 shows that N_n/N is greater for $n = 40$ at $p = 0.97$ than at either $p = 0.95$ or $p = 0.99$. This is generally true: Various n-mers go through a maximum in numerical abundance as p increases. Show that the extent of reaction at which this maximum occurs varies with n as follows: $p_{max} = (n - 1)/(n + 1)$. For a catalyzed AB reaction, extend this expression to give a function for the time required for an n-mer to reach its maximum numerical abundance. If $k_c = 2.47 \times 10^{-4}$ liter mol^{-1} sec^{-1} at $160.5°C$ for the polymerization of 12-hydroxystearic acid,[†] calculate the time at which 15-mers show their maximum abundance if the initial concentration of monomer is 3.0 M.

5. In the presence of a pyridine–cuprous chloride catalyst, the following polymerization occurs:

(monomer)

In an investigation to examine the mechanism of this reaction, the dimer $(n = 2)$

was used as a starting material. The composition of the mixture was studied as the reaction progressed and the accompanying results were obtained[‡]:

Percent of theoretical O_2 absorbed	Weight percent composition in reaction mixture			
	Monomer	Dimer	Trimer	Tetramer
9	1	69	15	9
12	1.5	68	24	9
20	3	38.5	23	9
35	6	26	21	11
60	11	4	4	1
80	1	0	0	0

[†] C. E. H. Bawn and M. B. Huglin, *Polymer* 3:257 (1962).
[‡] G. D. Cooper and A. Katchman, *Adv. Chem.* 91:660 (1969).

Plot a family of curves, each of different n, with composition as the y axis and O_2 absorbed as the x axis. Evaluate w_n by Eq. (5.30) for $n = 1, 2, 3$, and 4 and $0.1 \leqslant p \leqslant 0.9$ in increments of 0.1. Plot these results (w_n on y axis) on a separate graph drawn to the same scale as the experimental results. Compare your calculated curves with the experimental curves with respect to each of the following points: (1) coordinates used, (2) general shape of curves, and (3) labeling of curves.

6. The polymer described in the last problem is commercially called poly (phenylene oxide), which is not a proper name for a molecule with this structure. Propose a more correct name. Use the results of the last problem to criticize or defend the following proposition: The experimental data for dimer polymerization can be understood if it is assumed that one molecule of water *and* one molecule of monomer may split out in the condensation step. Steps involving incorporation of the monomer itself (with only water split out) also occur.

7. Taylor† carefully fractionated a sample of nylon-6,6 and determined the weight fraction of different n-mers in the resulting mixture. The following results were obtained:

n	$w_n \times 10^{-4}$	n	$w_n \times 10^{-4}$
12	6.5	311	15.2
35	19.6	334	14.1
58	29.4	357	13.0
81	33.0	380	11.5
104	35.4	403	11.0
127	36.5	426	9.1
150	33.0	449	7.2
173	27.6	472	6.5
196	25.2	495	4.9
219	22.9	518	4.3
242	19.4	541	3.9
265	18.5	564	3.3
288	16.8		

Evaluate $\bar{n}_w$ from these data; then use Eq. (5.32) to calculate the corresponding value of p. Calculate the theoretical weight fraction of n-mers using this value of p and a suitable array of n values. Plot your theoretical curve and the above data points on the same graph. Criticize or defend the following proposition: Although the fit of the data points is acceptable with this value of p, it appears that a slightly smaller value of p would give an even better fit.

---

† G. B. Taylor, *J. Am. Chem. Soc.* 69:638 (1947).

8. Paper chromatograms were developed for 50:50 blends of nylon-6,6 and nylon-6,10 after the mixtures had been heated at $290°C$ for various periods of time. The following observations[†] describe the chromatograms after the indicated times of heating:

0 hr—two spots with R_f values of individual polymers.
1/4 hr—two distinct spots, but closer together than those of 0 hr.
1/2 hr—spots are linked together.
3/4 hr—one long, diffuse spot.
1½ hr—one compact spot, intermediate R_f value.

On the basis of these observations, criticize or defend the following proposition: The fact that the separate spots fuse into a single spot of intermediate R_f value proves that block copolymers form between the two species within the blend upon heating.

9. Reimschuessel and Dege[‡] polymerized caprolactam in sealed tubes containing about 0.0205 mol H_2O per mole caprolactam. In addition, acetic acid (V), sebacic acid (S), hexamethylene diamine (H), and trimesic acid (T) were introduced as additives into separate runs. The following table lists (all data per mole caprolactam) the amounts of additive present and the analysis for end groups in various runs:

Additives	Moles additive	−COOH (mEq)	−NH₂ (mEq)
None	−	5.40	4.99
V	0.0205	19.8	2.3
S	0.0102	21.1	2.3
H	0.0102	1.24	19.7
T	0.0067	22.0	2.5

Neglecting end group effects, calculate $\bar{M}_n$ for each of these polymers from the end group data. Are the trends in molecular weight qualitatively what would be expected in terms of the role of the additive in the reaction mixture? Explain briefly.

10. In the study described in the last problem, caprolactam was polymerized for 24 hr at $225°C$ in sealed tubes containing various amounts of water. $\bar{M}_n$ and $\bar{M}_w$ were measured for the resulting mixture by osmometry and light scattering, respectively, and the following results were obtained[§]:

[†] C. W. Ayers, *J. Appl. Chem.* 4:444 (1954).
[‡] H. K. Reimschuessel and G. J. Dege, *J. Polym. Sci.* A-1:2343 (1971).
[§] Ibid.

Moles H_2O ($\times 10^3$)/mole caprolactam	$\bar{M}_n \times 10^{-3}$	$\bar{M}_w \times 10^{-3}$
49.3	13.4	20.0
34.0	16.4	25.6
25.6	17.9	29.8
20.5	19.4	36.6

Use the molecular weight ratio to calculate the apparent extent of reaction of the caprolactam in these systems. Is the variation in p qualitatively consistent with your expectations of the effect of increased water content in the system? Plot p versus moisture content and estimate by extrapolation the equilibrium moisture content of nylon-6 at $255°C$. Does the apparent equilibrium moisture content of this polymer seem consistent with the value given in Sec. 5.6 for nylon-6,6 at $290°C$?

11. At $270°C$ adipic acid decomposes[†] to the extent of 0.31 mol % after 1.5 hr. Suppose an initially equimolar mixture of adipic acid and diol achieves a value of p = 0.990 after 1.5 hr. Compare the expected and observed values of $\bar{n}_n$ in this experiment. Criticize or defend the following proposition: The difference between the observed and expected values would be even greater than calculated above if, instead of the extent of reaction being measured analytically, the value of p expected (neglecting decomposition) after 1.5 hr were calculated by an appropriate kinetic equation.

12. The Carothers equation can also be used as the basis for an estimate of the extent of reaction at gelation. Consider the value implied for each of the parameters in the Carothers equation at the threshold of gelation and derive a relationship between p_c and $\bar{f}$ on the basis of this consideration. Compare the predictions of the equation you have derived with those of Eq. (5.47) for a mixture containing 2 mol A_3, 7 mol AA, and 10 mol BB. Criticize or defend the following proposition: The Carothers equation gives a higher value for p_c than Eq. (5.47) because the former is based on the fraction of reactive groups that have reacted and hence considers wasted loops that the latter disregards.

13. For monomers of the AB type, reactions (5.CC) and (5.DD) become AB → ab and 2AB → AbaB, respectively. If k_r and k_l are the respective rate constants for these reactions, derive an expression which gives the ring to linear ratio in the product as a function of AB concentration and the two rate constants. Criticize or defend the following proposition[‡]: To obtain a test of Eq. (5.47) without the complications of intramolecular condensations, a series of otherwise identical polymeriztion reactions could be carried out on monomer mixtures at different concentrations. By

---

[†]V. V. Korshak and S. V. Vinogradova, *Polyesters*, Pergamon, Oxford, 1965.
[‡]W. H. Stockmayer and L. L. Weil, in *Advancing Fronts in Chemistry*, S. B. Twiss (Ed.), Reinhold, New York, 1945.

measuring the values of p_c obtained at different concentrations and extrapolating to zero concentration, an uncomplicated test of gelation theory is obtained.

Bibliography

1. Allcock, H. R., and Lampe, F. W., *Contemporary Polymer Chemistry*, Prentice-Hall, Englewood Cliffs, N.J., 1981.
2. Flory, P. J., *Principles of Polymer Chemistry*, Cornell University Press, Ithaca, N.Y., 1953.
3. Frisch, K. C., and Reegen, S. L. (Eds.), *Ring Opening Polymerization*, Marcel Dekker, New York, 1969.
4. Odian, G., *Principles of Polymerization*, 2nd ed., Wiley, New York, 1981.
5. Solomon, D. H. (Ed.), *Step Growth Polymerizations*, Marcel Dekker, New York, 1972.

6

Addition or Chain-Growth Polymerization

I remember snow, Soft as feathers, Sharp as thumb tacks, . . .
And ice, like vinyl, on the streets,
Cold as silver, White as sheets.

I Remember, words and music by Stephen Sondheim

6.1 Introduction

We indicated in Chap. 1 that the category of addition polymers is best character-ized by the mechanism of the polymerization reaction rather than by the addition reaction itself. This is known to be a chain mechanism, so in the case of addition polymers we have chain reactions producing chain molecules. One of the things to bear in mind is the double and different use of the word *chain* in this discussion. The word *chain* continues to offer the best description of the large polymer molecules. A chain reaction, on the other hand, describes a whole series of successive events triggered by some initial occurrence. We sometimes encounter this description of highway accidents in which one traffic mishap on a fogbound highway results in a pileup of colliding vehicles that can extend for miles. In nuclear reactors a cascade of fission reactions occurs which is initiated by the capture of the first neutron. In both of these examples some initiating event is required. This is also true in chain-growth polymerization.

In the above examples the size of the chain can be measured by considering the number of automobile collisions that result from the first accident, or the number of fission reactions which follow from the first neutron capture. When we think about the number of monomers that react as a result of a single initia-tion step, we are led directly to the degree of polymerization of the resulting molecule. In this way the chain mechanism and the properties of the polymer chains are directly related.

Chain reactions do not go on forever. The fog may clear and the improved visibility ends the succession of accidents. Neutron-scavenging control rods may be inserted to shut down a nuclear reactor. The chemical reactions which terminate polymer chain reactions are also an important part of the polymerization mechanism. Killing off the reactive intermediate that keeps the chain going is the essence of these termination reactions. Some unusual polymers can be formed without this termination; these are called living polymers.

The kind of reaction which produces a dead polymer from a growing chain depends on the nature of the reactive intermediate. These intermediates may be free radicals, anions, or cations. We shall devote most of this chapter to a discussion of the free-radical mechanism, since it readily lends itself to a very general treatment. The discussion of ionic intermediates is not as easily generalized.

In this chapter we deal exclusively with homopolymers. The important case of copolymers formed by the chain mechanism is taken up in the next chapter. The case of copolymerization offers an excellent framework for the comparison of chemical reactivities between different monomer molecules. Accordingly, we defer this topic until Chap. 7, although it is also pertinent to the differences in the homopolymerization reactions of different monomers.

6.2 Chain-Growth and Step-Growth Polymerizations: Some Comparisons

Our primary purpose in this section is to point out some of the similarities and differences between step-growth and chain-growth polymerizations. In so doing we shall also have the opportunity to indicate some of the different types of chain-growth polymerization systems.

In Chap. 5 we saw that step-growth polymerizations occur, one step at a time, through a series of relatively simple organic reactions. By treating the reactivity of functional groups as independent of the size of the molecule carrying the group, the entire course of the polymerization is described by the conversion of these groups to their condensation products. Two consequences of this are that both high yield and high molecular weight require extensive reaction to occur. By contrast, chain-growth polymerization occurs by introducing an active growth center into a monomer, followed by the addition of monomers to that center by a chain-type kinetic mechanism. The active center is ultimately killed off by a termination step. The (average) degree of polymerization that characterizes the system depends on the frequency of addition steps relative to the termination steps. Thus high molecular weight polymer is achieved almost immediately. The only thing that is accomplished by allowing the reaction to proceed somewhat further is an increased yield of polymer. The molecular weight of the product is relatively unaffected. This simple picture tends to

break down at high extents of conversion. For this reason we shall focus attention in this chapter on low conversions to polymer, except in those cases where noted otherwise.

Step-growth polymerizations can be schematically represented by one of the individual reaction steps $\curvearrowright A + B \curvearrowright \longrightarrow \curvearrowright ab \curvearrowright$ with the realization that the species so connected can be any molecules containing A and B groups. Chain-growth polymerization, by contrast, requires at least three distinctly different kinds of reactions to describe the mechanism. These three types of reactions will be discussed in the following sections in considerable detail. For now our purpose is to introduce some vocabulary rather than develop any of these beyond mere definitions. The principal steps in the chain growth mechanism are the following:

1. Initiation. An active species I^* is formed by the decomposition of an initiator molecule I:

$$I \longrightarrow I^* \tag{6.A}$$

2. Propagation. The initiator fragment reacts with a monomer M to begin the conversion to polymer; the center of activity is retained in the adduct. Monomers continue to add in some way until molecules are formed with degree of polymerization n:

$$I^* + M \longrightarrow IM^* \xrightarrow{M} IMM^* \longrightarrow \longrightarrow \longrightarrow IM_n^* \tag{6.B}$$

If n is large enough, the initiator fragment—an end group—need not be written explicitly.

3. Termination. By some reaction, generally involving two polymers containing active centers, the growth center is deactivated, resulting in dead polymer:

$$M_n^* + M_{n'}^* \longrightarrow \text{dead polymer} \tag{6.C}$$

Elsewhere in this chapter we shall see that other reactions—notably, chain transfer and chain inhibition—also need to be considered to give a more fully developed picture of chain-growth polymerization, but we shall omit these for the time being. Much of the argumentation of this chapter is based on the kinetics of these three mechanistic steps. We shall describe the rates of the three general kinds of reactions by the notation R_i, R_p, and R_t for initiation, propagation, and termination, respectively.

In the last chapter we presented arguments supporting the idea that reactivity is independent of molecular size. Although the chemical reactions are certainly different in this chapter and the last, we shall continue to maintain this position

for addition polymerization also. For step-growth polymerization this assumption simplified the discussion tremendously and at the same time needed careful qualification. We recall that the equal reactivity premise is valid only after an initial size dependence for smaller molecules. The same variability applies to the propagation step of addition polymerizations for short-chain oligomers, although things soon level off and the assumption of equal reactivity holds. We are thus enabled to treat all propagation steps by the single rate constant k_p. Since the total polymer may be the product of hundreds of such steps, no serious error is made in neglecting the variation that occurs in the first few steps.

In Sec. 5.3 we rationalized that, say, the first 50% of a step-growth reaction might be different from the second 50% because the reaction causes dramatic changes in the polarity of the reaction mixture. We shall see that, under certain circumstances, the rate of addition polymerization accelerates as the extent of conversion to polymer increases due to a composition-dependent effect on termination. In spite of these deviations from the assumption of equal reactivity at all extents of reaction, we continue to make this assumption because of the simplification it allows. We then seek to explain the deviations from this ideal or to find experimental conditions—low conversions to polymer—under which the assumptions apply. This kind of thing is common in chemistry: Most discussions of gases begin with the ideal gas law and describe real gases as deviating from the ideal at high pressures and approaching the ideal as pressure approaches zero.

The active centers that characterize addition polymerization are of two types: free radicals and ions. Throughout most of this chapter we shall focus attention on the free-radical species, since these lend themselves most readily to generalization. Ionic polymerizations not only proceed through different kinds of intermediates but, as a consequence, yield quite different polymers. Depending on the charge of the intermediate, ionic polymerizations are classified as anionic or cationic. These two types of polymerization are discussed in Secs. 6.10 and 6.11, respectively.

In the last chapter we saw that two reactive groups in a molecule are the norm for the formation of linear step-growth polymers. A pair of monofunctional reactants might undergo essentially the same reaction, but no polymer is produced because no additional functional groups remain to react. On the other hand, if a molecule contains more than two reactive groups, then branched or cross-linked products result from step-growth polymerizations. By comparison, a wide variety of unsaturated monomers undergo chain-growth polymerization. A single kind of monomer suffices—more than one yields a copolymer—and only one double bond per monomer may result in branching or crosslinking. For example, the 1,3-addition reaction of butadiene results in a chain which has a substituent vinyl group capable of branch formation. Divinyl benzene is an example of a bifunctional monomer which is used as a crosslinking agent in

chain-growth polymerizations. We shall be primarily concerned with various alkenes as the monomers of interest; however, the carbon–oxygen double bond in aldehydes and ketones can also serve as the unsaturation required for addition polymerization. The polymerization of alkenes yields a carbon atom backbone, while the carbonyl group introduces carbon and oxygen atoms into the backbone. The inadequacy of backbone composition as a basis for distinguishing between addition and condensation polymers is now apparent from both sides. We saw in the last chapter that condensation polymers could be formed with exclusively carbon backbones.

It might be noted that most (not all) alkenes are polymerizable by the chain mechanism involving free-radical intermediates, whereas the carbonyl group is generally not polymerized by the free-radical mechanism. Carbonyl groups and some carbon–carbon double bonds are polymerized by ionic mechanisms. Monomers display far more specificity where the ionic mechanism is involved than with the free-radical mechanism. For example, acrylamide will polymerize through an anionic intermediate but not a cationic one, N-vinyl pyrrolidones by cationic but not anionic intermediates, and halogenated olefins by neither ionic species. In all of these cases free-radical polymerization is possible.

The initiators which are used in addition polymerizations are sometimes called catalysts, although strictly speaking this is a misnomer. A true catalyst is recoverable at the end of the reaction, chemically unchanged. This is not true of the initiator molecules in addition polymerizations. Monomer and polymer are the initial and final states of the polymerization process, and these govern the thermodynamics of the reaction; the nature and concentration of the intermediates in the process, on the other hand, determine the rate. This makes initiator and catalyst synonyms for the same material: The former term stresses the effect of the reagent on the intermediate, and the latter its effect on the rate. The term *catalyst* is particularly common in the language of ionic polymerizations, but this terminology should not obscure the importance of the initiation step in the overall polymerization mechanism.

In the next three sections we consider initiation, termination, and propagation steps in the free-radical mechanism for addition polymerization. One should bear in mind that two additional steps, inhibition and chain transfer, are being ignored at this point. We shall take up these latter topics in Sec. 6.8.

6.3 Initiation

In this section we discuss the initiation step of free-radical polymerization. This discussion is centered around initiators and their decomposition behavior. The first requirement for an initiator is that it be a source of free radicals. In addition, the radicals must be produced at an acceptable rate at convenient temperatures; have the required solubility behavior; transfer their activity to

monomers efficiently; be amenable to analysis, preparation, purification, and so on. Some of the most widely used initiator systems are listed below, and Table 6.1 illustrates their behavior by typical reactions:

1. Organic peroxides or hydroperoxides.
2. Azo compounds.
3. Redox systems.
4. Thermal or light energy.

Peroxides and hydroperoxides are useful as initiators because of the low dissociation energy of the O—O bond in these compounds. This very property makes the range of possible compounds somewhat limited because of the instability of these reagents. In the case of azo compounds the homolysis is driven by the liberation of the very stable N_2 molecule, despite the relatively high dissociation energy of the C—N bond. The redox systems listed in Table 6.1 have the advantage of water solubility, although redox systems which operate in organic solvents are also available. One advantage of the redox reactions as a source of free radicals is the fact that these reactions often proceed faster and at lower temperatures than the thermal homolysis of the peroxide and azo compounds.

The initiation reactions shown under the heading of electromagnetic radiation in Table 6.1 indicate two possibilities of a large number of examples that might be cited. One mode of photochemical initiation shown involves the direct excitation of the monomer with subseqent bond rupture. The second example cited is the photolytic fragmentation of initiators such as alkyl halides and ketones. Because of the specificity of light absorption, photochemical initiators include a wider variety of compounds than those which decompose thermally. Photosensitizers can also be used to absorb and transfer radiation energy to either monomer or initiator molecules. Finally we note that high-energy radiation such as x rays and γ rays and particulate radiation such as α or β particles also produce free radicals. These latter sources of radiant energy are nonselective and produce a wider array of initiating species. Even though these high-energy radiations produce both ionic and free-radical species, the polymerizations that are so initiated follow the free-radical mechanism almost exclusively, except at very low temperatures, where ionic intermediates become more stable. We shall not deal further with these high-energy sources of initiating radicals, but we shall return to light as a photochemical initiator because of its role in the evaluation of kinetic rate constants.

All of the reactions listed in Table 6.1 produce free radicals, so we are presented with a number of alternatives for initiating a polymerization reaction. Our next concern is in the fate of these radicals or, stated in terms of our interest in polymers, the efficiency with which these radicals initiate polymerization. Since these free radicals are relatively reactive species, there are a variety of

Table 6.1 Some Examples of Initiation Reactions

1. Organic peroxides or hydroperoxides:

$$\phi-\overset{\overset{O}{\|}}{C}-\overset{\overset{O}{\|}}{C}-O-O-\overset{\overset{O}{\|}}{C}-C-\phi \longrightarrow 2\phi-\overset{\overset{O}{\|}}{C}-O\cdot$$

Benzoyl peroxide

$$\phi-\underset{\underset{CH_3}{|}}{\overset{\overset{CH_3}{|}}{C}}-O-O-H \longrightarrow \phi-\underset{\underset{CH_3}{|}}{\overset{\overset{CH_3}{|}}{C}}\cdot \; + \; \cdot OH$$

Cumyl hydroperoxide

2. Azo compounds:

$$CH_3-\underset{\underset{CN}{|}}{\overset{\overset{CH_3}{|}}{C}}-N=N-\underset{\underset{CN}{|}}{\overset{\overset{CH_3}{|}}{C}}-CH_3 \longrightarrow 2\,CH_3-\underset{\underset{CN}{|}}{\overset{\overset{CH_3}{|}}{C}}\cdot \; + N_2$$

2,2'-Azobisisobutyronitrile (AIBN)

3. Redox systems:

$$H_2O_2 + Fe^{2+} \longrightarrow OH^- + Fe^{3+} + \cdot OH$$

$$S_2O_8{}^{2-} + Fe^{2+} \longrightarrow SO_4{}^{2-} + Fe^{3+} + SO_4{}^-\cdot$$

4. Electromagnetic radiation:

$$\phi-CH=CH_2 \overset{h\nu}{\longrightarrow} \phi-CH=\overset{\cdot}{C}H + \overset{\cdot}{H} \quad \text{or} \quad \phi\cdot + \cdot CH=CH_2$$

$$\phi-\overset{\overset{O}{\|}}{C}-\underset{\underset{}{}}{\overset{\overset{OH}{|}}{C}}H\phi \overset{h\nu}{\longrightarrow} \phi-\overset{\overset{O}{\|}}{C}\cdot + \cdot \overset{\overset{OH}{|}}{C}H\phi$$

Benzoin

processes they can undergo as alternatives to adding to monomers to commence the formation of polymer.

In discussing mechanism (5.F) in the last chapter we noted that the entrapment of two reactive species in the same solvent cage may be considered a transition state in the reaction of these species. Reactions such as the thermal homolysis of peroxides and azo compounds result in the formation of two radicals already trapped together in a cage that promotes direct recombination, as with the 2-cyanopropyl radicals from 2,2'-azobisisobutyronitrile (AIBN),

$$
2(CH_3)_2\overset{\underset{\textstyle CN}{|}}{C}\cdot
\begin{cases}
(CH_3)_2-\overset{\underset{\textstyle |}{|}}{\underset{\textstyle C}{C}}-\overset{\overset{\textstyle CN}{|}}{\underset{\textstyle |}{C}}-(CH_3)_2 \\[2em]
(CH_3)_2C{=}C{=}N-\overset{\overset{\textstyle CN}{|}}{C}(CH_3)_2
\end{cases}
\tag{6.D}
$$

or the recombination of degradation products of the initial radicals, as with acetoxy radicals from acetyl peroxide,

$$
2CH_3\overset{\overset{\textstyle O}{\|}}{C}-O\cdot
\begin{cases}
CH_3-\overset{\overset{\textstyle O}{\|}}{C}-OCH_3 + CO_2 \\[1.5em]
CH_3CH_3 + 2CO_2
\end{cases}
\tag{6.E}
$$

In both of these examples, initiator is consumed, but no polymerization is started.

Once the radicals diffuse out of the solvent cage, reaction with monomer is the most probable reaction in bulk polymerizations, since monomers are the species most likely to be encountered. Reaction with polymer radicals or initiator molecules cannot be ruled out, but these are less important because of the lower concentration of the latter species. In the presence of solvent, reactions between the initiator radical and the solvent may effectively compete with polymer initiation. This depends very much on the specific chemicals involved. For example, carbon tetrachloride is quite reactive toward radicals because of the resonance stabilization of the solvent radical produced [I]:

$$
\overset{\underset{\textstyle Cl}{|}}{Cl-\overset{\cdot}{C}-Cl} \;\longleftrightarrow\; \overset{\underset{\textstyle Cl}{|}}{\overset{\cdot}{C}l{=}C-Cl} \;\longleftrightarrow\; \overset{\underset{\textstyle \overset{\cdot}{C}l}{|}}{Cl-\overset{\|}{C}-Cl} \;\longleftrightarrow\; \overset{\underset{\textstyle Cl}{|}}{Cl-\overset{}{C}{=}Cl\cdot}
$$

$$[I]$$

While this reaction with solvent continues to provide free radicals, these may be less reactive species than the original initiator fragments. We shall have more to say about the transfer of free-radical functionality to solvent in Sec. 6.8.

The significant thing about these and numerous other side reactions that could be described for specific systems is the fact that they lower the efficiency of the initiator in promoting polymerization. To quantify this concept we define the initiator efficiency f to be the following fraction:

$$f = \frac{\text{Radicals incorporated into polymer}}{\text{radicals formed by initiator}} \tag{6.1}$$

The initiator efficiency is not an exclusive property of the initiator alone, but depends on the conditions of the polymerization experiment, including the solvent. In many experimental situations, f lies in the range 0.3-0.8. The efficiency should be regarded as an empirical parameter whose value is determined experimentally. Several methods are used for the evaluation of initiator efficiency, the best being the direct analysis for initiator fragments as end groups compared to the amount of initiator consumed, with proper allowances for stoichiometry. As an end group method, this procedure is difficult in addition polymers, where molecular weights are higher than in condensation polymers. Research with isotopically labeled initiators is particularly useful in this application. Since this quantity is so dependent on the conditions of the experiment, it should be monitored for each system studied.

Scavengers such as diphenylpicrylhydrazyl radicals [II] react with other radicals and thus provide an indirect method for analysis of the number of free radicals in a system:

$$\phi_2\text{N}-\overset{\bullet}{\text{N}}\text{---}\langle\bigcirc\rangle\text{---NO}_2 + \text{R}\cdot \longrightarrow \text{adduct} \tag{6.F}$$

[II]

The diphenylpicrylhydrazyl radical itself is readily followed spectrophotometrically, since it loses an intense purple color on reacting. Unfortunately this reaction is not always quantitative.

We recall some of the ideas of kinetics from the summary given in Sec. 5.2 and recognize that the rates of initiator decomposition can be developed in terms of the reactions listed in the Table 6.1. Using the change in initiator radical concentration $d[\text{I}\cdot]/dt$ to monitor the rates, we write the following:

1. For peroxides and azo compounds

$$\frac{d[I\cdot]}{dt} = 2k_d[I] \tag{6.2}$$

where k_d is the rate constant for the homolytic decomposition of the initiator and $[I]$ is the concentration of the initiator. The factor 2 appears because of the stoichiometry in these particular reactions.

2. For redox systems

$$\frac{d[I\cdot]}{dt} = k[Ox][Red] \tag{6.3}$$

where the bracketed terms describe the concentrations of oxidizing and reducing agents and k is the rate constant for the particular reactants.

3. For photochemical initiation

$$\frac{d[I\cdot]}{dt} = 2\phi' I_{abs} \tag{6.4}$$

where I_{abs} is the intensity of the light absorbed and the constant ϕ' is called the quantum yield. The factor 2 is again included for reasons of stoichiometry.

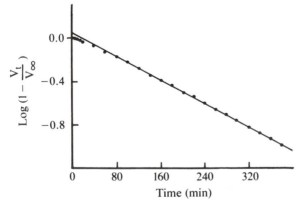

Figure 6.1 Volume of nitrogen evolved from the decomposition of AIBN at $77°C$ plotted according to the first-order rate law as discussed in Example 6.1. [Reprinted with permission from L. M. Arnett, *J. Am. Chem. Soc.* 74:2027 (1952), copyright 1952 by the American Chemical Society.]

Since $\frac{1}{2}\, d[I\cdot]/dt = -d[I]/dt$ in the case of the azo initiators, Eq. (6.2) can also be written as $-d[I]/dt = k_d[I]$ or, by integration, $\ln([I]/[I]_0) = -k_d t$, where $[I]_0$ is the initiator concentration at $t = 0$. Figure 6.1 shows a test of this relationship for AIBN in xylene at $77°C$. Except for a short induction period, the data points fall on a straight line. The evaluation of k_d from these data is presented in the following example.

Example 6.1

The deomposition of AIBN in xylene at $77°C$ was studied[†] by measuring the volume of N_2 evolved as a function of time. The volumes obtained at time t and $t = \infty$, are V_t and V_∞, respectively. Show that the manner of plotting used in Fig. 6.1 is consistent with the integrated first-order rate law and evaluate k_d.

Solution

The ratio $[I]/[I]_0$ gives the fraction of initiator remaining at time t. The volume of N_2 evolved is the following:

1. $V_0 = 0$ at $t = 0$ when no decomposition has occurred.
2. V_∞ at $t = \infty$ when complete decomposition has occurred.
3. V_t at time t when some fraction of initiator has decomposed.

Following the logic of Sec. 4.6, the fraction decomposed at t is given by $(V_t - V_0)/(V_\infty - V_0)$ and the fraction remaining at t is $1 - (V_t - V_0)/(V_\infty - V_0) = (V_\infty - V_t)/(V_\infty - V_0)$. Since $V_0 = 0$, this becomes $(V_\infty - V_t)/V_\infty$ or $[I]/[I]_0 = 1 - V_t/V_\infty$. Therefore a plot of $\ln(1 - V_t/V_\infty)$ versus t is predicted to be linear with slope $-k_d$. If logarithms to base 10 are used, the slope equals $-k_d/2.303$.

From Fig. 6.1,

$$\text{Slope} = \frac{-0.4 - (-0.8)}{160 - 320} = -2.5 \times 10^{-3} \text{ min}^{-1} = \frac{-k_d}{2.303}$$

$$k_d = 5.8 \times 10^{-3} \text{ min}^{-1}$$

•

Next we assume that a fraction f of these initiator fragments actually reacts with monomer to transfer the radical functionality to monomer:

$$I\cdot + M \xrightarrow{f} IM\cdot \qquad (6.G)$$

[†] L. M. Arnett, *J. Am. Chem. Soc.* 74:2027 (1952).

As indicated in the last section, we regard the reactivity of the species $IM_n\cdot$ to be the same regardless of the value of n. Accordingly, all subsequent additions to $IM\cdot$ in reaction (6.G) are propagation steps and reaction (6.G) represents the initiation of polymerization. Although it is premature at this point, we disregard end groups and represent the polymeric radicals of whatever size by the symbol $M\cdot$. Accordingly, we write the following for the initiation of polymer radicals:

1. By peroxides and azo compounds,

$$\frac{d[M\cdot]}{dt} = 2fk_d[I] \tag{6.5}$$

2. By redox systems,

$$\frac{d[M\cdot]}{dt} = fk[Ox][Red] \tag{6.6}$$

3. By photochemical initiation,

$$\frac{d[M\cdot]}{dt} = 2f\phi'I_{abs} = 2\phi I_{abs} \tag{6.7}$$

where we have combined the factors f and ϕ' into a composite quantum yield ϕ, since both of the separate factors are measures of efficiency.

Any one of these expressions gives the rate of initiation R_i for the particular catalytic system employed. We shall focus attention on the homolytic decomposition of a single initiator as the mode of initiation throughout most of this chapter, since this reaction typifies the most widely used free-radical initiators. Appropriate expressions for initiation which follows Eq. (6.6) are readily derived.

An important application of photochemical initiation is in the determination of the rate constants which appear in the overall analysis of the chain-growth mechanism. Although we shall take up the details of this method in Sec. 6.6, it is worthwhile to develop Eq. (6.7) somewhat further at this point. It is not possible to give a detailed treatment of light absorption here. Instead, we summarize some pertinent relationships and refer the reader who desires more information to textbooks of physical or analytical chemistry. The following results will be useful:

1. The intensity of light transmitted (subscript t) through a sample I_t depends on the intensity of the incident (subscript 0) light I_0, the thickness l of the sample, and the concentration [c] of the absorbing species,

$$I_t = I_0 e^{-\alpha[c]l} \tag{6.8}$$

where the proportionality constant α is called the absorption coefficient and is a property of the absorber.

2. The absorbance A as measured by spectrophotometers is defined as

$$A = \log_{10}\left(\frac{I_0}{I_t}\right) \tag{6.9}$$

The variation in absorbance with wavelength reflects the wavelength dependence of α.

3. Since I_{abs} equals the difference $I_0 - I_t$,

$$I_{abs} = I_0(1 - e^{-\alpha[c]l}) \tag{6.10}$$

If the exponent in Eq. (6.10) is small—which means dilute solutions in practice, since most absorption experiments are done where α is large—then the exponential can be expanded, $e^x \cong 1 + x + \cdots$, with only the leading terms retained to give

$$I_{abs} = I_0(\alpha[c]l) \tag{6.11}$$

4. Substituting this result into Eq. (6.7) gives

$$\frac{d[M\cdot]}{dt} = 2\phi I_0 \alpha[c]l \tag{6.12}$$

where [c] is the concentration of monomer or initiator for the two reactions shown in Table 6.1.

Note that Eqs. (6.5) and (6.12) are both first-order rate laws, although the physical significance of the proportionality factors is quite different in the two cases. The rate constants shown in Eqs. (6.5) and (6.6) show a temperature dependence described by the Arrhenius equation:

$$k = Ae^{-E^*/RT} \tag{6.13}$$

where E^* is the activation energy, which is interpreted as the height of the energy barrier to a reaction as discussed in Sec. 2.6. Activation energies are evaluated from experiments in which rate constants are measured at different temperatures. Taking logarithms of both sides of Eq. (6.13) gives $\ln k = \ln A - E^*/RT$. Therefore E^* is obtained from the slope of a plot of $\ln k$ against $1/T$. As usual, T is in degrees Kelvin and R and E^* are in the same energy units.

Table 6.2   Rate Constants (at Temperature Given) and Activation Energies for Some Initiator Decomposition Reactions

Initiator	Solvent	T (°C)	k_d (sec^{-1})	E_d^* (kJ mol^{-1})
2,2'-Azobisisobutyronitrile	Benzene	70	3.17×10^{-5}	123.4
	CCl$_4$	40	2.15×10^{-7}	128.4
	Toluene	100	1.60×10^{-3}	121.3
t-Butyl peroxide	Benzene	100	8.8×10^{-7}	146.9
Benzoyl peroxide	Benzene	70	1.48×10^{-5}	123.8
	Cumene	60	1.45×10^{-6}	120.5
t-Butyl hydroperoxide	Benzene	169	2.0×10^{-5}	170.7

Source: Data from J. C. Masson in Ref. 3.

Since E* is positive according to this picture, the form of the Arrhenius equation assures that k gets larger as T increases. This means that a larger proportion of molecules have sufficient energy to surmount the energy barrier at higher temperatures. This of course assumes that thermal energy is the source of E*, something that is not the case in photoinitiated reactions. The effective first-order rate constants k and $I_0 \alpha l$—for thermal initiation and photoinitiation, respectively—do not show the same temperature dependence. The former follows the Arrhenius equation, while the latter cluster of terms in Eq. (6.12) is essentially independent of T.

The activation energies for the decomposition (subscript d) reaction of several different initiators in various solvents are shown in Table 6.2. Also listed are values of k_d for these systems at the temperature shown. The Arrhenius equation can be used in the form $\ln(k_{d,1}/k_{d,2}) = -(E^*/R)(1/T_1 - 1/T_2)$ to evaluate k_d values for these systems at temperatures different from those given in Table 6.2.

6.4   Termination

The formation of initiator radicals is not the only process that determines the concentration of free radicals in a polymerization system. Polymer propagation itself does not change the radical concentration; it merely changes one radical to another. Termination steps also occur, however, and these remove radicals from the system. We shall discuss combination and disproportionation reactions as modes of termination.

Termination by combination results in the simultaneous destruction of two radicals by direct coupling:

$$M_n \cdot + \cdot M_{n'} \longrightarrow M_{n+n'} \tag{6.H}$$

The degree of polymerization values in the two combining radicals can have any value, and the molecular weight of the product molecule will be considerably higher on the average than the radicals so terminated. The polymeric product molecule contains two initiator fragments per molecule by this mode of termination.

Termination by disproportionation comes about when an atom, usually hydrogen, is transferred from one polymer radical to another:

$$\underset{\underset{X}{\overset{H}{\mid}}}{M_{n-1}-CH_2-\overset{H}{\underset{\mid}{C}}}\cdot + \cdot\underset{\underset{X}{\overset{H}{\mid}}}{\overset{H}{\underset{\mid}{C}}-CH_2-M_{n'-1}} \longrightarrow M_{n-1}-CH_2CH_2X + CHX=CH-M_{n'-1}$$
$$\text{(6.I)}$$

This mode of termination produces a negligible effect on the molecular weight of the reacting species, but it does produce a terminal unsaturation in one of the dead polymer molecules. Each polymer molecule contains one initiator fragment when termination occurs by disproportionation.

A kinetic analysis of the two modes of termination is quite straightforward, since each mode of termination involves a bimolecular reaction between two radicals. Accordingly, we write the following:

1. For general termination,

$$R_t = \frac{-d[M\cdot]}{dt} = 2k_t[M\cdot]^2 \tag{6.14}$$

    where R_t and k_t are the rate and rate constant for termination (subscript t) and the factor 2 enters because two radicals are lost for each termination step.

2. The polymer radical concentration in Eq. (6.14) represents the total concentration of all such species, regardless of their degree of polymerization; that is,

$$[M\cdot] = \underset{\text{all } n}{\Sigma} [M_n\cdot] \tag{6.15}$$

3. For combination,

$$R_t = \frac{-d[M\cdot]}{dt} = 2k_{t,c}[M\cdot]^2 \tag{6.16}$$

    where the subscript c specifically indicates termination by combination.

4. For disproportionation,

$$R_t = \frac{-d[M\cdot]}{dt} = 2k_{t,d}[M\cdot]^2 \tag{6.17}$$

where the subscript d specifically indicates termination by disproportionation.

5. In the event that the two modes of termination are not distinguished, Eq. (6.14) represents the sum of Eqs. (6.16) and (6.17), or

$$k_t = k_{t,c} + k_{t,d} \tag{6.18}$$

Combination and disproportionation are competitive processes and do not occur to the same extent for all polymers. For example, at 60°C termination is virtually 100% by combination for polyacrylonitrile and 100% by disproportionation for poly(vinyl acetate). For polystyrene and poly(methyl methacrylate), both reactions contribute to termination, although each in different proportions. Each of the rate constants for termination individually follows the Arrhenius equation, so the relative amounts of termination by the two modes is given by

$$\frac{\text{Termination by combination}}{\text{termination by disproportionation}} = \frac{k_{t,c}}{k_{t,d}} = \frac{A_{t,c}\,e^{-E_{t,c}{}^*/RT}}{A_{t,d}\,e^{-E_{t,d}{}^*/RT}}$$

$$= \frac{A_{t,c}}{A_{t,d}} \exp\left(\frac{-(E_{t,c}{}^* - E_{t,d}{}^*)}{RT}\right) \tag{6.19}$$

Since the disproportionation reaction requires bond breaking, which is not required for combination, $E_{t,d}{}^*$ is expected to be greater than $E_{t,c}{}^*$. This causes the exponential to be large at low temperatures, making combination the preferred mode of termination under these circumstances. Note that at higher temperatures this bias in favor of one mode of termination over another decreases as the difference in activation energies becomes smaller relative to the thermal energy RT. The experimental results on modes of termination cited above make it apparent that this qualitative argument must be applied cautiously. The actual determination of the partitioning between the two modes of termination is best accomplished by analysis of end groups, using the difference in end group distribution noted above.

Table 6.3 lists the activation energies for termination (these are overall values, not identified as to mode) of several different radicals. The rate constants for termination at 60°C are also given. We shall see in Sec. 6.6 how these constants are determined.

Table 6.3 Rate Constants at $60°C$ and Activation Energies for Some Termination Reactions

Monomer	E_t^* (kJ mol^{-1})	$k_{t, 60°} \times 10^{-7}$ (liter mol^{-1} sec^{-1})
Acrylonitrile	15.5	78.2
Methyl acrylate	22.2	0.95
Methyl methacrylate	11.9	2.55
Styrene	8.0	6.0
Vinyl acetate	21.9	2.9
2-Vinyl pyridine	21.0	3.3

Source: Data from R. Korus and K. F. O'Driscoll in Ref. 3.

The assumption that k values are constant over the entire duration of the reaction breaks down for termination reactions in bulk polymerizations. Here, as in Sec. 5.2, we can consider the termination process—whether by combination or disproportionation—to depend on the rates at which polymer molecules can diffuse into (characterized by k_i) or out of (characterized by k_o) the same solvent cage and the rate at which chemical reaction between them (characterized by k_r) occurs in that cage. In Chap. 5 we saw that two limiting cases of Eq. (5.8) could be readily identified:

1. Rate of diffusion > rate of reaction [Eq. (5.9)]:

$$k_t = \frac{k_i}{k_o} k_r \qquad (6.20)$$

2. This situation seems highly probable for *step-growth* polymerization because of the high activation energy of many condensation reactions. The constants for the diffusion-dependent steps, which might be functions of molecular size or the extent of the reaction, cancel out.
3. Rate of reaction > rate of diffusion [Eq. (5.10)]:

$$k_t = k_i \qquad (6.21)$$

4. This situation is expected to apply to radical termination, especially by combination, because of the high reactivity of the trapped radicals. Only one constant appears which depends on the diffusion of the polymer radicals, so it cannot cancel out and may be the source of a dependence of the rate constant on the extent of reaction or degree of polymerization.

Figure 6.2 shows how the percent conversion of methyl methacrylate to polymer varies with time. These experiments were carried out in benzene at

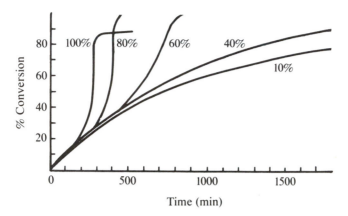

Figure 6.2 Acceleration of the polymerization rate for methyl methacrylate at the concentrations shown in benzene at 50°C. [Reprinted from G. V. Schulz and G. Haborth, *Makromol. Chem.* 1:106 (1948).]

50°C. The different curves correspond to different concentrations of monomer. Up to about 40% monomer, the conversion varies smoothly with time, gradually slowing down at higher conversions owing to the depletion of monomer. At high concentrations, however, the polymerization starts to show an acceleration between 20 and 40% conversion. This behavior, called the Trommsdorff effect, is attributed to a decrease in the rate of termination with increasing conversion. This, in turn, is due to the increase in viscosity which has an adverse effect on k_t through Eq. (6.21). Considerations of this sort are important in bulk polymerizations where high conversion is the objective, but this complication is something we wish to avoid. Hence we shall be mainly concerned with solution polymerization and/or low degrees of conversion where k_t may be justifiably treated as a true constant. We shall see in Sec. 6.8 that the introduction of solvent is accompanied by some complications of its own, but we shall ignore this possibility for now.

Polymer propagation steps do not change the total radical concentration, so we recognize that the two opposing processes, initiation and termination, will eventually reach a point of balance. This condition is called the stationary state and is characterized by a constant concentration of free radicals. Under stationary-state conditions (subscript s) the rate of initiation equals the rate of termination. Using Eq. (6.2) for the rate of initiation (that is, two radicals produced per initiator molecule) and Eq. (6.14) for termination, we write

$$2fk_d[I] = 2k_t[M\cdot]_s^2 \qquad (6.22)$$

or

$$[M\cdot]_s = \left(\frac{fk_d}{k_t}\right)^{\frac{1}{2}} [I]^{\frac{1}{2}} \qquad (6.23)$$

This important equation shows that the stationary-state free-radical concentration increases with $[I]^{\frac{1}{2}}$ and varies directly with $k_d^{\frac{1}{2}}$ and inversely with $k_t^{\frac{1}{2}}$. The concentration of free radicals determines the rate at which polymer forms and the eventual molecular weight of the polymer, since each radical is a growth site. We shall examine these aspects of Eq. (6.23) in the next section. We conclude this section with a numerical example which concerns the stationary-state radical concentration for a typical system.

Example 6.2

For an initiator concentration which is constant at $[I]_0$, the *non-stationary-state* radical concentration varies with time according to the following expression:

$$\frac{[M\cdot]}{[M\cdot]_s} = \frac{\exp[(16fk_d k_t [I]_0)^{\frac{1}{2}} t] - 1}{\exp[(16fk_d k_c [I]_0)^{\frac{1}{2}} t] + 1}$$

Calculate $[M\cdot]_s$ and the time required for the free-radical concentration to reach 99% of this value using the following as typical values for constants and concentrations: $k_d = 1.0 \times 10^{-4}$ sec^{-1}, $k_t = 3 \times 10^7$ liter mol^{-1} sec^{-1}, $f = \frac{1}{2}$, and $[I]_0 = 10^{-3}$ M. Comment on the assumption $[I] = [I]_0$ that is made in deriving the non-stationary-state equation.

Solution

Use Eq. (6.23) to evaluate $[M\cdot]_s$ for the system under consideration:

$$[M\cdot]_s = \left(\frac{fk_d}{k_t} [I]_0\right)^{\frac{1}{2}} = \left(\frac{(\frac{1}{2})(1.0 \times 10^{-4})(10^{-3})}{3 \times 10^7}\right)^{\frac{1}{2}} = (1.67 \times 10^{-15})^{\frac{1}{2}}$$

$$= 4.08 \times 10^{-8} \text{ mol liter}^{-1}$$

This low level of concentration is typical of free-radical polymerizations. Next we inquire how long it will take the free-radical concentration to reach $0.99[M\cdot]_s$, or 4.04×10^{-8} mol liter^{-1} in this case. Let $a = (16fk_d k_t [I]_0)^{\frac{1}{2}}$ and rearrange the expression given to solve for t when $[M\cdot]/[M\cdot]_s = 0.99$: $0.99(e^{at} + 1) = e^{at} - 1$, or $1 + 0.99 = e^{at}(1 - 0.99)$. Therefore the product $at = \ln(1.99/0.01) = \ln 199 = 5.29$, and $a = [16(\frac{1}{2})(1.0 \times 10^{-4})(3 \times 10^7)(10^{-3})]^{\frac{1}{2}} = 4.90$ sec^{-1}. Hence $t = 5.29/4.90 = 1.08$ sec. This short period is also typical of the time required to reach the stationary state.

The assumption that $[I] = [I]_0$ may be assessed by examining the integrated form of Eq. (6.2) for this system and calculating the ratio $[I]/[I]_0$ after 1.08 sec:

$$\ln\left(\frac{[I]}{[I]_0}\right) = -k_d t = -(1.0 \times 10^{-4})(1.08) = -1.08 \times 10^{-4}$$

$$\frac{[I]}{[I]_0} = 0.99989$$

Over the time required to reach the stationary state, the initiator concentration is essentially unchanged. As a matter of fact, it would take about 100 sec for $[I]$ to reach $0.99[I]_0$ and about 8.5 min to reach $0.95[I]_0$, so the assumption that $[I] = [I]_0$ is entirely justified over the short time involved.

•

6.5 Propagation

The propagation of polymer chains is easy to consider under stationary-state conditions. As the preceding example illustrates, the stationary state is reached very rapidly, so we lose only a brief period at the start of the reaction by restricting ourselves to the stationary state. Of course, the stationary-state approximation breaks down at the end of the reaction also, when the radical concentration drops toward zero. We shall restrict our attention to relatively low conversion to polymer, however, to avoid the complications of the Tromms-dorff effect. Therefore deviations from the stationary state at long times need not concern us.

Consideration of reaction (6.B) leads to

$$\frac{-d[M]}{dt} = k_p [M] [M \cdot] \tag{6.24}$$

as the expression for the rate at which monomer is converted to polymer. In writing this expression, we assume the following:

1. The radical concentration has the stationary-stage value given by Eq. (6.23).
2. k_p is a constant independent of the size of the growing chain and the extent of conversion to polymer.
3. The rate at which monomer is consumed is equal to the rate of polymer formation R_p:

$$\frac{-d[M]}{dt} = \frac{d[polymer]}{dt} = R_p \tag{6.25}$$

Combining Eqs. (6.23) and (6.24) yields

$$R_p = k_p [M] \left(\frac{fk_d}{k_t} \right)^{1/2} [I]^{1/2} = k_{app} [M] [I]^{1/2} \tag{6.26}$$

in which the second form reminds us that an experimental study of the rate of polymerization yields a single constant (subscript app) which the mechanism reveals to be a composite of three different rate constants. Equation (6.26) shows that the rate of polymerization is first order in monomer and half order in initiator and depends on the rate constants for each of the three types of steps—initiation, propagation, and termination—that make up the chain mechanism. Since the concentrations change with time, it is important to realize that Eq. (6.26) gives an instantaneous rate of polymerization at the concentrations considered. The equation can be applied to the initial concentrations of monomer and initiator in a reaction mixture only to describe the initial rate of polymerization. Unless stated otherwise, we shall assume the initial conditions apply when we use this result.

The initial rate of polymerization is a measurable quantity. The amount of polymer formed after various times in the early stages of the reaction can be determined directly by precipitating the polymer and weighing. Alternatively, some property such as the volume of the system (or the density, the refractive index, or the viscosity) can be measured. Using an analysis like that followed in Sec. 4.6 or Example 6.1, we can relate the values of the property measured at t, t = 0, and t = ∞ to the fraction of monomer converted to polymer. If the rate of polymerization is measured under known and essentially constant concentrations of monomer and initiator, then the cluster of contants $(fk_p^2 k_d/k_t)^{1/2}$ can be evaluated from the experiment. As noted above, f is best investigated by end group analysis. Even with the factor f excluded, experiments on the rate of polymerization still leave us with a single numerical value for a cluster of three constants: one equation with three unknowns. Two other measurable relationships among these unknowns must be found if the individual constants are to be resolved. In anticipation of this development, we list values of k_p and the corresponding activation energies for several common monomers in Table 6.4.

Equation (6.26) is an important result which can be expressed in several alternate forms:

1. The variation in monomer concentration may be taken into account by writing the equation in the integrated form and treating the initiator concentration as constant at $[I]_0$ over the interval considered:

$$\ln \left(\frac{[M]}{[M]_0} \right) = - \left(\frac{fk_p^2 k_d}{k_t} [I]_0 \right)^{1/2} t \tag{6.27}$$

where $[M] = [M]_0$ at t = 0.

Table 6.4 Rate Constants at 60°C and Activation Energies for
Some Propagation Reactions

Monomer	E_p^* (kJ mol^{-1})	$k_{p,60°} \times 10^{-3}$ (liter mol^{-1} sec^{-1})
Acrylonitrile	16.2	1.96
Methyl acrylate	29.7	2.09
Methyl methacrylate	26.4	0.515
Styrene	26.0	0.165
Vinyl acetate	18.0	2.30
2-Vinyl pyridine	33.0	0.186

Source: Data from R. Korus and K. F. O'Driscoll in Ref. 3.

2. Instead of using $2fk_d[I]$ for the rate of initiation, we can simply write this latter quantity as R_i, in which case the stationary-state radical concentration is

$$[M\cdot]_s = \left(\frac{R_i}{2k_t}\right)^{\frac{1}{2}}$$
(6.28)

and the rate of polymerization becomes

$$R_p = \left(\frac{k_p^2}{2k_t}\right)^{\frac{1}{2}} R_i^{\frac{1}{2}} [M]$$
(6.29)

If the rate of initiation is investigated independently, the rate of polymerization measures a cluster of k_p and k_t.

3. Alternatively, Equations (6.6) and (6.7) can be used as expressions for R_i in Eq. (6.29) to describe redox or photoinitiated polymerization.

Figure 6.3 shows some data which constitute a test of Eq. (6.26). In Fig. 6.3a, R_p and [M] are plotted on a log–log scale for a constant level of redox initiator. The slope of this line, which indicates the order of the polymerization with respect to monomer, is unity, showing that the polymerization of methyl methacrylate is first order in monomer. Figure 6.3b is a similar plot of the initial rate of polymerization—which essentially maintains the monomer at constant concentration—versus initiator concentration for several different monomer–initiator combinations. Each of the lines has a slope of ½, indicating a half-order dependence on [I] as predicted by Eq. (6.26).

The apparent rate constant in Eq. (6.26) follows the Arrhenius equation and yields an apparent activation energy:

$$\ln k_{app} = \ln A_{app} - \frac{E_{app}^*}{RT}$$
(6.30)

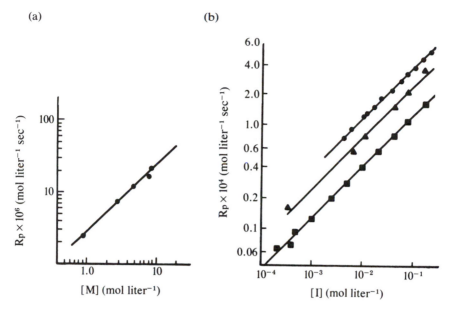

Figure 6.3 Log–log plots of R_p versus concentration which verify the order of the kinetics with respect to the constituent varied. (a) Monomer (methyl methacrylate) concentration varied at constant initiator concentration. [Data from T. Sugimura and Y. Minoura, *J. Polym. Sci.* A-1:2735 (1966).] (b) Initiator concentration varied: AIBN in methy methacrylate (○), benzoyl peroxide in styrene (□), and benzoyl peroxide in methyl methacrylate (△). (From P. J. Flory, *Principles of Polymer Chemistry*, copyright 1953 by Cornell University, used with permission.)

The mechanistic analysis of the rate of polymerization and the fact that the separate constants individually follow the Arrhenius equation means that

$$\ln k_{app} = \ln k_p \left(\frac{k_d}{k_t}\right)^{1/2} = \ln A_p \left(\frac{A_d}{A_t}\right)^{1/2} - \frac{E_p{}^* + E_d{}^*/2 - E_t{}^*/2}{RT} \tag{6.31}$$

This enables us to identify the apparent activation energy in Eq. (6.30) with the difference in E* values for the various steps:

$$E_{app}{}^* = E_p{}^* + \frac{E_d{}^*}{2} - \frac{E_t{}^*}{2} \tag{6.32}$$

Equation (6.32) allows us to conveniently assess the effect of temperature variation on the rate of polymerization. This effect is considered in the following example.

Example 6.3

Using typical activation energies out of Tables 6.2–6.4, estimate the percent change in the rate of polymerization with a 1°C change in temperature at 50°C for thermally initiated and photinitiated polymerization.

Solution

Write Eq. (6.26) in the form

$$\ln R_p = \ln k_{app} + \ln[M] + \tfrac{1}{2}\ln[I]$$

Taking the derivative, treating [M] and [I] as constants with respect to T while k is a function of T:

$$d \ln R_p = \frac{dR_p}{R_p} = d \ln k_{app}$$

Expand $d \ln k_{app}$ by means of the Arrhenius equation via Eq. (6.31):

$$\frac{dR_p}{R_p} = d \ln A_{app} - d\left(\frac{E_{app}{}^*}{RT}\right) = \frac{E_{app}{}^*}{RT^2}\, dT$$

Substitute Eq. (6.32) for $E_{app}{}^*$:

$$\frac{dR_p}{R_p} = \frac{E_p{}^* + E_d{}^*/2 - E_t{}^*/2}{RT^2}\, dT$$

Finally we recognize that a 1°C temperature variation can be approximated as dT and that $(dR_p/R_p) \times 100$ gives the approximate percent change in the rate of polymerization. Taking average values of E* from the appropriate tables, we obtain $E_d{}^* = 145$, $E_t{}^* = 16.8$, and $E_p{}^* = 24.9$ kJ mol^{-1}. For thermally initiated polymerization

$$\frac{dR_p}{R_p} = \frac{(24.9 + 145/2 - 16.8/2)(10^3)(1)}{(8.314)(323)^2} = 0.103$$

or 10.3% per degree Celsius.

For photoinitiation there is no activation energy for the initiator decomposition; hence

$$\frac{dR_p}{R_p} = \frac{(24.9 - 16.8/2)(10^3)(1)}{(8.314)(323)^2} = 1.90 \times 10^{-2}$$

or 1.90% per degree Celsius.

Note that the initiator decomposition makes the largest contribution to E^*; therefore photoinitiated processes display a considerably lower temperature dependence for the rate of polymerization.

•

Suppose we consider the ratio

$$R_p/R_i = \frac{-d[M]/dt}{-d[I]/dt}$$

under conditions where an initiator yields one radical, where $f = 1$ and where the final polymer contains one initiator fragment per molecule. For this set of conditions the ratio gives the number of monomer molecules polymerized per chain initiated, which is the degree of polymerization. A more general development of this idea is based on a quantity called the kinetic chain length $\bar{\nu}$. The kinetic chain length is defined as the ratio of the number of propagation steps to the rate of initiation, regardless of the mode of termination:

$$\bar{\nu} = \frac{R_p}{R_i} = \frac{R_p}{R_t} \tag{6.33}$$

where the second form of this expression uses the stationary-state condition $R_i = R_t$. The significance of the kinetic chain length is seen in the following statements:

1. For termination by disproportionation

$$\bar{\nu} = \bar{n}_n \tag{6.34}$$

2. For termination by combination

$$\bar{\nu} = \tfrac{1}{2}\bar{n}_n \tag{6.35}$$

3. $\bar{\nu}$ is an average quantity—indicated by the overbar—since not all kinetic chains are identical any more than all molecular chains are.

Using Eqs. (6.26) and (6.17) for R_p and R_t, respectively, we write

$$\bar{\nu} = \frac{k_p[M\cdot][M]}{2k_t[M\cdot]^2} = \frac{k_p[M]}{2k_t[M\cdot]} \tag{6.36}$$

This may be combined with Eq. (6.23) to give the stationary-state value for $\bar{\nu}$:

$$\bar{\nu} = \frac{k_p[M]}{2k_t(fk_d[I]/k_t)^{1/2}} = \frac{k_p[M]}{2(fk_tk_d[I])^{1/2}} \tag{6.37}$$

As with the rate of polymerization, we see from Eq. (6.37) that the kinetic chain length depends on the monomer and initiator concentrations and on the constants for the three different kinds of kinetic processes that constitute the mechanism. When the initial monomer and initiator concentrations are used, Eq. (6.37) describes the initial polymer formed. The initial degree of polymerization is a measurable quantity, so Eq. (6.37) provides a second functional relationship, different from Eq. (6.26), between experimentally available quantities—$\bar{n}_n$, [M], and [I]—and theoretically important parameters—k_p, k_t, and k_d. Note that the mode of termination which establishes the connection between $\bar{\nu}$ and $\bar{n}_n$ and the value of f are both accessible through end group characterization. Thus we have a second equation with three unknowns; one more and the evaluation of the individual kinetic constants from experimental results will be feasible.

There are several additional points about Eq. (6.37) that are worthy of comment. First it must be recalled that we have intentionally ignored any kinetic factors other than initiation, propagation, and termination. We shall see in Sec. 6.8 that another process, chain transfer, has significant effects on the molecular weight of a polymer. The result we have obtained, therefore, is properly designated as the kinetic chain length without transfer. A second observation is that $\bar{\nu}$ depends not only on the nature and concentration of the monomer, but also on the nature and concentration of the catalyst. The latter determines the number of different sites competing for the addition of monomer, so it is not surprising that $\bar{\nu}$ is decreased by increases in either k_d or [I]. Finally, we observe that both k_p and k_t are properties of a particular monomer. The relative molecular weight that a specific monomer tends toward—all other things being equal—is characterized by the ratio $k_p/k_t^{1/2}$ for a monomer. Using the values in Tables 6.3 and 6.4, we see that $k_p/k_t^{1/2}$ equals 0.678 for methyl acrylate and 0.0213 for styrene at 60°C. The kinetic chain length for poly(methyl acrylate) is thus expected to be about 32 times greater than for polystyrene if the two are prepared with the same initiator (k_d) and the same concentrations ([M] and [I]). Extension of this type of comparison to the degree of polymerization requires that the two polymers compared show the same proportion of the modes of termination. Thus for vinyl acetate (subscript V) relative to acrylonitrile (subscript A) at 60°C, with the same provisos as above, $\bar{\nu}_V/\bar{\nu}_A = 6$ while $\bar{n}_{n,V}/\bar{n}_{n,A} = 3$ because of differences in the mode of termination for the two.

The proviso "all other things being equal" in discussing the last point clearly applies to temperature as well, since the kinetic constants are highly sensitive to temperature. To evaluate the effect of temperature variation on the molecular weight of an addition polymer, we follow the same sort of logic as was used in Example 6.3:

1. Take logarithms of Eq. (6.37):

$$\ln \bar{\nu} = \ln k_p \, (k_t k_d)^{-\frac{1}{2}} + \ln \left(\frac{[M]}{2 \, (f[I])^{\frac{1}{2}}} \right) \tag{6.38}$$

2. Differentiate with respect to T, assuming the temperature dependence of the concentrations is negligible compared to that of the rate constants:

$$\frac{d\bar{\nu}}{\bar{\nu}} = d \ln k_p - \frac{1}{2} d \ln (k_t k_d) \tag{6.39}$$

3. By the Arrhenius equation $d \ln k = -d \, (E^*/RT) = (E^*/RT^2) \, dT$; therefore

$$\frac{d\bar{\nu}}{\bar{\nu}} = \frac{E_p{}^* - E_t{}^*/2 - E_d{}^*/2}{RT^2} \, dT \tag{6.40}$$

It is interesting to compare the application of this result to thermally initiated and photoinitiated polymerizations as we did in Example 6.3. Again using the average values of the constants from Tables 6.2–6.4 and using T = 50°C, we calculate that $\bar{\nu}$ *decreases* about 6.5% per degree Celsius for thermal initiation and *increases* about 2% per degree for photoinitiation. It is clearly the large activation energy of the initiator dissociation which makes the difference in the two cases. This term is omitted in the case of photoinitiation, where the temperature increase produces a bigger effect on propagation than on termination. On the other hand, for thermal initiation an increase in temperature produces a large increase in the number of growth centers, with the attendant reduction of the average kinetic chain length.

Photoinitiation is not as important as thermal initiation in the overall picture of free-radical chain-growth polymerization. The foregoing discussion reveals, however, that the contrast between the two modes of initiation does provide insight into and confirmation of various aspects of addition polymerization. The most important application of photoinitiated polymerization is in providing a third experimental relationship among the kinetic parameters of the chain mechanism. We shall consider this in the next section.

6.6 Radical Lifetime

In the preceding section we observed that both the rate of polymerization and the degree of polymerization under stationary-state conditions can be interpreted to yield some cluster of the constants k_p, k_t, and k_d. The situation is summarized diagramatically in Fig. 6.4. The circles at the two bottom corners

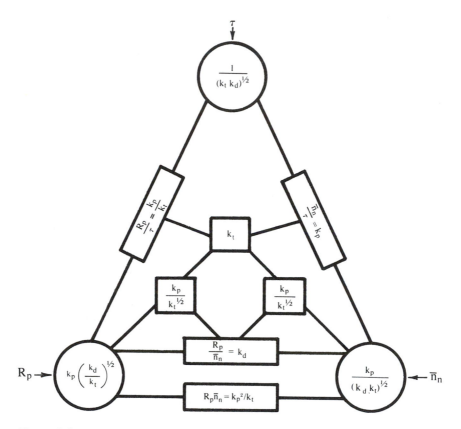

Figure 6.4 Schematic relationship between various experimental quantities $(R_p, \bar{n}_n,$ and $\tau)$ and the rate constants $(k_d, k_p,$ and $k_t)$ derived therefrom.

of the triangle indicate the particular grouping of constants obtainable from the measurement of R_p or $\bar{n}_n$, as shown. By combining these two sources of data in the manner suggested in the boxes situated along the lines connecting these circles, k_d can be evaluated, as well as the ratio k_p^2/k_t. Using this stationary-state data, however, it is not possible to further resolve the propagation and termination constants. Another relationship is needed to do this. A quantity called the radical lifetime $\bar{\tau}$ supplies the additional relationship and enables us to move off the base of Fig. 6.4.

To arrive at an expression for the radical lifetime, we return to Eq. (6.24), which may be interpreted as follows:

1. $d[M]/dt$ gives the rate at which monomers enter polymer molecules. This, in turn, is given by the product of the number of growth sites, $[M\cdot]$, and

the rate at which monomers add to each growth site. On the basis of Eq. (6.24), the rate at which monomers add to a radical is given by $k_p[M]$.

2. If $k_p[M]$ gives the number of monomers added per unit time, then $1/k_p[M]$ equals the time elapsed per monomer addition.

3. If we multiply the time elapsed per monomer added to a radical by the number of monomers in the average chain, then we obtain the time during which the radical exists. This is the definition of the radical lifetime. The number of monomers in a polymer chain is, of course, the degree of polymerization. Therefore we write

$$\bar{\tau} = \frac{\bar{n}_n}{k_p[M]} \tag{6.41}$$

4. The degree of polymerization in Eq. (6.41) can be replaced with the kinetic chain length, and the resulting expression simplified. To proceed, however, we must choose between the possibilities described by Eqs. (6.34) and (6.35). Assuming termination by disproportionation, we replace $\bar{n}_n$ by $\bar{\nu}$, using Eq. (6.37):

$$\bar{\tau} = \frac{k_p[M]}{2(fk_t k_d[I])^{\frac{1}{2}}} \frac{1}{k_p[M]} = \frac{1}{2(fk_t k_d[I])^{\frac{1}{2}}} \tag{6.42}$$

5. The radical lifetime is an average quantity, indicated by the overbar.

We shall see presently that the lifetime of a radical can be measured. When such an experiment is conducted with a known concentration of initiator, then the cluster of constants $(k_t k_d)^{-\frac{1}{2}}$ can be evaluated. This is indicated at the apex of the triangle in Fig. 6.4.

There are several things about Fig. 6.4 that should be pointed out:

1. In going from the experimental quantities R_p, $\bar{n}_n$, and $\bar{\tau}$ to the associated clusters of constants, it has been assumed that the monomer and initiator concentrations are known and essentially constant. In addition, the efficiency factor f has been left out, the assumption being that still another type of experiment has established its value.

2. By following the lines connecting two sources of circled information, the boxed result in the perimeter of the triangle may be established. Thus k_p is evaluated from τ and $\bar{n}_n$.

3. Here k_p can be combined with one of the various k_p/k_t ratios to permit the evaluation of k_t.

We can use the constants tabulated elsewhere in the chapter to get an idea of a typical radical lifetime. Choosing 10^{-3} M AIBN as the initiator ($k_d = 0.85 \times 10^{-5}$ sec^{-1} at 60°C) and vinyl acetate as the monomer (terminates entirely by

disproportionation, k_t = 2.9 X 10^7 liter mol^{-1} sec^{-1} at 60°C). Taking f = 1 for the purpose of calculation, we find τ = ½[(1.0)(2.9 X 10^7)(0.85 X 10^{-5}) $(10^{-3})]^{-½}$ = 1.01 sec. This figure contrasts sharply with the times required to obtain high molecular weight molecules in step-growth polymerizations.

Since the radical lifetime provides the final piece of information needed to independently evaluate the three primary kinetic constants—remember, we are still neglecting chain transfer—the next order of business is a consideration of the measurement of $\bar{\tau}$.

A widely used technique for measuring radical lifetime is based on photo-initiated polymerization using a light source which blinks on and off at regular intervals. In practice, a rotating opaque disk with a wedge sliced out of it is interposed between the light and the reaction vessel. Thus the system is in darkness when the solid part of the disk is in the light path and is illuminated when the notch passes. With this device, called a rotating sector, the relative lengths of the light and dark periods can be controlled by the area of the notch, and the frequency of the flickering can be controlled by the velocity of rotation of the disk. For simplicity, we shall consider light and dark periods of equal duration, although this need not be the case.

When results are compared for polymerization experiments carried out at different frequencies of blinking, it is found that the rate depends on that frequency. To see how this comes about, we must examine the variation of radical concentration under non-stationary-state conditions. This consideration dictates the choice of photoinitiated polymerization, since in the latter it is almost possible to turn on or off—with the blink of a light—the source of free radicals. The qualifying *almost* in the previous sentence is actually the focus of our attention, since a short but finite amount of time is required for the radical concentration to reach $[M \cdot]_s$ and a short but finite amount of time is required for it to drop back to zero after the light goes out.

Suppose the radical concentration begins at zero when the light is first turned on at t = 0 (unprimed t represents time in the light; primed t, time in the dark). The radical concentration then increases toward the stationary-state value during the time of illumination. We have already encountered in Example 6.2 the expression which describes the approach of $[M \cdot]$ to $[M \cdot]_s$. The equation is readily derived, but we leave it for Problem 6 at the end of the chapter and simply write the final result:

$$[M \cdot] = [M \cdot]_s \frac{\exp (8R_i k_t)^{½} t - 1}{\exp (8R_i k_t)^{½} t + 1} \tag{6.43}$$

When $R_i = 2fk_d [I]$, this expression is identical to that given in the example. For photochemical initiation, either Eq. (6.7) or (6.12) can be used for R_i. The result obtained with Eq. (6.12) is particularly informative, since it explicitly

shows the dependence of the radical growth in the system on the incident light intensity I_0. Likewise, the stationary-state free-radical concentration itself depends on the light intensity. Equating Eqs. (6.12) and (6.14), we obtain

$$[M\cdot]_s = \left(\frac{\phi\alpha l[c]}{k_t}\right)^{1/2}\sqrt{I_0} \qquad (6.44)$$

where [c] is the concentration of the light-absorbing species. Figure 6.5a shows schematically how the actual radical concentration approaches the stationary-state value.

Next we assume that the light is turned off after the radical concentration has reached $[M\cdot]_s$. Even in the dark the radicals will continue to undergo termination according to Eq. (6.14). Since this occurs without replacement owing to the darkness, the radical concentration decreases with time according to the integrated form of Eq. (6.14):

$$\frac{1}{[M\cdot]} = 2k_t t' + \text{const.} \qquad (6.45)$$

The constant of integration in this expression could be evaluated by remembering that $[M\cdot] = [M\cdot]_s$ at $t' = 0$ (prime representing darkness). A qualitative sketch of the decay of the radical concentration is shown in Fig. 6.5b.

If the light source is switched on and off and held for long periods of equal duration in either light or darkness, then the radical concentration in the system will consist of an alternation between the situation described in Figs. 6.5a and b. Because we have specified that the duration of each phase is long, the net behavior is essentially a series of plateaus in which the illumination is either I_0 or zero and the radical concentration is either $[M\cdot]_s$ or zero, with brief transitions in between. This is illustrated in Fig. 6.5c. The concentration of radicals is consistent with I_0, but is present only half of the time; hence the rate of polymerization is only half what it would be for the same illumination operating continuously.

Now suppose we consider another extreme of the same type of experiment. This time we alternate extremely short light and dark periods of equal duration. If we again start with $[M\cdot] = 0$ when the light goes on, we expect the type of behavior shown in Fig. 6.5d, where the radical buildup is interrupted by the extinction of the light before it reaches $[M\cdot]_s$. We could use Eq. (6.43) to evaluate the maximum radical concentration achieved, $[M\cdot]_{max}$, if that were the desired information. The decay of the radical concentration in the dark commences from $[M\cdot]_{max}$ and follows Eq. (6.45). This time the constant integration is evaluated from the fact that $[M\cdot] = [M\cdot]_{max}$ at $t' = 0$. During the dark phase the radical concentration drops to a nonzero value $[M\cdot]_{min}$ before the

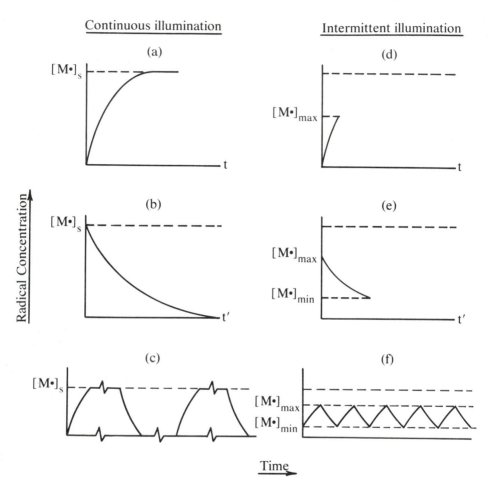

Figure 6.5 Schematic illustrations showing the variation of [M·] with time under conditions of continuous illumination (a–c) and intermittent illumination (d–f). The time in darkness is indicated as t'.

light switches back on again. This is qualitatively sketched in Fig. 6.5e and could be assessed quantitatively via Eq. (6.45). With rapid blinking, the radical concentration alternates between $[M\cdot]_{max}$ and $[M\cdot]_{min}$, as seen in Fig. 6.5f. Although the light is on half the time in this experiment, just as it was in the case of the slower blinking, the concentration of free radicals is always lower, since the time of illumination is never sufficient to reach $[M\cdot]_s$.

As the frequency of blinking increases, the range between $[M\cdot]_{max}$ and $[M\cdot]_{min}$ will narrow, approaching a plateau value below $[M\cdot]_s$ in the limit of

very fast flashing. The radical concentration in this case is consistent with an intensity of illumination $I_0/2$, since the blinking is too fast to cause a perceptible fluctuation in radical concentration, but only half the radiation energy is deposited in the system as would be the case for continuous illumination.

With these ideas as background, we can now consider the formation of polymer in experiments conducted under different conditions of illumination. The rate of polymerization is proportional to the concentration of radicals present in a system. For photoinitiated polymerization this quantity can be changed by varying the frequency of blinking. Thus the rate of polymerization can be measured in a system that is constant with respect to reagents, concentrations, and temperature, but which differs in conditions of illumination. On the basis of the foregoing discussion, we conclude the following for light of intensity I_0:

1. For continuous illumination (i.e., stationary state)

$$(R_p)_{cont} \propto \sqrt{I_0}$$

2. For intermittent illumination with very slow blinking

$$(R_p)_{slow} \propto \tfrac{1}{2} \sqrt{I_0}$$

3. For intermittent illumination with very fast blinking

$$(R_p)_{fast} \propto \sqrt{I_0/2}$$

Thus if we were to compare the rate of polymerization with intermittent illumination relative to that with continuous illumination, but under otherwise identical conditions, we would observe the following limits for equal periods of light and dark:

1. Slow blinking:

$$\frac{(R_p)_{slow}}{(R_p)_{cont}} = \tfrac{1}{2} \qquad\qquad (6.46)$$

2. Fast blinking:

$$\frac{(R_p)_{fast}}{(R_p)_{cont}} = \frac{1}{\sqrt{2}} \qquad\qquad (6.47)$$

If the dark period is m times longer than the lit period, then the fraction of time that the sample is illuminated is $1/(1 + m)$ (m = 1 for the case considered above).

For this general case the limits become the following:

3. Slow blinking:

$$\frac{(R_p)_{slow}}{(R_p)_{cont}} = \frac{1}{m + 1} \tag{6.48}$$

4. Fast blinking:

$$\frac{(R_p)_{slow}}{(R_p)_{cont}} = \frac{1}{(m + 1)^{1/2}} \tag{6.49}$$

Throughout the foregoing discussion, we have used terms like *slow* and *fast blinking* and *long* and *short periods of illumination*. The question that should have been stirring throughout this discussion is, How long a period of time is a lengthy illumination? or How slow is slow? The answer is that times are judged long or short compared to the lifetime of the radical. The radical, after all, is the species that "sees" the light; it is on *its* terms that intervals are described as long or short. Accordingly, we might ask how the rate of polymerization under intermittent illumination might compare with $(R_p)_{cont}$ when the time of illumination is on the order of $\bar{\tau}$.

The rate of polymerization under conditions where the period of illumination is comparable to $\bar{\tau}$ is obtained by integrating Eq. (6.24) in the following form:

$$\int dM = k_p[M]\left\{\int [M\cdot]\ dt + \int [M\cdot]\ dt'\right\} \tag{6.50}$$

The left-hand side of this expression gives the amount of monomer converted to polymer in one light–dark cycle. The first term on the right is integrated over one light period, using Eq. (6.43) for $[M\cdot]$, while the second term is integrated over one dark period, using Eq. (6.45) for $[M\cdot]$. The integration is rather messy, so we forgo the details and simply present the final result. Assuming equal periods for light and dark and expressing the result relative to the stationary-state rate, we integrate Eq. (6.50) to obtain the following result:

$$\frac{(R_p)_{blink}}{(R_p)_{cont}} = \frac{1}{2}\left[1 + \frac{\bar{\tau}}{t}\ \ln\left(\frac{[M\cdot]_{max}/[M\cdot]_{min} + [M\cdot]_{max}/[M\cdot]_s}{1 + [M\cdot]_{max}/[M\cdot]_s}\right)\right] \tag{6.51}$$

In the more general case where the dark period is m times longer than the light period, the factor $\frac{1}{2}$ in Eq. (6.51) is replaced by $1/(1 + m)$. This result is used by evaluating $[M\cdot]_{max}$, $[M\cdot]_{min}$, and $[M\cdot]_s$ using Eqs. (6.43), (6.45), and (6.44), respectively, for the conditions of the experiment. Then the ratio $(R_p)_{blink}/(R_p)_{cont}$ is evaluated for different values of the ratio τ/t. The results

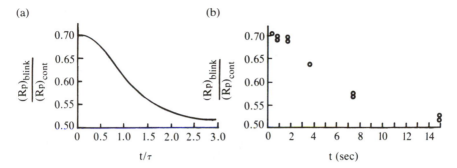

Figure 6.6   Plot of $(R_p)_{blink}/(R_p)_{cont}$ versus time: (a) theoretical curves with time expressed in units of τ and (b) experimental points for methyl acrylate. [Redrawn from Ref. 2.]

are plotted to yield a master curve such as that shown in Fig. 6.6a. Figure 6.6b shows some experimental points measured for the polymerization of methyl acrylate. By plotting the master curve and the data points on the same scale and then sliding the master curve along the ordinate until the points fall on the curve, it is possible to identify the time at which $t/\bar{\tau}$ equals unity. Applying this method to the data in Fig. 6.6 gives a value of $\bar{\tau}$ equal to about 6.5 sec for the system studied.

The superpositioning of experimental and theoretical curves to evaluate a characteristic time is reminiscent of the time–temperature superpositioning described in Sec. 4.10. This parallel is even more apparent if the theoretical curve is drawn on a logarithmic scale, in which case the distance by which the curve has to be shifted measures log $\bar{\tau}$. Note that the limiting values of the ordinate in Fig. 6.6 correspond to the limits described in Eqs. (6.46) and (6.47). Because this method effectively averages over both the buildup and the decay phases of radical concentration, it affords an experimentally less demanding method for the determination of $\bar{\tau}$ than alternative methods which utilize either the buildup or the decay portions of the non-stationary-state free-radical concentration.

To gain some additional familiarity with the concept of radical lifetime and to see how this quantity can be used to determine the absolute value of a kinetic constant, consider the following example.

Example 6.4

The polymerization of ethylene at 130°C and 1500 atm was studied using different concentrations of the initiator, 1-t-butylazo-1-phenoxycyclohexane.

The rate of initiation was measured directly and radical lifetimes were determined using the rotating sector method. The following results were obtained.[†]

Run	$\bar{\tau}$ (sec)	$R_i \times 10^9$ (mol liter^{-1} sec^{-1})
5	0.73	2.35
6	0.93	1.59
8	0.32	12.75
12	0.50	5.00
13	0.29	14.95

Demonstrate that the variations in the rate of initiation and $\bar{\tau}$ are consistent with free-radical kinetics and evaluate k_t.

Solution

Since the rate of initiation is measured, we can substitute R_i for the terms $(2fk_d[I])^{1/2}$ in Eq. (6.42) to give

$$\bar{\tau} = \frac{1}{(2k_t R_i)^{1/2}} \quad \text{or} \quad k_t = \frac{1}{2\bar{\tau}^2 R_i}$$

If the data follow the kinetic scheme presented here, the values of k_t calculated for the different runs should be constant:

Run	$k_t \times 10^{-8}$ (liter mol^{-1} sec^{-1})
5	3.99
6	3.64
8	3.83
12	4.00
13	3.98
Average	3.89

Even though the rates of initiation span almost a 10-fold range, the values of k_t show a standard deviation of only 4%, which is excellent in view of experimental errors. Note that the rotating sector method can be used in high-pressure experiments and other unusual situations, a characteristic it shares with many optical methods in chemistry.

[†] T. Takahashi and P. Ehrlich, *Polym. Prepr.* 22:203 (1981).

•

In the next section we shall examine the distribution of molecular weights for polymerization which follows the chain-growth mechanism.

6.7 Distribution of Molecular Weights

Until this point in the chapter we have intentionally avoided making any differentiation among radicals on the basis of the degree of polymerization of the radical. Now we seek a description of the molecular weight distribution of addition polymer molecules. Toward this end it becomes necessary to consider radicals of different n values. We begin by writing a kinetic expression for the concentration of radicals of degree of polymerization n, which we designate $[M_n \cdot]$. This rate law will be the sum of three contributions:

1. An increase which occurs by addition of monomer to the radical $M_{n-1} \cdot$.
2. A decrease which occurs by addition of a monomer to the radical $M_n \cdot$.
3. A decrease which occurs by the termination of $M_n \cdot$ with some other undifferentiated radical $M \cdot$.

The change in $[M_n \cdot]$ under stationary-state conditions equals zero for all values of n; hence we write

$$d[M_n \cdot] = k_p [M] [M_{n-1} \cdot] - k_p [M] [M_n \cdot] - 2k_t [M_n \cdot] [M \cdot] = 0 \qquad (6.52)$$

which can be rearranged to

$$\frac{[M_n \cdot]}{[M_{n-1} \cdot]} = \frac{k_p [M]}{k_p [M] + 2k_t [M \cdot]} \qquad (6.53)$$

Dividing the numerator and denominator of Eq. (6.53) by $2k_t [M \cdot]$ and recalling the definition of $\bar{\nu}$ provided by Eq. (6.36) enables us to express this result more succinctly as

$$\frac{[M_n \cdot]}{[M_{n-1} \cdot]} = \frac{\bar{\nu}}{1 + \bar{\nu}} \qquad (6.54)$$

Next let us consider the following sequence of multiplications:

$$\frac{[M_n \cdot]}{[M_{n-1} \cdot]} \frac{[M_{n-1} \cdot]}{[M_{n-2} \cdot]} \frac{[M_{n-2} \cdot]}{[M_{n-3} \cdot]} \cdots \frac{[M_{n-(n-2)} \cdot]}{[M_{n-(n-1)} \cdot]} = \frac{[M_n \cdot]}{[M_1 \cdot]} \qquad (6.55)$$

This shows that the number of n-mer radicals relative to the number of the smallest radicals is given by multiplying the ratio $[M_n \cdot]/[M_{n-1} \cdot]$ by $n - 2$ analogous ratios. Since each of the individual ratios is given by $\bar{\nu}/(1 + \bar{\nu})$, we can write Eq. (6.55) as

$$\frac{[M_n \cdot]}{[M_1 \cdot]} = \frac{[M_n \cdot]}{[M_{n-1} \cdot]} \left(\frac{\bar{\nu}}{1 + \bar{\nu}}\right)^{n-2} \tag{6.56}$$

or

$$[M_{n-1} \cdot] = [M_1 \cdot] \left(\frac{\bar{\nu}}{1 + \bar{\nu}}\right)^{(n-1)-1} \tag{6.57}$$

Since it is more convenient to focus attention on n-mers than $(n - 1)$-mers, the corresponding expression for the n-mer is written by analogy:

$$[M_n \cdot] = [M_1 \cdot] \left(\frac{\bar{\nu}}{1 + \bar{\nu}}\right)^{n-1} \tag{6.58}$$

Dividing both sides of Eq. (6.58) by $[M \cdot]$, the total radical concentration, gives the number fraction of n-mer radicals in the total radical population. This ratio is the same as the number of n-mers N_n in the sample containing a total of N (no subscript) polymer molecules:

$$\frac{N_n}{N} = \frac{[M_n \cdot]}{[M \cdot]} = \frac{[M_1 \cdot]}{[M \cdot]} \left(\frac{\bar{\nu}}{1 + \bar{\nu}}\right)^{n-1} \tag{6.59}$$

The ratio $[M_1 \cdot]/[M \cdot]$ in Eq. (6.59) can be eliminated by reexamining Eq. (6.52) explicitly for the M_1 radical:

1. Write Eq. (6.52) for $M_1 \cdot$, remembering in this case that the leading term describes initiation:

$$d[M_1 \cdot] = R_i - k_p [M] [M_1 \cdot] - 2k_t [M_1 \cdot] [M \cdot] = 0 \tag{6.60}$$

2. Rearrange under stationary-state conditions:

$$[M_1 \cdot] = \frac{R_i}{k_p [M] + 2k_t [M \cdot]} \tag{6.61}$$

The total radical concentration under stationary-state conditions can be similarly obtained:

3. Write Eq. (6.22) using the same notation for initiation as in (6.60):

$$d[M \cdot] = R_i - 2k_t [M \cdot]^2 = 0 \tag{6.62}$$

4. Rearrange under stationary-state conditions:

$$[M\cdot] = \frac{R_i}{2k_t[M\cdot]} \tag{6.63}$$

5. Take the ratio of Eq. (6.61) to Eq. (6.63):

$$\frac{[M_1\cdot]}{[M\cdot]} = \frac{2k_t[M\cdot]}{k_p[M] + 2k_t[M\cdot]} = \frac{1}{1+\bar{\nu}} \tag{6.64}$$

Combining Eq. (6.64) with Eq. (6.59) gives

$$\frac{N_n}{N} = \frac{1}{1+\bar{\nu}} \left(\frac{\bar{\nu}}{1+\bar{\nu}}\right)^{n-1} = \frac{1}{\bar{\nu}} \left(\frac{\bar{\nu}}{1+\bar{\nu}}\right)^n \tag{6.65}$$

This expression gives the number fraction or mole fraction of n-mers in the polymer and is thus equivalent to Eq. (5.25) for step-growth polymerization.

The kinetic chain length $\bar{\nu}$ may also be viewed as merely a cluster of kinetic constants and concentrations which was introduced into Eq. (6.54) to simplify the notation. As an alternative, suppose we define for the purposes of this chapter a fraction p such that

$$p = \frac{\bar{\nu}}{1+\bar{\nu}} = \frac{k_p[M]}{k_p[M] + 2k_t[M\cdot]} \tag{6.66}$$

It follows from this definition that $1/(1 + \bar{\nu}) = 1 - p$, so Eq. (6.65) can be rewritten as

$$\frac{N_n}{N} = (1 - p)p^{n-1} \tag{6.67}$$

This change of notation now expresses Eq. (6.65) in exactly the same form as its equivalent in Sec. 5.4. Several similarities and differences should be noted in order to take full advantage of the parallel between this result and the corresponding material for condensation polymers from Chap. 5:

1. In Chap. 5, p was defined as the fraction (or probability) of functional groups that had reacted at a certain point in the polymerization. According to the current definition provided by Eq. (6.66), p is the fraction (or probability) of propagation steps among the combined total of propagation and termination steps. The quantity 1 - p is therefore the fraction (or

probability) of termination steps. An addition polymer of degree of polymerization n has undergone n - 1 propagation steps and one termination step. Therefore it makes sense to describe its probability in the form of Eq. (6.67).

2. It is apparent from Eq. (6.66) that $p \to 1$ as $\bar{\nu} \to \infty$; hence those same conditions which favor the formation of a high molecular weight polymer also indicate p values close to unity.

3. In Chap. 5 all molecules—whether monomer or n-mers of any n—carry functional groups; hence the fraction described by Eq. (5.24) applies to the entire reaction mixture. Equation (6.67), by contrast, applies only to the radical population. Since the radicals eventually end up as polymers, the equation also describes the polymer produced. Unreacted monomer is specifically excluded, however.

4. Only one additional stipulation needs to be made before adapting the results that follow from Eq. (5.24) to addition polymers. The mode of termination must be specified to occur by disproportionation to use the results of Sec. 5.4 in this chapter, since termination by combination obviously changes the particle size distribution. We shall return to the case of termination by combination presently.

5. For termination by disproportionation (subscript d), we note that $p = k_p[M]/(k_p[M] + 2k_{t,d}[M\cdot])$, and by analogy with Eqs. (5.28), (5.32), and (5.33),

$$(\bar{n}_n)_d = \frac{1}{1 - p} \tag{6.68}$$

$$(\bar{n}_w)_d = \frac{1 + p}{1 - p} \tag{6.69}$$

$$\left(\frac{\bar{n}_w}{\bar{n}_n}\right)_d = 1 + p \to 2 \quad \text{as} \quad p \to 1 \tag{6.70}$$

Because of Eq. (6.66), $(\bar{n}_n)_d$ can also be written as $1 + \bar{\nu} \cong \bar{\nu}$ for large $\bar{\nu}$, which is the result already obtained in Eq. (6.34). Figures 5.5 and 5.6 also describe the distribution by number and the distribution by weight of addition polymers, if the provisos enumerated above are applied.

To deal with the case of termination by combination, it is convenient to write some reactions by which an n-mer might be formed. Table 6.5 lists several specific chemical reactions and the corresponding rate expressions as well as the general form for the combination of an (n - m)-mer and an m-mer. On the assumption that all $k_{t,c}$ values are the same, we can write the total rate of change of $[M_n\cdot]$:

Table 6.5 Some Free Radical Combination Reactions Which Yield n-mers and Their Rate Laws

Reaction	Rate law
$M_{n-1} \cdot + M_1 \cdot \longrightarrow M_n$	$\dfrac{d[M_n]}{dt} = k_{t,c}[M_{n-1} \cdot][M_1 \cdot]$
$M_{n-2} \cdot + M_2 \cdot \longrightarrow M_n$	$\dfrac{d[M_n]}{dt} = k_{t,c}[M_{n-2} \cdot][M_2 \cdot]$
$M_{n-3} \cdot + M_3 \cdot \longrightarrow M_n$	$\dfrac{d[M_n]}{dt} = k_{t,c}[M_{n-3} \cdot][M_3 \cdot]$
$\vdots$	$\vdots$
$M_{n-m} \cdot + M_m \cdot \longrightarrow M_n$	$\dfrac{d[M_n]}{dt} = k_{t,c}[M_{n-m} \cdot][M_m \cdot]$

$$\left(\frac{d[M_n \cdot]}{dt}\right)_{tot} = k_{t,c} \sum_{m=1}^{n-1} [M_{n-m} \cdot][M_m \cdot] \tag{6.71}$$

For the special case of $n - m = m = n/2$, $k_{t,c}$ is only half as large as for the combination of dissimilar radicals, but this effect is offset in Eq. (6.71) by the fact that the sumation counts all combinations twice except that where $n - m = m = n/2$.

The fraction of n-mers formed by combination may be evaluated by dividing $d[M_n \cdot]/dt$ by $\Sigma_n d[M_n \cdot]/dt$. Assuming that termination occurs exclusively by combination, then

$$\sum_n \frac{d[M_n \cdot]}{dt} = k_{t,c}[M \cdot]^2 \tag{6.72}$$

and the number fraction of n-mers formed by combination (subscript c) is

$$\left(\frac{N_n}{N}\right)_c = \frac{d[M_n \cdot]/dt}{\Sigma_n d[M_n \cdot]/dt} = \frac{k_{t,c} \sum_{m=1}^{n-1} [M_{n-m} \cdot][M_m \cdot]}{k_{t,c}[M \cdot]^2} \tag{6.73}$$

Equation (6.67) can be used to relate $[M_{n-m} \cdot]$ and $[M_m \cdot]$ to the total radical concentration:

$$[M_{n-m} \cdot] = (1 - p)p^{(n-m)-1}[M \cdot] \tag{6.74}$$

and

$$[M_m \cdot] = (1 - p) p^{m-1} [M \cdot] \tag{6.75}$$

Therefore

$$\left(\frac{N_n}{N}\right)_c = \frac{k_{t,c} \sum\limits_{m=1}^{n-1} (1 - p) p^{n-m-1} [M \cdot] (1 - p) p^{m-1} [M \cdot]}{k_{t,c} [M \cdot]^2}$$

$$= \sum_{m=1}^{n-1} (1 - p)^2 p^{n-2} \tag{6.76}$$

The index m drops out of the last summation; we compensate for this by multiplying the final result by $n - 1$ in recognition of the fact that the summation adds up $n - 1$ identical terms. Accordingly, the desired result is obtained:

$$\left(\frac{N_n}{N}\right)_c = (n - 1)(1 - p)^2 p^{n-2} \tag{6.77}$$

This expression is plotted in Fig. 6.7 for several large values of p. Although it shows a number distribution of polymers terminated by combination, the distribution looks quite different from Fig. 5.5, which describes the number distribution for termination by disproportionation. In the latter N_n/N decreases monotonically with increasing n. With combination, however, the curves go through a maximum which reflects the fact that the combination of two very small or two very large radicals is a less probable event than a more random combination.

Expressions for the various averages are readily derived from Eq. (6.77) by procedures identical to those used in Sec. 5.4. We only quote the final results for the case where termination occurs exclusively by combination:

$$(\bar{n}_n)_c = \frac{2}{1 - p} \tag{6.78}$$

$$(\bar{n}_w)_c = \frac{2 + p}{1 - p} \tag{6.79}$$

$$\left(\frac{\bar{n}_w}{\bar{n}_n}\right)_c = \frac{2 + p}{2} \tag{6.80}$$

These various expressions differ from their analogs in the case of termination by disproportionation by the appearance of occasional 2's. These terms arise

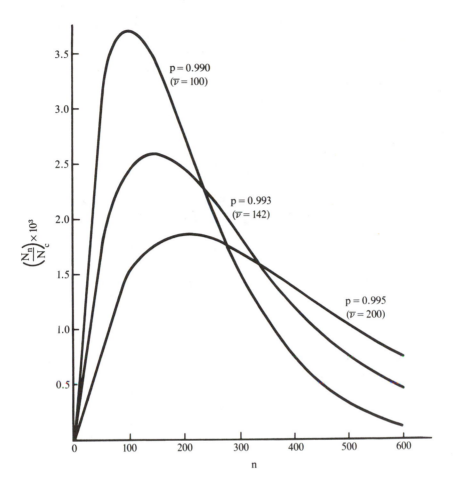

Figure 6.7 Number fraction of n-mers as a function of n for termination by combination. Drawn according to Eq. (6.77) for values of p indicated.

precisely because two chains are combined in this mode of termination. Again using Eq. (6.66), we note that $(\bar{n}_n)_c = 2(1 + \bar{\nu}) \cong 2\bar{\nu}$ for large $\bar{\nu}$, a result which was already given as Eq. (6.35).

One rather different result that arises from the case of termination by combination is seen by examining the limit of Eq. (6.80) for large values of p:

$$\frac{\bar{n}_w}{\bar{n}_n} \longrightarrow \frac{2+1}{2} = 1.5 \quad \text{as} \quad p \longrightarrow 1 \tag{6.81}$$

This contrasts with a limiting ratio of 2 for the case of termination by disproportionation. Since $\bar{M}_n$ and $\bar{M}_w$ can be measured, this difference is potentially a method for determining the mode of termination in a polymer system. In most instances, however, termination occurs by some proportion of both modes. Although general expressions exist for the various averages and their ratio when both modes of termination are operative, molecular weight data are generally not sufficiently precise to allow the proportions of termination modes to be determined in this way.

Throughout this section we have used mostly p and $\bar{\nu}$ to describe the distribution of molecular weights. It should be remembered that these quantities are defined in terms of various concentrations and therefore change as the reactions proceed. Accordingly, the results presented here are most simply applied at the start of the polymerization reaction when the initial concentrations of monomer and initiator can be used to evaluate p or $\bar{\nu}$. The termination constants are known to decrease with the extent of conversion of monomer to polymer, and this effect also complicates the picture at high conversions. Note, also, that chain transfer has been excluded from consideration in this section, as elsewhere in the chapter. We shall consider chain transfer reactions in the next section.

6.8 Chain Transfer

The three-step mechanism for free-radical polymerization represented by reactions (6.A)-(6.C) does not tell the whole story. Another type of free-radical reaction, called chain transfer, may also occur. This is unfortunate in the sense that it complicates the neat picture presented until now. On the other hand, this additional reaction can be turned into an asset in actual polymer practice. One of the consequences of chain transfer reactions is a lowering of the kinetic chain length and hence the molecular weight of the polymer without necessarily affecting the rate of polymerization.

Chain transfer arises when a hydrogen or some other atom, say, X in general, is transferred from some molecule in the system to the polymer radical. This terminates the growth of the original radical but replaces it with a new one: the fragment of the species from which X was extracted. These latter molecules will be designated by attaching the letter X to their symbol in this discussion. Thus if chain transfer involves an initiator molecule, we represent the latter IX in this section. Chain transfer can occur with any molecule in the system. The following reactions specifically describe transfer to initiator, monomer, solvent, and polymer molecules:

1. Transfer to initiator, IX:

$$M_n\cdot + IX \rightarrow M_nX + I\cdot \tag{6.J}$$

2. Transfer to monomer, MX:

$$M_n\cdot + MX \to M_nX + M\cdot \tag{6.K}$$

3. Transfer to solvent, SX:

$$M_n\cdot + SX \to M_nX + S\cdot \tag{6.L}$$

4. Transfer to polymer, M_mX:

$$M_n\cdot + M_mX \to M_nX + M_m\cdot \tag{6.M}$$

5. General, transfer to RX:

$$M_n\cdot + RX \to M_nX + R\cdot \tag{6.N}$$

It is apparent from these reactions how chain transfer lowers the molecular weight of a chain-growth polymer. The effect of chain transfer on the rate of polymerization depends on the rate at which the new radicals reinitiate polymerization:

$$R\cdot + M \xrightarrow{k_R} RM\cdot \xrightarrow{k_p} \longrightarrow RM_n\cdot \tag{6.O}$$

If the rate constant k_R is comparable to k_p, the substitution of a polymer radical with a new radical has little or no effect on the rate of polymerization. If $k_R \ll k_p$, the rate of polymerization will be decreased by chain transfer.

The kinetic chain length has a slightly different definition in the presence of chain transfer. Instead of being simply the ratio R_p/R_t, it is redefined to be the rate of propagation relative to the rates of *all* other steps that compete with propagation; specifically, termination and transfer (subscript tr):

$$\bar{\nu}_{tr} = \frac{R_p}{R_t + R_{tr}} \tag{6.82}$$

The transfer reactions follow second-order kinetics, the general rate law being

$$R_{tr} = k_{tr}[M_n\cdot][RX] \tag{6.83}$$

where k_{tr} is the rate constant for chain transfer to a specific compound RX. Since chain transfer can occur with several different molecules in the reaction mixture, Eq. (6.82) becomes

$$\bar{\nu}_{tr} = \frac{k_p [M\cdot][M]}{\left\{2k_t[M\cdot]^2 + k_{t,\,IX}[M\cdot][IX] + k_{tr,\,MX}[M\cdot][MX] \atop + k_{tr,\,SX}[M\cdot][SX] + k_{tr,\,M_nX}[M\cdot][M_nX]\right\}}$$

$$= \frac{k_p[M]}{2k_t[M\cdot] + \Sigma k_{tr,\,RX}[RX]} \tag{6.84}$$

where the summation is over all pertinent RX species. It is instructive to examine the reciprocal of this quantity:

$$\frac{1}{\bar{\nu}_{tr}} = \frac{2k_t[M\cdot]}{k_p[M]} + \frac{\Sigma k_{tr,\,RX}[RX]}{k_p[M]} \tag{6.85}$$

Since the first term on the right-hand side is the reciprocal of the kinetic chain length in the absence of transfer, this becomes

$$\frac{1}{\bar{\nu}_{tr}} = \frac{1}{\bar{\nu}} + \frac{\Sigma k_{tr,\,RX}[RX]}{k_p[M]} \tag{6.86}$$

This notation is simplified still further by defining the ratio of constants

$$\frac{k_{tr,\,RX}}{k_p} = C_{RX} \tag{6.87}$$

which is called the chain transfer constant for the monomer in question to molecule RX:

$$\frac{1}{\bar{\nu}_{tr}} = \frac{1}{\bar{\nu}} + \sum_{\text{all RX}} C_{RX} \frac{[RX]}{[M]} \tag{6.88}$$

It is apparent from this expression that the larger the sum of chain transfer terms becomes, the smaller will be $\bar{\nu}_{tr}$.

The magnitude of the individual terms in the summation depends on both the specific chain transfer constants and the concentrations of the reactants under consideration. The former are characteristics of the system and hence quantities over which we have little control; the latter can often be adjusted to study a particular effect. For example, chain transfer constants are generally obtained under conditions of low conversion to polymer where the concentration of polymer is low enough to ignore the transfer to polymer. We shall return below to the case of high conversions where this is not true.

If an experimental system is investigated in which only one molecule is significantly involved in transfer, then the chain transfer constant to that

material is particularly easy to obtain. If we assume that species SX is the only molecule to which transfer occurs, Eq. (6.88) becomes

$$\frac{1}{\bar{\nu}_{tr}} = \frac{1}{\bar{\nu}} + C_{SX}\,\frac{[SX]}{[M]} \tag{6.89}$$

This suggests that polymerizations should be conducted at different ratios of [SX]/[M] and the molecular weight measured for each. Equation (6.89) shows that a plot of $1/\bar{\nu}_{tr}$ versus [SX]/[M] should be a straight line of slope C_{SX}. Figure 6.8 shows this type of plot for the polymerization of styrene at 100°C in the presence of four different solvents. The fact that all show a common intercept as required by Eq. (6.89) shows that the rate of initiation is unaffected by the nature of the solvent. The following example examines chain transfer constants evaluated in this situation.

Example 6.5

Estimate the chain transfer constants for styrene to isopropylbenzene, ethylbenzene, toluene, and benzene from the data presented in Fig. 6.8. Comment

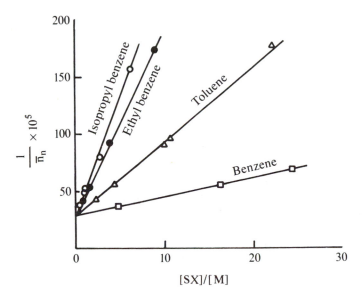

Figure 6.8 Effect of chain transfer to solvent according to Eq. (6.89) for polystyrene at 100°C. Solvents used were ethyl benzene (●), isopropyl benzene (○), toluene (△), and benzene (□). [Data from R. A. Gregg and F. R. Mayo, *Discuss. Faraday Soc.* 2:328 (1947).]

on the relative magnitude of these constants in terms of the structure of the solvent molecules.

Solution

The chain transfer constants are given by Eq. (6.89) as the slopes of the lines in Fig. 6.8. These are estimated to be as follow:

SX	i-C_3H_7	C_2H_5	CH_3	H
$C_{SX} \times 10^4$	2.08	1.38	0.55	0.16

The relative magnitudes of these constants are consistent with the general rule that benzylic hydrogens are more readily abstracted than those attached directly to the ring. The reactivity of the benzylic hydrogens themselves follows the order tertiary > secondary > primary, which is a well-established order in organic chemistry. The benzylic radical resulting from hydrogen abstraction is resonance stabilized. For toluene, as an example,

In certain commercial processes it is essential to regulate the molecular weight of the addition polymer either for ease of processing or because low molecular weight products are desirable for particular applications such as lubricants or plasticizers. In such cases the solvent or chain transfer agent is chosen and its concentration selected to produce the desired value of $\bar{v}_{tr}$. Certain mercaptans have especially large chain transfer constants for many common monomers and are especially useful for molecular weight regulation. For example, styrene has a chain transfer constant to n-butyl mercaptan equal to 21 at 60°C. This is about 10^7 times larger than the chain transfer constant to benzene at the same temperature.

Chain transfer to initiator or monomer cannot always be ignored. It may be possible, however, to evaluate the transfer constants to these substances by investigating a polymerization without added solvent or in the presence of a solvent for which C_{SX} is known to be negligibly small. In this case the transfer constants C_{IX} and C_{MX} can be determined from experiments in which $\bar{v}_{tr}$ (via

$\bar{M}_n$ and $\bar{n}_n$) and R_p are measured. The following steps outline a procedure for accomplishing this:

1. By Eqs. (6.14), (6.24), and (6.33),

$$\frac{1}{\bar{\nu}} = \frac{R_t}{R_p} = \frac{k_t R_p}{k_p [M]^2} \tag{6.90}$$

2. Solving Eq. (6.26) for [IX], we have

$$[IX] = \frac{k_t R_p^2}{f k_d k_p^2 [M]^2} \tag{6.91}$$

3. Substitute Eqs. (6.90) and (6.91) into (6.88) to obtain

$$\frac{1}{\bar{\nu}_{tr}} = \frac{k_t R_p}{k_p [M]^2} + C_{MX} + C_{IX} \frac{k_t R_p^2}{f k_d k_p^2 [M]^3} \tag{6.92}$$

4. Plot $1/\bar{\nu}_{tr}$ versus R_p. If the plot is linear, $C_{IX} = 0$ and the intercept gives the value of C_{MX}. If the plot is nonlinear, the term containing R_p^2 will become insignificant before the term which is first order in R_p as $R_p \to 0$, assuring that the plot becomes linear if examined at sufficiently low R_p values. In this case too the intercept gives C_{MX}.
5. Once the value of C_{MX} has been obtained, Eq. (6.92) can be rearranged to

$$\left(\frac{1}{\bar{\nu}_{tr}} - C_{MX}\right) \frac{1}{R_p} = \frac{k_t}{k_p [M]^2} + C_{IX} \frac{k_t R_p}{f k_d k_p^2 [M]^3} \tag{6.93}$$

If the left-hand side is plotted against R_p, the slope yields the value of C_{IX}, assuming all other kinetic constants are known.

Fairly extensive tables of chain transfer constants have been assembled on the basis of investigations of this sort. For example, the values of C_{MX} for acrylamide at 60°C is 6×10^{-5}, and that for vinyl chloride at 30°C is 6.3×10^{-4}. Likewise, for methyl methacrylate at 60°C, C_{IX} is 0.02 to benzoyl peroxide and 1.27 to t-butyl hydroperoxide.

As noted above, chain transfer to polymer does not interfere with the determination of other transfer constants, since the latter are evaluated at low conversions. In polymer synthesis, however, high conversions are desirable and extensive chain transfer can have a dramatic effect on the properties of the product. This comes about since chain transfer to polymer introduces branching into the product:

$$\underset{\underset{X}{|}}{\overset{\overset{Y}{|}}{\sim CH_2-C\sim}} + M\cdot \rightarrow MX + \underset{\underset{\cdot}{|}}{\overset{\overset{Y}{|}}{\sim CH_2-C\sim}} \overset{M}{\rightarrow \rightarrow \rightarrow} \underset{\underset{(M)_n}{}}{\overset{\overset{Y}{|}}{\sim CH_2-C\sim}} \qquad (6.P)$$

A moment's reflection reveals that the effect on $\bar{\nu}$ of transfer to polymer is different from the effects discussed above inasmuch as the overall degree of polymerization is not decreased by such transfers. Although transfer to polymer is shown in one version of Eq. (6.84), the present discussion suggests that this particular transfer is not pertinent to the effect described. Investigation of chain transfer to polymer is best handled by examining the extent of branching in the product. We shall not pursue the matter of evaluating the transfer constants, but shall consider instead two specific examples of transfer to polymer.

Remember from Sec. 1.3 that graft copolymers have polymeric side chains which differ in the nature of the repeat unit from the backbone. These can be prepared by introducing a prepolymerized sample of the backbone polymer into a reactive mixture—i.e., one containing a source of free radicals—of the side-chain monomer. As an example, consider introducing polybutadiene into a reactive mixture of styrene:

$$\sim CH_2-CH=CH-CH_2\sim + \underset{\underset{\phi}{|}}{\overset{\overset{H}{|}}{\sim CH_2-C\cdot}} \rightarrow \underset{\underset{\phi}{|}}{\sim CH_2-CH_2} + \underset{\underset{\cdot}{}}{\overset{\overset{H}{|}}{\sim C-CH=CH-CH_2\sim}}$$

$$\begin{array}{l}\text{styrene} \\ \rightarrow \rightarrow \rightarrow\end{array} \quad \overset{\overset{H}{|}}{\underset{\underset{\underset{H-C-\phi}{|}}{\overset{|}{CH_2}}}{\sim C-CH=CH-CH_2\sim}} \Big)_n \qquad (6.Q)$$

This procedure is used commercially to modify the properties of polybutadiene.

A second example of chain transfer to polymer is provided by the case of polyethylene. In this case the polymer product contains mainly ethyl and butyl side chains. At high conversions such side chains may occur as often as once every 15 backbone repeat units on the average. These short side chains are thought to arise from transfer reactions with methylene hydrogens along the same polymer chain. This process is called backbiting and reminds us of the stability of rings of certain sizes and the freedom of rotation around unsubstituted—beyond hydrogen—bonds:

$$
\begin{array}{c}
\sim CH \quad CH_2 \\
\quad \\
CH_2 \quad CH_2 \\
\quad \\
CH_2
\end{array}
\quad\longrightarrow\quad
\begin{array}{c}
\sim \overset{\cdot}{CH} \quad CH_3 \\
\quad \\
CH_2 \quad CH_2 \\
\quad \\
CH_2
\end{array}
\quad
\xrightarrow{CH_2=CH_2}
\quad
\begin{array}{c}
CH_2 \quad H \quad CH_2 \\
\quad \\
CH_2 \quad \quad CH \\
\quad \\
CH - CH_2
\end{array}
$$

$$
\Big\downarrow \; n \; CH_2=CH_2
$$

$$
H-\overset{?}{C}-CH_2CH_2CH_2CH_3
$$
$$
\underset{\overline{}}{|}
$$
$$
CH_2
$$
$$
|
$$
$$
CH_2
$$
$$
\underline{}{}_n
$$

$$
H-\overset{?}{C}-CH_2CH_3
$$
$$
|
$$
$$
CH_2
$$
$$
|
$$
$$
H-\overset{\cdot}{C}-CH_2CH_3
$$

$$
\Big\downarrow
$$

etc. (6.R)

We conclude this section by noting an extreme case of chain transfer, a reaction which produces radicals of such low reactivity that polymerization is effectively suppressed. Reagents that accomplish this are added to commercial monomers to prevent their premature polymerization during storage. These substances are called either retarders or inhibitors, depending on the degree of protection they afford. Such chemicals must be removed from monomers prior to use, and failure to achieve complete purification can considerably affect the polymerization reaction.

Inhibitors and retarders differ in the extent to which they interfere with polymerization, and not in their essential activity. An inhibitor is defined as a substance which blocks polymerization completely until it is either removed or consumed. Thus failure to totally eliminate an inhibitor from purified monomer will result in an induction period in which the inhibitor is first converted to an inert form before polymerization can begin. A retarder is less efficient and merely slows down the polymerization process by competing for radicals.

Benzoquinone [III] is widely used as an inhibitor:

$$
\begin{array}{c}
O \\
\| \\
\bigcirc \\
\| \\
O
\end{array}
+ \cdot CH_2{-}R
\quad
\nearrow \quad CH_2{=}R + HO-\bigcirc-O\cdot
$$
$$
\searrow \quad CH_3{-}R + \left[O{=}\bigcirc{=}O \right]
$$

[III]

Resonance forms

$$
\longrightarrow
\begin{array}{c}
\text{Inert} \\
\text{products}
\end{array}
$$

(6.S)

The resulting radical is stabilized by electron delocalization and eventually reacts with either another inhibitor radical by combination (dimerization) or disproportionation or with an initiator or other radical.

Molecular oxygen contains two unpaired electrons and has the distinction of being capable of both initiating and inhibiting polymerization. It functions in the latter capacity by forming the relatively unreactive peroxy radical:

$$\cdot O-O\cdot + M\cdot \rightarrow M-O-O\cdot \tag{6.T}$$

Inhibitors are characterized by inhibition constants which are defined as the ratio of the rate constant for transfer to inhibitor to the propagation constant for the monomer in analogy with Eq. (6.87) for chain transfer constants. For styrene at $50°C$ the inhibition constant of p-benzoquinone is 518, and that for O_2 is 1.5×10^4. *The Polymer Handbook* (Ref. 3) is an excellent source for these and most other rate constants discussed in this chapter.

6.9 Techniques of Polymerization: Emulsion Polymerization

A number of chain-growth polymers are commercially produced on a high tonnage basis, and the technology of polymerization deserves some comment even though it is not our main emphasis in this volume. Many important monomers polymerize with the evolution of large amounts of heat. This can result in temperature increases, increased kinetic constants, and accelerated reaction rates—in short, to runaway reactions—unless the heat is dissipated. Many important monomers are toxic, carcinogenic, or both and hence must be processed carefully for the safety of workers in the industry and users of the products. These facts, plus the diversity of monomers employed and the assortment of end uses for polymer products, make the choice of a polymerization technique a highly specific matter. In this section we shall discuss some of the possibilities.

Several polymerization techniques are in widespread usage. Our discussion is biased in favor of methods that reveal additional aspects of addition polymerization and not on the relative importance of the methods in industrial practice. We shall discuss four polymerization techniques: bulk, solution, suspension, and emulsion polymerization.

Bulk and solution polymerizations are more or less self-explanatory, since they operate under the conditions we have assumed throughout most of this chapter. A bulk polymerization may be conducted with as few as two components: monomer and initiator. Production polymerization reactions are carried out to high conversions which produces several consequences we have mentioned previously:

1. The rate process for termination is hindered through the Trommsdorff effect.
2. Because of the decrease in R_t, R_p and $\bar{\nu}$ increase as the percent conversion increases.
3. This produces even greater increases in viscosity, with the attendant increase in the difficulty of heat removal and processing.
4. Very strong stirring equipment is needed for mixing because of the high viscosity, and long tubular reactors with low cross-sectional area are needed for heat exchange.
5. Use of chain transfer agents may be indicated to regulate $\bar{\nu}$ and thus avoid some of the difficulties mentioned in items (3) and (4).

Solution polymerization is the name given to the technique of polymer formation in the presence of a solvent. The polymer may not have the same solubility in the solvent as the monomer, so the system may become heterogeneous with the formation of polymer. Although there is an increase in viscosity as polymer concentration in the system increases, the effect is mitigated by the presence of solvent, and autoacceleration through the Trommsdorff mechanism is less troublesome than in bulk polymerizations. This permits easier stirring and better heat exchange than in the bulk. The solvent is ordinarily chosen to show low chain transfer so that high molecular weight polymer is obtained. Solvent recovery must be considered as part of the overall process and this, of course, adds to the cost of this method.

A technique called suspension polymerization is sometimes employed with water-insoluble monomers. In this procedure the monomer is suspended by agitation as small drops in an aqueous medium. The diameters of the drops are in the range of micrometers to millimeters in this technique. Coalescence of the monomer drops is prevented by stirring and by addition of water-soluble polymers like gelatin or by suspending clay particles like kaolin in the mixture. Oil-soluble initiators are used so that each monomer drop behaves as a miniature bulk polymerization system. The advantage of suspension polymerization is the ease of heat removal, and a disadvantage is the need to separate the polymer from the suspending medium and wash it free of additives. Suspension polymerization is also called pearl polymerization because of the appearance of the polymer produced.

The fourth and most interesting of the polymerization techniques we shall consider is called emulsion polymerization. It is important to distinguish between suspension and emulsion polymerization, since there is a superficial resemblance between the two and their terminology has potential for confusion: A suspension of oil drops in water is called an emulsion. Water-insoluble monomers are used in the emulsion process also, and the polymerization is carried out in the presence of water; however, the following significant differences also exist:

1. Water-soluble initiators are used.
2. Polymerization occurs in particles whose dimensions are in the nanometer size range, perhaps 10^3 times smaller than the particles in suspension polymerization.
3. Emulsifying agents which are soaps or detergents play a central role in the emulsion polymerization process.
4. The kinetics of polymerization are considerably different from those we have considered elsewhere in this chapter.

These differences have been extensively researched and their origin is quite well understood. We shall present only an abbreviated discussion of the topic, however.

Before we examine the polymerization process itself, it is essential to understand the behavior of the emulsifier molecules. This class of substances is characterized by molecules which possess a polar or ionic group or head and a hydrocarbon chain or tail. The latter is often in the 10–20 carbon atom size range. Dodecyl sulfate ions, from sodium dodecyl sulfate, are typical ionic emulsifiers. These molecules have the following properties which are pertinent to the present discussion:

1. They are water soluble because of the polar head, if one assumes that the tail is not too long.
2. They lower the free energy of an interface between water and some other phase—for example, with air or hydrocarbon—by concentrating in the surface region. At the surface they may be visualized as having their polar heads in the water and their hydrophobic "water-fearing" tails in the other phase. This is called adsorption; such materials are said to be surface active and are also called surfactants. The role of these materials in washing and laundering applications is a consequence of their adsorption behavior.
3. At low concentrations surfactant molecules adsorbed at the surface are in equilibrium with other molecules in solution. Above a threshold concentration, called the critical micelle concentration (cmc, for short), another equilibrium must be considered. This additional equilibrium is that between individual molecules in solution and clusters of emulsifier molecules known as micelles.
4. When micelles are formed just above the cmc, they are spherical aggregates in which surfactant molecules are clustered, tails together, to form a spherical particle. At higher concentrations the amount of excess surfactant is such that the micelles acquire a rod shape or, eventually, even a layer structure.

Figure 6.9 represents schematically the formation of a micelle by the association of n surfactant molecules. The cutaway view of the spherical micelle shows the hydrocarbon interior of these particles. Incidentally, it is this sort of reversible

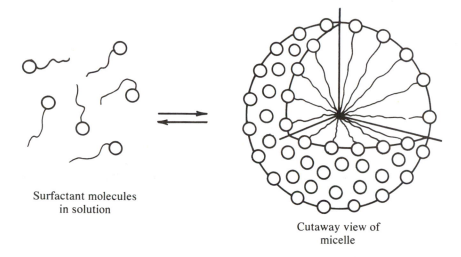

Surfactant molecules
in solution

Cutaway view of
micelle

Figure 6.9 Schematic illustration of the micellization process. Cutaway view of spherical micelle shows hydrocarbon interior with polar heads on surface.

association between molecules that was once thought to hold polymers together before covalent structures were accepted.

With this picture in mind, let us consider what happens when monomer is stirred into a surfactant solution—which also contains a water-soluble initiator— above the cmc.

The surfactant is initially distributed through three different locations: dissolved as individual molecules or ions in the aqueous phase, at the surface of the monomer drops, and as micelles. The latter category holds most of the surfactant. Likewise, the monomer is located in three places. Some monomer is present as individual molecules dissolved in the water. Some monomer diffuses into the oily interior of the micelle, where its concentration is much greater than in the aqueous phase. This process is called solubilization. The third site of monomer is in the dispersed droplets themselves. Most of the monomer is located in the latter, since these drops are much larger, although far less abundant, than the micelles. Figure 6.10 is a schematic illustration of this state of affairs during emulsion polymerization.

Polymerization begins in the aqueous phase with the decomposition of the initiator. The free radicals produced initiate polymerization by reacting with the monomers dissolved in the water. The resulting polymer radicals grow very slowly because of the low concentration of monomer, but as they grow they acquire surface active properties and eventually enter micelles. There is a possibility that they become adsorbed at the oil-water interface of the monomer

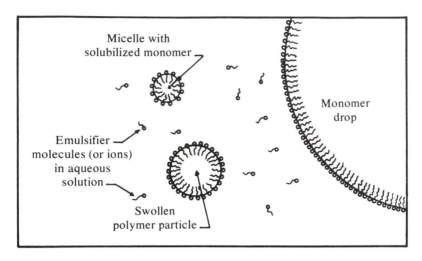

Figure 6.10 Schematic representation of the distribution of surfactant in an emulsion polymerization. Note the relative sizes of suspended particles. [From J. W. Vanderhoff, E. B. Bradford, H. L. Tarkowski, J. B. Shaffer, and R. M. Wiley, *Adv. Chem.* 34:32 (1962).]

drop, but this fate is inconsequential because the micelles are numerically so much more abundant. Once a radical enters a micelle, it polymerizes rapidly, since the concentration of solubilized monomer in the micelle is high. As this overall process continues, there is a general flux of monomer from the droplets to the micelles. The latter expand at the expense of the former. In fact, a micelle that is the locus of polymer growth is rapidly disrupted by the formation of polymer, so that it is no longer appropriate to even consider it a micelle. It is more accurate to consider it a polymer particle swollen by unreacted monomer, with the whole thing covered by adsorbed surfactants. There is a redistribution of surfactant as the polymerization progresses, with more adsorbed on polymer and less on monomer drops. The number of micelles also decreases, since this is the primary reservoir for surfactant.

It takes an initial period—stage one of the process—to get these various processes balanced out; but, after this first stage, polymerization progresses according to a straightforward pattern. We shall continue discussing only the stage which is analogous to a stationary state. The general flux of free radicals is from the aqueous phase into the micelle-swollen polymer particles, of which there are N* per unit volume in a particular experiment. The entry of the first radical into a micelle initiates the formation of a polymer chain; the arrival of a second radical terminates chain growth. A third radical entering a given particle will initiate a second chain, the the fourth radical terminates its growth, and so

on. Within one of these particles, then, there is either zero or one growing chain at any time. Since the entry of radicals into particles is random, there is a 50:50 chance that any one of the $N*$ micelle-swollen polymer particles is the site of polymer growth at any given moment. In other words, the rate of polymerization is proportional to half the concentration of such particles, $N*/2$. The rate of polymerization in the second stage of emulsion polymerization, R_p*, also depends on the concentration of monomer in the micelle $[M*]$; so we can write

$$R_p* = k_p [M*] \left[\frac{N*}{2} \right] \tag{6.94}$$

The kinetic chain length in emulsion polymerization $\bar{\nu}*$ is also readily derived by considering what happens in a single particle. The rate of propagation within a single particle is given by $k_p [M*]$, while the rate of termination is identical to the rate at which radicals enter the micelle-swollen polymer particle. This, in turn, is equal to the rate of initiation of radicals divided by $N*$ under second-stage reaction conditions. Therefore the kinetic chain length is

$$\bar{\nu}* = \bar{n}_n* = \frac{k_p [M*]}{R_i/N*} = \frac{k_p [M*] [N*]}{2fk_d [I]} \tag{6.95}$$

This also gives $\bar{n}_n*$, because the small terminating radical has essentially no effect on the degree of polymerization.

It is informative to compare Eqs. (6.94) and (6.95) with Eqs. (6.26) and (6.37), their counterparts for bulk polymerization. In the latter $R_p \propto [M] [I]^{1/2}$, while $\bar{\nu} \propto [M] [I]^{-1/2}$. Thus, if an initiator concentration were chosen to produce a predetermined value of $\bar{\nu}$, the rate would also be locked in to a particular value. Increasing $[I]$ to increase R_p would have a deleterious effect on the molecular weight of the product. By contrast, in emulsion polymerization, the rate of polymerization is independent of the initiator concentration. Both R_p* and $\bar{\nu}*$ depend on the number of micelle-swollen polymer particles in the second stage of the reaction, but this is primarily governed by the nature and concentration of the surfactant, the temperature, and the concentration of electrolytes which affect the cmc. In emulsion polymerization it is possible to achieve both high rates of polymerization and high degrees of polymerization without being caught in the inverse dependence on $[I]$ that characterizes bulk polymerization. The contrast in these behaviors is illustrated quantitatively by the following example:

Example 6.6

In an emulsion polymerization experiment at 60°C the number of micelles per unit volume is 5.0×10^{15} liter^{-1} and the monomer concentration in the micelle

is 5 M. Calculate R_p^* and $\bar{\nu}^*$ for this experiment, assuming the value of k_p from Table 6.4. Compare the calculated quantities for the emulsion polymerization with R_p and $\bar{\nu}$ for the bulk polymerization of styrene, assuming the same temperature, the same concentration of monomer, and the absence of chain transfer. In both cases take the rate of initiation to be 1.0×10^{-9} mol liter^{-1} sec^{-1}. What would be the effect on R_p and $\bar{\nu}$ in each of these cases if R_i were decreased by a factor of 4?

Solution

Assemble the pertinent equations. For emulsion polymerization, we take Eqs. (6.94), $R_p^* = k_p[M^*][N^*/2]$, and (6.95), $\bar{\nu}^* = k_p[M^*][N^*]/R_i = 2R_p^*/R_i$. For bulk polymerization, with the data given, we take Eqs. (6.29), $R_p = k_p(R_i/2k_t)^{1/2}[M]$, and (6.33), $\bar{\nu} = R_p/R_i$. From Tables 6.3 and 6.4, the appropriate constants at 60°C are $k_p = 165$ liter mol^{-1} sec^{-1} and $k_t = 2.9 \times 10^7$ liter mol^{-1} sec^{-1}. Direct substitution leads to the following results:

$$R_p^* = 165\,(5)\left[\frac{1}{2}\left(5 \times 10^{15}\,\frac{\text{micelles}}{\text{liter}} \times \frac{1\text{ molecule}}{\text{micelle}} \times \frac{1\text{ mol}}{6.02 \times 10^{23}\text{ molecules}}\right)\right]$$

$$= 3.41 \times 10^{-6}\text{ mol liter}^{-1}\text{ sec}^{-1}$$

and

$$R_p = 165\,(5)\left(\frac{1.0 \times 10^{-9}}{2\,(2.9 \times 10^7)}\right)^{1/2} = 3.41 \times 10^{-6}\text{ mol liter}^{-1}\text{ sec}^{-1}$$

In this example the number of micelles per unit volume is exactly twice the stationary-state free-radical concentration; hence the rates are identical. Although the numbers were chosen in this example to produce this result, neither N^* nor M^* are unreasonable values in actual emulsion polymerizations.

$$\bar{\nu}^* = \frac{2R_p^*}{R_i} = \frac{2\,(3.41 \times 10^{-6})}{(1.0 \times 10^{-9})} = 6820$$

$$\bar{\nu} = \frac{R_p}{R_i} = \frac{3.41 \times 10^{-6}}{1.0 \times 10^{-9}} = 3410$$

The difference of a factor of 2 between these values comes about because the conditions were chosen to give the same rates. Since a given micelle-swollen polymer particle is active only half of the time, it must produce chains which are twice as long to polymerize at the same rate as the bulk case. Reducing R_i by 1/4 produces the following effects on the calculated quantities:

R_p^*—no effect, since R_i does not appear in Eq. (6.94); therefore $R_p^* = 3.41 \times 10^{-6}$ mol liter^{-1} sec^{-1}.

R_p—rate reduced by ½, since $R_p \propto R_i^{1/2}$ in Eq. (6.29); therefore $R_p = 1.71 \times 10^{-6}$ mol liter^{-1} sec^{-1}.

$\bar{\nu}^*$—increased by a factor of 4, since $\bar{\nu}^* \propto R_i^{-1}$ in Eq. (6.95); therefore $\bar{\nu}^* = 27{,}300$.

$\bar{\nu}$—increased by a factor of 2, since $\bar{\nu} \propto R_p/R_i \propto 4 \times$ ½ in Eq. (6.33); therefore $\bar{\nu} = 6820$.

•

Emulsion polymerization also has the advantages of good heat transfer and low viscosity, which follow from the presence of the aqueous phase. The resulting aqueous dispersion of polymer is called a latex. The polymer can be subsequently separated from the aqueous portion of the latex or the latter can be used directly in eventual applications. For example, in coatings applications—such as paints, paper coatings, floor polishes—soft polymer particles coalesce into a continuous film with the evaporation of water after the latex has been applied to the substrate.

There is a great deal more that could be said about emulsion polymerization or, for that matter, about free-radical polymerization in general. We shall conclude our discussion of the free-radical aspect of chain-growth polymerization at this point, however. This is not the end of chain-growth polymerization, however. There are four additional topics to be considered:

1. Addition polymerization through anionic active species. This is discussed in the next section.
2. Chain-growth polymerization through cationic active species. This is taken up in Sec. 6.11.
3. Chain-growth copolymerization. This topic is considered in Chap. 7.
4. Stereoregular polymerization. This is also taken up in Chap. 7.

Ionic polymerizations, whether anionic or cationic, should not be judged to be unimportant merely because our treatment of them is limited to two sections in this text. Although there are certain parallels between polymerizations which occur via free-radical and ionic intermediates, there are also numerous differences. An important difference lies in the more specific chemistry of the ionic mechanism. While the free-radical mechanism is readily discussed in general terms, this is much more difficult in the ionic case. This is one of the reasons why only relatively short sections have been allotted to anionic and cationic polymerizations. The body of available information regarding these topics is extensive enough to warrant a far more elaborate treatment, but space limitations and the more specific character of the material are the reasons for the curtailed treatment.

Stereoregular polymerizations strongly resemble anionic polymerizations. We discuss these in greater detail in Chap. 7 because of their microstructure rather than the ionic intermediates involved in their formation.

6.10 Anionic Polymerization

Both modes of ionic polymerization are described by the same vocabulary as the corresponding steps in the free-radical mechanism for chain-growth polymerization. However, initiation, propagation, transfer, and termination are quite different than in the free-radical case and, in fact, different in many ways between anionic and cationic mechanisms. Our comments on the ionic mechanisms will touch many of the same points as the free-radical discussion, although in a far more abbreviated form.

The kinds of vinyl monomers which undergo anionic polymerization are those with electron-withdrawing substituents such as the nitrile, carboxyl, and phenyl groups. We represent the catalysts as AB in this discussion; these are substances which break into a cation (A^+) and an anion (B^-) under the conditions of the reaction. In anionic polymerization it is the basic anion which adds across the double bond of the monomer to form the active center for polymerization:

$$
\begin{array}{c} R \\ / \\ CH_2=C \\ \backslash \\ H \end{array} + AB \longrightarrow \begin{array}{c} R \\ | \\ B-CH_2-C^{\ominus}\ A^{\oplus} \\ | \\ H \end{array} \tag{6.U}
$$

The electron-withdrawing R group helps stabilize this anion. Chemically, these initiator ions are species like hydroxide, cyanide, amide, alkyl, or aryl ions, the latter two from organometallics such as alkyl or aryl compounds of lithium or aluminum. There is an inverse relationship between the electron-withdrawing tendency of a substituent and the required basicity of the catalyst. Substituents which are strongly electron withdrawing like nitrile can be polymerized with relatively weak bases like hydroxide ion, while phenyl requires a stronger base like the amide ion. Note that the base adds to the more positive carbon atom away from the electron-withdrawing substituent.

The reaction medium plays a very important role in all ionic polymerizations. Likewise, the nature of the ionic partner to the active center—called the counterion or gegenion—has a large effect also. This is true because the nature of the counterion, the polarity of the solvent, and the possibility of specific solvent–ion interactions determines the average distance of separation between the ions in solution. It is not difficult to visualize a whole spectrum of possibilities, from completely separated ions to an ion pair of partially solvated ions to an ion pair of unsolvated ions. The distance between the centers of the ions is different in

each of these situations, and this has a large effect on the ease of monomer addition to the active site.

Once the polymerization has been initiated by the addition of a catalyst to the monomer, propagation occurs by the entry of successive monomers between the ions of the active species:

$$B-CH_2-\overset{\overset{\displaystyle R}{|}}{\underset{\underset{\displaystyle H}{|}}{C}}{}^{\ominus}A^{\oplus} + (n-1)M \longrightarrow BM_n{}^{\ominus}A^{\oplus} \tag{6.V}$$

It is apparent why the distance of separation between the ions—hence the nature of the medium—is highly influential at this step.

In ionic polymerizations termination by combination does not occur, since all of the polymer ions have the same charge. In addition, there are solvents such as dioxane and tetrahydrofuran in which chain transfer reactions are unimportant for anionic polymers. Therefore it is possible for these reactions to continue without transfer or termination until all monomer has reacted. Evidence for this comes from the fact that the polymerization can be reactivated if a second batch of monomer is added after the initial reaction has gone to completion. In this case the molecular weight of the polymer increases, since no new growth centers are initiated. Because of this absence of termination, such polymers are called living polymers.

While living polymers can be prepared, there are some substances like water, alcohols, and carbon dioxide which are highly effective in terminating chain growth:

$$BM_n{}^{\ominus}A^{\oplus} + H_2O \longrightarrow BM_nH + AOH \tag{6.W}$$

$$BM_n{}^{\ominus}A^{\oplus} + CO_2 \longrightarrow BM_nCOO^{\ominus}A^{\oplus} \xrightarrow{HCl} BM_nCOOH + ACl \tag{6.X}$$

In practice, it is very difficult to completely exclude water and CO_2, so chain termination is often induced by these reactions.

An interesting situation is obtained when the catalyst-solvent system is such that the initiator is essentially 100% dissociated before monomer is added and no termination or transfer reactions occur. In this case all chain initiation occurs rapidly when monomer is added, since no time-dependent initiator breakdown is required. If the initial concentration of catalyst is $[AB]_0$, then chain growth starts simultaneously at $[B^-]_0$ centers per unit volume. The rate of polymerization is given by the analog of Eq. (6.24):

$$R_p = \frac{-d[M]}{dt} = k_p[B^-]_0[M] \tag{6.96}$$

which integrates to

$$[M] = [M]_0 e^{-k_p [B^-]_0 t} \qquad (6.97)$$

if $[M] = [M]_0$ at $t = 0$. Since no termination occurs, the kinetic chain length at any point during the reaction is given by the amount of monomer reacted at that point, $[M]_0 - [M]$, divided by the number of chain-growth centers:

$$\bar{\nu} = \frac{[M]_0 - [M]}{[B^-]_0} \qquad (6.98)$$

Combining Eqs. (6.97) and (6.98) gives

$$\bar{\nu} = \frac{[M]_0}{[B^-]_0} (1 - e^{-k_p [B^-]_0 t}) \qquad (6.99)$$

which approaches $[M]_0/[B]_0$ as $t \to \infty$.

The first living polymer studied in detail was polystyrene polymerized with sodium naphthalenide in tetrahydrofuran at low temperatures:

1. The catalyst is prepared by the reaction of sodium metal with naphthalene and results in the formation of a radical ion:

$$(6.Y)$$

2. These green radical ions react with styrene to produce the red styrene radical anions:

$$(6.Z)$$

3. The latter undergo radical combination to form the dianion, which subsequently polymerizes:

$$(6.AA)$$

In this case the degree of polymerization is $2\bar{\nu}$ because of the combination step.

The molecular weight distribution for a polymer like that described above is remarkably narrow compared to free-radical polymerization or even to ionic polymerization in which transfer or termination occurs. The sharpness arises from the nearly simultaneous initiation of all chains and the fact that all active centers grow as long as monomer is present. The following steps outline a quantitative treatment of this effect:

1. The first monomer addition to the active center occurs by the reaction

$$B-M_1^{\ominus}A^{\oplus} + M \longrightarrow B-M_2^{\ominus}A^{\oplus} \tag{6.BB}$$

for which the rate law is

$$- \frac{d[BM_1^-]}{dt} = k_p [BM_1^-] [M] \tag{6.100}$$

2. Substitution of Eq. (6.97) into Eq. (6.100) yields

$$\frac{d[BM_1^-]}{[BM_1^-]} = - k_p [M]_0 \exp(- k_p [B^-]_0 t) \, dt \tag{6.101}$$

Since $[BM_1^-] = [B]_0$ at $t = 0$, Eq. (6.101) can be integrated to

$$[BM_1^-] = [B^-]_0 \exp\left(- \frac{[M]_0}{[B^-]_0} (1 - e^{-k_p[B^-]_0 t})\right) \tag{6.102}$$

Consideration of Eq. (6.99) permits Eq. (6.102) to be written as

$$[BM_1^-] = [B^-]_0 \, e^{-\bar{\nu}} \tag{6.103}$$

3. The species $BM_2^{\ominus}$ is formed by reaction (6.BB) and lost by

$$BM_2^{\ominus}A^{\oplus} + M \longrightarrow BM_3^{\ominus}A^{\oplus} \tag{6.CC}$$

Therefore the expression for $d[BM_2^-]/dt$ is given by

$$\frac{d[BM_2^-]}{dt} = k_p [BM_1^-] [M] - k_p [BM_2^-] [M] \tag{6.104}$$

which becomes

$$\frac{d[BM_2^-]}{dt} = k_p[M]_0 e^{-k_p[B^-]_0 t} ([B^-]_0 e^{-\bar{\nu}} - [BM_2^-]) \tag{6.105}$$

by substitution of Eqs. (6.97) and (6.103). Differentiation of Eq. (6.99) with respect to t shows that $d\bar{\nu} = k_p[M]_0 e^{-k_p[B^-]_0 t}$ dt. Therefore Eq. (6.105) can be written as

$$\frac{d[BM_2^-]}{d\bar{\nu}} + [BM_2^-] = [B^-]_0 e^{-\bar{\nu}} \tag{6.106}$$

4. This is a standard differential equation for which the solution is given by

$$[BM_2^-] = [B^-]_0 \bar{\nu} e^{-\bar{\nu}} \tag{6.107}$$

We shall examine the solution to this equation in an example below. From the ratio of Eq. (6.106) to Eq. (6.103), note that $[BM_2^-]/[BM_1^-] = \bar{\nu}$.
5. The same sequence of steps outlined in items (3) and (4) can be followed to give the concentrations of n-mer anions resulting from n − 1 additions to the original active site:

$$[BM_n^-] = \frac{1}{(n-1)!} [B^-]_0 \bar{\nu}^{n-1} e^{-\bar{\nu}} \tag{6.108}$$

The expression shows that one extra factor $\bar{\nu}$ enters for each addition and that each successive expression is divided by an extra integer which counts the number of additions to the original growth center.
6. Since the total number of polymer chains is proportional to $[B^-]_0$, the number N_n of n-mers in a population of N polymer molecules is given by

$$\frac{N_n}{N} = \frac{[BM_n^-]}{[B^-]_0} = \frac{1}{(n-1)!} \bar{\nu}^{n-1} e^{-\bar{\nu}} \tag{6.109}$$

Inasmuch as $\bar{\nu}$ is the mean values for n, Eq. (6.109) shows that the distribution for the degree of polymerization follows the Poisson function, Eq. (1.38).

The integration of Eq. (6.106) is central to the kinetic proof that living polymers follow Poisson statistics. The solution of this differential equation is illustrated in the following example.

Example 6.7

Show that Eq. (6.107) is a solution to Eq. (6.106). Also consider the extension of the method to $[BM_3^-]$ and $[BM_4^-]$.

Solution

The integrating factor for Eq. (6.106) is $e^{\bar{\nu}}$; therefore multiply the equation through by this factor: $(d[BM_2^-]/d\bar{\nu})\,e^{\bar{\nu}} + [BM_2^-]\,e^{\bar{\nu}} = [B^-]_0$. Recognize that the left-hand side of the equation is $d([BM_2^-]\,e^{\bar{\nu}})/d\bar{\nu}$, so the expression becomes

$$d([BM_2^-]\,e^{\bar{\nu}}) = [B^-]_0\,d\bar{\nu}$$

This equation integrates to

$$[BM_2^-]\,e^{\bar{\nu}} = [B^-]_0\,\bar{\nu} + \text{const.}$$

where the constant of integration is found to be zero from the fact that at $t = 0$ both $\bar{\nu}$ and $[BM_2^-] = 0$. Rearrangement converts this to $[BM_2^-] = [B^-]_0\,\bar{\nu}e^{-\bar{\nu}}$ which is Eq. (6.107).

The same set of relationships beginning from Eq. (6.104) can be written for $[BM_3^-]$ and $[BM_4^-]$ and solved by the same procedure. Some key steps are set forth below:

For $[BM_3^-]$

For $[BM_4^-]$

$$\frac{d[BM_3^-]}{dt} = k_p[M]\left\{[BM_2^-] - [BM_3^-]\right\}$$

$$\frac{d[BM_4^-]}{dt} = k_p[M]\left([BM_3^-] - [BM_4^-]\right)$$

$$\frac{d[BM_3^-]}{d\bar{\nu}} + [BM_3^-] = [B^-]_0\,\bar{\nu}\,e^{-\bar{\nu}}$$

$$\frac{d[BM_4^-]}{d\bar{\nu}} + [BM_4^-] = [BM_3^-]$$

$$e^{\bar{\nu}}\left(\frac{d[BM_3^-]}{d\bar{\nu}} + [BM_3^-]\right) = [B^-]_0\,\bar{\nu}$$

$$e^{\bar{\nu}}\left\{\frac{d[BM_4^-]}{d\bar{\nu}} + [BM_4^-]\right\} = \frac{1}{2}\bar{\nu}^2\,[B^-]_0$$

$$\int d(e^{\bar{\nu}}\,[BM_3^-]) = [B^-]_0\int\bar{\nu}\,d\bar{\nu}$$

$$\int d(e^{\bar{\nu}}\,[BM_4^-]) = \frac{1}{2}[B^-]_0\int\bar{\nu}^2\,d\bar{\nu}$$

$$e^{\bar{\nu}}\,[BM_3^-] = [B^-]_0\,\frac{1}{2}\,\bar{\nu}^2$$

$$e^{\bar{\nu}}\,[BM_4^-] = \frac{1}{3}\frac{1}{2}\,[B^-]_0\,\bar{\nu}^3$$

$$[BM_3^-] = \frac{1}{2}\,\bar{\nu}^2\,e^{-\bar{\nu}}\,[B^-]_0$$

$$[BM_4^-] = \frac{1}{3!}\,\bar{\nu}^3\,[B^-]_0$$

These examples show that the power of $\bar{\nu}$ and the factorial are one less than the degree of polymerization for the ion under consideration.

•

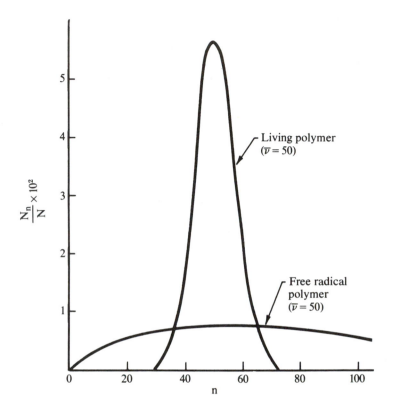

Figure 6.11 Comparison of the number distribution of n-mers for polymers prepared from anionic and free-radical active centers, both with $\bar{\nu} = 50$.

That the Poisson distribution results in a narrower distribution of molecular weights than is obtained with termination is shown by Fig. 6.11. Here N_n/N is plotted as a function of n for $\bar{\nu} = 50$, for living polymers as given by Eq. (6.109), and for conventional free-radical polymerization as given by Eq. (6.77). This same point is made by considering the ratio $\bar{M}_w/\bar{M}_n$ for the case of living polymers. This ratio may be shown to equal

$$\frac{\bar{n}_w}{\bar{n}_n} = 1 + \frac{\bar{\nu}}{(\bar{\nu} + 1)^2} \tag{6.110}$$

which is equal to 1.02 for $\bar{\nu} = 50$ and approaches unity as $\bar{\nu} \to \infty$ in contrast to a ratio of 1.5 (for combination) or 2 (for disproportionation) for the free-radical mechanism.

6.11 Cationic Polymerization

Just as anionic polymerizations show certain parallels with the free-radical mechanism, so too can cationic polymerization be discussed in terms of the same broad outline. There are some differences from the anionic systems, however, so the fact that both proceed through ionic intermediates should not be overextended.

The principal differences between cationic and anionic polymerizations center around the following points:

1. A single catalyst is often not sufficient in cationic polymerizations; frequently a cocatalyst is required.
2. Chain transfer is far more important than in the anionic case, so we do not encounter living polymers in cationic systems.
3. Total dissociation of the initiator system is rare, so the simplifications we explored for this condition in the anionic case cannot be used.

We shall consider these points below. The mechanism for cationic polymerization continues to include initiation, propagation, transfer, and termination steps, and the rate of polymerization and the kinetic chain length are the principal quantities of interest.

In cationic polymerization the active species is the ion which is formed by the addition of a proton from the initiator system to a monomer. For vinyl monomers the type of substituents which promote this type of polymerization are those which are electron supplying, like alkyl, 1,1-dialkyl, aryl, and alkoxy. Isobutylene and α-methyl styrene are examples of monomers which have been polymerized via cationic intermediates.

The catalysts for cationic polymerization are either protonic acids or Lewis acids, such as H_2SO_4 and $HClO_4$ or BF_3, $AlCl_3$, and $TiCl_4$:

1. The Lewis acids must be used with a protonic cocatalyst such as water or methanol which generates protons through the following kinds of equilibria:

$$BF_3 + H_2O \rightleftarrows F_3BOH^- + H^+$$

$$AlCl_3 + H_2O \rightleftarrows Cl_3AlOH^- + H^+ \tag{6.DD}$$

$$TiCl_4 + CH_3OH \rightleftarrows Cl_4TiOCH_3^- + H^+$$

With insufficient catalyst these equilibria lie too far to the left, while excess cocatalyst destroys the catalyst and/or terminates the chain. The optimum proportion of catalyst and cocatalyst varies with the system employed and also with the solvent for a specific system.

2. Protonic acids dissociate to some extent in the nonaqueous reaction mixtures to produce an equilibrium concentration of protons:

$$H_2SO_4 \rightleftarrows H^+ + HSO_4^-$$

$$HClO_4 \rightleftarrows H^+ + ClO_4^-$$

(6.EE)

3. The general formula for the initiator species can be written H^+B^-, where the degree of separation or ion pairing depends on the polarity of the medium and the possibility of specific solvation interactions. If we represent the equilibrium constant for the reactions in (6.DD) and (6.EE) by K, the initiator concentration can be written as

$$[H^+B^-] = K[\text{Lewis acid}][\text{cocatalyst}]$$

(6.111)

or

$$[H^+B^-] = K[\text{protonic acid}]$$

(6.112)

The proton adds to the more negative carbon atom in the olefin to initiate chain growth:

$$CH_2=C \overset{R}{\underset{H}{\diagup}} + H^+B^- \longrightarrow CH_3 - \overset{R}{\underset{H}{\overset{|}{C}}}{}^{\oplus}B^{\ominus}$$

(6.FF)

The electron-releasing R group helps stabilize this cation. As with anionic polymerization, the separation of the ions and hence the ease of monomer insertion depends on the reaction medium. The propagation reaction may be written as

$$CH_3 - \overset{R}{\underset{H}{\overset{|}{C}}}{}^{\oplus}B^{\ominus} + (n-1)M \longrightarrow M_n{}^{\oplus}B^{\ominus}$$

(6.GG)

Aldehydes can also polymerize in this manner, the corresponding reactions for $O=CH_2$ being

$$H^+B^- + O=CH_2 \longrightarrow H-O-\overset{H}{\underset{H}{\overset{|}{C}}}{}^{\oplus}B^{\ominus} \overset{O=CH_2}{\longrightarrow \rightarrow \rightarrow} H(O-CH_2)_n{}^{\oplus}B^{\ominus}$$

(6.HH)

Formaldehyde polymerizes even without added catalyst, but it is possible that traces of formic acid act as adventitious catalysts in this system.

One of the side reactions that can complicate cationic polymerization is the possibility of the ionic repeat unit undergoing the well-known carbonium ion rearrangement during the polymerization. The following example illustrates this situation.

Example 6.8

It has been observed[†] that poly(1,1-dimethyl propane) is the product when 3-methylbutene-1 is polymerized with $AlCl_3$ in ethyl chloride at $-130°C$. Write structural formulas for the expected repeat units and those observed and propose an explanation.

Solution

The structures expected and found are sketched here:

Expected

$$\left[\begin{array}{c} CH_2-CH \\ | \\ CH \\ / \quad \backslash \\ CH_3 \quad CH_3 \end{array} \right]_n$$

Found

$$\left[\begin{array}{c} CH_3 \\ | \\ CH_2-CH_2-C \\ | \\ CH_3 \end{array} \right]_n$$

The conversion of the cationic intermediate of the monomer to the cation of the product occurs by a hydride shift between adjacent carbons:

$$-C-C^{\oplus} \longrightarrow -C-C^{\oplus} \quad H \longrightarrow C-C-C^{\oplus}$$

This is a well-known reaction in organic chemistry which is favored by the greater stability of the tertiary cation compared to the secondary ion.

[†] J. P. Kennedy and R. M. Thomas, *Makromol. Chem.* 53:28 (1962).

The extent of such reactions depends on the monomer structure as well as the temperature and the solvent.

As intimated above, termination occurs in these systems by reactions with water or other proton sources:

$$M_n^{\oplus} B^{\ominus} + H_2O \longrightarrow M_n OH + HB \tag{6.II}$$

Chain transfer reactions to monomer and/or solvent also occur and lower the kinetic chain length without affecting the rate of polymerization:

$$M_n^{\oplus} B^{\ominus} + M \longrightarrow M_n + M^{\oplus} B^{\ominus} \tag{6.JJ}$$

A detailed kinetic analysis depends on the specifics of the initiation and termination–transfer steps. We shall illustrate only one combination; other possibilities are done similarly:

1. For initiation involving a cocatalyst, say, water,

$$R_i = k_i [H^+ B^-] [M] = k_i K [\text{Lewis acid}] [H_2O] [M] \tag{6.113}$$

2. For termination by transfer to monomer,

$$R_{tr} = k_{tr} [M_n^+] [M] \tag{6.114}$$

3. Under stationary-state conditions, $R_i = R_{tr}$; therefore

$$[M_n^+]_s = \frac{k_i K [\text{Lewis acid}] [H_2O]}{k_{tr}} \tag{6.115}$$

4. The rate of polymerization is given by $k_p [M] [M_n^+]$; hence

$$R_p = \frac{k_p k_i K [\text{Lewis acid}][H_2O] [M]}{k_{tr}} \tag{6.116}$$

5. Even though the catalyst may be only partially converted to $H^+ B^-$, the concentration of these ions may be on the order of 10^5 times greater than the concentration of free radicals in the corresponding stationary state of the radical mechanism. Likewise, k_p for ionic polymerization is on the order of 100 times larger than the sum of the constants for all termination and transfer steps. By contrast, $k_p / k_t^{1/2}$, which is pertinent for the radical mechanism, is typically on the order of 10^{-1}. These comparisons illustrate that ionic polymerizations occur very fast even at low temperatures.

6. Applying the Arrhenius equation to Eq. (6.116) shows that the apparent activation energy for the overall rate of polymerization is given by

$$E_{app}^* = E_i^* + E_p^* - E_{tr}^* \qquad (6.117)$$

In some cases E_{tr}^* may be larger than $E_i^* + E_p^*$, which leads to the unusual situation where the rate of the polymerization reaction decreases with increasing temperature. The specifics here depend on the reaction system, including the solvent.

7. The kinetic chain length is given by the ratio R_p/R_{tr} or

$$\bar{\nu} = \frac{k_p}{k_{tr}} = \frac{1}{C_M} \qquad (6.118)$$

where C_M is the chain transfer constant to monomer. These constants tend to be about two orders of magnitude larger than their counterparts for the radical mechanism at the temperatures of the respective reactions. There continues to be a significant dependence on the solvent and the nature of the catalyst as far as chain transfer is concerned.

Problems

1. The efficiency of AIBN in initiating polymerization at $60°C$ was determined by Bevington et al.[†] by the following strategy. They measured R_p and $\bar{\nu}$ and calculated $R_i = R_p/\bar{\nu}$. The constant k_d was measured directly in the system, and from this quantity and the measured ratio $R_p/\bar{\nu}$ the fraction f could be determined. The following results were obtained for different concentrations of initiator:

[I] (g liter^{-1})	$R_p/\bar{\nu} \times 10^8$ (mol liter^{-1} sec^{-1})
0.0556	0.377
0.250	1.57
0.250	1.72
1.00	6.77
1.50	10.9
2.50	17.1

Using $k_d = 0.0388$ hr^{-1}, evaluate f from these data.

2. AIBN was synthesized using ^{14}C-labeled reagents and the tagged compound was used to initiate polymerization of methyl methacrylate and styrene.

[†] J. C. Bevington, J. H. Bradbury, and G. M. Burnett, *J. Polym. Sci.* 12:469 (1954).

Samples of initiator and polymers containing initiator fragments were burned to CO_2. The radioactivity of uniform (in sample size and treatment) CO_2 samples was measured in counts per minute (cpm) by a suitable Geiger counter. A general formula for the poly(methyl methacrylate) with its initiator fragments is $(C_5H_8O_2)_n(C_4H_6N)_m$, where n is the degree of polymerization for the polymer and m is either 1 or 2, depending on the mode of termination. The specific activity measured in the CO_2 resulting from combustion of the polymer relative to that produced by the initiator is

$$\frac{\text{Activity of C in polymer}}{\text{activity of C in initiator}} = \frac{4m}{5n + 4m} \cong \frac{4m}{5n}$$

From the ratio of activities and measured values of n, the average number of initiator fragments per polymer can be determined. Carry out a similar argument for the ratio of activities for polystyrene and evaluate the average number of initiator fragments per molecule for each polymer from the following data[†]:

Methyl methacrylate		Styrene	
$\bar{M}_n$	Counts per minute	$\bar{M}_n$	Counts per minute
444,000	20.6	383,000	25.5
312,000	30.1	117,000	86.5
298,000	29.0	114,000	89.5
147,000	60.5	104,000	96.4
124,000	76.5	101,000	113.5
91,300	103.4		
89,400	104.6		

For both sets of data, the radioactivity from the labeled initiator gives 96,500 cpm when converted to CO_2.

3. In the same research described in Example 6.4, the authors[‡] measured the following rates of polymerization:

Run number	$R_p \times 10^4$ (mol liter^{-1} sec^{-1})
5	3.40
6	2.24
8	6.50
12	5.48
13	7.59

They also reported a k_p value of 1.2×10^4 liter mol^{-1} sec^{-1}, but the concentrations of monomer in each run were not given. Use these value of

[†] J. C. Bevington, H. W. Melville, and R. P. Taylor, *J. Polym. Sci.* 12:449 (1954).
[‡] T. Takahashi and P. Ehrlich, *Polym. Prepr.* 22:203 (1981).

R_p and k_p and the values of $\bar{\tau}$ and k_t given in Example 6.4 to evaluate [M] for each run. As a double check, evaluate [M] from these values of R_p (and k_p) and the values of R_i and k_t given in the example.

4. In a series of experiments at 60°C, the rate of polymerization of styrene agitated in water containing persulfate initiator was measured[†] for different concentrations of sodium dodecyl sulfate emulsifier. The following results were obtained:

[surfactant] $\times$ 10^3 (mol liter^{-1})	R_p (% min^{-1})
3.8	~0.08
4.8	~0.08
5.00	~0.08
6.00	0.40
6.25	0.75
6.8	0.77
7.2	0.79
7.8	0.80

Propose an explanation for this variation in R_p. Over the same range of surfactant concentrations, R_p for vinyl acetate is roughly constant at about 1.1% min^{-1}. Criticize or defend the following proposition: Since vinyl acetate is about 70 times more soluble in water than styrene at 30°C, the former undergoes solution polymerization in the aqueous phase while styrene displays true emulsion polymerization under suitable conditions.

5. Arnett[‡] initiated the polymerization of methyl methacrylate in benzene at 77°C with AIBN and measured the initial rates of polymerization for the concentrations listed:

[M] (mol liter^{-1})	$[I]_0 \times 10^4$ (mol liter^{-1})	$R_p \times 10^3$ (mol liter^{-1} min^{-1})
9.04	2.35	11.61
8.63	2.06	10.20
7.19	2.55	9.92
6.13	2.28	7.75
4.96	3.13	7.31
4.75	1.92	5.62
4.22	2.30	5.20
4.17	5.81	7.81
3.26	2.45	4.29
2.07	2.11	2.49

[†] S. Okamura and T. Motoyama, *J. Polym. Sci.* 58:221 (1962).
[‡] L. M. Arnett, *J. Am. Chem. Soc.* 74:2027 (1952).

Use these data to evaluate the cluster of constants $(fk_d/k_t)^{1/2} k_p$ at this temperature. Evaluate $k_p/k_t^{1/2}$ using Arnett's finding that $f = 1.0$ and assuming the k_d value determined in Example 6.1 for AIBN at $77°C$ in xylene also applies in benzene.

6. Since propagation does not change the number of radicals in the system, the rate of change of the radical concentration under non-stationary-state conditions is given by $d[M\cdot]/dt = 2k_d f[I] - 2k_t[M\cdot]^2$. For the short induction period before the stationary-state concentration of radicals is reached, $[I]$ may be regarded as a constant at the initial value $[I]_0$. Making this assumption and using a table of integrals, show by integration that this equation yields Eq. (6.43), the relationship used in Example 6.2, for the nonstationary value of $[M\cdot]$ at time t.

7. The lifetime of polystyrene radicals at $50°C$ was measured[†] by the rotating sector method as a function of the extent of conversion to polymer. The following results were obtained:

Percent conversion	$\bar{\tau}$ (sec)
0	2.29
32.7	1.80
36.3	9.1
39.5	13.9
43.8	18.8

Propose an explanation for the variation observed.

8. The equations derived in Sec. 6.7 are based on the assumption that termination occurs exclusively by either disproportionation or combination. This is usually not the case: Some proportion of each is the more common case. If A equals the fraction of termination occurring by disproportionation, we can write $\bar{n}_n = A[1/1 - p] + (1 - A)[2/(1 - p)]$ and $\bar{n}_w/\bar{n}_n = A(1 + p) + (1 - A)[(2 + p)/2]$. From measurements of $\bar{n}_n$ and $\bar{n}_w/\bar{n}_n$ it is possible *in principle* to evaluate A and p. May and Smith[‡] have done this for a number of polystyrene samples. A selection of their data for which this approach seems feasible is presented:

$\bar{n}_n$	$\bar{n}_w/\bar{n}_n$
1129	1.60
924	1.67
674	1.73
609	1.74

[†] M. S. Matheson, E. E. Auer, E. B. Bevilacqua, and J. E. Hart, *J. Am. Chem. Soc.* 73:1700 (1951).

[‡] J. A. May, Jr., and W. B. Smith, *J. Phys. Chem.* 72:216 (1968).

Since p is very close to unity, it is adequate to assume this value and evaluate A from $\bar{n}_w/\bar{n}_n$ and then use the value of A so obtained to evaluate a better value of p from $\bar{n}_n$.

9. In the research described in the last problem, the authors[†] determined the following distribution of molecular weights by a chromatographic procedure:

n	$w_n/w_{tot} \times 10^4$	n	$w_n/w_{tot} \times 10^4$
100	3.25	800	6.88
200	5.50	900	6.10
300	6.80	1200	4.20
400	7.45	1500	2.90
500	7.91	2000	1.20
600	7.82	2500	0.50
700	7.18	3000	0.20

They asserted that the points are described by the expression $w_n/w_{tot} = An(1-p)^2 p^{n-1} + 0.5(1-A)(n)(n-1)(1-p)^3 p^{n-2}$, with A = 0.65 and p = 0.99754. In this expression w_n/w_{tot} is the weight fraction n-mer. Calculate some representative points for this function and plot the theoretical and experimental points on the same graph. On the basis of Eq. (5.30) and the notation of the last problem, extract from the expression given the weight fraction n-mer resulting from termination by combination.

10. Palit and Das[‡] measured $\bar{\nu}_{tr}$ at 60°C for different values of the ratio [SX]/[M] and evaluated C_{SX} and $\bar{\nu}$ for vinyl acetate undergoing chain transfer with various solvents. Some of their measured and derived results are tabulated below (the same concentrations of AIBN and monomer were used in each run):

Solvent	$\bar{\nu}$	$\bar{\nu}_{tr}$	[SX]/[M]	$C_{SX} \times 10^4$
t-Butyl alcohol	6580	3709	—	0.46
Methyl isobutyl ketone	6670	510	0.492	—
Diethyl ketone	6670	—	0.583	114.4
Chloroform	—	93	0.772	125.2

Assuming that no other transfer reactions occur, calculate the values missing from the table. Criticize or defend the following proposition: The $\bar{\nu}$ values obtained from the limit [SX]/[M] → 0 show that the AIBN initiates polymerization identically in all solvents.

[†] J. A. May, Jr., and W. B. Smith, *J. Phys. Chem.* 72:216 (1968).

[‡] S. R. Palit and S. K. Das, *Proc. R. Soc. London* 226A:82 (1954).

11. Gregg and Mayo[†] studied the chain transfer between styrene and carbon tetrachloride at 60 and 100°C. A sample of their data is given below for each of these temperatures:

At 60°C		At 100°C	
$[CCl_4]/[styrene]$	$\bar{\nu}_{tr}^{-1} \times 10^5$	$[CCl_4]/[styrene]$	$\bar{\nu}_{tr}^{-1} \times 10^5$
0.00614	16.1	0.00582	36.3
0.0267	35.9	0.0222	68.4
0.0393	49.8	0.0416	109
0.0704	74.8	0.0496	124
0.1000	106	0.0892	217
0.1643	156		
0.2595	242		
0.3045	289		

Evaluate the chain transfer constant (assuming that no other transfer reactions occur) at each temperature. By means of an Arrhenius analysis, estimate $E_{tr}^* - E_p^*$ for this reaction. Are the values of $\bar{\nu}$ in the limit of no transfer in the order expected for thermal polymerization? Explain.

12. Soum and Fontanille[‡] prepared a living polymer of 2-vinyl pyridine using benzyl picolyl magnesium as the initiator. The values of $\bar{M}_n$ were measured experimentally for polymers prepared with different concentrations of initiator and different initial concentrations of monomer. The results are given below; calculate the theoretical molecular weights expected if polymerization proceeds completely from 100% predissociated initiator and compare the theoretical and experimental values:

$[I] \times 10^4$ (mol liter^{-1})	$[M]_0 \times 10^2$ (mol liter^{-1})	$\bar{M}_n$ (g mol^{-1})
4.8	8.2	20,000
3.7	8.5	25,000
1.7	7.1	46,000
4.8	7.1	17,000
5.8	7.3	14,000
1.5	15.0	115,000

13. Ionic polymers may exist as undissociated, unsolvated ion pairs; undissociated ion pairs solvated to some extent; solvated ions dissociated to some extent; or some combination of these. The propagation rate constant k_p and the dissociation equilibrium constant K of the lithium salt of anionic

---

[†] R. A. Gregg and F. R. Mayo, *J. Am. Chem. Soc.* 70:2373 (1948).
[‡] A. Soum and M. Fontanille, in *Anionic Polymerization*, J. E. McGrath (Ed.), ACS Symposium Series, Vol. 166 (1981).

polystyrene were measured in tetrahydrofuran containing various amounts of dioxane. The dielectric constant of the mixed solvents was also determined as a measure of polarity: The lower the dielectric constant, the less polar the medium. The following are some of the results obtained[†]:

Volume percent dioxane	Dielectric constant	k_p (liter mol^{-1} sec^{-1})	K
0	7.39	65,000	$\sim 2 \times 10^{-7}$
26.0	5.95	257	9×10^{-9}
47.6	4.81	~ 80	1.8×10^{-10}
69.4	3.75	6.1	Too small
84.8	2.93	~ 1.5	to measure
100.0	2.20	~ 0.9	reliably

Criticize or defend the following proposition: The decrease in K with increasing dioxane content is consistent with the diminished solvent properties for ions of substances with low dielectric constant. The decrease in k_p is quantitatively explained by this decrease in free ion concentration, since k_p eventually approaches a limiting value for unsolvated ion pairs.

14. The following table shows the values of ΔH°_{298} for the gas phase reactions $X(g) + H^+(g) \rightarrow HX^+(g)$ where X is an olefin.

X	XH^+	ΔH°_{298} (kJ mol^{-1})
$CH_2=CH_2$	$CH_2CH_3^+$	-640
$CH_3CH=CH_2$	$CH_3CH_2CH_2^+$	-690
$CH_3CH=CH_2$	$CH_3C^+HCH_3$	-757
$CH_3CH_2CH=CH_2$	$CH_3CH_2CH_2CH_2^+$	-682
$CH_3CH=CHCH_3$	$CH_3CH_2C^+HCH_3$	-782
$(CH_3)_2C=CH_2$	$(CH_3)_2CHCH_2^+$	-695

Use these data[‡] to quantitatively comment on each of the following points:

(a) The cation is stabilized by electron-supplying alkyl substituents.
(b) The carbonium ion rearrangement of n-propyl ions to i-propyl ions is energetically favored.
(c) With the supplementary information that $\Delta H^\circ_{f,298}$ of 1-butene and cis-2-butene are +1.6 and -5.8 kJ mol^{-1}, respectively, evaluate the ΔH_{298} for the rearrangement n-butyl to sec-butyl ions and compare with the corresponding isomerization for the propyl cations.

[†] M. van Beylen, D. N. Bhattacharyya, J. Smid, and M. Szwarc, *J. Phys. Chem.* 70:157 (1966).
[‡] P. H. Plesch (Ed.), *Cationic Polymerization*, Macmillan, New York, 1963.

(d) Of the monomers shown, only isobutene undergoes cationic polymerization to any significant extent. Criticize or defend the following proposition: The above data explain this fact by showing that this is the only monomer of those listed which combines a sufficiently negative ΔH for protonation with the freedom from interfering isomerization reactions.

Bibliography

1. Allcock, H. R., and Lampe, F. W., *Contemporary Polymer Chemistry*, Prentice-Hall, Englewood Cliffs, N.J., 1981.
2. Bagdasan'yan, K. S., *Theory of Free Radical Polymerization*, Israel Program for Scientific Translations, Jerusalem, 1968.
3. Brandup, J., and Immergut, E. H., and McDowell, W. (Eds.), *Polymer Handbook*, 2nd ed., Wiley, New York, 1975.
4. Ham, G. E. (Ed.), *Vinyl Polymerization*, Marcel Dekker, New York, 1967.
5. North, A. M., *The Kinetics of Free Radical Polymerization*, Pergamon, New York, 1966.
6. Odian, G., *Principles of Polymerization*, 2nd ed., Wiley, New York, 1981.
7. Smith, D. A. (Ed.), *Addition Polymers: Formation and Characterization*, Plenum Press, New York, 1968.

7

Polymers with Microstructure
Copolymers and Stereoregular Polymers

In a Rolls or a van
Wrapped in mink or Saran
Any way that you can
Come back to me!

Come Back to Me, words by Alan Jay Lerner, music by Burton Lane

7.1 Introduction

All polymer molecules have unique features of one sort or another at the level of individual repeat units. Occasional head-to-head or tail-to-tail orientations, random branching, and the distinctiveness of chain ends are all examples of such details. In this chapter we shall focus attention on two other situations which introduce variation in structure into polymers at the level of the repeat unit: the presence of two different monomers or the regulation of configuration of successive repeat units. In the former case copolymers are produced, and in the latter polymers with differences in tacticity. Although the products are quite different materials, their microstructure can be discussed in very similar terms. Hence it is convenient to discuss the two topics in the same chapter.

In the discussion of these combined topics, we use statistics extensively because the description of microstructure requires this kind of approach. This is the basis for merging a discussion of copolymers and stereoregular polymers into a single chapter. In other respects these two classes of materials and the processes which produce them are very different and their description leads us into some rather diverse areas.

The formation of copolymers involves the reaction of (at least) two kinds of monomers. This means that each must be capable of undergoing the same propagation reaction, but is is apparent that quite a range of reactivities is compatible with this broad requirement. We shall examine such things as the polarity of monomers, the degree of resonance stabilization they possess and the steric

423

hindrance they experience in an attempt to understand these differences in reactivity. There are few types of reactions for which chemists are successful in explaining all examples with general concepts such as these. Polymerization reactions are no exception. Even for the specific case of free-radical copolymerization, we shall see that reactivity involves the interplay of all these considerations.

To achieve any sort of pattern in configuration among successive repeat units in a polymer chain, the tendency toward random addition must be overcome. Although temperature effects are pertinent here—remember that high temperature is the great randomizer—real success in regulating the pattern of successive additions involves the use of catalysts which "pin down" both the monomer and the growing chain so that their reaction is biased in favor of one mode of addition or another. We shall discuss the Ziegler–Natta catalysts which accomplish this and shall discover these to be complicated systems for which no single mechanism is entirely satisfactory.

For both copolymers and stereoregular polymers, experimental methods for characterizing the products often involve spectroscopy. We shall see that nuclear magnetic resonance (NMR) spectra are particularly well suited for the study of tacticity. This method is also used for the analysis of copolymers.

In spite of the assortment of things discussed in this chapter, there are also a variety of topics that could be included but which are not owing to space limitations. We do not discuss copolymers formed by the step-growth mechanism, for example, or the use of Ziegler–Natta catalysts to regulate geometrical isomerism in, say, butadiene polymerization. Some other important omissions are noted in passing in the body of the chapter.

7.2 Copolymer Composition

We begin our discussion of copolymers by considering the free-radical polymerization of a mixture of two monomers, M_1 and M_2. This is already a narrow view of the entire field of copolymers, since more than two repeat units can be present in copolymers and, in addition, mechanisms other than free-radical chain growth can be responsible for copolymer formation. The essential features of the problem are introduced by this simpler special case, so we shall restrict our attention to this system.

The polymerization mechanism continues to include initiation, termination, and propagation steps. This time, however, there are four distinctly different propagation reactions:

$$-M_1\cdot + M_1 \xrightarrow{\ k_{11}\ } -M_1 M_1\cdot \tag{7.A}$$

$$-M_1\cdot + M_2 \xrightarrow{\ k_{12}\ } -M_1 M_2\cdot \tag{7.B}$$

$$-M_2 \cdot + M_1 \xrightarrow{k_{21}} -M_2 M_1 \cdot \qquad (7.C)$$

$$-M_2 \cdot + M_2 \xrightarrow{k_{22}} -M_2 M_2 \cdot \qquad (7.D)$$

Each of these reactions is characterized by a propagation constant which is labeled by a two-digit subscript: The first number identifies the terminal repeat unit in the growing radical, and the second identifies the adding monomer. The rate laws governing these four reactions are

$$R_{p,11} = k_{11} [M_1 \cdot] [M_1] \qquad (7.1)$$

$$R_{p,12} = k_{12} [M_1 \cdot] [M_2] \qquad (7.2)$$

$$R_{p,21} = k_{21} [M_2 \cdot] [M_1] \qquad (7.3)$$

$$R_{p,22} = k_{22} [M_2 \cdot] [M_2] \qquad (7.4)$$

In writing Eqs. (7.1)–(7.4) we make the customary assumption that the kinetic constants are independent of the size of the radical and we indicate the concentration of all radicals, whatever their chain length, ending with the M_1 repeat unit by the notation $[M_1 \cdot]$. This formalism therefore assumes that only the nature of the radical chain end influences the rate constant for propagation. We refer to this as the terminal control mechanism. If we wished to consider the effect of the next-to-last repeat unit in the radical, each of these reactions and the associated rate laws would be replaced by two alternatives. Thus reaction (7.A) becomes

$$-M_1 M_1 \cdot + M_1 \xrightarrow{k_{111}} -M_1 M_1 M_1 \cdot \qquad (7.E)$$

$$-M_2 M_1 \cdot + M_1 \xrightarrow{k_{211}} -M_2 M_1 M_1 \cdot \qquad (7.F)$$

and Eq. (7.1) becomes

$$R_{p,111} = k_{111} [M_1 M_1 \cdot] [M_1] \qquad (7.5)$$

$$R_{p,211} = k_{211} [M_2 M_1 \cdot] [M_1] \qquad (7.6)$$

when the effect of the next-to-last, or penultimate, unit is considered. For now we shall restrict ourselves to the simpler case where only the terminal unit determines behavior, although systems in which the penultimate effect is important are well known.

It is the magnitude of the various k values in Eqs. (7.1)-(7.4) that describes the intrinsic kinetic differences between the various modes of addition, and the k's plus the concentrations of the different species determine the rates at which the four kinds of additions occur. It is the proportion of different steps which determines the composition of the copolymer produced.

Monomer M_1 is converted to polymer by reactions (7.A) and (7.C); therefore the rate at which this occurs is the sum of $R_{p,11}$ and $R_{p,21}$:

$$- \frac{d[M_1]}{dt} = k_{11} [M_1 \cdot] [M_1] + k_{21} [M_2 \cdot] [M_1] \tag{7.7}$$

Likewise, reactions (7.B) and (7.D) convert M_2 to polymer, and the rate at which this occurs is the sum of $R_{p,12}$ and $R_{p,22}$:

$$- \frac{d[M_2]}{dt} = k_{12} [M_1 \cdot] [M_2] + k_{22} [M_2 \cdot] [M_2] \tag{7.8}$$

The ratio of Eqs. (7.7) and (7.8) gives the relative rates of the two monomer additions and, hence, the ratio of the two kinds of repeat units in the copolymer:

$$\frac{d[M_1]}{d[M_2]} = \frac{k_{11} [M_1 \cdot] [M_1] + k_{21} [M_2 \cdot] [M_1]}{k_{12} [M_1 \cdot] [M_2] + k_{22} [M_2 \cdot] [M_2]} \tag{7.9}$$

We saw in the last chapter that the stationary-state approximation is applicable to free-radical homopolymerizations, and the same is true of copolymerizations. Of course, it takes a brief time for the stationary-state radical concentration to be reached, but this period is insignificant compared to the total duration of a polymerization reaction. If the total concentration of radicals is constant, this means that the rate of crossover between the different types of terminal units is also equal, or that $R_{p,21} = R_{p,12}$:

$$k_{12} [M_1 \cdot] [M_2] = k_{21} [M_2 \cdot] [M_1] \tag{7.10}$$

or

$$\frac{[M_1 \cdot]}{[M_2 \cdot]} = \frac{k_{21} [M_1]}{k_{12} [M_2]} \tag{7.11}$$

Combining Eqs. (7.9) and (7.11) yields the important copolymer composition equation:

$$\frac{d[M_1]}{d[M_2]} = \frac{[M_1]}{[M_2]} \frac{(k_{11}/k_{12})[M_1] + [M_2]}{(k_{22}/k_{21})[M_2] + [M_1]} \tag{7.12}$$

Although there are a total of four different rate constants for propagation, Eq. (7.12) shows that the relationship between the relative amounts of the two monomers incorporated into the polymer and the composition of the monomer feedstock involves only two ratios of different pairs of these constants. Accordingly, we simplify the notation by defining

$$r_1 = \frac{k_{11}}{k_{12}} \tag{7.13}$$

and

$$r_2 = \frac{k_{22}}{k_{21}} \tag{7.14}$$

With these substitutions, Eq. (7.12) becomes

$$\frac{d[M_1]}{d[M_2]} = \frac{[M_1]}{[M_2]} \frac{r_1[M_1] + [M_2]}{r_2[M_2] + [M_1]} = \frac{1 + r_1[M_1]/[M_2]}{1 + r_2[M_2]/[M_1]} \tag{7.15}$$

Mayo and collaborators were among the earliest workers to clarify the relationship between copolymer and monomer solution compositions.

The ratio $d[M_1]/d[M_2]$ is the same as the ratio of the numbers of each kind of repeat unit in the polymer formed from the solution containing M_1 and M_2 at concentrations $[M_1]$ and $[M_2]$, respectively. Henceforth we shall designate this ratio as n_1/n_2. Since the composition of the monomer solution changes as the reaction progresses, Eq. (7.15) applies to the feedstock as prepared only during the initial stages of the polymerization. Subsequently, the instantaneous concentrations in the prevailing mixture apply unless monomer is added continuously to replace that which has reacted and maintain the original composition of the feedstock. We shall assume that it is the initial product formed that we describe when we use Eq. (7.15) so as to remove uncertainty as to the monomer concentrations.

As an alternative to the ratios n_1/n_2 and $[M_1]/[M_2]$ in Eq. (7.15), it is convenient to describe the composition of both the polymer and the feedstock in terms of the mole fraction of each monomer. Defining F_i as the mole fraction of the ith component in the polymer and f_i as the mole fraction of component i in the monomer solution, we observe that

$$F_1 = 1 - F_2 = \frac{d[M_1]}{d[M_1] + d[M_2]} \tag{7.16}$$

and

$$f_1 = 1 - f_2 = \frac{[M_1]}{[M_1] + [M_2]} \tag{7.17}$$

Combining Eqs. (7.15) and (7.16) into (7.17) yields

$$F_1 = \frac{r_1 f_1^2 + f_1 f_2}{r_1 f_1^2 + 2f_1 f_2 + r_2 f_2^2} \tag{7.18}$$

This equation relates the composition of the copolymer formed to the instantaneous composition of the feedstock and to the parameters r_1 and r_2 which characterize the specific system. Figure 7.1 shows a plot of F_1 versus f_1 —the mole fractions of component 1 in the copolymer and monomer mixture, respectively—for several arbitrary values of the parameters r_1 and r_2. Inspection of Fig. 7.1 brings out the following points:

1. If $r_1 = r_2 = 1$, the copolymer and the feed mixture have the same composition at all times. In this case Eq. (7.18) becomes

$$F_1 = \frac{f_1(f_1 + f_2)}{(f_1 + f_2)^2} = f_1$$

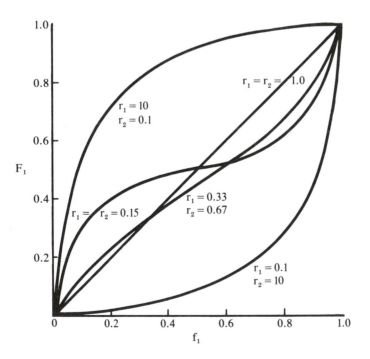

Figure 7.1 Mole fraction of component 1 in copolymers (F_1) and feedstock (f_1) for various values of r_1 and r_2.

2. If $r_1 = r_2$, the copolymer and the feed mixture have the same composition at $f = 0.5$. In this case Eq. (7.18) becomes $F_1 = (r + 1)/2(r + 1) = 0.5$.
3. If $r_1 = r_2$, with both values less than unity, the copolymer is richer in component 1 than the feed mixture for $f_1 < 0.5$, and richer in component 2 than the feed mixture for $f_1 > 0.5$.
4. If $r_1 = r_2$, with both values greater than unity, an S-shaped curve passing through the point $(0.5, 0.5)$ would also result, but in this case reflected across the $45°$ line compared to item (3).
5. If $r_1 \neq r_2$, with both values less than unity, the copolymer starts out richer in monomer 1 than the feed mixture and then crosses the $45°$ line, and is richer in component 2 beyond this crossover point. At the crossover point the copolymer and feed mixture have the same composition. The monomer ratio at this point is conveniently solved from Eq. (7.15):

$$\left(\frac{[M_1]}{[M_2]} \right)_{cross} = \frac{1 - r_2}{1 - r_1} \tag{7.19}$$

For the case of $r_1 = 0.33$ and $r_2 = 0.67$ shown in Fig. 7.1, $[M_1]/[M_2]$ equals 0.5 and $f_1 = 0.33$. This mathematical analysis shows that a comparable result is possible with both r_1 and r_2 greater than unity, but is not possible for $r_1 > 1$ and $r_2 < 1$.
6. When $r_1 = 1/r_2$, the copolymer composition curve will be either convex or concave when viewed from the F_1 axis, depending on whether r_1 is greater or less than unity. The further removed from unity r_1 is, the farther the composition curve will be displaced from the $45°$ line. This situation is called ideal copolymerization. The example below explores the origin of this terminology.

There is a parallel between the composition of a copolymer produced from a certain feed and the composition of a vapor in equilibrium with a two-component liquid mixture. The following example illustrates this parallel when the liquid mixture is an ideal solution and the vapor is an ideal gas.

Example 7.1

An ideal gas obeys Dalton's law; that is, the total pressure is the sum of the partial pressures of the components. An ideal solution obeys Raoult's law; that is, the partial pressure of the ith component in a solution is equal to the mole fraction of that component in the solution times the vapor pressure of pure component i. Use these relationships to relate the mole fraction of component 1 in the equilibrium vapor to its mole fraction in a two-component solution and relate the result to the ideal case of the copolymer composition equation.

Solution

We define F_1 to be the mole fraction of component 1 in the vapor phase and f_1 to be its mole fraction in the liquid solution. Here p_1 and p_2 are the vapor pressures of components 1 and 2 in equilibrium with an ideal solution and p_1^0 and p_2^0 are the vapor pressures of the two pure liquids. By Dalton's law, $p_{tot} = p_1 + p_2$ and $F_1 = p_1/p_{tot}$, since these are ideal gases and p is proportional to the number of moles. By Raoult's law, $p_1 = f_1 p_1^0$, $p_2 = f_2 p_2^0$, and $p_{tot} = f_1 p_1^0 + f_2 p_2^0$. Combining the two gives

$$F_1 = \frac{f_1 p_1^0}{f_1 p_1^0 + f_2 p_2^0} = \frac{f_1 (p_1^0/p_2^0)}{f_1 (p_1^0/p_2^0) + f_2}$$

Now examine Eq. (7.18) for the case of $r_1 = 1/r_2$:

$$F_1 = \frac{r_1 f_1^2 + f_1 f_2}{r_1 f_1^2 + 2 f_1 f_2 + (1/r_1) f_2^2} = \frac{r_1 f_1 (r_1 f_1 + f_2)}{(r_1 f_1 + f_2)^2} = \frac{r_1 f_1}{r_1 f_1 + f_2}$$

This is identical to the ideal liquid–vapor equilibrium if r_1 is identified with p_1^0/p_2^0.

The vapor pressure ratio measures the intrinsic tendency of component 1 to enter the vapor phase relative to component 2. Likewise, r_1 measures the tendency of M_1 to add to $M_1\cdot$ relative to M_2 adding to $M_1\cdot$. In this sense there is a certain parallel, but it is based on $M_1\cdot$ as a reference radical and hence appears to be less general than the vapor pressure ratio. Note, however, that $r_1 = 1/r_2$ means $k_{11}/k_{12} = k_{21}/k_{22}$. In this case the ratio of rate constants for monomer 1 relative to monomer 2 is the same regardless of the reference radical examined. This shows the parallelism to be exact.

•

Because of this parallel with liquid–vapor equilibrium, copolymers for which $r_1 = 1/r_2$ are said to be ideal. For those nonideal cases in which the copolymer and feedstock happen to have the same composition, the reaction is called an azeotropic polymerization. Just as in the case of azeotropic distillation, the composition of the reaction mixture does not change as copolymer is formed if the composition corresponds to the azeotrope. The proportion of the two monomers at this point is given by Eq. (7.19).

In this section we have seen that the copolymer composition depends to a large extent on the four propagation constants, although it is sufficient to consider these in terms of the two ratios r_1 and r_2. In the next section we shall examine these ratios in somewhat greater detail.

7.3 Reactivity Ratios

The parameters r_1 and r_2 are the vehicles by which the nature of the reactants enter the copolymer composition equation. We shall call these radical reactivity ratios, although similarly defined ratios also describe copolymerizations that involve ionic intermediates. There are several important things to note about radical reactivity ratios:

1. The single subscript used to label r is the index of the radical.
2. r_1 is the ratio of two propagation constants involving radical 1: The ratio always compares the propagation constant for the same monomer adding to the radical relative to the propagation constant for the addition of the other monomer. Thus if $r_1 > 1$, $M_1\cdot$ adds M_1 in preference to M_2; if $r_1 < 1$, $M_1\cdot$ adds M_2 in preference to M_1.
3. Although r_1 is descriptive of radical $M_1\cdot$, it also depends on the identity of the "other"; the pair of parameters r_1 and r_2 are both required to characterize a particular system and the product $r_1 r_2$ is used to quantify this by a single parameter.
4. The reciprocal of a radical reactivity ratio is sometimes used to quantitatively express the reactivity of monomer M_2 by comparing its rate of addition to radical $M_1\cdot$ relative to the rate of M_1 adding to $M_1\cdot$.
5. As the ratio of two rate constants, the radical reactivity ratio follows the Arrhenius equation with an apparent activation energy equal to the difference in the activation energies for the individual constants. Thus for r_1, $E_{app}{}^* = E_{p,11}{}^* - E_{p,12}{}^*$. Since the activation energies for propagation are not large to begin with, their difference is even smaller. Accordingly, the temperature dependence of r is relatively small.

The reactivity ratios of a copolymerization system are the fundamental parameters in terms of which the system is described. Since the copolymer composition equation relates the compositions of the product and the feedstock, it is clear that values of r can be evaluated from experimental data in which the corresponding compositions are measured. We shall consider this evaluation procedure in Sec. 7.7, where it will be found that this approach is not as free of ambiguity as might be desired. For now we shall simply assume that we know the desired r values for a system; in fact, extensive tabulations of such values exist. An especially convenient source of this information is the *Polymer Handbook* (Ref. 4). Table 7.1 lists some typical r values at 60°C.

Although Table 7.1 is rather arbitrarily assembled, note that it contains no systems for which r_1 and r_2 are *both* greater than unity. Indeed, such systems are very rare. We can understand this by recognizing that, at least in the extreme case of very large r's, these monomers would tend to simultaneously homopolymerize. Because of this preference toward homopolymerization, any copolymer that does form in systems with r_1 and r_2 both greater than unity will

Table 7.1 Values of Reactivity Ratios r_1 and r_2 and the Product $r_1 r_2$ for a Few Copolymers at 60°C

M_1	M_2	r_1	r_2	$r_1 r_2$
Acrylonitrile	Methyl vinyl ketone	0.61	1.78	1.09
	Methyl methacrylate	0.13	1.16	0.15
	α-Methyl styrene	0.04	0.20	0.008
	Vinyl acetate	4.05	0.061	0.25
Methyl methacrylate	Styrene	0.46	0.52	0.24
	Methacrylic acid	1.18	0.63	0.74
	Vinyl acetate	20	0.015	0.30
	Vinylidene chloride	2.53	0.24	0.61
Styrene	Vinyl acetate	55	0.01	0.55
	Vinyl chloride	17	0.02	0.34
	Vinylidene chloride	1.85	0.085	0.16
	2-Vinyl pyridine	0.55	1.14	0.63
Vinyl acetate	1-Butene	2.0	0.34	0.68
	Isobutylene	2.15	0.31	0.67
	Vinyl chloride	0.23	1.68	0.39
	Vinylidene chloride	0.05	6.7	0.34

Source: L. J. Young in Ref. 4.

be a block-type polymer with very long sequences of a single repeat unit. Since such systems are only infrequently encountered, we shall not consider them further.

Table 7.1 also lists the product $r_1 r_2$ for the systems included. These products lie in the range between zero and unity, and it is instructive to consider the character of the copolymer produced toward each of these extremes.

In the extreme case where $r_1 r_2 = 0$ because both r_1 *and* r_2 equal zero, the copolymer adds monomers with perfect alternation. This is apparent from the definition of r, which compares the addition of the same monomer to the other monomer for a particular radical. If both r's are zero, there is no tendency for a radical to add a monomer of the same kind as the growing end, whichever species is the terminal unit. When only one of the r's is zero, say r_1, then alternation occurs whenever the radical ends with an $M_1\cdot$ unit. There is thus a tendency toward alternation in this case, although it is less pronounced than in the case where both r's are zero. Accordingly, we find increasing tendency toward alternation as $r_1 \to 0$ and $r_2 \to 0$, or, more succinctly, as the product $r_1 r_2 \to 0$.

At the other end of the commonly encountered range we find the product $r_1 r_2 \to 1$. As noted above, this limit corresponds to ideal copolymerization and means the two monomers have the same relative tendency to add to both radicals. Thus if $r_1 = 10$, monomer 1 is 10 times as likely to add to $M_1 \cdot$ than monomer 2. At the same time $r_2 = 0.1$, which also means that monomer 1 is 10 times as likely to add to $M_2 \cdot$ than monomer 2. In this case the radicals exert the same influence, so the monomers add at random in a proportion governed by the specific values of the r's.

Recognition of these differences in behavior points out an important limitation on the copolymer composition equation. The equation describes the overall composition of the copolymer, but gives no information whatsoever about the distribution of the different kinds of repeat units within the polymer. While the overall composition is an important property of the copolymer, the details of the microstructural arrangement is also a significant feature of the molecule. It is possible that copolymers with the same overall composition have very different properties because of differences in microstructure. Reviewing the three categories presented in Chap. 1, we see the following:

1. Alternating structures [I] are promoted by $r_1 \to 0$ and $r_2 \to 0$:

$$M_1 M_2 M_1 M_2 M_1 M_2 M_1 M_2 M_1 M_2 M_1 M_2 M_1 M_2 M_1 M_2 M_1 M_2 M_1 M_2$$

[I]

2. Random structures [II] are promoted by $r_1 r_2 \to 1$:

$$M_1 M_2 M_2 M_2 M_1 M_1 M_2 M_1 M_1 M_2 M_1 M_2 M_1 M_2 M_1 M_2 M_2 M_1 M_1 M_2$$

[II]

3. Block structures [III] are promoted by $r_1 r_2 > 1$:

$$M_1 M_1 M_1 M_1 M_1 M_1 M_1 M_1 M_1 M_1 M_2 M_2 M_2 M_2 M_2 M_2 M_2 M_2 M_2 M_2$$

[III]

Each of these polymers has a 50:50 proportion of the two components, but the products probably differ in properties. As examples of such differences, we note the following:

4. Alternating copolymers, while relatively rare, are characterized by combining the properties of the two monomers along with structural regularity. We saw in Chap. 4 that a very high degree of regularity—extending all the

way to stereoregularity in the configuration of the repeat units—is required for crystallinity to develop in polymers. Although they fall short of perfect regularity, copolymers with alternating structures are capable of interactions which order separate chains, and this, in turn, can affect their solubility, chemical reactivity, and mechanical properties.

5. Random copolymers tend to average the properties of the constituent monomers in proportion to the relative abundance of the two comonomers.

6. Block copolymers are closer to blends of homopolymers in properties, but without the latter's tendency to undergo phase separation. As a matter of fact, diblock copolymers can be used as surfactants to bind immiscible homopolymer blends together and thus improve their mechanical properties. Block copolymers are generally prepared by sequential addition of monomers to living polymers, rather than by depending on the improbable $r_1 r_2 > 1$ criterion in monomers.

We shall return to the topic of copolymer properties in Sec. 7.8.

Returning to the data of Table 7.1, it is apparent that there is a good deal of variability among the r values displayed by various systems. We have already seen the effect this produces on the overall copolymer composition; we shall return to the matter of microstructure in Sec. 7.6. First, however, let us consider the obvious question. What factors in the molecular structure of two monomers govern the kinetics of the different addition steps? This question is considered in the few next sections; for now we look for a way to systematize the data as the first step toward an answer.

We noted above that the product $r_1 r_2$ can be used to locate a copolymer along an axis between alternating and random structures. It is by means of this product that some values from Table 7.1, supplemented by other results for additional systems, have been organized in Fig. 7.2. Figure 7.2 has been constructed according to the following general principles:

1. Various monomers are listed along the base of a triangle.
2. The triangle is subdivided into an array of diamonds by lines drawn parallel to the two sides of the triangle.
3. The spacing of the lines is such that each monomer along the base serves as a label for a row of diamonds.
4. Each diamond marks the intersection of two such rows and therefore corresponds to two comonomers.
5. The $r_1 r_2$ product for the various systems is the number entered in each diamond.
6. The individual monomers have been arranged in such a way as to achieve to the greatest extent possible values of $r_1 r_2$ that approach zero toward the apex of the triangle and values of $r_1 r_2$ which approach unity toward the base of the triangle.

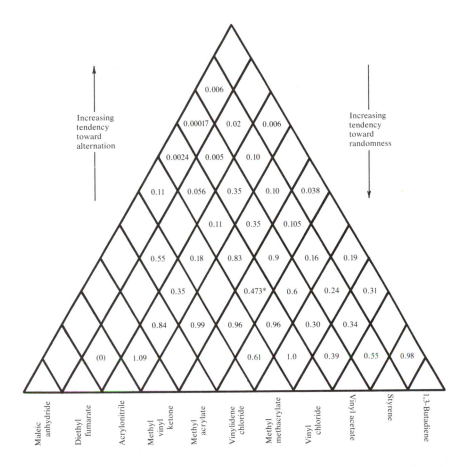

Figure 7.2 The product r_1r_2 for copolymers whose components define the intersection where the numbers appear. See the text for details of placement. The value marked * is determined in Example 7.5. Other values are from Ref. 4.

Before proceeding with a discussion of this display, it is important to acknowledge that the criteria for monomer placement can be met only in part. For one thing, there are combinations for which data are not readily available. Incidentally, not all of the r_1r_2 values in Fig. 7.2 were measured at the same temperature, but, as noted above, temperature effects are expected to be relatively unimportant. Also, there are outright exceptions to the pattern sought: Generalizations about chemical reactions always seem to be plagued by these. In spite of some reversals of ranking, the predominant trend moving upward from the base along any row of diamonds is a decrease in r_1r_2 values.

From the geometry of this triangular display, it follows immediately—if one overlooks the exceptions—that the more widely separated a pair of comonomers are in Fig. 7.2, the greater is their tendency toward alternation. Conversely, the closer they are together, the greater their tendency toward randomness. We recognize a parallel here to the notion that widely separated elements in the periodic table will produce more polar bonds than those which are closer together and vice versa.

This is a purely empirical and qualitative trend. The next order of business is to seek an explanation for its origin in terms of molecular structure. If we focus attention on the electron-withdrawing or electron-donating attributes of the substituent(s) on the double bond, we find that the substituents of monomers which are located toward the right-hand corner of the triangle in Fig. 7.2 are recognized as electron donors in organic chemistry. Likewise, the substituents in monomers located toward the left-hand corner of the triangle are electron acceptors. The demarcation between the two regions of behavior is indicated in Fig. 7.2 by reversing the direction of the lettering at this point. Pushing this point of view somewhat further, we conclude that the sequence acetoxy < phenyl < vinyl is the order of increase in electron-donating tendency. Chloro < carbonyl < nitrile is the order of increase in electron-withdrawing tendency. The positions of diethyl fumarate and vinylidene chloride relative to their monosubstituted analogs indicates that "more is better" with respect to these substituent effects. The location of methyl methacrylate relative to methyl acrylate also indicates additivity, this time with partial compensation of opposing effects.

Any discussion based on reactivity ratios is kinetic in origin and therefore reflects the mechanism or, more specifically, the transition state of a reaction. The transition state for the addition of a vinyl monomer to a growing radical involves the formation of a partial bond between the two species, with a corresponding reduction of the double-bond character of the vinyl group in the monomer:

$$
\begin{array}{c}
-\text{C}-\text{C}\cdot + \text{C}=\text{C} \\
\big| \phantom{-\text{C}\cdot + \text{C}} \big| \\
\text{X} \phantom{-\text{C}\cdot + } \text{Y}
\end{array}
\quad\longrightarrow\quad
\left[
\begin{array}{c}
-\text{C}-\text{C}\text{---}\overset{..}{\underset{}{\text{C}}}\overset{=}{=}\text{C} \\
\big| \phantom{-\text{C}\text{---}} \big| \\
\text{X} \phantom{-\text{C}\text{---}\text{C}} \text{Y}
\end{array}
\right]
\quad\longrightarrow\quad \text{product}
$$

Transition State

(7.G)

If substituent X is an electron donor and Y an electron acceptor, then the partial bond in the transition state is stabilized by a resonance form [IV] which attributes a certain polarity to the emerging bond:

$$\left[\begin{array}{c} \overset{\delta-}{} \ \overset{\delta+}{} \\ -C-C\cdots C{=\!=}C \\ \underset{X}{|} \ \overset{\smile}{\bullet} \ \underset{Y}{|} \end{array} \right]$$

[IV]

The contribution of this polar structure to the bonding lowers the energy of the transition state. This may be viewed as a lower activation energy for the addition step and thus a factor which promotes this particular reaction. The effect is clearly larger the greater the difference in the donor–acceptor properties of X and Y. The transition state for the successive addition of the same monomer (whether X or Y substituted) is structure [V]:

$$\left[\begin{array}{c} -C-C\cdots C{=\!=}C \\ \underset{X}{|} \ \overset{\smile}{\bullet} \ \underset{X}{|} \end{array} \right]$$

[V]

This involves a more uniform distribution of charge because of the identical substituents and thus lacks the stabilizing effect of the polar resonance form. The activation energy for this mode of addition is greater than that for alternation, at least when X and Y are sufficiently different.

Although we use the term *resonance* in describing the effect of polarity in stabilizing the transition state in alternating copolymers, the emphasis of the foregoing is definitely on polarity rather than resonance per se. It turns out, however, that resonance plays an important role in its own right in free-radical polymerization, even if polarity effects are ignored. In the next section we examine some evidence for this and consider the origin of this behavior.

7.4 Resonance and Reactivity

The tendency toward alternation is not the only pattern in terms of which copolymerization can be discussed. The activities of radicals and monomers may also be examined as a source of insight into copolymer formation. The reactivity of radical 1 copolymerizing with monomer 2 is measured by the rate constant k_{12}. The absolute value of this constant can be determined from copolymerization data (r_1) and studies yielding absolute homopolymerization constants (k_{11}):

$$k_{12} = \frac{k_{11}}{r_1} \tag{7.20}$$

Table 7.2 lists a few cross-propagation constants calculated by Eq. (7.20). Far more extensive tabulations than this have been prepared by correlating copolymerization and homopolymerization data for additional systems. Examination of Table 7.2 shows that the general order of increasing *radical* activity is

styrene $<$ acrylonitrile $<$ methyl acrylate $<$ vinyl acetate

An additional observation is that any one of these species shows the reverse order of reactivity for the corresponding monomers. As *monomers*, the order of reactivity in Table 7.2 is

styrene $>$ acrylonitrile $>$ methyl acrylate $>$ vinyl acetate

These and similar rankings based on more extensive comparisons are summarized in terms of substituents in Table 7.3.

An important pattern to recognize among the substituents listed in Table 7.3 is this: Those which have a double bond conjugated with the double bond in the olefin are the species which are more stable as radicals and more reactive as monomers. The inverse relationship between the stability of monomers and radicals arises precisely because monomers gain (or lose) stability by converting to the radical: The greater the gain (or loss), the greater (or less) the incentive for the monomer to react. It is important to realize that the ability to form conjugated structures is associated with a substituent whether it is in a monomer or a radical. Conjugation allows greater electron delocalization, which, in turn, lowers the energy of the system that possesses this feature.

Comparison of the range of k_{12} along horizontal rows and vertical columns in Table 7.2 suggests that resonance stabilization produces a bigger effect in the radical than in the monomer. After all, the right- and left-hand columns in Table 7.2 (various radicals) differ by factors of 100–1000, while the top and

Table 7.2   Values of the Cross-Propagation Constants k_{12} for Four Monomer-Radical Combinations

Monomer	Radical			
	Styrene	Acrylonitrile	Methyl acrylate	Vinyl acetate
Styrene	145	49,000	14,000	230,000
Acrylonitrile	435	1,960	2,510	46,000
Methyl acrylate	203	1,310	2,090	23,000
Vinyl acetate	2.9	230	230	2,300

Source: Ref. 4.

Table 7.3 List of Some Substituents Ranked in Terms of Their Effects on Monomer and Radical Reactivity

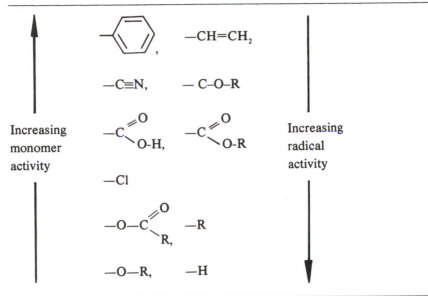

bottom rows (various monomers) differ only by factors of 50–100. In order to examine this effect in more detail, consider the addition reaction of monomer M to a reactant radical R· to form a product radical P·. What distinguishes these species is the presence or absence of resonance stabilization (subscript rs). If the latter is operative, we must also consider which species benefit from its presence. There are four possibilities:

1. Unstabilized monomer converts stabilized radical to unstabilized radical:

$$R_{rs} \cdot + M \rightarrow P \cdot \tag{7.H}$$

There is an overall loss of resonance stabilization in this reaction. Since it is a radical which suffers the loss, the effect is larger than in the reaction in which . . . .

2. Stabilized monomer converts stabilized radical to another stabilized radical:

$$R_{rs} \cdot + M_{rs} \rightarrow P_{rs} \cdot \tag{7.I}$$

Here too there is an overall loss of resonance stabilization, but it is monomer stabilization which is lost, and this is energetically less costly than reaction (7.H).

3. Unstabilized monomer converts unstabilized radical to another unstabilized radical:

$$R\cdot + M \rightarrow P\cdot \qquad (7.J)$$

This reaction suffers none of the reduction in resonance stabilization that is present in reactions (7.H) and (7.I). It is energetically more favored than both of these, but not as much as the reaction in which . . . .

4. Stabilized monomer converts unstabilized radical to stabilized radical:

$$R\cdot + M_{rs} \rightarrow P_{rs}\cdot \qquad (7.K)$$

This reaction converts the less effective resonance stabilization of a monomer to a more effective form of radical stabilization. This is the most favorable of the four reaction possibilities.

In summary, we can rank these reactions in terms of their propagation constants as follows:

$$R_{rs}\cdot + M < R_{rs}\cdot + M_{rs} < R\cdot + M < R\cdot + M_{rs}$$

Systems from Table 7.2 which correspond to these situations are the following:

Radical	Styrene		styrene		vinyl acetate		vinyl acetate
	+	$<$	+	$<$	+	$<$	+
Monomer	vinyl acetate		styrene		vinyl acetate		styrene

Note that this inquiry into copolymer propagation rates also increases our understanding of the differences in free-radical homopolymerization rates. It will be recalled that in Sec. 6.1 a discussion of this aspect of homopolymerization was deferred until copolymerization was introduced. The trends under consideration enable us to make some sense out of the rate constants for propagation in free-radical homopolymerization as well. For example, in Table 6.4 we see that k_p values at 60°C for vinyl acetate and styrene are 2300 and 165 liter mol^{-1} sec^{-1}, respectively. The relative magnitude of these constants can be understod in terms of the sequence above.

Resonance stabilization energies are generally assessed from thermodynamic data. If we define ϵ_i to be the resonance stabilization energy of species i, then the heat of formation of that species will be less by an amount ϵ_i than for an otherwise equivalent molecule without resonance. Likewise, the ΔH for a reaction which is influenced by resonance effects is less by an amount $\Delta\epsilon$ (Δ is the usual difference: products minus reactants) than the ΔH for a reaction which is otherwise identical except for resonance effects:

$$\Delta H_{rs} = \Delta H_{no\ rs} - \Delta \epsilon \qquad (7.21)$$

Thus if we consider the homopolymerization of ethylene (no resonance possibilities),

$$-CH_2-CH_2\cdot + CH_2=CH_2 \longrightarrow -CH_2\,CH_2\,CH_2\,CH_2\cdot$$

$$\Delta H_{no\ rs} = -88.7 \text{ kJ mol}^{-1} \qquad (7.L)$$

as a reference reaction, and compare it with the homopolymerization of styrene (resonance effects present),

$$-CH_2-\underset{\underset{\phi}{|}}{CH}\cdot + CH_2=\underset{\underset{\phi}{|}}{CH} \longrightarrow -CH_2-\underset{\underset{\phi}{|}}{CH}-CH_2-\underset{\underset{\phi}{|}}{CH}\cdot$$

$$\Delta H_{rs} = -69.9 \text{ kJ mol}^{-1} \qquad (7.M)$$

we find a value of $\Delta \epsilon = -19 \text{ kJ mol}^{-1}$, according to Eq. (7.21). Reaction (7.M) is a specific example of the general reaction (7.I), and the negative value of $\Delta \epsilon$ in this example indicates the overall loss of resonance stabilization which is characteristic of (7.I).

Although it is not universally true that the activation energies of reactions parallel their heats of reaction, this is approximately true for the kind of addition reaction we are discussing. Accordingly, we can estimate $E^* = k\ \Delta H$, with k an appropriate proportionality constant. If we consider the difference between two activation energies by combining this idea with Eq. (7.21), the contribution of the nonstabilized reference reaction drops out of Eq. (7.21) and we obtain

$$E_{11}{}^* - E_{12}{}^* = k[-\Delta\epsilon_{11} - (-\Delta\epsilon_{12})]$$

$$= -(\epsilon_{P_1\cdot} - \epsilon_{R_1\cdot} - \epsilon_{M_1}) + (\epsilon_{P_2\cdot} - \epsilon_{R_1\cdot} - \epsilon_{M_2}) \qquad (7.22)$$

In writing the second version of this, the proportionality constant has been set equal to unity as a simplification. Note that the resonance stabilization energy of the reference radical $R_1\cdot$ also cancels out of this expression.

The temperature dependence of the reactivity ratio r_1 also involves the $E_{11}{}^* - E_{12}{}^*$ difference through the Arrhenius equation; hence

$$r_1 \propto \exp\left(\frac{\epsilon_{P_1\cdot} - \epsilon_{M_1}}{RT}\right) \exp\left(\frac{-(\epsilon_{P_2\cdot} - \epsilon_{M_2})}{RT}\right) \qquad (7.23)$$

An analogous expression can be written for r_2:

$$r_2 \propto \exp\left(\frac{\epsilon_{P_2}. - \epsilon_{M_2}}{RT}\right) \exp\left(\frac{-(\epsilon_{P_1}. - \epsilon_{M_1})}{RT}\right) \tag{7.24}$$

According to this formalism, the following applies:

1. The reactivity ratios are proportional to the product of two exponential numbers.
2. Each exponential involves the difference between the resonance stabilization energy of the radical and monomer of a particular species.
3. The positive exponent is associated with the same species as identifies the r (i.e., for r_1, $M_1 \rightarrow P_1 \cdot$), while the negative exponent is associated with the other species (for r_1, $M_2 \rightarrow P_2 \cdot$).

We might be hard pressed to estimate the individual resonance stabilization energies in Eqs. (7.23) and (7.24), but the qualitative application of these ideas is not difficult. Consider once again the styrene–vinyl acetate system:

1. Define styrene to be monomer 1 and vinyl acetate to be monomer 2.
2. The difference in resonance stabilization energy $\epsilon_{P_1}. - \epsilon_{M_1} > 1$, since styrene is resonance stabilized and the effect is larger for the radical than the monomer.
3. The difference $\epsilon_{P_2}. - \epsilon_{M_2} \cong 0$, since neither the radical nor the monomer of vinyl acetate shows appreciable stabilization.
4. Therefore, according to Eqs. (7.23) and (7.24), $r_1 > 1$ while $r_2 < 1$.
5. The experimental values for this system are $r_1 = 55$ and $r_2 = 0.01$.

Although this approach does correctly rank the parameters r_1 and r_2 for the styrene–vinyl acetate system, this conclusion was already reached qualitatively above, using the same concepts and without any mathematical manipulations. One point that the quantitative derivation makes clear is that explanations of copolymer behavior based exclusively on resonance concepts fail to describe the full picture. All that we need to do is examine the product $r_1 r_2$ as given by Eqs. (7.23) and (7.24), and the shortcoming becomes apparent. According to these relationships, the product $r_1 r_2$ always equals unity, yet we saw in the last section that experimental $r_1 r_2$ values generally lie between zero and unity. We also saw that polarity effects could be invoked to rationalize the $r_1 r_2$ product.

The situation may be summarized as follows:

1. If resonance effects *alone* are considered, it is possible to make some sense of the ranking of various propagation constants.
2. In this case only random microstructure is predicted.
3. If polarity effects *alone* are considered, it is possible to make some sense out of the tendency toward alternation.
4. In this case homopolymerization is unexplained.

The way out of this dilemma is easily stated, although not easily acted upon. It is not adequate to consider *any one* of these approaches for the explanation of something as complicated as these reactions. Polarity effects and resonance are *both* operative, and, if these still fall short of explaining all observations, there is another old standby to fall back on: steric effects.

Resonance, polarity, and steric considerations are all believed to play an important role in copolymerization chemistry, just as in other areas of organic chemistry. Things are obviously simplified if only one of these is considered; but it must be remembered that doing this necessarily reveals only one facet of the problem. Nevertheless, there are times, particularly before launching an experimental investigation of a new system, when some guidelines are very useful. The following example illustrates this point.

Example 7.2

It is proposed to polymerize the vinyl group of the hemin molecule with other vinyl comonomers to prepare model compounds to be used in hemoglobin research. Considering hemin and styrene to be species 1 and 2, respectively, use the resonance concept to rank the reactivity ratios r_1 and r_2.

Solution

Hemin is the complex between protoporphyrin and iron in the +3 oxidation state. Iron is in the +2 state in the heme of hemoglobin. The molecule has the following structure:

It is apparent from the size of the conjugated system here that numerous resonance possibilities exist in this species in both the radical and the molecular form. Styrene also has resonance structures in both forms. On the principle that these effects are larger for radicals than monomers, we conclude that the difference $\epsilon_P - \epsilon_M > 0$ for both hemin and styrene. On the principle that greater resonance effects result from greater delocalization, we expect the difference to be larger for hemin than for styrene. According to Eq. (7.23), $r_1 \propto e^{larger} e^{-smaller} > 1$. According to Eq. (7.24), $r_2 \propto e^{smaller} e^{-larger} < 1$. Experimentally, the values for these parameters turn out to be $r_1 = 65$ and $r_2 = 0.18$.

•

In the next section we shall consider an attempt to combine both resonance and polarity effects.

7.5 The Price-Alfrey Equation

In the last two sections we have considered—separately—the effects of resonance and polarity on copolymerization. While these concepts provide some insights into various observations, it is artificial to consider either one of them operating exclusively. In fact, resonance and polarity features are both active in most molecules. A method for merging their contributions is clearly desirable.

Another troublesome aspect of the reactivity ratios is the fact that they must be determined and reported as a pair. It would clearly simplify things if it were possible to specify one or two general parameters for each monomer which would correctly represent its contribution to all reactivity ratios. Combined with the analogous parameters for its comonomer, the values r_1 and r_2 could then be evaluated. This situation parallels the standard potential of electrochemical cells which we are able to describe as the sum of potential contributions from each of the electrodes that comprise the cell. With x possible electrodes, there are $x(x - 1)/2$ possible electrode combinations. If x = 50, there are 1225 possible cells, but these can be described by only 50 electrode potentials. A dramatic data reduction is accomplished by this device. Precisely the same proliferation of combinations exists for monomer combinations. It would simplify things if a method were available for data reduction such as that used in electrochemistry.

An approach to copolymerization has been advanced by Price and Alfrey which attempts to both combine resonance and polarity considerations and accomplish the data reduction strategy of the last paragraph. It should be conceded at the outset that the Price-Alfrey method is only semiquantitative in its success. Its greatest usefulness is probably in providing some orientation to a new system before launching an experimental investigation.

The Price–Alfrey approach begins by defining three parameters—P, Q, and e— for each of the comonomers in a reaction system. We shall see presently that the parameter P is rapidly eliminated from the theory. As a result, the Price– Alfrey system is also called the Q–e scheme for copolymerization.

For the reaction of radical i with monomer j, Price and Alfrey assume that the cross-propagation rate constant can be written as

$$k_{ij} = P_i Q_j e^{-e_i e_j} \tag{7.25}$$

In this equation P and Q are parameters that describe the reactivity of the radical and monomer of the designated species, and the values of e measure the polarity of the two components without distinguishing between monomer and radical.

From Eqs. (7.13) and (7.14), the reactivity ratios can be written

$$r_1 = \frac{Q_1}{Q_2} \exp\left[-e_1(e_1 - e_2)\right] \tag{7.26}$$

and

$$r_2 = \frac{Q_2}{Q_1} \exp\left[-e_2(e_2 - e_1)\right] \tag{7.27}$$

Finally, Eqs. (7.26) and (7.27) can be combined to give

$$r_1 r_2 = \exp\left[(e_1 - e_2)(e_2 - e_1)\right] = \exp\left[-(e_1 - e_2)^2\right] \tag{7.28}$$

That these expressions do combine resonance and polarity effects can be seen as follows:

1. If molecules 1 and 2 differ widely in polarity, then the exponent in Eq. (7.28) will be large and the exponential will be small. We saw in Sec. 7.3 that alternation is favored by large differences in polarity and is described by small values of $r_1 r_2$.
2. If molecules 1 and 2 are identical in polarity, then $e_1 - e_2$ in Eqs. (7.26) and (7.27) is zero and $r_1 = Q_1/Q_2$ and $r_2 = Q_2/Q_1$. Comparing these limits with Eqs. (7.23) and (7.24) leads to the result

$$Q_j = \exp\left(\frac{(\epsilon_{P_j^-} - \epsilon_{M_j})}{RT}\right) \tag{7.29}$$

3. This last identification makes the Q's strictly a matter of resonance, whereas the general concept of "reactivity" also includes steric effects. The effects

of polarity are explicitly handled by the e's and are not therefore lumped together with these other concepts.

4. There are no inherent restrictions on Q and e; hence the individual reactivity ratios can take on a wide range of values.

An advantage of the Price-Alfrey system is that each monomer is characterized by its own values for Q and e, which are assumed to be independent of the nature of the comonomer. Thus if Q and e values were available for all monomers, then these could be combined at will to generate the parameters r_1 and r_2 which define copolymer composition and microstructure. This feature makes data reduction and predictions about new systems feasible. The only problem is that Q and e cannot be evaluated independently for a particular monomer any more than the potential of a single electrode can be measured. In electrochemistry we get around the latter problem by assigning to the hydrogen electrode under standard conditions a potential contribution of 0.0000 V. The choice of this electrode and this assignment of potential are arbitrary, but, once accepted, they permit other electrode potentials to be evaluated relative to the standard. A similar procedure is followed for the Q and e values. The reference monomer is chosen to be styrene and the parameters are assigned the following values: $Q_{sty} = 1.0$ and $e_{sty} = -0.8$. In their original work Price and Alfrey assigned styrene an e value of -1.0, but this was revised to the present value, which gives better agreement with experimental reactivity ratios.

This last statement shows that expressing Q and e values relative to the values for styrene is not identical to the procedure for electrode potentials. In the latter, cell potentials are calculated correctly from electrode potentials, regardless of the value assigned to the reference. The Price-Alfrey system is only semiquantitative; values are assigned which give the best average fit to the largest number of monomers. This is accomplished more effectively by assigning styrene an e value of -0.8. In this respect the Q-e values are like bond dissociation energies in which the properties of a variety of different compounds are considered to find the strength of an "average" bond. Both bond energies and Q-e values fall short of expectations when specific effects are present which separate a particular system from the "average."

Table 7.4 lists the Q and e values for an assortment of common monomers. The extremes in the column of e values in Table 7.4—which are listed in order—quantify the range of donor-acceptor properties which is used as the basis for ranking in Fig. 7.2. The Q values perform a similar ranking with respect to resonance effects. The eight different Q-e combinations in Table 7.4 allow the estimation of r_1 and r_2 values for 28 different copolymers. Of course, in these systems Q and e values were assigned to give the best fit to r values which had already been measured. As an illustration of the predictive values of the Q-e scheme, consider the following example:

Table 7.4   Values of the Price–Alfrey Q
and e Values for a Few Common Monomers

Monomer	Q	e
Acrylonitrile	0.60	1.20
Methyl vinyl ketone	1.0	0.7
Methyl acrylate	0.42	0.60
Methyl methacrylate	0.74	0.40
Vinyl chloride	0.044	0.20
Vinyl acetate	0.026	-0.22
Styrene (standard)	1.0	-0.8
Butadiene	2.39	-1.05

Source: L. J. Young in Ref. 4.

Example 7.3

Reactivity ratios for the N-vinylphthalimide (molecule 1)–styrene (molecule 2) system were measured, and found[†] to be $r_1 = 0.075$ and $r_2 = 8.3$. Use these values to estimate values of Q and e for N-vinylphthalimide; then estimate the parameters r_1 and r_2 for system in which molecule 2 is vinyl acetate.

Solution

Since styrene is used as the standard in the Price–Alfrey system, $Q_2 = 1.0$ and $e_2 = -0.8$. Use these values and the experimental r_1 and r_2 values for the styrene–N-vinylphthalimide system to evaluate Q_1 and e_1: Using Eq. (7.28), we have $r_1 r_2 = e^{-(e_1 - e_2)^2} = 0.075(8.3) = 0.623$, $\ln 0.623 = -[e_1 - (-0.8)]^2$, and $e_1 + 0.8 = \pm 0.688$. The phthalimide substituent is expected to be more electronegative than phenyl, so we choose the negative root; therefore $e_1 = -0.688 - 0.8 = -1.49$. Using Eq. (7.26), we have $r_1 = (Q_1/Q_2) e^{-e_1(e_1 - e_2)}$, $0.075 = (Q_1/1) e^{-(-1.49)(-0.688)}$, $Q_1 = 0.21$. Next we let vinyl acetate be monomer 2 and obtain the Q and e values for this monomer from Table 7.4: $Q_2 = 0.026$ and $e_2 = -0.22$. Using the Q and e values for N-vinylphthalimide calculated above, we find from Eq. (7.26), $r_1 = (Q_1/Q_2) e^{-e_1(e_1 - e_2)} = (0.21/0.026) e^{-(-1.49)[-1.49 - (-0.22)]} = 1.22$, and, from Eq. (7.27), $r_2 = (Q_2/Q_1) e^{-e_2(e_2 - e_1)} = (0.026/0.21) e^{-(-0.22)[-0.22 - (-1.49)]} = 0.16$. The experimental values of the reactivity ratios in this system are $r_1 = 2.4$ and $r_2 = 0.07$. Working backward from *these* data gives $Q_1 = 0.50$ and $e_1 = -1.56$ for N-vinylphthalimide.

●

This example also illustrates that "best-fit" values of Q and e are not absolute.

[†] Data from Ref. 4.

7.6 A Closer Look at Microstructure

In Sec. 7.3 we noted that variations in the $r_1 r_2$ product led to differences in the microstructure of the polymer, even when the overall composition of two compared systems is the same. Structures [I]–[III] are examples of this situation. In this section we shall take a closer look at this variation, using the approach which is best suited for this kind of detail: statistics.

Suppose we define as p_{ij} the probability that a unit of type i is followed in the polymer by a unit of type j, where both i and j can be either 1 or 2. Since an i unit must be followed by either an i or a j, the fraction of ij sequences out of all possible sequences defines p_{ij}:

$$p_{ij} = \frac{\text{number of ij sequences}}{\text{number of ij sequences} + \text{number of ii sequences}} \tag{7.30}$$

This equation can also be written in terms of the propagation rates of the different types of addition steps which generate the sequences:

$$p_{ij} = \frac{R_{ij}}{R_{ij} + R_{ii}} = \frac{k_{ij}[M_i \cdot][M_j]}{k_{ij}[M_i \cdot][M_j] + k_{ii}[M_i \cdot][M_i]} \tag{7.31}$$

For the various possible combinations in a copolymer, Eq. (7.31) becomes

$$p_{11} = \frac{k_{11}[M_1 \cdot][M_1]}{k_{11}[M_1 \cdot][M_1] + k_{12}[M_1 \cdot][M_2]} = \frac{r_1[M_1]}{r_1[M_1] + [M_2]} \tag{7.32}$$

$$p_{12} = \frac{[M_2]}{r_1[M_1] + [M_2]} \tag{7.33}$$

$$p_{22} = \frac{k_{22}[M_2 \cdot][M_2]}{k_{22}[M_2 \cdot][M_2] + k_{21}[M_2 \cdot][M_1]} = \frac{r_2[M_2]}{r_2[M_2] + [M_1]} \tag{7.34}$$

$$p_{21} = \frac{[M_1]}{r_2[M_2] + [M_1]} \tag{7.35}$$

Note that $p_{11} + p_{12} = p_{22} + p_{21} = 1$. In writing these expressions we make the assumption that only the terminal unit of the radical influences the addition of the next monomer. This same assumption was made in deriving the copolymer composition equation. We shall have more to say below about this so-called terminal assumption.

Next let us consider the probability of finding a sequence of repeat units in a copolymer which is exactly ν_1 units of M_1 in length. This may be represented as $M_2 (M_1)_{\nu_1} M_2$. Working from left to right in this sequence, we note the following:

1. If the addition of monomer M_1 to a radical ending with M_2 occurs L times in a sample, then there will be a total of L sequences, of unspecified length, of M_1 units in the sample.
2. If $\nu_1 - 1$ consecutive M_1 monomers add to radicals capped by M_1 units, the total number of such sequences is expressed in terms of p_{11} to be $Lp_{11}^{\nu_1-1}$.
3. If the sequence contains exactly ν_1 units of type M_1, then the next step must be the addition of an M_2 unit. The probability of such an addition is given by p_{12}, and the number of such sequences is $Lp_{11}^{\nu_1-1}p_{12}$.
4. Note that we use the symbol ν_i to indicate the number of M_i units in a particular sequence. This should be distinguished from n_i, which gives the total number of M_i units in the copolymer without regard to their distribution in various sequences.

Since L equals the total number of M_1 sequences of any length, the fraction of sequences of length ν_1, ϕ_{ν_1}, is given by

$$\phi_{\nu_1} = p_{11}^{\nu_1-1} p_{12} \tag{7.36}$$

The similarity of this derivation to those in Secs. 5.4 and 6.7 should be apparent. Substitution of the probabilities given by Eqs. (7.32) and (7.33) leads to

$$\phi_{\nu_1} = \left(\frac{r_1[M_1]}{r_1[M_1] + [M_2]}\right)^{\nu_1-1} \left(\frac{[M_2]}{r_1[M_1] + [M_2]}\right) \tag{7.37}$$

A similar result can be written for ϕ_{ν_2}. These expressions give the fraction of sequences of specified length in terms of the reactivity ratios of the copolymer system and the composition of the feedstock. Figure 7.3 illustrates by means of a bar graph how ϕ_{ν_1} varies with ν_1 for two polymer systems prepared from equimolar solutions of monomers. The shaded bars in Fig. 7.3 describe the system for which $r_1r_2 = 0.03$, and the unshaded bars describe $r_1r_2 = 0.30$. Table 7.5 shows the effect of variations in the composition of the feedstock for the system $r_1r_2 = 1$. The following observations can be made concerning Fig. 7.3 and Table 7.5:

1. In all situations, the fraction ϕ_{ν_1} decreases with increasing ν_1.
2. Figure 7.3 shows that for $r_1r_2 = 0.03$, about 85% of the M_1 units are sandwiched between two M_2's. We have already concluded that low values of the r_1r_2 product indicate a tendency toward alternation.
3. Figure 7.3 also shows that the proportion of alternating M_1 units decreases and the fraction of longer sequences increases as r_1r_2 increases. The 50 mol % entry in Table 7.5 shows that the distribution of sequence lengths gets flatter and broader for $r_1r_2 = 1$, the random case.

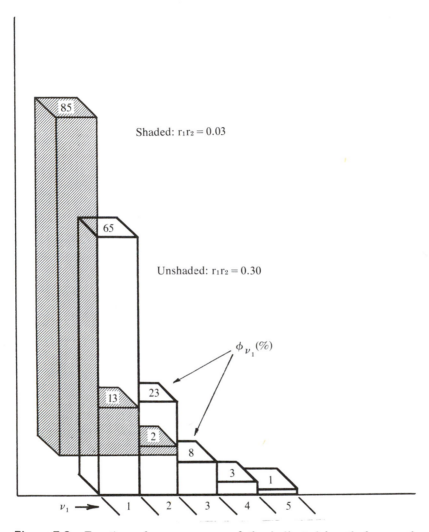

Figure 7.3 Fraction of n_1 sequences of the indicated length for copolymers prepared from equimolar feedstocks with $r_1 r_2 = 0.03$ (shaded) and $r_1 r_2 = 0.30$ (unshaded). [Data from C. Tosi, *Adv. Polym. Sci.* 5:451 (1968).]

4.   Table 7.5 also shows that increasing the percentage of M_1 in the monomer solution flattens and broadens the distribution of sequence lengths. Similar results are observed for lower values of $r_1 r_2$, but the broadening is less pronounced when the tendency toward alternation is high.

Table 7.5 Fraction of M_1 Sequences of Length n_1 for Copolymers Prepared from Feedstocks of Different Composition for a System with $r_1 r_2 = 1$

$f_1 \times 100$	ν_1											
	1	2	3	4	5	6	7	8	9	10	11	12
10	90.00	9.00	0.90	0.09								
20	80.00	16.00	3.20	0.64	0.13							
30	70.00	21.00	6.30	1.89	0.57	0.17	0.05					
40	60.00	24.00	9.60	3.84	1.54	0.62	0.25	0.10	0.04			
50	50.00	25.00	12.50	6.25	3.13	1.56	0.78	0.39	0.20	0.10	0.05	
60	40.00	24.00	14.40	8.64	5.18	3.11	1.87	1.12	0.67	0.40	0.24	0.14
70	30.00	21.00	14.70	10.29	7.20	5.04	3.53	2.47	1.73	1.21	0.85	0.59
80	20.00	16.00	12.80	10.24	8.19	6.55	5.24	4.19	3.36	2.68	2.15	1.72
90	10.00	9.00	8.10	7.29	6.56	5.90	5.31	4.78	4.30	3.87	3.59	3.23

Source: C. Tosi, *Adv. Polym. Sci.* 5:451 (1968).

Next we consider the average value of the sequence length for M_1, $\bar{\nu}_1$. Combining Eqs. (1.9) and (7.36) gives

$$\bar{\nu}_1 = \frac{\sum_{\nu_1=1}^{\infty} \nu_1 \phi_{\nu_1}}{\sum_{\nu_1=1}^{\infty} \phi_{\nu_1}} = \frac{\sum_{\nu_1=1}^{\infty} \nu_1 p_{11}^{\nu_1-1} p_{12}}{\sum_{\nu_1=1}^{\infty} p_{11}^{\nu_1-1} p_{12}} \tag{7.38}$$

Simplifying this result involves the same infinite series that we examined in connection with Eq. (5.27); therefore we can write immediately

$$\bar{\nu}_1 = \frac{1}{1 - p_{11}} = \frac{1}{p_{12}} \tag{7.39}$$

By combining Eqs. (7.33) and (7.39), we obtain

$$\bar{\nu}_1 = 1 + r_1 \frac{[M_1]}{[M_2]} \tag{7.40}$$

A value for $\bar{\nu}_2$ is obtained by similar operations:

$$\bar{\nu}_2 = 1 + r_2 \frac{[M_2]}{[M_1]} \tag{7.41}$$

The following example demonstrates the use of some of these relationships pertaining to microstructure.

Example 7.4

The hemoglobin molecule contains four heme units. It is proposed to synthesize a hemin (molecule 1)-styrene (molecule 2) copolymer such that $\bar{\nu}_1 = 4$ in an attempt to test some theory concerning hemoglobin. As noted in Example 7.2, $r_1 = 65$ and $r_2 = 0.18$ for this system. What should be the proportion of monomers to obtain this average hemin sequence length? What is the average styrene sequence length at this composition? Does this system seem like a suitable model if the four hemin clusters are to be treated as isolated from one another in the theory being tested? Also evaluate ϕ_{ν_1} for several values of ν_1 bracketing $\bar{\nu}_1$ to get an idea of the distribution of these values.

Solution

Use Eq. (7.40) to evaluate $[M_1]/[M_2]$ for $r_1 = 65$ and $\bar{\nu}_1 = 4$:

$$\frac{[M_1]}{[M_2]} = \frac{\bar{\nu}_1 - 1}{r_1} = \frac{4 - 1}{65} = 0.046 \quad \text{and} \quad \frac{[M_2]}{[M_1]} = 21.7$$

Use this ratio of concentrations in Eq. (7.41) to evaluate $\bar{\nu}_2$:

$$\bar{\nu}_2 = 1 + r_2 \frac{[M_2]}{[M_1]} = 1 + 0.18\,(21.7) = 4.9$$

The number of styrene units in an average sequence is a little larger than the length of the average hemin sequence. It is not unreasonable to describe the hemin clusters as isolated, on the average, in this molecule. The product $r_1 r_2 = 11.7$ in this system, which also indicates a tendency toward block formation. Use Eq. (7.37) with $[M_1]/[M_2] = 0.046$ and the r_1 and r_2 values to evaluate ϕ_ν:

$$\phi_\nu = \left(\frac{65\,(0.046)}{65\,(0.046) + 1} \right)^{\nu-1} \left(\frac{1}{65\,(0.046) + 1} \right)$$

$$= \left(\frac{2.99}{3.99} \right)^{\nu-1} \left(\frac{1}{3.99} \right) = (0.749)^{\nu-1}\,(0.251)$$

Solving for several values of ν, we obtain the following:

ν	1	2	3	4	5	6
ϕ_ν	0.251	0.188	0.140	0.105	0.079	0.059

The distribution of sequence lengths is very broad.

•

For the systems represented in Fig. 7.3 and the equimolar case in Table 7.5, the average sequence lengths are $\bar{\nu}_1 = 1.173$ for $r_1 r_2 = 0.03$, $\bar{\nu}_1 = 1.548$ for $r_1 r_2 = 0.30$, and $\bar{\nu}_1 = 2.000$ for $r_1 r_2 = 1.0$.

Equations (7.40) and (7.41) suggest a second method, in addition to the copolymer composition equation, for the experimental determination of reactivity ratios. If the average sequence length can be determined for a feed-stock of known composition, then r_1 and r_2 can be evaluated. We shall return to this possibility in the next section. In anticipation of applying this idea, let us review the assumptions and limitation to which Eqs. (7.40) and (7.41) are subject:

1. The instantaneous monomer concentration must be used. Except at the azeotrope, this changes as the conversion of monomers to polymer progresses. As in Sec. 7.2, we assume that either the initial conditions apply (little change has taken place) or that monomers are continuously being added (replacement of reacted monomer).

2. The kinetic analysis described by Eqs. (7.32) and (7.33) assumes that no repeat unit in the radical other than the terminal unit influences the addition. The next-to-last unit in the radical as well as those still farther from the growing end are assumed to have no effect.

3. Item (2) requires that each event in the addition process be independent of all others. We have consistently assumed this throughout this chapter, beginning with the copolymer composition equation. Until now we have said nothing about testing this assumption. Consideration of copolymer sequence lengths offers this possibility.

We have suggested above that both the copolymer composition equation and the average sequence length offer possibilities for experimental evaluation of the reactivity ratios. Note that in so doing we are finding parameters which fit experimental results to the predictions of a model. Nothing about this tests the model itself. It could be argued that obtaining the same values for r_1 and r_2 from the fitting of composition and microstructure data would validate the model. It is not likely, however, that both types of data would be available and of sufficient quality to make this unambiguous. We shall examine the experimental side of this in the next section.

Statistical considerations make it possible to test the assumption of independent additions. Let us approach this topic by considering an easier problem: coin tossing. Under conditions where two events are purely random—as in tossing a fair coin—the probability of a specific sequence of outcomes is given by the product of the probabilities of the individual events. The probability of tossing a head followed by a head—indicated HH—is given by

$$p_{HH} = p_H p_H \tag{7.42}$$

If the events are not independent, provision must be made for this, so we define a quantity called the conditional probability. For the probability of a head *given the prior event* of a head, this is written $p_{H/H}$, where the first quantity in the subscript is the event under consideration and that following the slash mark is the prior condition. Thus $p_{T/H}$ is the probability of a tail following a head. If the events are independent, $p_{H/H} = p_H$; if not, then $p_{H/H}$ must be evaluated as a separate quantity. If the coin being tossed were biased, that is, if successive events are not independent, Eq. (7.42) becomes

$$p_{HH} = p_{H/H} p_H \tag{7.43}$$

We recall that the fraction of times a particular outcome occurs is used to estimate probabilities. Therefore we could evaluate $p_{H/H}$ by counting the number of times N_H the first toss yielded a head and the number of times N_{HH} two tosses yielded a head followed by a head and write

$$p_{H/H} = \frac{p_{HH}}{p_H} = \frac{N_{HH}}{N_H} \tag{7.44}$$

This procedure is readily extended to three tosses. For a fair coin the probability of three heads is the cube of the probability of tossing a single head:

$$p_{HHH} = p_H p_H p_H \tag{7.45}$$

If the coin is biased, conditional probabilities must be introduced:

$$p_{HHH} = p_{H/HH} p_{H/H} p_H \tag{7.46}$$

Using Eq. (7.44) to eliminate $p_{H/H}$ from the last result gives

$$p_{HHH} = p_{H/HH} \left(\frac{p_{HH}}{p_H} \right) p_H \tag{7.47}$$

or

$$p_{H/HH} = \frac{p_{HHH}}{p_{HH}} = \frac{N_{HHH}}{N_{HH}} \tag{7.48}$$

If we were testing whether a coin were biased or not, we would use ideas like these as the basis for a test. We could count, for example, HHH and HH sequences and divide them according to Eq. (7.48). If $p_{H/HH} \neq p_H$, we would be suspicious!

A similar logic can be applied to copolymers. The story is a bit more complicated to tell, so we only outline the method. If penultimate effects operate, then the probabilities p_{11}, p_{12}, and so on, defined by Eqs. (7.32)-(7.35) should be replaced by conditional probabilities. As a matter of fact, the kind of conditional probabilities needed must be based on the two preceding events. Thus reactions (7.E) and (7.F) are two of the appropriate reactions, and the corresponding probabilities are $p_{1/11}$ and $p_{1/21}$. Rather than work out all of the possibilities in detail, we summarize the penultimate model as follows:

1. A total of eight different reactions are involved, since each reaction like (7.A) is replaced by a pair of reactions like (7.E) and (7.F).
2. There are eight different rate laws and rate constants associated with these reactions. Equation (7.1), for example, is replaced by Eqs. (7.5) and (7.6).
3. The eight rate constants are clustered in four ratios which define new reactivity ratios. Thus r_1 as defined by Eq. (7.13) is replaced by $r_1' = k_{111}/k_{112}$ and $r_1'' = k_{211}/k_{212}$ while r_2 [Eq. (7.14)] is replaced by $r_2' = k_{222}/k_{221}$ and $r_2'' = k_{122}/k_{121}$.

4. The probability p_{11} as given by Eq. (7.32) is replaced by the conditional probability $p_{1/11}$, which is defined as

$$p_{1/11} = \frac{k_{111} [M_1 M_1 \cdot] [M_1]}{k_{111} [M_1 M_1 \cdot] [M_1] + k_{112} [M_1 M_1 \cdot] [M_2]} = \frac{r_1' [M_1] / [M_2]}{1 + r_1' [M_1] / [M_2]}$$

(7.49)

There are eight of these conditional probabilities, each associated with the reactions described in item (1).

5. The probability p_{11} can be written as the ratio $N_{M_1 M_1} / N_{M_1}$ using Eq. (7.44). This is replaced by $p_{1/11}$, which is given by the ratio $N_{M_1 M_1 M_1} / N_{M_1 M_1}$ according to Eq. (7.48).

6. Equation (7.32) shows that p_{11} is constant for a particular copolymer if the terminal model applies; therefore the ratio $N_{M_1 M_1} / N_{M_1}$ also equals this constant. Equation (7.49) shows that $p_{1/11}$ is constant for a particular copolymer if the penultimate model applies; therefore the ratio $N_{M_1 M_1 M_1} / N_{M_1 M_1}$ also equals this constant, but the ratio $N_{M_1 M_1} / N_{M_1}$ does not have the same value.

These observations suggest how the terminal mechanism can be proved to apply to a copolymerization reaction if experiments exist which permit the number of sequences of a particular length to be determined. If this is possible, we should count the number of M_1's (this is given by the copolymer composition) and the number of $M_1 M_1$ and $M_1 M_1 M_1$ sequences. Specified sequences, of any definite composition, of two units are called dyads; those of three units, triads; those of four units, tetrads; those of five units, pentads; and so on. Next we examine the ratio $N_{M_1 M_1} / N_{M_1}$ and $N_{M_1 M_1 M_1} / N_{M_1 M_1}$. If these are the same, then the mechanism is shown to have terminal control; if not, it *may* be penultimate control. To prove the penultimate model it would also be necessary to count the number of M_1 tetrads. If the tetrad/triad ratio were the same as the triad/dyad ratio, the penultimate model is proved.

This situation can be generalized. If the ratios do not become constant until the ratio of pentads to tetrads is considered, then the unit before the next to last—called the antepenultimate unit—plays a role in the addition. This situation has been observed for propylene oxide–maleic anhydride copolymers.

The foregoing discussion has been conducted in terms of M_1 sequences. Additional relationships of the sort we have been considering also exist for dyads, triads, and so forth, of different types of specific composition. Thus an ability to investigate microstructure experimentally allows some rather subtle mechanistic effects to be studied. In the next section we shall see how such information is obtained.

7.7 Copolymer Composition and Microstructure: Experimental Aspects

As we have already seen, it is the reactivity ratios of a particular copolymer system that determines both the composition and microstructure of the polymer. Thus it is important to have reliable values for these parameters. At the same time it suggests that experimental studies of composition and microstructure can be used to evaluate the various r's.

Evaluation of reactivity ratios from the copolymer composition equation requires only composition data—that is, analytical chemistry—and has been the method most widely used to evaluate r_1 and r_2. As noted in the last section, this method assumes terminal control and seeks the best fit of the data to that model. It offers no means for testing the model and, as we shall see, is subject to enough uncertainty to make even self-consistency difficult to achieve.

Microstructure studies, by contrast, offer both a means to evaluate the reactivity ratios and also to test the model. The capability to investigate this type of structural detail was virtually nonexistent until the advent of modern instrumentation and even now is limited to sequences of modest length.

In this section we shall use the evaluation of reactivity ratios as the unifying theme; the experimental methods constitute the new material introduced.

The copolymer composition equation relates the r's to either the ratio [Eq. (7.15)] or the mole fraction [Eq. (7.18)] of the monomers in the feedstock and repeat units in the copolymer. To use this equation to evaluate r_1 and r_2, the composition of a copolymer resulting from a feedstock of known composition must be measured. The composition of the feedstock itself must be known also, but we assume this poses no problems. The copolymer specimen must be obtained by proper sampling procedures, and purified of extraneous materials. Remember that monomers, initiators, and possibly solvents are involved in these reactions also, even though we have been focusing attention on the copolymer alone. The proportions of the two kinds of repeat unit in the copolymer is then determined by either chemical or physical methods. Elemental analysis has been the chemical method most widely used, although analysis for functional groups is also employed.

If we represent the percent by weight of an element in monomer i as $(\%)_i$ and the molecular weight of the monomer (repeat unit) by $M_{0,i}$, then the mass of that element in a sample is

$$\text{Element mass} = n_1 M_{0,1}(\%)_1 + n_2 M_{0,2}(\%)_2 \qquad (7.50)$$

The total mass of the sample is

$$\text{Sample mass} = n_1 M_{0,1} + n_2 M_{0,2} \qquad (7.51)$$

since n_i represents the total number of M_i units in the copolymer. Therefore the percent by weight of the element in the copolymer is

$$\text{Percent} = \frac{n_1 M_{0,1}(\%)_1 + n_2 M_{0,2}(\%)_2}{n_1 M_{0,1} + n_2 M_{0,2}} \times 100 \tag{7.52}$$

Dividing numerator and denomintor by $n_1 + n_2$ and remembering that $F_1 + F_2 = 1$ gives

$$\text{Percent} = \frac{F_1 M_{0,1}(\%)_1 + (1 - F_1) M_{0,2}(\%)_2}{F_1 M_{0,1} + (1 - F_1) M_{0,2}} \times 100 \tag{7.53}$$

which gives F_1 in terms of elemental composition and known parameters.

Since the copolymer equation involves both r_1 and r_2 as unknowns, at least two polymers prepared from different feedstocks must be analyzed. It is preferable to use more than this minimum number of observations and to rearrange the copolymer composition equation into a linear form so that graphical methods can be employed to evaluate the r's. Several ways to linearize the equation exist:

1. Rearrange Eq. (7.18) to give

$$\frac{f_1(1 - 2F_1)}{F_1(1 - f_1)} = r_1 \left(\frac{f_1^2 (F_1 - 1)}{F_1(1 - f_1)^2} \right) + r_2 \tag{7.54}$$

This is the equation of a straight line, so r_1 and r_2 can be evaluated from the slope and intercept of an appropriate plot.

2. In terms of ratios rather than fractions, Eq. (7.54) may be written as

$$\frac{[M_1]/[M_2]}{n_1/n_2} \left(\frac{n_1}{n_2} - 1 \right) = r_1 \frac{([M_1]/[M_2])^2}{n_1/n_2} - r_2 \tag{7.55}$$

This expression is also of the form $y = mx + b$ if $x = ([M_1][M_2]^{-1})^2 / (n_1/n_2)$ and $y = [M_1][M_2]^{-1}/(n_1/n_2)(n_1/n_2 - 1)$, so the slope and intercept yield r_1 and $-r_2$, respectively. This type of analysis is known as a Finemann–Ross plot.

3. This last expression can be rearranged in several additional ways which yield linear plots:

$$\frac{y}{x} = - r_2 \frac{1}{x} + r_1 \tag{7.56}$$

$$x = \frac{1}{r_1} y + \frac{r_2}{r_1} \tag{7.57}$$

$$\frac{x}{y} = \frac{r_1}{r_2} \frac{1}{y} + \frac{1}{r_1} \tag{7.58}$$

Each of these last forms weigh the errors in various data points differently, so some may be more suitable than others, depending on the precision of the data. Ideally all should yield the same values of the reactivity ratios.

The following example illustrates the use of Eq. (7.54) to evaluate r_1 and r_2.

Example 7.5

The data in Table 7.6 list the mole fraction of methyl acrylate in the feedstock and in the copolymer for the methyl acrylate (M_1)-vinyl chloride (M_2) system. Use Eq. (7.54) as the basis for the graphical determination of the reactivity ratios which describe this system.

Solution

We calculate the variables to be used as ordinate and abscissa for the data in Table 7.6 using Eq. (7.54):

Ordinate = $f_1(1 - 2F_1)/F_1(1 - f_1)$	Abscissa = $f_1{}^2(F_1 - 1)/F_1(1 - f_1)^2$
0.0217	-0.0083
-0.1036	-0.0143
-0.2087	-0.0316
-0.3832	-0.0486
-0.6127	-0.0832
-0.9668	-0.1315
-2.8102	-0.2792
-6.4061	-0.7349

Least-squares analysis of these values gives a slope r_1 = 8.929 and an intercept r_2 = 0.053. Figure 7.4 shows these data plotted according to Eq. (7.54). The line is drawn with the least-squares slope and intercept. The last point on the left in Fig. 7.4, which this line passes through, corresponds to F_1 = 0.983 and f_1 = 0.867. Because the functional form plotted involves the small differences $F_1 - 1$ and $1 - f_1$, this point is also subject to the largest error. This illustrates the value of having alternate methods for analyzing the data. The authors of this research carried out several different analyses of this same data; the values they obtained for r_1 and r_2 averaged over the various methods were r_1 = 9.616 ± 0.603 and r_2 = 0.0853 ± 0.0239. The standard deviations of about 6 and 28% in r_1 and r_2 analyzed *from the same data* indicate the hazards of this method for determining r values.

•

Table 7.6   Values of F_1 as a Function of f_1 for the Methyl Acrylate (M_1)–Vinyl Chloride (M_2) System (Data used in Example 7.5)

f_1	F_1	f_1	F_1
0.075	0.441	0.421	0.864
0.154	0.699	0.521	0.900
0.237	0.753	0.744	0.968
0.326	0.828	0.867	0.983

Source: Data from E. L. Chapin, G. Ham, and R. Fordyce, *J. Am. Chem. Soc.* 70:538 (1948).

In spite of the compounding of errors to which it is subject, the foregoing method was the best procedure for measuring reactivity ratios until the analysis of microstructure became feasible. Let us now consider this development.

Most of the experimental information concerning copolymer microstructure has been obtained by physical methods based on modern instrumental methods. Techniques such as ultraviolet (UV), visible, and infrared (IR) spectroscopy, NMR spectroscopy, and mass spectroscopy have all been used to good advantage in this type of research. Advances in instrumentation and computer interfacing combine to make these physical methods particularly suitable to answer the question we pose: With what frequency do particular sequences of repeat units occur in a copolymer.

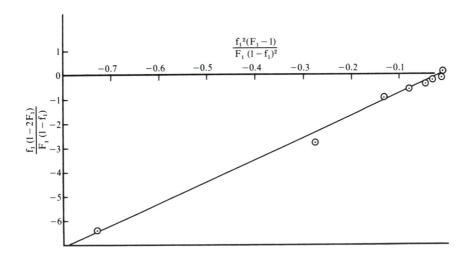

Figure 7.4   Finemann–Ross plot of the data in Table 7.6 and Example 7.5.

The choice of the best method for answering this question is governed by the specific nature of the system under investigation. Few general principles exist beyond the importance of analyzing a representative sample of suitable purity. Our approach is to consider some specific examples. In view of the diversity of physical methods available and the number of copolymer combinations which exist, a few examples barely touch the subject. They will suffice to illustrate the concepts involved, however.

Spectroscopic techniques based on the absorption of UV or visible radiation depend on the excitation of an electron from one quantum state to another. References in physical and/or analytical chemistry should be consulted for additional details, but the present summary is sufficient for our purposes:

1. The excitation energy ΔE measures the spacing between the final (subscript f) and initial (subscript i) quantum states of the electron:

$$\Delta E = E_f - E_i \qquad (7.59)$$

This difference is positive for absorbed energy.

2. The energy absorbed is proportional to the frequency of the radiation, with Planck's constant ($h = 6.63 \times 10^{-34}$ J sec) the factor of proportionality:

$$\Delta E = h\nu = h\frac{c}{\lambda} \qquad (7.60)$$

In the second version of this equation c is the speed of light, and λ the wavelength of the radiation.

3. The more widely separated two states are in energy, the shorter the wavelength of the radiation absorbed. Since vibrational quantum states are more closely spaced than electronic states, the former transitions are induced by IR radiation.

4. Different light-absorbing groups, called chromophores, absorb characteristic wavelengths, opening the possibility of qualitative analysis based on the location of an absorption peak.

5. If there is no band overlap in a spectrum, the absorbance at a characteristic wavelength is proportional to the concentration of chromophores present. This is the basis of quantitative analysis using spectra. With band overlap, things are more complicated but still possible.

6. The proportionality between the concentration of chromophores and the measured absorbance [Eqs. (6.8) and (6.9)] requires calibration. With copolymers this is accomplished by chemical analysis for an element or functional group that characterizes the chromophore, or, better yet, by the use of isotopically labeled monomers.

An elegant example of a system investigated by UV–visible spectroscopy is the copolymer of styrene (molecule 1) and 1-chloro-1,3-butadiene (molecule 2). These molecules quantitatively degrade with the loss of HCl upon heating in base solution. This restores 1,3-unsaturation to the butadiene repeat unit:

$$-(CHCl-CH=CH-CH_2)_n-CH_2-\underset{\phi}{CH}- \xrightarrow{-nHCl} -(CH=CH-CH=CH)_n-CH_2-\underset{\phi}{CH}-$$

$$(7.N)$$

It is these conjugated double bonds that are the chromophores of interest in this system. What makes this particularly useful is the fact that the absorption maximum for this chromophore is displaced to longer wavelengths the more conjugated bonds there are in a sequence. Qualitatively, this can be understood in terms of a one-dimensional particle-in-a-box model for which the energy level spacing is inversely proportional to the square of the length of the box. In this case the latter increases with the length of the conjugated polyene system. This in turn depends on the number of consecutive butadiene repeat units in the copolymer. For an isolated butadiene molecule dehydrohalogenation produces one pair of conjugated double bonds; two adjacent butadienes, four conjugated double bonds; three adjacent butadienes, six conjugated double bonds; and so on. Sequences of these increasing lengths are expected to absorb at progressively longer wavelengths.

The spectrum shown in Fig. 7.5 shows the appropriate portion of the spectrum for a copolymer prepared from a feedstock for which $f_1 = 0.153$. It turns out that each polyene produces a set of three bands: The dyad is identified with the peaks at $\lambda = 298$, 312, and 327 nm; the triad, with $\lambda = 347$, 367, and 388 nm; and the tetrad with $\lambda = 412$ and 437 nm. Apparently one of the tetrad bands overlaps that of the triad and is not resolved. Likewise, only one band (at 473 nm) is observed for the pentad. The identification of these features can be confirmed with model compounds and the location and *relative* intensities of the peaks has been shown to be independent of copolymer composition.

Once these features have been identified, the spectra can be interpreted in terms of the numbers of dyads, triads, tetrads, and pentads (?) of the butadiene units and compared with predicted sequences of various lengths. Further consideration of this system is left for Problems 4–6 at the end of the chapter.

Nuclear magnetic resonance (NMR) spectroscopy is another physical technique which is especially useful for microstructure studies. Because of the sensitivity of this technique to an atom's environment in a molecule, NMR is useful for a variety of microstructural investigations: We shall consider the application to copolymers now and to questions of stereoregularity in Sec. 7.11.

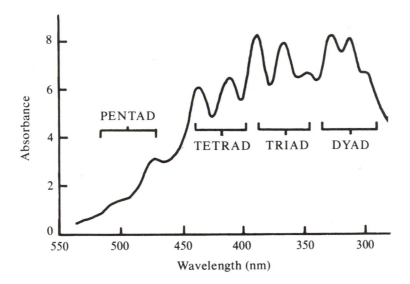

Figure 7.5 Ultraviolet–visible spectrum of dehydrohalogenated copolymers of styrene-1-chloro-1,3-butadiene. [Redrawn with permission from A. Winston and P. Wichacheewa, *Macromolecules* 6:200 (1973), copyright 1973 by the American Chemical Society.]

Nuclear magnetic resonance has become such an importnat technique in organic chemistry that contemporary textbooks in the subject discuss its principles quite thoroughly, as do texts in physical and analytical chemistry. We note only a few pertinent highlights of the method:

1. Certain nuclei—we shall only discuss ^1H, but ^{13}C is becoming increasingly important—possess magnetic moments and show two quantum states in a magnetic field.

2. If energy of the proper frequency is supplied, a transition between these quantum states occurs with the absorption of an amount of energy equal to the separation of the states. The frequency of the absorbed radiation lies in the radio-frequency range and depends on the local magnetic field at the atom in question.

3. In practice, NMR spectrometers vary the magnetic field strength and measure the value at which a constant radio frequency is absorbed.

4. The electrons in a molecule also have magnetic moments and set up secondary magnetic fields which partly screen each atom from the applied field. Thus atoms in different chemical environments display resonance at slightly different magnetic fields.

5. The displacement δ of individual resonances from that of a standard are small and are measured in parts per million (ppm) relative to the applied field. These chemical shifts are characteristic of a proton in a specific environment.

6. The interaction between nuclei splits resonances into multiple peaks, the number and relative intensity of which also assist in qualitative identification of the proton responsible for the absorption.

7. Splitting is most commonly caused by the interaction of protons on adjacent carbons with the proton of interest. If there are m equivalent hydrogens on an adjacent carbon, the proton of interest produces m + 1 peaks by this coupling.

8. More distant coupling is revealed in high magnetic fields. Unresolved fine structures in a field of one strength may be resolved at higher fields where more subtle long-range influences can be probed.

The use of NMR spectroscopy to characterize copolymer microstructure takes advantage of this last ability to discern environmental effects which extend over the length of several repeat units. This capability is extremely valuable in analyzing the stereoregularity of a polymer, and we shall have more to say about it in that context in Sec. 7.11.

For the case of copolymers, suppose we consider the various triads of repeat units. There are six possibilities: $M_1M_1M_1$, $M_1M_1M_2$, $M_2M_1M_2$, $M_2M_2M_2$, $M_2M_2M_1$, and $M_1M_2M_1$. These can be divided into two groups of three, depending on the identity of the central unit. Thus the center of a triad can be bracketed by two monomers identical to itself, different from itself, or by one of each. In each of these cases the central repeat unit is in a different environment, and a characteristic proton in that repeat unit will resonate at a different location, depending on the effect of that environment.

As a specific example, consider the ester methoxy group in poly(methyl methacrylate). The hydrogens in this group are magnetically equivalent and hence produce a single resonance at $\delta = 3.74$ ppm. Now suppose we look for this same resonance feature in the copolymer of methyl methacrylate (M_1) and acrylonitrile (M_2). Figure 7.6 shows the 60-MHz spectrum of several of these copolymers in the neighborhood of the methoxy resonance. Three resonance peaks rather than one are observed. Figure 7.6 also lists the methyl methacrylate content of each of these polymers. As the methyl methacrylate content decreases, the peak on the right decreases and that on the left increases. We therefore identify the right-hand peak with the $M_1M_1M_1$ sequence, the left-hand peak to $M_2M_1M_2$, and the peak in the center to $M_1M_1M_2$. The $M_1M_1M_1$ peak occurs at the same location as in the methyl methacrylate homopolymer. The areas under the three peaks give the relative proportions of

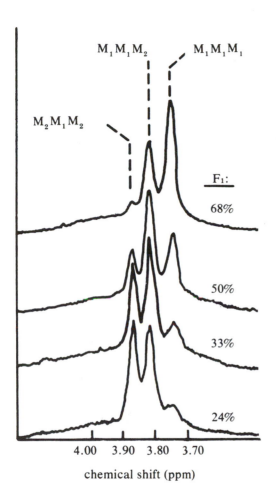

Figure 7.6 Chemical shift (from hexamethyldisiloxane) for acrylonitrile–methyl methacrylate copolymers of the indicated methyl methacylate (M_1) content. Methoxyl resonances are labeled as to the triad source. [From R. Chujo, H. Ubara, and A. Nishioka, *Polym. J.* 3:670 (1972).]

the three sequences. In the following example we consider some results on dyad sequences determined by comparable procedures in vinylidene chloride–isobutylene copolymers.

Example 7.6

The mole fractions of various dyads in the vinylidine chloride (M_1)–isobutylene (M_2) system were determined† by NMR spectroscopy. A selection of the values obtained are listed below, as well as the compositions of the feedstocks from which the copolymers were prepared:

	Mole fraction of dyads		
f_1	11	12	22
0.584	0.68	0.29	–
0.505	0.61	0.36	–
0.471	0.59	0.38	–
0.130	–	0.67	0.08
0.121	–	0.66	0.10
0.083	–	0.64	0.17

Assuming terminal control, evaluate r_1 from each of the first three sets of data, and r_2 from each of the last three.

Solution

Equations (7.32) and (7.34) provide the method for evaluating the r's from the data given. We recognize that a 12 dyad can come about from 1 adding to 2 as well as from 2 adding to 1; therefore we use half the number of 12 dyads as a measure of the number of additions of monomer 2 to chain end 1. Accordingly, by Eq. (7.30),

$$p_{11} = \frac{N_{11}}{N_{11} + (1/2)\,N_{12}} = \frac{2N_{11}}{2N_{11} + N_{12}} \quad \text{and} \quad p_{22} = \frac{2N_{22}}{2N_{22} + N_{12}}$$

Since $[M_1]/[M_2] = f_1/(1 - f_1)$, Eq. (7.31) can be written

$$p_{11} = \frac{r_1\,[f_1/(1 - f_1)]}{1 + r_1\,[f_1/(1 - f_1)]} \quad \text{and} \quad p_{22} = \frac{r_2\,[(1 - f_1)/f_1]}{1 + r_2\,[(1 - f_1)/f_1]}$$

---

† J. B. Kinsinger, T. Fischer, and C. W. Wilson, *Polym. Lett.* 5:285 (1967).

From		From	
$$\dfrac{2N_{11}}{2N_{11} + N_{12}} = \dfrac{r_1\,[f_1/(1-f_1)]}{1 + r_1\,[f_1/(1-f_1)]}$$		$$\dfrac{2N_{22}}{2N_{22} + N_{12}} = \dfrac{r_2\,[(1-f_1)/f_1]}{1 + r_2\,[(1-f_1)/f_1]}$$	
f_1	r_1	f_1	r_2
0.584	3.33	0.130	0.036
0.505	3.32	0.121	0.042
0.471	3.48	0.083	0.048
Average	3.38	Average	0.042

Particularly when r values are close to zero, this method for evaluating small r's is superior to the graphical analysis of composition data (compare Example 7.5 and Fig. 7.4).

By making measurements at higher magnetic fields, it is possible to resolve spectral features arising from still longer sequences. As a matter of fact, the authors of the research described in the last example were able to measure the fractions of tetrads of different composition in the same vinylidene chloride-isobutylene copolymer. Based on the longer sequences, they concluded that the penultimate model describes this system better than the terminal model, although the shortcomings of the latter are not evident in the example. Problems 7 and 8 at the end of the chapter also refer to this system.

Before examining NMR applications to problems of stereoregularity, let us conclude the discussion of copolymers by considering some copolymer applications.

7.8 Applications of Copolymers

Any list of copolymer applications would be as extensive as the applications of polymers in general. We shall only consider a few items in which the two-component nature of the copolymer plays an explicit role in determining the properties of the polymer. In addition, we shall examine several additional concerns which come up when applications technology is considered.

The properties of a copolymer can be viewed as hybrids of the properties of the separate homopolymers. Because of this, a good deal of refinement can be introduced into these properties by the use of copolymers. The situation is analogous to the use of pure liquids or binary solutions as solvents. The number of binary combinations, $n(n-1)/2$ as noted above, greatly exceeds the number of pure liquids, and any one of these combinations can be prepared over a range of compositions. Just as mixed solvents offer a wider range of properties than

pure liquids in, say, developing a chromatogram, so it is with copolymers. The analogy extends a bit further. Problems of limited miscibility can make certain solution compositions unattainable. Likewise, some copolymer compositions are not achievable because of the reactivity ratios for a system.

Even the limitations imposed by reactivity ratios—while real enough—are not as severe as our presentation suggests. The reason for this is that the same pair of comonomers may have very different r values when polymerized via anionic or cationic intermediates rather than free radicals. We have elected to concentrate on the latter, but the former may also be used to synthesize copolymers in some instances. Figure 7.7 illustrates this situation for the styrene (monomer 1)-methyl methacrylate system, showing how the copolymer composition F_1 varies with feed composition f_1 for the polymer produced by the three different mechanisms. This indicates that an equimolar feedstock produces a copolymer which is mostly styrene in the cationic case, mostly methyl methacrylate in the anionic case, and roughly equimolar in the free-radical case. As we saw in Chap. 6, ionic polymerizations tend to be highly selective, so the extra flexibility this adds to copolymer applications is somewhat limited. Another alternative is three-component systems; we shall not pursue this possibility.

There are a number of physical properties of copolymers that are directly related to the multicomponent character of these materials and which are

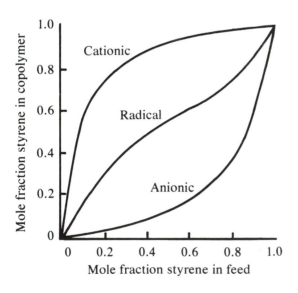

Figure 7.7 F_1 versus f_1 for styrene (M_1)-methyl methacrylate (M_2) copolymers prepared by the mechanisms indicated. [From D. C. Pepper, *Q. Review London* 8:88 (1954).]

utilized in various application situations. Solubility is perhaps the most easily visualized of these because of the familiar dictum: Like dissolves like. Thus copolymers may vary in their solubility in certain solvents, depending on their composition. For example, styrene–methyl methacrylate copolymers show increases in solubility in ester solvents as the methyl methacrylate content increases. This consideration is not merely of interest in dissolving polymers, but also affects such things as wet strength, swelling, and plasticizer compatibility.

In Chap. 4 we discussed the crystallizability of polymers and the importance of this property on the mechanical behavior of the bulk sample. Following the logic that leads to Eq. (4.17), the presence of a comonomer lowers T_m for a polymer. Carrying this further, we can compare a copolymer to an alloy in which each component lowers the melting point of the other until a minimum-melting eutectic is produced. Similar trends exist in copolymers.

Copolymers can be used to introduce a mixture of chemical functionalities into a polymer. Acidic and basic substituents can be introduced, for example, through comonomers like acrylic acid and vinyl pyridine. The resulting co-polymers show interesting amphoteric behavior, reversing their charge in solution with changes of pH.

Finally, the dielectric properties of a nonpolar polymer are modified by inclusion of even small amounts of a polar comonomer. In coatings applications the presence of polar repeat units in an otherwise nonpolar polymer reduces the tendency for static buildup during manufacture, printing, and ultimate use. On the other hand, in dielectric applications this increases the power loss and must be kept to a minimum, even to the exclusion of polar initiator fragments.

Table 7.7 lists the common names and the comonomers for several addition copolymers that are widely used as elastomers, fibers, or films.

Table 7.7   A Few Common Copolymers and Their Constituents

Common name or acronym for copolymer	Comonomers
ABS	Acrylonitrile, butadiene, styrene
Acrylan	Acrylonitrile, vinyl acetate
Buna N	Butadiene, acrylonitrile
Buna S (SBR)	Styrene, butadiene
Butyl rubber	Isoprene, isobutylene
SAN	Styrene, acrylonitrile
Saran	Vinyl chloride, vinylidene chloride
Vinoflax	Vinyl chloride, vinyl isobutyl ether
Vinylite	Vinyl chloride, vinyl acetate
Vinyon	Vinyl chloride, acrylonitrile

In any application of a copolymer the rate of formation of the product, its molecular weight, and the uniformity of its composition during manufacture are also important considerations. While the composition of a copolymer depends only on the relative rates of the various propagation steps, the rate of formation and the molecular weight depend on the initiation and termination rates as well. We shall not discuss these points in any detail, but merely indicate that the situation parallels the presentation of these items for homopolymers as given in Chap. 6. The following can be shown:

1. Termination steps can involve either similar or dissimilar radicals:

$$R_t = k_{t,11}[M_1\cdot]^2 + k_{t,12}[M_1\cdot][M_2\cdot] + k_{t,22}[M_2\cdot]^2 \qquad (7.61)$$

2. Under stationary-state conditions $R_i = R_t$, and with the use of Eqs. (6.5), (7.10), and (7.61), the stationary-state concentration of the various radicals can be found.

3. For copolymers, $R_p \propto [M]\sqrt{[I]}$ in analogy with Eq. (6.26), and $\bar{n}_n \propto [M]/\sqrt{[I]}$ in analogy with Eq. (6.37). In the copolymer case, however, $[M] = [M_1] + [M_2]$, and the proportionality "constants" are functions of the comonomer concentration ratio $[M_1]/[M_2]$, as well as the various kinetic constants.

Throughout this chapter we have assumed low conversions or monomer replacement, so that variation of the feed composition with the reaction could be neglected. Industrially, the former is impractical and the latter may be undesirable because of the added cost it entails. Figure 7.8 illustrates how the average composition of the copolymer drifts with the extent of reaction for the styrene–methyl methacrylate system. Although the instantaneous value of F follows f to either zero or unity, the average composition is seen to hold up better. For this system the reactivity ratios are very similar, so this is not a very stringent test of variability. In the context of applications, however, it raises the question as to just how much variability can be tolerated in a copolymer without deleterious effects on the properties of the product. Such considerations may reveal that the replacement of monomer through continuous addition is worth the added cost.

There are numerous other aspects of copolymer chemistry that might be taken up which we pass over because of space limitations. These include the following:

1. Systems of more than two components.
2. Complications arising from other types of chain isomerism, such as those mentioned in Sec. 1.6.
3. Copolymerization by ionic mechanisms.
4. Divinyl monomers and crosslinking.
5. Copolymers by step-growth processes.

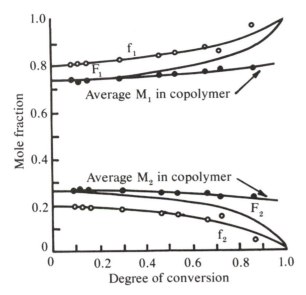

Figure 7.8 Mole fractions styrene (M_1) and methyl methacrylate (M_2) in feedstock (f) and copolymers (F) as a function of the extent of polymerization. Average copolymer compositions are also shown. [From V. E. Meyer and R. K. S. Chan, *Polym. Prepr.* 8:209(1967), used with permission.]

Instead of devoting more space to copolymers, we turn next to stereoregular polymers, in which many of the descriptions of microstructure developed in Sec. 7.6 can also find application.

7.9 Stereoregular Polymers

We introduced the concept of stereoregularity in Sec. 1.6. Figure 1.2 illustrates isotactic, syndiotactic, and atactic structures of a vinyl polymer in which successive repeat units along the fully extended chain lie, respectively, on the same side, alternating sides, or at random with respect to the backbone. It is important to appreciate the fact that these different structures—different configurations—have their origin in the bonding of the polymer, and no amount of rotation around bonds—changes in conformation—will convert one structure into another.

Our discussion of stereoregularity in this chapter is primarily concerned with polymers of monosubstituted ethylene repeat units. We shall represent these by

In this representation the X indicates the substituent; other bonds involve only hydrogens. This formalism also applies to 1,1-disubstituted ethylenes in which the substituents are different. With these symbols, the isotactic, syndiotactic, and atactic structures shown in Fig. 1.2 are represented by structures [VI] – [VIII], respectively:

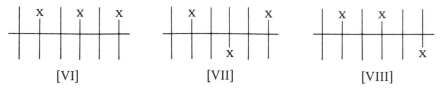

[VI]   [VII]   [VIII]

The carbon atoms carrying the substituents are not truly asymmetric, since the two chain sections—while generally of different length—are locally the same on either side of any carbon atom, except near the ends of the chain. As usual, we ignore any uniqueness associated with chain ends.

There are several topics pertaining to stereoregularity which we shall not cover to simplify the presentation:

1. Stereoregular copolymers. We shall restrict our discussion to stereoregular homopolymers.
2. Complications arising from other types of isomerism. Positional and geometrical isomerism, also described in Sec. 1.6, will be excluded for simplicity. In actual polymers these are not always so easily ignored.
3. Polymerization of 1,2-disubstituted ethylenes. Since these introduce two different "asymmetric" carbons into the polymer backbone (second substituent Y), they have the potential to display ditacticity. Our attention to these is limited to the illustration of some terminology which is derived from carbohydrate nomenclature (structures [IX]-[XII]):

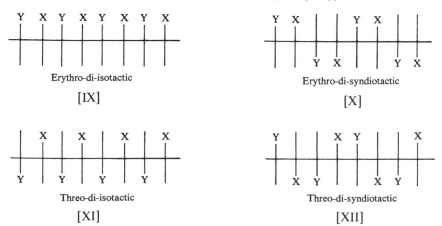

Erythro-di-isotactic   Erythro-di-syndiotactic

[IX]   [X]

Threo-di-isotactic   Threo-di-syndiotactic

[XI]   [XII]

The successive repeat units in strucutres [VI]-[VIII] are of two different kinds. If they were labeled M_1 and M_2, we would find that, as far as microstructure is concerned, isotactic polymers are formally the same as homopolymers, syndiotactic polymers are formally the same as alternating copolymers, and atactic polymers are formally the same as random copolymers. The analog of block copolymers, stereoblock polymers, also exist. Instead of using M_1 and M_2 to differentiate between the two kinds of repeat units, we shall use the letters D and L as we did in Chap. 1.

The statistical nature of polymers and polymerization reactions has been illustrated at many points throughout this volume. It continues to be important in the discussion of stereoregularity. Thus it is generally more accurate to describe a polymer as, say, predominately isotactic rather than perfectly isotactic. More quantitatively, we need to be able to describe a polymer in terms of the percentages of isotactic, syndiotactic, and atactic sequences.

Certain bulk properties of polymers also reflect differences in stereoregularity. We saw in Chap. 4 that crystallinity is virtually impossible unless a high degree of stereoregularity is present in a polymer. Since crystallinity plays such an important part in determining the mechanical properties of polymers, stereoregularity manifests itself in these other behaviors also. These gross, bulk properties provide qualitative evidence for differences in stereoregularity, but, as with copolymers, it is the microstructural detail that quantitatively characterizes the tacticity of a polymer. We shall examine the statistics of this situation in the next section, and the application of NMR to the problem in Sec. 7.11.

The comparison between stereoregular polymers and copolymers can be extended still further. We can write chemical equations for propagation reactions leading to products which differ in configuration along with the associated rate laws. We do this without specifying anything—at least for now—about the mechanism. There are several things that need to be defined to do this:

1. These are addition polymerizations in which chain growth is propagated through an active center. The latter could be a free radical or an ion; we shall see that coordinate intermediates is the more usual case.
2. The active-center chain end is open to front or rear attack in general; hence the configuration of a repeat unit is not fixed until the next unit attaches to the growing chain.
3. The reactivity of a growing chain is assumed to be independent of chain length. In representing this schematically, we shall designate active species, of whatever chain length, as either DM* or LM*. The M* indicates the terminal active center, and the D or L, the penultimate units of fixed configuration. From a kinetic point of view, we ignore what lies further back along the chain.
4. As in Chap. 6, the monomer is represented by M.

With these definitions in mind, we can write

$$-DM^* + M \left\langle \begin{array}{c} DDM^* \\ DLM^* \end{array} \right.$$

or   (7.0)

$$-LM^* + M \left\langle \begin{array}{c} LLM^* \\ LDM^* \end{array} \right.$$

What is significant about these reactions is that only two possibilities exist: addition with the same configuration ($D \rightarrow DD$ or $L \rightarrow LL$) or addition with the opposite configuration ($D \rightarrow DL$ or $L \rightarrow LD$). We shall designate these isotactic (subscript i) or syndiotactic (subscript s) additions, respectively, and shall define the rate constants for the two steps k_i and k_s. Therefore the rates of isotactic and syndiotactic propagation become

$$R_{p,i} = k_i [M^*] [M] \tag{7.62}$$

and

$$R_{p,s} = k_s [M^*] [M] \tag{7.63}$$

and, since the concentration dependencies are identical, the relative rates of the two processes is given by the ratio of the rate constants. This same ratio also gives the relative number of dyads having the same or different configurations:

$$\frac{R_{p,i}}{R_{p,s}} = \frac{k_i}{k_s} = \frac{\text{number dyads with same configuration}}{\text{number dyads with different configurations}} \tag{7.64}$$

The Arrhenius equation enables us to expand on this still further:

$$\frac{\text{iso dyads}}{\text{syndio dyads}} = \frac{A_i}{A_s} e^{-(E_i^* - E_s^*)/RT} \tag{7.65}$$

The main conclusion we wish to draw from this line of development is that the difference between E_i^* and E_s^* could vary widely, depending on the nature of the active center.

If the active center in a polymerization is a free radical unencumbered by interaction with any surrounding species, we would expect $E_i^* - E_s^*$ to be small.

Experiment confirms this expectation; for vinyl chloride it is on the order of 1.3 kJ mol^{-1}. (This is actually $\Delta H_i^* - \Delta H_s^*$ derived from an analysis based on the Eyring method: $k_i/k_s = e^{(\Delta S_i^* - \Delta S_s^*)/R} \; e^{-(\Delta H_i^* - \Delta H_s^*)/RT}$.) Thus at the temperatures usually encountered in free-radical polymerizations ($\sim 60°C$), the exponential in Eq. (7.65) is small and the proportions of isotactic and syndiotactic dyads are roughly equal. This is the case for polyvinyl chloride, for which $k_i/k_s = 0.63$ at 60°C. The preference for syndiotactic additions is greater than this (that is, $E_i^* - E_s^*$ is larger) in some systems, apparently because there is less repulsion between substituents when they are staggered in the transition state. In all cases, whatever difference in activation energies exists manifests itself in product composition to a greater extent at low temperatures. At high temperatures small differences in E* value are leveled out by the high average thermal energy available.

The foregoing remarks refer explicitly to free-radical polymerization. If the active center is some kind of associated species—an ion pair or a coordination complex—then predictions based on unencumbered intermediates are irrelevant. It turns out that the Ziegler-Natta catalysts—which won their discoverers the Nobel Prize—apparently operate in this way. The active center of the chain coordinates with the catalyst in such a way as to block one mode of addition. High levels of stereoregularity are achieved in this case. Although these substances also initiate the polymerization, the term *catalyst* is especially appropriate in the present context, since the activation energy for one mode of addition is dramatically altered relative to the other by these materials. We shall discuss the chemical makeup of Ziegler-Natta catalysts and some ideas about how they work in Sec. 7.12. For now it is sufficient to recognize that these catalysts introduce a real bias into Eq. (7.65) and thereby favor one pattern of addition.

In the next section we shall take up the statistical description of various possible sequences.

7.10 A Statistical Description of Stereoregularity

Since it is unlikely that a polymer will possess perfect stereoregularity, it is desirable to quantitatively assess this property both to describe the polymer and to evaluate the effectiveness of various catalysts in this regard. In discussing tacticity in terms of microstructure, it has become conventional to designate a dyad as meso if the repeat units have the same configuration, and as racemic if the configuration is reversed. This terminology is derived from the stereochemistry of small molecules; its basis is seen by focusing attention on the methylene group in the backbone of the vinyl polymer. This methylene lies in a plane of symmetry in the isotactic molecule [XIII],

$$
\begin{array}{ccc}
\text{X} & \text{H} & \text{X} \\
| & | & | \\
-\!\!\!\!- & \text{C} & -\!\!\!\!- \\
| & | & | \\
 & \text{H} &
\end{array}
$$

[XIII]

and thereby defines a meso (subscript m) structure as far as the dyad is concerned. Considering only the dyad, we see that these two methylene protons are in different environments. Therefore each will show a different chemical shift in an NMR spectrum. In addition, each proton splits the resonance of the other into a doublet, so a quartet of peaks appears in the spectrum. Still considering only the dyad, we see that the methylene is a syndiotactic grouping [XIV] contains two protons in identical environments:

$$
\begin{array}{ccc}
\text{X} & \text{H} & \\
| & | & | \\
-\!\!\!\!- & \text{C} & -\!\!\!\!- \\
| & | & | \\
 & \text{H} & \text{X}
\end{array}
$$

[XIV]

These protons show a single chemical shift in the NMR spectrum. This is called a racemic (subscript r) structure, since it contains equal amounts of D and L character. In the next section we shall discuss the NMR spectra of stereoregular polymers in more detail.

If we define p_m and p_r as the probability of addition occurring in the meso and racemic modes, respectively, then $p_m + p_r = 1$, since there are only two possibilities. The probability p_m is the analog of p_{ij} for copolymers; hence, by analogy with Eq. (7.30), this equals the fraction of isotactic dyads among all dyads. In terms of the kinetic approach of the last section, p_m is equal to the rate of an iso addition divided by the combined rates of iso and syndio additions:

$$
p_m = \frac{k_i}{k_i + k_s} \tag{7.66}
$$

This expression is the equivalent of Eq. (7.31) for copolymers.

The system of notation we have defined can readily be extended to sequences of greater length. Table 7.8 illustrates how either m or r dyads can be bracketed

Table 7.8 The Splitting of Meso and Racemic Dyads into Six Tetrads

Dyad	Triad

by two additional repeat units to form a tetrad. Each of the outer units is either m or r with respect to the unit it is attached to, so the meso dyad generates three tetrads. Note that the tetrads mmr and rmm are equivalent and are not distinguished. A similar set of tetrads is generated from the r dyad.

The same system of notation can be extended further by focusing attention on the backbone substituents rather than on the methylenes. Consider bracketing a center *substituent* with a pair of monomers in which the substituents have either the same or opposite configurations as the central substituent. Thus the two bracketing units are either m or r with respect to the central unit and the probabilities of the resulting triads are obtained from the probabilities of the respective m or r additions. The following possibilities exist:

1. An isotactic triad [XV] is generated by two successive meso additions:

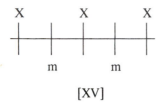

[XV]

The probability of the isotactic triad is

$$p_i = p_m{}^2 \tag{7.67}$$

2. A syndiotactic triad [XVI] is generated by two successive racemic additions:

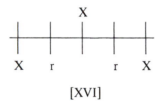

[XVI]

The probability of the syndiotactic triad is given by $p_r{}^2$, which becomes

$$p_s = (1 - p_m)^2 \tag{7.68}$$

since $p_r = 1 - p_m$.

3. A heterotactic traid [XVII] is generated by mr and rm sequences of additions:

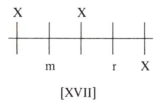

[XVII]

The probability of a heterotactic (subscript h) triad is

$$p_h = 2p_m(1 - p_m) \tag{7.69}$$

The factor 2 comes about because this particular sequence can be generated in two different orders.

These triads can also be bracketed by two more units to generate 10 different pentads following the pattern established in Table 7.8. It is left for the reader to verify this number by generating the various structures.

The probabilities of the various dyad, triad, and other sequences that we have examined have all been described by a single probability parameter p_m. When we used the same kind of statistics for copolymers, we called the situation one of terminal control. We are considering similar statistics here, but the idea that the stereochemistry is controlled by the terminal unit is inappropriate. The active center of the chain end governs the *chemistry* of the addition, but not the *stereochemistry*. Neither the terminal unit nor any other repeat unit considered alone has any stereochemistry. Equations (7.62) and (7.63) merely state that an addition must be of one kind or another, but that the rates are not necessarily identical.

A mechanism in which the stereochemistry of the growing chain does exert an influence on the addition might exist, but at least two repeat units in the chain are required to define any such stereochemistry. Therefore this possibility is equivalent to the penultimate mechanism in copolymers. In this case the addition would be described in terms of conditional probabilities, just as Eq. (7.49) does for copolymers. Thus the probability of an isotactic triad controlled by the stereochemistry of the growing chain would be represented by the reaction

$$(7.\text{P})$$

and described by the probability

$$P_{control} = p_m p_{m/m} \qquad (7.70)$$

where $p_{m/m}$, a conditional probability, is the probability of an m addition, given the fact of a prior m addition. As with copolymers, triads must be considered in order to *test* whether the simple statistics apply. Still longer sequences need to be examined to test whether stereochemical contol is exerted by the chain. Although such situations are known, we shall limit our discussion to the simple case where the single probability p_m is sufficient to describe the various additions. The latter, incidentally, may be called zero-order Markov (or Bernoulli) statistics to avoid the vocabulary of terminal control. The case where the addition is influenced by whether the last linkage in the chain is m or r is said to follow a first-order Markov process.

The number of m or r linkages in an "n-ad" is $n - 1$. Thus dyads are characterized by a single linkage (either m or r), triads by two linkages (either mm,

mr, or rr), and so forth. The m and r notation thus reduces by 1 the order of the description from what is obtained when the repeat units themselves are described. For this reason the terminal control mechanism for copolymers is a first-order Markov process and the penultimate model is a second-order Markov process. Note that the compound probabilities which describe the probability of an n-ad in terms of p_m are also of order $n - 1$. In the following example we calculate the probability of various triads on the basis of zero-order Markov statistics.

Example 7.7

Use zero-order Markov statistics to evaluate the probability of isotactic, syndiotactic, and heterotactic triads for the series of p_m values spaced at intervals of 0.1. Plot and comment on the results.

Solution

Evaluate Eqs. (7.67)-(7.69) for p_m between zero and unity:

p_m	$p_m{}^2$	$(1 - p_m)^2$	$2p_m(1 - p_m)$
0	0	1	0
0.1	0.01	0.81	0.18
0.2	0.04	0.64	0.32
0.3	0.09	0.49	0.42
0.4	0.16	0.36	0.48
0.5	0.25	0.25	0.50
0.6	0.36	0.16	0.48
0.7	0.49	0.09	0.42
0.8	0.64	0.04	0.32
0.9	0.81	0.01	0.18
1.0	1.0	0	0

These results are plotted in Fig. 7.9. The following observations can be made from these calculations:

1. The probabilities give the fractions of the three different types of triads in the polymer.
2. If the fractions of triads could be measured, they either would or would not lie on a single vertical line in Fig. 7.9. If they did occur at a single value of p_m, this would not only give the value of p_m (which could be obtained from the fraction of one kind of triad), but would also prove the statistics assumed. If the fractions were not consistent with a single p_m value, higher-order Markov statistics are indicated.

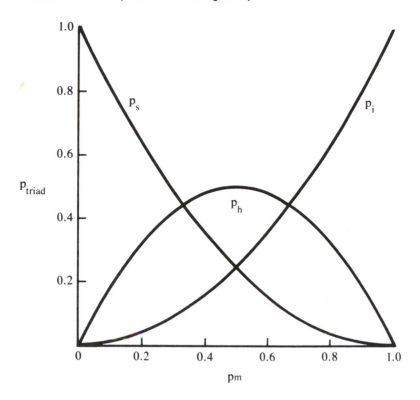

Figure 7.9 Fractions of iso, syndio, and hetero triads as a function of p_m, calculated assuming zero-order Markov (Bernoulli) statistics in Example 7.7.

3. The fraction of isotactic sequences increases as p_m increases, as required by the definition of these quantities.
4. The fraction of syndiotactic sequences increases as $p_m \to 0$, which corresponds to $p_r \to 1$.
5. The fraction of heterotactic triads is a maximum at $p_m = p_r = 0.5$ and drops to zero at either extreme.
6. For an atactic polymer the proportions of isotactic, syndiotactic, and heterotactic traids are 0.25:0.25:0.50.

To investigate the triads by NMR, the resonances associated with the chain substituent are examined, since structures [XV]-[XVII] show that it is these that experience different environments in the various triads. If dyad information is sufficient, the resonances of the methylenes in the chain backbone are measured. Structures [XIII] and [XIV] show that these serve as probes of the environment in dyads.

In the next section we shall examine in more detail how this type of NMR data is interpreted.

7.11 Measuring Stereoregularity by Nuclear Magnetic Resonance

It is not the purpose of this book to discuss in detail the contributions of NMR spectroscopy to the determination of molecular structure. This is a specialized field in itself and a great deal has been written on the subject. In this section we shall consider only the application of NMR to the elucidation of stereo-regularity in polymers. Numerous other applications of this powerful technique have also been made in polymer chemistry, including the study of positional and geometrical isomerism (Sec. 1.6), copolymers (Sec. 7.7), and helix-coil transitions (Sec. 1.11). We shall also make no attempt to compare the NMR spectra of various different polymers; instead, we shall examine only the NMR spectra of different poly(methyl methacrylate) preparations to illustrate the capabilities of the method, using the first system that was investigated by this technique as the example.

Figure 7.10 shows the 60-MHz spectra of poly(methyl methacrylate) pre-pared with different catalysts so that predominately isotactic, syndiotactic, and atactic products are formed. The three spectra in Fig. 7.10 are identified in terms of this predominant character. It is apparent that the spectra are quite different, especially in the range of δ values between about 1 and 2 ppm. Since the atactic polymer has the least regular structure, we concentrate on the other two to make the assignment of the spectral features to the various protons.

Several observations from the last section provide the basis of interpreting these spectra:

1. The hydrogens of the methylene group in the backbone of the poly(methyl methacrylate) produce a single peak in a racemic dyad, as illustrated by structure [XVI].

2. The same group of hydrogens in a meso dyad [XIII] produce a quartet of peaks: two different chemical shifts, each split into two by the two hydrogens in the methylene.

3. The peaks centered at $\delta = 1.84$ ppm—a singlet in the syndiotactic and a quartet in the isotactic polymers—are thus identified with these protons. This provides an unambiguous identification of the predominant stereo-regularity of these samples.

4. The features that occur near $\delta = 1.0$ ppm are associated with the protons of the α-methyl group. The location of this peak depends on the configura-tions of the nearest neighbors.

5. Working from the methylene assignments, we see that the peak at $\delta = 1.22$ ppm in the isotactic polymer arises from the methyl in the center of an

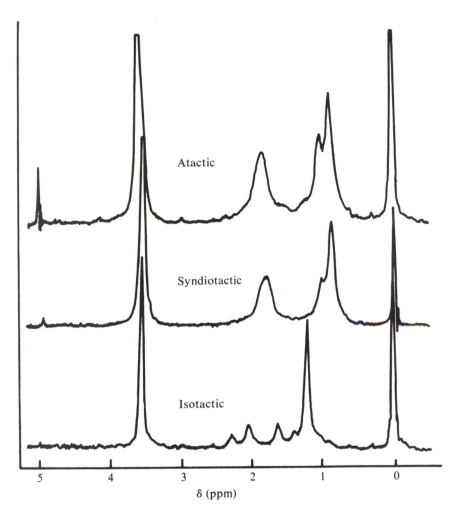

Figure 7.10 Nuclear magnetic resonance spectra of three poly(methyl methacrylate samples. Curves are labeled according to the preominant tacticity of samples. [From D. W. McCall and W. P. Slichter, in *Newer Methods of Polymer Characterization*, B. Ke (Ed.), Interscience, New York, 1964, used with permission.]

isotactic triad, the peak at δ = 0.87 ppm from a syndiotactic triad, and the peak at δ = 1.02 ppm from a homotactic triad.

6. The peak at δ = 3.5 ppm is due to the ester methoxy group.

Once these assignments are made, the areas under the various peaks can be measured to determine the various fractions:

1. The area under the methylene peaks is proportional to the dyad concentration: The singlet gives the racemic dyads and the quartet gives the meso dyads.

2. The area under one of the methyl peaks is proportional to the concentration of the corresponding triad.

3. It is apparent that it is not particularly easy to determine the exact areas of these features when the various contributions occur together to any significant extent. This is clear from the atactic spectrum, in which slight shoulders on both the methylene and methyl peaks are the only evidence of meso methylenes and iso methyls.

The spectra shown in Fig. 7.10 were early attempts at this kind of experiment, and the measurement of peak areas in this case was a rather subjective affair. We shall continue with an analysis of these spectra—subjectivity and all— even though improved instrumentation has resulted in much improved spectra. One development that has produced better resolution in spectra is the use of higher magnetic fields. As the magnetic field increases, the chemical shifts for the various features are displaced proportionately. The splitting caused by spin–spin coupling, on the other hand, is unaffected. This can produce a considerable sharpening of the NMR spectrum. Figure 7.11 demonstrates this with the methylene portion of the spectra of predominately isotactic and predominately syndiotactic poly(methyl methacrylate) measured at 220 MHz. The contrast of these 220-MHz spectra with those measured at 60 MHz (see Fig. 7.10) reveals greatly improved resolution. In fact, additional fine structure is revealed which is more confusing than helpful at this point. We shall return to this additional fine structure below. Other procedures such as spin decoupling, isotopic substitution, computerized stripping of superimposed spectra, and ^{13}C-NMR also offer methods for identifying and quantifying NMR spectra.

Table 7.9 lists the estimated fractions of dyads of types m and r and the fractions of triads of types i, s, and h. These fractions represent the area under a specific peak (or four peaks in the case of the meso dyads) divided by the total area under all of the peaks in either the dyad or triad category. As expected for the sample labeled isotactic, 89% of the triads are of type i and 87% of the dyads are of type m. Likewise, in the sample labeled syndiotactic, 68% of the triads are s and 83% of the dyads are r.

The sample labeled atactic in Fig. 7.10 was prepared by a free-radical mechanism and, hence, is expected to follow zero-order Markov statistics. As a test of this, we examine Fig. 7.9 to see whether the values of p_i, p_s, and p_h, which are given by the fractions in Table 7.9, agree with a single set of p_m values. When this is done, it is apparent that these proportions are consistent with this type

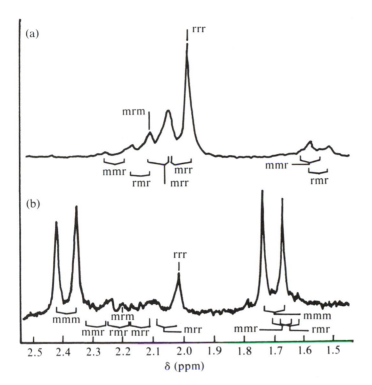

Figure 7.11 Methylene proton portion of the 220-MHz NMR spectrum of poly(methyl methacrylate): (a) predominately syndiotactic and (b) predominately isotactic. [From F. A. Bovey, *High Resolution NMR of Macromolecules*, Academic, New York, 1972, used with permission.]

Table 7.9 The Fractions of Meso and Racemic Dyads and Iso, Syndio, and Hetero Triads for the Data in Fig. 7.10

	Dyads		Triads		
Sample	Meso	Racemic	Iso	Syndio	Hetero
Atactic	0.22	0.78	0.07	0.55	0.38
Syndiotactic	0.17	0.83	0.04	0.68	0.28
Isotactic	0.87	0.13	0.89	0.04	0.07

Source: D. W. McCall and W. P. Slichter, in *Newer Methods of Polymer Characterization*, B. Ke (Ed.), Interscience, New York, 1964.

of statistics within experimental error and that $p_m \cong 0.25$ for poly(methyl methacrylate). Under the conditions of this polymerization, the free-radical mechanism is biased in favor of syndiotactic additions over isotactic additions by about 3:1, according to Eq. (6.66). Presumably this is due to steric effects involving the two substituents on the α-carbon.

With this kind of information it is not difficult to evaluate the average lengths of isotactic and syndiotactic sequences in a polymer. As a step toward this objective, we define the following:

1. The number of isotactic sequences containing n_i iso repeat units is N_{n_i}.
2. The number of syndiotactic sequences containing n_s syndio repeat units is N_{n_s}.
3. Since isotactic and syndiotactic sequences must alternate, it follows that

$$N_{n_i} = N_{n_s} \tag{7.71}$$

4. The number of iso triads in a sequence of n_i iso repeat units is $n_i - 1$, and the number of syndio triads in a sequence of n_s syndio repeat units is $n_s - 1$. We can verify these relationships by examining a specific chain segment [XVIII]:

-DDLDLDLDLD*DDDDDDDDL-

[XVIII]

In this example both the iso and syndio sequences consist of eight repeat units, with seven triads in each. The repeat unit marked * is counted as part of each type of triad, but is itself the center of a hetero triad.

5. The number of racemic dyads in a sequence is the same as the number of syndiotactic units n_s. The number of meso dyads in a sequence is the same as the number of iso units n_i. These can also be verified from structure [XVIII] above.

With these definitions in mind, we can immediately write expressions for the ratio of the total number ν of iso triads to the total number of syndio triads:

$$\frac{\nu_i}{\nu_s} = \frac{\Sigma N_{n_i}(n_i - 1)}{\Sigma N_{n_s}(n_s - 1)} = \frac{\Sigma N_{n_i}(n_i) - \Sigma N_{n_i}}{\Sigma N_{n_s}(n_s) - \Sigma N_{n_s}} \tag{7.72}$$

In this equation the summations are over all values of n of the specified type. Also remember that the ν's and n's in this discussion (with subscript i or s) are defined differently from the ν's and n's defined earlier in the chapter for copolymers (where they had subscript 1 or 2). Using Eq. (7.71) and remembering

the definition of an average provided by Eq. (1.9), we see that Eq. (7.72) becomes

$$\frac{\nu_i}{\nu_s} = \frac{\bar{n}_i - 1}{\bar{n}_s - 1} \tag{7.73}$$

where the overbar indicates the average length of the indicated sequence.

A similar result can be written for the ratio of the total number (ν) of dyads of the two types (m and r), using item (5) above:

$$\frac{\nu_m}{\nu_r} = \frac{\Sigma N_{n_i}(n_i)}{\Sigma N_{n_s}(n_s)} = \frac{\bar{n}_i}{\bar{n}_s} \tag{7.74}$$

Equations (7.73) and (7.74) can be solved simultaneously for $\bar{n}_i$ and $\bar{n}_s$ in terms of the total number of dyads and triads:

$$\bar{n}_i = \frac{1 - \nu_i/\nu_s}{1 - (\nu_i/\nu_s)(\nu_r/\nu_m)} \tag{7.75}$$

and

$$\bar{n}_s = \frac{1 - \nu_i/\nu_s}{(\nu_m/\nu_t) - (\nu_i/\nu_s)} \tag{7.76}$$

Use of these relationships is illustrated in the following example:

Example 7.8

Use the dyad and triad fractions in Table 7.9 to calculate the average lengths of isotactic and syndiotactic sequences for the polymers of Fig. 7.10. Comment on the results.

Solution

Since the total numbers of dyads and triads always occur as ratios in Eqs. (7.73) and (7.74), both the numerators and denominators of these ratios can be divided by the total number of dyads or triads to convert these total numbers into fractions: That is, $\nu_i/\nu_s = (\nu_i/\nu_{tot})/\nu_s/\nu_{tot} = p_i/p_s$. Thus the fractions in Table 7.9 can be substituted for the ν's in Eqs. (7.73) and (7.74). The values of $\bar{n}_i$ and $\bar{n}_s$ so calculated for the three polymers are the following:

	$\bar{n}_i$	$\bar{n}_s$
Atactic	1.59	5.64
Syndiotactic	1.32	6.45
Isotactic	9.14	1.37

This type of analysis adds nothing new to the picture already presented by the dyad and triad probabilities. It is somewhat easier to visualize an average sequence, however, although it must be remembered that the latter implies nothing about the distribution of sequence lengths.

•

We conclude this section by returning to Fig. 7.11 for a closer look at the fine structure revealed at higher magnetic fields. What we are observing in these spectra are the methylene resonances as influenced by tetrad sequences. Table 7.8 shows that the meso dyad can be the center of three possible tetrads. Each of these is split into two, giving six spectral features. Likewise, the racemic dyad can be the center of three tetrads, one of which is further split into a doublet while the other two produce singlets in the spectrum. We need not be concerned with the various splitting patterns or with the assignment of the possible tetrads to the different features of the spectrum. It is sufficient to note that once identified, the fraction of the possible tetrads in a polymer can be evaluated by the same procedure used above. The analysis of these longer sequences is particularly important for systems following higher-order Markov statistics. The resonances of the α-methyl protons as influenced by pentad sequences can also be measured at high magnetic fields.

In the next section we shall examine the catalysts which are able to introduce this kind of regularity into polymers.

7.12 Ziegler–Natta Catalysts

For the purposes of this discussion, stereoregulating catalysts will be taken to mean Ziegler–Natta catalysts. This is a somewhat restrictive view of the situation, since there are other catalysts—phenyl magnesium bromide is a Grignard reagent—which can produce stereoregularity; the Ziegler–Natta catalysts are also used to produce polymers—unbranched polyethylene to name one—which lack stereoregularity. Ziegler–Natta catalysts are the most widely used and best-understood stereoregulating systems, so the loss of generality in this approach is not of great consequence.

The fundamental Ziegler–Natta recipe consists of two components: the halide or some other compound of a transition metal from among the group IVB to VIIIB elements and an organometallic compound of a representative metal from groups IA to IIIA. Some of the transition metal compounds that have been

studied include $TiCl_4$, $TiCl_3$, VCl_4, VCl_3, $ZrCl_4$, $CrCl_3$, $MoCl_5$, and $CuCl$. Some of the representative element organometallics include $(C_2H_5)_3Al$, $(C_2H_5)_2Mg$, C_4H_9Li and $(C_2H_5)_2Zn$. These are only a few of the possible compounds, so the number of *combinations* is very large.

The individual components of the Ziegler–Natta system can separately account for the initiation of some form of polymerization reaction, but not for the fact of stereoregularity. For example, butyl lithium can initiate anionic polymerization (see Sec. 6.10) and $TiCl_4$ can initiate cationic polymerization (see Sec. 6.11). In combination, still another mechanism for polymerization, coordination polymerization, is indicated. When the two components of the Ziegler–Natta system are present together, complicated exchange reactions are possible. Often the catalyst must age to attain maximum effectiveness; presumably this allows these exchange reactions to occur. Some possible exchange equilibria are

$$2\,Al\,(C_2H_5)_3 \rightleftarrows Al_2(C_2H_5)_6 \rightleftarrows [Al(C_2H_5)_2]^+ [Al\,(C_2H_5)_4]^- \tag{7.Q}$$

$$Ti\,Cl_4 + [Al(C_2H_5)_2]^+ \rightleftarrows C_2H_5TiCl_3 + [Al\,(C_2H_5)\,Cl]^+$$

The organotitanium halide can then reduce to $TiCl_3$:

$$C_2H_5TiCl_3 \rightarrow Ti\,Cl_3 + C_2H_5\cdot \tag{7.R}$$

Among other possible reactions, these free radicals can initiate ordinary free-radical polymerization. The Ziegler–Natta systems are thus seen to encompass several mechanisms for the initiation of polymerization. Neither ionic nor free-radical mechanisms account for stereoregularity, however, so we must look further for the mechanism whereby the Ziegler–Natta systems produce this interesting effect.

The stereoregulating capability of Ziegler–Natta catalysts is believed to depend on a coordination mechanism in which both the growing polymer chain and the monomer coordinate with the catalyst. The addition then occurs by insertion of the monomer between the growing chain and the catalyst by a concerted mechanism [XIX]:

[XIX]

Since the coordination almost certainly involves the transition metal atom, there is a resemblance here to anionic polymerization. The coordination is an important aspect of the present picture, since it is this feature which allows the catalyst to serve as a template for stereoregulation.

The assortment of combinations of components is not the only variable to consider in describing Ziegler–Natta catalysts. Some other variables include the following:

1. Catalyst solubility. Polymerization systems may consist of one or two phases. Titanium-based catalysts are the most common of the heterogeneous systems; vanadium-based catalysts are the most common homogeneous systems. Since the catalyst functions as a template for the formation of a stereoregular product, it follows that the more extreme orienting effect of a solid surface (i.e., heterogeneous catalysts) are required for those monomers which interact only weakly with the catalyst. The latter are nonpolar monomers. Polar monomers interact more strongly with catalysts, and dissolved catalysts are able to exert sufficient control for stereoregularity.

2. Crystal structure of solids. The α-crystal form of $TiCl_3$ is an excellent catalyst and has been investigated extensively. In this particular crystal form of $TiCl_3$, the titanium ions are located in an octahedral environment of chloride ions. It is believed that the stereoactive titanium ions in this crystal are located at the edges of the crystal, where chloride ion vacancies in the coordination sphere allow coordination with the monomer molecules.

3. Tacticity of products. Most solid catalysts produce isotactic products. This is probably because of the highly orienting effect of the solid surface, as noted in item (1). The preferred isotactic configuration produced at these surfaces is largely governed by steric and electrostatic interactions between the monomer and the ligands of the transition metal. Syndiotacticity is mostly produced by soluble catalysts. Syndiotactic polymerizations are carried out at low temperatures, and even the catalyst must be prepared at low temperatures; otherwise specificity is lost. With polar monomers syndiotacticity is also promoted by polar reaction media. Apparently the polar solvent molecules compete with monomer for coordination sites, and thus indicate more loosely coordinated reactive species.

4. Rate of polymerization. The rate of polymerization for homogeneous systems closely resembles anionic polymerization. For heterogeneous systems the concentration of alkylated transition metal sites on the surface appears in the rate law. The latter depends on the particle size of the solid catalyst and may be complicated by sites of various degrees of activity. There is sometimes an inverse relationship between the degree of stereoregularity produced by a catalyst and the rate at which polymerization occurs.

The catalysts under consideration both initiate the polymerization and regulate the polymer formed. There is general agreement that the mechanism by which these materials exert their regulatory role involves coordination of monomer with the transition metal atom, but proposed details beyond this are almost as numerous and specific as the catalysts themselves. We shall return to a description of two specific mechanisms below. The general picture postulates an interaction between monomer and catalyst such that a complex is formed between the π electrons of the olefin and the d orbitals of the transition metal. Figure 7.12 shows that the overlap between the filled orbitals of the monomer can overlap with vacant $d_{x^2-y^2}$ orbitals of the metal. Alternatively, hybrid orbitals may be involved on the metal. There is a precedent for such bonding in simple model compounds. It is known, for example, that Pt^{2+} complexes with ethylene by forming a dsp^2 hybrid-π sigma bond and a dp hybrid-π^* pi bond. A crucial consideration in this coordination is maximizing the overlap of the orbitals involved. Titanium III ions seem ideally suited for this function; higher effective nuclear charge on the metal results in less spatial extension of d orbitals and diminished overlap.

Many mechanisms have been proposed that develop this picture more specifically. These are often so specific that they cannot be generalized beyond the systems for which they are proposed. Two schemes that do allow some generalization are presented here. Although they share certain common features, these mechanisms are distinguished by the fact that one—the monometallic model—does not include any participation by the representative metal in the mechanism. The second—the bimetallic model—does assume the involvement of both metals in the mechanism.

The monometallic mechanism is illustrated in Fig. 7.13a. It involves the monomer coordinating with an alkylated titanium atom. The insertion of the monomer into the titanium–carbon bond propagates the chain. As shown in

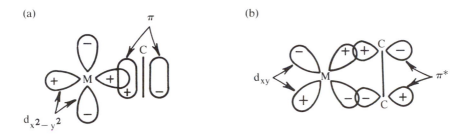

Figure 7.12 Orbital overlap between transition metal and olefin: (a) $d_{x^2-y^2}$ and π and (b) d_{xy} and π^*.

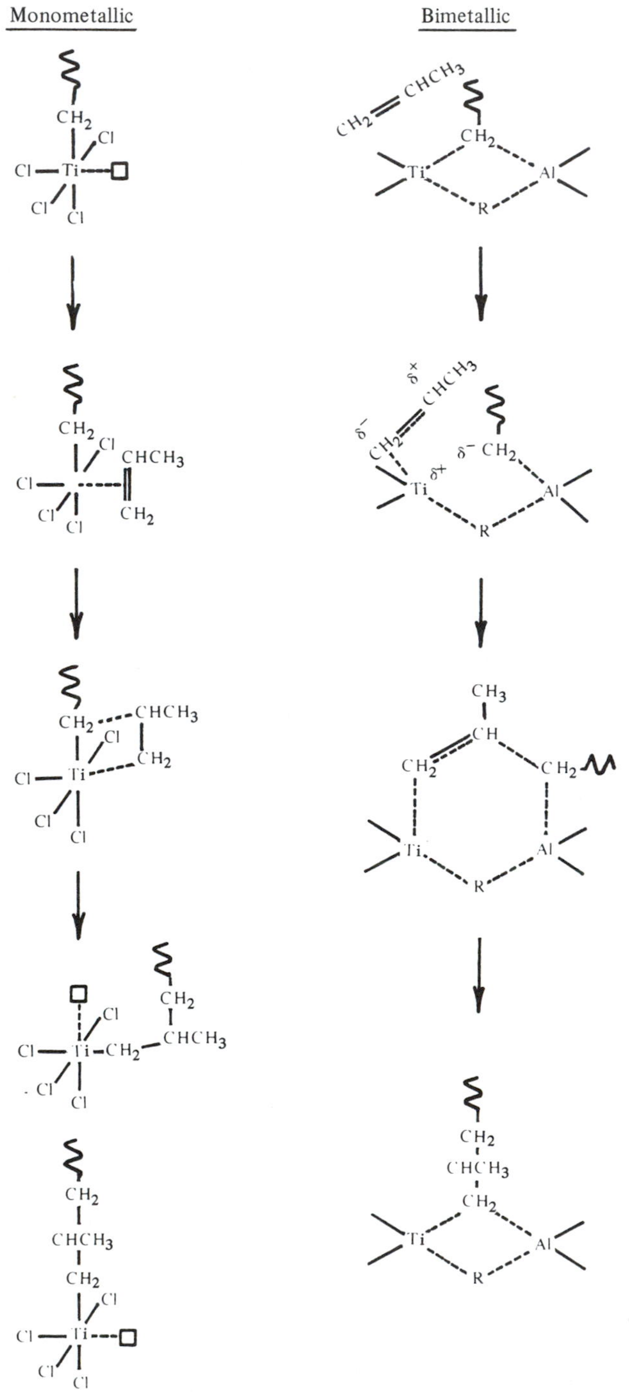

Figure 7.13 (a) The monometallic mechanism. The square indicates a vacant ligand site. (b) The bimetallic mechanism.

Fig. 7.13, this shifts the vacancy—represented by the square—in the coordination sphere of the titanium to a different site. Syndiotactic regulation occurs if the next addition takes place via this newly created vacancy. In this case the monomer and the growing chain occupy alternating coordination sites in successive steps. For the more common isotactic growth the polymer chain must migrate back to its original position.

Figure 7.14a illustrates the insertion of a propylene monomer into an edge vacancy in a crystal adjacent to an alkylated titanium atom. In Fig. 7.14b a cross-sectional view of the same site shows how the preferential orientation of the coordinated monomer is dictated by constraints imposed by the protuberances on the crystal surface.

The bimetallic mechanism is illustrated in Fig. 7.13b; the bimetallic active center is the distinguishing feature of this mechanism. The precise distribution of halides and alkyls is not spelled out because of the exchanges described by reaction (7.Q). An alkyl bridge is assumed based on observations of other organometallic compounds. The pi coordination of the olefin with the titanium is followed by insertion of the monomer into the bridge to propagate the reaction.

At present it is not possible to determine which of these mechanisms or their variations most accurately represents the behavior of Ziegler–Natta catalysts. In view of the number of variables in these catalyzed polymerizations, both mechanisms may be valid, each for different specific systems. In the following example the termination step of coordination polymerizations is considered.

Example 7.9

Polypropylene polymerized with triethyl aluminum and titanium trichloride has been found to contain various kinds of chain ends. Both terminal vinylidene unsaturation and aluminum-bound chain ends have been identified. Propose two termination reactions which can account for these observations. Do the termination reactions allow any discrimination between the monometallic and bimetallic propagation mechanisms?

Solution

A reaction analogous to the alkylation step of reaction (7.Q) can account for the association of an aluminum species with chain ends:

$$
\text{Ti–CH}_2\text{–C}\negmedspace\sim + \text{Al(C}_2\text{H}_5)_2^+ \;\rightarrow\; \text{Ti---CH}_2\text{–C}\negmedspace\sim \;\rightarrow\; \text{Ti} + \sim\negmedspace\text{C–CH}_2\text{–Al}^+\text{–C}_2\text{H}_5
$$

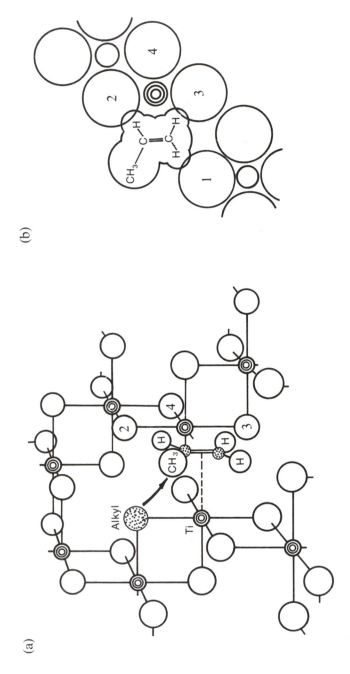

Figure 7.14 (a) The insertion of a propylene molecule into a site vacancy in the Ziegler–Natta catalyst. (b) The cross section shows the origin of stereoregulation. [From P. Cosee, *Tetrahedron Lett.* 17:17 (1960).]

The transfer of a tertiary hydrogen between the polymer chain and a monomer can account for the vinylidene group in the polymer:

$$
\begin{array}{c}
\text{X} \\
/ \\
\text{CH}_2\!\!=\!\!\text{C}---\text{H}
\end{array}
$$

$$
\underset{\text{X}}{\overset{\text{H}}{\text{Ti}-\text{CH}_2-\text{C}}}\!\!\sim + \underset{\text{X}}{\overset{\text{H}}{\text{CH}_2=\text{C}}} \;\rightarrow\; \text{Ti}---\text{CH}_2-\underset{\text{X}}{\overset{\text{H}}{\text{C}}}\!\!\sim \;\rightarrow\; \text{Ti}-\text{CH}_2-\text{CH}_2\text{X} + \underset{}{\overset{\text{X}}{\wedge\text{C}=\text{CH}_2}}
$$

These reactions appear equally feasible for titanium in either the monometallic or bimetallic intermediate. Thus they account for the different types of end groups in the polymer, but do not differentiate between propagation intermediates.

•

In the commercial process for the production of polypropylene by Ziegler–Natta catalysts, hydrogen is added to terminate the reaction, so neither of these reactions is pertinent to this process.

7.13 Preview of Things to Come

In part 2 of this book we have focused attention on some classes of reactions which produce polymers and on some properties of the resulting products. In the final three chapters we shall consider some of the methods that are used to characterize the polymeric products of these syntheses.

In the concluding chapters we again consider assemblies of molecules—this time, polymers surrounded by solvent molecules which are comparable in size to the repeat units of the polymer. Generally speaking, our efforts are directed toward solutions which are relatively dilute with respect to the polymeric solute. The reason for this is the same reason that dilute solutions are widely considered in discussions of ionic or low molecular weight solutes, namely, solute–solute interactions are either negligible or at least minimal under these conditions.

We shall discuss three types of phenomena for polymer solutions: thermodynamic properties in Chap. 8, frictional properties in Chap. 9, and light-scattering properties in Chap. 10. A common feature of virtually all phenomena in these areas is that they all depend on the molecular weight of the solute. Thus observations of these properties can be interpreted to yield values for M; we shall use this capability as a unifying theme throughout these chapters.

In Chap. 8 we discuss the thermodynamics of polymer solutions, specifically with respect to phase separation and osmotic pressure. We shall devote considerable attention to statistical models to describe both the entropy and the enthalpy of mixtures. Of particular interest is the idea that the thermodynamic

nonideality of polymer solutions can be used as a quantitative measure of either the volume of polymer molecules or the strength of solute–solvent interactions. We shall also briefly consider charged polymers in Chap. 8; this is the only place where polymeric electrolytes are discussed in this text.

At first glance, the contents of Chap. 9 read like a catchall for unrelated topics. In it we examine the intrinsic viscosity of polymer solutions, the diffusion coefficient, the sedimentation coefficient, sedimentation equilibrium, and gel permeation chromatography. While all of these techniques can be related in one way or another to the molecular weight of the polymer, the more fundamental unifying principle which connects these topics is their common dependence on the spatial extension of the molecules. The radius of gyration is the parameter of interest in this context, and the intrinsic viscosity in particular can be interpreted to give a value for this important quantity. The experimental techniques discussed in Chap. 9 have been used extensively in the study of biopolymers.

Chemistry students are generally more familiar with the absorption of light by molecules than with more classical optical phenomena such as refraction and scattering. In Chap. 10 we attempt to remedy this situation by detailed development of those aspects of light scattering that are especially pertinent to the study of polymers. The discussion of this topic brings together certain basic ideas from electromagnetism, optics, and solution thermodynamics. Even though it is explicitly through an expression for osmotic pressure that the molecular weight enters the light-scattering equations, these two techniques give different averages for M. This makes these two methods complementary rather than redundant, and, in combination, they provide information concerning the width of the molecular weight distribution.

Although the emphasis in these last chapters is certainly on the polymeric solute, the experimental methods described herein also measure the interactions of these solutes with various solvents. Such interactions include the hydration of proteins at one extreme and the exclusion of poor solvents from random coils at the other. In between, good solvents are imbibed into the polymer domain to various degrees to expand coil dimensions. Such quantities as the Flory–Huggins interaction parameter, the Θ temperature, and the coil expansion factor are among the ways such interactions are quantified in the following chapters.

Problems

1. Write structural formulas for maleic anhydride (M_1) and stilbene (M_2). Neither of these monomers homopolymerize to any significant extent, presumably owing to steric effects. These monomers form a copolymer,

however, with $r_1 = r_2 = 0.03$.[†] Criticize or defend the following proposition: The strong tendency toward alternation in this copolymer suggests that polarity effects offset the steric hindrance and permit copolymerization of these monomers.

2. Styrene and methyl methacylate have been used as comonomers in many investigations of copolymerization. Use the following list[‡] of r_1 values for each of these copolymerizing with the monomers listed below to rank the latter with respect to reactivity:

M_2	Styrene as M_1	Methyl methacrylate as M_1
Acrylonitrile	0.41	1.35
Allyl acetate	90	23
1,2-Dichloropropene-2	5	5.5
Methacrylonitrile	0.30	0.67
Vinyl chloride	17	12.5
Vinylidene chloride	1.85	2.53
2-Vinyl pyridine	0.55	0.395

To the extent that the data allow, suggest where these substituents might be positioned in Table 7.3.

3. The following reactivity ratios[§] describe the polymerization of acrylonitrile (M_1) with the monomers listed:

M_2	r_1	r_2
Butadiene	0.02	0.3
Methyl methacrylate	0.15	1.22
Styrene	0.04	0.40
Vinyl acetate	4.2	0.05
Vinyl chloride	2.7	0.04

Use the Q and e values listed in Table 7.4 for each of the comonomers to give five independent estimates of Q and e for acrylonitrile. Compare the average of these four with the values given for acrylonitrile in Table 7.4.

[†] F. M. Lewis and F. R. Mayo, *J. Am. Chem. Soc.* 70:1533 (1948).
[‡] Ref. 4.
[§] Ref. 4.

4. As part of the research described in Fig. 7.5, Winston and Wichacheewa†
    measured the percentages of carbon and chlorine in copolymers of styrene
    (molecule 1) and 1-chloro-1,3-butadiene (molecule 2) prepared from
    various feedstocks. A portion of their data is given below:

f_1	Percent C	Percent Cl
0.892	81.80	10.88
0.649	71.34	20.14
0.324	64.95	27.92
0.153	58.69	34.79

Use these data to calculate F_1, the mole fraction of styrene in these
copolymers.

5. Additional data from the research of the last problem‡ yield the following
    pairs of f_1, F_1 values (remember that styrene is component 1 in the styrene-
    1-chloro-1,3-butadiene system):

f_1	F_1	f_1	F_1
0.947	0.829	0.448	0.362
0.861	0.688	0.247	0.207
0.698	0.515	0.221	0.200
0.602	0.452		

Use the form suggested by Eq. (7.54) to prepare a graph based on these
data and evaluate r_1 and r_2.

6. The reactivity ratios for the styrene (M_1)–1-chloro-1,3-butadiene (M_2)
    system were found to be $r_1 = 0.26$ and $r_2 = 1.02$ by the authors of the
    research described in the last two problems§ using the results of all their
    measurements. Use these r values and the feed compositions listed below
    to calculate the fraction expected in the copolymer of 1-chlorobutadiene
    sequences of lengths $\nu = 2$, 3, or 4. From these calculated results, evaluate
    the ratios N_{222}/N_{22} and N_{2222}/N_{222}. Copolymers prepared from these
    feedstocks were dehydrohalogenated to yield the polyenes like that whose
    spectrum is shown in Fig. 7.5. The specific absorbance at the indicated
    wavelengths was measured for 1% solutions of the products after HCl
    elimination:

---

† A. Winston and P. Wichacheewa, *Macromolecules* 6:200 (1973).
‡ Ibid.
§ Ibid.

f_1	Specific absorbance		
	$\lambda = 312$ nm	$\lambda = 367$ nm	$\lambda = 412$ nm
0.829	74	13	–
0.734	71	19	–
0.551	154	77	20
0.490	151	78	42

As noted in Sec. 7.7, these different wavelengths correspond to absorbance by sequences of different lengths. Compare the appropriate absorbance ratios with the theoretical sequence length ratios calculated above and comment briefly on the results.

7. Use the values determined in Example 7.6 for the vinylidene chloride (M_1)– isobutylene (M_2) system[†] to calculate F_1 for various values of f_1 according to the terminal mechanism. Prepare a plot of the results. On the same graph, plot the following experimentally measured values of f_1 and F_1:

f_1	F_1	f_1	F_1
0.548	0.83	0.225	0.66
0.471	0.79	0.206	0.64
0.391	0.74	0.159	0.61
0.318	0.71	0.126	0.58
0.288	0.70	0.083	0.52

Comment on the quality of the fit.

8. Some additional dyad fractions from the research cited in the last problem[‡] are reported at intermediate feedstock concentrations (M_1 = vinylidene chloride; M_2 = isobutylene):

f_1	Mole fraction of dyads		
	11	12	22
0.418	0.55	0.43	0.03
0.353	0.48	0.49	0.04
0.317	0.44	0.52	0.04
0.247	0.38	0.58	0.04
0.213	0.34	0.62	0.04
0.198	0.32	0.64	0.05

Still assuming terminal control, evaluate r_1 and r_2 from these data. Criticize or defend the following proposition: The copolymer composition equation does not provide a very sensitive test for the terminal control mechanism.

[†] J. B. Kinsinger, T. Fischer, and C. W. Wilson, *Polym. Lett.* 5:285 (1967).
[‡] Ibid.

Dyad fractions are more sensitive, but must be examined over a wide range of compositions to provide a valid test.

9. Fox and Schnecko[†] carried out the free-radical polymerization of methyl methacrylate between -40 and $250°C$. By analysis of the α-methyl peaks in the NMR spectra of the products, they determined the following values of α, the probability of an isotactic placement in the products prepared at the different temperatures:

T(°C)	α	T (°C)	α
250	0.36	0	0.20
150	0.33	-20	0.18
100	0.27	-40	0.14
95	0.27		
60	0.24		
30	0.22		

Evaluate $E_i{}^* - E_s{}^*$ by means of an Arrhenius plot of these data using $\alpha/(1 - \alpha)$ as a measure of k_i/k_s. Briefly justify this last relationship.

10. As apparent from structure [XVIII], a hetero triad occurs at each interface between iso and syndio triads. The total number of hetero triads, therefore, equals the total number of sequences of all other types: $\nu_h = \Sigma N_{n_i} + \Sigma N_{n_s}$. Use this relationship and Eq. (7.71) to derive the expression $p_h = \nu_h/(\nu_h + \nu_i + \nu_s) = 2/\bar{n}_i + \bar{n}_s$. Criticize or defend the following proposition: The sequence DL_ is already two-thirds of the way to becoming a hetero triad, while the sequence DD_ is two-thirds of the way toward an iso triad. This means that the fraction of heterotactic triads is larger when the average length of syndio sequences is greater than the average length of iso sequences.

11. Randall[‡] used ^{13}C-NMR to study the methylene spectrum of polystyrene. In 1,2,4-trichlorobenzene at 120°C, nine resonances were observed. These were assumed to arise from a combination of tetrads and hexads. Using m and r notation, extend Table 7.8 to include all 20 possible hexads. Criticize or defend the following proposition: Assuming that none of the resonances are obscured by overlap, there is only one way that nine methylene resonances can be produced, namely, by one of the tetrads to be split into hexads while the remaining tetrads remain unsplit.

12. In the research described in the preceding problem, Randall[‡] was able to assign the five peaks associated with tetrads in the ^{13}C-NMR spectrum on the basis of their relative intensities, assuming zero-order Markov (or Bernoulli) statistics with $p_m = 0.575$. The five tetrad intensities and their chemical shifts from TMS are as follows:

[†] T. G. Fox and H. W. Schnecko, *Polymer* 3:575 (1962).

[‡] J. C. Randall, *J. Polym. Sci. Polym. Phys. Ed.* 13:889 (1975).

$^{13}C\ \delta_{TMS}$ (ppm)	Relative area under peak
45.38	0.10
44.94	0.28
44.25	0.13
43.77	0.19
42.84	0.09

The remaining 21% of the peak area is distributed among the remaining hexad features. Use the value of p_m given to calculate the probabilities of the unsplit tetrads (see Problem 11) and on this basis assign the features listed above to the appropriate tetrads. Which of the tetrads appears to be split into hexads?

13. The fraction of sequences of the length indicated below have been measured[†] for a copolymer system at different feed ratios:

$[M_1]/[M_2]$	$P(M_1)$	$P(M_1M_1)$	$P(M_1M_1M_1)$
3	0.168	0.0643	0.0149
4	0.189	0.0563	0.0161
9	0.388	0.225	0.107
19	0.592	0.425	0.278

From appropriate ratios of these sequence lengths, what conclusions can be drawn concerning terminal versus penultimate control of addition?

14. The following are experimental tacticity fractions of polymers prepared from different monomers and with various catalysts. On the basis of Fig. 7.9, decide whether these preparations are adequately described (remember to make some allowance for experimental error) by a single parameter p_m or whether some other type of statistical description is required:

Catalyst	Solvent	Temperature (°C)	Fraction of polymer		
			Iso	Hetero	Syndio
Methyl methacrylate[‡]					
Thermal	Toluene	60	8	33	59
n-Butyl lithium	Toluene	−78	78	16	6
n-Butyl lithium	Methyl isobutyrate	−78	21	31	48
α-Methyl styrene[§]					
$TiCl_4$	Toluene	−78	—	19	81
$Et_3Al/TiCl_4$	Benzene	25	3	35	62
n-Butyl lithium	Cyclohexane	4	—	31	69

[†] K. Ito and Y. Yamashita, *J. Polym. Sci.* 3A:2165 (1965).

[‡] K. Hatada, K. Ota, and H. Yuki, *Polym. Lett.* 5:225 (1967).

[§] S. Brownstein, S. Bywater, and O. J. Worsfold, *Makromol. Chem.* 48:127 (1961).

On the basis of these observations, criticize or defend the following proposition: Regardless of the monomer used, zero-order Markov (Bernoulli) statistics apply to all free radical, anionic, and cationic polymerizations, but not to Ziegler–Natta catalyzed systems.

15. Replacing one of the alkyl groups in R_3Al with a halogen increases the stereospecificity of the Ziegler–Natta catalyst in the order $I > Br > Cl > R$. Replacement of a second alkyl by halogen decreases specificity. Criticize or defend the following proposition on the basis of these observations: The observed result of halogen substitution is consistent with the effect on the ease of alkylation produced by substituents of different electronegativity. This evidence thus adds credence to the monometallic mechanism, even though the observation involves the organometallic.

16. The weight percent propylene in ethylene–propylene copolymers for different Ziegler–Natta catalysts was measured† for the initial polymer produced from identical feedstocks. The following results were obtained:

Catalyst components	Weight percent propylene	Catalyst components	Weight percent propylene
VCl_4 plus		$Al(i\text{-}Bu)_3$ plus	
$Al(i\text{-}Bu)_3$	4.5	$HfCl_4$	0.7
CH_3TiCl_3	4.5	$ZrCl_4$	0.8
$Zn(C_2H_5)_2$	4.5	$VOCl_3$	2.4
$Zn(n\text{-}Bu)_2$	4.5		

Interpret these results in terms of the relative influence of the two components of the catalyst on the product found.

Bibliography

1. Alfrey, T., Jr., Bohrer, J. J., and Mark, H., *Copolymerization*, Interscience, New York, 1952.

2. Bovey, F. A., *High Resolution NMR of Macromolecules*, Academic, New York, 1972.

3. Bovey, F. A., and Winslow, F. H., *Macromolecules: An Introduction to Polymer Science*, Academic, New York, 1979.

4. Brandup, J., Immergut, E. H., and McDowell, W. (Eds.), *Polymer Handbook*, 2nd ed., Wiley, New York, 1975.

5. Ham, G. E. (Ed.), *Copolymerization*, Interscience, New York, 1964.

6. Koenig, J. L., *Chemical Microstructure of Polymer Chains*, Wiley, New York, 1980.

7. Odian, G., *Principles of Polymerization*, 2nd ed., Wiley, New York, 1981.

---

† F. J. Karol and W. L. Carrick, *J. Am. Chem. Soc.* 83:585 (1960).

SOME PROPERTIES OF POLYMER SOLUTIONS AND THEIR RELATION TO POLYMER CHARACTERIZATION

8

The Thermodynamics of
Polymer Solutions

See plastic nature working to this end
The single atoms each to other tend
Attract, attracted to, the next in place
Formed and impelled its neighbor to embrace.

An Essay On Man, Alexander Pope

8.1 Introduction

The title of this chapter is somewhat misleading. In one sense it is too broad, in another sense too restrictive. We shall really discuss in detail only the phase separation and osmostic pressure of polymer solutions; a variety of other thermodynamic phenomena are ignored. In this regard the chapter title would better read "Some aspects of . . . ." Throughout this volume only a small part of what might be said about any topic is actually presented, so this modifying phrase is taken to be understood and is omitted.

In another sense the title is too restrictive, implying that only pure, phenomenological thermodynamics are discussed herein. Actually, this is far from true. Both thermodynamics and statistical thermodynamics comprise the contents of the chapter, with the second making the larger contribution. But the term *statistical* is omitted from the title, as it is too intimidating.

The phenomena we discuss, phase separation and osmotic pressure, are developed with particular attention to their applications in polymer characterization. Phase separation can be used to fractionate polydisperse polymer specimens into samples in which the molecular weight distribution is more narrow. Osmotic pressure experiments can be used to provide absolute values for the number average molecular weight of a polymer. Alternative methods for both fractionation and molecular weight determination exist, but the methods discussed in this chapter occupy a place of prominence among the alternatives, both historically and in contemporary practice.

Throughout this book we have emphasized fundamental concepts, and looking at the statistical basis for the phenomena we consider is the way this point of view is maintained in this chapter. All theories are based on models which only approximate the physical reality. To the extent that a model is successful, however, it represents at least some features of the actual system in a manageable way. This makes the study of such models valuable, even if the fully developed theory falls short of perfect success in quantitatively describing nature.

We shall devote a considerable portion of this chapter to discussing the thermodynamics of mixing according to the Flory-Huggins theory. Other important concepts we discuss in less detail include the cohesive energy density, the Flory-Krigbaum theory, and a brief look at charged polymers.

This is really the first time in this text that we have explicitly considered charged polymers as such. Even in the case of polymers prepared by ionic mechanisms, we tend to ignore the charge associated with the chain end as an insignificant end effect. Polymers in which ionizable functional groups occur throughout the molecule also exist, and these may carry quite high charges. Many biopolymers fall into this category. Even though we consider only a single aspect of charged polymers in this chapter, a number of pertinent considerations are clearly revealed.

8.2 Classical and Statistical Thermodynamics

In this chapter we shall consider some thermodynamic properties of solutions in which a polymer is the solute and some low molecular weight species is the solvent. Our special interest is in the application of solution thermodynamics to problems of phase equilibrium.

An important fact to remember about the field of thermodynamics is that it is blind to details concerning the structure of matter. Thermodynamics are concerned with observable, measurable quantities and the relationships between them, although there is a danger of losing sight of this fact in the somewhat abstract mathematical formalism of the subject. In discussing elasticity in Chap. 3, we took the position that entropy is often more intelligible from a statistical, atomistic point of view than from a purely phenomenological perspective. It is the latter that is *pure* thermodynamics; the former is the approach of statistical thermodynamics. In this chapter, too, we shall make extensive use of the statistical point of view to understand the molecular origin of certain phenomena.

The treatment of heat capacity in physical chemistry provides an excellent and familiar example of the relationship between pure and statistical thermodynamics. Heat capacity is defined experimentally and is measured by determining the heat required to change the temperature of a sample in, say,

a constant-pressure experiment. Numerous thermodynamic equations exist which relate the heat capacity to other thermodynamic quantities such as ΔH and ΔS. An alternative approach to heat capacity is to account for the storage of energy in molecules in terms of the various translational, rotational, and vibrational storage modes. *Doing* thermodynamics does not require so much as a knowledge that molecules exist, and much less how they store energy; *understanding* thermodynamics benefits considerably from the molecular point of view.

The drawback of the statistical approach is that it depends on a model, and models are bound to oversimplify. Nevertheless, we can learn a great deal from the attempt to evaluate thermodynamic properties from molecular models, even if the effort falls short of quantitative success.

There is probably no area of science that is as rich in mathematical relationships as thermodynamics. This makes thermodynamics very powerful, but such an abundance of riches can also be intimidating to the beginner. This chapter assumes that the reader is familiar with basic chemical and statistical thermodynamics at the level that these topics are treated in physical chemistry textbooks. In spite of this premise, a brief review of some pertinent relationships will be a useful way to get started.

Notation frequently poses problems in science, and this chapter is an example of such a situation. Our problem at present is that we have too many things to count: They cannot all be designated n. We have consistently used n to designate the degree of polymerization and shall continue with this notation. In thermodynamics n is widely used to designate the number of moles. Since we deal with (at least) two-component systems in this chapter, any count of the number of moles will always carry a subscript to indicate the component under consideration. We shall use the subscript 1 to designate the solvent, and 2 to designate the solute. The degree of polymerization is represented by n without a subscript.

To describe the state of a two-component system at equilibrium, we must specify the number of moles n_1 and n_2 of each component, as well as—ordinarily— the pressure p and the absolute temperature T. It is the Gibbs free energy that provides the most familiar access to a discussion of equilibrium. The increment in G associated with increments in the independent variables mentioned above is given by the equation

$$dG = V\,dp - S\,dT + \sum_{i=1,2} \mu_i\,dn_i \tag{8.1}$$

where μ_i is the chemical potential of component i. An important aspect of thermodynamics is the fact that the state variables (in the present context, this applies especially to the internal energy U, the enthalpy H, and the Gibbs

free energy G) can be expanded as partial derivatives of fundamental variables. Hence we can also write

$$dG = \left(\frac{\partial G}{\partial p}\right)_{T,n_1,n_2} dp + \left(\frac{\partial G}{\partial T}\right)_{p,n_1,n_2} dT + \left(\frac{\partial G}{\partial n_1}\right)_{p,T,n_2} dn_1$$

$$+ \left(\frac{\partial G}{\partial n_2}\right)_{p,T,n_1} dn_2 \qquad (8.2)$$

Comparing Eqs. (8.1) and (8.2) gives

$$V = \left(\frac{\partial G}{\partial p}\right)_{T,n_1,n_2} \qquad (8.3)$$

$$S = -\left(\frac{\partial G}{\partial T}\right)_{p,n_1,n_2} \qquad (8.4)$$

and

$$\mu_i = \left(\frac{\partial G}{\partial n_i}\right)_{p,T,n_{j\neq i}} \qquad (8.5)$$

The chemical potential is an example of a partial molar quantity: μ_i is the partial molar Gibbs free energy with respect to component i. Other partial molar quantities exist and share the following features:

1. We may define, say, partial molar volume, enthalpy, or entropy by analogy with Eq. (8.5):

$$\bar{Y}_i = \left(\frac{\partial Y}{\partial n_i}\right)_{p,T,n_{j\neq i}} \qquad (8.6)$$

where $Y = V$, H, or S, respectively. Except for the partial molar Gibbs free energy, we shall use the notation $\bar{Y}_i$ to signify a partial molar quantity, where Y stands for the symbol of the appropriate variable.

2. Partial molar quantities have "per mole" units, and for Y_i this is understood to mean "per mole of component i." The value of this coefficient depends on the overall composition of the mixture. Thus $\bar{V}_{H_2O}$ is not the same for a water-alcohol mixture that is 10% water as for one that is 90% water.

3. For a pure component the partial molar quantity is identical to the molar (superscript °) value of the pure substance. Thus for pure component i

$$\mu_i = G_i^{\circ} \qquad (8.7)$$

4. A useful feature of the partial molar properties is that the property of a mixture (subscript mix) can be written as the sum of the mole-weighted contributions of the partial molar properties of the components:

$$Y_{mix} = n_1 \bar{Y}_1 + n_2 \bar{Y}_2 \qquad (8.8)$$

In this expression n_1 and n_2 are the numbers of moles of components 1 and 2 in the mixture under consideration.

5. To express the value of property Y_{mix} on a per mole basis, it is necessary to divide Eq. (8.8) by the total number of moles, $n_1 + n_2$. The mole fraction x_i of component i is written

$$x_i = \frac{n_i}{\Sigma_{i=1,2} \, n_i} \qquad (8.9)$$

therefore

$$\frac{Y_{mix}}{n_1 + n_2} = x_1 \bar{Y}_1 + x_2 \bar{Y}_2 \qquad (8.10)$$

6. Relationships which exist between ordinary thermodynamic variables also apply to the corresponding partial molar quantities. Two such relationships are

$$\mu_i = \bar{H}_i - T\bar{S}_i \qquad (8.11)$$

and

$$\bar{V}_i = \left(\frac{\partial \mu_i}{\partial p} \right)_{T, n_{j \neq i}} \qquad (8.12)$$

As noted above, all of the partial molar quantities are concentration dependent. It is convenient to define a thermodynamic concentration called the activity a_i in terms of which the chemical potential is correctly given by the relationship

$$\mu_i = \mu_i^{\theta} + RT \ln a_i \qquad (8.13)$$

The quantity μ_i^{θ} is called the standard state (superscript θ) value of μ_i; it is the value of μ_i when $a_i = 1$. Neither μ_i nor G (nor U, H, etc.) can be measured

absolutely; we deal with differences in these quantities and the standard state value disappears when differences are taken. Although the standard state is defined differently in various situations, we shall generally take the pure component (superscript $^\circ$) as the standard state, so $\mu_1^\theta = \mu_1^\circ = G_1^\circ$. Equation (8.13) defines activity in terms of the difference $\mu_i - \mu_i^\circ$; an obvious question is how a_i relates to the concentration of the solution as prepared or as determined by analytical chemistry.

There are two ways to arrive at the relationship between a_i and the concentration expressed as, say, a mole fraction. One is purely thermodynamic and involves experimental observations; the other involves a model and is based on a statistical approach. We shall examine both.

The first point in developing the thermodynamic method is the observation that for equilbrium between two phases—say, α and β—the chemical potential must be equal in both phases for all components:

$$\mu_i^\alpha = \mu_i^\beta \tag{8.14}$$

This becomes apparent if we consider the increment in G associated with transferring a small number of moles of component i from phase α to phase β at constant pressure and temperature. For equilibrium, $dG = 0 = dG^\alpha + dG^\beta$, and for each phase $dG = \Sigma_i \mu_i \, dn_i$. Since $dn_i^\alpha = -dn_i^\beta$, it follows that

$$dG = 0 = dG^\alpha + dG^\beta = \sum_i \mu_i^\alpha \, dn_i^\alpha + \sum_i \mu_i^\beta \, dn_i^\beta = \sum_i (\mu_i^\alpha - \mu_i^\beta) \, dn_i^\alpha \tag{8.15}$$

from which Eq. (8.14) is obtained.

Next we apply this result to liquid–vapor equilibrium. The following steps outline the argument:

1. Equation (8.14) applies to both components in the liquid and the vapor:

$$\mu_i^l = \mu_i^v \tag{8.16}$$

   The equality also holds if we take the partial derivative of both sides of Eq. (8.16) with respect to p.

2. We use Eq. (8.13) to take the partial with respect to p of μ_i^l:

$$\left(\frac{\partial \mu_i^l}{\partial p}\right)_{T,n_1,n_2} = RT \frac{\partial \ln a_i}{\partial p} \tag{8.17}$$

3. We use Eq. (8.12) to evaluate the partial with respect to p of μ_i^v:

$$\left(\frac{\partial \mu_i^v}{\partial p}\right)_{T,n_1,n_2} = \bar{V}_1^v = \frac{RT}{p_i} \tag{8.18}$$

where the second version treats the vapor as an ideal gas, an assumption we can make without loss of generality concerning the solution.

4. Equations (8.16)–(8.18) can be combined to give

$$\partial \ln a_i = \frac{\partial p_i}{p_i} = \partial \ln p_i \qquad (8.19)$$

5. Equation (8.19) can be integrated using the convention that $a_i = 1$ for the pure component, which has the vapor pressure p_i°:

$$a_i = \frac{p_i}{p_i^\circ} \qquad (8.20)$$

6. Finally, we note that Raoult's law (see Example 7.1) is a limiting law that is observed to apply in all solutions in the limit $x_i \to 1$:

$$\frac{p_i}{p_i^\circ} = x_i \qquad (8.21)$$

7. Thus it is always observed in the limit of the pure component that

$$x_i = a_i \qquad (8.22)$$

A solution which obeys Raoult's law over the full range of compositions is called an ideal solution (see Example 7.1). Equation (8.22) describes the relationship between activity and mole fraction for ideal solutions. In the case of nonideal solutions, the nonideality may be taken into account by introducing an activity coefficient as a factor of proportionality into Eq. (8.22).

A second way of dealing with the relationship between a_i and the experimental concentration requires the use of a statistical model. We assume that the system consists of N_1 molecules of type 1 and N_2 molecules of type 2. In addition, it is assumed that the molecules, while distinguishable, are identical to one another in size and interaction energy. That is, we can replace a molecule of type 1 in the mixture by one of type 2 and both ΔV and ΔH are zero for the process. Now we consider the placement of these molecules in the $N_1 + N_2 = N$ sites of a three-dimensional lattice. The total number of arrangements of the N molecules is given by N!, but since interchanging any of the 1's or 2's makes no difference, we divide by the number of ways of doing the latter—N_1! and N_2!, respectively—to obtain the total number of different ways the system can come about. This is called the thermodynamic probabilty Ω of the system, and we saw in Sec. 3.3 that Ω is the basis for the statistical calculation of entropy. For this specific model

$$\Omega = \frac{N!}{N_1!N_2!} \tag{8.23}$$

The thermodynamic probability is converted to an entropy through the Boltzmann equation [Eq. (3.20)] so we can write for the entropy of the mixture (subscript mix)

$$S_{mix} = k \ln \Omega = k(\ln N! - \ln N_1! - \ln N_2!) \tag{8.24}$$

For large y's, we can use Sterling's approximation

$$\ln y! \cong y \ln y - y \tag{8.25}$$

in terms of which Eq. (8.24) becomes

$$S_{mix} = -k\left[N_1 \ln\left(\frac{N_1}{N}\right) + N_2 \ln\left(\frac{N_2}{N}\right)\right] \tag{8.26}$$

Multiplying and dividing the right-hand side of this expression by N converts the N_i's to mole fractions, and, if N is taken to be N_A, Avogadro's number, kN_A becomes R. Accordingly, we write for 1 mol of mixture

$$S_{mix} = -R(x_1 \ln x_1 + x_2 \ln x_2) \tag{8.27}$$

If either one of the mole fractions in Eq. (8.27) is unity, that is, for a pure component, S_{mix} becomes either S_1 or S_2. For the mixing *process* $1 + 2 \rightarrow$ mix,

$$\Delta S_m = S_{mix} - S_1 - S_2 = -R \sum_{i=1,2} (x_i \ln x_i) \tag{8.28}$$

Although the right-hand sides of Eqs. (8.27) and (8.28) are the same, the former applies to the mixture (subscript mix), while the latter applies to the mixing process (subscript m). The fact that these are identical emphasizes that in Eq. (8.27) we have calculated only that part of the total entropy of the mixture which arises from the mixing process itself. This is called the configurational entropy and is our only concern in mixing problems. The possibility that this mixing may involve other entropy effects—such as an entropy of solvation—is postponed until Sec. 8.12.

The model system for which this value of ΔS_m has been calculated is one for which ΔH_m has been specified to equal zero. Therefore, since $\Delta G = \Delta H - T \Delta S$, it follows that

$$\Delta G_m = RT \sum_{i=1,2} (x_i \ln x_i) \tag{8.29}$$

for this system. We can also use Eqs. (8.7), (8.10), and (8.13) to write ΔG_m in terms of chemical potentials:

$$\Delta G_m = G_{mix} - G_1^\circ - G_2^\circ = RT \sum_{i=1,2} (x_i \ln a_i) \tag{8.30}$$

Comparing Eqs. (8.29) and (8.30) also leads to the conclusion expressed by Eq. (8.22): $a_i = x_i$. Again we emphasize that this result applies only to ideal solutions, but the statistical approach gives us additional insights into the molecular properties associated with ideality in solutions:

1. The energy of interaction between a pair of solvent molecules, a pair of solute molecules, and a solvent-solute pair must be the same so that the criterion that $\Delta H_m = 0$ is met. Such a mixing process is said to be athermal.
2. The solvent and solute molecules must be the same size so that the criterion $\Delta V_m = 0$ is met.
3. Solutions can deviate from ideality because they fail to meet either one or both of these criteria. In reference to polymers in solutions of low molecular weight solvents, it is apparent that nonideality is present because of a failure to meet criterion (2), whether the mixing is athermal or not.

In the next section we shall examine the mixing process for molecules which differ greatly in size, building on the principles reviewed in this section. The reader who desires additional review of these ideas will find this material discussed in detail in textbooks of physical chemistry.

8.3 The Flory-Huggins Theory: The Entropy of Mixing

We concluded the last section with the observation that a polymer solution is expected to be nonideal on the grounds of entropy considerations alone. A nonzero value for ΔH_m would exacerbate the situation even further. We therefore begin our discussion of this problem by assuming a polymer-solvent system which shows athermal mixing. In the next section we shall extend the theory to include systems for which $\Delta H_m \neq 0$. The theory we shall examine in the next few sections was developed independently by Flory and Huggins and is known as the Flory-Huggins theory.

We assume that the mixture contains N_1 solvent molecules, each of which occupies a single site in the lattice we propose to fill. The system also contains N_2 polymer molecules, each of which occupies n lattice sites. The polymer molecule is thus defined to occupy a volume n times larger than the solvent molecules. Strictly speaking, this is the definition of n in the derivation which follows. We shall adopt the attitude that the repeat units in the polymer are equal to solvent molecules in volume, however, so a polymer of degree of

polymerization n will be larger than a solvent molecule by the factor n. In the event that the repeat unit and the solvent are not identical in volume, the chain can be hypothetically divided into n segments which *are* equal in volume to the solvent, and this n will be proportional to the degree of polymerization. The total number of sites in the lattice is $N_1 + nN_2 = N$.

We assign an index number to each of the polymer molecules and pick up the analysis of the problem after i polymer molecules have already been placed on an otherwise empty lattice. Our first question, then, concerns the number of ways the (i + 1)th polymer molecule can be placed in the lattice. The polymer is to be positioned one repeat unit at a time, so it is an easy matter to count the number of available positions for the first segment of the (i + 1)th molecule. Since the total lattice consists of N sites and ni of these are already occupied, the first segment of the (i + 1)th molecule can be placed on any one of the N – ni remaining sites.

There is nothing unique about the placement of this isolated segment to distinguish it from the placement of a small molecule on a lattice filled to the same extent. The polymeric nature of the solute shows up in the placement of the second segment: This must be positioned in a site adjacent to the first, since the units are covalently bonded together. No such limitation exists for independent small molecules. To handle this development we assume that each site on the lattice has z neighboring sites and we call z the coordination number of the lattice. It might appear that the need for this parameter introduces into the model a quantity which would be difficult to evaluate in any eventual test of the model. It turns out, however, that the z's cancel out of the final result for ΔS_m, so we need not worry about this eventuality.

If the molecule under consideration were being placed on an empty lattice, the second segment could go into any one of the z sites adjacent to the first. However, ni of the sites are already filled, so there is a chance that one of the z sites in the coordination sphere of the first segment is already occupied. To deal with this possibility, we assume that the fraction of vacant sites on the lattice *as a whole* also applies in the immediate vicinity of the segment positioned above. This fraction is (N – ni)/N, so the number of possible locations for segment 2 of the (i + 1)th molecule is z(N – ni)/N.

The logic that leads us to this last result also limits the applicability of the ensuing derivation. Applying the fraction of total lattice sites vacant to the immediate vicinity of the first segment makes the model descriptive of a relatively concentrated solution. This is somewhat novel in itself, since theories of solutions more commonly assume dilute conditions. More to the point, the model is unrealistic for dilute solutions where the site occupancy within the domain of a dissolved polymer coil is greater than that for the solution as a whole. We shall return to a model more appropriate for dilute solutions below. For now we continue with the case of the more concentrated solution, realizing

that with a significant fraction of the lattice sites already filled, we can treat the fraction of available sites as constant during the placement of a specific polymer molecule.

The third segment of molecule i + 1 would have z − 1 available sites on an otherwise empty lattice: The first segment, after all, already occupies a site adjacent to segment 2. For a lattice which already contains ni segments—if we consider those from molecule i + 1 as insignificant—the number of sites available for segment 3 is $(z - 1)(N - ni)/N$. This same number also applies to all subsequent segments. Therefore the number of ways ω_{i+1} that the $(i + 1)$th molecule can be placed on a lattice already containing i molecules is given by

$$\omega_{i+1} = (N - ni) z \left(\frac{N - ni}{N}\right) \left[(z - 1)\left(\frac{N - ni}{N}\right)\right]^{n-2}$$

$$= z(z - 1)^{n-2} N \left(\frac{N - ni}{N}\right)^n \qquad (8.31)$$

or for the ith molecule

$$\omega_i = z(z - 1)^{n-2} N \left(\frac{N - n(i - 1)}{N}\right)^n \qquad (8.32)$$

The total number of ways of placing N polymer molecules in the lattice is given by the product of factors like these, one for each molecule:

$$\Omega = \frac{\omega_1 \omega_2 \cdots \omega_i \cdots \omega_{N_2}}{N_2!} = \frac{1}{N_2!} \prod_{i=1}^{N_2} \omega_i \qquad (8.33)$$

In writing Eq. (8.33), we have divided by $N_2!$, since the polymer molecules are interchangeable. Equation (8.33) gives the thermodynamic probabilty for the system according to this model, since there is only one way to place the solvent molecules once the polymer molecules have been positioned on the lattice.

Application of the Boltzmann equation to Eq. (8.33) gives the entropy of the mixture according to this model for concentrated solutions:

$$S_{mix} = k \ln \left(\frac{1}{N_2!}\right) \prod_{i=1}^{N_2} \omega_i \qquad (8.34)$$

Substitution of Eq. (8.32) for ω_i in the above expression yields the desired result. A fair amount of algebra is required to convert this result into a usable form. We outline the tidying up in the following steps; the reader can supply the intervening steps:

1. Substitute Eq. (8.32) for ω in Ω:

$$\Omega = \frac{z^{N_2}(z-1)^{N_2(n-2)}}{N_2!\,N^{N_2(n-1)}} \prod_{i=1}^{N_2} [N - n(i-1)]^n$$

2. Examine the product of terms in this expression, replacing N by nN/n:

$$\prod_{i=1}^{N_2} [N - n(i-1)]^n = n^{nN_2} \prod_{i=1}^{N_2} \left(\frac{N}{n} + 1 - i\right)^n$$

3. Write out a few terms of the product:

$$\Pi = \left(\frac{N}{n} + 1 - 1\right)^n \left(\frac{N}{n} + 1 - 2\right)^n \left(\frac{N}{n} + 1 - 3\right)^n \cdots \left(\frac{N}{n} + 1 - N_2\right)^n$$

4. This is the same as

$$\Pi = \left(\frac{(N/n)!}{(N/n - N_2)!}\right)^n$$

5. Substitute items (2) and (3) into item (1):

$$\Omega = \frac{z^{N_2}(z-1)^{N_2(n-2)}\,n^{nN_2}}{N_2!\,N^{N_2(n-1)}} \left(\frac{(N/n)!}{(N/n - N_2)!}\right)^n$$

6. Apply Sterling's approximation to the logarithm of Ω and simplify:

$$\frac{S_{mix}}{k} = -N_2 \ln\left(\frac{nN_2}{N}\right) - N_1 \ln\left(\frac{N_1}{N}\right)$$

$$+ N_2 [\ln z + (n-2) \ln (z-1) + (1-n) + \ln n]$$

7. Consider the above result for the case of $N_2 = 0$, that is, for pure solvent: The expression in item (6) is proportional to the entropy of the pure solvent: $S_1 = 0$.
8. If $N_1 = 0$, that is, for pure polymer, the expression in item (6) gives

$$S_2 = kN_2 [\ln z + (n-2) \ln (z-1) + (1-n) + \ln n]$$

This is the entropy of the disordered polymer when the latter entirely fills the lattice.

The entropy of the mixture minus the entropy of the pure components—that is, item (6) minus items (7) and (8)—gives ΔS_m according to this model:

$$\Delta S_m = -k \left[N_1 \ln \left(\frac{N_1}{N} \right) + N_2 \ln \left(\frac{nN_2}{N} \right) \right] \tag{8.35}$$

Multiplying and dividing the right-hand side by $N_1 + N_2$ and letting this sum equal Avogadro's number of molecules gives ΔS_m per mole of solution:

$$\Delta S_m = -R \left[x_1 \ln \left(\frac{N_1}{N} \right) + x_2 \ln \left(\frac{nN_2}{N} \right) \right] \tag{8.36}$$

This result should be compared with Eq. (8.28) for the case of the ideal mixture. It is reassuring to note that for $n = 1$, Eq. (8.36) reduces to Eq. (8.28). Next let us consider whether a change of notation will clarify Eq. (8.36) still more. Recognizing that the solvent, the repeat unit, and the lattice site all have the same volume, we see that N_1/N is the volume fraction occupied by the solvent in the mixture and nN_2/N is the volume fraction of the polymer. Letting ϕ_i be the volume fraction of component i, we see that Eq. (8.36) becomes

$$\Delta S_m = -R (x_1 \ln \phi_1 + x_2 \ln \phi_2) \tag{8.37}$$

This model then leads us through a thicket of statistical and algebraic detail to the satisfying conclusion that going from small solute molecules to polymeric solutes only requires the replacement of mole fractions with volume fractions within the logarithms. Note that the mole fraction weighting factors are unaffected.

Since the system was defined to be athermal, $\Delta G_m = -T \Delta S_m$, so

$$\Delta G_m = RT (x_1 \ln \phi_1 + x_2 \ln \phi_2) \tag{8.38}$$

Since the ϕ's are fractions, the logarithms in Eq. (8.38) are less than unity and ΔG_m is negative for all concentrations. In the case of athermal mixtures entropy considerations alone are sufficient to account for polymer–solvent miscibility at all concentrations. Exactly the same is true for ideal solutions. As a matter of fact, it is possible to regard the expressions for ΔS_m and ΔG_m for ideal solutions as special cases of Eqs. (8.37) and (8.38) for the situation where n happens to equal unity. The following example compares values for ΔS_m for ideal and Flory–Huggins solutions to examine quantitatively the effect of variations in n on the entropy of mixing.

Example 8.1

Evaluate ΔS_m for ideal solutions and for athermal solutions of polymers having n values of 50, 100, and 500 by solving Eqs. (8.28) and (8.38) at regular intervals of mole fraction. Compare these calculated quantities by preparing a suitable plot of the results.

Solution

We express the calculated entropies of mixing in units of R. For ideal solutions the values of ΔS_m are evaluated directly from Eq. (8.28):

x_2	0.1	0.2	0.3	0.4	0.5	0.6	0.7	0.8	0.9
x_2	0.9	0.8	0.7	0.6	0.5	0.4	0.3	0.2	0.1
$\Delta S_m/R$	0.325	0.500	0.611	0.673	0.693	0.673	0.611	0.500	0.325

For polymer solutions we seek the relationship between mole fraction and volume fraction. Since $\phi_2/\phi_1 = nN_2/N_1$, $N_2/N_1 = (1/n)(\phi_2/\phi_1)$. Also,

$$x_2 = \frac{N_2}{N_1 + N_2} = \frac{N_2/N_1}{1 + N_2/N_1} = \frac{(1/n)\phi_2/\phi_1}{1 + (1/n)\phi_2/\phi_1}$$

Remember that both x and ϕ are fractional concentrations, so $x_1 + x_2 = 1$ and $\phi_1 + \phi_2 = 1$. Therefore

$$x_2 = \frac{(1/n)\phi_2/(1 - \phi_2)}{1 + (1/n)\phi_2/(1 - \phi_2)}$$

from which we find

$$\phi_2 = \frac{x_2}{(1/n) + x_2(1 - 1/n)}$$

This result enables us to convert mole fractions to volume fractions. Table 8.1 lists the corresponding values of ϕ_i and x_i for n = 50, 100, and 500 as needed for the evaluation of ΔS_m. With x_i's and the corresponding ϕ_i's available, the required values of $\Delta S_{mix}/R$ are calculated by Eq. (8.38):

n = 50	1.71	2.10	2.19	2.14	1.97	1.75	1.49	1.06	0.62
n = 100	2.25	2.62	2.65	2.53	2.31	1.99	1.66	1.25	0.69
n = 500	3.63	3.87	3.77	3.49	3.11	2.66	2.11	1.52	0.85

Table 8.1 Values of ϕ Corresponding to the Indicated Mole Fractions for Values of $n = 50$, 100, and 500 (These Quantities are Used in Example 8.1.)

x_2	0.1	0.2	0.3	0.4	0.5	0.6	0.7	0.8	0.9
x_1	0.9	0.8	0.7	0.6	0.5	0.4	0.3	0.2	0.1
For n = 50									
ϕ_2	0.847	0.926	0.955	0.971	0.980	0.987	0.992	0.995	0.998
ϕ_1	0.153	0.074	0.045	0.029	0.020	0.013	0.008	0.005	0.002
For n = 100									
ϕ_2	0.917	0.962	0.977	0.985	0.990	0.993	0.996	0.998	0.999
ϕ_1	0.083	0.038	0.023	0.015	0.010	0.007	0.004	0.002	0.001
For n = 500									
ϕ_2	0.982	0.992	0.995	0.997	0.998	0.9987	0.9991	0.9995	0.9998
ϕ_1	0.018	0.008	0.005	0.003	0.002	0.0013	0.0009	0.0005	0.0002

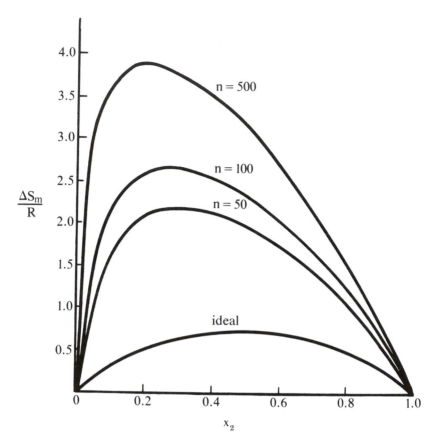

Figure 8.1   The entropy of mixing (in units of R) as a function of mole fraction solute for ideal mixing and for the Flory–Huggins lattice model with n = 50, 100, and 500. Values are calculated in Example 8.1.

A plot of these values is shown in Fig. 8.1. Note the increase in the entropy of mixing over the ideal value with increasing n value. Also note that the maximum occurs at decreasing mole fractions of polymer with increasing degree of polymerization.

•

Our ultimate goal is to develop expressions for the free energy of mixing according to the lattice model of the Flory-Huggins theory. The next step toward realization of this goal is to relax the restriction which limits the result to athermal solutions. In the next section we shall develop the theory for the enthalpy of mixing. Then we combine the equations for entropy and enthalpy to obtain an expression for ΔG_m according to the Flory-Huggins model.

8.4   The Flory–Huggins Theory: The Enthalpy of Mixing

In the liquid state molecules are in intimate contact, so the energetics of molecular interactions generally make a contribution to the overall picture of the mixing process. There are several aspects of the situation that we should be aware of before attempting to formulate a theory for ΔH_m:

1.   Our immediate goal is an expression for ΔH_m and we must remember that this is the *difference* in the enthalpies of the solution and the pure components. We need not worry about the absolute values of the enthalpies of the individual states.
2.   Enthalpies of mixing have their origin in the forces that operate between individual molecules. Intermolecular forces drop off rapidly with increasing distance of separation between molecules. This means that only nearest neighbors need be considered in the model.
3.   Until surface contact, the force between molecules is always one of attraction, although this attraction has different origins in different systems. London forces, dipole–dipole attractions, acid–base interactions, and hydrogen bonds are some of the types of attraction we have in mind. In the foregoing list, London forces are universal and also the weakest of the attractions listed. The interactions increase in strength and also in specificity in the order listed.
4.   Since London forces are universal and nonspecific, we shall eventually emphasize these, realizing that stronger forces may outweigh the London contribution in some systems. We anticipate the best results in nonpolar systems where London forces account for the interactions.

The lattice model that served as the basis for calculating ΔS_m in the last section continues to characterize the Flory-Huggins theory in the development of an expression for ΔH_m. Specifically, we are concerned with the change in enthalpy which occurs when one species is replaced by another in adjacent lattice sites. The situation can be represented in the notation of a chemical reaction:

$$(1,1) + (2,2) \rightarrow 2(1,2) \tag{8.A}$$

where 1 and 2 refer to solvent molecules and polymer *repeat units*, respectively, and the parentheses enclose the particular pair under consideration. Suppose we indicate the pairwise interaction energy between species i and j as w_{ij}; then for reaction (8.A),

$$\Delta w = 2w_{12} - w_{11} - w_{22} \tag{8.39}$$

The change in interaction energy per 1,2 pair is thus $\frac{1}{2}$ Δw. Next we must consider how this scales up for a large array of molecules, and particularly how to describe the concentration dependence of the result.

Each lattice site is defined to have z nearest neighbors, and ϕ_1 and ϕ_2, respectively, can be used to describe the fraction of sites which are occupied by solvent molecules and polymer segments. The following inventory of interactions can now be made for the mixture:

1. Each polymer segment (subscript 2) is surrounded, on the average, by $z\phi_2$ polymer segments and $z\phi_1$ solvent molecules.
2. The contribution to the energy of this polymer segment interacting with its neighbors is $z\phi_2 w_{22} + z\phi_1 w_{12}$.
3. Since the lattice consists of N sites of which $\phi_2 N$ are occupied by polymer segments, the contribution to the energy of all interactions of the sort described in item (2) is $(\frac{1}{2})$ $z\phi_2 N[(1 - \phi_1)w_{22} + \phi_1 w_{12}]$. The factor $\frac{1}{2}$ has been introduced, since all pairs are counted twice in calculating this total.
4. Each solvent molecule (subscript 1) is surrounded, on the average, by $z\phi_2$ polymer segments and $z\phi_1$ solvent molecules.
5. The contribution to the energy of this molecule interacting with its neighbors is $z\phi_2 w_{12} + z\phi_1 w_{11}$.
6. In parallel with item (3), the contribution of all solvent interactions to the total energy is $(\frac{1}{2})$ $z\phi_1 N[\phi_2 w_{12} + (1 - \phi_2)w_{11}]$.
7. When either pure component is considered, all w_{12} terms are zero, as well as the terms containing the volume fraction of the other component. Thus, for pure polymer, item (3) becomes $(\frac{1}{2})z\phi_2 Nw_{22}$, and for the pure solvent item (6) becomes $(\frac{1}{2})z\phi_1 Nw_{11}$.

The value of ΔH_m can be assembled from the foregoing by adding the energy contributions of items (3) and (6) and subtracting the contributions of the pure components given by item (7):

$$\Delta H_m = (\tfrac{1}{2})zN(2\phi_1\phi_2 w_{12} - \phi_1\phi_2 w_{11} - \phi_1\phi_2 w_{22}) \qquad (8.40)$$

or, substituting Eq. (8.39),

$$\Delta H_m = (\tfrac{1}{2})zN\phi_1\phi_2 \Delta w \qquad (8.41)$$

Since $\frac{1}{2}$ Δw is the change in interaction energy per 1,2 pair, it can be expressed as some multiple χ' of kT per pair or of RT per mole of pairs. It is also conventional to consolidate the lattice coordination number and χ' into a single parameter χ, since z and χ' are not measured separately. With this change of notation $\frac{1}{2}z$ Δw is replaced by its equivalent χRT, and Eq. (8.41) becomes

$$\Delta H_m = N\phi_1\phi_2\chi RT = N_1\phi_2\chi RT \qquad (8.42)$$

The quantity χ is called the Flory-Huggins interaction parameter: It is zero for athermal mixtures, positive for endothermic mixing, and negative for exothermic mixing. These differences in sign originate from Eq. (8.39) and reaction (8.A).

Until now we have been purposely vague about the quantity Δw. Since we shall use the notation of Eq. (8.42) from now on, it is convenient to say a few more things about Δw before we lose sight of it entirely. Reaction (8.A) is clear enough; what is unspecific are the conditions under which this "reaction" takes place. Several possibilities come to mind, and each imparts a slightly different meaning to the "energy" w:

1. Since the solvent molecules, the polymer segments, and the lattice sites are all assumed to be equal in volume, reaction (8.A) implies constant volume conditions. Under these conditions, ΔU is needed and what we have called Δw might be better viewed as the contribution to the internal energy of a pairwise interaction ΔU_{pair}, where the subscript reminds us that this is the contribution of a single pair formation by reaction A.
2. Especially for large values of Δw, there could be an additional entropy effect beyond that calculated in the last section which arises from the interaction of nearest neighbors. That is, reaction (8.A) might be characterized by both a ΔH_{pair} *and* a ΔS_{pair}. In this case Δw might be viewed as the pairwise contribution to a free energy ΔG_{pair} with

$$\Delta G_{pair} = \Delta H_{pair} - T \Delta S_{pair} \qquad (8.43)$$

3. In writing Eq. (8.41), we have clearly treated Δw as a contribution to enthalpy. This means we neglect volume changes (ΔH_{pair} versus ΔU_{pair}) and entropy changes beyond the configurational changes discussed in the last section (ΔG_{pair} versus ΔH_{pair}). In a subsequent development it is convenient to allow an additional entropy effect, and we shall return to Eq. (8.43) at that point.

The equations we have written until now in this section impose no restrictions on the species they describe or on the origin of the interaction energy. Volume and entropy effects associated with reaction (8.A) will be less if χ is not too large. Aside from this consideration, any of the intermolecular forces listed above could be responsible for the specific value of χ. The relationships for ΔS_m in the last section are based on a specific model and are subject to whatever limitations that imposes. There is nothing in the formalism for ΔH_m that we have developed until now that is obviously inapplicable to certain specific systems. In the next section we shall introduce another approximation

which will impose such a limitation, so it is important to recognize the point to which the generality extends.

In this section and the last, we have examined the lattice model of the Flory-Huggins theory for general expressions relating ΔH_m and ΔS_m to the composition of the mixture. The separate components can therefore be put together to give an expression for ΔG_m as a function of temperature and composition:

$$\Delta G_m = \Delta H_m - T\,\Delta S_m = RT[N\phi_1\phi_2\chi - (x_1 \ln \phi_1 + x_2 \ln \phi_2)] \qquad (8.44)$$

The contribution of ΔS_m to this expression is given by Eq. (8.37) and is always positive; hence it contributes to a negative value of ΔG_m. The ΔH_m contribution, on the other hand, can be either positive or negative. If ΔH_m, as given by Eq. (8.41), is negative, this also contributes to a negative value for ΔG_m. If ΔH_m is positive, however, ΔS_m and ΔH_m make opposing contributions to ΔG_m, and the net effect depends on the magnitudes of the quantities involved. We shall return to an examination of this possibility in Sec. 8.6.

There are several reasons for devoting so much attention to the foregoing derivation of ΔG_m. For one thing, a theory like this gives the reader a "feel" for the origin of the various effects, which is valuable even if the theory is somewhat oversimplified. This feature is of considerable value in a textbook such as this. In addition, theories help us both in the interpretation and the prediction of experimental observations. We shall apply the ideas of the Flory-Huggins theory to the interpretation of experiments in later sections. For the present let us consider the predictive aspect. Suppose, for example, that a researcher is seeking a solvent for a particular polymer such that the two would show athermal mixing. The foregoing theory suggests that the search should focus on solvents for which the interaction with polymer molecules and among its own molecules are similar to polymer–polymer interactions. In this sense, Eq. (8.40) quantifies the rule of thumb "like dissolves like"; with no opposition from enthalpy, entropy considerations guarantee mixing. In the next section we shall develop this idea still further, although for a more limited range of systems.

8.5 Cohesive Energy Density

It is not particularly difficult to find macroscopic measures of interactions between small molecules of the same type, that is, quantities which are proportional to w_{11} and w_{22} in Eq. (8.40). Among the possibilities, we consider the change in internal energy ΔU_v for the vaporization process for component i. This can be related to w_{ii} in terms of the lattice model by the expression

$$\Delta U_{v,i} = \tfrac{1}{2}zNw_{ii} \qquad (8.45)$$

where N equals Avogadro's number when ΔU_v is written per mole.

A more troublesome quantity is the interaction w_{12} between different species. It seems reasonable to expect this quantity to be some sort of hybrid of the w's for the separate components, at least provided that the mixing does not open the possibility for some specific interactions that is nonexistent among molecules of the pure components. We therefore postulate that no such possibility exists and consider for the remainder of this section only nonspecific interactions. In terms of the types of intermolecular forces enumerated in the last section, London and dipole–dipole attractions are certainly less specific than either acid–base or hydrogen bond interactions. The first two types are essentially physical interactions, depending on the distribution of charge in the interacting species. London forces are proportional to the polarizability of the molecules; dipole–dipole forces are proportional to their permanent dipole moments. These molecular parameters are also discussed in Sec. 10.3. While different molecules differ in the specific values of their polarizability and dipole moment, the notion that some sort of average value of these properties applies to the mixed interaction seems quite plausible. By contrast, the acidity or basicity of a molecule is the consequence of either empty or filled orbitals, and describing their interaction in terms of an "average orbital occupancy" is meaningless.

The application of these ideas to the mixing of low molecular weight liquids has been the object of extensive research. As a result of these investigations, the appropriate kind of "hybridization" of individual molecular properties in w_{12} is found to be the geometrical mean:

$$w_{12} \cong \sqrt{w_{11} w_{22}} \tag{8.46}$$

As argued above, this result is found to work best for substances in which both the 1,1 and 2,2 forces are either London or dipole–dipole. Even the case of one molecule with a permanent dipole moment interacting with a molecule which has only polarizability and no permanent dipole moment—such species interact by permanent dipole-induced dipole attraction—is not satisfactorily approximated by Eq. (8.46). In this context the "like dissolves like" rule means "like" with respect to the origin of intermolecular forces.

An advantage of expressing w_{12} as a geometrical mean is that Eq. (8.39) can now be written in an especially simple form. Combining Eqs. (8.39) and (8.46) gives

$$-\Delta w = w_{11} + w_{22} - 2\sqrt{w_{11} w_{22}} = (\sqrt{w_{11}} - \sqrt{w_{22}})^2 \tag{8.47}$$

As noted above, these homogeneous interaction energies can be represented by Eq. (8.45); therefore combining (8.45) and (8.47) yields

$$\Delta w \propto [(\Delta U_{v,1})^{1/2} - (\Delta U_{v,2})^{1/2}]^2 \tag{8.48}$$

where we have changed the sign in keeping with the convention that attractive energies such as w are negative while ΔU_v is positive.

Since we are explicitly interested in the difference in the sizes of solvent and solute molecules, it is more appropriate to express the values of ΔU_v on a "per unit volume" basis rather than on a molar basis. Accordingly, in Eq. (8.41) we replace the total number of sites N by the total volume of the mixture V and write

$$\Delta H_m = zV\phi_1\phi_2 \left[\left(\frac{\Delta U_{v,1}}{V_1^\circ}\right)^{1/2} - \left(\frac{\Delta U_{v,2}}{V_2^\circ}\right)^{1/2} \right]^2 \qquad (8.49)$$

The quantity $\Delta U_{v,i}/V_i^\circ$ is the internal energy of vaporization per unit volume and is called the cohesive energy density (CED) of component i. The square root of the CED is generally given the symbol δ_i for component i.

The special appeal of this approach is that it allows the heat of mixing to be estimated in terms of a single parameter assigned to each component. This considerably simplifies the characterization of mixing, since m components (with m δ values) can be combined into $m(m-1)/2$ binary mixtures, so a considerable data reduction follows from tabulating δ's instead of ΔH_m's. Table 8.2 is a list of CED and δ values for several common solvents, as well as estimated δ values for several common polymers.

Table 8.2 Values of the Cohesive Energy Density (CED) for Some Common Solvents and the Solubility Parameter δ for These Solvents and Some Common Polymers

Solvents	CED (cal cm^{-3})	δ [(cal cm^{-3})$^{1/2}$]	Polymers	δ [(cal cm^{-3})$^{1/2}$]
Cyclohexane	67.2	8.2	Polyisobutylene	7.5–8.0
CCl$_4$	74.0	8.6	Polyethylene	7.7–8.2
Toluene	79.2	8.9	Natural rubber	8.1–8.5
Benzene	84.6	9.2	Polystyrene	9.1–9.4
Methyl acetate	92.2	9.6	Poly(ethylene	
Acetone	98.0	9.9	phthalate)	9.3–9.9
Cyclohexanone	98.0	9.9	Polyacrylonitrile	12.0–14.0
Acetic acid	102.0	10.1	Nylon-6,6	13.5–15.0
Furfural	125.4	11.2		
Cyclohexanol	130.0	11.4		
Methanol	210.3	14.5		
Water	547.6	23.4		

Source: H. Burrell, in J. Brandrup and E. H. Immergut (Eds.), *Polymer Handbook*, 2nd ed., Wiley, New York, 1975.

There are several additional comments regarding Eq. (8.49) which should be noted:

1. The development of Eq. (8.49) is based on the approximation given by Eq. (8.46), and the latter is applicable only under the special circumstances cited above.

2. Athermal mixing is expected in the case of $\delta_1 = \delta_2$. Since polymers generally decompose before evaporating, the definition $\delta = (\Delta U_v / V^\circ)$ is not useful for polymers. There are noncalorimetric methods for identifying athermal solutions, however, so the δ value of a polymer is equated to that of the solvent for such a system to estimate the CED for the polymer. The fact that a range of δ values is shown for the polymers in Table 8.2 indicates the margin of uncertainty associated with this approach.

3. In addition to thermodynamic applications, δ_2 values have also been related to the glass transition temperature of a polymer, and the difference $\delta_2 - \delta_1$ to the viscosity of polymer solutions. The "best" values of δ have been analyzed into group contributions, the sum of which can be used to estimate δ_2 for polymers which have not been characterized experimentally.

4. Equation (8.49) accounts only for endothermic mixing. It is not too surprising that we are thus led to associate exothermic ΔH_m values with more specifically chemical interactions between solvent and solute as opposed to the purely physical interactions we have been describing in this approximation.

5. It is only the contribution of ΔH_m to ΔG_m that we are discussing here, but we see the effect of this contribution—in the systems for which the approximation is valid—is that a solvent becomes less suitable to dissolve a polymer the greater the difference is between their δ values. At best, when $\delta_1 = \delta_2$, the solvent effect is neutral. Cases for which a favorable specific interaction between solvent and polymer actually promotes solution are characterized by negative values of ΔH_m and are therefore beyond the capabilities of this model.

Note that the development of ΔH_m in terms of the cohesive energy density is an optional extension of the Flory–Huggins theory in which χ can be either positive or negative. As a matter of fact, the principles outlined in this section were first developed and have been extensively tested for solutions of low molecular weight compounds. As might be expected, a number of variations on these basic ideas have been examined; the interested reader will find additional details in references on the theory of solutions. As far as polymers are concerned, the greatest value of this approach is that it offers useful predictive guidelines for pairing solvents and polymers to give athermal mixing. This extends to such considerations as the compatibility of plasticizers with bulk polymers and the swelling of cross-linked polymers in contact with solvents, as well as the formation of polymer solutions.

We conclude this section with a numerical example illustrating the evaluation of δ for a low molecular weight solvent.

Example 8.2

For benzene $\Delta H_v = 33.90$ kJ mol^{-1} and $\rho = 0.879$ g cm^{-3} at 25°C. Use these data to evaluate the cohesive energy density and δ for benzene at 25°C. Why is the cohesive energy density defined in terms of ΔU_v rather than ΔH_v?

Solution

The vaporization process describes the phase change liquid (l) → gas (g), so from the definition of ΔH we can write

$$\Delta U_v = \Delta H_v - P\Delta V_v = \Delta H_v - P(V_g - V_1) \cong \Delta H_v - PV_g \cong \Delta H_v - RT$$

For benzene at 25°C this becomes $\Delta U_v = 33,900 - 8.314 \,(298) = 31,400$ J mol^{-1}. The molar volume of a compound is given by $V° = $ (molecular weight)/ (density). For benzene at 25°C, this becomes $V° = 78.0/0.879 = 88.7$ cm^3 mol^{-1}. The cohesive energy density is simply the ratio $\Delta U_v/V°$, but in evaluating this numerically, the question of units arises. By convention, these are usually expressed in calories per cubic centimeter, so we write

$$\text{CED} = \frac{31,400 \text{ J}}{\text{mol}} \times \frac{1 \text{ cal}}{4.184 \text{ J}} \times \frac{1 \text{ mol}}{88.7 \text{ cm}^3} = 84.6 \text{ cal cm}^{-3}$$

Therefore $\delta = (84.6)^{\frac{1}{2}} = 9.2 \,(\text{cal cm}^{-3})^{\frac{1}{2}}$. The unit $(\text{cal cm}^{-3})^{\frac{1}{2}}$ is often called the Hildebrand (symbol H) in honor of Hildebrand, who made extensive contributions to the theory of solutions.

The vaporization process requires energy both to overcome intermolecular attractions and to push back the surroundings to make room for the vapor. The quantity ΔU_v measures the former, while ΔH_v takes both into account. In connection with the mixing process, it is the contribution of intermolecular forces which we seek to evaluate, so ΔU_v is a more suitable measure of this quantity.

•

8.6 Phase Separation

The criterion for phase equilibrium is given by Eq. (8.14) to be the equality of chemical potential in the phases in question for each of the components in the mixture. In Sec. 8.8 we shall use this idea to discuss the osmotic pressure of a

polymer solution. First, however, let us consider the separation of a polymer-solvent mixture into two immiscible solutions. We shall begin this discussion by examining the general thermodynamics of such a situation. Next we shall use the Flory–Huggins model as the basis for a theoretical approach to the topic. Finally, we shall consider the application of these ideas to the practical problem of polymer fractionation with respect to molecular weight.

Let us suppose that both ΔS_m and ΔH_m, and therefore ΔG_m, are known as functions of concentration for a particular mixture. We shall eventually assume the latter is given by Eq. (8.44), but the following general remarks are independent of any model. When ΔS_m is positive and ΔH_m is negative, then ΔG_m is negative and the solution process occurs spontaneously. Such a situation is seen to apply over the full range of concentrations in Fig. 8.2a. Now let us consider the significance of the tangent to the curve drawn in Fig. 8.2a. From Eqs. (8.7) and (8.10), it follows that

$$\Delta G_m = x_1 \Delta\mu_1 + x_2 \Delta\mu_2 = (1 - x_2) \Delta\mu_1 + x_2 \Delta\mu_2 \tag{8.50}$$

where $\Delta\mu_i = \mu_i - \mu_i^\circ$. Rearranging, we obtain

$$\Delta G_m = \Delta\mu_1 + x_2(\Delta\mu_2 - \Delta\mu_1) \tag{8.51}$$

This is the expression of a straight line having a slope equal to $\Delta\mu_2 - \Delta\mu_1$ and an intercept at $x_2 = 0$ equal to $\Delta\mu_1$. By rewriting Eq. (8.50) in terms of x_1, we get the equation

$$\Delta G_m = \Delta\mu_2 + x_1(\Delta\mu_1 - \Delta\mu_2) = \Delta\mu_2 - x_1(\Delta\mu_2 - \Delta\mu_1) \tag{8.52}$$

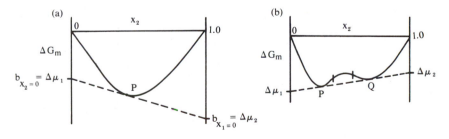

Figure 8.2 Schematic illustrations of ΔG_m versus x_2 showing how $\mu_i - \mu_i^\circ$ may be determined by the tangent drawn at any point. (a) The polymer–solvent system forms a single solution at all compositions. (b) Compositions between the two minima separate into equilibrium phases P and Q.

Both Eqs. (8.51) and (8.52) are seen to describe the line drawn tangent to the curve at point P in Fig. 8.2a, and they give the value of $\Delta\mu_i$ for each component for any mixture at which the tangent is drawn.

Next suppose ΔS_m and ΔH_m are both positive. In this case these two partially offset one another, and a plot of ΔG_m resembling that shown in Fig. 8.2b may result. We are particularly interested in the two minima in this curve and the hump between them. A common tangent can always be drawn to two such minima so the above discussion shows that the minima at points P and Q in Fig. 8.2b each have the same values of $\Delta\mu_1$ and $\Delta\mu_2$. Since $\Delta\mu_i$ is simply the difference between μ_i and its value for the pure component, the chemical potential for each component is seen to have the same value for both solution P and solution Q in Fig. 8.2b.

Now we turn our attention to the hump that separates these two minima. Between points P and Q, the plot of ΔG_m versus x_2 goes through a maximum and the values of ΔG_m along that portion of the curve are larger than the corresponding values on the line which is tangent to the two minima. This means that the mixtures whose compositions lie between points P and Q can achieve a lower free energy by separating into the two solutions at which the minima are located. Since these phases have the same chemical potential for both components, the phases at P and Q are in equilibrium, and any mixture which would otherwise fall between P and Q in composition will separate into phases with compositions $x_{2,P}$ and $x_{2,Q}$.

Remember that the hump which causes the instability with respect to phase separation arises from an unfavorable ΔH_m; considerations of configurational entropy alone favor mixing. Since ΔS_m is multiplied by T in the evaluation of ΔG_m, we anticipate that as the temperature increases, curves like that shown in Fig. 8.2b will gradually smooth out and eventually pass over to the form shown in Fig. 8.2a. The temperature at which the wiggles in the curve finally vanish will be a critical temperature for this particular phase separation. We shall presently turn to the Flory-Huggins theory for some mathematical descriptions of this critical point. The following example reminds us of a similar problem encountered elsewhere in physical chemistry.

Example 8.3

The van der Waals equation of state for 1 mol of a nonideal gas contains two constants a and b which are characteristic of a particular gas:

$$\left(p + \frac{a}{V^2}\right)(V - b) = RT$$

These constants can be related to the coordinates of the critical point of the gas: p_c, V_c, and T_c. Outline the strategy by which the van der Waals a and b

constants are related to p_c, V_c, and T_c. Point out the parallel between the van der Waals problem and the problem of separation into immiscible solutions, and suggest how the van der Waals strategy might be applied to the problem at hand.

Solution

Multiplied out, the van der Waals equation is cubic in volume,

$$pV^3 - (pb + RT)V^2 + aV - ab = 0$$

and, in general, there are three values of V which satisfy the equation for a given pressure and temperature. Of course, this contradicts the behavior of actual gases. Above a certain temperature, however, two of the roots of the van der Waals equation become imaginary, and above this temperature only one real solution to the equation exists. This mathematically unique temperature and the physically unique critical temperature are taken to be the same.

To identify this crossover point in the mathematical behavior, we recognize that the maximum and minimum that occur in the three-root region and the inflection point between them merge into a common point at the transition in behavior. The maximum and minimum are identified by the criterion $(\partial V/\partial p)_T = 0$, and the inflection point by the criterion $(\partial^2 V/\partial p^2)_T = 0$. Carrying out these two differentiations and identifying the volume and temperature that appear in the results as V_c and T_c gives two equations which are functions of a and b and the critical coordinates. These simultaneous equations can be solved for a and b to give $a = 27(RT_c)^2/64p_c$ and $b = RT_c/8p_c$.

For the phase separation problem, the maximum and minima in Fig. 8.2b and the inflection points between them must also merge into a common point at the critical temperature for the two-phase region. This is the mathematical criterion for the "smoothing out of wiggles," as the critical point was described above.

•

The procedure outlined in this example needs only one modification to be applicable to the critical point for solution miscibility. In Fig. 8.2b we observe that there are two inflection points in the two-phase region between P and Q. There is only one such inflection point in the "two-phase" region of the van der Waals equation. The presence of the extra inflection point means that still another criterion must be added to describe the critical point: The two inflection points must also merge with each other as well as with the maximum and the minima.

Before turning to the actual implementation of this strategy for characterizing the critical point, it seems worthwhile to summarize the procedure, since

the necessary steps have been described for an analogous problem above, and not for the specific problem of interest. In summary, we can state the following:

1. Figure 8.2b is a schematic representation of ΔG_m versus x_2 for a system which shows a miscibility gap. Any attempt to prepare a mixture between P and Q in composition will result in separation into the two phases P and Q at equilibrium.

2. There exists a critical temperature for this behavior above which Fig. 8.2b changes over to Fig. 8.2a, which describes miscibility in all proportions.

3. The mathematical behavior of the critical point is characterized by the two minima, the maximum, and the two inflection points all merging into a common point so that the entire function displays the smooth features seen *outside* the PQ region.

4. In Fig. 8.2b the minima and the maximum are described by $(\partial \Delta G_m / \partial x_2)_T = 0$ and the inflection points by $(\partial^2 \Delta G_m / \partial x_2^2)_T = 0$. If the second derivative describes the inflection point, the third derivative describes the *displacement* of the inflection point. Hence at the point where the two inflection points merge, $(\partial^3 \Delta G_m / \partial x^3)_T = 0$.

Since the Flory–Huggins theory provides us with an analytical expression for ΔG_m in Eq. (8.44), it is not difficult to carry out the differentiations indicated above to consider the critical point for miscibility in terms of the Flory–Huggins model. While not difficult, the mathematical manipulations do take up too much space to include them in detail. Accordingly, we indicate only some intermediate points in the derivation. We begin by recalling that $(\partial \Delta G_m / \partial n_i)_{p,T} = \Delta \mu_i$, so by differentiating Eq. (8.44) with respect to either N_1 or N_2, we obtain

$$\mu_1 - \mu_1^\circ = RT \left[\ln (1 - \phi_2) + \phi_2 \left(1 - \frac{1}{n} \right) + \chi \phi_2^2 \right] \qquad (8.53)$$

and

$$\mu_2 - \mu_2^\circ = RT \left[\ln \phi_2 + (1 - \phi_2)(1 - n) + \chi n (1 - \phi_2)^2 \right] \qquad (8.54)$$

These differentiations are easily accomplished by noting that $N_1 = V\phi_1$ and recalling that $\phi_2 = nN_2/(N_1 + nN_2)$. Of course, n, χ, and RT are constant in the differentiation.

Since $\Delta \mu_1$ and $\Delta \mu_2$ are both obtained from differentiating the same expression for ΔG_m, it makes no difference which of these we work with further. In addition, it makes no difference whether we differentiate with respect to x_2 or x_1, since $dx_1 = -dx_2$ and we are setting the results equal to zero. Furthermore, since higher derivatives will be set equal to zero, we can differentiate with respect to volume fraction instead of mole fraction. This is because $\partial / \partial x =$

$(\partial\phi/\partial x)(\partial/\partial\phi)$ and the factor arising from the $\partial\phi/\partial x$ operation will cancel when the result is set equal to zero. With these considerations in mind, the criterion for the critical point can be written as

$$\frac{\partial^2 \Delta G_m}{\partial x_2^{~2}} = 0 = \frac{\partial \Delta \mu_1}{\partial \phi_2} = -\frac{1}{1 - \phi_{2,c}} + 1 - \frac{1}{n} + 2\chi_c \phi_{2,c} \qquad (8.55)$$

and

$$\frac{\partial^3 \Delta G_m}{\partial x_2^{~3}} = 0 = \frac{\partial^2 \Delta \mu_1}{\partial \phi_2^{~2}} = -\frac{1}{(1 - \phi_{2,c})^2} + 2\chi_c \qquad (8.56)$$

In these expressions χ_c is the critical (subscript c) value of χ which marks the threshold at which immiscibility sets in, and $1 - \phi_{2,c}$ or $\phi_{1,c}$ is the volume fraction of the solvent in the solution at this point. Rearranging Eq. (8.56), we obtain

$$\chi_c = \frac{1}{2\phi_{1,c}^{~2}} \qquad (8.57)$$

which, on substitution back into Eq. (8.55), gives

$$-\frac{1}{\phi_{1,c}} + 1 - \frac{1}{n} + \frac{1 - \phi_{1,c}}{\phi_{1,c}^{~2}} = 0 \qquad (8.58)$$

Solving Eq. (8.58) for $\phi_{1,c}$ yields

$$\phi_{1,c} = \frac{1}{1 + n^{-\frac{1}{2}}} \qquad (8.59)$$

which shows that the critical solution is almost pure solvent for polymers of very large n. Combining Eqs. (8.57) and (8.59) gives

$$\chi_c = \frac{1}{2}(1 + n^{-\frac{1}{2}})^2 \qquad (8.60)$$

For a polymer of infinite chain length, $\chi_c = 0.500$, and for n values of 10^4, 10^3, and 10^2 respectively, χ_c equals 0.510, 0.532, and 0.605.

Experimental results describing limited mutual solubility are usually presented as phase diagrams in which the compositions of the phases in equilibrium with each other at a given temperature are mapped for various temperatures. As noted above, the chemical potentials are the same in the equilibrium phases, so Eqs. (8.53) and (8.54) offer a method for calculating such

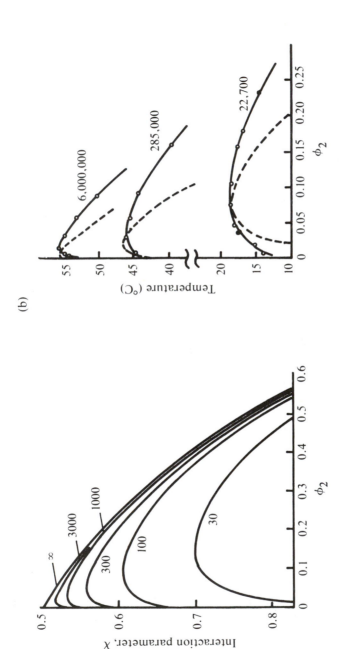

Figure 8.3 Volume fraction polymer in equilibrium phases for chains of different length. (a) Theoretical curves drawn for the indicated value of n, with the interaction parameter as the ordinate. Note that χ increases downward. (Redrawn from Ref. 6.) (b) Experimental curves for the molecular weights indicated, with temperature as the ordinate. [Reprinted with permission from A. R. Shultz and P. J. Flory, *J. Am. Chem. Soc.* 74:4760 (1952), copyright 1952 by the American Chemical Society.]

a phase diagram. The procedure is somewhat involved, so we dispense with the details and merely show the results of such calculations. Figure 8.3a shows the fraction of polymer in the equilibrium phases plotted against χ for polymers with the indicated values of n. Increasing values of χ are toward the bottom of Fig. 8.3a, so the curves have the same shape as would be obtained if temperature were the ordinate. Several features of these theoretical phase diagrams are noteworthy:

1. The miscibility gap becomes progressively more lopsided as n increases. This means that $\phi_{2,c}$ occurs at lower concentrations and that the tie line coordinates—particularly for the more dilute phase—are lower for large n.
2. For the case of n $\rightarrow \infty$, the limiting values of $\phi_{2,c}$ and χ_c are shown to be 0 and 0.5, as required by Eqs. (8.59) and (8.60), respectively.
3. Increasing positive values of χ—moving downward in Fig. 8.3—correspond to more endothermic values of ΔH_m. Interpreting the latter in terms of Eq. (8.49) means that systems for which $\delta_2 - \delta_1$ is large might show a miscibility gap for a given n, while complete miscibility is obtained for the same polymer in a solvent for which $\delta_2 - \delta_1$ is smaller. Decreasing the solvent "goodness" by the addition of a less suitable solvent may induce phase separation, at least for those molecules of large n.
4. If the poorer solvent is added incrementally to a system which is polydisperse with respect to molecular weight, the phase separation affects molecules of larger n, while shorter chains are more uniformly distributed. These ideas constitute the basis for one method of polymer fractionation. We shall develop this topic in more detail in the next section.

The curves shown in Fig. 8.3a are theoretical phase diagrams based on the Flory-Huggins model. Comparing theoretical predictions with experimental data thus provides our first opportunity to test the model. Figure 8.3b shows phase diagrams for polyisobutylene samples with the molecular weights indicated in diisobutyl ketone. Temperature variation is used to change the solvent "goodness." The broken lines in Fig. 8.3b are theoretical. We observe that the theory is qualitatively accurate, but that there is considerable discrepancy in quantitative detail. In particular, the experimental curves are considerably broader than predicted by theory. Generally speaking, the theory is more successful in accounting for χ_c than for $\phi_{2,c}$. It should be noted, however, that critical phenomena are extremely sensitive to small variations in a model—remember that at a critical point it only takes an infinitesimal variation to push the system into different regions of phase behavior—so Fig. 8.3b is not the best way to test the Flory-Huggins theory.

A far more satisfactory test of the Flory-Huggins theory is based on the chemical potential. According to Eqs. (8.13) and (8.20),

$$\mu_1 - \mu_1^{\theta} = RT \ln \left(\frac{p_1}{p_1^{\circ}} \right)$$ (8.61)

in which the vapor pressures are measurable quantities. According to the Flory–Huggins theory $\mu_1 - \mu_1^{\circ}$ is given by Eq. (8.53). Combining these two results gives

$$RT \ln \left(\frac{p_1}{p_1^{\circ}} \right) = RT \left[\ln \phi_1 + \left(1 - \frac{1}{n} \right) \phi_2 + \chi \phi_2^2 \right]$$ (8.62)

which suggests that a plot of $\ln (p_1/p_1^{\circ}) - \ln \phi_1 - (1 - 1/n) \phi_2$ versus ϕ_2^2 should give a straight line of slope χ. Figure 8.4 is such a plot for several different polymer–solvent systems. In view of the complexity of the phenomena

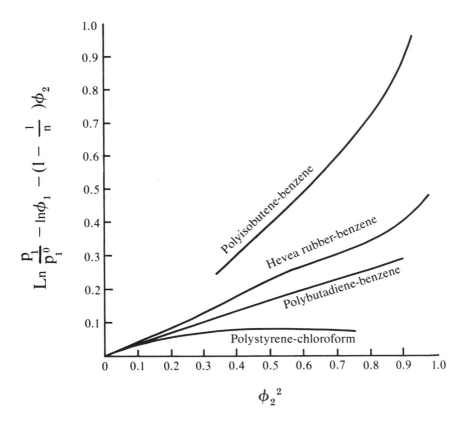

Figure 8.4 Experimental test of Flory–Huggins theory by Eq. (8.62) for the systems indicated. (From Ref. 3, used with permission.)

involved, the ability of the theoretical function to describe experimental results is remarkably good. It is interesting to note that the system for which the poorest agreement is observed—polystyrene in chloroform—is one in which it is easy to rationalize an acid–base-type of interaction between the hydrogen in $CHCl_3$ and the electrons in the phenyl groups. A chemical type interaction such as this is expected to show more complicated concentration effects than the Flory–Huggins model predicts. In the discussion of osmotic pressure in Sec. 8.8 we shall see that the latter experiment can also be used to measure $\Delta\mu_1$ and thereby test the Flory–Huggins theory.

8.7 Polymer Fractionation

Except for the living polymers described in Sec. 6.10, a high degree of poly-dispersity is expected in synthetic polymers. In most scientific research a narrow distribution of chain lengths must be present in the samples used to test theories. Similar though generally less stringent requirements also apply in applications technology. In both situations fractionation of the polymer sample into cuts of narrower molecular weight distribution is indicated. A variety of techniques exist for doing this, and the method of choice depends in part on the motive for doing the fractionation in the first place. If the objective of a fractionation experiment is the analysis of a molecular weight distribution, then it is not necessary to obtain large quantities of the separated product. For this type of application gel permeation chromatography has become the standard approach. We shall discuss this analytical method in Chap. 9.

For preparative purposes batch fractionation is often employed. Although fractional crystallization may be included in a list of batch fractionation methods, we shall consider only those methods based on the phase separation of polymer solutions: fractional precipitation and coacervate extraction. The general principles for these methods were presented in the last section. In this section we shall develop these ideas more fully with the objective of obtaining a more narrow distribution of molecular weights from a polydisperse system. Note that the final product of fractionation still contains a distribution of chain lengths; however, the ratio $\bar{M}_w/\bar{M}_n$ is smaller than for the unfractionated sample.

Figure 8.3b shows that phase separation in polymer mixtures results in two solution phases which are both dilute with respect to solute. Even the relatively more concentrated phase is only 10–20% by volume in polymer, while the more dilute phase is nearly pure solvent. The important thing to remember from both the theoretical and experimental curves of Fig. 8.3 is that both of the phases which separate contain some polymer. If it is the polymer-rich or precipitated phase that is subjected to further work-up, the method is called fractional precipitation. If the polymer-poor phase is the focus of attention, the method

is called coacervate extraction, since coacervation is the general term for this sort of phase separation.

The general procedure consists of first dissolving the polymer in a good solvent and then incrementally decreasing the solvent goodness by either lowering the temperature or adding a poor solvent as a precipitant. The high molecular weight polymers in a distribution are most affected by this reduction in solvent goodness; therefore it is this portion of the polymer that is most unevenly distributed between the two phases. We can use the Flory–Huggins theory to estimate the magnitude of this effect.

We shall identify the equilibrium phases by the labels P and Q in keeping with the notation established in Fig. 8.2b. The criterion for equilibrium is that for any n-mer, $\mu_{2,n}{}^P = \mu_{2,n}{}^Q$; using Eq. (8.54), we write

$$\ln \phi_{n,P} + \phi_{1,P}(1-n) + n\chi\phi_{1,P}{}^2 = \ln \phi_{n,Q} + \phi_{1,Q}(1-n) + n\chi\phi_{1,Q}{}^2$$

$$(8.63)$$

Rearranging, we obtain

$$\ln \frac{\phi_{n,P}}{\phi_{n,Q}} = n\left[\left(\frac{1}{n}-1\right)(\phi_{1,Q}-\phi_{1,P}) + \chi(\phi_{1,Q}{}^2 - \phi_{1,P}{}^2)\right] = nA$$

$$(8.64)$$

where the quantity inside the brackets is represented by A. If the (small) term $1/n$ inside the bracketed factor is treated as $1/\bar{n}$ for a polydisperse system and is independent of n, then A is constant. Strictly speaking, $\bar{n}_P \neq \bar{n}_Q$, and the first term in the brackets should not be factored as shown in Eq. (8.64), but this complication does not alter the essential conclusion that the bracketed term is constant. Therefore Eq. (8.64) can be written as

$$\frac{\phi_{n,P}}{\phi_{n,Q}} = e^{An}$$

$$(8.65)$$

which shows that the greater n becomes, the greater is the dissimilarity in the concentration of n-mer in the two equilibrium phases. This qualitative conclusion is certainly correct and independent of any model used to derive it.

Next we consider the fraction of n-mer in a specific phase. The volume of n-mer in each of the phases is given by $\phi_{n,P}V_P$ and $\phi_{n,Q}V_Q$; therefore the fraction of polymer in the two phases is

$$f_{n,P} = \frac{\phi_{n,P}V_P}{\phi_{n,P}V_P + \phi_{n,Q}V_Q} = \frac{V_P e^{An}}{V_P e^{An} + V_Q} = \frac{Re^{An}}{1 + Re^{An}}$$

$$(8.66)$$

and

$$f_{n,Q} = \frac{\phi_{n,Q}V_Q}{\phi_{n,P}V_P + \phi_{n,Q}V_Q} = \frac{V_Q}{V_P e^{An} + V_Q} = \frac{1}{1 + Re^{An}}$$

$$(8.67)$$

where Eq. (8.65) was used to eliminate $\phi_{n,P}/\phi_{n,Q}$ and $R = V_P/V_Q$. Now consider the case for which n is large and R is small. In this situation Re^{An} is large and $f_{n,P} > f_{n,Q}$. This indicates that the concentration of n-mer in phase P, for large values of n, will be maximized if the volume of the P phase is kept to a minimum compared to phase Q. Stated simply, the longer chains are concentrated in one of the phases, while the shorter chains are relatively unaffected by the phase separation and distribute themselves evenly and in proportion to the volume of the phases. As a practical matter, then, keeping the volume of the dilute phase as large as possible allows the optimum partitioning of the largest molecules into the smaller but more discriminating concentrated phase.

The relative amounts of various n-mers in a fractionated sample is examined numerically in the following example.

Example 8.4

Evaluate A in Eq. (8.64), assuming the polymer with n = 200 is divided equally between the two phases and taking the ratio of phase volumes R to be 10^{-1}, 10^{-2}, and 10^{-3}. Use these A values to evaluate the relative amounts of n-mer in the two phases for polymers with n = 100, 400, 600, and 800. Comment on the significance of the numerical results.

Solution

The ratio of Eq. (8.66) to Eq. (8.67) gives the ratio of the concentrations of n-mers in phases P and Q: $f_{n,P}/f_{n,Q} = Re^{An}$. Taking this ratio to be unity for n = 200 gives $Re^{A(200)} = 1$, which is readily solved for A using the R values given. Once these A values are obtained, $f_{n,P}/f_{n,Q}$ can be evaluated for the required n values. For the phase volume ratios under consideration, the corresponding values of A are listed below; also tabulated are the ratios $f_{n,P}/f_{n,Q}$ for the various n's:

	$R = V_P/V_Q$:	10^{-1}	10^{-2}	10^{-3}
	$A_{n=200}$:	1.15×10^{-2}	2.30×10^{-2}	3.45×10^{-2}
	100	0.32	0.099	0.032
	400	9.95	99.0	985
$\dfrac{f_{n,P}}{f_{n,Q}}$ for n =	600	99.9	9850	9.77×10^5
	800	990	9.80×10^5	9.69×10^8

Note that phase P is more concentrated in polymer, although smaller in volume than phase Q. It is apparent from these values that the combination of large n's

and large differences in the volumes of the separated phases gives rise to the most efficient segregation of polymer between phases. Also note that the requirement that 200-mers be evenly distributed between the phases creates the situation in which 100-mers are present in higher proportion in the more dilute phase.

•

Figure 8.5 illustrates the sort of separation this approach predicts. Curve A in Fig. 8.5 shows the weight fraction of various n-mers plotted as a function of n. Comparison with Fig. 6.7 shows that the distribution is typical of those obtained in random polymerization. Curve B shows the distribution of molecular weights in the more dilute phase—the coacervate extract—calculated for the volumes of the two phases in the proportion 100:1. The distribution in the concentrated phase is shown as curve C; it is given by the difference between curves A and B.

In practice, such a fractionation experiment could be carried out by either lowering the temperature or adding a poor solvent. In either case good temperature control during the experiment is important. Note that the addition of a poor solvent converts the system to one containing three components, so it is apparent that the two-component Flory–Huggins model is at best only qualitatively descriptive of the situation. A more accurate description would require a

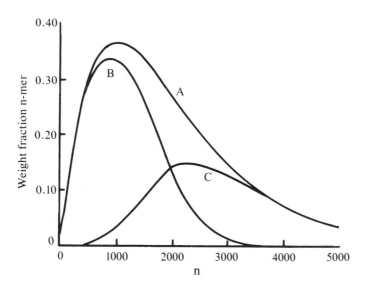

Figure 8.5 Theoretical plots of weight fraction n-mers versus n for unfractionated polymer (A), the dilute phase (B), and the concentrated phase (C) (drawn with $R = 10^{-2}$). (Adapted from Ref. 1.)

triangular phase diagram. The onset of precipitation is marked by the appearance of turbidity. In keeping with the principle outlined above, only a small volume of the precipitated phase is allowed to form. Then the sample is allowed to stand undisturbed until the two phases can be physically separated. This step can require quite a long wait.

This procedure is then repeated by decreasing the solvent goodness even further by another decrease in temperature or addition of precipitant. In this manner a set of fractions such as those shown in Fig. 8.6 are obtained from the initial distribution. In Fig. 8.6 curve A again represents the initial distribution. Eight fractions are obtained by precipitating successive portions of the polymer of progressively lower molecular weights as shown, until the dilute phase contains only the lowest molecular weight fraction as a residue. The curves in Fig. 8.6 are calculated for a phase volume ratio of 1000:1.

Figure 8.6 shows that the individual fractions still contain a considerable range of chain lengths, with this effect becoming less pronounced in the later cuts. In addition, there is a definite overlap among the fractions. Nevertheless, the approach results in a sharpening of the distribution of molecular weights in a sample. The method is time-consuming and involves large quantities of solvent, which must be subsequently removed to obtain pure polymer.

With these remarks, we conclude our discussion of phase separation. In the remainder of the chapter we consider another thermodynamic property of

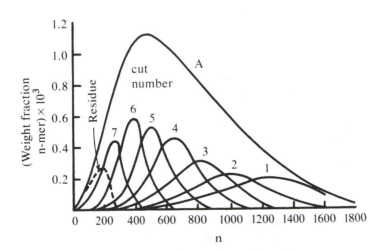

Figure 8.6 Effect of successive fractionations: weight fraction n-mer versus n for seven successive precipitates and final residue (calculated for $R = 10^{-3}$). [Adapted from G. V. Schulz, *Z. Phys. Chem.* B46:137 (1940); B47:155 (1940).]

polymer solutions, namely the osmotic pressure. As in the previous sections, we shall combine pure thermodynamics as well as statistical models to understand osmostic pressure more fully and to maximize our ability to interpret experimental osmometry.

8.8 Osmotic Pressure

Osmotic pressure is one of four closely related properties of solutions that are collectively known as colligative properties. In all four, a difference in the behavior of the solution and the pure solvent is related to the thermodynamic activity of the solvent in the solution. In ideal solutions the activity equals the mole fraction, and the mole fractions of the solvent (subscript 1) and the solute (subscript 2) add up to unity in two-component systems. Therefore the colligative properties can easily be related to the mole fraction of the solute in an ideal solution. The following review of the other three colligative properties indicates the similarity which underlies the analysis of all the colligative properties:

1. Vapor pressure lowering. Equation (8.20) shows that for any component in a binary liquid solution $a_i = p_i/p_i^\circ$. For an ideal solution, this becomes

$$p_1 = x_1 p_1^\circ = (1 - x_2) p_1^\circ \tag{8.68}$$

or

$$p_1^\circ - p_1 = \Delta p = x_2 p_1^\circ \tag{8.69}$$

This is an expression of Raoult's law which we have used previously.

2. Freezing point depression. A solute which does not form solid solutions with the solvent and is therefore excluded from the solid phase lowers the freezing point of the solvent. It is the chemical potential of the solvent which is lowered by the solute, so the pure solvent reaches the same (lower) value at a lower temperature. At equilibrium

$$\ln a_1 = - \frac{\Delta H_f^\circ}{R} \left(\frac{1}{T_f} - \frac{1}{T_f^\circ} \right) \tag{8.70}$$

where ΔH_f° is the heat of fusion of the solvent, T_f is the freezing point of the solution, and T_f° is the freezing point of the pure solvent. For an ideal solution the following relationships apply:

$$\ln a_1 = \ln x_1 = \ln (1 - x_2) \cong -x_2 \tag{8.71}$$

where the last approximation requires dilute conditions as well.

3. Boiling point elevation. A solute which does not enter the vapor phase to any significant extent raises the boiling point of the solvent. As above, the solute lowers the activity of the solvent, which, in turn, lowers the vapor pressure. Therefore the solution must be raised to a higher temperature before its vapor pressure reaches 1.0 atm. At equilibrium

$$\ln a_1 = \frac{\Delta H_v^\circ}{R} \left(\frac{1}{T_b} - \frac{1}{T_b^\circ} \right) \tag{8.72}$$

where ΔH_v° is the heat of vaporization of the solvent, T_b is the boiling point of the solution, and T_b° is the boiling point of the pure solvent. For ideal or ideal and dilute solutions, the approximations of Eq. (8.71) can be applied here as well.

For each of these phenomena in the limit of dilute solutions, some difference in the behavior of the solvent—Δp_1, ΔT_f, or ΔT_b—is proportional to the mole fraction of the solute. Since the solutions are already assumed to be dilute, we note that

$$x_2 = \frac{n_2}{n_1 + n_2} \cong \frac{n_2}{n_1} \tag{8.73}$$

under these conditions. The number of moles of solvent in a solution is generally known from its mass and molecular weight. Therefore measuring the colligative properties Δp_1, ΔT_f, and ΔT_b and using the dilute, ideal solution approximation provides a means for evaluating n_2, the number of moles of solute in the sample. If the solute is unknown with respect to molecular weight but its mass in the solution is known, then the colligative properties can be used to determine the molecular weight of the solute: $M_2 = m_2/n_2$.

One way to describe this situation is to say that the colligative properties provide a method for counting the number of solute molecules in a solution. In these ideal solutions this is done without regard to the chemical identity of the species. Therefore if the solute consists of several different components which we index i, then $n_2 = \Sigma_i n_{2,i}$ is the number of moles counted. Of course, the total mass of solute in this case is given by $m_2 = \Sigma_i n_{2,i} M_{2,i}$, so the molecular weight obtained for such a mixture is given by

$$\overline{M}_2 = \frac{\Sigma_i n_i M_i}{\Sigma_i n_i} \tag{8.74}$$

which is the definition of the number average molecular weight [Eq. (1.3)]. Measurements of the colligative properties are often used to evaluate the molecular weight of an unknown solute. The foregoing remarks show that the

molecular weight thus obtained is subject to whatever approximation arises from the assumption of ideality and is an average value over all species present. For something like NaCl in water, for example, $\overline{M} \cong 58/2 = 29$, since two moles of ions are counted for every mole of solute weighed into solution.

As noted above, all of the colligative properties are very similar in their thermodynamics if not their experimental behavior. This similarity also extends to an application like molecular weight determination and the kind of average obtained for nonhomogeneous samples. All of these statements are also true of osmotic pressure. In the remainder of this section we describe osmotic pressure experiments in general and examine the thermodynamic origin of this behavior.

In an osmotic pressure experiment two phases are brought to isothermal equilibrium: a solution and a pure solvent. Note that this statement also applies to the freezing and boiling equilibria described above. In an osmotic experiment the partitioning of the two phases is accomplished by means of a semipermeable membrane which, by definition, allows the solvent to pass but not the solute. Figure 8.7 is a schematic representation of this state of affairs. In order for the solvent to be in equilibrium on both sides of the membrane, the solution must be under a greater pressure than the pure solvent and it is this pressure difference that is designated as the osmotic pressure Π. If the solvent and solution are brought together at the same pressure in a suitable apparatus, an osmometer, then a hydrostatic pressure will develop on the solution side by the solvent diffusing through the membrane until the equilibrium criterion is satisfied.

Thermodynamics provide a straightforward method for quantifying this situation. The criterion for equilibrium is the equality of chemical potential in

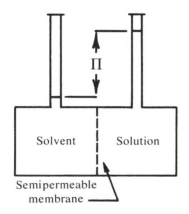

Figure 8.7 Schematic representation of an osmotic pressure experiment.

the equilibrium phases. Since one of the phases is pure solvent, the standard state, this means

$$\mu_1^\circ = \mu_1^{soln} = \mu_1^\circ + (\text{concentration effect}) + (\text{pressure effect}) \tag{8.75}$$

At equilibrium, these concentration and pressure effects must be equal and opposite for Eq. (8.75) to apply. Equation (8.13) describes the concentration dependence of μ_1, and Eq. (8.12) describes the pressure effect. Assembling these results, we write

$$\mu_1^\circ = \mu_1^\circ + RT \ln a_1 + \int_p^{p+\Pi} \bar{V}_1 \, dp \tag{8.76}$$

Remember that $\bar{V}_1$ is the partial molar volume of the solvent. Therefore a completely general relationship between Π and the solvent activity is given by

$$RT \ln a_1 = - \int_p^{p+\Pi} \bar{V}_1 \, dp \tag{8.77}$$

Next we seek a method for evaluating this integral. Since condensed phases show little volume change with changing pressure, $\bar{V}_1$ can be regarded as a constant and taken out of the integral. Then Eq. (8.77) is readily integrated to give

$$\ln a_1 = - \frac{\Pi \bar{V}_1}{RT} \tag{8.78}$$

Combining this result with Eq. (8.13) gives

$$\mu_1 = \mu_1^\circ - \Pi \bar{V}_1 \tag{8.79}$$

which could be used as the basis for testing the Flory–Huggins expression for μ_1.

Precisely the same substitutions and approximations that we wrote in Eq. (8.71) can be applied again to $\ln a_1$. Thus for the case of dilute, ideal solutions, Eq. (8.78) becomes

$$- \ln a_1 \cong x_2 \cong \frac{n_2}{n_1} = \frac{\Pi \bar{V}_1}{RT} \tag{8.80}$$

or

$$\Pi (n_1 \bar{V}_1) \cong \Pi V_1 \cong \Pi V = n_2 RT \tag{8.81}$$

where we use $\overline{V}_1 \cong V_1^\circ$ and $V_1 \cong V$ for dilute solutions. This solution–analog of the ideal gas law is called the van't Hoff equation and, like its counterpart, is a limiting law which applies perfectly in the limits of $\Pi \to 0$ or $n_2/V \to 0$. This relationship shows how the measurement of Π for solutions which are quantitatively characterized with respect to other pertinent factors can be used to measure n_2 and M_2. As with the other colligative properties, this latter value is a number average for nonhomogeneous samples. In the next section we shall examine osmometry as a method for molecular weight determination in more detail.

The approximations involved in going from Eq. (8.78) to Eq. (8.81) can be divided into three categories:

1. Those involving volume approximations. We replace $n_1\overline{V}_1$ by V_1 and identify the latter as the total volume of the solution. For dilute solutions these do not introduce serious errors and do simplify the notation somewhat.

2. Those involving series truncation. The quantity $\ln(1 - x_2)$ can be represented by the infinite series $-[x_2 + (1/2)x^2 + (1/3)x^3 + \cdots]$. Truncating this series after the first term is a valid approximation for dilute solutions and also simplifies the form of the equation. It is an optional step, however, and can be avoided or mitigated by simply retaining more terms in the series.

3. Those involving solution nonideality. This is the most serious approximation in polymer applications. As we have already seen, the large differences in molecular volume between polymeric solutes and low molecular weight solvents is a source of nonideality even for athermal mixtures.

In thermodynamics the formal way of dealing with nonideality is to introduce an activity coefficient γ into the relationship between activity and mole fraction:

$$a_i = \gamma_i x_i \tag{8.82}$$

such that γ_i is an empirical parameter which describes the nonideality. In view of the approximation mentioned in item (2) above, we introduce coefficients to deal with nonideality in the form

$$\frac{\Pi \overline{V}_1}{RT} = -\ln a_1 = x_2 + \frac{1}{2} B' x_2^2 + \frac{1}{3} C' x_2^3 + \cdots \tag{8.83}$$

where B' and C' are called the second and third virial coefficients. The first virial coefficient is unity, since the van't Hoff equation must be obtained in the limit $x_2 \to 0$.

In the next section we shall describe the use of Eq. (8.83) to determine the number average molecular weight of a polymer, and in subsequent sections we shall examine models which offer interpretations of the second virial coefficient.

Before doing this, however, it is informative to compare the sensitivity of the four colligative properties in the determination of molecular weight. In the following example this is done by making the appropriate numerical calculations.

Example 8.5

Calculate Δp_1, ΔT_f, ΔT_b, and Π for solutions which are 1% by weight in benzene of solutes for which $M = 10^2$ and 10^5. Assume that these solutions are adequately described by dilute ideal solution expressions. Consult a handbook for the physical properties of benzene. Comment on the significance of the results with respect to the feasibility of these various methods for the determination of M for solutes of high and low molecular weight.

Solution

We use the approximation $-\ln a_1 \cong x_2 \cong n_2/n_1$ as the basis for all of these calculations. Since a 1% solution contains 1.0 g solute per 99 g solvent,

$$\frac{n_2}{n_1} = \frac{1 \text{ g solute} \times (1 \text{ mol solute})/(M_2 \text{ g})}{99 \text{ g solvent} \times (1 \text{ mol benzene})/(78 \text{ g})} = \frac{0.788}{M_2}$$

This becomes 7.88×10^{-3} for $M_2 = 10^2$ and 7.88×10^{-6} for $M_2 = 10^5$. By Eq. (8.68), $p_1 = (n_2/n_1) p_1^\circ$. For benzene at $25°C$, $p_1^\circ = 95.9$ Torr; therefore

$$\Delta p_1 = 95.9 \left(\frac{n_2}{n_1}\right) \text{ Torr}$$

By Eqs. (8.70) and (8.71), $-x_2 = -(\Delta H_f^\circ/R)(1/T_f - 1/T_f^\circ)$; therefore $T_f^\circ - T_f \cong (n_2/n_1)(RT_f^{\circ 2}/\Delta H_f^\circ)$.
   For benzene $\Delta H_f^\circ = 10.6$ kJ mol^{-1} and $T_f^\circ = 5.5°C = 278.7$ K; hence

$$T_f^\circ - T_f = \left(\frac{n_2}{n_1}\right) (60.9) \text{ K}$$

By Eqs. (8.71) and (8.72), $-x_2 = (\Delta H_v^\circ/R)(1/T_b - 1/T_b^\circ)$; therefore $T_b^\circ - T_b \cong (n_2/n_1)(RT_b^{\circ 2}/\Delta H_v^\circ)$.
   For benezene $\Delta H_v^\circ = 30.8$ kJ mol^{-1} and $T_b^\circ = 80.1°C = 353.3$ K; hence

$$T_b^\circ - T_b = -33.7 \left(\frac{n_2}{n_1}\right) \text{ K}$$

By Eq. (8.80), $\Pi = (n_2 RT/n_1 \bar{V}_1) \cong (n_2/n_1)(RT/V_1^\circ)$. For benzene $\rho = 0.879$ g cm^{-3} and $V_1^\circ = 88.7$ cm^3 mol^{-1} at $25°C$; hence

$$\Pi = \frac{(82.1 \text{ cm}^3 \text{ atm mol}^{-1} \text{ K}^{-1})(298 \text{ K})}{88.7 \text{ cm}^3 \text{ mol}^{-1}} \frac{n_2}{n_1} = 276 \left(\frac{n_2}{n_1} \right) \text{ atm}$$

	$M_2 = 10^2$	$M_2 = 10^5$
n_2/n_1	7.88×10^{-3}	7.88×10^{-6}
Δp_1 at 25°C (Torr)	0.756	7.56×10^{-4}
$T_f^\circ - T_f^{'}$ (K)	0.480	4.80×10^{-4}
$T_b - T_b^\circ$ (K)	0.266	2.66×10^{-4}
Π (atm)	2.17	2.17×10^{-3}
Π (Torr)	1.65×10^3	1.65
Π (mm soln)	2.52×10^4	25.2

In these unit conversions on Π, we have used the facts that 1 atm = 760 Torr and the ratio of densities $\rho_{Hg}/\rho_{soln} = 13.5/\rho_{soln}$ converts from Torr to millimeters of solution. These numerical examples show that experiments in which Δp_1, ΔT_f, or ΔT_b are measured are perfectly feasible for solutes of molecular weight 100, but call for unattainable sensitivity for polymeric solutes of M = 10^5. By contrast, osmometry produces so much larger an effect that this method is awkward (at least for 1% concentration) for a low molecular weight solute, but is entirely feasible with the polymer.

•

8.9 Molecular Weight by Osmometry

Our primary objective in this section is the discussion of practical osmometry, particularly with the goal of determining the molecular weight of a polymeric solute. We shall be concerned, therefore, with the design and operation of osmometers, with the question of units, and with circumventing the problem of nonideality. The key to these points is contained in the last section, but the details deserve additional comment.

First, we consider the experimental aspects of osmometry. The semipermeable membrane is the basis for an osmotic pressure experiment and is probably its most troublesome feature in practice. The membrane material must display the required selectivity in permeability—passing solvent and retaining solute— but a membrane that works for one system may not work for another. A wide variety of materials have been used as membranes, with cellophane, poly(vinyl alcohol), polyurethanes, and various animal membranes as typical examples. The membrane must be thin enough for the solvent to pass at a reasonable rate, yet sturdy enough to withstand the pressure difference which can be

considerable. Above all, the membrane must be free from imperfections which allow solute to leak across the barrier and thus invalidate the experiment.

Osmometers are designed to allow maximum contact of the two liquids with the membrane and to provide sufficient support for the fragile membrane at the same time. Since the experiments are conducted under isothermal conditions, good temperature control is also required. Finally, it must be possible to accurately determine the difference in liquid column heights, since this is the quantity which is actually measured. Both of the osmometers shown in Fig. 8.8 meet these requirements, although they are but two of many designs that have been employed. In the osmometer of Fig. 8.8a the inner compartment contains the solution, while the solvent is placed in the outer vessel. The membrane is clamped across the wide opening at the bottom of the solution chamber and the pressure is measured in the capillary neck of the latter. The entire apparatus can be submerged in a constant-temperature bath for equilibration. In the apparatus shown in Fig. 8.8b, the solvent and solution are placed in the grooves—shown in Fig. 8.8c—of opposing faces of the central block and the membrane is sandwiched between them. The assembly is sealed and placed in a thermostat. The osmotic pressure is measured by the difference in liquid levels in the vertical capillaries.

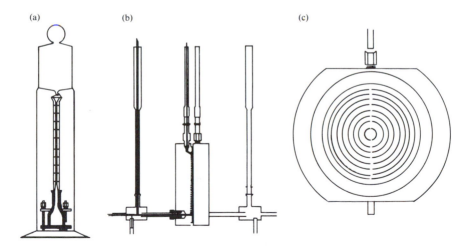

Figure 8.8 Two osmometer designs: (a) solution (inner chamber) separated from solvent by clamped membrane. [Reprinted with permission from D. M. French and R. H. Ewert, *Anal. Chem.* 19:165 (1947), copyright 1947 by the American Chemical Society.] (b) Solution and solvent in grooved faces shown in detail in (c). [Reprinted with permission from R. M. Fuoss and D. J. Mead, *J. Phys. Chem.* 47:59 (1943), copyright 1943 by the American Chemical Society.]

In both of these pieces of apparatus, isothermal operation and optimum membrane area are obtained. Good temperature control is essential not only to provide a value for T in the equations, but also because the capillary attached to a larger reservoir behaves like a thermometer, with the column height varying with temperature fluctuations. The contact area must be maximized to speed up an otherwise slow equilibration process. Various practical strategies for "presetting" the osmometer to an approximate Π value have been developed, and these also accelerate the equilibration process.

Since capillary tubing is involved in osmotic experiments, there are several points pertaining to this feature that should be noted. First, tubes that are carefully matched in diameter should be used so that no correction for surface tension effects need be considered. Next it should be appreciated that an equilibrium osmotic pressure can develop in a capillary tube with a minimum flow of solvent, and therefore the measured value of Π applies to the solution as prepared. The pressure, of course, is independent of the cross-sectional area of the liquid column, but if too much solvent transfer were involved, then the effects of dilution would also have to be considered. Now let us examine the practical units that are used to express the concentration of solutions in these experiments.

In molecular weight determinations it is conventional to dissolve a measured mass of polymer m_2 into a volumetric flask and dilute to the mark with an appropriate solvent. We shall use the symbol c_2 to designate concentrations in mass per volume units. In practice, 100-ml volumetric flasks are often used, in which case c_2 is expressed in grams per 100 ml or grams per deciliter. Even though these are not SI units, they are encountered often enough in the literature to be regarded as conventional solution units in polymer chemistry.

These mass per volume concentration units can be written as

$$c_2 = \frac{n_2 M}{n_1 \bar{V}_1 + n_2 \bar{V}_2} \tag{8.84}$$

where the $\bar{V}_i$'s are the partial molar volumes and may be expressed in whatever volumes unit is desired. We continue to use M (no subscript) for the (average) molecular weight of the polymer.

We are most often concerned with solutions which are dilute with respect to polymer. This means that $n_2 \bar{V}_2 \ll n_1 \bar{V}_1$ and that $n_2 \ll n_1$. Because the volume of the polymer molecule is so much greater than that of the solvent, the disparity in the number of moles is even more extreme than for the same approximation applied to compounds of comparably sized molecules. Since the approximation $x_2 = n_2/(n_1 + n_2) \cong n_2/n_1$ applies to dilute solutions, we can write

$$c_2 \cong \frac{n_2}{n_1} \frac{M}{\bar{V}_1} \cong x_2 \frac{M}{\bar{V}_1} \cong x_2 \frac{M}{V_1^{\circ}} \tag{8.85}$$

where the last relationship is justified because partial molar quantities approach molar quantities in the limit of the pure components.

Equation (8.85) can be solved for x_2 and the latter substituted into Eq. (8.83) to give an expression for Π which applies to dilute—but not necessarily ideal—solutions:

$$\Pi = RT \left(\frac{c_2}{M} + \frac{B'V_1^{\,\circ}}{2M^2} c_2^2 + \cdots \right) \tag{8.86}$$

Upon rearrangement, this becomes

$$\frac{\Pi}{RTc_2} = \frac{1}{M} + \frac{B'V_1^{\,\circ}}{2M^2} c_2 + \cdots = \frac{1}{M} + Bc_2 + \cdots \tag{8.87}$$

where the cluster of parameters $B'V_1^{\,\circ}/2M^2$ has been replaced by B, which is the second virial coefficient in these units. In writing Eq. (8.87), we have truncated the series in Eq. (8.83) after the second term. This restricts Eq. (8.87) to solutions which are sufficiently dilute that the term which is third order in c_2 is insignificant but at the same time does make allowance for the first manifestation of nonideality. The second virial coefficient embodies the latter.

What makes Eq. (8.87) especially important is that it is the equation of a straight line. It predicts that a plot of Π/RTc_2 versus c_2 will be linear for dilute solutions and that the slope and intercept of the plot will have the following significance:

$$\text{Intercept} = \left(\frac{\Pi}{RTc_2} \right)_0 = \frac{1}{M} \tag{8.88}$$

$$\text{Slope} = B \tag{8.89}$$

Figure 8.9 is a plot of osmotic pressure data for a nitrocellulose sample in three different solvents analyzed according to Eq. (8.87). As required by Eq. (8.88), all show a common intercept corresponding to a molecular weight of 1.11×10^5; the various systems show different deviations from ideality, however, as evidenced by the range of slopes in Fig. 8.9.

In the next sections we shall discuss the second virial coefficient in terms of several different models. Before turning to these models, however, it is important to summarize the foregoing conclusions:

1. The molecular weight analysis presented above is a purely thermodynamic result and is independent of any model. The procedure requires dilute solutions, but is not based on the assumption of ideality, even though Eq. (8.88) is a variation of the van't Hoff equation.

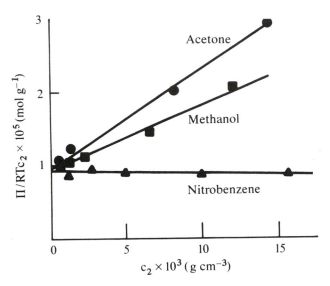

Figure 8.9 Osmotic pressure data plotted as Π/RTc_2 versus concentration for nitrocellulose in three different solvents. [Data from A. Dobry, *J. Chem. Phys.* 32:50 (1935).]

2. The solute molecular weight enters the van't Hoff equation as the factor of proportionality between the number of solute particles that the osmotic pressure counts and the mass of solute which is known from the preparation of the solution. The molecular weight that is obtained from measurements on polydisperse systems is a number average quantity.

3. Osmotic pressure experiments provide absolute values for $\bar{M}_n$: Neither a model nor independent calibration is required to use this method. Experimental errors can arise, of course, and we note particularly the effect of impurities. Polymers which dissociate into ions can also be confusing. We shall return to this topic in Sec. 8.13; for now we assume that the polymers under consideration are nonelectrolytes.

4. The ratio Π/c_2 is called the reduced osmotic pressure—and can be plotted with or without the RT—and the zero-intercept value (subscript 0) is the limiting value of the reduced osmotic pressure. Quite an assortment of different pressure units are used in the literature in reporting Π values, and the units of R in Eq. (8.88) must be reconciled with these pressure (as well as concentration) units.

The following example illustrates the sort of deciphering that is associated with the use of diverse units.

Example 8.6

The limiting reduced osmotic pressure for a sample of polystyrene in chloro-
benzene at 25°C is reported† to be 257 cm. Account for these units for $(\Pi/c_2)_0$
and evaluate $\bar{M}_n$ for the polymer.

Solution

The authors of this research selected units which could be written with maximum simplification to report their results. We must replace the factors which
have canceled out. Assuming that the cgs system of units was used throughout,
we note that possible units for Π are grams per square centimeter and possible
unts for c_2 are grams per cubic centimeter, which yield the required units for
Π/c_2: Note that these units of Π must be multiplied by the gravitational
constant to give Π in dynes per square centimeter:

$$257 \text{ cm} \times \frac{980 \text{ cm}}{\sec^2} = 2.52 \times 10^5 \frac{\text{dyne cm}^{-2}}{\text{g cm}^{-3}} = 2.52 \times 10^5 \frac{\text{erg}}{\text{g}}$$

$\bar{M}_n$ may be evaluated from Eq. (8.88) by using $R = 8.314 \times 10^7 \text{ erg K}^{-1} \text{ mol}^{-1}$:

$$\bar{M}_n = \frac{RT}{(\Pi/c_2)_0} = \frac{(8.314 \times 10^7 \text{ erg K}^{-1} \text{ mol}^{-1})(298 \text{ K})}{2.52 \times 10^5 \text{ erg g}^{-1}}$$

$$= 9.84 \times 10^4 \text{ g mol}^{-1}$$

●

The matter of units also complicates the second virial coefficient, which we
consider next.

8.10 Excluded Volume

Considering our approach to the topic, we might be tempted to believe that the
second virial coefficient is small for a polymer, depending as it does in Eq. (8.87)
on $\bar{M}^{-2}$. Experimental results such as those shown in Fig. 8.9, however, reveal
that plots of Π/c_2 versus c_2 have slopes quite different from zero. In fact, any
inquiry into the significance of the second virial coefficient must seek to explain
both positive and negative slopes in such plots. The form of the virial expansion
suggests that the molecular origin of the second virial coefficient is the pairwise
interaction of solute particles—B' is the coefficient of the x_2^2 term in Eq.
(8.83)—with the higher coefficients arising from the third- or higher-order

---

†J. Leonard and H. Daoust, *J. Phys. Chem.* 69:1174 (1965).

interactions. We shall restrict our discussion to the second virial coefficient: Things are complicated enough at this level of nonideality without looking for still more complex interactions!

It is convenient to begin by backtracking to a discussion of ΔS_m for an athermal mixture. We shall consider a *dilute* solution containing N_2 solute molecules, each of which has an excluded volume u. The excluded volume of a particle is that volume for which the center of mass of a second particle is excluded from entering. Although we assume no specific geometry for the molecules at this time, Fig. 8.10 shows how the excluded volume is defined for two spheres of radius a. The two spheres are in surface contact when their centers are separated by a distance 2a. The excluded volume for the pair has the volume $(4/3)\pi(2a)^3$, or eight times the volume of one sphere. This volume is indicated by the broken line in Fig. 8.10. Since this volume is associated with the interaction of two spheres, the excluded volume per sphere is

$$u_{sphere} = 4\,(\text{volume of sphere}) \tag{8.90}$$

Regardless of the particle geometry, the excluded volume exceeds the actual volume of the molecules by a factor which depends on the shape of the particles.

To arrive at an expression for ΔS_m, we follow a series of steps which parallel—for a different model—the development of the Flory–Huggins model for ΔS_m:

1. The number of ways of placing the first solute molecule ω_1 is proportional to the total volume of the solution V:

$$\omega_1 = KV$$

In this model we start with pure solvent so an undiminished value of V can be used for the first solute molecule.

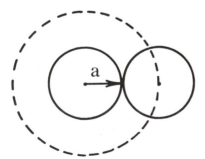

Figure 8.10 Excluded volume for two spheres (dotted surface) as determined by the distance of closest approach.

2. The number of ways of placing the second molecule is proportional to the volume remaining unoccupied after the first solute molecule is placed:

$$\omega_2 = K(V - u)$$

3. The number of ways of placing the ith molecule is therefore

$$\omega_i = K[V - (i - 1)u]$$

4. The total number of ways Ω of placing N_2 solute molecules is

$$\Omega = \frac{1}{N_2!} \prod_{i=1}^{N_2} K[V - (i - 1)u]$$

   where the factor $N_2!$ enters because of the interchangeability of the particles.

5. By changing the limits of the product, the latter can be written as

$$\prod_{i=1}^{N_2} K[V - (i - 1)u] = \prod_{i=0}^{N_2-1} K(V - iu) = \prod_{i=0}^{N_2-1} KV \left(1 - \frac{iu}{V}\right)$$

6. Since the entropy of the mixture S_{mix} is proportional to $\ln \Omega$ by the Boltzmann equation, we write

$$\frac{S_{mix}}{k} = N_2 \ln (KV) - \ln N_2! + \sum_{i=0}^{N_2-1} \ln \left(1 - \frac{iu}{V}\right)$$

   The summation in the last term results from taking the logarithm of the product.

7. The ratio iu/V is small for all values of i; hence the last term can be approximated by the series expansion of the logarithm $[\ln (1 - y) \cong -y]$:

$$\frac{S_{mix}}{k} = N_2 \ln KV - \ln N_2! - \frac{u}{V} \sum_{i=0}^{N_2-1} i$$

8. Since the sum of integers from 0 to y is $y(y + 1)/2$, the summation in the last expression can be written as

$$\sum_{i=0}^{N_2-1} i = \frac{(N_2 - 1)(N_2)}{2} \cong \frac{1}{2} N_2^2$$

   where the approximation applies since N_2 is large.

9. We can shift from numbers of molecules to numbers of moles of solute by dividing by Avogadro's number N_A and changing k to R. Also, recalling that $V = \Sigma_i n_i \bar{V}_i$, we write

$$\frac{S_{mix}}{R} = n_2 \ln K + n_2 \ln (n_1 \bar{V}_1 + n_2 \bar{V}_2) - \frac{1}{N_A} \ln N_2!$$

$$- \frac{1}{2} u N_A \frac{n_2^2}{n_1 \bar{V}_1 + n_2 \bar{V}_2}$$

10. By separately letting n_2 and n_1 equal zero, we obtain the configurational entropy for the pure solvent ($S_1 = 0$) and the pure solute ($S_2 = n_2 \ln K + n_2 \ln n_2 \bar{V}_2 - (1/N_A) \ln N_2! - u N_A n_2 / 2 \bar{V}_2$).

Subtracting the entropy contributions of the pure components from S_{mix} gives the entropy of mixing according to the present model:

$$\frac{\Delta S_m}{R} = n_2 \ln \left(\frac{n_1 \bar{V}_1 + n_2 \bar{V}_2}{n_2 \bar{V}_2} \right) - \frac{1}{2} u N_A \frac{n_2^2}{n_1 \bar{V}_1 + n_2 \bar{V}_2}$$

$$+ \frac{1}{2} u \frac{N_A n_2}{\bar{V}_2} \qquad (8.91)$$

This is converted to a partial molar quantity by differentiation:

$$\bar{S}_1 - S_1^\circ = \frac{\partial}{\partial n_1} (\Delta S_m) = R \left[\frac{n_2 \bar{V}_1}{n_1 \bar{V}_1 + n_2 \bar{V}_2} + \frac{1}{2} N_A \bar{V}_1 u \right.$$

$$\left. + \frac{1}{2} N_A \bar{V}_1 u \left(\frac{n_2}{n_1 \bar{V}_1 + n_2 \bar{V}_2} \right)^2 \right] \qquad (8.92)$$

Since the solution is athermal by hypothesis, $\Delta \mu_1 = - T \Delta \bar{S}_1$, or

$$\Delta \mu_1 = - RT \left[\frac{n_2 \bar{V}_1}{V} + \frac{1}{2} N_A \bar{V}_1 u \left(\frac{n_2}{V} \right)^2 \right] \qquad (8.93)$$

Upon incorporation of Eq. (8.85), this becomes

$$\Delta \mu_1 = - RT \bar{V}_1 \left(\frac{c_2}{M} + \frac{1}{2} \frac{N_A u}{M^2} c_2^2 \right) \qquad (8.94)$$

Equation (8.79) permits this result to be directly converted to an expression for osmotic pressure:

$$\Pi = RT \left(\frac{c_2}{M} + \frac{1}{2} \frac{N_A u}{M^2} c_2^2 \right)$$ (8.95)

or

$$\frac{\Pi}{RTc_2} = \frac{1}{M} + \frac{1}{2} \frac{N_A u}{M^2} c_2$$ (8.96)

Comparing Eqs. (8.87) and (8.96) shows that

$$B = \frac{1}{2} \frac{N_A u}{M^2}$$ (8.97)

according to this model.

Equation (8.97) shows that the second virial coefficient is a measure of the excluded volume of the solute according to the model we have considered. From the assumption that solute molecules come into surface contact in defining the excluded volume, it is apparent that this concept is easier to apply to, say, compact protein molecules in which hydrogen bonding and disulfide bridges maintain the tertiary structure (see Sec. 1.4) than to random coils. We shall return to the latter presently, but for now let us consider the application of Eq. (8.97) to a globular protein. This is the objective of the following example.

Example 8.7

The bovine serum albumin molecule is known to be nearly spherical and uncharged in a solution of pH 5.37. A plot of Π/c_2 versus c_2 for this polymer at 25°C is linear and has an intercept corresponding to $M = 69,000$. The slope of the line is 1.37×10^{-3} Torr liter^2 g^{-2}. Use this slope to estimate the radius of this spherical molecule.

Solution

We convert the units to SI units; then we use Eqs. (8.97) and (8.90) to calculate the volume of the sphere:

$$\text{Slope} = \frac{1.37 \times 10^{-3} \text{ Torr liter}^2}{g^2} \times \frac{1.01 \times 10^5 \text{ N m}^{-2}}{760 \text{ Torr}}$$

$$\times \left(\frac{(10^{-1} \text{ m})^3}{1 \text{ liter}} \right)^2 \times \left(\frac{10^3 \text{ g}}{1 \text{ kg}} \right)^2 = 0.182 \text{ N m}^4 \text{ kg}^{-2}$$

Since the plot is one of Π/c_2, not Π/RTc_2, versus c_2, we must divide the slope by RT to get B:

$$B = \frac{0.182 \ N \ m^4 \ kg^{-2}}{(8.314 \ J \ K^{-1} \ mol^{-1})(298 \ K)} = 7.35 \times 10^{-5} \ m^3 \ kg^{-2} \ mol$$

From Eq. (8.97),

$$u = \frac{2BM^2}{N_A} = \frac{2(7.35 \times 10^{-5} \ m^3 \ kg^{-2} \ mol)(69 \ kg \ mol^{-1})^2}{6.02 \times 10^{23} \ mol^{-1}}$$

$$= 1.16 \times 10^{-24} \ m^3$$

From Eq. (8.90), $V_{sphere} = (1/4) \ u = 2.90 \times 10^{-25} \ m^3$. Since $V_{sphere} = (4/3) \pi r^3$, we can readily convert this to the radius of the sphere: $r = 4.11 \times 10^{-9} \ m = 4.11 \ nm = 41 \ Å$.

•

We saw in Chap. 1 that the random coil is characterized by a spherical domain for which the radius of gyration is a convenient size measure. As a tentative approach to extending the excluded volume concept to random coils, therefore, we write for the volume of the coil domain (subscript d) $V_d = (4/3) \pi r_g^3$, and combining this result with Eq. (8.90), we obtain

$$u_{coil} = 4 \left(\frac{4}{3} \pi r_g^3 \right) \tag{8.98}$$

from which B can be calculated via Eq. (8.97). As noted above, the excluded volume concept appears to be more applicable to particles with a molecular structure which is more compact than the random coil. The following argument, however, shows how the excluded volume model can be extended to random coils as well.

Equations (1.48) and (1.59) show that r_g varies as $n^{1/2}$ for the random coil. This means that the volume of the spherical domain of the coil varies with the degree of polymerization as $n^{3/2}$. We shall disregard all other numerical coefficients in the remainder of this paragraph to argue that a spherical excluded volume of radius r_g is at least plausible for chains of large n. That portion of the domain which is actually occupied by chain segments is proportional to n, so the volume fraction of segments in the domain ϕ^* is proportional to $n/n^{3/2}$ or $n^{-1/2}$. For large n this is very small. The fraction of the coil domain which is not occupied by chain segments is $1 - \phi^*$ and is proportional to $1 - n^{-1/2}$. The number of sites into which segments of a *second* chain might be placed within the domain of the first is proportional to this "empty" volume. If the

second chain also consists of n segments, the number of ways of placing the second varies with n as $(1 - n^{-1/2})^n$. This is approximately equal to $e^{-n^{1/2}}$, which shows that the probability of a second coil overlapping the domain of the first decreases as n increases.

The above argument shows that complete overlap of coil domains is improbable for large n and hence gives plausibility to the excluded volume concept as applied to random coils. More importantly, however, it introduces the notion that coil interpenetration must be discussed in terms of probability. For hard spheres the probability of interpenetration is zero, but for random coils the boundaries of the domain are "softer" and the probability for interpenetration must be analyzed in more detail. One method for doing this will be discussed in the next section. Before turning to this, however, we note that the Flory-Huggins theory can also be used to yield a value for the second virial coefficient.

To use the Flory-Huggins theory as a source for understanding the second virial coefficient, we return to Eq. (8.53), which gives an expression for $\mu_1 - \mu_1^\circ$. Combining this result with Eq. (8.79) gives

$$-\Pi\bar{V}_1 = RT\left[\ln(1 - \phi_2) + \phi_2\left(1 - \frac{1}{n}\right) + \chi\phi_2^2\right] \tag{8.99}$$

At the same level of approximation as produced Eq. (8.85), we note

$$\phi_2 = \frac{n_2\bar{V}_2}{n_1\bar{V}_1 + n_2\bar{V}_2} \cong \frac{n_2}{n_1}\frac{\bar{V}_2}{\bar{V}_1}$$

so

$$\phi_2 \cong x_2\frac{\bar{V}_2}{\bar{V}_1} \cong c_2\frac{\bar{V}_2}{M} \tag{8.100}$$

Expanding the logarithm in Eq. (8.99) as $-\phi_2 - (1/2)\phi_2^2$ and combining with Eq. (8.100) yields

$$\frac{\Pi\bar{V}_1}{RT} = \left[\frac{\bar{V}_2}{M} - \left(1 - \frac{1}{n}\right)\frac{\bar{V}_2}{M}\right]c_2 + \left(\frac{1}{2} - \chi\right)\left(\frac{\bar{V}_2}{M}\right)^2 c_2^2 + \cdots \tag{8.101}$$

In terms of the reduced osmostic pressure, this becomes

$$\frac{\Pi}{RTc_2} = \frac{\bar{V}_2}{nM\bar{V}_1} + \frac{(1/2 - \chi)}{\bar{V}_1}\left(\frac{\bar{V}_2}{M}\right)^2 c_2 = \frac{1}{M} + \frac{(1/2 - \chi)}{\bar{V}_1}\left(\frac{\bar{V}_2}{M}\right)^2 c_2 \tag{8.102}$$

where the last simplification recalls the definition of n as $\bar{V}_2/\bar{V}_1$. Comparing this result with Eq. (8.87) shows that, according to the Flory-Huggins theory,

$$B = \left(\frac{1}{2} - \chi\right) \frac{\bar{V}_2{}^2}{\bar{V}_1 M^2} \tag{8.103}$$

If we further compare this result with Eq. (8.97), we can write

$$u_{coil} = \frac{2(1/2 - \chi)\bar{V}_2{}^2}{N_A \bar{V}_1} \tag{8.104}$$

which gives us two totally different expressions for the excluded volume of the random coil: Eqs. (8.98) and (8.104). In the next section we shall consider the reconciliation of these divergent relationships.

8.11 The Flory-Krigbaum Theory

We begin our attempt to reconcile these two expressions for the excluded volume of the random coil by reviewing some ideas about random coils from Chap. 1:

1. We saw in Sec. 1.11 that coil dimensions are affected by interactions between chain segments and solvent. Both the coil expansion factor α defined by Eq. (1.63) and the interaction parameter χ are pertinent to describing this situation.
2. Use of random flight statistics to derive r_g for the coil assumes the individual segments exclude no volume from one another. While physically unrealistic, this assumption makes the derivation mathematically manageable. Neglecting this volume exclusion means that coil dimensions are underestimated by the random fight model, but this effect can be offset by applying the result to a solvent in which polymer–polymer contacts are somewhat favored over polymer–solvent contacts.
3. Conditions in which the effects of item (2) exactly compensate are called Θ conditions. The expansion factor α gives the ratio of coil dimensions under non-Θ conditions to those under Θ conditions.

To apply these ideas to solution nonideality, we consider a theory developed by Flory and Krigbaum. This is only one of several approaches to the problem, but it is one which can be readily outlined in terms of material we have already developed. We shall only sketch the highlights of the Flory-Krigbaum theory, since the details are complicated and might actually obscure the principal ideas.

We consider the approach of two polymer coils in dilute solution until the two coil domains overlap to some extent, as shown in Fig. 8.11. According to

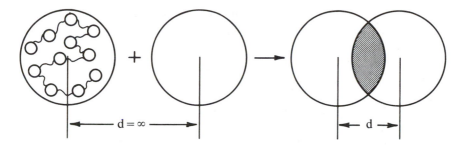

Figure 8.11 The coil domains of two polymer molecules. The schematic shows by shading how the region of overlap increases as the distance between centers decreases.

this theory, we shall consider the coil domain as a solution of chain segments in which the volume fraction is uniformly ϕ^*. Within the shaded area in Fig. 8.11, therefore, the segment volume fraction is $2\phi^*$. It is apparent that the volume of the lens-shaped overlap region V_{lens} is a function of the distance of separation of the coil centers.

The objective of the Flory-Krigbaum theory is to find a quantitative expression for the placement probability $\Omega(d)$ of the two coils as a function of their separation d. There are three stages to the derivation:

1. The statistical problem. The relative probability associated with the placement of the two coils such that $d = \infty$ is unity: The solution is so dilute that we can place the polymer molecules anywhere. It is at smaller d's that the placement probability drops off because of a generally unfavorable ΔG associated with the overlap. We assume that the decrease in probability is described by the Boltzmann factor and write

$$\Omega(d) = \Omega(d = \infty)\, e^{-\Delta G/kT} = e^{-\Delta G/kT} \qquad (8.105)$$

2. The thermodynamic problem. There is a ΔG associated with the change in ϕ^* in the overlap region. This, in turn, is the product of two factors:

$$\Delta G = (\text{concentration effect})(\text{geometrical effect}) \qquad (8.106)$$

The concentration effect involves the ΔG's of mixing for the solutions of concentrations ϕ^* and $2\phi^*$. We shall return to this presently.

3. The geometrical problem. This involves evaluating the "geometrical effect" in item (2). It requires calculation of the volume of the overlapping regions as a function of d and the coil dimensions, say, r_g. The mathematics of this step are tedious and add little to the polymer aspects of the theory.

Of the three stages outlined above, only the calculation of ΔG for the process in Fig. 8.11 warrants further discussion. Recalling that the volume of the shaded region is V_{lens}, we write

$$\Delta G = V_{lens} \, \Delta G_{m, 2\phi*} - 2V_{lens} \, \Delta G_{m, \phi*} \tag{8.107}$$

where the ΔG_m's are the free energies of mixing per unit volume of the segment solutions of the indicated concentrations. Factoring out V_{lens} and identifying it as the geometrical effect in Eq. (8.106), we describe the concentration effect in the lens-shaped volume when the coils interpenetrate:

$$\Delta G_{lens} = \Delta G_{m, 2\phi*} - 2 \, \Delta G_{m, \phi*} \tag{8.108}$$

Next we use the Flory–Huggins theory to evalute ΔG_m by Eq. (8.44). As noted above, the volume fraction occupied by polymer segments within the coil domain is small, so the logarithms in Eq. (8.44) can be approximated by the leading terms of a series expansion. Within the coil $N_2 = 1$ and $N_1 = (1 - \phi*) \, V_d N_A / \bar{V}_1$, where V_d is the volume of the coil domain. When all of these considertions are taken into account, Eq. (8.108) becomes

$$\Delta G_{lens} = 2kT \left(\frac{1}{2} - \chi \right) \frac{\bar{V}_2^{\,2}}{\bar{V}_1 V_d} \tag{8.109}$$

The volume of the coil domain which appears in Eq. (8.109) can also be grouped with related quantities as a factor in the geometrical part of the problem. Our primary interest is in the appearance of the factor $1/2 - \chi$ in the expression for ΔG_{lens} and—via Eqs. (8.105) and (8.106)—the idea that

$$\ln \left(\frac{\Omega(d)}{\Omega(d = \infty)} \right) \propto \left(\frac{1}{2} - \chi \right) \times f(d, r_g) \tag{8.110}$$

where the geometrical aspect of the problem is described by some function of the separation and the radii of gyration of the coils. While the precise details of such a function depend on the specific parameters used in the theory, it is sufficient to note that the probability undergoes a relatively rapid decrease from unity (for unhindered placement) to zero (for the excluded volume effect) as d decreases. This is sketched qualitatively in Fig. 8.12. For comparison, the broken line in Fig. 8.12 shows how the hard-sphere model would appear in the same representation. The more gradual transition in probability according to the Flory–Krigbaum model means the polymer coil is a relatively "soft" sphere but has a definite excluded volume nevertheless. Furthermore, the value of d at which the probability drops off is in the neighborhood of $2r_g$.

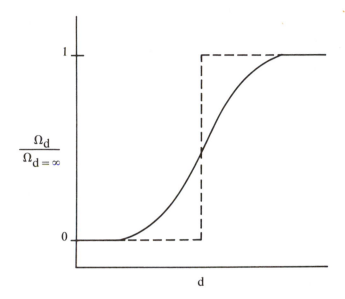

Figure 8.12 Schematic illustration showing with the solid line how the probability of placement varies with the distance of separation between the centers of the coils. The broken line is the equivalent result for hard spheres.

Converting the probability function described above into an excluded volume is accomplished by integrating the probability $1 - \Omega(d)$ of exclusion over a spherical volume encompassing all values of d:

$$u = \int_0^\infty [1 - \Omega(d)] \, 4\pi d^2 \, d(d) \tag{8.111}$$

The full Flory-Krigbaum theory results in the following expression for the excluded volume:

$$u = \frac{2\bar{V}_2{}^2}{\bar{V}_1 N_A} \left(\frac{1}{2} - \chi\right)\left(1 - \frac{y}{2!\, 2^{3/2}} + \frac{y^2}{3!\, 3^{3/2}} + \cdots\right) \tag{8.112}$$

where

$$y = \frac{2\bar{V}_2{}^2}{\bar{V}_1 N_A} \left(\frac{1}{2} - \chi\right)\left(\frac{3}{4\pi r_g}\right)^{3/2} \tag{8.113}$$

The complicated form of the final result makes it clear why we have skipped over the details of the Flory-Krigbaum derivation!

If we disregard the term in brackets in Eq. (8.112)—which we shall call $f(y)$ and which is close to unity for values of χ that are not too different from $1/2$—this expression gives the same result for the excluded volume of the coil as given by Eq. (8.104) from a comparison of Eqs. (8.97) and (8.103). We may regard $f(y)$ as a correction factor which is required for Eq. (8.104) to be valid as the difference $1/2 - \chi$ increases. In the next section we shall discuss the implications of Eq. (8.112) in greater detail.

Our primary interest in the Flory–Krigbaum theory is in the conclusion that the second virial coefficient and the excluded volume depend on solvent–solute interactions and not exclusively on the size of the polymer molecule itself. It is entirely reasonable that this should be the case in light of the discussion in Sec. 1.11 on the expansion or contraction of the coil depending on the solvent. The present discussion incorporates these ideas into a consideration of solution nonideality.

The parameter α which we introduced in Sec. 1.11 to measure the expansion which arises from solvent being imbibed into the coil domain can also be used to describe the second virial coefficient and excluded volume. We shall see in Sec. 9.7 that the difference $1/2 - \chi$ is proportional to $\alpha^5 - \alpha^3$. When the fully developed function of α is used to replace $1/2 - \chi$ in Eq. (8.112), the excluded volume according to the Flory–Krigbaum theory becomes

$$u = \left(\frac{4}{3} \pi r_g^{\,3} \right) 2 \left(\frac{4}{3} \pi \right)^{1/2} (\alpha^2 - 1) \left(1 - \frac{z}{2!\,2^{3/2}} + \frac{z^2}{3!\,3^{3/2}} + \cdots \right) \tag{8.114}$$

where $z = 2(\alpha^2 - 1)$. What is noteworthy about this form of the theory is that the excluded volume is proportional to the volume of a sphere whose radius is r_g, as proposed in Eq. (8.98). Depending, then, on the particular algebraic form we examine, the Flory–Krigbaum theory ties together several ideas that have been in apparent opposition until now.

8.12 Theta Conditions

So far we have avoided mentioning one of the most interesting features of Eq. (8.112): the fact that u and therefore B equal zero when $\chi = 1/2$. From an examination of Eq. (8.114), it is also apparent that this corresponds to the condition $\alpha = 1$. We shall presently introduce still another way of describing the condition $B = u = 0$, but first we summarize the information we have accumulated until now pertaining to this condition:

1. When $B = 0$, the solution behaves ideally, at least through second-order effects. This means that deviations from ideality might be observed at still higher concentrations, but that the van't Hoff equation applies at least in dilute solutions for systems with $B = 0$.

2. B = 0 when χ = 1/2, a condition we have already seen [Eq. (8.60)], corresponds to a critical value of χ for a copolymer of infinite molecular weight. For finite molecular weights this condition is not quite a threshold for precipitation, but is close to it. Polymer–polymer contacts are sufficiently favored over polymer–solvent contacts that a chain of infinite length would undergo phase separation.

3. We can imagine the coil tightening up as this point is approached from "better" conditions. This is not a shrinking to the vanishing point as suggested by the u = 0 criterion, but a contraction to the point where intramolecular exclusion effects are offset by shrinkage.

4. This was precisely the point where α was defined to equal unity, so it is entirely appropriate that inter- and intramolecular exclusion volumes are effectively zero under the same conditions.

5. In Chap. 1 we referred to these as Θ conditions, and we shall examine the significance of this term presently. Note that Θ conditions for a polymer solution are analogous to the Boyle temperature of a gas: Each behaves ideally under its respective conditions.

6. It is interesting to note that for a van der Waals gas, the second virial coefficient equals b – a/RT, and this equals zero at the Boyle temperature. This shows that the excluded volume (the van der Waals b term) and the intermolecular attractions (the a term) cancel out at the Boyle temperature. This kind of compensation is also typical of Θ conditions.

Many presentations of the second virial coefficient of polymer solutions contain different expressions for the quantities we have discussed. The difference lies in the fact that the factor $\psi(1 - \Theta/T)$ appears in place of $1/2 - \chi$. There are several attitudes we can take toward this difference. For one thing, we can regard the discrepancy as nothing more than different notation:

$$\frac{1}{2} - \chi \equiv \psi\left(1 - \frac{\Theta}{T}\right) \tag{8.115}$$

Based on this point of view, we can draw several conclusions:

1. ψ must be dimensionless.

2. Θ must have the units of temperature.

3. The conditions $\chi < 1/2$ and $\chi > 1/2$ correspond to positive and negative B values, respectively.

4. The same positive and negative B values must result from $T > \Theta$ and $T < \Theta$, respectively.

5. This alternative notation is especially suited to describe changes in solvent goodness which arise from temperature variations for a fixed system. By contrast, the χ notation is more descriptive of different solvents at a single temperature.

As a device for describing the effect of temperature on solution nonideality, it is entirely suitable to think of Eq. (8.115) as offering an alternate notation which accomplishes the desired effect with ψ and Θ as adjustable parameters. We note, however, that the left-hand side of Eq. (8.115) contains only one such parameter, χ, while the right-hand side contains two: ψ and Θ. Does this additional parameter have any physical significance?

To answer this question, we recall that χ is defined in Sec. 8.4 as a measure of the "energy" change associated with reaction (8.A), the solution process. In this context we acknowledged that there could very well be an entropy effect also associated with that process: a ΔS arising from some sort of "bonding" between solvent and polymer which is not included in the configurational entropy. If such an entropy effect is operative, then the quantity Δw on which the definition of χ is based would be better viewed as a free-energy change, as in Eq. (8.43). Although we discussed this possibility earlier, we have not acted upon it until now. Some manipulations on $1/2 - \chi$ provide a strategy for introducing this effect, which we shall call a solvation entropy ΔS_s:

1. We multiply $1/2 - \chi$ by RT to obtain

$$\left(\frac{1}{2} - \chi \right) RT = - \left(\chi RT - \frac{1}{2} RT \right)$$

2. With the absorbed coordination number, χRT is taken to be the value of ΔH for reaction (8.A); therefore we write

$$\left(\frac{1}{2} - \chi \right) RT = - \left(\Delta H - \frac{1}{2} RT \right)$$

3. The quantity in parentheses on the right-hand side is reminiscent of the expression $\Delta H - T \Delta S$, with the quantity $1/2R$ a contribution from the configurational entropy of the Flory–Huggins theory. Since our objective is to incorporate a solvation entropy into the discussion, we add the latter— in units of R for convenience—to $1/2R$:

$$\left(\frac{1}{2} - \chi \right) RT = - \left[\Delta H - \left(\frac{1}{2} + \Delta S_s \right) RT \right]$$

4. We define $1/2 + \Delta S_s$ by the parameter ψ and factor:

$$\left(\frac{1}{2} - \chi \right) RT = - \psi RT \left(\frac{\Delta H}{(1/2 + \Delta S_s) RT} - 1 \right)$$

5. We define Θ to be $\Delta H/(1/2 + \Delta S_s)R$. Since the numerator of this definition is an enthalpy and the denominator an entropy, Θ must have the units of a temperature:

$$\left(\frac{1}{2} - \chi \right) RT = - \psi RT \left(\frac{\Theta}{T} - 1 \right) = \psi RT \left(1 - \frac{\Theta}{T} \right)$$

6. By reversing step (1), we obtain Eq. (8.115), which we had set out to interpret.

The "new" parameter ψ, then, is a measure of the solvation entropy—except for the term $1/2$—which we can regard as a measure of ΔS of reaction (8.A), just as χ or $\psi \Theta R$ measures ΔH for that reaction.

We can therefore include ψ and Θ along with χ and α as parameters which are measured by the second virial coefficient. The usefulness of having these alternative formulations for the second virial coefficient lies in the flexibility they give us for describing the chemical potential of a polymer solution. Thus we can say that the chemical potential behaves ideally in a solvent for which $\chi = 1/2$ or at a temperature equal to Θ. From a practical point of view, this means that the slope of a Π/c_2-versus-c_2 plot is horizontal under Θ conditions. Therefore a correct value of M can be evaluated from the van't Hoff equation using data measured at a single concentration in the dilute range without any need to extrapolate to $c_2 = 0$. Of course, it may take extensive preliminary experimentation to establish what the Θ conditions are, as well as to determine the range of acceptable concentrations within which this ideality is observed.

Figure 8.13 shows the reduced osmotic pressure for solutions of polyisobutylene in benzene plotted against c_2 at several different temperatures. The

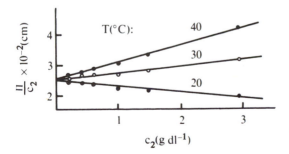

Figure 8.13 Π/c_2 versus c_2 for polyiosbutylene in benzene at the temperatures shown. [Reprinted with permission from W. R. Krigbaum and P. J. Flory, *J. Am. Chem. Soc.* 75:5254 (1953), copyright 1953 by the American Chemical Society.]

Table 8.3 Theta Temperatures for a Few Polymer–Solvent Systems

Polymer	Solvent	Θ (°C)
Poly(dimethyl siloxane)	Cyclohexane	-81
Poly(methyl methacrylate)	Toluene	-65
Polyisobutene	Toluene	-13
Poly(dimethyl siloxane)	2-Butanone	20
Poly(methyl methacrylate)	2-Ethylbutyraldehyde	22
Polyisobutene	Benzene	24
Polystyrene	Decalin	31
Poly(methyl methacrylate)	Amyl acetate	41
Polystyrene	Ethyl cyclohexane	70

Source: H. G. Elias et al., in J. Bandrup and E. H. Immergut (Eds.), *Polymer Handbook*, 2nd ed., Wiley, New York, 1975.

Θ temperature for this system is identified as the temperature at which this plot shows a horizontal slope. For the system shown in Fig. 8.13, Θ lies between 20 and 30°C; additional measurements in this range indicate this temperature to be about 24°C. Table 8.3 lists, in increasing order, the Θ temperatures for a few polymer–solvent systems. The following observations are pertinent to the entries in Table 8.3:

1. Although the systems listed were chosen arbitrarily, it is apparent that Θ values span a wide range.
2. A variety of systems exist for which Θ is in the vicinity of room temperature.
3. The nature of the solvent is as important as the polymer in determining Θ, as is apparent from the wide range of Θ values for poly(dimethyl siloxane) in different solvents.

We shall have occasion to refer to Θ conditions in the next two chapters as well, so the ideas of the past few sections have applications beyond merely describing the nonideality of osmostic pressure experiments.

In this chapter, and throughout this entire book for that matter, we have been concerned with uncharged polymers. We shall conclude this chapter with a brief discussion of the osmotic pressure of charged polymer molecules.

8.13 Osmotic Pressure of Charged Polymers

In this section we briefly consider the osmotic pressure of polymers which carry an electric charge in solution. These include synthetic polymers with ionizable functional groups such as $-NH_2$ and $-COOH$, as well as biopolymers such as proteins and nucleic acids. In this discussion we shall restrict our consideration

to solutions which are relatively dilute in polymer, thereby focusing attention on only the first or van't Hoff term of the concentration dependence of Π. This emphasis is strictly a matter of convenience; theories which deal with solution nonideality in charged systems are also available.

We shall be interested in determining the effect of electrolytes of low molecular weight on the osmotic properties of these polymer solutions. To further simplify the discussion, we shall not attempt to formulate the relationships of this section in general terms for electrolytes of different charge types—2:1, 2:2, 3:1, 3:2, and so on—but shall consider the added electrolyte to be of the 1:1 type. We also assume that these electrolytes have no effect on the state of charge of the polymer itself; that is, for a polymer such as, say, poly(vinyl pyridine) in aqueous HCl or NaOH, the state of charge would depend on the pH through the water equilibrium and the reaction

$$(8.B)$$

We shall refer to electrolytes which do not affect the charge of the polymer as indifferent electrolytes. In the situation illustrated by reaction (8.B), the HCl and NaOH are clearly not indifferent. We shall also assume that the indifferent electrolytes are 100% ionized in the polymer solution.

We continue to designate the solvent (usually water) as component 1, the polymer as component 2, and the indifferent electrolyte MX as component 3. We arbitrarily designate the polymer to be a cation with a relative charge of $+z$, having associated with it the same anion as is present in MX. Accordingly, we designate the polymer PX_z and represent its dissociation by

$$PX_z \rightarrow P^{+z} + zX^- \qquad (8.C)$$

We consider this system in an osmotic pressure experiment based on a membrane which is permeable to all components except the polymeric ion P^{+z}; that is, solvent molecules, M^+, and X^- can pass through the membrane freely to establish the osmotic equilibrium, and only the polymer is restrained. It does not matter whether pure solvent or a salt solution is introduced across the membrane from the polymer solution or whether the latter initially contains salt or not. At equilibrium both sides of the osmometer contain solvent, M^+, and X^- in such proportions as to satisfy the constaints imposed by electroneutrality and equilibrium conditions.

It is conventional to use molality—moles of solute per kilogram of solvent (symbol m)—as the concentration unit in electrolyte thermodynamics. Accordingly, we shall represent the concentrations of both the indifferent electrolyte and the polymer in these units in this section: m_3 and m_2, respectively. In the same dilute (with respect to polymer) approximation that we have used elsewhere in this chapter, m_2 is related to the mass volume^{-1} system of units c_2 by

$$m_2 = \frac{1000\,\overline{V}_1}{M_1 M}\, c_2 \tag{8.116}$$

Neglecting the higher-order terms, we can write the osmotic pressure for this three-component system in terms of the van't Hoff equation:

$$\Pi\,\frac{1000\,\overline{V}_1}{M_1} = mRT \tag{8.117}$$

where m (no subscript) is the number of moles of pressure-producing solute per kilogram of solvent and $1000\,\overline{V}_1/M_1$ is the volume of that amount of solvent. Next we examine the molal concentration of "pressure-producing solute." Considering the definition of the system, this appears, at first thought, to describe only the polymer. We shall see presently, however, that the indifferent electrolyte is generally not uniformly distributed on both sides of the membrane, so any asymmetry in its concentration will also contribute to m in Eq. (8.117). If we designate the phase containing the polyelectrolyte as the α phase, and that from which the polymer is excluded as the β phase, we can write

$$m = m_{2,\alpha} + m_{M^+,\alpha} + m_{X^-,\alpha} - m_{M^+,\beta} - m_{X^-,\beta} \tag{8.118}$$

If the concentrations of M^+ and X^- were both the same in the α and β phases, $m = m_2$; this formalism allows for the more general case where the α and β concentrations are not equal.

There are two additional concepts which we can invoke to simplify Eq. (8.118): electroneutrality and a less familiar principle called Donnan equilibrium. Some relationships pertaining to these are developed below:

1. Electroneutrality requires simply that the same amounts of positive and negative charge be present in a solution. Applying this separately to both the α and β phases, we obtain

$$zm_{P,\alpha} + m_{M^+,\alpha} = m_{X^-,\alpha} \tag{8.119}$$

and

$$m_{M^+,\beta} = m_{X^-,\beta} \tag{8.120}$$

2. Donnan equilibrium arises from applying the phase equilibrium criterion to the indifferent electrolyte: $\mu_3{}^{\alpha} = \mu_3{}^{\beta}$. From Eq. (8.13) this is $\mu_3 = \mu_3{}^{\theta} + RT \ln a_{3,\beta}$ for the side under normal pressure. From Eq. (8.76), $\mu_3 = \mu_3{}^{\theta} + RT \ln a_{3,\alpha} + \Pi \bar{V}_3$ for the side containing the polymer. If the polymer concentration is low, the term $\Pi \bar{V}_3$ is insignificant compared to the terms arising from the electrolyte itself; therefore

$$a_{3,\alpha} = a_{3,\beta} \tag{8.121}$$

3. The activity of an electrolyte is given by the product of the activities of the individual ions. Applied to MX, this means that Eq. (8.121) becomes

$$a_{M^+,\alpha} \, a_{X^-,\alpha} = a_{M^+,\beta} \, a_{X^-,\beta} \tag{8.122}$$

4. The activity of an ion is related to its molality through the mean activity coefficient $\gamma_{\pm}$; therefore

$$(m_{M^+,\alpha})(m_{X^-,\alpha}) \, \gamma_{\pm}{}^2 = (m_{M^+,\beta})(m_{X^-,\beta}) \, \gamma_{\pm}{}^2 \tag{8.123}$$

For low concentrations of electrolyte, $\gamma_{\pm} \to 1$. Since we are not concerned with electrolyte nonideality per se, we shall assume this last condition and write

$$m_{M^+,\alpha} \, m_{X^-,\alpha} = m_{M^+,\beta} \, m_{X^-,\beta} \tag{8.124}$$

5. This last result describes the Donnan equilibrium condition as it applies to the system under consideration. Like other ionic equilibrium expressions, it requires the equality of *ion products* in equilibrium solutions.

Equations (8.119), (8.120), and (8.124) provide some additional relationships which can be substituted into Eq. (8.118). When all of these results are combined, we obtain

$$m = m_{2,\alpha} (1 + z) + 2m_{M^+,\alpha} - 2 \left[m_{M^+,\alpha} (zm_{2,\alpha} + m_{M^+,\alpha}) \right]^{1/2} \tag{8.125}$$

Since all of the concentrations now apply to the phase containing the polymer, the α subscript has become redundant and will be dropped. Combining Eqs. (8.117) and (8.125) yields the desired result:

$$\Pi = \frac{M_1 RT}{1000 \bar{V}_1} \left[(1 + z) m_2 + 2m_{M^+} - 2m_{M^+} \left(1 + \frac{zm_2}{m_{M^+}} \right)^{1/2} \right] \tag{8.126}$$

Note that if $z = 0$, the entire quantity in the brackets would equal m_2, and Eq. (8.126) would be the van't Hoff equation applied to an ordinary polymer.

The effect of the charge as well as that of the indifferent electrolyte, then, is contained in the term in brackets. A numerical calculation is probably the easiest way to examine this effect. This is illustrated in the following example.

Example 8.8

Evaluate the bracketed quantity in Eq. (8.126) for m_{M^+} between 10^{-6} and 10^{-1} mol kg^{-1} at intervals differing by powers of 10. Do this for polyelectrolytes with the following charge-concentration $(z-m_2)$ combinations: $50\text{-}10^{-6}$, $50\text{-}10^{-5}$, $100\text{-}10^{-5}$, and $100\text{-}10^{-4}$. Comment on the results.

Solution

The factors zm_2 and $(z + 1)m_2$ in combination with the different values for m_{M^+} determine the magnitude of this term. We begin by tabulating these:

z	m_2 (mol kg^{-1})	zm_2 (mol kg^{-1})	$(z + 1)m_2$ (mol kg^{-1})
50	10^{-6}	5.0×10^{-5}	5.1×10^{-5}
50	10^{-5}	5.0×10^{-4}	5.1×10^{-4}
100	10^{-5}	1.00×10^{-3}	1.01×10^{-3}
100	10^{-4}	1.00×10^{-2}	1.01×10^{-2}

This particular combination of parameters enables us to do the following:

1. Vary m_2 at constant z.
2. Vary z at constant m_2.
3. Span a 200-fold variation in zm_2, the charge concentration contributed by polymer.
4. Use approximations developed for dilute polymer solutions. Note that the values for m_{M^+} satisfy the requirement for dilution at 10^{-6} m but not at 10^{-1} m.

Table 8.4 lists numerical values for the quantity in brackets for the required ranges of parameters.

The trends that we note in these calculations are the following:

1. The bracketed term approaches the value of m_2 as the concentration of indifferent electrolyte increases.
2. This is true in all cases, but is most evident in the first column of calculated quantities in Table 8.4.
3. The bracketed term approaches the value of $(z + 1)m_2$ as the concentration of indifferent electrolyte decreases.

Table 8.4 Numerical Values for the Bracketed Term in Eq. (8.126) for the Parameters m_2, z, and m_{M^+} Shown (Values Discussed in Example 8.8)

m_M^+ (mol kg^{-1}):				
m_2 (mol kg^{-1}):	10^{-6}	10^{-5}	10^{-5}	10^{-4}
$(z+1) m_2$ (mol kg^{-1}):	5.1×10^{-5}	5.1×10^{-4}	1.01×10^{-3}	1.01×10^{-2}
	Bracketed term in Eq. (8.126) (mol kg^{-1})			
10^{-6}	3.87×10^{-5}	4.67×10^{-4}	9.49×10^{-4}	9.90×10^{-3}
10^{-5}	2.20×10^{-5}	3.87×10^{-4}	8.29×10^{-4}	9.49×10^{-3}
10^{-4}	6.05×10^{-6}	2.20×10^{-4}	5.47×10^{-4}	8.29×10^{-3}
10^{-3}	1.61×10^{-6}	6.05×10^{-5}	1.82×10^{-4}	5.47×10^{-3}
10^{-2}	1.06×10^{-6}	1.61×10^{-5}	3.38×10^{-5}	1.82×10^{-3}
10^{-1}	1.01×10^{-6}	1.06×10^{-5}	1.25×10^{-5}	3.38×10^{-4}

4. This is true in all cases, but is most evident in the last column of Table 8.4. Of course, for large values of z, $z \cong z + 1$, and it is not readily apparent that it is $(z + 1)m_2$ rather than zm_2 that is the correct limit. The value given is correct, however.

•

The generalizations demonstrated in the preceding example can be proved without resorting to numbers, but we have looked at enough algebra in this chapter already without adding these manipulations as well!

Combining Eqs. (8.116) and (8.126) with the conclusions from the example, we obtain the following:

1. For the case of $m_{M^+} \rightarrow 0$,

$$\frac{\Pi}{RTc_2} = \frac{z + 1}{M} \tag{8.127}$$

2. For the other extreme in which the concentration of the indifferent electo-lyte is high,

$$\frac{\Pi}{RTc_2} = \frac{1}{M} \tag{8.128}$$

3. These results show more clearly than Eq. (8.126)– of which they are special cases–the effect of charge and indifferent electrolyte concentration on the osmotic pressure of the solution. In terms of the determination of molecular weight of a polyelectrolyte by osmometry . . . .

4. A correct value of the molecular weight is obtained for the charged polymer by the van't Hoff equation, provided that a large excess of indifferent electrolyte is present. These high concentrations are described as swamping electrolyte conditions.

5. A value for the molecular weight which is low by a factor $z + 1$ is obtained for salt-free solutions if the experimental results are analyzed as if the polymer were uncharged.

6. Again, Eq. (8.126) describes the situation for known values of z and inter-mediate concentrations of electrolyte.

What makes the latter items particularly important is the fact that the charge and electrolyte content of an unknown polymer may not be known; hence it is important to design an osmotic pressure experiment correctly for such a system. It is often easier to add swamping amounts of electrolyte than to totally eliminate all traces of electrolyte. Under the former conditions a true molecular weight is obtained. Trouble arises only when the experimenter is indifferent toward indifferent electrolyte; this sort of carelessness can be the source of much confusion.

The algebra and arithmetic of this section have led us to the correct conclusions, but the underlying physical reality may be obscured by all these manipulations. The situation is simply this: In the complete absence of indifferent electrolyte, the polymer ion *and* its counterions are restricted to the same phase. The polymer is held back by the membrane, the counterions are held back by electroneutrality. The polymer side of the membrane, therefore, contains z + 1 mol of solute particles for every molar mass of polymer introduced. Since osmotic pressure gives a number average molecular weight, Eq. (8.127) is the logical result. With a sufficiently large excess of indifferent electrolyte, the small ions effectively "swamp out" the charge contributed by the polymer, and the latter behaves as if it were uncharged.

Problems

1. In the derivation of Eq. (8.40) it is assumed that each polymer segment is surrounded by z sites which are occupied at random by either solvent molecules or polymer segments. Actually, this is true of only z − 2 of the sites in the coordination sphere−z − 1 for chain ends−since two of the sites are occupied by polymer segments which are covalently bound to other polymer segments. Criticize or defend the following proposition concerning this effect: The kinds of physical interactions that we identify as London or dipole–dipole attractions can also operate between segments which are covalently bonded together, so the w_{22} contribution continues to be valid. A slight error in counting is made−to allow for simplification of the resulting function−but this is a tolerable approximation in concentrated solutions. In dilute solutions the approximation introduces more error, but the model is in trouble in such solutions anyhow, so another approximation makes little difference.

2. By combining Eqs. (8.42), (8.49), and (8.60), show that $V_1^{\circ}(\delta_2 - \delta_1)^2 = (1/2)RT_c$, where T_c is the critical temperature for phase separation. For polystyrene with $M \cong 3 \times 10^6$, Shultz and Flory[†] observed T_c values of 68 and 84°C, respectively, for cyclohexanone and cyclohexanol. Values of V_1° for these solvents are abut 108 and 106 cm³ mol⁻¹, respectively, and δ_1 values are listed in Table 8.2. Use each of these T_c values to form separate estimates of δ_2 for polystyrene and compare the calculated values with each other and with the value for δ_2 from Table 8.2. Briefly comment on the agreement or lack thereof for the calculated and accepted δ's in terms of the assumptions inherent in this method. Criticize or defend the following proposition for systems where use of the above relationship is justified: Polymer will be miscible in all proportions in low molecular weight solvents from which they differ in δ value by about 3 or less.

[†] A. R. Shultz and P. J. Flory, *J. Am. Chem. Soc.* 75:5681 (1953).

3. The accompanying sketch qualitatively describes the phase diagram for the
    system nylon-6,6, water, phenol for $T \geqslant 70^\circ C$.† In this figure the broken
    lines are the lines whose terminals indicate the concentrations of the three
    components in the two equilibrium phases. Consult a physical chemistry
    textbook for the information as to how such concentrations are read. In
    the two-phase region, both phases contain nylon, but the water-rich phase
    contains the nylon at a lower concentration. On this phase diagram or a
    facsimile, draw arrows which trace the following procedure:

    a. Water is added to solution A until the system separates into solutions
        B and C.
    b. Solution C is removed and water is added to solution B until the system
        separates into D and E.
    c. Phenol is added to solution C until a solution A' results which is close
        to A in concentration.

    Would you expect solutions C and E or B and D to contain the higher
    molecular weight polymer? Briefly explain. Outline a strategy for nylon
    fractionation based on steps (a)–(c). Some steps may be repeated as needed.

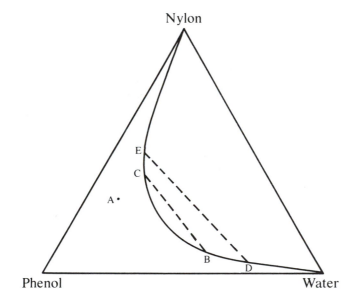

† G. B. Taylor, *J. Am. Chem. Soc.* 69:638 (1947).

4. Following the procedure described in the last problem, Taylor† fractionated nylon-6,6 into a series of cuts which he characterized by viscosity. These cuts were then refractionated. The following is a list of the intrinsic viscosities $[\eta]$ —which increase as $M^{0.72}$ in this system—of the various fractions:

	$[\eta]$ for the concentrated phase				$[\eta]$ for the dilute phase		
Cut number	First fraction	Cut number	Second fraction	Cut number	First fraction	Cut number	Second fraction
1	1.40	1.1	2.02	1	0.97	1.1	1.37
		1.2	1.80			1.2	1.24
		1.3	1.55			1.3	1.04
		1.4	1.17			1.4	0.65
		1.5	0.71			1.5	0.15
		1.6	0.30				
2	1.27	2.1	1.67	2	0.88	2.1	1.11
		2.2	1.61			2.2	0.93
		2.3	1.45			2.3	0.58
		2.4	1.30			2.4	0.37
		2.5	0.85				
		2.6	0.38				
3	1.01	3.1	1.28	3	0.74	3.1	0.94
		3.2	1.06			3.2	0.81
		3.3	0.68			3.3	0.58
		3.4	0.58			3.4	0.36
4	0.75	4.1	0.90	4	0.59	4.1	0.71
		4.2	0.75			4.2	0.62
		4.3	0.54			4.3	0.46
		4.4	0.46			4.4	0.36
5	0.57			5	0.44		
6	0.38			6	0.30		
7	0.23			7	0.18		

On the basis of these observations, criticize or defend the following proposition: In the first fractionation, the average molecular weight of the polymer in the concentrated phase is about 50% higher than that in the dilute phase, and between the first and last cut of either phase there is more than a 10-fold range in M. In the second fractionation the lower cut numbers fractionate into a narrower range of M's because they are more nearly pure samples of the high molecular weight polymer.

---

† G. B. Taylor, *J. Am. Chem. Soc.* 69:638 (1947).

5. In some IUPAC-sponsored research[†], samples of the same polystyrene preparation were distributed among different laboratories for characterization. The following molecular weights were obtained for one particular sample by osmotic pressure experiments using the solvents, membranes, and temperatures listed below:

Laboratory	T (°C)	Solvent[a]	Membrane	$M \times 10^{-3}$ (g mol^{-1})
I	27.5	T	Gel cellophane	79
I	27.5	MEK	Gel cellophane	78
H	25	MEK	Gel cellophane	98
M	25	B	Denitrated collodion	77.5
M	25	MEK	Denitrated collodion	77.5
U	28	B	Denitrated collodion	83
U	28	MEK	Denitrated collodion	100
O	25	B	Regenerated cellulose	75
T	26	MEK	Bacterial cellulose	78

[a] T = toluene, B = benzene, and MEK = methyl ethyl ketone.

Based on these observations, criticize or defend the following proposition: Except for two of the values in MEK, the reproducibility of M determinations is quite acceptable. MEK consistently gives erroneous values, probably because the solutions are highly nonideal for this solvent, which is more polar than either benzene or toluene.

6. The osmotic pressure of polystyrene fractions in toluene and methyl ethyl ketone was measured[‡] at 25°C and the following results were obtained:

	Toluene		Methyl ethyl ketone	
Fraction number	$c_2 \times 10^3$ (g cm^{-3})	Π (g cm^{-2})	$c_2 \times 10^3$ (g cm^{-3})	Π (g cm^{-2})
I	4.27	0.22	2.67	0.04
	6.97	0.58	6.12	0.14
	9.00	1.00	8.91	0.31
	10.96	1.53		

[†] The International Union of Pure and Applied Chemistry, *J. Polym. Sci.* 10:129 (1953).
[‡] C. Bawn, R. Freeman, and A. Kamaliddin, *Trans. Faraday Soc.* 46:862 (1950).

Fraction number	Toluene		Methyl ethyl ketone	
	$c_2 \times 10^3$ (g cm^{-3})	Π (g cm^{-2})	$c_2 \times 10^3$ (g cm^{-3})	Π (g cm^{-2})
II	1.55	0.16	3.93	0.40
	2.56	0.28	8.08	0.95
	2.93	0.32	10.13	1.30
	3.80	0.47		
	5.38	0.77		
	7.80	1.36		
	8.68	1.60		
III	1.75	0.31	1.41	0.23
	2.85	0.53	2.90	0.48
	4.35	0.88	6.24	1.11
	6.50	1.49	8.57	1.63
	8.85	2.36		
V	1.65	0.61	2.49	0.87
	2.97	1.16	4.21	1.53
	4.80	2.00	5.56	2.03
	7.66	3.52	7.76	2.96

From plots of Π/c_2 versus c_2, evaluate M for each of the four polymer fractions. Do the data collected from the two different solvents conform to expectations with respect to slope and intercept values?

7. The osmotic pressure of solutions of polystyrene in cyclohexane was measured[†] at several different temperatures, and the following results were obtained:

Fraction	T = 24°C	
	c_2 (g cm^{-3})	$\Pi/RTc_2 \times 10^6$ (mol g^{-1})
II	0.0976	8.0
	0.182	6.0
	0.259	8.7

[†] W. R. Krigbaum and D. O. Geymer, *J. Am. Chem. Soc.* 81: 1859 (1959).

		$T = 34°C$
Fraction	c_2 (g cm^{-3})	$\Pi/RTc_2 \times 10^6$ (mol g^{-1})
II	0.0081	13.3
	0.0201	14.2
	0.0964	14.2
	0.180	18.7
	0.257	26.2
III	0.0156	2.46
	0.0482	2.24
	0.0911	3.42
	0.126	4.96
	0.139	6.05
		$T = 44°C$
Fraction	c_2 (g cm^{-3})	$\Pi/RTc_2 \times 10^6$ (mol g^{-1})
II	0.0959	18.6
	0.178	28.1
	0.255	40.0
III	0.0478	5.50
	0.125	11.0
	0.138	13.2

Plot all of these data on a single graph as Π/RTc_2 versus c_2, connecting the points so as to present as coherent a display of the results as possible. Evaluate the molecular weights of the two polystyrene fractions. Criticize or defend the following propostion: These data show that the Θ temperature for this system is about 34°C. As expected, the range of concentrations which are adequately described by the first two terms of the virial equation is less for sample III than for sample II. Above this range other contributions to nonideality contribute positive deviations from the two-term osmotic pressure equation. It would be interesting to see how this last effect appears for sample III at 24°C, but this measurement was probably impossible to carry out owing to phase separation.

8. Krigbaum† measured the second virial coefficient of polystyrene in cyclohexane at several different temperatures. The observed values of B as well as some pertinent volumes at those temperatures are listed below:

---

† W. R. Krigbaum, *J. Am. Chem. Soc.* 76:3758 (1954).

T (K)	303	313	323
$B \times 10^5$ (cm^3 mol g^{-2})	-4.55	4.45	9.01
$\overline{V}_1$ (cm^3 mol^{-1})	109.5	110.9	112.3
$\overline{V}_2/M$ (cm^3 g^{-1})	0.930	0.935	0.940

Use these data to estimate the Θ temperature for this system by graphical interpolation. By combining Eqs. (8.97), (8.112), and (8.115), the product $\psi f(y)$ can be evaluated at each temperature once Θ has been determined. Since $f(y) = 1.00$ at $T = \Theta$, ψ can be evaluated by a graphical interpolation of $\psi f(y)$ versus T. Use this procedure to estimate ψ from these data. Based on these values of Θ and ψ, estimate χ at $25°C$ for polystyrene in cyclohexane.

9. By combining Eqs. (8.60) and (8.115), the following relationship is obtained between the critical temperature T_c for phase separation and the degree of polymerization:

$$\frac{1}{T_c} = \frac{1}{\Theta} - \frac{1}{\Theta \psi}\left(\frac{1}{n^{1/2}} + \frac{1}{2n}\right)$$

Derive this relationship and explain the graphical method it suggests for evaluating Θ and ψ. The critical temperatures for precipitation for the data shown in Fig. 8.3b are the following:

M (g mol^{-1})	22,700	285,000	6,000,000
T_c ($°C$)	18.2	45.9	56.2

Use the graphical method outlined above to evaluate Θ and ψ and, from these, χ for polyisobutylene in diisobutylketone.

10. Shultz and Flory† measured the critical temperature for precipitation for polystyrene fractions of different molecular weight in cyclohexane. The following results were obtained:

	A	B	C	D
T_c ($°C$)	19.4	23.6	27.5	31.1
$\overline{n}$	419	856	2,404	12,210

Use the method described in Problem 9 to obtain values of Θ and ψ from these data. How do the values of these parameters compare with the values obtained for the same system from osmotic pressure data in Problem 8?

---

† A. R. Shultz and P. J. Flory, *J. Am. Chem. Soc.* 74:4760 (1952).

Compare the two methods in terms of the assumptions involved, the supplementary information required, and the sensitivity to experimental error.

11. The following values of Π/c_2 versus c_2 were measured[†] at $25°C$ in 0.15 m NaCl solutions for bovine serum albumin:

At pH 5.37				
c_2 (g liter^{-1})	8.8	17.5	27.5	56.3
Π/c_2 (Torr liter g^{-1})	0.280	0.284	0.308	0.348
At pH 7.00				
c_2 (g liter^{-1})	16.9	29.4	50.6	56.9
Π/c_2 (Torr liter g^{-1})	0.312	0.336	0.384	0.399

At pH 5.37 this molecule is known to possess no net charge. The data under these conditions is the basis for Example 8.7. Verify the values of M and the slope given in that example. At pH 7.00 this polymer is known to have a charge of about -12. Based on a comparison of the above data at pH 5.37 and 7.00, what conclusion can be drawn about the swamping electrolyte concentration relative to 0.15 m NaCl? Criticize or defend the following proposition: The excluded volume of the bovine serum albumin molecule is larger at pH 7.00 than at pH 5.37 because the atmosphere of counterions contributes to the excluded volume for a charged polymer. If this is the case, the slope at pH 7.00 is expected to show a smaller intercept and a slope closer to that at pH 5.37 if the salt concentration is lowered significantly. (Hint: Consult a physical chemistry textbook for the concentration dependence of the thickness of the ion atmosphere according to the Debye-Huckel theory.)

Bibliography

1. Flory, P. J., *Principles of Polymer Chemistry*, Cornell Univesity Press, Ithaca, N.Y., 1953.
2. Hiemenz, P. C., *Principles of Colloid and Surface Chemistry*, Marcel Dekker, New York, 1977.
3. Morawitz, H., *Macromolecules in Solution*, Interscience, New York, 1965.
4. Richards, E. G., *An Introduction to the Physical Properties of Large Molecules in Solution*, Cambridge University Press, Cambridge, 1980.
5. Tanford, C., *Physical Chemistry of Macromolecules*, Wiley, New York, 1961.
6. Tompa, H., *Polymer Solutions*, Butterworths, London, 1956.

[†] G. Scatchard, A. C. Batchelder, and A. Brown, *J. Am. Chem. Soc.* 68:2320 (1946).

9

Frictional Properties of
Polymers in Solution

I give you the end of a golden string,
Only wind it into a ball
It will lead you in at Heaven's gate,
Built in Jerusalem's wall.

Jerusalem, William Blake

9.1 Introduction

This chapter contains one of the more diverse assortments of topics of any
chapter in the volume. In it we discuss the viscosity of polymer solutions,
especially the intrinsic viscosity; the diffusion and sedimentation behavior of
polymers, including the equilibrium between the two; and the analysis of
polymers by gel permeation chromatography (GPC). At first glance these seem
to be rather unrelated topics, but features they all share are a dependence on the
spatial extension of the molecules in solution and applicability to molecular
weight determination.

In addition to an array of experimental methods, we also consider a more
diverse assortment of polymeric systems than has been true in other chapters.
Besides synthetic polymer solutions, we also consider aqueous protein solutions.
The former polymers are well represented by the random coil model; the latter
are approximated by rigid ellipsoids or spheres. For random coils changes in the
goodness of the solvent affects coil dimensions. For aqueous proteins the
solvent–solute interaction results in various degrees of hydration, which also
changes the size of the molecules. Hence the methods we discuss are all potential sources of information about these interactions between polymers and their
solvent environments.

Both the intrinsic viscosity and GPC behavior of random coils are related to
the radius of gyration as the appropriate size parameter. We shall see how the
radius of gyration can be determined from solution viscosity data for these

systems, and how changes in the coil dimensions with changing conditions can be measured. We emphasize biopolymers in discussing diffusion-sedimentation behavior and discover that the axial ratio of the ellipsoidal particles is the kind of size information we obtain from these studies.

Another parameter that plays an important role in unifying viscosity, diffusion, and sedimentation is the friction factor. This proportionality factor between velocity and the force of frictional resistance was introduced in Chap. 2, and its role in interrelating the topics of this chapter is reflected in the title of the chapter.

Some exceptionally gifted researchers have made contributions to the topics we discuss here, and their impact on these and other areas of inquiry have earned them special recognition. Specifically, we note that Einstein, Svedberg, Staudinger, and Flory have all been awarded the Nobel Prize, and Stokes and Poiseuille have been honored by having units named after them. These are but a few of the "superstars" whose work we encounter in this chapter.

9.2 Viscous Forces on Rigid Spheres

In Sec. 2.2 we saw that the coefficient of viscosity is defined as the factor of proportionality between the shearing force per unit area $\sigma_s = F_s/A$ and the velocity gradient dv/dy within a liquid [Eq. (2.2)]:

$$\frac{F_s}{A} = \eta \frac{dv}{dy} \tag{9.1}$$

This definition is general; it applies equally to pure liquids and to solutions, to situations in which η is constant and those in which it is a function of the velocity gradient or rate of shear. Our interest in this chapter is the application of this concept to polymer solutions—dilute solutions at that—and we shall limit our attention to cases in which the rate of shear is low enough that the Newtonian limiting behavior applies and η is independent of dv/dy. In the next section we shall consider the measurement and interpretation of viscosity under these conditions, but, rather than turning to this immediately, we begin by considering the frictional force arising from viscosity as it affects rigid spheres submerged in a flowing liquid. The purpose of starting this way is twofold. First, it will review some of the general ideas about viscosity from Chap. 2, and, second, it will introduce two important relationships, due to Stokes and Einstein, respectively, which will recur often in this chapter.

The shearing force that is part of the definition of viscosity can also be analyzed in terms of Newton's second law and written as

$$F = ma = m \frac{dv}{dt} \tag{9.2}$$

When Eqs. (9.1) and (9.2) are combined, we obtain

$$\frac{m}{A}\frac{dv}{dt} = \eta\frac{dv}{dy} \tag{9.3}$$

If it were complete, Eq. (9.3) would be a differential equation whose solution would give v, the velocity of the flowing liquid, as a function of time and position within the sample. Equation (9.3) does not tell the whole story, however. Other forces are also operative: External forces of, say, gravitational or mechanical origin are responsible for the motion in the first place, and pressure forces are associated with the velocity gradient. In general, things are not limited to one direction in space, but the forces, gradients, and velocities have x, y, and z components. It is possible to bring these considerations together in a very general form by adding together all of the forces acting on a volume element of liquid, including the viscous force defined by Eq. (9.1), and using the net force and the mass of the volume element in Eq. (9.2). The resulting expression is called the equation of motion and is the cornerstone of fluid mechanics. The equation of motion is generally written in terms of vector operators and takes on a variety of forms, depending on the system of coordinates and the vector identities that have been employed. If the equation of motion is complicated even in the writing, things are even worse in the solving! Accordingly, we defer solutions to the experts in this area and merely note that a full-fledged hydrodynamic analysis of a flow problem begins with the equation of motion, which, in turn, is F = ma applied to a volume element subject to complex forces.

As with any differential equation, an important part of solving the equation of motion is defining the boundary conditions. For the purpose of our discussion, an important boundary condition is that no slippage occur at the boundary between a moving fluid and a rigid wall. This is best understood by imagining the liquid to consist of an array of thin slices which follow the contours of the wall. The nonslip condition means that the layer which is adjacent to a stationary wall has a velocity of zero also, with successive layers away from the wall possessing larger increments of velocity. At a sufficiently large distance from the wall, a net velocity is attained which is unperturbed by the presence of the surface. In Sec. 9.4, we shall apply this idea to the flow of a liquid through a capillary tube. For now we qualitatively consider two classic problems involving the effect of rigid spheres on the flow behavior of a liquid.

The first of these problems involves relative motion between a rigid sphere and a liquid as analyzed by Stokes in 1850. The results apply equally to liquid flowing past a stationary sphere with a steady-state (subscript s) velocity v_s or to a sphere moving through a stationary liquid with a velocity $-v_s$; the relative motion is the same in both cases. If the relative motion is in the vertical direction, we may visualize the slices of liquid described above as consisting of

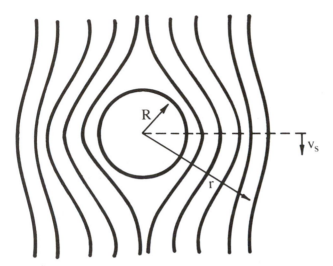

Figure 9.1 Distortion of flow streamlines around a spherical particle of radius R. The relative velocity in the plane containing the center of the sphere equals v_s as $r \to \infty$.

a bundle of layers, some of which are shown schematically in Fig. 9.1. In the plane containing the center of the sphere a limiting velocity v_s is reached as the distance r from the center of the sphere becomes large. This would be the observable settling velocity of such a spherical particle, for example. In an infinitesimally thin layer adjacent to the surface, the tangential component of velocity would be that of the solid sphere. This is the nonslip condition as it applies to this problem. This means that a velocity gradient exists that is described in terms of the distance from the center of the sphere and the component of velocity which is perpendicular to r. By considering the units of the velocity gradient and the units of some physically pertinent variables, it seems plausible that dv/dr is proportional to v_s/R, where R is the radius of the sphere. (In this chapter our problem is too many r's; be attentive to the specific radius this symbol represents in different contexts.) Formal analysis of the problem via the equation of motion verifies this argument from dimension and provides the necessary proportionality factors as well.

From Eq. (9.1) we see that the viscous force associated with this motion equals $[\eta(dv/dr)]$ (area), where the pertinent area is proportional to the surface of the sphere and varies as R^2. This qualitative argument suggests that the viscous force opposing the relative motion of the liquid and the sphere is proportional to $[\eta(v_s/R)](R^2)$. The complete solution to this problem reveals that both pressure and shear forces arising from the motion are proportional to $\eta R v_s$, and the total force of viscous resistance is given by

$$F_{vis} = 6\pi\eta Rv_s \qquad (9.4)$$

which is Stokes' law for rigid spheres. In Chap. 2 we found it convenient to call the factor of proportionality between the force of resistance and the steady-state velocity the friction factor f. We emphasize that the viscosity in Eq. (9.4) is that of the medium surrounding the sphere by labeling it with the subscript 0, and write

$$f = 6\pi\eta_0 R \qquad (9.5)$$

for spherical particles of radius R.

A somewhat similar problem arises in describing the viscosity of a suspension of spherical particles. This problem was analyzed by Einstein in 1906, with some corrections appearing in 1911. As we did with Stokes' law, we shall only present qualitative arguments which give plausibility to the final form. The fact that it took Einstein 5 years to work out the "bugs" in this theory is an indication of the complexity of the formal analysis. Derivations of both the Stokes and Einstein equations which do not require vector calculus have been presented by Lauffer [Ref. 3]. The latter derivations are at about the same level of difficulty as most of the mathematics in this book. We shall only hint at the direction of Lauffer's derivation, however, since our interest in rigid spheres is marginal, at best.

Figure 2.1 served as the basis for our initial analysis of viscosity, and we return to this representation now with the stipulation that the volume of fluid sandwiched between the two plates is a unit of volume. This unit is defined by a unit of contact area with the walls and a unit of separation between the two walls. Next we consider a shearing force acting on this cube of fluid to induce a unit velocity gradient. According to Eq. (2.6), the rate of energy dissipation per unit volume from viscous forces dW/dt is proportional to the square of the velocity gradient, with η_0 (pure liquid, subscript 0) the factor of proportionality:

$$\frac{dW}{dt} = \eta_0 \left(\frac{dv}{dy}\right)^2 \qquad (9.6)$$

Thus, to maintain a unit gradient, a volume rate of energy dissipation equal to η_0 is required.

Next we consider replacing the sandwiched fluid with the same liquid in which solid spheres are suspended at a volume fraction ϕ. Since we are examining a unit volume of liquid—a suspension of spheres in this case—the total volume of the spheres is also ϕ. We begin by considering the velocity gradient if the velocity of the top surface is to have the same value as in the case of the

pure liquid. Being solid, the suspended spheres contribute nothing to the velocity gradient. As far as the gradient is concerned, the spheres might as well be allowed to settle to the bottom and then be fused to the lower, stationary wall. The equivalency of the suspended spheres and a uniform layer of the same volume are illustrated schematicaly in Figs. 9.2a and b, respectively. Since the unit volume has a unit cross-sectional area, a volume ϕ fused to the base will raise the stationary surface by a distance ϕ and leave a liquid of thickness $1 - \phi$ to develop the gradient. These dimensions are also shown in Fig. 9.2b. If the velocity of the top layer is required to be the same in this case as for the pure solvent, then the gradient *in the liquid* need only be the fraction $1/(1 - \phi)$ of that for the pure liquid. Of course, since ϕ is less than unity, this "fraction" is grater than unity.

Now we return to consider the energy that must be dissipated in a unit volume *of suspension* to produce a unit gradient, as we did above with the pure solvent. The same "fraction" applied to the shearing force will produce the unit gradient, and the same "fraction" also describes the volume rate of energy dissipation compared to the situation described above for pure solvent. Since the latter was η_0, we write for the suspension, in the case of $dv/dy = 1$,

$$\frac{dW}{dt} = \eta = \frac{1}{1 - \phi}\, \eta_0 \qquad (9.7)$$

Again, since $\phi < 1$, $\eta > \eta_0$.

This is only one of the contributions to the total volume rate of energy dissipation; a second term which arises from explicit consideration of the individual spheres must also be taken into account. This second effect can be shown to equal $1.5\,\phi\eta_0/(1 - \phi)^2$; therefore the fully theory gives a value for η, the viscosity of the suspension:

$$\eta = \frac{\eta_0}{1 - \phi} + \frac{1.5\,\phi\eta_0}{(1 - \phi)^2} = \eta_0\,\frac{1 + 0.5\,\phi}{(1 - \phi)^2} \qquad (9.8)$$

One additional assumption which underlies the derivation of the second term in Eq. (9.8) is that ϕ is small. This being the case, $1/(1 - \phi)^2$ can be replaced by the leading terms of the series expansion $(1 + \phi + \phi^2 + \cdots)^2$ to give

$$\eta = \eta_0 \left(1 + \frac{1}{2}\,\phi\right)(1 + \phi + \cdots)^2 = \eta_0(1 + 2.5\,\phi + 4\,\phi^2 + \cdots) \qquad (9.9)$$

This is Einstein's famous viscosity equation:

1. η is the viscosity of the suspension as a whole; η_0 is the viscosity of the solvent, and ϕ is the volume fraction occupied by the spheres.

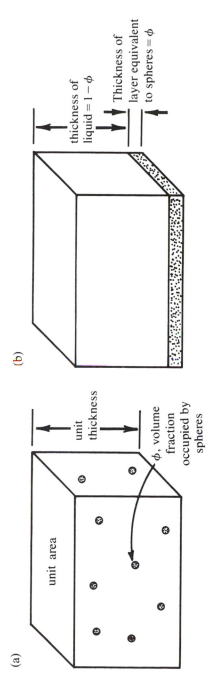

Figure 9.2 (a) Schematic representation of a unit cube containing a suspension of spherical particles at volume fraction ϕ. (b) The volume equivalent to the spheres in (a) is fused to the base, leaving $1 - \phi$ as the thickness of liquid.

2. The validity is limited to small values of ϕ, so Eq. (9.9) is generally truncated after the first two terms on the right-hand side.
3. By describing the concentration dependence of an observable property as a power series, Eq. (9.9) plays a comparable role for viscosity as Eq. (8.83) does for osmotic pressure.
4. The volume fraction emerges from the Einstein derivation at the natural concentration unit to describe viscosity. This parallels the way volume fraction arises as a natural thermodynamic concentration unit in the Flory–Huggins theory as seen in Sec. 8.3.

Both the Stokes and Einstein equations have certain features in common which arise from the hydrodynamic origins they share:

1. The liquid medium is assumed to be continuous. This makes the results suspect when applied to spheres which are so small that the molecular nature of the solvent cannot be ignored.
2. Both relationships have been repeatedly verified for a variety of systems and for spheres with a wide range of diameters. Despite item (1), both Eqs. (9.4) and (9.9) have often been applied to individual molecules, for which they work surprisingly well.
3. The spherical geometry assumed in the Stokes and Einstein derivations gives the highly symmetrical boundary conditions favored by theoreticians. For ellipsoids of revolution having an axial ratio a/b, friction factors have been derived by F. Perrin, and the coefficient of the first-order term in Eq. (9.9) has been derived by Simha. In both cases the calculated quantities increase as the axial ratio increases above unity. For spheres, a/b = 1.
4. In the derivation of both Eqs. (9.4) and (9.9), the disturbance of the flow streamlines is assumed to be produced by a single particle. This is the origin of the limitation to dilute solutions in the Einstein theory, where the net effect of an array of spheres is treated as the sum of the individual nonoverlapping disturbances. When more than one sphere is involved, the same limitation applies to Stokes' law also. In both cases contributions from the walls of the container are also assumed to be absent.

We shall make further use of the Stokes' equation later in this chapter; for the present, viscosity is our primary concern, and the Einstein equation is our point of departure.

9.3 The Intrinsic Viscosity of Rigid Molecules

One thing that is apparent at the outset is that polymer molecules in solution are very different species from the rigid spheres upon which the Einstein theory is based. On the other hand, we saw in the last chapter that the random coil contributes an excluded volume to the second virial coefficient that is at least

proportional to the volume of a sphere of radius r_g [see Eq. (8.114)]. Whether this thermodynamic conclusion applies to hydrodynamic behavior or not, and, if so, under what conditions, will be discussed in Sec. 9.6. By contrast, a compact protein molecule held in a spherical tertiary structure by hydrogen bonds and disulfide linkages may be reasonably approximated as a rigid sphere. Accordingly, we begin this section by considering the applicability of the Einstein equation to the latter systems. Then we shall consider the extension of these ideas to other rigid particles of nonspherical geometry. Again we emphasize that proteins, because of the intramolecular crosslinking they possess, are the kind of polymers that are best approximated by models of rigid bodies.

As in osmotic pressure experiments, polymer concentrations are usually expressed in mass volume^{-1} units rather than in the volume fraction units indicated by the Einstein equation. For dilute solutions, however, Eq. (8.100) shows that ϕ can be approximated as $c_2 \bar{V}_2/M$, where c_2 is the mass of polymer per volume of solution, $\bar{V}_2$ is the partial molar volume of the polymer in solution, and M is the molecular weight of the polymer. Substituting this relationship for ϕ in Eq. (9.9) gives

$$\eta = \eta_0 \left[1 + 2.5 \, \frac{\bar{V}_2}{M} c_2 + 4 \left(\frac{\bar{V}_2}{M} \right)^2 c_2{}^2 + \cdots \right] \tag{9.10}$$

a form which is reminiscent of Eq. (8.86) for osmotic pressure. A comparison of these results may momentarily suggest that Eq. (9.10) can also be used to evaluate the molecular weight of a polymer—at least for rigid-sphere molecules. Further reflection, however, shows that the first-order term in the osmotic pressure expression is proportional to M^{-1}, while the corresponding term in Eq. (9.10) involves $\bar{V}_2/M$, which is simply the reciprocal of the density of the polymer *as it exists in solution*. While this density is not as fundamental as the molecular weight in the characterization of a polymer, it turns out that the ability to measure this quantity through viscosity is an important application of viscosity studies. As indicated above, the most plausible systems on which to apply Eq. (9.10) are solutions of those proteins which are known to possess spherical molecules. Such molecules are often extensively hydrated in aqueous solutions, and the degree of solvation is reflected in $\bar{V}_2$. We shall presently consider an example which illustrates how the extent of protein hydration can be extracted from viscosity data.

Before turning to this, however, it is useful to introduce some additional vocabulary that is often employed in discussing solution viscosity. Equation (9.10) is a special case of the general function

$$\eta = A + Bc_2 + Cc_2{}^2 + \cdots \tag{9.11}$$

in which the coefficients A, B, and so on, are determined simply by fitting experimental data. Since the viscosity of the solution must approach that of

the solvent as the solute concentration goes to zero, it is apparent that $A = \eta_0$; the Einstein equation tells what the coefficient B must be for spherical particles. Working with the more general form for the moment, we can define the following quantities and symbols in terms of Eq. (9.11):

1. The relative viscosity η/η_0. This is obtained from Eq. (9.11) simply by dividing both sides of the equation through by the viscosity of the solvent ($B' = B/\eta_0$, etc.):

$$\eta_r = \frac{\eta}{\eta_0} = 1 + B'c_2 + C'c_2{}^2 + \cdots \tag{9.12}$$

2. The specific viscosity, $\eta/\eta_0 - 1$. This is obtained by subtracting unity from both sides of Eq. (9.12):

$$\eta_{sp} = \frac{\eta}{\eta_0} - 1 = B'c_2 + C'c_2{}^2 + \cdots \tag{9.13}$$

Since $\eta/\eta_0 - 1 = (\eta - \eta_0)/\eta_0$, the specific viscosity describes the excess viscosity of the solution above the viscosity of the solvent relative to the latter.

3. The reduced viscosity $(1/c_2)(\eta/\eta_0 - 1)$. This is obtained by dividing both sides of Eq. (9.13) by c_2; in this sense it is analogous to the reduced osmotic pressure:

$$\eta_{red} = \frac{1}{c_2}\left(\frac{\eta}{\eta_0} - 1\right) = B' + C'c_2 + \cdots \tag{9.14}$$

The reduced viscosity expresses the specific viscosity per unit of solute concentration.

4. The intrinsic viscosity $\lim_{c_2 \to 0} \eta_{red}$. Equation (9.14) shows that η_{red} varies linearly with c_2, at least for dilute solutions. In graphical terms, the intrinsic viscosity—B' in Eq. (9.14)—is the intercept in a plot of η_{red} versus c_2. Since the intrinsic viscosity is a limiting value at infinite dilution, it is a parameter which directly reflects the molecular properties of the solute. For this last reason intrinsic viscosity is the form of greatest usefulness in discussing solution viscosity. Intrinsic viscosity is given the symbol $[\eta]$ and has units of concentration^{-1}.

5. The inherent viscosity $(1/c_2) \ln(\eta/\eta_0)$. A plot of inherent viscosity versus concentration also extrapolates to $[\eta]$ in the limit of $c_2 \to 0$. That this is the case is readily seen by combining Eq. (9.12) with the definition of the inherent viscosity and then expanding the logarithm:

$$\eta_{inh} = \frac{1}{c_2} \ln (1 + B'c_2 + C'c_2{}^2 + \cdots) \cong \frac{B'c_2 + C'c_2{}^2 + \cdots}{c_2} \qquad (9.15)$$

In the polymer literature each of the five quantities listed above is encountered frequently. Complicating things still further is the fact that a variety of concentration units are used in actual practice. In addition, IUPAC terminology is different from the common names listed above. By way of summary, Table 9.1 lists the common and IUPAC names for these quantities and their definitions. Note that when ϕ is used as the unit of concentration, $[\eta] = 2.5$ for spheres according to the Einstein equation.

With this terminology in mind, we can restate the objective of this section as the interpretation of the intrinsic viscosities of solutions of rigid molecules. If the solute molecules are known to be spherical, comparison of Eqs. (9.10) and (9.14) shows that the intrinsic viscosity for such systems is given by

$$[\eta] = 2.5 \, \frac{\overline{V}_2}{M} \qquad (9.16)$$

Table 9.1 Summary of Names and Definitions of the Various Functions of η, η_0, and c_2 in Which Solution Viscosities are Frequently Discussed

Symbol	Definition	Common name	IUPAC name
η_r	$\dfrac{\eta}{\eta_0}$	Relative viscosity	Viscosity ratio
η_{sp}	$\dfrac{\eta}{\eta_0} - 1$	Specific viscosity	—
η_{red}	$\dfrac{1}{c_2}\left(\dfrac{\eta}{\eta_0} - 1\right)$	Reduced viscosity	Viscosity number
$[\eta]$	$\lim\limits_{c_2 \to 0} \dfrac{1}{c_2}\left(\dfrac{\eta}{\eta_0} - 1\right)$	Intrinsic viscosity	Limiting viscosity number
η_{inh}	$\dfrac{1}{c_2} \ln\left(\dfrac{\eta}{\eta_0}\right)$ $\left(\lim\limits_{c_2 \to 0} \eta_{inh} = [\eta] \text{ also}\right)$	Inherent viscosity	Logarithmic viscosity number

We take the position that $[\eta]$ is known from experiment and assume that M is also known from some other type of experiment, say, osmostic pressure. Assuming that these data are available enables us to evaluate $\bar{V}_2$, and we consider next what this quantity tells us about the state of solvation of the polymer.

If the solute molecule is solvated, then any bound (subscript b) solvent (subscript 1) must be added to the volume of the unsolvated solute (subscript 2); that is,

$$\text{Volume of solvated particle} = V_{2,dry} + V_{1,b} \tag{9.17}$$

where $V_{2,dry}$ is the volume of the unsolvated solute molecule. If we further define $m_{1,b}/m_2$ as the mass of solvent bound per mass of solute and represent the densities of the solvent and solute as ρ_1 and ρ_2, respectively, then Eq. (9.17) becomes

$$\text{Volume of solvated particle} = V_{2,dry} \left(1 + \frac{V_{1,b}}{V_{2,dry}} \right)$$

$$= V_{2,dry} \left(1 + \frac{m_{1,b}}{m_2} \frac{\rho_2}{\rho_1} \right) \tag{9.18}$$

An alternative in writing this last result would be to use $\rho_{1,b}$, since the density of the bound solvent is expected to be different from that in bulk. Rather than pursue this, however, we shall assume instead that $\rho_{1,b} = \rho_1$ and, on this basis, seek to evaluate $m_{1,b}/m_2$ from intrinsic viscosity data. The factor in parentheses in Eq. (9.18) corrects the volume of any amount of solute for the effects of solvation. Applied to the volume of 1 mol of dry polymer, we obtain for Eq. (9.16)

$$[\eta] = 2.5 \left(1 + \frac{m_{1,b}}{m_2} \frac{\rho_2}{\rho_1} \right) \bigg/ \rho_2 \tag{9.19}$$

where $1/\rho_2 = (V_2^{\circ}/M)_{dry}$. In the following example we see how this result can be applied to data collected from solutions of spherical protein molecules.

Example 9.1

The serum albumin molecule is known to have an approximately spherical shape (see Example 8.7) and is found[†] to have an intrinsic viscosity in aqueous buffer solutions of 3.7 cm^3 g^{-1}. Using $\rho_2 = 1.34$ g cm^{-3} as the density of the

---

[†] C. Tanford, *Adv. Protein Chem.* 23:121 (1968).

unsolvated polymer, estimate the quantity of water bound by this protein under the conditions of the experiment.

Solution

The units of $[\eta]$ reveal the concentration units in this experiment to be grams of protein per cubic centimer of solution. Dividing this concentration unit by the density of the unsolvated protein converts these concentration units to volume fractions:

$$\frac{\text{g protein}}{\text{cm}^3 \text{ solution}} \times \frac{\text{cm}^3 \text{ protein}}{\text{g protein}} = \frac{\text{cm}^3 \text{ protein}}{\text{cm}^3 \text{ solution}}$$

Since c_2 rather than ϕ is used in the determination of $[\eta]$, the factor $1/\rho_2$ as well as any contribution of solvation increases $[\eta]$ above the value 2.5; thus we can use Eq. (9.19) directly to interpret this data:

$$\left(1 + \frac{m_{1,b}}{m_2} \frac{\rho_2}{\rho_1}\right) = \frac{[\eta] \rho_2}{2.5} = \frac{3.7 (1.34)}{2.5} = 1.98$$

Therefore

$$\frac{m_{1,b}}{m_2} = 0.98 \frac{\rho_1}{\rho_2} = \frac{0.98 (1.00)}{1.34} = 0.73 \text{ g water (g protein)}^{-1}$$

•

In the last section we noted that Simha and others have derived theoretical expressions for η_{sp}/ϕ for rigid ellipsoids of revolution. Solving the equation of motion for this case is even more involved than for spherical particles, so we simply present the final result. Several comments are necessary to appreciate these results:

1. The totally symmetrical sphere is characterized by a single size parameter: its radius. Ellipsoids of revolution are used to approximate the shape of unsymmetrical bodies. Ellipsoids of revolution are characterized by two size parameters.
2. The ellipsoid of revolution is swept out by rotating an ellipse along its major or minor axis. When the major axis is the axis of rotation, the resulting rodlike figure is said to be prolate; when the minor axis is the axis of rotation, the disklike figure is said to be oblate.
3. We designate the length of the ellipsoid along the axis of rotation as 2a and the equatorial diameter as 2b to define the axial ratio a/b which characterizes the ellipticity of the particle. By this definition, a/b > 1 corresponds to prolate ellipsoids, and a/b < 1 to oblate ellipsoids.

4. The viscosity of a suspension of ellipsoids depends on the orientation of the particle with respect to the flow streamlines. The ellipsoidal particle causes more disruption of the flow when it is perpendicular to the stream-lines than when it is aligned with them; the viscosity in the former case is greater than in the latter. For small particles the randomizing effect of Brownian motion is assumed to override any tendency to assume a preferred orientation in the flow.

Based on these ideas, the intrinsic viscosity (in ϕ concentration units) has been evaluated for ellipsoids of revolution. Figure 9.3 shows $[\eta]$ versus a/b for oblate and prolate ellipsoids according to the Simha theory. Note that the intrinsic viscosity of serum albumin from Example 9.1–3.7(1.34) = 4.96 in volume fraction units—is also consistent with, say, a nonsolvated oblate ellipsoid of axial ratio about 5.

For solutions of rigid particles, then, the intrinsic viscosity exceeds 2.5 as a result of some combination of the following effects:

$$\frac{[\eta]}{2.5} = \text{(effect of concentration units)}\,\text{(effect of solvation)}\,\text{(effect of ellipticity)}$$

$$(9.20)$$

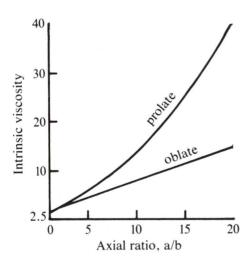

Figure 9.3 Intrinsic viscosity according to the Simha theory in terms of the axial ratio for prolate and oblate ellipsoids of revolution.

These factors having the following significance:

1. The first term reflects the fact that, in practice, volume fraction is not the concentration unit ordinarily used. Even for nonsolvated spheres, some factors will modify the Einstein 2.5 term merely as a result of reconciling practical concentration units with ϕ [Eq. (9.16)]. We assume that this factor poses no particular problem in evaluation.
2. The second term allows for solvation, which effectively increases the volume fraction of the particles to a larger value than that calculated on the basis of dry solute. Equation (9.18) shows how this can be quantified.
3. The effect of ellipticity also increases $[\eta]$ above the 2.5 value obtained for spheres. Analytical functions as well as graphical representations like Fig. 9.3 are available to describe this effect in terms of the axial ratios of the particles. In principle, therefore, a/b values for nonsolvated, rigid particles can be estimated from experimental $[\eta]$ values.
4. It is a frustrating aspect of Eq. (9.20) that the observed intrinsic viscosities contain the effects of ellipticity and solvation such that the two cannot be resolved by viscosity experiments alone. That is, for any value of $[\eta]$, there is a whole array of solvation-ellipticity values which are consistent with the observed intrinsic viscosity.

This state of affairs is summarized in Fig. 9.4a, which plots contours for different values of $[\eta]$ in terms of compatible combinations of $m_{1,b}/m_2$ and a/b. For the aqueous serum albumin described in Example 9.1 as an illustration, any solvation-ellipticity combination which corresponds roughly to $[\eta] = 5$ is possible for this system. Data from some other source are needed to pin down a more specific characterization.

We shall see in Sec. 9.10 that sedimentation and diffusion data yield experimental friction factors which may also be described—by the ratio of the experimental f to f_0, the friction factor of a sphere of the same mass—as contours in solvation-ellipticity plots. The two different kinds of contours differ in detailed shape, as illustrated in Fig. 9.4b, so the location at which they cross provides the desired characterization. For the hypothetical system shown in Fig. 9.4b, the axial ratio is about 2.5 and the protein is hydrated to the extent of about 1.0 g water (g polymer)$^{-1}$.

In all of these derivations concerning rigid bodies, no other walls are considered except the particle surfaces. Before we turn to the question of the intrinsic viscosity of flexible polymers, let us consider the relationship between the viscosity of a fluid and the geometry and dimensions of the container in which it is measured.

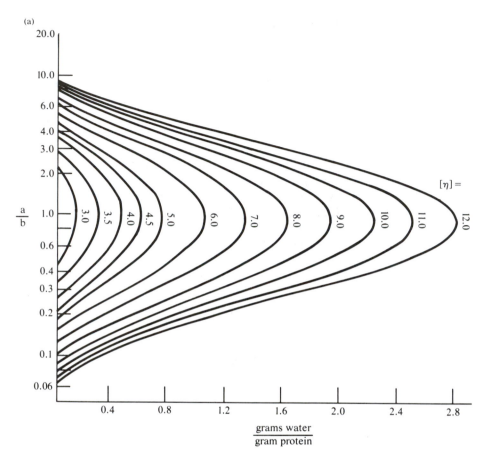

(a)

$\frac{a}{b}$

$[\eta] =$

3.0 3.5 4.0 4.5 5.0 6.0 7.0 8.0 9.0 10.0 11.0 12.0

$$\frac{\text{grams water}}{\text{gram protein}}$$

Figure 9.4 (a) For aqueous protein solutions, variation of $[\eta]$ with the axial ratio and extent of hydration. [Redrawn with permission from J. L. Oncley, *Ann. N.Y. Acad. Sci.* 41:121 (1941).] (b) Intrinsic viscosity and friction factor contours in terms of the axial ratio and extent of hydration. The crossover gives unambiguous characterization. (From Ref. 2, p. 106.)

9.4 The Poiseuille Equation and Capillary Viscometers

We defined the equation of motion as a general expression of Newton's second law applied to a volume element of fluid subject to forces arising from pressure, viscosity, and external mechanical sources. Although we shall not attempt to use this result in its most general sense, it is informative to consider the equation of motion as it applies to a specific problem: the flow of liquid through a capillary. This consideration provides not only a better appreciation of the equation of

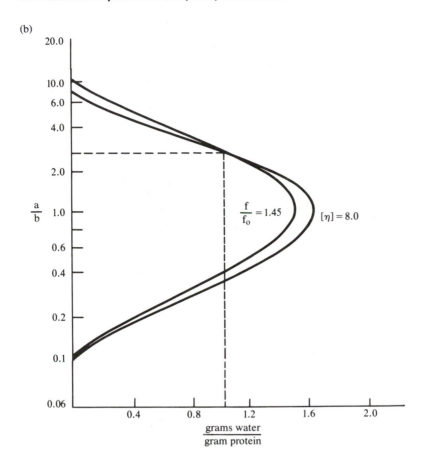

Figure 9.4 *(continued)*

motion, but also serves as the basis for an important technique for measuring solution viscosity. We shall examine the derivation first and then discuss its application to experiment.

Figure 9.5a shows a portion of a cylindrical capillary of radius R and length l. We measure the general distance from the center axis of the liquid in the capillary in terms of the variable r and consider specifically the cylindrical shell of thickness dr designated by the broken line in Fig. 9.5a. In general, gravitational, pressure, and viscous forces act on such a volume element, with the viscous forces depending on the velocity gradient in the liquid. Our first task, then, is to examine how the velocity of flow in a cylindrical shell such as this varies with the radius of the shell.

(a)   (b)

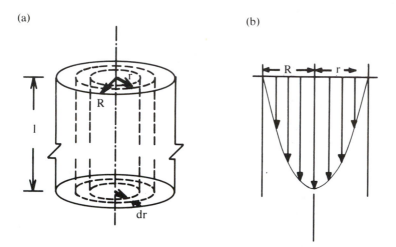

Figure 9.5 (a) Portion of a cylinder of radius R and length 1 showing (by broken lines) section of thickness dr. (b) Profile of flow velocity in the cylinder. (From Ref. 2, pp. 52 and 54.)

The net viscous force acting on this volume element is given by the difference between the frictional forces acting on the outer and inner surfaces of the shell:

$$F_{vis,net} = (F_{vis})_{out} - (F_{vis})_{in} = 2\pi(r + dr)\, l\eta \left(\frac{dv}{dr}\right)_{r+dr} - 2\pi r\, l\eta \left(\frac{dv}{dr}\right)_r$$

$$(9.21)$$

where the length times the circumference of the surface describes the appropriate area in Eq. (9.1). The relationship between the velocity gradient at the two locations is given by

$$\left(\frac{dv}{dr}\right)_{r+dr} = \left(\frac{dv}{dr}\right)_r + \left(\frac{d^2v}{dr^2}\right)\, dr$$

$$(9.22)$$

provided that dr is small. Combining Eqs. (9.21) and (9.22) and retaining only those terms which are first order in dr gives

$$F_{vis,net} = 2\pi\eta l \left[r\left(\frac{d^2v}{dr^2}\right) dr + \left(\frac{dv}{dr}\right) dr \right] = 2\pi\eta l \frac{d}{dr}\left(r\frac{dv}{dr} \right)$$

$$(9.23)$$

Under stationary-state conditions of flow, that is, when no further acceleration occurs, this force is balanced by gravitational and pressure forces. For

simplicity, we assume that the capillary is oriented vertically so that gravity operates downward and, for generality, we assume that an additional mechanical pressure Δp exists between the two ends of the capillary. Under these conditions the net gravitational and mechanical forces acting on the volume element equal

$$F_{grav,mech,net} = (2\pi \, lr \, dr)\rho g + (2\pi r \, dr)\Delta p \tag{9.24}$$

where $2\pi lr \, dr$ is the volume of the element and $2\pi r \, dr$ is its cross-sectional area. Under the stationary-state conditions we seek to describe, Eqs. (9.24) and (9.23) are equal and the following relationship applies to the volume element:

$$\eta \frac{d}{dr}\left(r \frac{dv}{dr}\right) = \left(\rho g + \frac{\Delta p}{l}\right) r \, dr \tag{9.25}$$

Integration converts this to

$$\eta r \frac{dv}{dr} = \frac{1}{2}\left(\rho g + \frac{\Delta p}{l}\right) r^2 \tag{9.26}$$

where the fact that $r(dv/dr) = 0$ at $r = 0$ is used to eliminate the integration constant. Note that the velocity gradient is directly proportional to the radial position in the fluid: It is zero at the axis and has a maximum value at $r = R$.

Equation (9.26) can be integrated again to give v as a function of r:

$$\int dv = \frac{\rho g + \Delta p/l}{2\eta} \int r \, dr \tag{9.27}$$

Because of the nonslip condition at the wall, $v = 0$ when $r = R$, and the constant of integration can be evaluated to give

$$v = \frac{\rho g + \Delta p/l}{4\eta} (r^2 - R^2) \tag{9.28}$$

This result describes a parabolic velocity profile, as sketched in Fig. 9.5b.

Equation (9.28) describes the velocity with which a cylindrical shell of liquid moves through a capillary under stationary-state conditions. This velocity times the cross-sectional area of the shell gives the incremental volume of liquid dV which is delivered from the capillary in an interval of time Δt. The total volume delivered in this interval ΔV is obtained by integrating this product over all values of r:

$$\frac{\Delta V}{\Delta t} = \frac{2\pi (\rho g + \Delta p/l)}{4\eta} \int_0^R (r^2 - R^2) r \, dr = \frac{(\rho g l + \Delta p)\pi R^4}{8\eta l} \tag{9.29}$$

This result is called the Poiseuille equation, after Poiseuille, who discovered this fourth-power dependence of flow rate on radius in 1844. The poise unit of viscosity is also named after this researcher. The following example illustrates the use of the Poiseuille equation in the area where it was first applied.

Example 9.2

Poiseuille was a physician–physiologist interested in the flow of blood through blood vessels in the body. Estimate the viscosity of blood from the fact that blood passes through the aorta of a healthy adult at rest at a rate of about 84 cm^3 sec^{-1}, with a pressure drop of about 0.98 mmHg m^{-1}. Use 9 mm as the radius of the aorta for a typical human.

Solution

The pumping action of the heart rather than gravity is responsible for blood flow; hence the term ρgl can be set equal to zero in Eq. (9.29) and the result solved for η:

$$\eta = \frac{\Delta p \pi R^4}{8l \Delta V/\Delta t}$$

The units must be expressed in a common system, with the pressure gradient requiring the most modification:

$$\frac{\Delta P}{l} = \frac{0.98 \text{ mmHg}}{m} \times \frac{133.3 \text{ N m}^{-2}}{1 \text{ mmHg}} = 131 \text{ kg m}^{-2} \text{ sec}^{-2}$$

Therefore

$$\eta = \frac{(131 \text{ kg m}^{-2} \text{ sec}^{-2}) \pi (9 \times 10^{-3} \text{ m})^4}{8 (84 \times 10^{-6} \text{ m}^3 \text{ sec}^{-1})} = 4.0 \times 10^{-3} \text{ kg m}^{-1} \text{ sec}^{-1}$$

At 37°C the viscosity of water is about 0.69 × 10^{-3} kg m^{-1} sec^{-1}; the difference between this figure and the viscosity of blood is due to the dissolved solutes in the serum and the suspended cells in the blood. The latter are roughly oblate ellipsoids of revolution in shape.

•

The Poiseuille equation provides a method for measuring η by observing the time required for a liquid to flow through a capillary. The apparatus shown in Fig. 9.6 is an example of one of many different instruments designed to use this relationship. In such an experiment the time required for the meniscus to drop

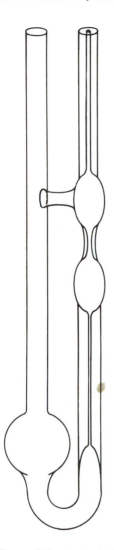

Figure 9.6 A typical capillary viscometer. (From Ref. 2, p. 55.)

the distance between the lines etched at opposite ends of the top bulb is measured. This corresponds to the drainage of a fixed volume of liquid through a capillary of constant R and l. The weight of the liquid is the driving force for the flow in this case, so the Δp term in Eq. (9.29) is zero and the observed flow time equals

$$\Delta t = \left(\frac{8 \, \Delta V}{\pi g R^4}\right) \frac{\eta}{\rho} \tag{9.30}$$

or

$$\eta = A\rho \, \Delta t \tag{9.31}$$

where A represents a cluster of factors which are constant for a particular apparatus. The constant A need not be evaluated in terms of the geometry of the apparatus, but can be eliminated from Eq. (9.31) by measuring both a known (subscript 2) and an unknown (subscript 1) liquid in the same instrument:

$$\eta_1 = \frac{\rho_1}{\rho_2} \frac{\Delta t_1}{\Delta t_2} \, \eta_2 \tag{9.32}$$

In more precise work an additional term is added to Eq. (9.31) which corrects for effects arising at the ends of the tube. This correction—which is often negligible—can be incorporated by writing

$$\eta = A\rho \, \Delta t - B \, \frac{\rho}{\Delta t} \tag{9.33}$$

where $B = \Delta V/8\pi$. As above, both A and B can be treated as instrument constants and evaluated by measuring *two* liquids which are known with respect to η and ρ, and then solving a pair of simultaneous equations for A and B.

The concentric cylinder viscometer described in Sec. 2.3, as well as numerous other possible instruments, can also be used to measure solution viscosity. The apparatus shown in Fig. 9.6 and its variations are the most widely used for this purpose, however. One limitation of this method is the fact that the velocity gradient is not constant, but varies with r in this type of instrument, as noted in connection with Eq. (9.26). Since we are not considering shear-dependent viscosity in this chapter, we shall ignore this limitation.

Experiments based on the Poiseuille equation make intrinsic viscosity an easily measured parameter to characterize a polymer. In the next section we consider how this property can be related to the molecular weight of a polymer.

9.5 The Mark–Houwink Equation

The viscosity of a polymer solution is one of its most distinctive properties. Only a minimum amount of research is needed to establish the fact that $[\eta]$ increases with M for those polymers which interact with the solvent to form a random coil in solution. In the next section we shall consider the theoretical foundations for the molecular weight dependence of $[\eta]$, but for now we approach this topic from a purely empirical point of view.

We saw in Sec. 2.9 that the viscosity of a bulk polymer is proportional to M^a, where a is either 1.0 or 3.4, depending on whether the polymer is below or above, respectively, the critical chain length for entanglement. For solutions, a similar result is obtained, only it is $[\eta]$ rather than η itself which is proportional to M^a:

$$[\eta] = kM^a \tag{9.34}$$

This relationship with a = 1 was first proposed by Staudinger, but in this more general form it is known as the Mark-Houwink equation. The constants k and a are called the Mark-Houwink coefficients for a system. The numerical values of these constants depend on both the nature of the polymer and the nature of the solvent, as well as the temperature. Extensive tabulations of k and a are available; Table 9.2 shows a few examples. Note that the units of k are the same as those of $[\eta]$, and hence literature values of k can show the same diversity of units as c_2, the polymer concentration.

Table 9.3 lists the intrinsic viscosity for a number of poly(caprolactam) samples of different molecular weight. The M values listed are number average figures based on both end group analysis and osmotic pressure experiments. The values of $[\eta]$ were measured in *m*-cresol at 25°C. In the following example we consider the evaluation of the Mark-Houwink coefficients from these data.

Table 9.2 Values for the Mark-Houwink Coefficients for a Selection of Polymer-Solvent Systems at the Temperatures Noted

Polymer	Solvent	T (°C)	$k \times 10^5$ (dl/g)	a
Polyisobutene	Benzene	40	43	0.60
Polystyrene	Toluene	25	17	0.69
Poly(vinyl alcohol)	Water	25	20	0.76
Poly(vinyl chloride)	Chlorobenzene	30	71.2	0.59
Polyacrylonitrile	Dimethyl formamide	50	30	0.752
Poly(methyl methacrylate)	Chloroform	25	4.8	0.80
Poly(ethylene terephthalate)	*m*-Cresol	25	0.77	0.95
Poly(ε-caprolactam)	*m*-Cresol	25	320	0.62

Source: M. Kurata, M. Iwama, and K. Kamada, in J. Bandrup and E. H. Immergut (Eds.), *Polymer Handbook*, 2nd ed., Wiley, New York, 1975.

Table 9.3   Intrinsic Viscosity as a Function of
Molecular Weight for Samples of Poly(caprolactam)[a]

$[\eta]$ (dl g^{-1})	M $\times$ 10^{-3}	$[\eta]/M^{0.69} \times 10^3$
0.34	3.24	1.29
0.36	3.50	1.29
0.43	4.46	1.30
0.48	5.52	1.26
0.54	6.11	1.32
0.61	7.69	1.27
0.69	8.6	1.33
0.71	8.8	1.35
0.74	9.63	1.32
0.79	10.3	1.35
0.88	12.2	1.33
0.87	13.0	1.26
0.89	13.92	1.23
0.95	14.1	1.30
0.94	15.8	1.19
0.98	17.4	1.16
1.06	17.5	1.25
1.10	17.6	1.29
1.17	19.5	1.28
1.25	21.6	1.28
1.11	22.2	1.11
1.40	25.0	1.29
1.59	30.8	1.27

[a]These data are used to evaluate k and a in Example 9.3
and Fig. 9.7. Values of k are calculated for each measurement
by the relationship k = $[\eta]/M^{0.69}$.
Source: Data from H. K. Reimschussel and G. J. Dege,
J. Polym. Sci. 9:2343 (1971).

Example 9.3

Evaluate the Mark-Houwink coefficients for poly(caprolactam) in *m*-cresol at
25°C from the data in Table 9.3.

Solution

By taking the logarithm of both sides of Eq. (9.34), the Mark-Houwink equation
is transformed into the equation of a straight line:

$$\ln [\eta] = \ln k + a \ln M$$

This form suggests that a plot of ln $[\eta]$ versus ln M is a straight line whose slope
equals a and whose intercept equals ln k. The data in Table 9.3 are plotted

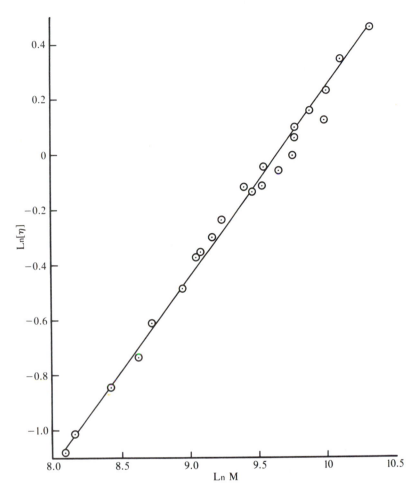

Figure 9.7 Plot of ln [η] versus ln M for the data in Table 9.3. An analysis of the Mark–Houwink coefficients from these data is presented in Example 9.3.

according to this representation in Fig. 9.7. The points† are observed to define a straight line over the 20-fold range of M values studied. The slope of this line is 0.69, which gives the Mark–Houwink a coefficient. Because of the high values of M, evaluation of the intercept at ln M = 0 would involve an extrapolation over too great a range. Accordingly, the graphically determined a value is used to evaluate k analytically for each measurement. The individual k values so

---

† H. K. Reimschuessel and G. J. Dege, *J. Polym. Sci.* 9:2343 (1971).

determined are shown in Table 9.3; the average value is 1.27×10^{-3} g dl^{-1} For these data the Mark–Houwink equation becomes

$$[\eta] = 1.27 \times 10^{-3} \, M^{0.69}$$

Note that this can be rearranged to give M directly in terms of the observed value of $[\eta]$:

$$M = \left\{ (1.27 \times 10^{-3})^{-1} \, [\eta] \right\}^{1/0.69} = 1.58 \times 10^4 \, [\eta]^{1.45}$$

Since viscometer drainage times are typically on the order of a few hundred seconds, intrinsic viscosity experiments provide a rapid method for evaluating the molecular weight of a polymer. A limitation of the method is that the Mark–Houwink coefficients must be established for the particular system under consideration by calibration with samples of known molecular weight. The speed with which intrinsic viscosity determinations can be made offsets the need for prior calibration, especially when a particular polymer is going to be characterized routinely by this method.

Even fractionated polymer samples are generally polydisperse, which means that the molecular weight determined from intrinsic viscosity experiments is an average value. The average obtained is the viscosity average as defined by Eqs. (1.20) and (2.40) as seen by the following argument:

1. The experimental (subscript ex) intrinsic viscosity is proportional to some average—the nature of which we seek to verify—value of M raised to the power a, according to Eq. (9.34),

$$[\eta]_{ex} = k\bar{M}^a \tag{9.35}$$

2. Since $[\eta]_{ex}$ is a limiting value as $c_2 \to 0$, the concentration effect it contains can be written as

$$[\eta]_{ex} = \frac{(\eta_{sp})_{ex}}{c_{ex}} \tag{9.36}$$

3. For a polydisperse system containing molecules in different molecular weight categories which we index i, we can write $(\eta_{sp})_{ex} = \Sigma_i \eta_{sp,i}$ and $c_{ex} = \Sigma_i c_i$, with

$$\eta_{sp,i} = c_i [\eta]_i = c_i k M_i{}^a \tag{9.37}$$

4. Combining the results in items (1)–(3), we obtain

$$\bar{M}^a = \frac{\Sigma_i c_i M_i{}^a}{\Sigma_i c_i} = \frac{\Sigma_i w_i M_i{}^a}{\Sigma_i w_i} \tag{9.38}$$

since $c_i = w_i / V$.

5. Since $w_i = n_i M_i$, Eq. (9.38) can be written as

$$\bar{M}_v = \left(\frac{\Sigma_i N_i M_i^{1+a}}{\Sigma_i N_i M_i} \right)^{1/a} \tag{9.39}$$

which is the viscosity (subscript v) average molecular weight as defined in Chap. 2.

In general, $\bar{M}_n < \bar{M}_v < \bar{M}_w$, and $\bar{M}_v = \bar{M}_w$ if a = 1. On the basis of this last observation, it can be argued that the Mark–Houwink coefficients should be evaluated using weight average rather than number average molecular weights as calibration standards. We shall see in Chap. 10 how $\bar{M}_w$ values can be obtained from light-scattering experiments.

It is apparent from an examination of Table 9.2 that the Mark–Houwink a coefficients fall roughly in the range 0.5–1.0. We conclude this section with some qualitative ideas about the origin of these two limiting values for a. We consider a polymer molecule consisting of n repeat units, and two different representations of its interaction with solvent.

According to one point of view, the entire domain of the coil is unperturbed by the flow. The coil in this case behaves effectively like a rigid body whose volume is proportional to r_g^3, where r_g is the radius of gyration of the coil. Such a coil is said to be nondraining, since the interior of its domain is unaffected by the flow. We anticipate using Eq. (1.58) to describe the molecular weight dependence of r_g. In view of this, we replace r_g^3 by $(\overline{r_g^2})^{3/2}$ and attach a subscript 0 to the latter as a reminder that, under Θ conditions, solvent and excluded-volume effects cancel to give a true value. With these ideas in mind, the volume fraction of the nondraining coil is written

$$\phi \propto (\overline{r_{g,0}^2})^{3/2} \frac{c_2}{M} \tag{9.40}$$

since the number of these spheres is proportional to c_2/M. Applying the Einstein equation and the idea that $\overline{r_{g,0}^2} \propto n$, we obtain

$$\left(\frac{\eta}{\eta_0} - 1 \right) \propto n^{3/2} \left(\frac{c_2}{n} \right) \tag{9.41}$$

or

$$[\eta] = \frac{1}{c_2} \left(\frac{\eta}{\eta_0} - 1 \right) \propto M^{1/2} \tag{9.42}$$

An alternative point of view assumes that each repeat unit of the polymer chain offers hydrodynamic resistance to the flow such that ζ—the friction factor per repeat unit—is applicable to each of the n units. This situation is called the free-draining coil. The free-draining coil is the model upon which the Debye viscosity equation is based in Chap. 2. Accordingly, we use Eq. (2.53) to give the contribution of a single polymer chain to the rate of energy dissipation:

$$\frac{dW}{dt} \propto (n\zeta \overline{r_{g,0}^2}) \left(\frac{dv}{dy}\right)^2 \tag{9.43}$$

Once again the subscript 0 is attached to $\overline{r_g^2}$ in anticipation of using Eq. (1.58) to give the molecular weight dependence of the latter. Since $dW/dt = \eta(dv/dy)^2$ by Eq. (2.6), this result gives the contribution to viscosity of a single molecule in the absence of interactions with other molecules. A solution containing $N_A c_2/M$ molecules per volume will therefore display an excess viscosity given by

$$\frac{\eta - \eta_0}{\eta_0} \propto \left(\frac{n\zeta \overline{r_{g,0}^2}}{\eta_0}\right) \left(\frac{N_A c_2}{M}\right) \tag{9.44}$$

Recalling that $M \propto n$ and $\overline{r_{g,0}^2} \propto n$, we obtain

$$\frac{\eta - \eta_0}{\eta_0} \propto nc_2 \tag{9.45}$$

or

$$[\eta] \propto M^1 \tag{9.46}$$

Equations (9.42) and (9.46) reveal that the range of a values in the Mark-Houwink equation is traceable to differences in the permeability of the coil to the flow streamlines. It is apparent that the extremes of the nondraining and free-draining polymer molecule bracket the range of intermediate permeabilities for the coil. In the next section we examine how these ideas can be refined still further.

9.6 To Drain or Not to Drain?

The emphasis in the foregoing analysis was explaining the range of exponents which describe the dependence of $[\eta]$ on M. In this section we retreat a bit from this objective and reexamine Eqs. (9.40) and (9.44) without regard to the molecular weight dependence of the individual factors:

1. In writing Eqs. (9.40) and (9.44) we did not worry about numerical coefficients and we continue to disregard these.

2. Equation (9.40) treats the nondraining coil as a rigid sphere and shows that in this limit $[\eta] \propto (\overline{r_{g,0}^2})^{3/2}/n$.
3. Equation (9.44) treats the free-draining molecule as an assembly of independent hydrodynamic units and shows that in this limit $[\eta] \propto (n\zeta/\eta_0)(\overline{r_{g,0}^2}/n)$.
4. These results arise from considering the same polymer molecule under different conditions of permeability to the streamlines of solvent flow.
5. As special cases, both Eqs. (9.40) and (9.44) must be limits of a *single* function which also describes the intermediate situations in which the interior of the coil domain is partially penetrated by the pattern of flow.

Taking the attitude described in item (5) toward the previously developed equations for $[\eta]$ is an important step. What this enables us to do is write a *general* expression for $[\eta]$ in solutions of flexible polymers:

$$[\eta] \propto \frac{(\overline{r_{g,0}^2})^{3/2}}{n} \times f\left(\frac{n\zeta}{\eta_0 (\overline{r_{g,0}^2})^{1/2}}\right) \qquad (9.47)$$

where $f[n\zeta/\eta_0 (\overline{r_{g,0}^2})^{1/2}]$ is some function—as yet unknown—of the indicated variables. Of this function, the following observations can be made:

1. The cluster of variables—which we designate X for simplicity—is dimensionless. Since ζ is the friction factor per repeat unit, it can at least be approximated by Eq. (9.5): $\zeta = 6\pi\eta_0 b$, where b is the radius of a repeat unit and η_0 is the viscosity of the medium. Thus ζ/η_0 has the units of length, as does $(\overline{r_{g,0}^2})^{1/2}$, making this particular cluster dimensionless.
2. The function $f(X)$ approaches a constant value for nondraining coils to generate Eq. (9.40), and approaches some constant times $n\zeta/\eta_0 (\overline{r_{g,0}^2})^{1/2}$ for free-draining coils to generate Eq. (9.44).
3. A detailed hydrodynamic theory has been developed by Kirkwood and Riseman which indeed reduces to the limits predicted above.
4. The details of the Kirkwood-Riseman theory are sufficiently involved that we shall not consider the derivation of this theory. We shall, however, examine in somewhat greater detail the cluster of variables we have designated by X as a measure of the permeability of the molecule to the flowing solvent.

Rather than discuss the penetration of the flow streamlines into the molecular domain of a polymer in terms of viscosity, we shall do this for the overall friction factor of the molecule instead. The latter is a similar but somewhat simpler situation to examine. For a free-draining polymer molecule, the net friction factor f is related to the segmental friction factor ζ by

$$f = n\zeta = n6\pi\eta_0 b \qquad (9.48)$$

The premise which underlies Eq. (9.48) is that all segments experience the flow as if it were undisturbed by other segments in the chain. This is the free-draining limit for the friction factor and, of course, assumes no interference from neighboring polymer molecules either. This result can be criticized on two grounds: applying Stokes' law to individual polymer segments and ignoring that there is ordinarily overlap of the hydrodynamic effect caused by neighboring units. The latter is the more serious of the two. The Kirkwood method of dealing with this overlap effect is to modify Eq. (9.48) as follows:

$$f = \frac{n 6 \pi \eta_0 b}{1 + (b/n) \sum_{i=1}^{n} \sum_{j=1}^{n} \overline{r_{ij}^{-1}}} \tag{9.49}$$

where $\overline{r_{ij}^{-1}}$ is the average value of the reciprocal distance between units i and j in the chain. As the value of this distance becomes larger, its reciprocal becomes smaller and the interference between units i and j decreases. The summations in Eq. (9.49) are to be carried out for each pair of i–j units, excluding i = j.

To evaluate r_{ij}^{-1} we return to the concepts of Sec. 1.10 and, using Eq. (1.44) for $P(n, r)$, write

$$\overline{r^{-1}} = \int_0^\infty r^{-1} P(n, r) \, dr = \left(\frac{6}{\pi \overline{r_0^2}} \right)^{1/2} \tag{9.50}$$

This integral is a gamma function and is readily solved using a table of integrals. In writing the last result nl_0^2 has been replaced by $\overline{r_0^2}$, the mean-square coil dimensions under Θ conditions. Equation (9.49) involves $\overline{r_{ij}^{-1}}$, not $\overline{r^{-1}}$, so we note that $\overline{r_0^2} = nl_0^2$ and replace n by $|i - j|$, the number of units separating units i and j, to obtain

$$\overline{r_{ij}^{-1}} = \left(\frac{6}{\pi |i - j| l_0^2} \right)^{1/2} \tag{9.51}$$

Substituting this value into Eq. (9.49), replacing the sums by integrals, and performing the indicated operations eventually yields

$$f = \frac{n 6 \pi \eta_0 b}{1 + (8/3)(6/\pi)^{1/2} (b/l_0) n^{1/2}} \tag{9.52}$$

This expression can be cast into a more useful form by making the following modifications on the second term in the denominator:

1. We begin by ignoring all purely numerical coefficients. In the final result we shall represent all such factors by the numerical constant K.

2. Multiply the numerator and denominator by $n^{1/2}$ and replace $n^{1/2}l_0$ by $(6\overline{r_{g,0}^2})^{1/2}$:

$$\frac{b}{l_0} n^{1/2} \propto \frac{bn}{(\overline{r_{g,0}^2})^{1/2}}$$

3. Replace b, the radius of the repeat unit, by $\zeta/6\pi\eta_0$:

$$\frac{bn}{(\overline{r_{g,0}^2})^{1/2}} \propto \frac{\zeta n}{\eta_0 (\overline{r_{g,0}^2})^{1/2}} \propto X$$

where X again represents the same cluster of constants used to describe $[\eta]$.

Substituting back into Eq. (9.52) gives

$$f = \frac{n6\pi\eta_0 b}{1 + Kn\zeta/[\eta_0 (\overline{r_{g,0}^2})^{1/2}]} = \frac{n\zeta}{1 + KX} \tag{9.53}$$

This result shows that the friction factor of the coil equals $n\zeta$ when $KX \ll 1$, and equals a numerical factor times $\eta_0 (\overline{r_{g,0}^2})^{1/2}$ when $KX \gg 1$. The first of these limits is the free-draining case; the second is Stokes' law for a nondraining sphere. The denominator in Eq. (9.53) can be regarded as a function $f'(X)$ which corrects the friction factor for variable coil permeability just as $f(X)$ in Eq. (9.47) corrects $[\eta]$ for this effect. The two correction functions $f(X)$ and $f'(X)$ are different, but the point we wish to illustrate is that the cluster of constants represented by X does describe the perturbation of the coil interior by the flow and illustrates that the condition $X \to 0$ corresponds to free-draining coils, and $X \to \infty$ to nondraining coils. Both of these conclusions are consistent with the anticipated interpretation of Eq. (9.47) for $[\eta]$.

From the definition of this permeability parameter [Eq. (9.47)], we see that X depends on $n/(\overline{r_{g,0}^2})^{1/2}$ or $n^{1/2}$, since $(\overline{r_{g,0}^2})^{1/2} \propto n^{1/2}$. Remember that the subscript 0 in all the preceding relationships specifies Θ conditions where the intramolecular volume exclusion effects and solvent effects on the random coil dimensions cancel. Since $X \to 0$ corresponds to the free-draining limit, and $X \to \infty$ to the nondraining limit, we see that the transition between these two extremes is predicted to occur with increasing n. Thus for short chains each segment experiences the flow of the solvent as if it were an isolated unit. For chains of intermediate length those segments near the perimeter of the coil partially shield those near the center from the flow. Finally, for very long chains the shielding is so effective that the solvent in the core of the molecular domain is immobilized.

We now have two entirely different contexts in which we speak of a coil as "unperturbed," with the term having a different meaning in each case:

1. In this section an unperturbed coil refers to the condition of immobilized solvent in the interior of the molecular domain. This is a hydrodynamic criterion and leads to Eq. (9.42).
2. In earlier chapters an unperturbed coil referred to molecular dimensions as predicted by random flight statistics. We saw in the last chapter that this thermodynamic criterion is met under Θ conditions.
3. To use the proportionality between $\overline{r_{g,0}^2}$ and n in this section, we have stipulated unperturbed dimensions—in the sense of item (2)—as indicated by the subscript 0 on the various size parameters.

In the nondraining limit of Eq. (9.47), the coils are unperturbed in both senses of the word: nondraining *and* Θ conditions. To emphasize the latter we attach the subscript Θ to $[\eta]$ when these conditions are met. Thus for high polymers under Θ conditions

$$[\eta]_\Theta = \frac{\Phi\,(\overline{r_{g,0}^2})^{3/2}}{M} = \Phi\left(\frac{\overline{r_{g,0}^2}}{M}\right)^{3/2} M^{1/2} \tag{9.54}$$

where Φ is the proportionality constant. Since $\overline{r_0^2}/n$ is a constant according to Eq. (1.61), this result predicts the value of 1/2 for the Mark–Houwink a coefficient under Θ conditions. Figure 9.8 shows the experimental intrinsic viscosities of four different polymer–solvent–temperature combinations corresponding to Θ conditions plotted against molecular weight on log–log coordinates. All of the data display lines of slope 1/2 in this representation in accord with Eq. (9.54).

The fully developed Kirkwood–Riseman theory predicts that in the limit of molecular weights above about 10^4 the function $f(X)$ in Eq. (9.47) reaches a limiting value of 5.32×10^{24} for $[\eta]$ expressed in cubic centimeters per gram and r_g in centimeters. A somewhat revised version of this theory gives 4.22×10^{24} for this limit. Because of the relationship between $\overline{r_{g,0}^2}$ and $\overline{r_0^2}$ [Eq. (1.59)], these values are decreased by $6^{3/2} = 14.7$ when written in terms of the end-to-end distance instead of the radius of gyration. We shall see in Chap. 10 that the radius of gyration of a polymer can be independently evaluated from light-scattering experiments. When $(\overline{r_{g,0}^2})$ values so determined are combined with measured M and $[\eta]$ values, an empirical value for Φ is found to be about 3.1×10^{24} —in the same units cited above—for a variety of systems under Θ conditions. Although there is some uncertainty as to its exact value, it is important to note that Φ is a universal parameter which permits the ratio $(\overline{r_{g,0}^2})^{3/2}/M$ to be evaluated from $[\eta]$ or—if the molecular weight is also known—allows the evaluation of $\overline{r_{g,0}^2}$ itself. Note that the precise numerical

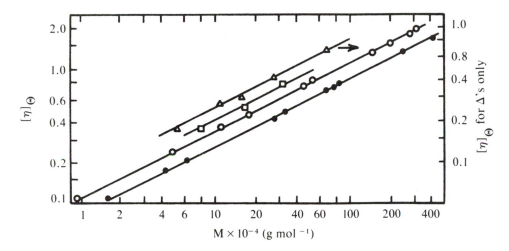

Figure 9.8 Log–log plot of $[\eta]_\Theta$ versus M for four different polymer–solvent–temperature combinations corresponding to Θ conditions. All lines have a slope of $1/2$ as required by Eq. (9.54). (Reprinted with permission from Ref. 1.)

value used for Φ depends on the system of units used, the choice of size parameters, and whether theoretical or empirical values are selected. The application of Eq. (9.54) is illustrated in the following example.

Example 9.4

Under Θ conditions occurring near room temperature, $[\eta] = 0.83$ dl g^{-1} for a polystyrene sample of molecular weight 10^6.† Use this information to evaluate $\overline{r_{g,0}^2}$ and $\overline{r_0^2}$ for polystyrene under these conditions. For polystyrene in ethylcyclohexane, $\Theta = 70°C$ and the corresponding calculation shows that $(\overline{r_0^2}/M)^{1/2} = 0.071$ nm. Based on these two calculated results, criticize or defend the following proposition: The discrepancy in calculated $(\overline{r_0^2}/M)^{1/2}$ values must arise from the uncertainty in Φ, since this ratio should be a constant for polystyrene, independent of the nature of the solvent.

Solution

Since these are Θ conditions, use Eq. (9.54) and the empirical value of Φ to evaluate $\overline{r_{g,0}^2}/M$:

---

† See Fig. 9.8.

$$(\overline{r_{g,0}^2}/M)^{3/2} = \frac{[\eta]}{\Phi M^{1/2}} = \frac{0.83 \text{ dl/g} \times 100 \text{ cm}^3/\text{dl}}{(3.1 \times 10^{24})(10^6)^{1/2}} = 2.68 \times 10^{-26} \text{ cm}^3$$

and

$$(\overline{r_{g,0}^2}/M)^{1/2} = (2.68 \times 10^{-26})^{1/3} = 3.0 \times 10^{-9} \text{ cm}$$

Since $M = 10^6$, $(\overline{r_{g,0}^2})^{1/2} = 30 \times 10^{-9}$ m = 30 nm. Since $\overline{r_0^2} = 6\overline{r_{g,0}^2}$, $(\overline{r_0^2})^{1/2} = 6^{1/2}(30) = 73.5$ nm. This figure was given in Example 1.6 as the rms end-to-end distance of an unperturbed polystyrene molecule of $M = 10^6$. Therefore for $\Theta \cong 25°C$, $(\overline{r_0^2}/M)^{1/2} = 73.5/(10^6)^{1/2} = 0.074$ nm, while for $\Theta = 70°C$, $(\overline{r_0^2}/M)^{1/2} = 0.071$ nm. As stated in the proposition, these values should be independent of the nature of the solvent. In both cases the ratios were evaluated using the empirical value of Φ, which, in turn, is based on a range of solvents. The above discrepancy in $(\overline{r_0^2}/M)^{1/2}$ is comparable to the range of empirical Φ values calculated from different solvents under Θ conditions. However, the calculated ratios are typical of a *systematic* trend to show slightly lower values of $(\overline{r_0^2}/M)^{1/2}$ at higher values of Θ. Therefore, before simply attributing this to the uncertainty in an empirical constant, let us examine the significance of this ratio in terms of Eq. (1.16). According to the latter,

$$\frac{\overline{r_0^2}}{n} \propto (\text{bond length})(\text{effect of bond angle})(\text{effect of hindered rotation})$$

The first two of these factors are constant for a given polymer, as are the absolute heights of the barriers to free rotation. The absolute barrier heights, however, are not as important as these heights *relative to RT*, since it is the relative height that determines the probability of overcoming the barrier. Therefore at higher temperatures the barriers to rotation are less effective and the coil can snuggle into a tighter coil. The slight decrease in $(\overline{r_0^2}/M)^{1/2}$ with increasing Θ is apparently a real effect.

•

Next we consider the situation of a coil which is unperturbed in the hydrodynamic sense of being effectively nondraining, yet having dimensions which are perturbed away from those under Θ conditions. As far as the hydrodynamics are concerned, a polymer coil can be expanded above its random flight dimensions and still be nondraining. In this case, what is needed is to correct the coil dimension parameters by multiplying with the coil expansion factor α, defined by Eq. (1.63). Under non-Θ conditions (no subscript), $r_g = \alpha(r_g)_0$; therefore under these conditions we write

$$[\eta] = \Phi\alpha^3 \frac{(\overline{r_{g,0}^2})^{3/2}}{M} = \frac{\Phi r_g^3}{M} = \Phi\alpha^3 \left(\frac{\overline{r_{g,0}^2}}{M}\right)^{3/2} M^{1/2} \qquad (9.55)$$

Comparing this result with Eq. (9.54) shows that

$$\alpha^3 = \frac{[\eta]}{[\eta]_\Theta} \tag{9.56}$$

for nondraining polymers. This provides a method for evaluating the coil expansion factor for a specific system.

In addition to the thermodynamic interpretation of Θ conditions that we discussed in Chap. 8, we have now established the hydrodynamic significance of these conditions as the point at which $[\eta] \propto M^{0.5}$. Summarizing in terms of the Mark-Houwink a coefficient we have the following:

1. For nondraining coils, a = 0.5 identifies the Θ condition.
2. As discussed in connection with Eq. (9.47), the Kirkwood-Riseman theory predicts that a = 1 in the free-draining limit. This limit is expected for small values of n, however, and does not explain a > 0.5 for high molecular weight polymers.
3. Under better-than-Θ conditions, $[\eta] > [\eta]_\Theta$, according to Eq. (9.56). Experimental a values are greater than 0.5 under these conditions also.

Next we shall examine the molecular weight dependence of the coil expansion factor α to see if the latter can explain the observations of a's greater than 0.5.

9.7 The Coil Expansion Factor

The coil expansion factor α has been examined by Flory and other workers, and is discussed in detail in Ref. 1. We shall not consider the full details of this topic, but shall be content with examining the dependence of α on the molecular weight of the polymer and on the energetics of the interaction between the polymer and the solvent. Figure 1.9 shows how a coil expands from its dimensions under Θ conditions as a result of imbibing a good solvent. Although coil shrinkage is also possible owing to solvent exclusion, we recognize that the range of the effect in this direction is limited. Before a solvent gets too poor, high molecular weight polymers will begin to precipitate out of solution; therefore we concentrate on the expansion aspect of the factor α.

The key to understanding α is the realization that two opposing forces operate on the coil to establish its equilibrium dimensions:

1. The molecule tends to imbibe a good solvent to dilute the chain segments within the coil domain. This is analogous to the solvent passing through the membrane in an osmotic pressure experiment, with the hypothetical boundary of the coil functioning as an imaginary membrane.

2. The expansion of the coil domain produces an elastic restoring force which opposes the expansion by tending to restore the molecule to its most probable conformation.
3. Since these forces work in opposition, the actual coil dimensions describe the point of balance between the two.

Our strategy in proceeding, therefore, is to write separate expressions for the forces cited in items (1) and (2), and then set them equal to each other as required by item (3). Since we have discussed osmotic effects in Chap. 8 and elastic forces in Chap. 3, we shall invoke certain concepts and relationships from these chapters in this discussion. In this derivation we continue to omit numerical coefficients and some of the less pertinent parameters (although we retain $\overline{V}_1$ for the sake of Problem 5 at the end of the chapter), and focus attention on the relationship between α, M, and the interaction parameter χ.

We begin by writing the volume of the spherical coil domain as $V_d = (4/3)\pi r^3$, where $r = \alpha(\overline{r_0^2})^{1/2}$. For simplicity we temporarily write this as $r = \alpha r_0$; we return to $(\overline{r_0^2})^{1/2}$ below to introduce the molecular weight dependence of the latter. Accordingly, $V_d = (4/3)\pi\alpha^3 r_0^3$, and the increment in domain volume associated with an incremental increase in r_0 is

$$dV_d \propto \alpha^3 r_0^2 \, dr_0 \tag{9.57}$$

Only a fraction of the chain segments will be present in this spherical shell, but whatever their number is, it will increase with the degree of polymerization n. Therefore, in the volume element associated with the expansion of the coil, the volume fraction of chain segments ϕ^* is proportional to n/dV_d, or $\phi^* \propto n/\alpha^3 r_0^2 \, dr_0$.

The change in free energy for the imbibed solvent is given by $\Delta G = (\mu_1 - \mu_1^\circ) \, dn_1$, where $dn_1 \propto dV_d(1 - \phi^*)/\overline{V}_1$, or $\alpha^3(1 - \phi^*)r_0^2 \, dr_0/\overline{V}_1$. Remembering that a force times the distance through which it operates gives an energy—in this case $F_{os} \, d\alpha = \Delta G$—we write the following for the osmotic force (subscript os) associated with the imbibed solvent:

$$F_{os} = \frac{\partial}{\partial\alpha} \left[(\mu_1 - \mu_1^\circ) \, dn_1 \right] \propto \frac{\partial}{\partial\alpha} \left(\frac{(\mu_1 - \mu_1^\circ)\alpha^3 (1 - \phi^*)r_0^2 \, dr_0}{\overline{V}_1} \right) \tag{9.58}$$

Expanding the logarithm in Eq. (8.53) $[\ln(1 - \phi^*) \cong -\phi^* - 1/2 \, \phi^{*2}]$ and considering the limit as $n \to \infty$ gives $\mu_1 - \mu_1^\circ \propto (1/2 - \chi) \, \phi^{*2}$ for this situation. Substituting into Eq. (9.58) and retaining no terms higher than second order in ϕ^*, we obtain

$$F_{os} \propto \frac{\partial}{\partial\alpha} \left[\frac{1/2 - \chi}{\overline{V}_1} \left(\frac{n}{\alpha^3 r_0^2 \, dr_0} \right)^2 \alpha^3 r_0^2 \, dr_0 \right] \tag{9.59}$$

Carrying out the indicated differentiation and reintroducing $(\overline{r_0^2})^{1/2}$ for r_0, this becomes

$$F_{os} \propto \frac{(1/2 - \chi)\, n^2}{\alpha^4 \overline{V}_1\, (\overline{r_0^2})\, d\,(\overline{r_0^2})^{1/2}} \tag{9.60}$$

Since $\overline{r_0^2} \propto n$, we obtain for the molecular weight dependence of F_{os}

$$F_{os} \propto \frac{(1/2 - \chi)\, n^2}{\overline{V}_1\, \alpha^4\, n^{3/2}} \propto \frac{(1/2 - \chi)\, n^{1/2}}{\overline{V}_1\, \alpha^4} \tag{9.61}$$

As noted above, this force is counterbalanced by an entropy-based, elastic force which prevents the molecule from uncoiling. Entropy elasticity was discussed in Sec. 3.4, where the elastic (subscript el) force between crosslinks is given by Eq. (3.19) to be

$$F_{el} \propto \frac{\partial\, \Delta S}{\partial L} \propto \frac{1}{L_0}\, \frac{\partial\, \Delta S}{\partial \alpha} \tag{9.62}$$

where ΔS is the entropy of stretching, L is the stretched length of the sample, and L_0 the unstretched length. The ratio L/L_0 is close enough to the definition of α in the present context to allow this expression from Chap. 3 to be used here.

We might be tempted to equate the forces given by Eqs. (9.61) and (3.38) and solve for α from the resulting expression. However, Eq. (3.38) is not suitable for the present problem, since it was derived for a cross-linked polymer stretched in one direction with no volume change. We are concerned with a single, un-cross-linked molecule whose volume changes in a spherically symmetrical way. The precursor to Eq. (3.36) in a more general derivation than that presented in Chap. 3 is

$$\Delta S \propto \frac{1}{2}\, [\alpha_x^2 + \alpha_y^2 + \alpha_z^2 - 3 - \ln\,(\alpha_x \alpha_y \alpha_z)] \tag{9.63}$$

where the α's are the relative elongations in the $x, y,$ and z directions. Equation (3.36) is obtained from Eq. (9.63) by letting $\alpha_x = \alpha_y = \alpha_z^{-1/2} = \alpha^{-1/2}$. If we allow each of the components of α to be identical, as appropriate for symmetrical expansion, we obtain from Eq. (9.63)

$$\Delta S \propto \frac{3}{2}\, [(\alpha^2 - 1) - \ln \alpha] \tag{9.64}$$

From this expression we obtain

$$F_{el} \propto \frac{\partial \, \Delta S}{\partial \alpha} \propto \alpha - \frac{1}{\alpha} \tag{9.65}$$

instead of Eq. (3.38).

Equating Eqs. (9.61) and (9.65) yields the desired result:

$$\frac{(1/2 - \chi) \, n^{1/2}}{\overline{V}_1 \alpha^4} \propto \alpha - \frac{1}{\alpha} \tag{9.66}$$

or

$$\alpha^5 - \alpha^3 \propto \frac{(1/2 - \chi) \, n^{1/2}}{\overline{V}_1} \tag{9.67}$$

This expression was anticipated in the last chapter in writing Eq. (8.114). Equation (8.115) can also be substituted into this expression to eliminate $1/2 - \chi$ and give

$$\alpha^5 - \alpha^3 \propto \psi \left(1 - \frac{\Theta}{T} \right) M^{1/2} \tag{9.68}$$

In the fully developed theory, the proportionality factors in Eqs. (9.67) and (9.68) are also evaluated.

Our primary objective in undertaking this examination of the coil expansion factor was to see whether the molecular weight dependence of α could account for the fact that the Mark–Houwink a coefficient is generally greater than 0.5 for $T \neq \Theta$. More precisely, it is generally observed that $0.5 \leqslant a \leqslant 0.8$. This objective is met by combining Eqs. (9.55) and (9.68):

1. At $T = \Theta$ (or $\chi = 1/2$), $\alpha = 1$ and $[\eta] \propto M^{1/2}$, as required by Eq. (9.54).
2. For good solvents where $\alpha > 1$, $\alpha^5 > \alpha^3$, so $\alpha^5 \propto M^{1/2}$ or $\alpha \propto M^{0.1}$. Because of the factor α^3 in Eq. (9.55), an additional factor $M^{0.3}$ enters the latter to give an overall proportionality between $[\eta]$ and $M^{0.8}$.
3. For conditions of intermediate solvent goodness, α shows a dependence on M which is intermediate between the limits described in items (1) and (2) with the corresponding intermediate values for the Mark–Houwink a coefficient.

What is especially significant about Eq. (9.68) is the observation that the coil expansion factor α definitely increases with M for good solvents, meaning that—all other things being equal—longer polymer chains expand above their Θ dimensions more than shorter chains. Even though the dependence of α on

M is a small fractional order, this can amount to a considerable effect over a wide range of M values. For example, in the limit described in item (2), where $\alpha \propto M^{0.1}$, two samples of the same polymer showing a 1000-fold range of M will differ in α—as well as $(\overline{r^2}/M)^{1/2}$ and $(\overline{r_g^2}/M)^{1/2}$—by a factor of 2. This result assumes the same (favorable) interaction between polymer and solvent and is a consequence of the difference in chain length alone.

Equation (9.68) requires that $(\alpha^5 - \alpha^3)/M^{1/2}$ be a constant for a particular system if the molecular weight dependence of α has been correctly assessed. While not absolutely constant, experimental values of such a ratio show only a slight variation over a 1000-fold change in M. This indicates that the functional dependence of α on M predicted above is correct, at least to a first approximation. Additional efforts have been directed toward refining this prediction still further, but we shall not pursue these.

This concludes our discussion of the viscosity of polymer solutions per se, although various aspects of the viscous resistance to particle motion continue to appear in the remainder of the chapter. We began this chapter by discussing the intrinsic viscosity and the friction factor for rigid spheres. Now that we have developed the intrinsic viscosity well beyond that first introduction, we shall do the same (more or less) for the friction factor. We turn to this in the next section, considering the relationship between the friction factor and diffusion.

9.8 The Diffusion Coefficient and the Friction Factor

We begin our discussion of diffusion by assuming that a gradient of solute concentration exists in a solution. We designate this dc/dx, without specifying the units of concentration. The second law of thermodynamics generalizes the familiar experience that such a gradient is unstable in the absence of some sustaining force: The molecules will diffuse throughout the space available to them until a uniform concentration is reached. We imagine a cross section of area A perpendicular to the concentration gradient, and consider the flow of material across that surface. The rate per unit area at which matter diffuses through this cross section is called the flux J and is directly proportional to the concentration gradient:

$$J = - D \frac{dc}{dx} \tag{9.69}$$

In this expression, called Fick's first law, the proportionality constant D is the diffusion coefficient of the solute. Since $J = (1/A)(dQ/dt)$ and $c = Q/V$, where Q signifies the quantity of solute in unspecified units, it follows that D has the units length^2 time^{-1}, or m^2 sec^{-1} in the SI system. The minus sign in Eq. (9.69)

is introduced in recognition of the fact that the direction of the flow is that of decreasing concentration. Rewriting Eq. (9.69) in terms of Q, we obtain

$$\frac{1}{A}\frac{dQ}{dt} = -\frac{D}{V}\frac{dQ}{dx} \tag{9.70}$$

which is formally the same sort of differential equation as the corresponding expression for viscosity, Eq. (9.3). The solution to Eq. (9.70) gives the amount of solute as a function of location and time so as to satisfy the boundary conditions of the problem. In general, both the gradient and the flux have x, y, and z components, and solving the three-dimensional analog of Eq. (9.70) for various boundary conditions can become quite complicated. For our purposes it suffices to consider only the one-dimensional case. The formal similarity of the defining equations for D and η makes the study of both of these—along with charge and heat transport—part of the general topic of transport phenomena. In Eq. (9.3) it is momentum, not matter per se, that is transported across the reference surface in response to a gradient.

As with viscosity, we must consider a volume element and the flux of solute in and out of that element. Figure 9.9a schematically represents three regions of an apparatus containing a concentration gradient. The end compartments contain the solute at two different concentrations c_1 and c_2, with $c_2 > c_1$. The center region, which in practice could be something like a porous plug, is the volume element of cross section A and thickness dx, along which the gradient exists. The arrows in Fig. 9.9a represent the flux of solute from the more concentrated solution to the less concentrated one. The change in the amount of material dQ in the center volume element can be developed in two different ways:

1. In terms of fluxes,

$$dQ = Q_{in} - Q_{out} = (J_{in} - J_{out})\,A\,dt \tag{9.71}$$

2. In terms of a concentration change dc in the element of volume A dx,

$$dQ = dc\,(A\,dx) \tag{9.72}$$

3. The expressions for dQ in items (1) and (2) can be equated and the difference in fluxes in Eq. (9.71) eliminated by substituting Eq. (9.69):

$$-D\left[\left(\frac{dc}{dx}\right)_x - \left(\frac{dc}{dx}\right)_{x+dx}\right]A\,dt = A\,dx\,dc \tag{9.73}$$

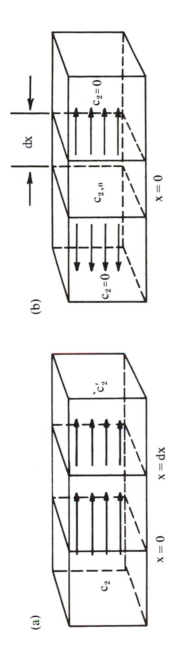

Figure 9.9 Schematic of diffusion with respect to a volume element of thickness dx located at $x = 0$: (a) $c_2 > c_1$ and (b) $c = c_0$ in volume element. The flux is shown by arrows.

4.   Since

$$\left[\left(\frac{dc}{dx} \right)_{x+dx} - \left(\frac{dc}{dx} \right)_{x} \right] = \frac{d^2 c}{dx^2} \ dx$$

this last result can be written as

$$\frac{dc}{dt} = D \frac{d^2 c}{dx^2} \qquad\qquad (9.74)$$

Equation (9.74) is a one-dimensional version of Fick's second law. We shall presently consider a statistical approach to solving this equation. If c is measured as a function of x and t in an experiment which corresponds to the boundary conditions of the mathematical solution to Eq. (9.74), then D can be evaluated for the solute. We shall consider this below also.

Before pursuing the diffusion process any further, let us examine the diffusion coefficient itself in greater detail. Specifically, we seek a relationship between D and the friction factor of the solute. In general, an increment of energy is associated with a force and an increment of distance. In the present context the driving force behind diffusion (subscript diff) is associated with an increment in the chemical potential of the solute and an increment in distance dx:

$$F_{diff} = - \frac{1}{N_A} \frac{d\mu_2}{dx} \qquad\qquad (9.75)$$

We divide by Avogadro's number to convert the partial molar Gibbs free energy to a molecular quantity, and the minus sign enters because the force and the gradient are in opposing directions. Recalling the definition of chemical potential [Eq. (8.13)], we write $\mu_2 = \mu_2{}^\theta + RT \ln a_2 = \mu_2{}^\theta + RT \ln \gamma_2 c$, where a_2 and γ_2 are the activity and activity coefficient, respectively, of the solute. In dilute solutions $\gamma_2 \rightarrow 1$ and $d\mu_2/dx = RT \, (d \ln c/dx)$; therefore

$$F_{diff} = - \frac{RT}{N_A} \frac{d \ln c}{dx} = - \frac{kT}{c} \frac{dc}{dx} \qquad\qquad (9.76)$$

Under stationary-state flow conditions, F_{diff} equals the force of viscous resistance experienced by the particle. The latter, in turn, equals the friction factor times the stationary velocity v_s; therefore

$$- \frac{kT}{c} \frac{dc}{dx} = fv_s \qquad\qquad (9.77)$$

The product cv_s can be written

$$\frac{Q}{V} \frac{dx}{dt} = \frac{Q}{A\,dx} \frac{dx}{dt} = \frac{1}{A} \frac{dQ}{dt}$$

which defines the flux. Accordingly, Eq. (9.77) becomes

$$J = -\frac{kT}{f} \frac{dc}{dx} \tag{9.78}$$

Comparing Eqs. (9.69) and (9.78) gives

$$D = \frac{kT}{f} \tag{9.79}$$

We shall see in Sec. 9.9 that D is a measurable quantity; hence Eq. (9.79) provides a method for the determination of an experimental friction factor as well. Note that no assumptions are made regarding the shape of the solute particles in deriving Eq. (9.79), and the assumption of ideality can be satisfied by extrapolating experimental results to $c = 0$, where $\gamma = 1$.

In contrast with Eq. (9.79), prior theoretical discussions of the friction factor for various particles were based on some assumed structure or geometry for the molecule:

1. Random coils. Equation (9.53) gives the Kirkwood–Riseman expression for the friction factor of a random coil. In the free-draining limit, the segmental friction factor can, in turn, be evaluated from f. In the nondraining limit the radius of gyration can be determined. We have already discussed ζ in Chap. 2 and $(\overline{r_g^2})^{1/2}$ in this chapter and again in Chapter 10, so we shall not examine the information provided by D for the random coil any further.

2. Rigid, unsolvated spheres. Stokes' law, Eq. (9.5), provides a relationship between f and the radius of the particle. Since this structure is a reasonable model for some protein molecules, experimental D values can be interpreted, via f, to yield values of R for such systems. Note that this application can also yield a value for M, since $M = N_A \rho_2 [(4/3)\pi R^3]$, where ρ_2 is the density of the unsolvated material.

3. Rigid particles other than unsolvated spheres. It is easy to conclude qualitatively that either solvation or ellipticity (or both) produces a friction factor which is larger than that obtained for a nonsolvated sphere of the same mass. This conclusion is illustrated in Fig. 9.10, which shows the swelling of a sphere due to solvation and also the spherical excluded volume that an ellipsoidal particle requires to rotate through all possible orientations.

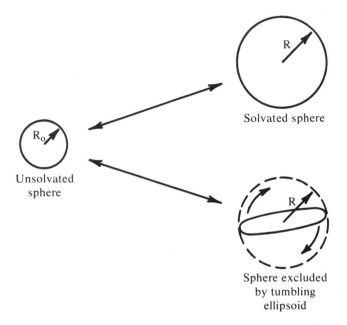

Figure 9.10 Schematic relationship between the radius R_0 of an unsolvated sphere and the effective radius R of a solvated sphere or of a spherical volume excluded by an ellipsoidal particle rotating through all directions.

Since f is a measurable quantity for, say, a protein, and since the latter can be considered to fall into category (3) in general, the friction factor provides some information regarding the ellipticity and/or solvation of the molecule. In the following discussion we attach the subscript 0 to both the friction factor and the associated radius of a nonsolvated spherical particle and use f and R without subscripts to signify these quantities in the general case. Because of Stokes' law, we write

$$\frac{f}{f_0} = \frac{R}{R_0} \tag{9.80}$$

where $R/R_0 > 1$. It is evident from Fig. 9.10 that the amount by which R/R_0 or f/f_0 exceeds unity increases with increasing solvation, ellipticity, or both.

The dependence of f/f_0 on solvation and ellipticity has been worked out in detail. Since the situation parallels the way $[\eta]$ exceeds its value for nonsolvated spheres, we shall not elaborate on the details, but merely summarize the conclusions:

1. The quantitative analysis of this problem results in a set of contours in terms of the axial ratio a/b and the solvation $m_{1,b}/m_2$ for constant values of f/f_0.
2. The general shape of the contours resembles the corresponding curves for $[\eta]$, which are shown in Fig. 9.4a. The contours for f/f_0 and $[\eta]$ differ in quantitative detail, however.
3. Figure 9.4b shows a theoretical f/f_0 contour for a value of this ratio equal to 1.45. As noted in the discussion of this figure in Sec. 9.3, the intersection of the f/f_0 and $[\eta]$ contours permits the state of solvation and ellipticity of such a protein molecule to be characterized uniquely.

The foregoing discussion is incomplete, since it offers an interpretation of f/f_0 while presenting, through Eq. (9.79), a method for evaluating f alone. What is needed to be able to take advantage of this approach is a value for the friction factor of a nonsolvated spherical particle having the same mass as the species under consideration. Fortunately, this reference value f_0 can be calculated if the molecular weight of the solute is known, since we are comparing f and f_0 for particles of the same mass. The mass of a molecule divided by its (unsolvated) density gives the volume of an equivalent spherical particle. From this volume R_0 can be evaluated, and this, in turn, allows f_0 to be calculated from Stokes' law. The following example illustrates this computational procedure.

Example 9.5

The hemoglobin molecule has a molecular weight of 62,300 and is observed to have a diffusion coefficient of 6.9×10^{-11} m^2 sec^{-1} in water at 20°C. Calculate the ratio f/f_0 which is consistent with these data, using 1.34 g cm^{-3} as the density of the dry polymer and 10^{-3} kg m^{-1} sec^{-1} as the viscosity of water at 20°C.

Solution

The friction factor of the actual particle is given by Eq. (9.79):

$$f = \frac{kT}{D} = \frac{(1.38 \times 10^{-23} \text{ J K}^{-1})(293 \text{ K})}{6.9 \times 10^{-11} \text{ m}^2 \text{ sec}^{-1}} = 5.89 \times 10^{-11} \text{ kg sec}^{-1}$$

If the hemoglobin molecule were a nonsolvated sphere, we could write for its volume

$$\left(\frac{62{,}300 \text{ g}}{\text{mol}} \times \frac{\text{cm}^3}{1.34 \text{ g}} \right) \left(\frac{1 \text{ mol}}{6.02 \times 10^{23} \text{ molecules}} \right) = \frac{4}{3} \pi R_0{}^3$$

from which $R_0 = 2.64 \times 10^{-7}$ cm $= 2.64 \times 10^{-9}$ m. By Stokes' law, $f_0 = 6\pi\eta R_0 = 6\pi(10^{-3} \text{ kg m}^{-1} \text{ sec}^{-1})(2.64 \times 10^{-9} \text{ m}) = 4.98 \times 10^{-11} \text{ kg sec}^{-1}$. Therefore $f/f_0 = 5.89 \times 10^{-11}/4.98 \times 10^{-11} = 1.18$. If this value had equaled unity, the particle would be proved to be a nonsolvated sphere.

•

All that can be concluded from the data given in the preceding example is that the particle is not an unsolvated sphere. However, when an appropriate display of contours is examined for f/f_0 (e.g., Ref. 2), the latter is found to be consistent with an unsolvated particle of axial ratio about 4:1 or with a spherical particle hydrated to the extent of about 0.48 g water (g polymer)$^{-1}$. Of course, there are a number of combinations of these variables which are also possible, and some additional experimental data—such as the intrinsic viscosity—are needed to select that combination which is consistent with all experimental observations.

In the next section we return to Eq. (9.74) and a consideration of the experimental determination of D.

9.9 The Diffusion Coefficient: Experimental Aspects

There is an intimate connection at the molecular level between diffusion and random flight statistics. The diffusing particle, after all, is displaced by random collisions with the surrounding solvent molecules, travels a short distance, experiences another collision which changes its direction, and so on. Such a zigzagged path is called Brownian motion when observed microscopically, describes diffusion when considered in terms of net displacement, and defines a three-dimensional random walk in statistical language. Accordingly, we propose to describe the net displacement of the solute in, say, the x direction as the result of a ν-step random walk, in which the number of steps is directly proportional to time:

$$\nu = Kt \tag{9.81}$$

When we discussed random walk statistics in Chap. 1, we used n to represent the number of steps in the process and then identified this quantity as the number of repeat units in the polymer chain. We continue to reserve n as the symbol for the degree of polymerization, so the number of diffusion steps is represented by ν in this section.

Next suppose we apply this idea to describe the spreading by diffusion of an infinitesimally thin solution layer sandwiched between two portions of pure solvent. Figure 9.9b represents this situation. In Fig. 9.9b the two end compartments contain solvent, with the central portion containing solution at some concentration c_0. The practical implementation of such an experiment is not our primary concern, although we can imagine successive portions of the various liquids being carefully layered in a tube, or the solution filling the pores of a plug between two portions of solvent. Of concern to us is the description of the process by which the solute diffuses from a narrow band at $x = 0$ to a condition where it exists at uniform concentration throughout the vessel. Any other factors which might disturb the solute concentration are assumed to be absent. In practice, this implies excellent temperature control, since convection currents would totally invalidate experimental results. The mathematics developed in Chap. 1 in reference to the one-dimensional random walk is what we propose to use to describe this diffusion process.

Equation (1.34) gives the probability that a molecule will be displaced a distance x after ν steps of length 1. Adapting this to the present problem, we use Eq. (9.81) to replace ν and write

$$P(x, t)\, dx = (2\pi K t\, l^2)^{-1/2} \, \exp\left(-\frac{x^2}{2\, K t\, l^2}\right) dx \qquad (9.82)$$

The ratio $c(x, t)/c_0$ would, in practice, be the logical measure of $P(x, t)$, so we write

$$c(x, t)\, dx = c_0\, P(x, t)\, dx = c_0\, (2\pi K t\, l^2)^{-1/2} \, \exp\left(-\frac{x^2}{2\, K t\, l^2}\right) dx \qquad (9.83)$$

The proposal we wish to examine is that Eq. (9.83) is a solution to Eq. (9.74) for the case of material initially present at $x = 0$ at a concentration c_0.

Several features of Eq. (9.83) assure us that this expression has the correct form:

1. $P(x, t)\, dx$ has the familiar bell shape of a normal distribution function [Eq. (1.39)], the width of which is measured by the standard deviation σ. In Eq. (9.83), t takes the place of σ. It makes sense that the distribution of matter depends in this way on time, with the width increasing with t.
2. Equation (1.34) is a normalized expression which means that, integrated over all values of x, it equals unity. Accordingly, adding together the

concentration of solute in a series of slices through the apparatus—that is, $\int c\, dx = \int c_0 P(x, t)\, dx$—always accounts for all of the solute, whatever its distribution.

3. The combined consideration of items (1) and (2) indicates that the profile of c throughout the apparatus gets flatter as it gets broader. At $t = 0$ the initial concentration is sharply peaked at c_0 at $x = 0$; at $t = \infty$, c is uniform from one end of the apparatus to the other.

All of these points are entirely reasonable. What is confusing about Eq. (9.83) is the step length and the undetermined constant K. Fortunately, they can both be eliminated in a single step. If Eq. (9.83) is a solution to Eq. (9.74), then both sides of the latter must be equal when the indicated operations are performed on Eq. (9.83). Carrying out these operations yields the following:

1. The first derivative of c with respect to t is

$$\left(\frac{\partial c}{\partial t}\right)_x = c_0 \, (2\pi\, Kt\, l^2)^{-1/2} \; e^{-x^2/2Ktl^2} \left[\frac{x^2}{2\, Kt^2\, l^2} - \frac{1}{2t}\right]$$

2. The second derivative of c with respect to x is obtained from

$$\left(\frac{\partial c}{\partial x}\right)_t = c_0 \, (2\pi\, Kt\, l^2)^{-1/2} \; e^{-x^2/2Ktl^2} \left(-\frac{x}{Kt\, l^2}\right)$$

followed by

$$\left(\frac{\partial^2 c}{\partial x^2}\right)_t = c_0 \, (2\pi\, Kt\, l^2)^{-1/2} \; e^{-x^2/2Ktl^2} \left[\left(-\frac{x}{Kt\, l^2}\right)^2 - \frac{1}{Kt\, l^2}\right]$$

3. Combining items (1) and (2) as required by Fick's second law yields

$$D = \frac{(\partial c/\partial t)_x}{(\partial^2 c/\partial x^2)_t} = \frac{x^2/2\, Kt^2 l^2 - 1/2t}{x^2/K^2 t^2 l^4 - 1/Ktl^2} = \frac{K^2 t^2 l^4}{2\, Kt^2 l^2} = \frac{Kl^2}{2}$$

These manipulations show that Eq. (9.83) satisfies Fick's second law with

$$K = \frac{2D}{l^2} \tag{9.84}$$

so the solution becomes

$$c(x, t)\, dx = c_0 (4\pi Dt)^{-1/2} \exp\left(-\frac{x^2}{4Dt}\right) dx \tag{9.85}$$

Figure 9.11 is a plot of this function at two different times for a solute with a diffusion coefficient arbitrarily selected to be 5×10^{-11} m^2 sec^{-1}.

We can imagine measuring experimental curves equivalent to those in Fig. 9.11 by, say, scanning the length of the diffusion apparatus by some optical method for analysis after a known diffusion time. Such results are then interpreted by rewriting Eq. (9.85) in the form of the normal distribution function, $P(z)\, dz$. This is accomplished by defining a parameter z such that

$$z = \frac{x}{(2Dt)^{1/2}} \tag{9.86}$$

in terms of which Eq. (9.85) becomes

$$\frac{c}{c_0}\, dx = \frac{1}{(2\pi)^{1/2}} \exp\left(-\frac{z^2}{2}\right) dz = P(z)\, dz \tag{9.87}$$

The inflection point of this function—where the second derivative changes sign—occurs at $z = 1$; hence the experimental analogs of Fig. 9.11 are examined for the location of their inflection points (subscript infl). The distance through which the material has diffused at this point is therefore given by

$$x_{infl} = (2Dt)^{1/2} \tag{9.88}$$

from which D can be evaluated if t is known. If Fig. 9.11 had been determined experimentally, such a procedure would be used by observing that $x_{infl} \cong 10^{-2}$ m for the curve at $t = 10^6$ sec. From this information, $D = (10^{-2}$ m$)^2 / 2(10^6$ sec$) = 5 \times 10^{-11}$ m^2 sec^{-1}.

The experiment we have just described is not very satisfactory from a practical point of view, since it is very difficult to deposit a thin layer of solution between two bulk portions of solvent without some mixing. An experimentally more convenient method consists of layering equal volumes of solvent and solution so that a sharp boundary exists between them at $x = 0$, with $c = c_0$ for

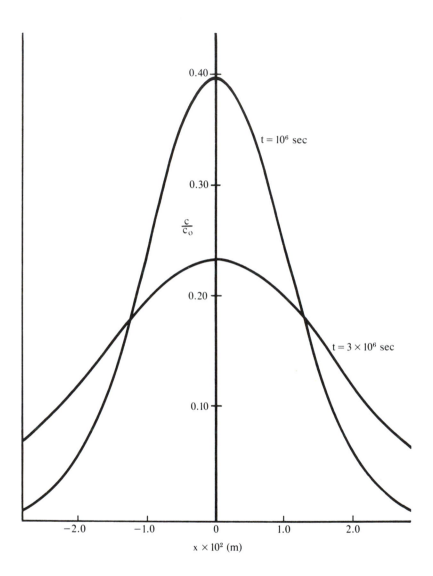

Figure 9.11 Variation of c/c_0 with x for one-dimensional diffusion [calculated from Eq. (9.85) with $D = 5 \times 10^{-11}$ m^2 sec^{-1}].

$x < 0$ and $c = 0$ for $x > 0$. This procedure also depends on layering liquids, but it is easier to do this with bulk quantities of material than to sandwich a thin layer between two bulk portions of solvent as required above.

At the outset of such an experiment, the gradient of concentration is zero on either side of the boundary and sharply peaked near $x = 0$. As diffusion occurs, the boundary becomes progressively less sharp. For $x < 0$, $c < c_0$, and for $x > 0$, $c > 0$, with both the magnitude and spatial extension of the effect increasing with time. Figure 9.12a shows how the overall concentration profile varies with time, and Fig. 9.12b shows the gradient dc/dx at corresponding times. Formal solution of Fick's second law with these boundary conditions proves that the shape of the gradient curves in this situation are identical to the curves in Fig. 9.11. Measuring the concentration gradient poses no serious experimental difficulty, so this method can also be analyzed by Eq. (9.87).

We are aware of the bending of a light ray when light passes from a medium of one refractive index to another. If light passes through a layer of optically homogeneous material and then reenters the original medium, then the emerging ray is bent back to the original direction. On the other hand, if there is a gradient of refractive index perpendicular to the direction of the light beam, the ray continues to experience additional bending as it passes through the medium with the gradient. Thus, when it reenters the surrounding medium, the emerging ray will make an angle with the incident beam. The size of this angle depends on

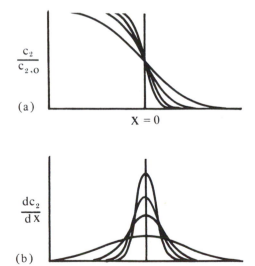

Figure 9.12 Progress of diffusion with time in an experiment like that shown in Fig. 9.9a, with the initial conditions given by $c = c_0$ for $x < 0$ and $c = 0$ for $x > 0$: (a) c/c_0 versus x and (b) dc/dx versus x.

the thickness of the sample and on the magnitude of the refractive index gradient. With the use of a diaphragm, the bent portion of the light beam can be excluded from a photographic plate or a photoelectric device. The magnitude of the resulting light attenuation is directly proportional to the refractive index gradient in the specimen. Such measurements can be made along the length of an apparatus to record the profile of the refractive index gradient. Procedures based on this concept are said to employ Schlieren optics.

The refractive indices of a solvent and a dilute solution, $\tilde{n}_0$ and $\tilde{n}$, respectively, can be related through

$$\tilde{n} = \tilde{n}_0 + \frac{d\tilde{n}}{dc} \; c \tag{9.89}$$

Over a reasonable range of c values, $d\tilde{n}/dc$ is constant; hence

$$\frac{d\tilde{n}}{dx} = \left(\frac{d\tilde{n}}{dc}\right)\frac{dc}{dx} = (\text{constant})\frac{dc}{dx}$$

This shows that Schlieren optics provide a means for directly monitoring concentration gradients. The value of the diffusion coefficient which is consistent with the variation of $d\tilde{n}/dx$ with x and t can be determined from the normal distribution function. Methods that avoid the difficulty associated with locating the inflection point have been developed, and it can be shown that the area under a Schlieren peak divided by its maximum height equals $(4\pi Dt)^{1/2}$. Since there are no unknown proportionality factors in this expression, D can be determined from Schlieren spectra measured at known times.

The standard procedure is to measure D at several different initial concentrations, using the procedure just described, and then extrapolating the results to c = 0. We symbolize the resulting limiting value $D°$. This value can be interpreted in terms of Eq. (9.79), which is derived by assuming $\gamma \to 1$ and therefore requires extreme dilution. It is apparent from Eqs. (9.79) and (9.5) that $D°$ depends on the ratio T/η_0, as well as on the properties of the solute itself. In order to reduce experimental (subscript ex) values of $D°$ to some standard condition (subscript s), it is conventional to write

$$D_s° = \frac{T_s}{T_{ex}} \; \frac{\eta_{0,ex}}{\eta_{0,s}} \; D_{ex}° \tag{9.90}$$

Note that this method of standardizing D values makes no allowance for the possibility that a molecule may change size, shape, or solvation with changes in temperature. In the next section we shall survey the behavior of polymeric materials in an ultracentrifuge. We shall see that diffusion coefficients can be

estimated from such experiments also and provide a valuable adjunct to the rate at which molecules settle in the ultracentrifuge.

9.10 Sedimentation Velocity and Sedimentation Equilibrium

In discussing diffusion in the last section we explicitly assumed the absence of forces which would complicate the boundary broadening in a diffusion experiment. Even though the force of gravity is always present in earthbound experiments, this is of negligible importance on a molecular scale, except over very large distances, such as in the earth's atmosphere. The equivalent to this force is not negligible in an ultracentrifuge, however, where accelerations, say, 10^5 times larger than gravity are readily achieved.

In ultracentrifugation a solution is placed in the cavity of a rotor which is spun at very high speeds in an instrument which allows vibration-free, constant-temperature operation. Two sidewalls of the sample compartment follow the radial lines of the rotor to prevent gradients of concentration from developing perpendicular to the radial direction of the acceleration. The sides of the sample space which are perpendicular to the axis of rotation are transparent, and an important part of the instrument is the optical system which allows a beam of light to pass through the rotating sample. The light which passes through the system can be analyzed by spectrophotometry or, most commonly, by Schlieren optics to monitor the progress of the molecules as they migrate in response to the radial acceleration.

Since the radial acceleration functions simply as an amplified gravitational acceleration, the particles settle toward the "bottom"—that is, toward the circumference of the rotor—if the particle density is greater than that of the supporting medium. A distance r from the axis of rotation, the radial acceleration is given by $\omega^2 r$, where ω is the angular velocity in radians per second. The midpoint of an ultracentrifuge cell is typically about 6.5 cm from the axis of rotation, so at 10,000, 20,000, and 40,000 rpm, respectively, the accelerations are 7.13×10^4, 2.85×10^5, and 1.14×10^6 m sec^{-2} or 7.27×10^3, 2.91×10^4, and 1.16×10^5 times the acceleration of gravity (g's).

The ultracentrifuge has been used extensively, especially for the study of biopolymers, and can be used in several different experimental modes to yield information about polymeric solutes. Of the possible procedures, we shall consider only sedimentation velocity and sedimentation equilibrium. We shall discuss these in turn, beginning with an examination of the forces which operate on a particle setting under stationary-state conditions.

Three kinds of forces must be considered:

1. The force of a molecule subject to radial acceleration is given by Newton's second law:

$$F_{accel} = \frac{M}{N_A} \omega^2 r \tag{9.91}$$

2. A buoyant force is given by the product of the volume V of the particle, the density ρ of the solution, and the radial acceleration:

$$F_{buoy} = V\rho \omega^2 r = \frac{M}{N_A \rho_2} \rho \omega^2 r \tag{9.92}$$

3. A force of viscous resistance is proportional to the stationary-stage velocity v_s according to Stokes' law:

$$F_{vis} = f v_s \tag{9.93}$$

4. The stationary-state velocity is rapidly achieved, and v_s corresponds to the force in item (1) equalling the opposing forces in items (2) and (3):

$$\frac{M}{N_A} \omega^2 r = \frac{M}{N_A} \frac{\rho}{\rho_2} \omega^2 r + f v_s \tag{9.94}$$

Since the velocity is dr/dt, Eq. (9.94) can also be written as

$$\frac{M}{f N_A} \left(1 - \frac{\rho}{\rho_2}\right) \omega^2 r = \frac{dr}{dt} \tag{9.95}$$

The stationary-state velocity per unit acceleration is a parameter which characterizes the settling particle and is called the sedimentation coefficient s:

$$s = \frac{dr/dt}{\omega^2 r} = \frac{M}{f N_A} \left(1 - \frac{\rho}{\rho_2}\right) \tag{9.96}$$

As a velocity divided by an accleration, s has units of time, and 10^{-13} sec—the order of magnitude for a typical solute—is called a svedberg (symbol S) in honor of Svedberg, a pioneer worker in this field.

Equation (9.95) can be integrated to give

$$\ln r = \frac{M}{f N_A} \left(1 + \frac{\rho}{\rho_2}\right) \omega^2 t + const. = s\omega^2 t + const. \tag{9.97}$$

which shows that the radial location of a species at various times is described by a straight line in a plot of $\ln r$ versus t. The slope of this line equals $\omega^2 s$. Many biopolymers are simply identified in terms of their sedimentation coefficient if

insufficient information is available to further resolve the cluster of constants which define s.

In a solution of molecules of uniform molecular weight, all particles settle with the same value of v_s. If diffusion is ignored, a sharp boundary forms between the "top" portion of the cell, which has been swept free of solute, and the "bottom," which still contains solute. Figure 9.13a shows schematically how the concentration profile varies with time under these conditions. It is apparent that the Schlieren optical system described in the last section is ideally suited for measuring the displacement of this boundary with time. Since the velocity of the boundary and that of the particles are the same, the sedimentation coefficient is readily measured.

As with the diffusion coefficient, sedimentation coefficients are frequently corrected for concentration dependence and reduced to standard conditions:

1. The concentration dependence of s is eliminated by making measurements at several different concentrations and then extrapolating to zero concentration. The limiting value is given by the symbol $s°$. This is the sedimentation analog of $D°$.

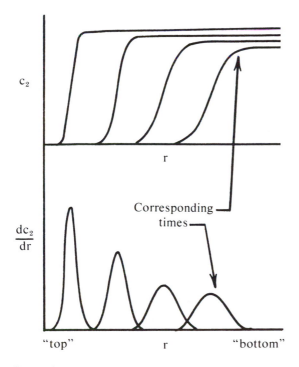

Figure 9.13 Location of sedimentation boundary after various times in an ultracentrifuge: (a) c versus r and (b) dc/dr versus r.

2. The temperature of an experiment affects s° through Eqs. (9.5) and (9.96), from which it is evident that s° is proportional to $(1 - \rho/\rho_2)/\eta_0$, all of which are temperature dependent.

3. In view of item (2), experimental (subscript ex) values of s° are reduced to standard (subscript s) temperature conditions by the expression

$$s_s^{\,\circ} = \frac{(1 - \rho/\rho_2)_s}{(1 - \rho/\rho_2)_{ex}} \frac{\eta_{0,ex}}{\eta_{0,s}} s_{ex}^{\,\circ} \tag{9.98}$$

This is the sedimentation analog of Eq. (9.90) for D°.

Diffusion effects are not absent during a sedimentation experiment as assumed above. As a matter of fact, the longer the experiment proceeds, the more diffuse the boundary between solvent and solution becomes owing to this effect. Figure 9.13b schematically illustrates the effect of this broadening in the Schlieren traces of the concentration profiles shown in Fig. 9.13a. The widths of these peaks depend on D, although correction must be made for the acceleration, which distorts the shape of the peaks from the ideal form indicated in Fig. 9.13b. Nevertheless, in principle, a study of sedimentation velocity permits the sedimentation coefficient to be evaluated unambiguously and the diffusion coefficient to be at least estimated. The proviso "in principle" is added, since s and D vary differently with M, and it may be only for an optimum range of M's that both s and D can be determined from a single experiment with sufficient accuracy.

Once a sedimentation coefficient has been measured, there are several ways in which it can be used:

1. As noted above, $s_s^{\,\circ}$ can be used directly to characterize a solute. This practice is widely followed with biopolymers.

2. The particle can be assumed to be spherical, in which case M/N_A can be replaced by $(4/3)\pi R^3 \rho_2$, and f by $6\pi\eta_0 R$. In this case the radius can be evaluated from the sedimentation coefficient: $s = 2R^2(\rho_2 - \rho)/9\eta_0$. Then, working in reverse, we can evaluate M and f from R. These quantities are called, respectively, the mass, friction factor, and radius of an equivalent sphere, a hypothetical spherical particle which settles at the same rate as the actual molecule.

3. If we assume that the densities of the solute and solution are known from separate experiments, the ratio M/f can be evaluated: $M/f = N_A s/(1 - \rho/\rho_2)$. By Eq. (9.79), f can be replaced by kT/D, with no assumptions regarding the shape of the particle. Therefore, if D is measured from the Schlieren traces of the sedimentation or in an independent experiment, s can be interpreted to give a value for M:

$$M = \frac{N_A kT}{(1 - \rho/\rho_2)} \frac{s}{D}$$

4. For polydisperse systems the value of M obtained from the values of $s°$ and $D°$—or, better yet, the value of the s/D ratio extrapolated to $c = 0$—is an average value. Different kinds of average are obtained, depending on the method used to define the "average" location of the boundary. The weight average is the type obtained in the usual analysis.

If a sedimentation experiment is carried out long enough, a state of equilibrium is eventually reached between sedimentation and diffusion. Under these conditions material will pass through a cross section perpendicular to the radius in both directions at equal rates: "downward" owing to the centrifugal field, and "upward" owing to the concentration gradient. It is easy to write expressions for the two fluxes which describe this situation:

1. The flux due to sedimentation is equal to the concentration of solute times v_s as given by Eq. (9.96):

$$J_{sed} = cv_s = c\omega^2 rs \tag{9.99}$$

2. The flux due to diffusion is given by Fick's first law:

$$J_{diff} = -D \frac{dc}{dr} \tag{9.100}$$

3. Under conditions of sedimentation equilibrium, the sum of these two fluxes equals zero, or

$$D \frac{dc}{dr} = c\omega^2 sr \tag{9.101}$$

which can be integrated to give

$$\ln c = \frac{s}{2D} \omega^2 r^2 + \text{const.} \tag{9.102}$$

Using Eqs. (9.96) and (9.79) to replace s and f, respectively, we obtain

$$\ln c = \frac{M(1 - \rho/\rho_2)}{2N_A kT} \omega^2 r^2 + \text{const.} \tag{9.103}$$

Note that this expression is equivalent to the barometric formula which gives the variation of atmospheric pressure ($\propto c$) with elevation ($\propto r$). A first-order dependence on the distance variable holds in the barometric equation, since the acceleration is constant in this case.

If the solute concentration is measured at different values of r in a sedimentation equilibrium experiment, Eq. (9.103) predicts that a plot of ln c versus r^2 is linear with a slope equal to $M(1 - \rho/\rho_2)\omega^2/2RT$. This type of experiment, therefore, allows M to be determined without either measuring D or assuming anything about the shape of the particle. For a polydisperse sample with a continuous distribution of molecular weights, several different kinds of average values of M—starting with $\bar{M}_w$ and including $\bar{M}_z$ [Eq. (1.19)] and still higher averages—can be extracted from this kind of data, particularly under Θ conditions. Protein preparations which are more commonly studied by this method are likely to contain several distinctly different molecular weight species rather than a broad distribution of molecular weights. In the case of discrete molecular weights it may be possible to resolve the data in such a way as to obtain M values for the individual components. The following example considers such a case.

Example 9.6

A preparation of reduced and carboxymethylated protein particles in water reached sedimentation equilibrium after 40 hr at 12,590 rpm. When the data are plotted as suggested by Eq. (9.103), two distinctly linear portions are observed in the graph. The following pairs of points are taken from these two regions:

	Region I		Region II	
Recorder displacement ($\propto$ c)	2.51	3.09	3.89	6.61
r (cm)	6.58	6.65	6.69	6.79

Calculate the molecular weight of the two fractions indicated to be present in this preparation, taking the density of each to be 1.37 g cm^{-3} and the density of the medium to be unity.

Solution

With only two points to work with in each region, it is preferable to integrate Eq. (9.101) between definite limits c_1 at r_1 and c_2 at r_2, thereby evaluating the integration constant in Eq. (9.103):

$$\ln\left(\frac{c_1}{c_2}\right) = \frac{M(1 - \rho/\rho_2)\omega^2}{2RT}(r_1^2 - r_2^2)$$

Since the factor $(1 - \rho/\rho_2)\omega^2/2RT$ is common to both fractions, it can be evaluated separately:

$$\frac{\left(1 - \dfrac{1}{1.37}\right)\left(\dfrac{12{,}590 \text{ rev}}{\min} \times \dfrac{2\pi \text{ rad}}{1 \text{ rev}} \times \dfrac{1 \text{ min}}{60 \text{ sec}}\right)^2}{2(8.314 \text{ J K}^{-1})(298 \text{ K})} = 94.7 \text{ mol kg}^{-1} \text{ m}^{-2}$$

Then for region I,

$$M = \frac{\ln\left(\dfrac{3.09}{2.51}\right)}{94.7 \dfrac{\text{mol}}{\text{kg m}^2}\, [(6.65)^2 - (6.58)^2]\, 10^{-4} \text{ m}^2} = 23.7 \text{ kg mol}^{-1}$$

$$= 23{,}700 \text{ g mol}^{-1}$$

For region II,

$$M = \frac{\ln\left(\dfrac{6.61}{3.89}\right)}{94.7\, [(6.79)^2 - (6.69)^2]\, 10^{-4}} = 41.5 \text{ kg mol}^{-1} = 41{,}500 \text{ g mol}^{-1}$$

These values are in agreement with the values obtained from the slopes of the two linear portions of the appropriate graph which includes many more points.

•

We have emphasized biopolymers in this discussion of the ultracentrifuge and in the discussion of diffusion in the preceding sections, because these two complementary experimental approaches have been most widely applied to this type of polymer. Remember that from the combination of the two phenomena, it is possible to evaluate M, f, and the ratio f/f_0. From the latter, various possible combinations of ellipticity and solvation can be deduced. Although these methods can also be applied to synthetic polymers to determine M, they are less widely used, because the following complications are more severe with the synthetic polymers:

1. Nonideality requires that especially dilute solutions be used.
2. Polydispersity obscures the nature of the average obtained, although the possibility of extracting more than one kind of average from the same data under optimum conditions partially offsets this.
3. The factor $1 - \rho/\rho_2$ cannot be too close to zero, nor can the refractive index of the polymer and the solvent be too similar. These additional considerations limit the choice of solvents for a synthetic polymer, while their values are optimal for aqueous protein solutions.

All of the experimental procedures we have discussed in this chapter yield, among other things, information concerning the average molecular weight of a

polymer sample. We conclude this chapter with still another technique for characterizing a polymer in this regard: gel permeation chromatography.

9.11 Gel Permeation Chromatography

Gel permeation chromatography (GPC) is one of several different types of liquid chromatography which separate a mixture of solutes by passing a solution through an appropriate column. As the mobile phase passes through the column packing, different solute species are retained to various degrees by their interaction with the stationary phase. Surface adsorption, liquid partitioning, and ion exchange describe the interactions which serve as the basis for other types of liquid chromatography. In GPC the columns are packed with porous particles, and the separation occurs because molecules of different size penetrate the pores of the stationary phase to various degrees. The method is somewhat like a reverse sieving operation at the molecular level. The largest molecules are excluded from the pores to the greatest extent and, hence, are the first to emerge from the column. Progressively smaller molecules permeate the porous stationary phase to increasing extents and are eluted sequentially. The eluted material is monitored for solute by a suitable detector, and an instrumental trace of the detector output provides distinct peaks for well-resolved mixtures and broad peaks for a continuous distribution of molecular sizes. With suitable calibration, this can be translated into a quantitative characterization of the sample. In our discussion of this material we shall consider the "safe" empirical approach to this calibration, partially successful approaches to theoretical calibration, and briefly some aspects of the instrumentation and technique associated with GPC.

What is essentially the same method is known by several different names— and their acronyms—by workers in different fields who are interested in this technique. As a matter of vocabulary, we review some alternatives in terminology:

1. The term *size exclusion chromatography* (SEC) is the general name for this method of separation, emphasizing—as it does—the mechanism for the separation (size exclusion) rather than directing attention to one of several column packing materials (the gels). This is the name of preference among analytical chemists who are concerned with both the theory and practice of the method in its broadest sense.

2. Gel filtration chromatography (GFC) is the name used to describe this method of separation in the biochemical literature. Under this heading, the method is primarily applied to aqueous solutions of solutes of biological origin.

3. Gel permeation chromatography (GPC) is the term found most widely in the polymer literature. In this context it is used most widely as an analytical

technique for determining the molecular weight and molecular weight distribution of a synthetic polymer sample.

4. All three of these names are also modified by the term *high-performance*, or its acronym, to give HPSEC, HPGFC, and HPGPC. The additional feature implied by this terminology is the increased speed of efficient separations due to rapid flow through the column under the influence of relatively high applied pressures.

To use GPC for molecular weight determination, we must measure the volume of solvent that passes through the column before a polymer of particular molecular weight is eluted. This quantity is called the retention volume V_R. Figure 9.14 shows schematically the relationship between M and V_R; it is an

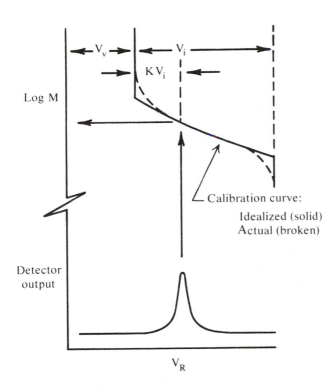

Figure 9.14 Calibration curve for GPC as log M versus the retention volume V_R, showing how the location of the detector signal can be used to evaluate M. Also shown are the void volume V_v and the internal volume V_i in relation to V_R, and KV_i as a fraction of V_i.

experimental fact that such calibration curves are approximately linear over about two orders of magnitude in M. In practice, the column is calibrated by constructing such a curve with standards of known molecular weight. We shall see presently that the "size" dependence of V_R is more a matter of spatial extension than molecular weight; hence the calibration should be carried out with standards of the same type as the unknown sample and under conditions which duplicate the treatment of the unknown. This sort of individual calibration is tedious and costly and thereby detracts from one of the principal advantages of GPC: speed.

To circumvent this need for calibration as well as to better understand the separation process itself, considerable effort has been directed toward developing the theoretical basis for the separation of molecules in terms of their size. Although partially successful, there are enough complications in the theoretical approach that calibration is still the safest procedure. If a calibration plot such as Fig. 9.14 is available and a detector output indicates a polymer emerging from the column at a particular value of V_R, then the molecular weight of that polymer is readily determined from the calibration, as indicated in Fig. 9.14.

Polydisperse polymers do not yield sharp peaks in the detector output as indicated in Fig. 9.14. Instead, broad bands are produced which reflect the polydispersity of synthetic polymers. Assuming that suitable calibration data are available, we can construct molecular weight distributions from this kind of experimental data. An indication of how this is done is provided in the following example.

Example 9.7

A broad chromatogram is subdivided into 20 slices, each 1 mm wide, and these are indexed from i = 1 to 20. The height h of the curve above a horizontal base line is carefully measured for each slice. The molecular weight of the ith slice is assigned from independent calibration via the retention volume. Columns 2–4 in Table 9.4 list h_i, $V_{R,i}$, and M_i values, respectively, for a particular chromatogram. Explain the significance and/or use of the remaining columns in Table 9.4 for the determination of a molecular weight distribution from this data.

Solution

The basic premise of this method is that the magnitude of the detector output, as measured by h_i for a particular fraction, is proportional to the weight of that component in the sample. In this sense the chromatogram itself presents a kind of picture of the molecular weight distribution. The following column entries provide additional quantification of this distribution, however.

Table 9.4 Data for the Analysis of the Gel Permeation Chromatogram of a Polydisperse Polymer Used in Example 9.7

(1) i	(2) h_i (mm)	(3) $V_{R,i}$ (ml)	(4) $M_i \times 10^{-6}$ (g mol^{-1})	(5) $\sum_{i=1}^{N} h_i$ (mm)	(6) $h_i/M_i \times 10^6$	(7) $h_i M_i \times 10^{-6}$	(8) A_i (mm^2)	(9) A_i/A_{tot}
21	0.0	20	4.709	545.0	0.0	0.0	545.0	1.0
20	0.0	21	3.302	545.0	0.0	0.0	545.0	1.0
19	0.8	22	2.327	545.0	0.34	1.86	544.6	0.999
18	3.5	23	1.640	544.2	2.13	5.74	542.5	0.995
17	16.8	24	1.1555	540.7	14.54	19.40	532.3	0.977
16	42.4	25	0.8142	523.9	52.08	34.52	502.7	0.922
15	67.9	26	0.5738	481.5	118.2	38.90	447.6	0.821
14	81.5	27	0.4003	413.7	203.6	32.62	373.0	0.684
13	81.4	28	0.2821	322.2	288.6	22.96	291.5	0.535
12	71.0	29	0.1988	250.8	357.1	14.12	215.3	0.395
11	57.0	30	0.1401	179.8	406.8	7.98	151.3	0.278
10	43.0	31	0.09872	122.8	435.6	4.24	101.3	0.186
9	30.0	32	0.06887	79.8	435.6	2.07	64.8	0.119
8	19.0	33	0.04853	49.8	391.5	0.92	40.3	0.074
7	12.2	34	0.03420	30.8	356.7	0.42	24.7	0.045
6	9.0	35	0.02410	18.6	373.4	0.22	14.1	0.026
5	4.0	36	0.01698	9.6	235.6	0.07	7.6	0.014
4	2.6	37	0.01197	5.6	217.2	0.03	4.3	0.008
3	2.0	38	0.00843	3.0	237.1	0.02	2.0	0.004
2	1.0	39	0.00588	1.0	170.0	0.01	0.5	0.001
1	0.0	40	0.00414	0.0	0.0	0.0	0.0	0.0

Source: Reprinted with permission from Ref. 6.

Column 5. $\Sigma_{i=1}^{N} h_i$ is proportional to the cumulative weight of all polymers in all categories up to the Nth.

Column 6. h_i/M_i is proportional to the weight of material in the ith class w_i divided by M_i, that is, to the number of moles in that class n_i. Therefore $\bar{M}_n$ can be evaluated as follows:

$$\bar{M}_n = \frac{\Sigma_i n_i M_i}{\Sigma_i n_i} = \frac{\Sigma_i (h_i/M_i)(M_i)}{\Sigma_i (h_i/M_i)} = \frac{\Sigma_i h_i}{\Sigma_i (h_i/M_i)} = \frac{545}{4.29 \times 10^{-3}}$$

$$= 127,000 \text{ g mol}^{-1}$$

Column 7. $h_i \times M_i$ is proportional to $w_i M_i$, and $\bar{M}_w$ is evaluated as follows:

$$\bar{M}_w = \frac{\Sigma_i w_i M_i}{\Sigma_i w_i} = \frac{\Sigma_i h_i M_i}{\Sigma_i h_i} = \frac{1.86 \times 10^6}{545} = 341,000 \text{ g mol}^{-1}$$

Column 8. $A_i = \Sigma_{i=1}^{N} [h_i + 1/2 (h_{i+1} - h_i)]$. Adding $1/2 (h_{i+1} - h_i)$ to h_i gives the height of the midpoint of each slice, and, since each slice is 1 mm wide, the summation gives the area under the curve up to the Nth class.

Column 9. A_i/A_{tot} gives that fraction of the area under the entire curve which has accumulated up to the Nth class. Since the curve is a weight distribution, this is equal to the weight fraction of material in the sample having $M < M_i$.

A plot of the last entry versus M gives the integrated form of the distribution function. The more familiar distribution function in terms of weight fraction versus M is given by the derivative of this cumulative curve. It can be obtained from the digitized data by some additional manipulations, as discussed in Ref. 6.

•

A detailed examination of the correlation between V_R and M is discussed in references on analytical chemistry such as Ref. 6. We shall only outline the problem, with particular emphasis on those aspects which overlap other topics in this book. To consider the origin of the calibration curve, we begin by picturing a narrow band of polymer solution being introduced at the top of a solvent-filled column. The volume of this solvent can be subdivided into two categories: the stagnant solvent in the pores (subscript i for internal) and the interstitial liquid in the voids (subscript v) between the packing particles:

$$V_{solvent} = V_v + V_i \tag{9.104}$$

The entire interstitial volume must pass through the column before any polymer emerges. Then the first polymer that does appear is the one with the highest

molecular weight. This solute has spent all its time in the voids—not the pores—of the packing and passes through the column with the velocity of the solvent.

Progressively smaller molecules have access to successively larger fractions of the internal volume. Therefore, as V_i emerges, consecutive fractions of the polymer come with it. Thus we can write the retention volume for a particular molecule weight fraction as

$$V_R = V_v + KV_i \qquad (9.105)$$

where K is a function of both the pore size and the molecular size and indicates what fraction of the internal volume is accessible to that particular solute. The relationships between V_R, V_v, V_i, and KV_i are also shown in Fig. 9.14. When K = 0, the solute is totally excluded from the pores; when K = 1, it totally penetrates the pores.

It is instructive to consider a simple model for the significance of the constant K in Eq. (9.105). For simplicity, we assume a spherical solute molecule of radius R and a cylindrical pore of radius a and length 1. As seen in Fig. 9.15a, an excluded volume effect prevents the center of the spherical solute molecule from approaching any closer than a distance R from the walls of the pore. This effectively decreases the volume accessible to the solute to a smaller cylinder of radius a – R. In this accessible cylinder the concentration of the solute is the same as in the interstitial fluid outside the pore. The excluded volume—that shell of thickness R around the walls of the pore—is devoid of solute. Hence the average concentration of solute in the pore *as a whole* is less than that outside the pore. The fraction of the external concentration in the pore is given by the ratio of the accessible volume to the actual volume of the cylindrical pore: $\pi(a - R)^2 l / \pi a^2 l$. This fraction gives K for the case of spherical solute molecules in cylindrical cavities. If we assume that the pore is long enough to neglect end effects, we have

$$K = \frac{(a - R)^2}{a^2} = \left(1 - \frac{R}{a}\right)^2 \qquad (9.106)$$

Note that the fraction is zero when R = a, and unity when R = 0.

This simple model illustrates how the fraction K and, through it, V_R are influenced by the dimensions of both the solute molecules and the pores. For solute particles of other shapes in pores of different geometry, theoretical expressions for K are quantitatively different, but typically involve the ratio of solute to pore dimensions.

The extension of these ideas to random coils can proceed along two lines. In one analysis the coil domain is visualized as a sphere, as in the case above, with r_g taking the place of R. Alternatively, statistical methods can be employed

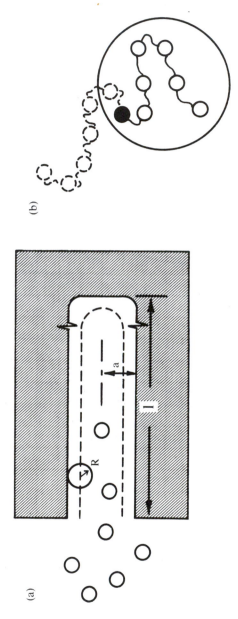

Figure 9.15 Schematic illustration of size exclusion in a cylindrical pore: (a) for spherical particles of radius R and (b) for a flexible chain, showing allowed (solid) and forbidden (broken) conformations of polymer.

to consider those conformations of a random chain which are excluded for a coil confined to a pore. This latter situation is illustrated in Fig. 9.15b. Figure 9.15b represents by solid and broken lines two conformations of the same chain, with the filled-in repeat unit being held in a fixed position. If the molecule were in bulk solution, both conformations would be possible. In a pore, represented by the enclosing circle in Fig. 9.15b, the broken line conformation is impossible. This is equivalent to a decrease in entropy for the coil in the pore, and the effect can be translated into an equilibrium constant between the solute in the pore and in the bulk solution. The factor K in Eq. (9.105) is just such a constant—the distribution coefficient—and can be evaluated by this approach for pores of different shape.

Figure 9.16 shows the theoretical predictions for K versus r_g/a compared with experimental findings. The solid line is drawn according to the statistical theory. The experimental points correspond to the same porous beads used as the stationary phase with their pore size analyzed by two different experimental procedures: mercury penetration (circles in Fig. 9.16, $\bar{a} = 21$ nm) and gas adsorption (squares in Fig. 9.16, $\bar{a} = 41$ nm). We can draw several conclusions from an examination of Fig. 9.16:

1. The characterization of the solid is also a source of discrepancy: The polymers are not the only source of difficulty!
2. Despite item (1), the fact that one set of experimental points agrees reasonably well with theory verifies the basic soundness of this approach.
3. Since K represents the fraction of V_i at which a particular molecular weight fraction emerges from the column, and since $\ln M \propto \ln r_g$, we see that this model correctly accounts for the form of the calibration curve shown in Fig. 9.14.

An interesting outgrowth of these considerations is the idea that $\ln r_g$ versus K or V_R should describe a universal calibration curve in a particular column for random coil polymers. This conclusion is justified by examining Eq. (9.55), in which the product $[\eta]M$ is seen to be proportional to $(\overline{r_g^2})^{3/2}$, with $r_g = \alpha(\overline{r_{g,0}^2})^{1/2}$. This suggests that $\ln r_g$ in the theoretical calibration curve can be replaced by $\ln[\eta]M$. The product $[\eta]M$ is called the hydrodynamic volume, and Fig. 9.17 shows that the calibration curves for a variety of polymer types merge into a single curve when the product $[\eta]M$, rather than M alone, is used as the basis for the calibration.

Although theoretical considerations still fall short of the goal of giving absolute values of M from V_R, they do suggest ways in which empirical calibrations can be extended beyond the stringent requirement of identical polymers under identical conditions being the only acceptable calibrations.

We conclude this section with a few remarks about GPC instrumentation. As with other areas of chromatography, this field is served by a variety of

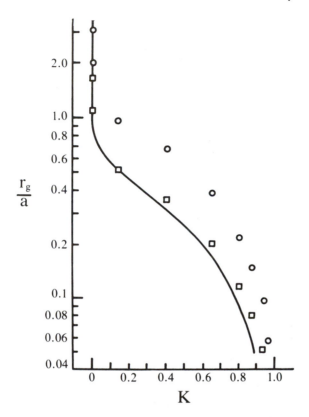

Figure 9.16 Comparison of theory with experiment for r_g/a versus K. The solid line is drawn according to the theory for flexible chains in a cylindrical pore. Experimental points show some data, with pore dimensions determined by mercury penetration (circles, $\overline{a}$ = 21 nm) and gas adsorption (squares, $\overline{a}$ = 41 nm). [From W. W. Yau and C. P. Malone, *Polym. Prepr.* 12:797 (1971), used with permission.]

commercial instruments, with "new and improved" models appearing regularly. It is therefore inappropriate to attempt to summarize the state of the art in a general discussion such as this. We can, however, consider some of the components of a GPC apparatus. While these have already and will continue to evolve through successive generations of instruments, we may reasonably expect that their basic functions will continue to be pertinent. Some key features of any GPC instrument are high-pressure liquid pumps, precise sample injection, sensitve detectors, and efficient columns.

A constant, reproducible flow rate on the order of 3–10 ml min^{-1} at a pressure of several thousand pounds per square inch makes the pump the most

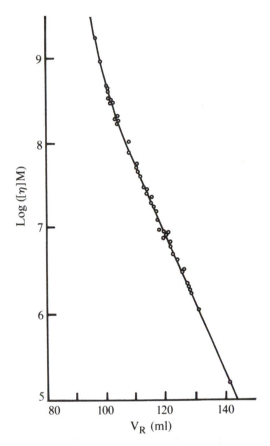

Figure 9.17 Plot of log [η] M versus retention volume for various polymers, showing how different systems are represented by a single calibration curve when data are represented in this manner. The polymers used include linear and branched polystyrene, poly(methyl methacrylate), poly(vinyl chloride), poly(phenyl siloxane), polybutadiene, and branched, block, and graft copolymers of styrene and methyl methacrylate. [From Z. Grubisec, P. Rempp, and H. Benoit, *Polym. Lett.* 5:753 (1967), used with permission of Wiley.]

important part of the solvent-monitoring system. Good sample injection capability is crucial because of the desirability of maintaining as narrow a band of solute as possible in the column. A certain amount of band broadening inevitably occurs during passage through the column, but the sample injection procedure should not aggravate this. The detector output is proportional to the solute concentration in the column effluent. Therefore versatility, sensitivity, and accuracy are all important requirements in the detection system. The

difference in refractive index between the solvent and the solution is probably the most widely used method of detection. Various spectrophotometric methods are also employed.

The column is the site of actual separation, and, of the variables involved in its preparation, the packing material is the most important. Historically, cross-linked polystyrene gel particles were used as the stationary phase. Being only semirigid, these are of limited utility in modern high-performance, high-pressure equipment. These and other cross-linked polymers in the form of spheres with diameters in the range 10-100 μm are commercially available with a range of pore sizes, and are still widely used when pressures less than about 3000 psi are sufficient. More recently, silica particles with dimensions in the 5-10 μm range have also become available with a range of pore dimensions. Being rigid, these can withstand higher pressures. Silica particles are also available with various organic functional groups chemically bonded to their surface to minimize adsorption. If adsorption occurs, then both size exclusion and adsorption mechanisms are involved and the experiment is no longer governed by size exclusion principles alone.

Finally, we note that the size and shape of the particles of the packing, the packing technique, and column dimensions and configuration are additional factors which influence a GPC experiment. In addition, the flow rate, the sample size, the sample concentration, the solvent, and the temperature must all be optimized. Details concerning these considerations are found in analytical chemistry references, as well as in the technical literature of instrument manufacturers.

Problems

1. The intrinsic viscosity of poly(γ-benzyl-L-glutamate) (M_0 = 219) shows such a strong molecular weight dependence in dimethyl formamide that the polymer was suspected to exist as a helix which approximates a prolate ellipsoid of revolution in its hydrodynamic behavior[†]:

$M \times 10^{-3}$ (g mol^{-1})	21.4	66.5	130	208	347
$[\eta]$ (dl g^{-1})	0.107	0.451	1.32	3.27	7.20

Using 1.32 g cm^{-3} as the density of the polymer, estimate the axial ratio for these molecules, using Simha's equation:

$$\frac{\eta_{sp}}{\phi} = \frac{p^2}{15[\ln(2p) - 3/2]} + \frac{p^2}{5[\ln(2p) - 1/2]} + \frac{14}{15} \cong 175\left(\frac{p}{50}\right)^{1.8}$$

(for large p = a/b)

[†] P. Doty, J. H. Bradbury, and A. M. Holtzer, *J. Am. Chem. Soc.* 78:947 (1956).

For the α-helix, the length per residue is about 1.5 Å. Use this figure with the molecular weight to estimate the length 2a of the particle. Use the estimated a/b ratios to calculate the diameter 2b of the helix, which should be approximately constant if this interpretation is correct. Comment on the results.

2. Fox and Flory[†] used experimental molecular weights, intrinsic viscosities, and rms end-to-end distances from light scattering to evaluate the constant Φ in Eq. (9.55). For polystyrene in the solvents and at the temperatures noted, the following results were assembled:

Solvent	T (°C)	$M \times 10^{-3}$ (g mol^{-1})	$[\eta]$ (dl g^{-1})	r_{rms} (Å)
Methyl ethyl ketone	22	1760	1.65	1070
	22	1620	1.61	1015
	67	1620	1.50	980
	22	1320	1.40	900
	25	980	1.21	840
	22	940	1.17	750
	22	520	0.77	545
	25	318	0.60	475
	22	230	0.53	400
Dichloroethane	22	1780	2.60	1410
	22	1620	2.78	1335
	67	1620	2.83	1295
	22	562	1.42	760
	22	520	1.38	680
Toluene	22	1620	3.45	1290
	67	1620	3.42	1280

Evaluate Φ for each set of data and compare the average with the value given in the text.

3. Under Θ conditions for the polystyrene–cyclohexane system, intrinsic viscosities were measures[‡] for polymers of different molecular weights:

M (g mol^{-1})	320,000	16,000	10,400	8,370	3,990
$[\eta]$ (dl g^{-1})	0.473	0.107	0.089	0.078	0.056

Use these data to evaluate $(\overline{r_{g,0}^2})^{1/2}$ for these polymers. To what step length l_0 in Eq. (1.62) do these values of $r_{g,0}$ correspond?

[†] T. G. Fox, Jr., and P. J. Flory, *J. Am. Chem. Soc.* 73:1915 (1951).
[‡] W. R. Krigbaum, L. Mandelkern, and P. J. Flory, *J. Polym. Sci.* 9:381 (1952).

4. The intrinsic viscosity of polystyrene in benzene at $25°C$ was measured[†]
    for polymers with the following molecular weights:

M (g mol^{-1})	[η] (dl g^{-1})	M (g mol^{-1})	[η] (dl g^{-1})
6,970,000	11.75	277,000	1.07
4,240,000	8.15	63,800	0.358
2,530,000	5.54	63,100	0.356
838,000	2.43	43,200	0.268
784,000	2.32	16,050	0.136
676,000	2.07	10,430	0.106
335,000	1.23	8,370	0.0932
		3,990	0.0608

Evaluate $(\overline{r_g^2})^{1/2}$ for these polymers from these data. Use the value of l_0
determined in the last problem to calculate $(\overline{r_{g,0}^2})^{1/2}$ and, from the ratio
of the two radii, evaluate α for each fraction.

5. According to Eq. (8.42), the parameter χ is proportional to the energy of
    interaction per 1–2 pair of molecules. To allow for solvent molecules of
    various sizes, we can write this as $\chi \propto n_1^* \, \Delta w$, where n_1^* is the number of
    segments *in the solvent molecule* and equals $\overline{V}_1/V_1^*$, with V_1^* the volume
    per mole of solvent segments. Use this concept to criticize or defend the
    following proposition: According to Eq. (9.67), $\alpha^5 - \alpha^3$ is proportional to
    $(1/2 - \chi)/\overline{V}_1$ or to $(2\overline{V}_1)^{-1} - (z \, \Delta w/2RTV_1^*)$. For a particular polymer in
    a homologous series of solvent molecules of different sizes, the second of
    these terms should be constant, while the first term decreases with increas-
    ing $\overline{V}_1$. This shows that $\alpha = 1$ for a polymer molecule dissolved in a
    "solvent" of other polymer molecules identical to itself. That is, that
    unperturbed coil dimensions apply to bulk polymer.

6. Mandelkern and Flory[‡] have assembled the sedimentation coefficients and
    intrinsic viscosities for polymers of various molecular weights. As shown
    by the following data, the quantity $s_0 [\eta]^{1/3}/M^{2/3}$ is constant for a particu-
    lar system:

[†] W. R. Krigbaum and P. J. Flory, *J. Polym. Sci.* 11:37 (1953).
[‡] L. Mandelkern and P. J. Flory, *J. Chem. Phys.* 20:212 (1952).

| Polystyrene molecular weight | $s_0 [\eta]^{1/3}/M^{2/3} \times 10^{17}$ | | Cellulose acetate molecular weight | $s_0 [\eta]^{1/3}/$ $M^{2/3} \times 10^{17}$ |
	In methyl ethyl ketone	In toluene		In acetone
1,240,000	25.3	13.6	194,000	54.0
830,000	25.0	13.4	130,000	59.5
519,000	26.6	13.4	53,000	59.0
254,000	25.6	13.9	11,000	50.3

Propose a brief explanation why this should be the case.

7. Protein molecules extracted from *Escherichia coli* ribosomes were examined by viscosity, sedimentation, and diffusion experiments for characterization with respect to molecular weight, hydration, and ellipticity. These data[†] are examined in this and the following problem. Use Fig. 9.4a to estimate the axial ratio of the molecules, assuming a solvation of 0.26 g water (g protein)$^{-1}$. At $20°C$, $[\eta] = 27.7$ cm^3 g^{-1} and $\rho_2 = 1.36$ for aqueous solutions of this polymer.

8. For the protein described in the last problem[‡], M = 24,000 and $D° = 5.59 \times 10^{-11}$ m^2 sec^{-1}. Estimate f/f_0 from these data, taking $\rho = 1.00$ and $\eta_0 = 0.01$ P for the medium. The experimental ratio f/f_0 is the product of contributions from solvation and ellipticity: $(f/f_0)_{ex} = (f/f_0)_{solv}(f/f_0)_{ellip} = (R/R_0)_{solv}(f/f_0)_{ellip} = [1 + (m_{1,b}/m_2)(\rho_2/\rho_1)]^{1/3}(f/f_0)_{ellip}$. Briefly justify this expansion of the $(f/f_0)_{solv}$ factor. Assuming these particles were solvated to the extent of 0.26 g water (g protein)$^{-1}$, calculate $(f/f_0)_{ellip}$. For prolate ellipsoids of revolution (b/a < 1), Perrin has derived the following expression:

$$\left(\frac{f}{f_0}\right)_{ellip} = \left[1 - \left(\frac{b}{a}\right)^2\right]^{1/2} \Big/ \left(\frac{b}{a}\right)^{2/3} \ln\left(\frac{1 + [1 - (b/a)^2]^{1/2}}{b/a}\right)$$

Verify—or revise, if necessary—the axial ratio estimated in the last problem for this protein.

9. The sedimentation boundary of an enzyme preparation in an aqueous buffer at $20.6°C$ was measured after various times in an ultracentrifuge at 56,050 rpm. The following results were obtained[§]:

[†] K-P. Wong and H. H. Paradies, *Biochem. Biophys. Res. Commun.* 61:178 (1974).
[‡] Ibid.
[§] Data of P. Modrich, quoted by I. Tinoco, Jr., K. Sauer, and J. C. Wang, *Physical Chemistry*, Prentice-Hall, Englewood Cliffs, N.J. (1978).

Time (min)	r (cm)
0	5.9110
20	6.0217
40	6.1141
60	6.2068
80	6.3040
100	6.4047
120	6.5133
140	6.6141

Determine the sedimentation coefficient of this enzyme and, from this, its molecuar weight, assuming that $f = 8.24 \times 10^{-11}$ kg sec^{-1} and taking the factor $1 - \rho/\rho_2$ to be 0.256.

10. The sedimentation and diffusion coefficients for three different preparations of poly(methyl methacrylate) were measured[†] in n-butyl chloride at $35.6°C$ ($= \Theta$) and in acetone at $20°C$ ($> \Theta$) and the following results were obtained:

Solvent	Preparation	$s°$ (S)	$D° \times 10^7$ (cm^2 sec^{-1})	M
n-Butyl chloride	1	15.7	7.18	1.97×10^5
	2	39.1	2.91	—
	3	86.0	1.15	—
Acetone	1	20.3	6.80	—
	2	46.1	2.20	—
	3	88.5	0.92	6.35×10^6

Use the molecular weights given to evaluate the factor $1 - \rho/\rho_2$ for each of the systems; then use these factors to evaluate M for the other fractions. Compare the molecular weights obtained in the two solvents.

11. Both preparative and analytical GPC were employed to analyze a standard (NBS 706) polystyrene sample. Fractions were collected from the preparative column, the solvent was evaporated away, and the weight of each polymer fraction was obtained. The molecular weights of each fraction were obtained using an analytical gel permeation chromatograph calibrated in terms of both $\overline{M}_n$ and $\overline{M}_w$. The following data were obtained[‡]:

[†] H. Lütje and G. Meyerhoff, *Makromol. Chem.* 68:180 (1963).
[‡] Y. Kato, T. Kametani, K. Furukawa, and T. Hashimoto, *J. Polym. Sci. Polym. Phys. Ed.* 13:1695 (1975).

Fraction number	Mass polymer (mg)	$M_n \times 10^{-4}$ (g mol^{-1})	$M_w \times 10^{-4}$ (g mol^{-1})
6	2	109	111
7	8	90.8	92.5
8	20	76.7	78.0
9	42	62.3	63.5
10	64	51.5	52.5
11	84	41.7	42.5
12	102	34.7	35.4
13	110	28.7	29.3
14	110	23.3	23.8
15	96	18.9	19.4
16	86	15.9	16.3
17	68	13.0	13.3
18	54	11.0	11.2
19	42	9.14	9.35
20	30	7.52	7.68
21	28	6.16	6.28
22	18	5.12	5.22
23	12	4.09	4.18
24	8	3.33	3.40
25	6	2.63	2.69
26	5	2.01	2.06
27	4	1.52	1.56
28	3	1.13	1.16
29	2	0.83	0.85
30	1	0.59	0.61

Calculate $\bar{M}_n$ and $\bar{M}_w$ and the ratio $\bar{M}_w/\bar{M}_n$ for the original polymer. Also evaluate the ratio $\bar{M}_w/\bar{M}_n$ for the individual fractions. Comment on the significance of $\bar{M}_w/\bar{M}_n$ for both the fractionated and unfractionated polymer.

12. Use the model for the size exclusion of a spherical solute molecule in a cylindrical capillary to calculate K_{GPC} for a selection of R/a values which are compatible with Fig. 9.16. Plot your values on a photocopy or tracing of Fig. 9.16. On the basis of the comparison between these calculated points and the line in Fig. 9.16 drawn on the basis of a statistical consideration of chain exclusion, criticize or defend the following proposition: There is not much difference between the K values calculated by the equivalent sphere and statistical models. The discrepancy between various experimental methods for evaluating $\bar{a}$ is much greater than the differences arising from different models. Even for random coil molecules the simple equivalent sphere model is acceptable for qualitative discussions of V_R.

13. Three polystyrene samples of narrow molecular weight distribution were investigated† for their retention in GPC columns in which the average particle size of the packing was varied. In all instances the peaks were well resolved. The following results were obtained:

Mean dimension of particles in packing (μm)		Molecular weight of polystyrene		
		411,000	51,000	2,030
120	t_R (min)	160	200	230
	V_R (ml)	160	200	230
44	t_R (min)	9	12	15
	V_R (ml)	4.5	6.0	7.5
6	t_R (sec)	20	2.7	32
	V_R (ml)	0.33	0.43	0.53

It is also known that smaller particles in the packing material are more difficult to pack homogeneously, and that higher pressures are required to pump liquid through them. On the basis of these observations, criticize or defend the following proposition: Between the largest and smallest particle sizes in these packing materials there is a 7.5-fold decrease in retention time. This increase in rate with undiminished resolution shows that still smaller packing particles should be developed, perhaps using something like emulsion polymerization to prepare small, cross-linked polymer gels.

Bibliography

1. Flory, P. J., *Principles of Polymer Chemistry*, Cornell University Press, Ithaca, N.Y., 1953.
2. Hiemenz, P. C., *Principles of Colloid and Surface Chemistry*, Marcel Dekker, New York, 1977.
3. Lauffer, M. A., *J. Chem. Educ.* 58:250 (1981).
4. Richards, E. G., *An Introduction to the Physical Properties of Large Molecules in Solution*, Cambridge University Press, Cambridge, 1980.
5. Tanford, C., *Physical Chemistry of Macromolecules*, Wiley, New York, 1961.
6. Yau, W. W., Kirkland, J. J., and Bly, D. D., *Modern Size Exclusion Liquid Chromatography*, Wiley, New York, 1979.

---

† E. P. Otocka, *Acc. Chem. Res.* 6:348 (1973).

10
Light Scattering by Polymer Solutions

Nature, and Nature's Laws lay hid in Night.
God said, *Let Newton be!* and All was Light.

Epitaph for Isaac Newton, Alexander Pope

10.1 Introduction

This chapter is the narrowest in scope of any chapter in this book. In it we discuss a single experimental procedure and its interpretation. It is appropriate to examine light scattering in considerable detail, since the theory underlying this method is relatively unfamiliar to students and the interpretation yields information concerning a variety of polymer parameters.

There are really only two major conclusions presented in the chapter, and even these can be consolidated into a single analysis when applied to experimental data. First, we shall develop the Rayleigh theory for the scattering of light by molecules whose linear dimensions are small compared to the wavelength of the light. For visible light Rayleigh scattering applies to gases and low molecular weight liquids, and we discuss these applications as part of the process for gaining understanding of this powerful technique. Next we derive the Debye theory for scattering by particles whose dimensions are no longer insignificant compared to the wavelength of light. This theory corrects Rayleigh scattering for interference effects and therefore includes the assumptions and limitations of Rayleigh scattering, plus some added features of its own.

Although we take a while before eventually casting these theories in forms which are directly applicable to polymers, the final results are highly practical. Throughout the chapter the presentation is aimed toward these eventual applications. We begin by comparing and contrasting the turbidity of solutions which scatter light with the absorbance of solutions which absorb light. We describe the experiments whereby scattering data are collected, and discuss the extrapolation procedures that must be followed to match experimental results with

theoretical models. Although we develop the various stages stepwise, we conclude by describing the Zimm method for combining all extrapolations in a single graphical method. Through these manipulations of light-scattering data, absolute values of the molecular weight, the second virial coefficient, and the radius of gyration can all be determined. Thus a single procedure can be used to evaluate several different parameters which would otherwise entail more than one kind of experiment. In contrast with osmometry, light scattering is rapid and free from complications associated with finding a suitable membrane. In contrast with viscometry and GPC, light scattering is absolute and does not require prior calibration. In spite of these advantages, light scattering has some limitations of its own which we shall discover as the theory unfolds. Leaving their respective limitations aside, we see that light scattering and osmometry complement each other, since each gives a different kind of molecular weight average; therefore, taken together, they provide information about the width of the molecular weight distribution.

This chapter is the only place in this volume that we encounter electrical units. Certain equations in electrostatics differ by the factor 4π, depending on whether they are written for SI or cgs units. To help clarify this situation, the chapter contains an appendix on electrical units which may be helpful, particularly when references based on other units are consulted.

10.2 The Intensity of Scattered Light and Turbidity

Chemistry students are familiar with spectrophotometry, the qualitative and quantitative uses of which are widespread in contemporary chemistry. The various features of absorption spectra are due to the absorption of radiation to promote a particle from one quantized energy state to another. The scattering phenomena we discuss in this chapter are of totally different origin: classical not quantum physics. However, because of the relatively greater familiarity of absorption spectra, a comparison between absorption and scattering is an appropriate place to begin our discussion.

We begin with a consideration of notation, defining I_0 as the intensity of light incident (subscript 0) upon a sample and I_t as the intensity of the light transmitted (subscript t) through a sample of thickness x. There are two different mechanisms that can account for the fact that $I_t < I_0$:

1. When the energy of the light matches the spacing of quantum states, some light is absorbed. In this case $I_0 - I_t = I_{abs}$.
2. When light interacts with the electrons in molecules in a nonquantized fashion, some of the incident light is redistributed in all directions, that is, scattered. As a result of this redistribution, $I_t < I_0$, with $I_0 - I_t = I_{sca}$.
3. It items (1) and (2), the effects are considered separately, although they may both occur together. When absorption is the primary interest, it is

rarely necessary to consider the accompanying scattering, since the latter contributes far less than absorption to the attenuation of intensity. When scattering is the primary interest, it is generally investigated in a portion of the spectrum which is free from absorption peaks. We shall always assume the latter situation, although scattering theories for absorbing particles are also available.

Figure 10.1 schematically illustrates the relationship between I_0, I_t, and I_s.

In spectrophotometry the absorbance per unit length of path through the sample ϵ is defined as

$$\epsilon = - \ln\left(\frac{I_t}{I_0}\right) \tag{10.1}$$

In the absence of absorption, scattering alone is responsible for any attenuation; therefore

$$\epsilon = - \ln\left(\frac{I_0 - I_s}{I_0}\right) = - \ln\left(1 - \frac{I_s}{I_0}\right) \cong \frac{I_s}{I_0} \tag{10.2}$$

where the last approximation is justified since the scattering intensity is ordinarily quite small. It does not make much sense to call this quantity "absorbance" when no absorption is involved; the scattering equivalent is called turbidity and given the symbol τ. These ideas enable us to write $(I_t/I_0)_{abs} = \exp(- \epsilon x)$,

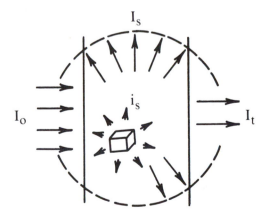

Figure 10.1 Relationships between I_0, I_t, and I_s. The light scattered per unit volume i_s is also shown.

$(I_t/I_0)_{sca} = \exp(-\tau x)$ and $(I_t/I_0)_{net} = \exp[-(\epsilon + \tau)x]$ for the situations itemized above. The turbidity of a specimen, then, is experimentally equivalent to
absorption, except it is a small $(\tau \ll \epsilon)$, classical effect which is generally ignored
in absorption studies. To prevent it from being overshadowed by absorption,
we consider nonabsorbing systems for which $\epsilon = 0$.

Like ϵ, τ is the product of two contributions: the concentration N/V of the
centers responsible for the effect and the contribution per particle to the
attenuation. It may help us to become oriented with the latter to think of
the scattering centers as opaque spheres of radius R. These project opaque cross
sections of area πR^2 in the light path. The actual cross section is then multiplied
by the scattering efficiency factor Q_{sca}, so that $\pi R^2 Q_{sca}$ gives the *optical* cross
section of the particle. The fact that the actual particle may not be opaque or
spherical is taken into account by Q_{sca}. Thus we can think of the turbidity
as the product of three factors:

$$\tau = (\pi R^2) Q_{sca} \left(\frac{N}{V} \right) \tag{10.3}$$

At this point it is instructive to examine the units of each of the terms in
Eq. (10.3):

1. τ has units of length^{-1}. By analogy with Eq. (10.1), it is the "absorbance"
    per unit path length.
2. πR^2 has the units of length^2, since it is an area. This is the case regardless
    of the geometry of the actual particle.
3. N/V has units of length^{-3}, since it is a concentration.

It is apparent from these considerations that Q_{sca} is dimensionless. It is also
clear that neither πR^2 nor N/V have anything to do with the wavelength λ of
the light used in the experiment. In spite of this, τ is wavelength dependent,
showing a broad, smooth variation with λ—as opposed to sharp peaks—for nonabsorbing particles. What this means is that the wavelength dependence of τ
enters Eq. (10.3) through Q_{sca}. Furthermore, since Q_{sca} is dimensionless, λ
must enter Q_{sca} in the form of a ratio, with some other variable having units of
length. A fairly obvious choice for the latter is R, since Q_{sca} is a property of
the scattering center.

An important aspect of the realization that Q_{sca} can be represented by
$f(R/\lambda)$ is the fact that the function has the same value for any particles with the
same R/λ ratio. Thus x rays interacting with atoms and microwaves interacting
with fog drops have about the same R/λ ratio as polymer molecules interacting
with visible light. As far as the size dependence of Q_{sca} is concerned, all of these
systems are described by the same value of Q_{sca}, provided that the wavelength
of the illumination is scaled to make the R/λ ratio the same in all cases.

The purpose of these qualitative remarks is to show that turbidity experiments are potential sources of information concerning both the spatial extension ($\propto R$) of the scatterers and their molecular weight (since $N/V \propto c_2/M$). We anticipated these conclusions earlier in this volume by noting elsewhere that light scattering provides absolute values for the radius of gyration and the weight average molecular weight of a polymer.

In developing these ideas quantitatively, we shall derive expressions for the light scattered by a volume element in the scattering medium. The symbol i_s is used to represent this quantity; its physical significance is also shown in Fig. 10.1. [Our problem with notation in this chapter is too many i's!] Before actually deriving this, let us examine the relationship between i_s and I_s or, more exactly, between i_s/I_0 and I_s/I_0.

In order to do this, we anticipate the form of the expression for i_s/I_0. Equation (10.31) will show that i_s/I_0 can be written as the product of two terms: an optical-molecular factor we symbolize as R_ϕ and a geometrical factor $1 + \cos^2 \phi_x/r^2$, where r is the distance from the scattering molecule and ϕ_x is the angle between the x axis and a specific line of sight. The unscattered—that is, incident and transmitted—light beam in Fig. 10.1 is assumed to travel in the x direction. Accordingly, the total scattered intensity I_s is equal to the summation *over all angles* of the scattering per unit volume, i_s. The factor R_ϕ does not affect this summation and can be factored out. For the present we are only concerned with the summation:

$$\frac{I_s}{I_0} = \sum_{\substack{all \\ angles}} \frac{i_s}{I_0} = \int_0^\pi \frac{i_s}{I_0} 2\pi r \sin \phi_x \, (r \, d\phi_x) \tag{10.4}$$

The justification for replacing the summation with this integral is seen by examining Fig. 10.2. An element of area on the surface of a sphere of radius r has a circumference $2\pi r \sin \phi_x$ and a thickness $r \, d\phi_x$. Integration of these increments of area gives the light scattered at all angles. As noted above, $i_s/I_0 = R_\phi (1 + \cos^2 \phi_x)/r^2$. Substituting this into Eq. (10.4) and canceling the r^2's gives

$$\frac{I_s}{I_0} = 2\pi R_\phi \int_0^\pi (1 + \cos^2 \phi_x) \sin \phi_x \, d\phi_x \tag{10.5}$$

This standard integral is readily evaluated to give the numerical factor 8/3; therefore

$$\tau = \frac{I_s}{I_0} = \frac{16\pi}{3} R_\phi \tag{10.6}$$

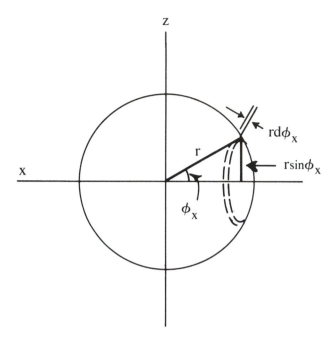

Figure 10.2   Definition of an element of area for the purpose of integrating $i_s(r, \phi_x)$ over all angles to evalute I_s. (Reprinted from Ref. 2, p. 178.)

In Sec. 10.5 we shall consider the derivation of i_s/I_0 and the factor R_ϕ which appears in Eq. (10.6). First, however, it is worthwhile to review some basic ideas about light itself.

10.3   Electric Fields and Their Interaction with Matter

The scattering of visible light by polymer solutions is our primary interest in this chapter. However, since Q_{sca} is a function of the ratio R/λ, as we saw in the last section, the phenomena we discuss are applicable to the entire range of the electromagnetic spectrum. Accordingly, a general review of the properties of this radiation and its interactions with matter is worthwhile before a specific consideration of scattering.

In this discussion we define the x direction to be the direction of propagation of the light waves. This means that the yz plane contains the oscillating electrical and magnetic fields which carry the energy of the radiation. Only the electric field concerns us in scattering. Since the oscillation is periodic in both time t and location x, the electric field can be represented by the equation

$$E = E_0 \cos \left[2\pi \left(\nu t + \frac{x}{\lambda} \right) \right] \tag{10.7}$$

in which the following applies:

1. ν is the frequency and at a fixed location—say, x = 0—one full wave is traced out in a time $1/\nu$.
2. λ is the wavelength and at a fixed time—say, t = 0—one full wave is traced out over a distance λ.
3. E_0 is the maximum amplitude of the field, since the cosine factor which modifies it oscillates between −1 and +1.
4. E oscillates in sign, as described by Eq. (10.7), yet the wave manifests itself with an intensity which is always positive. This suggests that E^2 rather than E itself be used as a measure of light intensity.
5. E causes a particle of charge q to experience a force and hence a displacement. Both the force and the displacement are proportional to E; therefore the energy of the field–charge interaction—the product of the force and the displacement—is proportional to E^2.
6. The combination of items (4) and (5) leads to the important conclusion that light intensity is the measure of the flux of energy through a surface perpendicular to the direction of propagation—the yz plane in our convention—and this is proportional to E^2.

Under vacuum, the velocity of propagation c of an electromagnetic wave is 3.0×10^8 m sec^{-1}, and this is related to the frequency and wavelength by

$$c = \nu \lambda_0 \tag{10.8}$$

In a medium of refractive index ñ, both c and λ_0 are decreased to $1/ñ$ of their value under vacuum: $v = c/ñ$ and $\lambda = \lambda_0/ñ$:

$$v = \nu \lambda \tag{10.9}$$

Note that c and λ_0 are used as symbols for velocity and wavelength, respectively, under vacuum, and that v and λ signify their counterparts in some medium. In addition, we observe that the frequency ν is not affected by the passage from one medium to another. As the light passes through a substance, its electric field interacts with the electrons of that substance, inducing oscillations in them at the same frequency as the frequency of the original field. Since no change in frequency is involved in going from vacuum into matter, it is the wavelength that must also adjust along with velocity. Figure 10.3 illustrates this by showing successive crests of electromagnetic waves traveling in the direction of the arrow as they pass from vacuum into a substance of refractive index ñ. At the surface

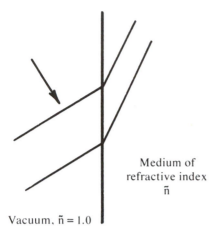

Medium of
refractive index
ñ

Vacuum, ñ = 1.0

Figure 10.3 Schematic illustration showing the bending of light and the decrease in wavelength as the radiation passes from a vacuum to a medium of refractive index ñ.

between the two media, there is a continuity in E with respect to frequency. That is, from whichever side of the surface it is viewed, the trace of E on the boundary is the same. To achieve this "fit," the wave front bends, as shown in Fig. 10.3, and the spacing between the wave crests, the wavelength, decreases. Both the bending of light and the change in wavelength are required by the continuity of E at the interface.

As our discussion of scattering proceeds, we shall examine the coupling between the oscillating electrical field of light and the electrons of the scatterer in detail. First, it is useful to consider the interaction of an electric field with matter, as this manifests itself in the dielectric behavior of a substance. This will not only introduce us to the field–matter interaction, but will also provide some relationships which will be useful later.

For this purpose we compare a parallel plate capacitor under vacuum and one containing a dielectric, as shown in Figs. 10.4a and b, respectively. The plates of the capacitor carry equal but opposite charges ±Q which can be described as ±σA, where σ is the surface charge density and A is the area of the plates. In this case, the field between the plates is given by

$$E_0 = \frac{\sigma}{\epsilon_0} \tag{10.10}$$

where ϵ_0 is the permittivity of vacuum, 8.85×10^{-12} C^2 J^{-1} m^{-1}. If the space between the plates is filled with a dielectric as shown in Fig. 10.4b, the field is decreased to a value

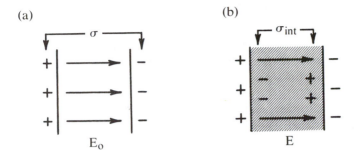

Figure 10.4 Parallel-plate capacitor with surface charge density σ. (a) The field is E_0 with no dielectric present. (b) The field is reduced to E by a dielectric which acquires a surface charge of its own, σ_{int}.

$$E = \frac{\sigma}{\epsilon} \tag{10.11}$$

where the ratio $\epsilon/\epsilon_0 = \epsilon_r$ is called the relative dielectric constant of the medium and is measured as the ratio of the capacitances of the apparatus with and without the substance. Sometimes the term *relative permittivity* and the symbol K_r are used instead of ϵ_r to describe this ratio.

Since $\epsilon > \epsilon_0$, we seek to explain the smaller field in the presence of the dielectric in terms of molecular properties and the way in which they are affected by the electric field. An easy way to visualize the effect is to picture an opposing surface charge—indicated as σ_{int} in Fig. 10.4b—accumulating on the dielectric. This partially offsets the charge on the capacitor plates to a net charge density $\sigma - \sigma_{int}$ so that E_0 becomes E and is given by

$$E = \frac{\sigma - \sigma_{int}}{\epsilon_0} \tag{10.12}$$

Next we can eliminate σ from Eqs. (10.11) and (10.12) to obtain

$$\sigma_{int} = \epsilon_0 (\epsilon_r - 1) E \tag{10.13}$$

Now let us examine the molecular origin of σ_{int}. Molecular polarity may be the result of either a permanent dipole moment μ or an induced dipole moment μ_{ind}, where the latter arises from the distortion of the charge distribution in a molecule due to an electric field. We saw in Chap. 8 that each of these types of polarity are sources of intermolecular attraction. In the present discussion we assume that no permanent dipoles are present and note that the induced dipole moment is proportional to the net field strength at the molecule:

$$\mu_{ind} = \alpha E_{net} \tag{10.14}$$

where α is called the polarizability of the molecule. It is apparent from this definition that the larger α is, the greater is the induced dipole moment per unit field. Thus α is aptly named, since it measures the ability of a molecule to undergo polarization.

Next we consider the net field at the molecule. This turns out to be the sum of two effects: the macroscopic field given by Eq. (10.12) plus a local field that is associated with the charge on the surface of the cavity surrounding the molecule of interest. The latter may be shown to equal $(1/3)(\sigma_{int}/\epsilon_0)$. Hence the net field at the molecule is

$$E_{net} = E + \frac{1}{3}\frac{\sigma_{int}}{\epsilon_0} = \frac{\sigma_{int}}{\epsilon_0(\epsilon_r - 1)} + \frac{1}{3}\frac{\sigma_{int}}{\epsilon_0} = \frac{\sigma_{int}}{3\epsilon_0}\left(\frac{\epsilon_r + 2}{\epsilon_r - 1}\right) \tag{10.15}$$

All that remains to be done is to connect σ_{int} with μ_{ind}. This is done by the observation that $\sigma_{int} = Q_{int}/A$, where Q_{int} is the charge on the surface of the dielectric. Next, Q_{int}/A can be written $(Q_{int}/V)d$, where V and d are the volume and the thickness of the dielectric, respectively. The product of a charge and the distance that separates it from its opposite charge defines a dipole moment. This means that σ_{int} has the significance of a net dipole moment per unit volume for the sample. If only induced dipole moments operate, as we have specified, this net dipole per unit volume is simply the product of the number of molecules per unit volume $\rho N_A/M$ and the induced dipole moment of each: $\sigma_{int} = (\rho N_A/M)\mu_{ind}$. Combining this idea with Eqs. (10.14) and (10.15) gives

$$\sigma_{int} = \frac{\rho N_A \alpha}{M}\left[\frac{\sigma_{int}}{3\epsilon_0}\left(\frac{\epsilon_r + 2}{\epsilon_r - 1}\right)\right] \tag{10.16}$$

or

$$\frac{1}{3}\frac{\rho N_A \alpha}{M\epsilon_0} = \frac{\epsilon_r - 1}{\epsilon_r + 2} \tag{10.17}$$

This result, called the Clausius–Mosotti equation, gives the relationship between the relative dielectric constant of a substance and its polarizability, and thus enables us to express the latter in terms of measurable quantities. The following additional comments will connect these ideas with the electric field associated with electromagnetic radiation:

1. If molecules have permanent dipole moments and can orient themselves with respect to the field, then Eq. (10.17) must be modified by inclusion of a term associated with μ.

2. In discussing capacitance, we implied nothing about the frequency of the field. The results given are general, applying equally to the frequencies of alternating current and of the electric field in electromagnetic radiation.
3. At the high frequencies of visible light, any permanent dipoles present cannot respond rapidly enough to contribute to the dielectric behavior; hence Eq. (10.17) applies to polar molecules also under the influence of light.
4. Under the same conditions, Maxwell's theory of radiation shows that the refractive index and the relative dielectric constant are simply related by

$$\epsilon_r = \tilde{n}^2 \tag{10.18}$$

Both ϵ_r and $\tilde{n}$ are clearly frequency dependent, since the foregoing argument shows that various effects contribute to the polarity of a molecule at different frequencies.

Equations (10.17) and (10.18) show that both the relative dielectric constant and the refractive index of a substance are measurable properties of matter that quantify the interaction between matter and electric fields of whatever origin. The polarizability is the molecular parameter which is pertinent to this interaction. We shall see in the next section that α also plays an important role in the theory of light scattering. The following example illustrates the use of Eq. (10.17) to evaluate α and considers one aspect of the applicability of this quantity to light scattering.

Example 10.1

The refractive index of CCl_4 at $20°C$ and 589 nm, the D line of the sodium spectrum, is 1.4607. At this temperature the density of this compound is 1.59 g cm^{-3}. Use this information to calculate α for CCl_4. Criticize or defend the following proposition: The prediction that $Q_{sca} = f(R/\lambda)$ may have been premature. The consideration of Eq. (10.3) which led to this conclusion could just as well predict $Q_{sca} = f(\alpha^{1/3}/\lambda)$.

Solution

According to Eq. (10.18), $\epsilon_r = \tilde{n}^2 = (1.4607)^2 = 2.1336$ at $\nu = 5.09 \times 10^{14}$ Hz (589 nm). The number of CCl_4 molecules at $20°C$ is given by

$$\frac{\rho N_A}{M} = \frac{(1.59 \text{ g cm}^{-3})(6.02 \times 10^{23} \text{ molecules mol}^{-1})}{153.8 \text{ g mol}^{-1}}$$

$$= 6.245 \times 10^{21} \text{ molecules cm}^{-3}$$

Therefore, by Eq. (10.17),

$$\alpha = \frac{3\epsilon_0 \, (\epsilon_r - 1)/(\epsilon_r + 2)}{\rho N_A / M} = \frac{3 \, (8.85 \times 10^{-12} \; J^{-2} \; C^{-2} \; m^{-1})(0.2742)}{6.245 \times 10^{21} \; molecules \; cm^{-3}}$$

$$= 1.166 \times 10^{-39} \; C^2 \; J^{-1} \; m^2 \; molecule^{-1}$$

This is the correct value for α in SI units, as can be verified by examining the units in Eq. (10.14): αE_{net} has units $(C^2 \; J^{-1} \; m^2)(V \; m^{-1}) = C \; m$, which is correct for μ_{ind}. In this instance cgs units are also informative; to effect the transformation we divide by $4\pi\epsilon_0$ as described in the appendix, Sec. 10.12:

$$\alpha_{cgs} = \frac{\alpha}{4\pi\epsilon_0} = \frac{1.166 \times 10^{-39} \; C^2 \; J^{-1} \; m^2 \; molecule^{-1}}{1.113 \times 10^{-10} \; C^2 \; J^{-1} \; m^{-1}}$$

$$= 1.05 \times 10^{-29} \; m^3 \; molecule^{-1} = 10.5 \; \mathring{A}^3 \; molecule^{-1}$$

Expressed in these units, α resembles a molecular volume; hence $\alpha^{1/3}$ has units of length and is thus a length which characterizes the interaction between a molecule and the field. In view of this, it may be true that $Q_{sca} = f(\alpha^{1/3}/\lambda)$, since all we can say about Q_{sca} is that it is dimensionless.

•

With this as background, we are finally in a good position to look at the scattering process itself.

10.4   Light Scattering by an Isolated Molecule

Ordinarily, light-scattering experiments are conducted with unpolarized light, but the following discussion is more easily visualized in terms of light which is linearly polarized. For light propagating in the x direction, the electric field lies in the yz plane and may be resolved into y and z components. Polarizing filters have different absorption coefficients in perpendicular directions, and hence absorb light in which the field oscillates in one direction and pass the perpendicular component. For convenience, we speak of vertically and horizontally polarized light when the oscillations are parallel to the z and y axes, respectively.

To quantify the interaction between a molecule and an incident ray of light, we imagine the molecule situated at the origin of a coordinate system as shown in Fig. 10.5 and consider its interaction with vertically polarized light traveling in the x direction. Since it is the mobile charge in the molecule that couples with the light, we represent the molecule as a quantity of charge q. As discussed in the last section, the oscillating field induces an oscillation in the charge q,

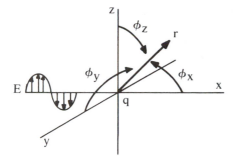

Figure 10.5 Definition of the variables used to describe the electric field produced by the oscillation of the charge q under the influence of vertically polarized light. (Reprinted from Ref. 2, p. 164.)

and it is a basic fact of electromagnetism that this situation is a source of radiation. On a molecular scale this molecule behaves like an antenna, sending out a signal along a line of sight represented by the arrow in Fig. 10.5. The radial distance from the radiating dipole is r and the angles between the line of sight and the coordinate axes are designated ϕ_x, ϕ_y, and ϕ_z, respectively.

Since the charge becomes coupled with the oscillating field, q undergoes a periodic acceleration which we represent by a_p. Next we borrow a relationship from electromagnetic theory to describe the field produced by an oscillating dipole such as the molecule we have described:

$$E = \frac{q a_p \sin \phi_z}{4\pi \epsilon_0 c^2 r} \tag{10.19}$$

Since we have introduced this relationship without proof, it will be helpful to consider its components with respect to their plausibility:

1. It is reasonable that the field produced by the accelerating charge should be proportional to the magnitude of both the charge and the acceleration.

2. The energy that is radiated spreads out over a solid angle and is therefore proportional to r^{-2}. We saw in Sec. 10.2 that this energy is proportional to E^2; hence E varies with r^{-1}.

3. The factor $4\pi\epsilon_0$ arises from the choice of SI units. Since a_p has units time^{-2}, the acceleration is divided by c^2 to convert the units of the denominator to length^2, as required by the definition of the field.

4. The $\sin \phi_z$ factor shows that the field produced by the oscillator is maximum in the xy plane, zero along the z axis, and symmetrical with respect to the z axis. This geometry is consistent with the vertical polarization of the field which is driving the dipole and producing the field described by Eq. (10.19).

Next we look for a substitution for a_p, the acceleration experienced by the charge. A convenient device for doing this originates from considering the oscillating dipole produced by the driving field. Since $\mu = \alpha E$,we can describe the periodic (subscript p) dipole moment of a molecule by

$$\mu_p = \alpha E_0 \cos(2\pi\nu t) \tag{10.20}$$

where we have used Eq. (10.7) with x = 0 to describe the incident field. Since the dipole moment also equals the charge times the distance of charge separation and it is the latter that is periodic, we identify the displaced charge q as αE_0 and $\xi = \cos(2\pi\nu t)$ as the separation. Therefore the acceleration is

$$a_p = \frac{d^2\xi}{dt^2} = -4\pi^2\nu^2 \cos(2\pi\nu t) \tag{10.21}$$

Substituting these results into Eq. (10.19) gives

$$E = \frac{(\alpha E_0)[-4\pi^2\nu^2 \cos(2\pi\nu t)] \sin\phi_z}{4\pi\epsilon_0 c^2 r} \tag{10.22}$$

for the field produced by the oscillating dipole.

As noted previously, the intensity of light is proportional to the square of the field; hence the intensity of the light radiated by the dipole is given by

$$i_v \propto \frac{\pi^2\nu^4\alpha^2 E_0^2 \cos^2(2\pi\nu t) \sin^2\phi_z}{\epsilon_0^2 c^4 r^2} \tag{10.23}$$

while the intensity of the driving radiation incident (subscript 0) on the molecule is given by squaring Eq. (10.7):

$$I_{0,v} \propto E_0^2 \cos^2(2\pi\nu t) \tag{10.24}$$

The subscript v is attached to both of these intensities as a reminder that the foregoing analysis is based on the assumption of vertical polarization for the incident light beam. The ratio of these intensities gives the fraction of light scattered per molecule by vertically polarized light:

$$\frac{i_v}{I_{0,v}} = \frac{\pi^2\nu^4\alpha^2 \sin^2\phi_z}{\epsilon_0^2 c^4 r^2} = \frac{\pi^2\alpha^2 \sin^2\phi_z}{\epsilon_0^2 \lambda_0^4 r^2} \tag{10.25}$$

Since the vertically polarized light postulated in the derivation of Eq. (10.25) involves both the z component of the electric field and the angle ϕ_z, it is

apparent that the corresponding expression for horizontally (subscript h) polarized light—which involves the y component of the field—is identical to Eq. (10.25), except for the $\sin^2 \phi$ term:

$$\frac{i_h}{I_{0,h}} = \frac{\pi^2 \alpha^2 \sin^2 \phi_y}{\epsilon_0^2 \lambda_0^4 r^2} \qquad (10.26)$$

Note that this also involves the assumption of isotropic molecules, which have the same polarizability in all directions. Unpolarized light consists of equal amounts of vertical and horizontal polarization, so the fraction of light scattered in the unpolarized (subscript u) case is given by

$$\frac{i_u}{I_{0,u}} = \frac{1/2\,(i_v + i_h)}{I_{0,u}} = \frac{1}{2} \frac{\pi^2 \alpha^2 \,(\sin^2 \phi_z + \sin^2 \phi_y)}{\epsilon_0^2 \lambda_0^4 r^2} \qquad (10.27)$$

It is awkward to use two different angles to describe the intensity of light scattered along a particular line of sight, but this situation is easily remedied by referring back to Fig. 10.5. It is apparent from Fig. 10.5 that $r \cos \phi$ is the projection of r along either the x, y, or z axis, depending on the choice of ϕ. We therefore see that

$$r^2 \,(\cos^2 \phi_x + \cos^2 \phi_y + \cos^2 \phi_z) = r^2 \qquad (10.28)$$

Replacing $\cos^2 \phi_y$ by $1 - \sin^2 \phi_y$ and $\cos^2 \phi_z$ by $1 - \sin^2 \phi_z$ leads to the relationship

$$\sin^2 \phi_y + \sin^2 \phi_z = 1 + \cos^2 \phi_x \qquad (10.29)$$

in terms of which Eq. (10.27) becomes

$$\frac{i_u}{I_{0,u}} = \frac{1}{2} \frac{\pi^2 \alpha^2}{\epsilon_0^2 \lambda_0^4 r^2} \,(1 + \cos^2 \phi_x) \qquad (10.30)$$

The $\sin^2 \phi$ terms in Eqs. (10.25) and (10.26) arise from the consideration of polarized light. The light scattered by polarized incident light is also polarized in the same direction, so the term $1 + \cos^2 \phi_x$ in Eq. (10.30) describes the overall polarization of the scattered light. Before we lose sight of the individual contributions to this, it will be helpful to consider this polarization somewhat further. This is done in the following example.

Example 10.2

Describe the angular dependence of the vertically and horizontally polarized light scattered by a molecule and their resultant by considering the intensity as a vector anchored at the origin whose length in various directions is given by the trigonometric terms in Eqs. (10.25), (10.26), and (10.30).

Solution

The intensity of the vertically polarized scattered light is proportional to $\sin^2 \phi_z$ which, in polar coordinates, is described by a figure 8-shaped curve centered at the origin and having maximum values of ±1 at $\phi_z = 90°$. Because ϕ_z is symmetrical with respect to the z axis, this component of scattered light is described in three dimensions by a doughnut-shaped surface—in which the hole has shrunk to a point—centered symmetrically in the xy plane.

The horizontal component is identical except for its orientation in space. In the horizontal case the "doughnut" lies in the xz plane.

Since the intensities are additive, we can make the following statements about their resultant:

1. Along the z axis, $\sin^2 \phi_z$ contributes nothing but $\sin^2 \phi_y = 1$, so the net scattered light is horizontally polarized.
2. Along the y axis, $\sin^2 \phi_y$ contributes nothing but $\sin^2 \phi_z = 1$, so the net scattered light is vertically polarized.
3. Along the x axis, both $\sin^2 \phi_y$ and $\sin^2 \phi_z$ equal unity. The light consists of equal amounts of horizontally and vertically polarized components, that is, it is unpolarized, and has twice the intensity observed in the perpendicular directions.

This situation is summarized by the term $1 + \cos^2 \phi_x$ in Eq. (10.30), which we now consider. The following table lists some values for this factor for various values of ϕ_x:

ϕ_x (°)	0	15	30	45	60	75	90
$1 + \cos^2 \phi_x$	2.00	1.93	1.75	1.50	1.25	1.07	1.00

These are plotted in Fig. 10.6, which shows the net intensity envelope in the xy plane as a solid line and represents the horizontally and vertically polarized contributions to the resultant by the broken lines. Since ϕ_x is symmetrical with respect to the x axis, the three-dimensional scattering pattern is generated by rotating the solid contour around the x axis.

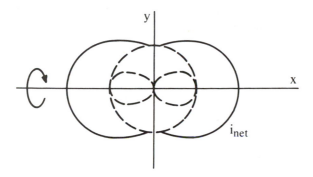

Figure 10.6 Two-dimensional representation of i_v and i_h (broken lines) and their resultant i_{total} (solid line) for scattering by a molecule situated at the origin and illuminated by unpolarized light along the x axis. The intensity in any direction is proportional to the length of the radius vector at that angle. (Reprinted from Ref. 2, p. 168.)

From now on we shall describe the scattered light by Eq. (10.30) exclusively, rather than considering the separate components. We shall also consider ϕ_x only in the xy plane, in which case we use the symbol θ to describe this angle. By convention, the incident light approaches the scattering dipole from $\theta = 180°$, and the transmitted light leaves the sample at $\theta = 0°$.

The equations we have developed in this section describe scattering originating from a point charge located at the origin. Since the charge is, in fact, the electron cloud of a molecule, we must consider the implications of treating a molecule as if it had no spatial extension. The way this consideration explicitly enters the relationships of this section is through the assumption that the dipole "sees" the same field throughout [see Eq. (10.20)]. In visible light, the wavelength is on the order of 500 nm. If the linear dimension of a molecule is small compared to this—as is the case for low molecular weight compounds—then the field is approximately uniform over the dimensions of the molecule and the assumption is valid. For polymer molecules, on the other hand, the dimensions of the molecule may not be insignificant compared to the wavelength of the radiation. In larger particles different parts of the molecule experience significantly different fields, oscillate independently, and produce light which interferes with itself. One way to circumvent this complication appears to be in working with radiation of longer wavelength, but the latter (e.g., infrared) can be absorbed in more ways, so this approach simply trades one source of difficulty for another. It turns out, however, that extrapolating i_s measured at different angles to $\theta = 0°$ also eliminates this interference effect. In Sec. 10.10 we shall discover how the additional complexity of interference can

be interpreted to yield additional information about the scattering source. Before considering this, however, let us further examine the implications of Eq. (10.30) for molecules which are small compared to the wavelength of light.

10.5   Rayleigh Scattering

The scattering formula we have derived in Eq. (10.30) applies to an isolated molecule whose dimensions are small compared to λ. A gas molecule satisfies this description, so we expect that this relationship applies to gases. Since the molecules are far apart in that case, each molecule behaves as an independent scattering center. Therefore the light scattered per unit volume of gas is given simply by the number of molecules per unit volume $\rho N_A/M$ times $i_u/I_{0,u}$. Using the symbol i_s/I_0 for the light scattered per unit volume (unpolarized light is assumed, and the subscript u is dropped), we obtain

$$\frac{i_s}{I_0} = \frac{\rho N_A}{M} \frac{1}{2} \frac{\pi^2 \alpha^2}{\epsilon_0^2 \lambda_0^4} \frac{1 + \cos^2 \theta}{r^2} \tag{10.31}$$

This is precisely the same quantity we discussed in Sec. 10.2 and illustrated in Fig. 10.1. As anticipated in that section, i_s/I_0 can be written as the product $R_\phi(1 + \cos^2 \theta)/r^2$. In the xy plane, R_ϕ becomes R_θ:

$$R_\theta = \frac{1}{2} \frac{\pi^2 \alpha^2 \rho N_A}{\epsilon_0^2 \lambda_0^4 M} \tag{10.32}$$

which is called the Rayleigh ratio and is given for a gas by Eq. (10.32). By combining Eqs. (10.6) and (10.32), we obtain

$$\tau = \frac{8 \pi^3 \alpha^2 \rho N_A}{3 M \epsilon_0^2 \lambda_0^4} \tag{10.33}$$

as the turbidity of a gas. Recall from Sec. 10.2 that the turbidity is the analog of absorbance, the light attenuation per unit path through a substance, in the absence of absorption. Equation (10.33) shows that this quantity does the following:

1. It increases with the concentration of scattering centers, which is equivalent to Beer's law.
2. It depends on the nature of the molecules through the factor α^2, where α measures the ability of the molecules to be polarized by an electric field.
3. It varies with the wavelength of the incident radiation as λ_0^{-4}.

The application of this last result to light scattered by the earth's atmosphere is especially interesting. The wavelengths of light at the red and blue ends of the

visible spectrum differ in λ_0 by about a factor of 2, which means that bluish light is scattered about 16 times as much as reddish light. If there were no atmosphere around the earth, the sky would look black, except along direct lines of sight toward the sun or other stars. Instead, we see blue sky overhead where the earth's atmosphere scatters light toward an earthbound observer. This situation is illustrated in Fig. 10.7. Figure 10.7 also shows why sunsets have a reddish hue. In that case the light we see is essentially transmitted light from which the blues have been more effectively removed by scattering. Several additional comments about the color of the sky are pertinent to this discussion:

1. We are explicitly excluding absorption effects: Light-absorbing pollutants modify this description.
2. Skylight is polarized to an extent that depends on the angle between I_0 and I_s in Figs. 10.6 and 10.7.
3. Water drops condensed in the atmosphere have much larger dimensions than gas molecules; hence they are subject to the interference phenomena mentioned at the end of the last section. This alters the color of the scattered light. Smoke and dust particles are also larger and may absorb as well.

The scattering relationships we have considered in this section were published by Lord Rayleigh in 1871 as an explanation for the color and polarization of skylight. Light scattering by small, nonabsorbing particles is known as Rayleigh scattering and is characterized by the λ^{-4} dependence on wavelength. The factor R_θ is known as the Rayleigh ratio of a substance; additional expressions for this quantity in terms of other variables will be developed below.

Before attempting further development of Eq. (10.33), it is useful to examine the dimensions of the various terms in that expression in SI units:

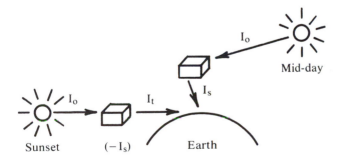

Figure 10.7 Schematic of light reaching an earthbound observer from different regions of the sky.

1. τ is expressed per length of path and has units of meter^{-1}.
2. $\rho N_A / M$ is a concentration and has units of meter^{-3}.
3. λ_0 is the wavelength, so λ_0^{-4} has units of meter^{-4}.
4. The equation is dimensionally consistent if α^2 / ϵ_0^2 has units of meter^6. Example 10.2 shows that $\alpha / 4\pi\epsilon_0$ has units of meter^3, so the relationship has the proper dimensions.

For the moment, let us define $\alpha / 4\pi\epsilon_0$ as α_V, where the subscript V reminds us that the polarizability is being expressed in volume units. Multiplying the numerator and denominator of Eq. (10.33) by $(4\pi)^2$ gives

$$\tau = \frac{128\pi^5}{3} \frac{\alpha_V^2}{\lambda_0^4} \frac{\rho N_A}{M} \tag{10.34}$$

If we think of α_V as a parameter which is proportional to the cube of some kind of molecular "radius" R, then $\alpha_V^2 \propto R^6$ and $\alpha_V^2 / \lambda_0^4 \propto R^2 (R/\lambda)^4$, which is the form predicted by Eq. (10.3).

The Clausius–Mosotti equation with $\bar{n}^2$ written for ϵ_r can be used to eliminate α from Eq. (10.33). For gases the refractive index is close to unity, so the factor $\bar{n}^2 - 1/\bar{n}^2 + 2$ is approximately

$$\frac{(\bar{n} + 1)(\bar{n} - 1)}{\bar{n}^2 + 2} \cong \frac{2}{3} (\bar{n} - 1)$$

Therefore Eq. (10.17) becomes

$$\alpha \cong \frac{2\epsilon_0 M}{\rho N_A} (\bar{n} - 1) \tag{10.35}$$

and Eq. (10.33) is given by

$$\tau = \frac{8\pi^3}{3\lambda_0^4 \epsilon_0^2} \left(\frac{2M\epsilon_0}{\rho N_A} (\bar{n} - 1) \right)^2 \frac{\rho N_A}{M} = \frac{32\pi^3}{3\lambda_0^4} (\bar{n} - 1)^2 \frac{M}{\rho N_A} \tag{10.36}$$

An important historic application of this relationship was the determination of Avogadro's number from measurements of light scattered by the atmosphere (see Problem 3).

Next let us consider the light scattered by liquids of low molecular weight compounds. We are actually not directly interested in this quantity per se, but in scattering by solutions—polymer solutions eventually, but for now solutions of small solute molecules. The solvent in such a solution does scatter, but, in practice, the intensity of light scattered by pure solvent is measured and subtracted as a "blank" correction from the scattering by the solution.

Therefore we do not need a theory for scattering by pure liquids to be able to deal with solutions experimentally. The theory for scattering by homogeneous liquids is somewhat simpler to visualize than that for solutions, and the same principles are involved for each. Accordingly, we shall develop the results for pure liquids up to a point and then apply the result to solutions by drawing the appropriate analogy.

The first thing to realize about scattering by liquids is that individual molecules can no longer be viewed as independent scatterers. If a liquid were perfectly uniform in density at the molecular level, its molecules could always be paired in such a way that the light scattered by each member of a pair would be exactly out of phase with the other, resulting in destructive interference. No net scattering results in this case. The second thing to realize, however, is that density is *not* perfectly uniform at the molecular level.

Molecules are in continuous random motion, and as a result of this, small volume elements within the liquid continuously experience compression or rarefaction such that the local density deviates from the macroscopic average value. If we represent by $\delta\rho$ the difference in density between one such domain and the average, then it is apparent that, averaged over all such fluctuations, $\overline{\delta\rho} = 0$: Equal contributions of positive and negative δ's occur. However, if we consider the average value of $\delta\rho^2$, this quantity has a nonzero value. Of these domains of density fluctuation, the following statements can be made:

1. The domain has slightly different properties than its surroundings and can be considered a scattering center itself.
2. Their random, fluctuating nature prevents these domains from destructively interfering with each other's scattered light.
3. The domain is small compared to the wavelength of visible light, so Eq. (10.33) describes the scattering, provided that we can find appropriate values for the concentration and polarizability of these domains.

In the next section we shall pursue the scattering by fluctuations in density. In the case of solutions of small molecules, it is the fluctuations in the solute concentration that plays the equivalent role, so we shall eventually replace $\delta\rho$ by δc_2. First, however, we must describe the polarizability of a density fluctuation and evaluate $\delta\rho$ itself.

10.6 Fluctuations and Rayleigh Scattering

We define the concentration of fluctuation domains at any instant by the symbol N*. In addition, we assume that the polarizability associated with one of these domains differs from the macroscopic average value for the substance

by $\delta\alpha$. As with $\delta\rho$, we expect $\overline{\delta\alpha}$ to equal zero, but $\overline{\delta\alpha^2}$ to have a nonzero value. Accordingly, we can write an expression which is equivalent to Eq. (10.33) for the fluctuations:

$$\tau = \frac{8\pi^3}{3\epsilon_0^2\lambda^4} \ \overline{\delta\alpha^2} \ N^* \tag{10.37}$$

Note that the wavelength in vacuum has been replaced by the wavelength in the medium, since it is the latter that drives the oscillations in the fluctuation domain. Now what can we say about $\overline{\delta\alpha^2}$?

The Clausius–Mosotti equation relates the polarizability of a substance to

$$\frac{\epsilon_r - 1}{\epsilon_r + 2} = \frac{\epsilon - \epsilon_0}{\epsilon + 2\epsilon_0}$$

where ϵ is the permittivity of the substance and ϵ_0 is the permittivity of the vacuum that surrounds it. Applied to a fluctuation, we replace ϵ by $\epsilon + \delta\epsilon$ and use ϵ itself rather than ϵ_0 to describe the surroundings. Therefore $(\epsilon - \epsilon_0)/(\epsilon + 2\epsilon_0)$ becomes $\delta\epsilon/(3\epsilon + \delta\epsilon)$, and if we assume that $\delta\epsilon$ is small compared to 3ϵ, Eq. (10.17) becomes

$$\delta\alpha = \frac{\epsilon_0}{N^*} \ \frac{\delta\epsilon}{\epsilon} \tag{10.38}$$

In addition, we can treat the δ's as differential quantities and write

$$\delta\epsilon = \frac{d\epsilon}{d\rho} \ \delta\rho \tag{10.39}$$

Since $\epsilon = \epsilon_r\epsilon_0$ and $\epsilon_r = \tilde{n}^2$ for optical frequencies, Eq. (10.39) becomes

$$\delta\epsilon = \epsilon_0 2\tilde{n} \ \frac{d\tilde{n}}{d\rho} \ \delta\rho \tag{10.40}$$

Combining Eqs. (10.38) and (10.40) gives

$$\overline{\delta\alpha^2} = \frac{\epsilon_0^2}{N^{*2}} \left(\frac{\epsilon_0 2\tilde{n} \ d\tilde{n}/d\rho}{\epsilon} \right)^2 \overline{\delta\rho^2} \tag{10.41}$$

where the method of averaging is based on the considerations presented above. Recognizing that $\lambda^{-4}(\epsilon_0/\epsilon)^2 = \lambda^{-4} \ n^{-4} = \lambda_0^{-4}$, Eq. (10.41) can be substituted into Eq. (10.37) to yield

$$\tau = \frac{8\pi^3}{3\lambda_0^4} \frac{1}{N^*} \left(2\tilde{n} \frac{d\tilde{n}}{d\rho}\right)^2 \overline{\delta\rho^2} \tag{10.42}$$

Next we consider how to evaluate the factor $\overline{\delta\rho^2}$. We recognize that there is a local variation in the Gibbs free energy associated with a fluctuation in density, and examine how this value of G can be related to the value at equilibrium, G_0. We shall use the subscript 0 to indicate the equilibrium value of free energy and other thermodynamic quantities. For small deviations from the equilibrium value, G can be expanded about G_0 in terms of a Taylor series:

$$G = G_0 + \left(\frac{\partial G}{\partial \rho}\right)_0 \delta\rho + \frac{1}{2!} \left(\frac{\partial^2 G}{\partial \rho^2}\right)_0 \delta\rho^2 + \cdots \tag{10.43}$$

The quantity $G - G_0 = \delta G$ is the change in G associated with the fluctuation, and the term $(\partial G/\partial \rho)_0 \, \delta\rho = 0$ because of the cancellation of positive and negative density fluctuations. Therefore we obtain

$$\delta G \cong \frac{1}{2} \left(\frac{\partial^2 G}{\partial \rho^2}\right)_0 \delta\rho^2 \tag{10.44}$$

Now we evaluate the probability of a fluctuation $\delta\rho$ in terms of a Boltzmann factor:

$$P(\delta\rho) = A \exp\left(-\frac{\delta G}{kT}\right) \tag{10.45}$$

where normalization requires that the constant A be given by

$$\left[\int \exp\left(-\frac{\delta G}{kT}\right) d\,\delta\rho\right]^{-1}$$

integrated over all fluctuations, that is, from $\delta\rho = 0$ to ∞. Using the definition of an average provided by Eq. (1.9), we write

$$\overline{\delta\rho^2} = \frac{\int_0^\infty \delta\rho^2 \, e^{-\delta G/kT} \, d\,\delta\rho}{\int_0^\infty e^{-\delta G/kT} \, d\delta\rho}$$

$$= \frac{\int_0^\infty \delta\rho^2 \exp\left[-(1/2)(\partial^2 G/\partial \rho^2)_0 \, \delta\rho^2/kT\right] d\delta\rho}{\int_0^\infty \exp\left[-(1/2)(\partial^2 G/\partial \rho^2)_0 \, \delta\rho^2/kT\right] d\delta\rho} \tag{10.46}$$

By letting $y = \delta\rho$ and $a = (\partial^2 G/\partial\rho^2)_0/2kT$, these rather formidable-looking integrals are recognized as gamma functions: $\int y^m e^{-ay^2} dy$. Using tabulated values for these integrals for $m = 0$ and $m = 2$, we obtain

$$\overline{\delta\rho^2} = \frac{kT}{(\partial^2 G/\partial\rho^2)_0} \tag{10.47}$$

Substitution of this result into Eq. (10.42) yields

$$\tau = \frac{32\pi^3}{3\lambda_0^4} \frac{1}{N^*} \left(\bar{n} \frac{d\bar{n}}{d\rho}\right)^2 \frac{kT}{(\partial^2 G/\partial\rho^2)_0} \tag{10.48}$$

As it stands, Eq. (10.48) is not an encouraging-looking result, but it is actually very close to a highly useful form.

Rather than continuing to discuss the scattering of pure liquids at the theoretical level, let us consider an example to illustrate the application of these ideas.

Example 10.3

By an assortment of thermodynamic manipulations, the quantities $d\bar{n}/d\rho$ and $[N^*(\partial^2 G/\partial\rho^2)_0]^{-1}$ can be eliminated from Eq. (10.48) and replaced by the measurable quantities α, β, and $d\bar{n}/dT$: the coefficients of thermal expansion, isothermal compressibility, and the temperature coefficient of refractive index, respectively. With these substitutions, Eq. (10.48) becomes

$$\tau = \frac{32\pi^3}{3\lambda_0^4} \frac{kT\beta}{\alpha^2} \bar{n} \frac{d\bar{n}}{dT}^2$$

For benzene†, $\alpha = 1.21 \times 10^{-3}$ deg^{-1}, $\beta = 9.5 \times 10^{-10}$ m^2 N^{-1}, and $d\bar{n}/dT = 6.38 \times 10^{-4}$ deg^{-1}. At 23°C, $\bar{n} = 1.503$ for benzene at $\lambda_0 = 546$ nm. Use these data to evaluate τ for benzene under these conditions.

Solution

Direct substitution into the equation given provides the required value for τ:

$$\tau = \frac{32\pi^3(1.38 \times 10^{-23} \text{ J K}^{-1})(296 \text{ K})(9.5 \times 10^{-10} \text{ m}^2 \text{ N}^{-1})}{3(546 \times 10^{-9} \text{ m})^4(1.21 \times 10^{-3} \text{ deg}^{-1})^2}$$

$$= 9.07 \times 10^{-3} \text{ m}^{-1} \qquad (\text{since J} = \text{N m})$$

●

---

† M. Kerker, *The Scattering of Light and Other Electromagnetic Radirion*, Academic, New York, 1969.

Experimentally the Rayleigh ratio for benzene at $90°$ has been observed to equal about 1.58×10^{-3} m^{-1} under the conditions described in this example. By Eq. (10.6), $\tau = (16\pi/3)\, R_\theta$ so the value of R_θ corresponding to this calculated turbidity is $R_{\theta,\text{calc}} = 5.41 \times 10^{-4}$ m^{-1}. The ratio between the observed value of R_θ and that calculated in the example is called the Cabannes factor and equals about 2.9 in this case.

The origin of this enhanced scattering is understood to originate from the anisotropy of the scattering centers. In terms of the analysis we have presented, this amounts to some scrambling of horizontally and vertically polarized components of light. This means that the light scattered at $90°$ is not totally polarized as suggested by Fig. 10.6. By using polarizing filters with the detection system, the intensities of the horizontally and vertically polarized components of the scattered light can be measured. The ratio of $i_{s,h}(\theta)$ to $i_{s,v}(\theta)$ is called the depolarization ratio and is given the symbol $\rho_u(\theta)$ (u unpolarized incident light; θ, angle at which measured). The Cabannes factor which corrects for this effect can be evaluated from the measured depolarization ratio. Applied to turbidity, the Cabannes factor C as a function of $\rho_u(90)$ is $C_\tau = [6 + 3\rho_u(90)] / [6 - 7\rho_u(90)]$ and applied to the Rayleigh ratio at $90°$, $C_{R\,90} = [6 + 6\rho_u(90)] / [6 - 7\rho_u(90)]$. Note that for the isotropic scatterers we have assumed, $\rho_u(90) = 0$, so the Cabannes factor equals unity. Corrections of this sort are also required for scattering by solutions. When the turbidity value calculated in Example 10.3 is multiplied by the appropriate Cabannes factor, the calculated and experimental results agree very well. We shall not pursue this correction any further, but shall continue to assume isotropic scatterers; additional details concerning the measurement and use of the Cabannes factor will be found in Ref. 3.

As noted at the end of the last section, it is fluctuations in concentration δc_2 rather than density which act as the scattering centers of interest for solutions of small molecules. There is nothing in the forgoing theory that prevents us from placing $\delta \rho$ by δc_2, the solute concentration in mass volume^{-1} units. Therefore we write for a solution of small molecules

$$\tau = \frac{32\pi^3}{3\lambda_0^4}\ \frac{1}{N^*}\ \left(\tilde{n}\ \frac{d\tilde{n}}{dc_2}\right)^2\ \frac{kT}{(\partial^2 G/\partial c_2^2)_0} \tag{10.49}$$

In the next section, we consider the application of Eq. (10.49) to scattering from fluctuations in concentration.

10.7   Light Scattering by Solutions

We saw in Example 10.3 that Eq. (10.48) for the turbidity of pure liquids could be converted to a usable expression by suitable thermodynamic manipulations. The corresponding relationship for solutions can also be transformed into the following useful form:

$$\tau = \frac{32\pi^3}{3\lambda_0^4} \, kT \left(\bar{n} \, \frac{d\bar{n}}{dc_2} \right)^2 \frac{c_2}{(\partial\Pi/\partial c_2)_0} \tag{10.50}$$

where $(\partial\Pi/\partial c_2)_0$ describes the concentration dependence of the equilibrium (subscript 0) osmotic pressure. The following outline summarizes the steps involved in this transformation:

1. From the definition of the partial molar quantities [Eq. (8.8)] we write $\delta G = \mu_1 \delta n_1 + \mu_2 \delta n_2$ and $\delta V = \bar{V}_1 \delta n_1 + \bar{V}_2 \delta n_2$, where the δ's refer to an individual fluctuation. The changes in concentration in a fluctuation arise from changes in the number of moles of solvent (subscript 1) and solute (subscript 2), not because of volume changes. Hence $\delta V = 0$; therefore $\delta n_1 = -(\bar{V}_2/\bar{V}_1)\delta n_2$ and $\delta G = [\mu_2 - (\bar{V}_2/\bar{V}_1)\mu_1]\delta n_2$.

2. Since c_2 is expressed as mass volume^{-1}, N* can be related to the δc_2 by the relationship $\delta c_2/M = N^*\delta n_2$. Therefore $\delta G = [\mu_2 - (\bar{V}_2/\bar{V}_1)\mu_1](\delta c_2/MN^*)$ or $\partial G/\partial c_2 = [\mu_2 - (\bar{V}_1/\bar{V}_2)\mu_1]/MN^*$.

3. Differentiating item (2) again with respect to c_2 gives

$$\frac{\partial^2 G}{\partial c_2^2} = \left(\frac{d\mu_2}{dc_2} - \frac{\bar{V}_2}{\bar{V}_1} \frac{d\mu_1}{dc_2} \right) \Big/ MN^*$$

The Gibbs–Duhem equation also follows from the definition of partial molar quantities: $n_1 d\mu_1 + n_2 d\mu_2 = 0$. With the Gibbs–Duhem equation, $\partial^2 G/\partial c_2^2$ becomes

$$-\left(\frac{n_1}{n_2} + \frac{\bar{V}_2}{\bar{V}_1} \right) \frac{(\partial\mu_1/\partial c_2)_0}{MN^*}$$

4. The factor $[N^*(\partial^2 G/\partial c_2^2)_0]^{-1}$ in Eq. (10.49) can therefore be written

$$\left[-\left(\frac{n_1}{n_2} + \frac{\bar{V}_2}{\bar{V}_1} \right) \frac{(\partial\mu_1/\partial c_2)_0}{M} \right]^{-1} \quad \text{or} \quad \left(-\frac{n_1\bar{V}_1 + n_2\bar{V}_2}{n_2 M} \frac{(\partial\mu_1/\partial c_2)_0}{\bar{V}_1} \right)^{-1}$$

Since $c_2 = n_2 M/(n_1\bar{V}_1 + n_2\bar{V}_2)$, this is more concisely written as

$$\left(-\frac{1}{c_2} \frac{(\partial\mu_1/\partial c_2)_0}{\bar{V}_1} \right)^{-1}$$

5. Equation (8.79) gives $\mu_1 = \mu_1{}^0 - \Pi\bar{V}_1$ at equilibrium, so $(\partial\mu_1/\partial c_2)_0 = -\bar{V}_1(\partial\Pi/\partial c_2)_0$, and the factor in item (4) becomes

$$\left[\frac{1}{c_2} \left(\frac{\partial\Pi}{\partial c_2} \right)_0 \right]^{-1}$$

Substituting this result into Eq. (10.49) gives Eq. (10.50).

Although Eq. (10.50) is still plagued by remnants of the Taylor series expansion about the equilibrium point in the form of the factor $(\partial\Pi/\partial c_2)_0$, we are now in a position to evaluate the latter quantity explicitly. Equation (8.87) gives an expression for the equilibrium osmotic pressure as a function of concentration: $\Pi = RT(c_2/M + Bc_2^2 + \cdots)$. Therefore

$$\left(\frac{\partial\Pi}{\partial c_2}\right)_0 = RT\left(\frac{1}{M} + 2Bc_2 + \cdots\right) \tag{10.51}$$

in terms of which Eq. (10.50) becomes

$$\tau = \frac{32\pi^3}{3\lambda_0^4 N_A} \frac{(\tilde{n}\, d\tilde{n}/dc_2)^2\, c_2}{(1/M + 2Bc_2)} \tag{10.52}$$

This is the result toward which we have been working. Representing the following cluster of constants by the symbol H, we have

$$H = \frac{32\pi^3\, (\tilde{n}\, d\tilde{n}/dc_2)^2}{3\lambda_0^4\, N_A} \tag{10.53}$$

Eq. (10.52) can be written

$$\frac{Hc_2}{\tau} = \frac{1}{M} + 2Bc_2 \tag{10.54}$$

This is the equation of a straight line and indicates that a plot of Hc_2/τ versus c_2 has the following properties:

$$\text{slope} = 2B \tag{10.55}$$

and

$$\text{intercept} = \frac{1}{M} \tag{10.56}$$

Thus we have finally established how light scattering can be used to measure the molecular weight of a solute. The concentration dependence of τ enters Eq. (10.54) through an expression for osmotic pressure, and this surprising connection deserves some additional comments:

1. In Chap. 8 we saw how the equilibrium osmotic pressure of a solution is related to ΔG for the mixing process whereby the solution is formed. Any difference in the concentration of the solution involves a change in ΔG_{mix}

and would be reflected by a change in Π. Since the occurrence of a fluctuation in concentration depends on the value of the associated δG, the fluctuation can also be expressed in terms of an equivalent $\delta \Pi$.

2. The value of B given by Eq. (10.55) has exactly the same significance that we discussed for the second virial coefficient in Chap. 8.

3. The limiting value of Hc_2/τ is proportional to $1/M$, which shows that τ/c_2 increases with increasing M, at least if interference effects are ignored. This is just the opposite of the molecular weight dependence of the colligative properties and makes light-scattering experiments ideally suited for polymeric solutes. We shall discuss in Sec. 10.10 the implications of interference effects for solute particles whose dimensions are comparable to λ.

4. For polydisperse systems the molecular weight obtained from light scattering is a weight average value, rather than the number average value obtained from an osmotic pressure experiment. This is an unexpected result in view of the role of Eq. (10.51) in relating τ to M.

It is easy to show that a light-scattering experiment "sees" a polydisperse system in such a way as to average the different molecular weights in terms of their mass rather than their number. We only need to consider the leading term of Eq. (10.54) to see the origin of this effect. For a polydisperse system we write $Hc_{ex}/\tau_{ex} = 1/M$, with $c_{ex} = \Sigma c_i$ and $\tau_{ex} = \Sigma \tau_i$, where the summations extend over all molecular weight categories. For any one molecular weight category it is also true that $Hc_i/\tau_i = 1/M_i$. Combining these various results gives

$$\overline{M} = \frac{\tau_{ex}}{Hc_{ex}} = \frac{\Sigma \tau_i}{H\Sigma c_i} = \frac{H\Sigma c_i M_i}{H\Sigma c_i} = \frac{\Sigma m_i M_i}{\Sigma m_i} \tag{10.57}$$

which is the weight average as defined by Eq. (1.12).

Since the development of a method for polymer characterization has been spread over several sections and since the literature contains several variations in the manner data is displayed, a summary of some pertinent definitions and relationships will be helpful at this point:

1. The relative intensity of light scattered per unit volume i_s/I_0 is a function of factors of three different origins: geometrical (r and θ), optical (λ, $\tilde{n}$, and $d\tilde{n}/dc_2$), and thermodynamic (c_2, M, and B).

2. The Rayleigh ratio combines the intensity factors with those associated with the geometry of the experiment:

$$R_\theta = \frac{i_s}{I_0} \frac{r^2}{1 + \cos^2 \theta} \tag{10.58}$$

Thus R_θ is a constant in any particular experiment where Rayleigh scattering is obtained, since the entire angular dependence of the light intensity is correctly contained in the $1 + \cos^2 \theta$ term.

3. As the attenuation of the incident beam per unit path through the solution, the turbidity is larger than the Rayleigh ratio by the factor $16\pi/3$, since τ is obtained by integrating R_θ over a spherical surface. Thus, if Eq. (10.54) is written in terms of R_θ rather than τ, the proportionality constant H must also be decreased by $16\pi/3$, in which case the constant is represented by the symbol K:

$$K = \frac{2\pi^2 \, (\tilde{n} \, d\tilde{n}/dc_2)^2}{\lambda_0^4 \, N_A} \tag{10.59}$$

and

$$\frac{Kc_2}{R_\theta} = \frac{1}{M} + 2Bc_2 \tag{10.60}$$

4. If turbidity itself is the quantity which is used to characterize a polymer solution, then Eqs. (10.53) and (10.54) are used as derived above:

$$H = \frac{32\pi^3 \, (\tilde{n} \, d\tilde{n}/dc_2)^2}{3\lambda_0^4 \, N_A} \tag{10.53}$$

and

$$\frac{Hc_2}{\tau} = \frac{1}{M} + 2Bc_2 \tag{10.54}$$

Figure 10.8 shows two sets of data plotted according to these conventions, after correction for the effect of interference. In Fig. 10.8a, Hc_2/τ is plotted against c_2 for three different fractions of polystyrene in methyl ethyl ketone. Figure 10.8b shows Kc_2/R_θ versus c_2 for solutions of polystyrene in cyclohexane at five different temperatures. These results are discussed further in the following example.

Example 10.4

In both parts of Fig. 10.8, c_2 is expressed in grams per cubic centimeter, with H (or K) and τ (or R_θ) in cgs units also. Verify that the units of the intercept are appropriate units for M^{-1}. Evaluate $\overline{M}_w$ and B for the data in Fig. 10.8a, and $\overline{M}_w$ and Θ from the data in Fig. 10.8b. (K is independent of temperature over the range of T's studied in Fig. 10.8b.)

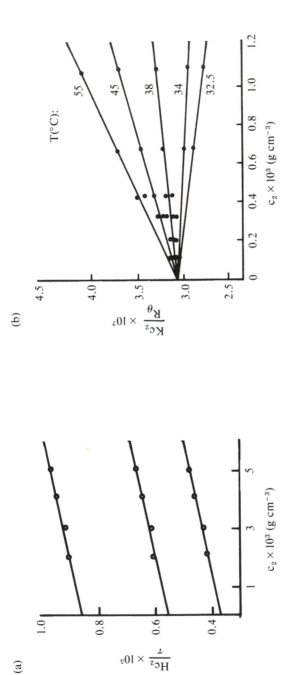

Figure 10.8 Light-scattering data plotted to give slope–intercept values which can be interpreted in terms of M and B. (a) Polystyrene in methyl ethyl ketone. [From B. A. Brice, M. Halwer, and R. Speiser, *J. Opt. Soc. Am.* 40:768 (1950), used with permission.] (b) Polystyrene in cyclohexane at temperatures indicated. Units of ordinates are given in Example 10.4. [Reprinted with permission from W. R. Krigbaum and D. K. Carpenter, *J. Phys. Chem.* 59:1166 (1955), copyright 1955 by the American Chemical Society.]

Solution

Since c_2 has units of grams per cubic centimeter, $\tilde{n}\ d\tilde{n}/dc_2$ has units of cubic centimeters per gram, because $\tilde{n}$ is dimensionless. This means that H (and K) has the cgs units centimeter² mole gram⁻². τ (and R_θ) has units of centimeter⁻¹, hence Hc_2/τ (and Kc_2/R_θ) has the units moles per gram, which are appropriate for the interpretation of this quantity given by Eq. (10.56). The three lines in Fig. 10.8a have the following intercepts, and the corresponding molecular weights are simply the reciprocals of these values:

$(Hc_2/\tau)_{c=0} \times 10^5$ (mol g⁻¹)	0.862	0.556	0.370
$\bar{M}_w$ (g mol⁻¹)	116,000	180,000	270,000

Each fraction in Fig. 10.8a has the same slope, and therefore the same value of B is expected. The slope of the lines is approximately (13 × 10⁻⁷ mol g⁻¹)/ (6 × 10⁻³ g cm⁻³) = 2.2 × 10⁻⁴ cm³ g⁻² mol and B = 1.1 × 10⁻⁴ cm³ g⁻² mol. These are also the correct cgs units for B, as seen by comparison with, say, Eq. (8.97). In Fig. 10.8b the lines for the different temperatures meet at a common intercept of about 3.1 × 10⁻⁷ mol g⁻¹, which means that $\bar{M}_w$ = 3.2 × 10⁶ g mol⁻¹ for the polystyrene sample under investigation. The slopes and B values for the various temperatures are tabulated below:

T (°C)	55	45	38	34	32.5
T (K)	328.2	318.2	311.2	307.2	305.7
Slope × 10⁵ (cm³ g⁻² mol)	10.76	6.24	2.26	−1.68	−3.06
B × 10⁵ (cm³ g⁻² mol)	5.38	3.12	1.13	−0.84	−1.53

The Θ temperature is that value of T for which B = 0; therefore Θ can be determined from the results above by graphical interpolation. Although there is some scatter in such a graph, the best value for the temperature at which B = 0 appears to be 308.4 K, which agrees well with values determined for this system by other methods.

Figure 10.8 and Example 10.4 show that light-scattering experiments can be interpreted to yield much the same information as obtained from an osmotic pressure experiment. The fact that the latter yields $\bar{M}_n$ while light scattering gives $\bar{M}_w$ prevents the two methods from being redundant, however. When we discuss interference phenomena in Sec. 10.10, we shall see that the radius of gyration can also be obtained from scattering experiments. Now that the utility of light-scattering experiments in polymer chemistry is well established, let us consider the experimental aspects of this topic.

10.8   Experimental Aspects of Light Scattering

We initiated our discussion of light scattering at the beginning of this chapter by comparing turbidity with the (more familiar) absorbance of a solution. The comparison of these two quantities is also a useful place to begin a consideration of the experimental aspects of light scattering. The components of a spectrophotometer and a light-scattering photometer are largely identical, the primary difference being that absorbance is always measured at $\theta = 0°$, while it is advantageous to measure scattering at various different angles. There are several reasons for using the Rayleigh ratio evaluated at various angles in reporting scattering results:

1.   Since i_s is a small quantity, it is better to measure it directly and report it as R_θ rather than by difference, as in the case when turbidity is reported.
2.   If experimental values of R_θ are observed to be independent of θ, then Rayleigh scattering is established and Eq. (10.60) can be applied to the data with confidence.
3.   If the experimental values of R_θ vary with θ, then the effects of interference are demonstrated. We shall see later in this section that these effects can be eliminated by extrapolating R_θ values to $\theta = 0°$, at which limit Eq. (10.60) also applies.
4.   We shall see in subsequent sections that measuring R_θ as a function of θ can be used to evaluate the radius of gyration of the scattering molecules, thereby providing more information about the polymer in addition to M and B.

As a result of these considerations, the primary difference between a spectrophotometer and a light-scattering photometer is the fact that the photodetector is mounted on an arm which pivots at the sample so that intensity measurements can be made at various angles.

Figure 10.9a is a schematic top view of such a photometer, and Fig. 10.9b is a cutaway view of a commercially available instrument which operates on the principles we describe here. The apparatus shown is the Brice-Phoenix universal scattering photometer. Like spectrophotometers, these light-scattering photometers consist of a light source, a sample cell, and a detector, as well as filtering and collimating systems for both the incident and scattered light. The interior of the photometer is painted black and the transmitted beam is absorbed in a light trap to prevent stray light from reaching the detector.

In contrast to spectrophotometry, light-scattering experiments are generally conducted at constant wavelength. Mercury vapor lamps are the most widely used light sources, since the strong lines at 436 and 546 nm are readily isolated by filters to allow monochromatic illumination. Polarizing filters are also included for both the incident and scattered beams so that depolarization can

(a)

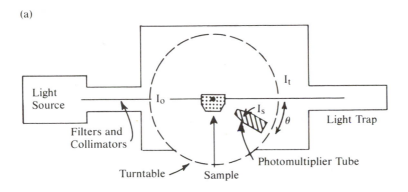

(b)

Figure 10.9 Light-scattering photometers. (a) Schematic top view showing movable photodetector. (Reprinted from Ref. 2, p. 176.) (b) Cutaway photograph of commercial light-scattering instrument, the Brice-Phoenix Universal Scattering Photometer. (Photo courtesy of the Virtis Co., Gardiner, New York.)

be studied. The usual assortment of lenses and slits assures that the beams are properly collimated, but these details need not concern us here.

The scattering of laser light is also of considerable importance in contemporary light-scattering practice. As a source of high-intensity monochromatic light, the laser source makes it possible to carry out light-scattering experiments on samples of greatly reduced volume. This means that laser light scattering can be used as a detection system in GPC and also aids in minimizing the scatter produced by dust particles, gas bubbles, and so on, which can invalidate measurements made on larger volumes of solution. Incidentally, as a detection system in GPC, light scattering offers sensitivity to both concentration changes and differences in molecular weight. Combined with refractive index measurements—to supply values for $d\bar{n}/dc_2$ —laser light-scattering detectors provide data which allow for calibration with respect to molecular weight, as well as measure the relative concentrations of the various components.

A variety of cell designs have been successfully employed in light-scattering experiments. Cylindrical cells offer symmetry with respect to viewing angle, but care must be exercised in their use because of reflection from cell walls. This difficulty is not encountered when intensity measurements are made normal to planar cell windows. Cells of octagonal cross section have planar viewing surfaces at θ = 0, 45, 90, 135, and 180°; these are especially convenient for light-scattering experiments.

The solutions must be carefully prepared so as to be free of dust particles and other extraneous scatterers. Filtration through sintered glass or centrifugation is widely used to clarify solutions of particles which would compete with polymeric solutes. This concern for cleanliness also extends to glassware, especially scattering cells. A fingerprint on the viewing window is disastrous!

Photomultipliers are used to measure the intensity of the scattered light. The output is compared to that of a second photocell located in the light trap which measures the intensity of the incident beam. In this way the ratio i_s/I_0 is measured directly with built-in compensation for any variations in the source. When filters are used for measuring depolarization, their effect on the sensitivity of the photomultiplier and its output must also be considered. Instrument calibration can be accomplished using well-characterized polymer solutions, dispersions of colloidal silica, or opalescent glass as standards.

The Rayleigh ratio is not the only optical measurement that must be made in order to interpret light-scattering experiments. In addition, the factor $\bar{n}\, d\bar{n}/dc_2$ must also be accurately measured. The refractive index itself is easily determined at the temperature and wavelength of the experiment and requires no further comment. The refractive index gradient $d\bar{n}/dc_2$, by contrast, presents more of a challenge. Although nominally the slope of a plot of the solution refractive index versus c_2, the gradient is not determined in this manner, since acceptable precision could not be achieved by differences: The solutions

involved are too dilute and the solvent and solute are often not too different in their respective refractive indices.

Instead, the difference between the refractive index of the solvent and that of a solution can be measured directly by one of several designs of differential refractometer. It is apparent that solute and solvent must differ in refractive index, otherwise the solute and its concentration fluctuations are effectively "invisible" in the solvent. The intrinsic difference in the ñ values of the constituents as well as the concentration of the solution, therefore, determine the small differences in refractive index that must be accurately measured. What is sought is an optical measurement that compares the light interacting with both the solvent and the solution to produce an effect which is sensitive to the refractive index difference. One instrument design for a differential refractometer passes light through a divided cell, the two halves of which contain solvent and solution, respectively. The amount by which the light beam deviates in passing through such a cell is measured by a position-sensitive photodetector. Alternatively, another design uses a split beam to reflect light off the surface of another divided cell, so that part of the beam is reflected by the solvent and part by the solution. The two reflected beams are compared by a dual-element photodetector, and the difference in intensity is related to the difference in ñ. By amplification of the small signals involved, each of these methods is capable of measuring down to 10^{-7} refractive index units under optimum conditions. The usefulness of $d\tilde{n}/dc_2$ as the basis for a detection system in liquid chromatography has contributed to the development of instrumentation in this area.

As noted at the beginning of this section, extrapolation of R_θ to $\theta = 0°$ is one way to correct light-scattering data for interference effects. Interference becomes troublesome for particles whose dimensions are larger than about $\lambda/20$, because light scattered from one portion of the molecule interferes with that scattered by another portion. This situation is shown schematically in Fig. 10.10, which shows the incident light in phase as it passes the surface AA', but shows different phase relationships in the scattered light at the (distant) surface BB'. It is typical of interference phenomena (think of the colors displayed by a soap film) to show different intensities depending on the angle of observation. Thus in Fig. 10.10 the light scattered at the smaller value of θ remains more nearly in phase than that scattered at the larger angle, where significant destructive interference occurs. This situation is generally true and constitutes the basis for extrapolating to $\theta = 0°$ to eliminate interference effects. The data in Fig. 10.8b were "corrected" in this way.

In the last sentence the word *corrected* is in quotation marks to emphasize the point that the observation *is* corrct: Rather, we are not in a position to deal with the information it tells us, namely, that these particles do not have negligible dimensions compared to λ. Until now we have lacked theory for dealing with this fact. It should be recognized that, by circumventing the

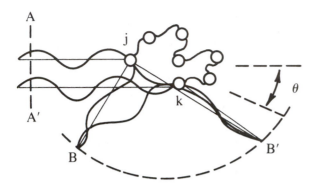

Figure 10.10 Interference of light rays scattered by segments j and k in a polymer chain. Destructive interference increases with increasing θ.

interference effect by extrapolation, we are wasting potentially usable information. Interference occurs in Fig. 10.10 because the light scattered at site j and that scattered at k travel different distances between AA′ and BB′. The wavelength of the light is the "yardstick" that measures this difference in distance traveled. Of course, the latter difference is somehow related to the dimensions of the scattering molecule taken as a whole. Therefore, to bypass this effect by extrapolation is to ignore light as a probe of molecular dimensions. In the next section we shall examine the theory of interference per se, and in Sec. 10.10 we apply the resulting theory to the interpretation of light-scattering data.

10.9 Optical Interference

The Rayleigh scattering theory which culminates in Eq. (10.60) as its most pertinent form for our purposes is based on the explicit assumption that interference effects are absent. The objective of the present section is to correct the Rayleigh theory to allow for interference effects. There are several assumptions-limitations that are implied by our approach:

1. We assume that the Rayleigh theory can be corrected by subdividing the actual solute particle into an array of scattering sites which, considered individually, obey the Rayleigh theory. It can be shown that this approach is a valid approximation so long as $(4\pi R/\lambda)(\tilde{n}_2/\tilde{n}_1 - 1) \ll 1$, where R is the radius of the overall molecule and $\tilde{n}_2$ and $\tilde{n}_1$ are the refractive indices of the solute and solvent, respectively. This shows that the validity of the model involves a trade-off: To apply to progressively larger molecules, the difference in refractive index between solute and solvent must decrease. This particular limitation is not especially severe in polymer applications

of light scattering, but can be a real problem in applications to still larger colloidal particles.

2. We assume that the observed interference is the cumulative effect of the contributions of the individual polymer molecules and that solute-solute interactions do not enter the picture. This effectively limits the model to dilute solutions. This restriction is not particularly troublesome, since our development of the Rayleigh theory also assumes dilute solutions.

3. We assume that there exists a function which we represent by $P(\theta)$—in recognition of the fact that it is angle dependent—which can be multiplied by the scattered intensity as predicted by the Rayleigh theory to give the correct value for i_s, even in the presence of interference. That is,

$$P(\theta) = \frac{i_{s,actual}}{i_{s,Rayleigh}} \tag{10.61}$$

4. Based on considerations we have encountered earlier in this chapter, we can anticipate two limiting cases of this function: $P(\theta)$ approaches unity both in the limit of small particles and in the limit of small angles of observation. Interference is absent in both of these cases.

Equation (10.61) suggests how Eq. (10.60), which is valid in the absence of interference, should be corrected. Using Eq. (10.58) for $R_{\theta,Rayleigh}$, Eq. (10.60) is corrected for interference by dividing both sides of the equation by $P(\theta)$:

$$\frac{Kc_2 (1 + \cos^2 \theta) I_0}{i_{s,Rayleigh} P(\theta) r^2} = \frac{1}{P(\theta)} \left(\frac{1}{M} + 2Bc_2 \right) \tag{10.62}$$

Since $P(\theta) i_{s,Rayleigh}$ gives the *observed* scattering intensity at θ—the value that is actually used in the evaluation of R_θ—Eq. (10.62) can be simply written as

$$\frac{Kc_2}{R_\theta} = \frac{1}{P(\theta)} \left(\frac{1}{M} + 2Bc_2 \right) \tag{10.63}$$

Our objective now becomes finding an expression for $P(\theta)$.

The theory for $P(\theta)$ proceeds through three stages:

1. We must describe the light scattered with interference in terms of phase differences that develop as the waves pass through a molecule consisting of multiple scattering sites.

2. We must find a way to describe these phase differences in terms of the distances traveled through the array of scattering sites, since this is how the size of the molecule enters the theory.

3. We must average the effect described in item (2) so that no assumed orientation of the scattering sites remains in the final result.

We shall take up these various steps in the following paragraphs.

Figure 10.10 shows that interference is the result of phase differences which arise as the incident light—initially in phase—is scattered by different sites in the molecule. In principle, any particle can be subdivided into a number of hypothetical sites whose isolated behavior is described by the Rayleigh theory. For polymers it is convenient to use the repeat unit of the polymer as the individual scattering site. Thus a polymer whose degree of polymerization is n consists of n independent scattering sites. We can describe each of these by an index number and describe the electric field scattered by the jth using Eq. (10.7), including a phase angle δ_j which is characteristic of the jth repeat unit:

$$E_j = E_0 \cos (2\pi\nu t + \delta_j) \tag{10.64}$$

The net field produced by n such sites is given by

$$E_{net} = \sum_{j=1}^{n} E_j = \sum_{j=1}^{n} E_0 \cos (2\pi\nu t + \delta_j) \tag{10.65}$$

A very useful way to simplify Eq. (10.65) involves the complex number e^{iy} in which $i = \sqrt{-1}$; e^{iy} equals cos y + i sin y. Therefore cos y is given by the real part of e^{iy}. Since exponential numbers are easy to manipulate, we can gain useful insight into the nature of the cosine term in Eq. (10.65) by working with this identity. Remembering that only the real part of the expression concerns us, we can write Eq. (10.65) as

$$E_{0,net} \, e^{i(2\pi\nu t + \delta_{net})} = \sum_{j=1}^{n} E_0 \, e^{i(2\pi\nu t + \delta_j)} \tag{10.66}$$

Dividing both sides by $e^{i2\pi\nu t}$ gives

$$E_{0,net} \, e^{i\delta_{net}} = \sum_{j=1}^{n} E_0 \, e^{i\delta_j} \tag{10.67}$$

It is the net intensity, not the electric field, which concerns us. We previously used the fact that intensity is proportional to E^2 to evaluate i_s. Using complex numbers to represent E requires one slight modification of this procedure. In the present case we must multiply E by its complex conjugate—obtained by replacing $\sqrt{-1}$ by $-\sqrt{-1}$—to evaluate intensity:

$$i_s \propto \left(\sum_{j=1}^{n} E_0 \, e^{i\delta_j} \right) \left(\sum_{k=1}^{n} E_0 \, e^{-i\delta_k} \right) \tag{10.68}$$

where we have introduced k as the index simply to distinguish between the two summations. Equation (10.68) can be written

$$i_s \propto E_0^2 \sum_{j=1}^{n} \sum_{k=1}^{n} e^{i(\delta_j - \delta_k)} \tag{10.69}$$

For every term $\delta_j - \delta_k$ in this double summation, there is a term $\delta_k - \delta_j$ which equals $-(\delta_j - \delta_k)$; therefore Eq. (10.69) is equivalent to

$$i_s \propto \frac{1}{2} E_0^2 \sum_{j=1}^{n} \sum_{k=1}^{n} \left(e^{i(\delta_j - \delta_k)} + e^{-i(\delta_j - \delta_k)} \right) \tag{10.70}$$

By using this form, we can take advantage of the identity that $\cos y = 1/2(e^{iy} + e^{-iy})$, from which it follows that

$$i_s \propto E_0^2 \sum_{j=1}^{n} \sum_{k=1}^{n} \cos (\delta_j - \delta_k) \tag{10.71}$$

For Rayleigh scattering, $\delta_j - \delta_k = 0$—there are no phase differences—and each of the cosine terms in Eq. (10.71) equals unity. In this case, which corresponds to $i_{s,\,Rayleigh}$, the right-hand side of Eq. (10.71) equals $E_0^2 n^2$, and we can write

$$\frac{i_s}{i_{s,\,Rayleigh}} = P(\theta) = \frac{1}{n^2} \sum_j \sum_k \cos (\delta_j - \delta_k) \tag{10.72}$$

This expression formalizes the anticipated conclusion that it is the *difference* in phase between light scattered by different segments that is responsible for the interference effect we seek to analyze. Equation (10.72) completes the first of the three stages in the development of $P(\theta)$.

In the next stage of the derivation we replace the difference in phase angles by the difference in the distance traveled by the light in reaching the observer via site j compared with site k. We shall develop this connection for the specific geometry of Fig. 10.11, but in the third stage of the derivation an average for all possible geometries is performed. Hence we need not worry about the specific model for which the second stage is obtained.

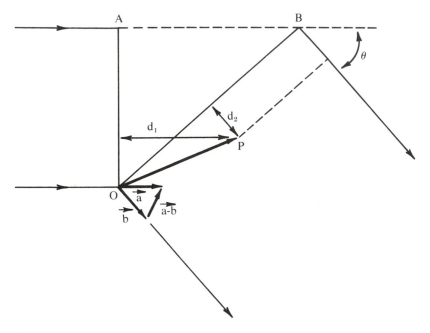

Figure 10.11   Definition of variables required to describe interference of light scattered from points O and P.

Equation (10.72) shows that it is the difference in phase between light scattered by two sites that determines the interference. Hence in Fig. 10.11 we construct two surfaces OA and OB which are perpendicular to the incident and scattered light, respectively. These lines might represent wave crests and, since they meet at O, any segment situated at O is defined to have a phase angle of zero. Suppose we consider a second segment situated at P. To reach a distant observer in the scattering direction, light scattered by a segment located at P must travel a distance d_1 farther in the incident direction compared to a scatterer located at O. In addition, light scattered by a segment at P would have to travel a distance d_2 *less* than light scattered from O to reach the observer. The difference in the distance traveled by light scattered from this pair of sites is therefore given by $d_1 - d_2$. These distances are illustrated in Fig. 10.11. Since there is no phase angle for a scattering segment located at O, the difference in phase angles between O and P can be written

$$\delta_O - \delta_P = \frac{2\pi (d_1 - d_2)}{\lambda} \qquad\qquad (10.73)$$

Next we look for a relationship between $d_1 - d_2$ and the geometry of the experiment.

The easiest way to proceed is to use vectors to describe this part of the problem. We represent the distance between the pair of scattering sites by the vector $\overrightarrow{OP}$ the length of which is simply r. To express d_1 and d_2 in terms of $\overrightarrow{OP}$ we construct the unit vectors $\vec{a}$ and $\vec{b}$ which are parallel to the incident and scattered directions, respectively. The projection of $\overrightarrow{OP}$ into direction $\vec{a}$, given by the dot product of these two vectors, equals d_1. Likewise, the projection of $\overrightarrow{OP}$ into direction $\vec{b}$ gives d_2. Therefore we can write

$$d_1 - d_2 = \overrightarrow{OP} \cdot \vec{a} - \overrightarrow{OP} \cdot \vec{b} = \overrightarrow{OP} \cdot (\vec{a} - \vec{b}) \tag{10.74}$$

The vector $\vec{a} - \vec{b}$ is also shown in Fig. 10.11 and may be described as the product of a unit vector $\vec{c}$ in the direction $\vec{a} - \vec{b}$ times the scalar length of $\vec{a} - \vec{b}$. This length is easily evaluated, since $\vec{a}$ and $\vec{b}$ are both of unit length and are separated by the angle θ. Therefore a perpendicular to $\vec{a} - \vec{b}$ through O bisects both the angle θ and the length of $\vec{a} - \vec{b}$. The length of $\vec{a} - \vec{b}$ is therefore $2 \sin(\theta/2)$ and we can describe $\vec{a} - \vec{b}$ as

$$\vec{a} - \vec{b} = 2 \sin\left(\frac{\theta}{2}\right) \vec{c} \tag{10.75}$$

Combining Eqs. (10.73)–(10.75) gives

$$\delta_O - \delta_P = \frac{4\pi}{\lambda} \sin\left(\frac{\theta}{2}\right) (\overrightarrow{OP} \cdot \vec{c}) = s(\overrightarrow{OP} \cdot \vec{c}) \tag{10.76}$$

where the scalar quantity s is defined by

$$s = \frac{4\pi}{\lambda} \sin\left(\frac{\theta}{2}\right) \tag{10.77}$$

The phase difference $\delta_O - \delta_P$ given by Eq. (10.76) describes a particular geometrical arrangement between scatterers. This difference can be substituted for the phase difference $\delta_j - \delta_k$ between an arbitrary pair of segments in Eq. (10.72), provided that a suitable averaging is carried out to allow for all possible orientations between segments j and k. We shall take up this averaging in the final stage of the derivation. For now we simply anticipate the average by using an overbar and substitute Eq. (10.76) for $\delta_j - \delta_k$ in Eq. (10.72):

$$P(\theta) = \frac{1}{n^2} \Sigma \Sigma \overline{\cos(s[\overrightarrow{r_{jk}} \cdot \vec{c}])} \tag{10.78}$$

In writing this last result, we have replaced $\overrightarrow{OP}$ by the vector $\overrightarrow{r_{jk}}$ between the generalized pair of segments j and k.

In the final stage of this involved derivation, we have to free Eq. (10.78) from the dependence it contains on the geometry of Fig. 10.11. The problem lies in the dot product of the vector $\overrightarrow{r_{jk}}$ –which replaces $\overrightarrow{OP}$ in Fig. 10.11 – and $\overrightarrow{c}$, the unit vector in the direction $\overrightarrow{a} - \overrightarrow{b}$ in Fig. 10.11. In Fig. 10.11 these two vectors have a specific orientation with respect to each other, but between an arbitrary pair of segments the vectors are separated by a general angle we call γ. The dot product $\overrightarrow{r_{jk}} \cdot \overrightarrow{c}$ then becomes $r_{jk} \cos \gamma$, where r_{jk} is the (scalar) distance between j and k, since the vector $\overrightarrow{c}$ has unit length. With this substitution, Eq. (10.78) becomes

$$P(\theta) = \frac{1}{n^2} \sum_j \sum_k \overline{\cos (sr_{jk} \cos \gamma)} \tag{10.79}$$

Now we consider how the averaging implied by the overbar is carried out. What this involves is multiplying $\cos (sr_{jk} \cos \gamma)$ by $P(\gamma) \, d\gamma$ –the probability that a particular angle is between γ and $\gamma + d\gamma$ –and then integrating the result over all values of γ in keeping with the customary definition of an average quantity.

We can describe the function $P(\gamma) \, d\gamma$ in terms of the same geometrical arrangement shown in Fig. 10.2, with γ replacing ϕ_x. That is, $P(\gamma) \, d\gamma$ is proportional to the area traced on the surface of a sphere by angles in the range γ to $\gamma + d\gamma$:

$$P(\gamma) \, d\gamma = A r \sin \gamma \, (r \, d\gamma) \tag{10.80}$$

where $A = (\int_0^\pi r^2 \sin \gamma \, d\gamma)^{-1}$ satisfies the normalization criterion. Using this as the expression for $P(\gamma) \, d\gamma$ gives

$$P(\theta) = \frac{1}{n^2} \frac{\int_0^\pi \Sigma_j \Sigma_k \cos (sr_{jk} \cos \gamma) \sin \gamma \, d\gamma}{\int_0^\pi \sin \gamma \, d\gamma} \tag{10.81}$$

This unattractive integral is readily solved by introducing a change of variable in the numerator. If we let $y = sr_{jk} \cos \gamma$, then $dy = -sr_{jk} \sin \gamma \, d\gamma$. The corresponding limits for y –after dividing the range of integration in half and multiplying the integral by 2 –are $y = sr_{jk}$ for $\gamma = 0$ and $y = 0$ for $\gamma = \pi/2$. With this change of variable, Eq. (10.81) becomes

$$P(\theta) = \frac{(-2/sr_{jk}) \int_{y=sr_{jk}}^{0} \Sigma_j \Sigma_k \cos y \, dy}{n^2 \, 2 \int_{\gamma=0}^{\pi/2} \sin \gamma \, d\gamma} = \frac{(2/sr_{jk}) \Sigma_j \Sigma_k \sin (sr_{jk})}{2}$$

$$= \Sigma_j \Sigma_k \frac{\sin (sr_{jk})}{sr_{jk}} \tag{10.82}$$

This is the result we have sought, although it needs a bit of additional manipulation to make its usefulness evident. The derivation we have followed in this section was developed by Debye in the context of x-ray scattering by the individual atoms of small molecules. Since $s \propto \lambda^{-1}$, this function again emphasizes the idea that R/λ ratios rather than absolute distances themselves are the pertinent quantities in the discussion of optical phenomena. We shall call this theory and its subsequent developments the Debye scattering theory. (Remember that Chap. 2 contains the Debye viscosity theory.) In the next section we examine how Eq. (10.82) can be converted into a practical form.

10.10   The Radius of Gyration

Equation (10.82) is a correct but unwieldy form of the Debye scattering theory. The result benefits considerably from some additional manipulation which converts it into a useful form. Toward this end we assume that the quantity sr_{jk} is not too large, in which case $\sin (sr_{jk})$ can be expanded as a power series. Retaining only the first two terms of the series, we obtain

$$P(\theta) = \frac{1}{n^2} \Sigma_j \Sigma_k \frac{\sin (sr_{jk})}{sr_{jk}} = \frac{1}{n^2} \Sigma_j \Sigma_k \frac{sr_{jk} - (sr_{jk})^3/3!}{sr_{jk}}$$

$$= \frac{1}{n^2} \Sigma_j \Sigma_k \left(1 - \frac{(sr_{jk})^2}{6} \right) \tag{10.83}$$

Since $\Sigma_j \Sigma_k 1 = n^2$, this becomes

$$P(\theta) = 1 - \frac{s^2}{6n^2} \Sigma_j \Sigma_k r_{jk}^2 \tag{10.84}$$

At this point it is useful to compare the result we have obtained with the expected behavior of $P(\theta)$ that we anticipated from the definition of this quantity:

1. The smaller the overall dimensions of a molecule, the smaller will be the r_{jk} value for any pair of sites in the molecule. As the values of r_{jk} approach zero, $P(\theta) \to 1$, as required.
2. Equation (10.77) shows that $s \propto \sin(\theta/2)$; therefore $\sin(\theta/2)$ and s approach zero as $\theta \to 0$. This means that $P(\theta) \to 1$ in this limit also, as required.
3. Since the product sr_{jk} appears in Eq. (10.84), there is a trade-off possibility between particle size and the angle of observation. That is, the Debye scattering theory applies with the same level of accuracy to larger molecules at smaller angles and to smaller molecules at larger angles.

Since Eq. (10.63) contains $1/P(\theta)$, since we have already assumed sr_{jk} to be small, and since $1/(1-y) \cong 1+y$, Eq. (10.84) can be rewritten as

$$\frac{1}{P(\theta)} = 1 + \frac{s^2}{6n^2} \sum_j \sum_k r_{jk}^2 \tag{10.85}$$

At this point we return to Chap. 1 to connect Eq. (10.85) with the radius of gyration. Although we have not encountered the form $\sum_j \sum_k r_{jk}^2$ explicitly before, a moment's reflection will convince us that it is identical to the bracketed quantity in Eq. (1.54):

$$\sum_{j=1}^{n} \sum_{k=1}^{n} r_{jk}^2 = \sum_{j=1}^{n} \left(\sum_{k=1}^{j} P(r)r^2 + \sum_{k=1}^{n-j} P(r)r^2 \right) \tag{10.86}$$

To clarify this identification we note the following:

1. The bracketed summations span the full range of k values from 1 to n and hence can be combined into a single summation.
2. The probability $P(r)$ times r^2 gives a particular value of r^2 which, in the context of Eq. (10.86), is equivalent to r_{jk}^2.
3. Therefore it follows from Eq. (1.54) that

$$\overline{r_g^2} = \frac{1}{2n^2} \sum_j \sum_k r_{jk}^2 \tag{10.87}$$

in the present notation.

Combining Eqs. (10.84) and (10.87) gives

$$\frac{1}{P(\theta)} = 1 + \frac{s^2}{3}\overline{r_g^2} = 1 + \frac{16\pi^2}{3\lambda^2}\overline{r_g^2}\sin^2\left(\frac{\theta}{2}\right) \tag{10.88}$$

Finally, we can incorporate this result into Eq. (10.63) to obtain

$$\left(\frac{Kc_2}{R_\theta}\right)_{c=0} = \frac{1}{M}\left[1 + \frac{16\pi^2}{3\lambda^2}\,\overline{r_g^2}\sin^2\left(\frac{\theta}{2}\right)\right] \tag{10.89}$$

where the concentration term on the right-hand side of Eq. (10.63) has been omitted, since the Debye scattering theory applies in the limit of $c_2 \to 0$. We see that there is a certain analogy in the way the Rayleigh and the Debye theories must be "corrected":

1. In applying the Rayleigh theory to large polymer molecules, we had to extrapolate results measured at different θ's to $\theta = 0$ to eliminate the interference effect.
2. In applying the Debye theory to concentrated solutions, we must extrapolate the results measured at different concentrations to $c_2 = 0$ to eliminate the effects of solute–solute interactions.
3. Experimentally, R_θ is measured at a series of different c_2's and θ's, which makes the extrapolations of Kc_2/R_θ at constant θ to $c_2 = 0$ and of Kc_2/R_θ at constant c_2 to $\theta = 0$ equally feasible. In the next section we shall examine a specific graphical technique which combines these two extrapolations in a single procedure.

Assuming that concentration effects have been eliminated by extrapolating Kc_2/R_θ to $c_2 = 0$ (subscript $c = 0$), we see that Eq. (10.89) is the equation of a straight line if $(Kc_2/R_\theta)_{c=0}$ is plotted against $\sin^2(\theta/2)$. The characteristic parameters of the line have the following significance:

$$\text{Slope} = \frac{16\pi^2}{3\lambda^2 M}\,\overline{r_g^2} \tag{10.90}$$

$$\text{Intercept} = \frac{1}{M} \tag{10.91}$$

or

$$\overline{r_g^2} = \frac{3\lambda^2}{16\pi^2}\left(\frac{\text{slope}}{\text{intercept}}\right)_{c=0} \tag{10.92}$$

Figure 10.12 shows data for cellulose nitrate in acetone measured at $\lambda_0 = 436$ nm, and plotted in the manner suggested in Eq. (10.89). The following example completes the analysis of these data.

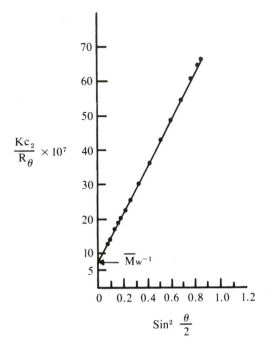

$\frac{Kc_2}{R_\theta} \times 10^7$

Figure 10.12 Light-scattering data in the limit of $c_2 = 0$ plotted according to Eq. (10.89) for cellulose nitrate in acetone. [Data from H. Benoit, A. M. Holtzer, and P. Doty, *J. Phys. Chem.* 58:635 (1954).]

Example 10.5

Interpret the slope and intercept values of the line in Fig. 10.12 in terms of the molecular weight and radius of gyration of cellulose nitrate in this solution. At 436 nm the refractive index of acetone is 1.359.

Solution

Examination of the graph shows that the line is characterized by the following parameters:

Slope $= 6.78 \times 10^{-6}$ mol g^{-1}

Intercept $= 7.87 \times 10^{-7}$ mol g^{-1}

The units of these quantities are determined solely by the ordinate, since the abscissa is dimensionless. Example 10.4 verifies that moles per gram are the units for Kc_2/R_θ. The molecular weight is given by the inverse of the intercept:

$$\bar{M}_w = 1.27 \times 10^6 \text{ g mol}^{-1}$$

Since the intercept corresponds to the Rayleigh limit also (i.e., $\theta = 0°$), Eq. (10.57) demonstrates this to be the weight average value of M.

In acetone $\lambda = \lambda_0/\bar{n} = 436/1.359 = 321$ nm; therefore

$$\overline{r_g^2} = \frac{3\lambda^2 M \text{ (slope)}_{c=0}}{16\pi^2}$$

$$= \frac{3 (321 \text{ nm})^2 (1.27 \times 10^6 \text{ g mol}^{-1})(6.78 \times 10^{-6} \text{ mol g}^{-1})}{16\pi^2}$$

$$\overline{r_g^2} = 1.69 \times 10^4 \text{ nm}^2 \qquad \text{or} \qquad (\overline{r_g^2})^{1/2} = r_{g,rms} = 130 \text{ nm}$$

This distance parameter which characterizes the polymer is about 40% of λ at $\lambda_0 = 436$ nm.

For a polydisperse system not only M but also the radius of gyration will be an average value. Let us next consider the type of average that is obtained for this quantity from light scattering. For this purpose it is sufficient to consider the kind of average obtained for $P(\theta)$: Eq. (10.84) shows that the weighting factor used to average $P(\theta)$ is also that for r_g^2. We proceed as in Sec. 10.7, where we considered the same question for M:

1. We begin with Eq. (10.89)—that is, we neglect solute–solute interactions— and write

$$\frac{Kc_2}{R_\theta} = \frac{1}{\bar{M}_w} \frac{1}{\bar{P}(\theta)} \qquad (10.93)$$

This result uses the already established fact that $M = \bar{M}_w$ when the molecular weight is determined by light scattering for a polydisperse system.

2. Equation (10.93) can be written for both the individual components and the mixture as a whole, yielding $Kc_i M_i P_i(\theta) = R_{\theta,i}$ and $Kc_{ex}\bar{M}_w\bar{P}(\theta) = R_{\theta,ex}$, respectively.

3. Since $c_{ex} = \Sigma_i c_i$ and $R_{\theta,ex} = \Sigma_i R_{\theta,i}$, these results can be combined to give

$$\bar{M}_w \bar{P}(\theta) = \frac{R_{\theta,ex}}{Kc_{ex}} = \frac{K\Sigma_i M_i c_i P_i(\theta)}{K\Sigma_i c_i} \qquad (10.94)$$

4. Because $c_i \propto m_i$, $\bar{M}_w = \Sigma_i c_i M_i / \Sigma_i c_i$; therefore $(\Sigma_i c_i M_i / \Sigma_i c_i)\bar{P}(\theta) = \Sigma_i c_i M_i P_i(\theta)/\Sigma_i c_i$ or

$$\bar{P}(\theta) = \frac{\Sigma_i c_i M_i P_i(\theta)}{\Sigma_i c_i M_i} = \frac{\Sigma_i m_i M_i P_i(\theta)}{\Sigma_i m_i M_i} \tag{10.95}$$

5. Finally, $m_i = n_i M_i$; therefore

$$\bar{P}(\theta) = \frac{\Sigma_i n_i M_i^2 P_i(\theta)}{\Sigma_i n_i M_i^2} \tag{10.96}$$

which shows that the weighting factor for each of the categories averaged is $m_i M_i = n_i M_i^2$. This kind of average is known as a z average. The similarly defined z-average molecular weight, $\bar{M}_z = \Sigma_i n_i M_i^3 / \Sigma_i n_i M_i^2$, is given as Eq. (1.19) in Chap. 1.

Although we presented the derivation for $P(\theta)$ in terms of a random coil, the result is applicable to particles of other geometries—for example, rigid spheres or ellipsoids—provided that the particles fall in the size range where the Debye theory is applicable. The radius of gyration thus obtained is an exact measure of this parameter for the particle in question, regardless of its shape, although its relationship to the physical dimensions of the scatterer does depend on the geometry of the particle. Relationships between the radius of gyration and the dimensions of bodies of various geometries are derived in elementary physics textbooks. Several of these are listed in Table 10.1 for some geometries that are encountered among polymeric solutes. Thus if a particle is known to possess some specific geometry, the radius of gyration can be translated into a geometrical particle dimension through these relationships. It should be emphasized, however, that this type of conversion merely helps us picture the molecule; r_g itself is an equally valid way to describe its dimensions.

In Example 10.5 we extracted both the molecular weight and the radius of gyration from light-scattering data. There may be circumstances, however, when nothing more than the dimensions of the molecule are sought. In this case a simple alternative to the analysis discussed above can be followed. This technique is called the dissymmetry method and involves measuring the ratio of intensities scattered at $45°$ and $135°$. The ratio of these intensities is called the dissymmetry ratio z:

$$z = \frac{i_{s,45°}}{i_{s,135°}} \tag{10.97}$$

This parameter should also be extrapolated to $c_2 = 0$, so the amount of experimental data required in this approach is not significantly less than in the method

Table 10.1 Relationships Between the Radius of Gyration and Geometrical Dimensions for Some Bodies Having Shapes Pertinent to Polymers

Geometry	Definition of parameters	Radius of gyration through the center of gravity
Random coil	$\overline{r^2}$: mean-square end-to-end distance ($\propto n$)	$r_g^2 = \dfrac{\overline{r^2}}{6}$
Sphere	R: radius of sphere	$r_g^2 = \dfrac{2}{5} R^2$
Thin rod	L: length of rod (approximation for prolate ellipsoids for which $a/b \gg 1$)	$r_g^2 = \dfrac{L^2}{12}$
Cylindrical disk	R: radius of disk (approximation for oblate ellipsoids for which $a/b \ll 1$)	$r_g^2 = \dfrac{1}{2} R^2$

described above. The advantage of working with this quantity, however, is that it can yield a particle dimension with very little calculation through the use of published tables and graphs. To see how this is possible, consider the following points:

1. The factor $1 + \cos^2 \theta$ in R_θ has the same value at $\theta = 45°$ and $135°$; hence the ratio $i_{s,45°}/i_{s,135°}$ is the same as $R_{45°}/R_{135°}$ [Eq. (10.58)].
2. Equation (10.63) shows that in the limit of $c_2 = 0$ this ratio also equals the ratio $P(45°)/P(135°)$.
3. By Eq. (10.88), this becomes

$$z = \frac{P(45°)}{P(135°)} = \frac{1 + (16\pi^2/3)(r_g/\lambda)^2 \sin^2 67.5}{1 + (16\pi^2/3)(r_g/\lambda)^2 \sin^2 22.5} \tag{10.98}$$

and a master curve can be drawn which shows z for different ratios r_g/λ. Of course, this result is subject to the limitations of Eq. (10.88). Especially pertinent is the idea that the particles should not be too large if Eq. (10.98) is used, since large angles are involved.

4. Taking the concept of a master curve a step further, the relationships in Table 10.1 can also be incorporated into such curves so that graphs of z versus a characteristic dimension relative to λ are plotted for various geometries.

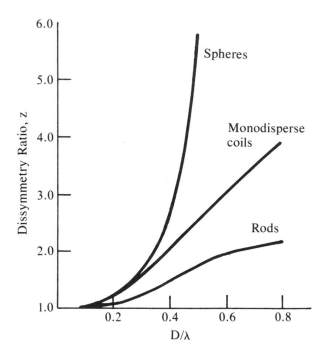

Figure 10.13   Variation of the dissymmetry ratio z with a characteristic dimension D (relative to λ) for spheres, random coils, and rods. (Data from Ref. 4.)

Figure 10.13 shows such plots of z versus D/λ, where D is r_{rms} for random coils,
R for spheres and disks, and L for rods. More detailed theories permit these
curves to be extended to larger values of r_g/λ than is justified by consideration
of Eq. (10.97) alone. In the following example we illustrate an application of
this simple method for estimating particle dimensions.

Example 10.6

Poly(γ-benzyl-L-glutamate) is known to possess a helical structure in certain
solvents. As part of an investigation† of this molecule, a fractionated sample
was examined in chloroform ($CHCl_3$) and chloroform saturated ($\sim$0.5%) with
dimethyl formamide (DMF). The following results were obtained:

---

† P. Doty, J. H. Bradbury, and A. M. Holtzer, *J. Am. Chem. Soc.* 78:947 (1956).

	$CHCl_3$	$CHCl_3$ + DMF
$\bar{M}_w$ (g mol^{-1})	144,000	73,000
$(z)_{c=0}$ at 436 nm	1.30	1.11

Taking ñ = 1.446 as the refractive index in both media, estimate the length of the helix in these two situations. Propose a possible interpretation of the results.

Solution

The helix can be approximated as a rod; therefore values of L/λ which are consistent with the observed dissymmetries can be read from Fig. 10.13 or equivalent sources. Also, $\lambda = \lambda_0/ñ = 436/1.446 = 302$ nm in each of these systems. In view of these considerations, the following results are obtained:

	$CHCl_3$	$CHCl_3$ + DMF
L/λ	0.31	0.19
L (nm)	94	57

The observed molecular weight suggests that this polymer associates into a "dimer" in $CHCl_3$, but that this aggregation is effectively blocked by small amounts of DMF. The particle lengths are not quite in the 2:1 ratio indicative of end-to-end association, but the increase in length is sufficiently large to make such a mechanism worthy of additional study.

•

Until now we have looked at various aspects of light scattering under several limiting conditions, specifically, $c_2 = 0$, $\theta = 0$, or both. Actual measurements, however, are made at finite values of both c_2 and θ. In the next section we shall consider a method of treating experimental data that consolidates all of the various extrapolations into one graphical procedure.

10.11 Zimm Plots

If we substitute Eq. (10.88) into Eq. (10.63) we obtain

$$\frac{Kc_2}{R_\theta} = \left(\frac{1}{M} + 2Bc_2\right)\left[1 + \frac{16\pi^2 \overline{r_g^2}}{3\lambda^2} \sin^2\left(\frac{\theta}{2}\right)\right] \qquad (10.99)$$

and, since R_θ is measured as a function of both c_2 and θ, this relationship brings together all of the experimental variables and molecular parameters that are

related through light scattering. In the development of this chapter we have looked at special cases of this general function:

1. In the limit of $c_2 = 0$ *and* $\theta = 0°$, $(Kc_2/R_\theta)_{\theta=c=0} = 1/\bar{M}_w$.

2. In the limit of $\theta = 0°$, $(Kc_2/R_\theta)_{\theta=0} = 1/\bar{M}_w + 2Bc_2$.

3. In the limit of $c_2 = 0$, $(Kc_2/R_\theta)_{c=0} = (1/\bar{M}_w) [1 + (16\pi^2/3) (r_g/\lambda)^2 \sin^2 (\theta/2)]$.

The assumptions made in deriving the two factors in Eq. (10.99) restrict the validity of the respective parts of the general expression to these limiting cases. A method for extrapolating experimental data to the limits itemized above has been devloped by Zimm, and the resulting graph—called a Zimm plot—has become a standard way of representing light-scattering data.

There is really nothing in this method that we have not already considered, one aspect at a time, in this chapter. The Zimm plot simply brings it all together in a single analysis.

We shall presently construct a Zimm plot in detail in an example. In anticipation of this, we label each of the paragraphs describing Zimm's procedure for ease of cross-referencing in the example.

1. The method consists of plotting Kc_2/R_θ as the ordinate and $\sin^2 (\theta/2) + kc_2$ as the abscissa, where k is a number which is chosen to give a good distribution of data points in the graph. We assign to this constant reciprocal concentration units so that kc_2 can be added to the dimensionless $\sin^2 (\theta/2)$. When a suitable scale has been selected, the experimental points are spread over a large area in the graph.

2. Next the points at constant values of c_2 are connected. Likewise, points at constant values of θ are also connected. This produces a grid of intersecting lines that may be straight but which are not necessarily so.

3. On each of the lines that has been drawn, a mark is made at the value of the abscissa corresponding to one of the limiting cases above. Specifically, along the line where $c_2 = c^*$, a mark is placed where the abscissa has the value kc^*. Since the abscissa is $\sin^2 (\theta/2) + kc_2$, this mark corresponds to $\theta = 0°$ for this concentration. Similarly, along a line for which $\theta = \theta^*$, a mark is placed where the abscissa equals $\sin^2 (\theta^*/2)$. This corresponds to the $c_2 = 0$ limit at this angle.

4. When each of the constant c_2 and constant θ lines has been marked off in this way, the various derived points can be connected. One of these groups of points gives values of Kc_2/R_θ for a range of c's at $\theta = 0°$. The other group of points gives Kc_2/R_θ for a range of θ's at $c_2 = 0$. The two lines thus derived should meet at a common intercept which equals $1/\bar{M}_w$, and their slopes are given by Eqs. (10.55) and (10.90), respectively.

Table 10.2 Values for Kc_2/R_θ at the Indicated Values of θ and c_2 for Solutions of Polystyrene in Benzene at 546 nm

$c_2 \times 10^3$	θ (deg)							
(g cm^{-3})	30	37.5	45	60	75	90	105	120
2.00	3.18	3.26	3.25	3.45	3.56	3.72	3.78	4.01
1.50	2.73	2.76	2.81	2.94	3.08	3.27	3.40	3.57
1.00	2.29	2.33	2.37	2.53	2.66	2.85	2.96	3.12
0.75	2.10	2.14	2.17	2.32	2.47	2.64	2.79	2.93
0.50	1.92	1.95	1.98	2.16	2.33	2.51	2.66	2.79

Source: Data from D. Margerison and G. C. East, *An Introduction to Polymer Chemistry*, Pergamon, Oxford, 1967.

For a sample of polystyrene in benzene, experimental values of Kc_2/R_θ are entered in the body of Table 10.2. The values are placed at the intersection of rows and columns labeled c_2 and θ, respectively. In the following example these values are used to construct a Zimm plot.

Example 10.7

Prepare a Zimm plot using the data in Table 10.2 and evaluate M, B, and $\overline{r_g^2}$ for this solution of polystyrene in benzene. The effective wavelength in the medium is $\lambda_0/\tilde{n} = 546/1.501 = 364$ nm for this experiment.

Solution

We follow the procedure outlined above, cross-referencing the individual steps with the labels introduced above.

1. The abscissa values for the Zimm plot are given by $\sin^2 (\theta/2) + kc_2$. For these data, $k = 100$ cm^3 g^{-1} gives a good array of points. Table 10.3 shows the values of the abscissa in the same format used in Table 10.2.
2. Figure 10.14a shows a plot of these data with the lines drawn between points of constant c_2. Likewise, Fig. 10.14b shows the same data with lines drawn between points of constant θ.
3. When $\theta = 0°$, $\sin^2 (\theta/2) = 0$ and the abscissa in Fig. 10.14a is simply kc_2. For the solution with $c_2 = 0.5 \times 10^{-3}$ g cm^{-3}, $kc_2 = 0.05$, and this point is marked by an x on the line through the points at this concentration. The x's on other lines in Fig. 10.14a correspond to the $\theta = 0°$ limit for each concentration. When $c_2 = 0$, the abscissa in Fig. 10.14b becomes $\sin^2 (\theta/2)$. For the measurements at $\theta = 30°$, $\sin^2 15 = 0.067$, and this

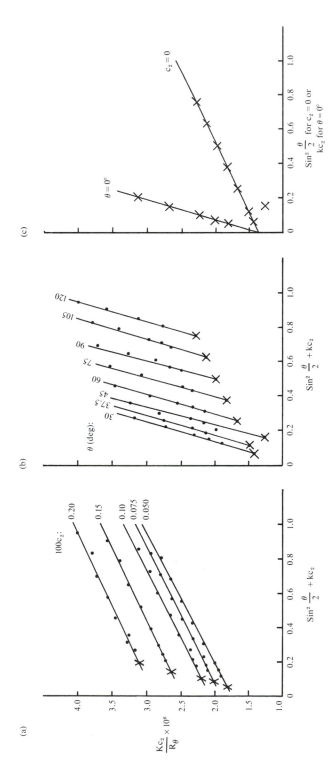

Figure 10.14 Construction of a Zimm plot from the data of Tables 10.2 and 10.3: (a) extrapolation to $\theta = 0°$, (b) extrapolation to $c_2 = 0$, and (c) derived lines.

Table 10.3 Values of $\sin^2(\theta/2) + 100\,c_2$ for the Data in Table 10.2 (used as Abscissa Coordinates in the Construction of Fig. 10.14)

$c_2 \times 10^3$	θ (deg)							
(g cm^{-3})	30	37.5	45	60	75	90	105	120
2.00	0.267	0.303	0.346	0.450	0.571	0.700	0.829	0.950
1.50	0.217	0.253	0.296	0.400	0.521	0.650	0.779	0.900
1.00	0.167	0.203	0.246	0.350	0.471	0.600	0.729	0.850
0.75	0.142	0.178	0.221	0.325	0.446	0.575	0.704	0.825
0.50	0.117	0.153	0.196	0.300	0.421	0.550	0.679	0.800

   point is marked by an x on the line through the points at 30°. The x's on
   the other lines in Fig. 10.14b correspond to the $c_2 = 0$ limit for each angle.
4.   Figure 10.14c shows the two limiting lines defined in Figs. 10.14a and b.
   The derived lines in Fig. 10.14c have the following properties: common
   intercept = 1.35×10^{-6} mol g^{-1}, slope $c_2 = 0$ line = 1.25×10^{-6} mol
   g^{-1}, slope $\theta = 0°$ line = 9.0×10^{-6} mol g^{-1} (as drawn), slope $\theta = 0°$ line =
   9.0×10^{-4} cm^3 mol g^{-2} (corrected for k). Using Eqs. (10.56) and (10.91),
   we have

$$\bar{M}_w = \text{intercept}^{-1} = 7.41 \times 10^5 \text{ g mol}^{-1}$$

Using Eq. (10.55), we have

$$B = \frac{(\text{slope})_{\theta=0}}{2} = \frac{9.0 \times 10^{-4}}{2} = 4.5 \times 10^{-4} \text{ cm}^3 \text{ mol g}^{-2}$$

Using Eq. (10.92), we have

$$\overline{r_g^2} = \frac{3\lambda^2}{16\pi^2}\left(\frac{\text{slope}}{\text{intercept}}\right)_{c=0} = \frac{3(364)^2}{16\pi^2}\frac{1.25 \times 10^{-6}}{1.35 \times 10^{-6}} = 2330 \text{ nm}^2$$

$$(\overline{r_g^2})^{1/2} = r_{g,\text{rms}} = 48.2 \text{ nm}$$

   The objective of the Zimm plot is to conduct all extrapolations on a single
graph. The three-stage development of Fig. 10.14 is not typical, but it is
intended to clarify the discussion.
   Not all Zimm plots show the same grid of essentially parallel straight lines
found in Fig. 10.14. In some cases there is considerable curvature, and quite
a bit of "interpretation" is required to extract the molecular parameters from
the data. In this connection we note that the reciprocal of Eq. (10.83)

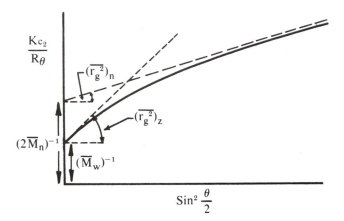

Figure 10.15 Schematic showing alternative limits in the plot of Kc_2/R_θ versus $\sin^2 (\theta/2)$ and their interpretation. [Reprinted with permission from H. Benoit, A. M. Holtzer, and P. Doty, *J. Phys. Chem.* 58:635 (1954), copyright 1954 by the American Chemical Society.]

approaches an asymptotic value for large values of sr_{jk}. Instead of Eq. (10.88), the reciprocal of $P(\theta)$ for *large* values of s, r_{jk}, or both (indicated by a prime) is given by

$$\frac{1}{P'(\theta)} = \frac{8\pi^2}{\lambda^2} \; (\overline{r_g^2})_n \sin^2 \left(\frac{\theta}{2}\right) \tag{10.100}$$

In addition, the intercept obtained by extrapolating this asymptote back to $\sin^2 (\theta/2) = 0$ equals $(2\overline{M}_n)^{-1}$. Note that both $\overline{M}$ and r_g^2 are *number* averages when this asymptotic limit is used. This is illustrated schematically in Fig. 10.15 and indicates that even more information pertaining to polymer characterization can be extracted from an analysis of the curvature in Zimm plots.

10.12 Appendix: Electrical Units

At one time or another, all of us have tangled with problems of units, but generally these decrease in severity and frequency with experience. Advanced students juggle kilograms and grams, centimeters and angstroms, joules and calories, and rarely fumble in the process. Electrical units are sometimes more troublesome.

In 1785 Coulomb showed that the force between two charges q_1 and q_2 separated by a distance r is proportional to $q_1 q_2/r^2$, a result we know as

Coulomb's law. This relationship poses no particular difficulties as a qualitative statement; the problem arises when we attempt to calculate something with it, since the proportionality constant depends on the choice of units. In the cgs system of units, the electrostatic unit of charge is defined to produce a force of 1 dyne when two such charges are separated by a distance of 1 cm. In the cgs system the proportionality factor in Coulomb's law is unity and is dimensionless. For charges under vacuum we write

$$F_{cgs} = \frac{q_1 q_2}{r^2} \tag{10.101}$$

By contrast, in SI units, the coulomb (C) is the unit of charge and is defined as an ampere second (A sec). To reconcile this with newtons and meters, the units of F and r, respectively, a proportionality constant that is numerically different from unity and which has definite units is required. For charges under vacuum we write

$$F_{SI} = \frac{1}{4\pi\epsilon_0} \frac{q_1 q_2}{r^2} \tag{10.102}$$

where ϵ_0, the permittivity of vacuum, is 8.854×10^{-12} C^2 N^{-1} m^{-2} (or C^2 J^{-1} m^{-1} or kg^{-1} m^{-3} sec^2) and $1/4\pi\epsilon_0$ is 8.988×10^9 N m^2 C^{-2}.

In a medium where the relative dielectric constant is ϵ_r, the force between fixed chages at a definite separation is decreased by the dimensionless factor ϵ_r. This is true regardless of the system of units and is incorporated into Eqs. (10.101) and (10.102) by dividing the right-hand side of each by ϵ_r.

So far, so good. The situation is really no different, say, than the ideal gas law, in which the gas constant is numerically different and has different units depending on the units chosen for p and V. The unit change in Example 10.1 is analogous to changing the gas constant from liter-atmospheres to calories; it is apparent that one system is physically more meaningful than another in specific problems. Several considerations interfere with this straightforward parallel, however, and cause confusion:

1. The fact that the proportionality factor in Eq. (10.101) is numerically equal to unity and is dimensionless makes us tend to forget that any such factor is needed.
2. The fact that the proportionality constant in Eq. (10.102) is not written as, say, k but also includes the factor $(4\pi)^{-1}$ is a recognition of the fact that 4π arises frequently in equations from geometrical considerations and can be conveniently eliminated by this device.

3. In media other than vacuum, the product $\epsilon_0 \epsilon_r$ is sometimes written ϵ (no subscript), where ϵ is the permittivity of the medium. Thus in a specific substance, $F_{cgs} \propto 1/\epsilon_r$ and $F_{SI} \propto 1/4\pi\epsilon_r\epsilon_0 = 1/4\pi\epsilon$.
4. In calculation, however, ϵ_r and ϵ are quite different. We must remember that ϵ_r is dimensionless, while ϵ is the product of ϵ_r and ϵ_0, with the latter having definite units.
5. Since the factor 4π is introduced into Eq. (10.102) to encourage cancellation, we frequently find expressions for the same quantity differing by this factor, depending on the system of units used by the author. For example, the Clausius–Mosotti equation [Eq. (10.17)] is written $(4\pi/3)(\rho N_A \alpha/M) = (\epsilon_r - 1)/(\epsilon_r + 2)$ in the cgs system.

Since an electric field E in space is defined as the force experienced by a unit test charge q_t (strictly, in the limit of $q_t \to 0$), it follows that the field produced by q_1 is obtained by letting $q_2 = q_t = 1$ in Coulomb's law:

$$E_{cgs} = \frac{q_1}{\epsilon_r r^2} \tag{10.103}$$

and

$$E_{SI} = \frac{q_1}{4\pi\epsilon_0 \epsilon_r r^2} = \frac{q_1}{4\pi\epsilon r^2} \tag{10.104}$$

Even when we discuss the electric field of light without reference to any particular charge, we must be aware of these differences. When that field interacts with a charge, as in light scattering, we will be in trouble unless a self-consistent set of units has been employed.

Problems

1. The geometry of Fig. 10.3 leads to a result known as Snell's law, which relates the refractive index of the medium to the angles formed by two wave fronts with the interface. Defining θ_0 and θ, respectively, as the angles between the phase boundary and the wave front under vacuum and in the medium of refractive index ñ, show that Snell's law requires ñ $= \sin\theta_0/\sin\theta$.
2. Use the expression given in Example 10.3 to evaluate the compressibility of CCl_4 from the fact[†] that $R_{90} = 5.38 \times 10^{-4}$ m^{-1} at room temperature for $\lambda_0 = 546$ nm. The depolarization ratio $\rho_u(90)$ is 0.042 for this liquid and ñ $= 1.460$, $\alpha = 1.21 \times 10^{-3}$ deg^{-1}, and dñ/dT $= 58.6 \times 10^{-5}$ deg^{-1}. Compare the light-scattering value with a literature value for β_{CCl_4}; be sure to cite the reference consulted.

[†] Data from Ref. 3.

3. Fowle[†] measured the turbidity of air at Mt. Wilson, California, on a clear day in 1913. Values of τx for dry air at different wavelengths are tabulated below, where x is essentially the thickness of the atmosphere corrected to standard temperature and pressure (STP) conditions:

λ (μm)	τx	λ (μm)	τx
0.3504	0.459	0.5026	0.122
0.3600	0.423	0.5348	0.108
0.3709	0.377	0.5742	0.100
0.3838	0.338	0.5980	0.091
0.3974	0.285	0.6238	0.074
0.4127	0.245	0.6530	0.064
0.4307	0.213	0.6858	0.0419
0.4516	0.174	0.7222	0.0304
0.4753	0.147	0.7644	0.0212

Prepare a log–log plot of τx versus λ and evaluate the slope as a test of the Rayleigh theory applied to air. The factor $M/\rho N_A$ in Eq. (10.36) becomes $6.55 \times 10^5/N_0$, where N_0 is the number of gas molecules per cubic centimeter at STP and the numerical factor is the thickness of the atmosphere corrected to STP conditions. Use a selection of the above data to determine several estimates of N_0, and from the average, calculate Avogadro's number. The average value of $\tilde{n} - 1$ is 2.97×10^{-4} over the range of wavelengths which are most useful for the evaluation of N_A.

4. Bhatnagar and Biswas[‡] measured the turbidity at 436 nm of a *single* sample of poly(methyl methacrylate) in several solvents, including acetone and methyl ethyl ketone (MEK):

	$d\tilde{n}/dc_2$	$(Hc_2/\tau)_{c=0}$	M
In acetone	0.13914	1.75×10^{-6}	571,000
In MEK	0.01385	2.60×10^{-8}	38,400,000

Working with different samples of the same polymer, other researchers have published conflicting values for the refractive index gradient in these solvents:

	$d\tilde{n}/dc_2$ in acetone	$d\tilde{n}/dc_2$ in MEK
Cohn and Schuele §	0.137	0.113
Tremblay et al.#	0.107	0.093

[†] F. E. Fowle, *Astrophys. J.* 40:435 (1914).

[‡] H. L. Bhatnagar and A. B. Biswas, *J. Polym. Sci.* 13:461 (1954).

[§] E. S. Cohn and E. M. Schuele, *J. Polym. Sci.* 14:309 (1954).

[#] R. Tremblay, Y. Sicotte, and M. Rinfret, *J. Polym. Sci.* 14:310 (1954).

Using the original Hc_2/τ values, recalculate M using the various refractive index gradients. On the basis of self-consistency, estimate the molecular weight of this polymer and select the best value of $d\tilde{n}/dc_2$ in each solvent. Criticize or defend the following proposition: Since the extension of the Debye theory to large particles requires that the difference between ñ for solute and solvent be small, this difference should routinely be minimized for best results.

5. Various amounts of either ethanol or hexane were added to polystyrene solutions in benzene and τ was measured for several concentrations of polymer. The following results were obtained† (c_2 in g liter^{-1}; Hc_2/τ in mol g^{-1}):

Pure benezene	
c_2	$Hc_2/\tau \times 10^6$
0.73	1.47
1.21	1.82
2.00	2.38

Benzene + percent ethanol			
20%		28.5%	
c_2	$Hc_2/\tau \times 10^6$	c_2	$Hc_2/\tau \times 10^6$
0.52	0.75	0.52	0.51
1.06	0.87	1.00	0.56
1.60	1.00	—	—

Benzene + percent hexane			
50%		58%	
c_2	$Hc_2/\tau \times 10^6$	c_2	$Hc_2/\tau \times 10^6$
0.28	0.95	0.28	0.89
0.66	1.00	0.66	0.89
0.88	1.06	0.88	0.91

Evaluate M and B for each of the five runs on this polymer sample and comment on the following points:

a. What is the significance of the runs for which $B \cong 0$?

b. What is the significance of the difference in the amount of the two diluents needed to produce the $B = 0$ condition?

c. What is the significance of the different behavior with respect to M for the two diluents?

---

† E. D. Kunst, *Rec. Trav. Chim. Pays Bas* 69:125 (1950).

6. Zimm[†] has reported the intensity of scattered light at various angles of observation for polystyrene in toluene at a concentration of 2×10^{-4} g cm^{-3}. The following results were obtained (values marked * were estimated and not measured):

θ (deg)	i_s (arbitrary units)
0	4.29*
25.8	3.49
36.9	2.89
53.0	2.18
66.4	1.74
90.0	1.22
113.6	0.952
143.1	0.763
180	0.70*

Draw a plot in polar coordinates of the scattering envelope in the xy plane. How would the envelope of a Rayleigh scatterer compare with this plot? By interpolation, evaluate i_{45}, i_{135}, and z. Use Fig. 10.13 to estimate the value of r_{rms} to which this dissymmetry ratio corresponds if λ (in toluene) is 364 nm. What are some practical and theoretical objections to this procedure for estimating r_{rms}?

7. The effect of adenosine triphosphate (ATP) on the muscle protein myosin was studied by light scattering in an attempt to resolve conflicting interpretations of viscosity and ultracentrifuge data. The controversy hinged on whether the myosin dissociated or changed molecular shape by interaction with ATP. Blum and Morales[‡] reported the following values of $(Kc_2/R_\theta)_{c=0}$ versus $\sin^2(\theta/2)$ for myosin in 0.6 M KCl at pH 7.0:

$\sin^2(\theta/2)$	0.15	0.21	0.29	0.37	0.50	0.85
$Hc_2/\tau \times 10^7$						
(Before ATP)	0.9	1.1	1.5	1.8	2.2	2.7
(During ATP)	1.9	2.8	3.7	4.6	6.0	6.8

Which of the two models for the mode of ATP interaction with myosin do these data support? Explain your answer by quantitative interpretation of the light-scattering data.

[†] B. H. Zimm, *J. Chem. Phys.* 16:1093 (1948).
[‡] J. J. Blum and M. F. Morales, *Arch. Biochem. Biophys.* 43:208 (1953).

8. Aggregation of fibrinogen molecules is involved in the clotting of blood. To learn something about the mechanism of this process, Steiner and Laki[†] used light scattering to evaluate M and the length of these rod-shaped molecules as a function of time after a change from stable conditions. The stable molecule has a molecular weight of 540,000 g mol^{-1} and a length of 840 Å. The following table shows the average molecular weight and average length at several times for two different conditions of pH and ionic strength μ:

pH = 8.40 and μ = 0.35 M			pH = 6.35 and μ = 0.48 M		
t (sec)	M × 10^{-6} (g mol^{-1})	length (Å)	t (sec)	M × 10^{-6} (g mol^{-1})	length (Å)
650	1.10	1300	900	1.10	1100
1150	1.65	1600	1000	2.0	1200
1670	2.20	1900			
2350	3.30	2200			

Criticize or defend the following proposition: The apparent degree of aggregation x at various times can be obtained in terms of either the molecular weight *or* the length. The ratio of the values of x based on M to that based on length equals unity for exclusively end-to-end aggregation and increases from unity as the proportion of edge-to-edge aggregation increases. In the higher pH–lower μ experiment there is considerably less end-to-end aggregation in the early stages of the process than in the lower pH-higher μ experiment.

9. Zimm plots at 546 nm were prepared for a particular polystyrene at two temperatures and in three solvents. The following summarizes the various slopes and intercepts obtained[‡]:

		T = 22°C	
Solvent	Intercept	$\left(\dfrac{\text{slope}}{\text{intercept}}\right)_{c=0}$	$\left(\dfrac{\text{slope}}{\text{intercept}}\right)_{\theta=0}$
Methyl ethyl ketone	0.896	0.608	260
Dichloroethane	1.61	1.16	900
Toluene	3.22	1.14	1060

[†] R. F. Steiner and K. Laki, *Arch. Biochem. Biophys.* 34:24 (1951).
[‡] P. Outer, C. I. Carr, and B. H. Zimm, *J. Chem. Phys.* 18:830 (1950).

Solvent	Intercept	$\left(\dfrac{\text{slope}}{\text{intercept}}\right)_{c=0}$	$\left(\dfrac{\text{slope}}{\text{intercept}}\right)_{\theta=0}$
		$T = 67°C$	
Methyl ethyl ketone	0.840	0.551	230
Dichlorethane	1.50	1.05	870
Toluene	2.80	1.09	800

The slope–intercept ratios have units of cubic centimeters per gram, and the intercepts are $c/R_{\theta,v}$, where the subscript v indicates vertically polarized light. In this case $K_v = 4\pi^2\,(\tilde{n}\,d\tilde{n}/dc_2)^2/\lambda_0^4 N_A$. The following values of $\tilde{n}$ and $d\tilde{n}/dc_2$ can be used to evaluate K_v:

	$T = 20°\,C \cong 22°C$		$T = 67°C$	
	$\tilde{n}$	$d\tilde{n}/dc_2$	$\tilde{n}$	$d\tilde{n}/dc_2$
Methyl ethyl ketone	1.378	0.221	1.359	0.230
Dichloroethane	1.444	0.158	1.423	0.167
Toluene	1.496	0.108	1.472	0.118

Evaluate M, $(\overline{r_g^2})^{1/2}$, and B from each piece of pertinent data and comment on the following points:

a. Agreement between M values.
b. Correlation of $(\overline{r_g^2})^{1/2}$ and B with solvent "goodness."

10. For polystyrene in butanone at 67°C the following values of $Kc_2/R_\theta \times 10^6$ were measured† at the indicated concentrations and angles. Construct a Zimm plot from the data below using $k = 100$ cm^3 g^{-1} for the graphing constant. Evaluate M, B, and $(\overline{r_g^2})^{1/2}$ from the results. In this experiment $\lambda_0 = 546$ nm and $\tilde{n} = 1.359$ for butanone at this temperature.

	c_2 (g cm^{-3})	
θ (deg)	1.9×10^{-3}	3.8×10^{-4}
25.8	—	1.48
36.9	1.84	1.50
53.0	1.93	1.58
66.4	1.98	1.62
90.0	2.10	1.74
113.6	2.23	1.87
143.1	2.34	1.98

---

† B. H. Zimm, *J. Chem. Phys.* 16:1093 (1948).

11. Benoit et al.† prepared a mixture of two different fractions of cellulose nitrate and determined the molecular weight of the mixture by light scattering. The mixture was 25.8% by weight fraction A and 74.2% fraction B, where the individual fractions have the following properties:

Fraction	$\bar{M}_n$	$\bar{M}_w$
A	635,000	1,270,000
B	199,000	400,000

Calculate $\bar{M}_n$ and $\bar{M}_w$ for the mixture on the basis of this information concerning the components and their proportions. The following light-scattering data for the mixture allow $\bar{M}_n$ and $\bar{M}_w$ to be evaluated by the procedure shown in Fig. 10.15:

θ (deg)	$(Kc_2/R_\theta) \times 10^7$ $(mol\ g^{-1})$	θ (deg)	$(Kc_2/R_\theta) \times 10^7$ $(mol\ g^{-1})$
30	19.6	80	42.2
35	21.7	90	48.7
40	24.4	100	53.0
45	26.4	110	57.0
50	28.3	120	61.3
55	31.0	130	64.9
60	33.0	140	67.4
70	37.4		

Evaluate $\bar{M}_n$ and $\bar{M}_w$ from the light-scattering data and compare the values with those calculated from the preparation of the mixture.

Bibliography

1. Dover, S. D., in *An Introduction to the Physical Properties of Large Molecules in Solution*, by E. G. Richards, Cambridge University Press, Cambridge, 1980.
2. Hiemenz, P. C., *Principles of Colloid and Surface Chemistry*, Marcel Dekker, New York, 1977.
3. Kerker, M., *The Scattering of Light and Other Electromagnetic Radiation*, Academic, New York, 1969.
4. Stacey, K. A., *Light Scattering in Physical Chemistry*, Butterworths, London, 1956.
5. Tanford, C., *The Physical Chemistry of Macromolecules*, Wiley, New York, 1961.

† H. Benoit, A. M. Holtzer, and P. Doty, *J. Phys. Chem.* 58:635 (1954).

Index to Tables

Subject Index

Encyclopedia of the City

Encyclopedia of the City

Edited by
Roger W. Caves

Routledge
Taylor & Francis Group

LONDON AND NEW YORK

First published 2005
by Routledge
2 Park Square, Milton Park, Abingdon, Oxon, OX14 4RN

Simultaneously published in the USA and Canada
by Routledge
270 Madison Avenue, New York, NY 10016, USA

Routledge is an imprint of the Taylor & Francis Group

© 2005 Taylor & Francis Group

Typeset in Times New Roman by
Newgen Imaging Systems (P) Ltd, India
Printed and bound in Great Britain by
T.J. International Ltd, Padstow, Cornwall

British Library Cataloguing in Publication Data
A catalogue record for this book is available from the British Library

Library of Congress Cataloging in Publication Data
Encyclopedia of the City / edited by Roger W. Caves.
p. cm.
Includes bibliographical references and index.
1. Cities and towns–Encyclopedias.
2. Sociology, Urban–Encyclopedias.
3. City planning–Encyclopedias.
I. Caves, Roger W.

HT108.5.E63 2005

307.76′03–dc22   2004051142

ISBN 0–415–25225–3 (alk. paper)

To my wife Carol who has encouraged me in everything I do
and who has been there for everything I've done

and

To my late parents, Raymond and Helen Caves,
for everything they did for me and
for their love and encouragement

Contents

Editors

Volume editor

Roger W. Caves
San Diego State University, US

Consultant editors

Carl Abbott
Portland State University, US

David Amborski
Ryerson University, Canada

Eugenie L. Birch
University of Pennsylvania, US

Lineu Castello
*Formerly of the Universidade Federal do Rio
Grande do Sul, Brazil*

Richard Huggins
Oxford Brookes University, UK

Andrew Kirby
Arizona State University West, US

Richard LeGates
San Francisco State University, US

Yuichi Takeuchi
*Institute of Behavioural Science,
Japan*

Alma H. Young
Wayne State University, US

Georges Yves Kervern
Sorbonne, France

Contributors

Carl Abbott
School of Urban Studies and Planning
Portland State University, US

Silvio Abreu
Faculty of Architecture
Universidade Federal do Rio Grande do Sul, Brazil

Luis Ainstein
School of Architecture, Design and Urbanism
Buenos Aires University, Argentina

Stuart C. Aitken
Department of Geography
San Diego State University, US

Arturo Almandoz
Department de Planificacíon Urbana
Universidad Simón Bolívar, Venezuela

David Amborski
School of Urban and Regional Planning
Ryerson University, Canada

R. Jerome Anderson
School of Architecture, Planning and Landscape
University of Newcastle, UK

Leandro M.V. Andrade
Faculty of Architecture
Universidade Federal do Rio Grande do Sul, Brazil

Jerry Anthony
Graduate Program in Urban and Regional Planning
University of Iowa, US

Ari-Veikko Anttiroiko
Department of Local Government Studies
University of Tampere, Finland

Bruce Appleyard
School of Urban Studies and Planning
Portland State University, US

Ernesto G. Arias
College of Architecture and Planning
University of Colorado at Boulder, US

Karl-Olov Arnstberg
Department of Ethnology
Stockholm University, Sweden

Deborah A. Auger
School of Urban Affairs and Public Policy
University of Delaware, US

Susan E. Baer
School of Public Administration and Urban Studies
San Diego State University, US

Adrian R. Bailey
School of Geography and Environmental Science
University of Birmingham, UK

Victoria Basolo
Department of Urban and Regional Planning
University of California, Irvine, US

Robert A. Beauregard
Milano Graduate School of Management and Urban Policy
New School University, US

Luigi Biocca
Italian National Research Council, Italy

John Blewitt
School of Lifelong Education and Development
University of Bradford, UK

Scott A. Bollens
Department of Urban and Regional Planning
University of California, Irvine, US

Elinor Bradley
Smart Communities Program
Industrie Canada, Canada

Rae Bridgman
Department of City Planning
University of Manitoba, Canada

Ray Bromley
Geography and Planning Department
University of Buffalo, State University of
New York, US

A. Lex Brown
School of Environmental Planning
Griffith University, Australia

John R. Bryson
School of Geography, Earth and
Environmental Sciences
University of Birmingham, UK

Trudi E. Bunting
School of Planning
University of Waterloo, Canada

Nico Calavita
Graduate City Planning Program
San Diego State University, US

Matthew Carmona
Bartlett School of Planning
University College London, UK

Iára Regina Castello
Faculty of Architecture
Universidade Federal do
Rio Grande do Sul, Brasil

Lineu Castello
Faculty of Architecture
Universidade Federal do Rio Grande do Sul, Brasil

Roger W. Caves
Graduate City Planning Program
San Diego State University, US

Mittie Olion Chandler
Levin College of Urban Affairs
Cleveland State University, US

Carla Chifos
School of Planning
University of Cincinnati, US

Jeffrey M. Chusid
University of Texas at Austin, US

Tara Lynne Clapp
Department of Community and Regional Planning
Iowa State University, US

Anthony J. 'Tony' Costello
College of Architecture and Planning
Ball State University, US

Roman Cybriwsky
Geography and Urban Studies
Temple University, US

Richard Dagenhart
College of Architecture
Georgia Institute of Technology, US

Joe T. Darden
Geography and Urban Affairs
Michigan State University, United States

Joaquín Farinós Dasí
Department of Geography
Universidad de Valencia-Estudios General, Spain

Alex Deffner
Department of Planning and
Regional Development
University of Thessaly, Greece

Bożena Degórska
Polish Academy of Science, Poland

Vicente del Rio
Department of City and Regional Planning
California Polytechnic State University,
San Luis Obispo, US

Davide Deriu
Bartlett School of Planning
University College London, UK

Frans M. Dieleman
Faculty of Geographical Sciences
Utrecht University, the Netherlands

D. Gregg Doyle
Department of City and Regional Planning
California Polytechnic State University,
San Luis Obispo, US

Michael Dudley
Institute of Urban Studies
University of Winnipeg, Canada

Montserrat Pareja Eastaway
Departament de Teoria Econòmica
Universitat de Barcelona, Spain

Dimitris Economou
Department of Planning and
Regional Development
University of Thessaly, Greece

Alessandra Faggian
School of Business
University of Reading, UK

James A. Fawcett
Sea Grant Program
University of Southern California, US

Daniel Felsenstein
Department of Geography
Hebrew University, Israel

Pierre Filion
School of Urban Planning
University of Waterloo, Canada

Raphaël Fischler
School of Urban Planning
McGill University, Canada

Ann Forsyth
Design Center for American Urban Landscape
University of Minnesota, US

Andrea I. Frank
Department of City and Regional Planning
Cardiff University, UK

Lance Freeman
Graduate School of Architecture, Planning and
Preservation
Columbia University, US

Robert Freestone
Planning and Urban Development Program
University of New South Wales, Australia

Steven P. French
Center for Geographic Information Systems
Georgia Institute of Technology, US

Jan Marie Fritz
School of Planning
University of Cincinnati, US

Duncan Fuller
Division of Geography
University of Northumbria, UK

Catherine C. Galley
College of Architecture
Texas Tech University, US

George C. Galster
College of Urban, Labor and
Metropolitan Affairs
Wayne State University, US

Daniel Garr
Urban and Regional Planning Department
San José State University, US

Edward G. Goetz
Humphrey Institute of Public Affairs
University of Minnesota, US

Rubén C. Lois González
Department of Geography
Universidade de Santiago de Compostela, Spain

Jamie Gough
Division of Geography
Northumbria University, UK

Jill Grant
School of Planning
Dalhousie University, Canada

Louise S. Gresham
Epidemiology Division
County of San Diego, US

Jennifer E. Gress
Department of Urban and
Regional Planning
University of California, Irvine, US

Michael Gunder
School of Planning
University of Auckland, New Zealand

Stephen Gurman
Smart Communities Program
Industrie Canada, Canada

Penny Gurstein
School of Community and
Regional Planning
University of British Columbia, Canada

Darrene L. Hackler
Public and International Affairs
George Mason University, US

Emilio Haddad
Faculdade de Arquitetura e Urbanismo
Universidade de São Paulo, Brasil

Anna L. Haines
Center for Land Use Education
University of Wisconsin at Stevens Point, US

John S. Hall
School of Public Affairs
Arizona State University, US

Sir Peter Hall
Institute of Community Studies
London, UK

Tim Hall
Geography and Environmental Management
Research Unit
University of Gloucestershire, UK

Pierre Hamel
INRS Urbanisation, Culture et Société
Montréal, Canada

Tigran Hasic
Division of Urban Studies
Royal Institute of Technology, Sweden

Amy Helling
Andrew Yound School of Policy Studies
Georgia State University, US

Sue Hendler
Department of Urban and Regional Planning
University of Buffalo, State University of
New York, US

Daniel Baldwin Hess
School of Urban and Regional Planning
Queen's University, Canada

Harrison Higgins
Department of Urban and Regional Planning
Florida State University, US

Phil Hubbard
Department of Human Geography
Loughborough University, UK

Richard Huggins
Department of Politics
Oxford Brookes University, UK

Lori M. Hunter
Institute of Behavioral Science
University of Colorado, US

David L. Imbroscio
Department of Political Science
University of Louisville, US

Yosef Jabareen
Department of Urban Studies and Planning
MIT, US

A.J. Jacobs
School of Planning
University of Cincinnati, US

Curtis W. Johnson
Citistates
St. Paul, Minnesota, US

Sanda Kaufman
Levin College of Urban Affairs
Cleveland State University, US

Loraleigh Keashly
College of Urban, Labor and Metropolitan Affairs
Wayne State University, US

W. Dennis Keating
Levin College of Urban Affairs
Cleveland State University, US

Robert Kerstein
Department of Government, History and Sociology
University of Tampa, US

Joochul Kim
School of Planning and Landscape Architecture
Arizona State University, US

Andrew Kirby
Department of Social and Behavioral Sciences
Arizona State University, West, US

Christopher Klemek
Department of History
University of Pennsylvania, US

Kimberly Knowles-Yánez
Urban Studies Department
California State University, San Marcos, US

Rob Konings
OTB Research Institute for Housing,
Urban and Mobility Studies
Delft University of Technology,
the Netherlands

Tomoko Kuroda
Department of Human Environmental Sciences
Mukogawa Women's University, Japan

Konstantinos Lalenis
Department of Planning and Regional Development
University of Thessaly, Greece

Diona Lambiri
Centre for Spatial and Real Estate Economics
University of Reading, UK

Larissa Larsen
School of Natural Resources and Environment
University of Michigan, US

Richard W. Lee
Fehr and Peers Associates
Lafayette, California, US

Richard LeGates
Urban Studies Program
San Francisco State University, US

Suzanne H. Crowhurst Lennard
International Making Cities Livable Conferences
Carmel, California, US

Myron A. Levine
Political Science Department
Albion College, US

Miles Lewis
Faculty of Architecture, Building and Planning
University of Melbourne, Australia

Christopher Long
School of Architecture
University of Texas, US

Erik Louw
OTB Research Institute for Housing,
Urban and Mobility Studies
Delft University of Technology, the Netherlands

Thomas S. Lyons
School of Urban and Public Affairs
University of Louisville, US

Kees Maat
OTB Research Institute for Housing,
Urban and Mobility Studies
Delft University of Technology, the Netherlands

Heather MacDonald
Urban and Regional Planning
University of Iowa, US

Thomas Maloutas
Department of Planning and
Urban Development
University of Thessaly, Greece

Peter Marris
Department of Sociology
Yale University, US

Stephen Marshall
Bartlett School of Planning
University College London, UK

Judith A. Martin
Urban Studies Program
University of Minnesota, US

Benjamin Mason
University of Illinois, US

Doreen J. Mattingly
Department of Women's Studies
San Diego State University, US

Gary A. Mattson
Department of Community and
Regional Planning
Iowa State University, US

W. Arthur Mehrhoff
Local and Urban Affairs
St. Cloud State University, US

Julia Ellen Melkers
College of Urban and Public Affairs
University of Illinois at Chicago, US

Ali Modarres
Edmund C. Brown Institute for Public Affairs
California State University,
Los Angeles, US

Daniel J. Monti Jr
Department of Sociology
Boston University, US

GianPiero Moretti
School of Urban Planning
McGill University, Canada

Annalisa Morini
Italian National Research Council, Italy

Ian Morley
Graduate School of Design
Ming Chuan University, Taiwan ROC

Gabriel Moser
Institut de Psychologie
Université René Descartes – Paris V, France

Elizabeth J. Mueller
Community and Regional Planning Program
University of Texas, US

Vincent Nadin
Centre for Environment and Planning
University of the West of England, UK

A. Ramprasad Naidu
Department of Anthropology
University of Pune, India

Corinne Nativel
School of Earth Sciences and Geography
Kingston University, UK

Michael Neuman
Department of Landscape Architecture and
Urban Planning
Texas A&M University, US

Mai Nguyen
Department of Urban and
Regional Planning
University of California, Irvine, US

Phil O'Keefe
Division of Environmental Management
Northumbria University, UK

Paul Ong
School of Public Policy and
Social Research
University of California, Los Angeles, US

Takashi Onishi
Department of Urban Engineering
University of Tokyo, Japan

A. Sule Özüekren
Faculty of Architecture
Istanbul Technical University, Turkey

Ronan Paddison
Department of Geography
University of Glasgow, Scotland

Urbano Fra Paleo
Department of Geography and Spatial Planning
University of Extremadura, Spain

Ayse Pamuk
Urban Studies Program
San Francisco State University, US

Kenneth Pearlman
City and Regional Planning
Ohio State University, US

Jesús M. González Pérez
Department of Earth Sciences
Universitat de les Illes Balears, Spain

Claire Poitras
INRS Urbanisation, Culture et Société
Montréal, Canada

Patricia Baron Pollak
Cornell University, US

Rita Pomposini
Italian National Research Council, Italy

Gheorghe Popescu
Faculty of Economics
Babeş-Bolyai University, Romania

Frank J. Popper
Urban Studies Department
Rutgers University, US

Hugo Priemus
OTB Research Institute for Housing,
Urban and Mobility Studies
Delft University of Technology, the Netherlands

David C. Prosperi
Department of Urban and Regional Planning
Florida Atlantic University, US

Jeffrey A. Raffel
Graduate School of Urban Affairs and
Public Policy
University of Delaware, US

Samina Raja
Department of Urban and Regional Planning
University of Buffalo, State University of
New York, US

Laxmi Ramasubramanian
College of Urban Planning and Public Affairs
University of Illinois at Chicago, US

Edward Ramsamy
Department of African Studies
Rutgers University, US

Peter Reed
School of Urban Planning
McGill University, Canada

Roger Richman
Public Administration and Urban Studies
Old Dominion University, US

Caroline Rodenburg
Department of Spatial Economics
Free University, the Netherlands

Arie Romein
OTB Research Institute for Housing,
Urban and Mobility Studies
Delft University of Technology, the Netherlands

Kevin Romig
Department of Geography
Arizona State University, US

Bernard H. Ross
School of Public Affairs
American University, US

Beverly A. Sandalack
Faculty of Environmental Design
University of Calgary, Canada

Gary Sands
Department of Geography and Urban Planning
Wayne State University, US

Sheila Sarkar
California Institute of Transportation Safety
San Diego State University, US

Andrew Seidel
College of Architecture
Texas A&M University, US

Siddhartha Sen
City and Regional Planning Program
Morgan State University, US

Ellen Shoshkes
Bloustein School of Planning and Public Policy
Rutgers University, US

William J. Siembieda
Department of City and Regional Planning
California Polytechnic State University,
San Luis Obispo, US

Carlos Nunes Silva
Department of Geography
University of Lisbon, Portugal

Chris Silver
Department of Urban and Regional Planning
University of Illinois, US

Glenn L. Silverstein
Graduate School of Urban Affairs and
Public Policy
University of Delaware, US

Ian Skelton
Department of City Planning
University of Manitoba, Canada

Paola Somma
Town and Regional Planning
University of Venice, Italy

Daphne Spain
Department of Urban and Environmental Planning
University of Virginia, US

Gregory D. Squires
George Washington University, US

Sumeeta Srinivasan
Division of Engineering and Applied Sciences
Harvard University, US

Michael Steiner
Institute of Technology and Regional Policy
Karl-Franzens University, Austria

Donovan Storey
School of People, Environment and Planning
Massey University, New Zealand

Frederic Stout
Urban Studies
Stanford University, US

Elaine Stratford
Department of Geography and
Environmental Sciences
University of Tasmania, Australia

Aki Suwa
Environmental Management
Nagoya Sangyo University, Japan

Erik Swyngedouw
School of Geography and the Environment
Oxford University, UK

Yuichi Takeuchi
Institute for Behavioural Science, Japan

Mark Tewdwr-Jones
Bartlett School of Planning
University College London, UK

Michelle Thompson-Fawcett
Department of Geography
University of Otago, New Zealand

Wendy Tinsley
Architectural Resources Group
San Francisco, US

Alison Todes
School of Architecture, Planning and Housing
University of Natal, South Africa

Simon Ungar
Department of Psychology
University of Surrey, UK

Antonella Valitutti
Faculty of Architecture
Università degli
Studi di Roma La Sapienza, Italy

Ronald van Kempen
Geographical Sciences
Utrecht University, the Netherlands

David van Vliet
Department of City Planning
University of Manitoba, Canada

Jan van Weesep
Utrecht University, the Netherlands

David P. Varady
School of Planning
University of Cincinnati, US

María Teresa Vázquez Castillo
Pitzer College, California, US

Avis C. Vidal
Department of Geography and
Urban Planning
Wayne State University, US

Rainer vom Hofe
School of Planning
University of Cincinnati, US

Ron Vreeker
Department of Spatial Economics
Free University, the Netherlands

Carole C. Walker
Center for Urban Policy Research
Rutgers University, US

R. Alan Walks
Department of Geography
University of Toronto, Canada

Robert J. Waste
Department of Public Policy and Administration
California State University,
Sacramento, US

Philip Rodney Webb
Graduate Institute of the Liberal Arts
Emory University, US

Steven M. Webber
Ryerson University, Canada

Grzegorz Weclawowicz
Polish Academy of Science, Poland

Ana Maria Cabanillas Whitaker
Urban and Regional Planning Department
California State Polytechnic University,
Pomona, US

Catherine White
Division of Geography
University of Northumbria, UK

Patricia A. Wilson
Community and Regional Planning Program
University of Texas at Austin, US

Murat Cemal Yalcintan
City and Regional Planning
Mimar Sinan University, Turkey

Alma H. Young
College of Urban, Labor and Metropolitan Affairs
Wayne State University, US

Evelyn Zellerer
Criminal Justice Administration
San Diego State University, US

Wil Zonneveld
OTB Research Institute for Housing,
Urban and Mobility Studies
Delft University of Technology,
the Netherlands

Introduction

The term 'city' means anything and everything. Things we hear and see in the arts, films and books often determine our views. It is indeed a complex organism. We can examine the form of a city, its evolution, its people, cultures, public spaces, governance, institutions, environment, economics, etc. As such, it encompasses many things and gives rise to broad definitions. Acknowledging there is 'no neat definition of a city', Chant (1999: ix, quoting Hammond 1972: 8) offers the following:

> The city may be defined initially as a community whose members live in close proximity under a single government and in a unified complex of buildings, often surrounded by a wall. Since, however, this definition would also cover many villages, military camps, religious communities and the like, the city may further be described as a community in which a considerable number of the population pursue their main activities within the city, in non-rural occupations. But other communities, such as a monastery or small factory surrounded by the dwellings of its workmen, might be similarly characterized. A third characterization may therefore be that the city is a community which extends at least its influence and preferably its control over an area wider than that simply necessary to maintain its self-sufficiency.

Gallion and Eisner (1983: 5) suggest the following:

> The word 'city' implies a concentration of people in a geographic area who can support themselves from the city's economic activities on a fairly permanent basis. The city can be a center of industry, exchange, education, and government or involve all these activities. These diverse areas of opportunity attract people from rural areas to cities.

In the classic *The City in History*, Mumford (1961) discusses the vastness of what might be considered a city. In rejecting a single definition of a city, Mumford (*ibid.*: 3) wrote 'no single definition will apply to all of its manifestations and no single description will cover all its transformations, from the embryonic social nucleus to the complex forms of its maturity and the corporeal disintegration of its old age'. Nevertheless, we keep trying to do so.

A number of scholars define 'city' according to their discipline. As Passoneau (1963: 9) has suggested:

> To an economist, a city is a large, complex, input–output device. To a sociologist, a city is distinguished from a village by its higher degree of social differentiation and by the wider opportunity it offers by fruitful interaction between diverse individuals. To a political scientist, a group of compact, contiguous, but separately governed suburbs might not be a city, while a sprawling series of communities under a single government might be a city with a distinctive personality.

This belief may provide reasons why the problems of a city persist. Constantinos Doxiades (1970: 5) acknowledged as much when he commented:

> Man often continues to see the city through the eyes of experts in separate disciplines dealing with single aspects of city life. People speak of the transportation problem and try to solve it through the transportation engineers or the transportation economists only, in many cases without even bringing these two professions together . . . . To continue to deal with them separately by isolating its parts is like refusing to see that man himself is a single organism which cannot be looked at separately as body or senses or mind.

Park (1925: 1) has observed that a city could not be viewed as purely a physical entity:

> The city is, rather, a state of mind, a body of customs, and traditions, and of the organized attitudes and sentiments that inhere in these customs and are transmitted with this tradition. The city is not, in other words, merely a physical mechanism and an artificial construction. It is involved in the vital processes of the people who compose it; it is a product of nature, and particularly of human nature.

If it is not a physical entity, then how should we view a city? Mumford (1961) likened a city to 'a theater of social action'. Saarinen (1943: ix) viewed the city as 'an open book in which to read aims and ambitions', while Gutkind (1962: 81) offered an intriguing analogy comparing a city to a power station:

> Cities are the power stations of our technical mass civilization. In these giant containers, ideas and habits, technical skills and inventions are transformed into new energy, spreading over vast areas and connected by the invisible bonds of similar pursuits and interests. The power lines and the pylons are symbolic media of this process.

Early development of the city

Early man was nomadic in nature. Individuals grouped together in areas near natural resources like water and fertile grounds. The ability to locate at these locations allowed the people to sustain themselves. When the resources were depleted, man moved to another location to take advantage of the new resources. They located on elevated sites and other areas that allowed them to see if anyone (e.g. outsiders or invaders) were getting close to them. Developing at such strategic locations allowed them to control the countryside. Agricultural occupations dominated early settlements. Distance was limited by how far an individual could travel. Population density was not a concern of the times.

Walls soon became commonplace as individuals continued to group together. Constructed of various stones and later bricks, walls were created for defensive purposes. They stood as barriers against invaders. They also functioned as boundaries of ancient cities. On many occasions, the walls were extended to accommodate the growing city. As various types of weaponry were developed, the construction materials changed. Many segments of ancient walls can still be found around the world.

The evolution of the city continued during the Greek and Roman times. Order started being placed on land uses within the walls. For example, the meeting place and market place had specific locations with the cities. Trading started occurring between towns. Forms of government were created. The development of roads, aqueducts and sanitation sewers followed.

As civilization advanced, problems started to surface. It is often said that citizen participation was valued in early cities. While this may be true, it must be noted that it was the participation of some people, not all people. Women, slaves or individuals foreign to the city were not allowed to participate in civic matters. Only certain people in the city were afforded the opportunity to participate. As such, a type of segregation or discrimination among classes existed. A rigid class system was therefore clearly evident in many early populations. Noblemen or religious leaders were the leaders of early cities. They were clearly in charge. No one attempted to challenge their leadership. The 'average' man or woman was unable to rise to a leadership level.

The population of many cities continued to grow. Crowded conditions within the walls led to various health problems, increases in pollution, crime, and other issues in search of better living conditions.

Concerns over living conditions led to the creation of early rules, regulations and codes designed to protect an individual's health, safety and welfare – commonly known today as an early version of the 'police power' in the United States. For example, the Code of Hammurabi called for strict punishment to builders that constructed faulty buildings. If the building collapsed and killed the owner of the building, then the individual that constructed the building would be put to death. If the building owner's son were fatally injured from the falling building, then the son of the building's builder would be put to death. Early regulations prevented individuals from building structures that would hang over a street for fear the structure might collapse and injure someone. Other regulations existed in many areas that sought to restrict traffic (horse, cart or some other mode of transportation) at certain times of the day. Moreover, early city regulations placed restrictions on the width of carts due to the narrow nature of the early roads. It was not uncommon to see horse-drawn wagons getting stuck between two buildings.

Early cities remained small in size for a number of years. Monuments to various individuals were common fixtures in the cities, as were temples honouring various gods. In fact, it could be said that monuments and temples dominated many early cities.

The living conditions found in early cities were not conducive to good health. People threw their wastes into the streets. Animals travelled in the narrow streets depositing their wastes. Combined with the fact that there was no drainage, health concerns plagued early cities.

The seeds of change were planted. In order to facilitate pedestrian travel within the cities, the narrow and confusing street patterns had to be changed. Hippodamus sought to correct the confusion and plan cities by creating designs based on geometric layouts during the fifth century BC in Greece. His ideas have been credited as the start of the 'gridiron' street system and are still discussed today.

The Greek influence on the design of cities was clearly evident in surrounding areas. However, the small nature of the Greek city was soon to pass. New areas soon opened up. New cities were created with large monuments and temples. Cities and their leaders competed to see who could build the biggest monuments and temples. More people began migrating to the cities. Increased overcrowding and unsanitary living conditions soon occurred. The possibilities of fires became a harsh reality in densely populated areas. The leaders of the city, in an attempt to avoid these problems, moved to the surrounding areas, leaving the unhealthy living conditions to the ordinary people.

As time passed, new towns and cities were created. New empires were formed. A rigid class system became evident in the cities. Rulers trying to make earlier buildings seem insignificant continued to construct bigger and more elaborate buildings. Life revolved around the rulers. Although improvements in water and drainage systems were made, living conditions in the cities suffered. While the rulers lived in large and luxurious buildings, the average person suffered. Extreme congestion and overcrowding occurred in most cities. Instead of opening up land outside the cities to meet the needs of the people, the rulers kept the lands for themselves.

After the fall of the Greek and Roman Empires, many cities had been destroyed. People started returning to the rural areas for a better life. As time passed, people started joining together once again for safety purposes or some other common purposes. A period of feudalism soon surfaced.

So-called 'city-states' were created by rulers. The rulers, as their seats of power, built castles. People began moving closer to the castles for protection. The economy remained primarily agriculturally based. Within the castle, streets remained narrow. Market places and public spaces continued to be found.

The number of cities continued to increase during the Middle Ages. They remained rather small due to the size of the castles and the availability of natural resources like water. Having only one source of water exacerbated matters.

Poor living conditions continued to plague the people within the castles. Density started increasing as more people sought jobs and shelter within the confines of the castles. Cart traffic increased in the already narrow streets, causing congestion problems. People continued to dispose of their waste in the streets. Combined with poor air quality, environmental and health problems soon emerged as major concerns.

The city continued to evolve. The invention of gunpowder and other weapons signalled the eventual downfall of castle walls. As the walls came down, populations started to decentralize. Designers and planners were brought in to put order into the cities. New street patterns were created. The arts took on increased importance. Changes within the cities continued during the Renaissance and the Baroque

periods. Although many changes had taken place in the cities, the living conditions of the people continued to deteriorate.

Populations continued to move as people sought more freedom and better lives. People left countries in the hope of finding better places to live. They took their experiences and lessons from their old cities to their new areas. Various plans and development patterns were followed in these new settlements. Some cities adopted simple development schemes while others adopted new and elaborate development ideas. Additional changes were on the horizon.

The Industrial Revolution

The dawn of the Industrial Revolution signalled a number of changes in cities. The transition from a traditional agricultural and rural economy to an urban and industrial economy and the resulting changes from the invention of new technologies and production processes made dramatic changes in cities. Many of the changes altered the construction of urban space.

The transition to an industrial economy represented a 'process' of change, not an overnight change. Moreover, a single change did not trigger the transition. Instead, it was the result of various changes in agriculture, technology, demography and other areas. Some countries were quick to embrace the changes while, due to political and other factors, other countries took longer to embrace the changes. The impacts of the technological changes clearly prompted various changes in settlement patterns.

Technological innovations in machinery offered new and improved means of doing things. New discoveries and innovations in agriculture, textiles, coal and iron led to the establishment of factories and assembly plants. People started migrating from rural areas to cities in search of jobs. Many rural areas were hurt by this migration. Some rural areas were devastated. Concomitantly, immigrants from different countries ventured to many parts of the newly developing industrial world in search of jobs. With improvements in transportation, additional land was opened for development. Factory towns were starting to occur along transportation corridors. Many of these towns were prompted by the presence of certain resources. The factories were there to extract materials like coal and other raw materials so they could be used for other products. Many of them thrived for a number of years. However, once the resource was depleted, the settlement became a thing of the past. People left and migrated to other areas in search of employment.

The advent of the changes prompted rivalries between areas. Once compact cities were opening up to more development and to more people. Once isolated areas could now be reached by new modes of transportation. New markets opened up.

The growth of many cities did not occur without a number of drawbacks. Population densities continued to increase. Industries had a cheap labour force. Factory workers lived in poorly built units. With limited incomes, overcrowding became a common occurrence in industrial cities around the world. Basic sanitation facilities were lacking. The poor living conditions led to a number of health problems among the workers.

The role of transportation

As previously mentioned, walking represented the mode of transportation in early cities. Since early cities were compact in scale, it did not take very long to get from one place to another in the city. Distance was limited to how far an individual could walk. As time progressed, the use of horse-drawn 'cars' allowed individuals to cover more territory and to make new areas accessible to the public.

The advent and presence of streetcars helped shape the commercial and residential development of many cities. Their placement hastened the transformation of some agricultural lands into commercial and residential areas. Streetcar routes became the primary locations of development. As a result, these areas were in high demand. They wanted to take advantage of prime locations made accessible by transportation. A market for land speculation was created. Ultimately, the streetcars contributed, as Sam Bass Warner Jr (1962) has written, to the creation of 'streetcar suburbs'. Their development has also led scholars

to conclude that the presence and introduction of streetcar suburbs contributed to the increased separation of various socio-economic groups.

The advent of motorized transportation continued to contribute to the spreading out of many cities. The opening up of new land for development would lessen the amount of congestion that was common in the existing cities. Land started being consumed in great amounts as new land was opened for urban expansion. Unfortunately, some areas expanded with little or no regard to the impacts on the natural environment.

The importance of transportation to an area's development varies by city and by country. If you ever wonder how important transportation is to a city's development, consider several examples. Hall (1989: 119–20) claims that modern London is largely a creation of its transportation system. Bullock (1999) notes that the railway system increased accessibility throughout Berlin in the 1980s. In the following passage, Bartholomew (1999: 342) describes the impact of the railway on the old city of Shahjahanabad during the building of New Delhi:

> Making just as great an impact on the city as the military, came the railway, demolishing one of the city's old gates and a segment of the palace complex itself as it was laid out from west to east across the northern edge of the city, crossing the river at a new bridge close to the palace. The western edge of the city was hemmed in by further rail developments. Then a cordon sanitaire was cleared next to the walls along the southern boundary of the city to separate it from New Delhi, which was laid out to the south with hardly any organic links to the old city. The result of all these developments was that the city, which was bounded on its eastern side by the river, became hemmed in and compressed on its other three sides.

City form

A number of ideas about city form have surfaced over the years. Camillo Sitte and Benjamin Ward Richardson raised concerns over artistic principles and city planning and public health and city planning. Daniel Burnham, through the City Beautiful movement, embraced using principles of art, civic improvement and landscape architecture to reorder public space. Social utopians like Ebenezer Howard and Arturo Soria y Mata advocated proposals for garden cities with green belts and the linear city respectively. Le Corbusier offered his version of a city with a large skyscraper surrounded by open spaces. An additional proposal called for the creation of a city that incorporated rows of tall buildings radiating outwards through landscaped space. In the United States, Frank Lloyd Wright spoke against dense urban environments. Under his 'Broadacre City' proposal, each citizen would be provided with a minimum of one acre each with access to transportation linking people together. The role of government would be greatly reduced under his proposed scheme. Ultimately, these ideas and others provided fruitful discussions on the future form of cities. The ideas were applauded by some and ridiculed by others. Nevertheless, their views and proposals created even more discussion on the need to reorganize the urban form.

The structure and patterns of a city's growth have been examined by a number of scholars. A concentric zone model was proposed by Burgess, a multiple-nuclei model was advocated by Harris and Ullman, the sectoral model was presented by Hoyt, while the central place theory was offered by Christaller. These and other alternative models of growth have provided keen insights into how cities have developed.

Looking into the future

In order to understand the future of the 'city', it is imperative that we examine the past and present time of a city. We cannot understand a city at one particular time in its history. As Lynch (1960: 1) has so eloquently noted, 'like a piece of architecture, the city is a construction in space, but one of vast scale, a thing perceived only in the course of long spans of time'.

We have seen where and why cities were created. We have seen the factors that influenced their development. We can point to the importance of land and other natural resources in pre-industrial

settlements, the importance of technological innovations to the restructuring of urban space, the importance and implications of new forms of urban space, and to the importance of competition between cities.

No one can foresee the future of any city. We can, however, be sure of some things. The world's population will continue to increase, faster in some areas and slower in other areas. Oucho (2001) indicates that projections suggest a world population increase from 5.7 billion people in 1995 to 8.9 billion in 2050. Some of this growth will occur in rural areas. Some of the population will migrate from rural areas to urban areas and others from urban areas to rural areas. Others will migrate from one urban area to another urban area.

The exodus of people from rural areas to urban areas suggests a number of potential problems. Some people lack the skills necessary to find a job. This suggests a potential problem in finding affordable housing. These problems could lead to additional problems such as crime, drugs, vandalism, etc. The ability to respond to and to 'solve' these problems represents a serious challenge to our cities.

Urban area growth will be dramatic. A United Nations report compiled in 1999 (United Nations 2001) estimated that the world's urban population in 2000 would reach 2.9 billion persons. The report also indicated that the world's urban population is set to increase by some 2 billion persons to 4.9 billion by the year 2030. This increase represents virtually all of the world's expected population growth.

The growth is no more evident than in the Asia-Pacific region. According to the 'Cities of Asia' website (http://whc.unesco.org/events/asiaciti.htm), in 1970 only eight cities in the Asia-Pacific region had populations of more than 5 million inhabitants. That number has increased to more than thirty. By 2020, the website indicates, more than half of the urban areas on the planet will be located in Asia, holding more than a third of the world's population. Problems such as poverty, overcrowding, environmental degradation, etc. are bound to accompany this growth in such large cities as Seoul, South Korea; Bombay, India; and Jakarta, Indonesia.

We are currently in an age where our cities and businesses have the ability to create and disseminate information globally. This age is generally referred to as 'the new global economy' or 'knowledge-based economy', an economy based on information as a commodity. Cities will continue to change. As Graham and Marvin (1996: 124) have suggested, 'no longer can we understand cities primarily as centres for the manufacturing and the exchange and production of physical goods and commodities – as most were during the last hundred years'. Today, the production and transmission of information or knowledge are prime factors enabling cities and regions to innovate and grow. Distance no longer remains as a major impediment or obstacle to an area's growth. Markets can be anywhere. They can be thousands of miles from the home business or industry. Today, instead of a California business having markets in Arizona, Texas, Washington, Michigan, Florida and Massachusetts, changes in technology enable a California business to reach potential markets in such countries as the United Kingdom, Brazil, Italy, Sweden, China, South Africa and Australia. Unfortunately, many countries have been unable to adapt as quickly to participate in the new economy. Some scholars have speculated that the collapse of the Soviet Union may have been hastened due to the structural inability of its economic system to adapt industrial processes to a changing global economy.

Today, workers, in a number of professions, can take advantage of new information technologies and conduct work or business from their home or another off-site location, a practice known as 'telecommuting'. Cities might be able to realize a number of potential benefits from telecommuting including a reduction in the number of vehicles on the roads. The reduction also helps to reduce air pollution problems, a growing problem in many cities of the world.

Technological advances have contributed to even more changes in cities and countries around the world. Whereas earlier cities prospered because of their location, they have had to reassess their roles, functions and capabilities in relation to the new technologies. Manuel Castells (1989: 1), in *The Informational City*, provides an eloquent statement about the potential of the new changes in technology:

A technological revolution of historic proportions is transforming the fundamental dimensions of human life: time and space. New scientific discoveries and industrial innovations are extending the productive capacity of working hours while superseding spatial distance in all realms of social activity.

The unfolding promise of information technology opens up unlimited horizons of creativity and communication, inviting us to the exploration of new domains of experience, from our inner selves to the outer universe, challenging our societies to engage in a process of structural change.

As such, the new economy is more than simply an area of 'high tech'. It involves examining new business practices and markets.

Advances in the field of information and communications technologies will continue to influence the future of the 'city' (Cairncross 2001; Castells 1989, 1996; Graham and Marvin 1996, 2001; and Mitchell 1996). Clark (2000: 151) has noted that 'the communication city of the twenty-first century will likely further weaken the arguments for a compact urban form with a dominant center'. However, cities will remain. As Aurigi and Graham (2000: 489) point out:

Clearly, electronic networks *can* substitute for some physical travel and face-to-face encounters, as phone banking and online shopping demonstrate. But it does not follow that cities will somehow 'vanish' with the growth of the online realm.

The long-term impact of electronic networks on cities is still being debated. We have certainly witnessed a number of changes in how cities conduct government business (Castells 1989, 1996; Castells and Hall 1994; Caves and Walshok 1999; Graham and Marvin 1996, 2001; Mälkiä *et al.* 2004; Moss 1987). Today, cities are providing increasing amounts of government information via their internet websites. The minutes of city council meetings are posted on city internet sites. If an individual was unable to attend a public meeting, some communities offer the opportunity to download various meetings in video format for viewing in the comfort of your own home. Other communities use the internet to promote their jurisdictions for economic development. Social service agencies, hospitals, schools, etc. use the internet to provide citizens with a wealth of information on their services. Communities all over the world continue to develop and implement programmes using information and communications technologies. For example, programmes such as Smart Communities, Digital Cities, TeleCities, etc. have been created in the United States, Canada, the United Kingdom and elsewhere.

While the internet does offer the opportunity to provide information and services to people without regard to their location, some people may still lack the ability to take advantage of this opportunity. A 'digital divide' exists among various individuals and groups, between the 'information haves' and the 'information have-nots' (Norris 2001; Schiller 1996; Warschauer 2003). The divide could be categorized along racial, economic, ethnic, age, gender or education lines. It has been seen in rich countries vs. poor countries or developed vs. non-developed countries. Moreover, some areas lack community technology centres or facilities that might enable these individuals and groups to take advantage of the information or services. The ability of cities and groups to bridge this 'digital divide' represents a global challenge. This challenge has been discussed worldwide by such organizations as the World Bank and the United Nations.

Concluding comments

As this encyclopedia was being assembled, a number of activities and events occurred that could have drastic impacts on cities. These impacts could be either direct impacts or indirect impacts. For example, after the terrorist attacks on the World Trade Center towers and the devastating consequences of what is now commonly known as '9/11', cities around the world are on heightened security alerts for possible terrorism attacks. No country is immune from such atrocities. Phrases such as 'code yellow' and 'code orange' security alerts are commonly heard in everyday life. Concerns over security have risen to the top of many public policy agendas. Security issues surrounding the home, the workplace and public spaces have taken on increasingly important dimensions in family life, communities, states and nations.

Whether they occur in cities in the developing or developed world, conflicts or wars have the potential of devastating parts of or entire towns or cities. The hopes of many individuals and families are shattered. Homes and businesses are destroyed. Hospitals and government buildings are destroyed, as are various systems or modes of transportation.

These events lead to even more problems. With housing destroyed, spontaneous settlements lacking the infrastructure necessary to sustain a given population have occurred. Living conditions suffer, health problems may surface, and tensions heighten. Many families are separated. Critically needed human services are hindered or completely stopped. People are forced to move in search of new employment opportunities. Lacking sufficient funds for housing and a lack of employment may intensify a lack of jobs or housing problems in their new area.

The economic base of a city or country can be drastically impacted by armed conflicts or wars. The economic base may essentially be destroyed and a new one will need to be created. Unfortunately, if the residents' employment skills are devoted or tied to one industry, they may need to be retrained to meet the needs of new industries. New workers will be attracted to the new opportunities while the earlier labour force may be forced into unemployment.

The partial or complete destruction of an area's transportation infrastructure presents a formidable challenge to areas devastated by armed conflicts or war. Many areas lack the resources necessary to make repairs or to rebuild roads, railroads, etc. They need time to make the necessary repairs or to engage in reconstruction efforts. Many will need to seek the assistance of such groups as the World Bank, the United Nations or other organizations to help in the various reconstruction efforts.

Cities continue to be a product of the interaction of many things. As Abrams (1965: 16) has suggested, 'a city, even an American city, is the pulsating product of the human hand and mind, reflecting man's history, his struggle for freedom, his creativity, his genius – and his selfishness and errors'.

What is meant by the word 'future'? Some areas have developed plans for the future and use something like five or ten years from now. It really represents a hard word to define. It has a way of alarming people. All we know is that it represents a time that has not yet happened. People are leery of the unknown. In fact, while some people look forward to the future, other people will be frightened by it. To them, not knowing what the future holds concerns them. They like the status quo. They will actively resist change, or, as Schon (1971: 32) indicates, practise 'dynamic conservation' – 'a tendency to fight to remain the same'.

It is important that we think about the future. Reacting to problems as they occur may simply exacerbate the problems. We need people and futurist thinkers like Jules Verne (1996), Alvin Toffler (1970, 1980, 1990) and Nicholas Negroponte (1995). To many, discussing the future is venturing into a realm of science fiction, pure speculation. This is what attracts many people to Verne. Written approximately 100 years before the period it discusses, in *Paris in the Twentieth Century*, Verne (1996) describes a Paris he envisions in the 1960s where technological progress is more important than culture. In *being digital*, Negroponte (1995: 231) hails the power of being digital:

The access, the mobility, and the ability to effectuate change are what will make the future so different from the present. The information superhighway may be mostly hype today, but it is an understatement about tomorrow. It will exist beyond peoples' wildest predictions. As children appropriate a global information resource, and as they discover that only adults need learner's permits, we are bound to find new hope and dignity in places where very little existed before.

This encyclopedia is designed to help individuals better understand the city. It is not an end. As such, additional individuals, concepts, issues, etc. will surface that will give rise to their being included in later editions. In order to truly understand the complexity and fascinating nature of the study of cities, we must continue to study the past and the present, and project into the future.

There is no question that the structure and form of the world's cities have changed over time. Many former compact cities are experiencing varying degrees of peripheral expansion. Cities that once possessed a single downtown business area now have multiple business areas scattered throughout the city. New cities are developing on the periphery of other cities. New patterns of land development are emerging. Environmental concerns over haphazard development are surfacing everywhere.

Population growth continues to be an issue in many parts of the world. Some areas will be better prepared to deal with it than others. Concerns over the continuing expansion of cities remain. It poses numerous dilemmas. Gutkind (1962: 96) offered the following questions that must be answered: 'But where

and how can the population increase be absorbed? Can the cities of tomorrow remain the centers of attraction and culture, or will they become mere refuse dumps for human beings?'

Handling population increases and city expansion varies greatly by city, country, and region of the world. Many areas are ill-equipped to provide guidance. Some countries make decisions at the national level while other countries have decentralized the ability to make certain decisions. In some areas, economic development remains the main activity of a government. Such governments are apparently willing to accept environmental problems in the name of jobs and economic development.

The original compact nature of a city has changed dramatically over the years. Many cities lack a defined shape. They grow in different directions. Some scholars question whether the city still exists. Arrango (1970: 65) acknowledged this when he mentioned more than thirty years ago that 'the typical American city is not, in fact, a city; it is a pan urban region, a conglomeration of cities, suburbs, and semi-urbanized areas loosely set together, among which there is little social unity'. This idea suggests the possibility and probability that many problems facing cities are no longer purely local problems. Instead, many of the problems are regional in nature. Air and water pollution does not respect any boundary. Many of a city's problems require regional cooperation and regional actions.

City expansion remains a critical issue facing many cities around the world. Many areas have seen the creation of squatter settlements at the periphery of large cities. These areas have generally lacked the critical infrastructure needed to meet the needs of the residents. With many of the inhabitants lacking employment, the areas have become breeding grounds for many problems.

Differences will continue to exist between cities. European cities are more compact than their American counterparts. The population in many parts of the world has accepted mass transit. The United States, on the other hand, is still predominantly an automobile society. Unfortunately, as Glazer (1970: 5) points out, 'no modern society with any degree of consumer freedom has figured out how to control the impact of the automobile'. Many US cities are trying to offer mass transit alternatives to the automobile. Some of these cities are seeing increased numbers of people taking advantage of the alternative transportation.

In the film *Field of Dreams*, a voice is heard saying 'If you build it, they will come.' Although the voice was talking about building a baseball field, the same is true about cities. Cities are like magnets. They attract people. People have flocked to the city through the ages. There is nothing to suggest this will not hold true for the future.

References

Abrams, C. (1965) *The City is the Frontier*, New York: Harper & Row.

Arrango, J. (1970) *The Urbanization of the Earth*, Boston, MA: Beacon Press.

Aurigi, A. and Graham, S. (2000) 'Cyberspace and the city: the "virtual city" in Europe', in G. Bridge and S. Watson (eds) *A Companion to the City*, Oxford: Blackwell, pp. 489–502.

Bartholomew, M. (1999) 'Colonial India', in D. Goodman and C. Chant (eds) *European Cities and Technology: Industrial to Post-Industrial City*, London: Routledge, pp. 326–54.

Bullock, N. (1999) 'A short history of everyday Berlin, 1871–1989', in D. Goodman and C. Chant (eds) *European Cities and Technology: Industrial to Post-Industrial City*, London: Routledge, pp. 225–56.

Cairncross, F. (2001) *The Death of Distance: How the Communications Revolution is Changing our Lives*, Boston, MA: Harvard Business School Press.

Castells, M. (1989) *The Informational City*, Oxford: Blackwell.

—— (1996) *The Rise of the Network Society*, Oxford: Blackwell.

Castells, M. and Hall, P. (1994) *Technopoles of the World: The Making of 21st Century Industrial Complexes*, London: Routledge.

Caves, R.W. and Walshok, M.G. (1999) 'Adopting innovations in information technology', *Cities* 16: 3–12.

Chant, C. (1999) 'Introduction', in D. Goodman and C. Chant (eds) *European Cities and Technology: Industrial to Post-Industrial City*, London: Routledge, pp. vii–x.

Clark, W.A.V. (2000) 'Monocentric to policentric: new urban forms and old paradigms', in G. Bridge and S. Watson (eds) *A Companion to the City*, Oxford: Blackwell, pp. 141–54.

Doxiades, C.A. (1970) *Emergence and Growth of an Urban Region: The Developing Urban Detroit Area*, vol. 3, *A Concept for Future Development*, Detroit: The Detroit Edison Company.

Gallion, A.B. and Eisner, S. (1983) *The Urban Pattern*, 4th edn, New York: Van Nostrand Reinhold.

Glazer, N. (1970) *Cities in Trouble*, Chicago: Quadrangle Books.

Graham, S. and Marvin, S. (1996) *Telecommunications and the City: Electronic Spaces, Urban Places*, London: Routledge.

Graham, S. and Marvin, S. (2001) *Splintering Urbanism: Networked Infrastructures, Technological Mobilities and the Urban Condition*, London: Routledge.

Gutkind, E.A. (1962) *The Twilight of Cities*, New York: Free Press of Glencoe.

Hall, P. (1989) *London 2001*, London: Unwin Hyman.

Hammond, M. (1972) *The City in the Ancient World*, Cambridge, MA: Harvard University Press.

Lynch, K. (1960) *The Image of the City*, Cambridge, MA: MIT Press.

Mälkiä, M., Anttiroiko, A.-V. and Savolainen, R. (2004) *eTransformation in Governance: New Directions in Government and Politics*, Hershey, PA: Idea Group Publishing.

Mitchell, W. (1996) *City of Bits: Place, Space and the Infobahn*, Cambridge, MA: MIT Press.

Moss, M.L. (1987) 'Telecommunications, world cities, and urban policy', *Urban Studies* 24: 534–46.

Mumford, L. (1961) *The City in History*, New York: Harcourt, Brace & World.

Negroponte, N. (1995) *Being Digital*, New York: Vintage Books.

Norris, P. (2001) *Digital Divide: Civic Engagement, Information Poverty, and the Internet Worldwide*, New York: Cambridge University Press.

Oucho, J.O. (2001) 'Urban population trends', *Habitat Debate* 7(2): np.

Park, R.E. (1925) 'The city: suggestions for the investigation of human behavior in the urban environment', in R.E. Park and E.W. Burgess (eds) *The City*, Chicago: University of Chicago Press, pp. 1–46.

Passoneau, J.R. (1963) 'Emergence of city form', in W.Z. Hirsh (ed.) *Urban Life and Form*, New York: Holt, Rinehart & Winston, pp. 9–29.

Saarinen, E. (1943) *The City: Its Growth, Its Decay, Its Future*, New York: Reinhold Publishing Corporation.

Schiller, H.I. (1996) *Information Inequality: The Deepening Social Crisis in America*, London: Routledge.

Schon, D.A. (1971) *Beyond the Stable State*, New York: W.W. Norton & Co.

Toffler, A. (1970) *Future Shock*, New York: Random House.

—— (1980) *The Third Wave*, New York: Morrow.

—— (1990) *Powershift: Knowledge, Wealth, and Violence at the Edge of the 21st Century*, New York: Bantam Books.

United Nations (2001) *World Urbanization Prospects: The 1999 Revision*, New York: Population Division, Department of Economic and Social Affairs, United Nations.

Verne, J. (1996) *Paris in the Twentieth Century*, New York: Random House.

Warner, S.B., Jr (1962) *Streetcar Suburbs: The Process of Growth in Boston, 1870–1900*, Cambridge, MA: Harvard University Press.

Warschauer, M. (2003) *Technology and Social Exclusion: Rethinking the Digital Divide*, Cambridge, MA: MIT Press.

A

ABERCROMBIE, LESLIE PATRICK

b. 6 June 1879, Ashton-upon-Mersey,
England; d. 23 March 1957,
Aston Tirrold, Berkshire, England

Key works

(1933) *Town and Country Planning*, London: Home
University Library.
(1945) *Greater London Plan 1944*, London: His Majesty's
Stationery Office.

Patrick Abercrombie may appear a late developer:
he was 63 when his *County of London Plan*
(co-authored with J.H. Forshaw, chief architect-
planner of the London County Council) appeared
in 1943, 66 on publication of his even more celebrated *Greater London Plan* in 1945; he was
knighted that year, retired from academia the year
after, and died thirteen years later. But he was
already the most celebrated British planner of his
generation. Articled to an architect aged 18, lacking an academic degree, he had been appointed
research fellow in town planning – and first editor
of the *Town Planning Review* – in the new School of
Civic Design at the University of Liverpool in
1912; when Professor Stanley Adshead moved
to University College London (UCL) in 1915,
Abercrombie succeeded him in the Liverpool chair;
when Adshead retired in 1935, Abercrombie
followed him to UCL at the ripe age of 56.

Starting with his successful competition entry
for the Dublin Plan in 1913, he became a highly
successful planning consultant; in the 1920s
and 1930s, he developed a specialty in regional
planning – for the Doncaster region, for East
Kent and Suffolk, for Sheffield, and for Bath

and Bristol; he also acted as consultant on
Manchester's Garden City at Wythenshawe, where
Barry Parker was the architect. He became chair-
man of the Council for the Preservation of Rural
England, which he helped found in 1926, and was
on the Council of the Town and Country Planning
Association.

Abercrombie followed Patrick Geddes's principle of survey before plan; he used experienced
local collaborators to perform the survey work;
his skill was to synthesize complex planning concepts in memorable cartoon-like diagrams. After
Raymond Unwin's death in the United States, in
1940, he was Britain's pre-eminent planner, natural
choice for the London plans.

He was confident in his concept of planning:
'Planning simply means proposing to do, and
then doing, certain things in an orderly, pre-
meditated, related and rational way, having in view
some definite end that is expected to be beneficial.' He argued that 'Town and Country Planning seeks to proffer a guiding hand to the trend of
natural evolution as a result of a careful study
of the place itself, and its external relationships.
The result is to be more than a piece of skilful
engineering or satisfactory hygiene or successful
economics: it should be a social organism and
a work of art.'

Those qualities were evident in his London
plans – especially the *Greater London Plan*, which
proposed that London's physical growth should be
stopped by a green belt and that over a million
people should move out to new and expanded
towns beyond it. Rather remarkably, the Plan's
key elements were implemented over the following
quarter century, powerfully shaping the wider
London region as we know it today. Within
London, he was less successful: large parts of east

and south London were rebuilt according to his prescriptions, with a mixture of high- and low-rise development at a density of 136 people per acre, but the process remained incomplete and the bold highway plans – five orbital highways, at least two of them expressways, and high-capacity radials – fell victim to political opposition and a change in Zeitgeist. Despite that, of few other planners can Sir Christopher Wren's obituary seem so applicable: *si monumentum requiris, circumspice*.

Further reading

Dix, G. (1978) 'Little plans and noble diagrams', *Town Planning Review* 49: 329–52.

Manno, A. (1980) *Patrick Abercrombie: A Chronological Bibliography with Annotations and Biographical Details*, Leeds: Planning Research Unit, Leeds Polytechnic.

SIR PETER HALL

ABU-LUGHOD, JANET

b. 3 August 1928, Newark, New Jersey

Key works

(1989) *Before European Hegemony: The World System A.D.*, New York: Oxford University Press.

(1999) *New York, Chicago, Los Angeles: America's Global Cities*, Minneapolis: University of Minnesota Press.

Abu-Lughod's distinguished career as academic, author and practitioner devoted to processes of social change has spanned fifty years and has included positions in both the United States and the Middle East. Upon receiving her BA from the University of Chicago, Janet held positions as director of research for the American Society of Planning Officials (1950–52), research associate at the University of Pennsylvania, consultant and author for the American Council to Improve Our Neighborhoods (1954–57), and assistant professor at the American University in Cairo. Returning to the United States, Janet completed her Ph.D. at the University of Massachusetts and assumed a position as professor of sociology at Northwestern University. Upon retirement from Northwestern, she moved to New York to direct REALM, Urban Research Center at the New School for Social Research. She retired again and has held

visiting and short-term teaching appointments at Bosphorous University in Istanbul and on the International Honors Program at the University of Cairo.

Abu-Lughod has received over a dozen prestigious national government and philanthropic fellowships and grants to write and practise in the areas of demography, urban planning, urban sociology, economic and social development, world systems, and urbanization in the United States, the Middle East and the Third World. Her numerous publications include books, several hundred reviews, short pieces, limited circulation reports and monographs. Many of her articles have been reprinted in various collections of readings.

Abu-Lughod approaches the economic and social development of global cities not only with a scholarly understanding of the patterns and developments but also with a commitment to seeing and acting on possibilities for constructive social change. The span of her work goes from historical studies of medieval cities, to the analysis of the diffusion of global economies to cities in both the Western and the Arab world, to micro-level studies of territoriality and social change.

Further reading

(1991) *Changing Cities: Urban Sociology*, New York: HarperCollins.

(1994) *From Urban Village to East Village: The Battle for New York's Lower East Side*, Oxford: Blackwell.

(1999) *Sociology for the 21st Century: Continuities and Cutting Edges*, Chicago: University of Chicago Press.

DAVID C. PROSPERI

ACCESSIBLE CITY

Accessibility is more or less used internationally with the same meaning as that given to universal design (UD): the main difference is probably that UD is widely devoted to every field, from products to components and built environments, while accessibility is directly related to the building of environments in the appropriate way, to be easily used by everybody. Even if these concepts were also expressed in other countries, the term UD was first used in the United States by Ron Mace in 1985, who then affirmed that UD is an approach to design that incorporates products as well as

building features which, to the greatest possible extent, can be used by everyone.

Accessibility, a term principally adopted in Europe, in many countries is also related to technical parameters, including not only a space usable by everybody but also any effort to improve safety, comfort and easy use of the environment. Accessibility was first applied to public buildings, then housing, residential buildings and related surroundings, while for transport means and departure/arrival areas, the concept firstly involved air transport, then railways and, yet not sufficiently, bus systems and maritime transport. Finally, the concept is applied to every built environment, including open-air areas such as public parks or archaeological areas.

At a legislative level, measures were addressed by the United States through the Americans with Disabilities Act of 1990, a law focusing on all the building aspects, products and design that is based on the concept of respecting human rights. The United Kingdom established technical rules among the principal builders while other industrialized countries focused on laws on public buildings (schools, hospitals, offices).

Accessible city is a newer concept: planning an accessible city in the first half of the 1990s meant rehabilitation works to facilitate people in visiting specific tourist areas, especially of the historic centres, and allowing them to access public offices. Then, it spread to all the elements of a city, especially in new planning: pathways and pedestrian areas, bus stops, public parking, green areas, street furniture.

The accessible planning concept evolved too, introducing first only barrier-free elements, such as lifts, ramps and wide entrances, but subsequently also tools to overcome sensorial impairments (e.g. hearing traffic lights, pedestrian pathways for people with visual impairment), and in the late 1990s, a barrier was considered to be any element which could make uncomfortable the use of the environment (e.g. poor public street lighting).

Since design must face many considerations, in order to include all possible needs of end-users, the integration of disciplines is essential, from human and psychological subjects to technical and economic competencies. Design is, therefore, a sensible approach to allow the built environment and technical elements to satisfy daily life exigencies. Because users' needs change during their lifespan,

accessibility must also guarantee flexible use, which characterizes the continuous evolution of the concept.

Further reading

Christophersen, J. (2002) *Universal Design: 17 Ways of Thinking and Teaching*, Oslo: Husbanken.
Preiser, W.F.E. and Ostroff, E. (2001) *Universal Design Handbook*, New York: McGraw-Hill.

ANNALISA MORINI

ACROPOLIS

An acropolis (meaning 'city on a hill') is a settlement in ancient Greek cities that evolves into a religious sanctuary. The high site provides natural defences as a CITADEL and the mountain itself contributes to its religious significance by the presence of sacred springs. If the city was attacked, citizens could congregate in the acropolis where all the important buildings were. In Greek colonies, the acropolis is often a fortified citadel. As the city grew and expanded beyond this high area, the site became the most sacred space in the city.

The Acropolis in Athens is the best-known example. It achieved its form in the fifth century BC and is currently an archaeological site. It is on a hill 300 feet high with easy access only from the west, which made it into a natural fortress. It contains a temple to the city goddess, Athena, called the Parthenon (447–432 BC), an elaborate entrance to the religious precinct called the Propylaea (437–432 BC), and other temples such as Erechtheum (420–393 BC) and the Temple of Athena Nike (426 BC). These buildings, sited in accordance with topography and respect for traditional siting, are considered models of Greek architecture due to their outstanding designs. The Acropolis is located south of the AGORA, the multi-purpose city centre.

Further reading

Hurwit, J.M. (2001) *The Athenian Acropolis*, Cambridge: Cambridge University Press.
Morris, A.E.J. (1994) 'Greek urban form components', in *History of Urban Form*, 3rd edn, New York: John Wiley & Sons, pp. 41–50.

ANA MARIA CABANILLAS WHITAKER

ADAMS, THOMAS

b. 10 September 1871, Corstorphine, Scotland; d. 24 March 1940, Henleys Down, England

Key works

(1917) *Rural Planning and Development: A Study of Rural Conditions and Problems in Canada*, Ottawa: Commission of Conservation of Canada.
(1929) *The Graphic Regional Plan*, New York: Regional Plan of New York and Its Environs.
(1931) *The Building of the City*, New York: Regional Plan of New York and Its Environs.
(1935) *Outline of Town and City Planning: A Review of Past Efforts and Modern Aims*, New York: Russell Sage Foundation.

Planning advocate, leader of town planning in Great Britain and Canada, and pioneer of REGIONAL PLANNING in the United States, Thomas Adams began a first career in journalism in London in 1900. Between 1903 and 1906, he became familiar with the GARDEN CITY movement while serving as the first manager of Letchworth. In England, he established a successful practice as town planner by designing low-density residential developments known as garden suburbs. In 1909, Adams was appointed as Town Planning Adviser to the Local Government Board, where he remained up until 1914. In 1914, he was invited to Canada to work for the Commission of Conservation, whose members were seeking ways to provide better housing for the growing population of industrial cities. Adams brought with him the garden city ideas; in addition, he advocated more planning regulations and institutions at the local level. He prepared comprehensive plans for Canadian resource communities and garden suburbs in which he included the ideas of balance and inter-dependence between city and country. In 1923, he moved to the United States to become director of the Regional Plan of New York, a position he held until 1930. Funded by the Russell Sage Foundation, this project involved the physical and social survey of an area of 5,000 square miles with nearly 9 million people. Published in 1929, the plan anticipated the region's basic transportation and infra-structure needs for the next three decades. Adams took an active role in creating planning institutions in Great Britain, Canada and the United States. In addition to being a prolific lecturer and writer, Adams held various teaching positions.

Further reading

Simpson, M. (1985) *Thomas Adams and the Modern Planning Movement: Britain, Canada and the United States, 1900–1940*, London and New York: Mansell.

CLAIRE POITRAS

ADAPTIVE REUSE

Adaptive reuse, also known as recycling and conversions, refers to the reuse of a building by adapting it to accommodate a new use or uses. Gaining impetus from both the HISTORIC PRESERVATION movement and proven economic feasibility, adaptive reuse has prevented the demolition of thousands of buildings and has allowed them to become critical components of URBAN REGENERATION.

Since the 1970s, a select number of adaptive reuse projects, from the Ghiradelli Chocolate Factory/Ghiradelli Square Shopping Complex (San Francisco) to the Bankside Power Station/Tate Modern Art Museum (London), have gained international recognition. However, the greatest value of the adaptive reuse movement is represented by the hundreds of abandoned schools, factories, warehouses, military posts and hotels of local historic value that have been creatively adapted for use as affordable housing, office buildings, commercial complexes, as well as civic, educational, cultural and recreational centres.

Most adaptive reuse projects, especially in the private sector, are dependent on their economic feasibility. This is typically determined by the building's existing physical configuration and condition – especially the need to mitigate hazardous materials such as asbestos or lead-based paints; the viability of its location and immediate context; and the cost of compliance with current ZONING, building, fire and/or HISTORIC DISTRICT codes and ordinances.

Successful adaptive reuse projects require the knowledge and skills of creative architects and engineers, code consultants, construction managers, contractors and skilled tradespersons.

Further reading

Thompson, E.K. (ed.) (1977) *Recycling Buildings*, New York: McGraw-Hill.

ANTHONY J. 'TONY' COSTELLO

ADDAMS, JANE

b. 6 September 1860, Cedarville, Illinois;
d. 21 May 1935, Chicago, Illinois

Key works

(1909) *Twenty Years at Hull-House*, New York: Macmillan.
(1910) *The Spirit of Youth and the City Streets*, New York: Macmillan.

Jane Addams, the first American woman to win a Nobel Peace Prize, is remembered as a clinical sociologist, social worker, peace activist and urban reformer. In 1889, Addams and her good friend Ellen Gates Starr established a settlement house in the decaying Hull Mansion in Chicago. Hull-House, as it was called, had many aims, not the least of which was to give privileged, educated young people contact with the real life of the majority of the population. The core Hull-House residents, an important group in the development of urban sociology, were well-educated women bound together by their commitment to progressive causes such as labour unions, the National Consumers League and the suffrage movement. During the next 45 years, Jane Addams would travel widely, but Hull-House remained her home.

Hull-House, a national symbol of the settlement house movement, was a centre of activities for the ethnically diverse, impoverished immigrants in the Nineteenth Ward of Chicago. Within five years, some forty clubs were based in the settlement house and over 2,000 people came into the facility each week. Hull-House operated a day nursery, hosted meetings of four women's unions, established a labour museum, ran a coffee house, and held economic conferences bringing together businessmen and workers. The Working People's Social Science Club held weekly meetings, and a college extension programme offered evening courses for neighbourhood residents. A few University of Chicago courses were available there and the Chicago Public Library had a branch reading room on the premises.

Addams challenged the competency of male city administrators. She criticized their civic housekeeping skills, questioned their willingness to meet social needs and thought they deprived American citizens of genuine democracy. Nearly every major reform proposal in Chicago (1895–1930) had Jane Addams's name attached in some way. Her involvement in major issues – such as factory inspection, child labour laws, improvements in welfare procedures, recognition of labour unions, compulsory school attendance and labour disputes – catapulted her to national prominence. Intellectuals, including Beatrice Webb and Sidney Webb, came from around the world to Chicago to meet Addams and her colleagues.

Hull-House was a base for promoting political, economic and social reform as well as systematic investigation. In 1895, *Hull-House Maps and Papers* was published. This ground-breaking document, dealing with tenement conditions, sweatshops and child labour, was the first systematic attempt to describe immigrant communities in an American city.

Addams was a prolific writer whose most successful book was her moving autobiography, *Twenty Years at Hull-House* (1910). Her years of work and writing in the interest of peace and democracy earned her the Nobel Peace Prize in 1931. When she died in 1935, Addams was America's best-known female public figure.

Further reading

Elshtain, J. (ed.) (2002) *The Jane Addams Reader*, New York: Basic Books.

JAN MARIE FRITZ

ADVOCACY PLANNING

Advocacy planning is one vision of how urban planning might be organized, proposed by lawyer/planner Paul Davidoff. Davidoff attacked city planning commissions as 'non-responsible' and 'vestigial' institutions which produced 'unitary' plans focusing only on physical development. He deplored the idea that urban planners should serve as mere technicians in such a limited and dull process.

Rather, Davidoff called for planners to represent different constituencies and different points of view in a political planning process, much as lawyers advocate for the interests of their clients. Ideally, Davidoff would have liked to see multiple and competing urban plans addressing both physical and social issues, developed by advocate planners representing different interest groups, presented directly to local elected officials.

The conventional view of urban planning calls for planners to develop a single unitary plan ostensibly promoting the public interest. The planner's role is seen as largely technical – developing the one best plan to reach the public interest. Technique is paramount; goals and values secondary. This model assumes that there is a common public interest and that planners have the knowledge and technical skills to define it and propose the best way to achieve it.

Davidoff himself worked tirelessly to bring racially integrated low- and moderate-income housing into affluent white suburbs. In his practice he encountered neighbourhood organizations, environmentalists, real estate developers and civil rights groups with radically different ideas about what the plans for the future of a community should be. Davidoff certainly did not want to let one set of planners working for the city fathers of the smug, affluent, exclusionary communities where he was trying to get racially integrated low- and moderate-income housing built define the public interest. Their concept of what was good for their communities was to keep out low-income and minority people. The technical subdivision, ZONING, building code and other tools they proposed did not advance racial integration or social equity.

Advocacy was one cornerstone of Davidoff's approach. In a courtroom, skilled lawyers advocate for the interests of their clients. In courts of law, advocacy makes the lawyers of both plaintiffs and defendants work hard and do quality work because anything they say will be analysed and may be challenged by an opposing advocate. Advocacy exposes judges to a full range of choices for deciding a case. Davidoff argued that if the same kind of vigorous advocacy of competing claims happened in the urban planning process, weaknesses in proposed plans would be exposed and alternative courses of action would come to light. Davidoff argued that plans were (and should be) political documents and that the practice of planning was inherently political.

Pluralism was the other cornerstone of Davidoff's approach. In a democracy, multiple points of view are encouraged and even minority points of view protected. Davidoff was concerned that as central bureaucratic government control grew, unpopular views and views held by a minority of citizens would not be heard. For him, planning to protect local, specialized interests was essential. The style of advocacy planning that Davidoff proposed would acknowledge, indeed further, the engagement of many different groups in urban planning.

Davidoff believed that planning could not be done from a position of value neutrality. He argued that prescriptions themselves depend on desired objectives. In the advocacy planning model, goals and ideals take precedence over technical issues. Distributive justice is a paramount value.

Davidoff believed that many different organizations might develop their own plans: chambers of commerce, real estate boards, labour organizations, pro- and anti-civil rights groups, and anti-poverty groups. Davidoff was particularly interested in effective advocacy on behalf of low-income and minority communities and other disenfranchised groups. He believed that if advocacy planners were available to these groups, they could change plans formulated by elites to better serve their interests.

Davidoff argued that his model of advocacy planning would improve urban planning practice in several ways. The public would be better informed of alternative choices, because competing groups would forcefully argue in favour of their plans. Advocacy planning would require planning agencies to do high-quality work in order to have their plans prevail. It would force critics of establishment plans to produce superior plans themselves, rather than merely critiquing plans they did not like.

Davidoff's lucid and revolutionary approach achieved critical acclaim among urban planners. It generated criticism from conservative physical planners who were comfortable with unitary physical planning before city planning commissions and threatened by multiple plans coming from multiple sources. Radicals attacked advocacy planning on the theory that it could co-opt community anger away from political mobilization and protest that could achieve change into mere plan-making.

Davidoff's message spoke to the generation of socially conscious planners in the 1960s and 1970s. Some styled themselves advocacy planners and set forth to advocate on behalf of groups whose voices they felt were not heard in the planning process. In a few cases, funding was made available for advocate planners. But city planning commissions did not wither away, unitary plans are still the order of the day, and most urban planning today remains physical planning. Nonetheless, planners continue to be inspired by the idea of advocacy planning, and alternative plans are sometimes

developed by neighbourhood organizations and other groups.

In recent decades, the idea of advocacy planning has blended into the idea of EQUITY PLANNING. Equity planners may not consciously take the side of one group, but their planning goals are informed by concern for different segments of the population.

Further reading

Blecker, E. (1971) *Advocacy Planning for Urban Development*, New York: Irvington.

Cloward, R.A. and Piven, F.F. (1965) 'Whom does the advocate planner serve', in *The Politics of Turmoil*, New York: Vintage.

Davidoff, P. (1965) 'Advocacy and pluralism in planning', *Journal of the American Institute of Planners* XXI(4).

Rothblatt, D. (1982) *Planning the Metropolis: The Multiple Advocacy Approach*, New York: Praeger.

SEE ALSO: community conflict resolution; community goal setting; community visioning; direct democracy; social inequality; social justice; social polarization

RICHARD LEGATES

AGORA

An agora (derived from the Greek word meaning 'to meet') is a central open space in classical Greek cities (900–338 BC) which best represents the response of city form to accommodate the social and political order of the POLIS. The agora was used for many functions. It was a place of assembly where all the citizens except for women and slaves could convene and make democratic decisions. It accommodated commercial, educational, religious and social activities such as theatrical and gymnastic performances and special celebrations. The agora was surrounded by buildings used for special functions such as civic meetings in the bouleterion (council chambers) or the tholos (a circular committee meeting building), a variety of activities in the stoas (long rectangular colonnaded buildings), and religious activities (temples and shrines).

The agora in the city of Athens (currently an archaeological city) was an open, irregular area surrounded by buildings at different angles. In classical times, it was crossed by the Panathenaic Way, a processional path towards the ACROPOLIS. The Athenian agora was changed several times until in Roman times it became too crowded and a new Roman forum was built. In Greek colonies, such as Miletus and Priene in Asia Minor, there were one or more agoras, but they were rectangular, open spaces surrounded by colonnades, or stoas.

Further reading

Camp, J.M. (1986) *The Athenian Agora: Excavations in the Heart of Classical Athens*, London: Thames & Hudson.

Morris, A.E.J. (1994) *History of Urban Form*, 3rd edn, New York: John Wiley & Sons, pp. 41–50.

ANA MARIA CABANILLAS WHITAKER

AIR RIGHTS

An air right or an air space parcel is a real property right providing for the legal ownership of a space above a parcel of land or within a building. As a real property interest, it is defined by survey and can be conveyed by deed. The property itself may be conjectural since it cannot be touched but it can be exactly delineated by the horizontal limits within which it lies by establishing a datum, i.e. a recognizable height marker, with respect to the underlying property. The slice of air space can also be described as horizontal property to differentiate it from other real properties that extend in principle from the centre of the earth 'to the heavens above'. In the system of common law, the legal description of the property right can be established by a unique contract signed and sealed with the owner of the underlying piece of property. Such contracts were at one time uniquely tailored to a specific situation, but since the adoption of enabling legislation – known in the United States as Horizontal Property Acts, in Canada as the Air Space Act – such contracts have become largely standardized. This facilitates their negotiability in business transactions.

As a normal facet of real property, air rights are a commodity. A landowner can sell them separately, just like mineral rights, user rights or development rights. As real estate becomes scarce in urban areas, air rights become more valuable. By allowing multiple use of a parcel, air rights can be developed to create space in an already developed area, especially where land use is extensive, such as traffic infrastructure, piers on the waterfront, etc.

For example, a developer may negotiate for control of air rights above an urban railway and then construct buildings above the tracks.

The most widespread applications of air rights are condominiums. In most cases, the unit owners only own the air rights, the space within the unit. The homeowner association owns the common elements – the walls, floors, ceilings, etc. Within a unit, the owner enjoys the usual bundle of real property rights such as possession, control, exclusion and disposition. The homeowner association places restrictions on these ownership rights, mostly in the form of easements that are appurtenant to the property.

Before the advent of condominiums, the most prevalent application of air rights involved railroads. These air rights were established when cities became densely developed and stronger construction materials, particularly concrete and steel to support superstructures, became widely available. Many railroad companies made extra money by selling or leasing the air rights above the tracks, a common practice in the early twentieth century. Some famous examples of buildings constructed in the air space of railroad property are the Merchandise Mart and the Post Office in Chicago, and Park Avenue and the office building erected on top of Grand Central Terminal in New York. The development of air rights is still on the rise; the Mass Turnpike in Boston will support new park space, and derelict docklands will provide air space to construct Manhattan's new Westside stadium complex.

Further reading

Whyte, W.H. (1988) *City: Rediscovering the Center,* New York: Doubleday, pp. 277–9.

JAN VAN WEESEP

ALBERTI, LEON BATTISTA

b. 14 February 1404, Genoa, Italy;
d. 25 April 1472, Rome, Italy

Key works

Descriptio urbis Romae (1443–49).
De re aedificatoria (1452, published 1485).

Italian humanist scholar. Although he was educated in the classical texts and studied law at the University of Bologna, his training in mathematics was the foundation for his studies of perspective and architecture.

Alberti represents the ideal Renaissance man who comprises the knowledge of his time. In 1432, he was appointed as secretary in the Papal Chancery, and in the late 1440s he began to work as an architect, designing both civil and religious buildings like the Palazzo Rucellai, the church of San Francesco (Rimini) and the façade of Santa Maria Novella (Florence). Alberti is more a theoretical than a practical architect, but he managed to keep a balance. After studying the text on architecture of Vitruvius, he writes *De re aedificatoria* (*On the Art of Building*), the first treatise on architecture of the Renaissance, the result of his fascination with Roman and Greek architecture. He proposes that use of the nine basic ideal geometric shapes be harmonically combined. He anticipates the principle of street hierarchy, with wide and straight MAIN STREETS connected to secondary streets, and buildings with equal height.

Further reading

Westfall, C.W. (1974) *In This Most Perfect Paradise: Alberti, Nicholas V, and the Invention of Conscious Urban Planning in Rome, 1447–1455,* University Park: Pennsylvania State University Press.

URBANO FRA PALEO

ALEXANDER, CHRISTOPHER

b. 4 October 1936, Vienna, Austria

Key works

(1964) *Notes on the Synthesis of Form,* Cambridge, MA: Harvard University Press.
(1979) *The Timeless Way to Building,* New York: Oxford University Press.

In the scenery of contemporary architecture, Christopher Alexander occupies a singular position, marked by his critical attitude to modern architecture, and the profound coherence between a theoretical reflection seated in a challenging epistemological eclecticism, and his vigorous architectural production.

After studying mathematics and architecture in Cambridge, England, he studied in the United States, obtaining his Ph.D. at the University of Harvard. Since 1963, he has held the post of

professor at the University of California in Berkeley, being also founder and director of the Center for Environmental Structure.

A pioneer in understanding the possibilities of the systemic and cybernetic approach – and, by extension, of the use of computers – as a project tool, in the early 1960s, his work is recognized principally by the *patterns* notion, social-spatial entities that are integrated as language that involves relationships of form, order and structure. This approach already appears sketched in his seminal work, *Notes on the Synthesis of Form* (1964) – a conceptual and mathematical essay on design processes – and is consolidated in the trilogy *Pattern Language* (1977), *The Timeless Way to Building* (1979) and *The Oregon Experiment* (1975).

Another essential aspect of his thought and work is the attempt to understand users' concrete demands and their role in design decisions. This perspective is present in the University of Oregon Plan – whose cooperation process between users and architects, objectifying the construction of an organic order, is described in full detail in *The Oregon Experiment* – and in the Mexicali housing project (1976), in which low-cost housing was produced with participation from dwellers, as is detailed in *The Production of Houses* (1985).

In *A New Theory of Urban Design* (1987), Alexander suggests the idea of *growing whole*, as a condition of organic growth that governs complex relationships between nature and human ingenuity, between space and behaviour, and between parts and the whole. *The Nature of Order* (2002) concludes his comprehensive theory in the search for an understanding of the generating processes of living structures.

Works and projects that stand out in his vast production are the Eishin School (Japan, 1985–87), the Shelter for the Homeless (California, 1990) and the Mary Rose Museum (the story of which is told in the homonymous book of 1993). However, the reach of his work goes beyond architecture, finding significant interest in the area of object-oriented programming, for the application of original principles of the pattern language idea.

Further reading

Gabriel, R.P. (1996) *Patterns of Software: Tales from the Software Community*, New York: Oxford University Press.
King, I.F. (1993) *Christopher Alexander and Contemporary Architecture*, Tokyo: a + u Publishing.

LEANDRO M.V. ANDRADE

ALINSKY, SAUL D.

b. 30 January 1909, Chicago, Illinois;
d. 12 June 1972, Carmel, California

Key works

(1946) *Reveille for Radicals*, New York: Vintage.
(1971) *Rules for Radicals: A Primer for Realistic Radicals*, New York: Vintage.

Saul Alinsky, the renowned community organizer and self-styled radical and rabble-rouser, helped develop the *direct action* approach of the Industrial Areas Foundation that influenced workplace and neighbourhood organizing efforts in American cities for generations that followed. Alinsky grew up in the poor, Russian-Jewish immigrant section of Chicago. He came to believe that the poor needed to adopt pragmatic, power tactics as the 'haves' of American society were never going to voluntarily cede privileges to the 'have-nots' and the 'have-little-want-mores'.

Alinsky's first noteworthy success came in 1939 in the predominantly Catholic and impoverished Back of the Yards area that bordered the city's stockyards. In the 1960s, he helped African-American residents on the city's South Side to form The Woodlawn Organization (TWO). TWO aimed fair pricing campaigns at GHETTO merchants (where TWO used scales to weigh purchases and dramatically publicize just which shops cheated customers). The threat of protests forced Chicago's downtown department stores to adopt fair hiring and promotion practices. TWO also fought against expansion plans by the University of Chicago that would have displaced large numbers of neighbouring, poor, African-American residents. In the mid-1960s, Alinsky organized a similar campaign for racial justice in Rochester, New York, a campaign that was largely aimed at the city's dominant employer, the Eastman Kodak Corporation. At the time of his death, Alinsky was expanding his organizing efforts to help the middle class combat the power of corporate America.

Under the 'Alinsky method', the organizer works with a community's natural indigenous leaders to discover just what is a source of grievance in a community. The organizer then polarizes the situation, rubbing raw the wounds in order to mobilize community members in a protest action that will disorient and intimidate the target of the

action. Once the target of an action is picked, the Alinsky-style organizer freezes it, not allowing the target to shift the blame for unjust social conditions to others. The organizer seeks to sustain the organization through a succession of small, quick victories.

Alinsky believed in the possibilities of building on neighbourhood progressive organizations, especially labour unions and churches, which attracted the membership of a community. He sought to identify the natural indigenous leaders of a community who would eventually oust the professional organizer. Alinsky wanted the poor to understand that they could only depend on themselves in the fight for social justice. He had little patience for critics, especially those self-styled radicals of the 1960s and early 1970s who charged that his tactics were too conservative, that he focused on small, discrete projects rather than systems change. For Alinsky, the true radical was one who pursued and achieved pragmatic change, not one who espoused the rhetoric of radical restructuring.

Further reading

Horwitt, S. (1989) *Let Them Call Me Rebel: Saul Alinsky, His Life and Legacy*, New York: Knopf.

MYRON A. LEVINE

ALLEY

An alley is a laneway in the interior of a block giving access from the street to the rear of buildings or lots. The alley has existed in virtually every CULTURE and every era, and its form and function vary greatly. It can be rectilinear or curvilinear, lead directly from one street to another, or be organized in a 'T' or 'H' or other pattern. Unlike a STREET, an alley is generally well defined by buildings, walls or fences. Narrower than the street, it provides a high level of privacy and of enclosure, and can be either privately or publicly owned. It provides an unstructured setting for local COMMUNITY interaction, and often houses uses and activities, from cheap housing to garbage removal, which are often seen as undesirable on the more public side of properties.

The alley originated as an efficient way to subdivide land: it provided maximum frontage and access, and organized the delivery of services and utilities. But it was vilified by nineteenth-century social reformers because of the BLIGHT and associated health and safety problems that plagued rear laneways in poorer urban areas. The alley ceased to be a common form of subdivision in the 1930s, but it re-emerged with the advent of the NEW URBANISM. In older neighbourhoods, the alley largely remains a hidden resource, waiting to be used for infill housing, play and recreation, or urban agriculture.

Further reading

Clay, G. (1978) *Alleys: A Hidden Resource*, Louisville, KY: Grady Clay & Co.

PETER REED

ALONSO, WILLIAM

b. 29 January 1933, Buenos Aires, Argentina; d. 11 February 1999, Boston, Massachusetts, United States

Key works

(1964) *Location and Land Use*, Cambridge, MA: Harvard University Press.
(ed.) (1987) *Population in an Interactive World*, Cambridge, MA: Harvard University Press.

A regional and urban planning scholar, William Alonso was one of the first to create the basis for a generalized theory of inter- and intra-regional location. After a BA in architecture (1954) and an MA in city planning (1956), both from Harvard, Alonso completed his Ph.D. in regional science (1960), the first Ph.D. ever awarded in this field, at the University of Pennsylvania. On the subject of his doctoral thesis was based his first book, *Location and Land Use* (1964), one of the first books on location patterns in metropolitan areas.

One of his main contributions to urban economics research was the development of the well-known Alonso model. Extending the Von Thünen model on agricultural land use and rent, Alonso developed a model that deals with land use, land rent and location in and around the central business district of a metropolitan area. The model revolves around a mono-centric urban form, and explains the location patterns of households and businesses in metropolitan areas, giving great importance to accessibility and transport costs

as the main determinants of land use and rents in urban areas. Furthermore, Alonso's research covered the areas of MIGRATION and policy-oriented studies of demographic change.

In addition to his academic work at universities such as Harvard, California at Berkeley and Yale, Alonso served as an adviser and consultant to the United Nations, the World Bank and many international foundations.

DIONA LAMBIRI

ALTERNATIVE DISPUTE RESOLUTION (ADR)

Whenever people come together, there will be conflict. Cities bring people together and, hence, create opportunities for conflict. Familiar sources of conflicts in urban settings include public policy formulation, housing development, environmental concerns, labour issues, racial and class tensions, crime, and simple neighbourhood living. As inevitable as these conflicts are, it is not inevitable that they will be destructive. The resultant processes and outcomes are heavily dependent upon the ways in which the conflict is managed.

'ADR' is a catchall phrase to refer to a variety of dispute resolution mechanisms that are distinguished by not involving the formal adversarial method of litigation. These processes are voluntary and non-violent in nature. Processes that fall under this rubric include negotiation, conciliation, facilitation, mediation, fact-finding, mini-trials and arbitration. One way to think about these processes is to arrange them along a continuum defined by the level of involvement and control of the disputing parties and that of one or more neutral third parties. Thus, *negotiation* would be at one end as this process involves only the disputants who develop their own agreement through constructive dialogue. *Arbitration* would be at the other end where the third party makes a decision that in some forms is binding on the parties. *Mediation* would be midway on this continuum as it involves a third party facilitating the development of an agreement by the parties to resolve their difficulties.

The interest in ADR has been fuelled by overburdened court systems whose delays and costs result in limited access to quick and efficient justice for many people. In addition, there is recognition that for many disputes, particularly those involving ongoing relationships, litigation and adjudication are not appropriate. In order for a dispute to be handled in the courts, it needs to be packaged as a matter of rights and that one party is clearly wrong. In addition, judgments are limited mainly to fines and sentences. For many disputes, such as neighbour, family and labour, this packaging does not capture the complexity of the issues and the judgments do not meet the need for workable solutions that will move the relationship forward (assuming the relationship is to be preserved). Though not intentionally, the litigation process often creates further hostilities and makes more enemies than it does mend and rebuild relationships. The weakening of traditional conflict resolvers who were provided by extended families, churches and civic groups has fuelled this reliance on the legal system as a mechanism for handling disputes. Increasing mobility, urbanization and increasing anonymity of citizens have contributed to the breakdown of strong social ties and a sense of community and connection. In essence, the role of the broader community in providing guidance regarding appropriate behaviours and helping its members deal with their disputes has been diminished, reducing the social controls on hostile and disputing behaviours. While the legal system frames issues as rights and solutions that are win-lose in nature, ADR processes reflect the view that many disputes are about fundamental interests or needs that are not met and concern hurt feelings and disappointments regarding unmet (and often unspoken) expectations within the context of the relationship. By viewing disputes in this way, these processes seek to help surface those interests and help devise solutions that will address the important needs of *all* the parties to the dispute. Thus, the perspective that many disputes can and should be addressed in a win-win fashion underlies much of the ADR practice.

An ADR movement that is particularly noteworthy is community mediation in the United States. This movement focuses on providing a variety of low-cost, informal and effective processes for handling a variety of disputes. In the late 1960s and early 1970s, a few innovative programmes were developed to deal with crimes such as harassment, minor assault and fraudulent financial dealings in a more effective manner. The field grew rapidly during the late 1970s and early

1980s through funding from federal, state and local governments as well as foundations. As of the late 1990s, every state was providing support and funding for a statewide system of community mediation centres. The range of services offered by such centres has expanded over time to include minor criminal and civil case processing, school-based dispute resolution, divorce/custody dispute resolution, intergroup dispute resolution, public policy dispute resolution mechanisms and victim-offender reconciliation efforts. There is accumulating evidence that ADR processes produce greater disputant satisfaction, longer-lasting agreements, and are more cost effective than litigation. Unfortunately, despite such evidence, many community mediation centres struggle to remain open, hampered by uncertain funding and limited public awareness of these alternative approaches and their effectiveness.

In addition to these non-profit efforts, there has been increasing movement within government and for-profit businesses towards handling disputes in a more consensual manner. This interest has manifested itself in the design and implementation of dispute resolution systems within organizations that provide a variety of entry points and mechanisms for dealing with disputes involving their employees. For example, the US Postal Service (USPS) in the late 1990s introduced mediation as an optional mechanism for dealing with some disputes that were being brought to their Equal Employment Opportunity offices. In doing this, the USPS recognized that these disputes involved people who had ongoing working relationships that needed to be preserved. By providing a more consensual form of handling the dispute, it was believed that relationships would be improved and the culture of the organization enhanced. A tall order, but the current evidence is that it is having a significant effect on employees' satisfaction with the resolutions reached and with the company overall.

While ADR has not yet reached its full potential for enhancing the quality of relationships for all people, it is meeting a clear need to increase the possibilities for dealing with conflict constructively, non-violently and humanely.

Further reading

Deutsch, M. and Coleman, P.T. (2000) *The Handbook of Conflict Resolution*, San Francisco: Jossey-Bass.

Merry, S.E. and Milner, N. (eds) (1993) *The Possibility of Popular Justice: A Case Study of Community Mediation in the U.S.*, Ann Arbor: University of Michigan Press.

LORALEIGH KEASHLY

ALTSHULER, ALAN A.

b. 9 March 1936, Brooklyn, New York

Key works

(1965) *The City Planning Process: A Political Analysis*, Ithaca, NY: Cornell University Press.
(2003) *Mega-Projects: The Changing Politics of Urban Public Investment*, Washington, DC: The Brookings Institution Press (with D. Luberoff).

Urban planning scholar and practitioner, Altshuler is an internationally renowned expert in the fields of urban politics, land use planning and transportation policy. He received his BA from Cornell University (1957), and his MA and Ph.D., both in political science, from the University of Chicago (1959, 1961), where he worked under the direction of Edward C. BANFIELD and Herbert J. Storing. Since 1988, Altshuler has been the Ruth and Frank Stanton Professor of Urban Policy and Planning at Harvard University, director of the Taubman Center for State and Local Government, and, until 1998, director of the Ford Foundation Program on Innovations in American Government. Previously, he served as dean of the Graduate School of Public Administration, New York University (1983–88), and professor at MIT (1966–71, 1975–83) and Cornell University (1962–66).

Altshuler has also held various public appointments in the Commonwealth of Massachusetts. He chaired Governor Francis W. Sargent's Task Force on Transportation (1969–70), which recommended new transportation policy initiatives for Greater Boston. As a gubernatorial appointee (1970–71), he implemented a US$3.5 million programme to develop a multi-modal transportation plan for metropolitan Boston, and the Central Artery project was first envisioned while he was Massachusetts' Secretary of Transportation and Construction (1971–74).

His research has brought to the fore the political dimension and policy issues of urban and transportation planning. In his first book, *The City Planning Process* – a case study of the political

dilemmas of plan implementation – he analysed the actions of planning officials in Minneapolis/ St. Paul, Minnesota, during the 1950s. Observing that planners lacked power and political skills, he challenged the legitimacy and effectiveness of comprehensive planning. In the 1990s, his examination of innovation in American government revealed some fundamental changes in public management – increased emphasis on outcome accountability, efficiency, effectiveness and public satisfaction – as well as a number of challenges, opportunities and dilemmas.

At the beginning of the twenty-first century, his thought-provoking perspective is even more relevant. Focusing on the politics of mega-projects over the past half-century, he remarks that in an era of shrinking budgets, increased requirement for local consensus and involvement of more interest groups in the decision-making process, planning has become more political. Also, although he does not exclude the possibility of future large public investment projects, he claims that they will be vastly more expensive than their predecessors and that enormous mega-projects such as Boston's Big Dig are unlikely.

Further reading

Altshuler, A.A. and Behn, R.D. (eds) (1997) *Innovation in American Government: Challenges, Opportunities, and Dilemmas*, Washington, DC: The Brookings Institution Press.
Altshuler, A.A. and Gómez-Ibáñez, J.A. (1993) *Regulation for Revenue: The Political Economy of Land Use Exactions*, Washington, DC: The Brookings Institution Press; Cambridge, MA: Lincoln Institute of Land Policy.

CATHERINE C. GALLEY

ANNEXATION

Annexation occurs when a municipality extends its boundaries outwards, absorbing neighbouring territory. Many of America's largest cities achieved their present size as a result of annexations carried out at an earlier point in time. In 1854, Philadelphia expanded from 2 square miles to 136 square miles. In 1898, New York City, then only 40 square miles in size, consolidated with Brooklyn, capturing over 240 new square miles of territory and 1 million new residents. Los Angeles pursued an aggressive policy of annexation and more

than tripled its size in just ten years, growing from 108 square miles in 1915 to 415 square miles in 1925.

The rules that govern municipal annexations vary by country. Under Britain's unitary system, the national government has the authority to establish local governments and determine their territorial jurisdiction and powers. In the United States, state law, not national law, regulates and sets the procedures for annexation. In France, efforts to induce mergers among the country's 36,000 historically rooted communes (many of quite small population size) have more often than not proved futile.

In many cases, the residents of rapidly growing and underdeveloped outlying areas may see annexation as a means of gaining public water, road paving and other urban services. In the United States, however, a watershed event occurred when Brookline, Massachusetts, in 1873, refused incorporation into the city of Boston, marking the beginning of a new era of suburban resistance to annexation. By the latter half of the twentieth century, large-scale annexation largely proved impossible in the Northeast and the Midwest. In the SUNBELT and the Pacific Rim states, though, more relaxed state annexation statutes facilitated the dynamic growth of a number of major cities, including Albuquerque, Austin, Charlotte, Fort Worth, Fresno, Houston, Oklahoma City, Phoenix, Portland, San Antonio, San Diego, San Jose and Tulsa.

In the United States, state laws make it difficult for a jurisdiction to swallow up a legally incorporated neighbour; consequently, most contemporary annexations are of unincorporated areas, that is areas that have not formed their own legally recognized municipal governments. Cities annex relatively sparsely populated tracts of land in order to promote economic growth. Houston and Denver each annexed sizeable acreage outside their old city border in order to build a new international airport.

Except for relatively minor boundary adjustments, American state law tends to require dual referenda for an annexation; citizens in the annexing jurisdiction and the area to be annexed must both give their consent for a proposed consolidation to proceed. This requirement acts as a great deterrent to annexation. Baltimore, Boston, Chicago, Cleveland, Detroit, Los Angeles, Milwaukee, New York, Philadelphia and other

'landlocked' cities fully surrounded by already-incorporated suburban municipalities can no longer use annexation as a tool for growth.

Further reading

Jackson, K.T. (1985) *Crabgrass Frontier: The Sub-urbanization of the United States*, New York: Oxford University Press, ch. 8.
Miller, J. (1993) 'Annexations and boundary changes in the 1980s and 1990–1991', in *Municipal Year Book 1993*, Washington, DC: International City/County Management Association.

BERNARD H. ROSS AND MYRON A. LEVINE

APPLEYARD, DONALD

b. 26 July 1928, Southgate, Middlesex, England; d. 23 September 1982, Athens, Greece

Key works

(1963) *The View from the Road*, Cambridge, MA: MIT Press (with J.R. Myer).
(1974) 'The Berkeley Environmental Simulation Project: its use in environmental impact assessment', in T.G. Dickert and K.R. Domeny (eds) *Environmental Impact Assessment: Guidelines and Commentary*, Berkeley: University of California (with Kenneth Craik).
(1981) *Livable Streets*, Berkeley: University of California Press.

Donald Appleyard's diverse writings (which include over 100 articles/reports and 11 books) explored URBAN DESIGN, ENVIRONMENTAL PERCEP-TION, cognition and the symbolic meanings in the environment. He studied city planning at MIT under his mentor Kevin LYNCH. After teaching at MIT for six years, he left in 1967 to become professor of urban design at the College of Environmental Design at the University of California at Berkeley. There he produced a wide range of research and wrote some seminal works defining the emerging field of environment and behaviour research.

Appleyard produced a number of significant studies related to transportation and STREETS. He was particularly concerned with streets as livable environments in their own right and established that residents exhibited identifiable patterns of environmental perception and experience, depending on their residential proximity to streets with heavy motor traffic. Those living on heavily used roads knew fewer neighbours than did those on quieter streets, were less satisfied with their situations and even used their homes differently, to the extent that they avoided using rooms that were nearer these streets.

One of his key contributions to environment and behaviour studies involved a series of elaborate simulation experiments undertaken in partnership with Berkeley psychologist Kenneth Craik. Utilizing computer-controlled television cameras and tabletop miniature models, Appleyard and Craik prepared filmed simulations that would give participants the experience of travelling through and over a variety of landscapes. The experiments at the Berkeley Environmental Simulation Laboratory provided insights into how people perceive environments cognitively; demonstrated a new means of undertaking ENVIRONMENTAL IMPACT ASSESSMENTS and participatory planning processes; and paved the way for the computer-generated simulations used today.

His approach to analysing the built environment was highly multidisciplinary, encompassing not only psychology but also history, politics, economics and sociology.

Later in his career he also explored the realm of environmental symbolism, including the meanings of home, trees, parks and other environments. He proposed that patterns of symbols in these environments served to communicate messages to others.

Appleyard's ability to communicate to a diverse range of audiences (including the layperson) resulted in his work finding wide practical application. He was also much in demand internationally: he lectured at over forty universities and acted in a professional capacity in architecture and planning firms in the United Kingdom, Italy and the United States.

Donald Appleyard's work continues to be highly influential and respected more than twenty years after his tragic death in a traffic accident.

Further reading

Anthony, K. (1983) 'Major themes in the work of Donald Appleyard', *Environment and Behavior* 15(4): 411–18.
Cuff, D. (1984) 'The writings of Donald Appleyard', *Places* 1(3): 75–83.

MICHAEL DUDLEY

ARCHITECTURAL REVIEW BOARD

Architectural review boards (ARBs) are bodies of citizens whose purpose is to review proposed architectural changes and additions to buildings and properties within a city or a specific historical district. Usually, three to ten members, including architects, planners, real estate agents, lawyers, district residents and other laypersons, are elected or appointed, often every three years, to constitute these ARBs. Those boards which preside over historic districts are given the legal means to govern the architectural characteristics of new structures, as well as renovations of existing structures, including roofing and siding materials, colour, and window styles, in order to preserve the character of the district. Most ARBs preside over entire cities and work with city councils to regulate the architectural standards for their city. This relationship has, at times, created tension between the two bodies, especially as a result of the governing bodies' capacity to overturn board decisions, but these strains tend to be the exception, not the rule.

While some legal battles have ended in restrictions on the boards' authority, including the clarification of ambiguous regulations, many cities have found these boards to be helpful in maintaining the aesthetic value and continuity of their city. Dissension has also arisen regarding the creation of boards on the basis that they create too great an obstacle to the fulfilment of home designers', builders' or owners' dreams, as well as with respect to the philosophical question concerning the government's role in private property interests. Yet these too have tended to be minority positions.

ANDREW SEIDEL

B

BACK-TO-THE-CITY MOVEMENT

Towards the end of the twentieth century, cities across the world – New York, Chicago, Toronto, Vancouver, London and Melbourne, to name only a few – experienced GENTRIFICATION as 'yuppies' and other young urban 'homesteaders' began to buy and renovate houses in certain 'rediscovered' inner-city neighbourhoods. The phrase 'back-to-the-city movement' is often used synonymously with the term 'gentrification'. Both refer to the reinvestment in, and upgrading of, declining and undervalued inner-city neighbourhoods that have suddenly become fashionable. But there is a difference.

The back-to-the-city terminology emphasizes the return of suburbanites to the city, a process of reinvasion that may be adding new wealth to the city, even if the cost is the displacement of the urban poor from properties that can now command higher market value. Many critics, however, argue that the extent of a back-to-the-city movement is overstated as there are few permanent returnees from the suburbs. Instead, neighbourhood revival is primarily the product of 'incumbent upgrading' by existing neighbourhood residents and the housing moves of persons who previously resided in other parts of the city having discovered the value of living in a trendy 'new' neighbourhood. Gentrification is in part the result of changes in family demography, where, due to divorce and delayed marriages, growing numbers of singles and young marrieds have found city life attractive, resulting in an increased demand for urban property. To the extent that the back-to-the-city movement is overstated, then gentrification does not really add all that much to the wealth and tax base of a city; where the back-to-the-city movement

is limited, gentrification does not draw a wealth of new taxpayers and taxable resources to the city.

While the back-to-the-city movement contributed to some degree to the inner-city gentrification and urban revival of the late twentieth century, on the whole the back-to-the-city movement represents only a small countertrend to the continuing exodus of wealthy and middle-class families to the suburbs. For every suburbanite who moves back to the city, it is estimated that three city residents leave for the suburbs. The urban crisis of many cities continues despite the small but significant increase in the number of returnees from suburbia. Where gentrification leads to displacement and gentrifiers demand new services for their neighbourhoods, the back-to-the-city movement exacerbates the housing and service problems of the inner-city poor.

Further reading

Laska, S.B. and Spain, D. (eds) (1980) *Back to the City*, New York: Pergamon Press.
Smith, N. and Williams, P. (eds) (1986) *Gentrification of the City*, Boston, MA: Allen & Unwin.

MYRON A. LEVINE

BACON, EDMUND N.

b. 2 May 1910, Philadelphia, Pennsylvania

Key work

(1967) *Design of Cities*, New York: Viking Press (revised 1977).

American architect, urban designer, planner, educator (University of Pennsylvania) and prolific writer, Edmund N. Bacon studied architecture at

Cornell University and at the Cranbrook Academy of Art under Eliel Saarinen in the 1930s. Bacon is recognized in public, private and academic sectors for his focus over several decades on the city of Philadelphia, Pennsylvania, and his role in its rebirth, most notably as the visionary executive director of the Philadelphia Planning Commission from 1949–70. Bacon was a critic of the ahistorical processes of post-war planning, and of the failure of planners and designers to develop a viable concept of the contemporary city. His landmark text, *Design of Cities*, is a seminal work on urban design that richly combines text and drawings to illustrate the relationship between historical and modern principles and practices of urban planning, applied particularly to Philadelphia.

His national and international honours include Emeritus Fellow of the American Institute of Architects, the 1990 International Union of Architects' Sir Patrick Abercrombie Prize for Town Planning, the 1989 Chicago Architecture Award, the 1976 R.S. Reynolds Memorial Award for Community Architecture and the 1971 Distinguished Service Award from the American Institute of Planners.

Further reading

Guinther, J. (1996) *Direction of Cities*, foreword by E. Bacon, New York: Viking Press.

BEVERLY A. SANDALACK

BALANCED COMMUNITY

Prior to the INDUSTRIAL REVOLUTION, villages, towns and even the few cities that existed were generally well balanced in their activities and functions, social composition, and relationship to the surrounding environment. There were exceptions, such as imperial Rome, a city that exploited its power and trade opportunities offered by the Mediterranean to reach far and wide to bring goods of all sorts to the city, allowing it to grow to a gargantuan size while supporting a large, idle population.

However, the great majority of settlements were in an organic relationship with their countryside, contained a wide array of activities commensurate to their size, and integrated housing for all classes. They were in this sense balanced communities.

With the Industrial Revolution and its accompanying urbanization, all organic limits to city growth were thrown off and so was the balance of human activities. London of the late nineteenth century came to exemplify the industrial metropolis run amok, with the countryside out of reach for the great majority of residents, extreme segregation dividing the social classes, work and living spaces separated, and overcrowding limiting the urban masses' access to sunlight and open space.

In reaction to the mass congestion of the London conurbation, Ebenezer HOWARD, with his GARDEN CITY and social city proposal, sought to reintroduce – as urban historian Lewis Mumford had noted – the ancient Greek idea of a natural limit to the growth of the city. In so doing, Howard joined together the advantages of town and country. Howard's garden city – and at the regional level, the social city – would contain all the essential functions of an urban community, while limiting urban expansion with a green belt. In addition, people working in the garden city would also live there. In short, the garden city was to be self-contained and socially balanced.

Howard's idea of self-containment and social balance was extended to the British new towns programme of the post-Second World War period. A 1946 report called for new towns to be 'self-contained and balanced communities...places where residential growth was carefully geared to the provision of new jobs'.

The garden city and the earlier British new towns were attempts to accommodate growth on the metropolitan fringe as an alternative to unplanned suburban expansion. They were generally successful, but they remain the exception, not the rule, of suburban expansion. In Third World countries, the metropolitan periphery is overrun with spontaneous, illegal settlements of poor immigrants from the countryside, lacking even the most basic infrastructure, let alone jobs and adequate housing. In developed countries, suburban SPRAWL is endemic and, especially in the United States, the result is degradation at the centre of the metropolis, environmental degradation at the periphery, and class and racial segregation throughout.

The problems associated with sprawl have led to a not-so-new strategy in the United States, called SMART GROWTH. With its call for geographic containment of development, mixed land

uses, social equity and regional fair-share housing, and environmental conservation, smart growth is in essence a strategy for the creation of balanced communities. It remains to be seen, however, whether smart growth carries enough force to shake the entrenched and institutionalized interests of a balkanized metropolis and counter the ideological trend towards weaker government intervention.

Further reading

Calthorpe, P. and Fulton, W. (2001) *The Regional City*, Washington, DC: Island Press.
Clawson, M. and Hall, P. (1971) *Planning and Urban Growth*, Baltimore: Johns Hopkins University Press.
Mumford, L. (1961) *The City in History*, New York: Harcourt, Brace & World.

NICO CALAVITA

BANFIELD, EDWARD C.

b. 19 November 1916, Bloomfield, Connecticut; d. 30 September 1999, East Montpelier, Vermont

Key works

(1958) *The Moral Basis of a Backward Society*, Glencoe, IL: Free Press; Chicago, IL: Research Center in Economic Development and Cultural Change, University of Chicago.
(1970) *The Unheavenly City: The Nature and Future of our Urban Crisis*, Boston, MA: Little, Brown & Co. (revised 1974).

Controversial and influential expert on urban affairs, Edward C. Banfield was a political scientist who, through the publication of sixteen books and numerous articles on urban politics, URBAN PLANNING and civic culture, contributed to shape American conservatism in the latter half of the twentieth century.

After graduating from Connecticut State College at Storrs – now the University of Connecticut – in 1938, he became a journalist for the *Rockville Journal*. In subsequent years, he worked for the US Forest Service, the New Hampshire Farm Bureau federation and the US Farm Security Administration. In 1947, interested in understanding the failure of the New Deal's agricultural experiments, he started studying the politics and economics of planning with Edward

Shils and Rexford G. TUGWELL, and received his Ph.D. from the University of Chicago in 1952. Appointed professor of political science, he developed close friendships with his colleagues Leo Strauss and Milton Friedman. In 1959, he moved to Harvard University where he stayed until his death, with the exception of a four-year appointment at the University of Pennsylvania (1972–76).

Questioning the adequacy of his contemporaries' analyses of urban affairs, Banfield viewed city governments and urban problems as political processes determined by culture. His early investigations focused on the concept of class culture. He observed that various social classes have distinctive orientations towards providing for the future – the upper class being the most future-oriented and the lower class being the most present-minded – which ultimately influence their physical environment.

Although he acknowledged the existence of an URBAN CRISIS, he downplayed its significance, noticing that the quality of life in American cities had improved over the last century. He challenged the idea that racism was the cause of urban African-American POVERTY, provided a merciless image of the urban underclass and did not believe that national economic planning was the solution. *The Unheavenly City* inflamed public debates and stirred up waves of criticism from liberals and radicals who hated Banfield's conservative views on urban poverty and attacks on GREAT SOCIETY programmes. The controversies were invigorated by the fact that under President Richard M. Nixon, Banfield headed the Presidential Task Force on Urban Affairs, evaluated the MODEL CITIES programme and served as an adviser with the Commission on Intergovernmental Relations.

While Banfield deliberately shocked, his provocative arguments succeeded in leading people to redefine the nature of problems and solutions. In the early twenty-first century, his ideas about the primacy of CULTURE and SOCIAL CAPITAL are echoed in the work of many scholars.

Further reading

Banfield, E.C. (1985) *Here the People Rule: Selected Essays*, New York: Plenum Press.
Banfield, E.C. and Wilson, J.Q. (1963) *City Politics*, Cambridge, MA: Harvard University Press.

CATHERINE C. GALLEY

BARNETT, JONATHAN
b. 6 January 1937, Boston, Massachusetts

Key works

(1969–71) Director, Urban Design Group, New York City Planning Department.
(1974) *Urban Design as Public Policy: Practical Methods for Improving Cities*, New York: Architectural Record Books.

The architect, planner and urban designer Jonathan Barnett was educated at Yale and Cambridge Universities. Barnett served as the first director of New York City's Urban Design Group, created in 1969 by Mayor John Lindsay to advise the city's planning commission. As director, Barnett oversaw a group of architects tasked with achieving design excellence through advising developers, addressing urban redevelopment, preserving neighbourhoods and creating more open space. The group advanced several important land use policies in New York, including encouraging developers to provide plazas and arcades through zoning incentives, establishing special districts to encourage the provision of theatres in new office buildings in Manhattan's theatre district, and the introduction of mixed-use buildings. From 1971–98, Barnett directed the graduate programme in urban design at the City College of New York, where he was professor of architecture. In 1998, he became professor of city and regional planning at the University of Pennsylvania. Barnett is the author of several books, including *Urban Design as Public Policy* (1974); *The Architect as Developer* (1976), with John C. Portman Jr; *Introduction to Urban Design* (1982); *The Elusive City* (1986); and *Redesigning Cities* (2003).

Further reading

Barnett, J. (1986) *The Elusive City: Five Centuries of Design, Ambition and Miscalculation*, New York: Harper & Row.

HARRISON HIGGINS

BAROQUE CITY

A city can be defined as baroque when it has been designed to create dramatic movement, tension and monumentality and which departs from established rules and proportions. The baroque movement started in Italy at the end of the sixteenth century and extended into the eighteenth century into other European countries and Latin America.

The typical elements of the baroque urban form were based on the elements of the Renaissance such as the straight STREET, but they were carried further by using them to connect key points, and by introducing the diagonal street within regular orthogonal grids. Other concepts included the creation of heights through the use of natural topography or stairs. These design elements are illustrated in the city of Rome. The city was partially remodelled under several popes, but it was Sixtus V (1585–90) and his architect Domenico Fontana who established a better system of movement within the city by connecting key shrines with wider, straight streets to accommodate the large number of pilgrims. Placing obelisks at each location designated the key points in this system.

In France, the centralized power of the kings permitted the construction of the royal palace of Versailles that also illustrates baroque elements of design. This palace served as a model to other cities including Washington DC, which was designed by Pierre L'ENFANT starting in 1791. He used the topographic elements of the site to create a dramatic setting for the nation's capital, which reflects the building hierarchy. He connected the Capitol, the President's house and the Supreme Court with streets in order to represent the three governmental powers. He created vistas and public spaces as the settings for monuments. L'Enfant, who grew up by Versailles, where his father worked, was also familiar with other European cities that used similar concepts. These included St Petersburg, the Russian city created by Peter the Great in 1703, and Karlsruhe (1715–81), the German Versailles, which had thirty-two radial streets originating from a central palace. Baroque elements were also present to enhance the city of Paris with wide streets such as the original Champs-Elysées two hundred years before Haussmann redesigned its BOULEVARDS.

Designed public spaces include pedestrian, residential and traffic squares. The Capitol Piazza or Campidoglio in Rome (1567–1664), redesigned by Michelangelo as a complete unit, and St Peter's square in Vatican City (1655–67), designed by Gian Lorenzo Bernini, with an emphasis on monumentality and massiveness, are

examples of pedestrian squares. The enclosed Renaissance residential squares were developed in France and England. In Paris, the Place des Vosges (1605–12), originally the Place Royale, is a rectangular, landscaped square surrounded by residential buildings and arcades with identical façades. In London, Covent Garden (1630) was a rectangular square enclosed by buildings. Traffic squares, such as the Place des Victories in Paris (1687), were designed to facilitate traffic circulation within a city.

Further reading

Bacon, E. (1974) *The Design of Cities*, New York: Viking Press, pp. 131–55.

Kostof, S. (1992) *The City Shaped*, Boston, MA: Little, Brown & Co., pp. 209–48.

ANA MARIA CABANILLAS WHITAKER

BARRIO

A term that in its Arabic original means a peripheral part of a city, or a village separate yet close to a major one, in modern Spanish language 'barrio' generally means each area of a city, usually differentiated by functional (residential, commercial, industrial, etc.), social, morphological or architectural features. Partly as the legacy of Spain's colonial planning in Latin America, the barrio is often articulated around a landmark activity or space, such as a plaza, a market, a park or any other node. In this respect, the barrios were often established on the basis of parochial SUB-DIVISIONS within the colonial cities. From the mid-nineteenth century onwards, the barrio was the design and planning unit associated with the plans of 'ensanche' (urban expansion) of the biggest Spanish cities, such as in Carlos María de Castro's for Madrid (1860) and Idefonso Cerdá's for Barcelona (1860), which were followed by proposals for Bilbao and San Sebastián. This use of the barrio as a unit of the 'ensanche' took place as well in Buenos Aires, Mexico City and Havana, among other cities of the expanding economies of Latin America, where by the 1880s the penetration of foreign investment paved the way for adding new areas to the crowded centres of the previously untouched colonial 'dameros' (checkerboards). It is worth noticing that in Mexico, the equivalent to 'barrio' was 'colonia', though it came to connote working-class or uncontrolled areas later in the twentieth century. Also in Brazil, the Portuguese word 'bairro' was used from the late nineteenth century in Rio de Janeiro and São Paulo, where the new developments of private initiative promoted new architectural styles contrasting with those of the ancient city, thus acquiring an aesthetic value and expression.

Throughout the twentieth century, the planning of the modern barrios in many contexts of Latin America incorporated some of the features of the NEIGHBOURHOOD unit, both in the versions by Clarence PERRY or by the Congrès Internationaux d'Architecture Moderne (CIAM), and especially through José Luis Sert's interpretation. A school, a church, a community venue, a shopping centre and other services were often included within the limits of the new areas, which were of both low and high density. These were targeted at BOURGEOIS and middle-class groups that wanted to leave the downtown areas, crowded with migrants and administrative and business activities. At the same time, from a sociological perspective, the barrios can be said to have acquired social signification as community-based, functional and territorial units of the metropolis, as they were early described by Robert PARK, Ernest Burgess, Roderick McKenzie and other members the CHICAGO SCHOOL OF SOCIOLOGY. They were later revisited as essential components of urban diversity and health in the works of Jane JACOBS and Suzanne Keller. This community and segregational sense has been strongly associated with the use of the word 'barrio' in the context of big North American cities such as New York and Los Angeles, where it usually conveys the sense of a GHETTO or a COMMUNITY of Latinos that enrich the ethnical and cultural variety of metropolitan globalization.

In Spain and most of Latin America, the barrio generally is a consolidated district of the inner urban areas. However, the term can also be used to refer to SUBURBS that are peripheral barrios of high-rise buildings for the working classes. (Such suburbs are different, however, to those in Britain and North America, which are usually out of the metropolitan perimeter, made up of single-family houses and widespread services.) There is also a suburban meaning related to illegal, squatter or makeshift settlements, as in the 'bairros clandestinos' (clandestine areas) in Portugal, and the 'barriadas' (slum areas) in Peru. Perhaps the most dramatic case in this respect are cities in Venezuela,

where the word 'barrio' – different from 'urbanización' (urban development) – has become associated with the countless shanty towns and self-help housing areas that have featured in this country's urbanization over the last century.

Further reading

Keller, S. (1968) *The Urban Neighborhood: A Sociological Perspective*, New York: Random House.
Potter, R. and Lloyd-Evans, S. (1998) *The City in the Developing World*, Harlow: Longman.

ARTURO ALMANDOZ

BARTHES, ROLAND

b. 12 November 1915, Cherbough, Manche, France; d. 23 March 1980, Paris, France

Key works

(1957) *Mythologies*, Paris: Éditions du Seuil.
(1982) *Empire of Signs*, trans. R. Howard, New York: Hill & Wang.

A literary theorist and critic who was principally known for developing and extending the field of semiotics through his analysis of a variety of sign systems predominantly derived from BOURGEOIS, occidental popular culture. A central concern of Barthes's work was the detailed revelation of the ways in which sign systems naturalize certain ideologies, especially that of the universalism of Western bourgeois, occidental culture. His most popular book was *Mythologies* (1957), a series of short journalistic essays on a variety of popular cultural texts including advertising, film, newspapers, magazines, guidebooks and cuisine. In considering the reader and intertextuality in his analysis, Barthes prefigured many post-structural concerns with the fluidity of meaning. Although producing little work directly concerned with the city, his ideas were influential among scholars of a variety of disciplines concerned with unmasking the symbolic and ideological meanings of the built landscape. His influence was most explicitly felt among semioticians working on the urban landscape, although his influence can be traced in the burgeoning post-structural approaches developed by 'new' cultural geographers in the 1980s and 1990s.

Further reading

Culler, J. (1983) *Roland Barthes*, London: Fontana.
Gottdiener, M. and Lagopoulos, A.P. (eds) (1986) *The City and the Sign: An Introduction to Urban Semiotics*, New York: Columbia University Press.

TIM HALL

BARTHOLOMEW, HARLAND

b. 14 September 1889, Stoneham, Massachusetts; d. 2 December 1989, Florida

Harland Bartholomew was a founding member of the American City Planning Institute in 1917 who headed one of the largest planning consulting firms in the United States. After two years of study in civil engineering at Rutgers University, he gained a position with consultants George B. Ford and J.P. Goodrich in 1911 which involved preparing a plan for Newark, New Jersey. Three years later, the city plan commission hired him as the first full-time city planner in the nation. He moved in 1916 to St Louis to take a similar position with their City Planning Commission, replacing the renowned landscape architect George Kessler, who had designed the fairgrounds for the city's 1902 Louisiana Purchase Exposition. St Louis remained Bartholomew's home for the next seventy years and the home office of a multi-city consulting firm that produced plans, zoning ordinances and other specialized studies for over 500 cities and towns in the United States and Canada. Bartholomew relied upon extensive socio-economic data and detailed engineering drawings for road improvements, with less emphasis upon landscape design and aesthetic concerns that had been so central to the previous generation of City Beautiful plans. Projecting future land use needs based upon a variety of variables was also a trademark of Bartholomew's plans, and this data served as the basis for his book *Land Uses in American Cities* (1955).

Bartholomew was influential in devising federal urban policy, first through his contribution to shaping urban redevelopment strategies during the 1930s that led to the urban renewal programme under the Housing Act of 1949. As a member of Franklin D. Roosevelt's Interregional Highway Committee in the early 1940s, he helped to devise the Interstate Highway System. Bartholomew received an honorary B.Sc. in civil engineering from Rutgers in 1921 and an honorary doctorate from the same institution in 1952.

Further reading

Johnston, N.J. (1973) 'Harland Bartholomew: precedent for the profession', *Journal of the American Institute of Planners* 40: 115–24.

Lovelace, E. (1993) *Harland Bartholomew: His Contributions to American Urban Planning*, Champaign: University of Illinois.

CHRIS SILVER

BAUER, CATHERINE

b. 11 May 1905, Elizabeth, New Jersey;
d. 22 November 1964, Marin Coast,
north of San Francisco

Key work

(1934) *Modern Housing*, Boston, MA: Houghton Mifflin.

Catherine Bauer was a housing reformer, pioneer, advocate and educator whose life and career are intertwined with the US public housing movement of the 1930s and 1940s. Soon after graduating from Vassar – a prestigious women's college in the United States – in 1926 with an English degree, Bauer travelled to Europe and began her lifetime inquiry into housing and urban development policy and her lifelong commitment to MODERNISM. Upon returning to the United States, she described the housing achievements she had observed first-hand in Europe and lessons for her home nation in *Modern Housing* (1934). The book captured the attention of New Deal policy-makers wanting to boost the economy through housing construction.

Bauer was particularly enamoured with the idea of bringing about social change through strategic interventions in housing policy and architectural design. As a modernist, she advocated large-scale multi-family housing like the housing developments she had seen in Sweden and Germany. She became an active participant of the Regional Planning Association of America, whose members included Lewis MUMFORD, Clarence STEIN and Edith Elmer Wood.

She quickly established herself as a housing expert, and began arguing for a strong role for federal intervention in housing, especially for low-income households. Through her work as the executive secretary of the Labor Housing Conference – a group advocating multi-family housing with childcare centres and recreational facilities – she travelled across the United States to help build support among labour groups for the passage of important national housing legislation. She played an active role in drafting Senator Robert Wagner's housing bill that was passed in 1937, after a lengthy struggle between construction unions, builders and low-income housing advocates. The US Housing Act of 1937 provided funding for the construction of housing for low-income households by local public housing authorities. She continued her campaign for expanded public housing programmes that led to the passage of the Housing Act of 1949. She became disenchanted with the way in which public housing actually evolved. Title I of the Act provided federal funding for local 'slum clearance and urban redevelopment' programmes, ultimately destroying many low-income city neighbourhoods without replacing these areas with housing that low-income households could afford. Urban REDEVELOPMENT was later known as urban renewal under the Housing Act of 1954.

Academic work followed her activist years that had centred on housing legislation. She was the first woman to serve on the Graduate School of Design faculty at Harvard and at the Department of City and Regional Planning at the University of California, Berkeley. At Berkeley, she was a supporter of offering a Ph.D., and a strong advocate of policy-oriented urban research on campus, her involvement in this issue leading to the establishment of the Institute of Urban and Regional Research in 1963. Late in life, she was involved in international development issues.

Further reading

Oberlander, H.P. and Newbrun, E. (1999) *The Houser: The Life and Work of Catherine Bauer*, Vancouver: University of British Columbia Press.

SEE ALSO: modernism; Mumford, Lewis

AYSE PAMUK

BELL, DANIEL

b. 10 May 1919, New York, NY

Key works

(1973) *The Coming of Post-Industrial Society: A Venture in Social Forecasting*, Harmondsworth: Penguin.

(1976) *The Cultural Contradictions of Capitalism*, London: Heinemann.

An American sociologist best known for his account of the 'information society' and for coining the term 'post-industrial society', Bell's first career was in journalism. He was managing editor of *The New Leader* (1941–45), labour editor of *Fortune* (1948–58) and co-founder of *The Public Interest* magazine. He obtained his doctorate from Columbia University (1960) and was professor of sociology at Columbia (1959–69) and at Harvard from 1968.

Bell's contribution to understanding the development of the city comes in his classic account of the sociology of the service economy. In *The Coming of Post-Industrial Society*, he argues that industrial societies are being transformed into information-led service-oriented societies. This new society with its post-industrial cities is characterized by a heightened presence and significance of information; a shift from manufacturing to service occupations; increased importance of new knowledge-based industries; and the rise of new technical elites. This account has been criticized for overstating the development of a new society when all that is occurring is an extended division of labour. This work influenced Manuel CASTELLS's work on informational capitalism and Saskia SASSEN's account of global cities.

Further reading

Liebowitz, N. (1985) *Daniel Bell and the Agony of Modern Liberalism*, Westport, CT: Greenwood Press.
Webster, F. (2002) *Theories of the Information Society*, London: Routledge.

JOHN R. BRYSON

BELL, WENDELL

b. 27 September 1924, Chicago, Illinois

Key works

(1985 [1955]) *Social Area Analysis: Theory, Illustrative Application and Computational Procedures*, Palo Alto: Stanford University Press (with E. Shevky).
(1997) *Foundations of Futures Studies: Human Science for a New Era*, vol. 1, *History, Purposes and Knowledge*, New Brunswick, NJ: Transaction Publishers.

(1997) *Foundations of Futures Studies: Human Science for a New Era*, vol. 2, *Values, Objectivity, and the Good Society*, New Brunswick, NJ: Transaction Publishers.

Professor emeritus of sociology at Yale University. His areas of specialization include futures studies, sociology, social class, race and family life. Upon completing his Ph.D. in sociology from UCLA in 1952, Bell embarked on a forty-three-year career that is noteworthy in three separate spheres. First, working at Stanford, Northwestern and Yale, Bell studied issues of nationalism and social change in the Caribbean and served as president of the Caribbean Studies Association. Secondly, Bell pioneered (with SHEVKY) during the mid-1950s the analytical tool – social area analysis – that revolutionized quantitative urban studies in geography, political science and sociology for over three decades. Finally, over the last quarter of the twentieth century, Bell worked as a founder and leading advocate of the field of futures studies, dedicating himself to developing the epistemology of critical realism as the framework with which to obtain knowledge of the future. Served as consultant to the Commission of National Security for the Twenty-first Century.

DAVID C. PROSPERI

BENCHMARKING

In surveying, a benchmark is a point of known elevation that serves as a reference to measure other elevations. Since the 1970s, however, the term has been captured as a tool in management by which quality or excellence is gauged and improved in organizations, operations or activities. Benchmarking requires that participants identify, understand and continually adapt their practices such that they provide efficient and effective service to stakeholders. Various approaches to benchmarking exist, but each asserts the need to (a) know one's organizational and operational CULTURE and practices; (b) understand how similar and contrasting organizations work to produce quality outcomes; (c) compare the different practices among such organizations; (d) identify opportunities for improvement in the practices of the home organization; (e) take steps to close the gap between existing and desired practices; and (f) monitor and adjust practices for continual improvement.

The origins of benchmarking can be traced to Frederick Winslow Taylor (1856–1915), after whom the analysis of time-and-motion in production lines, or Taylorism, is named. By scrutinizing how long it took each worker to complete a step, and then by adjusting equipment, Taylor thought he could anticipate average production rates in optimum conditions, and if rates were tied to wages, he surmised that workers would have an incentive to exceed the average. Attempting to humanize such industrial processes, others – such as the industrial psychologist Lillian Moller Gilbreth (1878–1972) – pioneered research into indirect incentives, job satisfaction, wage planning, stress and time management.

Individual companies may focus their benchmarking efforts on refining processes related to the production of goods or services in which they specialize and for which they anticipate sales and profits. But benchmarking has become widely used in government and non-government (or not-for-profit) organizations. It has developed from a means to measure production of goods to a means to measure production of services, including policy implementation, asset and risk management, and sustainability outcomes. Thus, benchmarking is now common practice in the management of towns and cities, a public sector activity in which town planning and STRATEGIC PLANNING are especially implicated. Municipal governments in charge of city management often emphasize best value, equity, and quality and relevance of service delivery to community members.

Two cases illustrate different motivations for benchmarking in the management of towns and cities. The City of Atlanta's Public Works Department has, for example, measured the road it maintains (1,900 lane miles in total), and the proportion that is paved (98.8 per cent of all lane miles). This internal measure has been compared to similar data gathered by other cities in the United States in order to outline two further facts – that these other cities maintain between 1,400 and 2,000 lane miles, and that the City of Atlanta spends US$5,651 per lane mile, second only to Reno (US$7,860). Alone, these findings have limited utility. But when the governors and people of the City of Atlanta ask 'Is this what we want to be spending on road maintenance? Can we spend less and achieve the same results? Should we be spending more on other City infrastructure or on public transport?' or like questions, the findings

have greater utility. The benchmark provides information about 'where we're at', and prompts the City's people to deliberate about 'where we'd like to be'. In turn, a series of ethical and political questions arise about sustainable transport, the distribution of resources and so on (see SUSTAINABLE URBAN DEVELOPMENT).

Secondly, it is clear that international, national and sub-national policies on quality assurance can prompt individual town and city administrations to embrace the practice of benchmarking. In the state of Tasmania, Australia, for example, since 2001 all 29 local governments have been required to adhere to a system of performance measures, gauging up to 50 key performance indicators or results areas that benchmark the excellence of GOVERNANCE, management and finance, regulation, infrastructure and utilities, and community development. Once benchmarks are established, local governments are expected to make continual improvements to practice over time and demonstrate to their communities that such improvements are being made.

A local government area taking in the central business district and inner SUBURBS of the state's capital, the City of Hobart has commenced a number of benchmarking exercises. Noteworthy among them, and illustrative of the power of international, national and sub-national trends, is a plan to reduce Hobart City Council's greenhouse gas emissions by 70 per cent over five years, and to encourage the community of the City to reduce its greenhouse gas emissions by 20 per cent over the same period. This plan has required the City's elected representatives and staff to identify existing emission levels, activities which contribute to them, means to reduce them, and a range of social, cultural, institutional, organizational and economic practices that will support such reductions. Moreover, the City's involvement in the international Cities for Climate Protection programme has drawn links from the local to the global. In short, benchmarking is becoming integral to better SUSTAINABLE URBAN DEVELOPMENT practices, and here its efficacy is clear.

Notwithstanding these strengths of benchmarking and related techniques, like many tools it can be misapplied – for example by not accounting for the needs of minority groups, SOCIAL JUSTICE and equity issues, or the environment. It can be applied piecemeal – by being under-resourced by city managers and elected representatives such that

ongoing monitoring and improvement are not guaranteed. It can be applied punitively – in ways that penalize staff or other participants if improvements are not made in ways anticipated. Finally, a tendency to focus on quantification and objective measures – such as road miles – often marginalizes those parts of city life, or life in general, which are best measured qualitatively – trust, conviviality or quality of life, for instance. If such challenges are addressed, however, benchmarking may aid reflection on city management and its improvement for better collective outcomes.

Further reading

Hart, M. (1999) *Guide to Sustainable Community Indicators*, North Andover, MA: Hart Environmental Data.
O'Reagan, S. and Keegan, R. (2000) 'Benchmarking explained', *Benchmarking in Europe – Working Together to Build Competitiveness*, Public Sector Information and the European Union.

ELAINE STRATFORD

BENEVOLO, LEONARDO

b. 25 September 1923, Orta San Giulio, Novara, Italy

Key works

(1967) *The Origins of Modern Town Planning* (*Le origini dell'urbanistica moderna*), Cambridge, MA: MIT Press.
(1980) *The History of the City* (*Storia della città*), Cambridge, MA: MIT Press.

Leonardo Benevolo has written extensively on the history of the city and architecture, especially modern architecture. His major books have been translated into several languages. He has received literary awards such as the Lugano Libera Stampa Prize in 1964, the Paris Medaille de l'Histoire de l'Art de l'Académie d'Architecture in 1985 and the Capri Prize in 1989. In addition to his academic activities, Benevolo has worked as a planning and urban design consultant for many Italian cities.

Benevolo's concept of the 'neo-conservative' city is an important contribution to the understanding of the evolution of cities. The crushing of the working class during the 1848 revolutions in Europe led to the creation of conservative regimes that adopted a technocratic model of planning at the service of the now victorious bourgeoisie.

Instead of laissez-faire, the state limits the complete freedom of the private sector through building and planning regulations, while at the same time carrying out huge public works. Thus, the respective roles of the public and private sectors were established and made possible the restructuring of several European cities – including Paris, Vienna, Barcelona and Florence – during the second half of the nineteenth century. The new cities reflected the values and interests of the bourgeoisie through the elimination of slums, the creation of boulevards for fashionable society to see and be seen, parks, and major public buildings.

NICO CALAVITA

BENJAMIN, WALTER

b. 15 July 1892, Berlin, Germany;
d. 26 September 1940, Portbou, Spain

Key works

(1999) *The Arcades Project* (*Das Passagen-Werk*), trans. Howard Eiland and Kevin McLaughlin, Cambridge, MA: Harvard University Press.

Known for an idiosyncratic style and original approach to subject matter, Benjamin, a German essayist, critic and theorist, writes on topics ranging from literature, art, film, photography, history and cities. His understanding of the city primarily derives from the two metropolises in which he lived for substantial periods: Berlin and Paris.

An unsystematic thinker, Benjamin's views of cities are found throughout essays and books. In a style of essay called *Denkbilder*, or 'thought images', he writes cityscapes that reflect particular localities – sights, sounds, smells and tastes. In the posthumously published, unfinished work *The Arcades Project*, he uses a fragmentary style to write about the rise of modern European urban culture.

According to Benjamin, although the city is the locus for the rise of middle-class consumer culture and capitalism, it is also the site for the rise of the industrial working class. While it is a place of excitement where one can lose oneself, the city can also have a stultifying character. On the one hand, the modern city symbolizes technological progress which creates fascinating structures; on the other, it gives rise to alienation. In Benjamin's work,

cities encapsulate the fragmentation of modern culture.

Further reading

Gilloch, G. (1996) *Myth and Metropolis: Walter Benjamin and the City*, Cambridge: Polity Press.

PHILIP RODNEY WEBB

BERRY, BRIAN J.L.

b. 16 February 1934, Sedgley, Staffordshire, England

Key works

(1967) *Geography of Market Centers and Retail Distribution*, Englewood Cliffs, NJ: Prentice Hall.
(1973) *The Human Consequences of Urbanization*, New York: St. Martin's Press.
(1991) *Long-Wave Rhythms in Economic Development and Political Behavior*, Baltimore: Johns Hopkins University Press.

An extraordinarily prolific author, Brian J.L. Berry has made major contributions to economic and urban geography and to URBAN STUDIES and planning. Throughout his career he has been an intellectual innovator, opening up new fields of study, pioneering quantitative analysis, and then moving on to still newer fields. He completed his first degree in economics and geography at University College London in 1955, then moved to the United States to study geography at the University of Washington, completing his Ph.D. in 1958. He worked at the University of Chicago till 1976, acquiring dual citizenship (US–UK) in 1965. Subsequently, he has worked at Harvard (1976–81), Carnegie-Mellon (1981–86) and at the University of Texas at Dallas since 1986. In the 1960s, Berry's research on urban systems and regional development in the United States and India fuelled geography's 'quantitative revolution', and for more than two decades he was the most prolific and widely cited author in geography. Among his many honours, he was elected to the National Academy of Sciences (NAS) in 1975, he served on the NAS Council from 1999 till 2002, and he is a fellow of the British Academy and the American Academy of Arts and Sciences. He has worked extensively with the Ford Foundation, USAID, the US Department of Commerce, the US Department of Housing and Urban Development, the World Bank, the Brookings Institution and the Urban Institute, including extended international development assignments with India, Chile, Brazil, Indonesia and Sri Lanka.

In the 1960s, Berry's work focused on the structure of central place systems, retail location and consumer behaviour, industrial location, urban spatial structure, internal migration, commuting, and the analysis of highway impacts and commodity flows. In the 1970s and early 1980s, he focused on commuter fields, racial discrimination in housing markets, the definition and spread of metropolitan areas, and the phenomenon of counter-urbanization – metropolitan cores losing population while outer metropolitan fringes, smaller metropolitan areas and some non-metropolitan areas gain population. In the 1980s, he began major work on long-wave cycles in economic development, and he linked these with social and political changes. As his historical interests grew, he also wrote on the rise and fall of utopian communities, and he developed a passion for his British family genealogy. Berry was very influential in US urban policy in the 1970s, increasing attention to metropolitan growth and sprawl, and he has played a major role ever since in assessing needs and standards for population and transportation data collection in the United States.

RAY BROMLEY

BLIGHT

Blight is an official policy designation assigned to declining urban areas that are regarded as threats to local health, safety, public welfare and morals. Cities in the United States often declare blighted status after determining that urban renewal strategies are the most appropriate means to encourage the private investment necessary to reverse deteriorating downtown conditions. Formal proof of blight is incorporated into the local decision-making process through a set of measures based on structural, economic, social, environmental and planning indicators. Empirical confirmation is necessary because state redevelopment laws cite the existence of blight as a precondition for granting local governments the authority to implement a variety of tools including EMINENT DOMAIN and TAX INCREMENT FINANCING programmes. Objective assessments became essential following the 1949 Housing Act stipulation that

federal funding for SLUM clearance and redevelopment would only be made available to certified blighted districts. This requirement was subsequently incorporated into COMMUNITY DEVELOPMENT BLOCK GRANTS and urban development block grant programmes complemented by a strong preventative element that is included to identify areas with the potential to experience blight.

Initial definitions in the 1940s resulted in blight becoming synonymous with the dire structural circumstances found in neglected inner-city slums. Blight emerged as a concept representing the physical decay in aging, poorly maintained and dilapidated TENEMENTS located in districts that also experienced a disproportionate amount of abandonment. In these NEIGHBOURHOODS, local building and fire code violations, inadequate sanitary facilities, poor maintenance, and industrial environmental hazards contributed to unsafe living conditions. Blighted buildings are characterized as dilapidated, meaning they are in a severe state of building disrepair not suited for human habitation and unlikely to be rehabilitated to acceptable standards due to physical and economic constraints. Blight also applies to deteriorated buildings with leaky ceilings, wall cracks, broken windows and inoperable doors.

Blight has also been associated with improper or inadequate planning leading to residential overcrowding and poorly configured central city built environments. The resulting densities and building placements provide inadequate access to natural light levels, a lack of proper air ventilation, fire and safety risks, and unsatisfactory access to parks and open space. Growth in the absence of institutionalized planning regulations permitted inappropriate land use combinations leading to blight caused by conflicts between residential and industrial activities exposing the population to dangerous environmental hazards.

MODERNISM emerged as a central guiding principle used by planners and politicians to designate blighted situations. Beyond the deteriorated and unhealthy conditions constituting blight, modernists equated decline with chaotic development patterns, unattractive neighbourhoods without public amenities, and land use mixes that overall did not impose immediate threats to residential well-being. These assertions influenced public deliberations by politicizing the belief that distressed districts did not conform to the contemporary vision of order, separated land uses with a focus on isolating residential development, and an aesthetically pleasing built form.

Blight status has also been used to encompass situations identified as obstacles to modernist large-scale redevelopment plans. Narrow streets and confusing road patterns create inefficient traffic networks that are unable to sustain proposed commercial and residential development. Additionally, inadequate infrastructure capacities cannot supply water, sewer and electricity services necessary to support new growth. Small and irregularly shaped lots defined during initial downtown development qualify as blight because the resulting fragmented landownership has made it difficult in the past for local officials to gain political consensus. Consequently, cities use eminent domain to assemble tracts of land into large sites that are unencumbered by outdated development patterns and uncooperative property owners.

Long-standing controversies questioned the relevance of applying modernist planning standards to inner-city cores as the basis for approving and funding slum clearance projects undertaken in the public interest. Proponents asserted that addressing these physical conditions would reduce dysfunctional social behaviour, high crime rates, health threats and poverty. Opponents claimed that blight designations made by middle-class professionals ignored vital and established neighbourhoods by using suburban sensibilities as the basis for passing value judgments on inner cities. Neighbourhood preservationists argued that inappropriate blight designations stigmatized immigrant and minority communities and reinforced the incorrect assertion that betterment could only be achieved by demolishing the offending districts and relocating residents in new redevelopment projects. Advocates such as Jane JACOBS praised the high densities and people-oriented urban design that modernists viewed as problematic, while Herbert GANS used the West End of Boston to document rich urban villages found in blighted districts. These critiques led to a fundamental re-evaluation of the social element modernists chose to discount in their blight definitions. Though cities continued to utilize blighted status as a precursor for renewal activities, protests led to greater public attention being placed on preservation and revitalization options in addition to clearance strategies.

During the 1980s, blight designations moved beyond traditional inner-city problems to address aging suburbs with the potential for serious

economic decline due to emerging shifts associated with SUBURBANIZATION and globalization. Blight has been adapted to structurally sound and properly serviced districts impeded by obsolete building configurations that are unable to accommodate the highest and best use in a changing market. Suburban blight also emphasizes the existence of declining property values that threaten the local tax base necessary to finance adequate service delivery. High local unemployment rates or the lack of proximity to jobs is also interpreted as sufficient cause to declare blighted status.

Further reading

Teaford, J.C. (1990) *The Rough Road to Renaissance: Urban Revitalization in America, 1940–1985*, Baltimore: Johns Hopkins University Press.

SEE ALSO: eminent domain; Gans, Herbert; Jacobs, Jane; redevelopment; slum; tenements; urban decline

STEVEN M. WEBBER

BLUMENFELD, HANS

b. 18 October 1892, Osnabruck, Germany;
d. 30 January 1988, Toronto, Ontario, Canada

Key works

(1967) *The Modern Metropolis, Its Origins, Growth, Characteristics, and Planning: Selected Essays by Hans Blumenfeld*, ed. Paul Speiregen, Cambridge, MA: MIT Press.
(1987) *Life Begins at 65: The Not Entirely Candid Autobiography of a Drifter*, Montreal: Harvest House.

Born in Germany, it was expected that Blumenfeld would be destined for a position in the family businesses of banking and 'merchant princes'. Blumenfeld rejected this expectation and sought his own path by pursuing his interest in architecture. His education began in 1911 at several institutions including Darmstadt Polytechnical Institute, the Technical University of Munich, and Karlsruhe. Interrupted by the advent of the First World War, Blumenfeld volunteered for service in the German army, serving four years as a private. This experience, plus the loss of his brother in the war, influenced his work to promote peace throughout the rest of his life.

Following the War, Blumenfeld continued his education in Munich and earned his degree in 1921. He spent three years working in the United States and then returned to Hamburg in 1927. After working in Vienna for three years, in 1930 he worked in the Soviet Union. Hoping to work in an atmosphere of social purpose and creativity, Blumenfeld became a member of the Russian State Planning Institute in the cities of Moscow and Gorki, working on various city plans from 1933 to 1935. In 1937, when his German passport expired, he was compelled to leave and went to New York City via France, arriving in 1938.

In New York, he worked on housing projects and was active on the Citizens Housing and Planning Council of New York, where he worked with notable planning/housing specialists including Charles Abrams. In 1941, he accepted a research position with the Philadelphia Housing Association. In 1944, he became a US citizen and started working for the City Planning Commission of Philadelphia, where he worked with Edmund BACON. In 1952, when Pennsylvania required its public servants to sign a loyalty oath, Blumenfeld resigned in protest. He continued to work as a private consultant until he moved to Toronto in 1955.

He became assistant director of the newly created Metropolitan Toronto Planning Board and made significant contributions to this body until he left in 1961 to lecture at the University of Toronto. Blumenfeld served as a consultant to a number of cities around the world.

Blumenfeld published many articles stressing the economic and social effects of planning, and emphasizing the interdependence of physical, social and economic renewal. Through his writing, Blumenfeld displays the ability to explain complex subjects in basic and understandable language. Two other hallmarks of his life were advocating peace and the establishment of a foundation towards this end in his brother's name. Blumenfeld has been recognized by professional organizations in the United States (Distinguished Service Award of the American Institute of Planners) and in Canada (Fellow of the Canadian Institute of Planners).

Further reading

Speiregen, P. (ed.) (1979) *Metropolis and Beyond: Selected Essays by Hans Blumenfeld*, New York: Wiley.

DAVID AMBORSKI

BONUS ZONING

In countries such as the United States and Canada, where land use regulation is generally reactive and non-discretionary, various techniques have given municipalities greater range and flexibility in controlling development. One such technique is bonus zoning, also known as incentive zoning.

First adopted in New York City in 1961, bonus zoning entails a quid pro quo between officials and developers. In exchange for providing a desired public benefit, builders are given development rights that go beyond those specified in the ZONING code. Desired public amenities can be of various sorts: open spaces (plazas), through-block connections, direct access to the subway, day-care centres, affordable housing, cultural facilities, preservation of historic structures, etc. Benefits accruing to the developer generally take the form of added floor space in the project, that is a higher DENSITY or a greater height than normally allowed.

Municipalities have seen in incentive zoning a means to finance public facilities that they cannot afford. But the multiplication of 'bonusable' amenities has enabled developers to erect structures far in excess of desired densities. In many cases, the benefits to the developers have far surpassed the benefits to the public. Cities have therefore reduced the number of possible bonuses, made them much more directly related to developers' contributions, and generally defined these contributions in the most explicit terms possible. To a certain extent, bonus zoning has reintegrated the system of non-discretionary regulation that it had come to supplement at first.

Further reading

Lassar, T.J. (1989) *Carrots and Sticks: New Zoning Downtown*, Washington, DC: The Urban Land Institute.
Whyte, W.H. (1988) 'The rise and fall of incentive zoning', in *City: Rediscovering the Center*, New York: Doubleday, Anchor Books.

RAPHAËL FISCHLER

BOOSTERISM

Americans have been indefatigable promoters of their towns and cities. During the nineteenth century, fierce competition for economic success among newly founded cities led to an outpouring of booster literature. In books, pamphlets, magazine articles and editorials, businessmen and publicists extolled the virtues of their communities. Boosterism served to inspire local pride and was intended to attract settlement and investment. Booster literature typically inventoried the visible signs of growth, cited statistics on population and trade, and looked to local geography for reasons why their town was most likely to succeed. In part because professional publicists readily transferred loyalty from one town to another, they became easy targets for writers such as Charles Dickens (in *Martin Chuzzlewit*) and Sinclair Lewis (in *Babbitt*).

In the twentieth century, the overblown rhetoric of the frontier journalism gave way to the promotional efforts of economic development professionals in chambers of commerce and development agencies. The new boosterism uses the language of economics to stress the presence of industrial clusters, advantages in the local supply of labour, cost of living and amenities. Twentieth-century boosters also turned to professional sports franchises and to major events such as world fairs and the Olympic Games as techniques to attract attention and demonstrate status as a 'big league city'.

Further reading

Glaab, C. (1968) 'Historical perspectives on urban development schemes', in L. Schnore (ed.) *Social Science and the City*, New York: Praeger.
Hamer, D. (1990) *New Towns in the New World: Images of Nineteenth Century Urban Frontiers*, New York: Columbia University Press.

CARL ABBOTT

BOOTH, CHARLES

b. 30 March 1840, Liverpool, England;
d. 23 November 1916, Thringstone, Leicestershire, England

Key works

(1889) *Labour and Life of the People*, London: Macmillan.
(1902–03) *Life and Labour of the People of London*, 17 vols, London: Macmillan.

Charles Booth took over the running of his father's company at 22 years of age and became an energetic leader, adding glove manufacturing to the company's expanding shipping interests. In the

1860s, he became interested in the philosophy of Auguste Comte, particularly with Comte's idea that scientific industrialists would take over social leadership from church ministers. Booth campaigned unsuccessfully in Toxteth, Liverpool, for the Liberal Party in the parliamentary election of 1865, but house-to-house canvassing shocked Booth and provided him with first-hand evidence of squalor and poverty. In 1885, he became agitated at a claim made by H.H. Hyndman, leader of the Social Democratic Foundation, that 25 per cent of the population of London lived in abject poverty. Inspired by Comte, Booth investigated the incidence of poverty in the city's East End, and the result of Booth's investigations, *Labour and Life of the People*, was published in 1889. The investigations revealed that the situation was even worse than that suggested by Hyndman; 35 per cent were living in abject poverty. He extended his research to cover the remainder of London and employed an army of researchers in an effort to obtain comprehensive and reliable survey material. The results of the work were gathered over a twelve-year period from 1891 in 17 volumes under the title of *Life and Labour of the People of London* and organized into three sections: poverty, industry and religious influences. Significant outputs of the inquiry were the maps of London, coloured street by street to indicate levels of poverty and wealth. Booth argued that the state should assume responsibility for those living in poverty, and first suggested the introduction of old-age pensions as an example of 'limited socialism'. Booth was not a socialist and he feared the onset of socialist revolution if the state did not act to alleviate the poverty of the working classes. Some of his researchers became socialists as they proceeded with the research, but Booth became more conservative and, unhappy with the 1906 Liberal Government's relationship to the trade unions, renounced his support for the Party and joined the Conservatives. In 1893, he served on the Royal Commission on the Aged Poor. He was made a Privy Councillor in 1904, and in 1907 he served with Beatrice Webb on the Royal Commission on the Poor Law.

Further reading

Fried, A. and Elman, R. (eds) (1969) *Charles Booth's London: A Portrait of the Poor at the Turn of the Century, Drawn from His 'Life and Labour of the People in London'*, London: Hutchinson.

Norman-Butler, B. (1972) *Victorian Aspirations: The Life and Labour of Charles and Mary Booth*, London: Allen & Unwin.

O'Day, R. and Englander, D. (1993) *Mr Charles Booth's Inquiry: Life and Labour of the People in London Reconsidered*, London: Hambledon Press.

MARK TEWDWR-JONES

BORDERLESS SOCIETY

Labour, industries and capital move across borders, which is historically not so new. Furthermore, even outside the sphere of economics, international organizations, such as the United Nations, worldwide sports competitions, such as the Olympic Games, and the building up of global telecommunications networks, which are means by which people and nations have been carrying out mutual exchanges, have a history exceeding well over a century. However, since the 1970s, when multinational companies appeared on the world economy and began to develop movements that cross over borders on a global basis and perform economic activities that go beyond the limits of nation-states, a new form of global system has clearly come into view. The world economy has begun to shift to globalization from internationalization. When internationalization was advancing, goods and services were traded across borders, but trading was strictly regulated by the nation-states that held sovereignty. When globalization is taking place, goods and services are still provided in the same way as before, as they cross over borders when businesses are being developed. But production and trading are carried out by oligopolistic global companies that have established worldwide networks subject to lenient regulations of the nations involved. Time will come when multinational companies will set up head offices, branch offices and plants throughout the world and establish their own networks and promote cooperation with other related firms through diversified businesses. Consequently, companies will carry on business and make decisions as if borders have disappeared.

Kenichi Ohmae named the society which enabled companies to cross over borders and produce and distribute goods freely 'the borderless world' after experiencing for himself changes in the processes of corporate activities as an

entrepreneur. Since the 1980s, when rapid progress took place in information communications technology based mainly on the INTERNET, the phenomenon of a borderless world has become even more distinct in the area of economic activities. Multinational companies carry out activities on a global basis; clothing and cars, demand for which is generated around the world at the same time, have appeared and spread, and markets that handle these items have been formed globally; and in particular, integration and reorganization of commercial and financial functions progressed on a global basis in the 1990s. As corporate activities spread, the North American Free Trade Agreement and the General Agreement on Tariffs and Trade were set up and formed regional alliances for free trade among nations. Moreover, regional integration, such as in the form of the European Union, which went further than establishing a simple cooperation among nations to unifying their currencies, made progress. The term 'borderless' is now used not only in relation to economic aspects but also often in relation to politics, culture and learning; the term may be used when activities going beyond the boundaries of conventional frameworks and areas advance.

Further reading

Ohmae, K. (1990) *The Borderless World: Power and Strategy in the Interlinked Economy*, New York: Harper Business.
—— (1995) *The End of the Nation-State*, New York: Free Press.

YUICHI TAKEUCHI

BOSS

The term 'boss' refers to the head of a strong, often hierarchically organized, political organization. Political bosses reward their friends and punish their enemies. In the United States, bosses led the political party 'machines' that were a prominent feature of many cities for much of the nineteenth and early twentieth centuries. Bossism was an important aspect of big East Coast and Midwestern cities, where, for instance, the leaders of New York's Tammany Hall and Chicago's regular Democratic Party organization dominated municipal affairs for decades. Boss-led party organizations were also found in smaller cities and across

the United States, in Albany, Cincinnati, Jersey City, Kansas City, Memphis, New Orleans, San Antonio, San Francisco and Tampa.

The boss phenomenon is international in scope. In France, Gaston DeFerre was the mayor of Marseilles for four decades; Pierre Mauroy similarly governed Lille for a long duration. Both bosses gained great influence as a result of the positions they held simultaneously in both local and national government. In France, smaller communities have been traditionally governed by influential local leaders, or *notables*. In Italy, traditional family heads have traditionally been powerful local figures. In Russia, the power of local and regional bosses is so considerable that it imposes severe limits on the ability of the central government in Moscow to administer the country's affairs. Latin America, too, has witnessed a great many local party strongmen who have headed clientelist organizations.

Boss-driven organizations often arose as a response to the weakness of the formal municipal governments which could not provide much-needed services or govern effectively. Party organizations delivered job patronage and provided housing services, emergency assistance and a whole wide range of favours that the party faithful required. In what was essentially a brokerage operation, the boss delivered services and exchanged favours for political support; alternatively, party leaders sought those resources (i.e. the promise of jobs from a business owner) that could be exchanged for votes. The party bosses quite often worked with and brokered the demands of business leaders. The local boss represented someone with whom business leaders could deal, someone who had the power to 'get things done' and ensure that the necessary permissions and other important concessions were granted. Of course, party bosses often enriched themselves in the process.

In the United States and other countries, the power of party bosses declined as levels of affluence and education increased and the government assumed new welfare state responsibilities; citizens became less willing to trade their votes for the favours that party bosses could deliver. In the United States, the GOOD GOVERNMENT MOVEMENT instituted DIRECT DEMOCRACY and other reforms that undermined the power of party bosses. A few cities like Chicago retain elements of boss-run politics; but, on the whole, the strong local boss in the United States is an anachronism.

Further reading

Allswang, J.H. (1977) *Bosses, Machines, and Urban Voters: An American Symbiosis*, New York: Kennikat.

Callow, A.B. (ed.) (1976) *The City Boss in America: An Interpretive Reader*, New York: Oxford University Press.

BERNARD H. ROSS
MYRON A. LEVINE

BOULEVARD

A boulevard is a wide, landscaped street, within a city, with several lanes of traffic and pedestrian walkways, in a variety of combinations. The design of boulevards involves precise geometry to create pleasant recreational spaces through extensive landscaping, the development of vistas and the creation of a clear city structure. Boulevards often become an important axis in a city, used for public ceremonies.

The first boulevards ('boulevard' is derived from a French word, 'boulevart', meaning large bastion, which in turn was a corruption of the Nordic word 'bulvirke' or 'bulwark', or palisade) developed in Paris, France, in 1670, when Louis XIV ordered the demolition of the city walls no longer needed for defence. Over time, the ramparts were levelled to form a landscaped three-and-a-half-mile-long street network for carriages and pedestrians, used as a promenade by the wealthy and the aristocracy. These boulevards were not used as traffic arteries and had few interruptions, except at the city gates, but served as a boundary between city and country. Many other European cities converted their ramparts into boulevards, including Vienna, which created a ring road with public spaces and cultural uses. Other cities used the boulevards to connect older city centres to newer developments, such as El Paseo del Prado in Madrid.

In the 1850s, Baron Georges-Eugène HAUSSMANN in Paris expanded the concept of the boulevard to its ultimate form. With the intent of facilitating army access and controlling civil unrest, he demolished the heart of the medieval city and created a system of remodelled and new boulevards (e.g. the Champs-Elysées and the boulevards around the Place de l'Etoile). The surrounding buildings with consistently designed façades provided fashionable housing to a new middle class. Restaurants and shops along the walkways contributed to the festive nature of this social space. These boulevards also accommodated new infrastructure systems. Both the trees and the buildings framed vistas, which focused on key city monuments such as the Avenue de l'Opera.

The United States first embraced the boulevard in plans for new cities, such as Washington DC, and later as the basis for landscaped residential streets (for example in Kansas City). It also became an important component of the CITY BEAUTIFUL MOVEMENT. The boulevard lost its popularity as streets became engineering projects intended to facilitate traffic movement. In Latin America, elegant boulevards developed as part of the modernization of the colonial cities. These include El Prado in Havana, Cuba, La Reforma in Mexico City and the Avenida Nueve de Julio in Buenos Aires, Argentina, which is defined as the widest boulevard in the world.

Jacobs, MacDonald and Rofe (2002) provide a description and classification of different boulevards throughout the world, and specific guidelines for the development of new boulevards.

Further reading

Jacobs, A.B., MacDonald, E. and Rofe, Y. (2002) *The Boulevard Book*, Cambridge, MA: MIT Press.

Kostof, S. (1992) 'Boulevards and avenues', in *The City Shaped*, Boston, MA: Little, Brown & Co., pp. 249–61.

Wilson, W.H. (1989) *The City Beautiful Movement*, Baltimore: Johns Hopkins University Press.

ANA MARIA CABANILLAS WHITAKER

BOURGEOIS

The bourgeois (a term possibly derived from the French word 'bourgeoisie', which refers to the middle classes of cities who were engaged in trading) are a social group. They received notoriety through the work of Karl Marx, who named them as an exploiter class of people characteristic of the capitalist system developed under the Industrial Revolution in Britain. Given the slight differences in the general definitions of the bourgeois, certain traits are nonetheless commonly noted, such as the concern for, or dependency on, industrial and/ or commercial enterprise, and belonging to the middle class of society. However, like other social classes, it is necessary to understand that the bourgeois have not been the same for all of history. They have, like other social groups, evolved.

The term 'bourgeois' in the modern sense is usually referred to in the context of the capitalist

economic system. In such a system, it is the bourgeois who control work patterns, own property, and sit in powerful managerial and government positions, although sometimes the bourgeois are simply known to be the owners of the means of production, that is machinery, patents and factories. In Marxist thinking, for example, the bourgeois are both property-owning and exploitative in nature, but despite such descriptive differences, common elements are nonetheless evident in the range of meanings given, principally in that the term 'bourgeois' implies some form of access to the means of production. That is the bourgeois are widely perceived to have access to the means of production, and those who do not are labelled as the labouring classes or proletariat (in Marxist thought). It is with the work of Friedrich Engels and Karl Marx that attention was first given to the particular relationship between those who do and those who do not own the means of production, and their social conditions as a consequence. Engels, for example, in his 1844 study of Manchester, wrote about the poor living conditions of labouring people, while Marx argued that the poor living conditions of the workers were a consequence of the factory-owning bourgeois who kept wages low so as to ensure maximum profits and maintain their economic and social position as owners of industry.

In the modern sense, a number of bourgeois-associated terms are commonly used to describe those persons who aspire to or conform to the standards and principles of the middle class. Often, where a person has a preoccupation with materialism and middle-class respectability, the terms 'bourgeois mentality', 'bourgeois culture' and 'bourgeois lifestyle' can be noted. Such terms may often be employed in a derogatory fashion to label persons as money-oriented, of a shallow character or without a cultivated sense of taste. Furthermore, people who are perceived as exploiters in small-sized workplaces, or small-scale property owners, can be understood to be 'petty bourgeois'. Small-scale exploiters thus are petty bourgeois and more powerful exploiters are labelled bourgeois or merely 'capitalists'.

Further reading

Marx, K. and McCellan, D. (1996) *Capital*, Oxford and New York: Oxford University Press.
Rose, D. (1988) *Social Stratification and Economic Change*, London: Hutchinson.

IAN MORLEY

BOURNE, LARRY S.

b. 24 December 1939, London, Ontario, Canada

Key works

(ed.) (1982) *Internal Structure of the City: Readings on Structure, Growth and Policy*, New York: Oxford University Press.
(1992) 'Self-fulfilling prophecies? Decentralization, inner city decline, and the quality of urban life', *Journal of the American Planning Association* 58(4): 529–34.

Larry Bourne, whose academic career has been based in geography/planning at the University of Toronto, is a leading expert on Canadian urban issues. His interest lies primarily with North American cities but also includes substantial international work. A distinctive character of Bourne's work is that it generally brings together statistically substantiated trends with theoretically based interpretation and policy implications. His writings embrace a diversity of scales and topics: globalization; urban systems; urban physical form; land use; population growth, migration, and social and demographic change; housing; and modes of regulation, models of governance and the changing role of the state. Herein, he concerns himself most with documenting change and reflecting on its impacts on people, places and planning. His research is best represented in over seventy articles in refereed professional journals. His other writings include an equal number of book chapters and numerous books (written and edited) and research monographs. Bourne is particularly recognized for his contribution to understanding URBAN FORM, especially in relation to the CENTRAL CITY.

TRUDI E. BUNTING

BOYER, M. CHRISTINE

b. 17 January 1939, New York, NY

Key works

(1983) *Dreaming the Rational City: The Myth of American City Planning*, Cambridge, MA: MIT Press.
(1994) *The City of Collective Memory: Its Historical Imagery and Architectural Entertainments*, Cambridge, MA: MIT Press.

M. Christine Boyer is an eclectic urban historian whose eloquently written books on the city and

the profession of city planning emphasize the relationship between the visual perception of urban space and its production, and how the nature of this relationship has changed over the past two centuries. Her writings also criticize the American planning profession's imposition of an abstract but visually rational order on cities that not only failed to recognize humanistic influences on the production of urban space, but instead served the interests of the capitalist economy. In recent years, Boyer has drawn on her earlier career in computer science to explore the impact that information technology and virtual reality is having on the way we perceive and think, and what our relationship with electronically mediated environments portends for the future of urban public spaces.

Boyer obtained both her MA and Ph.D. in city planning from MIT, and served as professor and chair of the City and Regional Planning Program at the Pratt Institute. She is currently the William R. McKenan Jr Professor of Architecture and Urbanism at the School of Architecture at Princeton University.

Further reading

Yen, L. (2000) 'M. Christine Boyer and recent debates over virtual public space', *Critical Planning* 7: 51–61.

MICHAEL DUDLEY

BRAUDEL, FERNAND

b. 24 August 1902, Luméville-en-Ornois, France; d. 27 November 1985, Haute-Savoie, France

Key works

(1972–73) *The Mediterranean and the Mediterranean World in the Age of Philip II*, New York: Harper & Row.
(1981–84) *Civilization and Capitalism*, New York: Harper & Row.

The work of Fernand Braudel reflects a revolutionary change of focus in the craft and analysis of history. No longer do individuals or short-lived events occupy the stage of understanding that past. Rather, his work signals a shift to the analysis of world systems at the very time the Mediterranean, once the hub of Western ancient and medieval history, was superseded by the nations of the Atlantic

seaboard and their exploitation of trade, native populations and of ensuing colonial empires. Braudel enables us to view the past through the more contemporary lens of globalism in all its facets.

An understanding of this approach incorporates on the macro level long-term cycles, political and economic rivalries, the development of regions and their peripheries, money flow, and fluctuations in climate, as well as more micro considerations such as crop yields, patterns of consumption, and the fabrication of housing. Thus, we can view history as a process with an underlying base of demographics and interrelated markets mediated by technological and cultural change, emerging national economies, and their relationships to the development of international capitalism.

Further reading

Braudel, F. (1981) *The Structures of Everyday Life: Civilization and Capitalism, 15th–18th Century*, vol. 1, New York: Harper & Row.

DANIEL GARR

BRIGGS, ASA

b. 7 May 1921, Keighley, England

Key works

(1954) *Victorian People*, London: Odhams Press.
(1961–95) *A History of Broadcasting in the United Kingdom*, vols 1–5, Oxford: Oxford University Press.
(1963) *Victorian Cities*, London: Penguin Books.
(1994) *A Social History of England*, London: Penguin Books.

Asa Briggs achieved international recognition during his long and prolific career for examining many aspects of modern British history. A graduate of the universities of Cambridge and London, Briggs spent the War years serving in the Intelligence Corps. In 1945, he took up a post as a fellow at Worcester College, Oxford, where he began to carve himself a distinguished academic career in the field of nineteenth- and twentieth-century British social and cultural history. This resulted, in 1954, in *Victorian People* being published and, in 1955, in him being awarded a professorship in history at the University of Leeds. Briggs's career continued to develop at Leeds, where he continued to publish, and in 1961 he became the first

academic to be appointed to the newly formed University of Sussex. During this period at the university he completed the widely regarded *Victorian Cities*, which examined the urban growth, civic development and changing nature of a number of large-sized provincial British cities, and after six years as a professor in Sussex he received promotion to the post of vice-chancellor. Despite achieving this lofty academic height, Briggs's career was far from complete and was continuing to gain further momentum. In 1976, for example, he became provost of Worcester College, Oxford, and in 1979 he became the chancellor of the Open University, a major non-residential institution that educates its students through the mediums of television, radio and the INTERNET. Under Briggs's leadership, the Open University has become a major academic force in the United Kingdom. In 1976, due to his distinguished contribution to historical study and higher education, the British Government awarded Briggs a peerage with the title Baron Briggs of Lewes, East Sussex. During his career, Briggs has held prominent posts such as the presidencies of the Victorian Society, the Social History Society and the British Association for Local History, and has been awarded numerous honorary degrees in Britain and elsewhere.

Briggs has published prolifically, often to critical success, and in so doing has arguably become the most influential and important historian of British broadcasting, particularly the history of the British Broadcasting Corporation (BBC), as well as an authority on modern British history. Briggs's work on British broadcasting includes a major book series which chronicles the birth of the BBC and its evolution through to its modern-day form, while his work on nineteenth- and twentieth-century Britain covers the rise of industrial cities such as Manchester, Leeds and Birmingham, the political changes occurring in British society at a local and national level, economic developments, the British Empire, and the cultural transition of day-to-day British life. Currently, the British Library holds approximately 180 pieces of Briggs's work.

IAN MORLEY

BROWNFIELDS

URBAN development is a process under constant evolution. As a consequence, functions performed in certain city quarters also vary, and usually accompany the modifications that emerge from changes in society's lifestyles. As functions change, land uses that used to be prevalent in certain city locations are also subject to experiencing change. This would be the case with the vast tracts of land occupied by central railway stations, DOCKLANDS and, in general, by most of the production land uses. In fact, one of the most dramatic changes in urban land uses, which marked indelibly the twentieth-century lifestyle, occurred with the location of industries in MANUFACTURING CITIES. It is well acknowledged that change in INDUSTRIALIZATION goes together with the cyclical alterations that take place in the technologies of production and transportation. Sometimes, these changes imply drastic adjustments in the locational patterns of industry, which are primarily manifested by an outward migration of factories from their original CENTRAL CITY locations. As industries move out, usually in search of larger and more accessible sites offered in the CONURBATION areas, post-industrial physical patterns soon begin to rule in the vacated areas. These patterns are characterized by the presence of buildings and sites left vacant in prime inner-city locations, due to the succession of changes faced by the functions formerly performed there. As a result, the urban scenario drawn from this situation is a cumulative environmental obsolescence that generates what are today known in the literature as 'brownfields', a worrying picture of BLIGHT, commonly associated with the post-industrial landscapes of most urban environments. The phenomenon is not only confined to advanced economies, and the question of land left vacant in central locations is extremely relevant for the organization of all cities that have experienced changes in their industrial cycles of development. Even in THIRD WORLD CITIES the problem cannot be minimized because, there, the continued presence of underused land in centrally located areas, usually well provided with urban infrastructure, is a dismal hazard to the city's economy. Moreover, the need to avoid longer journeys traversing across the city, attributable to the expansion of the urban boundaries, is a matter of paramount importance for the less mobile population of those cities.

As an immediate follow-up comes the inevitable question of what to do or how to deal with the vacant land, in order to restore the proper running

of the cities. From the mid-1980s onwards, a broad range of alternatives started to occupy the discussions of the planning circles. Among some of the alternative suggestions that have been issued, one indicates an URBAN REVITALIZATION policy for the blighted areas as a viable choice. The argument runs that it is always more economical to serve a compact city than to extend costly services to the urban sprawls. HISTORIC PRESERVATION of the industrial buildings is another suitable consideration. After all, as observed by Kevin LYNCH, a renowned writer on city subjects, the images of factories in an urban landscape, and their meaningful reuse as PLACES marking a city's identity, should be kept as a symbolic paradigm of twentieth-century industrial cultural heritage. Also, an ADAPTIVE REUSE of the sites and warehouses can be reckoned, even if at the cost of an exhaustive examination of the socio-economic benefits that reuse strategies will eventually bring to the brownfield sites. Finally, a mixed strategy, combining some of the previous alternatives in order to promote the sustainable reuse of the declining industrial areas, might also be thought of.

The convenience of reusing brownfield areas, as an alternative to extending urban growth, and the corresponding search for planning strategies capable of revitalizing the abandoned areas are new concepts, and still dominate many of the planning discussions of the 2000s. Those in favour of recycling ground their support on the necessity of reducing pressure on exurban expansion. The argument runs that the reintroduction of local economic activities is often advised as a basic requisite to balance physical, administrative and economic problems. Of course, such argument allows for the phenomenon of vacant inner-city industrial sites being regarded as a source of opportunities for effective land policies, rather than as a cause for cumulative problems. Contrarily, those who are against it argue that the upgrading of the existing infrastructure services, together with the cleaning up of the polluted brownfield sites, demand such a large number of operations that, in the end, the initiative becomes largely cost-prohibitive. This unveils another important facet of the topic: it is necessary to weigh the potential opportunities for redeveloping an urban brownfield against the constraints posed by its implementation. Reuse opportunities and constraints, as well as the need for an environmental clean-up, are all items that must be carefully assessed. Also, the

allocation of liabilities caused by polluted sites is a matter of concern, since it may incur serious risks for the redevelopment policy.

As a final point, it is realistic to appreciate that the twenty-first century started with a worried acknowledgement of the hazardous consequences of industrial culture for soil, water and air. It was also recognized that this pressed mankind to struggle against the obnoxious effects of the industrial age, as a path to improve the environmental sustainability of life on the urbanized planet. However, since the individual conditions of brownfields vary considerably, almost on a site-to-site basis, it is reasonable to accept that a good first step would be to avoid the unrestricted standardization of patterns, and their use as solutions in all cases. A rational approach to the problem would certainly advocate a piecemeal process as the most advisable practice.

Further reading

Barnett, J. (1996) *The Fractured Metropolis: Improving the New City, Restoring the Old City, Reshaping the Region*, New York: IconEditions/HarperCollins.

Castello, L. (2000) 'Technological districts or commercial districts: alternatives for the reuse of industrial brownfields', *LAC Papers*, Cambridge, MA: Lincoln Institute of Land Policy, available online at www.lincolninst.edu/main.html

Wright, J.G. (1997) *Risks and Rewards of Brownfield Redevelopment*, Cambridge, MA: Lincoln Institute of Land Policy.

LINEU CASTELLO

BURNHAM, DANIEL HUDSON

b. 4 September 1846, Henderson, New York, United States; d. 1 June 1912, Heidelberg, Germany

Key works

(1908) Union Station, Washington, DC.
(1909) Plan of Chicago, Chicago, IL.

Born in rural upstate New York, by age 27, Daniel Hudson Burnham was well on his way to becoming one of the great architects of the twentieth century. At age 9, his family moved from Henderson, New York, to Chicago, where he was exposed to a rapidly growing urban complex. After attending high school there, the young Burnham travelled

west to Nevada to prospect for gold, having been rejected for undergraduate study by both Harvard and Yale; institutions that later lauded him with honorary degrees. Returning to Chicago in 1872, he began work for the architectural firm of Carter, Drake and Wright. In 1873, he formed the legendary architectural partnership Burnham and Root, with John Wellborn Root. Together they designed some of the earliest modern skyscrapers. Aided by the recent inventions of electric lighting, the elevator and telephones, they advanced building principles of the day using a cast-iron skeleton to create the early skyscraper.

With the death of Root in 1891, the firm, renamed D.H. Burnham and Co., continued its successes. In 1893, Burnham extended his reach into city design. Responding to a request to coordinate the design of the Columbian Exposition of 1893, he became chief of the Columbian Exposition Construction Department. In that project, he was instrumental in guiding the work of more than a dozen major architectural firms to create a national exposition celebrating the 400th anniversary of Columbus's discovery of America. Designed in the *beaux arts* style, the Exposition reflected the academic tradition of many of its architects who had studied in Paris at the ÉCOLE DES BEAUX-ARTS.

The Columbian Exposition also marked the beginning of the CITY BEAUTIFUL MOVEMENT in the United States, with Burnham as one of its major proponents. The movement argued that American cities would become as culturally prominent as European cities through the use of the *beaux arts* style and that beautiful cities would promote civic loyalty. That aesthetic judgment would find its expression in further designs by Burnham such as the Flatiron Building in New York (1902) and Union Station in Washington DC.

Following his success with the Columbian Exposition, Burnham was commissioned in 1909 to develop the Plan of Chicago, one of the first comprehensive land use plans in the United States. With his associate, Edward Bennett, the Plan of Chicago protected lakeside properties from development, concentrated mixed-use developments, provided for a comprehensive transit system and identified sensitive natural areas that should be preserved. In its wake, he created similar plans for San Francisco, Cleveland and Manila.

He died in 1912 while travelling in Germany.

Further reading

Burnham, D. and Bennett, E. (1909) *Plan of the City of Chicago*, Chicago: Commercial Club of Chicago.
Moody, D. (1915) *Wacker's Manual of the Plan of Chicago*, Chicago: Chicago Plan Commission.

JAMES A. FAWCETT

BUSTEE

The word 'bustee' is a distortion of the Bengali word 'basati', which means habitation, a residence or a colony. Bustees are the predominant type of housing for the urban poor in Kolkata (Calcutta). These are legal 'slums' and should be differentiated from illegal squatter settlements along canals and railroad tracks, under bridges, or on pavements. Unlike squatters, bustee dwellers have housing rights and cannot be evicted. About 30–35 per cent of Kolkata's population lives in bustees. They have existed as long as the city has, but their phenomenal growth began in the mid-nineteenth century. Bustees are scattered throughout the city. Kolkata's indispensable labour and petty traders such as domestic servants and maids, rickshaw pullers, street vendors, small shop owners, scavengers, factory workers and artisans make bustees their home. Clearly, bustee dwellers are not marginal city residents, but play a vital role in the functions of the city. Bustees are generally rented huts made of permanent and semi-permanent materials such as bricks, concrete, asbestos, clay tiles, tin and mud. Although the British did little to improve bustees before independence in 1947, Kolkata's local authorities were the first in India to recognize bustees as legitimate housing for the poor. This was primarily because of the role played by local leftist parties in the 1950s. These leftist parties started building a political base among bustee dwellers, and by the late 1950s the principal leftist party at that time, the Communist Party of India (CPI), led a mass movement against proposed legislation for bustee clearance. The then ruling Congress Party, who were initially opposed to CPI's activities, soon realized the futility of slum clearance and adopted a bustee improvement plan as early as the mid-1960s. The subsequent involvement of the World Bank in 1971 saw the launching of the most successful slum improvement programme in the developing world. From deplorable living conditions, bustees were

transformed to more habitable dwellings with the provision of paved roads, sanitary and drainage facilities, electricity, and drinking water. The Communist Party of India (Marxist) (CPI(M)) (which was formed by a faction of the CPI and which came to power in the late 1970s) passed the Thika Tenancy (Acquisition and Regulation Act) of 1981, thereby increasing the security of the bustee dwellers. Prior to the passing of the Act, the land in the bustees was owned by landlords who leased the land to intermediary developers known as Thika Tenants. These tenants in turn constructed huts and rented them to the inhabitants. The landlords paid taxes on the land and collected revenue for their property from the intermediary developers. After the Act was passed in 1981, the local government abolished the landlord system by purchasing and acquiring the land and renting it directly to the Thika Tenants. Although CPI(M) still has a stronghold, most political parties now have a base in the bustees. With Kolkata's growing housing shortage and rising real estate prices, land speculation has become common in bustees. Individual owners and speculators have also erected illegal structures, thereby violating building codes for profit.

Further reading

Pugh, C. (1990) *Housing and Urbanization: A Study of India*, New Delhi: Sage.
Sen, S. (1998) 'On the origins and reasons behind non-profit involvement and non-involvement in low income housing in urban India', *Cities* 15(4): 257–68.

SIDDHARTA SEN

C

CALTHORPE, PETER A.

b. 25 June 1949, London, England

Key works

Northwest Landing, DuPont, Washington.
The Crossings Neighborhood, Mountain View, California.

American architect, urban designer, land use planner, author and one of the leading proponents of the NEW URBANISM, Peter Calthorpe has been named one of twenty-five 'innovators on the cutting edge' by *Newsweek* magazine for his work on redefining the models of urban and suburban growth in America. After attending Antioch College and Yale's Graduate School of Architecture, he joined various architectural firms. Starting practice in 1972, he has had a long and honoured career in the planning and architecture fields, combining his experience in both disciplines to develop an environmental approach to community development and urban design. Since forming Calthorpe Associates in 1983 (www.calthorpe.com), based in Berkeley, California, his work has further diversified to major projects in urban, new town and suburban settings in the United States and abroad.

Peter Calthorpe has published technical papers, articles and a number of books. In the early 1980s, he co-authored the book *Sustainable Communities: A New Design Synthesis for Cities, Suburbs, and Towns* with Sym Van der Ryn. His book *The Next American Metropolis: Ecology, Community, and the American Dream*, a landmark text of the new urbanism, provides key principles and techniques for regional and local planning. His latest book, *The Regional City: Planning for the End of Sprawl*, is co-written with William Fulton.

Peter Calthorpe lectures extensively throughout the United States, Europe, Australia and South America, and teaches at various universities in the United States. He has received numerous honours and awards. He was also selected to represent the United States in an exchange with Russia on city and regional planning issues, and was appointed to the President's Council for Sustainable Development. Most recently, he provided direction for the US Department of Housing and Urban Development's Empowerment Zone and Consolidated Planning Programs, which emphasize the important relationship between land use, transportation and community design. One of the pioneers of the new urbanism movement, he was the founder of the Congress for the New Urbanism and was its first board president.

Peter Calthorpe's design philosophy focuses on creating communities that are ecologically sustainable, diverse, mixed-use and pedestrian-friendly. Calthorpe Associates place special emphasis on fostering neighbourhoods that provide a range of housing in close proximity to shopping, jobs, recreation and transit – walkable communities that offer realistic housing and transportation choices. Projects range from urban infill and redevelopment plans to new towns and regional growth strategies. Calthorpe has been a pioneer in finding acceptability for progressive urban planning ideas such as the URBAN VILLAGE, pedestrian pockets and TRANSIT-ORIENTED DEVELOPMENT. Calthorpe's current work extends his vision from neighbourhood to regional design (Twin City Project).

Further reading

Calthorpe, P. (1993) *The Next American Metropolis: Ecology, Community, and the American Dream*, New York: Princeton Architectural Press.

Van der Ryn, S. and Calthorpe, P. (1986) *Sustainable Communities: A New Design Synthesis for Cities, Suburbs, and Towns*, San Francisco: Sierra Club Books.

TIGRAN HASIC

CALVINO, ITALO

b. 15 October 1923, Santiago de las Vegas, Cuba; d. 19 September 1985, Siena, Italy

Key works

(1974) *The Invisible Cities*, New York: Harcourt Brace Jovanovich.
(1981) *If on a Winter's Night a Traveler*, New York: Harcourt Brace Jovanovich.

Journalist, essayist and writer, Calvino was born in Cuba, of Italian parents. He grew up in Italy and graduated with a degree in literature from the University of Turin (1947). His experience in the Italian resistance during the Second World War marked his first efforts as a fictionist, but his acclamation as a writer of fantastic stories came in the 1950s, in particular for the trilogy formed by *Il visconte dimezzato* (1952), *Il barone rampante* (1957) and *Il cavaliere inesistente* (1959).

Some of his titles mark his innovative experimentation. In *Marcovaldo* (1963), Calvino drives his character in unexpected directions in the daily life of the modern city. In *Cosmicomiche* (1965), he tells the adventures of Qfwfq, strange hero who, after testifying to the Big Bang, turns into the chronicler of the evolution of the Universe. In *Il castello dell destini incrociati* (1973), in a hypertextual game, a pack of tarot cards played on a table structures the double narrative – in a castle and in a tavern – of speechless characters.

The amazing work of Calvino cannot be understood without considering *The Invisible Cities* (1974), one of his masterpieces, in which he recounts imaginary dialogues between Marco Polo and the Mongolian emperor Kublai Kahn, when the legendary Venetian navigator of the thirteenth century, in the powerful emperor's service, is entrusted with visiting the numerous cities of the kingdom. Each city represents a brief but intense symbolic allegory, narrated to the Lord of the Mongols.

In *If on a Winter's Night a Traveler* (1981), another definitive masterpiece, the hypertextuality returns in a not well-published book, whose second chapter begins a different work, sending the reader/character on a search that takes in ten different incomplete romances, ending with a loving conclusion.

In 1984, Calvino was invited to give the Charles Eliot Norton Poetry Lectures for the 1985/86 academic year at Harvard University. He decided to write on those essential values that literature should preserve in the new millennium. His unexpected death left uncompleted the task that he had proposed. Five of the planned lectures are gathered in *Six Memos for the Next Millennium*, published posthumously, and discuss 'lightness', 'quickness', 'exactitude', 'visibility' and 'multiplicity' respectively. In the sixth and last, Calvino was to write about 'consistency'.

LEANDRO M.V. ANDRADE

CAPACITY-BUILDING

The term 'community development' indexes a series of initiatives from the 1950s in which the active participation of members of local communities in social and economic development was encouraged via national and sub-national plans. The term 'capacity-building' appears from the 1990s in literature on community development and sustainable development, and refers to strategies and practices by which people realize their potential and contribute to individual and collective outcomes for a better quality of life. Applicable to rural contexts, capacity-building is nevertheless often championed in URBAN settings because of the resources available to urban populations and governments.

Community involvement is an expression of active CITIZENSHIP, which is different from passive citizenship – itself characterized by voting, paying taxes, obeying the law and protecting personal rights. Active citizens prioritize collective responsibilities and common goals, asking whether, how and to what extent we should care for the future; how much we are prepared to sacrifice for future generations; and to what degree non-humans and non-citizens are due consideration. In urban environments, with their large ecological footprints, such questions are imperative and demanding. They require advanced capacities – such as civic and ecological literacy or an understanding of

governmental and environmental processes – to engage in responsible and tolerant social debate (see CIVIL SOCIETY).

These capacities are best fostered when individual health and well-being are engendered, and city governors are increasingly aware that leadership is important to optimize such outcomes. Thus, in Australia, the City of South Sydney's strategic plan includes a local food policy to cultivate HUMAN CAPITAL via better dietary habits, knowledge of nutrition and food preparation; improve the quality of food available to the COMMUNITY; ensure that the Council's role in direct food services is appropriate to the needs of the community; and support environmentally sustainable food production and delivery (South Sydney City Council, 1995).

City governments also have a role in building social capacities, often referred to as 'SOCIAL CAPITAL' – the glue that strengthens the social stability of communities and the lubricant that enhances cooperation and trust for mutual benefit. Social capital enhances effective participation in decision-making and increasingly – in the United Kingdom, Canada, the United States, New Zealand or Australia, for example – is formally fostered via partnerships between councils and communities, or among groups of councils. In such partnerships, social learning is often emphasized.

Change in the operations of Marrickville Council in the relatively poor inner-west of Sydney, Australia, illustrates this growing tendency to foster human and social capacities for wider community development and sustainability outcomes. In a series of partnerships with community and industry, specific attention has been paid to social capital building, sustainability and citizenship. The Council's programme focused on strategic thinking and innovation; communication and citizen participation; performance management and teamwork developed through a management leadership programme; and a staff recognition scheme. It reported an overall increase in equity; involvement in consultation; low levels of conflict; and changes in organizational CULTURE. In 2001, it won an Australian Local Government Excellence Award for this work.

Further reading

Myers, G. and Macnaghten, P. (1998) 'Rhetorics of environmental sustainability: commonplaces and places', *Environment and Planning A* 30(2): 333–53.

United Nations (1992) *Agenda 21, Chapter 37*, available online at http://habitat.igc.org/agenda21/a21-37.htm, accessed August 2004.

ELAINE STRATFORD

CAPITAL ACCUMULATION

Accumulation is an objective, complex and permanent process, having a law character. Its aim is creating new fixed and working capitals, broadening and modernizing the existing ones, growing the material basis of social-cultural activities, as well as constituting the necessary resources for reserve and insurance.

In a general sense, accumulation means gathering up, collection, progressive increasing of some goods that allow continual enlargement and diversification of production and economic growth.

In an economic sense, accumulation refers to creating possibilities for continuing production to a higher level. In a strict sense, accumulation means transforming a part of the net income into investment goods needed for production rise. In a broad sense, accumulation means using all the economic resources for enlarging production (a part of the net income, the depreciation (amortization) fund, external credits, etc.). The depreciation fund, designed for replacing obsolete fixed capital, is also used for expanding and perfecting fixed capital, allowing the input and the use of high-performance fixed assets with higher returns. Accumulation could be either productive when the investment goods are used in the production process, or unproductive when these are used for social-cultural activities. Depending on the formation level, accumulation could be individual, group or national, and considering the nature of ownership, accumulation could be private or public (state). The interaction between accumulation and socio-economic development is expressed through the accumulation law according to which 'accumulation is society's most progressive function', the foundation for continuing production at a higher level, which in turn creates new development possibilities. The accumulation's dimensions and dynamics depend upon the net national product (NNP), the accumulation rate, the depreciation fund, and on the external credits volume. The accumulation fund depends upon the size of NNP

used and on the ratio between consumption and savings. The accumulated wealth embraces firms' reproducible assets in the form of both fixed and working capital goods, reproducible state assets including roads, bridges, channels, etc. (except capital goods of an 'offensive' military nature), and households' durable goods.

Capital accumulation represents the process of augmenting a firm or a country's capital. Increasing capital through accumulation could be realized (a) by maintaining unchanged the capital inner structure, or (b) by modifying the capital composition. The sources for accumulation are (a) the profit obtained; (b) the fixed capital depreciation fund; (c) capital contributions; and (d) credits. Into a stock company, accumulation is realized through (a) new (supplementary) capital contributions, mostly money, from stockholders, and more rarely through capital assets following a merger process; (b) incorporating reserves into capital; or (c) debt-equity swap. The supplementary capital contributions are made either by (a) existing stockholders (closed companies), or (b) new stockholders (open companies). Transforming reserves into capital is done (a) through raising the value of existing shares (without increasing their number), or (b) by issuing new shares (of an equal value, or different).

Primitive (primary) capital accumulation represents the formation and raising of capital through extra-economic means, often by force, a process widespread at the beginning of the capitalist age, gradually reduced once the rule of law was enforced, but still present in various degrees nowadays. Primary accumulation was realized in Western European countries during the Mercantilism Age (sixteenth to seventeenth centuries).

Further reading

Hicks, J.R. (1965) *Capital and Growth*, New York: Oxford University Press.

Marx, K. and Engels, F. (1965) *Opere*, vol. 23, Bucharest: Editura Politica, pp. 576–777.

GHEORGHE POPESCU

CAPITAL FACILITY PLANNING

Within the United States, capital facility planning provides a framework for knowing which public project activities will be provided, in what quantity, at what cost, over what time period, in what manner, and for whom.

Capital facility planning is a systematic process for recommending the replacement or the construction of capital projects in the community. The two key elements for a successful capital facility planning process are the adoption of an updated comprehensive plan and an extensive financial investment strategy for big-ticket community projects. Capital facility planning is built upon the continuous assessment of the demand for public improvements within the community. Consequently, an updated comprehensive plan provides the knowledge base to anticipate the types of public infrastructure improvements required with respect to the population being served.

Capital facility planning is the umbrella process for both the Capital Improvement Plan (CIP) and the Capital Budget. Both are programme-planning components for determining the financial priority of public projects. The CIP is a schedule for the placement and timing of capital projects. It notes when a public project needs to be replaced or built. The CIP should not be confused with the Capital Budget, which is part of a city's annual operating budget. The CIP is a seven-year schedule for all present and future capital facility projects that is required by the comprehensive plan. The CIP can be viewed as a multi-year capital budget that itemizes the timing and financing of capital facility projects.

In contrast, the Capital Budget establishes a yearly priority review of replacement needs of existing infrastructures as well as future capital facility projects. Derived from the CIP, the Capital Budget is a multistage process by which a planning staff considers each public project, then costs each of them out, and finally determines the method of payment for each listed project. Ideally, the Capital Budget is viewed as part of the city's budget document, providing information pertaining to the financial aspects of all capital facility projects. The Capital Budget, therefore, helps public officials to decide on an annual basis the priority rankings for all big-ticket item expenditures by being the financial arm of the capital facilities plan.

Together, both capital facilities planning components aid the planner in his/her effort to undertake capital debt financing and to efficiently allocate public services. Consequently, several interrelated factors, some which are socio-economic and others which are fiscal, determine the parameters of a capital improvement plan. These interrelated

factors include the fiscal capacity – the capacity to raise revenues from the tax base – of a community, the policy skills of the planning staff for obtaining federal aid, and the willingness of its citizen tax-payer to assume additional community debt. Therefore, capital facilities planning cannot be thought of in simple engineering terms. The objective of the capital facilities plan-making process is to anticipate costs for improvement of the overall quality of life within a city while maintaining the viability of that community's economic base.

SEE ALSO: fiscal impact analysis

GARY A. MATTSON

CAPITALIST CITY

Scholars continue to ponder the nature of the relationship between the growth of certain types of cities and the different stages of capitalist development. According to Marx and ENGELS, the form and function of the city are a direct consequence of economic and social organization. Both the relations between different towns and the living conditions within a specific location can be explained as responses to the mode of production of a given historical moment. After them, the debate focused on those aspects of the city that reflect the interplay between the accumulation process and class struggle (see CASTELLS; LEFEBVRE; HARVEY). If there is no unanimity between the various interpretations, there is general agreement on some points.

It is acknowledged that the main elements defining the city are the alienation of assets and resources to private hands, their use for profit to meet the market's interests rather than individual needs, and the physical separation between the working class and those with capital.

There is also agreement on the fact that the capitalist city is not a stable entity with features defined once and for all. As capitalism evolves from one stage to another, it produces a geographical and social setting suited to its own accumulation dynamics. Urban settlements played a key role in the emergence of capitalism but, at the same time, they were strongly shaped by its needs. The search for scale economies was at the heart of the rapid growth of urban centres; on the other hand, urban development was made possible and necessary by the demands of industry.

The main features of this period were the concentration of workers in the factories and factories concentrated within the towns, with a consequent UNEVEN ECONOMIC DEVELOPMENT, and spatial differentiation.

In the transition to the corporate accumulation stage, manufacturing industries started moving out of the city centres and were replaced by various types of services. Successive waves of decentralization became the rule. The previous commercial areas were transformed into central business districts. Empty industrial areas were abandoned and the workers' quarters were often turned into GHETTOS. The SUBURBS started developing, with the settlement of industrial complexes and commercial centres and specialized and segregated residential areas.

In the interpretation of how concrete spatial features are determined by capitalist logic, the issue of land value is of significant importance. When land became a commodity, the capital-holders started realizing surplus values, not only through the sale of industrial products, but through real estate investments, deciding the various uses of the land according to the highest profit criterion. The growing importance of financial motivations with respect to industrial ones is the main characteristic of the capitalist city in the post-industrial period.

Further reading

Gordon, M.D. (1984) 'Capitalist development and the history of American cities', in W.T. Tabb and L. Sawers (eds) *Marxism and the Metropolis*, New York and Oxford: Oxford University Press.
Harvey, D.W. (1985) *The Urbanization of Capital*, Oxford: Blackwell.

PAOLA SOMMA

CARTESIAN CITY

An early URBAN form, the result of a fixed plan adopted for a new settlement or urban expansion, with an orthogonal structure where STREETS are drawn at right angles to each other as in the Cartesian coordinate system, resulting in regular blocks. It responds to a rational conception of the world as a form of spatial organization and not a result of organic growth.

The Cartesian plan is simple in the sense that it is easy to design, and distributes land equitably, the same principles that led to the adoption of the township and range system for land division west of the Appalachians. However, some drawbacks emerge as a result of the lack of flexibility of this structure. The adaptation of the regular pattern to an uneven terrain is problematical, resulting in local steep slopes or variations from the planned structure, as in the case of San Francisco. Distances are increased and, although it favours communication in the two main directions, it hinders communication in diagonal directions.

The Cartesian plan has also been called the Hippodamic plan, since Hippodamus of Miletus is considered to be the author of the plan for his home POLIS after its destruction by the Persian army. It was reconstructed following the standard plan used in the foundation of the Mediterranean Sea colonies. Although there are earlier examples of Cartesian-like plans in the valley of the Indo dating from 3000 BC, such as Mohenjo Daro or Harappa, the settlement of Greek colonies in the eighth century BC extended the plan's implementation in a wide area. The Roman Empire resumed the diffusion of the model throughout southern Europe and northern Africa with planned walled military camps organized around two principal orthogonal axes, the *cardus* and *decumanus*, crossing in the forum, or city centre. They later became consolidated as commercial towns, which in some cases kept this structure in their HISTORIC DISTRICTS.

In the Middle Ages, as well as the bastide towns of the thirteenth century in Aquitaine, France, some new cities were founded and planned according to a regular orthogonal scheme in the German expansion area to the East.

Between the sixteenth and eighteenth centuries, other geometrical forms were developed associating URBANISM and architecture. The colonization and settlement of South and North America which began in the sixteenth century was an opportunity to implement the Cartesian plan in territories without an urban system, although it was not always adopted.

The Industrial Revolution initiated a phase of rapid urban growth accommodated in SUBURBS planned following a grid pattern. The nineteenth and twentieth centuries saw the application of the Cartesian plan in the New World, as a response to the rapid economic development and the arrival of immigrants from Europe and rural areas, as well as the emergence of metropolises like New York and Los Angeles. The plan has since been modified with diagonal avenues over the underlying structure to overcome the obstacles to long-distance communication, such as in Washington and Barcelona.

Further reading

Morris, A.E.J. (1994) *History of Urban Form before the Industrial Revolution*, New York: Longman.

URBANO FRA PALEO

CASTELLS, MANUEL

b. 9 February 1942, Barcelona, Spain

Key works

(1977) *The Urban Question: A Marxist Approach* (*La question urbaine*), trans. Alan Sheridan, London: Edward Arnold.
(1996–98) *The Information Age: Economy, Society and Culture*, vols I–III, Oxford: Blackwell.

Manuel Castells is a Spanish-born academic who became one of the leading neo-Marxist urban sociologists in the 1970s and who later, in the 1990s, gained international renown after publishing the trilogy *The Information Age*.

As a student activist against Franco's dictatorship, Castells went into exile in the early 1960s. He headed for Paris. Castells graduated from the Sorbonne in 1964 and received his Ph.D. from the University of Paris in 1967. His doctoral dissertation discussed the impact of technological factors on firms' locational decisions in metropolitan Paris.

Castells started his academic career as a researcher and later became an associate professor in the Research Centre of Social Movements of the University of Paris. His academic career took a major turn in 1979 when he was appointed as professor of sociology and planning at the University of California, Berkeley. It was to become his intellectual home for the rest of his productive academic life. Castells has taught and carried out research in several universities in America, Europe and Asia. After gaining a worldwide reputation, his writings and speeches have attracted a lot of public and professional attention. He has

also served as an expert in many national and international organizations.

Castells has published some twenty books and co-authored or edited a further fifteen. His early works were critical urban analyses. His first book, *The Urban Question*, published originally in French in 1972, had a greater influence on the formation of urban political economy than any other single work at that time. Another important theme in Castells's writings is technology and its impact on society. *The Informational City*, published in 1989, turned out to be one of the most important macro-theoretical works on the impact of technological development on urban-regional processes.

The culmination of Castells's work, *The Information Age* (1996–98), outlines the impact of the emerging informational mode of development on economics, politics and culture. The core message of the trilogy is that there is a growing tension between global networks of informational capitalism ('the Net') and the everyday lives of ordinary people ('the Self'). In fact, this tension pervades all his works.

Castells has rightly been characterized as the Karl Marx of the information age. Yet he has never been an orthodox Marxist scholar. Rather, he is politically a social democrat and intellectually some kind of analytical neo-Marxist, whose mission has been to connect human social experience to the critical structural analysis of informational capitalism. This is how Castells prepared the ground for the silent revolution emerging from emancipatory identity-building processes of people and their communities.

ARI-VEIKKO ANTTIROIKO

CATEGORICAL GRANT PROGRAMME

Grants-in-aid (also known as *intergovernmental transfers*) have become an increasingly important part of the revenue sources of local governments across the world. Britain, France, Germany, Sweden and the Netherlands are only a few of the countries where the financial and technical support of the central government helps sub-national governments to combat important problems. In countries such as Japan, central government assistance provides the bulk of municipal spending monies.

As seen in the United States, there are two major types of grants-in-aid: categorical grants and block grants. *Categorical grants* are designed for very narrow and specific objectives as determined by Congress; they have accompanying rules and guidelines that constrain the recipient government in the use of grant funds. A local government that receives a categorical grant to upgrade its police communications equipment, for instance, cannot use that grant for any other police or non-police purposes. In contrast, a *block grant* has fewer accompanying rules and regulations and allows the recipient government greater discretion in its use of funds within a designated functional area.

There are two further variants of grants. *Formula grants* are distributed to sub-national jurisdictions in accordance with such statistical criteria as the population, per capita income, tax effort, or the number of senior citizens or school-age children in a jurisdiction. Recipient governments are entitled to the grants by virtue of the law; they do not have to submit to elaborate proposals and compete for funds. *Project grants*, in contrast, are not automatically allocated on the basis of a formula; instead, eligible jurisdictions must apply and compete for the programme money.

The resulting array of grants can be very confusing; local governments may not even be aware of all of the available assistance programmes or of the different application and reporting procedures for each programme. This confusion and inefficiency has led to the call for grant simplification and rationalization. In Britain, Margaret Thatcher in 1987 combined the urban development grant, the urban regeneration grant and the derelict land grant programmes into a new city grant with a simplified grant application process.

In the United States, most federal aid programmes take the form of categorical grants. But Richard Nixon in the late 1960s and 1970s introduced what was then a relatively new form of federal aid, the block grant, in order to add a new degree of sub-national flexibility to the intergovernmental assistance system.

Yet Nixon did not entirely succeed in grant consolidation. Important constituencies continued to fight for new programmes of categorical assistance. When Nixon entered office, there were

306 federal grant programmes available to states and localities. By the time Ronald Reagan took office in 1981, the figure had grown to 539. Reagan cut the number to 492; yet, under Presidents Bush and Clinton, the pressures for growth reasserted themselves, and the number of grant programmes exceeded 600. By the mid-1990s, the grant system was worth over US$220 billion and exceeded 15 per cent of federal outlays.

Further reading

Walker, D.B. (2000) *The Rebirth of Federalism: Slouching Toward Washington*, 2nd edn, New York: Seven Bridges Press/Chatham House.

BERNARD H. ROSS AND MYRON A. LEVINE

CENTRAL CITY

'Central city' is a term that is used to denote the municipality containing the central business district within a larger metropolitan area. The term is used most widely in the US context, although since the 1970s it has also become relatively common in Canada and, to a lesser extent, Europe and Australia. Within the United States, the central city is strictly defined by the US Census Bureau as most populated municipality within an urban area containing at least 50,000 people with a population density exceeding 1,000 persons per square mile. In the case of (consolidated or standard) METRO-POLITAN STATISTICAL AREAS, there may in fact be more than one central city municipality within a metropolitan region. Outside the United States, there are no official census definitions of what constitutes a central city. However, when used in other national contexts, the term often conveys or implies a definition similar to that of the US Census Bureau.

The central city is typically the oldest, most densely populated municipality within an urban region. Often (but not always), it contains the largest concentrations of the poor, visible minorities, tenants and low-rent housing, as well as the largest concentration of high-level financial and business services activities and headquarters. Frequently, the central city is the locus of what is referred to as the URBAN UNDERCLASS and of the GHETTO. Since the end of the Second World War, many of the functions previously belonging to central cities, including industrial, office and retail

functions, have been dispersed to nearby SUBURBS, some of which have in turn grown into edge cities. This decentralization, coupled with the flight of many middle-class families from central cities during the post-war period, is cited as the main cause of urban fiscal stress and URBAN DECLINE. Decentralization from central cities has been most acute in the United States, but similar problems have beset central cities in Canada, Australia and Europe. On the other hand, since the 1970s, many central cities have witnessed REDEVELOPMENT and reinvestment. This, in some cities, is related to the process known as GENTRIFICATION, which in turn is cited as a cause of the displacement of the poor, new immigrants and non-whites from central-city neighbourhoods.

The term 'central city' is closely related to, and often used interchangeably with, the term 'inner city'. However, the latter does not usually connote any explicit political structure or overtly defined boundaries. Instead, the concept of the inner city conveys a less technical and more implicitly socio-cultural set of (usually negative) meanings, and is frequently associated with social deterioration, blighted commercial districts and dilapidated residential neighbourhoods, high crime rates, and racial segregation. 'Inner city' is the more common term outside the United States.

Further reading

Ley, D. (1996) *The New Middle Class and the Remaking of the Central City*, Oxford: Oxford University Press.
Marshall, H.M. and Stahura, J.M. (1982) 'The growth and decline of American central cities', *Journal of Urban Affairs* 4(2): 55–66.

R. ALAN WALKS

CENTRAL PLACE THEORY

Central place theory was first introduced in 1933 by geographer Walter Christaller to explain the spatial distribution of cities across the landscape. In *Central Places in Southern Germany*, Christaller argued that the primary purpose of a settlement or market centre is to provide goods and services for the population of the surrounding area. Central place theory uses the basic concepts of *threshold* and *range*. The location of any place is determined by its threshold, or minimum market area necessary for the goods and services offered to be economically viable, i.e. to bring a firm selling goods

and services into existence and to keep it in business. Christaller suggested that each place seeks to expand its market area until the range, or maximum distance consumers will travel to purchase a commodity or service, is reached. Under ideal conditions, market centres of the same size and function would be equidistant from each other.

Christaller's theory assumes the ideal condition of a uniform homogeneous plane of equal population density and purchasing power. In this regard, central place theory is similar to the location theories of Weber and Von Thünen, where locations are assumed to be on a Euclidean, isotropic plane with similar consumer purchasing power in all directions. Christaller suggests that goods and services can be categorized along a continuum from low order or basic (such as groceries) to high order or highly specialized (such as sub-specialties of medical care or specialized appliances or automobiles). Each item or service has its own optimal market area that can be expressed as the radius of a circle. To ensure that the entire landscape area is served, the circles of market area must overlap. The resulting pattern therefore can be described using geometric shapes of circles, hexagons and triangles.

Since movement across the landscape is posited to be uniformly easy in any direction, transportation costs vary by distance. Delivered price of a good or service is composed of store price plus transportation cost. As a result of increasing transportation costs, a consumer living further from a market pays more for the same goods and services as a consumer living close to the market. In theory, consumers will act rationally to minimize costs by shopping at the nearest location offering the desired good or service. Places with a lesser range or smaller market area provide the lower-order goods and services that are purchased most frequently.

Market centres providing highly specialized goods and services are called higher-order places and are larger and fewer in number than lower-order places. Since quantity purchased, or demand, is a function of price, higher-price goods and services will have lower demand than lower-priced goods. A market providing more expensive goods and services will therefore require a greater range than one providing lower-cost items. Higher-order places offer goods that are the least sensitive to changes in cost, have a greater range and can take advantage of ECONOMIES OF SCALE. Since the quantity and specialization of goods and services offered determines the size of the place, a place with more specialized goods and services will be larger and consumers will travel further and pay the associated transportation costs to acquire the goods and services offered.

It is useful to keep in mind that Christaller's theory rests on certain assumptions that render the model inapplicable to realistic situations. These include:

1   the landscape is a flat, unbounded, homogeneous plain with similar and evenly distributed resources – there is no geographic differentiation;

2   political and administrative boundaries do not exist to distort the even spacing and development of settlements;

3   there are no external economies or diseconomies to interfere with the market;

4   the population is evenly distributed across the market plane with no solely residential centres;

5   many small sellers offer the same product – there is no product differentiation;

6   all consumers have the same purchasing power;

7   transportation costs are equal in all directions and vary in proportion to distance;

8   the buyer pays the transportation cost of the product or service; and

9   there is no accommodation for innovation or entrepreneurship.

German economist August Lösch made extensions and modifications of Christaller's central place theory. In his book *The Spatial Organization of the Economy* (1940), Lösch began with the smallest scale of economic activity, farms, which were regularly distributed across the landscape in a triangular lattice pattern. Lösch proposed a consumer model based on an administrative and manufacturing structure as opposed to Christaller's service centres. Despite the underlying unrealistic assumptions, central place theory was a breakthrough in helping to structure thinking about differentiation in the development of communities and has been nonetheless useful in considering the location of trade and service activity and the location of the provision of distinctive goods and services. The conceptualization of a hierarchical arrangement of market communities

also permits a particular consideration of the impact on social networks as the economic activities and movement of people are modified according to the hierarchical level of services provided. Central place theory has served as a foundation for a large body of empirical research on the external structure of cities and is applicable to urban and regional economic development issues concerning the location and viability of economic activity.

Further reading

Berry, B.J.L. and Harris, C.D. (1970) 'Walter Christaller: an appreciation', *Geographical Review* LX(1): 116–19.

Christaller, W. (1966 [1933]) *Central Places in Southern Germany* (*Die zentralen Orte in Süddeutschland*), trans. C.W. Baskin, Englewood Cliffs, NJ: Prentice Hall.

—— (1972) 'How I discovered the theory of central places: a report about the origin of central places', in P.W. English and R.C. Mayfield (eds) *Man, Space and Environment*, Oxford: Oxford University Press, pp. 601–10.

Lösch, A. (1954 [1940]) *Economics of Location*, trans. William Woglom from 2nd rev. edn, New Haven: Yale University Press.

PATRICIA BARON POLLAK

CENTRALIZATION

In a broad sense, centralization is a principle of organization where decision-making is concentrated within a superior group that must control subordinate groups in order to facilitate the activities of the whole. The degree of centralization depends on the scale, extent of divergence, and amount of territory belonging to an organization. It is also a characteristic of a hierarchical power structure, since the activity of the whole is controlled by and dependent on the central system. Such hierarchies can be seen in the management of corporate enterprises and in the granting of official permissions by governments to such enterprises. In a more narrow sense, centralization in a political organization is the relationship between the national government and the local governments. In addition, URBAN centralization describes both the flow of a population into major cities as well as the philosophy that affirms the importance of the metropolis as an ideal for city planning.

The activities of multinational corporations and financial firms affect world markets, within which competition has been increasing ever since the end of the Cold War. In addition, the increase in both border disputes and borderless terrorism has made coordination among intelligence agencies a requirement for national security. Consequently, advanced nations with a modern state-structure, regardless of whether they are federations, must consider the degree to which government and society are centralized, and the effects such centralization will have on world economics, international relations and the culture inherent in a centralized bureaucracy.

These important considerations follow from the growth of the modern state: on the one hand, the state-structure is associated with the vertical growth of the power hierarchy; on the other hand, it is associated with the horizontal growth resulting from territorial expansion. As a result, the dominant structure of a state is one which consists of a 'compulsory branch' – the military, police force, intelligence agencies and judiciary system – and an 'administrative branch' – the cabinet, assembly and the bureaucracy – the existence of which is guaranteed by the compulsory branch. Even though centralization varies from institution to institution, the idea of a control centre is necessary, at the very least, in the military, so far as we understand the notion of a state in the twenty-first century.

The degree of centralization is less evident in the world of economics as it is in the military. Domestic enterprises have largely been emancipated by deregulation in America and Europe and privatization of state ownership in China. Nevertheless, the economic branch of the central government should act as a buffer for society, insuring it against economic risks through national economic management. Such management is necessary to minimize uncertainty and maximize the overall profit of a nation.

A strong military force is also known to be a deterrent to the use of weapons, thus affecting international relations. Weapons and weapons technology are dispersed across borders, both as defensive measures and as a means of war and terrorism. The state must coordinate its military technology, strategy and intelligence in order to maintain national and international security. This type of centralization necessarily implies a degradation of personal privacy and liberties.

As a representative system, bureaucracies have a tendency to operate impersonally, more like

a machine than a part of humanity. However, unlike a machine, bureaucracies, left unchecked, have a tendency to grow and become motivated out of self-interest. It is in this very human need for self-preservation that the bureaucracy becomes ineffective. Since the 1980s, there has been an effort to decentralize the bureaucracies of economic management and social welfare in Japan, while at the same time centralizing the bureaucracies of law and social order. The democratic administration of the bureaucracy is vital to the strength of the nation. This, in turn, requires frequent and focused interaction between the state and society.

Finally, the theme of centralization directly concerns modern city planning. In the nineteenth century, many states intended to promote industrialization and the optimal use and production of resources by letting URBANIZATION take its own course. The influx of people to large cities directly resulted in the outflow of people to the SUBURBS later in the twentieth century, due to a lack of urban facilities. A major theme of modern city planning was how to optimize the balance between these two competing flows in the population. The ideal and appropriate size, function and structure of a city were examined by theory and practice on the scale of state and city. Centralization was practised with the interests of urban function and economic growth in mind. The results affirmed the notion of a metropolis and attached great importance to city REDEVELOPMENT: the ideal city should utilize the high density of the population and capital accumulation in the most social, economic and culturally beneficial way by fulfilling all urban functions.

However, as more and more large cities grew in size and the small to medium-sized cities declined, this view of centralization was reconsidered. The idea of a GLOBAL CITY, that includes the interaction of cities both within a nation as well as between nations, and takes into consideration not only geographic but also political boundaries, has become the focus of URBAN PLANNING since the 1980s.

Further reading

Beetham, D. (1985) *Max Weber and the Theory of Modern Politics*, Cambridge: Polity Press.
Benevolo, L. (1971) *The Origins of Modern Town Planning*, Cambridge, MA: MIT Press.
Gilbert, F. (ed.) (1975) *The Historical Writings of Otto Hintze*, New York: Oxford University Press.
Wallerstein, I. (1974) *The Modern World System*, vol. 1, New York: Academic Press.

TOMOKO KURODA

CERDÀ, ILDEFONS

b. 23 December 1815, Centelles (Barcelona), Spain; d. 21 August 1876, Caldas de Besaya (Santander), Spain

Key works

(1859) Barcelona Extension Plan.
(1867) *Teoría General de la Urbanización* (*General Theory of Urbanization*), Madrid: Imprenta Espanola.

Spanish engineer, liberal politician and urbanist, Cerdà was a founder of modern town planning as science, coined the new word 'URBANIZATION' and achieved the outstanding Barcelona Extension Plan. After studying mathematics and architecture in Barcelona (1831–32), he later secured a job as a road, channel and dock-building engineer in Madrid (1841), and served as a civil engineer and officer in several public commissions. During the 1850s, he dealt with Barcelona's urban and social condition, producing a Topographic Plan of Barcelona and Outskirts (1854–55), a first Extension project (1855) and a working-class *Statistical Monograph* (1856). As a free extension to the city was authorized in 1858, Cerdà was entitled to have his project appointed by central government, regardless of the contest organized by the municipality.

Cerdà's plan of 1859 is a regular, rectangular grid extending for over 5 miles on Barcelona's coastal plain, integrating the old city and existing SUBURBS in an isotropic layout based on a square city block 113 metres in length, with 45-degree truncated corners (to ensure easier turns), setting up a unique pattern of octagonal squares. The road network had 20-metre-wide streets and 30–50-metre-wide avenues (three of them crossing the grid diagonally), foreseeing urban transport by rail. The grid's neutrality allowed a combinatory system of public facilities, open areas and special blocks for public buildings. Morphology anticipated open city models, with generally 2 rows of buildings per block, 4 stories high and 28 metres deep, with interior gardens. A later project review (1863) introduced more extensive railway lines in

separate levels, and combinations of 2, 4 or 6 blocks forming 'railway block units', like modern movement 'superblocks'.

The plan's main features are its optimism and unlimited possibilities for expansion; its mathematical, geometric nature; the healthy role of urban parks, squares and block gardens; the concern for technical advances, especially in transport; and the systematic, scientific approach to urban functions, with two priorities, traffic and hygiene.

Despite Cerdà's initial supervision, the plan's execution faced changes of policy, land speculation, construction pressure and adjustments to increasing demographic growth: with a projected density of 250 persons per hectare, by 1890 there were 1,400 per hectare, and over 2,000 by 1925. The blocks were enclosed on all sides, buildings were made higher and deeper, and the inner gardens were privatized. However, the structure remains a powerful formal matrix defining the overall character of Barcelona.

As a synthesis of his work in town planning, Cerdà published *Teoría General de la Urbanización* (*General Theory of Urbanization*) in 1867, at that time the most advanced text on the subject. He also published an urban reform project for Madrid (1861), and was elected to the Spanish Congress and at different times to city and provincial assemblies.

Further reading

Fundació Catalana per a la Recerca (1996) *CERDÀ Ciudad y Territorio* (*City and Territory*), Barcelona: Electa.

SILVIO ABREU

CHARTER OF ATHENS

The Charter of Athens summarizes and codifies the doctrine of the 'functional city' postulated by the International Congress of Modern Architecture movement, founded in 1928, in La Sarraz (Switzerland), and usually known by its French acronym CIAM (Congrès Internationaux d'Architecture Moderne), with a total of eleven congresses, all held in Europe between 1928 and 1959. The Charter of Athens corresponds to the conclusions of the 4th CIAM, held in Athens in 1933. It was first published, anonymously, in the Netherlands in 1935, and with small changes in

1941–42 in France, during the German occupation, again anonymously, and in 1957 under the name of LE CORBUSIER, including this time a personal commentary. It is the 1957 Le Corbusier version that is considered to be *the* Charter of Athens.

Starting from a critique of the contemporary city, its disorder and inefficiency, based on what was believed to be a scientific analysis, it ends with proposals for a model city of the future, the 'functional city', establishing a link between architectural forms and positive urban social change. The Charter is structured in three parts: an introduction ('Généralités') and a conclusion ('Points de doctrine') and five chapters in between, one for each of the main functions of a city – housing, leisure, work, transportation – and a small fifth chapter entitled 'Historic patrimony of cities'. The city would be organized as a large green area from which would emerge high-rise buildings, separated from nearby buildings, leaving most of the area free of constructions. Each of those four functions and its sub-types (high-density housing, low-density housing, services, industries, etc.) should be spatially separated, which would be done through zoning. Car traffic would pass through independent and specialized streets and not in front of the buildings, as was usual in traditional urban streets.

A central aim of the CIAM movement was the creation of a more egalitarian and rational society through better urban design, reflecting a clear socialist influence in its first years, replaced by an approach with fewer political connotations in the post-war period. The Charter of Athens is considered to be the manifesto of the modern movement in URBANISM and proved to be the most important reference for generations of urban planners – and still continues to be so in different parts of the world – even though not all of them shared exactly the same principles. The exile of several CIAM members due to political reasons, before and during the Second World War, helped the spread of the Charter of Athens's principles to other parts of the world, with different adaptations to local circumstances. In Europe and North America, its influence continues to be felt, as the new urban utopias – the Charter of NEW URBANISM and the 'NEW CHARTER OF ATHENS' – are essentially reactions against CIAM's discourse on urbanism.

Further reading

Le Corbusier (1971 [1957]) *La Charte d'Athènes*, Paris: Éditions du Seuil.

Mumford, E. (2000) *The CIAM Discourse on Urbanism, 1928–1960*, Cambridge, MA: MIT Press.

SEE ALSO: Costa, Lucio; Gropius, Walter; Mies van der Rohe, Ludwig; radiant city; utopia

CARLOS NUNES SILVA

CHICAGO SCHOOL OF SOCIOLOGY

Conceived in 1892, the same year in which the university was founded, the Department of Sociology at the University of Chicago rose to international prominence as the epicentre of advanced sociological thought between 1915 and 1935. The significance of the department to the emerging field of sociology was the faculty's insistence on integrating research into the traditional theoretical underpinnings of the sociological movement and its linking of sociological inquiry to current issues such as URBANISM. Chicago and other growing American cities were the laboratories from which they generated their data and about which they constructed influential theories regarding human ecology.

Albion Woodbury Small chaired the department from 1892 to his retirement three decades later, and is primarily remembered for building the department at an administrative level. It was a colleague in the department, William Isaac Thomas, who made the seminal contribution to the department's approach to sociology in *The Polish Peasant in Europe and America* (1918–19). Written with Florian Znaniecki, Thomas's study of Polish immigrants incorporated a large amount of family letters, newspaper articles, public records and other personal materials from Poles living in both the United States and Poland. In addition to new research methods, Thomas introduced the concept of social disorganization as a cause of social problems. Thomas also wrote extensively about the emerging field of social psychology; specifically, he expounded on his interest in human attitudes and values. He developed a theory that attitudes and values are interrelated and that men respond to objects, not through an instinctual response, but through a conscious understanding of them as perceived by the human mind.

In 1916, Robert Ezra Park and Ernest Watson Burgess, the latter a former student of the department, were hired as faculty. These two men, working closely with one another during their tenure in the department, significantly expanded Thomas's tradition of research-oriented scholarship complementing theory. Park was the elder of the two, and a man who, Burgess claimed, 'lived and slept research' (Bulmer 1984: 94). The pair co-authored the seminal sociological textbook of the era, *Introduction to the Science of Sociology* (1921).

Park and Burgess developed, during the 1920s, the human ecology approach to understanding social organization patterns in cities. Burgess's famous piece, *The Growth of the City* (1924), viewed the city as a valuable research laboratory due to its inherent intensification of daily life from which a heightened understanding of human behaviour could be gleaned. Cities were conceived of by Burgess as expanding outwards from a central core, 'by a series of concentric circles, which may be numbered to designate both the successive zones of urban extension and the types of areas differentiated in the process of expansion'. Popularly referred to as the 'concentric zone theory', Burgess's construct posited that people were sorted into 'natural economic and cultural groupings' – a process, he believed, which defined a city's physical form and character. Burgess identified mobility as essential to understanding the change and growth that occur in a city, as mobility could be numerically quantified and yielded the (human) pulse of a city.

Other research studies on the city, conducted by both faculty and students during the period of Park and Burgess's influence, and that utilized empiricism to attain theory, include Frederick Thrasher's *The Gang* (1927); Louis Wirth's *The Ghetto* (1928); and Harvey Zorbaugh's *The Gold Coast and the Slum* (1929). Wirth's theoretical essay 'Urbanism as a way of life' (1938) examined the psychological and behavioural consequences of living in cities and greatly influenced future urban policy initiatives.

The Chicago School influenced successive generations of urban sociologists, including William Julius Wilson, author of *The Truly Disadvantaged: The Inner City, the Underclass, and Public Policy* (1987). Richard Wright, the noted black novelist, stated in an introduction to St Clair Drake and Horace Cayton's *Black Metropolis* (1962) that many of his famous novels, including *Uncle Tom's Children* (1938) and *Native Son* (1940), were, in part, inspired by the scientific research of the

Chicago School whose writings enabled him to understand the 'urban Negro's body and soul'.

Further reading

Bogue, D.J. (ed.) (1974) *The Basic Writings of Ernest W. Burgess*, Chicago: University of Chicago Press.

Bulmer, M. (1984) *The Chicago School of Sociology: Institutionalization, Diversity, and the Rise of Sociological Research*, Chicago: University of Chicago Press.

Faris, R.E.L. (1967) *Chicago Sociology, 1920–1932*, San Francisco: Chandler Publishing.

Kurtz, L.R. (1984) *Evaluating Chicago Sociology: A Guide to the Literature, with an Annotated Bibliography*, Chicago: University of Chicago Press.

Smith, M.P. (1979) *The City and Social Theory*, New York: St. Martin's Press.

BENJAMIN MASON AND CHRIS SILVER

CHOAY, FRANÇOISE

b. 29 March 1925, Paris, France

Key works

(1960) *Le Corbusier*, New York: Braziller.

(1969) *The Modern City: Planning in the XIXth Century*, New York: Braziller.

(1997 [1980]) *The Rule and the Model* (*La règle et le modèle*), Cambridge, MA: MIT Press.

(2001 [1992]) *The Invention of the Historic Monument* (*L'Allégorie du patrimoine*), trans. Lauren M. O'Connell, Cambridge: Cambridge University Press.

Former professor and director at the Institut français d'urbanisme (Paris), teacher at MIT, Princeton, Cornell, Louvain and Milano, Françoise Choay is the author of numerous books, essays and articles on the history of architecture and urban planning, as well as the co-editor of a French dictionary of urban and regional planning. Her erudite work on the history of modern urban planning and historic preservation is methodologically related to that of Michel Foucault. She shows the emergence of those singular, yet taken-for-granted concepts and textual forms that underlie modern planning and preservation and thereby shape our relationship to the city.

Choay demonstrates the genealogical importance of utopias and architectural treatises, starting in the fifteenth century, as foundational texts in the emergence of modern urban planning discourse, and she denounces the abusive scientific pretensions of modern planning which hide moral, utopian and even mythical elements inherited from the past. Likewise, she traces the social construction of the concept of historic monument in the past two centuries and the institutionalization of the practice of historic preservation as a narcissistic response to a sense of loss of identity. Throughout her work, she pleads for the recovery of a 'competence to build' that is grounded in the bodily (and not just visual) interaction of people and their environment and in the expression of their own needs.

RAPHAËL FISCHLER

CHRISTALLER, WALTER

b. 21 April 1893, Berneck, Germany;
d. 9 March 1969, Königstein, Germany

Key work

(1966 [1933]) *Central Places in Southern Germany* (*Die Zentralen Orte in Süddeutschland*), trans. C.W. Baskin, Englewood Cliffs, NJ: Prentice Hall.

A German geographer, Christaller laid the foundations of CENTRAL PLACE THEORY in his pioneer work, published in 1933. Being primarily concerned with the urban space, he worked on the role of towns as geographic-economic units, besides analysing the relationships between towns belonging to the same region. Unlike most of the scholars dealing with economic science at that time, he was fully aware of the fact that economic relationships and economic events were closely related to space, since it was in space that they, necessarily, expressed themselves.

As a child, Christaller used to play with an atlas, a point that denotes his early concern with regional studies and spatial location. As a scholar, he maintained the habit of observing particular spatial distributions. Investigating his home surroundings, the flat Bavarian landscape of south Germany, Christaller noticed that towns of a certain size were roughly equidistant. By examining and defining the functions of the cities, the settlement structures and the size of the HINTERLAND, he found it possible to model a pattern of settlement locations, using geometric shapes. His major aim was to set up a purely deductive and coherent theory to explain the size, number and spatial

arrangement of towns, in the belief that there was a principle governing the distribution.

From the basic assumption that there can be identified a central order in human community life, meaning that a certain amount of land and an adequate human DENSITY leads to the creation of a town, Christaller went further into the analysis of its functional characteristics. He assumed that the major function of a town was to be the centre of its rural surroundings and the mediator of local commerce with the outside world or, in a more general sense, to be the market area of a region, a condition that leads to the concepts of centrality, central place and tributary area.

His masterwork, a comprehensive coverage of geographic theory and empirical work, remained relatively unknown for almost thirty years, due to its late translation into English. The theoretical approach, actually the most inventive part, explains the main function and the spatial arrangement of economic entities – the market cities – and the relationships they establish over their area of influence – the region. Actually, being a geographer, Christaller derived an economic theory that explains patterns of URBANIZATION and the establishment of market areas for different goods and services.

The theory also offers insights into why specific goods and services are or are not present in a particular community. Moreover, and more importantly, it specifically recognizes that no community's trade sector can be viewed in isolation.

Further reading

Berry, B.J.L. and Pred, A. (1965) *Central Place Studies: A Bibliography of Theory and Applications*, Philadelphia, PA: Regional Science Research Institute.

IÁRA REGINA CASTELLO

CITADEL

A citadel (from the French 'citadelle', or small city) is a fortified settlement or stronghold, sometimes located on higher ground, built to protect its people from attack. In some cases, the citadel is the city itself with one system of defence. In others, the citadel is a separate, fortified unit next to a walled SUBURB developed by those who had remained next to their protective stronghold. In this case, the citadel becomes an administrative centre, or a religious centre, to the residential town. In the citadel, the functions of the police and the army were developed, as well as the army barracks. It also served as a granary for the food supplies of the city. The citadel may also be located within the fabric of the city as a fortified unit. Lewis Mumford states that the citadel has the marks of a sacred enclosure originally built to honour the gods which was much stronger than was required for protection. Over time, this strength was helpful as protection against enemies.

Citadels have been found in every culture. Early evidence was found in the Harappan cities (2154–1864 BC), such as Mohenjo-Daro (today located in Pakistan) which had a citadel separate from the city. In mainland Greece, the citadel in Mycenae (from 3000 BC) was surrounded by walls, and, during its peak period (2200–1600 BC), was occupied by the royal families and maybe craftsman. The ACROPOLIS in Athens was originally a citadel where people could find refuge from attack. In medieval times, citadels became heavily fortified. Kingston-upon-Hull in England, a city with a harbour, built an extensive citadel in 1680 after two previous fortifications failed to protect its strategic site. This new citadel was almost half the size of the old walled town. During the Renaissance, elegant citadels were first built next to the city, such as the one in Mannheim in Germany, which was then rebuilt in 1697 as a continuous defensive system that included the city grid. A baroque palace replaced it in 1720. In the eighteenth century, a massive star-shaped citadel was built in Barcelona, Spain, as ordered by Philip V. The citadel was demolished after a plan for the city was prepared in 1854.

In the United States, in Pensacola, Florida, the British built a massive citadel during their occupation of the city (1778) on the site of the original Spanish plaza adjacent to the water. In Latin America, Saqsaywaman was identified as a citadel high above the Inca city of Cuzco. However, it may have had other uses beyond the protection of the city.

Further reading

Morris, A.E.J. (1994) *History of Urban Form*, 3rd edn, New York: John Wiley & Sons (Mohenjo-Daro, pp. 30–3; Mycenae, pp. 38–40; Kingston-upon-Hull, pp. 126–8; Mannheim, p. 235; Barcelona, p. 299; Pensacola, p. 324; Cuzco, p. 311).

Mumford, L. (1961) *The City in History*, New York: Harcourt, Brace & World, pp. 37, 65, 101.

ANA MARIA CABANILLAS WHITAKER

CITIES AND FILM

Moving pictures were the quintessential urban art form of the early twentieth century, appealing to unsophisticated city dwellers throughout the world and replacing variety theatre as the most popular form of entertainment until the advent of television. Yet despite this urban base – perhaps even because of it – movies often have had relatively little to do with the cities in which they are watched. Indeed, it is little exaggeration to suggest that they have been influential in promoting an anti-urban bias within popular culture and have done much to establish the norms of consumerism that include suburbanization at their core.

The film industry began in various places at the end of the nineteenth century, but the technologies and the distribution process were consolidated in the United States. The moving pictures were enormously popular as escapist entertainment, precisely because they were simple, cheap and silent; no language skills were necessary in order to enjoy the images of everyday life (mastered in Chaplin's early productions) or the more escapist fare that portrayed fantasies of the Wild West, pirates or Robin Hood, as Hall (1998) shows at length.

In contrast to these popular products, two early films stand out as powerful statements of anti-URBANISM. The first of these is Fritz Lang's *Metropolis*, which was released in 1926. It is a tale of two cities, in fact, one the modernist nightmare of the assembly line and the enormous clock, that enforces a tyranny upon the exhausted workers and causes them to make fatal mistakes; and the other, a bucolic realm to be found in penthouse gardens above the rooftops, where rich adolescents cavort in a mindlessly naive hedonism. The film works on a number of levels, so to speak, notably as a critique of monopolistic capitalism and of a brutalist urbanism, but also as a precursor of the collective society that was about to descend upon Germany. Curiously, given his sympathies, Lang was offered control of the German film industry by Goebbels, but fled to Hollywood, where he made movies such as *While the City Sleeps*, a characteristic study of urban deviance.

In some ways, we see echoes of Lang's work in a 1939 documentary funded by the Depression-era Works Progress Administration (later renamed Works Projects Administration), written and narrated by Lewis Mumford, and scored by Aaron Copland. Organized in discrete parts, the film shows a number of idyllic rural scenes before plunging us into another industrial landscape, this time a real vista of smoke, coal and trash. Once more, grimed and tired workers are shown to be the hapless victims of industrial rapacity, living unfulfilled lives from which there is no escape. Or is there? The documentary goes on to show the potential of suburbanization, in which new, clean homes are on offer and both men and women can develop fulfilling, albeit separate, lives, as workers and domestic partners, enjoying a version of Ebenezer Howard's new-town existence.

Popular Hollywood movies have tended to play out these themes over the years. The city proper is often a place of deviance at best (often typified by nightclubs and jazz – as in *Cabaret* – but also by the sordid realities that result from high-density living – as in *Rear Window*) or savage crime, at worst (*Taxi Driver*). It is a complex place with complicated conflicts, as in *West Side Story*. Or it is the manifestation of evil, and the larger the city, the more likely it is that it is hell on earth; this has been portrayed routinely in Hollywood's films about New York, from *Ghostbusters* to the *Devil's Advocate*. Here, the premise is that all that is bad about urban life is the perfect setting for the planet's darker forces. Quite how King Kong and Godzilla fit into this story would require a much longer entry; indeed, it would be possible to imagine a separate entry detailing just the representations of New York, which range from the apocalyptic (*Escape from New York*) to the hagiographic (*Manhattan*, *You Got Mail*) (see Sanders's veritable encyclopedia on this topic). Suffice it to say that whenever Hollywood wants to portray definitive catastrophe, it involves the destruction of New York – by aliens, by tidal waves or by nuclear war (*Independence Day*, *Armageddon*, *Planet of the Apes*).

In contrast, life in the SUBURBS has been treated much more like normalcy. The white picket fence was a Hollywood staple for several decades (notably from the 1930s to the 1950s), and although it is not remembered for this part of its

plot, it is symptomatic that one of the best-loved American movies – *It's a Wonderful Life* – has at its core a story about trying to save a Savings and Loan so that a suburban housing market can survive. The prototypical Anglo family, as portrayed in the movies, aspired to live on tree-lined streets, surrounded by versions of itself and served by the ubiquitous kid on a bike throwing newspapers. There was rarely a mention of the realities of suburban life, such as the restrictive covenants that specified race and religion; and the finances of the household rarely seemed to match the cost of the house they occupied. The home portrayed has become ever larger and more expensive, so that by the 1990s the 'typical' suburban home in the movies may be worth as much as a million dollars (*Father of the Bride*).

Of course, there is a strong counterweight to this tendency. This has manifested itself as a positive portrayal of urbanity (think Cary Grant in virtually any movie of the 1950s) contrasted with the unsophistication of the suburban dweller (*The Out-of-Towners*). Moreover, the picket fence has become a symbol of secrets and hypocrisy, which can be traced from *To Kill a Mockingbird* through to *Blue Velvet* and on to *Pleasantville*. The *Bad Lieutenant* robs everyone in sight to support his suburban lifestyle, although it drives him to heroin in the process; in *Copland*, the corrupt urban cops have created an entire suburban enclave for themselves. In *American Beauty*, the film begins with a long shot of the tree-lined street, and returns to this to express the hollowness of every relationship, both between the neighbours and within the families, who exist via voyeurism. Ultimately, Lester Burnham's efforts to change his life can only result in death, and it is not even immediately clear who his executioner is, so great is the rage directed against him.

The way in which Hollywood treats the city is not entirely surprising, given what we know about Los Angeles. The movie industry grew there at precisely the moment that it was laid out as a vast speculative suburb, based not on the individualism of the automobile, as one might expect, but upon collective values – the streetcar on the one hand, and a vast municipal water project on the other (which can be seen in Hollywood's version of itself, *Chinatown*). Although Los Angeles is at the core of a vast global entertainment industry, it remains a singularly un-urban place. And just as Hollywood has a fairly cynical view of itself (*The Player*), it

has, too, a fairly critical view of its own environs (*LA Story*). This is revealed most starkly in one of the most influential science fiction films, *Blade Runner*. Set in LA in 2019, it reveals a shattered city of perpetual rain in which remnants of a Pacific Rim culture are embraced by those who cannot leave for a new life in the 'offworld colonies'. This is, of course, the quintessential urban dystopia; it is balanced by the rather smug suburban alternative *Demolition Man*, in which the post-apocalyptic San Angeles is run as a vast, crime-free gated community. In reality, it is hard to imagine a less metropolitan film-setting than Southern California, especially if we include its neighbouring offspring, the pornography industry, based in the quintessential suburbs of the San Fernando Valley (*Boogie Nights*).

It is not entirely surprising that one of the corporate names most closely tied with the global entertainment media has diversified into real estate. The Disney Corporation, associated closely with rigorously clean cartoons and theme parks, has used its name to launch a community in the 'NEW URBANISM' style in Florida, named Celebration. The latter has proven to be more complicated than a movie version of a new community (such as *The Truman Show*), as residents have struggled for control of their own development, so that it is in Celebration that we see the most interesting collision of studio values and the real world of urban development.

Hollywood is very much in contrast to the film industries of other nations (notably Germany, France, Great Britain, Italy and Japan). These are based in some of the largest metropolitan areas, and all produce realistic urban dramas as a matter of course. Examples of work by Fassbinder (*Berlin Alexanderplatz*), Bertolucci (*Last Tango in Paris*) and Fellini (*Fellini's Roma*) are all dependent on the urban settings in which they are filmed. The apotheosis of this urban realism is *Trainspotting*, based on Irvin Welsh's novel of urban life in Scotland. This movie, which would never have been made in Hollywood because of its unrepentant and explicit drug content, draws on Edinburgh's bleak STREETS to show the collapse of family life amid the wreckage of 1970s public housing. It stands in contrast to the whimsical *Gregory's Girl*, based entirely in the nearby Scottish New Town of Cumbernauld, and one of the few films to worship the pastoral simplicity of such an explicitly planned location.

An exception to this trend is to be found in the other major producer of feature films, namely India. Although approximately 800 films are produced annually, most of them around Bombay (which also carries, in consequence, the name Bollywood), these are, for the most part, fantasies designed for a ravenous domestic population eager to devour escapist plots: the biggest hit of Hindi cinema in the past decade, titled *Hum Apke Hain Kaun?*, has been described as little more than a four-hour celebration of domestic consumption. Powerful urban dramas, such as *Salaam Bombay!*, which is an unflinching account of Bombay's extreme poverty and its many street children, who hover between work, drug abuse, crime and prostitution, are very much the exception, and are often more acclaimed in the West than they are domestically.

Another alternative to the Hollywood production line, and one of the more interesting developments in its own right, is the Japanese school of animated features, usually characterized as anime. Many of these productions have an urban backdrop, in which high-tech characters with a cyberpunk consciousness do battle with assorted enemies; they clearly share some common ground with the numerous computer games that have the same dystopic outlook, from *Metroid* to *State of Emergency*. To show that all things come full circle, one of the first productions of the twenty-first century is a new anime version of Lang's *Metropolis*.

Further reading

Banham, R. (1971) *Los Angeles: The Architecture of Four Ecologies*, New York: Harper & Row.
Hall, P.G. (1998) *Cities in Civilization*, New York: Phoenix Giant.
Khilhani, S. (1997) *The Idea of India*, New York: Farrah Strauss Giroux.
Sanders, J. (2001) *Celluloid Skyline*, New York: Knopf.

ANDREW KIRBY

CITIES AND THE ARTS

While it is not unknown for cultural events to take place in rural settings (operatic series are held annually at Glyndebourne, in England, and Santa Fe, in the United States, for example), it is more common to find them located in metropolitan areas. There are a number of good reasons for this.

Music, dance and the visual arts are all expensive and, as they tend to be a minority taste, they demand a relatively large audience in order to survive, like expensive restaurants or high-class jewellers. In that sense, they can be thought of as just another aspect of the marketing function of cities, in which higher-order goods are provided in large cities with large HINTERLANDS. From that standpoint, New York, London and a handful of other world cities serve as cultural capitals, and receive visitors from around the globe. However, the story is more complex than that, insofar as cultural attributes can define a city (such as Florence or Kyoto) and give it a cachet regardless of more conventional indicators such as economic prowess or sporting success. Consequently, we find a complicated link between URBAN development and CULTURE.

Historically, the relationship between cities and the arts was a product of the patronage of the urban ruling class. Even by the time of Rome's decline, cities had a reputation for sheltering culture from the barbarians beyond the gates. In medieval Europe, and in Edo Japan, the concentration of artistic activity in large cities again depended on an urban elite that drew upon the ruling class and the church, and whose patronage supported and defined what Adorno termed a 'high culture'. Cities existed as mercantile centres and as symbols of authority, but artistic adornments were unambiguous signals, both of wealth and of prestige. Cathedrals and palaces became repositories of the fine arts and settings for performances of all kinds.

The relation between culture, urban growth and nation building was cemented in the eighteenth and nineteenth centuries as competition within the world system accelerated. It became the norm to collect artefacts from one's colonies and to place them in new museums that showcased economic power and global reach. Artefacts were taken from China, Persia, India and Africa and displayed in Berlin, Paris and London. These 'national treasures' represented clear claims of national and racial superiority, but were also important in creating a broader public interest in culture, and, importantly, in its visibility and its display. This process was at its most marked in the nineteenth century, and Evans (2001) points out that it has become more common to disperse national cultural holdings to smaller centres for economic development purposes. Nonetheless, it is not uncommon to

strip-mine one nation's treasures to enhance another. The Nazi party was responsible for looting cultural artefacts from around Europe and bringing them to Berlin; in turn, many were removed in 1945 and taken to Moscow and New York, from whence they have only slowly begun to be returned to their former owners. Armies have sometimes been remarkably circumspect in attacking certain cities through the years, precisely because of their cultural treasures (the survival of Paris is a case in point), although there are more than enough examples of wilful destruction – such as the bombing of Belgrade in 1999 – to balance this generalization.

In addition to their role of housing artefacts, it is common for cities to be redesigned with the goal of showcasing certain elements and diminishing others. Engels commented on the manner in which the creation of elaborate façades could hide the poorest districts. Similarly, the work of Haussmann in the extensive redesign of nineteenth-century Paris has often been commented upon, by writers such as Benjamin and Harvey. In reality, the construction of tall façades of marble and iron (as Whitman portrayed it) was a general practice witnessed in many nineteenth- and early twentieth-century cities. This was the city-as-art at its most ambitious, and perhaps at its most pretentious; behind the façades of Barcelona, visitors commented, yellow fever was rife, but then façades can be seen, and sewage systems cannot. In the twentieth century, the hubris has intensified; rather than façadism, civic leaders have proposed elaborate reconstructions of whole cities. Hitler proposed a wholesale redesign of Berlin to reflect the cultural aspirations of the Reich (one never built), but the sterility of Brasília reflects the apotheosis of the artificial city, indicating that successful cities grow rather than being designed.

This notwithstanding, it is often hard to rationalize the manner in which individual centres have always created reputations for particular artistic accomplishments. Sixteenth-century London is remembered for its written word at a time when state censorship was powerful. Schorske (1981) shows how Vienna was a powerful cultural centre at the close of the nineteenth century, the home of varied intellectuals ranging from Klimt to Freud. It was also, in his words, a fin-de-siècle society, with a moribund economic class, a rigid nobility and profound anti-Semitism. New York in the 1950s produced a roiling intellectual elite that included expressionist poets, abstract painters and fabulous musicians at precisely the moment that the country demanded social conformity, sexual responsibility, racial apartheid and political intolerance, and was not afraid to make those demands public. Jacoby (1987) has, more recently, complained that the United States now contains no such thing as a public intellectual, a phenomenon that he attributes to the destruction of what he terms fragile urban habitats, where bohemians once thrived on low rents and cheap cafes.

How, in fact, does a city become a cultural centre? Peter Hall (1998) argues for the existence of a 'creative milieu'. This, he states, is a complex notion that can be constructed synthetically from a number of blueprints, drawing on a little Kuhn here, some Foucault there. Using a number of case studies, he teases out some important exemplars from the Western tradition. From fifth-century Athens, we see the importance of mercantilism and a social system that brought a large number of alien cultures to the city. From Florence in the Renaissance, we see again the importance of trade relations, but in this case it was channelled into expensive forms of artistic patronage. Fifteenth-century London suggests similar tensions between a relatively stable aristocracy and a vibrant bourgeoisie that demanded artistic entertainment, even while the state and religious leaders tried to limit performance and display. In Vienna between 1780 and 1910 we see again this tension between an aristocracy on the one hand and its many artists and intellectuals on the other. Berlin after the First World War took this to the ultimate stage, after the monarchy collapsed, after the economy collapsed, and everything had to be reconstructed – socially, politically, and especially artistically.

Is this an organic process, or can a creative milieu be manufactured? City elites have long believed in their ability to draw attention, and the crowds, to large semi-permanent displays; this extended from the Great Exhibition in London in 1851 to the World's Fair in New York approximately a century later. Many in smaller cities still believe in this culture principle, and despite the seeming paradox of using the funds of the many to subsidize the pleasures of the few, they have attempted to create a culture industry as a form of local economic development strategy. Such a strategy can take many forms. In some cities, it can consist of nothing more elaborate than the relaxation of land use controls so that functioning

arts districts in inner cities can prosper. In other cases, it involves a more elaborate commodification of a city's reputation so that it can be marketed more aggressively, via advertising on the one hand (increasingly web-based), but also via a coordination of hotel, airport and other infrastructural expansions. The 'I ♥ NY' campaign was a particularly successful example of this, and many other cities have followed suit, including New Orleans, Florence, Barcelona, Budapest, Cairo, Athens, Agra and Venice. And for the would-be traveller and cultural consumer who does not wish to travel, there is also of course Las Vegas, in which pastiches of many great cities are rammed together in a postmodern cultural chaos.

While it is clear that cities can serve to create art, and may even house leaders who have aspirations for their cities to be works of art, it remains a paradox that there is little high culture that fully embraces the city (popular culture is a somewhat different story). In part, this reflects the deep nostalgia that exists for bucolic images – is it not the case that Grant Woods's *American Gothic* is his country's best-known painting? Does anyone remember that Georgia O'Keefe painted New York before she painted New Mexico? This is, though, simply another way of stating the obvious: cities have not been the inspiration for much good art. What there is, is dazzling: the street scenes of Béraud, Caillebot, Hopper and Grosz, for instance; the vignettes by members of the Ashcan School in New York, such as Bellows and Sloan; the photographs of Riis and Stieglitz. Yet the urban experience has not been explored systematically by the visual arts, and, provocatively, it has been argued that URBANIZATION was reflected most clearly in the grand *opéra* of nineteenth-century Paris, where productions by Meyerbeer, Verdi and Bertin astonished audiences by their reflections on the contemporary urban condition. Quickly, however, this became a commonplace, and by 1920, Erich Wolfgang Korngold's opera proclaimed *Die Tote Stadt* (*The Dead City*).

The globalization of economies has had numerous implications for the culture industries, notably in terms of making expensive events accessible to larger audiences via telecommunications (a process we might characterize as 'the Three Tenors syndrome', for the performers who have sung together in larger and larger stadiums for ever-larger global audiences). One consequence is that while cities have routinely commodified themselves for mass events such as the Olympics, they have also tended to include cultural signifiers in their marketing (the ubiquity of the Sydney Opera House, the Pompidou Centre and the Vietnam Memorial as design symbols is indicative of the way in which cities can be known for more than their more obvious pop cultural attributes). Yet the attack on the World Trade Center in 2001 was more than the destruction of an urban symbol, it also involved the loss of significant art holdings, and it is likely that the circulation of major exhibitions from one GLOBAL CITY to another will cease as the risks of other attacks are taken into account.

Further reading

Evans, G. (2001) *Culture Planning: An Urban Renaissance?*, London: Routledge.

Gerhard, A. (1998) *The Urbanization of Opera*, Chicago: University of Chicago Press.

Hall, P.G. (1998) *Cities in Civilization*, London: Phoenix Giant.

Jacoby, R. (1987) *The Last Intellectuals: American Culture in the Age of Academe*, New York: Basic Books.

Schorske, C.E. (1981) *Fin-de-siècle Vienna: Politics and Culture*, New York: Vintage Books.

Vale, L.J. and Warner, S.B. (2001) *Imaging the City*, New Brunswick, NJ: Center for Urban Policy Research.

ANDREW KIRBY

CITIES IN LITERATURE

Since Hugo defined the city as a 'book of stone' and compared the urban architecture project with the drafting of a novel, the concept of the city as a literary text and the novel as a source of information has been widely accepted. This gave rise to innumerable attempts to identify and interpret the relationships between city and literature. The criteria adopted by some scholars are somehow similar to those described by MUMFORD in *The City in History*, and are aimed at going over the novel's evolution in parallel with the evolution of economy and society, in particular since the nineteenth century.

The initial assumption of this type of reading is that the realist novel and the modern city are two products of the same historical period. It was actually at the time when the outcomes of the INDUSTRIAL REVOLUTION were fully revealed that the former became an autonomous literary genre and the latter embodied characteristics so specific

as to make it a settlement model clearly different from the preceding ones. Since then, the city – as a prevailing form of the organization of space and of economic and social relations – has been a dominant feature of many novels.

The way in which cities were represented, however, changed dramatically through the various development stages of capitalist society. As the MERCANTILE CITY gradually shifted to the establishment of the INDUSTRIAL CITY, writers' attentions focused on the effects of production and technological transformations on the town environment. The reactions are contradictory. While some express optimism and confidence in a future that appears rich in opportunities, many criticize not only the new physical shape of the city but chiefly its social institutions.

Most novelists do not reject the industrial city as such, but they are revolted by the mixture of environmental and moral ugliness, symbolized by the image of 'poisonous smog'. The conditions of the workers and their families, the greed of the middle class, the consequences of the demolition and major restructuring of the city fabric are the actual protagonists of nineteenth-century novels.

In many cases, these provide information similar to that of a real sociological inquiry and can be considered as the literary equivalents of contemporary economic and social analyses. On the other hand, Zola believed that the novelist was as a scientist and explicitly declared that he wanted to depict the inevitable downfall of working-class families in the polluted atmosphere of most urban areas.

Human labour and its transformations within the new social environment is one of the major themes in the novels of many American authors of the same period, while some of them nourished contradictory feelings towards big cities. For example, Dreiser depicts the metropolis as a kind of magnet that attracts the mirage of the American dream and fosters hopes – often illusory – of social emancipation. While dismay is strongly felt for a world that proves difficult to understand, innovations stir some sense of admiration.

The crowd, the lights at night, the skyscrapers, the large new trade buildings, and in particular the train and the railway strike the imagination of the novelists who, in spite of the suffering brought to many individuals by the new social organization, are impressed by the 'beauty' of the new urban landscape and by the enlargement of individual liberty.

As a whole, however, the prevailing IMAGE OF THE CITY, particularly the big city, is increasingly associated with that of the danger and fear of getting lost in a grey and livid labyrinth. The disorientation and hostility towards the city that can be found in many works of fiction during the formation and development stages of the industrial city are widespread feelings, even in poetry.

For Poe and Sandburg, the city is chiefly a destructive force. Baudelaire defines it as sordid, even though he acknowledges the stimulating and fascinating spectacle for the creative observer, and Eliot uses the narrow streets of the miserable areas of a decayed town as a desolate background.

As the industrial city gradually becomes a business city and the urban environmental decay appears as an inexorable phenomenon, new metaphors become customary. The city is no longer seen as a labyrinth but as a jungle, a term that embodies and materializes a sense of uneasiness and powerlessness in the face of injustice and social exploitation.

The city as a synonym of corruption and social disintegration, as a place where money governs every single activity, is a recurrent theme, from Dos Passos to Fitzgerald, and the leitmotif of specific literary sub-genres such as hard-boiled detective fiction.

The end of the Second World War marks the end of the industrial city and the sad fate of its inhabitants, who become the protagonists of many novels. The destruction of a system based on factory work with its tragedies, yet with a strong sense of solidarity, is accompanied by the rupture of the town organization. Algren's post-industrial city is no longer seen as a unitary phenomenon but is made up of ethnically defined areas, separated by increasing distances.

The domination of technology, violence, and fragmentation become the basic characteristics of the subsequent corporate capitalist city together with fortuitousness and the impossibility of acknowledging a system of logic in the city layout or in the social relationships. For Auster, chance is the engine of events and the distinctive element of the city, which has now become an unknowable, unintelligible entity. In the past, besides the city's evils, the logic of the overall design could be understood. Nowadays, unbalance and contrasts are emphasized, but an organic design that would allow one to recognize and give a meaning to the single parts is lacking.

Meaninglessness and discontinuity are the main features of the world cities that have lost any interconnectedness. The road networks and transport facilities that were previously used to connect activities and population groups have evolved to become dividing barriers. It is not haphazard if the subject of borders within cities, symbolic divisions and real physical barriers is one of the main themes in many fictional works written in the last two decades of the nineteenth century.

Belleville, the most multi-ethnic district of Paris, is at the centre of the novels by Pennac, who makes it a kind of URBAN VILLAGE where ethnic and class solidarity is the only antidote to building speculation and city transformation into mere concentrations of financial opportunity locations. Doris Lessing's depiction of London is not an exaltation of the British MELTING POT but a clear-sighted and sad representation of those areas of the city inhabited by immigrants, far from the idea of world-city glamour.

The attention given to specific areas within large metropolitan settlements is accompanied by the appearance of new literary cities. While the planet is becoming more and more urbanized, these new entries no longer belong to Western countries only, but to other contexts and cultures. The novels by Isabel Allende and Amado, set in Santiago and Bahia, are a true yet fantastic background for the deep imbalance hidden behind the apparent assimilation of the European and North American culture and way of life.

The 'cities in literature' theme is open to investigations from many different viewpoints. For example, some scholars focus on a single city and follow its evolution through the literary texts, like urban historians read a city through its maps. In this type of exercise, some places eventually become the emblem of a determined type of city. Chicago is the archetype of the industrial city, Detroit that of the post-industrial city, Los Angeles is the postmodern city par excellence, and Belfast is the last divided European city.

There is an obvious risk of creating stereotypes; however, this often leads to highly interesting (re)constructions, because the city events resulting from other sources are the recurrent theme of the reasoning, but the narratives of the novels provide in turn the connecting element to frame them. This approach often proves rich in suggestions, since it does not merely provide a transformed representation of real facts and actual

situations, but proposes an original image of the city, giving creative and designing indications, as if a more detached and at the same time more selective perspective were better able to perceive and recompose the truly significant details.

'Cities of feeling' and 'cities of fact' are not separate entities but affect each other to such an extent that it is no longer possible to make a distinction between physical place and state of mind.

On the other hand, the so-called realistic novels of the nineteenth century also contain a strong metaphorical charge. Dickens and Balzac give a clear depiction of the social dislocation following the Industrial Revolution. However, they do not reproduce the material reality of London and Paris but a transformed reality.

After them, the evolution of the city from fictional background – more or less transfigured artistically – to actual protagonist is irreversible. The houses in Joyce's Dublin have human feelings, and the writers do not tell about the city but speak through it. As BENJAMIN says about Alfred Doblin, 'Berlin is his loudspeaker'.

Through his wanderings in the metropolitan space, Benjamin has gathered descriptive documents as opposed to denunciations of human living conditions, and he has explained with masterly skill how the writer does not describe the population and the physical city, but manages to evoke the former in the image of the latter.

The impossibility of defining the city in a univocal and unambiguous manner and the concept that the only way of getting close to its comprehension is through the dream are the intuitions from which CALVINO starts for his catalogue cities that do not exist in reality but can be surmised from an openly distorted viewpoint.

Whether it is expressed as a realistic representation or whether it suggests metaphors and allegories, urban space is the landscape that the writer must compete with. A significant number of writers are inextricably associated with a particular city. Not only are their works set in a specific city and their descriptions so accurate that we can walk along with the protagonists and find the addresses, the buildings, monuments and natural elements, but the image that emerges from the text affects the way in which we interpret the city to such an extent that we recognize these cities even when they have been entirely transformed and even destroyed.

Singer's Warsaw or Dostoyevsky's St Petersburg no longer exist, yet the strength of the literary work

is such that our perception and our understanding continue to be moulded by literary portraits whose models have been irretrievably destroyed.

Literature succeeds in placing on the map cities that we would hardly know about, owing to their small dimensions or their location in peripheral countries, if major writers had not made them the protagonists of their novels. In the novels, the urban landscape is a real, non-fictitious character, and the city reflected in the literary work becomes in turn one of the dimensions of the actual city.

Unlike the cities of statistics and of many geographical descriptions, literary cities are populated with human beings of flesh and blood, not with abstract numbers. And this is probably why literature is an instrument rich in stimulus for the architect and the designer, who remember that projects and buildings have no sense as such, but are meant to improve life and human activities.

The authors of the so-called utopian novels that foreshadow models based on a different and more advanced physical and social organization are explicitly aware of this, as well as the sad dystrophic visions of those who imagine the future as an exasperation of the worst elements of the present.

Further reading

Lehan, R. (1998) *The City in Literature*, Berkeley and Los Angeles: University of California Press.
Preston, P. and Simpson, H.P. (eds) (1994) *Writing the City*, London: Routledge.
Rotella C. (1998) *October Cities: The Redevelopment of Urban Literature*, Berkeley: University of California Press.
Wirth-Nesher, H. (1996) *City Codes*, Cambridge: Cambridge University Press.

PAOLA SOMMA

CITIES OF DIFFERENCE

From the beginnings of the modern INDUSTRIAL CITY, scholars have recognized the variety of experiences and processes that constitute URBAN life. Social and cultural difference and diversity are woven together in cities at varied scales and levels of intensity. Some early commentators, like Max WEBER (1864–1920) and Emile Durkheim (1858–1917), predicted the need for more specialization and bureaucracy to help manage the increased complexity of social and spatial difference.

Georg SIMMEL (1858–1922) argued that the creation of large-scale bureaucracy liberates individuals and groups in ways that support difference. In his famous 'Urbanism as a way of life', Louis WIRTH (1938) posited that large numbers account for individual variability, distinctive ways of life and the segmentation of human relations. He argued that density involves diversification and specialization. For most of these early writers, specialization, capitalism and the promotion of difference were either good or neutral, but for others, and especially Friedrich Engels (1820–95), they raised concerns about urban inequality and ghettoization. The growing industrial city spawned a desperate need for social welfare. The writings of these social critics influenced the famous Chicago School of Ecology from the 1920s onwards in its attempts to map and model the complexities of social and spatial difference.

Recent developments in urban theory suggest intensified interest in issues of political identity and difference. Many of these developments are a reaction to the universal rational modelling of the Chicago School where averages and norms often represented individual lives. Whereas the Chicago School looks at singular aspects of difference such as race and class where groups are mapped by one variable, contemporary theorists engage complex and multiple axes of difference. As Fincher and Jacobs (1998: 5) point out, 'we are racialized, classed *and* gendered (to select but three possible axes of difference)...[and] how these distinct attributes come into play in terms of life chances is by no means given'. Empirical studies focusing on everyday lives, articulated especially but not exclusively by feminists, suggest new ways of understanding the varied and multi-layered contexts of urban experience. In addition, the reassertion of space in urban social theory suggests new expressions of identity politics and spatial differentiation.

There is a move away from attempting to compartmentalize diversity and universalize our understanding of it from a white, male, middle-class perspective. Much of this change resonates with an intensification of political struggles by women, minorities and other oppressed groups. Feminist scholars, queer studies scholars, anti-racist scholars and scholars concerned with ability and disability raise academic consciousness about the ways in which difference structures cities and urban life. In addition, concerns about the social

construction and welfare of children and elderly people highlight issues of societal exclusion. Understanding social and spatial difference is important today because of increasing concern over the multiple realities, complex daily lives and varied experiences of people living and working in cities.

Within URBAN STUDIES, the study of difference generally takes two forms: (1) representations of difference, and (2) social polarization and its effects. Analysis of social polarization and spatial inequality in the last two decades has been influenced in large part by the post-Marxist political-economic explanations of David Harvey and others (see SOCIAL JUSTICE), who argue that inequality is a fundamental aspect of the capitalist city. Katherine Gibson (1998) argues that the urban literature on social polarization is now largely a discourse about class and the transformation of class relations. This literature approaches difference from an economic standpoint and recognizes that equity issues are now systemic and global, and largely beyond the control of any one city government. From the perspective of representation, theories of difference highlight gender, race, ability, sexuality, ethnicity and age as valued aspects of our multi-layered political identities. For over a century, for example, children in American cities have been seen as needing special places to grow and develop such as schools, parks and playgrounds. It was not until recently, however, that scholars began to highlight the ways that institutions such as the late-nineteenth-century American Playground Movement (APM) constructed children. For example, it is now generally accepted that Jacob Riis's famous *How the Other Half Lives* (1890) spawned projects and institutions (such as the APM) that were as much about displacing, disciplining and monitoring low-income children as they were about uncovering the seeming squalor of immigrant neighbourhoods. Gibson points out that the politics in studies of representations of difference do not focus solely on the emergence of new class structures, but are also enacted around a multiplicity of identities and activism. The latter discourse is embedded in psychoanalytic, post-structural and feminist theories that often draw from post-Marxist structural theories seeking economic and class-based interpretations.

Iris Young's (1990) exposé of the politics of urban difference and justice takes a different tact with a novel appeal to an explicit and public political imaginary that is emancipatory at the day-to-day and individual levels. For Young (1990: 318), urban residents may remain strangers, but they can acknowledge 'their contiguity in living and the contributions each make to others'. Like many of the Chicago ecologists, Young believes that large cities liberate people from conformist pressures, but she is sceptical of any kind of political value for difference at the scale of the community or local government. Rather, difference is constituted, experienced and politicized through the infinitely unique spatial and temporal distinctions that exist throughout all cities. Young (1990) wants to promote the promiscuous mingling of different peoples that Lewis MUMFORD (1895–1988) long ago identified as the essence of urbanity.

Further reading

Fincher, R. and Jacobs, J.M. (eds) (1998) *Cities of Difference*, New York: The Guilford Press.

Gibson, K. (1998) 'Social polarization and the politics of difference: discourses in collision or collusion?', in R. Fincher and J.M. Jacobs (eds) *Cities of Difference*, New York: The Guilford Press, pp. 301–16.

Wirth, L. (1938) 'Urbanism as a way of life', *American Journal of Sociology* 44: 1–24.

Young, I.M. (1990) *Justice and the Politics of Difference*, Princeton, NJ: Princeton University Press.

STUART C. AITKEN

CITISTATE

A citistate is a region consisting of one or more historic CENTRAL CITIES surrounded by cities and towns which have a shared identification, function as a single zone for trade, commerce and communication, and are characterized by social, economic and environmental interdependence.

The contemporary citistate may be seen as the rebirth of the form traceable to the city-states of antiquity, such as Athens, Rome and Carthage. Even more striking similarities emerge with medieval URBAN FORMS, such as the Hanseatic League, which at its peak comprised eighty city-states, exerting influence from Russia in the east to London in the west.

The rebirth of this regional urban form found new roots in the upheavals of the late twentieth century. The collapse of the Soviet empire, marked most dramatically by the 1989 fall of the Berlin

Wall, encouraged a wave of decentralizing forces, as though the ice of mega-nationalism melted and ethnic waters were freed to flow again. The population shift to metropolitan centres accelerated. Military clashes, while serious threats to prosperity and civility, seemed increasingly confined within regional theatres. Competition among major national powers began shifting from military to economic struggles. Trade barriers continued their steady decline.

At the same time, technology was transforming developed economies, globalizing every significant market. By the end of the twentieth century, communications and money moved instantly around the globe. Markets were born, matured, and withered rapidly, largely ignoring national borders. An international economy emerged, one organized around major metropolitan regions.

This reorganization transcends the mere joining of city and state. Its function and force signify more than conventional terminology suggests; terms such as 'metropolitan area' or 'metroplex' are not sufficiently descriptive. The European use of 'CONURBATION' is likewise a dreary, derivative name for what has become a powerful organizing principle of the new global century. Hence the logic of 'citistate', symbolizing the melding of forms into one driven by organic forces and defined by fundamental changes in the political economy. In short, the contemporary citistate is what the economy does. Its geography is variable and elastic – delineated by markets for newspapers, broadcast signals, employment commuting patterns, patronage of cultural and educational centres, and the reach of healthcare services. While maps may depict myriad political jurisdictional boundaries, the actual citistate is best seen from the air, most dramatically in lighted form from a satellite camera at night.

CURTIS W. JOHNSON

CITIZENSHIP

The idea of citizenship has been defined as the capacity of individuals to defend their rights in front of a governmental authority. Citizenship is consolidated and enlarged when individuals acquire rights and participate actively in governing. It can present varied natures: civic, social, political,

intercultural or, more recently, ecological – the right of each citizen to public heritage (historical, environmental or economic) and to be defended from those who, in an individual way, try to appropriate goods that belong or must belong to the community (see CULTURAL HERITAGE).

Its origins go back to classical Greece where it represents the capacity of a free man or woman living in the 'POLIS' to participate in government of the city-state. Since then, the appearance of modern states is a concept closely bound to the members of a city. The appearance of the nation-state as the main rightful form of political institution led to the extinction of another option of medieval urban representation. That would carry unconnected problems between elites and the masses because of the larger distance between them. Later, the BOURGEOIS revolutions did not change this model given that political sovereignty over a protected market inside borders was the favoured tool of the bourgeoisie. As a result of this history, there were doubts about the legitimacy of CIVIL SOCIETY sharing political space, which turned into rejection in periods of increased market forces.

In chronological order, following T.H. Marshall's classical analysis, the first rights of citizens, determined in the eighteenth century, were civil rights – liberty and property – against an oppressive and despotic state. That laid the foundations of liberalism. In the nineteenth century, industrial middle classes defined political rights: the right, initially partial, to vote and be voted to participate in government against an oligarchic state. Both civil and political rights laid the foundations for the liberal democracy of the twentieth century, although citizenship does not necessarily entail democracy. In the second half of the nineteenth century, working classes established social rights, strengthened in the twentieth century and included in the constitution and laws of many nations. In 1948, three rights (civil, political and social) were joined together in the Universal Declaration of Human Rights.

Thus, the citizen became a political subject, owner of a statute that grants him civil, social and political rights. Citizenship has progressively been extended as a result of the social and civil development of the democratic state. Civil progress enlarged citizenship to include women and young people, social progress generalized the welfare state, and political advance made possible the emergence of a new political culture characterized

by new mechanisms of representation and citizen participation, wider and more efficient.

At the beginning of the twenty-first century, citizenship faces challenges at both the national and the international level. At the state level, the defence of acquired rights, enlarging the rights, and extending the rights to the full population are challenges. Even so, its legal competency to promote public economic and social policies is reduced by economic globalization, informational revolution, supranational integration processes (e.g. the EU), territorial differentiation (localisms and regionalisms with cultural and political substrata encouraged by institutions) and social fracture (multiculturalism of social groups, like immigrants, with strong elements of a specific identity).

At the transnational level, most problematically, it is necessary to find solutions to the matter of globalization without democracy (e.g. the International Monetary Fund, the World Bank, the World Trade Organization). Globalization collides with the traditional foundations of citizenship: membership, participation, association, inclusion or exclusion, national identity, and guaranteed sovereign law. In this new context, the 'global citizen' is a 'deterritorialized citizen' afflicted with a loss of identity at both the state and the local level (diffused city instead of compact city). If citizenship has been conceived inside states, now it is recognized as a transnational matter. Its practice can be located within broader institutional structures that create multiple levels of citizenship both within nation-states and beyond them.

Two choices are possible: a passive citizenship, understood as the right to receive goods and services guaranteed through law, or an active citizenship, whereby one looks for direct participation in political decision-making (a field of interest to political theory). The first involves the risk that the market will be considered as an alternative to citizenship or the way to widen it (the neo-liberal view). A progressive degradation strategy of the public sphere ends up as unable to confer citizenship rights (overall social rights) on users. Deregulation, externalization and public expenditure reduction give new scope to the capital in a time of global competence intensification. As a result, there is pressure on public industries and services to privatize, and also on producers looking for new markets and profits in other parts of the world to relocate. This reduces the communities' capacity to control their local factors to attract and retain investments. As a result, in turn, labour rights are affected by deregulation strategies. Thus, consolidated social rights of workers, even though they do not reach the point of disappearing, are transformed in a matter of voluntary concession by the employer. These rights leave out the scope of citizenship, and a consolidated right (in the EU) is changed into privilege only for some. However, global competitiveness demands the cooperation and involvement of workers, an active labour citizenship inside the firm.

Active citizenship development is easier when a lot of cultural traits are shared inside a community. Therefore, there is a closed interconnection between active citizenship and membership that is developed first at the local level. Significant efforts are necessary to apply the subsidiarity principle inside states (see NEW FEDERALISM), multiplying and extending the power of the community to involve civil society and promote active citizenship (see GOVERNANCE). The development of a principle of subsidiarity also constitutes an alternative to solving new conflicts in a 'multi-level citizenship', the result of global migration fluxes, the movement of refugee populations, the formation of supranational bodies (e.g. the EU), the formation of new successor states and the codification of international human rights norms.

Further reading

Delanty, G. (2000) *Citizenship in a Global Age: Society, Culture, Politics*, Philadelphia, PA: Open University Press.

Faulks, K. (2000) *Citizenship*, New York: Routledge.

Holston, J. (ed.) (1998) *Cities and Citizenship*, Durham, NC: Duke University Press.

Marshall, T.H. (1950) *Citizenship and Social Class*, Cambridge: Cambridge University Press.

JOAQUÍN FARINÓS DASÍ

CITY

Cities are the biggest things human beings make that actually work. A permanent and densely settled place with boundaries that are administratively defined, a city is the accomplishment of a population whose members work primarily on non-agricultural tasks. Their accomplishment is evident in three ways. First, it is seen in the city's physical layout, buildings, infrastructure, and in

the sheer number of people who live and work there. Secondly, it is apparent in the distinctly URBAN way of life or CULTURE that the population creates to organize life inside that space. Thirdly, it is evident in the great number and variety of ways in which the city is tied to settlements dispersed over a much larger territory.

Though clearly the most dominant and recognizable features of the man-made world, cities are not universally admired, much less revered, for their accomplishments. They are, instead, respected as centres of economic and political power and sometimes as repositories of great cultural achievement. At the same time, cities also have been viewed for many centuries as showcases for social inequality, harbingers of social decay and moral depravity, and, more recently, technologically obsolete to the point of being irrelevant.

That is really quite a lot of weight for one man-made creation to carry, particularly in light of the fact that cities have not been around all that long. Indeed, cities as we know them today are a comparatively recent invention. For the better part of human history, including most of the time for which we have no written records, people did not live in cities or anything like them. That only happened in the last three centuries or so.

Small but relatively permanent settlements built on the back of sustained agricultural surpluses began to appear in scattered locations across semi-tropical parts of Asia and Africa only 12,000 years ago. Larger and more urban-like settlements made an appearance in the fertile river valleys of these same regions not quite 6,000 years later. But it was not until 2,000 years ago, a period that came after the glory days of the ancient Greek city-states but before the downfall of the Roman Empire, that cities of any appreciable size and stature arose.

There was nothing inevitable in the rise of cities generally and certainly not in the rise of any particular city. Rarely holding more than a small portion of a given society's population, certainly no more than 5 per cent of the total and usually not that much, the size and eminence of these early cities rose and fell along with the fate of the empire to which they were attached. Nothing like sustained urban growth makes an appearance until late medieval times, less than 1,000 years ago, and then only in response to renewed war-making and trade between still feudal kingdoms in Europe and Muslim societies in what we call today the Middle East.

Cities at this time were still very much centres of command and control and not yet places where much production took place. They housed a comparatively small number of elite families and a larger but still modest number of retainers, merchants and artisans that supported the thin layer of the population that lived relatively well, if not especially longer, compared to the overwhelming majority of people who grew crops and fought wars. It was several centuries later, which brings us to about 250 years ago, before there was enough agricultural surplus, manufacturing and trade to sustain many more and much larger cities.

When this threshold was finally reached, city building took off in a big and sustained way, first throughout most of Europe and North America and only later in other parts of the world. That is not to say that cities were not present or prominent in Asia, Africa and the rest of the Americas before the Industrial Revolution or even as recently as the nineteenth and twentieth centuries. It is only to suggest that cities on these other continents grew dramatically comparatively later and in response to continued colonization and, more recently yet, renewed war-making, trade and the introduction of large-scale manufacturing in these faraway places.

The so-called 'globalization' of the world's economy and the rise of the 'GLOBAL CITY' during the second half of the twentieth century were new only in the sense that the volume and speed of exchanges, friendly or otherwise, with 'less developed' countries picked up dramatically. Globalization involved little more than finding new city-based regimes to fight and trade with in parts of the world that our city-based regimes already knew but had not yet fully explored. The nature and timing of globalization were such, however, that the process of city building in these other parts of the world unfolded a lot faster. It also produced cities that look different from our own and play a much different role in their country's economic and social life.

Cities in more recently developed parts of the world are often referred to as 'primate cities' for reasons that frankly have never been made clear to me but which I am pretty sure have nothing to do with monkeys. They tend to be larger, sometimes much larger, than even the largest cities in more developed countries like the United States; and their number is growing. Back in 1950, there were some 43 cities in Europe and North America with

populations greater than 1 million, but only 36 in all of Africa, Asia and Latin America. Less than fifty years later, there were just over 100 such places in Europe and North America but well more than 200 in Africa, Asia and Latin America.

The number of people living in the rapidly growing urban centres of developing countries is staggering; and living conditions for most of these people are not good. Often lacking the most basic of amenities like decent housing and clean water, and with no promise of steady employment or a good education, their prospects are grim. To make matters worse, these countries do not have well-developed networks of small and medium-sized cities like there are in North America and Europe. There is one and maybe a second really large city (of the primate sort) and the rest of the country remains overwhelmingly rural or city-less. The primate city acts like a magnet that attracts a great deal of that country's wealth, energy and, of course, people. This makes it much harder to develop the HINTERLAND around these big places, much less the whole country.

The situation in countries that developed earlier is much different. Cities sit atop a pyramid of medium-sized and smaller places, and in different regions of the country there are cities that are tied socially and economically to a number of these smaller places. It is not just the cities that benefit and grow. Other communities also grow and their residents usually can find gainful employment. Furthermore, when people are dissatisfied or need to search for work there are a lot more places for them to explore than there are in less developed parts of the world.

The dramatic growth in so-called 'clean' or non-manufacturing industries since the end of the Second World War greatly expanded the number of professional, technical and service jobs in developed countries. Combined with advances in short-distance transportation and information technologies, people were no longer as tied to central cities for work as they once had been. People, capital and jobs were all more mobile; and the smaller and medium-sized cities and towns surrounding major cities attracted increasing shares of all of them.

Concern that cities had become technologically obsolete in light of these changes proved greatly overblown. Older cities in developed countries like the United States did not wither and die, as many people feared and some analysts and government leaders predicted. Cities remained important command and control centres even as they lost large numbers of manufacturing jobs and residents and a good portion of their tax base to smaller places located some distance from the CENTRAL CITY. Their leaders and residents were reminded that the city's first and most sustained growth had come as a trading centre, capital lending house, cultural showcase and surplus gobbler rather than as a manufacturer and surplus producer.

The picture we have of contemporary cities is clear enough. Downtown areas have many more tall buildings today, and they house more professional, technical and service industries or increasingly up-scale residents. There may be secondary outcroppings of commercial and institutional buildings away from the traditional downtown area, but most of the remaining parts of the city house smaller business districts and neighbourhoods filled with more low-income and minority residents. The metropolitan area surrounding the traditional central city may have several smaller cities of its own. These places, so-called edge cities on the periphery of metropolitan areas, and some of the surrounding SUBURBS have become more 'urban' with time, even to the point of acquiring some of the people and problems that were once considered the exclusive property of larger cities.

However fearful they may have been about the city's future as an engine for regional and at times even international economic growth and development, many persons remained absolutely convinced that American cities had hit their high watermark as culture-makers a long time ago. In the eyes of many observers and analysts, cities no longer were places that gave most of us important reasons to want to stay. The way of life or culture practised in cities did not seem whole or inviting, except perhaps to people who had only recently arrived or would have been unwanted most anywhere else. Further contributing to this sense of unease about our way of life were studies indicating that Americans were increasingly disinterested in visiting each other or joining groups that brought different people together to do important volunteer work in their community.

Exaggerated as these perceptions may have been, there was a great deal of precedent for them in American history. Its great energy notwithstanding, there was something decidedly confusing and off-putting about the culture of cities. Indeed, some serious-minded persons believed that cities

did not have one culture that everyone could recognize and embrace, only a number of smaller cultures that welcomed a few of us at a time and could not be united in a congenial way.

While cities have always had their detractors and boosters, what was especially noteworthy about these claims was that they were coming from people who normally do not find much to agree about. We might have expected more conservative types to rail against the perceived immorality of big-city life and to bemoan the decline of civic-mindedness among the people living there. At the end of the twentieth century, however, a chorus of more progressive thinkers joined them in raising alarms about our civic habits.

Upon closer inspection, what appears to be going on with our urban way of life is neither new nor particularly scary. There certainly are problems with the way city people organize and fill their days; but these problems have nothing to do with the ability of people to see right from wrong or to work together on matters of importance most of the time. It has everything to do with expecting them to come up with a way of being together that would satisfy either liberal or conservative people all of the time.

The hallmark of an urban way of life, at least in the United States, is that it promotes values and ways of getting along that both conservative *and* liberal persons would endorse, but not together. Historically, conservatives have liked to keep close tabs on who joined their club. They talked a lot about following rules, and they expected people to watch out for each other. Liberal types tended to be more welcoming and tolerant, and at the same time they preferred that you kept your nose out of their affairs.

The distinguishing feature of an urban way of life is that it does all of these things, at the same time. It embraces both exclusivity *and* inclusiveness in terms of who can belong, piety *and* tolerance in following rules, and public-regarding behaviour *as well as* privacy in the conduct of its members. More important, perhaps, what research we have on the civic culture of American cities seems to indicate that this mix of liberal and conservative ways works well, or at least a whole lot better than most persons would have guessed.

Further reading

Hall, P.G. (1998) *Cities in Civilization*, New York: Pantheon.
Light, I. (1983) *Cities in World Perspective*, New York: Macmillan.
Monti, D. (1999) *The American City*, London: Blackwell.
Paddison, R. (ed.) (2001) *Handbook of Urban Studies*, London: Sage.

DANIEL J. MONTI JR

CITY BEAUTIFUL MOVEMENT

The City Beautiful movement developed in the transition from the nineteenth to the twentieth century as part of the progressive social reform movement in North America, under the leadership of the upper-middle class concerned with the social deprivation and potential social unrest of those living in poor and unhealthy conditions in all major cities, namely immigrants, in a period (1860–1910) during which the total population grew, on average, by more than 10 million new inhabitants per year. At the same time, the City Beautiful movement aimed to achieve a cultural parity with the cities of Europe.

It was inspired by the *beaux arts* movement developed at the Paris ÉCOLE DES BEAUX-ARTS, founded in 1816, whose teaching was based on the imitation of classic models, postulating the necessity of order, dignity and harmony in architecture, and whose basic premise was that through reform of the urban landscape, together with policies in other sectors, it would be possible to control the problems caused by the growing urban population. It was based on the assumption that beauty could influence social behaviour in the sense that a beautiful city would educate its inhabitants to civic virtue, particularly the poor, in spite of the fact that, in practice, these city plans had the effect of removing the poor from the area that was supposed to socialize them. As such, the City Beautiful movement was clearly an environmental deterministic perspective influenced, as were many others at that time, by Darwin's ideas. A second influence came from earlier planners in the United States who had already argued in favour of the positive effects of city beauty as a social control device, such as landscape architect Frederick Law OLMSTED, author of the New York Central Park plan, whose designs are considered to have inspired the City Beautiful pioneers.

The City Beautiful movement also attempted through the monumental and classical forms of the European *beaux arts* style to revitalize the city

centre and increase property value, which explains why they designed mainly civic centres, grand BOULEVARDS and PARKS.

Daniel BURNHAM was one of the early proponents of the City Beautiful movement and the first to apply this monumental style in the United States, in the construction of the World's Columbian Exposition of 1893 in Chicago, in commemoration of the 400th anniversary of Columbus's arrival in America. This exposition introduced the notion of a monumental civic centre, as well as the concept of a comprehensive city plan. It is not only regarded as marking the birth of the City Beautiful movement, but also became its model. As director of the construction, Burnham contracted architects trained in the *beaux arts* principles to build the exposition with monumental and classical buildings, painted white and with similar decoration, through which it became known as the 'White City', in contrast to the grey suburban sprawl of Chicago at that time.

After the World's Columbian Exposition of 1893, the first application of the *beaux arts* architectural style, or City Beautiful doctrine, was the plan for Washington DC, considered the first attempt at city planning in the United States, known as the McMillan Plan of 1901–02, named after the senator who directed the plan commission. It was a redesign of the monumental centre, due to the fact that the original plan, designed by Pierre L'ENFANT for the capital city of the United States (1791), at the invitation of George Washington, had not been fully implemented. It was prepared to commemorate the city's centenary with the explicit objective of giving monumentality to Washington DC, in accordance with the growing international importance of the United States. The influence of the classical style of the *beaux arts* movement was present from the beginning of the plan's preparation when the technical commission, headed by Daniel H. Burnham, visited the major European cities, including Paris, Versailles, Vienna and Berlin. Ebenezer HOWARD'S GARDEN CITY model (1898), mixing the advantages of both city and country life, was another major influence on American urban planning during this period. The City Beautiful projects included usually monumental buildings, boulevards, parks, beautiful vistas, wide streets and large civic centres, all based on axial and cross-axial geometric patterns.

The McMillan Plan for Washington DC, completed in 1922, influenced subsequent beautification projects in other cities, from New York to San Francisco, such as Chicago (Burnham's Chicago Plan of 1909), Seattle, Cleveland, Denver, Kansas City and Portland, among others. Its influence was felt well into the twentieth century and its legacy in North America and Europe is still being felt today in debates about URBAN REGENERATION. By the 1920s, however, the City Beautiful movement was being questioned by all sorts of reformers guided by the new flag of efficiency. There was then a large consensus that much more than aesthetics was necessary in order to deal with the urban problems of transportation, housing, etc. Like *beaux arts* in Europe, modern movements in America defined themselves against this approach. It was time for the 'functional city' to replace the 'city beautiful'.

Further reading

Cullingworth, J.B. (1993) *The Political Culture of Planning: American Land Use Planning in Comparative Perspective*, London: Routledge.

Kostof, S. (1995) *A History of Architecture*, Oxford: Oxford University Press.

Ragon, M. (1991) *Histoire de l'architecture et de l'urbanisme modernes*, vol. 1, *Idéologies et pionniers, 1800–1910*, Paris: Éditions du Seuil.

SEE ALSO: classicism; progressive planners

CARLOS NUNES SILVA

CITY OF UR

The City of Ur (5500–300 BC) represents one of the first known cities whose urban form is characteristic of the city-states built by the Sumerian civilization. It was located in southern Mesopotamia (Iraq), by the River Euphrates, which provided water and access to the Persian Gulf, making the city an important trading centre.

The excavations of L. Wooley documented several important periods in the life of this city. During the Third Dynasty (2110–2015 BC), Ur was a prosperous capital with a population of 35,000 (250,000 in the city-state). It was an organic city with an oval layout located on a 'tell', a high, man-made platform built with the remains of previous buildings. A defensive wall on a rampart surrounded it. It had two harbours, a temenos (religious precinct), residential areas and cemeteries with royal tombs. The walled temenos (rectangular

in the excavated version of 600 BC) dominated the city and contained a ziggurat (a stepped, pyramidal temple), palaces, an open courtyard, and administrative buildings with access reserved for priests and the royal household. The residential areas, which grew over time without a preconceived form, had narrow, winding streets, irregular buildings sharing walls, a variety of housing sizes including some with central courtyards, temples, markets and no significant open space.

Further reading

Forte, M. and Siliotti, A. (eds) (1997) *Virtual Archaeology*, New York: Harry N. Abrams, 'Ur: the city of the flood', pp. 88–93.
Morris, A.E.J. (1994) *History of Urban Form*, 3rd edn, New York: John Wiley & Sons, pp. 7–8, 21–4.

ANA MARIA CABANILLAS WHITAKER

CITY TYPOLOGIES

A city type may be regarded as a representative or characteristic kind of city, which embodies the characteristics of a group of actual cities. A city type therefore implies a generalized, typical city – an abstract, idealized entity, of which each actual city is an example. A typology may refer to the system of recognition or classification of types, or may more loosely refer to an actual set of types.

Typologies allow generalizations to be made about cities. This effectively allows controlling for one variable while discussing diversity on other fronts, or setting one variable against another. For example, the recognition of WALKING CITIES allows comparison or contrast between different kinds of HISTORIC CITY or compact city. A type allows us to recognize a common characteristic in a group of cities, which may assist examination of a shared set of problems or opportunities – for example, the recognition of the post-communist city allows for diagnoses and solutions that may be distinct from both communist and non-communist cases. In general, typologies can be useful for recognizing emerging trends: the identification of a new type can help to encapsulate a new phenomenon, such as with edge cities, GLOBAL CITIES or 24-hour cities.

The extent to which city types – and their inherent generalizations – are meaningful and useful will depend on the purpose to which they are to be put, and their context of application. The historian may be most interested in the dynasty in the origin of a city, whereas the economist or town planner might be more concerned with its present economic function. The anthropologist or sociologist might be interested in the city's inhabitants, while the engineer or architect may be most interested in its physical form. Accordingly, there can be no single correct or definitive way of classifying cities or identifying city types, and a diversity of overlapping types and themes are both appropriate and inevitable.

An important theme for classification is the function of cities, which gives rise to recognition of such types as the INDUSTRIAL CITY, MERCANTILE CITY, PORT CITY, resort city and capital city. These labels refer to the ongoing purpose (or modus operandi) of cities; alternatively, types may refer to the original purpose (or *raison d'être*) of settlements, as with FORTIFIED CITIES, planned cities or towns, new towns, or company towns. A city type may refer to its inhabitants, based on themes such as ethnicity, race, religion, linguistic or CULTURAL IDENTITY, demography, POLITICAL ECONOMY, or socio-economic composition (e.g. working-class city). A city may be characterized according to its geographical position – with respect to climatic region (e.g. tropical city, temperate city), continental or national location, topographical situation (e.g. coastal city), or urban context (e.g. CENTRAL CITY, SATELLITE TOWN). A type may be used to describe the three-dimensional form of a city (e.g. WALLED CITY, high-rise city), or its means of construction (e.g. prefabricated city). City types may also refer to the two-dimensional pattern or plan of a settlement, such as with the LINEAR CITY, grid city, radial or radio-concentric city, or its articulation in MONO-CENTRIC, MULTI-CENTRIC or POLYCENTRIC forms (see also URBAN FORM). Some city labels may be metaphorical, as in the 'city of the dead' or NECROPOLIS.

A typological label may describe a city in terms of a period when it acquired its most familiar character: this may refer to the specific chronology of its origin or founding, or the era or period when it made its mark (e.g. the description of Florence as a RENAISSANCE CITY). Such a label might apply indefinitely. Alternatively, the label may refer to any city during a particular period (e.g. 'the MEDIEVAL CITY'). The type still exists, even if the cities themselves move on: yesterday's twentieth-century city becomes today's twenty-first-century city.

City types change over time as the cities themselves change in function or form. A city might be

founded as a religious city, but later be recognized more for being an industrial city; a city might change its function absolutely from being a provincial city to a national capital, or vice versa. However, the typological label may change while the city stays much the same. A city once considered 'modern' or 'new' may mature to the extent that it is recognized as a 'traditional' city. Conversely, the typological label may stay the same over time, while the kind of city implied changes. An industrial city of the past might have implied a Dickensian 'Coketown' of steam engines and mills, whereas the industrial city of today might imply a hi-tech 'Silicon Valley' of electronics factories and business parks.

As well as cities changing, the abstract types themselves might evolve over time, and generate new evolutionary branches or lineages. The industrial city might spawn the post-industrial city and the INFORMATIONAL CITY, TECHNO-CITY and CYBERCITY. The original type may be superseded by the new, or may continue in parallel – modern cities continue to co-exist with postmodern ones (see POSTMODERNISM).

City typologies may be used both in a descriptive sense, such as by geographers describing cities as they are or were, and in a prescriptive sense, as with urban planners envisioning optimal forms which suggest how cities could or should be. Among the latter cases are subjective terms such as HUMANE CITY, HEALTHY CITY, NON-SEXIST CITY and cosmopolitan city. In principle, any city might aspire to becoming any of these types. The general aspiration to a healthy city or linear city might crystallize into a specific manifestation or prescription in the plan for HYGEIA or *Ciudad Lineal*. There is a whole class of MODEL CITIES intended to represent exemplars for cities to follow in the future. These imply specific packages, such as Broadacre City or RADIANT CITY.

In general, among typological labels, it is possible to distinguish 'singular' types, which focus on a specific facet of a city, and 'composite' types, which bundle together several meanings, facets or connotations under a single label.

The singular type implies a single defining feature or characteristic, possession of which admits membership to the class. For example, possession of a port (among other things) qualifies a city as a port city. The singular type implies various degrees of specificity. First, a singular type could imply a single basis for distinction across a spectrum of mutually exclusive types, for example by

continental location (African, Asian, European, etc.). Secondly, a label might refer to a single polarity – a special case of a mutually exclusive set, with two members. Examples might include the planned city versus the unplanned city, or general law city versus charter city. Thirdly, a singular type might refer to a single feature that sets a particular city apart from the general ruck of cities – for example, capital city, DIVIDED CITY, world city. Here, the complement is not explicitly defined; this may be quite appropriate because the remainder cities do not have anything particular in common (the majority of cities are non-capital, undivided, non-world cities). Finally, a singular label might refer to a class which happens presently to have one member. Here, the label would effectively pinpoint a single city, a unique case; or the first or last city of a given type.

In contrast, a composite type uses a single descriptive label to bundle together a variety of characteristics that typically go together. These composite terms can act as a useful shorthand, where a single word can convey a multitude of connotations. For example, the modern city may imply the use of modern technology, the era of conception or construction, the style of building or planning philosophy. The COLONIAL CITY implies something about the creation or establishment of the city, its governance, possible connotations of planned form, and possible counterpoint to the existing 'native city'. The 'Victorian' city implies not just the characteristics of that era as a whole, but in a sense the residual connotations that we may nowadays associate with that era, such as certain kinds of Victorian technology or social values. Other composite types with multiple connotations would include the renaissance city, BAROQUE CITY and THIRD WORLD CITY.

Creating a typology is strongly concerned with bases for distinction and difference, and the relative positioning of boundaries, rather than necessarily with absolute values. The boundaries between different city types may be 'fuzzy' – where there is really a continuous spectrum or a series of shades of meaning, rather than a discrete set of mutually exclusive types. For example, there are subtle distinctions between sacred, religious and ecclesiastical cities, or between industrial and manufacturing cities, between communist and socialist cities, and so on. That said, even if a typological label can be precisely defined, the meaning is not necessarily significant. We could in

principle define 'the Asian city' in precise geographical terms, but whether or not the label is useful in characterizing cities as different as Jerusalem, Chandigarh and Vladivostok would depend on the context of application.

There is a problem of over-generalization, then, where a label is too loose to be meaningful. This may arise if a type is generated by negation, for example non-medieval city, or non-western city. The group of cities that the label stretches to cover is too heterogeneous. Over-generalization may also occur where a label implying a bundle of meanings and connotations is applied to a city to which only some of those meanings or connotations apply. Conversely, a label may be too narrow, reducing the multi-faceted nature of a city's character to a single aspect. Stereotyping can be seen as a combination of over-generalizing and over-particularizing attributes of a city, which might result in the 'pigeon-holing' of a city by just one or a few attributes. This is especially harmful if the stereotype is a negative one, for example a 'rustbelt' city.

As a result, the issue of city types is liable to be a contested area. Yet the use of types can be appropriate depending on the context of use. It may be a simplification to talk of 'western' cities and 'eastern' cities, but in context the meaning of the distinction might well be quite clear and appropriate. Cities like Istanbul or Tel Aviv could be uncontroversially regarded in geographical terms as 'eastern' from a European perspective or 'western' from an Asian one.

In general, there is a balance to be struck between having too few broad categories or too many narrow ones. In the former case, taken to the extreme, we have a single category, into which all actual cities are lumped. Clearly, this fails to differentiate and is not much use as a typological system. At the other extreme, we could have a multitude of finely defined categories, each of which contains a unique actual case (e.g. a 'Rio de Janeiro'). This is not much use as a typology either. Hence, although the use of types may carry the latent danger of over-generalization or stereotyping, we have to have some sort of collective or representative label, if we are to be able to generalize about cities at all, and not simply refer to each one by their own unique name.

Finally, types can distinguish cities from what are not cities: to be granted membership of a particular class of cities implies prior acceptance as a

city in the first place. Urban units that might not qualify as cities would include smaller settlements such as towns and villages; larger elements such as CONURBATIONS; and subunits such as SUBURBS or SUBDIVISIONS, or portions of divided cities. The very recognition of an edge city as a city, and not just a peripheral suburban accretion, is as significant in terms of defining what constitutes a city as it is in distinguishing different types of cities.

Further reading

King, A.D. (1994) 'Terminologies and types: making sense of some types of dwellings and cities', in K. Franck and L. Schneekloth (eds) *Ordering Space: Types in Architecture and Design*, New York: Van Nostrand Reinhold.

STEPHEN MARSHALL

CIVIL RIGHTS MOVEMENT

Historically, the CITY has been a contested space in the civilian struggle for social justice, spatial equality and CITIZENSHIP. The civil rights movement (CRM) was an effort to transform race-relations in the United States between the Second World War and the Vietnam War. On 1 December 1955, the burgeoning movement gained national visibility when Rosa Parks was arrested in Montgomery, Alabama, for refusing to give up her seat to a white patron on a public bus. Parks's arrest ignited a year-long boycott of segregated public transportation and marked the beginning of one of the most significant urban social movements in American history. The success of the Montgomery bus boycott was directly attributable to the activism of the Women's Political Council (WPC), which was formed in the mid-1940s after the local, all-white League of Women Voters refused to integrate. The WPC sought fair treatment of the city's black residents as well as the DESEGREGATION of public transportation and other facilities. The WPC and the newly formed Montgomery Improvement Association (MIA) successfully used Parks's arrest to challenge the SEGREGATION of public transportation. Martin Luther King Jr was elected to head the MIA and subsequently became one of the key national spokespersons of the CRM. King was influenced by Mahatma Gandhi's non-violent resistance strategies, as evidenced by the movement's boycotts, sit-ins and mass protests against desegregation.

Within a decade, mass non-cooperation overturned the foundations of *de jure* segregation and won voting rights for black citizens of the South.

By the mid-1960s, economic concerns rose to the forefront of the movement's agenda. In 1966, King and other leaders began to test their strategies in the urban centres of the North. They launched the Chicago Freedom Movement, which demanded better employment opportunities for blacks, equal treatment in mortgage lending, and the construction of low-income housing. In short, they sought to dismantle that city's segregated black GHETTO and transform it into a democratic urban space. The CRM also spawned the inter-racial, class-based Poor People's Campaign, which highlighted the need for radical economic reforms to address the persistence of widespread poverty in the United States. Trade union leader A. Philip Randolph and civil rights leader Bayard Rustin proposed a 'Freedom Budget' in 1966, calling for a ten-year, US$100 billion federal programme to address urban poverty.

While the CRM led to a number of significant victories in the South, such as bringing an end to legal segregation in American society, its marches, sit-ins and mass arrests proved ineffective in challenging the structures of racial inequality in the North. Some reasons for the impasse include the entrenched power of Chicago mayor Richard Daley's political machine and the ferocious opposition of white homeowners. However, the movement did succeed in placing economic justice on the national agenda; Lyndon Johnson's GREAT SOCIETY programme was a direct consequence of civil rights activism. While Johnson's 'WAR ON POVERTY' did succeed in addressing poverty on a small scale, the costs of the Vietnam War and a rising conservative tide in the nation eroded the gains of the initiative and left many racial issues unresolved. In conclusion, the colour line remains a persistent reality in American cities.

Further reading

Crawford, V.L., Rouse, J.A. and Woods, B. (eds) (1990) *Women in the Civil Rights Movement: Trailblazers and Torchbearers, 1941–1965*, Bloomington, IN: Indiana University Press.

Kelley, R.D.G. and Lewis, E. (2000) *To Make Our World New: A History of African Americans*, New York: Oxford University Press.

EDWARD RAMSAMY

CIVIL SOCIETY

There was a great deal of speculation at the end of the twentieth century that American civil society was not in great shape. Both its civic hardware (i.e. the organizations we join, our ties to other persons, the amount of time we volunteer) and software (i.e. the norms, values and customs that guide our civic habits) were not working as well as they once had, many analysts said. This may have come as a big shock to many persons, but it could not have been much of a surprise to students of cities who have been hearing and making this kind of noise about URBAN life for generations.

Cities are supposed to be places where lots of different persons alternately bump into each other or try to stay out of each other's way. The result in either case is that many people live *among* each other in cities but not necessarily *with* each other or especially well. Urban dwellers who cannot avoid persons or groups that are different, obnoxious or downright scary end up fighting with them, or so our theories about urban life tell us.

What appears to have happened, if these observers are correct, is that an urban malaise has finally spread through the rest of American society. Large chunks of national civic hardware are said to be obsolete or inactive, meaning that people are not as engaged in each other's lives as they once were. Furthermore, whatever understanding and standards for the common good people once held no longer seem relevant.

Contemporary critics of American civic habits with few exceptions made no connection between city life and America's crumbling civic INFRA-STRUCTURE; but such a connection ought to be clear enough to anyone who has spent much time writing about the way people live in cities. The problem is that this pinched view of urban life is not supported by much evidence, while there are a great many reasons to think that the civic hardware and software available in cities still works very well.

While their views are still very much in the minority, persons who have studied urban communities in something more than a superficial way point to the continuing viability of civic organizations and the vitality of the values and customs that guide the people who run them. Cities may lack the kind of folksy face-to-face intimacy that is thought to be a hallmark of life in small out-of-the-way

towns and villages. On the other hand, surveys of persons who live in cities consistently show that urban dwellers have social lives that are every bit as rich as those of their small-town counterparts. There also is no great difference between them when one compares both the number and the variety of voluntary associations to which they belong. In short, many of the supposed differences in the social lives of small-town and big-city people are more apparent than real.

An alternative point of view regarding the connection between cities and civil society is that city life is built on a foundation of good BOURGEOIS principles, meaning that big-city dwellers remain committed to being prosperous and behaving in more rather than less orderly ways. The civic society or community that they build with those principles is neither lost nor saved, or as contrived or conditional as many observers have asserted over the years. It is instead rather paradoxical in the sense that urban civic ties and values reflect an interesting mix of liberal *and* conservative ways of being in the world together.

This mix is apparent in the four main ways that urban dwellers make a civil society or 'do' COMMUNITY. There is *commercial communalism*, which entails the kind of community-building associated with business leaders. The finest examples of commercial communalism are found in the voluntary subscription campaigns that business leaders use to drum up support and money for projects they deem to be in the interest of the whole community. These campaigns have long been used to launch a variety of initiatives inside cities. Some are more obviously public in character, such as efforts to build a PARK or provide charity. Other efforts have a decidedly private and even self-serving edge to them, such as joint stock companies that would bring more jobs to their city (and more wealth to them).

Ethnic communalism entails the kind of community-building that one associates with ethnic enclaves. The quintessential strategies for doing this kind of community-building are the rotating credit association and mutual trade association, which involve loaning money and restricting trading privileges to one's fellow ethnic group members. Both take advantage of the trust and regard that fellow ethnic group members have for each other and are the ethnic versions of the voluntary subscription campaigns favoured by more established business leaders.

Consumer communalism is accomplished through the spending habits of individual shoppers and investors. Although predicated upon a lot of self-seeking behaviour, shopping and investing involve nearly all of us in pursuits that enforce certain kinds of disciplines and re-enforce values that are decidedly bourgeois in character. The central ritual in this kind of communal activity may be shopping; but it is the extension of credit to more and different kinds of persons that makes all of us act more alike than we might otherwise think necessary. Principles of voluntary subscription campaigns apply here as well, with persons giving up some portion of their wealth and circulating it among more people in order to get something they want.

Government communalism also seeks to build communities one person at a time, only as citizens who participate in public rituals like elections and voting. The most important customs associated with this kind of community-building, however, are paying taxes and getting favours. The act of paying taxes is akin to a mandatory subscription campaign, and the revenues that are generated are supposed to be used to serve a broader common interest.

City people and organizations, including businesses, may watch out for themselves. Sometimes explicitly but often without even knowing it, however, they help make a world in which different types of persons are able to live and work together, just like the ancient Greeks said they should. This is civil society.

Further reading

Lofland, L.H. (1998) *The Public Realm: Exploring the City's Quintessential Social Territory*, New York: Aldine de Gruyter.
Monti, D. (1999) *The American City*, London: Blackwell.
Zukin, S. (1995) *The Cultures of Cities*, Oxford: Blackwell.

DANIEL J. MONTI JR

CLASSICISM

Classicism refers to a 'style', 'historical period' or 'quality' of work in art, architecture and town planning, but also in literature, music and science. The term is best understood as a whole system of thought. In its purest form, *classicism is an aesthetic attitude dependent on principles based on*

the culture, art and literature of ancient Greece and Rome, and characterized by emphasis on form, simplicity, proportion and restrained emotion. It also places emphasis on the *clarity of structure, an explicit appeal to the intellect, and on perfection.*

Classicism is sometimes compared or contrasted to other styles, periods or qualities. The most often cited contrasts are those between the baroque or rococo and classicism as well as between classicism and romanticism. Classicism spars with other whole systems of thought for dominance in time and space. It never completely dominates nor is it totally subjugated. Well-known periods where classicism was in fashion include the Renaissance (*c.* 1400–1600), the neo-classical revival of the late eighteenth and early nineteenth centuries (*c.* 1750–1850), and in the early twentieth century (up to, say, 1940). Classical influences persist into the twenty-first century.

As a *style*, classicism can be seen in art, in architecture and in town plans. Common to all of these fields of endeavour is the notion that contemporary creative individuals use the art, literature and aesthetics created by ancient Greeks and Romans in copied, suggested or derived forms.

In art, the classical style can be understood in pictures that portray ancient themes. The best examples are from the late eighteenth and early nineteenth centuries seen in the works of Jacques-Louis David (*Oath of the Horatio*, 1784–85) and Jean-Auguste-Dominique Ingres (*Apotheosis of Homer*, 1827).

The classical style in architecture can be seen in the works and buildings from the period of academic classicism (1885–1920), including the World's Columbian Exposition of 1893, which is the icon of the CITY BEAUTIFUL MOVEMENT, and the New York City Public Library; from the neo-classical period in such buildings as the Capitol in Washington DC or the State Hermitage in St Petersburg; and from the Renaissance period in such ecclesiastical works as St Peter's in Rome.

Notable examples of the classical style associated with city, town and urban planners include the Mall in Washington DC, the ring road in Vienna (in terms of both its layout and its Greek- and Roman-inspired buildings reflecting the glory of the Habsburg Dynasty) and the layout of civic buildings by Shinkel in Berlin. In town plans, classical thinking is reflected, perhaps most succinctly, in J.B. Jackson's article about the hypothetical Optimo City, 'The almost perfect town', published in the journal *Landscape* in 1952.

Historically, classicism was 'dominant' (remember, it ebbs and flows) in a number of specific time periods. The first occurred during the Renaissance (*c.* 1400–1600). The rise of humanism (a philosophical outlook that emphasizes the intrinsic value, dignity and rationality of human beings as opposed to the medieval focus on sin and redemption) was based significantly on the ideological viewpoint that ancient Greek and Roman cultures reflected the positive virtues of 'civilized' behaviour and could serve as the foundation in the emergence of Western European culture as an ultimate civilization. In art, Botticell's *Birth of Venus* and Michelangelo's *David* represent idealized 'perfect' forms. In architecture, the Greek and Roman orders of architecture were revived and applied to ecclesiastical designs. Leon Battista ALBERTI wrote the first Renaissance treatise on architecture (1485), based on his reading of Vitruvius.

The second great period is termed the neo-classical, generally agreed to encompass the late seventeenth and early nineteenth centuries (*c.* 1680–1850). Neo-classicism replaced the frivolity and superficiality of the earlier baroque and rococo period. Following the archaeological rediscovery of Herculaneum and Pompeii, and like its previous rise in influence, the revival of antiquity was closely tied to political events, this time the American and French revolutions, in which parallels were drawn between ancient and modern forms of government. Knowledge, for the *philosophes*, was a transcendent and universal goal. Intellectual seriousness yielded both a cult of sensibility and a profound belief that the fine arts could – and should – spread knowledge and enlightenment. Structural clarity returned in painting, sculpture and architecture, and above all in music, which found the culmination of a remarkable century of its history in works of emotional depth and formal inventiveness.

The third great period of classical influence occurred in the late nineteenth and early twentieth centuries in Europe and the United States. Due largely to the influence of the ÉCOLE DES BEAUX-ARTS in France, neo-classicism never entirely vanished. In architecture, the works of Ludwig MIES VAN DER ROHE and Richard Meier insist on order beyond all else. The classical style and its qualities remain dominant in public buildings, particularly those that represent an ideology or culture, for

example in the official Aryan style of the architect Albert Speer.

As a *quality*, classicism in art refers to vanishing points and composition on the canvas. In architecture, the classical orders are the three Greek orders – the Doric, Ionic and Corinthian – and the two Roman additions to them – the Composite and the Tuscan. In virtually all the arts, in architecture and in other creative endeavours, there are two dominant qualities, derived from the ancient Greeks and Romans, which define a work as 'classical': (1) a sense of rational ordering and proportioning of forms yielding a regularity of form; and (2) a sense of conscious restraint in the handling of themes yielding beauty.

Further reading

Curl, J.S. (2003) *Classical Architecture: An Introduction to Its Vocabulary and Essential Terms Glossary*, New York: W.W. Norton & Co.

Greenhalgh, M. (1990) *What is Classicism?*, New York: St. Martin's Press.

DAVID C. PROSPERI

CLOSED CIRCUIT TELEVISION (CCTV)

CCTV has been one of the most significant developments in terms of the URBAN environment and crime reduction of the last twenty years. The extensive growth and deployment of this technology has facilitated a significant growth in state surveillance, a substantial rise in the variety of methods of advanced social monitoring and control, and a host of crime prevention measures throughout the world. Indeed, many commentators have noted that it is proving increasingly difficult to move through public space without having your movements surveyed and recorded. Specifically, CCTV offers a distinct approach to what is often called situational crime prevention in that the deployment of such surveillance technologies is seen as offering an opportunity to tackle specific crimes – such as car theft, shoplifting or violent assault – in specific places – such as car parks, SHOPPING MALLS or city centres – through the systematic management of PUBLIC SPACE.

CCTV technology is popular with politicians, many security experts and services (e.g. the police),

and the public alike as overt CCTV deployment offers apparently concrete signs of anti-crime measures. Furthermore, the use of CCTV appeals to a simplistic notion of crime detection, reduction and control, and appears to offer a relatively quick and inexpensive solution to crime control across a range of different environments. For proponents, CCTV offers the security, guardianship and protection of the citizen by the state and its agencies.

On the other hand, civil liberties groups, academics and some citizens stress the intrusive nature of CCTV and fear the intensification of state and security surveillance. Such critics question the appropriateness, justification and underlying attractions to the deployment of CCTV in public spaces, and not surprisingly popular and critical discourses alike draw parallels with Orwell's representation of state surveillance in his novel *Nineteen Eighty-Four*. Others argue that CCTV technology and its use works to further exclude and marginalize those already at the edges of society (e.g. the homeless) and effectively reinforces social divisions. Social geographers have drawn attention to the role of CCTV in terms of the structural transformation of public space in which public spaces cease to operate as public arenas for the social and political interaction of individuals and social groups and become simply spaces of mass, increasingly uniform (and global) consumerism.

More importantly, perhaps, some critics question the actual effectiveness of CCTV technologies in combating crime. Such critics highlight the fact that rather than reduce or prevent crime, CCTV deployment – which characteristically takes place in urban centres and areas of general business or economic significance (e.g. town centres) – is most effective at displacing crime to other localities nearby.

Whatever the verdict, we should note that the development of CCTV and related technologies continues. Recent developments have included systems of automatic car number recognition (which facilitates faster identification of vehicle owners who may be committing certain types of crime) and greater use of mobile CCTV units to counter charges of crime displacement and for surveying particular events and public gatherings (e.g. demonstrations, protests or large sporting events). Furthermore, the popularity of technological strategies for tackling crime remains high and the continued deployment of CCTV is

likely to remain a consistent feature of the contemporary city.

Further reading

Lyon, D. (1994) *The Electronic Eye: The Rise of the Surveillance Society*, Cambridge: Polity Press.

Norris, C. and Armstrong, G. (1999) *The Maximum Surveillance Society: The Rise of CCTV*, Oxford: Berg.

Staples, W.G. (1997) *The Culture of Surveillance: Discipline and Social Control in the United States*, New York: St. Martin's Press.

RICHARD HUGGINS

COASTAL MANAGEMENT

Inspired in part by the social activism of the period, during the late 1960s environmentalists in the United States, Europe and Oceania came to appreciate that the world's coastlines were a finite resource whose quality was rapidly being compromised by URBANIZATION, industrialization and pollution. In the United States, urbanization along the California coastline created areas where a virtual wall of homes prevented access to the state's public beaches (see VIEW CORRIDORS). In portions of the eastern United States, for decades industries had discharged industrial pollutants into fragile coastal wetlands (see WATERFRONT DEVELOPMENT). Along the coastline of the Gulf of Mexico, oil production had similarly degraded coastal wetlands. Environmental degradation was likewise experienced in other parts of the developed world where an emerging environmental consciousness inspired public policies to protect the coast.

In many ways, the Netherlands remains the home of coastal management, reclaiming shallow coastal polders for human habitation. Hundreds of years of experience by the Dutch have informed modern efforts at coastal management. Building on that base, coastal management efforts now span the globe from the United States, Europe and Oceania to the islands of the Philippines to the coastlines of China, Japan, Tanzania and Chile. The management processes are often similar throughout a wide variety of coastal resources.

In the United States, the states have responsibility for developing and operating coastal management programmes and there are as many styles of programmes as there are coastal states. The federal Coastal Zone Management Act of 1972 guides state programmes and provides funding, but the states retain responsibility for managing their own coastlines. Two states, California (1972) and Rhode Island (1971), preceded the federal government in establishing coastal management programmes.

While definitions of the 'coastal zone' differ from state to state, the US federal statute defines it as:

> the coastal waters (including the lands therein and thereunder) and the adjacent shorelands (including the waters therein and thereunder), strongly influenced by each other and in proximity to the shorelines of the several coastal states, and includes islands, transitional and intertidal areas, salt marshes, wetlands, and beaches. The zone extends, in Great Lakes waters, to the international boundary between the United States and Canada and, in other areas, seaward to the outer limit of State title and ownership under the Submerged Lands Act.
> (43 U.S.C. 1301 *et seq.*)

In some cases, the coastal zone subject to unique land use controls is defined primarily by its physical proximity to the sea or other major body of water. In other cases such as Hawaii, the entire state and its coastal waters are included. In still other cases, the coastal zone is defined by the boundaries of the waters that drain into coastal waters. Identifying the coastal zone is the first step to applying special land use management tools to those areas. Common objectives are to preserve environmental qualities, provide opportunities for public access to the coastline, limit development that would further degrade the coastline and plan development in such a way as to preserve environmental values in coastal waters. Achieving these objectives usually involves empowering a coastal management agency with land use controls specifically designed for the coastal environment. In some cases such as Oregon, a statewide comprehensive land use planning agency already exists and the coastal management statute merely provides additional guidance for a pre-existing agency. This is, however, the exception to the rule.

Exercising land use controls over highly valued coastal lands is often a contentious process. When completed, the plans are implemented by regulatory schemes that often overlay existing land use controls.

Where the coastline is urbanized, typically land use controls permit coastal management agencies

to condition any subsequent application for development. Thus, existing structures are permitted to remain unchanged until such time as a landowner seeks a change in use. Landowners have brought legal challenges to the exercise of such controls and cases have moved for years through US courts of law.

The most vigorous struggles over the appropriate extent of coastal land use controls occur at the edge of the urban coastline (see SPRAWL). At those points, the coastal ecosystem is most threatened, yet because there is little pre-existing development, those shorelands are often attractive to land developers.

Coastal management plans seek to balance use of the shoreline with reasonable use of its amenities. However, deciding what is reasonable is enormously difficult. For that reason, coastal management planning processes have been crafted to include wide opportunities for the public to influence the process. Acceptance of a broad interpretation of citizen participation has meant that coastal planning matters are unpopular with development interests, but the decisions of the process acquire legitimacy by virtue of the acceptance of wide public comment. Nevertheless, legal challenges to coastal agencies' decisions abound in the United States.

Because of the diversity of uses in the coastal zone, the discipline of coastal management requires attention to a wide variety of academic disciplines including urban/town planning, geology, terrestrial and marine biology, coastal engineering, resource management, recreation management and law. Coastal managers call on various aspects of these fields as they craft plans for wise use of coastal resources. Thematically, the notion of 'carrying capacity' dominated the field in the 1970s and 1980s: to what extent could the coastline support human activity? That notion has now evolved into a new term: 'sustainability' (see SUSTAINABLE URBAN DEVELOPMENT). That is, to what extent can environmental and human use co-exist without compromising either?

A corollary principle for coastal management is coastal dependency. There is a relatively narrow array of uses that must be located, by their nature, at the shoreline. Among them are coastal recreation, both commercial and recreational fishing, seaports and shipping, and naval uses (see DOCK-LANDS; PORT CITY). Many other uses make claims to use coastal resources and some of those claims are at least in part legitimate. The challenge for coastal managers is to balance the coastal-dependent nature of the use with the capacity of the resource to support that, and other, uses.

Thus, it is the nature of coastal management to craft plans and management policies for wise use of often sensitive coastal resources. And, those plans and policies must be crafted from a wide array of claims for use, many of which are only partially coastal-dependent.

Further reading

Beatley, T., Brower, D.J. and Schwab, A.K. (2002) *An Introduction to Coastal Zone Management*, 2nd edn, Washington, DC: Island Press.
Cicin-Sain, B. and Knecht, R. (1998) *Integrated Coastal and Ocean Management: Concepts and Practices*, Washington, DC: Island Press.

JAMES A. FAWCETT

COGNITIVE MAPS

The term 'cognitive map' refers to mental representations underlying our ability to remain oriented in our surroundings as we move about. It was coined by Edward Tolman to explain the behaviour of rats that appeared to learn the layout of a maze, rather than just sequences of responses that would lead to the goal. Subsequently, studies have considered other animals, including humans, asking how mental representations of our surroundings are acquired, structured and used.

Various research methods are used for investigating, or 'externalizing', cognitive maps: from navigation tasks or simple sketch mapping, to complex statistical procedures on people's estimates of distance or direction. Two main areas of research have used such methods. By externalizing people's cognitive maps to investigate their IMAGE OF THE CITY, researchers attempted to determine the 'legibility' of various cities, to inform and improve URBAN PLANNING (see Kevin LYNCH). In contrast, psychologists and behavioural geographers have investigated the structure, geometric properties and acquisition of people's mental representations of their surroundings.

Cognitive maps are acquired in two main ways: direct experience of an environment, and indirect experience from maps or verbal descriptions.

With direct learning, representations develop through a series of stages: first, a sequence of landmarks along a route is learned; later, the route is structured into segments; and finally, the various routes are integrated into a coherent representation of one's surroundings. When an environment is learned from a map, we are able to skip these stages and perceive the structure of the environment immediately. However, there are important differences in the resulting representations. Most significantly, they tend to retain a bias in respect of the map's orientation (i.e. it will usually be more difficult to use the information when facing south than when facing north).

The term 'cognitive map' suggests a 'cartographic map in the head', in the form of some kind of mental image that can be 'scanned by the mind's eye'. However, cognitive maps appear not to be structured in a uniform way, like the continuous sheet of a map, but have a hierarchical structure and include some systematic distortions, which are thought to reduce the amount of information contained in the mental representations, thus making it easier to store and retrieve the information, without producing significant limitations in spatial orientation ability. For instance, information about relative locations of districts in a city would be represented at a separate hierarchical level from information about the locations of places (e.g. shops) within the districts. Accordingly, people find it more difficult to make spatial judgments about pairs of locations across districts than between pairs of locations in the same district. Many other simplifying distortions in cognitive maps result from the schematization of spatial information, such as representing curvy roads as straighter or representing a side road as joining the main road at a right angle whatever the actual angle.

Further reading

Kitchin, R.M. and Blades, M. (2002) *The Cognition of Geographic Space*, London: I.B. Tauris.

Kitchin, R.M. and Freundschuh, S. (eds) (2000) *Cognitive Mapping: Past, Present and Future*, London: Routledge.

SEE ALSO: environmental perception

SIMON UNGAR

COHOUSING

Cohousing is a form of residential development combining individual and common ownership. Cohousing originated in Denmark in the late 1960s. The concept was popularized in the United States by a young Berkeley-based husband-and-wife team – Charles Durrett and Kathryn McCamant – who coined the English-language term 'cohousing' in their splendid and very influential book of the same name.

Residents of a cohousing development want to be part of a COMMUNITY, but also value privacy and individual homeownership. Cohousing encourages people to participate more actively in a community than in conventional single-family detached housing developments, apartment buildings or condominiums, but they retain more privacy and individual ownership than in a commune or Kibbutz. Residents can have the benefits of jointly owned common areas and shared cooking, childcare and other activities combined with the privacy of an individually owned housing unit. Cohousing appeals particularly to young couples with children, but other age groups and household configurations may also find it attractive.

Cohousing project residents own their own individual housing units. They also own an interest in a common house with a kitchen and dining room large enough to accommodate all the cohousing project's residents. Residents take turns cooking and usually eat their evening meal in the common house. They share ownership of facilities such as a cooperative store, day-care centre, community garden, bicycle repair shop, workshop, darkroom, guestroom, and recreational facilities for children, teenagers and adults.

Cohousing is not cheaper to build than comparable apartments or condominiums. Common areas add costs above what it would cost to build comparable individual units alone. Cohousing residents accept the trade-off between smaller individual units and more jointly owned shared space.

Cohousing developments vary in size, location and design. The residential units in most cohousing projects consist of moderate density, attached row houses clustered around COURTYARDS. Some include single-family detached houses. Some cohousing projects have been built by retrofitting other buildings such as schools or factories. Most units are owned as condominiums, but some are organized and financed as cooperatives. A few

include rental units. Most cohousing developments contain between 15 and 30 residences. Cohousing designs seek to foster a sense of community by including shared facilities, eliminating fences, being pedestrian-friendly, reducing the intrusiveness of automobiles, and including space for children and teenagers.

Cohousing projects involve residents in decision-making. Residents usually participate in designing and building the projects. They almost always manage the cohousing projects – usually in democratic community meetings.

Cohousing is the object of a robust worldwide housing reform movement today. While the absolute number of cohousing units remains a tiny percentage of all housing, cohousing attracts passionate followers who want the housing and social alternative that it affords. Most cohousing is now located in Denmark and the United States, but there are cohousing projects in existence or planned in Sweden, the Netherlands, Norway, Germany, France and other countries.

Further reading

Hanson, C. (1996) *The Cohousing Handbook: Building a Place for Community*, Point Roberts, WA: Hartley & Marks.
McCamant, K. and Durrett, C. (1994) *Cohousing: A Contemporary Approach to Housing Ourselves*, 2nd edn, Berkeley: Ten Speed Press.

RICHARD LEGATES

COLONIAL CITY

Colonization in any systematic sense is generally taken to originate in classical Greece, when population pressures caused individual cities to found settlements to the east (in the Aegean and on the coast of Asia Minor), and to the west, mainly in Sicily and Italy (the Achaean colonies). Roman colonies were not founded in otherwise alien countryside, but within conquered territory, with a view to maintaining control over it. In the medieval period, colonies founded in a similar way to dominate conquered territory, especially in the south of France, and in Wales, were known as *bastides*, and were generally more heavily fortified but less systematically planned.

From the sixteenth century, the European powers founded colonies of quite different types. In broad terms, the Spanish and Portuguese, who partitioned South America under the authority of the Pope, expected to absorb and exploit all the territory and to convert the populace to Catholicism. Their colonies therefore tended to be versions of Iberian cities. Other nations, and especially the British, tended to found colonies for specific purposes without (at first) aspiring to control the surrounding territory. They were 'factories' (trading depots) or forts designed to protect and control specific land or sea routes, but they were therefore not planned as cities.

Private and utopian settlements are an entirely distinct category, and draw upon diverse intellectual traditions. J.V. Andreae, commonly believed to be the founder of Rosicrucianism, in 1619 proposed a plan of concentric squares for his town of Christianopolis; eight years later, Tommaso Campanella, in *La Città del Sole*, proposed a circular plan for his Austrinopolis. Colonizers on the ground were, fortunately, a little more pragmatic, and James Oglethorpe's Savannah, of 1833, was successfully planned upon an ideal geometric basis, but one capable of indefinite extension, a grid within which public squares were interspersed in a diagonal pattern. At the opposite end of the scale were the German peasants who settled in South Australia as religious refugees, and recreated medieval *hufendorf* village plans.

When the British became broad-scale colonizers, almost despite themselves, they were the most systematic of all. The instructions issued in relation to settlements from Florida to Sydney were virtually identical, and required the administrator to choose 'townships' (meaning something more like parishes) which had so far as possible natural boundaries and some frontage to the coast. Within the township, the actual node of settlement was the 'town', located where possible on a navigable river, or on the coast, and containing town and pasture lots and reservations of land for future government requirements. All this assumed a rural society with farms and market towns, and in general these plans were effective for at least the first few decades.

Today, the colonial town is often threatened not merely by the usual pressures of development but by the perception that it challenges the nationhood of the host country. But that perception cannot be maintained. Tourists visit such towns, from Antigua in Guatemala to New Delhi in India. And where the tourist dollars flow, national identity and national pride soon follow.

MILES LEWIS

COMMUNITAS

Originally published in 1947 and re-issued as a best-selling paperback in 1960, *Communitas* by Percival and Paul Goodman became one of the most influential texts – Jane Jacobs's *The Death and Life of Great American Cities* was another – of an entire generation of urban activists and radical critics of modernist city planning.

The book was the collaborative effort of two extraordinary brothers. Percival (1904–89) was an architect, teacher and artist who authored *The Double E* (1973), a plea for the integration of ecology and economics. Paul (1911–72) was a poet, gestalt psychologist and philosophical anarchist who helped to define the 1960s counterculture. Among his books are *Growing Up Absurd* (1960), on the alienation of modern youth, and *The Community of Scholars* (1962), an unflattering dissection of American higher education.

Subtitled *Means of Livelihood and Ways of Life*, *Communitas* is divided into 'A manual of modern plans', which brilliantly reviews the conceptual history of twentieth-century planning, and 'Three community paradigms', a series of UTOPIAS, each of which proposes answers to the central question of the book: 'How to find the right relations between means and ends?'

Communitas offers few practical solutions for urban problems and laid out no plans that were ever put into effect. Still, its power as a moral beacon inspired and challenged countless future architects, planners and activists who saw humane possibilities beyond the sterility of corporate modernism and bulldozer-style urban renewal.

Further reading

Goodman, P. and Goodman, P. (1960) *Communitas*, New York: Vantage Books.

FREDERIC STOUT

COMMUNITY

The term 'community' is used to refer to a wide range of social groups, both actual and ideal. Linguistically, 'community' has its roots in the Latin 'communitas' (fellowship) and 'communis' (common). At the most general level, community is used to refer to associations of individuals who share a range of values and a way of life and

identify with the group. Beyond this, the concept of community has two distinct types of applications. First, it is used to describe existing groupings and associations. Secondly, it is used to communicate social ideals. Even when used to describe actually existing groups, the term has a range of meanings. In many, although not all, uses, 'community' implies face-to-face interactions and therefore geographic proximity and small groups. Within URBAN STUDIES, the term 'community' is often used as a description of the NEIGHBOURHOOD or local scale. In other areas of social life, the term 'community' is used to describe social aggregates as large as the nation or a race. In some cases, the term is used to describe groups that have no spatial interaction but share interests or practices, such as INTERNET communities. The term is applied to both ascriptive associations (such as nation and family) and those where membership is voluntary. When used to describe an ideal, 'community' implies the existence of solidarity, mutual concern and equality among group members. This concept of community is often used to critique existing social and political practices and to suggest ideal futures. The word's enduring power, as well as its ambiguity, comes from its positive and 'warmly persuasive' connotation; it is rarely used unfavourably and almost never opposed. Despite the 'warmly persuasive' nature of community, critics have pointed out the negative side of both the concept and its use. Feminists have been particularly critical of the regulatory nature of community and its ability to define and enforce norms of gender and sexuality. As a concept, community is necessarily exclusive; insiders are defined by characteristics that distinguish them from outsiders. In *Imagined Communities*, his well-known treatise on nationalism, Benedict Anderson explores the ways that the imagined community of the modern nation was constructed. While Anderson was concerned with nationalism, his insights are invaluable in discerning the diverse uses of the term in urban studies, politics and planning.

In the nineteenth century, the term 'community' was used to describe, and later idealize, the organic solidarity of pre-industrial villages. In 1887, Tönnies distinguished between *gemeinschaft*, the immediate and connected 'community' of pre-industrial villages, and *gesellschaft*, the more mediated, abstract and instrumental 'society' of modern, urban and industrialized life. His terms distinguished the rapidly emerging modern

industrial society from the collective way of life that lay in the recent past, and endowed the term 'community' with a particular anti-urban cast. In a similar vein, Marx used the term to refer to unalienated social groups, who enjoy relative equality within the group. In URBAN PLANNING, Ebenezer HOWARD and his followers invoked the term 'community' as an attribute of the more 'natural' socio-spatial organization of the countryside, and saw its restoration as one of the goals of enlightened planning.

Through the first half of the twentieth century, the term 'community' continued to be associated with traditional lifestyles and rural settlements. As the West grew ever more urban and relations were increasingly mediated, the term 'community' evoked an old sense of values and personal relations. This ideal vision of a rural way of life characterized by face-to-face interactions and small geographic scales is closely related to the more descriptive use of the term found in the Anglo-American 'community studies' tradition, which had its heyday in the 1950s. These intensive local studies sought to investigate a rural–urban continuum in social relations through empirical research. In both rural villages and urban neighbourhoods, community studies collected a wealth of information about the daily lives of residents and examined local variations in national tendencies. In the 1970s, many urban scholars rejected empirical studies in favour of a more critical, theoretical approach. Critics argued that community studies reified an ideology of community, pathologized urban life, and obscured the influences of structural processes on rural and urban life.

In the 1960s and 1970s, activists outside the university were changing the meaning of community. Individuals and organizations engaged in direct social action and local grass-roots organizing used the term 'community' to refer to poor urban people outside of and marginalized by the bureaucratic state. For these activists, community meant 'the people', especially the poor and oppressed, as opposed to the formal politics of the local state. The term 'community activist' is therefore often used to describe those working for SOCIAL JUSTICE. In a related usage, members of oppressed social groups often use the term 'community' to refer to their own group, such as the African-American community or the gay and lesbian community. This usage not only describes a social group; it also communicates a political ideal

and a politics of identity. When used as an expression of solidarity among members of oppressed social groups, the term emphasizes collective pride and shared experiences. Once again, we see the warmly persuasive quality of the word, as it is used to reflect oppressed social groups in a positive light and to emphasize the quality of relationships among group members. In this usage, locality does not necessarily play a significant role in identifying the group. Rather, the term 'community' appeals to a sort of shared identity that transcends geography and a common experience that unites strangers. The usage of 'community' by oppressed social groups suggests a form of nationalism, not unlike the horizontal comradeship central to the formation and maintenance of nation-states. It also can be used to downplay significant differences among group members by emphasizing one shared attribute (such as race) over many others that are not shared (such as sex, class, ability, age, etc.).

The term 'community' is particularly common, and perhaps particularly appropriate, in describing groups of recent immigrants. As in other oppressed social groups, the term is used as an ideal to create solidarity, emphasize shared struggles and communicate group pride. The term is also an apt description of many immigrant groups who often experience high levels of interdependence, as demonstrated by the importance of social networks to daily life. In addition, new immigrant groups are usually somewhat linguistically and culturally isolated and often share their workplace and neighbourhood with other immigrants from the same country. As a result, they are very likely to have a shared way of life distinct from the rest of society.

In many parts of the industrialized world, the empowering use of the term 'community' by oppressed and at times oppositional groups was accompanied by a shift in urban planning towards community participation. Many saw community participation as a means of developing effective democratic involvement in local planning and development. This usage of 'community' is found in the activism of Saul ALINSKY and his followers, and in the work of COMMUNITY DEVELOPMENT CORPORATIONS (CDCs). Clearly, this usage overlaps with the previous purely place-based usage; most CDCs serve urban neighbourhoods, and in most projects community participation means the involvement of local residents. As is often the

case, this usage of 'community' is both descriptive and idealized. While the term describes a group of people, it also provides political and moral legitimacy to those who use it; community activists and organizations derive some of their authority with residents and also with external institutions because they are perceived as acting on behalf of those they are connected to.

In Britain and the United States, the language and practice of community participation waned in the conservative 1980s. During the 1990s, the emphasis on community in urban policy was partially revived through an emphasis on community partnerships. Urban studies research from both North America and Britain has pointed out several differences between the contemporary emphasis on community partnerships and the community participation of the 1970s. Critics argue that community partnerships have tended to have top-down structures and rarely empower marginalized groups. Rather than aiming at increasing democratic participation, they have been more corporatist coalitions designed to maximize resources. In the eyes of some critics, the ideal of community partnerships has been used to legitimate privatization and government cutbacks and to pass the blame for failed national programmes and policies.

Despite these pessimistic analyses of community participation, the 1990s saw widespread interest in community among a wide range of people, including politicians, academics on the left and the right, planners, and developers. Some of the rhetorical and strategic appeal must be understood in terms of the reduction of central government power and the increase in local and regional political and economic autonomy. The term 'community' is often also mobilized in attempts to increase the power of local governments relative to state, provincial, regional or federal authorities. In industrialized nations, a second source of the revived interest in community is the continuing demographic shift from large cities to SUBURBS. Through its anti-urban implications and appeal to the moral virtue of small rural settlements, the ideal of community can be mobilized by suburbanites against the interest of large cities. A third motivation for people's renewed interest in community, at least in the United States, is a widely felt sense of fragmentation, isolation and disengagement, aptly depicted by Robert Putnam as 'bowling alone'. High rates of residential mobility, looser family ties, and the prevalence of individualism and consumerism have weakened social ties and left many nostalgic for the past SENSE OF COMMUNITY they recall or imagine. In urban studies, a large literature has analysed the role of the built environment in the loss of community, emphasizing the negative consequences of SPRAWL, architecture, ZONING, functional segregation and forbidding PUBLIC SPACE.

Prescriptions for recovering, reviving, renewing and recreating community are plentiful. The academic community has seen the emergence of a new school of thought called communitarianism. The term refers to a loose grouping of scholars who share an emphasis on the importance of community. In the 1980s, communitarians critiqued the individualism of liberal theory and argued that one's values are shaped by attachment to one's community. In the 1990s, a younger generation of scholars sought to apply communitarians' ideas in order to stop the erosion of communal life and bolster democratic institutions and local community. Since its founding in 1990, the journal *The Responsive Community* has been the site of both theoretical developments in communitarianism and practical policy proposals.

Concerns about the demise of community, and the vision of recovery, have influenced urban planning. Many planners and urban scholars in the NEW URBANISM tradition are strongly influenced by the communitarian theoretical work. Neo-traditional planners seek to develop dense, mixed-use, pedestrian-friendly developments, arguing that such spaces are more supportive of social interaction and public life, and therefore of community. Often called the new urbanism, this planning movement emphasizes design elements such as pedestrian walkways, public space, public transportation and affordable housing. Some argue that new urbanism shares with other large-scale residential developments the further commodification of an image of community. The criticism, like others before, has done nothing to erode the appeal of the term 'community'.

Further reading

Anderson, B. (1983) *Imagined Communities: Reflections on the Origin and Spread of Nationalism*, London: Verso.

Gordon, I. and Low, M. (1998) 'Community, locality, and urban research', *European Planning Studies* 6(1): 5–15.

Putnam, R.D. (2000) *Bowling Alone: The Collapse and Revival of American Community*, New York: Simon & Schuster.

Williams, R. (1985) 'Community', in *Keywords: A Vocabulary of Culture and Society*, rev. edn, New York: Oxford University Press, pp. 75–6.

DOREEN J. MATTINGLY

COMMUNITY CONFLICT RESOLUTION

The term 'community' encompasses groups that share interests rooted in a common culture, history or profession, in belonging to the same institutions, or living in the same space. These interests affect, and are affected by, public decisions such as those resulting from planning processes. Context is key to understanding both conflict dynamics and alternatives to its management. For example, interpersonal disputes within a professional community may require a different approach than conflict among segments of a place-based community sharing URBAN space. The focus here will be on *place-based* community conflicts.

Community conflicts range in complexity and intensity from low to high, with differing consequences for community members. At one end of this range are interpersonal disputes fairly limited in scope, about behaviours around shared spaces: noise management, gardening practices, trespassing, pet care or parking habits. Duration, contentiousness and scale increase in landlord–tenant disputes, youth groups and gang fights, disputes over externalities of industrial facilities, conflicts over the handling of neighbourhood eyesores, demolition of cherished schools or churches, and reuse of buildings and BROWNFIELDS. At the other end of this range are long-term, high-stakes conflicts about costly policy and planning decisions fraught with uncertainty and issues of justice, engulfing entire communities: the management of traffic patterns, the location of amenities or of locally unwanted facilities such as affordable housing, homeless shelters, jails and landfills, economic development policies, environmental remediation and the protection of sensitive environmental assets, and education policies. At the interpersonal level, conflict resolution is often possible. Community-wide conflicts resist resolution for long time periods. Intervention efforts aim to manage the situation and reduce the likelihood of destructive, escalatory moves and consequences while parties search for solutions.

Conflict management approaches and procedures differ in suitability to specific situations, goals sought, level of resources and specialized skills necessary to support them, and locus of responsibility (Deutsch and Coleman 2000). Some procedures, such as interpersonal mediation or variants of arbitration by respected community members, are as old as humanity. Others – facilitated negotiations and consensus-building (Susskind *et al.* 1999) – are of relatively recent vintage, a cultural outcrop of democratic societies that value and can support deliberative processes (see DIRECT DEMOCRACY). The former are still widely used for resolving the more tractable community disputes (see ALTERNATIVE DISPUTE RESOLUTION). The latter have evolved in response to challenges posed by participatory decision environments where individuals exercising civic and legal rights can hamper or prevent the implementation of public decisions, effectively entrenching the status quo by default although at least some segments of the community want change (Susskind and Cruikshank 1987). Therefore, conflict management processes enabling change are all the more important in the realm of public decisions, whose consequences may linger beyond the lifespan of the generation responsible for them.

What, then, are the options for assisting segments of place-based communities in finding mutually acceptable solutions to protracted, complex, long-term conflict? It is not unusual for such conflicts to be waged in several venues simultaneously. Since there are no clear, set procedures for engaging opponents, a frequent first move by an aggrieved party is filing a lawsuit. However, legal remedies are typically costly, and not the ideal forum for policy decisions, especially when heavily technical content is involved. If all active parties come to expect a negotiated solution to be more advantageous to their interests than a legal remedy, they will agree to negotiate directly or assisted by interveners – mediators, facilitators, or community members with widespread credibility and skills. Depending on the situation, the ensuing deliberative process can last for weeks to years, at times resolving dispute episodes while leaving the underlying conflict to fester.

Conflict resolution processes can vary considerably from one case to another, depending on

number of parties involved, complexity of issues, history of relationships, importance of stakes, and other factors. However, a convening step is common, aiming to identify who should be at the negotiation table and what their key interests are, what critical issues need to be discussed, what procedures are acceptable to participants, whether and what kind of intervention is necessary, and how likely it is to arrive at a negotiated solution. The parties at the table then proceed to acquaint themselves with each other and with the issues, and negotiate, exchanging information and engaging in problem-solving. A sometimes extended period of time is necessary for learning, building trust and exploring alternatives but can also undermine resolution: during the months of negotiation, the political and institutional context and even the parties at the table may change, making it necessary to backtrack and rebuild the trust and procedural agreements and to renew commitments. The negotiating parties may choose among a number of decision rules, including consensus. An important component of conflict management processes is evaluation, not only summative (at the end) but also formative, informing the parties as the process unfolds and enabling them to make the necessary mid-course corrections.

For deliberative processes to work in moving the community towards implementable change, civic capacity to negotiate competently on their own behalf is critical for the feuding parties. A community group has civic capacity if it is able to engage effectively in decision-making, internally as well as with other community groups and with the relevant political and institutional structure (Elliott and Kaufman 2003). Thus, internally, groups need to rally, formulate goals and strategies, and act in concert, by resolving dissension as it arises. Externally, the group needs to garner resources and information, identify allies in the community, and understand sufficiently the existing institutional and political processes and channels to work through them to their advantage rather than despite them. The civic capacity of community groups engaged in conflict is important to themselves and, paradoxically, to their opponents, since negotiating with a competent counterpart is preferable, especially when implementing the agreements. A party whose acquiescence is rooted in misunderstandings, feelings of intimidation, or perception of poor self-efficacy is more likely to default or otherwise

sabotage implementation than if agreeing to a solution because it appears to serve its interests.

Further reading

Deutsch, M. and Coleman, P.T. (eds) (2000) *The Handbook of Conflict Resolution*, San Francisco: Jossey-Bass.

Elliott, M. and Kaufman, S. (2003) 'Building civic capacity to manage environmental quality', *Environmental Practice* 5(3): 265–72.

Susskind, L. and Cruikshank, J. (1987) *Breaking the Impasse*, New York: Basic Books.

Susskind, L., McKearnan, S. and Thomas-Larner, J. (eds) (1999) *The Consensus Building Handbook: A Comprehensive Guide to Reaching Agreement*, Thousand Oaks, CA: Sage.

SANDA KAUFMAN

COMMUNITY DEVELOPMENT BLOCK GRANT

After more than a quarter of a century, the Community Development Block Grant (CDBG) programme is the single largest direct URBAN aid programme in the United States, providing assistance that allows local governments to expand housing, public services, minority entrepreneurship, and job training and related activities. CDBG monies are a major element in the affordable housing efforts of many communities. The CDBG programme dispenses aid to over 1,000 entitlement communities (with a population of 50,000 or greater) according to a formula that considers such factors as local population, poverty, growth lag, and the age and overcrowding of housing.

The CDBG programme was established in 1974 as part of the consolidation that merged a number of urban categorical grants – MODEL CITIES, urban renewal and five smaller urban programmes, including urban parks and sewer monies – into a single, more flexible block grant. President Richard Nixon saw the creation of the Community Development Block Grant as part of his NEW FEDERALISM initiative to give sub-national governments greater flexibility in the use of programme monies. But Democrats in Congress did not totally trust local governments to meet the needs of the poor. The result was a compromise. Overall, the number of regulations governing local use of aid money were greatly reduced and recipient governments gained considerable new programme discretion. Still, Congress added certain 'strings' or conditions

establishing programme priorities – that the money be used to eliminate SLUMS and BLIGHT, to aid low- and moderate-income families, and to meet urgent community needs.

Much of the history of the CDBG programme has reflected a continuing tension between the programme goals of increasing local discretion and ensuring that the needs of the poor are being met. Republican administrations generally relaxed the citizen participation requirement and loosened other grant rules and regulations; Democratic administrations generally tightened programme targeting. The Republican Richard Nixon and Gerald Ford administrations largely took a hands-off approach, deferring to the decisions made by local elected officials. Democrats saw such deference as an abdication of the social policy goals inherent in the programme. The Democratic Jimmy Carter administration required recipient governments to use at least 75 per cent (by the year 2002 the requirement stood at 70 per cent) of their community development funds on projects that served low- and moderate-income neighbour-hoods. The Republican Ronald Reagan administration soon returned to a philosophy of maximum programme DECENTRALIZATION. The number of pages specifying how local governments were to spend their CDBG funds was reduced from 52 to just 2. Local governments were seen as entitled to the programme monies as determined by the aid formula. The Reagan administration also virtually abandoned any notion of a federal review of, and veto over, local spending plans. Municipalities were only required to file bare-bones spending plans with the federal Department of Housing and Urban Development (HUD); they did not need to gain HUD's approval.

While many cities used their community development monies responsibly, problems occurred when cities spent money on programmes that were popular with the local electorate and the local growth coalition interests rather than on projects aimed at neighbourhoods in need. In some instances, local officials used CDBG funds to build new tennis courts in affluent areas of town and to help pay for such economic development projects as a new boat marina. Although Democratic administrations were less willing than Republicans to approve of such uses, over time great deference has been accorded to sub-national priorities. Municipal elected officials gained power as a result of CDBG decentralization.

In allocating their CDBG funds, cities tended to chose physical development over social service spending. There was also a tendency for local officials to spread community development monies widely, with each local council member acting to ensure that a portion of CDBG funds was spent on projects in his constituency. On the whole, CDBG spending was less redistributive than were the old categorical programmes that it had replaced. Compared to the categorical grants of the GREAT SOCIETY, the CDBG programme was also less well targeted as it allowed a greater spread of urban spending to suburban and non-urban communities.

The Reagan administration, hostile to the notion of continued federal urban aid, slashed the level of CDBG assistance from US$6.1 billion in 1980 to US$2.8 billion in 1990. For the financial year 2000/01, the CDBG programme was budgeted at US$5.1 billion.

Further reading

Dommel, P. (1982) *Decentralizing Urban Policy*, Washington, DC: The Brookings Institution Press.
Hays, A. (1995) *The Federal Government and Urban Housing*, 2nd edn, Albany, NY: State University of New York Press.
Kettl, D.F. (1987) *The Regulation of American Feder-alism*, Baltimore: Johns Hopkins University Press.

MYRON A. LEVINE

COMMUNITY DEVELOPMENT CORPORATION

A community development corporation (CDC) is a non-profit organization formed and controlled by community stakeholders with the purpose of revitalizing poor, often minority, communities. Most focus primarily on the provision of housing affordable to low- and moderate-income community residents, although they may also offer a range of economic development and social services. They are tax-exempt under Section 501(c)(3) of the United States Internal Revenue Code.

The earliest CDCs grew out of the CIVIL RIGHTS MOVEMENT and anti-poverty activism in low-income URBAN communities during the 1960s. Some began with church support and many received federal funding under the Equal Opportunity Act's Special Impact Program. During the

1970s, new CDCs often sprang from neighbourhood-based advocacy and protest against threats to NEIGHBOURHOODS, often from proposed highway projects or plans for expansion of universities and other large institutions. By the 1980s, the sharp fall in federal support for affordable housing was paralleled by an increase in funding from corporate and private philanthropy, resulting in a substantial net increase in the number of CDCs operating across the nation. During the 1990s, CDCs began working more comprehensively, often partnering with other local organizations to tackle a broad range of local needs.

While CDCs were initially concentrated in a handful of Northeastern and Midwestern cities, a 1989 survey of US cities of more than 100,000 people found CDCs operating in more than 95 per cent of them. Networks of CDCs exist in two-thirds of these cities. Among the best known are the Boston Housing Partnership, the Cleveland Housing Network and the Pittsburgh Partnership for Neighborhood Development. CDCs have also organized at the state level. By the late 1990s, 17 states and the District of Columbia had state-wide networks. These groups, whether city- or state-level, act as trade organizations, promoting the interests of the group in the policy-making process and serving as a conduit for funding from a variety of sources to individual CDCs.

The rise in the number and technical sophistication of CDCs has been facilitated by the support of a number of institutional backers. These 'intermediaries' include foundations, government-sponsored enterprises (Fannie Mae, for example), banks, training institutes, and information networks. Financial intermediaries, such as the Local Initiatives Support Corporation, package funds from for-profit investors (through tax credit programmes) and foundations and allocate them, along with technical assistance, to CDCs.

While it is difficult to measure the impact of the work of CDCs, due to the wide range of goals and contexts in which they operate, a 1998 census of CDCs estimated that they were responsible for the production of 550,000 units of affordable housing, the creation of 247,000 private sector jobs, the development of 71 million square feet of commercial and industrial real estate, and loans worth US$1.9 billion to 59,000 small and micro businesses. Qualitative studies underscore the difficult context for their work, the tensions that arise between their roles as neighbourhood advocates

and developers, and their longer-term success at stabilizing their neighbourhoods.

ELIZABETH J. MUELLER

COMMUNITY GOAL-SETTING

URBAN PLANNING and/or policy-making that follows a rational process begins with the selection of goals for the future, followed by an iterative process to identify, analyse and select among alternative means of achieving those goals. The selected plan is then the preferred means of attaining these. (Non-rational planning processes do not require that participants agree upon goals.) Goals are typically expressed in rather general terms, and should represent the aspirations of the client group for whom the plan is done. Thus, when a goal-oriented plan for the future of a community is prepared, the community members should normally define its aims. This is somewhat more complex than might at first appear.

As described above, 'COMMUNITY' is a synonym for the client group for whom the plan is being prepared. Thus, it might potentially be an URBAN region, a single municipality, a NEIGHBOURHOOD or a demographic sub-group. A good planning process will clearly define who is included, and who thus ought to have an opportunity for input into the planning process. In some cases, the community may be defined as all those with an interest in the outcome, sometimes referred to as 'stakeholders'. Different definitions of community will result in different goals, and thus different results, if the goals are met.

How the community is represented will also make a difference. Most processes devised to choose community goals do not involve every individual member of the community, but assume that one individual can represent a group. Yet formal elections for these representatives are almost never held. Thus, there is no guarantee that a Hispanic businesswoman 'represents' either Hispanics or the business community in any way.

Once the client group has been defined and representatives chosen or meetings thrown open for general participation, a variety of methods exist for ascertaining community aspirations. Different methods are appropriate under different circumstances, though normative theorists and consultants do not always acknowledge this. Theory

has shown that the process used to arrive at a decision affects the outcome. While legislative bodies typically follow strict procedural rules for making decisions, many planning processes are more informal. Some theorists argue for clearly mapping out the process in advance, so that participants can see how their input will be used, while others maintain that allowing the group to determine its own goal-setting and planning processes allows greater flexibility and thus more potential for creative solutions.

One very basic process attribute is the degree of agreement necessary in order for the group to make a decision, on goals or anything else. Planning processes commonly presume that decision will be by consensus, rather than by majority rule, for example. Complete consensus is difficult to obtain in a large or diverse group, posing a significant problem for public processes attempting to formulate community goals. This difficulty is one reason that community goals are commonly very general, as no agreement can be reached on highly specific aims. Other means of facilitating agreement, which may be pursued knowingly or inadvertently, include: selecting participants for their willingness to cooperate; limiting participation in some other manner; encouraging negotiation and compromise; using technical information or social pressure to persuade; and discouraging or ignoring dissent. Clearly, some of these techniques conflict with the spirit of community goal-setting. Small groups, groups that follow an organized hierarchy and groups with many shared values face far fewer problems agreeing upon goals. Thus, consensual goal-setting is usually easier for private sector firms, non-profit organizations and special interest groups than for public sector entities, particularly large, general-purpose governments.

Procedural techniques can help foster agreement within diverse communities. Some consultants specialize in these types of activities, running COMMUNITY VISIONING meetings or charrettes, or facilitating STRATEGIC PLANNING processes or discussions of goals. However, most communities will be able to identify a neutral person capable of leading such discussions as a volunteer, potentially using one or more of the many non-technical handbooks available. In fact, some advocates of community goal-setting pursue it primarily to develop indigenous community leaders and a social network oriented towards planning for the future.

Community goal-setting, like planning generally, can fail at implementation. Ascertaining whether a planning or policy-making process achieves the goals set by the community requires evaluation after sufficient time has passed for results to have been achieved. It is reasonable to compare the outcomes of a process to the goals originally established for that process, and usually *un*reasonable to expect it to achieve something different, imagined by the evaluators or others. Because goals are typically general, whether or not they have been achieved will commonly be determined at least in part by the evaluator's judgment, unless measurable objectives or unambiguous indicators of goal attainment were also created when the goals were originally formulated.

Further reading

Brooks, M.P. (2002) *Planning Theory for Practitioners*, Chicago: Planners Press.
Gans, H.J. (1991) 'The goal-oriented approach to planning', in *People, Plans and Policies*, New York: Columbia University Press, pp. 145–58.
Helling, A. (1998) 'Collaborative visioning: proceed with caution! Results from evaluating Atlanta's Vision 2020 project', *Journal of the American Planning Association* 64(3): 335–49.
Vogel, R.K. and Swanson, B.E. (1988) 'Setting agendas for community change: the community goal-setting strategy', *Journal of Urban Affairs* 10(1): 41–61.

AMY HELLING

COMMUNITY POLICING

Community policing is an approach to and philosophy of law enforcement that stresses the involvement of the police with the COMMUNITY they patrol and police. This approach emphasizes knowledge of, involvement with and increased accountability to the community and involves the use of more strategic, problem-solving, community-based approaches to crime detection, control and reduction. It is an approach which proponents maintain is an effective method of reducing crime and disorder and – for some more importantly – the fear of crime and promotes quality of life and community well-being through enhanced local democracy, community safety and crime prevention, and greater collaboration between the community and law enforcement agencies. In this sense it is, in some ways, an approach to law

enforcement which stresses the consensual ideal of policing based on idealized notions of community, policing, authority and civic life.

More pragmatically, this approach rests on the recognition – particularly in the light of the experiences of civil and urban unrest (e.g. Britain in the 1980s or the United States in the 1990s) – that in increasingly complex and diverse communities, in which more traditional forms of authority and order may have broken down, policing may be more effective (and less inflammatory) where law enforcement agencies work with the community and respond more directly to the community's concerns, issues and voices. It also rests on the realization that law enforcement agencies – whose clear-up and detection rates for crimes can be very low – cannot solve the problem of crime on their own – nor can detection or clear-up, in and of themselves, reduce public perceptions about and fear of crime and disorder. Community policing approaches include the strategic use of locally visible officers who are or become known to the community and – through the development of community knowledge and familiarity – enjoy the trust, confidence and support of local people. Such approaches may have considerable impact on considerations of operational policing and strategy including the level, timing and location of police patrols, the use of police vehicles, foot patrols, and crime and disorder priorities.

However, idealized notions of the effectiveness and value of community policing approaches can be challenged on a number of levels. For some, the reality of community policing is the extension of state control and surveillance through local police activity and operations. Furthermore, some question the narrowness and exclusivity of the notions of community that underpin this approach and highlight the absence of effective systems for consultation, accountability and democracy. Finally, although the community policing approach to law enforcement is seen as potentially highly valuable for effective policing, for improving police and community relations, and for enhancing community safety and crime prevention while reducing the fear of crime, it can – at the level of operational policing – conflict with more managerialist, target-driven approaches to policing and crime reduction which stress arrest and conviction rates and short-term responses to specific crimes or political objectives which can be seen to characterize police responses to certain crimes (e.g. street crime, or mugging, drug-related crime, burglary and car-theft) at certain times.

Further reading

Greene, J.R. and Mastrofski, S.D. (1991) *Community Policing: Rhetoric or Reality?*, New York: Praeger Press.

McLaughlin, E. (1994) *Community, Policing, and Accountability*, Aldershot: Gower.

Skogan, W.G. and Hartnet, S.M. (1997) *Community Policing: Chicago Style*, Oxford: Oxford University Press.

RICHARD HUGGINS

COMMUNITY POWER STRUCTURE

In the last half-century, scholars studying the governing politics – the community power structure – of cities have developed five separate explanations for *Who Governs?* (1961) or *Who Really Rules?* (1978) cities. Floyd Hunter, one of the earliest pioneers of community power research, developed the reputational approach to studying local decision-makers. In *Community Power Structure* (1953), Hunter studied Atlanta, Georgia, by asking a panel of knowledgeable Atlanta observers who exercised real power in Atlanta politics. In 1953, and again in a 1980s-era follow-up study, Hunter found the Atlanta community power structure to be a power elite drawn from the ranks of local corporations, banks and investment firms.

The power elite view was challenged by the pluralist view of Robert DAHL in his *Who Governs?* (1961) study of New Haven, Connecticut. Dahl interviewed observers of New Haven politics and studied key financial decisions by the mayor and city council. Dahl found New Haven to be a 'community of unequal', but more pluralist than elitist, since power was widely, if unevenly, shared. For example, Dahl found that low-income groups such as African-American residents of New Haven had influence in New Haven electoral politics even if they faced considerable discrimination in economic and social circles in New Haven.

Dahl's 1960s-era pluralist community power structure view of New Haven was challenged by sociologist G. William Domhoff in his 1970s re-analysis of Dahl's New Haven findings, *Who Really Rules?* (1978). Drawing on Dahl's interviews and further study, Domhoff concluded that,

like Atlanta, New Haven was ruled by a set of economic and political elites drawn from New Haven's 'higher circles'. Domhoff's 1970s-era 'higher circle' argument would in turn be supplanted by two, more recent elite community power structure theories, Harvey Molotch's GROWTH MACHINE theory in the 1980s and Clarence Stone's URBAN REGIME THEORY beginning in the late 1980s and continuing to the present day.

Further reading

Dahl, R. (1961) *Who Governs?*, New Haven: Yale University Press.

Hunter, F. (1953) *Community Power Structure*, Chapel Hill, NC: University of North Carolina Press.

Mills, C.W. (1956) *The Power Elite*, New York: Oxford University Press.

Waste, R.J. (1986) *Community Power*, Thousand Oaks, CA: Sage.

ROBERT J. WASTE

COMMUNITY VISIONING

Visioning is a process by which a community envisions the future it wants, and plans how to achieve it. Through public involvement, communities identify their purpose, core values and a desired vision of the future. The primary product of such an event is to guide subsequent planning efforts; it is the basis on which goals, objectives and policies are formulated. Community visioning focuses residents on possibilities rather than problems. It establishes a desired end state for a community and a vision of the future to strive towards. The visioning process is so important in many places that it is given its own event. A community may convene a special meeting, or a series of meetings, to develop a community vision. Depending on the detail desired, visioning may last four to six months and entail several community meetings, or last up to a year or more and involve numerous meetings and task forces.

Shipley and Newkirk (1998) see visioning as a way to return to the roots of planning when individuals had visions of place, such as LE CORBUSIER, Daniel BURNHAM, John NOLEN and Frank Lloyd WRIGHT. Rather than establish a vision of place through one individual's view, the visioning technique strives to incorporate broad public participation. In theory, a community vision occurs through a group process that tries to arrive at a consensus about the future of a place. Ideally, a community vision can direct and guide the creation of a comprehensive plan and from there the future growth and development of a place.

Many communities have used the technique to establish broader public participation on the direction a community should move in the future. In the United States, many cities in Oregon have used a visioning process, as have Chattanooga, Tennessee; Atlanta, Georgia; and Racine, Wisconsin. Several of these programmes are very similar. Probably no other state has been more involved in community visioning than the state of Oregon. The Oregon model has four basic steps: a community profile (where are we now?), a trend statement (where are we going?), a vision statement (where do we want to be?), and an action plan (how do we get there?). This model suggests that a target year be chosen that is at least ten but no more than twenty-five years in the future (Ames 1996). A simplified version of this model can be completed in six months or less, while a comprehensive version can take a year or more. The Missouri model focuses on the future of the community as a whole. The object is to focus on future possibilities rather than being limited by present or past problems. The centrepiece of the process is an 'Action Planning Workshop' that is devoted to formulating a vision and developing action plans to carry out the vision. The Arkansas model is similar to most of the others but tends to prompt participants about specific areas such as economic development, education, parks and recreation, etc. The Wisconsin model is similar to Missouri's since the visioning workshop is not focused on past trends or current problems. This model advocates for developing a community-wide vision and then developing visions based on thematic areas, such as housing, transportation and land use.

Community visioning has been used in many communities around the world, including, but not limited to, Curitiba, Brazil; Dublin, Ireland; Vancouver, Canada; Brisbane, Australia; and Kirklees Council, Yorkshire, England. In addition, a number of states in India have produced tourism vision statements. Lodi, Wisconsin, produced a vision in May 2000. It reads: 'In 2025, Lodi is a community that links the future with the past by recognizing the importance of history in growth and development. The center of our small town is a pedestrian-friendly main street that celebrates historical architecture, while our waterways and

surrounding vistas nourish the health and beauty of the valley.' Lodi's vision continues with statements in the following areas: land use and growth management, downtown revitalization, housing and historic preservation, community services and public works, business retention and expansion, natural resources, and parks and recreation. Visioning programmes have occurred in many places, whether in a small city of 2,000 people, a large city of several million or a large region encompassing many square miles.

Several studies have examined the experience of visioning programmes. These studies have identified many factors necessary for a successful visioning effort, such as impartial leadership, attention to detail, media attention and creating a process that is appropriate for the community. Other studies focused on the outcomes of visioning projects. Helling (1998) reports that the Atlanta metropolitan visioning project yielded few significant, immediate results, despite a US$4.4 million budget and an effective programme of community participation. The general ineffectiveness of this programme was caused by three factors: a focus on process rather than actions, a requirement for consensus without a method for achieving consensus, and little technical assistance from urban and regional planners. Chattanooga, in contrast, had a successful visioning process that resulted in over forty goal statements and specific activities and projects. The visioning process is considered a key catalyst that led to the city's rebirth. Translating a broad vision into specific actions remains a key challenge in visioning programmes. Despite the challenges, visioning programmes can galvanize a community into creating a new future.

Further reading

Ames, S.C. (1996) *A Guide to Community Visioning: Hands-on Information for Local Communities*, Portland, OR: Oregon Visions Project (rev. 1998).

Green, G., Haines, A. and Halebsky, S. (2000) *Building the Future: A Guide to Community Visioning*, Madison, WI: University of Wisconsin Extension Publications.

Helling, A. (1998) 'Collaborative visioning: proceed with caution! Results from evaluating Atlanta's Vision 2020 project', *Journal of the American Planning Association* 64(3): 335–49.

Shipley, R. and Newkirk, R. (1998) 'Visioning: did anybody see where it came from?', *Journal of Planning Literature* 12(4): 407–15.

ANNA L. HAINES

COMMUNITY-BASED FINANCING

Mainstream financial institutions are not charities; they lend money for low-risk projects at commercial interest rates. These lending practices are producing financial exclusion in which individuals and organizations are unable to access mainstream financial services. In extreme cases, the mainstream banks are withdrawing provision (see DISINVESTMENT) from poorer localities (see DIVIDED CITIES) leaving socially excluded communities behind who increasingly have limited access to affordable financial services. There are two processes at work here. First, the withdrawal of financial provision from parts of the city, and secondly, the exclusion of people and organizations from the mainstream financial system. Both processes are driven by bank restructuring as the financial institutions attempt to enhance their profitability by cherry-picking the wealthy and excluding the poor (see REDLINING; SOCIAL POLARIZATION).

These gaps in the availability of financial services have led to the creation of alternative forms of lending institution often organized by and for a local COMMUNITY. Such micro-credit institutions are being established on the understanding that the latent capacity of the poor for entrepreneurship would be encouraged by the availability of small-scale loans. This would make people more self-reliant and create employment opportunities. The most common form of micro-credit institution is the *credit union*. These are not-for-profit, financial cooperatives that offer low-cost financial services to their members. Credit unions are increasingly playing an important role in maintaining and enhancing community economic stability.

The first credit union was established in Heddesdorf (Germany) in 1869. Since then, the movement has developed into two distinct but related forms. In Europe credit unions developed into national cooperative banks, while in North America they became local, autonomous institutions whose membership was restricted to individuals who shared a common bond – living or working in a particular locality, working for a specific employer or following a particular occupation. The US model was often a response against loan sharks and moneylenders. This form of community finance institution has spread from North America to Europe.

Credit unions operate in 87 countries with a combined membership of over 85 million. They

are also playing an important role in the transition economies of Eastern Europe. In Poland, for example, in just four years, 220 credit unions were established with over 260,000 members. American community development credit unions are smaller than mainstream credit unions linked to occupations or employers. On average, each credit union has 1,300 members and US$2.4 million in assets. These institutions are professionally run financial institutions designed to become economically viable, but using a business model that highlights the social benefits that result from their lending activities and a lending policy that provides reasonably priced financial services to those who are excluded from other forms of banking provision.

Further reading

Arrossi, S., Bombarolo, F., Hardoy, J., Mitlin, D., Coscio, L. and Satterthwaite, D. (eds) (1994) *Funding Community Initiatives*, London: Earthscan Publications Ltd.

Guene, C. and Mayo, E. (eds) (2001) *Banking and Social Cohesion: Alternative Responses to a Global Market*, Charlbury: Jon Carpenter Publishing.

JOHN R. BRYSON

COMMUTERS

In the late nineteenth century, American railways began selling multiple-use tickets known as 'commutation tickets' because they commuted (exchanged) travellers' daily payments for weekly or monthly payments. In the twentieth century, the term 'commuter' was generalized to mean anyone who undertook the same journey regularly (particularly to and from work) regardless of his or her mode of transport.

Commuting is generally viewed as a consequence of the INDUSTRIAL REVOLUTION. Before 1800, most people lived and worked in the same place, whether that place was URBAN or (for the vast majority) rural. The Industrial Revolution brought specialization of work and workplaces, and removed most paid work from households and rural areas, relocating it to factories in urban areas.

Commuter railways initially allowed the wealthy to live in suburban areas with rural amenities while managing enterprises located in the urban

centre. At the end of the nineteenth century, electric streetcars (trams) enabled members of the middle class to commute from suburban residences. The proliferation of affordable public transport, roads and automobiles further democratized commuting during the twentieth century.

Commuting is much more than a transport phenomenon. Commuting patterns are used to define METROPOLITAN STATISTICAL AREAS by the US Census Bureau, and are integral to many other definitions of urban regions. These official definitions in turn fundamentally affect how metropolitan areas are studied, funded and governed.

Commuting has significant social and economic effects on communities. Living in one jurisdiction while working in another divides the time and loyalties of the commuter and may reduce political and civic participation at both ends. The commuter is likely to pay the majority of local taxes (e.g. property-based assessments) in the community of residence, while still requiring public services in the community where he or she works.

There are also considerable environmental consequences to commuting. The severity of these environmental impacts depends on the modes of transport used for commuting. The environmental costs of automobile commuting are generally greater than for public transport, due to higher per capita fuel use, pollution emissions and infrastructure costs. The twentieth century witnessed a growing dependence upon automobiles for commuting (by the 1990s, nine out of ten US commuters used automobiles, most driving alone). Intervening communities, i.e. where the commuter neither lives nor works, often bear the brunt of the environmental impacts of commuting.

This rise in automobile commuting has been both a cause and a consequence of the DECENTRALIZATION of employment in metropolitan areas. Almost universally, the magnitude of suburb-to-city commuting is declining relative to suburb-to-suburb and other forms of peripheral commuting. Traffic CONGESTION, once a city-centre phenomenon, is now common in many peripheral locales.

Solutions to problems posed by commuting are by no means limited to transportation infrastructure, particularly since faster forms of transport have historically led to longer commutes. URBAN PLANNING schemes and technology (e.g. TELECOMMUTING) that bring work and home closer together also have important roles to play.

Further reading

Downs, A. (1992) *Stuck in Traffic*, Washington, DC: The Brookings Institution Press.

RICHARD W. LEE

COMPACT DEVELOPMENT

Most meanings of compact development include medium to high built densities and continuous, contiguous development (apart from planned open spaces) to achieve sufficient thresholds for services and facilities and to enable efficient public transport, or movement on foot. Clear divisions between town and countryside are usually implicit. Two dominant forms of compact urban development have been proposed: concentrated dense cities where further development is contained through intensification within existing boundaries, and decentralized, interlinked compact settlements.

Compact development has been an important theme in URBAN PLANNING. Nineteenth-century utopian planning proposals such as HOWARD's GARDEN CITIES embodied decentrist approaches, while LE CORBUSIER's later vision of high-rise, high-density urban living represented a modernist version of centrist compact development. Despite differences in conception, both were reactions against market-driven SPRAWL. Both were criticized as anti-city in the 1960s by Jane JACOBS and the urbanists who argued for compact cities based on an appreciation of older (European) cities with their characteristic mix of uses, high-density living, urban vitality, sociability, pavement life and pedestrian orientation. This version of centrist compact development prevailed after the 1960s, but Dantzig and Saaty's 1975 high-tech compact city recalled modernist approaches.

The 1970s fiscal crisis focused attention on links between urban form, dependence on the motor car, energy use and pressures on resources. Several studies linked compact development to reduced infrastructure costs, lower travel costs, better public transport, reduced energy consumption, more efficient use of land and greater accessibility for the poor. Advocacy of sustainable urban development, following the 1987 Bruntland Commission and the 1993 Local Agenda 21, fuelled interest in compact development internationally. NEW URBANISM and design models such as

TRANSIT-ORIENTED DEVELOPMENT and traditional neighbourhood development embraced these ideas. Compaction became a common theme within urban policy, for example Australia's urban consolidation policy, Europe's compact cities and the United States' SMART GROWTH movement. Policies included URBAN REGENERATION, INFILL DEVELOPMENT, densification and REDEVELOPMENT within the existing fabric, particularly along public transport routes, and new compact settlements.

Compact development received less attention in developing countries, where rapid URBANIZATION and informality in the late twentieth century frequently underpinned 'spontaneous' forms of compact cities: dense, mixed and vital, but rather environmentally degraded. The lateral spread of settlements, however, became a concern. Compaction policies were adopted in some places previously characterized by sprawl, such as South Africa, and Brazil's Curitiba.

The merits of compact development were extensively debated from the 1970s. Critics questioned the claimed environmental, transport and cost benefits of compact development, and argued that compaction was contrary to market forces towards sprawl, the DECENTRALIZATION of work and residents' desires. Debates focused largely on developed-country contexts and centrist approaches, but attention shifted to the merits of centrist versus decentrist compact development in the 1990s.

Further reading

Breheny, M. (1996) 'Centrists, decentrists and compromisers: views on the future of urban form', in M. Jenks, E. Burton and K. Williams (eds) *The Compact City: A Sustainable Urban Form?*, London: E&FN Spon.

SEE ALSO: sprawl

ALISON TODES

CONDOMINIUM

The term 'condominium' denotes a system of real estate ownership combining a title to an individually owned unit – usually an apartment or townhouse – with an undivided interest in the building, the land and its improvements. Before attaining

legal status in the United States in the 1960s, this form of tenure was well established in European cities and in countries with a tradition of apartment living. In the nineteenth century, it was enshrined in the Napoleonic Civil Code (Article 664), providing the French with a legal basis for the combination of individual and joint ownership. This form of tenure gradually spread to other European and Latin American countries whose legal system was based on the Civil Code. American developers became acquainted with it in Puerto Rico and brought the concept to Florida, but to realize its growth potential in the United States, there was a need for enabling legislation. Such legislation would set standards for the title description, specifying what is allowed, thereby making transactions in the property market more transparent. At the same time, it had to deal with a variety of developer abuses by closing loopholes in condominium statutes (HUD 1975).

The condominium tenure gained a foothold in the United States after 1961, when the real estate industry convinced Congress to amend the National Housing Act. The amendment (Section 234) authorized the Federal Housing Administration to extend its mortgage insurance programme to cover individually owned units in multi-unit structures in those states where this tenure was legally recognized. Every state of the union adopted enabling legislation in the 1960s. It was promoted by developers who had seen the lucrative advantages of dealing in horizontal property. In the following years, the real estate industry made a major effort to familiarize its membership and the public with the advantages of this new tenure. In fact, the growth of the condominium sector was primarily supply-driven (Van Weesep 1986). By 1968, every state and the District of Columbia had passed a Horizontal Property Act.

Since then, the term 'condominium' – 'condo' for short – has become a household word in the United States. At first, condos were mainly found in resort-type communities in the South and the West. But the sector's explosive growth through new construction and the conversion of existing rental properties – from 85,000 units in 1970 to over 2 million in 1980 – and its spread across the country and to all large cities turned this foreign concept into a familiar form of home ownership (HUD 1980). To sustain the demand, the financial side of the industry – notably lawyers, mortgage bankers and realtors – popularized the

condominium by devising specific legal and financial instruments.

One of these, a model condominium declaration, was widely used early on, thanks to developers of apartment buildings and planned-unit developments. With a uniform legal title, the marketability of the units was greatly improved. Demand for this type of housing grew as the combined effect of overall economic growth, the shift in the spatial economy towards the South and the West, lifestyle change, and retirement migration to the Sunbelt. As part of a condominium association, people had access to the recreational amenities they desired but could rarely afford as individual owners.

Once the condominium system of ownership had proven its worth in the residential sector, it was applied to other forms of real estate where it was advantageous to combine individual ownership and joint control over additional facilities and services. Professional offices, boat marinas and recreational centres represent some forms of real estate that now make up the sector known as condominiums.

Further reading

HUD (1975) *HUD Condominium/Cooperative Study*, vol. 1, *National Evaluation*, HUD – PDR – 111–1, July 1975, Washington, DC: US Department of Housing and Urban Development.
—— (1980) *The Conversion of Rental Housing to Condominiums and Cooperatives: A National Study of Scope, Causes and Impacts*, HUD – PDR – 554(2), Washington, DC: US Department of Housing and Urban Development, Division of Policy Studies.
van Weesep, Jan (1986) 'The creation of a new housing sector: condominiums in the United States', *Housing Studies* 2: 122–33.

JAN VAN WEESEP

CONGESTION

Traffic congestion on URBAN road networks has been an increasingly problematic characteristic of cities since the 1950s. Congestion is defined as the time lost per vehicle per kilometre and is calculated by comparing average journey times for routes with those under congested conditions. Car ownership has expanded rapidly since the early 1950s, a period of population growth and economic expansion in the West; by 1990, the private car accounted for over 85 per cent of all journeys in the

United States and over 70 per cent in Western Europe. During this time, local expansions in road network capacity have almost universally been exceeded by demand, leading to conditions of frequent and extreme congestion. The SUB-URBANIZATION and dispersal of residential, leisure and commercial land uses in the West have made MASS TRANSIT systems less viable and attractive relative to the car. The problem is made worse by the concentration of daily COMMUTER trips into peak 'rush hour' periods.

While car traffic produces a range of negative impacts in itself, in the majority of cases these are exacerbated when traffic is congested. In addition to the inconvenience caused by congestion, it leads to increased levels of noise, pollution and greenhouse gas emission and a reduction in air quality, all of which contribute to a general reduction in quality of life. The economic impacts of congestion include costs related to both lost time and fuel and reductions in worker productivity. These costs are felt by individual commuters, by companies operating in congested areas, and by cities, regions and at a national level. For example, congestion has been cited as a factor hindering the ability of cities to attract businesses and residents, while at a national level the economic costs of congestion in the late 1990s have been estimated to have exceeded US$180 billion per annum in the United States and US$30 billion in the United Kingdom.

Central government policy in the early twenty-first century has become increasingly designed to alleviate the negative impacts of congestion rather than to reduce the numbers of cars on the road per se. Policy measures include expanding road capacity and encouraging better use of existing capacity, for example through the employment of INTELLIGENT TRANSPORT SYSTEMS. However, critics have argued that such approaches are insufficiently integrated, as they fail to reduce the demand for peak-hour car travel or provide practical alternatives to the car. Alternative solutions have advocated changing work practices, for example by promoting flexitime and working from home, and the promotion of more sustainable modes of travel such as cycling and car sharing, along with expanding and improving mass transit systems. While these offer some long-term hope, in some severely affected cities local authorities have sought to deter car travel to congested areas through the introduction of congestion charges.

Further reading

Brotchie, J., Batty, M., Blakely, E., Hall, P. and Newton, P. (eds) (1995) *Cities in Competition: Productive and Sustainable Cities for the 21st Century*, Melbourne: Longman Australia (chs 17–22).

Buchanan, C. (1963) *Traffic in Towns*, Harmondsworth: Penguin.

TIM HALL

CONURBATION

The term 'conurbation' applies to URBAN settlements that have grown together to form a large built-up cluster of urban SPRAWL, a network of merged urban communities possibly dominating a region that has arisen due to population growth and spatial expansion. The origin of the word is associated with the work of Scottish biologist/sociologist/URBAN PLANNING pioneer Patrick GEDDES, who first coined it circa 1915. Geddes, known today for being a founder of modern town and REGIONAL PLANNING, first applied the term while examining the urban development of many existing places during the nineteenth century and employed it as a means to suggest the possible emergence of large super-cities in the future, which he believed would be formed when existing cities grew so big that they would merge with neighbouring ones. Geddes, when he put forward the term, was thinking of existing British urban settlements such as Liverpool and Manchester and Glasgow and Edinburgh joining together, and due to Geddes's term being closely connected with urban growth and the spatial extent of settlements, it is often considered as a synonym of urban sprawl. In the modern context, conurbations exist all over the world and examples of large-sized conurbations include the West Midlands (Birmingham) region of England, the Cologne region in Germany and the Paris metropolitan area in France. Most modern, industrialized nations have conurbations of some form and these places may also include those settlements that are today known as MEGACITIES, world cities or GLOBAL CITIES, places such as London, Los Angeles, Tokyo and New York.

As noted previously, the origin of the word 'conurbation' is associated with the urban studies work of Patrick Geddes. In his work, such as that included in the classic book *Cities in Evolution*, Geddes applied the term in relation to urban

centres of national and international importance, centres of trade, finance, industry, political power, education and culture, that had grown in size to encompass formerly isolated nearby places and which Geddes predicted would continue to grow until neighbouring cities had adjoined with each other. If Glasgow in Scotland is taken as a modern example of a conurbation, this once spatially independent settlement grew in the nineteenth and twentieth centuries to such an extent that its sprawl joined with nearby places like Clydebank to the west and Paisley to the south, forming what is today an urban region. Some might argue that the Greater Glasgow area is a conurbation when using the term to mean a predominantly urban region including adjacent towns and SUBURBS, i.e. a metropolitan area. Thus, towns such as Hamilton and Motherwell which are not yet part of the Glasgow sprawl become incorporated into the conurbation as they are located close to the city and are within what is considered the city's metropolitan area. However, a metropolitan district usually combines a conurbation proper, a large built-up area, with peripheral districts or settlements which may not necessarily be urban in character yet are closely tied to the conurbation due to matters of commerce and employment.

Further reading

Geddes, P. (1915) *Cities in Evolution: An Introduction to the Town Planning Movement and the Study of Civics*, London: Williams & Norgate.

IAN MORLEY

CONVENTION CENTRES

A convention is a type of general or formal meeting. Its participants can be members of a legislative body, or of a social or economic group. A convention has set objectives and its aim is to provide information on a particular situation in order to deliberate and establish consent on policies. Its main temporal characteristics are limited duration as well as no determined frequency. The term 'convention' is widely used, particularly by American associations, to describe the traditional form of total membership meetings. The leading conference city for international association meetings has been, for eighteen years, Paris (249 meetings in 1997). The leading country is the United States (1,054 meetings in 1997), while the leading US city is Washington DC (100 meetings in 1997).

Conventions usually take place in convention centres, congress centres, convention hotels, public halls or arts centres (e.g. Bridgewater Hall, Manchester). Two of the most impressive convention hotels are Jumeirah Beach Hotel and Conference Centre, Dubai (by W.S. Atkins), and the Convention and Events Centre, Estel Hotel, Berlin. There are also some special cases of convention centres that are part of THEME PLACES: the most prominent example is Disneyland Paris, but another interesting example is the Corfu Grecotel Congress Centre and Resort in Greece. Exhibitions or trade fairs increasingly accompany conventions, and this is the main reason for the construction of combined convention and exhibition centres (e.g. Singapore International Convention and Exhibition Centre, Scottish Exhibition and Conference Centre in Glasgow, Sydney Convention and Exhibition Centre, Darling Harbour) or convention and trade fair centres (e.g. Congress Center Messe Frankfurt). Convention centres can also be combined with concerts (e.g. Le Palais de la Musique et des Congrès in Strasbourg, Corum Conference Centre in Montpelier). Furthermore, there are examples of cities that have a combined infrastructure, for example Birmingham (International Convention Centre, National Exhibition Centre, National Indoor Arena) and Manchester (G-Mex Centre/MICC, Bridgewater Hall).

Conventions contribute to urban development in various ways that can be transformed into planning goals: revitalization of declining cities, extension of the tourist season, comprehensive reconstruction of large depressed downtown areas, and improvement of the competitiveness of the city and its region. It can be argued that special cases exist, such as convention cities (Atlanta) and convention districts in some cities (Boston).

In some cases, the architectural result is impressive (e.g. Pacifico Yokohama by N. Sekkei, Valencia Congress Centre by Foster and Partners). However, in most cases, the originality of the architectural design is usually neglected, especially in comparison with other buildings in the same city, for example the Guggenheim Museum in comparison with the convention centre in Bilbao (since attending conferences is usually combined with visiting museums). In recent years, as in the case of FESTIVALS, the importance of TOURISM has

grown considerably. Conventions belong to incentive travel and particularly business tourism.

Further reading

Lawson, F. (2000) *Congress, Convention and Exhibition Facilities*, Oxford: Architectural Press.
Swarbrooke, J. and Horner, S. (2001) *Business Travel and Tourism*, Oxford: Butterworth-Heinemann.

ALEX DEFFNER

COOPERATIVE HOUSING

Cooperative housing refers to various living arrangements that contrast sharply with other housing in terms of resident control. Proponents have argued that it is collective resident control that enables the creation of innovative environments oriented to women's needs, income mixing, universal design, affordability, community development, cultural appropriateness and other issues. Cooperatives have been associated, both within and alongside public policy, with aspects of urban development such as expansion of the housing stock, provision of low-cost housing, transfer of under-invested stock from the privately rented sector, neighbourhood stabilization and transfer of rental stock from public housing.

Cooperatives in housing have been traced to eighteenth-century France, though it is recognized that the cooperative movement emerged in strength in Britain in the nineteenth century. It was in that setting that the Rochdale Principles, which underlie the broader cooperative movement, were recorded:

- democratic resident control on the basis of one member, one vote;
- open and voluntary membership;
- limited returns on investment and the return of surplus to members;
- open disclosure, active participation and continuing education;
- expansion of services to members and to the community; and
- cooperation among cooperatives.

Adhering to some or all of the Rochdale Principles, cooperatives have taken different forms in different circumstances, accounting for about 20 per cent of national housing stock in Scandinavia and under 1 per cent in North America and Britain. Common to all forms is the distinguishing feature that members own a share in the cooperative organization but not an individual dwelling unit; membership of the organization confers the right to occupy the unit. Members are usually engaged in management of aspects of their organizations, such as resident recruitment and training, building maintenance, and finance.

An early form is the building cooperative, incorporated for the purpose of producing housing for group members, who combine resources such as personal cash, sweat equity, government loans and grants, and money raised on lending markets. When all members' housing has been built and paid for, the cooperative 'privatizes' (dissolves) and members take freehold possession of their housing. Therefore, the term 'terminating cooperatives' is also used.

The co-ownership society is the form in which the cooperative is established by a third party such as a union, commercial developer or dedicated cooperative sponsor. Members buy a share equivalent to a portion of development costs and take a lease for the balance. Where property values rise, share values appreciate and thus members' capital also appreciates. Co-ownership societies were promoted in Britain in the 1970s for their emulation of owner occupation, a point which has drawn attention to cooperatives in the United States as well. However, success was limited and many societies privatized. This form has nevertheless been an important component of housing provision in Scandinavia.

In limited equity cooperatives (LECs) the by-laws of the organization include provisions restricting share resale values, thereby dampening speculative gains. Where the contribution of a non-profit sponsor such as a trade union, credit union or government lowers members' equity requirement, the term 'co-partnership cooperative' is used. Where members' equity has nominal value, the terms 'par value' or 'zero-equity cooperative' apply. LECs have been used in the transfer of housing across tenures, with public programmes providing differing levels of public financial and organizational support for renovations and repairs. In the United States, LECs have been formed in landlord-abandoned housing, and in both the United States and Britain they have been formed in public housing. Par value cooperatives were part of the SOCIAL HOUSING expansion in Canada, where they are also known as continuing

cooperatives, until the programme freeze in the early 1990s.

In tenant management cooperatives the tenant group handles management and maintenance on behalf of a landlord, such as a municipal or private non-profit corporation, which retains ownership. This form has been used in transitional situations in Britain, for example in local authority housing awaiting clearance or upgrading.

Secondary cooperatives are federations formed, typically on a regional basis, for purposes such as pooling resources, representing needs to government and providing technical assistance. In the widely adopted 'Swedish model' an umbrella organization takes the lead in recruiting members, forming new cooperatives and providing them with organizational, technical and sometimes financial support. Under this model, the secondary cooperative may also provide services on a fee basis to ongoing cooperatives. Regional secondary cooperatives may form a tertiary cooperative serving the nation. The development of upper-tier cooperatives has been recognized as an essential resource for the success of the individual cooperatives.

Collective resident control has been recognized as a key to the creation of residential environments different from those produced in social housing and by HOUSING MARKETS. In comparison with renters in the public and private sectors, cooperative members are clearly able to determine many aspects of the use and occupancy of their dwellings because they formally control these issues through the cooperative. Control at the level of the individual dwelling may not be as great as in owner occupation. However, in comparison with both renters and individual owner occupants, cooperative members may have greater influence on the local environment through their joint action.

Some commentators have argued that particular aspects of cooperatives – for example buffering members from housing markets, providing members with strength through collective organization, or altering conventional notions of the household – have enabled new types of relations among people to emerge. Others have argued that substantial and durable social change in cooperatives has been muted. The prospect of creating alternative environments raises significant questions around agency and structure, as discussed by GIDDENS and others. This refers to the ability of individuals in cooperatives to act in ways that contradict social structures such as race, gender, ability and class that mediate relations among people in conventional forms of housing.

Further reading

Birchall, J. (1988) *Building Communities the Co-operative Way*, London: Routledge & Kegan Paul.
Heskin, A. and Leavitt, J. (1995) *The Hidden History of Housing Cooperatives*, Davis, CA: Center for Cooperatives.
Open House International 17(2) (1992), theme issue on collective and cooperative housing.

IAN SKELTON

COSTA, LUCIO
b. 27 February 1902, Toulon, France;
d. 13 June 1998, Rio de Janeiro, Brazil

Key works
(1937–43) The Ministry of Education and Health Building, Rio de Janeiro.
(1957) Pilot Plan, Brasília.

Brazilian architect, author and creator of the planned new capital of the country, Brasília, Lucio Costa graduated in architecture from the National School of Fine Arts, in Rio, where he was later appointed director, introducing the vanguard of modernist architecture. Leader of the team of young architects commissioned to design the Ministry of Education and Health Building (1937–43), a breakthrough in modern architecture, he arranged the visit of LE CORBUSIER to obtain his assessment of the project. Also in the team was architect Oscar Niemeyer, with whom Costa designed the Brazilian Pavilion for the New York World's Fair (1939). In the 1950s, he was invited to advise on the UNESCO building, Paris, and to lecture at the Conference of Venice. At this time, his project for the Guinle Park apartments in Rio was granted the São Paulo Biennial Award. He was successively invited to several assignments, including to act as a consultant for the restoration of Florence, Italy, and to provide the initial layout for the Brazilian House at the Cité Universitaire in Paris. In the United States, he was invited to visit the Parsons School (New York) and Yale and Cornell Universities; and to write *The New Scientific and Technological Humanism* to celebrate the Massachusetts Institute of Technology centennial

in 1961. He also researched extensively on Brazil's colonial architecture, and designed the Missions Museum while at the Institute for Historical and Artistic Heritage.

In 1957, his project won the competition to design the Pilot Plan for the new capital. Located on a desert plateau in the country's geographic centre, Brasília's fundamentals are linked to the modernist URBANISM concepts brought out by the avant-garde manifesto of the Congrès Internationaux d'Architecture Moderne (CIAM) (1930s). CIAM postulated that architecture and urbanism had an important role in modernizing society, and were believed to become agents for social change and economic development. Showing acute inventiveness, Costa was able to merge CIAM's principles with Brazilian urban-architectural patterns, thus establishing a paragon. A paradigm of modern urbanism, Brasília was included in UNESCO's World Cultural Heritage List in 1987.

Also an author, Costa's theoretical writings are a legacy for architecture and urbanism scholars and show his aptitude for combining the Brazilian cultural past with the dynamics of the present. His last work was a book of memories he published assisted by his daughter, Maria Elisa, also an architect.

Among his prestigious international tributes, he was awarded a Doctor Honoris Causa by Harvard University (1960) and France's Légion d'Honneur (1970). He was a member of the American Institute of Architects, the Royal Institute of British Architects and France's Académie d'Architecture.

Further reading

Costa, Lucio (1962) *Sobre Arquitetura* (*On Architecture*), Porto Alegre: Centro dos Estudantes de Arquitetura.
—— (1995) *Registro de uma Vivência* (*Record of a Life*), São Paulo: Empresa das Artes.

LINEU CASTELLO

COUNCIL HOUSING

Most governments have directly provided many homes in response to housing shortages and the poor health and sanitary standards of some privately rented housing. This is particularly so in the United Kingdom, where from the end of the First World War many homes were provided by local authorities otherwise known as 'councils', hence the name 'council housing'. At its peak in the mid-1970s, one-third of all households in the United Kingdom (about 6 million) lived in council houses. In 2001, 2.9 million households lived in council houses and a further 1.9 million in other public housing. In the United States, public housing in total accounts for only about 1 per cent of the stock.

Between 1918 and 1939, a million council houses were built in the United Kingdom by councils (10 per cent of the housing stock) using subsidies from central government. Much of this was used to rehouse families from SLUM clearance areas. National standards outlined in the Tudor Walters Report guaranteed decent space and inside amenities. Estates often followed a GARDEN CITY layout of about twelve houses to the acre, generally on the edge of cities. While council housing expanded quickly, various experiments with rent controls among other factors resulted in the steady decline of private rented housing. The driving force for the high level of direct public intervention in the housing market was the Fabian socialists who followed the ideas of Henry GEORGE on community ownership of the development value of land.

After the Second World War, a renewed effort on building council housing made much use of system-built techniques, and from the late 1950s high-density, high- and medium-rise flats. As much as 70 per cent of housing in the major cities was high-rise. Much of this housing has since been demolished because of problems of design, construction quality and management. Estates of mainly semi-detached homes were also built at this time under the Parker Morris standards, which provided for relatively generous space and amenities.

The term 'council housing' has now been replaced in most contexts with 'SOCIAL HOUSING', a change that reflects the much wider range of providers, primarily not-for-profit housing associations. Since the 1960s, these associations have received subsidies from the national Housing Corporation to provide 'affordable housing'. The Cullingworth Report of 1969 emphasized the need for councils to take a much broader view of housing, and the emphasis is now on councils as enablers of the provision of all types of housing, including special provision for the elderly and people with disabilities. Local authorities prepare housing strategies in consultation with all households, examining provision across all tenures.

Since 1980, government has pursued a policy which gives most council house tenants a 'right to buy' at very favourable terms: originally 50 per cent and later 60 per cent of the market value. Some councils have sold most of their housing or handed it over to other providers. Nevertheless, councils remain the most significant providers of social housing in the United Kingdom.

Further reading

Malpass, P. and Murie, A. (1999) *Housing Policy and Practice*, 5th edn, London: Macmillan.

Ravetz, A. (2001) *Council Housing and Culture: The History of a Social Experiment*, London: Routledge.

VINCENT NADIN

COURTYARDS

Courtyards are common elements in both Eastern and Western building patterns and have been included as a typical and traditional building feature by both ancient and contemporary architects alike. Several courtyards can be found within one building – each representing a different use by its owners. Each maintains a different spatial relationship from another courtyard within the same complex. Reynolds (2002) suggests that within Hispanic cultures, when a building contains one courtyard, it serves as a formal showcase for the owners, while at the same time accommodating the informal activities that occur in and around the courtyard. In a building containing two or more courtyards, the first is accessed from the street and serves as the parlour or entry to the building rather than nutritional or functional reasons. The second courtyard, located in close proximity to the kitchen or laundry area, is utilized as a functional space as a garden, play area or work area. In Middle Eastern Islamic culture, a courtyard is part of a women's domain and is a private part of the building not meant for public viewing. In northern Chinese architecture, the central courtyard (*Siheyuan*) often represents as much as 40 per cent of the total area of the building complex, and according to Knapp (1989) is often larger than any of the structures which make up the house. In current times, contemporary uses of older courtyards have expanded to include outdoor offices, restaurants, waiting rooms and living spaces, while

new courtyard designs are often modified through the enclosure of the space with a glass roof creating an atrium.

Further reading

Knapp, R.G. (1989) *China's Vernacular Architecture: House Form and Culture*, Honolulu: University of Hawaii Press.

Reynolds, J.S. (2002) *Courtyards: Aesthetic, Social, and Thermal Delight*, New York: John Wiley & Sons.

WENDY TINSLEY AND ROGER W. CAVES

CRIME

The relationship between the city as a social space and the actuality, fear and perception of crime is central to any understanding of the nature, organization and experience of the contemporary built environment and URBAN landscape. The fear and perception of crime, criminality and the 'criminal classes' has been reflected in the social ordering of urban space in the history of URBAN PLANNING and development throughout the world. The construction of the city as a dangerous and unruly place of criminal activity, of threat, of licentiousness and of disorder is readily reflected in popular literature and visual media, from early modern to contemporary representations of urban space. For some, the combined effects of crime and the fear of crime have been deeply corrosive of the economic, social and political fabric of the city in general and more specifically the urban ideal of democratic and civic virtue. For others, the relationship between crime, disorder and the social, economic and political values and processes within the city is altogether more ambiguous in terms of its meaning and outcomes. Thus, for some, the occurrence and fear of crime impact on the individual and collective response to, and use and experience of, urban space, while for others it is the design, experience and interpretation of urban space and the built environment that shapes our senses of fear, crime and risk. A further approach suggests that the fear and perception of crime (and consequently the impact such concerns have on anti-crime strategies and actions) are more about social and personal anxieties and diffuse quality of life issues than they are about crime itself. In this reading, the fear of crime is a kind of displaced anxiety about a range

of other fears and concerns generated by contemporary urban living.

The history of architectural, development and planning interventions within the built environment represents a series of attempts to deal with various and different actual and imagined threats of crime from different and diverse dangerous groups. Thus, commentators have highlighted the city walls of pre-modern cities as crucial for generating actual and psychological security from certain fears and crimes and for protecting the city and the citizen from the outsider. However, with the rise of the modern, urban, industrial and increasingly politicized mass cities of the nineteenth century, concerns focused more directly on the unruly masses as possible threats to individual and collective security, order and safety. HAUSSMANN's reconstruction of Paris is frequently cited as one such intervention directly driven by concerns about social order and disorder – albeit of a particular social and political form.

In the twentieth century, the urban sociology of the CHICAGO SCHOOL made a significant contribution to the analysis of urban crime and disorder through a number of important studies of urban life. For example, Park, Burgess and Mackenzie's 1925 study of Chicago presented a significant analysis of the dynamics of urban change and the phenomena of crime and social order through the notion of *zonal theory*. This approach highlighted the tendency in European and American cities for continued expansion into large metropolitan districts that can be divided up into different and relatively segregated concentric zones in which affluence flows from the core to the periphery. At the core of this model of the city was the business or factory zone. Immediately surrounding this zone was the zone in transition – in which successive waves of migrants would establish themselves between the factories and work opportunities of the core and the increasingly affluent and stable residential, outer zones. Within the transitional zone, life could be characterized as disorderly and unstable and crime was common – an area of social disorganization.

Later twentieth-century commentators have focused on the role crime and the fear of crime appear to have played in increasing the role of surveillance, policing and security within the highly sanitized and standardized urban spaces of SHOPPING MALLS, retail parks and leisure spaces of the contemporary urban environment. Such developments appear to demonstrate the close interplay between individual and collective concerns about personal safety and security and the political and economic concerns that difference, disorder and social complexity may impact negatively upon commercial profit, property prices and patterns of urban consumption. The experience and the fear of crime coupled with the rise of crime-prevention technologies and strategies have led to the emergence of the 'anxious' or 'frightened' city in which the fear of crime, concerns about public safety and the threat of the 'other' have become a crucial influence on political, social and economic processes and issues and an important factor in influencing how individuals organize their private and public lives, such as work and leisure activities. Furthermore, such concerns play a central role in influencing urban planning, security and policing activities and broader sweeps of public and social policy such as approaches to URBAN REGENERATION and environment management. Indeed, a number of commentators highlight concerns about the degree to which responses to crime in the form of community safety and crime prevention initiatives and concerns now dominate many other areas of the governance of the city. Such influences result in periodic programmes of coordinated response to such issues in the form of campaigns to create 'safer cities' and 'safer communities', with an increased targeting of specific crimes and crime-related activities and new and increased levels of public order legislation. Interestingly, such responses tend to focus not simply on actual crime but also on more diffuse and difficult-to-define notions of order and anti-social behaviour, including begging, street-drinking, noise generation, protest and drug use.

If, once, the idea of the city represented a relatively utopic vision of cosmopolitan civic virtue and prosperity, such a vision is long gone. Contemporary concern with and anxiety about crime, order and social disorganization is frequently reflected in relatively dystopic visions of the contemporary and future city in all media forms, from popular journalism, film and television to literary and musical commentary on the contemporary urban environment. For many commentators, the notion of a vibrant space of individual and collective social and political engagement within the urban public sphere has been replaced by a commodified space of individual and collective fear and institutionalized anxiety of crime.

Further reading

Davis, M. (1990) *City of Quartz*, London: Verso.
Ellin, N. (1996) *Postmodern Urbanism*, Oxford: Blackwell.
Park, R.E., Burgess, E.W. and MacKenzie, R.D. (1925) *The City*, Chicago: University of Chicago Press.
Pile, S., Brook, C. and Mooney, G. (eds) (2000) *Unruly Cities?*, London: Routledge.
Sennett, R. (1996) *The Uses of Disorder: Personal Identity and City Life*, London: Faber & Faber.

RICHARD HUGGINS

CRITICAL COALITION

The critical coalition (CC) is an approach to identify those interest groups *effecting* or *affected by* the impacts of a planning policy or a development action. They represent the community's needs, attitudes and values. They may also share common interests due to overlaps in their perceived problems and objectives relative to the impacts from the locational decisions of such policy or action. Through their participation in the decision-making process, the CC provides not only specific knowledge as to problems and objectives from various interest groups, but also a political base for the implementation of future policy actions to reinforce, maintain or redirect growth in an URBAN setting.

The CC is a policy science concept initially utilized in housing policy and growth management. It was subsequently transferred to support URBAN DESIGN, urban redevelopment and physical planning applications addressing growth, change and location.

Justifying the concept

Growth management policies and actions are made by and affect people representing special interests, from institutions to individuals. Therefore, the importance of including the CC in growth management or physical planning is basically twofold. First, it affords any locational decision-support approach with cultural, social or behavioural input for the establishment of priorities and importance weights representative of a group of people living in the particular spatial level being planned (region, city, district, NEIGHBOURHOOD or block). Secondly, it provides through their participation in the process the political base for the implementation of future policy actions to reinforce, maintain or redirect existing or needed physical change in an urban setting.

It is this array of entities and their specific agendas, from providers to users, that when brought into the locational decision-making process, contribute both the inherent 'wickedness' of physical planning and design problems, for example lack of evaluation criteria, no solutions only resolutions, etc. (Rittel and Webber 1973), as well as the distributed tacit knowledge necessary to resolve them. Therefore, identifying and understanding objectives and problems which bring these interest groups together as stakeholders around a particular action is central to any sensitive growth management method or tool (see SUSCEPTIBILITY-TO-CHANGE).

Identifying the players

Five steps need to be considered to identify and formalize the CC for its integration into a growth management application. First, the interest groups are identified in terms of those affected by or effecting the policies and actions for new development. This initial list can be quickly developed from sources such as a city planning department, a neighbourhood agency, the chamber of commerce of a municipality, neighbourhood groups or other sources such as phone directories.

Next, the problems impacting the quality of life and the objectives perceived by each interest group to improve or maintain the quality of life are identified in terms of the physical urban systems (existing or needed) and their relationships to other non-physical processes of the existing study setting. This is simply accomplished through a traditional survey instrument, mail questionnaires or interviews.

Thirdly, to prioritize problems a cross-tabulation of all perceived problems versus all identified interest groups is performed to yield a 'problem importance' rating. The rating can be specified as a function of both the 'problem's occurrence', i.e. how often a particular problem is shared by other groups, and an interest group's 'problem intensity', which is the summation of all problems perceived by the group as impacting their quality of life.

Fourthly, the prioritization of objectives follows a similar process, with the output being the 'objective importance', again, as a function of the 'objective's occurrence' across entities and the 'objective perception' of each stakeholder group.

In this manner, we can identify those interest groups who share a particular problem as the summation ($\sum IG_n$) of all interest groups (IG_a, IG_b, IG_c, ..., IG_n) perceiving to be impacted by a particular problem P_i. Similarly, we can identify those groups sharing a particular objective O_j. Thus, we are able to compute the importance of a particular problem or objective. On the other hand, we are able to interpolate the importance of each group by observing the occurrence summations of problems (P_1, P_2, P_3, ..., P_i) and objectives (O_1, O_2, O_3, ..., O_j) of the group UG_n. Both are measures instrumental in establishing the membership of a particular interest group in the CC.

The final step represents the identification of the CC. Once the most important problems and objectives are identified, the establishment of the CC is simply accomplished by the summation of overlaps for each of the problems and objectives ($\sum O_{PO}$) of each group with all the other stakeholders. A list of the most important problems and objectives can now be developed from the information obtained in the third and fourth steps by giving a qualitative value to the priority factors of both the problems and the objectives, for example a 'most important', 'relatively important' or 'unimportant' value. It is important to note that such values cannot be totally defined by quantitative or qualitative methods alone; rather, their validity hinges on the appropriateness of each method with respect to the particular development actions, spatial level or cultural context of the growth management situation.

In this manner, the CC's concept assures sensitivity to importance weights assigned to variables utilized when applying it to a location decision-analysis in growth management, urban design or physical planning.

Applications of the concept

The CC has been utilized in the United States and internationally. In policy science applications, the concept has been utilized in housing policy in Baltimore, Maryland (Grigsby and Rosenburg 1975), in some neighbourhood revitalization efforts such as the Cole Neighborhood in Denver, Colorado (Arias 1996), and in the growth management of some city-centre redevelopment efforts such as the growth management plan of Detroit (Wallace *et al.* 1979). In Latin America, this concept has also been proposed as an approach for physical planning and

urban development (Arias and Martinez 1981), and has been utilized in urban planning by the Programa Nacional de Desarrollo Urbano Sostenible (PRODUS; see www.produs.ucr.ac.cr) at the University of Costa Rica, in the development of various *planes reguladores* (growth management plans) of municipalities as well as in the urban design of districts of CENTRAL CITIES in Costa Rica.

Further reading

Arias, E.G. (1996) 'Bottom-up neighborhood revitalization: participatory decision support approaches and tools', *Urban Studies Journal – Special Issue on Housing Markets, Neighborhood Dynamics and Societal Goals* 33(10): 1831–48.

Arias, E.G. and Martinez, L. (1981) 'A conceptual decision-making model for the physical planning of urban development in Third World nations', in W.G. Vogt and M.H. Mickle (eds) *Modeling and Simulation*, vol. 12, Pittsburgh, PA: University of Pittsburgh, pp. 1011–19.

Grigsby, W.G. and Rosenburg, L. (1975) *Urban Housing Policy*, New York: APS Publications.

Rittel, H. and Webber, M. (1973) 'Dilemmas in a general theory of planning', *Policy Sciences* 4: 155–69.

Wallace, D.A., McHarg, I., Roberts, W. and Todd, T. (1979) 'Downtown Detroit redevelopment plan: 1979', City of Detroit, MI: WMRT technical report, Philadelphia, PA.

ERNESTO G. ARIAS

CROSS-BORDER REGION

Today, when people, goods and services move freely, the direct and indirect consequences of such movement are forced upon regions and URBAN policies. Because a region of a country that is adjacent to a border (a 'border region') is geographically situated in the periphery of the country, it does not receive much attention from a political and economic standpoint, especially economically, and is very often placed in a weak position. Nevertheless, borderless activities have an impact on such regions or cities. There are a number of instances where neighbouring regions of two countries adjacent to a border take advantage of geographical conditions to strengthen their competitiveness and form a cooperative relationship. Such a borderless relationship is called a 'cross-border cooperation', and a region in which such a relationship is established is known as a 'cross-border region'.

Borderless cooperative relationships have existed for a long time; an early example would be

fishing and sailing in lakes, marshes and rivers along borders. As the globalization of economies advances and people, goods and services move more freely, policies regarding these regions and cities and their improvement are gradually becoming important for central governments, local autonomous governments, international relationships among governments and international organizations. The purposes of the policies are varied. They include guaranteeing the distribution of commodities and the daily movement of people that cross the border of adjacent regions; activating the interchange of people in regions that have a homogeneous culture and language although a border separates them; controlling the pollution of lakes, marshes and rivers that straddle multiple countries; and preventing pollutants spreading widely. To resolve economic, political and environmental issues, a complex cooperative relationship between the regions, nations and international organizations will be established. It is characteristic of these relationships that they are established among local municipal governments rather than among nation-states, or among mutual regions rather than administrative jurisdictions.

There are diversified instances of linkage; from an economic point of view, one is the economic linkage between cities along the US–Mexican border such as San Diego and Tijuana, and the other is the establishment of the European Spatial Development Perspective by the European Union, which is a more extensive alliance among communities and nations and transcends borders. On the other hand, the political power of nation-states is still strong in world mentalities and political conflicts occur very often in border regions. But the nation-state no longer circumscribes our social reality, as globalization studies have illustrated or world systems studies have theorized.

Further reading

Aykac, A. (1994) 'Transborder regionalisation: an analysis of transborder cooperation structures in Western Europe within the context of European integration and decentralization towards regional and local governments', Libertas Paper 13, Europaisches Institut GmbH, Sindelfingen, Germany.

Friedmann, J. and Weaver, C. (1979) *Territory and Function: The Evolution of Regional Planning*, Berkeley and Los Angeles: University of California Press.

YUICHI TAKEUCHI

CULLEN, GORDON

b. 9 August 1914, Bradford, England;
d. 11 August 1994

Key works

(1957) 'Downtown is for people', in *The Exploding Metropolis*, special issue of *Fortune Magazine*, New York: Doubleday (with J. Jacobs and I. Nairn).
(1961) *Townscape*, London: The Architectural Press.

English architect who started the important 'townscape movement' in the 1960s and 1970s in Britain. Cullen presented us with a new theory and methodology for URBAN visual analysis and design which is based on the psychology of perception, such as on the human need for visual stimulation and the notions of time and space. His best-known technique is the 'serial vision': the experience of the CITY as 'an uninterrupted sequence of views which would unfold themselves, like "stills" from a movie'. He also introduced design concepts that where to become popular, such as the use of bollards and of the floorscape in pedestrian realms.

Although Cullen's townscape owes much to his capacity as an artist for visually perceiving and graphically reproducing the physical context of the city within an evocative and socially engaged environment, it is not to be taken as merely picturesque. His work generated a school of thought and had a profound impact on URBAN DESIGN, particularly in Britain, where it still influences public planning and design guidelines. Cullen was a busy practitioner with a number of influential studies, master plans and urban design projects, such as the new town for Alcan Industries (1964–68), and the Isle of Dogs and Canary Wharf redevelopment projects in London.

Further reading

Broadbent, G. (1990) 'The Neo-Empiricists: Gordon Cullen', in *Emerging Concepts in Urban Space Design*, New York: Van Nostrand Reinhold.
Cullen, G. (1971) *The Concise Townscape*, London: The Architectural Press.
Gosling, D. (1996) *Gordon Cullen – Visions of Urban Design*, London: Academy.

VICENTE DEL RIO

CULTURAL HERITAGE

In the past, heritage conservation and progress were considered incompatible, but since the 1970s a new vision has arisen: cultural heritage has become a major impulse for social and economic progress. It is largely identified with 'built' heritage (e.g. heritage cities, cultural sites and MONUMENTS) but can also be considered in a broader sense to include cultural landscapes and intangible assets that shape local culture.

Because of the menace of excessive dependence on foreign forces to the social and economic vibrancy of local areas, many territories are increasingly adopting cultural markers (crafts, folklore, historical sites, landscape systems, local visual arts and emblematic architecture, literary references, regional languages, etc.) as key resources in their development strategies. This attempt to valorize a place through its CULTURAL IDENTITY, localizing economic control, is called 'culture economy'. This can be understood as strategies to transform CULTURE (local knowledge, own way of doing things) into resources available for spatial development. Such strategies take several forms.

The most easily recognizable are the construction of an identity and its projection to the outside world (typical in the context of URBAN REGEN-ERATION and the growth of cities, selling places for tourism and inward investment), and the commoditization of local and regional culture (culture and history are encapsulated in the different kinds of products to be sold). Others are already raising the self-confidence of local people and organizations in their own ability to achieve development and valorize local resources (including local culture that may have been disabled by a dominant culture during the construction and maintenance of a nation-state), and the normative capacity of every territory, as a global player, to choose alternative development trajectories where geographic identity and love of place guide action and policy (see GLOCALIZATION; STRATEGIC PLANNING). This idea is related to the concept of a 'development repertoire', a stock of all kinds of resources from which the owner selects according to each situation. This concept, in turn, gathers together the principles of endogeneity: the idea of local ownership of resources and the sense of choice to employ them to achieve local objectives (Ray 1999: 525).

The new European Spatial Development Perspective cultural heritage, which includes contemporary realizations, constitutes a form of regional development with positive effects for the unity and cohesion of EU territory, urban and rural (see URBAN–RURAL TENSION). In the case of rural areas, both those away from and those close to urban centres, but especially the former, one of their development strategies is to consider themselves as a part of the cultural and natural heritage. These two kinds of heritage are strictly related in different cultural landscapes, reflecting local identity outcomes of history and interactions between man and nature. Some of these landscapes are unique – veritable sanctuaries that not only have to be preserved but also need rehabilitation and creative planning; like many places of exceptional cultural value, they are often threatened by a slow but constant deterioration.

Further reading

Graham, B., Ashworth, G.J. and Tunbridge, J.E. (2000) *A Geography of Heritage: Power, Culture and Economy*, New York: Oxford University Press.
Kearns, G. and Philo, C. (1993) *Selling Places: The City as Cultural Capital, Past and Present*, Oxford: Pergamon Press.
Ray, C. (1999) 'Towards a meta-framework of endogenous development: repertoires, paths, democracy and rights', *Sociologia Ruralis* 39(4): 521–37.

JOAQUÍN FARINÓS DASÍ

CULTURAL IDENTITY

A highly subjective matter, closely related to space, cultural identity is a topic that is gaining greater relevance in URBAN STUDIES. Cultural identity has to do with the links that yield from an interaction between people and space in cities. Rooted in cultural and historic values, people establish connections to the urban space. These connections attribute a meaning to that space, activating people's awareness about their own cultural identity.

Twentieth-century globalization brought a renewed interest to the area. On the one hand, this was due to the growing concern about the influence of globalization on local cultures, and the likely diffusion of a 'global culture'. On the other hand, cultural identity became increasingly regarded as a

powerful tool for the preservation of values and traditions of singular urban regions.

In urban studies, cultural identity is approached through the interrelationship the concept has to space, and through the role it plays in the quality of urban life. As for the first point, it is acknowledged that cities have the power to act as hubs for the diffusion of cultural factors. Cultural identity is a collective manifestation of societal characteristics, and as such derives from social interactions. But it is in space, and particularly in PUBLIC SPACES, that urban social interaction takes place, and thus the process of cultural identity may mature. It follows a close interrelationship between urban space and cultural identity. Globalization also exerts an influence on this interrelationship. The dissemination of global practices occurs mainly in the multicultural public spaces of cities, but the morphological patterns these spaces display are normally the same in the GLOBAL CITIES. Therefore, social interaction will tend to become culturally standardized. Furthermore, the boundaries between what is public and what is private are growing almost undifferentiated in such spaces. Concepts such as the 'McDonaldization' of the environment, the 'mal' de mall, META-URBANISM and generic cities are often referred to by contemporary urban researchers, connoting an acculturation and spatial uniformity amid the globalized society.

As far as the quality of urban life is concerned, cultural identity may be regarded as creating a SENSE OF COMMUNITY, a quality highly sought after by physical and social planners alike. In a COMMUNITY, a collective cultural identity becomes responsible for the prevalent behaviour, implying the formation of a cultural group. This would be the case, for example, with gated CONDOMINIUMS, ETHNIC ENCLAVES and bohemian districts. Gains in the quality of urban life are experienced in those places, but there are also worries about the build-up of urban enclaves. The community's tendency is to become inward-looking, and to maintain an immobile identity, but in the long run this may produce a welcoming outcome, since the persistence of cultural values will secure the presence of meaning in space.

Spatial implications for planning may be noticed in both alternatives, encouraging more specialized studies.

Further reading

Carr, S., Francis, M., Rivlin, L. and Stone, A. (1995) *Public Space*, Cambridge: Cambridge University Press.
Zukin, S. (1995) *The Cultures of Cities*, Oxford: Blackwell.

LINEU CASTELLO

CULTURE

Culture has been the interdisciplinary object par excellence, so much so that, in recent years, a specific field (cultural studies) has been established. It is, however, very difficult, if not impossible, to define. By as early as 1952, Kroeber and Kluckhohn had found 160 definitions from the different social sciences. However, two of the multiple meanings of culture are that it refers to whole ways of life or systems of shared beliefs, as well as to the codes with which meaning is constructed, conveyed and understood, or, in geographical terms, 'maps of meaning' through which groups and individuals make sense of their social world. Time and space are crucial, in the sense that culture, like society, is temporally and spatially constituted. Culture is important in the process of determining particularity: when it is public (and not kept in the human mind), it is made available through social life by particular people, to particular people, and refers to particular times and places.

The importance of culture, as well as time, has already been recognized by 'classic' writers such as SITTE and MUMFORD. There is an ongoing debate on the existence of urban culture in the anthropological sense (see RAPOPORT), and this debate is connected to the concept of space as a theoretically important issue, which implies its relation to society (see CASTELLS; HARVEY; LEFEBVRE). An important breakthrough in cultural values is observed in POSTMODERNISM, and examples of its impact can be found in the establishment of various interrelationships: production–consumption, high culture–popular culture and global–local (see DISNEY ENVIRONMENTS; FESTIVALS; GLOCALIZATION; LUNA PARKS; THEME PLACES). Globalization has already been a characteristic of MODERNISM in the sense that similar plans are proposed in most parts of the world – a typical example being the attitude, and the influence, of LE CORBUSIER. Globalization implies that popular culture is largely concentrated in the cities. However, its end result is

not a global homogenization of culture, because of the existence, especially in GLOBAL CITIES, of what Hannerz calls 'creole cultures' ('systems of meaning and expression mapped onto structures of social relations'), in which the periphery can respond to the centre.

Other cultural issues raised by postmodernism are the recognition of multiculturalism, i.e. of the plurality of cultures (Zukin refers to 'the cultures of cities'), and the existence of multiple identities (this means that CULTURAL IDENTITY is also characterized by plurality). An example of an older issue that was addressed in a new light is SENSE OF PLACE. On the one hand, the spatial, as well as the temporal, constitution of identities is considered as a part of the construction of cultural identity. On the other hand, the attachment to the locality is considered as a reaction to the impact of globalization.

Examples of current cultural issues are: the growth of the economic importance of culture (Zukin even speaks of a 'symbolic economy of cities'), the necessity of cultural policy (this relates to the recognition of the work done by UNESCO, especially in the fields of cultural development and CULTURAL HERITAGE, and by the Council of Europe), and, more generally, the necessity of cultural GOVERNANCE.

Culture as an activity constitutes one of the four main types of leisure activity – the other three being sport (see STADIA), TOURISM and entertainment/social life. Leisure constitutes, according to the CHARTER OF ATHENS, one of the four main functions of URBAN PLANNING. The cultural dimension of cities has many aspects: the role of cultural activities in the function of cities (e.g. in URBAN REGENERATION), the impact of cultural infrastructure (i.e. culture as land use) in urban planning, the weight of culture in policy issues (e.g. cultural heritage), the effect of urban interventions in the cultures of cities, and the location and planning of cultural spaces. The aims of urban cultural development can be summarized in two factors: the improvement of cultural spaces, and the amelioration of the cultural level of the inhabitants, i.e. of their 'cultural capital', according to Bourdieu.

Most of these general issues constitute the relatively new field of cultural planning, which is compatible with sustainability (see SUSTAINABLE URBAN DEVELOPMENT). Cultural planning functions as an alternative to both traditional cultural policies and cultural policy-led urban regeneration strategies. The aim is to overcome the main strategic dilemmas of cultural policy, which are: city centre vs. periphery tensions and the risk of gentrification (spatial dilemma); consumption vs. production (economic development dilemma); and buildings (or property or capital development) vs. human networks or activity (cultural funding dilemma).

The importance of culture for cities can be confirmed in the variety of specific issues that have been dealt with by cultural planning: cultural industries (cities in the Ruhr area in Germany), creativity/innovation (Creative City Network), PUBLIC ART as an important element of URBAN DESIGN (Battery Park City), cultural quarters (London, Times Square; see DISNEY ENVIRONMENTS), WATERFRONT DEVELOPMENT (Baltimore), arts centres (several US, UK and French cities), cultural buildings (Guggenheim Museum in Bilbao), grand projects (Paris), mega events (Barcelona), place marketing (i.e. promotion of the IMAGE OF THE CITY), and cultural tourism (art cities in Belgium). As far as the city as a whole is concerned, in Europe the most important institution created by the European Commission is the European City of Culture (already in its eighteenth year), and Glasgow (in 1990) is considered the most successful example. However, as the example of New York shows, in order for a city to become a 'culture capital', it is not necessary to be part of an institution.

Further reading

Council for Cultural Co-operation (1995–98) *Culture and Neighbourhoods*, vol. 1, *Concepts and References*; vol. 2, *A Comparative Report*; vol. 3, *Talking About the Neighbourhood*; vol. 4, *Perspectives and Keywords*, Strasbourg: Council of Europe.

Evans, G. (2001) *Cultural Planning: An Urban Renaissance?*, London: Routledge.

Landry, C., Greene, L., Matarasso, F. and Bianchini, F. (1996) *The Art of Regeneration*, Stroud: Comedia.

Zukin, S. (1995) *The Cultures of Cities*, Oxford: Blackwell.

ALEX DEFFNER

CYBERCITY

A cybercity is a 'city' in which people meet in virtual space using networked computers with the city metaphor functioning as the interface. It

is a virtual space for computer-mediated communication and interaction serving various social and commercial purposes.

The concept of the cybercity has two origins: cybernetics on the one hand, and the concept of cyberspace on the other.

A father of cybernetics, the mathematician Norbert Wiener coined the term in the late 1940s. He understood cybernetics as the science of communication and control in living organisms and machines. It relates to general systems theory, as the complex automatic control systems are referred to as cybernetic systems. In this respect, the cybercity can be seen as a locus of computer-mediated communication and social interaction.

The other, more important dimension of the cybercity relates to the concept of cyberspace. Cyberspace was originally a concept used by William Gibson in his 1984 fantasy novel *Neuromancer*. It concerns the virtual space of networked computers, and the society that gathers around them. In the contemporary world, the INTERNET can be understood as the global cyberspace.

Building a cybercity is to build a website that is rich in content, or, in its most sophisticated form, a virtual reality model on the basis of a highly realistic image of a city. The cybercity is not only an 'artificial' creation. Rather, it creates new social practices and use value for various people, thereby providing a connection between global virtual space and people with their intentions and needs.

For obvious reasons, the concept of the cybercity is close to such concepts as the virtual city, the digital city, the INFORMATIONAL CITY and the TECHNO-CITY. The basic difference is that the cybercity is associated more with varieties of URBAN life, even with something radical and more unpredictable than the other conceptions of the city. One reason for this is that the 'cyber' prefix also appears in expressions referring to a more or less darker side of virtual space, such as 'cyber-terrorism', 'cyberpunk', 'cybersex' and 'cyborg'.

A cybercity can basically be defined in the same way as a virtual city. In a narrow sense, it is the whole set of the activities of the city government provided for citizens and stakeholders through the information networks. This is, however, a rather rarely applied definition. A cybercity is principally conceived in a broader sense as a web-based city formation in which people interact with each other and use services or buy products. Indeed, the essence of this city formation is the richness of a postmodern 'urban' life on the web. Accordingly, a cybercity can be seen as a set of socially and commercially motivated activities organized on the web around the city metaphor, including various expressions of urban life, from science fiction, computing and computer games to chat, cybersex, sports, e-commerce and electronic public services.

ARI-VEIKKO ANTTIROIKO

D

DAHL, ROBERT

b. 17 December 1915, Inwood, Iowa

Key works

(1961) *Who Governs?*, New Haven, CT: Yale University Press.
(1971) *Polyarchy*, New Haven, CT: Yale University Press.

Robert Dahl, Sterling Professor of Political Science at Yale University from the 1950s to the 1990s, is the most important scholar associated with the pluralist approach to describing and understanding both city and national power structures. In *Who Governs?* (1961), Dahl described New Haven, Connecticut – and by extension all American cities – as a 'republic of unequals'. For Dahl, cities are divided into two sectors: a stratum of *homo politicus*, or citizens generally but not exclusively of higher income and education who participate or at least actively follow local politics; and a stratum of *homo civicus* with no discernible active participation or interest in local politics.

Because the *homo politicus* stratum was in most cities a sizeable plurality but not a majority, Dahl referred to such cities as 'pluralist' rather than 'majoritarian', and referred to such systems as 'pluralistic' rather than 'democratic'. A decade later, in *Polyarchy* (1971), Dahl developed the term 'polyarchy' to distinguish between the rough plurality or pluralistic political systems found in American cities – and in many nations – and pure or fully developed democratic regimes.

While American cities share many, perhaps most, of the same characteristics as fully developed democracies – including free elections, freedom of assembly and freedom of the press – American cities still exhibit distinctly stratified political processes and outcomes. Particularly true at the time of his writing in the 1960s, Dahl noted that blacks, for example, had significant barriers to full participation in the economic or social life of American cities. Dahl argued that in contrast to the situation that the American blacks faced in the private socio-economic sphere, in local politics and government the barriers were less than in other spheres because lower-resourced groups could take advantage of 'slack resources'. By slack resources, Dahl meant that lower-resourced or traditionally less influential groups could attempt to outmanoeuvre traditionally more influential groups and achieve their ends via the use of effective leadership, better organization, protest action, or voter and/or neighbourhood mobilization strategies.

Recent scholars associated with URBAN REGIME THEORY have raised significant doubts about whether slack resources can be used effectively by traditionally less influential groups in many cities. Still, Dahl's description of American cities as pluralities or 'pluralist' political systems, and as 'polyarchies' or roughly democratic rather than fully democratic regimes, remains one of the leading explanations of political life in modern American cities.

Further reading

Dahl, R. (1986) 'Rethinking *Who Governs?*', in R. Waste (ed.) *Community Power*, Thousand Oaks, CA: Sage.
Judge, D. (1995) 'Pluralism', in D. Judge, G. Stoker and H. Wolman (eds) *Theories of Urban Politics*, Thousand Oaks, CA: Sage.

ROBERT J. WASTE

DAVIDOFF, PAUL

b. 14 February 1930, New York, NY;
d. 27 December 1984, New York, NY

Key works

(1965) 'Advocacy and pluralism in planning', *Journal of the American Institute of Planners* 31(4): 331–8.
(1970) 'Suburban action: advocate planning for an open society', *Journal of the American Institute of Planners* 36(1): 12–21 (with L. Davidoff and N.N. Gold).

Paul Davidoff is the father of 'advocacy planning' in the United States. Davidoff developed this idea as a counterweight to such programmes as URBAN renewal, which displaced the poor, especially minorities. Davidoff was a defender of those vulnerable populations and he challenged planners and their professional organization – the American Institute of Planners (AIP) – to change the ways cities and urban neighbourhoods were planned and redeveloped. Davidoff's argument was contained in his seminal 1965 AIP journal article entitled 'Advocacy and pluralism in planning'. In this article, he argued for advocacy on behalf of poor communities, social planning and greater citizen involvement in the planning process. His ideas inspired the creation of agencies to provide alternative planning resources to urban communities and were guides to many who became associated with advocacy and equity planning. Davidoff became a leading voice of Planners for Equal Opportunity, a voice for progressive planners often opposed to government policies and an alternative to the AIP.

Davidoff earned planning and law degrees from the University of Pennsylvania. He served as professor and director of the graduate programme in URBAN STUDIES at Hunter College, City University of New York, from 1965 to 1969. In 1968, Davidoff was the unsuccessful Democratic candidate for the US Congress from the 26th Congressional District in New York. In 1969, Davidoff (and Neil Gold) founded the Suburban Action Institute, which in 1980 became the Metropolitan Action Institute. Davidoff fought exclusionary and racially discriminatory ZONING and land use policies prevalent in most US suburban communities. He participated in lawsuits challenging these practices, as well as lawsuits against major employers in suburban areas abetting these policies. These lawsuits targeted SUBURBS like Mahwah and similar cities in New Jersey and Brookhaven and other suburbs in New York. Davidoff, along with other allies, sought to promote the opportunity for low- and moderate-income urban residents to be able to afford housing in the suburbs where jobs were available to them. This was in part a reaction to the 1968 report by the Kerner Commission, which investigated the causes of the urban riots and warned against two separate societies – largely white, affluent suburbs and increasingly poor CENTRAL CITIES where racial and ethnic minorities were concentrated.

Paul Davidoff died in 1984. In his memory, the American Planning Association created the Paul Davidoff award for outstanding books that promote justice and equity in URBAN PLANNING.

Further reading

Checkoway, B. (ed.) (1994) 'Paul Davidoff and advocacy planning in retrospect', *Journal of the American Planning Association* 60(2): 139–234.
Hartman, C. (1985) 'Paul Davidoff: an appreciation', reprinted in C. Hartman (ed.) (2002) *Between Eminence and Notoriety: Four Decades of Radical Urban Planning*, New Brunswick, NJ: Center for Urban Policy Research, Bloustein School of Planning and Public Policy, Rutgers University, pp. 335–56.

W. DENNIS KEATING

DE CERTEAU, MICHEL

b. 1925, Chambéry, France; d. 9 January 1986, Paris, France

Key works

(1984) *The Practice of Everyday Life*, trans. Steven Rendall, Berkeley: University of California Press.
(1986) *Heterologies: Discourse on the Other*, trans. Brian Masuumi, Minneapolis: University of Minnesota Press.

Known as the philosopher of everyday life, Michel de Certeau wrote numerous books and articles, including *Culture in the Plural*, *The Writing of History* and *The Mystic Fable*. He was widely regarded as a historian with a range of interests, which included travelogues of the sixteenth and seventeenth centuries, as well as contemporary URBAN life. Among this wide intellectual spectrum, de Certeau consistently focused on his search for the production and articulation of the 'other' and

the importance of the ordinary. His writing on urban spaces inspired theorists and designers and has reinvigorated discussions about the importance of pedestrians in humanizing cities. As a historian, he introduced alternative interpretations of history and historiography. He viewed history not as a form of truth, but as a method of imposing order. Despite his critique of post-colonial writing, de Certeau was also uncomfortable with nationalism and false hegemonies, and his work celebrated the importance of everyday life. Michel de Certeau taught at the University of California, San Diego from 1978 to 1984, but returned to France to teach at the Ecole des Hautes Etudes en Sciences Sociales in Paris.

Further reading

Buchanan, I. (2001) *Michel de Certeau: Cultural Theorist*, Thousand Oaks, CA: Sage.
Ward, G. (ed.) (1999) *The Certeau Reader*, Oxford: Blackwell.

ALI MODARRES

DECENTRALIZATION

The word 'decentralization' can be used in various fields and with different meanings. In the political system of a nation, decentralization means shifting the balance of power from central to local government. Decentralization can also mean that the population or industrial activities of a nation move from its central region to peripheral or rural regions in physical planning. More strictly, decentralization can be defined in a physical planning context: to move industrial functions or people from congested metropolitan regions to rural regions or SUBURBS through incentives or controls in order to avoid regional disparities, and to try to realize a balanced development throughout the nation.

Historically, decentralization policies were implemented in the United Kingdom before the Second World War to ease the concentration of the population and industries in London. After the war, the Japanese government implemented similar policies to alleviate similar problems in Tokyo. In both countries, the governments have often taken two kinds of policy measures to stimulate decentralization. First, the government offers incentives to those industries which are likely to move to or are newly located in rural regions, such as tax exemptions, subsidies, low interest rate loans and so on. Some of the incentives are given to local governments in the regions where industries are expected to be located, and others are given to the companies. Secondly, the government introduces regulations to restrict the operation or location of factories in congested metropolitan regions. For example, the new location or the expansion of factories and universities had been regulated in the central part of Tokyo for more than forty years until the regulations were abolished in 2002.

The background to these policies is that factories or universities attract young people with their jobs or higher education. And, of course, those facilities are at the centre of the industrial activities that stimulate local economies. Finally, the policies clearly imply that people should move from metropolitan areas to rural regions. In reality, however, although the factories or universities are decentralized, people are not easy to move. Although factories were the symbol of an advanced industrial nation in the past, they are not now. The tertiary sector in the form of advanced business activities attracts more labour force. The offices where the tertiary sector is operated are now in the metropolitan regions. As a result, people are often attracted to the metropolitan areas despite the decentralization policies mentioned above.

Another type of decentralization policy, the sub-centre policy, is introduced within a metropolitan region. Through such a policy, sub-centres are developed in the suburbs of the metropolitan region to gather business or administrative activities. The centre and sub-centres may form a hierarchical structure which may alleviate part of the congestion problem, although it cannot solve the over-concentration of the metropolitan area as a whole. The development of sub-centres following this idea was planned in many cities in the world, and some of them were put into practice.

Finally, we can draw lessons from the many attempts of decentralization policies to capture and make use of market mechanisms and social tendencies. Policies that try to change the movement of society must make use of the power of society. Eventually, decentralization policies may affect social trends only slightly.

TAKASHI ONISHI

DEFENSIBLE SPACE

The defensible space concept originates from the idea that architectural and environmental design plays a crucial part in increasing or reducing criminality. It was made popular in the early 1970s by the architect Oscar Newman, who was working in St Louis and had the opportunity to observe the decline of the PRUITT-IGOE housing complex, where CRIME and vandalism had contributed to create unacceptable living conditions.

On the basis of concepts previously developed by Jane JACOBS, in particular the assumption that people are more inclined to control a territory that they identify as their own, Newman argued that a good project should generate a strong sense of territoriality and suitable opportunities for active surveillance among the residents.

To transform this hypothesis into design guidelines, Newman proposed to identify which environmental and functional characteristics are most associated with crime occurrence. To this purpose, he compared two blocks of buildings situated on the opposite sides of the same New York street, i.e. a high-rise housing complex and an older settlement of row houses. The population was identical, with a majority of single-parent and welfare families, while the crime rate was very different. According to Newman, this difference was due to the fact that one of the settlements incorporated defensible space techniques while the other did not.

The next step was to develop a set of principles and tools aimed at enhancing the possibility of surveillance through an increased observability of the people's movement patterns and the identification of potential intruders.

This approach gained popularity. The defensible space concept entered into use and was put into practice in a variety of settings. A number of local authorities implemented programmes such as the CPTED (Crime Prevention Through Environmental Design) to redesign the physical layout of some popular housing quarters, especially through the creation of clear physical boundaries and the subdivision of public areas into several smaller parts assigned to the residents' control. 'Safety by design' has thus gained a place on the planning agenda, but has also undergone massive criticism.

The scientific value of the statistical procedures on the basis of which a quarter is defined insecure and the effectiveness of solutions that, in the best of cases, lead to a redistribution of the crimes in the surrounding neighbourhoods were questioned. In particular, some scholars believe that neglecting the mutual relations between physical and social variables is a dangerous signal of environmental determinism that leads one to ignore the fact that a specific architectural solution, for example high-rise buildings, can have different effects on different people. Moreover, critics claim that this approach, by presenting an image of widespread violence as an inevitable feature of URBAN life, has contributed to the rise of phenomena such as GATED COMMUNITIES, private armed guards and NEIGHBOURHOOD WATCH that do not provide answers to the problems of a dysfunctional society.

Further reading

Newman, O. (1972) *Defensible Space: Crime Prevention through Urban Design*, New York: Macmillan.
Poyner, B. (1983) *Design against Crime: Beyond Defensible Space*, London: Butterworths.

PAOLA SOMMA

DEMOGRAPHY

Demography is the field of study that inquires about the human population. In 1855, a Belgian scholar, Achille Guillard, originally defined demography as the natural and social history of the human species or the mathematical knowledge of populations, of their general changes, and of their physical, civil, intellectual and moral condition. Although academic scholars and professionals studying the elements of population often used such terms as 'demographic analysis' and 'population studies', the field of demography is commonly defined by its scope of use. Formal demography is narrow in its scope and usually restricted to the size, changes, distribution and structure of populations, while a broader notion of demography includes in its analysis additional population characteristics and their relationships with other attributes in society.

Population size refers to the number of people in a defined area (i.e. metes and bounds) or a geographical unit (i.e. neighbourhood, city, county, region, state or country). Population change measures the growth or decline of the total population

over time in any one of the specifically defined areas or the geographical units (i.e. change in total number of people over two time periods due to migration). Population distribution shows special or unique arrangements of people at a specified time period (i.e. total number of people by states or by region). Population structure presents sex and age compositions among the people under study.

Additional population characteristics include, among others, socio-economic, ethnic and geographic. Social characteristics consist of level of educational attainment, family structure, languages spoken at home, marital status, place of birth and life-cycle stages. Economic characteristics include family income, occupational status, employment status, home ownership and manufacturing activities. Ethnic characteristics look closely at race. Geographic characteristics show an urban and rural distinction.

Although many countries collect data for their own population count and information for vital statistics, the United Nations provides the most widely accepted demographic statistics for most nations in the world. One of the most established data sources is the *Demographic Yearbook*, which has been published since 1948, and it includes such information as population trends, economic characteristics and vital statistics.

In the United States, the main source of data for the field of demography is the decennial census, which began in 1790 as a requirement by the US Constitution for apportionment of the House of Representatives. Every ten years, the Bureau of the Census has enumerated the total population in the country and presented official reports on population gains and losses by each state. Because the number of state representatives is proportional to the total population number of the state, in accordance with the decennial population counts from the Bureau of the Census, each state may gain or lose congressional seats for the House of Representatives.

The Bureau of the Census also provides a special census between the decennial census periods for some states experiencing rapid population growth. Following the special census, state and local jurisdictions often redistrict or redesign congressional boundaries. In addition, in the United States, demographic data are carefully reviewed, analysed and often scrutinized by state and local governments, because most federal government funding given to state and local governments is based on the official census count.

Further reading

Shryock, H.S. and Siegel, J.S. (eds) (1975, 1979, 1980) *The Methods and Materials of Demography*, Washington, DC: US Government Printing Office.

JOOCHUL KIM

DENSITY

Density, a number of items per unit land area, is a crucial issue in both popular and professional debates about URBAN FORM and architecture. Systematic work on density makes a number of key distinctions between different measures of density and between density and related terms.

Residential population density and dwelling unit density are the most common measures used in research on the city. Population density is important for planning such facilities and services as transit, retail and recreational areas. Housing unit density is important both as an indirect measure of population density and also as a rough measure of the physical character of the space. In most countries, it is also easier to regulate the number of the dwelling units in an area, through mechanisms such as minimum and maximum lot sizes, than it is to regulate the number of people living in a household or city. This makes housing density a useful measure in practical terms. The density of many other items – such as jobs, pedestrians, automobiles, trees or shops – may be calculated for specific purposes but are less often used.

Perceived density is a person's subjective evaluation of population or built density, a measure that is related to environmental cues, culture and design. Perceived density may have little to do with the actual measured density. For example, high-rise housing units are often perceived as high density when in fact low-rise developments that cover more of the site may have more persons or dwellings per unit land area.

An elaborate terminology has developed to distinguish between density measures at different scales. The most confusion occurs in the measurement of area, as any unit – from the site to the metropolis – may be used. Different fields, countries and even cities also use the same term, such as

'gross density', to mean different things. Given these variations, site or net density is typically the number of dwelling units or persons per site area where the site is just the parcel of land being developed for housing. For detached housing this includes the driveway, and for apartments related internal circulation. In the United States, gross density often includes the residential site area and half the adjacent roads but no other uses. However, in some countries such as Australia, gross density may be closer to what would be called neighbourhood density in the United States, and includes local shops, parks, schools and roads but not such uses as regional parks or regional industrial land. City or metropolitan density includes as the base land calculation all the built-up land in some particular city or metropolitan area, although how much land is excluded as unbuilt is open to some interpretation. The area used in such city density calculations may be all land within some local government boundary, all land counted as urbanized or continuously built-up in a metropolitan area, or even all land within counties classed as URBAN. In addition, densities are often divided into low, medium and high density, but there is no agreement on the specific numerical cutoff points for such measures. A low density in Israel may be a high density in the United States.

Residential or population density figures also vary quite dramatically with the scale of density measurement and with household size. A site or net density of 20 units per hectare may amount to a neighbourhood density of 10 to 12 units and a metropolitan density of perhaps 5 to 8 dwelling units per hectare. Alternatively, a site density of 20 units per hectare would represent 50 people at a household size of 2.5, typical of urban areas in wealthy countries, but 100 persons per hectare with a household size of 5.0. The arrangement of this density is also important. For example, in terms of the practical viability of rail transportation systems, there is a great difference between a metropolitan area with a uniform density and one with significant high-density neighbourhoods or centres and very low-density areas in between.

'Density' is often confused with other terms. One of the most important confusions is with 'crowding', which is a perception of there being too many people in a space. While about perceptions, work in housing generally measures crowding in more objective terms such as there being more than a specified number of people per bedroom or per habitable room. However, because density is measured in terms of land area, the same housing unit may be either low density and crowded, as in a one-room farmhouse with several residents, or high density and uncrowded, as in the high-rise downtown apartment of an affluent individual in Hong Kong. A number of measures of building bulk and massing are also commonly used as proxies for density. These include building coverage (the ratio of the building footprint to the site), FLOOR AREA RATIO (the ratio of built areas on all floors to the site), building height and setbacks, and the proportion of detached housing units.

Different density measures are used to answer different substantive questions. In the area of SUSTAINABLE URBAN DEVELOPMENT, metropolitan or city density has been an important topic of debate because quite high densities are required before people will walk to conduct daily activities or to use transit, and higher-density developments use less land per person or per dwelling. Employment density is important in this light along with residential density. However, very low densities allow self-sufficiency in food and reduce the proportion of impervious surface, leading to vigorous debates about density patterns.

Further reading

Alexander, E. (1993) 'Density measures: a review and analysis', *Journal of Architectural and Planning Research* 10(3): 181–202.

Churchman, A. (1999) 'Disentangling the concept of density', *Journal of Planning Literature* 13(4): 389–411.

Newman, P. and Kenworthy, J. (2000) *Cities and Sustainability*, Washington, DC: Island Press.

Rapoport, A. (1984) 'Toward a redefinition of density', *Environment and Behavior* 7(2): 7–32.

ANN FORSYTH

DENSITY BONUS

A density bonus is a public policy tool used in land planning and development to encourage real estate developers to serve a public purpose. A density bonus allows a development to exceed the FLOOR AREA RATIO or the DENSITY of units permitted by planning and ZONING regulations in exchange for contributing to a public goal.

The density bonus is used widely in cities and counties throughout the United States. The public

goals served by the bonus include protection of the natural environment and historic sites – this approach often entails a TRANSFER OF DEVELOPMENT RIGHTS – promotion of cluster and INFILL DEVELOPMENTS, and production of affordable housing units. The density bonus is most commonly associated with this last goal.

Many localities use the density bonus to encourage developers to build housing units affordable to low- and moderate-income households. The bonus typically is part of an inclusionary zoning policy. The bonus provides developers with additional density to offset profits lost as a result of developing lower-income units.

The density bonus approach has been resisted by some cities. Opposition to this approach prompted the State of California to pass a Density Bonus Law in 1979. As amended in 1989, this Law mandates all cities and counties to adopt a density bonus ordinance providing a 25 per cent density increase, if 20 per cent of the housing units are affordable. In 2002, some localities in California still had not adopted such an ordinance.

Further reading

Mallach, A. (1984) *Inclusionary Housing Programs: Policies and Practices*, New Brunswick, NJ: Center for Urban Policy Research, Rutgers University.

VICTORIA BASOLO

DEPENDENCY THEORY

Dependency theory was primarily concerned with explaining Third World underdevelopment through its incorporation into the global capitalist economy. It was ascendant in the 1960s and 1970s at a time when the assumptions of liberal theories of development were under attack. Cities played an important role in dependency analysis, in terms of acting as critical nodes between internal and external economies. As such, post-colonial and IMPERIAL CITIES were not the result of indigenous processes but reflected the nature of their entry into, and articulation with, the global capitalist economy.

In dependency theory, cities are at the core of the twin processes of development and underdevelopment. Historically, colonialism was essential in creating (mostly PORT CITIES, which were established as sites of resource extraction, foreign ownership and colonial administration (see COLONIAL CITIES). In the post- or neo-colonial world system, foreign trade and investment, primarily through transnational corporations (TNCs), continue to be the major drivers of Third World urban development (and malaise). For dependency theorists, this has resulted in THIRD WORLD CITIES maintaining a peripheral role in the global economy, resulting in economic marginalization, income inequalities, social exclusion and urban primacy.

The Third World city is then a 'dependent metropolis'. These cities are not global decision-making 'cores', but peripheral sites of outsourced jobs and investment subordinate to the control held by the 'core' (or 'First World'). Foreign capital, and capitalists, essentially hold sway over urban development and form. This means that such things as environmental regulation or labour standards are negotiable to the extent that capital can be attracted and some benefits can be realized. But while these cities are peripheral in terms of global power, they do assert an unhealthy dominance vis-à-vis local capital and spatial development. Dependency writers assert that transnational capital is drawn almost entirely to the core metropolitan regions of a country, making them key centres of local capital accumulation. This results in a polarization between ('parasitic') cities and their rural HINTERLANDS.

There have been several criticisms made about the arguments put forward from dependency theory. Critics saw their analysis as static economic reductionism, and overstating of external processes. Several studies suggested that indigenous capital was as important in shaping URBAN FORM and PRIMACY as FOREIGN INVESTMENT. Rather than being impediments, cities are still the most important factors associated with economic development. Nevertheless, many aspects of urban development raised by dependency theory, such as economic concentration and ownership, the role of urban (often foreign) elites in policy, and the dependence on foreign capital (TNCs), remain highly relevant in URBAN STUDIES, as does analysis of the role of capitalism and the relationship between the state and the market. Even if some of its conclusions have been left wanting, dependency analysis has resulted in a more critical appreciation of the economic bases of cities, their role in development, and the relationship of Third World cities to global capitalism.

Further reading

Castells, M. (1977) *The Urban Question: A Marxist Approach*, London: Edward Arnold.
Friedmann, J. and Wulff, G. (1982) 'World city formation: an agenda for research and action', *International Journal of Urban and Regional Research* 6: 309–44.

SEE ALSO: imperial cities; Third World cities

DONOVAN STOREY

DESEGREGATION

Desegregation refers to the removal of barriers such as laws, customs or practices that separate groups of people, typically on the basis of race or religion. SEGREGATION in the areas of housing, education and the use of public facilities has been a common characteristic of URBAN areas in many societies. Because patterns of segregation often reflect deep social, cultural and religious divisions, movements to desegregate cities often involve momentous social change. The American CIVIL RIGHTS MOVEMENT and the South African anti-apartheid movement are two prominent examples.

Desegregation can be pursued in a passive way, through the elimination of barriers to the mixing of populations. Passive desegregation includes executive orders to open all schools to black enrolment, to the US judicial decision in *Shelley v. Kramer* finding racially restrictive deeds unconstitutional. Desegregation can also be more actively pursued through court-ordered busing programmes for school children or housing subsidy programmes designed to facilitate the movement of black public housing residents into predominantly white neighbourhoods.

Desegregation is typically measured by the INDEX OF DISSIMILARITY. This measure allows researchers to determine whether desegregation efforts are having an impact on the settlement patterns of various groups.

Cities across the world exhibit characteristics of racial, religious, ethnic and ECONOMIC POLARIZA-TION. Often, these patterns are supported by state action. The Jim Crow laws in the United States established a system of separation between whites and blacks that extended to separate drinking fountains, restaurants and entranceways into hotels and other public buildings. Separation of the races was also the intent of various racial

ZONING laws that proliferated in southern cities during the first half of the twentieth century. In South Africa, the Group Areas Act led to the designation of neighbourhoods for specific races and legalized the forced eviction of non-whites.

Desegregation movements can have a range of effects on urban areas. School desegregation has led to WHITE FLIGHT in many cities as whites attempt to avoid the integration of the schools. The US Supreme Court ruled in *Brown v. Board of Education* that segregated schools were unconstitutional, and that local and state governments should desegregate with, in their words, 'all deliberate speed'. However, the subsequent Supreme Court decision in *Milliken v. Bradley* ruled that desegregation could not be applied across municipal boundaries. This made white flight from the CENTRAL CITY a common reaction to school desegregation. School segregation is now more common across boundaries of local government than within them. In addition, the *Milliken* decision made fragmented local government another barrier to desegregating schools.

Housing desegregation is often limited when the percentage of minorities within a neighbourhood reaches what is called 'the tipping point'. The tipping point is the proportion of minority residents within a neighbourhood that triggers wholesale white flight from the area, resulting in re-segregation of the neighbourhood rather than desegregation. There are various estimates of what the tipping point is, and it may vary from place to place, but it is generated by the unwillingness of whites to tolerate as much diversity in residential neighbourhoods as is tolerated by non-whites.

Efforts to desegregate housing are sometimes confounded by the need for more affordable housing. Shortages of affordable housing induce governments to build as much housing as possible where land is inexpensive and opposition is minimal. This has meant that government-built housing has often reinforced patterns of segregation, as has been the case in post-apartheid South Africa, and in cities across the United States. Alternatively, strict adherence to desegregation objectives may slow down the provision of affordable housing to families who need it.

Desegregation programmes in both housing and education have shown benefits for the families or children they were meant to assist. In the United States, minority students have significantly increased school performance under desegregation

programmes. The gap in performance between white and black students narrowed considerably and graduation rates have increased greatly since desegregation of the nation's schools was begun. Housing desegregation programmes in the United States have also produced measurable benefits for minority families who were able to move away from disadvantaged central city neighbourhoods.

Despite documented benefits from desegregation programmes in housing and education, there are significant barriers to achieving lasting impacts. Decades of institutionalized support for the separation of races and ethnic or religious groups are not easily overcome. Prolonged segregation in housing and education is often accompanied by DISCRIMINATION in employment. These factors combine to impoverish the excluded group, making their integration in the housing market very difficult even after desegregation has lifted the legal barriers. Long-term progress in education has been limited by changing political and judicial support for desegregation. A range of US Supreme Court rulings during the 1980s and 1990s, for example, has allowed many local governments to end desegregation programmes. The result has been a reversal of the progress made during the 1960s and 1970s, and a slow re-segregation of schools.

Desegregation efforts are often historic social movements that aim to dismantle well-established, formal and informal systems of domination. Such change does not occur easily. Desegregation has typically involved violent and non-violent forms of protest, and has led to significant social reforms and sometimes, as in the case of South Africa, regime change. Desegregation of housing and schools in the United States has frequently required the intervention of the Supreme Court, and sometimes even the deployment of federal troops (as in the desegregation of public high schools in Little Rock, Arkansas, in 1958).

Further reading

Orfield, G. (1983) *Public School Desegregation in the United States*, Washington, DC: Joint Center for Political Studies.

Rubinowitz, L.S. and Rosenbaum, J.E. (2000) *Crossing the Class and Color Lines: From Public Housing to White Suburbia*, Chicago: University of Chicago Press.

EDWARD G. GOETZ

DEVELOPMENT AGREEMENTS

Development agreements in the United States arose as the nature of land development changed. Beginning in the 1970s, developers planned much larger developments than historically had been created. The overall scheme was divided into phases to be implemented over time according to the developer's schedule. A major problem faced by developers was the potential for change in land use regulations. If one of these multi-year developments were begun under a given land use regulatory framework, but then that framework were substantially changed to restrict development, the developer could face substantial economic loss. These multi-year, phased developments required a different regulatory framework from the then-existing land use planning regime.

The statutory framework that resulted was express authorization in law for agreements between developers and municipalities. Under these agreements, developers achieved legally binding guarantees that the land use regulatory scheme under which the approvals were given would not be changed over the life of the project. In return, the developers agreed to provide enumerated public facilities to the municipality. Developers achieved certainty and stability; municipalities received public facilities at the developers' expense.

As of 2001, thirteen states had enacted statutes permitting development agreements. One common requirement of these statutes is that developers' agreements be in accordance with the community's comprehensive plan. Some statutes require public hearings before the agreements are adopted by municipalities. Design standards must be specified in the agreements under most statutes. Review of the implementation of the agreements is also a common requirement. Some statutes require the agreements to specify the nature and type of public facilities to be provided by the developer. In some cases, the duration of the contract is limited to a fixed term.

Enabling legislation for development agreements helps to shield municipalities and developers from a number of legal challenges, such as the charge that the municipality has contracted away its police power. One open question is whether recent US Supreme Court cases that require (1) a 'rational nexus' between a legitimate state interest and a condition attached to a permit, and

(2) 'proportionality' in the scope of the requirement imposed on the developer apply to development agreements. If development agreements are viewed as regulatory in nature, the rationality/ proportionality tests may be applied to these agreements. If, on the other hand, development agreements are examined under traditional contract principles, developers and municipalities may be free to bargain for amenities unrelated to the developments in these agreements. This legal issue is not yet settled.

In the United Kingdom, development agreements arose in a much different context. Unlike the United States, where there is at least an implicit right to develop property, subject to existing land use regulations, the right to develop is not one of the rights in the United Kingdom's property rights 'bundle of sticks'. Developers must receive express permission for development from the relevant local authority. In exchange for this permission to develop, developers are often required to provide planning gain to the community. Planning gain is simply community benefits provided at the expense of developers. These benefits may include such things as affordable housing, restrictions on housing occupancy, restrictions on goods sold in retail outlets, transportation plans, open space, drainage areas, community facilities or development phasing.

These requirements are often embodied in formal agreements. They should be related to the associated developments and follow public policy. In practice, however, there is somewhat more flexibility than is suggested by the United Kingdom's planning acts, and there is certainly more flexibility than is found in the United States.

One study in Scotland found that more than 25 per cent of the development agreements entered into in a three-year period were for developments of less than ten dwellings. It is also possible, as acknowledged by a Supplementary Planning Guidance issued by a county council in Wales, that a single dwelling could cause sufficient externalities to require a development agreement. In comparison with US practice, in which development agreements typically apply to large-scale developments, these agreements may apply to rather small developments in the United Kingdom.

The United Kingdom has taken the concept of development agreements much further, however, than simply using them to acquire community amenities through the land use approval process. Some new public sector projects are being funded through agreements with private sector developers. Projects funded through this mechanism include hospitals, roads and schools. The private developer enters into an agreement with a public body to construct and operate the needed facility, and is then reimbursed by the government body over a period of years.

In Continental Europe, development agreements are used in conjunction with municipal powers of expropriation to assemble large tracts of land for commercial development. In exchange for use of municipal expropriation power to assemble small parcels with complicated ownership interests, developers agree to pay for the land thus assembled, either by direct payment or by giving the municipality a share of the equity in the project. This facilitates commercial development when otherwise it would be more difficult or impossible.

Development agreements have thus progressed from a device originally designed to preserve a right to develop a plan in a certain way in exchange for providing community facilities to an instrument through which public facilities are designed, financed, built and operated by private entities. Development agreements may thus be viewed as important instruments of public sector finance as well as land use control devices.

Further reading

Callies, D.L. and Tappendorf, J.A. (2001) 'Unconstitutional land development conditions and the development agreement solution: bargaining for public facilities after Nollan and Dolan', *Case Western Reserve Law Review* 51: 663–96.

Haywood, R.A. and Hartman, D. (2001) 'Legal basics for development agreements', *Texas Tech Law Review* 32: 955–78.

Scottish Executive Central Research Unit (2001) *The Use and Effectiveness of Planning Agreements*, Edinburgh: The Stationery Office.

Wenger, J.W. (1987) 'Moving toward the bargaining table: contract zoning, development agreements, and the theoretical foundations of government land use deals', *North Carolina Law Review* 65: 957–1038.

SEE ALSO: exactions; property rights

R. JEROME ANDERSON

DIGITAL DIVIDE

The term 'digital divide' refers to the gap in technology ownership and access between those who are affluent and those who are poor. The rapid development of information technology (IT) in the 1990s revolutionized the organization of work and life in advanced industrialized countries. Because of its high cost, however, the adoption and utilization of technology is highly uneven across the globe. According to the United Nations, only 2 per cent of the entire world (about 250 million people) has access to the INTERNET. As the spread of information technology accelerates and IT enters into different facets of modern life, the lack of access to technology by a significant percentage of the world population raises serious concerns about equity.

Inequities are well documented in the United States. In 2000, half of all households nationwide did not have access to the internet from home. Meanwhile, getting access to government services and participation in ordinary social and economic activities increasingly require access to information and communications technologies. Low access of low-income minority groups to the internet became a pressing public policy issue in the late 1990s in the United States, prompting the creation of federal grant programmes.

Despite federal efforts to introduce IT into disadvantaged communities in urban and rural America, the technology gap across income, gender, race, geography and disability still remains. Studies show that technology adoption tends to rise more slowly for low-income users; children in low-income families have significantly less access to the internet; rates of adoption for Hispanics and blacks are lower than for whites; Spanish-speaking households have far less access to technology; and unskilled workers have few opportunities to develop their technology skills.

Furthermore, the digital divide is not only a problem of access to hardware and software, but also a problem of access to training that can enable individuals and groups to use technology in problem-solving. Realizing the empowering potential of IT, non-profit organizations in the United States began forging links between community-building and community technology movements. Because of their long-time community-based work in disadvantaged inner-city neighbourhoods, COMMUNITY DEVELOPMENT CORPORATIONS (CDCs) are in a good position to integrate IT functionality into their traditional community-building and social service provision activities. Increasingly, CDCs are utilizing IT to increase their capacity to deliver services. Community mapping is a particularly powerful tool that involves the use of GEOGRAPHIC INFORMATION SYSTEMS (GIS) technology to map community assets in neighbourhoods to help residents identify resources in their immediate and larger communities.

Public funding and leadership are essential in bridging the digital divide. In the absence of government intervention to connect the 'haves' and the 'have-nots', the technology gap is likely to widen, resulting in a significant portion of the world's population living deprived of the benefits IT can provide.

Further reading

Graham, S. (2002) 'Bridging urban digital divides? Urban polarisation and information and communications technologies (ICTs)', *Urban Studies* 39(1): 33–56.

Schon, D.A., Sanyal, B. and Mitchell, W.J. (eds) (1999) *High Technology and Low-Income Communities: Prospects for the Positive Use of Advanced Information Technology*, Cambridge, MA: MIT Press.

SEE ALSO: capacity-building; cybercity

AYSE PAMUK

DIGITAL GOVERNMENT

Digital government (e-government) describes the application of information technology to government operations and public services, and to the development of new communication linkages between citizens and government. This rapidly changing field applies innovations in computers, and in database and information storage systems, GEOGRAPHIC INFORMATION SYSTEMS, INTERNET access, and advances in telecommunications technologies (e.g. wireless communications) to government. Digital government offers new opportunities for more direct and convenient citizen access to government, and for government provision of services directly to citizens. Exciting new possibilities for citizen participation in government enabled by digital government range from proposals for DIRECT DEMOCRACY through internet voting in elections or in local government referenda

on public issues, to the development of online community networks bringing together citizens, schools, non-profit organizations, the private sector and government in locally oriented online networks, to practical means for decentralizing certain public services to the neighbourhood level.

Within government, the digital revolution is changing government operations. New methods of gathering, storing and accessing information are challenging traditional agency organizations and procedures. The digital revolution is changing government from the inside, reordering traditional organization structures around the new information infrastructure.

Digital government is a powerful new technology raising new issues that must be addressed in democratic CIVIL SOCIETIES. The same digital revolution that will improve government service delivery raises concerns about information security and about citizen privacy in the digital age. Information security and privacy are important issues that must be addressed in the development of digital government.

Some of the most exciting aspects of digital government involve offering new paths for citizen–government interaction. Today, many local government web pages offer citizens direct access to information about public agency services. As interactive web-based communications are added to city webs, residents are able to fill out forms and apply for public services, and pay taxes and user fees online. Trips to local government offices to stand in lines to fill out forms or pay taxes will be a thing of the past. As high-speed internet connections (broadband access) become more widely available and the costs of access decrease, the experience of interacting with government will become more user-friendly, and governments will be able to serve individuals and households more directly, targeting public services less through government programmes and more to individuals' requirements. Before digital government becomes universal, however, it must address issues raised by the DIGITAL DIVIDE – a gap in access to the internet, and thus to digital government services, when people are sorted by income levels and education levels. People with higher incomes and educational levels almost universally have access to the internet, but people with low incomes and educational levels in many instances lack internet access, and so are unable to participate in digital democracy (e.g. internet voting) or benefit from internet access to

public services. Addressing the digital divide through public policy with the aim of making high-speed broadband access as available as telephone services throughout metropolitan areas should facilitate the development of digital government and the efficiencies and benefits it promises.

Further reading

Cohill, A. and Kavanaugh, A. (eds) (2000) *Community Networks: Lessons from Blacksburg, Virginia*, Boston, MA: Artech House.

Howard, M. (2001) 'e-government across the globe: how will "e" change government?', *Government Finance Review* 17(4): 8.

Susman, A.M. (2001) 'The good, the bad, and the ugly: e-government and the people's right to know', *Vital Speeches* 68(2): 38.

SEE ALSO: cybercity; informational city; Smart Community; techno-city; technopoles

ROGER RICHMAN

DIRECT DEMOCRACY

Direct or participatory democracy is allowing citizens to participate directly in managing their own affairs. Direct democracy first flourished in the Greek city-states, reaching its fullest expression in ancient Athens. There, the citizens participated (see CITIZENSHIP) directly in governance as members of the popular assembly and citizen participation was considered an honour and a civil duty, but it was often limited by class distinctions.

The theory and practice of direct democracy was the core of the work of many theorists, philosophers and politicians alike. The common characteristic of their work is their stress on participation. Here, reference will be made to the approaches of the three most important ones: Jean Jacques Rousseau, John Stuart Mill and G.D.H. Cole.

Rousseau is considered as the theorist par excellence of participation, and his work *The Social Contract* is vital for the theory of direct democracy. In his image for the ideal society he refers to a non-industrial city-state, made up of small, peasant proprietors, and he advocates a society of economic equality and economic independence. Rousseau's entire political theory about direct democracy hinges on the individual participation of each citizen in decision-making. In

his theory, participation also has a psychological effect on the participants, ensuring that there is a continuing interrelationship between the workings of institutions and the psychological qualities and attitudes of individuals interacting within them. Each citizen would be, as he puts it, 'excessively dependent on the republic' and equally subject to law. He also argues that in direct democracy, the feeling among individual citizens of 'belonging' to their community is increased. Rousseau's direct democracy makes two points clear: first, that participation should be in the making of decisions, and secondly, that it is a way of protecting private interests and ensuring good government. According to Rousseau's political theory, there is an interrelationship between the authority structures of institutions and the psychological qualities and attitudes of individuals. The theories of J.S. Mill and Cole reinforce Rousseau's arguments about the educative functions of direct democracy, but more interestingly, in them the related theory is lifted out of the context of a city-state of peasant proprietors into that of a modern political system.

John Stuart Mill considers 'good government' as one form of security against 'the sinister interests of the holders of power', provided that it sufficiently promotes the good management of the affairs of society. Fundamental is government, which has great influence acting on the human mind, and promotes the general mental advancement of the community, including advancement in intellect, in virtue, and in practical activity and efficiency. In Mill's arguments, there is the basic assertion that responsible social and political action depends largely on the sort of institutions within which the individual has, politically, to act. Like Rousseau, Mill sees these qualities being as much developed by direct democracy institutions as existing beforehand, and thus the political system has a self-sustaining character. For Mill, it is at the local level where the real educative effect occurs, since by participating at the local level the individual 'learns democracy'. Participation also aids the acceptance of decisions and Mill specifically points to the integrative function of participation. He says that through political discussion, the individual becomes consciously a member of that great community. In his theory, apart from the merits of participation as an educative device, Mill regards direct democracy as something very much wider than a set of 'institutional arrangements' at the national level and as expanding in all spheres of human expression. This wider view of direct democracy can also be found in the political theory of G.D.H. Cole.

In his book *Guild Socialism Restated*, Cole worked out a detailed scheme of how direct democracy might be organized and brought into being. The guild was to be the unit of organization on the production side. The Guild Socialist structure was organized vertically and horizontally, from the grass roots upwards, and was participatory at all levels and in all aspects. It provided for local communes for each town or country area, for regional communes bringing together both town and country, and for regional guilds and the National Commune, which would be a purely coordinating body neither functionally, historically nor structurally continuous with the existing state. Cole's social and political theory is built on Rousseau's argument that will, not force, is the basis of social and political organization. To translate their will into action in a way that does not infringe upon their individual freedom, Cole argues that men must participate in the organization and regulation of their associations. The democratic principle, Cole says, must be applied not only or mainly to some special sphere of social action known as 'politics', but to any and every form of social action. Like Mill, Cole argued that it was only by participation at the local level and in local associations that the individual could 'learn democracy'. For Cole, like Rousseau, there could be no equality of political power without a substantive measure of economic equality, and his theory provides some interesting indications of how the economic equality in Rousseau's ideal society of peasant proprietors might be achieved in a modern economy.

In the above theories of direct democracy, there is a central assertion about the stability of such a system; participation develops and fosters the very qualities necessary for it; and the more individuals participate, the better able they become to do so.

Finally, in examining direct democracy as a better alternative to representative democracy, besides its virtues, one has also to tackle some serious problems related to the very definition and assessment of participation, and the extent to which the paradigm of direct democracy can be replicated under conditions where representation is going to be widely necessary.

Further reading

Caves, R. (1992) *Land Use Planning: The Ballot Box Revolution*, Thousand Oaks, CA: Sage.

Cole, G.D.H. (1920) *Guild Socialism Restated*, London: Leonard Parsons.

Pateman, C. (1974) *Participation and Democratic Theory*, Cambridge: Cambridge University Press.

KONSTANTINOS LALENIS

DISCRIMINATION

The concept of discrimination refers to the ability to discern differences and act accordingly. One could, for example, say about a dog that he is good in discriminating between different scents. Usually, however, discrimination is applied to social contexts, where the concept alludes to unequal treatment of groups of basically equal status. This qualifier, *basically equal status*, is important, since individuals in all societies are differentiated by a number of criteria. Distinctions are made on the basis of age groupings, sex roles, family ties, education, and according to the division of wealth and labour. Most societies have distinct social classes and the discrepancy between those in superior and those in subordinated positions is clearly evident. Discrimination is one of several tools for the rich and powerful to maintain their wealth and positions. Discrimination is also a tool for people lower down on the ladder to raise themselves in the status hierarchy. The unaccepted behaviour is culturally determined, which is to say that it varies between different societies and different ages. In other words, it is important how the distinction is drawn between accepted and unaccepted strategic behaviour. Subsequently, what is labelled as discrimination should, as something unacceptable, be counteracted, fought and legislated against.

Discrimination is attributed to contexts like housing, employment, age, race and ethnicity. We talk about discrimination against women, disabled persons, homosexuals, religious, political and sexual minorities, people of colour, indigenous people, people favouring a deviating lifestyle, etc. It is a concept referring to a hegemonic group wielding power over weaker groups. Usually, but not always, it is about a majority oppressing minorities. As the concept connotes something unfair, unreasonable and arbitrary, there are, around the world, committees and different organizations that fight against discrimination. In Sweden, for example, there is even an *ombudsman* against ethnic discrimination.

The standardized scenario is like this: prejudice (something mental) leads to discrimination (acting), with segregation (social and physical environments) as a problematic result. Discrimination is, however, not the unavoidable consequence of prejudice. Prejudiced individuals do not always discriminate, and prejudice is not always the reason why people discriminate. Often, there are economic reasons behind discrimination – for example when restaurants deny access to a certain clientele. The owners are afraid of lowering the status of their establishment, with fewer clients and diminished incomes as a result. Neither are clearly segregated URBAN environments necessarily a result of discrimination. GHETTOS and SLUM areas are rather to be understood as what is left when people who had an opportunity to move elsewhere, for one reason or another, did so. Concepts such as 'WHITE FLIGHT' and 'GATED COMMUNITIES' say something about these processes, where people no longer wish to identify and make common cause with society as a whole. They prefer to live in COMMUNITIES made up of people like themselves. When they drop 'the presence of otherness', they could be said to discriminate.

In a way, discrimination is an instrument for the shaping of order, and it should come as no surprise that modern planning has its streaks of discrimination. ZONING is one example, SOCIAL HOUSING another. This idea of everything in its proper place has become obsolete in recent planning theory. The dense, mixed and multicultural city is preferred to the sprawling and divided city, and older European cities often serve as the prototypes for the cities of tomorrow. Nevertheless, discriminating processes are powerful. The urban landscape of today as well as of tomorrow, in its basic components, is the result of discriminatory thinking, with single-use physical objects like business towers, malls and single-family houses as a result.

In postmodern theory, discrimination as a human act is as negative as in modern ways of understanding society. Nevertheless, there is a twist in the conceptualizing, since *difference* is a concept with a positive ring to it. As Ruth Fincher and Jane M. JACOBS (1998) write: 'What happens to studies of housing, suburbia, the inner city, ghettos, gentrification, social polarization and urban social

movements when framed not by a theory of "the city" but by theories of difference?'

Further reading

Blakely, E.J. and Snyder, M.G. (1997) *Fortress America: Gated Communities in the United States*, Washington, DC: The Brookings Institution Press.
Body-Gendrot, S. (1997) *The Social Control of Cities? A Comparative Perspective*, Oxford: Blackwell.
Ellin, N. (ed.) (1997) *Architecture of Fear*, New York: Princeton Architectural Press.
Fincher, R. and Jacobs, J.M. (eds) (1998) *Cities of Difference*, New York: The Guilford Press.

KARL-OLOV ARNSTBERG

DISINVESTMENT

Disinvestment is an important contributing cause of the decline of lower-income neighbourhoods, often those in the older CENTRAL CITY but also older inner-ring SUBURBS and smaller rural COMMUNITIES. It represents the cumulative effect of many individual decisions not to invest further resources in the physical and social infrastructure of an area. Continued disinvestment may lead to abandonment. If enough structures are abandoned, the neighbourhood may be blighted, leading to further declines in property values in a self-reinforcing cycle. Crime, declining employment opportunities, a concentration of households in poverty, a decline in social institutions such as schools and civic or religious organizations, and a resulting loss of SOCIAL CAPITAL are all symptoms of disinvestment. Some have argued that social disinvestment (as residents stop taking care of their neighbourhood or participating in its institutions) may be a precursor to financial disinvestment.

Several groups of actors may decide to disinvest. Banks and other private firms may stop making loans or investments in particular neighbourhoods (see REDLINING). Loans and business investments may be seen as risky ventures in neighbourhoods where property values are falling. This becomes a self-fulfilling prophecy, as property owners and firms without access to credit or capital will be unable to maintain the value of their assets in the long term. Critics argue that disinvestment represents prejudice against particular types of neighbourhoods, rather than an economically motivated decision. Neighbourhoods with high concentrations of racial minority residents may be far more likely to experience disinvestment than predominately white neighbourhoods. Financial disinvestment makes it more likely that other groups of actors will decide to disinvest. Calvin Bradford and others have argued that disinvestment by private sources of capital is most problematic, because it leads residents of those neighbourhoods to rely instead on public sources of capital, creating a dual financial system. The abuse of government-insured mortgage programmes in the United States, by lenders who have incentives to make loans regardless of whether they will be repaid, has led to widespread mortgage defaults and thus abandonment and BLIGHT in particular neighbourhoods. Disinvestment by conventional banks is also seen as one of the reasons for the rise in predatory lending in those neighbourhoods.

Property owners may stop maintaining their buildings in response to perceived or actual declines in property values. If the market value of a building has declined, investing more money in maintaining it may result in further losses. Absentee landlords are most likely to disinvest, collecting rents but spending nothing on maintenance. However, homeowners may also disinvest if they lack access to credit for essential maintenance. City governments may disinvest by deciding to reduce maintenance of public infrastructure, responding to and reinforcing declines in property values and thus property taxes.

BROWNFIELD sites (obsolete, abandoned property that may be contaminated) represent a special form of disinvestment that accelerates neighbourhood decline, because of the environmental threats the sites pose and because of the complexity and expense of redeveloping them. Attracting new investment to brownfield sites is complicated by concerns about liability and about the profitability of investing in a blighted neighbourhood.

Geographer David HARVEY developed one theory of how and why disinvestment occurs. He argued that speculators make profits through a cycle of investment and disinvestment in particular neighbourhoods. Older neighbourhoods with many obsolete structures gradually undergo disinvestment that leads to their abandonment. Disinvestment during this period is a way for owners to maximize their profits – continuing to collect rents but not using the money to cover operating expenses, because the property would be almost

worthless on resale. At some point, the property indeed becomes worthless when it is uninhabitable. Widespread disinvestment results in neighbourhood-wide blight. Abandoned neighbourhoods are left to 'lie fallow' for a period until property values have dropped to the point where they become attractive targets for speculators. At that point, they are ripe for reinvestment, often resulting in GENTRIFICATION and the displacement of the remaining residents or businesses by more affluent households and firms.

Disinvestment poses difficult urban policy questions, because it is a self-reinforcing cycle. It is one cause of urban SPRAWL, and it also affects the well-being of residents left behind when investment leaves. Community development efforts have tried to reverse this cycle through a variety of strategies. One group of strategies attempts to revitalize neighbourhoods by providing spatially targeted incentives for business investment, such as enterprise zones or empowerment zones. Incentives are also used to attract INFILL DEVELOPMENT to abandoned sites by subsidizing private developers. Often, this strategy aims to attract middle-class residents back to neighbourhoods, and it may entail some gentrification disguised as 'income mixing'.

Another group of strategies seeks to reverse disinvestment by forcing banks to serve the markets they discriminate against, or by providing alternative sources of capital to neighbourhood firms and residents. Community reinvestment agreements with banks and the active enforcement of anti-discrimination laws have been effective ways to reverse disinvestment. Community development financial institutions (such as locally supported credit unions, venture capital funds for community development or banks that specialize in meeting the needs of a specific community) provide alternative sources of investment targeted at specific neighbourhoods, to counter disinvestment.

The first set of strategies grapples with the problem of reversing the market forces that undermine the economic feasibility of investing in a declining area. The second set of strategies addresses the political and social roots of disinvestment as well, by attempting to regulate private industry to prohibit discrimination and to set up alternatives to the private sector institutions that some argue have devastated many city neighbourhoods.

Further reading

Bradford, C. and Rubinowitz, L. (1975) 'The urban–suburban investment–disinvestment process: consequences for older neighborhoods', *Annals of the American Academy of Political and Social Science* 422: 77–86.

Harvey, D. (1978) 'The urban process under capitalism: a framework for analysis', *International Journal of Urban and Regional Research* 2(1): 101–31.

Squires, G.D. (ed.) (1992) *From Redlining to Reinvestment: Community Responses to Urban Disinvestment*, Philadelphia, PA: Temple University Press (see esp. chs 1 and 9).

HEATHER MacDONALD

DISNEY ENVIRONMENTS

Disney environments constitute the example par excellence of THEME PLACES. Disneyland was the first theme park (it opened in 1955 in Anaheim), although it was not a wholly original idea since it borrowed several ideas from De Efteling Park (Eindhoven). The next Disney park opened in 1971 near Orlando in Florida: it was the first phase of Walt Disney World (WDW) and the biggest private plan in North America. The global dimension of theme parks can be observed in the ease with which the same formula was diffused in Europe, i.e. the bigger British theme parks have been modelled on the popular Disney complexes in Florida and California. The first attempt by Disney for a large-scale theme park in Europe took place in France: Disneyland Paris opened in 1992 at Marne-la-Vallée.

Entries to the Disney parks in California and Florida are by far the highest, compared to theme parks all over the world, with over 14 million visiting Anaheim in 1995. By 1996, WDW had received 500 million visitors, managing to attract over 30 million a year. The attendance figures for Disney USA parks can only be compared with those for Disney parks in other countries: Tokyo had 16 million visitors in 1993. In Europe, the most popular theme park is Disneyland Paris, with over 11.7 million visitors in 1996.

The peculiarity of Disney parks lies in the combination of various characteristics: a unique attempt to make imagination safe, a will to express order, the existence of visual coherence, the absence of transport and pollution problems, and

an effort (one of the first) to apply globalization. Walt Disney had a utopian dream, similar to that of HOWARD, to rebuild the city as it should be, not as it actually was. The organization and scale of the parks is that of the GARDEN CITY.

The Walt Disney Corporation (WDC) had a particular policy towards the design of the parks: it showed a preference for 'entertainment architecture' and it used respectable (mostly postmodern) architects. The WDC has also been involved in URBAN PLANNING projects, such as the Times Square REDEVELOPMENT and the new town of Celebration. The first is an example of 'urban theming' and reflects the recent shift in urban economic development strategies towards leisure services. Disney was particularly involved in theatre renovation. Celebration is an example of 'NEW URBANISM'. Its idea as an experimental community prototype was first conceived for the second phase of WDW, i.e. EPCOT (Experimental Prototype Community of Tomorrow). Celebration is a pedestrian-oriented, mixed-use development built in Osceola County between 1991 and 1996, with 20,000 inhabitants and 8,000 housing units designed by top-name architects. The impact of Disney environments in urban development can be characteristically observed in that the establishment of WDW in Central Florida has contributed to the fact that this area has been the fastest-growing area of the state.

Further reading

Marling, K.A. (ed.) (1997) *Designing Disney's Theme Parks*, Paris and New York: Flammarion.
Ross, A. (1999) *The Celebration Chronicles*, New York: Ballantine Books.

SEE ALSO: culture

ALEX DEFFNER

DISPERSION

The term 'dispersed city' was originally used in the field of URBAN and economic geography in conjunction with CENTRAL PLACE THEORY. It defined an urban agglomeration, composed of a series of politically autonomous but geographically linked urban places, that together were able to provide high-order functions such

as would usually only be found in a larger, MONO-CENTRIC CITY. This usage is no longer popular, and the term 'dispersion' is now commonly used to describe outward types of urban development associated with post-Second World War suburban form in North America. It is sometimes (incorrectly) used synonymously with DECENTRALIZATION and SPRAWL.

Dispersed styles of URBAN FORM are characterized by decentralization of almost all types of activities from the core, by relatively low residential and employment densities, by plentiful open space, by relatively unfocused travel patterns and by near universal dependence on the car. This urban form produces many negative environmental impacts and requires costly infrastructure. It is also blamed for creating suburban gridlock and amplifying SOCIAL POLARIZATION. Policies that aim to reduce dispersion include: COMPACT DEVELOPMENT; GROWTH MANAGEMENT; MULTICENTRIC CITY form; neo-traditional planning; NEW URBANISM; SMART GROWTH; SUSTAINABLE URBAN DEVELOPMENT; TRANSIT-ORIENTED DEVELOPMENT; and URBAN GROWTH BOUNDARIES.

The major factors known to promote dispersed development are consumer preference for low-density, single-family housing, for privacy and green, rural-like landscapes, and for auto-based travel and auto-oriented communities. Widely perceived association of these types of living arrangements with healthy environments and community safety compounds this consumer predisposition. The other major factor that propels this form of development in the face of an increasingly strong critique is economic. Assuming land is available, the dispersed style of development is believed to be less costly for developers than more compact alternatives or central city REDEVELOPMENT – unless strong incentives are provided for alternative development initiatives. Other factors recognized as promoting dispersion include: central business district (CBD) and CENTRAL CITY decline; the downward FILTERING PROCESS; WHITE FLIGHT; and the growth of an URBAN UNDERCLASS.

Bunting and Filion (1999) distinguish between dispersed styles of urban development which prevail everywhere across suburban North America and 'dispersed cities' which are characterized by depleted CBDs so that features of dispersion prevail throughout the entire cityscape, rather than

being confined to outer, 'suburban' zones. Their dispersed city form is centred on a depleted CBD. This distinctive city type is more likely to be found among places with a mid-size population, a manufacturing economic base and/or a twentieth-century development history.

Further reading

Bunting, T. and Filion, P. (1999) 'Dispersed city form in Canada: a case of Kitchener CMA', *The Canadian Geographer* 43(2): 268–87.
Filion, P., Bunting, T. and Warriner, K. (1999) 'The entrenchment of urban dispersion: residential preferences and location patterns in the dispersed city', *Urban Studies* 36: 1317–47.

TRUDI E. BUNTING

DIVIDED CITIES

Inhabitants of cities may be spatially sorted (SEGREGATION) by residence on the basis of such individual characteristics as ethnic identity, race, income and age. Such spatial and social divisions in cities can be produced through self-selective sorting into ETHNIC ENCLAVES by individuals with like characteristics, economic determinants, racial discrimination by private and public institutions, and/or racial prejudice by private individuals.

African-American – white segregation persists at high levels in US cities, despite decades of national government efforts. According to the 2000 US Census, there is little change in residential segregation despite the growing ethnic diversity of the country. The typical white lives in a neighbourhood that is nearly 83 per cent white, 7 per cent African-American, 6 per cent Hispanic and 3 per cent Asian. The experience of minorities is very different. For example, the typical African-American lives in a neighbourhood that is 54 per cent African-American, 33 per cent white, 9 per cent Hispanic and 3 per cent Asian.

Measured by the INDEX OF DISSIMILARITY, African-American – white residential patterns in the top fifty metropolitan areas in the United States show slightly declining, but stubbornly high, segregation, both in CENTRAL CITIES and SUBURBS. In 1967, the Kerner Commission (National Commission on Civil Disorders) in the United States warned of 'two nations, one black, one white, separate and unequal'. There has been little change since 1970 in residential segregation of

African-Americans – in some smaller and newer metropolitan areas, their segregation from whites had declined slightly, but in the larger places where most African-Americans lived, segregation remained high and unchanging. Hispanics and Asians were less residentially segregated in most parts of the country than African-Americans. However, Hispanics and Asians lived in more isolated settings in 2000 than they did in 1990, with a smaller proportion of white residents in their neighbourhoods. In Britain's cities, the extent of ethnic concentration is less than in the United States, although the distribution of foreign-born population in London and other cities displays socio-spatial segregation.

Along with segregation, concentrated poverty has become an indelible part of the American metropolitan landscape. In 1990, 8.5 million people lived in census tracts having poverty rates in excess of 40 per cent, compared to 4.2 million people in 1970. Fully 33 per cent of the African-American poor in the United States live in these high-poverty neighbourhoods, almost 30 per cent of which are composed of 90 per cent African-American or more. The location of households and individuals in segregated, poverty-stricken neighbourhoods significantly influences the quality of their schools, the level of municipal services, tax burdens, access to work and the level of safety. As poverty concentrates, the physical fabric of neighbourhoods deteriorates, anti-social behaviours take over the streets, and communities are spatially and psychologically separated from good URBAN jobs, education and amenities. Because racial segregation concentrates poverty and systematically builds deprivation into the residential structure of African-American communities, it has been linked to the creation of an URBAN UNDERCLASS.

In many urban areas, the building of walls, separations, and the policing of boundaries are used as ways to organize difference amid violence and fear. Examples include GATED COMMUNITIES, the bisecting and cordoning-off of inner-city minority neighbourhoods by major urban highways, and the increased use of private security personnel and hardware to monitor urban space. There is much concern that such divisions of cities into rich–poor or by colour are increasingly closing off the common physical and psychological spaces needed for civic GOVERNANCE and society to continue. Peter Marcuse has described five types of

neighbourhoods in the divided, or quartered, city. A 'dominating city' consists of high-income enclaves and luxury buildings, a 'gentrified city' contains residences of professional and managerial groups, a 'suburban city' consists of single-family housing in the outer city, a 'tenement city' contains cheaper housing of lower-paid, blue-collar workers disproportionately African-American and ethnic minority, and 'abandoned cities' are places left for the poor, the unemployed and the excluded.

Increased economic globalization, according to Saskia SASSEN, is increasing socio-economic division and social segmentation in large, globally connected world cities as growth contributes to a bifurcation of the economy into low-wage service and high-wage advanced information jobs rather than an expansion of the middle class. The social segregation and economic inequality associated with the separate societies of divided cities can stimulate ETHNIC CONFLICT if the issues and grievances related to group-based inequalities are not effectively addressed in the political process. In the most extreme examples of urban divisions (such as Cold War Berlin, Jerusalem between 1948 and 1967, and contemporary Nicosia (Cyprus)), cities are physically divided into two parts by walls due to ideological disagreements and fights over competing political claims to the city.

Further reading

Feinstein, S., Gordon, I. and Harloe, M. (1992) *Divided Cities: Economic Restructuring and Social Change in London and New York*, New York: Blackwell.
Goldsmith, W.W. and Blakely, E.J. (1992) *Separate Societies: Poverty and Inequality in U.S. Cities*, Philadelphia, PA: Temple University Press.
Marcuse, P. (1997) 'Walls of fear and walls of support', in Nan Ellin (ed.) *Architecture of Fear*, New York: Princeton Architectural Press, pp. 101–14.
Massey, D.S. and Denton, N.A. (1993) *American Apartheid: Segregation and the Making of the Underclass*, Cambridge, MA: Harvard University Press.

SCOTT A. BOLLENS

DOCKLANDS

Generically, the term refers to those areas of a city devoted to commercial waterborne transportation and includes docks, wharves, piers, quays, dry docks and associated landside facilities. The term is more common in Britain and the British Commonwealth.

However, 'the Docklands' has come to refer to that portion of the City of London east of the city centre and approaching Greenwich. Historically, that portion of the city was reclaimed from marshland, Stepney Marsh, and since the early nineteenth century has been devoted to waterborne transportation.

In 1802, the West India Docks opened and were considered at that time to be the country's greatest civil engineering structure. 'The great stretch of docks from St. Katharine's by the Tower of London, past Surrey Commercial Docks, through the East and West Indies and the Royal Docks to Barking was the world's largest port', states Canary Wharf Group PLC, the current owner of a large portion of the Docklands.

While the docks at one time employed thousands, the advent of containerized cargo in the 1970s saw rapid disuse of the facilities as cargo operations moved to more favourable locations. Realizing the public consequences of disuse, Parliament in 1980 created the London Docklands Development Corporation (LDDC) charged with redeveloping the area (see REDEVELOPMENT). The LDDC was to oversee the restoration of buildings, bring appropriate industry to the area, improve aesthetics and assist in providing housing to encourage Londoners to live and work in the same area.

Early in the 1980s, one portion of the Docklands, the Isle of Dogs, was quickly designated an enterprise zone with tax incentives to encourage development. Following its designation, G. Ware Travelstead proposed developing a 10 million square feet office complex on a portion of the Docklands, Canary Wharf (named for its nineteenth-century role as the port of entry for goods from the Canary Islands). Unable to fund his plan, Travelstead sold his interest in Canary Wharf to the firm of Olympia & York of Toronto, Canada. By 1987, the firm reached an agreement with the LDDC for a 12.2 million square feet development at Canary Wharf. Four years later, the LDDC was able to celebrate both its 10th anniversary and the opening of the first phase of the Canary Wharf complex.

However, in 1992, Olympia & York filed for bankruptcy on the project, and over the next three years the LDDC and the company endeavoured to work out a scheme whereby the project could be completed. By 1995, new owners resumed construction. As part of their DEVELOPMENT AGREEMENT,

Olympia & York had agreed to jointly fund an extension of the Jubilee Line of the London Underground to Canary Wharf. By 1993, construction commenced and the line was completed to Canary Wharf in 1999.

By 1998, the series of projects at Canary Wharf were well underway by private developers and the LDDC closed its doors, the project seen as a success. At that time, with other projects on the drawing boards, the working population at Canary Wharf stood at 35,000.

JAMES A. FAWCETT

DOT DENSITY MAP

If someone presents data in a map, it is important to choose the proper type of map. There are a variety of different types of maps available, but by choosing the wrong type of map it is possible to completely misrepresent the data. The right type depends on the nature of the data, the audience and the purpose of the map.

The two main ways of representing data are by choropleth (the colour of an area represents its value) or proportional symbol maps. In most cases, the area is represented by its total or mean value, for instance the average population density of a region. But usually, the data is not equally spread over a region, but concentrated and dispersed. Therefore, if accurate point data is available, a dot location map will probably be the best. A dot map represents events as much as possible on the right location. This is possible as long as there is only one event per dot. If a dot represents more events, one will choose the location of the dot as a gravity or mean point. Dot maps can be a very useful method of identifying the location of surveyed points or events. They can help in spotting patterns that cannot be uncovered any other way.

An often-used example is a population map. A choroplete would show one colour for the average value of a region. A dot density map shows one dot for each, say, 25 inhabitants. Built-up areas are identified by a cluster of dots, while it is clear that the rest of the region is less populated. In doing this, one gets a more differentiated image of the region.

If one knows the real location of events and can associate points with each one, it is possible using mapping software to create a good and useful dot map. However, one needs to be very careful here. Most GIS (GEOGRAPHIC INFORMATION SYSTEMS) packages provide a dot density map. Here a random pattern of points is used to represent the value in a given area – the more points, the higher the value. Although such a map gives a nice picture, this type (which is really nothing to do with dots, but rather just a particular type of choropleth map) is very dangerous. It is too easy for the reader of the map to assume that each point on the map represents a real occurrence of an event. It does not; each one is just randomly placed in the area. Readers will regularly misinterpret random lines or clusters on the map as roads or towns. Even with experience, it is easy to fall into this trap.

In sum, if the data is available on a detailed level, a dot density map allows one to present a detailed picture. However, if the data is not available on a detailed level, a dot density map may induce misinterpretations.

Further reading
ESRI (1996) *Using ArcView GIS*, Redlands: ESRI.

KEES MAAT

DOWNTOWN

Downtown refers to the concentrated employment, shopping and recreation district that developed in American cities from the late nineteenth century to the mid-twentieth century. It is marked by a cluster of tall buildings, by cultural institutions and by the convergence of rail transit and bus lines. Although the following discussion focuses on the United States, downtown districts are a common feature of cities in Western capitalist societies. With variations arising from national histories and regimes of land management, US examples have their counterparts in Canada and New Zealand, in Britain and Peru.

The term 'downtown' originated in New York, where residents in the nineteenth century began to distinguish between uptown, midtown and downtown sections of Manhattan. Because downtown Manhattan was an early centre of wholesale and retail commerce, the term was adopted by many other cities to refer to comparable districts. Several large cities have distinctive terms, such as 'Center City' in Philadelphia and the 'Loop' in Chicago, but 'downtown' is the widely recognized generic term.

Downtowns evolved in the course of the nineteenth century from mixed financial, whole-saling and manufacturing districts that had appeared along waterfronts or around land transportation terminals. Key factors were the expanded buying power of the middle class in growing cities, the construction of streetcar systems that radiated from a single core, and new building technologies that allowed office sky-scrapers and massive department stores. The downtown of a large American city in the early twentieth century typically contained several massive department stores surrounded by smaller shops; bank headquarters; 10–20-storey office towers; government buildings; theatres and concert halls; and large hotels. Much of the effort of CITY BEAUTIFUL planning between 1900 and 1920 was devoted to improving transportation and providing public spaces in these growing and congested districts.

It is possible to describe a continuum of central business districts that vary in size with the population of their town or city. The small-town 'Main Street' district is downtown writ small, with a city hall or courthouse square, several blocks of banks and stores, and perhaps an outmoded hotel building. Downtown districts step up in size and complexity with the increased population of their city and HINTERLAND.

Public policy since the 1950s has tried to anchor and support the private functions of downtowns through urban renewal and targeted public investment. The older retail and office core is likely to contain one or more new mixed-use developments and FESTIVAL MARKETPLACES developed with public financial assistance. The core is likely to be ringed by such facilities as a convention centre, a sports arena and/or stadium, museums, a community college or university campus, and a hospital/medical complex.

Downtowns as residential districts tend to accommodate the poor and the affluent, but few members of the middle class. Skid row districts that served single working men in the early twentieth century are now likely to house pensioners in single room occupancy hotels and the homeless on streets and in shelters. The affluent live in high-rise apartment and condominium towers from the urban renewal era of the 1960s and 1970s and in newer residential districts of converted office and loft buildings and newly built imitations. Such newer districts proliferated in the 1990s. More affluent downtowners include both 'empty nesters' and younger professionals.

The transition from 'mainstream' employment to tourism was an important change in the function of downtowns in the 1980s and 1990s. As routine office work has been automated, moved to suburban locations or even displaced overseas, public officials and property owners have tried to attract conventions and recreational tourists with festival marketplaces, museums, PARKS, plazas, performing arts centres, sports facilities, upscale hotels, and supporting restaurants and clubs. They hope that the same mix will also attract suburbanites who no longer work downtown or shop there on a regular basis. Examples of cities that followed this strategy include Baltimore, Cleveland, San Antonio, Seattle and Denver.

At the start of the twenty-first century, American downtowns presented a mixed picture. In many middle-sized cities, downtown had failed to maintain a critical mass of private uses. Losing too much shopping and employment to SUBURBS and too many corporate headquarters to buyouts, downtowns struggle to survive on government offices and special attractions. In larger cities with booming economies, the picture is different. Professional and business services and information industries are supplanting routine back-office jobs. Economic strength, the associated downtown housing boom, and the emphasis on culture and recreation have resulted in lateral growth. In the 1950s, experts distinguished between a downtown core or central business district and a frame or fringe of lower-intensity uses (cheap housing, parking and automobile services, wholesaling, institutions). In cities that prospered during the 1980s and 1990s, downtowns expanded into adjacent fringe areas – south of Market Street in San Francisco; north of Burnside Street and east of the Willamette River in Portland, Oregon; west from the downtown core in Atlanta.

Further reading

Abbott, C. (1993) 'Five downtown strategies: policy discourse and downtown planning since 1945', *Journal of Policy History* 5(1): 5–27.

Fogelson, R. (2001) *Downtown: Its Rise and Fall, 1880–1950*, New Haven, CT: Yale University Press.

Ford, L. (1994) *Cities and Buildings: Skyscrapers, Skid Rows, and Suburbs*, Baltimore: Johns Hopkins University Press.

Frieden, B. and Sagalyn, L. (1989) *Downtown, Inc.: How America Rebuilds Its Cities*, Cambridge, MA: Harvard University Press.

CARL ABBOTT

DOXIADES, CONSTANTINOS

b. 14 May 1913, Stenemachos, Greece;
d. 28 June 1975, Athens, Greece

Key works

(1960) Master Plan, Islamabad.
(1968) *Ekistics: An Introduction to the Science of Human Settlements*, London: Hutchinson.

Constantinos Doxiades was a Greek city planner and founder of EKISTICS. Doxiades graduated in architecture from Athens Technical University and obtained his doctorate at Charlottenburg University, Berlin. Doxiades began his career as chief town planning officer for the Greater Athens area in 1937 and later became the first head of the Office of Regional and Town Planning, Ministry of Public Works. During the occupation of Greece in the Second World War, he took part in the resistance and published an underground magazine called *Regional Planning, Town Planning and Ekistics*. After the war, Doxiades played a considerable role in restoring Greece to a normal peacetime existence, first as under-secretary and director-general of the Ministry of Housing and Reconstruction (1945–48), and subsequently as coordinator of the Greek Recovery Programme and under-secretary of the Ministry of Coordination (1948–51). Over these years, as well as in the 1950s and 1960s, he also served as Greece's representative on many international forums and as a consultant to several countries and the United Nations.

In 1951 he founded Doxiades Associates, a private consulting firm, with a small group of architects and planners. The company grew rapidly until it had offices on five continents and undertook projects in forty countries throughout the world, such as Pakistan, Iran, Saudi Arabia, Ghana, Nigeria, Brazil, United States, France, Spain, Yugoslavia and Greece. These projects included the Plan of Islamabad, the new capital of Pakistan for 2 million people, an URBAN renewal plan for Eastwick, Pennsylvania, the National Physical Plan of Spain, and numerous other assignments in REGIONAL and URBAN PLANNING, architecture, public works, agriculture, housing, transportation, etc.

Doxiades' professional work grew in parallel to, and partly reflected, the development of his theoretical ideas. In a series of publications, he proposed a new science, ekistics, as an interdisciplinary effort to solve the mounting problems of contemporary cities and URBANIZATION. Doxiades believed that with proper planning, the cities of the world would eventually mature into a stable and attractive form he called the ECUMENOPOLIS. He worked for more than a decade to establish an international and interdisciplinary community of scholars who would complete research on cities and the best ways to manage them, and he created the Athens Technological Organization (1959) and the Athens Centre of Ekistics (1963). *Ekistics*, the journal he founded, continues to publish articles on a wide range of urban topics. Although his writings often do not follow strictly academic lines, many of his ideas have influenced the disciplines dealing with human settlements.

Doxiades received several awards and decorations, including the Order of the British Empire, the Sir Patrick Abercrombie Prize of the International Union of Architects, and the Royal Architectural Institute of Canada's Gold Medal.

Further reading

Doxiades, C. (1966) *Between Dystopia and Utopia*, Hartford, CT: Trinity College Press.
(1976) *Ekistics* 41(247): whole issue.

DIMITRIS ECONOMOU

DRUGS

The production, trafficking and use of illicit drugs (e.g. heroin, cocaine and LSD) have attracted significant attention in recent years. Following the prohibition of certain drug use in many developed and advanced economies in the early twentieth century, illegal production, distribution and use of illegal drugs has become viewed as an increasingly worrying phenomenon by political and moral leaders around the world. Such concern has resulted in sustained and systematic responses to illegal drug use at the local, regional, national and global levels. Illegal drug use is now seen as a major global problem in terms of production, trafficking

and use and in terms of the perceived relationships between such drug use, crime and public health concerns. Such concerns have been particularly heightened by the apparent rapid increase in illegal drug use in many countries throughout the 1980s and 1990s and with the rise of powerful, organized and transnational cartels in other regions such as South America.

Frequently, concern has been centred on the impact of illegal drug use on URBAN crime rates and perceptions of urban danger in terms of street crime (mugging), gun crime and sex-working. In terms of city and URBAN PLANNING, the impact of concerns and anxieties about illegal drug use can be seen at both macro and micro levels. At the macro level, significant political and policy attention has been focused on attempts to strategically address illegal drug production and trafficking in terms of cross-national anti-drug cooperation and initiatives through agencies such as the United Nations, the G8 and the European Union. At national levels, considerable attention has been focused on what are characterized as the (external) threats of trafficking, on domestic serious crime and on the links between illegal drug use and broader issues of social and economic regeneration and inclusion. At more micro or local levels, measures to combat illegal drug dealing and use in urban areas include the use of fixed and permanent barriers such as gated walkways, CCTV, certain forms of street lighting, low-level market disruption by security agencies and various other measures. In addition to the type of crime reduction and control measures outlined above, significant emphasis on harm reduction and minimization has impacted on the visibility of illegal drug use in urban areas including the use of needle disposal schemes and experiments with the development of officially sanctioned and supported public injecting spaces in some countries such as Australia and Switzerland. Public and political concerns with illegal drug use and other associated behaviours which are also seen as high risk – for example intravenous injecting drug use and the sharing of drug paraphernalia – have also resulted in high-profile public information and awareness campaigns in many cities around the world.

Symbolically, illegal drug use has played a significant role in the discursive construction of the city as dangerous, exotic and unruly, with emphasis placed on the 'opium den', 'shooting gallery' and, most recently, 'crack house'. But such 'dangerousness' is also accompanied by more ambiguous and celebratory notions that have attended the recreational use of illegal drugs such as marijuana, amphetamine and certain psychedelic drugs and, most recently, have seen the rise and celebration of the 'chemical generation' and the recreational use of Ecstasy. Furthermore, the cultural significance of illegal drug use has been an important element of how illegal drug use has been popularly understood. Historically, strong associations – often inspired by overt or implicitly racist discourses and attitudes – have been deployed to explain or characterize the use of certain illegal drugs by particular groups and to convey senses of threat and danger, and such discourses have contributed further to contemporary representations of urban street life and city living.

In recent years, commentators have stressed the apparent relationship between illegal drug production, distribution and consumption and the rise of transnational or global organized crime, arguing that over the last decade, the perceived threat of transnational organized crime has become a principal issue on the agenda of key international forums such as the G8, the United Nations and the Council of Europe. Castells, for example, in tracing the explosion of drug trafficking in Latin America since the 1970s, argues that the size of the global drugs trade is a clear example of the importance and extent of the globalization of criminal activity. Furthermore, Castells also explores the close relationship between illicit trade and legitimate global business, the transformation of economies and politics under the influence of 'narco-dollars', and the deep penetration of state institutions and social organization by the illegal drugs industry. In this argument, illicit and illegal drugs provide both an example of global flows of production, distribution and consumption and examples of emerging global responses to such problems. The production, trafficking and use of illicit drugs are defined as problems with global reach in terms of criminal justice, health, public policy, and social and political GOVERNANCE.

Given all of the above, it might well be argued that any serious analysis of the contemporary city – in a period of intense globalization – requires some attention to the social, political and economic impact of illegal drug use, whether that focuses on production, distribution, consumption or, in some cases, all three.

Further reading

Castells, M. (1998) *The Information Age: Economy, Society and Culture*, vol. 3, *End of Millennium*, Oxford: Blackwell.

Courtwright, D. (2001) *Forces of Habit: Drugs and the Making of the Modern World*, Cambridge, MA: Harvard University Press.

Gootenberg, P. (ed.) (1999) *Cocaine: Global Histories*, London: Routledge.

McAllister, W. (2000) *Drug Diplomacy in the Twentieth Century*, London: Routledge.

RICHARD HUGGINS

DUANY, ANDRÉS M.

b. 7 September 1949, New York, NY

Key works

(1981) Seaside, Florida.
(1988) Kentlands, Maryland.
(1992) *Towns and Town-making Principles*, New York: Rizzoli (with E. Plater-Zyberk).
(2001) *Suburban Nation: The Rise of Sprawl and the Decline of the American Dream*, New York: North Point Press (with E. Plater-Zyberk and J. Speck).

American architect, town planner and author, Andrés Duany received his undergraduate degree in architecture and URBAN PLANNING from Princeton University. After a year of study at the ÉCOLE DES BEAUX-ARTS in Paris, he also received an MA in architecture from the Yale School of Architecture. Furthermore, he has received honorary doctorates from two other universities, as well as various prestigious awards for architecture. He is a fellow of the American Institute of Architects. Duany is also an adjunct professor at the University of Miami, and has worked as a visiting professor at many other institutions. He teaches planning at the Harvard Graduate School of Design, as well as being the founder and member of the board of directors of the Congress for the New Urbanism.

Duany has been a founding partner of two very influential architecture firms: Arquitectonica, and Duany Plater-Zyberk & Company (www.dpz.com). As a principal of the latter firm, he has co-designed more than 200 NEIGHBOURHOODS, towns and cities. Duany has spearheaded a resurgence in neighbourhood-based design in the United States and abroad. He has been one of the pioneers and founders of the NEW URBANISM movement in the United States, which has been described by the *New York Times* as 'the most important phenomenon to emerge in American architecture in the post-Cold War era'. His planning and design philosophy has been guided by the idea that in order to create traditionally organized towns, current ZONING laws would have to be rewritten and the real estate development industry would need revising. He introduces methods of integrating planning with codes. He propagates and puts into practice principles of livable communities. The basic idea behind his thinking, which is also the heart of new urbanism, is the promotion, creation and restoration of diverse, walkable, compact, vibrant, mixed-use communities composed of the same components as conventional developments, but assembled in a more integrated fashion, in the form of complete communities. Duany has spent the last two decades travelling the world lecturing on the distinctions between traditional neighbourhood development and urban SPRAWL.

Duany is also an accomplished author. Together with Elizabeth PLATER-ZYBERK, he is the author of the influential book on the methodology of town planning *Towns and Town-making Principles*. They recently co-wrote the book *Suburban Nation*, together with Jeff Speck. Books due out soon include *The New Civic Art* and *The Smart Growth Manual*. Duany is also the author of *The Lexicon of the New Urbanism*. Currently, he is completing a new universal land use ordinance called the SmartCode, which will be distributed nationwide by the Municipal Code Corporation as an alternative to conventional (sprawl-oriented) ordinances.

TIGRAN HASIC

DUBOIS, WILLIAM EDWARD BURGHARDT (W.E.B.)

b. 23 February 1868, Great Barrington, Massachusetts, United States;
d. 27 August 1963, Accra, Ghana

Key works

(1899) *The Philadelphia Negro*, Philadelphia, PA: University of Pennsylvania Press.
(1982 [1903]) *The Souls of Black Folk*, New York: Penguin Books.

Civil rights activist and social scientist whose ideas on race still reverberate a century later, W.E.B. DuBois's most significant contribution to our understanding of the American city lay in his seminal work, *The Philadelphia Negro*, which was the first study to employ social science methodology to the study of race.

DuBois became the first African-American to earn a Ph.D. from Harvard University in 1896. After a brief stint as a professor at Wilberforce College, DuBois was commissioned by the University of Pennsylvania and the College Settlement Association in 1896 to undertake a study that would shed light on the deplorable conditions of the predominantly African-American seventh ward of Philadelphia. The sponsors expected and received a study thoroughly cataloguing and describing the social pathologies that were rampant in the seventh ward at the close of the nineteenth century. But in *The Philadelphia Negro*, DuBois delivered much more. For at that time, DuBois believed that social science, by using reason and empiricism to analyse social problems like racism, could show the short-sightedness of race-based discrimination. Thus, DuBois not only chronicled the social pathologies afflicting black Philadelphia, but traced their roots to centuries of oppression manifested in slavery and the relegation of Philadelphia's African-Americans to the most menial and ill-paying jobs. The hope was that by highlighting the structural obstacles faced by African-Americans, the enlightened elite would act to remove these barriers. At a time when hereditarian notions of race held sway, DuBois's linking of the plight of African-Americans to political and economic forces was revolutionary. Moreover, the methodology employed in *The Philadelphia Negro*, the description and mapping of social characteristics onto neighbourhood areas, was a forerunner to the types of studies done under the CHICAGO SCHOOL OF SOCIOLOGY.

Besides his eminent work as a social scientist, DuBois is probably best known as a civil rights activist who helped found one of the premier civil rights institutions, the National Association for the Advancement of Colored People, for helping to organize the first four Pan-African Congresses, for his writings on race and race relations as typified by *The Souls of Black Folk*, and as one of the progenitors of the Harlem Renaissance. Based in the predominantly black enclave of Harlem in New York City, this renaissance was predicated on the notion that by fostering and drawing attention to African-American talent in the arts, racial prejudice would recede.

In his later years, DuBois was a victim of the rabid anti-communism of the United States in the 1950s and became increasingly disenchanted with his native land, retiring to Ghana where he renounced his American citizenship and died on 27 August 1963.

Further reading

Lewis, D.L. (1993) *W.E.B. DuBois: Biography of a Race*, New York: Henry Holt.

LANCE FREEMAN

E

ÉCOLE DES BEAUX-ARTS

Located in the Latin Quarter in Paris, France, the École Nationale Supérieure des Beaux-Arts – or, concisely, the École des Beaux-Arts – is the cradle of the legendary *beaux arts* style in architecture and city planning, which flourished in France and the United States during the end of the nineteenth century and the first quarter of the twentieth century. The school is the descendant of the Académie Royale d'Architecture founded in 1671 by Louis XIV, and the Académie Royale de Peinture et de Sculpture established in 1648 by Cardinal Mazarin. It offered free instruction in architecture, drawing, engraving, painting and sculpture to students selected by competitive examination. Until Napoleon III's reforms of 1863, the school was held under the king's tutelage to ensure the training of artists that would work for the royalty.

During the nineteenth century, the École des Beaux-Arts was viewed as the ultimate in architectural education in the Western world. Its student body became increasingly cosmopolitan and many US architecture schools adopted a French-inspired curriculum. Richard Morris Hunt was the first American student admitted to the École des Beaux-Arts, and between 1846 and 1918, more than four hundred Americans studied architecture in Paris. Carrère and Hastings, McKim, Mead and White, Bernard Maybeck, Julia Morgan, H.H. Richardson, and Louis Sullivan are some of the most well-known American architects who studied in Paris. When they returned to the United States, they held influential positions and quickly disseminated the French artistic and architectural trends.

The *beaux arts* style is characterized by formal planning and a grandiose architectural style applied mainly to public and institutional buildings, and some large country estates. Its aesthetic emphasizes symmetry, monumental axes, radiating avenues, hierarchical sequences of spaces and neoclassical architecture. This style was embodied in the CITY BEAUTIFUL MOVEMENT, which started at the World's Columbian Exposition of 1893, in Chicago. Other famous North American examples include the New York Public Library, New York Grand Central Terminal, Boston Public Library, Toronto Union Station and the mall in Washington DC.

After the First World War, the *beaux arts* style was criticized and became obsolete – Frank Lloyd WRIGHT called it 'Frenchite pastry'. The school seemed increasingly disconnected from the emerging theories of the modern movement, as exemplified in the curriculum of the Bauhaus, another legendary institution. The revolts of May 1968 ended the three-century-old tradition of the prestigious Grands Prix de Rome and terminated the architectural section of the École des Beaux-Arts. The 1975 exhibition 'The Architecture of the École des Beaux-Arts', organized by Arthur Drexler in the MOMA in New York, initiated the rehabilitation of the *beaux arts* style. At the beginning of the twenty-first century, the consensus is to preserve what is left of *beaux arts* architecture and planning.

Further reading

Drexler, A. (ed.) (1976) *The Architecture of the École des Beaux-Arts*, New York: Museum of Modern Art.
Noffsinger, J.P. (1955) *The Influence of the École des Beaux-Arts on the Architects of the United States*, Washington, DC: Catholic University of America Press.

CATHERINE C. GALLEY

ECONOMIC CLUSTERS

Clusters have the discreet charm to obscure objects of desire. As condensed forms of economic cooperation and knowledge exchange, they are an old topic in a new setting for the analysis of URBAN systems, regional development and technology policy. The charm of these forms of economic activity resides in the idea that knowledge-intensive, interlinked activities which frequently are regionally concentrated create synergies, lead to innovation, increase productivity and hence result in economic advantages.

This growing interest in spatially condensed forms of cooperation and knowledge exchange has several roots, and is based both on new theoretical concepts as well as on new technological developments. The basic arguments for cluster development rely on Alfred Marshall's distinction made in 1890 that there are three principal factors for the development of clusters: labour-market effects, input–output dependency and knowledge spillovers. The GROWTH POLE approach by Perroux in the 1950s showed clear affinities to today's cluster debate. In the 1980s, industrial districts (with an emphasis on the Italian example) were at the centre of attention as new manifestations of the advantages of interfirm cooperation at a concentrated geographical scale. A similar focus was taken by the innovative milieu approach and the Californian school of economic geography analysing territorial innovative processes and the production–reproduction modalities of the competitive advantages of their complex socio-economic fabric. In the 1990s, the approach of the 'New Economic Geography' (NEG) put an emphasis on ECONOMIES OF SCALE emerging and/or leading to spatial concentrations under conditions of monopolistic competition and decreasing transaction costs.

An additional approach consists in the analysis of urban clustering, taking up questions already posed by CHRISTALLER, who asked why are there large and small towns with an irregular distribution, suggesting that the pattern of urban settlement is the result of different functions individual urban places may perform. In later studies, the role of urban clusters as a means of reducing the perception of spatial isolation was emphasized. As reasons for cluster development, three – often interrelated – processes were distinguished: natural bifurcation of existing population centres, simultaneous growth and eventual merging of adjacent quasi-urban localities, and deliberate planning actions. This research on the urban system and urban cluster development has also recently been given an impetus with the emergence of the NEG, where agglomeration effects within an economy are analysed as developing into a system of cities as a self-organizing process.

The empirical aspect of clusters is based on quite different analytical approaches and quantitative methods. These also depend on the dominant perspective: input–output relations, trade patterns, spillovers of different kinds (technological, labour market), or patterns of communication and cooperation. One of the fundamental interests is the identification of clusters. At a national (macro) level, clusters are conceived as broad industry groups linked within the overall macro-economic transaction networks of universally documented flows. Clusters at the industry (meso) level constitute the extended value chains of given end-market products involving best-practice benchmarking and studies of cluster-specific technology adoption and innovation processes. At the firm (micro) level, clusters are conceived as one enterprise or a few linked enterprises often restricted to a single visible collection of similar-sector firms, overlooking linkages that some of the members may have with regionally collocated firms from very different sectors and with research units. A special focus in recent years has become the knowledge-intensity of these linkages and the specific forms of learning resulting from these interactions.

The development of clusters as a challenge for policy raises the question of whether policy intervention in favour of clusters is legitimate at all. For someone with a strong belief in the allocative and dynamic power of markets, this is of course an economically incorrect question. For someone who believes in the possibility of the economic incorrectness of markets, there is ample legitimization of policy intervention. Supporters of active cluster development point to different phases of industrial policy with the cluster approach – after a first phase of 'backing losers' and a second of 'picking winners', there is a third stage with a more modest ambition combining the strong aspects of the previous phases ('backing winners'). They distinguish between two policy alternatives: knowledge intensification of existing clusters and the creation of new cooperative structures. Here the emphasis is on the institutional aspect of

clusters: They are regarded as an institution supporting the adoption of new technologies. They are a broad institutional concept responding to a change in the dominant form of production – a change from Fordist mass production to a post-Fordist system of flexible specialization calling for efficient means to coordinate small and medium-sized firms. The crucial change is that the production of scientific and technological knowledge is increasingly self-contained. To be able to continue, i.e. to be 'sustainable', the innovation process has to get and keep a systemic dimension, which derives from the fact that a local production system is made by a plurality of actors. This sense of belonging represents the base of an 'associative approach' or of an 'associative governance', that leads to the creation of clubs, forums, consortia and different institutional schemes of partnerships. Clusters then become institutions for knowledge management and organizational learning.

Further reading

Cooke, P. and Morgan, K. (1999) *The Associational Economy*, Oxford: Oxford University Press.

McCann, P. (ed.) (2002) *Industrial Location Economics*, Cheltenham: Edward Elgar.

Portnov, B.A. and Erell, E. (2001) *Urban Clustering: The Benefits and Drawbacks of Locations*, Aldershot: Ashgate.

Steiner, M. (ed.) (1998) *Clusters and Regional Specialisation: On Geography, Technology, and Networks*, European Research in Regional Science, vol. 8, London: Pion.

MICHAEL STEINER

ECONOMIC POLARIZATION

Economic polarization describes the distribution of economic wealth among a small number of groups showing a high degree of homogeneity within each group and great heterogeneity across all groups. Income in all its forms is usually the most accepted measure of economic wealth. Clustering of groups around distant poles on the basis of income or any other variable related to economic wealth points to the existence of, and allows measurement of the scope of, economic polarization.

Economic polarization is, in principle, distinct from inequality. Economic polarization shows the allocation of economic wealth in a small number of significantly sized groups of agents, whereas inequality describes a situation where wealth is appropriated by a few. In a particular context, for instance a neighbourhood, inequality can be reduced if the difference between families' earnings decreases, while economic polarization can be increased if families' earnings become attached to a few poles. In other words, income inequality between rich and poor can grow without any changes in the number of rich and poor groups. Difficulties in reaching a greater social cohesion are related mostly to reducing economic polarization rather than inequality.

The concept of economic polarization can be applied to different scenarios – from the global arena, where countries are the main actors involved, to the national or local level, where regions, neighbourhoods and individuals are the units considered. Economic polarization can be identified through analysis and comparison between and within actors.

The transformation of the world economy during the second half of the twentieth century has contributed to the concentration of high unemployment rates, low per capita incomes and high poverty rates in determined areas, leading to an increase in economic polarization. These global dynamics are also reflected in local settings which usually capture worldwide trends, embodying and reflecting new processes that consolidate the growing gap between the wealthiest and poorest groups. Hereby, changes in economic and political developments may contribute to the emergence of new polarization patterns within countries and cities.

Several studies have been carried out by international organizations such as the United Nations and the World Bank to estimate and assess the degree and consequences of economic polarization in the world. The North–South dialogue between the poorest and the richest nations aims to narrow the wealth gap.

From an URBAN point of view, forces driving polarization within the city have been deeply discussed since the early 1990s, along with the relationship between economic and SOCIAL POLARIZATION. Certainly, besides the income gap, additional facts must be taken into account in order to properly assess to what extent a pattern of social polarization corresponds only to the existence of economic polarization.

Further reading

Esteban, J. and Ray, D. (1994) 'On the measurement of polarization', *Econometrica* 62: 819–51.

Fainstein, S. (2002) 'The changing world economy and urban restructuring', in S. Fainstein and S. Campbell (eds) *Readings in Urban Theory*, Oxford: Blackwell.

MONTSERRAT PAREJA EASTAWAY

ECONOMIC RESTRUCTURING

The concept of economic restructuring refers to the rescaling and redistribution of economic activities across space. The reconfiguration of the world economy has profound implications for the productive capacities and competitiveness of cities and regions. Since Piore and Sabel's classic work (1984), it is commonly argued that the economic geography of post-Fordism and flexible specialization have generated deep spatial transformations, and that localities have been unequally affected by the reorganization and relocation of the factors of production. Broadly speaking, two sets of complementary perspectives are adopted: a first strand of research is constructed around the mapping and understanding of restructuring patterns, processes and outcomes; a second strand of work can be classed as 'normative' as it focuses on the public policy implications, i.e. on the question of how REGIONAL and URBAN PLANNING can mitigate the negative effects of economic restructuring and create positive impulses for territorial (re)development.

Economic restructuring is invariably associated with the forces of deindustrialization, globalization and technological development that have unfolded since the 1970s and led to 'creative destruction'. Intensified competition, market integration and the rapid spread of information and communications technologies have spurred the circulation of global capital and widened the set of locational choices. Multinational and transnational corporations have sought to capture higher returns through a larger spectrum of market segments and thus increasingly favoured economies of scope over the old ECONOMIES OF SCALE based on models of vertically integrated conglomerates. Following Doreen Massey's terminology, a new 'spatial division of labour' has surfaced. The downsizing of manufacturing activities has strongly hit the 'old' industrial areas and led to the familiar symptoms of plant closures, job losses and derelict land. In contrast to capital, land and labour remain relatively immobile so that URBAN DECLINE, human capital depreciation and social exclusion have ensued.

Moreover, the old distinction between manufacturing and services is seen as obsolete since both strands of activities are now part of a highly integrated production system. In this system, knowledge represents the most strategic resource, and learning the most important process. Many regional and URBAN scholars have focused their attention on systems of interfirm linkages and cluster development, particularly in sectors such as electronics and biotechnology. Innovation is at the heart of these new interactive processes and represents a source of comparative advantage. Some writers such as Michael Storper have highlighted the importance of 'untraded interdependencies' based on relational assets such as trust and reciprocity. California's Silicon Valley, the 'Third Italy' and the South East of England are among the often-cited examples of 'learning' regions and 'innovative milieux'.

Cities are of course major sites of structural changes. There is much evidence that globalization is changing the positions of cities, as a new world hierarchy of urban centres has emerged, according to Saskia SASSEN and others. The location of corporate headquarters in GLOBAL CITIES facilitates access to financial centres and large pools of sub-contractors, consumers and highly skilled workers. Allen Scott (1988) argues that an important outcome of restructuring is the revitalization of spatial agglomeration economies, resulting in the creation of new industrial spaces, or so-called 'Neo-Marshallian' landscapes. The tendency to agglomerate arises because of the increased fragmentation of the production process and the need to reduce transportation, contract enforcement and other transaction costs. Empirical case studies of industrial specialization have included the clothing and cultural industries in cities such as New York, Paris and Los Angeles.

Furthermore, Manuel CASTELLS (1989) holds that the network-bound NEW ECONOMY with its 'spaces of flows' and 'timeless time' has transformed the INDUSTRIAL CITY into an INFORMATIONAL CITY. Discreet buildings and campus-like environments, populated mainly by affluent

'knowledge' workers, have superseded the coal mining and iron foundries. This shift has raised concerns around the emergence of a dual city characterized by new forms of social and spatial SEGREGATION and the existence of an urban underclass.

David HARVEY's 'entrepreneurial city' illustrates the competitive twist favoured by urban administrations. To alleviate and pre-empt the effects of economic restructuring, these have implemented a plethora of area-based urban renewal strategies based on aggressive PLACE marketing, business development and 'quality of life' factors. The numerous attempts to encourage the creation of enterprise zones or TECHNOPOLES, i.e. new growth poles specializing in global high-tech industries, are striking examples of territorial specialization in urban areas. Yet technopoles are not a panacea for economic development. In some instances, they have proved useful tools to attract external investment and create business incubators, but have equally failed to foster self-sustained adjustment.

Additionally, the notions of democratic GOVERNANCE, local empowerment, CAPACITY-BUILDING and leadership have gained increasing prominence in the work on regeneration. Extended COMMUNITY involvement has been particularly promoted and has given rise to a proliferation of partnerships, often criticized for their project-based and overlapping character and their lack of strategic orientation.

In sum, economic restructuring is a complex process that has generated a vast and vibrant literature in economic geography, urban/regional planning and policy studies. However, this body of literature is by and large concerned with discussions surrounding analyses of 'hierarchies' and 'markets' and thus with the behaviour of firms, local governments and civil society actors, and tends to under-explore the social and cultural foundations of economic restructuring. Individual identities, consumption norms and lifestyles, formed through the transmission of knowledge, attitudes and values from one generation to the next, also affect the space economy. The focus on social reproduction illuminates the role played by individuals, households and national welfare states, but necessitates stronger theoretical and conceptual linkages between economic and welfare restructuring.

Further reading

Castells, M. (1989) *The Informational City: Information Technology, Economic Restructuring, and the Urban-Regional Process*, Oxford: Blackwell.

OECD (2001) *Cities and Regions in the New Learning Economy*, Paris: OECD.

Piore, M.J. and Sabel, C.F. (1984) *The Second Industrial Divide: Possibilities for Prosperity*, New York: Basic Books.

Scott, A.J. (1988) *Metropolis: From Division of Labor to Urban Form*, Berkeley and Los Angeles: University of California Press.

SEE ALSO: glocalization; labour markets; uneven economic development; urban regeneration

CORINNE NATIVEL

ECONOMIES OF SCALE

Economies of scale provide a fundamental explanation of the inexorable growth of cites and the concentration of people and economic activity within them. They are said to exist when average costs decrease with an increase in production. In the URBAN arena, economies of scale exist at both the *internal* scale of the firm or industry and at the *external* scale of the city or metropolitan area. Internal economies of scale relate to the cost savings accorded to the firm or industry from expanding production at a single location. When a firm improves a production process that allows it to simultaneously lower costs and increase output, it has attained economies of scale from its own efforts. These internal economies have implications for business location in the city. A firm experiencing large internal scale economies will concentrate production at a few select locations. The existence of a large-scale producer in turn will attract a concentration of related activity close by. Another source of internal scale economies in the city arises from the production of local public goods. Cities 'produce' civic buildings, urban monuments and urban infrastructure. These are lumpy, immobile and non-divisible investments, internal to the city. They create an urban advantage and act as the nuclei around which other related activities cluster.

External economies of scale refer to those production advantages arising from interactions between firms and industries and their locational association. They are the result of spatial

proximity and are commonly referred to as agglomeration economies of scale. A distinction is often drawn between two subsets of agglomeration economy: *localization economies* and *urbanization economies*. The former arise when firms benefit from spatial proximity with other similar firms. This concentration of similar firms in the city allows for labour-pooling across proximate firms, production linkages and the development of local input production, and benefits arising from information flows (between competitors, suppliers and clients). Localization economies generally encourage the development of a specialized economy. Urbanization economies relate to the benefits accruing from location in a large urban area irrespective of the composition of proximate economic activities. These include access to a large market, the existence of specialized services in the city and the potential for innovation through the flow of information and knowledge which is released randomly in the urban area, across both space and time. This type of scale economy will promote diversified city economies. Empirical studies have generally found stronger evidence of localization than urbanization economies. However, they have often concentrated solely on manufacturing activities which do not accurately reflect the range of activities in the modern city. Urban advantage in areas such as amenities, climate and education are less tradable and therefore tend to be overlooked. Where urbanization economies have been identified, they have been explained as resulting from savings in transportation costs rather than any productivity gain arising from location in a large city.

Further reading

Henderson, J.V. (1988) *Urban Development: Theory, Fact and Illusion*, New York: Oxford University Press (ch. 5).

DANIEL FELSENSTEIN

ECOVILLAGE

An ecovillage is a small-scale development that strives to produce the least possible negative impact on the natural environment through intentional physical design and resident behaviour choices. Although ecological sustainability is the core, and sometimes the only, objective, most ecovillages also have a strong community-building component as well. Common to most ecovillages is their intentionality. Such communities consciously choose ecologically based values to underpin their common objectives and guide their group efforts to achieve them. These objectives often include implementing an overall physical layout and use of space that is conducive to increased social contact as well as reduced use of natural resources; preferring construction materials and techniques that rely on renewable resources; reducing auto dependence; conserving water and energy sources; integrating organic agriculture or permaculture practices into the settlement; incorporating recycling and resource management into the everyday life of the COMMUNITY; preserving and, if necessary, restoring pre-existing ecosystems; and creating a mutually supportive social network among residents.

Underlying the actions that most ecovillages undertake as essential to environmentally responsible living are elements of a philosophy that has been drawn from eco-development theory, URBAN PLANNING theory, utopianism and sustainable development theory. Themes in this evolving hybrid philosophy include: size matters (small is better); physical layout matters (compact is better); economic relations matter (local is better); use of resources matters (small eco-footprint is better); environmental relations matter (awareness, sound management and eco-ethics are important); and people matter (SOCIAL CAPITAL, participation and empowerment are important). In addition to these themes, a tension between a 'backward-looking' and a 'forward-looking' approach is apparent. Many ecovillagers are looking for a simpler, low-tech, more traditional agricultural life, or in some cases a more traditional URBAN life, reminiscent of the idealized village or NEIGHBOURHOOD of a century ago. The idea is that this former way of life was less stressful on the environment and allowed for more social cohesion. On the other hand, some ecovillages do not find their models in the past, but are searching for new, more appropriate technologies to manage the land and the environment within and around a settlement. The belief is that innovation, as opposed to returning to tradition, is needed to replace our harmful practices.

Divergent types of places call themselves ecovillages, including housing developments, villages, neighbourhoods and even small towns, with

populations ranging from a handful to over 10,000 people. Examples of ecovillages exist in rural, suburban and urban settings and occur all over the world, built and maintained by private, public and volunteer initiatives. Ecovillages have been initiated as CO-HOUSING projects (such as EcoVillage at Ithaca, New York), utopian rural communities (such as The Farm, Tennessee), URBAN REVITALIZATION projects (for example Price Hill/Seminary Square Ecovillage, Cincinnati), greenfield developments (as at Ecovillage of Loudoun County, Virginia) and as the basis of development schemes in the Third World (as done through the Senegal Ecovillages Network).

Further reading

Barton, H. (ed.) (2002) *Sustainable Communities: The Potential for Eco-Neighborhoods*, London: Earthscan Publications Ltd.

SEE ALSO: sustainable urban development

CARLA CHIFOS

ECUMENOPOLIS

Ecumenopolis was the fifteenth level of ekistic units according to DOXIADES (see EKISTICS). It was also the most significant one because, being the uppermost echelon of the classification, it was the epitome of both the potential and the problems of human settlements. Although when Doxiades formulated his theories, during the 1960s, evolution had only attained the twelfth unit, MEGALOPOLIS, to be found mainly on the east coast of the United States, he considered that the creation of Ecumenopolis, that is the interconnection of all great cities of the ecumene in one system, would be a reality by the end of the twenty-first century. Doxiades thought that this outcome was inevitable, due to a combination of forces already operating in cities, but that the form Ecumenopolis would assume was open to several alternatives and would be satisfactory only if man would guide events, using rational and scientific planning. Major issues in this process should be the conservation of nature, transportation networks placed beneath the surface and the preservation inside Ecumenopolis of smaller units corresponding to lower levels of the overall URBAN organization. Ecumenopolis was a futuristic vision rather

than an elaborated theoretical concept; moreover, Doxiades' claims that the above planning principles had already been applied in some of his projects, for instance in Islamabad (Pakistan), do not seem that plausible today. However, and besides the poetic quality of this notion, it is also true that it captured some essential facets of the expanding scale of URBANIZATION which became more apparent in the era of globalization and the world wide web.

DIMITRIS ECONOMOU

EDUCATION AND THE CITY

Educating children is one of the most significant responsibilities of government, and improving the present education system is a serious concern of virtually every nation; education is global in nature, yet local in practice and though it varies from classroom to classroom, its success is similarly measured. Ultimately, the success or failure of URBAN schools signals the success or failure of the city itself. Despite the importance of education to the achievement of cities and nations, the field of urban education did not even begin until the 1970s, and it has focused almost exclusively on the United States and Great Britain. Thus far, research on urban education has documented the increasing problems of urban education while identifying few effective remedies or reforms. This entry highlights the characteristics and the structural problems experienced by most urban schools as well as the suggested reforms to overcome them.

City schools, whether public, religious or private, vary from suburban, small-town and rural schools in several fundamental ways that have serious consequences for administrators and teachers educating their pupils. While still the centres of modern civilization, cities have more and more become the epicentre of the poor, the disadvantaged and minorities. In developed nations, cities have been losing their middle-class population to the SUBURBS; in developing nations, cities have attracted those with few resources from rural areas. The result has been that cities are more dependent on quality schools while urban education systems are deteriorating. The greatest differences between city schools and other schools lie in their environmental and population demographics. Urban schools service a much larger

geographic area and a larger, more diverse population base; this racial and cultural diversity often creates serious conflicts and offers a severe challenge to city schools. While the poor predominate in cities, the economic diversity of urban populations remains very broad, and the difficult economic conditions urban inhabitants face are reflected in the schools their children attend. The cost of living in urban areas is comparatively more expensive, the urban tax base is often limited, and the wealthy and the politically powerful that reside in the city often send their children to private or parochial schools; therefore, many of those that have the capacity to positively affect urban education may not have the desire or initiative necessary to introduce essential reforms. City school buildings are often old and their deteriorating structures are badly in need of repair and technological updating; city school students suffer from low achievement, high rates of absenteeism, high dropout rates and growing rates of general disorder. Furthermore, the diversity and needs of the student body, the conflict inherent in cities and the resulting threat of violence make city schools less attractive places for educators to work at a time when cities must compete with other regions for the most skilled education professionals.

To service a growing and changing population, and to accommodate the most recent technological and pedagogical requirements, new schools need to be built or expand. Urban school systems, however, face much larger barriers to development than do their regional counterparts. Land to build or expand a public school within an urban area is difficult to find and commands a very high premium. Urban schools are typically housed in older buildings designed to accommodate a much smaller number of students. The floor plans are inflexible and not expandable. Moreover, the structures are not easily modified to accommodate the additional wiring and power requirements for the latest technology. Finally, changing the structure of an urban school requires overcoming the very long and arduous task of securing government approval even though the results of such a struggle are not certain or, in many cases, encouraging.

Yet urban schools have the potential to draw on the wealth of non-financial resources found in cities. Cities provide the business and cultural foundation for the surrounding region. Most urban areas have a variety of museums, galleries and architectural achievements to inspire the passion and creativity of students. Urban universities and research institutions are among the best in the world and their resources can be harnessed to enhance the educational possibilities of nearby schools. Finally, the managerial and creative skills of corporate and cultural professionals can be accessed to provide administrative assistance as well as educational alternatives. These rich educational resources may be available, but only a school able to make use of these resources can take full advantage of the urban potential.

City schools operate within an urban environment which creates structural barriers and limitations that inhibit exploitation of these non-financial resources and prevent city schools from satisfying the students' needs. The sheer size of most city schools creates bureaucratic, organizational and pedagogical difficulties that impede successful implementation of change. Many school organizations use 'tall' bureaucracies with many management levels to manage the urban school system. This type of bureaucracy is rife with inefficiency and notoriously slow decision-making.

The huge scale of urban schools presents additional challenges to teachers and principals. The sheer diversity of an urban school's student population presents significant challenges to the ability of teachers and school administrators to address the needs of individual students. For example, a typical city school might have immigrant students from many different countries speaking many different languages, particular health and nutritional concerns, various learning difficulties, cultural idiosyncrasies, family issues, and behavioural difficulties. Addressing these issues requires significant financial and staffing commitments from the local and regional authorities, but many times the government is unable to fully address the educational needs of its students. Local schools are reflections of the larger city around them. The financial resources of cities, like the schools themselves, are generally stretched very thin and cannot accommodate the needs of all inhabitants. Therefore, it is very difficult to increase the financial allotment given to urban schools. Schools are forced to address their needs within the budgetary and staffing limitations placed upon them; this leaves an increasing number of urban students unserved or underserved by the institution in the best position to help.

Clearly, urban schools face and must overcome significant challenges to their vital mission. To surmount these obstacles, researchers and school managers have considered and introduced many reforms to the governance structures, school organization and pedagogy of the education environment. These reforms have been advocated to change the structure of the institution, to improve the delivery of education information, to exploit the resources found in cities, to diminish the difficult realities of the urban environment and to create more effective classrooms.

School governance plays an important role in designing the most efficient and effective school system, establishing the strategy and strategic vision for the school system, creating a capable management structure for proper decision-making, initiating reforms, and building an effective education community. The ability of a governance structure to accomplish these goals is hampered by the huge scale most urban school organizations encompass. In the business world, economy of scale is generally considered a positive feature that helps add purchasing power and efficiency, but in the field of education, scale may not bring similar benefits. Scale is expensive; scale breeds bureaucracy; scale creates inefficiency; and scale geometrically increases the number and types of 'typical' problems found in every school. Therefore, the first alternative open to urban governance bodies is to reduce the scale of urban school systems. Smaller school districts or administrative decentralization provides more flexible and efficient structures to implement education reform. Once the size of districts and the size of schools become more manageable, the bureaucracy should shrink as well. Plans for 'site-based management' or flattening the decision-making structures, as per Chicago's local school councils, can also be adapted.

Education organization and governance reforms have been driven primarily by existing business models which stress market-driven and accountability solutions. Market-driven solutions include 'new' school structures that give parents increased choice and control over the school and the school environment, such as: choice/voucher programmes, management contracts, privately funded tuition scholarships, leadership changes, charter schools, support for parochial schools, and 'for profit' schools. Included in this solution model is an emphasis on accountability, outcomes assessment and increased public information.

Under many accountability systems, schools and districts are required to produce statistics, such as drop-out and graduation rates, achievement scores, and student–teacher ratios, to 'educate' parents on the success or failure of their schools and give parents tools to make intelligent comparisons to other schools. The theory behind this market-driven improvement strategy is that the competition or 'market' for students will bring efficiency and improvement to schools as it has been theorized to improve companies in the consumer market.

Some reformers focus on improving the school environment rather than the governance or decision-making structure. They argue that the school's leadership team should create a common strategic vision for all its teachers and staff and serve as a communications conduit between teachers and the executive management team. Under this approach, it is recognized that the people most directly affected by any instituted reform should be more than simply 'part of the process'. The entire school community must be given direct input before reforms are implemented and some method by which they can monitor and alter their progress. The leadership must also present the educational vision to the local community. Through effective communication between the school and the local community, the school gains much-needed parental support. True support, however, will not come easily. Parents will want a voice in the strategic plan, staffing decisions, school structure and curriculum development. In return, parents can provide classroom support, curriculum assistance, school endorsement and additional funding sources. Achieving a beneficial school–parent relationship can be the difference between a school and a community institution.

In many urban environments, schools also act as an integrated service delivery institution and community action centre. Schools are where the local populations directly touch the governing authority and where they can voice their opinions about issues that directly affect them. Researchers have found that this type of 'community-based' school which utilizes the enormous resources of the city, including the business community, nearby universities and other strong communities, can benefit from the business and research knowledge unavailable in the typical school district and the district will have the opportunity to realize some of the non-financial benefits described above.

While some reformers focus on improving the school environment, other reformers seek pedagogical and curriculum improvements. These reformers advocate using targeted programmatic investments, such as early childhood development, drop-out prevention, youth-to-work initiatives, compensatory education programmes and pre-schooling programmes (e.g. Head Start); they also suggest using more effective instructional methods, such as active learning, innovative evaluation scales and culturally focused programmes of study, to create a more productive learning environment. Reformers stress the probable linkage between education success and success in life; they also suggest introducing 'whole school' curricula to reflect a specialized programme of study found in many 'magnet' or job preparatory schools. These reformers believe that altering and improving the content and delivery of educational information will lead to greater information retention and a more meaningful reflection of the lives of urban students.

Use of the latest educational technology may offer improvements to the structural, governance and pedagogical environment as well. While technology has not proved to be the educational panacea it was once perceived as, computer technology has been shown to have the potential to improve communication among schools and districts, among teachers and students, and among different education groups. It can also offer additional classroom options. However, introducing the latest technology into urban schools will require additional infusions of cash and a great deal of technological skill. Furthermore, the age and physical plants of city schools add complications to computer installations. Some reform groups find that preliminary contact with the surrounding businesses and universities can overcome or mitigate these difficulties.

The paradoxes of education in cities are striking – cities are the centres of learning but many students are at risk of not learning enough to fully function in urban society; cities have great resources but these are often insufficient to create great schools; cities are at the centre of a nation's vitality and innovation but their schools are often mired in bureaucratic traditionalism; the world is becoming more urban but city schools are facing more failure. Even more unsettling, however, is the understanding that despite the great enthusiasm for education reform, in general, and for specific improvement initiatives,

there is little concrete evidence which suggests that a successful path to urban education improvement even exists. Perhaps the true test of urban education reform may not occur until the systemic causes underlying the need for education reform, such as POVERTY, SEGREGATION and violence, have been successfully overcome.

Further reading

Cibulka, J., Reed, R.J. and Wong, K.K. (eds) (1992) *The Politics of Urban Education in the United States*, Washington, DC: The Falmer Press.

Raffel, J.A., Boyd, W.L., Briggs, V.M., Jr, Eubanks, E.E. and Fernandez, R. (1992) 'Policy dilemmas in urban education: addressing the needs of poor, at-risk children', *Journal of Urban Affairs* 14(3/4): 263–89.

Ravitch, D. (ed.) (1998–2003) *Brookings Papers on Education Policy*, Washington, DC: The Brookings Institution Press.

Stone, C.N. (ed.) (1998) *Changing Urban Education*, Kansas: University Press of Kansas.

JEFFREY A. RAFFEL
GLENN L. SILVERSTEIN

EKISTICS

'Ekistics', a term which derives from the Greek 'οικίζω', meaning to settle down, was used by Constantinos DOXIADES to denote a new 'science of human settlements'. A major incentive for this was the emergence of increasingly large and complex CONURBATIONS, tending even to a worldwide city; however, ekistics aimed to encompass all scales of human habitation, and to examine them from every point of view (as opposed to disciplines like architecture, town planning or geography). The starting point was a two-way classification scheme: that of scale, comprising fifteen ekistic units (man, room, dwelling, dwelling group, small NEIGHBOUR-HOOD, neighbourhood, small town, town, large city, metropolis, conurbation, MEGALOPOLIS, urbanized region, urbanized continent, ECUMENOPOLIS), and five elements common to all settlements (nature, society, shells, networks and culture).

The theoretical postulates of Doxiades were less clear, consisting of the following: settlements are created to satisfy human needs; their development is a continuous process depending upon the dynamic balance of their elements; their form ultimately results from the combination of centripetal, linear and circumstantial forces. In fact, ekistics was more an attempt at the creation of

a wide knowledge basis for human settlements than a science with a rigorous methodological and conceptual framework and explicit hypotheses. However, it is also true that many of the ideas that Doxiades developed have influenced, or have been incorporated into, mainstream academic and popular thought, running from ecological-environmental studies and the use of information technology in planning to the NEW URBANISM movement.

Further reading

Doxiades, C. (1968) *Ekistics: An Introduction to the Science of Human Settlements*, New York: Oxford University Press.

DIMITRIS ECONOMOU

EMINENT DOMAIN

Eminent domain is recognized as a fundamental power of government, the power to take private property (most often land) for public use or for a public purpose. For centuries it has been recognized that when acting in the public welfare, government has the power to compel a property owner to sell his or her property, even against the owner's assertion of PROPERTY RIGHTS. The property taken must be for a public use, or valid public purpose; eminent domain does not include the power to take and transfer ownership of private property from one property owner to another private property owner unless a valid public purpose is forthcoming. The public purpose may be indirect, for example urban REDEVELOPMENT for economic revitalization and job creation, but it must exist and courts must be able to find the public purpose sufficient to sustain the use of eminent domain. In general, courts look to legislation authorizing cities to exercise the eminent domain power to determine if the city's condemnation of property serves a valid public purpose; in most circumstances a city's finding that the transaction serves a valid public purpose will be accepted by the courts. There are, however, instances where courts have denied cities use of the eminent domain power when a valid public purpose has not been established.

While the eminent domain power is inherent in government, it is given form by constitutions and by acts of legislative bodies. In the United States, for example, the Fifth Amendment to the Constitution includes a moral obligation of government to fairly compensate individuals for taking their property, 'nor shall private property be taken for public use, without just compensation'. As a result, in the United States, as in most countries, eminent domain can only be exercised if accompanied by the payment of fair value of the property taken. Disputes between property owners and the government over the value of property taken by eminent domain are settled in court.

Cities are granted the power of eminent domain by state legislatures. In granting the power, legislatures may place limits or constraints on it, limiting, for example, the public purposes for which cities may utilize the power. Legislatures may also grant the power to non-governmental parties, for example to cable television companies for the express limited purpose of wiring communities. In general, however, eminent domain is seen as a governmental power not delegated to private sector interests.

Further reading

McQuillin (2001) *The Law of Municipal Corporations*, St. Paul, MN: West Publishing (ch. 32, 'Eminent domain').

SEE ALSO: property rights

ROGER RICHMAN

EMPOWERMENT ZONE

The Empowerment Zone and Enterprise Communities Program was established as part of the Omnibus Budget Reconciliation Act of 1993 and grew out of earlier unsuccessful proposals for an enterprise zone programme at the federal level. Based primarily upon local-level incentives without federal financial support, the federal enterprise zone programme never gained essential political support. However, under the Clinton administration, the Empowerment Zone Program offered financial incentives that motivated cities to participate.

The Empowerment Zone Program was the legislative response to RIOTS that erupted in 1992 in Los Angeles after the controversial acquittal of police officers charged in the beating of an African-American motorist during an arrest. The riots brought renewed attention to the deteriorated

conditions in many CENTRAL CITY areas across the country which had not been focused upon since the major initiatives of the GREAT SOCIETY programme during the 1960s. The Empowerment Zone Program was ostensibly designed to empower cities with the flexibility to design their own strategies from the ground up with broad citizen participation.

The US Department of Housing and Urban Development (HUD) designated cities as empowerment zone (EZ) awardees based upon an extensive, competitive application process. The cities identified target areas that met the criteria set in the application. The Empowerment Zone Program resembled the MODEL CITIES programme of the 1960s in that it called for a comprehensive approach to address physical, economic and social urban ills in cities troubled with poverty, unemployment and general distress. Between 1994 and 2001, thirty-one EZs were named, which received various combinations of grants, loans, incentives and preferences from the federal government.

In 1994, six EZs and two supplemental EZs were named. The EZs received US$100 million in Title XX Social Services Block Grant (SSBG) funds, access to wage tax credits, increased tax deductions for businesses located in the zones and new tax-exempt bond financing to the six cities: Atlanta, Baltimore, Chicago, Detroit, New York and Philadelphia/Camden. One of the two supplemental zones, Cleveland, received US$3 million in Title XX SSBG funds, US$177 million in Economic Development Initiative (EDI) grant funds and Section 108 loan guarantee dollars, and tax-exempt bond financing. Los Angeles, the other supplemental zone, received US$450 million in EDI grant and Section 108 loan dollars.

Wage tax credits, formally known as empowerment zone employment credits, were available to EZs as a tool to encourage firms within the zones to hire local residents. EZ communities were given priority in consideration for other federal assistance programmes for activities proposed to take place within their boundaries. In addition, they were encouraged to request waivers for federal programme provisions that deterred implementation of their revitalization plans.

In 1998, the two supplemental zones were converted to full zones (with all benefits except the $100 million SSBG grant) and fifteen additional zones were added. These were: Santa Ana, CA; New Haven; Miami Dade County; Gary/East

Chicago, IN; Boston; Minneapolis; St. Louis/East St. Louis; Cumberland County, NJ; Cincinnati; Columbus; Ironton, OH/Huntington, WV; Columbia Sumter, SC; Knoxville; El Paso; and Norfolk/Portsmouth. These communities received EZ grants from HUD and access to all wage credits, deductions and other incentives given to the Round I awardees.

Eight additional EZ slots were approved under the 2000 Community Renewal Tax Relief Act. The programme attracted thirty-five applications. The selection process entailed a two-part process that weighed population and poverty criteria and the community's strategic plan. The Round III EZ communities were: Pulaski County, AR; Tucson; Fresno; Jacksonville; Syracuse; Yonkers; Oklahoma City; and San Antonio. The EZ communities in this round received no direct funding but were eligible for the tax incentive benefits associated with the programme.

The first EZ round was unique because funding was assured for a ten-year period, freeing the cities from the uncertainty of whether annual appropriations would be approved by Congress. Furthermore, the cities had discretion in determining programme designs, activities undertaken and strategies incorporated. However, each funded applicant was expected to incorporate four key principles in their revitalization plans: economic opportunity, sustainable community development, community-based partnerships and a strategic vision for change.

1   The economic development principle reflects the priority given to job creation and job placement for residents of the zone by the federal government. Some of the ways that EZ communities pursued this principle were loans to small businesses, commercial real estate development to attract new or expanded businesses, technical assistance, business training, and redevelopment of BROWNFIELDS.

2   Sustained community development of economic development and human and physical capital was pursued through childcare provision, health outreach and education, and substance abuse services, among other activities.

3   'Community-based partnerships' describes the expectation that broad participation of community members would occur, drawing in residents, community organizations,

government officials, social service providers and the private sector. Furthermore, the design proposed a process that engaged the residents consistently and created partnerships among all participants. EZ communities pursued these ends by establishing governing boards that included residents and other participants, such as elected officials, local business people and external corporate representatives. The cities varied considerably in their governance structures and programme implementation.

4   The strategic visions for change were expected to reveal a comprehensive plan for what the communities would become and how they would achieve those ends. All EZ communities envisioned economic and commercial activity as a major theme, and most saw a well-educated, trained and connected workforce, an attractive physical environment, and a healthy, nurturing and safe social environment as critical themes. EZ communities uniformly identified business development as a prime strategy to move them towards their long-term visions. Workforce development, i.e. preparing residents for employment and job placement, was a major strategy for most EZ sites. Overall, the EZ communities varied greatly in the combinations of approaches and outcomes.

HUD sponsored evaluations of the Empowerment Zone Program that were conducted by the Rockefeller Institute at the State University of New York and Abt Associates. Both evaluations used local researchers to assess the conduct and outcomes of EZ programmes.

Further reading

Gittell, M. (1995) 'Growing pains, politics beset empowerment zones', *Forum for Applied Research and Public Policy* 10: 107–11.

HUD (2001) *Interim Assessment of the Empowerment Zones and Enterprise Communities (EZ/EC) Program: A Progress Report*, Washington, DC: US Department of Housing and Urban Development.

Keating, W.D. and Krumholz, N. (eds) (1999) *Rebuilding Urban Neighborhoods*, Thousand Oaks, CA: Sage.

Thomas, J.M. (1995) 'Applying for empowerment zone designation: a tale of woe and triumph', *Economic Development Quarterly* 9(3): 212–24.

MITTIE OLION CHANDLER

ENGELS, FRIEDRICH

b. 28 November 1820, Barmen, Rhine Province, Prussia; d. 5 August 1895, London

Key works

(1914) *Revolution and Counter-Revolution, or Germany in 1848*, by K. Marx (originally written in the form of articles for the *New York Tribune* from 1851 to 1852), ed. E.M. Aveling, Chicago: Charles H. Kerr & Co.

(1955) *The Communist Manifesto, with selections from The Eighteenth Brumaire of Louis Bonaparte and Capital by Karl Marx*, ed. S.H. Beer, New York: Appleton-Century-Crofts (with K. Marx).

(1958 [1845]) *The Condition of the Working Class in England*, trans. and ed. W.O. Henderson and W.H. Chaloner, New York: Macmillan.

(1972 [1884]) *The Origin of Family, Private Property, and the State*, New York: Pathfinder Press.

'Engels was the finest scholar and teacher of the modern proletariat in the whole civilised world.' So wrote Vladimir Lenin in the autumn of 1895, on Friedrich Engels's death. Engels's life is no less dramatic and interesting than that of his lifelong friend and intellectual comrade, Karl Marx. Born to a well-off family in manufacturing, Engels received first-hand education in capitalism and the plight of the labour class by working for his father. After an initial interest in Hegelian teachings, Engels turned to alternative philosophical explanations of the state, the capitalist economy and the social structure. Through his acquaintance with Moses Hess, Engels became familiar with communism and its political economy and started an intellectual career that took him through revolutions, exiles and the production of some of the most exciting manuscripts of the last two centuries.

In 1842, Engels worked in Manchester at a company in which his family had shares. Engels first met Marx on the way to Manchester, but it was not until two years later that they began to work together more closely. In Manchester, Engels observed the living conditions of the working class and made the final intellectual journey to a lifelong commitment to battling capitalist ideologies and their social externalities. This experience resulted in an important book in 1845: *The Condition of the Working Class in England*.

By 1844, when Engels made a stop in Paris to see Marx, the two had already begun to publish and correspond with each other. In 1845, Engels

went to Brussels to be with Marx and they returned to Paris. The League of Communists authorized them in 1847 to prepare a manuscript on communist principles. By 1848, *The Communist Manifesto* was completed and translated into multiple languages. By the end of that year, Engels moved to Cologne with Marx, established a newspaper, *Neue Rheinische Zeitung*, and participated in the revolution of 1848. The revolution's failure put Marx and Engels on a new path. In 1849, the two met in London, and in need of money, Engels returned to Manchester to work in his previous place of employment, allowing him to provide financial assistance to Marx and help the communist cause. However, Engels continued to publish on various issues, including a series of articles on the 1848 revolution that appeared under Marx's name in the *New York Tribune* from 1851 to 1852. Engels and Marx remained in constant correspondence. In 1870, after returning to London, Engels spent most of the remaining years of Marx's life (until 1883) with him. From 1883 until his death in 1895, Engels edited and published the second and third volumes of Marx's *Das Kapital*.

With the death of Engels, the nineteenth century witnessed an intellectual diminishment and the loss of one of its great minds: a thinker, a champion and always a friend of the working class.

ALI MODARRES

ENVIRONMENT AND THE CITY

Efforts to integrate concepts of nature into URBAN theory (or for that matter, concepts of the CITY into theories of nature) have been very unsatisfying. The reasons for this are not hard to fathom, insofar as the city has long existed as a symbol of everything that is unnatural. Historically, cities seemed detached from rural places; they were places of government, places of exchange and commerce, places of change and urbanity, in contrast to their HINTERLANDS, which were marked by the more fixed rhythms of the seasons, agricultural practice and landownership. Consequently, there evolved from the first a dualism between city and country, with one being associated with the natural world and the other with all that is artificial and constructed, a created environment, as GIDDENS terms it. Nor was this a simple study of social and economic contrasts, as between, say, life in the

valley versus life in the uplands; this has long been a very emotional exercise in nostalgia, in which rural life has been defined as normal and idyllic, while urban life has been contrasted as abnormal, unhealthy and dangerous. Raymond Williams traced this trope back for several centuries, and it is possible to find laments from city dwellers regarding their detachment from nature that extend back prior to the INDUSTRIAL REVOLUTION. Shelley wrote of hell being a city much like London, which has echoes in biblical verses, such as Zephania 3: 1, that allude to the oppressing city that is filthy and polluted. Recent studies have also shown that there is a gendered element to this, with the natural world traditionally being characterized as fruitful and female, while the city is harsh, male and sterile. It is no coincidence that one of the most powerful, and enduring, silent movies was the social commentary of Lang's *Metropolis*, made in 1926, in which the concrete of the city is contrasted with the access to nature enjoyed on the penthouse gardens of the rich. Only materialists like Trotsky have lauded capitalism's victory of town over country.

This self-evident dualism is, however, based on two serious fallacies. The first is that there is something natural about rural life, while the second is that there is nothing natural about cities. The first of these is not immediately relevant to this argument, but has important implications for much urban theory – it is after all the basis of much early URBAN SOCIOLOGY (such as the work of Tönnies) that relations between individuals in cities are inorganic, in contrast to those to be found in rural communities, as Peter Saunders reminds us. Suffice it to say that there is little to be found in most rural areas that maintains traditional social relations and that has not been touched by capitalist development, so once more this must be an exercise in nostalgia. The question of whether there is anything natural about cities is more compelling. Advanced societies measure their progress in their ability to distance themselves from nature; the ability of their residents to isolate themselves from epidemic diseases, to ignore the extremes of cold and heat, and even to live in locations where the basics of life (such as potable water) have to be obtained from far away. Yet it is increasingly clear that this popular alienation ignores crucial aspects of environmental reality, which can be summarized as follows. First, there is the fact that most, if not all, cities contain

elaborate examples of what is more commonly thought of as the natural world. Secondly, there is the more contested assertion that policy-makers in many nations and their cities have created successful plans to improve the natural environment over the past fifty years, and have in many cases succeeded in bringing about a real improvement in air quality, to take a particular example. And thirdly, there is also a more belated recognition that the dystopic image of the postmodern city, a *Blade Runner* in the making, so to speak, ignores the complex ecological foundations of the city, which new research is at last beginning to comprehend.

First, there is the question of the heterogeneity of flora and fauna in most cities. This extends far beyond the 'pictures at 11' recognition that bears, parakeets, deer and coyotes range throughout the SUBURBS and can prove remarkably adaptable to urban conditions. While there is in any city typically a high population density that is sustained via residential uniformity and large amounts of concrete, there is also much interstitial land use that is in complete contrast. Globally, most cities have highly intensive agriculture within the city limits, ranging from poultry production to dairy farming, and including marijuana cultivation. Even the poorest and most densely settled places contain parks, which have often been self-consciously created and designed by individuals whose contributions to civic development remain clearly understood, such as Frederick OLMSTED. These spaces are rarely examples of pristine environment, but that is increasingly true of most rural areas too; true wilderness is at a premium everywhere. Rather more can be said for the many zoos that mark the world's largest cities; it is here that can be found some of the last remnants of the planet's most exotic species, animals that can no longer survive in the wild but have been bred successfully amid the concrete. In contrast, the use of animals for food from the Paris zoo in 1870 and the Berlin zoo in 1945 also indicates how tenuous the urbanity of the city's residents really is.

Secondly, there is the more contentious issue of whether cities are basically anti-nature, amounting to little more than 'a transgression of environmental common sense', as Davis (1998) claims. This is, of course, just the most recent expression of the popular dichotomy rural = good, urban = bad, which is currently manifesting itself once again in the cyberpunk genre, for instance (see POPULAR CULTURE AND CITIES). It is also connected to postmodern architectural thought (the urban style in which elements are whimsically juxtaposed), so that natural elements, such as trees and waterfalls, are brought into the atriums of public buildings, or lakes are created in the desert – Las Vegas is the most obvious, and the most often ridiculed, example of this tendency. In its most recent academic form, this dichotomy has been re-expressed and posed as a fundamental question of whether cities can be sustainable.

The environmental critique of INDUSTRIALIZATION and URBANIZATION traces a clear line from the existence of cities through to the manifestation of global warming. While there is some empirical correlative evidence for the latter, there is no causal evidence, yet urban growth has been strongly implicated in this equation, notably the phenomenon of 'urban SPRAWL'. Despite the rhetorical appeal and the self-evident righteousness of the anti-urban argument, it remains hard to quantify. Most serious analysts recognize that cities are highly efficient modes of human organization, packing large numbers of human beings into relatively small spaces, even when this involves lower-density sprawl. It is the case that the ecological footprint of cities in advanced societies – that is the impact that they have on the surrounding hinterland – can extend for hundreds of miles and more; but this is often the product of an affluent consumer society rather than of urbanization per se. Most telling is the fact that many Western cities now have better environmental conditions than they did fifty years ago, just as many of the fastest-growing settlements in developing nations are seeing a decline in their conditions. In Los Angeles and London, for example, air quality is demonstrably better now than it was at the midpoint of the twentieth century. In LA, the aggressive use of catalytic converters in automobiles has diminished the levels of ozone and thus the presence of smog. In London, the abandoning of coal for fuel and power generation has eliminated the presence of fog. In contrast, the rise of the petrol engine at the expense of walking and the bicycle throughout the developing world is contributing to reduced air quality and an increase in respiratory illness and death; but this reflects the weakness of public discourse with respect to transport policy, rather than the inherent drawbacks of city life, as the UN Habitat programme argues. Nor is this an academic argument; the popular environmental

critique has extended itself into many facets of public discourse in an insidious manner, linking, for instance, immigration to a supposed population explosion in the United States and Europe. By thus connecting urbanization to global warming, it serves to oppose urban growth in developed nations while condemning families in the evolving economies to a diminishing quality of life by staying in overpopulated rural areas. Castigation of urban growth is merely another contemporary facet of rural nostalgia, therefore.

There is a third context to the issue of whether there is a natural dimension to cities, and that is to be found in the field of URBAN ECOLOGY. To this point, most research in ecology has addressed itself to ecology *in cities*, simply measuring the complexity of groundcover, the extent of the urban heat island and so forth. Much of this work has, again, been driven by a nostalgic anti-urbanism and a concern to document the impacts of sprawl rather than an ideology-free science. In complete contrast, recent team research, such as that done by Pickett and colleagues (2001), calls instead for an integrated ecology *of cities* that evaluates the complex system integration that occurs in all urban areas. This research promises to lay the foundations for a new understanding of the created environment, and will also allow us to integrate the various aspects of urban analysis. For instance, it will be possible to investigate the impacts that new forms of urban GOVERNANCE – such as the PUBLIC–PRIVATE PARTNERSHIPS that define many master-planned communities – have on the way that landscaping is designed and resources like water are used. This in turn also serves to emphasize the role that such concerns should play in planning and economic development decision-making, where environmental impacts are defined very narrowly rather than being considered at the broadest level of quality of life impacts.

The realization that humans are natural too is deeply antithetical to Western culture, yet it is a basic key to rethinking the city and our relationship to nature. Simplistic critiques of sprawl on environmental grounds rather miss the point – that on the one hand urban populations are growing and thus demand more space, while on the other the genie cannot be put back into the bottle: we already know where urbanization is going to take place over the next century, and growth collars and GREEN BELTS serve only to drive up housing prices in inner cities, and to promote leapfrog developments that paradoxically increase commuter distances. The alternative is to recognize that cities can be efficient and can, moreover, sustain a complex natural life. The challenge is to understand this formation, to comprehend its strengths and weaknesses, and to develop public and private policy that informs the stewardship of communities.

Further reading

Davis, M. (1998) *Ecology of Fear: Los Angeles and the Imagination of Disaster*, New York: Metropolitan Books.

Pickett, S.T.A., Cadenasso, M.L., Gove, J.M., Nilon, C.H., Pouyat, R.V., Zipperer, W.C. and Costanza, R. (2001) 'Urban ecological systems: linking terrestrial ecological, physical and socioeconomic components of metropolitan areas', *Annual Reviews: Ecology and Systematics* 32: 127–57.

United Nations Centre for Human Settlements (Habitat) (2001) *Cities in a Globalizing World: Global Report on Human Settlements 2001*, London: Earthscan Publications Ltd.

Williams, R. (1973) *The City and the Country*, New York: Oxford University Press.

ANDREW KIRBY

ENVIRONMENTAL BEHAVIOUR

Environmental behaviour represents the totality of activities and actions undertaken by governmental and non-governmental organizations, and by all inhabitants of the earth, leading to the improvement or worsening of environmental performance. The effects of such activities directly influence environmental quality, and consequently the quality of human life, the physical and psychological health of humans, as well as the safety and functioning of the economy.

The activities, oriented mainly towards curbing pollution introduced into the environment, turned out to be insufficient for stopping the process of environmental degradation at the global level. The after-effects include global warming, ozone depletion, growing shortage of freshwater, etc. The currently observed hazards for human life quality and economic safety at the end of the twentieth century caused attention to be given to undertaking environmental action in all the domains of human activity. These ideas found their reflection in the concept of sustainable development, which was officially adopted

at the conference in Rio de Janeiro in 1992. The supreme goal of environmental activity is to improve human life quality by reconciling the growth and environmental needs of present and future generations. This requires not only the implementation of safeguards against the disadvantageous human impact on the environment through the limitation of pollution, or the reclamation of degraded areas, but also the preservation of natural resources appropriate with respect to their quality and quantity.

Environmental behaviour requires the collaboration of many domains of human activity, primarily the legal, technical and technological, market, economic and financial, political, institutional, social, academic, educational, and managerial, as well as socio-economic and spatial development planning.

The implementation of environmental laws and regulations and respect for them are meant to achieve environmental performance standards. The resulting legal system should comply with the international conventions concerning environmental protection and sustainable development, linking ecological and economic effectiveness, and social acceptance.

Technical manipulations may serve to prevent, reduce and smooth out the degrading environmental impacts from civilization-related factors. Within this area, environmental behaviour consists of the elaboration and implementation of the best available – environmental – technologies, processes and materials, alternative solutions, and modern pro-ecological infrastructure, in association with the advancement of research and development.

The economic and financial as well as market mechanisms consist of the creation of economic motivations for the observance of the legal and administrative requirements concerning environmental protection. These undertakings induce definite environmental behaviour through economic calculations, and create the possibility of minimizing the social cost of environmental protection. An important part of this is integrated systems management.

Environmental behaviour depends largely upon the level of education and the advancement of research, as well as upon access to environmental information, whose basis is constituted by the development of modern, integrated environmental monitoring systems.

The effectiveness of activities meant to improve environmental performance is to a large extent dependent upon the involvement and collaboration of governmental and non-governmental institutions, the business sector, politicians, scientists, public interest groups, etc. Of high importance is the partnership between the levels and branches of government and the self-governmental authorities, as well as citizen participation in decision-making.

BOŻENA DEGÓRSKA

ENVIRONMENTAL DESIGN

Environmental design seeks to create spaces that will enhance the natural, social, cultural and physical environment of particular areas. It has become increasingly allied to the need for producing a sustainable URBAN form. The relationship of people to place, their identification with specific NEIGHBOURHOODS and their use of particular spaces for social, political and cultural activities may be influenced by design and URBANISM as a way of life may be renewed through environmentally sensitive planning processes. Designers must understand social psychology, human behaviour and ecology. Cafe society, culturally vibrant street life, pedestrianized shopping precincts, crime prevention and community safety through natural surveillance, the construction of children's play areas, accessible street furniture and resting places for the elderly and infirm, and open public spaces that support the practice of social and political democracy are all aspects of good environmental design as they promote and support social sustainability. Some areas, buildings or MONUMENTS may have significant symbolic importance. Although mainly a historical curiosity and tourist attraction, Speaker's Corner in London's Hyde Park still signifies the liberal freedoms of speech and assembly. GATED COMMUNITIES, razor wire fences and 'vagrant proof' park benches signify and shape different social relationships and values. Environmental design is therefore about helping to fashion human experience through a created physical space.

The classic Western SUBURB has detached homes built to fairly low densities with high levels of embodied energy. They are usually energy inefficient and their inhabitants are frequently

socially atomized. Public transport is often non-existent. Although in some ways closer to the natural environment than the inner city, the impact of suburban lifestyles seriously impacts on the wider natural environment, most notably in terms of auto and energy use. Environmentally aware architectural design and planning can deal with these problems. Architects and engineers may exploit solar and wind power, choose environmentally sound building materials, recycle old brick and concrete as aggregate, install double or triple glazing, insulate effectively, use natural ventilation to provide thermal comfort and healthy air circulation, design roof gardens or even turf roofs, allow for the recycling of grey water, minimize noise pollution through effective sound absorption and be open to unconventional built forms. Without cooperation between architects, engineers and planners, a great deal of environmental design will never be seen.

On a bigger scale, the high-density compact city is sometimes presented as a possible solution to environmental problems: combat suburban SPRAWL by building at higher densities; ensure new developments encourage walking, cycling and social interaction; discourage auto use through congestion charging and reduced parking facilities; and construct minimally polluting and cheap and reliable public transit systems (particularly light rail and tram). Declining DOWNTOWN areas may be revitalized and long commutes to work may be avoided by planning for mixed-use developments. By contrast, others argue that large gardens reduce high temperatures, increase potential for domestic food production and composting, offer greater opportunities to collect and use rainwater, and allow more space for solar energy panels.

Environmental urban design may also aim to reduce greenhouse gas emissions through energy-efficient buildings, combined heat and power systems, water recycling, and waste minimization. The implications for politicians, architects, planners, urban designers and ordinary city dwellers are immense. Politicians will need to develop long-term strategies that impact upon voters' lifestyles and consumer preferences. Individuals' lifestyles will need to be less car dependent, less materially acquisitive, and perhaps more socially settled and community focused. Planners will need to be empowered to enforce the construction or conversion of properties to ensure settlement patterns have low energy requirements, offering greater

accessibility but less mobility. Toronto in Canada and Perth in Australia indicate that a completely auto-dependent city is not inevitable. Policies should make visible the environmental consequences of everyday living.

Environmental enhancement is frequently associated with improvements in people's quality of life. Parks and gardens, the city's 'urban lungs', provide healthy recreational areas that were also once viewed as a civilizing influence on new immigrants. OLMSTED's Central Park in New York is a prime example. The acquisition of land for nature reserves, the creation of town trails, widespread tree planting and urban agriculture (a tool for transforming urban organic waste into food and jobs, improving public health and land, and saving water and other natural resources) are ways urban dwellers may build mutually supportive social relationships and reconnect to the larger ecosystems on which they depend.

The recent development of 'URBAN VILLAGES' may offer environmental benefits, high-quality and affordable neighbourhoods, and mixed-use urban space with stable and diversely populated communities. Although criticized for its conservative vernacular architecture and limited potential for replication, the ideas of Léon KRIER and the Prince of Wales have been successfully realized in the new development of Poundbury in West Dorset, England. It is financially viable with highly marketable real estate and good-quality social housing. Similarly, the Crown Street development in the centre of Glasgow has demonstrated the significance of the urban village concept, transcending the monoculturalism of other experiments like the Disney Corporation's town of Celebration in Florida.

Environmental design may need to take cognizance of the comprehensive vision of sustainable land use promoted by the Regional Planning Association of America in the 1920s. The principal idea was that the urban economy and its communities should be rescaled to accord with the ecological limitations of a physical region. Regions should live within their means, avoiding the problems of increasing a city's ecological footprint by excessive resource use and the exportation of the city's waste and pollution. This vision offers an alternative design and development model focusing on the desirability of the human scale, the importance of proximity, and the pleasure and necessity of cultural and biological diversity.

Urban environments have hitherto been shaped by economic rather than social and environmental goals. Michael Hough (1995) argues that designers must ensure that urban development positively contributes to the environments they change. Natural processes need to become incorporated into human activities through the creation of multi-functional, productive and working spaces that integrate people, economic activity and the environment.

Further reading

Hough, M. (1995) *Cities and Natural Process*, London: Routledge.

Thomas, R. (ed.) (1996) *Environmental Design: An Introduction for Architects and Engineers*, London: Spon.

Williams, K., Burton, E. and Jenks, M. (eds) (2000) *Achieving Sustainable Urban Form*, London: Spon.

JOHN BLEWITT

ENVIRONMENTAL IMPACT ASSESSMENT

Environmental impact assessment (EIA) is a tool of environmental management that forms part of project approval and decision-making. It usually involves the production of a report, often called the environmental impact statement. The intent of the tool is to provide a specific, and documented, assessment of the environmental consequences of a proposed development while it is still in its planning stage. EIA has classically been applied to large infrastructure projects such as water impounds, roads, power generation or transmission facilities, mining operations, etc., and while it continues to be used for these purposes, the scale and nature of EIA applications have expanded over time. The practice has in some places been taken up at local authority level, as part of the process of development assessment of land use changes by municipalities, and also at an international level by multilateral institutions, including international banks, for the assessment of international development assistance projects. EIA is undertaken for private sector projects and generally also for projects where government is the proponent.

While EIA terminology varies between countries and jurisdictions, the origins can be traced to the National Environmental Policy Act of 1969 in the United States. The tool has now been adopted in most countries in the world. The procedures and format of EIA are, in most instances, defined by legislation, and similarities and differences in EIA practice and requirements in different jurisdictions are well documented. The tool has had remarkable longevity.

The technical steps of EIA include scoping the range of potential environmental effects on the physical and social systems in which the development will be located. This includes direct effects and higher-order effects. For example, a direct effect of a new freeway would be the dislocation of people from the route; a secondary effect could be the changes in land use induced by the new roadway. The magnitude of each of the potential effects is then predicted and evaluated for significance. The prediction stage of EIA remains problematic, for while the ability to predict physical changes such as air quality, or noise, has improved remarkably, the ability to predict ecological change, or social impact, is less robust. EIA also proposes strategies to mitigate significant environmental effects and ongoing monitoring activities that may be necessary if the project is approved. EIA has had an important role in the improvement of prognostic capabilities in the environmental field.

The theory of impact assessment requires that the need for the project should be demonstrated and that a range of alternative ways to meet the same objectives should be included in the analysis. In practice, need and alternatives are not always adequately pursued, particularly where the project is in the private sector. EIA is most often conducted by consultants employed by the proponent of the development. The resulting documentation is reviewed by either an environment agency or the authority responsible for development approval, though in some countries expert panels may undertake the review. In most jurisdictions, the requirement for EIA is separately legislated, though the tool can function as part of the administrative requirements of development approval.

EIA practice has developed over three decades. As initially implemented, the role of EIA was the provision of information on environmental effects of a development proposal to decision-makers. While this role remains, over the years it has variously expanded into an environmental design tool closely linked with the planning of the project

itself, and into a tool for public consultation. There are some tensions between perspectives of EIA as a technocratic tool and perspectives of it as a tool for increasing the transparency of decision-making. There is also increasing attention to the role of EIA in assessing the sustainability of development proposals.

The effectiveness of EIA has always been under scrutiny. Effectiveness appears to depend largely on the potential to test the EIA in the courts, or on the ability of third parties such as affected communities to use the process to participate in the assessment, either formally, or informally through public pressure. While EIA is practised in most developing as well as developed countries, there are concerns about the ineffectiveness of environmental assessment in the former associated with a lack of opportunity for community participation in and input into the assessment, lack of transparency in the process, and lack of enforcement of conditions recommended by the assessment. Despite this, EIA is still seen by environmental practitioners as an important environmental management tool. As well as its explicit role in mitigating environmental effects of projects, a broader view suggests that, over time, EIA has been effective in an educational role for many of the stakeholders and decision-makers in the development process and in promoting a commitment to environmental management.

There is growing concern that EIA of development activities on a project-by-project basis is inadequate. Cumulative effects of developments result in environmental degradation, despite EIA practice. The response is awareness that project-level EIA needs to be supplemented by environmental assessment at the levels of programme, plan and policy, too. Accordingly, the field of strategic environmental assessment (SEA) (programmatic assessment in the United States) has evolved to supplement what might now more appropriately be called project-based EIA. There is an emerging concept of tiering, where SEA at policy and programme levels is supplemented by EIA of projects that eventually arise from the frameworks provided by these policies and plans.

Different jurisdictions have different definitions of what constitutes 'environment'. In particular, concern that 'environment' in EIA is sometimes defined narrowly to incorporate only biophysical issues has led to the development of the tools of social impact assessment, health impact assessment, and so on in other fields. While there is no doubt that positive emphasis on such social dimensions within environmental assessment is appropriate, all of these tools are best considered as component dimensions of a holistic EIA process.

Further reading

Petts, J. (ed.) (1999) *The Handbook of Environmental Impact Assessment*, vols 1 and 2, Oxford: Blackwell Science.
Woods, C. (2002) *Environmental Impact Assessment: A Comparative Review*, Harlow: Longman.

A. LEX BROWN

ENVIRONMENTAL JUSTICE

Environmental justice is a social movement that seeks to protect those that have borne the brunt of environmental hazards. It seeks to prevent minority and lower-income communities from continuing to serve as dumpsites for industry and society. And, it seeks to protect workers from occupational exposures to pesticides, radiation and other hazardous chemicals. Within the United States and other countries of the North, the environmental justice movement organizes both inner-city and rural communities, as these have historically borne more environmental risk. In an international context, the citizens of some countries have borne the environmental risks of the global economy.

Minority and lower-income communities share a disproportionate exposure and proximity to toxic industry, hazardous waste and hazardous waste management sites. Several influential studies in the 1980s documented this co-location with its attendant higher risks. Some argue that the existing distribution of environmental risk and hazard is more strongly correlated with race than with any other community measure such as income. This is termed environmental racism.

As a social movement, environmental justice is a loose affiliation of groups that share common ideas and work towards similar ends. Minority and working-class communities have greater exposure to environmental hazards. Decisions about locations continue to be made that increase this exposure. Prevention and safety policies have not been equally enforced. Many workers have borne the health costs of an industrial organization that

benefits the rest of society. In an international context, countries of the South have had fewer environmental protections and the poor of these countries have had greater workplace exposures and greater hazards from incidents such as Bhopal.

Those communities and workers that benefit the least from industrialization are simultaneously those that bear its greatest harms. It is not merely the uneven distribution of environmental risk and harm that must be rejected, but also the willingness to subject anyone to the preventable harms of unregulated industrial development.

A diverse set of groups in the environmental justice movement aims to increase URBAN environmental quality, better regulate toxic industries, ensure the safe disposal of toxic hazards, protect industrial and agricultural workers, and reform decision-making processes.

In contrast to the environmental justice framework, according to normative economics individuals value risk and benefit differently, depending on their individual economic situation, risk tolerance and the rate at which they discount their future health for their current wages. An individual that values health more than the benefit received from a particular job or place to live is free to move or change occupation. The fact that an individual continues to suffer this risk means that the value of employment outweighs the value that the individual places on health. For a migrant or an unskilled worker, the freedom to choose another occupation is more likely to be the freedom to be destitute.

Opinions also differ as to the cause of the uneven distribution of environmental hazards. Some argue that if the co-location of these communities and environmental risks were racist, the location of toxic industries would have followed the location of minority communities. In many inner-city locations, toxic industrial locations predate the establishment of particular minority communities. The fact that the minority communities seem to have moved to the hazard makes the economic explanation sound more plausible than the racist one.

However, the environmental justice movement takes a systemic view of racism and injustice. Environmental racism is part and parcel of larger societal racism that includes lower incomes for minority groups and a less meaningful range of life choices. Minority individuals are less likely to be able to (or to want to) move to historically white enclaves where they may face everyday racism. The co-location of minority communities with environmental hazards is proof of systemic racism rather than a high toleration for poor health.

Historically, the movement draws on a tradition of worker protection and urban environmentalism. Expressed in the context of early INDUSTRIAL CITIES, this tradition is exemplified by the work of Alice Hamilton and Jane ADDAMS of the settlement house movement. These social reformers sought to improve sanitation and working conditions for the urban poor.

The modern environmental movement is often dated from Rachel Carson's *Silent Spring* and its systematic critique of a society killing itself with pesticides. The publicity surrounding this book made ecology a household word. Environmentalism became a widespread middle-class social movement. While *Silent Spring* expressed concern for workers, women and human health, early environmentalism of the 1960s and 1970s focused its efforts on the environment outside the cities.

As a social movement conscious of shared causes, environmental justice dates from the early 1980s. It was not until the publicity given to Love Canal and other communities poisoned by industrial waste that industrial toxics were understood as 'environmental'. Hazardous waste dumping and disposal was the issue around which those active in the environmental justice movement first organized themselves. Activists usually date the movement from a situation in Warren County, North Carolina, where an African-American community organized itself to fight both the illegal dumping of polychlorinated biphenyls (PCBs) and then a proposed PCB disposal facility. In concert with this community resistance, the term 'environmental justice' was first used to signify the connection between the CIVIL RIGHTS MOVEMENT and the environment. Since then, many communities and groups have organized in support of the causes of environmental justice.

The ideas of environmental justice have helped to organize resistance to the siting of new environmental harms in minority and lower-income communities. Environmental justice activists make strong arguments for better environmental management and the equitable enforcement of existing standards in these communities.

Further reading

Bullard, R.D. (1990) *Dumping in Dixie: Race, Class and Environmental Quality*, Boulder, CO: Westview.

Cole, L.W. and Foster, S.R. (2001) *From the Ground Up: Environmental Racism and the Rise of the Environmental Justice Movement*, New York: New York University.

Gottlieb, R. (1993) *Forcing the Spring*, Washington, DC: Island.

Krauss, C. (1993) 'Women and toxic waste protests: race, class and gender as resources of resistance', *Qualitative Sociology* 16(3): 247–62.

TARA LYNNE CLAPP

ENVIRONMENTAL PERCEPTION

Environmental perception is an area of knowledge dealing with the apprehension human beings experience when exposed to the stimuli presented by the environment that surrounds them. The basis for studies and research in environmental perception is that human interaction with the environment is guided by perceptive mechanisms; that is to say, in their daily lives people exercise a permanent acknowledgement of the environmental conditions, and they do so by using their perceptive processes.

Environmental perception starts with people's sensorial response to the external incentives they receive from the environment. Nevertheless, perceptions are not solely restricted to the sensorial field. Perception of the environment is a cumulative experience. It starts with the registering of the sensorial stimulations, which are communicated to the human brain via the five senses – sight, hearing, touch, smell and taste. But the process does not end in this reception alone. People not only get informed about the environmental conditions; they also share the experience of actually living in the environment. Both sensorial and experiential feelings will merge to inform people about the environment; and next, people will process an evaluation of the features presented by the environment, becoming eventually aware of how to better behave in it. So, perceptions allow people to understand their environment, and to communicate the information they acquire via their senses, straight to their cognitive level, to the intelligence. In other words, people will tend to adopt certain attitudes in their spatial behaviour, and these attitudes vary according to the environmental patterns they perceive in the environment.

The conceptual framework employed in environmental perception research is able to bring in contributions that help to identify the structural elements in an URBAN landscape. Furthermore, among the whole range of elements interspersed throughout the landscape, it is able to point out the ones that represent what really makes 'sense' for that specific society. It is usually assumed that there are tangible and intangible urban elements present in cities, which act upon people's perception, and consequently upon their behaviour. Two considerations may be derived from this assumption. One refers to the opportunities this assumption has to be used as a tool in URBAN DESIGN strategies. The other recommends its use as a likely tool for supporting sustainable development policies.

There is a growing acceptance that environmental data gathered by environmental perception techniques can be of great value in city planning operations. This is so because the sort of technical procedures employed in environmental perception research can provide planners with a better understanding about the people–environment interaction that is actually happening in the territory. Data collected by these means provide a more reliable account of what is genuinely valued in the environment, and hence provide better leads for diminishing the uncertainty content that accompanies the process of bringing out design guidelines. Therefore, it is believed that the planning objectives will be more accurately calibrated, and in conformity to what people effectively perceive as meaningful for the preservation of the CULTURAL IDENTITY of their environment.

Seen from another viewpoint, this sort of reasoning seems to be in accordance with what was understood as a revolutionary process, when advanced in the philosophical debates that took place at the end of the twentieth century. Theoretical postulations forwarded by leading postmodern thinkers of the time, such as Jean-François LYOTARD, questioned the legitimacy of scientific knowledge in the new postmodern condition, proposing that the world should be understood in terms of 'local narratives' rather than the 'grand narratives' instituted by science, which dictated the norms in MODERNISM. Within this reasoning, it is implicit that the local narratives implied in the theories of POSTMODERNISM can be brought to the fore by the revelation of the perceptions people share more intimately about the environment they

live in. These perceptions would help to decode the phenomenological elements symbolized in the environment, which are produced by the lively interaction performed between people and their urban settings. On this basis, it is believed that applying the findings of environmental perception research to the postulation of planning guidelines enables ENVIRONMENTAL DESIGN to relate more aptly to people's behaviour.

A similar reasoning applies to the use of environmental perception techniques in SUSTAINABLE URBAN DEVELOPMENT, a concept directly related to environmental sustainability. It seems fair to accept that it will be easier to preserve what the local populations perceive as local environmental assets. As a consequence, a proactive policy towards the conservation of non-renewable resources would tend to gain wider acceptance, since what would be proposed would result from what is effectively valued as significant in the perceptions people have about their environments.

Generally speaking, environmental perception techniques ultimately point to the affectionate and cognitive dimensions that arise from people–environment interactions, and the urban elements identified in such a procedure will be exactly the ones that are best understood by the community, and more accepted within the prevalent local societal codes. As an output, the inventory of elements which results from this type of investigation will necessarily comprise both socio-psychological and urban-architectural stimulations, furnishing a broad view of the overall ENVIRONMENTAL BEHAVIOUR. In the end, the overall likelihood is that environmental perception will be effective in providing useful guidelines for architects and planners, because the sort of information it discloses reveals precisely those elements of the city which should be strategically included in the environmental design and environmental sustainability proposals to be produced.

Further reading

Banerjee, T. and Southworth, M. (eds) (1991) *City Sense and City Design: Writings and Projects of Kevin Lynch*, Cambridge, MA: MIT Press.

Bonnes, M. and Secchiaroli, G. (1995) *Environmental Psychology: A Psycho-Social Introduction*, London: Sage.

Castello, L. (2001) 'When perception gets designed by the market', in *Honey, I Shrunk the Space – Planning in the Information Age*, Proceedings of the 37th International ISoCaRP Congress, Utrecht, 16–20 September.

Zube, E. (1980) *Environmental Evaluation: Perception and Public Policy*, Monterey, CA: Brooks/Cole.

LINEU CASTELLO

ENVIRONMENTAL SOCIOLOGY

Environmental sociology represents a sub-discipline of sociology, the systematic study of human behaviour in the social context. Emerging in the early 1970s, environmental sociology represents a relatively new area of inquiry, with its focus an extension of earlier sociology through inclusion of the physical context as related to social factors.

From its formative years, environmental sociology has drawn heavily from HUMAN ECOLOGY, URBAN SOCIOLOGY, rural sociology, as well as work by ecological and cultural anthropologists. Important linkages are also found with URBAN PLANNING and environmental psychology.

Reflecting this multidisciplinarity, within environmental sociology the 'environment' reflects both natural and built contexts. In both cases, researchers aim to examine how society affects the environment, as well as how the environment affects, and reflects, society. Topics of inquiry include public environmental opinion, environmentalism as a social movement, human-induced environmental decline, social response to natural disasters, and social dimensions of the built environment.

Regarding the built environment, sociologists have contributed more to our understanding of the social aspects of regional and urban development ('macro' level), as opposed to social dimensions of buildings and interiors ('micro' level). Still, at both analytical levels, two broad questions guide sociological consideration of the urban and built environments.

First, sociologists ask: *Why does the built environment take on a particular spatial arrangement?* Here, many dimensions of social processes combine with important aspects of the physical environment to shape the built context. As an example, consider the influence on land use patterns of interest groups with different levels of power. Indeed, social structure is reflected in urban spaces, as evidenced by the prominence of lower-income and minority communities in proximity to industrial urban sectors. Social structure vis-à-vis

the built environment could also be examined with regard to the nature and adequacy of housing stock for the poor, elderly and/or handicapped. Inequalities can also be considered on a global scale; variation in international allocation of power and influence plays a role in the shape of the built context within nation-states.

As another example of social influences on the built environment, consider the ways in which the built environment reflects the *cultural* context, therefore possessing social meaning. As an example, American SUBURBS reflect American culture's materialism, value of privacy, and automobile dependence.

The second broad sociological question regarding the built environment is: *What impact does the built environment have upon social processes?* Consider the ways in which the built environment *shapes* the context in which social activity takes place. As an example, the availability, and configuration, of public spaces has tremendous influence on levels, and patterns, of social interaction. As another example, drawing from classic work by Louis WIRTH on URBANISM, sociologists examine how social relationships vary along the rural–urban continuum. Wirth argued that with urbanism comes a decline in neighbourly relations, a loss of sense of 'community' and a lessening of social control as evidenced by higher crime levels.

Further reading

Ahrentzen, S. (2002) 'Socio-behavioral qualities of the built environment', in R.E. Dunlap and W. Michelson (eds) *The Handbook of Environmental Sociology*, Westport, CT: Greenwood Press.

Popenoe, D. and Michelson, W. (2002) 'Macro-environments and people: cities, suburbs, and metropolitan areas', in R.E. Dunlap and W. Michelson (eds) *The Handbook of Environmental Sociology*, Westport, CT: Greenwood Press.

LORI M. HUNTER

ENVISIONMENT AND DISCOVERY COLLABORATORY

The Envisionment and Discovery Collaboratory (EDC) is a physical-computational environment or system. It was initially developed around a conceptual framework to support participation and collaboration in city planning activities such as URBAN DESIGN. Its aim is to empower citizens to discover and envision ways to address collaboratively collective concerns or problems in city planning. To this end, it integrates interactive *physical games* and their respective *languages of objects* with *computational simulations* and *dynamic information systems* into two tightly linked spaces, which are called the *action* and *reflection* spaces. Their linkage is attained via the web (http://l3d.cs.colorado.edu/systems/EDC/).

Conceptual basis

The EDC design is based on a framework for creating shared understanding among stakeholders in collaborative planning. It is based on notions from city planning, design, computer science and cognitive science. Thus, in its development and evolution, the EDC:

- *introduces the notion of a common language of design* by integrating the use of *physical objects* – to support and encourage face-to-face interaction among the participants – with *virtual objects* – to provide computational support for the models underlying the simulations (Arias *et al.* 2001);
- *uses simulations* to engage in 'what if' games and to replace anticipation of the consequences of our assumptions by analysis (Repenning and Sumner 1995);
- *deals with a set of possible worlds* effectively (i.e. exploring design alternatives) to account for the fact that design is an argumentative process in which we do not prove a point but instead create an environment for a design dialogue (Simon 1996);
- *incorporates an emerging design* in a set of external memory structures and records the design process and the design rationale (Arias *et al.* 2001);
- *creates low-cost modifiable models* that assist stakeholders in creating shared understanding by having a 'conversation with the materials' (Schön 1983).

The framework provides the design process with decision-making support when different interests and opinions conflict, when alternative proposals compete for limited resources, and when it is considered important that different stakeholders need to be enabled and encouraged to

collaborate in the planning activity (see CRITICAL COALITION). Challenged by such real-world contexts, the participatory, experiential and interactive nature of the EDC framework allows users to:

- define the problem in a way that is amiable to solution;
- reduce areas of disagreement;
- suggest directions that are consistent with opposing positions; and
- determine what the different stakeholders are willing to do to resolve the problem as they perceive it.

Present form

Individuals using the EDC convene around a computationally enhanced table, which together with the Physical Languages of planning objects serve as the *action space*. Currently realized as a touch-sensitive surface, the action space allows users to manipulate a computational simulation projected on this surface by interacting with the physical languages of objects on the table. Thus, participants, for example, can construct and modify a physical and computational representation of their neighbourhood by placing and moving physical objects of the language that represent elements such as houses, parks, schools or bus stops which are developed for a particular application. Physical representations of problems are often extremely useful in describing complex phenomena and serve as an anchor for discussion (Arias *et al.* 2001). Thus, the 'physical languages of objects' integrated with computational simulations in the action space support the players to collaboratively construct descriptions of reality, evaluate them and, based on their assessments, come up with collaborative prescriptions to resolve them.

The action space table is flanked by a second computer which is driving another touch-sensitive whiteboard serving as the *reflection space*. The computers driving these spaces are linked via the web. Since much of the rich conversation around the action space takes place in a single session, and given that 'groups have no head', the reflection space uses web-based dynamic information systems to provide a flexible forum for capturing and presenting information to users during interactions with the EDC. Also, utilizing the web, neighbours who are not present in a session can fill out in a distributed manner a web-based planning survey

associated with the model being constructed in the action space, for example to provide descriptions, opinions, preferences or evaluations of the area being analysed or planned using the EDC. In this manner, like the collaboratively constructed problem in the action space, the information in the reflection space is a shared representation. This interplay of information creation and presentation within a design task creates a forum in which information is captured and evolves through design activity, as well as providing a way of framing the various issues and perspectives that surround the problem.

In this manner, the construction and presentation in the action space provides a shared physical context for a target city planning activity, while the externalized representation of the information surrounding the issues in the activity is presented in the reflection space. Thus, through the immersion of people *within* the representations of the problem-solving tasks being carried out and supported by the two spaces, the EDC creates an integrated human–computer system grounded in the physical world.

It is being developed at the Center for LifeLong Learning and Design of the College of Engineering and Urban Simulations Laboratory of the College of Architecture and Planning at the University of Colorado, with support from the National Science Foundation. Its development as a new generation of collaborative technology systems aims to address and overcome current limitations of human–computer interactions. It shifts the emphasis away from the computer screen as the focal point and creates an *immersive* environment in which stakeholders as users can incrementally create a *shared understanding* through their participation in collaborative city planning. It is an environment that is not restricted to the delivery of predigested information to individuals; rather, it provides opportunities and resources for design activities embedded in social debates and discussions where *all* stakeholders can actively contribute rather than being confined to passive consumer roles. As a computational research prototype, it represents what support technologies and collaborative approaches to the planning and design of cities will be like in the future.

Further reading

Arias, E.G., Eden, H., Fischer, G., Gorman, A. and Scharff, E. (2001) 'Transcending the individual

human mind: creating shared understanding through collaborative design', in J.M. Carroll (ed.) *Human-Computer Interaction in the New Millennium*, Boston, MA: Addison-Wesley, pp. 347–72.

Repenning, A. and Sumner, T. (1995) 'Agentsheets: a medium for creating domain-oriented visual programming languages', *IEEE Computer: Special Issue on Visual Programming* 28(3): 17–25.

Schön, D.A. (1983) *The Reflective Practitioner: How Professionals Think in Action*, New York: Basic Books.

Simon, H.A. (1996) *The Sciences of the Artificial*, 3rd edn, Cambridge, MA: MIT Press.

ERNESTO G. ARIAS

EQUITY PLANNING

Equity planning seeks to move resources, political power and participation towards lower-income, disadvantaged people in order to provide them with more choices than those available through existing political-economic and institutional arrangements. Equity planning is closely related to ADVOCACY PLANNING in that both approaches seek to serve the disadvantaged. However, while advocacy planning encourages the preparation of plans alternative to official plans – with advocacy planners preparing plans for disadvantaged communities – equity planning applies to official city planning agencies within local government.

The first example of equity planning took place in the 1970s in Cleveland, Ohio, where planning director Norman KRUMHOLZ and a core staff of progressive planners worked under three different administrations on behalf of 'those who had few, if any, choices'. Their efforts led to the enhancement of transit services for the transit-dependent population, the saving of lakefront parkland, 'fair share' low-income housing distribution plans for Cuyahoga County, improvements in public service delivery and progressive changes in Ohio's property law. Their contribution is significant at three levels. First, instead of attempting to serve an abstract public interest, they chose to serve those with few choices, thus establishing a framework that guided them in their planning efforts. Secondly, in deciding to serve those with the least political power, they chose a path strewn with political difficulties. Thirdly, they demonstrated that – even when advocating for the poor – planners can be effective by becoming assertive and taking risks.

An especially important example of equity planning took place in Chicago during the 1980s as a result of the 1982 election of Chicago's first black mayor, Harold Washington. The new mayor brought into City Hall a group of progressive planners and academics, including Robert Mier, a planning professor who had funded the Center for Urban Economic Development at the University of Illinois, Chicago. These planners included redistributive and social justice as explicit goals for the Washington administration. The emphasis of their Chicago Economic Development Plan was on generating jobs for Chicago residents, with a particular focus on the unemployed. They also implemented a 'linkage' programme, making developers receiving city incentives or allowed to build in profitable areas of the city contribute to a low-income housing trust fund, or assist the community-based development corporation building in declining areas of the city.

The goals of equity planning have been officially integrated in the code of ethics of the American Planning Association (APA). The fourth APA ethical principle urges planners to 'strive to expand choice and opportunity for all persons, recognize a special responsibility to plan for the needs of disadvantaged people, and urge changing policies, institutions, and decisions that restrict their choices and opportunities'.

Even with the official exhortation of the APA, equity is rarely addressed in official planning documents in American cities. Advocacy and equity planning were a response to the URBAN problems of the 1960s and 1970s. With the more recent weakening of cities in national electoral politics, the delegitimization of government intervention and a general shift to the right, equity planning has not prospered. But the problems of the disadvantaged persist, or have worsened. In this context, equity planning will find a home where local political conditions match individual planners' progressive politics, motivation and confidence. An organization called Planners' Network, with a 1999 membership of over 800 planners and academics concerned with economic and SOCIAL JUSTICE, has been in existence for more than a quarter of a century and continues to pursue the goals of advocacy and equity planning, publishes a journal, *Planners Network*, and holds annual conferences.

Further reading

Cleveland City Planning Commission (CPC) (1975) *Policy Planning Report*, Cleveland: CPC.

Krumholz, N. and Clavel, P. (1994) *Reinventing Cities: Equity Planners Tell Their Stories*, Philadelphia, PA: Temple University Press.

Krumholz, N. and Forester, J. (1990) *Making Equity Planning Work*, Philadelphia, PA: Temple University Press.

NICO CALAVITA

ETHNIC CONFLICT

Disagreements over the distribution of costs and benefits of URBAN living that are linked to ethnic or racial group identity. In these circumstances, an urban group perceives it is obtaining disproportionately low benefits (such as services, access to good education and jobs, amenities) and/or absorbing disproportionately high costs (such as urban pollution, ENVIRONMENTAL JUSTICE, mistreatment by police) that are associated with urban living.

The growing ethnic diversity of cities in the world is challenging the traditional assimilation, or MELTING POT, model of intergroup relations, premised on the integration of majority and minority ethnic groups into a common culture and social structure, for the most part approximately that of the dominant, or majority, group. The pluralism model of intergroup relations, in contrast, works to retain the distinct subcultures of ethnic groups within a common culture, creating not a melting pot but a multicultural mosaic of many distinct parts.

In most cities, ethnic conflicts focus on issues of urban service delivery, political representation and voice, community development, housing, police protection, and the siting of controversial facilities. These conflicts are commonly addressed within accepted political frameworks. Questions of what constitutes the public good are debated but largely within a sanctioned framework. It is the hope in a democracy that political coalition-building remains possible across ethnic groups and that diversity of interests and perspectives can defuse and moderate ethnic group-based conflict. Urban policy-making and planning seek to ameliorate urban conflict through an acceptable allocation of urban services and benefits across ethnic groups and their neighbourhoods.

When grievances cannot be effectively addressed in the political arena, anger and frustration can result in riots and violence. The United States experienced three consecutive summers of race riots as African-Americans took to the streets to express their anger over urban inequalities. African-Americans were frustrated over their relative poverty and their lack of full access to economic opportunity and political equality. A total of 329 rioting incidents took place across urban America from 1965 to 1968, with 220 deaths. In one year alone (the 'long hot summer' of 1967), over 500,000 people were arrested for rioting, over 8,000 were injured and more than 200 were killed. One of the more well-known riots was the 1965 Watts riot of Los Angeles, ignited when a crowd of African-Americans gathered to protest against police actions in the arrest of a young African-American youth. In 1967, a presidential commission (Kerner Commission on Civil Disorders) concluded that 'Segregation and poverty have created in the racial ghetto a destructive environment . . . White institutions created it, white institutions maintain it, and white society condones it.'

Violent urban uprisings have occurred in the United States since the 1960s. Major riots occurred in Miami in 1980 and 1989, the Crown Heights area of Brooklyn, New York, in 1991, Cincinnati in 2001, and racial conflict is associated with arson incidents in Detroit in the mid-1980s. In 1992, rioting hit Los Angeles a second time, again ignited by police mistreatment of a stopped African-American motorist. However, this time Latinos as well as African-Americans engaged in rioting and looting, and Korean as well as white establishment businesses were targeted. Indeed, some have called this America's first multi-ethnic riot. Along with continuing African-American–white tensions, economic decline in Los Angeles in the early 1990s placed ethnic and racial groups in conflict and competition. Deaths (53), injuries (2,383) and property damage (over US$700 million) were considerable, making it America's worst civil disorder of the twentieth century. Rioting violence since the 1980s has been targeted against civilians to a greater extent than in the riots of the 1960s, which aimed their anger more towards property and the police.

The potential for future urban ethnic conflict remains due to the continued existence of DIVIDED CITIES, in particular the spatial mismatch between

low-income, often minority residents and outer-suburban jobs, increased minority poverty concentration and social isolation, persistence of racial and economic SEGREGATION, and widening disparities between the revenue-producing abilities of higher-income, outer-suburban cities and lower-income, disproportionately minority inner-suburban and CENTRAL CITIES.

Ethnic conflict has occurred in Britain's cities. Minority ethnic groups are over-represented in declining INDUSTRIAL CITIES such as Birmingham, Manchester, Newcastle, Liverpool and Sheffield; increasing competition for a shrinking job base has led to ethnic conflict. Riots and conflict can take on a visible ethnic characteristic, such as violence in the city of Bradford that contains a highly segregated Pakistani population.

A deeper, and more polarizing, type of ethnic conflict is apparent when ethnic and political claims combine and impinge significantly on issues of GOVERNANCE and control in cities. In such circumstances, such as in Jerusalem (Israel/West Bank), Belfast (Northern Ireland), apartheid cities in South Africa, and Nicosia (Cyprus), a strong minority of the urban population may reject urban and societal institutions, making consensus regarding political power-sharing impossible. Whereas in most cities there is a belief maintained by all groups that the existing system of governance is properly configured and capable of producing fair outcomes, assuming adequate political participation and representation of minority interests, governance in these more deeply contested cities is often viewed by a substantial segment of the ethnic minority population as artificial, imposed or illegitimate. An ethnic minority will deeply mistrust the intrinsic capability of city government to respond to calls for equal, or group-based, treatment. A combustible mixture of political grievances can then combine to turn the attention of subordinated ethnic leaders from urban reform of the existing system to radical restructuring or, finally, to separation and autonomy.

Further reading

Farley, J.E. (2000) *Majority-Minority Relations*, 4th edn, Upper Saddle River, NJ: Prentice Hall.

National Advisory Commission on Civil Disorders (Kerner Commission) (1988) *The Kerner Report: The 1968 Report of the National Advisory Commission on Civil Disorders*, New York: Pantheon.

Pincus, F.L. and Ehrlich, H.J. (1994) *Race and Ethnic Conflict: Contending Views on Prejudice, Discrimination, and Ethnoviolence*, Boulder, CO: Westview.

SCOTT A. BOLLENS

ETHNIC ENCLAVES

Ethnic enclaves are NEIGHBOURHOODS that are identified with immigrants from other countries or their descendants. There may be persons living and working in the area whose ancestors come from other countries, but the place is usually known for one particular group. The best-known ethnic enclaves in US cities began appearing with the arrival of large numbers of Irish immigrants during the first third of the nineteenth century and continued forming throughout that century and well into the twentieth as successive waves of immigrants arrived in the United States. Members of other nationality groups formed their own enclaves after they arrived, or they took over an area being abandoned by a group that had preceded them. While remnants of some of those early enclaves remained intact at the end of the twentieth century, many had long ago passed into the hands of different immigrant populations.

Ethnic enclaves are supposed to be common in cities, because that is where so many immigrants settled upon their arrival. Although this kind of settlement has also been found in small towns some distance from major URBAN centres and even in contemporary SUBURBS, the connection between cities and ethnic enclaves remains strong.

In European cities, the traditional GHETTO or 'quarter' held certain kinds of persons (e.g. Jews, members of a particular family and their attendants or extended kin). These persons may have travelled freely throughout the city or even outside of their community, but their permanent homes and often their livelihoods as well were found in these compact neighbourhoods. The custom of forming such neighbourhoods was carried over into the United States.

Indeed, the practice of forming such compact and identifiable settlements inside cities was expanded to include other kinds of persons. Wealthy persons had Boston's Beacon Hill, the nineteenth-century version of 'GATED COMMUNITIES' on the edge of St. Louis, and Nob Hill in San Francisco. Gays and lesbians predominated

in the Castro and Mission districts of San Francisco.

How long a particular nationality group remained tied to and identified with a particular neighbourhood depended on many factors, a number of which the people in question had no control over. It is generally recognized that with the exception of black neighbourhoods, most ethnic enclaves were home to persons from several different nationality groups, and that the 'borders' between them were often quite porous. People moved in and out of these neighbourhoods with some regularity and not always because they acquired enough money to move to a better part of town.

Though improvements to their economic condition certainly played a part in the decision of many families and individuals to move, not everyone who 'made it' moved out. Most enclaves were able to hold on to persons from different rungs on the social class ladder for at least two generations. By then, the community would have established both a solid commercial base of small and medium-sized businesses and a network of voluntary organizations that included a number of religious institutions. These businesses and organizations gave ethnic enclaves their staying power in the wake of all the people moving in and out of the neighbourhood.

Further reading

Abrahamson, M. (1996) *Urban Enclaves: Identity and Place in America*, New York: St. Martin's Press.
Light, I. and Gold, S.J. (2000) *Ethnic Economies*, San Diego, CA: Academic Press.

DANIEL J. MONTI JR

ETHNO-CLASS RELATIONS

In a series of publications that began in 1958, Milton M. Gordon explored the relationship between ethnicity, class and assimilation in the United States. These sociological investigations metamorphosed into a series on the organizational theories of American social structure, which culminated in his proposal of a new conceptual framework, called ethclass. Even though he had discussed this matter in his famous 1964 book, *Assimilation in American Life*, Gordon formally argued for a new terminology in understanding the

relationship between class and ethnicity in *Human Nature, Class, and Ethnicity* (1978). He suggested that the term 'ethclass' be employed to refer to the sub-society (or social groups) created by 'the intersection of the vertical stratification of ethnicity with the horizontal stratification of social class'. Within this conceptual framework, he posed the questions of group identity, social participation and cultural behaviour. He proposed that while ethnic affiliation provides a sense of historical identification, ethclass provides the basis for participatory identification. The latter provides explanations of social participation patterns in society. Gordon's ethclass concept, while similar to Shevky and Bell's social area analysis (circa 1955), is a much richer theoretical proposal, since on the one hand it acknowledges the persistence of ethnicity in American society and on the other hand illustrates that solidarity of social classes may be viewed as fragmentary in a racially and ethnically conscious society. Furthermore, by its language, ethclass confirms contemporary understanding of the early industrial era, when ethnicity provided a safer alternative to social division by class. Immigrants found their flat division of ethnic nomenclature less restrictive than a vertically organized social structure. The outcome, as proposed by Gordon and others after him, has been a society that divides itself by both ethnicity and social class. This multidimensional social division manifests itself in how the members of an immigrant society choose (or are forced/manipulated) to participate in the social process. A number of scholars have operationalized ethclass in order to examine political participation patterns, educational achievement, the residential distribution of immigrants and other social indicators.

The word 'ethno-class' is also used to refer to cultural groups who are typically either immigrants or their descendents and who occupy a specific social class position within a society, typically low (how the term is defined by Ted Gurr and used by Oren Yiftachel). This terminology appears more often among those with an interest in politically marginalized populations, or more commonly in issues of international ethno-political conflicts. In this context, the term 'ethno-class relations' appears more politically charged than Gordon's conception of ethclass. Ethno-class groups are typically populations afforded a social and political position because of their ethnicity, within the framework of a national ethnic hierarchy. Based on

current authorship, the Turkish population in Germany, the Kurdish population in Turkey and Iraq, and the Mizrahi Jews in Israel are examples of such groups. Therefore, ethno-class can be viewed as a three-dimensional model of ethclass, where a political axis has been added to ethnicity and class. In this manner, a group's position in relation to the national political, social and cultural construct can be mapped against this structural model. However, it should be noted that ethno-class is typically used to denote a specific marginalized population within this three-dimensional construct. Gurr and others have used various forms of public protest as a way of measuring a specific ethno-class population's resistance to the state. Examples of this conceptual and methodological approach can be seen in a number of publications from various social science disciplines, including political science, planning and policy analysis, sociology, geography, ethnic studies, history, and international studies.

Further reading

Gordon, M.M. (1958) *Social Class in American Sociology*, Durham, NC: Duke University Press.
—— (1964) *Assimilation in American Life*, New York: Oxford University Press.
—— (1978) *Human Nature, Class, and Ethnicity*, New York: Oxford University Press.
Gurr, T.R. (1970) *Why Men Rebel*, Princeton, NJ: Princeton University Press.
Shevky, E. and Bell, W. (1955) *Social Area Analysis: Theory, Illustrative Application, and Computational Procedures*, Stanford, CA: Stanford University Press.
Yiftachel, O. (2000) 'Social control, urban planning and ethno-class relations: Mizrahi Jews in Israel's "development towns"', *International Journal of Urban and Regional Research* 24(2): 418–38.

ALI MODARRES

EUROPEAN SQUARE

Whether called 'AGORA', 'forum', 'piazza', 'plaza', 'Platz', 'platia', 'náměsti' or 'rynek', the main (market) square has been a distinguishing characteristic of European cities for over two thousand years.

Unlike the grand squares and plazas of ancient Chinese cities intended to create a stage for the display of power, the European multifunctional market squares were shaped by the needs of inhabitants and merchants to have a place for social life, trade, community festivals, for general assembly and to elect representatives. Indeed, the traditional European square was intimately connected to the development of democratic self-government.

The traditional European square is an URBAN space surrounded by an almost continuous wall of buildings, with small or angled entrances and exits, creating the feeling of an outdoor salon. Mixed-use shops/houses, complemented by important civic and religious edifices, enclose the square.

Most cities and towns are organized around one central square, but this is not the case everywhere. In Padova, Piazza delle Erbe, Piazza della Frutta and Piazza dei Signori are interconnected; in Verona, the city's heart is composed of Piazza delle Erbe and Piazza dei Signori; Antwerp's heart is Grote Markt, Handschoenmarkt and Groenplaats; and Salzburg's Alter Markt, Residenzplatz, Mozartplatz and Universitätsplatz adjoin one another. In Venice, there are large and small squares – 'campi' – in every NEIGHBOURHOOD throughout the city.

Every European square is unique. The greatest squares are fan-shaped in plan (Siena), trapezoidal (Venice's Piazza San Marco), rectangular (Ascoli Piceno), triangular (Tübingen's Markt), funnel-shaped (Telc), elliptical (Vigevano), oval (Verona), square (Salamanca's Plaza Mayor), square doughnut-shaped (Krakow's Rynek), or triangular doughnut-shaped (Olomouc), or they are broadened streets visually closed at each end (Landshut, Cirencester).

Regional differences in architectural forms and materials, predominant architectural styles, characteristics and placement of focal buildings are uniquely combined to imbue each square with its individual personality, reflecting the town's unique history, the character and values of its inhabitants. Annual FESTIVALS, such as Siena's Palio, reflecting the square's unique character, involve broad citizen participation and renew COMMUNITY pride.

If the city is the second most important invention of mankind (after language), as Lewis MUMFORD maintained, then the market square is the most important invention of European city-making. For over two thousand years, since the ancient Greeks first invented the agora (imitated by the Romans in the forum, and reinvented in the Middle Ages), the multifunctional market square stimulated development of participatory, representational self-government. By providing a setting that generated negotiation and dialogue, it became

clear that the well-being of all depended on their ability to work together for the common good.

On the market square the people debated and assembled to demand rights of self-government; here they met to elect representatives, and to swear them into office.

During the eleventh, twelfth and thirteenth centuries, hundreds of market squares were created as the centres of new European cities, from Spain to Sweden and from Belgium to Hungary. This democratic movement was so strong that the most prominent location on the square was selected for the new symbol of self-government, city hall. Indeed, when the Polish cities of Tarnów and Poznan were founded in the thirteenth century, city hall was considered such a potent attraction for new settlers that it was placed in the centre of the square.

Over their long history, most squares were adapted to express the values of the time. Salamanca's Plaza Mayor was a modest, irregularly shaped marketplace before it was transformed in 1728 into the regal architectural setting we see today. By 1890 it had become a Victorian-style park with trees and a fountain. By 1954 the park had disappeared and the plaza was used as a car park, but by 1985 the plaza had resumed its original multifunctional social character as an open space for social life, festivity and celebration.

Some medieval squares, such as Place Plumereau in Tours, or the Markt in Tübingen, present themselves today virtually unchanged in size, scale and form.

During the Renaissance, many squares were enhanced with porticoes, and cities were further beautified with architecturally significant urban spaces and structures. Padova's Piazza delle Erbe, Verona's Piazza delle Erbe and Venice's Piazza San Marco have retained their architectural heritage almost intact since the Renaissance.

Most European squares, however, are surrounded by buildings from distinctly different periods, which still fit together in scale, height, building materials and range of colours (e.g. Stadtplatz in Steyr).

As the medieval system of self-government gave way to autocratic rule, single-function rather than multifunctional squares were created. Michelangelo's Campidoglio is an impressive forecourt for the civic buildings that enclose it on three sides.

The baroque period saw the creation of many architecturally magnificent squares designed to impress the population with the power of the ruling prince or archbishop.

In Place des Vosges (1605), originally called Place Royale, and in the numerous London residential squares such as Grosvenor Square (1695) and Bedford Square (1775), the OPEN SPACE was intended for carriages, and later fenced in as a garden for residents' exclusive use.

In the twentieth century, most European squares were transformed into car parks. It was not until the 1970s that they were reclaimed for the pedestrian, their architectural beauty was restored, seating and outdoor cafes were encouraged, and farmers' markets and community festivals were revived to generate social life and foster community.

In hundreds of cities throughout Italy, France, Spain, Belgium, the Netherlands, Germany, Austria, the Czech Republic and Poland, the main square answers the yearning for the vitality of urban life, for social encounters in a beautiful setting. These social experiences, repeated over and over again, with small variations, contribute to a sense of belonging, and create community. Civic dialogue on the square provides an unparalleled school for social learning, exercise of responsibility and development of democratic decision-making.

In the best squares, inhabitants and visitors of all ages and diverse backgrounds feel at home. The genius of the European square is its multifunctionality and adaptability: varied shops, cafes, restaurants, businesses and private dwellings, civic, religious and institutional buildings clustered around the square provide multiple reasons for people to gather on a regular basis.

For over two thousand years the multifunctional square has been the essence of the European city, epitomizing the community's heritage and symbolizing its identity. The question for the twenty-first century will be whether this unique heritage of urban places can continue to generate the cooperative, entrepreneurial, cultural and democratically organized urban civilization that was a uniquely European invention.

Further reading

Suzanne H. Crowhurst Lennard (1911–36) *Cambridge Medieval History*, vols V, VI and VIII, Cambridge: Cambridge University Press.

Durant, W. (1960) *The Age of Faith*, New York: Simon & Schuster.

Gutkind, E.A. (1964) *International History of City Development*, New York: Free Press.

Mumford, L. (1961) *The City in History*, New York: Harcourt, Brace & World.

SUZANNE H. CROWHURST LENNARD

EXACTIONS

An exaction is a requirement placed upon a real estate developer to make a dedication or payment to the local government body as a condition of development approval. The rationale for imposing the condition is to offset the costs, defined broadly in economic terms, of the development to the municipality. Exactions may include dedications of land, impact fees, construction of public facilities or actions to mitigate environmental damage.

Historically, in the United States, as new SUB-DIVISIONS were developed, municipalities paid for the paving of roads and installation of sewer and water lines. The developer would simply show the streets and easements on the plan. As development progressed and streets or other INFRASTRUCTURE were needed, the municipality installed them at its expense. Subsequently, municipalities imposed benefit or betterment assessments on the property owners who directly benefited from the streets or infrastructure, thus shifting the cost from general municipal revenues.

By the 1970s and 1980s, communities in growing areas faced financial difficulties as population increase and rising personal incomes put burdens on existing infrastructure. At the same time, taxpayers resisted higher levies and even sought to limit or reduce existing taxes. Federal support to communities also declined. This situation of increasing demands yet steady or declining revenues forced local officials to identify new sources of funds.

Simultaneously, communities became concerned with the effects of growth. CONGESTION, SPRAWL and increasing demands on infrastructure of all types came to be viewed as 'EXTERNALITIES'. Municipal officials in many cases saw developers as intentionally or otherwise shifting a significant part of the burden of their developments to the public at large. Local governments sought ways to force developers to bear these costs. In economic terms, municipalities wanted developers to 'internalize' these externalities.

This combination of factors led to new thinking about both development and the financing of development in growing areas. Roadway widening, water and sewer lines, sewage treatment plants, schools, and in some cases police stations and libraries came to be viewed as costs of development which developers should internalize. Municipalities began to require developers to pay these costs through exactions, impact fees and LINKAGE FEES. 'Exactions' as a generic term includes impact and linkage fees, but more specifically means conveyance of fee title or easements to the community for park, recreation, flood plain, utility or other purposes, or for the construction of improvements necessitated by the development. Exactions may also include payments in lieu of those dedications or constructions. Impact fees include any payment to finance improvements required as a condition of approval of the development. Linkage fees are payments required in order to provide low-income housing or other social benefits to those who may be negatively affected by the development. While the developer is legally obligated to make these payments or other contributions, the cost thereof is generally passed to the ultimate purchasers or tenants of the development. In a more difficult market, developers may seek to recoup these costs by lowering the price they pay for the land they develop.

England, which is much more densely populated with less open space for development, has also developed mechanisms by which developers internalize costs their projects impose on the population at large. 'Planning loss' is the term used in England to describe the impact of development on the community. 'Planning gain' is the term used to describe the measures taken by developers, either voluntarily or by government requirement, to mitigate planning loss. While the terminology is different, the essential concepts are the same.

Municipalities are not free to impose exactions on developers at will. In the United States, the Constitution and constitutional case law provide the framework within which exactions may be imposed. Recent Supreme Court cases have enunciated the tests that municipalities must follow in designing exactions. The tests may be simply stated as follows: (1) the condition for which exactions are sought must be the result of the development in question; (2) the solution imposed

on the developer must be roughly proportional to the condition caused by the development.

A further limitation in the United States is that exactions must only be used to offset development externalities and cannot be used as general revenues. Such use will cause them to be characterized as taxes and stricken as void. In the United Kingdom and other European countries, developers sometimes forego legal challenges to excessive exactions, rendering the exactions more like taxes than fees.

It is possible for municipalities and developers to agree on the conditions for plan approval, including exactions. Developers and communities may negotiate the details of the development, the nature of the approval and the specifics of any impact fees or other exactions to be imposed on the developer. In order to overcome possible legal challenges to these agreements, statutes have been passed in a number of US states authorizing the use of such DEVELOPMENT AGREEMENTS and specifying requirements, such as public hearings, for their adoption. Under the Development Charges Act, municipalities in Ontario, Canada, may also enter into development agreements with developers. Municipalities are required to justify the exactions they impose and developers are given the right to a judicial determination if they disagree with the municipalities' calculations. In England, developers' agreements are governed by Section 106 of the Town and Country Planning Act 1990. Agreements entered into pursuant thereto are generally upheld if the requirements imposed are reasonable and material to the development in question.

Further reading

Altshuler, A.A. and Gómez-Ibáñez, J.A. (1993) *Regulation for Revenue: The Political Economy of Land Use Exactions*, Washington, DC: The Brookings Institution Press; Cambridge, MA: Lincoln Institute of Land Policy.

Faus, R.D. (2000) 'Practice commentary: exactions, impact fees and dedications – local government responses to Nollan/Dolan takings law issues', *Stetson Law Review* 29: 675–708.

Saxer, S. (1990) 'Planning gain, exactions and impact fees: a comparative study of planning law in England, Wales and the United States', *The Urban Lawyer* 32(1): 21–71.

SEE ALSO: development agreements; police power

R. JEROME ANDERSON

EXCLUSIONARY ZONING

Because ZONING is, by definition, the allocation of certain activities and building types to particular areas of the city, it can be said to be inherently exclusionary. The phrase 'exclusionary zoning', however, refers to the particular practice of so regulating the use of land as to exclude particular groups of people from a given territory, generally a residential area. In the United States in particular, the emergence of zoning at the turn of the twentieth century, its diffusion over the following decades and even its current application show the importance of exclusionary motivations in municipal regulation.

Exclusive residential areas are exactly that: they are places where common people are not welcome. To enforce this SEGREGATION, homeowners at first relied on restrictive covenants that prevented the sale of property to members of minority groups (Jews and blacks in particular) and required the erection of houses of a certain cost and quality. As private restrictions made way for public ones and as the latter were deemed unconstitutional if they segregated by race, ethnicity or religion, regulating built form became the means to keep out undesirables, with home value as a proxy for social value.

The most obvious form of exclusionary zoning is the prohibition against the erection of apartment buildings in single-family residential areas. Further regulations within areas of single-family homes pertain to development density (requiring large lots and large setbacks), unit size (setting minimum standards of floor area), building design (demanding expensive amenities and excluding manufactured homes) and the permitting process (imposing high fees and a slow review). SUBDIVISION regulations and construction codes may impose additional constraints (by imposing high standards for roads, materials, etc.). All these measures have exclusionary consequences, even if they are not adopted explicitly for the purpose of exclusion and are allegedly meant to protect the environment or to limit the burden on public infrastructure and services. They raise the cost of housing, thereby preventing poorer households from settling in the area; given the correlation between income and race, their effects are strongest on the members of minority groups.

Exclusionary zoning has been challenged in court, often successfully. In the Mount Laurel

cases, the New Jersey Supreme Court identified a pattern of exclusionary zoning in suburban municipalities and mandated a fairer distribution of affordable housing. But exclusionary zoning remains a reality, albeit in more discreet forms. Growth management, i.e. control over the volume and pace of development, has also enabled homeowners and their elected representatives to hide exclusionary motivations behind environmental and infrastructural concerns.

Further reading

Babcock, R.F. and Bosselman, F.P. (1973) *Exclusionary Zoning: Land Use Regulation and Housing in the 1970s*, New York: Praeger (Part I).

Danielson, M.N. (1976) *The Politics of Exclusion*, New York: Columbia University Press.

RAPHAËL FISCHLER

EXOPOLIS

One of several neologisms seeking to characterize the form (or formlessness) of URBAN areas that developed in the late twentieth century (others include 'technoburb', 'EXURBIA' and 'edge city'). Each of these terms emphasizes different aspects of postmodern metropolitan development; all convey the sense that many traditional distinctions between CITY, SUBURB and countryside have been eradicated.

California geographer Edward Soja coined 'exopolis' to describe Orange County, south of Los Angeles, which for Soja represented a paradigm of postmodern urban SPRAWL. Though it had over 2 million inhabitants in 1990, Orange County lacked the bustling DOWNTOWNS, impressive skylines or central places that have been the hallmarks of modern cities. Instead, it featured many diffuse sub-centres, loosely organized around SHOPPING MALLS and business parks.

These sub-centres cannot be called peripheral, since they are the highest order places in the new urban hierarchy. The foremost sub-centres possess what Soja terms 'nowness', a form of attractiveness that substitutes for centrality and supplants the old dichotomy between centre and periphery.

Exopolis is thus the city turned inside out: the urban periphery has many (though by no means all) of the characteristics that historically were the hallmarks of CENTRAL CITIES (e.g. white-collar jobs, diverse retail, and cultural displays).

Meanwhile, many central cities have lost these once-defining features, and are viewed by many as disadvantaged places lacking social and economic status.

Further reading

Soja, E.S. (1992) 'Inside exopolis: scenes from Orange County', in M. Sorkin (ed.) *Variations on a Theme Park*, New York: Noonday Press, pp. 94–122.

RICHARD W. LEE

EXPORT BASE THEORY

Export (or economic) base theory presents a simple and intuitive account of URBAN or regional growth. Central to the theory is the assumption that all economic activity in the urban area can be divided into a *basic* (i.e. primary or export) sector and a *non-basic* (i.e. secondary or service) sector. The relationship between the two sectors remains constant over time and the two sectors are intricately linked. The theory posits that external (or export) demand is the driving force behind urban growth. Income generated from exports by the basic sector is used to purchase goods and services from the non-basic sector. Firms and workers in the non-basic sector then use this income for further rounds of purchases and expenditures. This circular flow process sets a *multiplier effect* in motion: stimulating the basic sector triggers off an economy-wide effect much greater than the initial increase. The multiplier measures the magnitude of this ripple-through process and is contingent on the propensity to consume locally. Export base theory assumes this propensity to be constant across all rounds of spending and re-spending. Furthermore, it implies an absence of supply-side constraints with local capital and labour responding immediately to changes in demand. Finally, the theory posits the existence of a homogeneous export sector where earnings across all activities are roughly equivalent.

The roots of export base theory lie in the staple theory of economic development which promoted the idea of economic growth through the export of staple products to metropolitan markets. Export base theory was initially utilized by geographical and planning analysis in the 1930s. It was developed as a practical quantitative technique for identifying urban and regional self-sufficiency and

for population and employment projection. It was adopted in the 1950s by economists applying macroeconomic growth theory to the study of urban areas via the Keynesian regional income model. The links between these various sources were recognized by Charles TIEBOUT in a classic treatise. In recent years, the theory has re-emerged as both an economic rationale and a practical technique for urban economic development strategies such as industrial recruitment, targeting of 'critical' industries and identifying competitive advantage.

For all its intuitive appeal, export base theory has some serious conceptual and practical limitations. First, the bifurcation of the urban economy into basic and non-basic sectors is plagued with theoretical and empirical difficulties. Secondly, the emphasis on the driving role of exports in stimulating the urban economy overlooks the important role of endogenous growth in urban development and the increasing importance of non-basic sectors (services and amenities) in the modern city. Finally, by focusing solely on aggregate growth (in terms of income and employment change), the model ignores welfare and distributional issues. Thus, while not a general theory of urban growth, the export base model provides some partial and short-run insights for urban economic development applicable to small areas with relatively narrow economies.

Further reading

Tiebout, C.M. (1962) *The Community Economic Base Study*, Committee for Economic Development, Supplementary Paper No. 16, New York: Committee for Economic Development.

DANIEL FELSENSTEIN

EXPROPRIATION

'Expropriation' has been defined as a legally sanctioned acquisition of one's property by another, without the owner's consent. In effect, it is a compulsory acquisition of property rights. The basis of this power is typically a statutory grant of authority. The grant is normally given by provincial (state) or national government legislation. Expropriation is generally not exercised by prerogative.

Expropriation may be used in a variety of contexts. It has been used by governments, typically in the developing world, to acquire property owned by foreign citizens. Expropriation may also be used as a penalty in criminal proceedings. However, the most common use of expropriation, and the one of concern here, is the acquisition of interests in real estate for public purposes, normally in conjunction with urban development.

Expropriation statutes generally require some degree of connection with the public good in order to permit the use of the expropriation power. The test varies, depending on the wording of the statute. In some instances, there may be no expropriation unless there is a 'public necessity'. Other statutes require a 'public purpose' or perhaps only a 'public interest'.

Various uses of the expropriation power demonstrate these nuances of meaning. A basic use of expropriation is for the creation or widening of streets and highways. Assuming the proper grant of statutory authority, and assuming that the landowner and governing body cannot agree on a price for the property to be acquired, the governing body may exercise its powers of expropriation to compel the transfer of the needed real estate. The connection with the public good is apparent in the usual street or highway case.

Other uses of expropriation may be further removed from a 'public necessity'. Acquisition of property for SLUM clearance has generated controversy, especially when the properties acquired are later sold to private developers. Likewise, expropriations of private property for sports arenas have also been attacked. Many such expropriations, however, survive legal challenge on the grounds that there is sufficient public 'purpose' or 'interest' to permit the taking.

Expropriation may be effected in an indirect manner. Noise generated from a municipal airport may so reduce property values in the flight paths that owners may claim compensation on the grounds that their properties have been 'taken'. In the United States, regulations of land use, such as a prohibition on construction in a flood plain, have been interpreted as 'regulatory takings'.

Other countries, notably the United Kingdom, do not recognize regulatory expropriation. However, in the United Kingdom, owners whose properties lose value due to inclusion in a proposed development scheme may require the government

body to purchase their interests due to this 'planning blight'.

Typically, the compensation to be paid to the landowner whose property interest is expropriated is the market value of that interest. The goal is to put the landowner in the same position after the expropriation as before the expropriation.

Further reading

Erasmus, G.M. (ed.) (1990) *Compensation for Expropriation: A Comparative Study*, Oxford: Oxford University Press.

SEE ALSO: eminent domain

R. JEROME ANDERSON

EXTERNALITIES

The concept of externalities has its basis in economic theory. Consequently, anyone wishing to fully understand the technical aspects of externalities should refer to an economic textbook such as the public finance textbook *Economics of the Public Sector*, by Joseph Stiglitz.

Externalities represent a situation where market failure occurs. Consequently, the market failure, if left uncorrected, leads to an inefficient allocation of resources. In the most basic terms, externalities are unintended impacts on third parties during either the production or the consumption process. They occur either when producers produce a good or service, or when consumers consume a good or service. Externalities result in generating either 'spillover' benefits or spillover costs. The spillover results from a divergence between either private and social costs or private and social benefits. The result of either type of externality is economic inefficiency.

Perhaps the best-known example is the case of external costs generated by a firm that produces air or water pollution as part of its production process. If it emits either pollution into the air or effluents into a watercourse, it may be imposing social costs via the creation of an unhealthy environment resulting in medical problems. In the case of water pollution, downstream costs will be borne by water users in the form of either reduced water quality, with health impacts, or the expense of removing the pollutants from the water prior to use. In order to take corrective action, a policy action is required to correct the problem. The three traditional policy instruments used to address externalities are taxes, subsidies or regulation.

In URBAN settings, externalities are often generated across property lines or political jurisdictions. In terms of individual properties, land use and type and height of building can have an external effect on adjacent properties. Consequently, municipalities use the regulatory tool known as ZONING in an attempt to mitigate externalities in terms of land use. When externalities are generated across political boundaries, they are generally referred to as 'interjurisdictional spillovers'. This occurs when the decision of one political jurisdiction has spillover costs or benefits for the residents of another political jurisdiction. These spillovers may arise due to either land use decisions or service provision decisions. In the first case, a municipality may locate industrial properties, high traffic-generating land uses or a sanitary landfill at its political boundary, thereby having negative impacts upon properties in the adjacent jurisdiction. With regard to service provision, there are circumstances where services provided by one jurisdiction either provide or have the potential to provide spillover benefits to citizens of adjacent municipalities. These could result from the provision of parks, recreational facilities or a library. If residents of adjacent municipalities use these facilities, they will reap spillover benefits. In cases where a higher level of government deems it efficient to provide these services to residents of adjacent municipalities, it could provide a subsidy to ensure that adequate or optimal amounts of these services are provided. An alternative solution is that a payment could be negotiated between the service-providing and the service-receiving municipality. In the case of reorganization, the new government unit must be large enough to encompass the previous spillover benefits. This will ensure that the decision-makers will select the service level to meet the needs of all residents in the new larger jurisdiction. The negotiated solution could be achieved either through appropriate user charges for non-residents or through a system where the government unit representing the residents who receive spillover benefits agrees to some form of payment. Externalities and spillovers abound in an urban setting and need to be addressed in a variety of ways, i.e. with policy instruments to attempt to minimize the inefficiency generated by externalities.

Further reading

Coase, R. (1960) 'The problem of social cost', *Journal of Law and Economics*, October 1960.
Stiglitz, J.E. (2000) *Economics of the Public Sector*, 3rd edn, New York: W.W. Norton & Co.

DAVID AMBORSKI

EXURBIA

The loose zone surrounding a metropolitan area beyond the SUBURBS and within its commuting fringe, shaping an interface between the URBAN and rural landscapes. It holds an urban nature for its functional, economic and social interaction with the urban centre, due to its dominant residential character.

This least densely populated urban area adopts a dispersed settlement pattern, comprising low-density residential SUBDIVISIONS, exclusive estates, mobile homes, hobby farms, cottages and small businesses, inserted in farmland with associated services and small urban centres. This diverse mixture of land uses occurs as a result of the integration of heterogeneous spatial and economic components, following urban expansion. Urban growth has shown a process of extensification by SUBURBANIZATION, which ultimately led to the development of exurbs, and a major demographic shift. Exurbs progressively turn into suburbs as the process intensifies.

Exurbia is strongly linked to the implementation of transportation systems, and the increasing number and extension of roadways allowed to expand the limits of urban areas. Accessibility has been a key factor in exurban growth, letting other factors intervene. The out-migration is a result of people's preference for a better quality of life, suburban lifestyles, rural residences, larger homes, open spaces, cheaper land and housing prices, lower tax rates, lower crime rates, and a sense of community, all driving the development of livable communities far away from the problems of CENTRAL CITIES.

Relocation of residences in the periphery is concurrent with the decentralization of economic activity, since new populations attract service activities and retail businesses, increasing suburban employment. However, the installation of industrial facilities comes into collision with the driving forces of exurb development. Parallel to this, information technologies are increasingly facilitating teleworking and online shopping, reducing dependence on urban areas and travelling requirements. Distance and commuting time can be applied as measurable criteria to define the general location of exurbia. Although highly dependent on urban size, exurbia could be found at a distance from 6 to 60 miles or at a commuting time of half an hour for different city sizes.

Although there is a debate on the costs and benefits of urban SPRAWL regarding land use and environmental implications, there is an agreement on its social consequences, namely social stratification. Longer distances imply more expensive travelling, more affordable for the wealthier, which, subsequently, induces impoverished and older people to remain in inner cities. The enlargement of the commuting zone produces a higher automobile dependence, rising accessibility times and a growing population of COMMUTERS, resulting in more traffic CONGESTION and higher air pollution levels. Challenges shift to local governments, who face policy changes in land use, transportation, housing and public services, which translate into higher taxes to be borne by local long-term residents, who benefit instead from increased land prices.

Further reading

Davis, J.S., Nelson, A.C. and Dueker, K.J. (1994) 'The new 'burbs: the exurbs and their implications for planning policy', *Journal of the American Planning Association* 60(1): 45–59.
Nelson, A.C. and Dueker, J. (1990) 'The ex-urbanisation of America and its planning policy implications', *Journal of Planning Education and Research* 9: 91–100.

URBANO FRA PALEO

F

FAÇADE PRESERVATION

Since the late 1970s, the heritage movement has gained momentum in reaction to the tabula rasa approach of the modern movement in architecture and URBAN planning. This period witnessed an increased interest in preservation and ADAPTIVE REUSE of historical buildings. Façade preservation is a contextual approach to URBAN DESIGN and redevelopment involving the complete removal of the interiors of historic buildings and the construction of new buildings behind retained façades to accommodate contemporary interiors. When an entire building cannot be salvaged and reused for economic or technical reasons, façade preservation is considered an acceptable compromise mainly because it maintains the historical continuity of the street and the cityscape, while granting a new and economically productive use to strategically located pieces of land and buildings. In addition to technical problems, façade preservation poses aesthetical and architectural challenges related to the integration of different scales, materials and functions. This approach, also called 'façadism' (a term coined at the beginning of the 1980s for a form of preservation focused on street façades), has been criticized by advocates of architectural and urban preservation. According to its opponents – those in favour of the preservation of the entire building – façade preservation undermines architectural authenticity by transforming the cityscape into a stage set. The discontinuity between the exterior and the interior represents a loss of heritage value. For those who support it, this practice is sometimes the only answer to the secure economic preservation of historic buildings and streetscapes.

Further reading

Richards, J. (1994) *Façadism*, London and New York: Routledge.

CLAIRE POITRAS

FAIR SHARE HOUSING ALLOCATION

A fair share housing allocation is a policy adopted to encourage an equitable distribution of affordable housing among communities within a region. Although there are a number of reasons to implement a fair share housing policy, including jobs–housing imbalances and concomitant traffic CONGESTION, many of these types of policies emerged to combat EXCLUSIONARY ZONING by local communities. Exclusionary zoning contributed to unequal distributions of income groups across a region. In other words, lower-income households were concentrated in some communities and virtually absent in others.

The most widely known case of exclusionary zoning and subsequent fair share policy comes from New Jersey. Lawsuits brought against the township of Mount Laurel in the 1970s and 1980s resulted in rulings by the State Supreme Court castigating Mount Laurel's land use policy and instituting a fair share housing allocation for local jurisdictions in New Jersey. Other states including California and Massachusetts adopted legislation to encourage fair share housing. In addition, there are a number of regions, the Minneapolis/St. Paul area for example, that encourage the development of fair share housing goals. However, the majority

of states and regions in the United States have not adopted these policies.

Local autonomy and self-interest most likely account for much of the resistance to fair share housing policies. Even if these obstacles can be overcome, the development of a fair share housing allocation policy is difficult for several reasons. First, fair share housing allocation is a form of REGIONAL PLANNING and, as such, requires the identification of regional boundaries – a potentially difficult task, conceptually and politically. Secondly, the housing covered by a fair share policy may include publicly subsidized units only, lower-income units only or housing for all income groups. Thirdly, the form of the policy, voluntary or required, must be determined.

Fourthly, the allocation formula may be based on current needs only or extended with projections for future needs. Fifthly, an oversight agency must be identified to monitor fair share progress and enforce penalties, if they exist, for non-compliance.

The difficulties associated with fair share housing policies suggest that states may be the most effective initiators of this type of policy. In California, for example, fair share policy became law and was tied to existing law mandating housing elements as part of the general plan. Each incorporated jurisdiction is required to plan for their regional fair share of all housing types as identified by their regional council of governments. The fair share numbers are determined for existing and future needs with a model that uses US Census data and other data collected from local jurisdictions. The housing elements of each local jurisdiction must be approved by the state, otherwise a jurisdiction runs the risk of challenges to any proposed development in the community.

Further reading

Connerly, C.E. and Smith, M.T. (1996) 'Developing a fair share housing policy for Florida', *Journal of Land Use and Environmental Law* 12(1): 64–109.

Listokin, D. (1976) *Fair Share Housing Allocation*, New Brunswick, NJ: Center for Urban Policy Research, Rutgers University.

VICTORIA BASOLO

FAVELA

Originating in Rio de Janeiro, 'favela' is the commonly used word for SQUATTER SETTLEMENT, GECEKONDU or shanty town in Brazil. It originally meant a group of poorly built shacks of mud, wood and recycled materials, settled on land with no legal title; generally, no public services – piped water, sewage systems or electricity – were available. Typically, the 'favelados' (favela dwellers) occupy areas in the city that are well located and may be easily occupied; for example, government-controlled land – railroad rights-of-ways, leftovers from highway projects, or preservation areas – or private property left unoccupied over legal disputes are likely target areas for favelas. The favela may result from an informal and timely squatting process, from collective social movements, or from overnight invasions planned and led by strongmen, political or religious leaders, in which case they are also called 'invasions' – particularly when in non-urban areas.

While the act of squatting must be done quickly in order to preserve locational rights, upgrading a house and the favela usually results from a lengthy self-help process that largely depends on the needs and possibilities of the favelado, and on the collective capabilities of the community. Because the construction industry in Brazil is largely dependent on unskilled labour, this working experience is common in favelas. Community-based initiatives and self-help solutions are capable of incredible achievements – such as collective water tanks and complete piped water and sewage systems. At times, favela communities provide support to local politicians or parties in exchange for their influence in getting public services and utilities installed.

As the earlier favelados tend to achieve a higher standard of housing due to a gradual upgrading and betterment process, and as the settlements mature over time and consolidate better living conditions, the internal structure of favelas tends to reproduce the land-market rationale and social stratification of the formal city. There are 'poorer' and 'richer' areas, differentiated land values, commercial 'strips', apartment units built for rent, home industries, specialty shops, etc. Land values in a favela also reflect its locational assets within the city.

A good example is Rocinha, one of the oldest and largest favelas in Rio, with a population of over 80,000. It is very well located next to expensive residential districts and the beach, and enjoys a high degree of URBANIZATION. Its strong residents' association has been pivotal in building and managing various facilities: schools, nurseries, a waste disposal system, theatre groups, local

newspapers, computer training centres and even a website – many of the community-generated projects are supported by local politicians and governmental and non-governmental organizations. These conditions made of Rocinha a dense, popular neighbourhood with apartment buildings up to six storeys in height, and with real estate prices high in the informal market.

The first favela was started in the mid-nineteenth century in Rio de Janeiro by former slaves drafted by the army to fight a rebellion in the northern state of Bahia. Upon their return, housing was unavailable and they were allowed to settle with their families on a hill next to the military garrison, nicknaming their settlement after 'fava' (a bean common in the savannahs of Bahia). This temporary solution became permanent, as new shacks were added by other former slaves arriving from rural areas, and by low-income families evicted from inner-city SLUMS by the 1875 Public Works and Sanitation Plan. In the first decades of the twentieth century, favelas began to appear in larger cities due to rural–urban migration, the birth of INDUSTRIALIZATION, and new restrictive city codes and sanitation and beautification projects. They reflected the expansion of the capitalist mode of production and the realization that urban land was a valuable commodity.

While Brazil's developmental efforts of the 1950s resulted in growing rates of industrialization and rapid urbanization, they inflated the accelerating demand for affordable housing. Until the early 1960s, this shortage was largely ignored, favelas were taken for granted as a temporary stage in the life of a migrant, and the favelado was perceived as a 'marginal' to society who, once with a steady job, would move into the formal market for better housing (see FILTERING PROCESS). By then, the perception of the favelas changed and they started to be considered a 'social disease' and eyesores in the formal city – especially when occupying prime sites. National and local policies turned to their eradication and the eviction of favelados to new government-subsidized housing.

From 1964 to the late 1970s, this view was imposed by the controlling military regime through a strongly centralized and technocratic view of the housing problem. The creation of the BNH – National Housing Bank – and its affiliated housing agencies pursued mass production of low-income housing as a mere quantitative and affordability equation. In 1965, 417,000 of the total 3,750,000 population of Rio de Janeiro lived in 211 favelas, and in that decade alone more than 100,000 favelados were evicted to new housing projects in the outskirts of town, a solution replicated in major cities until the late 1970s. The favelados were 'suburbanized' and thus further marginalized.

Unable to respond to the housing demand, these programmes also ignored the assets of a favela as the best possible response to structural conditions set by political and socio-economic circumstances. By squatting, the favelado is able to match housing expenditures to cash inflow, enjoying a supportive social network, good accessibility, and proximity to job opportunities. The housing unit is flexible enough to allow for expansions to accommodate family growth and alternative sources of income generation such as rooms for rent or sweatshops.

With Brazil's return to democracy in the mid-1980s, low-income housing policies became more realistic, smaller scale and participatory, and the favela was recognized as a legitimate solution to be respected. In addressing matters of urban policy and PROPERTY RIGHTS, the new Constitution of 1988 protected squatters by subordinating private property to the 'social function of property' and by acknowledging a squatter's right to the land after five years of unrefuted occupation.

At the start of the twenty-first century, Brazil faces an enormous growing demand for housing, the continuous growth of existing favelas, and new ones emerging in small towns and increasingly farther into the periphery of larger cities. Data from 2000, for example, indicate that Rio still holds the largest relative percentage of favelados, with more than 20 per cent of its 5.5 million inhabitants residing in over 600 favelas. Nevertheless, most major cities have readdressed the housing problem by moving their ordinances and urban policies towards alternative social and income-generation programmes, innovative design solutions, upgrading existing settlements by small-scale interventions, installing public services/facilities without disrupting communities, and respecting the lengthy housing investments of the favelado.

Further reading

Fernandes, E. (2000) 'The legalisation of favelas in Brazil: problems and prospects', *Third World Planning Review* 22(2): 167–87.

Neuwith, R. (2000) 'Letter from Brazil', *The Nation*, 10 July, pp. 29–31.

Perlman, J. (1976) *The Myth of Marginality: Urban Policy and Politics in Rio de Janeiro*, Berkeley, CA: University of California Press.

VICENTE DEL RIO

FEMINIST THEORY

Feminist theory is the body of explanations about sexual difference and gender inequality coming out of feminism or the women's movement. These theories exist at a number of levels of abstraction, from the quite grounded and action-oriented to the highly conceptual. Feminist theories of relevance to the city also come from a variety of fields, leading to a disciplinary diversity that parallels the diversity of work in the area of GENDERED SPACES. However, with its close connections to the feminist social movement, feminist theory has always had at least some relation to action, or at least the potential for transformation. For many, feminist theory is only useful as it can inform feminist activism.

Feminist theory has had diverse explanations of the reasons for women's inequality and accounts of the characteristics of women. Anglo-American feminism was originally a modernist project. In its liberal form, modernist feminism advocated equal opportunities for women to be fully functioning, rational participants in modern political, economic and social life. In the socialist or Marxian form of modernism, feminist theory was concerned with economic equality and, particularly in Marxism, with revaluing women's traditional household work as economically important in reproducing labour. In contrast, an alternative strand of feminism called radical or cultural feminism has stressed sexual difference as the primary area of oppression and simultaneously valorized women's unique history, experiences, sexuality, capacities and common traits. Some eco-feminists have drawn on this radical tradition, emphasizing women's innate closeness to nature. While analysing multiple dimensions of women's lives – for example class and race – in modernist and radical feminism, gender has often been seen as essentially similar for all women and even as the model for other oppressions such as racism. In contrast with these more universalizing theories, a number of groups have analysed the specific situation of particular groups of women or the general diversity of women's lives and identities. US black feminists (sometimes called womanists), Third World women, women of colour and a number of other groups have articulated distinctive positions in theoretical terms. In the late 1980s and 1990s, a number of broadly postmodern feminist theories dealt even more critically with the category of women, pointing to the vast diversity of women's experiences and identities. Feminist theories influenced by POSTMODERNISM still use the category of 'women', but in a much more limited way, articulating a cluster of interests shared among particular women and emphasizing differences among women as well as their common concerns. This difference debate was highly influential in the 1990s, meaning that by the end of the twentieth century, few feminist theorists were making grand claims about women and many in fact became theorists of diversity, difference or multiculturalism. At the same time, gender has been increasingly incorporated into theories used in general work in particular subject areas. Overall, this has led to theorizing about gender that may not be recognizable as uniquely feminist.

Feminist theory has been used to understand the city in URBAN SOCIOLOGY, geography, planning and architecture. While there have been somewhat separate trajectories in each field, each has drawn from theoretical work from a variety of sources including literary criticism, political theory, economics, cultural studies, psychology, sociology and more interdisciplinary feminist analyses. However, those using feminist theory to understand the city have had to deal with the issue of space, at least as a setting for social action if not in a more fundamental role, creating a unique contribution to feminist theorizing from these spatially oriented fields. This theoretical diversity is evident in debates in such journals as *Gender, Place and Culture*.

Feminist theories have influenced work in several key debates about society and space, both explaining phenomena and contributing a normative vision. At the smallest scale has been an interest in the meaning or performance of the self in space, including issues of sexuality, violence and personal identity. Work on the division between public and private has examined forms of democratic participation, the location of violence against women, gendered access to public spaces, the character of the family and issues of economic power. The spatial manifestations of the gendered

division of labour have been examined from the home to the factory and the multinational corporation, including the role of care in the economy and the workplace. The issue of community and its relation to autonomy and diversity has been another theme, including the implications of class, ethnic, racial, cultural and other divisions among women for the forging of feminist agendas and the practice of feminist activism. While women have been a focus of feminist theory, this has often included their relation to men and at times feminist analyses of masculinity. Finally, feminist theory has helped understand gender dimensions of globalization and international inequality.

Gender analyses and feminist theory obviously touch on such a wide range of issues that it is extremely interesting that such analyses are not more pervasive. Instead, even in the twenty-first century, the explanatory power of gender is too frequently ignored.

Further reading

Butler, J. (1990) *Gender Trouble: Feminism and the Subversion of Identity*, New York: Routledge.
Colomina, B. (ed.) (1992) *Sexuality and Space*, New York: Princeton Architectural Press.
Spain, D. (2002) 'What happened to gender relations on the way from Chicago to Los Angeles?', *City and Community* 1: 155–69.
Young, I. (1997) *Intersecting Voices: Dilemmas of Gender, Political Philosophy, and Policy*, Princeton, NJ: Princeton University Press.

ANN FORSYTH

FESTIVAL MARKETPLACES

Festival marketplaces are generally found on restored waterfronts, combining the idea of a historical urban community with that of a street market. They are found in many places in the United States, such as South Street Seaport in New York City and Pier 39 in San Francisco, as well as in Canada, Europe, Australia and Japan. The festival marketplace was pioneered at Faneuil Hall Marketplace in Boston in 1976 and the firm James Rouse Enterprise Development Company built many, including the Bayside Marketplace in Miami and the Riverwalk in New Orleans. They were very successful in the 1980s and seen as a keystone for URBAN REGENERATION. Festival marketplaces are developed by PUBLIC–PRIVATE PARTNERSHIPS as a means of encouraging URBAN REVITALIZATION and are mainly found in WATERFRONT DEVELOPMENTS.

Festival marketplaces are characterized by the lack of anchor stores and the reliance on representations of historic life in the architecture, and in exhibitions and festivals. The spectacles are carefully orchestrated using such features as bandstands, cafes, plazas and promenades. They differ from PEDESTRIAN MALLS in that they open onto streets or the waterfront. Their similarity to the malls is in their presentation of public space as 'symbolic public life'. The local labour of the working class is aestheticized in the material culture of the marketplace side by side with a presentation of the lives of the wealthy and the well travelled. Kiosks and pushcarts invite interaction with the consumers while retail tenants are selected for the liminality of their products.

Festival marketplaces connect modernist ideas of the mall with the postmodern need for spectacle. They satisfy the postmodern need for a sense of their own history, for authenticity and tradition. The marketplaces have been constructed both as metaphors for a renewed way of urban life and as an escape from the realities of poverty and homelessness.

Goss (1996) suggests that festival marketplaces are a response to 'agorafilia', the collective desire for authentic relations of consumption and the spontaneous sociality of the marketplace. Yet in these marketplaces there is a tension between the presentation of lifestyles and the strategies of control by which they are presented. Groups of undesirable people are excluded and there is a general tension between the private ownership with its concern with profit and the creation of a new form of public space. There is a contradiction in the two ways in which festival marketplaces can be presented. They can be presented as egalitarian settings in which people buy real and essential things or as stages for the display of consumer capital. Goss (1996) suggests that they are important to public politics as well as commerce.

Further reading

Goss, J. (1996) 'Disquiet on the waterfront: reflections on nostalgia and utopia in the urban archetypes of festival marketplaces', *Urban Geography* 17(3): 221–47.

Goss, J. (1999) 'Modernity and postmodernity in the retail landscape', in K. Anderson and F. Gale (eds) *Cultural Geographies*, 2nd edn, Australia: Longman.

CATHERINE WHITE

FESTIVALS

Festivals are special and planned events that focus on a theme (see THEME PLACES), and they constitute one of the most common public forms of cultural celebration. They constitute typical cases of GLOCALIZATION, as well as the high CULTURE–low culture interrelationship. Festivals can contribute to the construction of identity as a reaction to the constraints and standardization of URBAN life (e.g. Daimonji in Kyoto). The temporal dimension of festivals is very important in the sense that they usually occur on an annual basis during the conventional May–August tourist period, and they consist of a combination of the past (folk) and the present (garden) as well as of temporal (music) or spatio-temporal arts (film).

An area that is especially rich in festivals is Mediterranean Europe, with Spain being the most typical example (Mystery of Elche, San Fermin in Pamplona). Festivals can act as a boost to the development of cities, especially those that host a variety of festivals, for example Chicago (Jazz, Blues, Gospel, Latin Music, Country Music) and Edinburgh (Hogmanay, Book, International, Film, Fringe, Jazz, and Military Tattoo). Some local authorities have developed an event-focused arts strategy (e.g. Bath and North-East Somerset Council), using events to deliver the cultural strategy (e.g. Brighton and Hove Council) or developing a specific event and festival strategy (e.g. Nova Scotia Department of Tourism and Culture). In recent years, something that has been recognized is the importance of TOURISM, as well as the necessity of planning and management. In relation to tourism, there are two main types of cities: established tourist resorts (Cannes), or cities that try to create a tourist industry through festivals (Bayreuth).

Festivals can be categorized according to various factors: size (Notting Hill Carnival), purpose, fame (Rio de Janeiro Carnival, rodeos) and type of activity. According to the last categorization, two of the most common festivals are garden (e.g. Glasgow) and arts. Arts festivals share a number of characteristics including intense artistic output and a clear time-specific programme delivered with a clear purpose and direction. Seven categories can be used for the classification of arts festivals: high-profile general celebrations of the arts, celebrations of a particular location, specific art form festivals, celebrations of work by a community of interest, calendar (cultural or religious), amateur, and commercial music. Examples of arts festivals are Adelaide and Charleston (Spoleto Festival of Performing Arts). The most popular arts festivals are music festivals, for example Glastonbury (rock) and Montreux (jazz). Classical music – and especially opera – festivals usually attract a particular type of audience, for example Salzburg and Glyndebourne. Another type of arts festival is the film festival, for example Berlin and Venice. Fast-growing types of festivals are those related to food and drink, for example Taste of Chicago and Oktoberfest in Munich.

Further reading

Falassi, A. (ed.) (1987) *Time Out of Time: Essays on the Festival*, Albuquerque: University of New Mexico Press.

Getz, D. (1997) *Event Management and Event Tourism*, Elmsford, NY: Cognizant Communication Corporation.

SEE ALSO: festival marketplaces

ALEX DEFFNER

FILTERING PROCESS

Housing analysts have identified filtering as one of the basic characteristics of local housing markets. During the relatively long life of a housing unit, a more or less gradual decline in quality and value will generally occur. The factors contributing to this decline may include deterioration of the physical structure, style or functional obsolescence, or changes in locational preferences. As the value of the structure declines, it becomes accessible to households with progressively lower incomes. This downward filtering of the housing stock is the process by which those households unable to afford new housing units are able to obtain their housing.

This conceptualization of filtering as a market process is widely accepted but has not been well defined empirically. Nevertheless, US policy-makers

have adopted the filtering model as a keystone of their housing strategies. The widely recognized phenomenon of filtering can be employed to increase housing opportunities for households at the lower end of the income distribution. If the process of value decline can be accelerated through the creation of excess supply at the higher cost levels, the price declines among the existing housing stock will occur more rapidly. By encouraging prices to decline more rapidly than other attributes of quality, the result will be lower prices for better-quality housing.

Public policies support filtering strategies both indirectly and directly. Building and housing codes restrict housing construction to relatively costly new units, limiting increases in the housing supply to middle- and higher-income households. New housing that is affordable by low-income households can only be provided with substantial public subsidies. While the United States provides some direct subsidies for SOCIAL HOUSING, the majority of the public subsidies are in the form of tax subsidies for homeowners, as well as mortgage insurance. The result is the continued production of new single-family homes for owner occupancy.

Filtering strategies are attractive to policy-makers for a number of reasons. First, the marginal cost of encouraging an excess supply of middle-income housing is less than that of direct subsidies. Indirect strategies may be more politically popular as well.

There are a number of difficulties with this strategy, however. Empirical studies have found that filtering strategies are relatively inefficient means for realizing housing improvement for lower-income households. Moreover, the ultimate consequence of filtering, the abandonment and demolition of poor-quality housing, can be problematic. The continued production of new housing at the suburban fringe has contributed to the widespread abandonment of the housing stock in many US CENTRAL CITIES, such as Detroit, Philadelphia and St. Louis. Not only does this contribute to poor living environments for those who remain in these neighbourhoods, it also results in capital losses for the inner-city property owners.

Further reading

Emmi, P. (1995) *Opportunity and Mobility in Urban Housing Markets*, New York: Pergamon Press.

Galster, G. and Rothenberg, J. (1991) 'Filtering in urban housing: a graphical analysis of a quality-segmented market', *Journal of Planning Education and Research* 11: 37–50.

GARY SANDS

FISCAL IMPACT ANALYSIS

Fiscal impact analysis assessment of how a particular project or programme will affect the costs and revenues of a local government. Most local governments derive revenue from property taxes, sales taxes, hotel and motel taxes, and business and other licences. They also derive user fees from some types of municipal services, such as water and sewer services and garbage collection. Governments use these revenues to provide a variety of services including police and fire protection, water and sewer services, libraries, parks and recreation, general administration, and sometimes social services. In the United States, public schools are often administered by a separate government agency, a school district that is supported by a combination of local property taxes and state funding.

Development projects can affect the amount and mix of revenues generated and costs incurred by the local government. A new housing development will add to the property tax base; a new shopping centre may capture new sales and thereby add to sales tax revenue as well as adding to the property tax base. However, new developments also require additional services from the local government. A new housing SUBDIVISION will provide new citizens who will need water and sewers as well as libraries and police protection. It may also house new students who may require additional teachers and classrooms. A new shopping centre will require additional police and fire protection as well as water and sewer services. Fiscal impact analysis seeks to determine the revenues a project will generate and how much it will cost to provide the additional services required. Most fiscal analyses attempt to project the costs and revenues for at least ten years into the future.

Estimating the projected revenue from a project should be fairly straightforward if one has a detailed description of the proposed project. The tax rates and user fees are applied to the proposed development. However, particular care must be given to estimating sales tax receipts. Many cities

have found that a new development on the edge of town merely takes sales from the existing DOWNTOWN. From the municipal perspective, the focus must be on the total sales in the community. If the new development will take sales from existing retail establishments, it will not be adding net new sales tax revenue.

Costs are often more difficult to estimate. The preferred type of fiscal impact analysis is based on true marginal costs; however, this is the most difficult type of fiscal impact analysis to perform. As an example, if there is existing capacity in the sewer plant, the marginal cost of treating the additional sewage from 100 new houses is relatively small. However, if the plant is operating at its full capacity, the new development will require a significant expansion or even a new plant. In this case, the marginal cost would be quite high. Therefore, the cost of providing the service depends on the state of the service delivery system. Assessing the true marginal cost of servicing a new development requires a detailed understanding of each of the services provided, the amount of existing capacity in the service delivery system and the costs of likely expansion plans for the system.

Since true marginal costs are often difficult to determine, many fiscal analyses are based on average costs. In this type of analysis, the costs per person, per dwelling unit or per square foot (for commercial and industrial uses) are used to estimate the costs associated with a proposed development. These per capita factors can be calculated using the city or county budget and detailed land use and population data. While they are not as accurate as marginal cost methods, they are much easier to implement.

While it is prudent for a local government to be aware of the costs and revenues associated with its development decisions, it is important not to let fiscal concerns completely dominate land use decisions. There are other factors such as equity, economic development and quality of life that must be considered in creating a desirable community. Local governments should balance fiscal concerns against their other goals when evaluating a proposed development.

Further reading

Burchell, R.W., Listokin, D. and Dolphin, W.R. (1985) *The New Practitioner's Guide to Fiscal Impact Analysis*, New Brunswick, NJ: Center for Urban Policy Research.

Robinson, S.G. (ed.) (1990) *Financing Growth: Who Benefits?, Who Pays?, and How Much?*, Chicago: Government Finance Research Center of the Government Finance Officers Association.

STEVEN P. FRENCH

FISCAL ZONING

For most US communities, few issues are more controversial than local ZONING and SUBDIVISION land use regulations. Within a metropolitan setting, each suburban municipality will likely devise a land use regulatory policy plan that is designed to improve its overall relocation competitiveness. Charles M. TIEBOUT had asserted that individual households and firms were the consumers of space who will act rationally when demanding ideal locations for homes and business sites. A dynamic COMMUNITY will respond to these preference demands by devising an ideal municipal tax-package bundle while allocating land for ideal site opportunities. Failure to do so will lead to citizens' dissatisfaction, causing an individual household or firm to comparison-shop and more than likely relocate to another municipality that better fits its public goods preferences.

If Tiebout's assertion is valid, then the blueprint for suburbia's SPRAWL began with the fiscal zoning policies during the 1960s that were tied to the property tax. In theory, zoning and subdivision controls are designed to prevent irrational and inefficient land use patterns. In practice, both land use regulatory systems have been used to protect a community's tax base and to sort out the provision of public services. Generally, each suburban community will attempt to maximize its own tax revenue base by restricting land uses to those activities that will produce high-income property revenues and a low demand for public services. Known as fiscal zoning, a community will establish a land use scheme which allocates an oversupply of high-priced residential housing, supplemented by commercial and light industrial parcels at the expense of moderate and lower-priced housing.

Fiscal zoning deals with the fiscal implications of residential and non-residential ratables of the tax base. As jurisdictions felt the fiscal strains in the 1970s, suburban communities began to change their land use patterns to offset municipal service delivery costs. On the one hand, William Fischel and others have argued that local property taxation,

in conjunction with zoning, produces an efficient location and fiscal policy scheme for firms and households. On the other hand, George Zodrow contends that fiscal zoning is much like an excise tax, causing distortions in local and regional housing and business land use choices. Yet fiscal zoning is a complicated process, involving housing type, housing density, housing size and composition, and the associated costs for public service delivery. Contemporary research has indicated that it is exceedingly difficult to identify the true motivations of a particular community's zoning and subdivision regulatory codes. Better predictors for the fiscal implications of municipal service costs are the demographic character of population growth, the development standards for site improvements, and the spatial arrangement of existing capital facilities and infrastructure placement.

Further reading

Fischel, W.A. (2001) 'Homeowners, municipal corporate governance, and the benefit view of the property tax', *National Tax Journal* 54: 157–73.
Zadrow, G.R. (2001) 'The property tax as a capital tax', *National Tax Journal* 54: 139–56.

SEE ALSO: exclusionary zoning; Tiebout hypothesis

GARY A. MATTSON

FLOOR AREA RATIO

The floor area ratio (FAR) is the ratio of total floor area of a building to land, and is often used as one of the regulations in city planning together with the building-to-land ratio. Larger buildings can be developed in the districts where higher FAR is designated. Different from the height regulation, FAR does not limit the height of a building. Therefore, higher buildings with more floors can be built on smaller areas of built-up land, keeping broader open spaces free. Generally, higher FAR is applied in commercial or business land use, whereas lower FAR is used in low-rise residential districts. Since the total floor area of a building increases when FAR increases, its change is used as an incentive to promote development of the district. Increased FAR also increases demands on URBAN facilities such as roads, railways, water supply, sewerage systems and energy supply;

thus, it must be carefully considered whether the capacity of this INFRASTRUCTURE is sufficient for when the increased FAR eventually results in greater demand. Supplementary policies to collect fees out of the profits yielded by the increase in FAR have been introduced in various cities to expand and increase urban facilities. Although FAR is seen as a more rational regulation than the height regulation because it can control the density of activities if the urban activities are more or less proportionate to floor area, FAR may have drawbacks for the skyline if the buildings in a city are not aesthetically arranged.

TAKASHI ONISHI

FORD, GEORGE B.
b. 24 June 1879, Clinton, Massachusetts;
d. 13 August 1930, New York City

Key works

(1913) 'The city scientific', in *Proceedings of the Fifth National Conference on City Planning, Chicago, Illinois, May 5–7, 1913*, Boston, MA: n.p.
1925, founded *City Planning* magazine.
(1931) *Building Height, Bulk, and Form*, Cambridge, MA: Harvard University Press.

A leading figure of the US and international city planning movement in its early days, George B. Ford was first and foremost an advocate of the City Efficient movement. Though trained as an architect (Harvard, 1899; MIT, 1901; ÉCOLE DES BEAUX-ARTS, 1907), he aimed to lessen the dominance of CITY BEAUTIFUL concepts in URBAN PLANNING and to put the new profession on a scientific footing. His ideal planner was a sober analyst of the city's functional needs, an expert who could find the right solution to the city's problems on the basis of thorough empirical analysis, in fact an exponent of scientific management in the municipal realm. Ford wrote the technical chapter of Benjamin Marsh's *Introduction to City Planning* (1909), the first textbook devoted to the field in the United States, and published an early textbook in French, *L'Urbanisme en pratique* (1920).

Ford was particularly interested in the intersection of housing, architecture and city planning: from his thesis in Paris on *A Tenement in a Large*

City (1907) to his last work, *Building Height, Bulk, and Form* (1931, published posthumously), he stressed the concrete effects of planning on the physical city. This pragmatism, which spoke to the businessman's perspective on planning, made him a respected consultant. He collaborated on the 1916 comprehensive ZONING code of New York City and on the Regional Plan of New York and Its Environs. He headed the reconstruction efforts of the American Red Cross and advised the French government on the rebuilding of several war-torn cities at the end of the First World War, work which he documented in *Out of the Ruins* (1919). He also advised the government of the Philippines on the planning of Manila and its region. At the time of his death, he was president of the National Conference on City Planning and of the American Planning Institute.

Further reading

Scott, M. (1969) *American City Planning Since 1890*, Berkeley, CA: University of California Press.

RAPHAËL FISCHLER

FOREIGN INVESTMENT

Foreign investment is considered a very important factor in competitiveness (along with research and development, infrastructure, HUMAN CAPITAL, institutions and SOCIAL CAPITAL). It is also a significant mechanism in order to integrate international markets, together with trade. The economic development of a region depends on its capacity to attract foreign investment, involving an increase of the capital stock (see CAPITAL ACCUMULATION) and productive capacity. Increasing production capacity can generate employment directly, whereas the strengthening of competitiveness does it indirectly and in the long term. For this reason, regional development policies tend to attract these investments, reducing incertitude and improving working conditions of the local companies by encouraging the endowment of tangible and intangible factors (see UNEVEN ECONOMIC DEVELOPMENT).

While the pure neo-classical model excludes the state in resource allocation, the Keynes focus recommends its participation, direct or indirect, in the creation of conditions to attract new industries

to a region. The regions are open economies so it is not possible to use mechanisms of economic regulation such as monetarism or rates of exchange, and the use of taxation policies is rare. For this reason, Keynesian theorists will defend an expansion of the export sector (see EXPORT BASE THEORY) or a careful regionalization of expenditure policy, justifying either action with the hypothesis that regional problems are due to a lack of investment. That is why incentives and public investments are needed in order to attract foreign capital. Export-sector expansion is based on the capture of external wealth by export activities – called basic sectors – and it might be the fundamental motor of regional development (export-led growth).

Since the 1980s, post-Keynesian focuses have been developed, retrieving the concept of 'uncertainty' in a wider analysis and emphasizing the unpredictability of structural change processes (chaos theory). In this new context, it is very important that the geographic aspect of investment be the determinant element of economic growth. The concept of 'investment (or DISINVESTMENT) series' is introduced at the regional level. Each series shapes different 'strata'. The evolution of these strata ends with the configuration of each locality.

It is necessary to integrate the enterprise investor into the local economy in order to obtain the maximum benefits of foreign investment in local development, encouraging relations with native enterprises and the diffusion of new technologies in them. If the foreign enterprise is not integrated and, as an element of its own global strategy, decides to reduce production and employment in a locality, negative effects might appear. For this reason, the results of local initiatives are considered more permanent in time than the results of exterior ones. However, local enterprises (usually small and medium-sized enterprises) often cannot participate in the global economy without partnerships with large players (multinational enterprises), but in such relationships the division of power is asymmetric.

Further reading

UNCTAD (1999) *Foreign Direct Investment and Development*, UNCTAD Series on Issues in International Investment Agreements, Geneva: UNCTAD.

JOAQUÍN FARINÓS DASÍ

FOREIGN TRADE ZONES

Foreign trade zones are places located within the United States but deemed to be outside US territory and the jurisdiction of the US Customs Service for purposes of tariffs on goods from foreign countries or domestic products for export. The purpose of these zones is to help American businesses to be competitive in the global economy by reducing tariff burdens on the importation of foreign inputs and on exported finished products. The international equivalent to a foreign trade zone is a free trade zone. In a free trade zone, duty-free imports are only those that are inputs to goods or services for export. Free trade zones often provide a wide array of support services to businesses operating within them, including day-care centres, banks, employment centres and post offices, among others.

The US version of this development tool was created in 1934 by the Foreign Trade Zones Act, which established the Foreign Trade Zones Board to oversee the approval of new trade zones and the operations, including rate setting, of all zones. Approval must also be received from the US Customs Service before a trade zone may begin operations.

Initially, virtually all foreign trade zones were located in major PORT CITIES. Towards the latter part of the twentieth century, however, new trade zones were being established in smaller, land-locked communities seeking to take advantage of their economic development potential. Foreign trade zones are typically operated by public or non-profit organizations. They act much like public utilities, having a public purpose and published rates. They are bonded and provide US Customs Service security. Customs procedures are streamlined for the convenience of participating businesses.

The typical general-purpose foreign trade zone consists of an industrial park with warehouse space. Participating businesses lease space from the trade zone operator. These businesses may engage in storing, processing, fabricating, exhibiting, repacking, breaking down and reshipping abroad goods from foreign countries while in the trade zone without paying US Customs duties or excise taxes. Applicable duties and taxes are deferred until such time as the goods are transferred to US Customs territory for domestic consumption. Businesses that will ultimately seek to transfer their goods into US Customs territory for consumption may choose between paying duties at the rates applicable for materials coming into the trade zone

from a foreign country and paying them at the rates for items that are transferred out, whichever is lower. Goods from a zone may be exported overseas without payment of duties or taxes. There is no limit on how long goods may remain in a trade zone. Foreign merchandise may not be sold at retail in a foreign trade zone, however.

A foreign trade zone may have a sub-zone. The latter is located outside of the general-purpose foreign trade zone boundaries and consists of a separate manufacturing facility where components from the trade zone are assembled into a finished product. This arrangement can help a business delay the payment of customs duties until the finished product is shipped.

Further reading

Hanks, G.F. and Alst, L.V. (1999) 'Foreign trade zones', *Management Accounting* 80(7): 20–3.
Lyons, T.S. and Hamlin, R.E. (2001) *Creating an Economic Development Action Plan: A Guide for Development Professionals*, Westport, CT: Praeger, pp. 139–41.

THOMAS S. LYONS

FORESTER, JOHN

b. 23 July 1948, Washington, DC

Key works

(1968) *Planning in the Face of Power*, Berkeley, CA: University of California Press.
(1993) *The Argumentative Turn in Planning and Policy Analysis*, Durham, NC: Duke University Press (ed. with F. Fischler).
(1999) *The Deliberative Practitioner*, Cambridge, MA: MIT Press.

Forester's work constitutes a substantial contribution to planning theory in the United States. His scholarship draws upon moral philosophy, oral history and ethnographic social science, as well as planning and policy studies. He received his Ph.D. from the University of California, at Berkeley, in 1977.

His work has enriched North American planning theory by incorporating theoretical discussions drawn not only from John Dewey's pragmatism but from European philosophical thought, particularly the critical theory of the Frankfurt School and discourse theories of micro-politics. His work has received international acknowledgement. *Planning*

in the Face of Power has appeared in Chinese and Italian editions. Response to power in planning has been Forester's major theme, explored through applied questions of daily ethics and conflict resolution practice.

Since the late 1980s, he has produced as teaching and research materials six collections of practice stories: oral histories of planners reflecting upon their specific practice. Drawing upon them, his book *The Deliberative Practitioner* has received great attention.

Forester has a major interest in planning education; he teaches at Cornell University, and has visiting appointments in other major universities. He has been active in the Association of Collegiate Schools of Planning and serves on the editorial advisory boards of several academic journals.

EMÍLIO HADDAD

FORTIFIED CITIES

A fortified city is a city with a system of permanent defence against attacks. In some cases, cities take advantage of natural features, such as rivers or mountains, to protect themselves. In other cases, they built fortifications to provide protection either as walls or as forts or other types of structures placed to offer the best protection. From the beginnings of cities, defence has been a preoccupation for most URBAN centres and fortification has been a strong determinant of URBAN FORM even after the built systems were no longer needed for protection. In addition, systems of fortifications changed as war technology changed and this in turn required adjustments within the cities.

From very simple palisades to elaborate structures, there are many types of fortified cities. Some of those identified by E.J. Morris as the cities with the most outstanding defensive systems could be used to illustrate different types of fortified cities. Constantinople (Istanbul) best represents a city that used its outstanding natural setting, at the tip of a peninsula, for defence and completed its system by building three successive walls inland, from Roman times to AD 413. Naarden, Holland, is a small, strongly fortified city of the seventeenth century with three systems of fortification: six formidable bastions, protected gun structures and a moat. It was demolished in the nineteenth century. Vienna, Austria, was fortified by the Romans in the first century AD due to its strategic site. It

also served as an important military centre during the crusades. In the fifteenth and sixteenth centuries, it resisted the Turkish attacks because of its strong fortifications. SUBURBS developed outside the walls and were destroyed during the sieges. After 1583, the suburbs were protected by another fortification system. Eventually, the defensive system was demolished in 1858 and the area was occupied by the Ringstrasse, a ring road.

The most famous designer of fortifications was Sebastian Le Pestre de Vauban (1633–1707), who worked under Louis XIV in France. He designed thirty new fortified towns, and the fortifications of three hundred existing towns. One of his projects was the city of Neuf Brisach (designed in 1698, built in 1708). It was located on the west bank of the Rhine. Vauban selected this strategic site as it was a meeting point of important roads and out of reach from artillery. It was to serve as a fortified garrison for the deployment of forces and protect the river. In this new town, Vauban used the French bastide model (a medieval planned city) of a grid around a central square, La place d'Armes. A ring road was built around the walls. The formidable fortification included 66-foot ramparts, a ditch and tower bastions.

Since the twentieth century, new advances in warfare technology make the development of permanent fortifications ineffective. However, defence is still a major concern of the modern city.

Further reading

Morris, A.E.J. (1994) *History of Urban Form*, 3rd edn, New York: John Wiley & Sons (Constantinople, pp. 89–91; Naarden, p. 167; Vienna, pp. 227–9; Vouban, pp. 214–17).
Scully, V. (1991) *Architecture*, New York: St. Martin's Press, pp. 275–311.

ANA MARIA CABANILLAS WHITAKER

FOUCAULT, MICHEL

b. 15 October 1926, Poitiers, France;
d. 25 June 1984, Paris, France

Key works

(1980) 'Questions on geography', in *Power/Knowledge*, trans. C. Gordon, New York: Pantheon, pp. 63–77.
(1982) 'The subject and power', in H. Dreyfus and P. Rabinow (eds) *Michel Foucault: Beyond Structuralism and Hermeneutics*, Brighton: Harvester, pp. 208–26.

French philosopher and political activist, internationally known for work on the history of systems of thought, Foucault's purpose is partly captured by the term 'power/knowledge', but is also always informed by his understanding of geography and of space, something acknowledged by him in interviews and writings.

Foucault was intrigued by problems of government (the conduct of conduct), including the functions of citizens and the POLIS. He was concerned less with given periods in time and more with problems and effects of discourses, i.e. how civic life and government are constituted (see CIVIL SOCIETY). Occupied with analysing historical statements as he found them in the archive (his 'archaeology of knowledge'), he was committed to the analysis of power-histories of the present (his 'genealogy of knowledge') and of the self (his 'ethics'). This interest became the means to interrogate the problems and effects of different attributes of government, the ethical and physical 'comportment' of the ruling class and its citizens. Thus, Foucault's interest in civic life and government became a strategy to deliberate on power, citizenship and URBAN life in modern as well as in past times and spaces, and profoundly affected how others understand the city and citizens. His insights into SURVEILLANCE, the oppositions between public and private, or the production of displacement and difference have been critical to later analyses of space/power/knowledge in the city.

Further reading

Dean, M. (1999) *Governmentality: Power and Rule in Modern Society*, London: Sage.
Soja, E. (1996) *Thirdspace: Journeys to Los Angeles and Other Real-and-Imagined Places*, Malden, MA: Blackwell.

SEE ALSO: de Certeau; Lefebvre

ELAINE STRATFORD

FRIEDMANN, JOHN

b. 1926, Vienna, Austria

Key works

(1973) *Retracking America: A Theory of Transactive Planning*, Garden City, NY: Anchor Press.

(1986) 'The world city hypothesis', *Development and Change* 17: 69–83.
(1987) *Planning in the Public Domain: From Action to Knowledge*, Princeton, NJ: Princeton University Press.
(1992) *Empowerment: The Politics of Alternative Development*, Cambridge, MA: Blackwell.

John Friedmann numbers among significant influences on his work and life his father, Robert Friedmann, a historian and philosopher, Karl Mannheim, Hannah Arendt, Martin Buber, Lewis MUMFORD and Harvey S. Perloff. Perloff was Friedmann's doctoral supervisor at the University of Chicago from 1949 to 1955. Friedmann then worked in Brazil, South Korea, the United States, Venezuela and Chile, for academic, government and non-government agencies, between 1955 and 1969, forging his international reputation in regional development planning, URBANIZATION studies and policy, socio-economic development, and planning theory, in which he is a central figure. Perloff brought in Friedmann to head the Program for Urban Planning in the Graduate School of Architecture and Planning at UCLA. He was at UCLA from 1969 to 1996 and head for a total of 14 years, and is now emeritus professor there. He was an honorary professor at Royal Melbourne Institute of Technology from 1996 to 1998 and at the University of Melbourne during 2000 and 2001. He is now an honorary professor at the School of Community and Regional Planning at the University of British Columbia, married to Leonie SANDERCOCK, a professor in that school.

Friedmann's most recent publication is *The Prospect of Cities* (2002), in which he examines patterns of global urbanization and issues of POVERTY, violence, democracy, citizenship and transnational MIGRATION, as well as the importance of the experience of everyday life. These important themes build on a long-standing concern with matters of social justice, life space, regional development and the WORLD CITY hypothesis. Chapter 7 provides a reflection on his career, and the work as a whole is both captivating for its real interest and important for the insights that Friedmann draws from fifty years' experience in planning and related disciplines.

Friedmann suggests that there has been a thread running through his work since the 1970s, which can be traced in *Retracking America, The Good Society, Planning in the Public Domain, Empowerment: The Politics of an Alternative*

Development, Cities for Citizens (with Mike Douglass) and The Prospect of Cities. This thread is underscored by a concern with radical planning as the mediation of theory and practice in social transformation. It has been central to an abiding project in Friedmann's work, namely to protect and enlarge democracy and the spaces of democracy. And it has been informed by particular understandings of the good society; by critical realism; and by what he calls an epistemology of social learning, or praxis. The notion of insurgent non-violent citizenship at the point between local and global social spaces is central to this thread.

In 1988, Friedmann received the Association of Collegiate Schools of Planning Distinguished Planning Educator Award, and international credits for his achievements include honorary doctorates from the University of Dortmund and the Catholic University of Chile. He has written 14 books, co-edited 11 others, and is the author of 150 chapters, articles and reviews.

Further reading

Friedmann, J. (1998) 'Planning theory revisited', European Planning Studies 6(3): 245–53.

ELAINE STRATFORD

FROSTBELT

Although no consensus exists as to exactly which cities and states constitute the frostbelt, this term generally refers to the cities and states in the Northeastern and Midwestern sections of the United States. It became a commonly used term when scholars began to employ the phrase 'SUNBELT' in the 1960s to refer to the Southeastern and Southwestern regions of the country.

Both the population and private sector employment in most major frostbelt cities declined in the decades following the Second World War. These declines were primarily the result of two trends: the SUBURBANIZATION of people and employment, and the movement of population and jobs to the sunbelt. Due to these trends, many cities in the frostbelt suffered considerable social distress, including high levels of unemployment and POVERTY. They also suffered from fiscal stress, having a difficult time raising enough revenue to provide adequate services to their populations. Unlike many sunbelt cities, most cities in the frostbelt were unable to expand their boundaries by annexing surrounding suburban territory due to restrictive state legislation and to the fact that they were generally surrounded by incorporated SUBURBS that could not legally be annexed.

Not all frostbelt cities suffered to the same extent, and some that had been losing population and jobs managed to gain in the 1990s. For example, New York City benefited from the strong national economy and increased its number of jobs during the decade. It also grew in population, primarily due to the large number of immigrants who moved to the city. Many other frostbelt cities, however, such as Pittsburgh and St. Louis, continued to lose population and jobs through the end of the twentieth century. Even in some of these, however, signs of revitalization surfaced. For example, GENTRIFICATION of some neighbourhoods near their DOWNTOWNS brought a middle-class population to certain areas and renovation of the older housing. Also, the downtowns in many frostbelt cities benefited from the construction of office buildings, CONVENTION CENTRES and other facilities during the 1980s and 1990s.

Many cities in the frostbelt had been governed by political machines that remained in office by providing favours and services to citizens and businesses in exchange for electoral and financial support. Most of these machines had weakened or been completely replaced by other modes of governance by the 1940s. By the 1980s, many frostbelt cities elected African-Americans to mayoral and to other public offices who in some cases strove to increase the hiring of minorities by city government and to increase services to minority neighbourhoods. In the 1990s, several large frostbelt cities, including Philadelphia and Chicago, elected mayors who emphasized a 'pragmatic' style that incorporated efforts to strengthen the economy, decrease crime rates, and work with neighbourhood groups.

Further reading

Bernard, R.M. (ed.) (1990) Snowbelt Cities: Metropolitan Politics in the Northeast and Midwest since World War II, Bloomington, IN: Indiana University Press.
Mollenkopf, J.H. (1983) The Contested City, Princeton, NJ: Princeton University Press.

ROBERT KERSTEIN

G

GANGS

Gangs composed of young persons, as distinct from organized criminal syndicates, arose in America by the middle of the nineteenth century and were a concern for city leaders from the time they first appeared. Frederick Thrasher's book entitled *The Gang* was the first serious piece of research on the subject, published in 1927. Since then, research on the subject has become common.

Though not easily summarized, there appears to be consensus on two points. First, gangs are in disorganized COMMUNITIES. These places have no stable core of conventional institutions and groups to guide their residents, or people living there have unclear and questionable values. Secondly, the only way to control gangs is to cut off the supply of members or break up the groups. This is to be accomplished by employing a variety of carrots (i.e. programmes that attract youngsters to conventional groups and styles of behaviour) and sticks (i.e. police harassment and incarceration).

Although local people sometimes help implement plans to discourage gang activities, it is usually outside agencies and experts that assume responsibility for fashioning intervention strategies and carrying them out. Most of these programmes have not been effective. In some cases, the use of repressive tactics actually emboldens gang members or makes gangs more attractive to young persons. Further adding to a sense of urgency on the part of many concerned persons is the fact that the number of gangs increased in the last half of the twentieth century. New minority populations arrived in URBAN areas, and their children formed gangs that both mimicked and diverged from gangs formed by earlier ethnic groups.

Some recent gangs are well connected to adults in their neighbourhoods. They have a 'gang tradition' that has lasted for several generations and is tied to older members of their own families (e.g. Hispanic and some ethnic Chinese gangs). Others are more independent and not so tied up with the ongoing routines of their community or its conventional adult-run groups (e.g. gangs composed of African-American youth and more recent Asian immigrants). Some, like many white working-class gangs rooted in older ETHNIC ENCLAVES, have a tradition of defending their neighbourhood against 'outsiders'. Others, like the drug-dealing gangs affiliated with the Crips or Bloods, are viewed more as predators than defenders of their community.

Other changes in gangs occurred in the late twentieth century. Girls are now creating their own gangs instead of being the female auxiliary to boy gangs. Gangs appear in SUBURBS and even some small towns located some distance from any large city. Some gangs are more deeply involved in serious illegal activities, use deadly force to solidify their control over an area and are more mobile.

It would seem that either the 'social disorganization' hypothesis is wrong or all of America now suffers from the same crippling disorders that once affected only inner-city SLUMS. Alternatively, gangs may not be so alien a creation as we think. They have many features in common with conventional groups, meet similar needs for their members and do not discourage members from becoming conventional adults. That is why so many youngsters 'mature out' of gangs as they age.

Gang members, like teenagers generally, have presented themselves increasingly in adult-like ways, despite not being prepared to take on most

adult responsibilities. In their stylized dress, ritualistic declarations of brotherhood, indecipherable graffiti and crude capitalization of home-grown entrepreneurs, contemporary youth gangs are a cruel parody of eighteenth- and nineteenth-century male fraternities.

Further reading

Klein, K., Maxson, C. and Miller, J. (1995) *The Modern Gang Reader*, Los Angeles, CA: Roxbury Press.
Monti, D. (1994) *Wannabe: Gangs in Suburbs and Schools*, Oxford: Blackwell.

DANIEL J. MONTI JR

GANS, HERBERT J.

b. 7 May 1927, Cologne, Germany

Key works

(1962) *The Urban Villagers: Group and Class in the Life of Italian-Americans*, Glencoe: Free Press.
(1967) *The Levittowners: Ways of Life and Politics in a New Suburban Community*, New York: Pantheon.

Born in Germany, Herbert J. Gans moved to the United States in 1940 and became an American citizen in 1945. Trained as a sociologist, he obtained a Ph.D. in sociology and planning. Formerly at the University of Pennsylvania and MIT, Gans moved to Columbia University in 1971, serving as the Robert S. Lynd Professor of Sociology in 1985.

Gans's interest in linking theoretical research to social problems and public policies was clear in two classics of URBAN SOCIOLOGY. In *The Urban Villagers*, Gans investigated Boston's West End, an inner-city district mainly occupied by workers' families of Italian origin, designated for demolition. He discovered that, despite the modest appearance of the buildings, the area had a strong vitality and people lived there by choice.

To depict similar areas, Gans used the term 'URBAN VILLAGE'. He challenged the idea that central areas with small, old buildings were synonymous with SLUMS. Gans focused on the inequality of urban renewal and changed the public perception of the problem, making it clear that tearing down buildings does not improve the living conditions in neighbourhoods where the main issue is the residents' low income.

The purpose of the second survey was to study the emergence and growth of a new large SUBURB and its effects on the individuals and families who moved there. Working in Levittown, Gans developed an interest for the suburbs that led him to revise WIRTH's theory on URBANISM.

According to Gans, the mere fact of living in a suburb does not automatically lead the individual to adopt a specific behaviour. On the contrary, people's way of life is largely determined by economic and cultural features.

In both surveys, Gans used 'participant observation'. He lived in the area and became a member of the community while simultaneously acting as a scientific observer. Gans reached the conclusion that architectural projects and urban plans have little impact on either behavioural patterns or people's choice of value. The fact that the city is essentially a social entity had a huge influence on all his subsequent work as a teacher, researcher and consultant for several civil rights and POVERTY and planning agencies. According to him, planning should be user-oriented and any policy aimed at improving poor living conditions should address the causes and not the symptoms. Since poverty and SEGREGATION are the main urban issues, only a firm and coordinated attack on them can have some positive outcome.

Further reading

Feinstein, S. (1970) 'People and plans', *Journal of American Planners Association* 36(2): 208–9.
Kornblum, W. (1996) 'The war against the poor', *American Journal of Sociology* 102(2): 627–30.

PAOLA SOMMA

GARDEN CITY

A type of city invented by British reformer Ebenezer HOWARD (1850–1928) in his classic 1898 book *Garden Cities of Tomorrow* (originally titled *To-morrow: A Peaceful Path to Real Reform*). Letchworth and Welwyn Garden City were built near London according to Howard's concept during his lifetime, and many other garden cities inspired by (though often deviating from) his model have since been built all over the world.

Howard was reacting to the appalling physical and social conditions in nineteenth-century English INDUSTRIAL CITIES. The INDUSTRIAL REVOLUTION had brought unprecedented population growth

coupled with air and water pollution, crowding, disease, filth and social exploitation.

Ebenezer Howard, a modest British stenographer with no formal training in planning or architecture, produced a lucid little book that hit exactly the right note in a world searching for a serious, but not too revolutionary, physical and social alternative to nineteenth-century industrial cities. In place of the polluted, grim and sprawling industrial CONURBATIONS of late nineteenth-century England, Howard argued in favour of 'social cities' of about 32,000 people on 1,000 acres of land, surrounded by a 5,000-acre GREEN BELT.

Howard argued that garden cities should be built far enough from existing cities that land could be purchased at agricultural values. Citizens would own the land in perpetuity, eventually generating abundant public revenue.

A central illustration in *Garden Cities of Tomorrow* is Howard's brilliant polemical diagram of a garden city 'town-country magnet' pulling people with the force of bright homes and gardens, no smoke, no SLUMS, and ultimately freedom and cooperation away from 'town magnets' and 'country magnets', which simultaneously attracted and repelled people because of their respective mix of good and bad features.

Howard provided illustrative schematic diagrams of possible physical layouts of his garden city – with a crystal palace shopping centre in the middle, circular boulevards, a peripheral railroad and then the green belt.

British philanthropists and reformers formed a garden city society and purchased a site for a garden city at Letchworth, about 40 miles from London, in 1902. Raymond Unwin and Barry Parker created a brilliant design for Letchworth. To the astonishment of its critics, Letchworth was actually built along garden city principles. When visitors to Letchworth saw attractive Unwin/Parker buildings, the garden city movement gained great credibility. Hampstead (really a garden SUBURB), Welwyn Garden City and other English examples followed, as did garden cities in Germany, France and even Australia. In the United States, Lewis MUMFORD championed Howard's ideas. Many American garden cities have been proposed and a few built.

Further reading

Buder, S. (1990) *Visionaries and Planners: The Garden City Movement and the Modern Community*, Oxford: Oxford University Press.

Fishman, T. (1977) *Urban Utopias in the Twentieth Century*, New York: Basic Books.

Hall, P. and Ward, C. (1998) *Sociable Cities: The Legacy of Ebenezer Howard*, New York: John Wiley & Sons.

Howard, E. (1898) *To-morrow: A Peaceful Path to Real Reform*, London: Swan Sonnenschein (reprinted in R.T. LeGates and F. Stout (eds) (1998) *Early Urban Planning: 1870–1940*, vol. 2, London: Routledge/Thoemmes).

RICHARD LEGATES

GARNIER, TONY

b. 13 August 1869, Lyon, France;
d. 19 January 1948, Bedoule, France

Key works

(1917) *Une Cité Industrielle, étude pour la construction des villes*, Paris: C. Massin & Cie ((1990) *Une Cité Industrielle, étude pour la construction des villes*, ed. R. Mariani, New York: Rizzoli International Publications).

(1920) *Les Grands Travaux de la Ville de Lyon* (*Civil Engineering in Lyon*), Paris: C. Massin.

French architect, URBAN designer and creator of the project for an industrial city (La Cité Industrielle, 1901–04), Tony Garnier graduated in 1889 from the École des Beaux-Arts of Lyon. In 1899, he won the Prix de Rome. It was during his stay in Rome that he began working on the project of an industrial city that became his main contribution to town planning. Drafted between 1901 and 1904 and finally published in 1917, this imaginary city of 35,000 inhabitants located in the vicinity of Lyon introduced innovations such as land use planning and standardized housing. Radically opposed to the academic principles of town planning as taught by the ÉCOLE DES BEAUX-ARTS, Garnier's ideas made him the forerunner of MODERNISM that later influenced planners and architects such as LE CORBUSIER. His design favoured rational, functional and well-balanced arrangements that included small buildings and detached cubic housing with terrace roofs. His modernity was also expressed through the use of reinforced concrete. In 1905, Garnier became municipal architect of Lyon where he pursued a successful architectural practice. He designed the livestock market and slaughterhouse (1909–28), the Gerland Olympic Stadium (1914–26), the Grange-Blanche hospital (1910–33) and the Quartier des États-Unis

(1919–34). In 1912, Garnier was named professor of architecture at the École régionale d'architecture in Lyon.

Further reading

Pawlowski, K.K. and Vechambre, J.-M. (1993) *Tony Garnier: pionnier de l'urbanisme du XXe siècle*, Lyon: Les Créations du Pélican.
Wiebenson, D. (1969) *Tony Garnier: The Cité Industrielle*, New York: G. Braziller.

CLAIRE POITRAS

GARREAU, JOEL

b. 1948

Key works

(1981) *The Nine Nations of North America*, Boston, MA: Houghton Mifflin.
(1988) *Edge City: Life on the New Frontier*, New York: Doubleday.

Joel Garreau has been a senior writer for the *Washington Post* for many years. He became world famous through his book *Edge City: Life on the New Frontier*. In this work he presents a five-part definition of Edge City. Edge City is any place that:

1 has 5 million square feet or more of leasable office space – the workplace of the information age;
2 has 600,000 square feet or more of leasable retail space;
3 has more jobs than homes;
4 is perceived by the population as one place;
5 was nothing like a 'city' as recently as 30 years ago.

In 1981 he published *The Nine Nations of North America*, in which he identified the key ingredients for a high-rise city.

Garreau is a student of global culture, values and change, whose current interests range from human networks and the transmission of ideas to the hypothesis that the 1990s set the stage for a social revolution to come. He is editor in charge of cultural revolution reporting at the *Washington Post*, and a principal of The Edge City Group, which is dedicated to the creation of more livable and profitable URBAN areas worldwide. At the School of Public Policy at George Mason University he is a senior fellow, leading two groups, one studying the future of universities and the other examining which global gateway city regions will be the winners and losers in the year 2020.

HUGO PRIEMUS

GATED COMMUNITIES

A gated community is a COMMUNITY living behind a gate and a fence. No clear definition and differentiation of various types of gated communities has been agreed upon. A controversial discussion concerns the notion of 'community'. Only a few empirical studies have been conducted on the social life and motivations of the people who choose to live in these settlements, as well as on the relations between the inhabitants and the society 'outside the gate'.

The main purpose of a gate, on a low-crime property, is not to deter or prevent crime but to provide the perception of security and exclusivity. In affluent residential neighbourhoods, privacy means exclusivity and mostly increases property values. The impact may be that crime shifts from one area to another. Gates and fencing works best on a stable property with non-criminal, mature residents. When the property houses criminal types, the impact on security may even be negative.

In the 1980s and 1990s, the construction of gated communities became a mass trend in the United States when more than 8 million people were found living in such communities. In many parts of the country, especially in the metropolitan areas of the SUNBELT states, these communities have changed the American URBAN landscape, and suburban society and its lifestyle as well. In the United States, these settlements are predominantly privately built and maintained. Residents live behind guarded or remote-controlled gates, walls or fences and adopt a variety of other defensive measures, such as privately organized NEIGHBOURHOOD WATCH schemes or professional security personnel. Gated communities reflect the progressive trends in US cities towards privatization of urban services and an increasing polarization, fragmentation and diminished solidarity within urban society.

Gated communities have developed into a global phenomenon. They are not only popular in

the United States; gated housing estates are being established in some countries of South America, Asia, South Africa, Europe and the Middle East.

Gated settlements have historical predecessors, such as the traditional Medina and Mellah, the walled city in medieval Europe, and some colonial towns. The recent development is shaped by global socio-economic changes, marketing strategies of developers, and the spreading of architectural concepts and lifestyles by international migration.

Further reading

Blakely, E.J. and Snyder, M.G. (1997) *Fortress America: Gated Communities in the United States*, Washington, DC: The Brookings Institution Press; Cambridge, MA: Lincoln Institute of Land Policy.

Lang, R.E. and Danielsen, K.A. (1997) 'Gated communities in America: walking out the world?', *Housing Policy Debate* 8(4): 867–99.

HUGO PRIEMUS

GECEKONDU

'Gecekondu' literally means 'built overnight' in Turkish and refers to unauthorized informal housing settlements. For many low-income households who cannot afford to purchase or rent formal housing, gecekondu neighbourhoods offer an affordable alternative for shelter. Like other informal housing settlements in the developing world, gecekondu settlements have four broad features: (1) lack of land tenure security; (2) lack of basic infrastructure such as piped drinking water inside dwelling units, sewerage and electricity; (3) predominance of physically sub-standard dwellings; and (4) locations that are not in compliance with land use regulations and are often not suitable for development (e.g. hillsides, wetlands, flood plains). Other examples of informal housing settlements are the FAVELAS in Brazil, katchi abadis in Pakistan, kampungs in Indonesia and bidonvilles in former French colonies.

Gecekondu settlements first appeared in Turkish cities in the 1940s and 1950s when a push for industrial growth was not accompanied by corresponding efforts to meet the housing needs of the rural migrants drawn to the industrial jobs being created in major cities. When newly arrived migrants could not find adequate and affordable housing in cities, they began constructing single-storey gecekondu dwellings on vacant government or private land near their jobs, without the necessary building permits. Much of this development took place using self-help construction methods with the assistance of family members and neighbours.

To address the massive unplanned and unauthorized development of gecekondu settlements, the Turkish parliament passed its first Gecekondu Act in 1966. The Act established a framework for regularization programmes that would provide infrastructure and services in gecekondu neighbourhoods. Some gecekondu neighbourhoods were selected for upgrading while others were slated for demolition. New areas were zoned as gecekondu prevention areas. These efforts, however, did not slow down the growth of gecekondu neighbourhoods. Ultimately, policy-makers could no longer ignore the gecekondu population as an important political constituency, and a series of amnesty laws were passed that enabled gecekondu dwellers to start the process of obtaining legal title to their land.

By the late 1980s and 1990s, gecekondu settlements were undergoing fundamental changes. The era of traditional gecekondu construction – a single-family house quickly built with communal help on public land – was coming to an end. The construction of multi-storey buildings (apartmankondu) began characterizing housing construction activities in gecekondu settlements. The late 1990s saw the commercialization of gecekondu construction, the proliferation of rental housing markets within gecekondu settlements and the densification of gecekondu settlements. By the late 1990s, more than half of the population in three of Turkey's largest cities lived in gecekondu settlements: 70 per cent in Ankara; 55 per cent in Istanbul; and 50 per cent in Izmir.

Further reading

Pamuk, A. (1996) 'Convergence trends in formal and informal housing markets: the case of Turkey', *Journal of Planning Education and Research* 16: 103–13.

United Nations Centre for Human Settlements (Habitat) (2001) *Cities in a Globalizing World: Global Report on Human Settlements 2001*, London: Earthscan Publications Ltd.

AYSE PAMUK

GEDDES, PATRICK

b. 2 October 1854, Ballater, Scotland;
d. 17 April 1932, Montpellier, France

Key works

(1905) 'Civics: as applied sociology, Part I and Part II',
Sociological Papers, London: Macmillan.
(1915) *Cities in Evolution: An Introduction to the Town
Planning Movement and to the Study of Civics*,
London: Williams & Norgate.

Lewis MUMFORD said of Patrick Geddes that his
was one of the truly seminal minds to emerge from
the nineteenth century. His vision transcended
three worlds: a waxing Victorian era into which he
was born, a disintegrating pre-First World War
world that he confronted in his maturity, and a
new world (his 'neotechnics era') towards which all
his creative efforts were bent. He belonged to
a generation of writers and thinkers who were
concerned about the social consequences of the
INDUSTRIAL REVOLUTION. In his time, Geddes's
influence spanned several continents and countries
including Europe, Australia and India.

Geddes took an inductive, comprehensive
approach to any city problem he undertook,
relating physical planning to the natural environment, integrating it with sociology and economics,
and always taking into account the people and
their ways of life. Inherent in all Geddes's ideas
was the belief that social processes and spatial
form are intimately related. A critic of the popular
utopian planning of his day, English GARDEN
CITIES and Parisian BOULEVARDS were crushed by
what Patrick ABERCROMBIE called Geddes's
'nightmare of complexity'. Geddes's willingness to
embrace the complexity of social and spatial relations earned him a lauded place in the contemporary spatial turn in the social sciences (see
SOCIAL JUSTICE).

In his concern for promoting novel and new
ideas that arose from specific planning contexts,
Geddes is often remembered for his polemics
rather than the whole of his ideas. His major
written work, *Cities in Evolution* (1915), although
resting on a good deal of substance, elucidates only
part of his planning discourse. Geddes's first major
concept was the 'Valley Section', a heuristic device
developed in the 1890s as a graphic expansion of
Le Play's place/work/folk scheme of social analysis.
In Edinburgh, he advocated SLUM revitalization

rather than removal. He established the world's
first 'sociological laboratory' in Edinburgh, with a
camera obscura so that 'the manifestation of the
region could be intimately revealed'.

Geddes was fond of neologisms; it is to him that
we attribute the terms 'MEGALOPOLIS' and 'CON-
URBATION'. Later, he elaborated 'neotechnics'
as a way of remaking a world apart from over-
commercialization and money dominance. In his
time, Geddes was criticized as a polymath and an
idealist, but today this criticism falls short in praise
of ideas that, in his own account, reflected euto-
pianism (good places) rather than utopianism (no
places).

Further reading

Meller, H. (1990) *Patrick Geddes: Social Evolutionist and
City Planner*, London and New York: Routledge.
Stalley, M. (ed.) (1972) *Patrick Geddes, Spokesman for
Man and the Environment*, New Brunswick, NJ:
Rutgers University Press.

STUART C. AITKEN

GEMEINSCHAFT

Defined by Ferdinand Tönnies (1957 [1887]) as
deep, horizontal social relations of pre-modern
society wherein people remain united, in spite of all
separating factors. Controls over individual behaviour are exerted through the informal disciplining
of family and neighbours. Place is valued as primary, in part because limited social mobility fosters parochialism and intensive, face-to-face local
social relations. Within this circumscribed space,
people interact frequently in highly personalized
ways. *Gemeinschaft* provides the founding principles for contemporary communitarian philosophies. The distinction between *gemeinschaft*
(COMMUNITY) and GESELLSCHAFT (society or
association) informs a large part of the last century's discussion and debate about what constitutes community. It heavily influenced social
theorists in the late nineteenth and early twentieth
centuries such as SIMMEL, Durkheim and WEBER.
One of the problems of this dualism is that
gemeinschaft is always historicized to an earlier
pre-modern period and is often romanticized as
what is lacking in contemporary communities. It is
de-politicized to an element of nostalgia that people try to regain and has become the ideological

backbone of large parts of neo-traditional URBANISM.

Further reading

Garber, J. (1995) 'Defining feminist community: place, choice and the urban politics of difference', in J.A. Garber and R.S. Turner (eds) *Gender in Urban Research*, Thousand Oaks, CA: Sage, pp. 24–43.

Tönnies, F. (1957 [1887]) *Community and Society*, trans. C. Loomis, East Lansing: Michigan State University Press.

STUART C. AITKEN

GENDER DISCRIMINATION

Gender discrimination, especially in employment, has been the topic of much research and policy in recent decades. In general terms, DISCRIMINATION describes situations where individuals are treated differently because they are members in different social groups. Gender discrimination refers specifically to the differential treatment of women. The task of defining gender discrimination is a complicated one, largely because of two different uses of the term 'gender'. Feminist scholarship makes a distinction between sex – biological difference between males and females – and gender – socially constructed difference between masculinity and femininity. The distinction between biological sex and socialized gender is central to late twentieth-century Western feminist thought. From this perspective, gender is the social organization of sexual difference; it is a process by which the world is defined in binary and hierarchical terms. Outside of feminist scholarship, however, gender and sex tend to be used interchangeably to refer to the biologically distinct categories of males and females. Thus, non-feminists often use the term 'gender' to refer to what feminists call sex.

Accordingly, the phrase 'gender discrimination' has two different meanings. For feminists, gender discrimination is different from sex discrimination. In feminist terms, sex discrimination means not hiring, or renting an apartment to, or educating, or giving a loan to someone solely because she is a woman. It is sex discrimination if a man similar in all attributes except sex would have received the job or promotion or education or loan that the woman did not receive. For feminists, gender discrimination encompasses much more than instances where similarly situated women and men are treated differently. Rather, the term 'gender discrimination' captures the myriad of ways that societal norms routinely assign women secondary status and discriminate against them. Social prohibitions against women working for pay, rules requiring a woman to have her husband's permission to access healthcare or enter into contracts, and assumptions that childbearing is the priority of all women are all forms of gender discrimination. Gender discrimination occurs in both public and private spheres, it affects women's political and social rights, and it is cumulative. From a feminist perspective, a woman who does not receive a job because her domestic responsibilities limit her ability to work overtime can be seen as a victim of gender discrimination, even if she is not a victim of sex discrimination. Mainstream non-feminist scholarship, however, tends to use the term 'gender discrimination' to refer to what feminists call sex discrimination. Economists, for example, distinguish between gender discrimination (not receiving the job because one is a woman) and individual tastes and preferences (not receiving the job because one prefers to not work overtime because one has primary responsibilities for children). In this definition, the social process feminists call gender discrimination disappears into the category of individual choice.

The feminist definition of gender has been particularly apparent in the United Nations Convention on the Elimination of All Forms of Discrimination Against Women (Women's Convention, or CEDAW). The treaty defines discrimination against women as any 'distinction, exclusion or restriction made on the basis of sex which has the effect or purpose of impairing or nullifying the recognition, enjoyment or exercise by women, irrespective of marital status, on the basis of equality between men and women, of human rights or fundamental freedoms in the political, economic, social, cultural, civil, or any other field'. In keeping with this broad definition, CEDAW addresses a diversity of issues affecting women, including prostitution, stereotyped concepts of men and women in educational materials, the right to maternal health, and the right to freely decide the number and spacing of children. The broad definition of gender discrimination in CEDAW has influenced policies in a number of countries, many in the global south, that have ratified the treaty and used it to inform national laws.

In the United States, sex discrimination is regulated in a manner similar to discrimination based on race, religion, national origin, physical disability or age. The Fifth and Fourteenth Amendments to the US Constitution limit the power of the federal and state governments to discriminate. Discrimination in the private sector is not directly constrained by the Constitution, but has become subject to a growing body of federal and state statutes, especially the Equal Pay Act (1963) and the Civil Rights Act (1964). Subsequent court decisions have clarified, and in some cases expanded, the definition of discrimination against women. For example, court decisions have defined sexual harassment as a form of sex discrimination. Employment discrimination in particular has been the focus of numerous court decisions as well as extensive academic research. Although 'gender discrimination' is the term used in most studies of employment discrimination against women, the main focus of research and policies has been what feminists call sex discrimination.

In addition to discrimination in employment, there are other forms of gender discrimination in city life. In many areas, gender discrimination overlaps with discrimination on the basis of race, ethnicity, sexual orientation, income, age or ability. Indeed, feminist scholars stress that gender is never independent of other dimensions of SOCIAL INEQUALITY. Access to housing, especially for female-headed households, is one important area of gender discrimination. Since female-headed households are over-represented among the poor, gender overlaps with POVERTY in unequal access to housing. Scholars have also pointed out ways gendered assumptions are imbedded in the architecture. For example, the separation of the kitchen from other living spaces exemplifies and contributes to the isolation and low status of housework. Issues of safety are another way that URBAN landscapes are gendered. The fear of sexual assault constricts many women's access to public space, and therefore constitutes a form of gender discrimination. Gender discrimination has also been noted and challenged in access to services such as medical care. Given the high concentration of women among the elderly, gender discrimination and age discrimination often intersect in health service provision. Finally, some feminist urban scholars have argued that the dominant planning trend of segregating land use, especially the physical separation of residence and employment, has meant gender discrimination is a pervasive characteristic of modern Western URBAN FORM. Feminist challenges to the distinction between sites of production and reproduction have articulated their vision of a NON-SEXIST CITY free of underlying discrimination.

Further reading

Acker, J. (1990) 'Hierarchies, jobs, bodies: a theory of gendered organizations', *Gender and Society* 4: 139–58.

McDowell, L. (1983) 'Towards an understanding of the gender division of urban space', *Environment and Planning D: Society and Space* 1: 15–30.

Tomasevski, K. (1998) 'Rights of women: from prohibition to elimination of discrimination', *International Social Science Journal* 50(158): 545–58.

DOREEN J. MATTINGLY

GENDER EQUITY PLANNING

In the 1960s and 1970s, ADVOCACY PLANNING and other movements for social equity in cities began to identify distinct social groups which not only could be identified by unique sets of political interests regarding the spatial and material outcomes of city development processes, but also could be organized into somewhat coherent (if overlapping) political constituencies. Among these was the largest disenfranchised group, women.

Resulting from several decades of research since, by the early twenty-first century there was a well-developed literature on differences between women and men in terms of paid employment, household responsibility, daily transportation, personal safety, housing, public services and civic participation. Based on this increasing social science documentation of persistent sex differences and the ways in which local and national policies contribute to enforcing and maintaining them, gender equity planning developed as a movement to account for and neutralize the gender bias in URBAN PLANNING, DESIGN and policy.

As feminist approaches themselves have developed more sophisticated analyses of the role of gender within social relations, gender planning has focused less on strict statistical differences between women's and men's behaviour or resources, and more on analyses of the nuances of social relations and expectations regarding gender as they apply to social activity. Of particular importance is the

feminist distinction between gender and sex: the latter is a generally biological distinction characterized by man/woman, while the former is a more complicated set of socially constructed behaviour patterns, characterized by masculine/feminine – and which are thereby learned and implicitly unlearnable. One goal of gender equity planning is to eliminate the assumption that feminine social roles are necessarily tied to female actors. In addition, the importance of viewing gender simultaneously with ethnicity, class, sexual orientation and other social categorizations has become prominent, as differences which had been viewed as universally applicable have been shown to be true primarily for white, Western, middle-class women and comparable men, but not necessarily for women in other minority groups or non-white-majority nations.

Examples of issues with which gender planning is concerned include limitations on the time or spatial range of women, which result from a gender expectation that they will be the primary caregiver for children or dependent elder family members. Metropolitan or resource area plans and policies must account for such expectations, and ensure that spatial and material outcomes both recognize the existence of such gender-based standards and work to eliminate them in policy.

For instance, transportation plans which emphasize the importance of direct home-to-work round-trip commutes may neglect the fact that one or more household workers are required to make other stops on the way home, to retrieve offspring from childcare or to acquire the evening meal. Thus, an increasing insistence on rail-based strategies – which are fixed-route by nature and provide speed and convenience advantages chiefly to those with longer, non-stop commutes – inadvertently may support not only workers with longer, uninterrupted journeys to work (more often male), but also the notion that one worker within the household should be encouraged to shirk travel for unpaid purposes while another adult takes care of household maintenance activities. Gender equity planning would make the role of caregiving central in analysing costs and benefits of a particular solution, and would strive to erase gender assumptions that support inequality.

Other gender differences relate not to planning outcomes but to social and political processes and relationships. For example, findings in psychology and ethics indicate that girls' moral development is different from that of boys – for example, boys tend to be more interested in the neutral application of abstract rules and standards, whereas girls are more likely to alter outcomes of games or other relationships in order to achieve more inclusive outcomes for all. Applied to participatory political processes, this could mean that judicial or electoral processes, which favour participant-blind outcomes and vote counting, are less effective at giving voice to community-development concerns such as the nurturing of social networks and edification of individual potential to contribute to the neighbourhood, city and society. Thus, an equitable process would both enable voices to be heard through a variety of mechanisms and highlight the importance of traditionally feminine concerns to the sustenance of communities and of everyday living.

In the late twentieth and early twenty-first centuries, gender planning became increasingly institutionalized, most often at the national or supranational level. In many countries within the European Union, for example, organizations and agencies underwent 'gender audits', required to receive EU grant funds, to ensure not only that issues of gender inequality were raised in their substantive focus areas – for example in international development or resource protection projects – but also that the structure and governance of each organization itself was consciously composed to reduce gender inequality. Such institutionalized legal standing for gender equality can be expected to increase. As women join men in equal numbers in the paid workforce and the political arena, issues which have traditionally been the primary concern of women – caregiving, community-building, urban crime and personal safety – will likely become more prominent as public issues, to be addressed through planning and policy.

Further reading

Greed, C. (1994) *Women and Planning: Creating Gendered Realities*, London: Routledge.

Hayden, D. (2002) *Redesigning the American Dream: The Future of Housing, Work, and Family Life*, rev. edn, New York: W.W. Norton & Co.

Sandercock, L. and Forsyth, A. (1992) 'A gender agenda: new directions for planning theory', *Journal of the American Planning Association* 58(1): 49–56.

D. GREGG DOYLE

GENDERED SPACES

'Gender and space' is one term for an inter-disciplinary area of study and activism that examines how sexual difference is practically and symbolically reflected, constructed and reinforced in space. Contributions to this area of analysis have come from the fields of human geography, planning, architecture, environmental psychology, anthropology, URBAN SOCIOLOGY and urban history. However, the area has never been purely academic and has always had an activist or practice-oriented component.

Starting in the 1970s, the area was known as 'women and environments', which became the title of a Canadian magazine founded in the 1970s. By the 1980s, as English-language scholarship and practice developed, specialists from a number of disciplines and professions dealt with the issue in sub-fields such as feminist planning, geography and gender, sexuality and space, and gender and place. The issue of gender and development, while frequently focusing on social and economic concerns, has often also dealt with space. This shift towards a number of sub-fields reflected larger changes in both FEMINIST THEORY and the feminist social movement. By the 1980s there was a more sophisticated theoretical understanding of gender, the cultural elaboration of sexual difference, with a small number of gender scholars starting to deal with issues of masculinity along with inequality between the sexes. Feminism began to include a growing appreciation of the diversity among women both within and across national cultures. The development of gay and lesbian, and later queer, activism, as well as the field of queer studies, also meant that there was a second significant social movement focused centrally on the issue of gender.

There are a number of spaces that have been traditionally segregated by sex, although few were exclusively the domain of men or women but rather the locations for different forms of inequality. Military bases, sports clubs and fields, PARKS, shopping areas, workplaces, educational institutions, social clubs, transportation facilities, the buildings of civic organizations, and the interiors of homes have all been formally or informally segregated by sex. One of the contributions of this area has been to examine just how this SEGREGATION has created or reinforced inequality.

The analysis of gendered spaces in the United States, Britain, Canada and Australasia has always been diverse in aims, scale of analysis and theoretical position. Some scholars have been concerned with a historical or broadly sociological understanding of the development of sex segregation or spatial inequality along gender lines. Others have developed a theoretical analysis of such issues as the body in relation to the city or of gendered differences in the meaning of space. Still others have been more concerned with change, either making proposals for more egalitarian environments or examining women's activism around spatial issues. The issue of scale has also been important – whether the space under scrutiny is the body, building, neighbourhood, city, region or globe – with people working from different disciplines bringing their own biases and expertise. While some very early research, criticism and activism considered women as a group, quite early on differences among women were explored, reflecting both practical needs and developments in feminist theory. At first, in the 1970s, specific groups attracting attention included mothers, single parents, poor and working-class women, and women in non-Western societies. By the 1980s, analyses became more complicated, or are least more specific, with more work about lesbians, women from non-dominant racial or ethnic groups, indigenous women, homeless women, women with disabilities and women in post-colonial societies.

Many involved in this area have used their knowledge to promote practical changes, including those working in the area of GENDER EQUITY PLANNING. In the 1970s and 1980s, feminist design professionals organized non-profit and activist groups. Groups such as the British design cooperative Matrix and the Canadian network Women Plan Toronto are well-known examples of groups formed in the 1980s. Using volunteers and government funding, they collaborated with other women's organizations and non-profits to build feminist facilities such as women's shelters, redesign the existing environment, and advocate for changes in spatial policies. Queer activism in the 1990s often used dramatic actions to highlight issues of violence in public spaces. Still other organizations were sponsored directly by governments. These included the women's initiatives started in British local governments in the early 1980s and projects initiated by the United Nations.

The status of women and the character of the spaces they occupy have undergone significant changes in recent decades. For example, when US feminism emerged in the 1960s and 1970s, many of the middle-class women drawn to the movement were like those described in Betty Friedan's *Feminine Mystique* – unhappy residents of the new postwar SUBURBS. Others had been involved with the CIVIL RIGHTS MOVEMENT that had at least one base in centre cities, a location where public housing was increasingly occupied by poor, single mothers. Key issues included inequality within domestic space, spatial isolation, and support for women seeking jobs and housing. As women increasingly entered the labour force, and as the gay and lesbian population emerged in some locations, the focus of this area also shifted. Work on GENTRIFICATION by women and gay men, on gender and transportation, and on the gendering of employment in the global economy reflects some of these actual changes.

The area of gender and space has raised awareness of spatial issues among scholars interested in gender. Among those areas more centrally concerned with space, the area has contributed a gender analysis. While many forms of inequality and difference exist in the city, gendered space is likely to remain an important dimension of inequality for some time.

Further reading

Bell, D. and Valentine, G. (eds) (1995) *Mapping Desire: Geographies of Sexualities*, London: Routledge.

McDowell, L. (1999) *Gender, Identity, and Place: Understanding Feminist Geographies*, Cambridge: Polity Press.

Rothschild, J. (ed.) (1999) *Design and Feminism: Re-Visioning Spaces, Places, and Everyday Things*, New Brunswick, NJ: Rutgers University Press.

Spain, D. (1992) *Gendered Spaces*, Chapel Hill: University of North Carolina Press.

ANN FORSYTH

GENERAL PLAN

Also known as a comprehensive plan, development plan, master plan, city, county or district plan; the term 'general plan' emphasizes the broad policy nature of these local territorial plans. The general plan has been one of the most important instruments in city and REGIONAL PLANNING since the early twentieth century. Following Kent (1964), the general plan is best conceived as the official statement of the governing body of a city or other territorial region which establishes its major policies concerning desired and desirable future development.

At minimum, a general plan covers all territory within a jurisdiction. Any unincorporated areas likely to be annexed during the time frame covered by the plan also should be included. A larger sphere of influence (e.g. a trade area or watershed) may also be examined in the general plan at a more cursory level. The time frame of the plan is typically twenty years or longer.

The general plan includes a series of maps showing a unified plan for the city or territory. General plans have historically emphasized physical development, but since the late twentieth century general plans typically also describe relationships between physical development policies and social, economic and environmental concerns.

The objectives and policies of the general plan are intended to underlie most land use decisions including decisions on development approvals and investments in infrastructure and public services; California law has deemed general plans to be 'the constitution' for all future development, meaning that all forms of development regulation must be subservient to and consistent with the goals, objectives and policies of the general plan.

The general plan provides a measure of predictability by informing citizens, land developers, decision-makers and other jurisdictions (superior, subordinate and neighbouring) of the basic principles and rules that guide development within the COMMUNITY. In addition to identifying a community's land use, circulation, environmental, economic and social goals and associated development policies, the general plan should also provide its citizens with opportunities to participate in shaping their community.

Thus, the general plan should serve to connect identified community values, visions and objectives with physical decisions such as land development and public works projects. It should encompass a statement of the current conditions in the community, and establish the community's long-term goals regarding development. The policy element of the plan then establishes the means and mechanisms for moving towards the desired long-term goals.

Without question, the comprehensiveness of the general plan conflicts with its need to be relevant and accessible. Many jurisdictions publish their general plans on the INTERNET, increasing the accessibility and portability of the document as well as simplifying updates. However, the general plan remains a demanding endeavour, requiring both a considerable commitment of resources and the active participation of the entire community to be successful.

Further reading

Kaiser, E.J. and Godschalk, D.R. (1995) 'Twentieth century land use planning: a stalwart family tree', *Journal of the American Planning Association* 61(3): 365–85.
Kent, T.J. (1964) *The Urban General Plan*, San Francisco: Chandler.

RICHARD W. LEE

GENTRIFICATION

The process of renewal and upgrading of NEIGH-BOURHOODS connected with the influx of middle- and upper-middle-class people into deteriorating areas of inner cities and resulting displacement of poorer residents. The distinction between revitalization and gentrification is important. While the 'revitalization' of inner-city neighbourhoods may be a desired outcome for local government agencies because it increases the tax base necessary for providing social services, long-time residents often resist and fight gentrification – the type of revitalization that results in the displacement of low-income, mostly minority residents by more affluent people.

Ruth Glass, a British urbanist, first coined the word 'gentrification' in 1964 while describing change in the West End of London. In the United States, the best-known examples of gentrified areas are Washington DC's Georgetown and Dupont Circle, New York's Lower East Side, and San Francisco's Castro and Mission neighbourhoods. London's Docklands is another well-known gentrified area.

Common features of gentrifying neighbourhoods include: (a) location close to DOWNTOWN, museums and other amenities; (b) architectural quality of buildings often marked for historic preservation; (c) dense, mixed-use and ethnic character of the area; (d) concentration of young upwardly mobile professionals (yuppies) who are mostly single and couples without children; (e) existence of old industrial buildings that can be converted into 'lofts'; and (f) visible local government and business community investments like urban furniture, pavements, landscaping, private security and other neighbourhood amenities.

General theories of neighbourhood change provide some insight into understanding the underlying causes of gentrification as well as its impact on inner-city neighbourhoods and residents. The most influential thinking goes back to research carried out by the CHICAGO SCHOOL OF SOCIOLOGY in the 1920s. Robert PARK, Ernest Burgess and others in this school conceptualized neighbourhood change as a process of 'invasion and succession' by different social groups. According to this theory, now known as the human ecology approach, cities develop through a process of competition for space where newcomers (poor immigrants in the case of 1920s Chicago) invade an area, and when crowded, higher-income groups locate further away from the central business district. Gentrification is the reverse process depicted by the 'invasion and succession' model of the Chicago School because higher- rather than lower-income groups 'invade' an area, displacing lower-income groups. Some scholars use the terms 'filtering-up' and 'reinvasion' to characterize the process of gentrification.

The emergence of trendy neighbourhoods in the 1980s amid declining inner cities attracted great attention in the media. A 'BACK-TO-THE-CITY MOVEMENT' theory quickly gained popularity to explain gentrification. According to this theory, children born to parents in the SUBURBS after the Second World War – the baby boomer generation – exhibited a strong desire to live in inner cities (unlike their parents, who had fled the ills of the inner city). These 'urban pioneers' were seen as transforming dilapidated neighbourhoods through their sweat equity, idealism and pro-city values. Careful empirical studies by demographers, however, showed that the magnitude of migration from suburbs to inner cities by the 'gentry' (a term borrowed from the British for the upper class) during the 1980 and 1990 period was too small to support the back-to-the-city movement theory. Instead, neighbourhoods were being upgraded by existing city residents who moved into gentrifying neighbourhoods from other parts of the city. This latter process is known as 'incumbent upgrading'.

While tastes and preferences are important in understanding the housing choice behaviour of different households, a predominant focus on urban lifestyle and consumption patterns falls short in explaining the complexity of gentrification. The strongest critique of the purely ecological and demographic approaches in explaining gentrification comes from scholars highlighting the role of public policy, history, economic restructuring, social movements and culture in shaping cities.

By describing city development as a product of capitalist production, David HARVEY, for example, emphasizes the role of money flows into and out of certain city neighbourhoods dictated by the needs of capital. Economic and spatial restructuring of cities are indeed closely connected. Deindustrialization and DECENTRALIZATION in the 1950s saw the relocation of most industries from CENTRAL CITIES to suburbs in search of cheaper land and lower taxes. Coupled with the flight of middle- and upper-class whites from inner cities in pursuit of homeownership in the suburbs (funded by federal mortgage subsidies and highways), this led to the gradual deterioration of inner cities. Neil Smith explains that capital returned back to selected inner-city neighbourhoods only to take advantage of the 'rent gap'. Both Sharon Zukin and Neil Smith attribute the transformation of New York's SoHo and Lower East Side neighbourhoods to investment and disinvestment cycles of capital.

Gentrification of neighbourhoods in major US cities is also linked to global population movements and the globalization of economic activities. On the demographic side, most inner cities in the United States would have lost greater numbers of people between 1990 and 2000 if there had not been significant additions to city populations through IMMIGRATION from abroad. On the economic side, the unprecedented technology boom of the late 1990s had important implications for cities like San Francisco and New York. The concentration of INTERNET firms in downtown areas and the strong demand for residential and office space resulted in gentrification of neighbourhoods that provided the right mix of attributes for the technology workforce, such as the Mission and SoMa (South of Market) neighbourhoods in San Francisco. A new wave of gentrification was set in motion with an influx of dot-com industry employees into these neighbourhoods. Not surprisingly, massive evictions and rent hikes

followed, fuelling intense anti-displacement efforts by low-income housing advocates and neighbourhood activists. While the collapse of the technology sector in 2002 eased intense development pressures city-wide to some extent, neighbourhoods like the Mission and SoMa are likely to prevail as sites of the next wave of gentrification in San Francisco.

Social movements can also gentrify neighbourhoods. Manuel CASTELLS describes how San Francisco's Castro neighbourhood was transformed from a working-class neighbourhood into an affluent gay community in the 1980s.

Further reading

Castells, M. (1983) *The City and the Grassroots*, Berkeley, CA: University of California Press.
Downs, A. (1981) *Neighborhoods and Urban Development*, Washington, DC: The Brookings Institution Press.
Hartman, C., Keating, W.D. and LeGates, R. (1982) *Displacement: How to Fight It*, Berkeley, CA: Legal Services Anti-Displacement Project.
Smith, N. (1996) *The New Urban Frontier: Gentrification and the Revanchist City*, London: Routledge.

SEE ALSO: back-to-the-city movement

AYSE PAMUK

GEOGRAPHIC INFORMATION SYSTEMS

A geographic information system (GIS) is a collection of computer hardware and software designed to display, store and manipulate information about spatially distributed phenomena. Since the mid-1980s, geographic information systems have become valuable tools that are used to support a variety of CITY and REGIONAL PLANNING functions. These systems allow planners to conduct types of analyses that were not possible using coloured pens and paper maps.

There are two basic ways in which a GIS can represent mapped data: vector and raster. A *vector* GIS represents a map as a set of discrete objects, usually points, lines and polygons (closed areas). Each mapped object is linked to a record in a relational database table. This database contains text and numeric information that describes the object on the map (e.g. the value of a land parcel or the traffic volume on a segment of road). Vector

representation is most appropriate when the map consists of discrete items, such as individual buildings or land parcels. A *raster* system divides an area into a set of regular grid cells, usually squares. Information about what lies within each cell is represented by a numeric value assigned to that cell. Raster representation is most appropriate for continuous variables, such as soil or vegetation types. Digital photographic images are another example of a raster structure. In a photograph, colours or grey shades are represented by cell values. Satellite imagery and aerial photography with grid cell sizes as small as 1 metre or less are now widely available for use in a GIS. Such images provide detailed representations of URBAN areas. Many modern GIS software packages allow the user to combine raster and vector information to provide the most complete depiction of spatial information.

In addition to standard database queries, the GIS can perform a number of spatial analyses. The most common use of a GIS is to display information that is stored in a database. This allows the user to see and understand spatial patterns that may not be apparent in a tabular database report. For example, a planner may display census data on household income to see if poverty is concentrated within a certain part of the city. However, a GIS is more than just a computerized mapping system because it allows the user to analyse the spatial relationships between sets of mapped objects. The most common type of spatial relationship is proximity. The GIS can determine how close one set of objects is to another set of objects. Using a GIS, a planner can easily determine how many housing units are located within half a mile of a transit station. Another type of spatial relationship is containment, or whether an object lies within the boundaries of an area feature. For example, a crime analyst may want to know how many crimes were committed within a particular police precinct, or a water resource planner may need to know the amount of production of all the wells that fall within a given watershed. The GIS can count or sum values associated with objects that are contained within a given area.

Most geographic information systems group similar types of information into layers. One of the most powerful functions of the GIS allows the user to overlay multiple layers to determine what features are coincident. This technique is widely used to conduct land suitability analysis. The purpose of this type of analysis is to find areas that are best suited for a particular use. If she is trying to locate a new landfill, the planner might want to find an area that is not in a flood plain, has impermeable soil, is not near housing, but is close to a main road. By performing spatial analysis on separate layers of data and then combining them in an overlay process, the planner can find one or more locations that meet all of these criteria simultaneously. If the criteria are correctly specified, this analysis will determine the locations that are most suitable for a new landfill.

Raster systems are especially powerful for doing this type of overlay analysis. Each type of data is represented on a separate layer. Since the grids on each layer are congruent, it is a simple matter to add, subtract or multiply the values in the corresponding grid cells on multiple layers. To estimate sedimentation, an environmental planner may want to identify areas that have steep slopes, little vegetation and erodible soils. Combining these layers to find the sources of sediment is quite straightforward in a raster GIS system.

Geographic information systems often rely on photogrammetric techniques to produce data. Satellite imagery or aerial photography can be used to determine the type and health of vegetation. Conversely, these techniques can be used to identify impervious surfaces that can be used in storm water modelling. At the regional scale, change detection compares aerial photographs of the same area at two points in time to identify new areas of URBANIZATION.

As can be seen from these examples, GIS technology cuts across a wide variety of disciplines. In fact, GIS can be useful in any application where location is important. Within a local government, geographic information systems not only support the planning department, but are also used by the public works department, the tax assessor, the police department and the fire department. Recognizing this fact, many cities are moving to an enterprise GIS model. In an enterprise GIS, each department owns and maintains the data layers that are necessary for its functions, but it shares them with other departments in a common system. In such a system, the tax assessor might maintain land parcel boundaries and assessed values; the public works department would maintain the streets, water and sewer layers; and the planning department would maintain ZONING and future land use layers. By sharing their data through

a common system, each department has access to more and better information. In the enterprise model, the geographic information system becomes a platform to integrate departments that may otherwise operate separately. It ensures that all of the departments are operating off a consistent base of information.

Further reading

DeMers, M.N. (2003) *Fundamentals of Geographic Information Systems*, 2nd edn, New York: John Wiley & Sons.

Huxhold, W.E. (1991) *An Introduction to Urban Geographic Information Systems*, New York: Oxford University Press.

Longley, P.A., Goodchild, M.F., Mcguire, D. and Rhind, D.W. (2001) *Geographic Information Systems and Science*, New York: John Wiley & Sons.

O'Sullivan, D. and Unwin, D.J. (2003) *Geographic Information Analysis*, New York: John Wiley & Sons.

STEVEN P. FRENCH

GEORGE, HENRY

b. 2 September 1839, Philadelphia, Pennsylvania; d. 27 October 1897, New York City

Key work

(1966) *Progress and Poverty*, New York: Robert Schalkenback Foundation.

Henry George was a social reformer who was one of the best-known people in America during his era. George was born in Philadelphia, undertook a significant amount of travel and was basically self-educated as he terminated his formal education before the age of 13. He drew his analysis from what he observed and what he read. George went to sea as a young boy, made a voyage around the world, and during his second left the ship in San Francisco and became a journeyman printer. George observed that as wealth was created, there were also those who were poor. While those of means enjoyed affluence and leisure, there were others who suffered degradation. George wanted to understand why these two conditions existed concurrently. This led him to reading and studying classical works in political economy, such as those by Adam Smith, David Ricardo and John Stuart Mill.

As a consequence of his studies, he began to examine the unequal distribution of wealth and the way in which wealth was produced from the production of goods and services. George asserted that working the land was a natural right of individuals, and consequently landlords were not entitled to the unearned increment or increased land value that reflected the 'economic rent' earned on the land. His solution to this was a single tax that would capture the unearned increment or increased land value via the implementation of this site value land tax. George felt that the revenue from a single site value land tax could finance government. What he was advocating is known as 'site value' taxation. This is a form of property tax where the tax is based only on the value of the land; the improvements, buildings, are not part of the tax base. It is often argued that this form of tax encourages the development of vacant sites in the CENTRAL CITY where markets are strong. George believed that all of the activities of local government could be financed via one tax, a site value property tax. George articulated his analysis and views in a number of essays and books. His best-known publication is *Progress and Poverty*, published in 1879, and still read today. This simply written book that he completed in San Francisco was initially published as an author's edition as he was unable to secure a publisher. It was later published in New York and England, and he gained international notoriety.

It is important to note that in 1896, as he continued to write and speak to convey his views, George ran for the office of mayor of New York City on a single tax platform and lost. However, it is notable that Theodore Roosevelt, who was also in the race, finished behind Henry George. During his life, communities in Delaware and Alabama were developed based on his single tax on land. This legacy continued through applications in a number of areas around the world, including Australia, New Zealand and Taiwan. His influence continues today among scholars analysing such topics as property rights, land value capture and sustainable development.

Further reading

Brown, H.J. (ed.) (1997) *Land Use and Taxation: Applying the Insights of Henry George*, Cambridge, MA: Lincoln Institute of Land Policy.

Wenzer, K.C. (ed.) (1997) *An Anthology of Henry George's Thought*, Rochester, NY: University of Rochester Press.

DAVID AMBORSKI

GESELLSCHAFT

Defined by Ferdinand Tönnies (1957 [1887]) as the predominant social relations of modern society wherein people are separated despite all uniting factors such as spatial proximity or common needs. Social relations are founded on rationality, efficiency and the contractual obligations of capitalist organization. In their separateness, individuals find COMMUNITY among common-interest groups rather than through neighbouring and local socialization. *Gesellschaft* provides a basis for contemporary philosophies on individualism. The distinction between GEMEINSCHAFT (community) and *gesellschaft* (society or association) informs a large part of the last century's discussion and debate about what constitutes community. It heavily influenced social theorists in the late nineteenth and early twentieth centuries such as SIMMEL, Durkheim and WEBER. Simmel's concept of URBAN *anomie* is directly derived from *gesellschaft*. Some contemporary cultural critics use the concept of *gesellschaft* to argue that urban POSTMODERNISM is about disengagement, withdrawal and solipsism.

Further reading

Tönnies, F. (1957 [1887]) *Community and Society*, trans. C. Loomis, East Lansing: Michigan State University Press.
Young, I.M. (1990) *Justice and the Politics of Difference*, Princeton, NJ: Princeton University Press.

STUART C. AITKEN

GHETTO

Ghetto is an area or contiguous areas of a CITY where more than 50 per cent of a racial, ethnic or religious group lives due to past and/or present DISCRIMINATION in housing. Louis WIRTH argued that the modern ghetto is of medieval URBAN European origin and that the term applied originally to European Jewish settlements in Venice which were segregated from the rest of the population. Wirth recognized that although the ghetto originated as a Jewish institution, there are ghettos in the United States that never had Jewish occupants although the forces that underlie the formation and development of the ghettos in the United States bear a close resemblance to those that contributed to the early Jewish ghettos. Wirth's ideas are expressed in Darden's *The Ghetto* (1981).

Ghetto is one of the most misunderstood spatial concepts. Some social scientists and members of the general public refer to a ghetto and SLUM as interchangeable. The ghetto is perceived as an area of POVERTY and social disorganization where members of minority groups live under conditions of involuntary SEGREGATION. A number of social scientists disagree with this perception. They argue that the ghetto is not necessarily an area of poverty, nor is it socially disorganized. The ghetto is not a homogeneous area of poor people. Instead, it is inhabited by all incomes and social classes. The ghetto is not synonymous with slum. It covers a wider geographic area than a slum, and the ghetto expands because of population growth, rather than because of deterioration.

The spatial expansion of the ghetto has been described historically within the context of the growth of cities. New groups of immigrants arrive in cities. Those with low income, education and occupational status settle close to the central business district. As these groups improve their socio-economic status, they move out spatially towards the periphery. Thus, residents on the periphery of the ghetto are more likely to be of higher socio-economic status than those residents living in the inner core.

Structurally, most ghettos can be generalized as forming sectors emanating from a core near the central business district. In the inner part of the ghetto is the slum, which is an area of extreme poverty. But outer parts of the ghetto are composed of middle-income families.

However, economic restructuring combined with civil rights legislation of the 1960s began to change the spatial structure of the ghetto. William Julius Wilson expressed these changes in the United States in a book entitled *The Truly Disadvantaged* (1987). According to Wilson, the ghetto was impacted by a decline of manufacturing jobs in the CENTRAL CITY and a relocation of those jobs to the SUBURBS. There was also an increase in low-wage service sector jobs. Such

wages were not sufficient to support a family. This trend led to high rates of unemployment especially among ghetto residents, and a reduced pool of male employable marriage partners. Marriage thus became less attractive to poor women in the ghetto and unwed childbearing increased, resulting in a higher rate of female-headed households. Those residents in the ghetto suffered the most from these economic changes due to their location in the central city.

These economic changes, according to Wilson, were followed by civil rights legislation (especially the Fair Housing Act) which provided middle-class black ghetto residents with opportunities for housing outside of the ghetto. The out-migration of middle-class black families from the ghetto left the poor residents of the ghetto even poorer as the institution of the ghetto weakened, resources declined, schools deteriorated and an expanded 'URBAN UNDERCLASS' within the ghetto emerged. Thus, the post-1960 ghetto has a larger slum portion or area of extreme poverty than what existed previously.

Massey and Denton argued in *American Apartheid* (1993) that the high level of black–white residential segregation is a necessary component to an understanding of the ghetto underclass and urban poverty. According to Massey and Denton, residential segregation created the structural conditions for the emergence of an oppositional culture in the ghetto that devalues work, schooling and marriage, and instead stresses attitudes and behaviours that are often hostile to success in the larger restructured economy.

Residential segregation, according to Massey and Denton, is the institutional apparatus that supports other racially discriminatory processes and binds them together into a coherent effective system of racial subordination. Until the black ghetto is dismantled as a basic institution of American urban life, progress ameliorating racial inequality in other arenas will be slow and incomplete.

In sum, two separate perspectives have evolved over time in our understanding of the ghetto. One perspective has placed the central emphasis on race and the other on class. Spatially, due to economic restructuring and persistent racial segregation caused by discrimination in housing, the ghetto today is increasingly the home of the 'ghetto poor'. However, Wacquant argues in 'Three pernicious premises in the study of the American ghetto'

(1997) that to say that they are ghettos because they are poor is to reverse social and historical causation. It is because they were and are ghettos that joblessness and misery are unusually acute and persistent in them – not the other way round. According to Wacquant, to call any area exhibiting a high degree of poverty a ghetto is not only arbitrary and empirically problematic, it robs the term of its historical meaning. It also obliterates its sociological import, thereby thwarting investigation of the criteria and processes whereby exclusion effectively operates in it. Such exclusion has been racial, involuntary and permanent for the purpose of perpetuating an inferior social structure.

Further reading

Darden, J.T. (1981) *The Ghetto: Readings with Interpretations*, Port Washington, NY: Kennikat Press.

Massey, D.S. and Denton, N. (1993) *American Apartheid: Segregation and the Making of the Underclass*, Cambridge, MA: Harvard University Press.

Wacquant, L.J.D. (1997) 'Three pernicious premises in the study of the American ghetto', *International Journal of Urban and Regional Research* 21(2): 341–53.

Wilson, J. (1987) *The Truly Disadvantaged: The Inner City, the Underclass and Public Policy*, Chicago: University of Chicago Press.

JOE T. DARDEN

GIDDENS, ANTHONY

b. 18 January 1938, Edmonton, North London, England

Key works

(1984) *The Constitution of Society*, Berkeley, CA: University of California Press.
(1990) *The Consequences of Modernity*, Stanford, CA: Stanford University Press.

Anthony Giddens is a British sociologist and author and editor of thirty-five books translated into twenty-nine languages. Twelve of the books concentrate solely on his own work. Giddens is agreed to be the most widely read and cited contemporary social theorist. He has academic appointments in approximately twenty different universities throughout the world and has received numerous honorary degrees. Currently, he is

serving as the director of the London School of Economics and Political Science.

Giddens's writings range from sociology, social theory and history of social thought to class and power structures, nationalism and identity. Giddens developed the 'theory of structuration', which is an attempt towards the understanding of relations between individuals and conditions surrounding them. Accordingly, life in a society is organized around a series of ongoing individual activities and practices, which at the same time reproduce structures. In other words, the moment of the production of individual action is also that of the structure's reproduction. Structures have no existence without the agent and its social activity.

In the mid-1980s, Giddens was one of the first authors to emphasize the 'process of globalization' and its impacts on our lifeworlds, and he has worked on the concept intensively since then. He explains globalization as a consequence of modernity. Accordingly, globalization is inevitable, unavoidable and irreversible, since not only is the process forced by the capitalist actors, some of whom are unprecedentedly strong, but also there remained no alternative. This new version of capitalism weakens the welfare state by enforcing flexible business structures. Consequently, new forms of class divisions are being developed, undermining the power balance between capital and labour. He proposes to reconstruct our existing institutions or develop new ones in accordance with globalization. Together with Ulrich Beck, Giddens developed the concept of 'reflexive modernization', which argues a continuous interplay between local happenings and global flows.

His latest contribution has been in the field of political thinking with the notion of the 'Third Way', which has had a major impact on the restructuring of left-of-centre politics throughout the world. Giddens argues that Third Way politics represents the renewal of social democracy in a world where the politics of the old left and right have become obsolete and those of the new right are inadequate and contradictory in terms of recent developments. It is simply a modernized social democracy – sustaining socialist values and applying them to the new era. A positive but still critical position is taken with respect to globalization. Third Way politics is required to help people in their negotiations with the contemporary revolutions. For this to be realized, a comprehensive

approach to social and economic policy should be developed combining social solidarity with the dynamism of the liberal economy. Partnerships between state and community and between public and private are crucial. In particular, such an approach looks for a new relationship between the individual and the community with an attempt to redefine rights and obligations.

Further reading

Cassel, P. (ed.) (1993) *The Giddens Reader*, Hong Kong: Stanford University Press.
Held, D. and Thompson, J. (eds) (2002) *Social Theory of Modern Societies: Anthony Giddens and His Critics*, Cambridge: Cambridge University Press.

MURAT CEMAL YALCINTAN

GLOBAL CITIES

Since the 1970s and 1980s, the economy has been undergoing a process of transformation, from being international to being global. The rapid development of information and telecommunications technology and infrastructure is bringing advances such as the worldwide spread of finance and capital markets, round-the-clock availability of financial transactions, and the expansion of global operations by multinational enterprises; economies are transcending the framework of nation-states; and cultural mobility is also becoming global. This process has advanced the unbounding of various economic and cultural functions, but instead of being spatially dispersed, those various functions have been concentrated in a number of cities around the world.

The reorganization of these various global functions has caused a metamorphosis of the metropolises that were the historical and traditional centres of international trade, providing them with a new strategic role. Globalization has brought great changes to the nature of these major URBAN spaces. Their various functions have expanded beyond the boundaries of nation-states; they have become ambiguous, their scale and scope having expanded dramatically, their spatial patterns becoming closely related to social interdependence. There are now more than 300 city regions around the world with populations of more than 1 million. Their political and administrative boundaries being practically meaningless, concepts such as MEGALOPOLIS or MEGACITY can no longer be

applied to such regions. The world's metropolises have been studied with respect to these changing circumstances and the new paradigm research has made remarkable progress since the late 1980s. These new cities have been given various labels, such as 'world city', 'INFORMATIONAL CITY', 'cosmopolis' and 'GLOBAL CITY'. After the proposal of the 'WORLD CITY HYPOTHESIS' by John FRIEDMANN and others, Saskia SASSEN turned her attention to their function as world economic nodes and pointed out that a new type of city had appeared. She called it the 'global city' and cited as examples New York, London and Tokyo. Subsequently, the world city is often referred to as the global city.

For a city to be regarded as global, it must be perceived from diverse aspects, including: (1) having networks that constitute or create an urban system within the world system, not just within one country's urban system; (2) being a spatial constituent on a global scale and integrated with an adjoining multi-level structure, whose position in the levels is not only strongly dependent on the scale of its economy but also largely related to its power of global dominance, which is concentrated in other areas of the city such as its cultural functions; (3) being perceived as an interface between the city and the national districts that constitute its central, semi-peripheral and peripheral regions; and (4) being perceived both in terms of large-scale urban space for the realization of higher-order mutual socio-economic actions, called 'the power of places', and in terms of highly developed 'computer space' in the city, called 'the power of flows'. However, the world city hypothesis summarized these in seven propositions. Of these, 'the place' and 'central administrative functions', which the global economy will develop, are strongly associated with the global city that Sassen is advocating.

Up to that point she had been focusing on economics related to organizations with authority over service economies, especially financing, insurance and real estate, with respect to global city functions that she had been treating with a vague definition such as the social regenerative functions possessed by this kind of metropolis, and she was now advocating a new paradigm for grasping these functions as extremely productive activities. She treats global cities as places for the accumulation of four new functions: (1) as command towers in the world economy; (2) as places for locating finance in a leading position in place of manufacturing and for locating the related service functions; (3) as places for leading industries and the innovative technologies to support them; and (4) as places for conducting these transactions. To give places such as these a centralized command structure with administrative power over production bases and service centres scattered geographically around the world means concentrating total administrative power over the world economy in, for example, the three cities mentioned above. However, if we look closely at global economic systems, we see that each of them has a different circuit, and if among the cities that are included in such a circuit there is also a city which has sophisticated specialized global city functions, then even if that city itself is not a global city, it will be placed in a specialized network and connected to a global city.

Sassen-style global cities represent the strategic space where global processes are materialized in the national domain and global dynamics work through national institutions. This means 'places' called sophisticated specialized network functions where there are transborder network functions that connect strategic places. It does not mean a single unified body. Global city functions are central administrative functions that include producers' service work, and even if this is not national work, it is apparently domestic work such as accounting, law and advertising, and includes functions that support the global activities of the city. Moreover, it also includes the informal or underground economy. Therefore, although they are not financial centres like New York and London, urban regions where finance functions are widely dispersed as in Los Angeles, where there is a multiracial mix, or where economic, cultural, entertainment and technical development functions, with military functions, are widespread are also global cities.

Further reading

Castells, M. (1989) *The Informational City: Information Technology, Economic Restructuring, and the Urban-Regional Process*, Oxford: Blackwell.

Sassen, S. (2001) *The Global City: New York, London, Tokyo*, 2nd edn, Princeton, NJ: Princeton University Press.

Soja, E.W. (2000) *Postmetropolis: Critical Studies of Cities and Regions*, Oxford: Blackwell.

YUICHI TAKEUCHI

GLOCALIZATION

'Glocalization' is a word coined from globalization and localization. The advance of globalization gave rise to simultaneously reactive and resistant local movements in various dimensions of nationalism, regionalism and transnational governance networks. This process, which is closely associated with URBANIZATION, signifies a situation in which local units smaller than nation-states are constrained by the need to redefine their own economic, geopolitical and cultural positions in the world. Globalization and localization are not unilateral processes. Globalization is driving localization and localization is driving globalization. In other words, while they are mutually interacting, both are progressing in a competitive relationship. This competitive process is called glocalization.

The term 'glocalization' is in fairly wide use today. In the sense of 'think globally, act locally', the term 'glocal management' has been in use in the business strategies of companies from an early stage. In particular, 'dochakuka', which is a business strategy of Japanese companies that are expanding overseas, is used without translation into English. Even among the NGOs that help the development of developing nations, the necessity for the bi-directional idea of 'think locally, act globally' is being advocated together with the idea of 'think globally, act locally'.

Among scholars, attempts are being made to use glocalization as an analytical concept. In sociology, cultural homogenization and heterogenization indicate processes that are proceeding simultaneously and are called glocalization. In a world that is being increasingly compressed, and in which there is rising universality of people's sense of values and awareness, moves to restore the traditions and identities possessed by nations and peoples are growing stronger. In other words, globalization does not obliterate local matters; globalization has become a compositional arrangement whereby local matters can float upwards.

The fact that globalization causes the importance of local matters to float upwards is also being pointed out in economics. Even though we are facing an age of instantaneous global money transfers through computer networks, 24 hours a day, anywhere in the world, such global transactions are actually being carried out in very limited regions confined to major cities such as New York, London and Tokyo, which are especially equipped with telecommunications networks and various kinds of associated service industries. That is to say, for globalization to advance, specific sites will be needed, and it is being shown that the provision of such sites will be a local process.

Further reading

Robertson, R. (1995) 'Glocalization: time-space and homogeneity-heterogeneity', in M. Featherstone, S. Lash and R. Robertson (eds) *Global Modernities*, London: Sage, pp. 25–44.

Swyngedouw, E. (1997) 'Neither global nor local: "glocalization" and the politics of scale', in K.R. Cox (ed.) *Spaces of Globalization*, London: Guilford Press, pp. 137–66.

YUICHI TAKEUCHI

GOOD GOVERNMENT MOVEMENT

The Good Government movement emerged in the late nineteenth century as a response to the civic corruption and patronage associated with BOSSES and machine politics. Advocates inspired by the national progressive movement sought to promote efficient municipal GOVERNANCE and a heightened sense of civic morality as the antidote to chaotic city life. This alliance defined its agenda during the First Annual Conference for Good City Government held in Philadelphia in 1894. The gathering resulted in the formation of the National Municipal League, an organization that subsequently became the National Civic League.

Efficiency objectives focused on reforming an overtly politicized and unqualified civic administration that threatened the viability of local economic growth. Changes occurred as Good Government supporters in local business communities successfully led efforts to apply a formal corporate management model to municipal administration. The resulting shift meant that city bureaucracies were staffed by permanently appointed municipal experts who were responsible for applying scientific principles to planning and service delivery decisions. A city manager position was created to serve as the municipal equivalent of a chief executive officer, with decision-making responsibilities that often surpassed those available to the mayor and council. Structural changes

reinforcing the desired separation of city bureaucracies from political oversight led to a proliferation of unaccountable special purpose bodies and
independent boards responsible for local service
provision.

Reforms also expanded municipal governing
responsibilities to ensure efficiency principles could
guide ongoing city development. This action was
taken in response to the frustration real estate
developers experienced over an inability to secure
inexpensive and on-demand servicing for their
sites. Profit motives and corruption associated with
privately owned transit, electricity and water companies produced haphazard infrastructure decisions that did not always respond to the needs of
local builders. In response, cities assumed ownership of these services to promote orderly growth
and cost-efficient services in accordance with real
estate industry preferences.

Electoral changes sought to maintain a democratic presence in city government while reducing
the relevance of local councillors. This strategy
hinged on eliminating the NEIGHBOURHOOD influences that gave bosses their local power. Good
Government leaders viewed these parochial COM-
MUNITY interests as being an impediment to their
URBAN efficiency objectives. Some cities sought
to take neighbourhoods out of government by
abolishing council elections based on geographically defined wards in favour of at-large city-wide
representation. The resulting structure disenfranchised ethnic and racial groups through the
dilution of neighbourhood voting blocs, making
it very difficult for lower-income and minority
citizens to win a council seat. Good Government
supporters regarded this change as a moral act that
would reduce urban social problems by integrating immigrants with the broader population.
Ultimately, this new approach combined with
bureaucratic reforms merely replaced political
machines and their supporters with local elites who
benefited from greater access to the municipal
decision-making process.

Further reading

Judd, D. and Swanstrom, T. (1994) *City Politics: Private
Power and Public Policy*, New York: HarperCollins.
Scott, M. (1995) *American City Planning Since
1890*, Chicago: American Planning Association,
pp. 110–82.

STEVEN M. WEBBER

GOODMAN, ROBERT

b. 9 October 1936, Brooklyn, New York

Key works

(1971) *After the Planners*, New York: Simon & Schuster.
(1979) *The Last Entrepreneurs: America's Regional Wars
for Jobs and Dollars*, Boston, MA: South End Press.
(1995) *The Luck Business: The Devastating Consequences
and Broken Promises of America's Gambling Explosion*, New York: Free Press.

Robert Goodman made his name in the 1960s as
the first president of Urban Planning Aid (UPA), a
non-profit organization of planners, architects
and engineers that provided technical help to
Boston-area communities. UPA campaigned against
expressway building and URBAN renewal, and
Goodman was a pioneer of 'ADVOCACY PLANNING'.
As an associate professor of architecture at MIT he
wrote *After the Planners*, a devastating critique of
top-down modernist planning in the 1950s and
1960s. In the 1970s, his research and writing
focused on regional economic development and
deindustrialization in the United States. This led to
The Last Entrepreneurs, a path-breaking analysis
of how state and local governments compete for
investment by giant oligopolistic corporations,
subsidizing big business with locational incentives
and infrastructure. In the 1990s, Goodman directed the Ford Foundation and Aspen Institute-
funded US Gambling Study. He authored *The
Luck Business*, revealing the negative impacts of
gambling on regional economies, government ethics, Native American communities, and the finances, families and health of gamblers. As a professor
at Hampshire College in Amherst, Massachusetts,
Goodman continues to research and write on
urban development issues.

RAY BROMLEY

GŌTŌ, SHINPEI

b. 25 July 1857, Muzusawa, Iwate
Prefecture, Japan; d. 13 April 1929,
Tokyo, Japan

One of the most important politicians and
administrators in Japanese national government
during a time of modernization and reform in
the late nineteenth and early twentieth centuries,

and influential mayor of Tokyo from 1920–23. Among his most significant achievements was the rebuilding of Tokyo following the Great Kantō Earthquake of 1 September 1923 that devastated the city.

Gōtō was born in 1857 in what is now Iwate Prefecture, Japan, to a samurai family and left home as a young man to study medicine at the Sukagawa Medical School in Fukushima Prefecture, from which he graduated in 1876. After a brief stint as director of the Aichi Hospital in Nagoya, Gōtō joined the Home Ministry in 1883. He was put in charge of its Health Bureau in 1892, responsible for improving public health and sanitation in Japan. In 1898, he was sent to the Japanese colony of Taiwan to be in charge of civil administration. He also served in Manchuria, another Japanese possession, and became president of the South Manchuria Railway Company in 1906. From 1908 to 1912, he was Japan's Minister of Communications as well as director of the National Railways Bureau and the Colonization Bureau. He became Home Minister in 1916 and Minister of Foreign Affairs in 1918, during which time he advocated Japan's expansion into Siberia while the Russian government was preoccupied with revolution.

In December 1920, Gōtō was coaxed out of retirement to serve as mayor of Tokyo. He had earned a reputation in his career as a capable administrator who could get things done, and was urgently needed in Tokyo to reform municipal government after a bribery and corruption scandal, and to improve the city's squalid living conditions. He proposed ambitious plans for the capital, paved and widened major streets, constructed sewer systems, and improved public education and welfare. In 1922, he founded the Tokyo Institute for Municipal Research as a vehicle for applying 'scientific methods' to URBAN problems and improving public administration. Funded initially through the will of Japanese business leader Yasuda Zenjirō, the Tokyo Institute was modelled after the New York Bureau of Municipal Research, a similar organization in the United States, and was assisted by American historian Charles Austin Beard, whom Gōtō brought to Tokyo as an adviser. Because of his progressive, science-based approach to city administration, Gōtō's biographers have referred to him as Japan's 'statesman of research'.

The earthquake of 1923 and its fires provided Gōtō with an opportunity to completely modernize Tokyo. Together with Beard, he put forward ambitious plans for wide and straight streets, new PARKS and improved housing. However, only a small part of what was proposed was actually built; as the plans were criticized as being too expensive, the city was reconstructed largely on foundations from before the disaster. Gōtō finished his public service by representing Japan in negotiations with Russia to normalize diplomatic and commercial relations.

Further reading

Hayase, Y. (1974) 'The career of Gōtō Shinpei: Japan's statesman of research, 1857–1929', unpublished Ph.D. thesis, Florida State University.

ROMAN CYBRIWSKY

GOTTMANN, JEAN

b. 10 October 1915, Kharkov, Ukraine;
d. 1 March 1994, Oxford, England

Key works

(1955) *Virginia at Mid-Century*, New York: Henry Holt.
(1961) *Megalopolis: The Urbanized Northeastern Seaboard of the United States*, New York: Twentieth Century Fund.

Jean Gottmann was born in the Ukraine and educated at the University of Paris. He taught geography at Johns Hopkins University, the Ecole des Hautes Etudes and the University of Oxford. He is best known for his writing on the United States. Gottmann established himself as a leading regional geographer with *L'Amerique* (1954) and *Virginia at Mid-Century* (1955). This introduction to the rapidly urbanizing Atlantic states led to a larger project under the sponsorship of the Twentieth Century Fund. In *Megalopolis* (1961), Gottmann appropriated an antique Greek term and applied it to the vast city region that was emerging from Washington north to Boston. Gottmann's book received wide attention in a nation concerned about the impacts of rapid SUBURBANIZATION after the Second World War. In the popular press, 'MEGA-LOPOLIS' was taken to mean a continuously developed urbanized region. Gottmann, however, was careful to point out that Megalopolis was a network

of large metropolitan regions that functioned in some ways as a single entity but that embraced substantial amounts of farms and forests. Residents in other parts of the world looked for incipient megalopolises around the Great Lakes, on the Pacific coast, in Western Europe, on Honshu, and elsewhere. In the 1980s, Gottmann turned to the growing importance of high-speed communication for reshaping URBAN systems, writing numerous essays to develop his ideas about 'transactional' cities whose primary economic function is the processing and distribution of information.

CARL ABBOTT

GOVERNANCE

Governance is the exercise of public authority. This means making and administering public policy, as well as making and acting on public decisions. Gradually over time in many countries and cultures, this power to steer society has been diffused from singular rulers to powerful governments and now to a broader base of stakeholders and constituents. The United States has a rich history of experimentation in city governance. In this splendid laboratory for observations of the range of local governance forms and reforms, we now find far more to governance than government. Government provides formal institutional structure while governance provides the process and participants for the exercise of public authority. Today, government plays a major role in governance along with thousands of citizens, other public institutions, private companies and non-profit organizations. Governance in twenty-first-century America is complex, despite the simplicity and flexibility of the US Constitution.

Although the Constitution does not mention city government, it establishes the context for modern governance in its preamble, which makes plain that governments and institutions do not own governance, the people do: 'We the people of the United States, in order to form a more perfect union . . . do ordain and establish this Constitution.'

Despite constitutional clarity on this point, the questions of who governs and who wields power in communities remain very debatable. In theory, citizens have the last word on public decisions. Yet public capacity for decision-making is limited according to many critics because public officials, staff members, interest group representatives and other 'experts' and 'elites' make many decisions with only minimal input from the general public. Public influence is important but episodic and varies enormously across the vast landscape of American local government and issues.

Most of America's 85,000 governments are local and 'semi-sovereign', meaning that they often share certain powers but are governed by independent, elected boards and commissions with significant autonomy and policy-making and taxing authority. Nearly half a million people are elected to serve America's many governments, most at the local level. Currently, the vast array of American local governments include approximately 20,000 general-purpose municipalities and counties, 17,000 townships, 31,000 special districts and 14,000 school districts. Legally, American local governments, according to the famous Dillon's Rule (named after Judge John Dillon, who ruled on the powers of municipal government in the nineteenth century), are 'creatures of state government' whose powers are expressly granted to them by the states.

Over time, many cities have gained significant discretionary authority via a process called HOME RULE, which is now available for municipalities in 48 states and for counties in 37 states. Most American big cities and some URBAN counties have been allowed by state guidelines to develop home rule charters that serve as mini-constitutions prescribing functions, purpose and rules of their operation.

There is ample diversity in the structure of American local government and numerous questions exist about the 'best' ways to govern. The National Civic League's 100-year-old *Model City Charter*, which has served city governments through seven editions, is being revised to address both basic and advanced questions about local government structure. Passionate debates about the 'best' ways to govern cities include such topics as: form of government; mayoral powers; election of mayor by and from council vs. direct, at-large, district and hybrid elections of council and mayor; non-partisan ballots; mayor-council compensation; and city manager accountability and political neutrality. Principal forms of American city government are:

- *mayor-council* – elected mayor acts as the chief executive officer (CEO) with varying administrative authority depending on the state;

- *council-manager* – manager is appointed by and responsible to an elected council and serves as the city's CEO;
- *commission* – elected commission performs both legislative and executive functions, with functional departmental administration divided among the commissioners; and
- *town meeting* – voters of a town meet annually or more frequently to set policy.

According to *The Municipal Year Book: 2000*, the council-manager form is used by about 50 per cent of cities while 43 per cent choose the mayor-council form. Among large US cities (populations of 250,000 or more), however, mayor-council is the form of choice.

City government structure is but one, albeit important, dimension of local governance. Promoted as recent innovations, PUBLIC–PRIVATE PARTNERSHIPS, privatization and community collaborations are derived from earlier traditions of self-reliance and reliance on community organization and participation. In his classic study of early American democracy, Alexis de Tocqueville found people of many communities working through various non-governmental institutions to solve community problems. These historical patterns combined with contemporary themes of distrust and cynicism associated with politics and government provide the platform for governance through networks of shared responsibility, neighbourhood organizations, PUBLIC–PRIVATE PARTNERSHIPS, and reliance on a dynamic mix of non-profit and 'faith based' organizations to deliver public services and identify community needs.

According to recent studies, including a 2001 national survey, many Americans who are dissatisfied with governments and politicians are still optimistic about potential contributions to community problem-solving. In every large community, hundreds – sometimes thousands – of non-profit organizations provide public services in health, social services, education, environment, criminal justice and every other major public policy area with combinations of public and private resources.

Among the most complicated modern local challenges is regional governance. To compete in the world economy and to cope with destructive social forces, urban regions – CITISTATES – must function as whole units. Yet many public problems such as air quality, transportation, land use, workforce development and emergency services span city borders and are unaddressed or approached in piecemeal fashion because of the lack of coherent regional governance, either formal or informal.

To deal with these and other challenges of changing times, such as globalism, technological change and devolution, will require informal and structural change, including increased collaborative leadership and community-building, to bring governments and people together for effective governance.

Further reading

Frug, G. (1999) *City Making: Building Communities Without Building Walls*, Princeton, NJ: Princeton University Press.

International City/County Management Association (ICMA) (2000) *The Municipal Year Book: 2000*, Washington, DC: ICMA.

Peirce, N., Johnson, C. and Hall, J. (1993) *Citistates*, Washington, DC: Seven Locks Press.

Svara, J. (1990) *Official Leadership in the City: Patterns of Conflict and Cooperation*, New York: Oxford University Press.

JOHN S. HALL

GRAFFITI

The impulse to make a mark on the environment is a long-standing one. 'Graffiti' is derived from the Greek term 'graphein', meaning to write, while also being the plural of the Italian word 'graffito', meaning scratch. Forms and styles of graffiti vary from the quick and simple 'tag' to the more elaborate (master)'pieces' and 'burners'. Many involved in 'doing graffiti' disassociate themselves with the term, preferring notions such as 'aerosol', 'freestyle', 'spraycan' and 'street influenced' *art* (graffiti is/as art), primarily to distinguish themselves from those whom they consider actually do 'graffiti' – 'taggers', 'bombers' or 'vandals' (graffiti is/as vandalism). 'Graffiti', therefore, is viewed by 'writers' as a narrow term/label that has been imposed from elsewhere, predominantly through the media. A range of aspects of popular URBAN/ 'street' culture are intrinsically linked with (and feed off) the graffiti scene, including hip hop, break-dancing and skateboarding.

Predominantly considered an underclass, subcultural, urban phenomenon, graffiti has become visualized as a growing urban 'problem' for many

cities in industrialized nations, spreading from the New York City subway system in the early 1970s to the rest of the United States and to Europe and other world regions. Outside of the 'appropriate' space of the art gallery, it raises questions about ownership and control of urban space(s), generating fear, dread and panic regarding the apparent impending loss of social order and civility to an invisible army of faceless perpetrators. As such, strategies to control graffiti include diversionary-based interventions that seek to include those involved in graffiti, law enforcement-based interventions, including partnership approaches, situational crime prevention interventions, and education campaigns.

Further reading

Castleman, C. (1982) *Getting Up: Subway Graffiti in New York*, Cambridge, MA: MIT Press.
Cresswell, T. (1996) *In Place/Out of Place: Geography, Ideology and Transgression*, Minneapolis: University of Minnesota Press.

DUNCAN FULLER

GREAT SOCIETY

Coined during a 1964 speech by President Lyndon B. Johnson at the University of Michigan, the phrase 'great society' came to represent his domestic agenda. As described in this speech, the Great Society would be a place of revitalized cities, with vibrant cultural and social institutions, embodying a SENSE OF COMMUNITY, where all children, regardless of race or income, would have access to quality education and economic opportunities, and where the splendour and abundance of our natural resources would be preserved for future generations. It would be 'a place where men are more concerned with the quality of their lives than with the quantity of their goods'.

Johnson's Great Society programme represents the greatest period of legislative activity since the New Deal. Of the 115 Great Society bills submitted to Congress, 78 passed and were signed into law. Three landmark bills were passed in the area of civil rights – the Civil Rights Act of 1964, the Voting Rights Act of 1965 and the Open Housing Act of 1968. Collectively, these laws ended legal DISCRIMINATION against minorities in, respectively, public accommodation, voting and private housing

markets. Forty programmes were established as President Johnson declared a national 'WAR ON POVERTY' aimed at eliminating poverty by improving living conditions and by giving the poor access to opportunities long denied them. Such programmes included food stamps, healthcare and job training. Another sixty bills focused on educational reforms, providing funding for new and better-equipped classrooms, minority scholarships and low-interest student loans. Quality pre-school for poor children was provided through the Head Start programme.

Many bills introduced new areas of federal action. Legislation was passed that focused the attention and resources of the federal government on environmental degradation, particularly the problems of air and water pollution. Federal funding for the arts was established through the creation of the National Endowment for the Arts and the Humanities (NEA). The NEA, in turn, spurred creation of arts councils in all fifty states. The Corporation for Public Broadcasting was established, helping to support the development of 350 public television and 699 public radio stations across the country by 2000.

Access to healthcare for the elderly was guaranteed through the creation of the Medicare programme. The Department of Housing and Urban Development was established to coordinate programmes aimed at improving living conditions in cities, especially for the poor. Legislation also supported major URBAN mass transportation projects and established important consumer protections.

Ultimately, this period of legislative innovation and optimism came to an end as the war in Vietnam expanded and became more controversial and expensive, and protest undermined Johnson's effectiveness in gaining support for this domestic agenda both in Congress and at the voting booth. In 1968, Johnson announced that he would not run for re-election.

The meaning and legacy of the Great Society remains hotly debated. Consensus exists on the scale of the effort, however. During this period, the federal government's role in domestic social programmes expanded dramatically. At the end of the Eisenhower administration there were only 45 such programmes; by 1969 there were 435. Federal social spending (excluding social security) rose from US$9.9 billion in 1960 to US$25.6 billion in 1968 – an increase of 158 per cent. Most of this

funding was not targeted at the poor but instead to programmes benefiting the middle class, such as education. Similarly, consensus exists on the progress made in overcoming discrimination in this era. Few would fail to credit Johnson with the gains made due to the passage of civil rights legislation. The number of black elected officials rose from 300 in 1964 to 9,000 in 1998. Great Society programmes arguably laid the foundation for improvements in life expectancy, infant mortality and educational attainment for African-Americans.

During this period, poverty was redefined as a national issue and responsibility for addressing it a federal responsibility. Supporters point out that poverty declined from 22.2 per cent of the population in 1963, when President Johnson took office, to 12.6 per cent in 1970. The foundation was laid for gains in educational attainment, critical to economic development. The share of the population completing high school and college also rose dramatically, from 41 per cent completing high school and 8 per cent completing college in the early 1960s, to 81 per cent and 24 per cent, respectively, in the mid-1990s. By 2000, Great Society programmes and their descendents provided financial assistance to nearly 60 per cent of full-time undergraduate students. These programmes helped establish the notion that access to higher education should be determined by ability and motivation rather than family wealth.

Access to health insurance also improved dramatically. In 1963, most elderly Americans had no health insurance. The poor had little access to medical treatment, apart from emergency care. High-quality care was available only to the richest residents of our largest cities. By 1998, 39 million elderly people were enrolled in Medicare, and Medicaid served a similar number of low-income people. The 1968 heart, stroke and cancer legislation helped create centres of medical excellence in most major cities. Legislation increasing funds for medical education helped double the number of doctors graduating from medical schools. Funding for anti-hunger programmes provided breakfast for millions of schoolchildren. Supporters credit these programmes collectively with improving life expectancy and infant mortality statistics, especially for the poor.

Conservative criticisms of the Great Society focus on impacts of anti-poverty programmes.

Critics note that welfare caseloads ballooned during this period, and that easy access to benefits that they argue collectively exceeded the value of the minimum wage undermined the motivation of the poor to leave welfare for work. Extended welfare dependence, coupled with the programmes' rules that prohibited welfare receipt by families with an able-bodied male in residence, are credited with undermining the 'family values' of the poor and for the sharp increase in female-headed households and illegitimacy among the poor since 1959 (see WAR ON POVERTY). Free market advocates objected to the overall rise in the regulatory powers of the government.

Further reading

Califano, J.A., Jr (1991) *Triumph and Tragedy of Lyndon Johnson: The White House Years*, New York: Simon & Schuster.

ELIZABETH J. MUELLER

GREEN BELT

Many governments have some form of planning policy to contain URBAN development, though the effectiveness of these policies varies considerably. In the United States, urban growth management policies are referred to as SMART GROWTH, although this has been applied in only a minority of states. In parts of Europe, particularly England and the Netherlands, the idea of urban containment is a long-standing and deeply embedded part of the policies and practice of city planning. The result is that the rate of conversion of land uses from rural to urban is (relatively) very low indeed. For example, in 1995 in England, 13,320 hectares of land changed to urban use, which is about 0.1 per cent of the land area, and only about half of this was a change from rural uses, the rest being from previously developed land. The net change from rural to urban uses in England is about 14,800 acres (6,000 hectares) per year (down from 8,000 hectares in 1987). In comparison, in Germany about 116,000 acres (47,000 hectares) of land in total is converted to urban uses each year. Because of different definitions, the figures are not directly comparable, but the increase in 'developed land' in the United States is over 1.67 million acres (600,000 hectares) per year.

One reason for these great differences in conversion rates is the importance and success of the main policy instrument for urban containment in England: the green belt. Similar approaches are used in the Netherlands, Ireland and other countries. In the United States, proposals for green belts made in the 1930s were soon abandoned, though they came back on the agenda in the 1990s.

The green belt is probably the most well-known and well-supported planning policy in the United Kingdom, although the policy objectives are less well understood. Green belts were established in England from 1955 to simply prevent the physical growth of large built-up areas; to prevent neighbouring cities and towns from merging; and, where relevant, to preserve the special character of a town. They had no place in the provision of recreation or environmental benefits until the 1990s. Green belts now cover 13 per cent of the land area of England, which is more than 3.7 million acres (1.5 million hectares). It should also be noted that planning authorities in England also operate strict development constraint in open countryside that is not designated as green belt.

The main feature of green belts in England is their permanence. Changes to boundaries are allowed in only exceptional circumstances. Reaching agreement on boundaries for green belts has been fraught with difficulty, and once agreed, the review of boundaries happens only in exceptional circumstances. Strategic planning instruments first indicate the approximate area of green belt in diagrammatic form. Local development plans then draw the precise boundaries, which generally follow physical features on the ground. Some development interests and landowners fiercely challenge the designation of green belts so as to retain the development value of their land interests, although the planning authority must ensure that adequate development land is also allocated in the plan. Needless to say, rural conservation interests and particularly the rural local authorities surrounding large cities are equally fierce in their defence of the green belt.

Once designated, development outside cities in green belt is allowed in only exceptional circumstances, for example for the renovation of historic buildings or the improvement of local amenities such as playing fields. There has been some relaxation in recent years to permit, for example, park-and-ride sites where it can be argued there are major sustainability benefits for the city. In other ways, urban containment policy is being strengthened with the introduction of a 'greenfield direction' requiring planning authorities to consult the national planning minister before allowing housing developments on any greenfield site over 12 acres (5 hectares). Another initiative is the *sequential test*, which requires developers to demonstrate that they have considered options for using previously developed sites, increasing densities in existing urban centres and exploiting the potential for conversions of existing non-residential properties.

During the 1990s, the effects of green belt policy in England were the subject of extensive evaluation. Research found that the policy had been very successful in checking unrestricted SPRAWL and in preventing towns from merging. Boundary changes to allow development had affected less than 0.3 per cent of green belts over an eight-year period. But green belt policy has been less successful in encouraging the regeneration of cities, especially in directing jobs onto sites within cities. Where there has been selective release of green belt land on the fringe of cities, it has mostly been to supply sites for employment growth on the grounds that inner-city sites will not provide adequate substitutes for certain uses seeking to locate on the city fringe. It is important to emphasize that green belt policy has not been about stopping development (which is how it is sometimes portrayed) but about the selection of uses that are allowed in the green belt and the redirection of many uses elsewhere.

Critics also argue that green belt has caused development to leap-frog to small towns beyond the outer boundary, thus creating new long-distance commuting into cities. Green belts, alongside other planning controls, have also affected house prices, although the effect varies considerably across the country. Criticisms that green belt has prevented the diversification of the rural economy by not allowing the change of use of farm buildings and that it has not paid attention to providing sport and recreation opportunities have led to a limited relaxation of controls. These criticisms have always had less force in Scotland where there has been much better integration of green belt, recreation and environmental objectives. Despite its critics, green belt policy is being more widely implemented at the turn of the century with, for example, plans or action on green belt designations around cities in Germany, California and Oregon.

Further reading

Elson, M.J., Walker, S. and Macdonald, R. (1993) *The Effectiveness of Green Belts*, London: HMSO.

VINCENT NADIN

GROPIUS, WALTER

b. 18 May 1883, Berlin, Germany;
d. 5 July 1969, Boston, Massachusetts,
United States

Key work

(1919) Bauhaus School of Architecture.

Walter Gropius, one of the leading architects of the twentieth century, founded the Bauhaus in Weimar in 1919. He believed that art should combine aesthetics with the needs of the modern industrial world. The guiding principle of the Bauhaus was that architecture should reflect the desires of society, sound engineering concepts, and commitment to forward-looking modern design. Characterized by the absence of ornament and elaborate façades, Bauhaus style introduced the concept of simplicity of shape and the extensive use of glass.

In 1928, Gropius returned to private practice but had ensured the reputation of the Bauhaus with a faculty led by Paul Klee, Wassily Kandinsky and Laszlo Moholy-Nagy (who later founded the Chicago Institute of Design based on Bauhaus principles). Gropius fled Germany and joined the architecture faculty at Harvard University in 1938. His fourteen years in Cambridge, Massachusetts, defined a generation of American architects with broad URBAN sensibilities. In 1943, Gropius and Martin Wagner presented their ideas on URBAN FORM and livability on a metropolitan scale, anticipating post-war urban expansion and inner-city decline.

After the war, Gropius formed the Architects Collaborative, a firm which won significant commissions, including the Harvard Graduate Center, the US Embassy in Athens and the Pan Am Building in New York.

Further reading

Giedion, S. (1992) *Walter Gropius*, New York: Dover.

DANIEL GARR

GROUP HOME

A group home is a single housekeeping unit occupied by a number of people who are unrelated by blood, marriage, adoption or consanguinity and are living together and sharing a common household. They are appropriate for both service-dependent as well as independent populations including those who are developmentally disabled, physically challenged, mentally ill, recovering from alcohol or substance abuse, elderly, and others for whom living alone is either socially, physically or economically not feasible or desirable.

The location of a group home becomes a COMMUNITY issue when ZONING or other land use regulations define family in such a way that a number of unrelated individuals are prohibited from living together in a single-family residence. In the United States, numerous state courts have concluded that when residents of a group home live together as the functional equivalent of a traditional family, the group may occupy a home in an area zoned for single-family residences. Federal courts have addressed the issue in cases related to the Fair Housing Amendments Act (FHAA) of 1988, which prohibits DISCRIMINATION on the basis of handicap. The FHAA's definition of handicapped includes those with physical or mental disabilities. Cases have extended this to include the elderly as well.

Further reading

Pollak, P.B. (1993) 'Zoning matters in a kinder gentler nation: balancing needs, rights and political realities for shared residences for the elderly', in K.H. Young (ed.) *Zoning and Planning Law Handbook*, New York: Clark Boardman Callaghan, pp. 491–516.
—— (1994) 'Rethinking zoning to accommodate the elderly in single family housing', *Journal of the American Planning Association* 60(4): 521–31.

PATRICIA BARON POLLAK

GROWTH MACHINE

The growth machine model attempts to explain URBAN political power in the United States. Growth machine theory maintains that growth is the essence of local politics, and that those

who benefit from development – landowners, developers, bankers, construction companies, etc. – are generally able to manipulate the planning process to foster growth and increase land use intensities. They legitimate their actions with 'value-free growth' ideology, claiming that everyone benefits from growth. A combination of factors distinct to the United States has encouraged coalitions of pro-growth elites in that country: a weak land use regulatory system, heavy dependence on local revenues, and significant local autonomy in land use decisions. In this context, those who would profit from more development are strongly motivated to influence the decision-making process. They develop alliances with elected officials to promote a good business climate and increase the demand for land consumption.

The growth machine model was introduced in 1976, when Harvey Molotch, a sociologist at the University of California, Santa Barbara, published the article 'The city as a growth machine', in which he made a strong case for a model of local politics based on the power of the growth machine. Molotch, in a series of articles and later in the book *Urban Fortunes*, written with John Logan, maintains that this type of growth comes inevitably at the expense of the quality of life of the general population, who as a result suffer from higher taxes, more traffic congestion, increased taxes and so forth. Social groups seeking to use their locale as a place to live and work, and not as a place to make money, will fight back, often under the banner of environmental protection. The theory stresses that the outcome of such conflict depends on local conditions and historical circumstances. Pro-growth elites, however, have many institutional advantages, like more staff and money to promote their agendas.

Empirical research is ambiguous on whether growth can be significantly altered by growth machine efforts to promote growth or efforts by residents to slow it down. It increasingly appears, however, that citizens' growth limitation efforts do not slow growth but create the political conditions that allow jurisdictions to exact more concessions from developers.

The undisputed success of the theory, however, lies in the way it highlights the centrality of growth in American urban politics. While the exact weight that should be given to growth is a point of contention (see URBAN REGIME THEORY), that it matters a great deal is not.

Further reading

Logan, J. and Molotch, H. (1987) *Urban Fortunes: The Political Economy of Place*, Berkeley, CA: University of California Press.

Logan, J., Whaley, R.B. and Crowder, K. (1997) 'The character and consequences of growth regimes: an assessment of twenty years of research', *Urban Affairs Review* 32(5): 603–30.

NICO CALAVITA

GROWTH MANAGEMENT

Integrated set of public programmes and policies aimed at shaping URBAN growth so as to maintain a livable balance between conservation and development, equilibrium between development and the capital facilities needed to accommodate it, and a sense of COMMUNITY. Growth management commonly goes beyond traditional land use planning, ZONING and SUBDIVISION controls in both the characteristics of development influenced (timing and financing, in particular) and the scope of government powers used. These contemporary land use programmes seek increasingly to consciously manage the process of urban growth, and thus tend to have greater impacts on urban growth patterns than more traditional land use planning efforts such as zoning and general plans that envisioned an end-state rather than managing a process. Increasingly today, in efforts to reduce SPRAWL, growth management is linking land use regulations to CAPITAL FACILITY PLANNING to shape growth in fiscally efficient ways.

Several different goals of growth management are evident – to shape the location of future growth, to preserve environmental qualities, to accommodate growth in an efficient way from the view of the city's budget, to phase or slow down development so it does not overwhelm the city's community character, to manage local growth in ways that recognize regional consequences. Growth management programmes usually emphasize one or two of these goals but they will differ across cities.

Whereas traditional planning controls were legally defended in the 1920s, growth management began more recently, in the late 1960s/early 1970s. At this time, three concerns – environmental, fiscal and psychological – over continued URBANIZATION catalysed new, more forceful ways of shaping and even limiting growth. A growing environmental

GROWTH POLES

consciousness in the 1970s highlighted the conflict between urbanization and fragile ecosystems such as wetlands and critical habitats and legitimated early growth limitation/management programmes. The concept of environmental carrying capacity was asserted by some including Godschalk and Parker (1975) – a limit beyond which human activity and growth will cause irreparable damage to the environment. A second rationale behind growth management was awareness of the fiscal, or budgetary, costs to government of unplanned and sprawling SUBURBANIZATION. In cases of low-density suburban growth, it was found that tax revenues often do not balance with needed new public expenditures that come with growth (costs of new sewer and water lines; new and expanded roads; schools). Thirdly, there was concern about the psychological costs of mass suburbanization and its impacts on preservation of lifestyle and small-town character.

Early growth management programmes sought to manage the rate and timing of growth by applying a yearly growth quota (Petaluma, California, in 1972; and Boulder, Colorado, in 1976) or to link growth to capital facility availability (Ramapo, New York, in 1969). Since these early programmes tended to restrict growth below historical rates, they were commonly referred to as growth control programmes. To this day, growth management is controversial. Some believe that it is a counter-productive and onerous intrusion into the growth process that creates regulatory barriers to community vitality. Others support it as a necessary means of achieving balance in the ways our cities and regions urbanize. Growth management programmes have generally been upheld by courts of law as an appropriate use of a local government's POLICE POWER if regulations and techniques used are logically connected to the city's stated growth management goals.

Other types of growth management programmes have tended to guide or shape growth without any implication of limiting it. One type of programme here are those anti-sprawl measures that demarcate URBAN GROWTH BOUNDARIES or urbanizing tiers, encouraging growth within boundaries through regulatory incentives or public financing of infrastructure, and discouraging growth outside through restrictive regulations, limited public funding of infrastructure, and public acquisition of OPEN SPACE. Examples of this type of programme can be found in Portland, Oregon;

San Diego, California; Montgomery County, Maryland; and Minneapolis/St. Paul, Minnesota.

About fifteen state governments in the United States also use growth management principles. States, in these cases, commonly do not directly regulate or manage growth, but provide standards or principles that must or should be incorporated into local general or comprehensive plans. Examples include Oregon, Florida, California (for its coastal region), New Jersey, Georgia and Maryland. Oregon and Florida require local comprehensive plans to address issues of urban sprawl, public facility availability and housing diversity. If local governments do not submit a local plan in compliance with state standards, penalties and state funding withdrawals can occur. Other states, such as New Jersey and Maryland, use a more incentive-based approach, targeting state funding to locally designed urban growth areas.

Contemporary growth management programmes are increasingly seeking balanced growth and are focusing on sprawl containment, the sharing of infrastructure funding responsibilities between public and private sectors, and efforts to plan for housing diversity and community and neighbourhood character. Growth management, in its integrated effort to address a full range of urban issues – land use, housing diversity, economic opportunity, fiscal efficiency, environmental protection – merges conceptually and operationally with the ideas of SMART GROWTH.

Further reading

Downs, A. and Godschalk, D. (1992) 'Growth management: Satan or savior?', *Journal of the American Planning Association* 58(4): 419–24.

Godschalk, D. and Parker, F.H. (1975) 'Carrying capacity: a key to environmental planning', *Journal of Soil and Water Conservation* 30(4): 160–5.

Kelly, E.D. (1993) *Managing Community Growth: Policies, Techniques, and Impacts*, Westport, CT: Praeger.

Porter, D.R. (1997) *Managing Growth in America's Communities*, Washington, DC: Island Press.

SCOTT A. BOLLENS

GROWTH POLES

The 'growth pole' concept was established in the 1960s, in the context of the development pole and unbalanced growth theory. It has served the purposes of, at least, three areas of study. First, it has

been used in research activities which attempted to achieve a general theory of regional development planning. Secondly, it has been undertaken as a planning instrument in the formulation of intra- and inter-regional planning strategies, regardless of the level of development attained by the area under analysis. Thirdly, it has also been a useful tool in applied research, as a hypothesis to be tested in the light of historical series of statistical information. Perhaps the strongest reason for the popularity of the growth pole notion is its claimed suitability as a reference point for the launching of public intervention strategies relating to the development of problem regions.

The notion was first formulated by the French economist François Perroux, as an attempt to overcome the difficulties brought about by the use of traditional tools in regional analysis. Looking at the real world, Perroux realized that growth does not appear everywhere at the same time but, rather, it manifests at certain points, with uneven intensities. Furthermore, it spreads through different channels, with variable outputs for the economy as a whole. The recognition that economic activities tend to concentrate in growth poles opened the way to the development of a theoretical framework, which aimed at explaining the entire process of structural change in the economy as well as in the social and even institutional systems. In this context, a growth pole is an economic and functional concentration, a field of forces occurring in an abstract and a-spatial economic space, defined by economic relations.

Jacques Boudeville, a French geographer, introduced the geographical dimension of the growth pole, through the definition of the polarization concept. The emergence of a growth pole is linked to the innovation process, expressed by the generation of entrepreneurial innovations, their diffusion and, most importantly, their adoption. Successful innovations establish the dynamics of the process of growth, giving rise to the emergence of dominant units and an increase of inequalities among entrepreneurs. These dominant industries may greatly affect the economy, by inducing growth and structural change in the economic space in which they are set. The cluster of dominant – or leading – and induced units, sharing the same portion of land and inducing further economic activity in their zone of influence, constitutes a growth pole. The two main characteristics of a geographical growth pole seem to be its internal mechanism of expansion and its positive relationship with the surrounding region. In this fashion, a great pole is a large heavy industrial centre, which can achieve self-sustained growth and, eventually, can diffuse this growth outwards into its polarized region.

Further reading

Boudeville, J.R. (1966) *Problems of Regional Economic Planning*, Edinburgh: Edinburgh University Press.

Perroux, F. (1964) *L'économie du XXème siècle* (*The Economics of the 20th Century*), Paris: Presses Universitaires de France.

IÁRA REGINA CASTELLO

GRUEN, VICTOR

b. 18 July 1903, Vienna, Austria;
d. 14 February 1980, Vienna, Austria

Key works

(1954) Northland Shopping Center, Detroit.
(1956) Southdale Shopping Center, outside Minneapolis.
(1960) *Shopping Towns USA*, New York: Van Nostrand Reinhold.
(1964) *The Heart of Our Cities. The Urban Crisis: Diagnosis and Cure*, New York: Simon & Schuster.
(1973) *Centers for the Urban Environment*, New York: Van Nostrand Reinhold.

Austrian-American architect, URBAN planner, author and inventor of the SHOPPING MALL, Victor Gruen (born Viktor David Grünbaum) received his architectural training at the Akademie der Bildenden Künste in Vienna. In 1932, he opened an architectural office in Vienna's city centre, specializing in the remodelling of shops and apartments.

After the German annexation of Austria in 1938, Gruen emigrated to the United States. He worked briefly for the Ivels Corporation and for Norman Bel Geddes before moving, in 1940, to Los Angeles. Gruen, who adopted the anglicized version of his name after 1943, quickly expanded his practice, and in 1949 he founded Victor Gruen Associates with Rudolf Baumfeld, another émigré architect from Vienna. The firm developed into one of the nation's leading architectural, planning and engineering firms with offices in Los Angeles, Washington and New York.

Gruen believed that architectural practice extended beyond designing isolated buildings. He called for the creation of entire environments that could fit within existing contexts but transform usual patterns of living. His breakthrough came in 1954 with the Northland Shopping Center in Detroit, a multifunctional shopping centre. Two years later, he designed the first fully enclosed shopping mall in Southdale, outside Minneapolis, which included an auditorium, a school and a skating rink, as well as a large variety of shops.

Gruen argued that such suburban malls could take the place of traditional civic centres, but by the later 1950s he became preoccupied with URBAN PLANNING issues. In the early 1960s, he produced designs for pedestrian malls in Fort Worth, Texas, and Fresno, California, and he wrote two books, *Shopping Towns USA* and *The Heart of Our Cities*, exploring his ideas for urban redevelopment. Gruen retired in 1967 and returned to Vienna, where he put together an extensive proposal for the renewal of Vienna's inner city. He also continued to write, publishing his final book, *Centers for the Urban Environment*, in 1973.

Gruen was a fellow of the American Institute of Architects and a frequent lecturer at US universities, including Harvard, Yale, MIT, Rice, Columbia and the University of Southern California. Although some of his ideas are now considered passé, he is still widely hailed for his contributions to twentieth-century URBAN DESIGN and theory.

CHRISTOPHER LONG

HABITAT 96

An international conference on human settlements (also known as Habitat II) convened by the United Nations (UN) in Istanbul (3–14 June 1996) to formally adopt global principles to solve human settlement problems in the world. It was preceded by a series of Habitat-related activities after the first Habitat conference in Vancouver (June 1976). Of particular importance is the Urban Management Programme – an interagency undertaking between the UN Development Programme, the UN Centre for Human Settlements (UNCHS) and the World Bank. Other milestones between 1976 and 1996 include the Global Shelter Strategy (UNCHS, 1988) and the Rio Declaration (UN, 1992).

The Habitat II conference marked the culmination of collaborative efforts worldwide over two decades to advance two principal goals: (1) shelter for all; and (2) sustainable human settlement development in an urbanizing world. To facilitate the global process, the UN asked governments to prepare official national reports describing shelter and human settlement conditions, and policies in place for improvements. In the United States, the national report was prepared under the leadership of the Department of Housing and Urban Development, the State Department, and the US Agency for International Development. To engage communities in the debate, an NGO (US Network for Habitat II) organized town meetings across the United States and brought core Habitat II themes – civic engagement, sustainability and equity – to local communities. Official meetings at the international level involved drafting and negotiating the contents of the principal Habitat II document – the Global Plan of Action (also known as the Habitat

Agenda). In a second document produced at the conference – the Istanbul Declaration on Human Settlements – governments agreed to address lack of basic infrastructure and services, unsustainable population changes, and increased vulnerability to disasters among other priorities.

In preparation for the Habitat II conference, the UN established – for the first time in its history – a forum for 'partners' to provide input into the Habitat Agenda. Eight groups of partners were identified including non-governmental and community-based organizations, national academies of science and technology, and business groups. The NGO forum was by far the largest among these partner activities. Further exchange of ideas occurred through a 'best practices' exhibition where innovative approaches to human settlement problems were showcased.

The key innovations of the Habitat II conference were: (1) the incorporation of views from civil society institutions into the official Habitat Agenda through the partners forum; (2) the emphasis on the role of local government agencies and NGOs in implementation; and (3) support of global partnerships and networks in finding solutions to human settlement problems.

Despite these innovations and the accomplishment of formulating the Habitat Agenda, two important constraints in implementation remain: (1) commitments made by governments are non-binding; and (2) the lack of an institutionalized arrangement to monitor and critically evaluate progress.

Further reading

Leaf, M. and Pamuk, A. (1997) 'Habitat II and the globalization of ideas', *Journal of Planning Education and Research* 17: 71–8.

United Nations Centre for Human Settlements (Habitat) (2001) *Cities in a Globalizing World: Global Report on Human Settlements 2001*, London: Earthscan Publications Ltd.

SEE ALSO: favela; gecekondu; informal sector

AYSE PAMUK

HALL, EDWARD T.

b. 16 May 1914, Webster Groves, Missouri

Key works

(1959) *The Silent Language*, Garden City, NY: Doubleday.
(1966) *The Hidden Dimension*, Garden City, NY: Doubleday.

The American anthropologist E.T. Hall developed the field of intercultural communication studies. Earning his doctorate at Columbia, Hall was influenced by the work of Franz Boaz, Ruth Benedict and Ralph Linton.

After army service overseas and a few years of teaching, Hall joined the Foreign Service Institute training programme in Washington as an anthropologist in 1950. Along with the linguist George Trager, he created a system for teaching future diplomats to see the link between CULTURE and communication. Hall argued that non-verbal communication is often more important than verbal communication in transmitting meaning.

In the 1960s, Hall published his theory of proxemics, the study of the human use of space. His work created a new field of research investigating the nature of personal and PUBLIC SPACE, and how it may differ between cultures. He suggested that cultures create invisible boundaries around space. In American culture, personal space involves a three-foot 'bubble' around the body; other cultures control territory according to their own patterns.

Hall's understanding that space conveys cultural values played a significant role in subsequent thinking about the nature and design of public and private space. He helped designers to recognize that space conveys opportunities for generating meaning.

Further reading

E.T. Hall (1992) *An Anthropology of Everyday Life*, New York: Doubleday.

JILL GRANT

HALL, SIR PETER GEOFFREY

b. 19 March 1932, Blackpool, England

Key works

(1966) *The World Cities*, London: Weidenfeld.
(1988) *Cities of Tomorrow: An Intellectual History of Urban Planning and Design in the Twentieth Century*, Oxford: Basil Blackwell.
(1998) *Cities in Civilization: Culture, Technology and Urban Order*, London: Weidenfeld & Nicholson.

Along with Manuel CASTELLS, a colleague with whom he has collaborated, Peter Hall is one of the most prolific and influential urbanists of the twentieth century. Three foundations define his work: a historical understanding, a geographical imagination and an evolutionary political consciousness. His earliest works were a translation of Von Thünen's classic *The Isolated State* (1966) and *The Industries of London* (his doctoral study), the first of eight books published within the first decade after his doctorate, and the precursors to over forty volumes in English, many of which have been subsequently translated into other languages. He established himself in the University of London in 1957, and almost immediately began to position himself as a public intellectual with much to say on URBAN and REGIONAL PLANNING issues, particularly those related to the capital and the South East of England. He published constantly in magazines such as *New Society*, which were influential during the first periods of Labour Party government, while also writing provocative and academically challenging books such as *London 2000* (1963). He was, additionally, active in a wide range of other academic enterprises, such as editing the journals *Regional Studies* and *Built Environment*. He took the chair of geography and planning at Reading in 1968 and built up a premier applied department. His subsequent research projects and publications pioneered a number of interests that have become established throughout the social sciences: an interest in world cities; the symbiotic roles that technology plays in the URBANIZATION process; the

issue of planning failures and urban sustainability; and the development of futures research.

Hall moved his chair to Berkeley in 1980, running the Institute for Urban and Regional Development and teaching in both geography and city and regional planning, while producing a new round of studies that focused on US themes, including *The Rise of the Gunbelt*, a study of military spending and urban and regional development, in 1991. He returned to the United Kingdom in 1992, as professor of planning at the Bartlett School, University College London. He was chairman of the Town and Country Planning Association (1995–99), and a member of the Office of the Deputy Prime Minister's Urban Task Force (1998–99). His book *Sociable Cities* was published in 1998 to launch the centenary of the Town and Country Planning Association, as was his magnum opus *Cities in Civilization*, a historical study of urban creativity. Hall has been recognized internationally for his academic and applied work, receiving numerous medals and nine honorary doctorates. He was knighted in 1998.

ANDREW KIRBY

HARRIS, BRITTON

b. 7 June 1914, East Orange, New Jersey

Harris is professor emeritus of city and regional planning at the University of Pennsylvania where he has spent his entire academic career. A pioneer in the advocacy of a scientific approach to URBAN and REGIONAL PLANNING, he foresaw the use of computers to support such practices. Harris is often credited with the original formulation of the concept of planning support systems.

Harris has been instrumental in creating the Graduate School of Fine Arts at the University of Pennsylvania and was one of the founding members of the Regional Science Association. Ever the intellectual innovator, Harris's earlier work focused on the need for applied location theory and the importance of human values and behaviour in planning. His tireless advocacy for better planning models helped pave the way for new advances in the scientific basis of such models. This work was popularized in the late 1980s as he sought to link GEOGRAPHIC INFORMATION SYSTEMS

and later subsequent advances in visual planning technologies to his previous work in transportation and land use planning. His collected works are stored in the Regional Science Archives at Cornell University.

Harris is a fellow of the American Institute of Certified Planners and has won numerous meritorious awards including the Association of Collegiate Schools of Planning Distinguished Educator Award of 2000 for lifetime contributions to the field of urban and regional planning.

DAVID C. PROSPERI

HARRIS, CHAUNCY

b. 1914, Logan, Utah; d. 26 December 2003, Chicago, Illinois

Key works

(1943) 'A functional classification of cities in the United States', *Geographical Review* 33: 86–99.

(1943) 'Suburbs', *American Journal of Sociology* 49: 1–13.

(1945) 'The nature of cities', *Annals of the American Academy of Political and Social Sciences* 242: 7–17 (with E. Ullman).

(1970) *Cities of the Soviet Union: Studies in Their Functions, Size, Density, and Growth*, Chicago: Rand McNally.

(1975) *Guide to Geographical Bibliographies and Reference Works in Russian or on the Soviet Union: Annotated List of 2660 Bibliographies or Reference Aids*, Chicago: Department of Geography, University of Chicago.

There are no names more familiar to students of URBAN STUDIES than the names of Ullman and Harris (see EDWARD ULLMAN). Almost every introductory course either presents or alludes to their 1945 'multiple nuclei' model in its review of the three models of American spatial structure. The article in which this model appeared, 'The nature of cities', has been frequently reprinted in various books and translated into many languages. The importance of Ullman and Harris's contribution was so great that, in 1997, *Urban Geography* dedicated a special issue to this article and its impact on urban studies.

Chauncy Harris, a classmate of Edward Ullman, received his Ph.D. from the University of Chicago in 1940. From 1939 to 1941, he was at Indiana University, followed by a two-year residence at the University of Nebraska. In 1943, he

returned to the University of Chicago and remained there until his retirement in 1984. During the Second World War, like his friend Ullman, Harris too served the country by working in various posts, including the Office of Geographer, Department of State (1942–43), and the Office of Strategic Services (1943–44).

Harris published for six decades, beginning in 1938, and was an avid student of cities and their SUBURBS. In 1943, he published two articles on this topic, and as late as 1997, his articles on 'The nature of cities' and 'Urban geography in the last half century' appeared in *Urban Geography*. In 1945, the third thread of his scholarly interest appeared in a piece on Russian cities, followed by one on the ethnic population of Soviet cities. On the Soviet topic alone, Chauncy Harris wrote over twenty articles and a number of books. His contribution to the field of Soviet studies was celebrated in a special volume, edited by George Demko and Roland Fuchs (1984). Harris was not only among the first few American geographers who specialized in the field of Soviet studies, but he also greatly promoted the field through a number of comprehensive publications, including *Guide to Geographical Bibliographies and Reference Works in Russian or on the Soviet Union* (1975). Six decades and three inter-related topics have made Chauncy Harris the perpetual agent of geographic advocacy, an urban studies enthusiast and a Russian specialist, who has consistently contributed to teaching and research on all three topics.

Further reading

Demko, G.J. and Fuchs, R.J. (eds) (1984) *Geographical Studies of the Soviet Union: Essays in Honor of Chauncy D. Harris*, University of Chicago, Department of Geography, Research Paper No. 211.

ALI MODARRES

HARVEY, DAVID

b. 31 October 1935, Gillingham, England

Key works

(1973) *Social Justice and the City*, London: Edward Arnold.
(1981) *Limits to Capital*, Oxford: Basil Blackwell.
(1989) *The Condition of Post-Modernity*, Oxford: Basil Blackwell.

David Harvey is one of the leading URBAN geographers whose intellectual influence has travelled widely across disciplines. With a Ph.D. in geography from Cambridge University, he embarked on a lifelong intellectual and political trajectory that would transform the ways in which urban theorists approach the capitalist city and urban activists seek urban social or political change. Already noted for the landmark publication in 1969 of *Explanation in Geography*, a book that transformed geography as a discipline, his epistemological and political attention soon turned to a more radical and Marxist understanding of the urban. This epistemological shift coincided with his transatlantic migration to the Johns Hopkins University, where he would teach Marxist urban theory for the next fifteen years or so. The confrontation between the deep injustices that had just come to the boil in rioting US cities and a rediscovery of the power of historical materialist Marxist analysis resulted in the publication of *Social Justice and the City* (*SJC*). While Manuel CASTELLS introduced Marxism into URBAN STUDIES via an Althusserian detour, Harvey's theorization of the city is deeply embedded in the original writings of Marx, combined with the radical urban theories and politics pioneered by Henri LEFEBVRE. *SJC* radically transformed urban theory and inspired generations of scholars and activists in search of more 'genuinely human geographies' of and for capitalist cities.

Harvey's intellectually most monumental work, *Limits to Capital*, was completed in 1981. He not only brought together brilliantly the core of Marx's analysis, he expanded and innovated Marxist theory, particularly with respect to the functioning of money and finance, and the 'spatial moment' in the unfolding of capitalist crisis formation. For Harvey, 'urban' represents a pivotal space in both the unfolding of capitalism and the reproduction of the contradictory and crisis-ridden dynamics that characterize capitalist societies. He explored the interweaving of cultural, social and political-economic processes in the development of capitalist URBANIZATION in the 1985 twin volumes *The Urbanization of Capital* and *Consciousness and the Urban Experience*.

His temporary return to the United Kingdom to take up the Halford McKinder Chair of Geography at Oxford University was followed by *The Condition of Post-Modernity*. This book was intended to be a theoretical and political

intervention in what was rapidly becoming a new terrain in urban studies and research. From a broadly Marxist perspective, Harvey argues convincingly how the condition of postmodernity reflects and embodies the contradictory cultural and political-economic dynamics of capitalism as they have exploded since the 1970s. For him, accelerating time-space compression and the ongoing 'annihilation of space by time' paralleled a radical overhaul of the cultural and socio-spatial experiences of space and time and, with it, the emergence of a new urban condition.

During the 1990s, Harvey turned to considering urban environmental questions in the context of enduring and pervasive (in)justices in *Justice, Nature, and the Geography of Difference*. In *Spaces of Hope* (2000), he returns explicitly to what has always guided his intellectual work, i.e. imagining alternative urban visions and charting the contours for a humanized URBANISM in the face of the disempowering, uneven development of capitalist forms of urbanization. Harvey currently lives in New York and teaches at the Graduate Center of the City University of New York. He has received several honorary doctorates from universities around the world.

Further reading

(1987) *The Urban Experience*, Oxford: Basil Blackwell.
(2002) *Spaces of Capital*, University Press: Edinburgh.

ERIK SWYNGEDOUW

HAUSSMANN, BARON GEORGES-EUGÈNE

b. 27 March 1809, Paris, France;
d. 11 January 1891, Paris, France

Key work

(1890) *Memoires*, Paris: V. Havard.

Baron Georges-Eugène Haussmann was prefect of the Seine under Emperor Napoleon III from 1853 to 1870. Haussmann transformed Paris from a thoroughly disorganized city into the first great city of the industrial age. What changed the face of Paris was a system of BOULEVARDS that was ruthlessly pushed through the old medieval fabric of narrow, filthy and twisting streets and alleys to

unify what had been separate parts of the city. It is true that the boulevard system was conceived as a mechanism for the easy deployment of troops and artillery – too many revolutions had succeeded in the recent past because of the medieval layout of Paris – but its main purpose was to help solve the traffic problem in a city in the midst of explosive growth and interconnect its landmark buildings.

The boulevards represented for Haussmann the circulatory system of the city. But Haussmann realized that Paris needed also what he called a respiratory or ventilation system. That was accomplished through the creation of promenades, like the Champs-Elysées, public gardens and, most importantly, suburban PARKS at the western and eastern edges of the city, the Bois de Boulogne and the Bois de Vincennes, the lungs that Paris so sorely lacked. The sewer and water distribution systems were upgraded and expanded as well on a grandiose scale.

All these public works needed financing, and the power of EMINENT DOMAIN. The creative assumption made by Napoleon and Haussmann was that, given the financial wealth of the city, money could be borrowed on a massive scale and repaid with rising tax revenues. While private property had been declared a basic human right with the revolution, an 1852 law made it possible for the executive to expropriate land for works of public utility.

According to BENEVOLO, it is with Haussmann's plan for Paris that modern city planning is born. Instead of 'laissez-faire', or the absolute control of land on the part of the state in pre-revolutionary France, a compromise is reached. The state limits the complete freedom of the private sector through building and planning regulations and the power of eminent domain, while at the same time carrying out huge public works that service private property. Property owners whose land was expropriated were handsomely compensated. Thus, the respective roles of the public and private sectors were established, making possible the restructuring of several European cities – including Paris, Vienna, Barcelona and Florence – during the second half of the nineteenth century. The new cities reflected the values and interests of the new class in power, the bourgeoisie, through the elimination of SLUMS, the creation of parks and the construction of major public buildings and boulevards, where fashionable society could live, see and be seen, and

shop. But no other city can match the huge scale, comprehensiveness and splendour of Haussmann's Paris.

Further reading

Choay, F. (1969) *The Modern City: Planning in the 19th Century*, New York: George Braziller.

NICO CALAVITA

HAYDEN, DOLORES

b. 15 March 1945, New York City

Key works

(1981) *The Grand Domestic Revolution: A History of Feminist Designs for American Homes, Neighborhoods, and Cities*, Cambridge, MA: MIT Press.
(1984) *Redesigning the American Dream: The Future of Housing, Work, and Family Life*, New York: Norton (rev. edn 2002).

Straddling the boundary between cultural history, architecture and physical planning, Dolores Hayden, URBAN historian and architect, has made ground-breaking contributions to the understanding of the social importance of urban space and to the history of the built environment in the United States.

After training in architecture at Harvard University, Hayden achieved recognition for her first book on the history of utopian communities, and then for her work on women's history. *The Grand Domestic Revolution* reclaimed a nineteenth- and early twentieth-century tradition of 'material feminism' – women's cooperative experiments that redefined household work and neighbourhood space. Her classic 1980 article, 'What would a non-sexist city be like?', and later book, *Redesigning the American Dream*, extended these historical precedents into prescriptions for designing and redeveloping contemporary urban areas. At a key period in the interdisciplinary study of gender and space, Hayden moved beyond the prevalent scholarship of critique to provide hopeful examples of URBAN FORMS better suited to new household types (see GENDERED SPACES). At a time when progressive planning had moved away from a concern with physical planning and URBAN DESIGN, Hayden staked out a courageous feminist position in physical planning and architecture that has been

inspiring to many practitioners, particularly in the United States and Europe.

In the 1980s, Hayden became interested in ethnic diversity and public memory. Hayden put her ideas into practice in the fields of HISTORIC PRESERVATION and PUBLIC ART, in 1982 founding the non-profit The Power of Place in Los Angeles and publishing a 1995 book on the experience. The Power of Place developed a number of innovative projects to preserve and commemorate women's, ethnic and labour history, including walking tours, public history workshops and public art installations.

Hayden was important in establishing a feminist URBAN PLANNING curriculum at UCLA, where she taught from 1978 to 1991 before moving to Yale University to teach in architecture and American studies. The UCLA curriculum was pioneering in a national sense and was publicized through a series of conferences that engaged a broad group of scholars and activists.

Hayden's writing, activism and practice have won her significant recognition as a public intellectual. In 1987, the American Planning Association awarded Hayden the Diana Donald Award for contributions to women and planning. She has also won fellowships from the Guggenheim and Rockefeller Foundations, book awards from the American Library Association, the Association of Collegiate Schools of Planning and the National Endowment for the Arts, and the Radcliffe Graduate Medal.

Further reading

Hayden, D. (1976) *Seven American Utopias: The Architecture of Communitarian Socialism, 1790–1975*, Cambridge, MA: MIT Press.
—— (1995) *The Power of Place: Urban Landscapes as Public History*, Cambridge, MA: MIT Press.

ANN FORSYTH

HEALTHY CITY

A healthy city is a municipality that continually improves on both a physical and a social level until environmental and pathological conditions are reached which establish an acceptable morbidity rate for the population. Its principal aim is to reach a high standard of living, a goal that goes beyond mere sanitation, through the collaboration of local government, citizens and various social sectors. At

the beginning of the third millennium, more than half of the world's population lived in cities. POVERTY and deteriorating levels of quality of life in inner cities have made certain URBAN areas insalubrious both physically and socially. Various factors, such as a complex social composition, a difference in quality of living conditions among various sectors, population density and the concentration of diverse activities within cities, have contributed to the degradation of the environment. The term 'healthy city' can be associated, on the one hand, with an increased awareness of ecology and threats to urban environmental well-being and, on the other, with a series of movements or groups supported by international organizations which seek to create healthy urban areas from an integrated perspective.

Historically, antecedents can be found in the early era of public health, the so-called sanitation era (1840–80), in the region where the INDUSTRIAL REVOLUTION began: Great Britain. The relocation of large segments of the rural population to INDUSTRIAL CITIES caused huge increases in population in these urban centres, which contributed to their transformation. The immigrant working class lived and worked in squalid conditions, which gave way to an alarming increase in mortality rates and the spread of infectious diseases. The concept of public hygiene was developed as a reaction to the threat that INDUSTRIALIZATION and rapid URBANIZATION posed to public health. In 1844, in Exeter, England, the first Association for Health in the Cities was formed. Its purpose was to inform the public of the results of studies done which analysed the deplorable living conditions of much of the working population, and to put pressure on the government to introduce legislative reforms.

During the new public health era, initiated in 1974, it was demonstrated that an improvement in the general health of the population was not so much a factor of improved treatments, but rather was produced by an investment in prevention and therapy, and by the acknowledgement of a need for a change in environmental politics. Diverse initiatives by the World Health Organization (WHO) in the realm of ecology gave a new impulse to the arena of public health. The Healthy Cities Project arose out of this debate. Although its basic principles were established in 1984 with the celebration of the Congress of Toronto, the idea was further developed by the European Regional Office of the WHO (1986) as an attempt to develop programmes

and legislation that promoted health and improved standards of living on a local level, and whose goal was 'Health for all in the year 2000'. By the beginning of the year 2000, there were more than 3,000 municipalities worldwide participating in the movement. In Europe, more than twenty-five countries have formed national networks. In some areas, this movement joined ranks with other initiatives relevant to the urban environment such as Sustainable Cities and Local Agenda 21.

Further reading

Ashton, J., Grey, P. and Barnard, K. (1986) 'Healthy Cities: WHO's new public health initiative', *Health Promotion* 1(3): 319–24.
Hancock, T. and Duhl, L. (1988) *Promoting Health in the Urban Context*, Copenhagen: FADL Publishers.

JESÚS M. GONZÁLEZ PÉREZ
RUBÉN C. LOIS GONZÁLEZ

HETEROTOPIA

The term 'heterotopia' is derived from a combination of 'hetero', meaning other or different, and 'topo', meaning place.

Ideas about UTOPIA (the good place/no place) and counter-spaces have antecedents in long-standing debates about whether the world is exactly as we perceive it, or is the way it appears to our reason, alongside concerns about how the world should or could be. Immanuel Kant (1724–1804), for example, distinguished between things in and of themselves and things as they appear to be. Frederick Nietzsche (1844–1900) maintained that when people explain the world around them, it is rather their interpretation of it that is presented, because for him, there are no objective truths. Understanding time and space to be fragmented, discontinuous and functions of position was also characteristic of Michel FOUCAULT.

'Of other spaces' (1986) summarizes Foucault's idea about heterotopias – other/counter-spaces, spaces of difference. Asserting that modern existence is about simultaneity, juxtaposition, the near and far, Foucault also maintained that oppositions such as discipline and deviance, URBAN and rural, or public and private govern space, whose unsettling constituted his (incomplete) development of the idea, which has six basic characteristics.

First, all CULTURES constitute heterotopias: crisis spaces, the spaces of the sacred and profane, menstrual huts, military colleges, honeymoon suites, boarding schools. Secondly, heterotopias have precise functions as other-spaces that nevertheless are neither stable nor continuous. For Foucault, then, the displacement of urban cemeteries to suburban sites represents a destabilization of the understanding of city life and the PRIVATIZATION of death. Thirdly, heterotopias put side by side in any one PLACE several heterogeneous spaces. Libraries and Oriental gardens exemplify this idea. The city, too, illustrates this unpredictable adjacency to which Foucault alludes: for example, a group of English architects, Archigram, has designed numerous virtual WALKING CITIES, plug-in cities, instant cities and inflatable cities, responding to urban spaces as nodes of communication and consumption. Fourthly, heterotopias are constituted through other-times; museums, libraries and archives being places where time accretes; city streets during carnivals being places where time is precarious. Fifthly, heterotopias are inside power/knowledge: counter-spaces appear accessible but are subject to regulation or exclusion. SHOPPING MALLS provide one instance of such disciplinary technologies. Mass private spaces furnish a series of civic functions but exclude or marginalize the uncivil – youth, vagrants, the impoverished. Finally, heterotopias are relational – creating spaces of illusion and hyper-reality. In this regard, heterotopias paradoxically represent limit and tension, and were used by Foucault to illustrate the impossibility of absolutely differentiated and contested spaces.

Further reading

Foucault, M. (1986) 'Of other spaces', *Diacritics* 16: 22–7.
Soja, E. (1996) *Thirdspace: Journeys to Los Angeles and Other Real-and-Imagined Places*, Malden, MA: Blackwell, pp. 145–63.

ELAINE STRATFORD

HIGH-OCCUPANCY VEHICLE LANES

High-occupancy vehicle (HOV) lanes have been conceived to maximize the relative person-carrying capacity of the roadway rather than the absolute number of vehicles, by providing priority movement of high-occupancy vehicles. They represent a metropolitan-area strategy for improving road network efficiency by increasing mobility, reducing traffic CONGESTION in peak hours, absorbing increasing traffic and scaling down infrastructure development. They help to reduce environmental impact by reducing fuel consumption and pollutant emissions. Potential users of such lanes include car pools, van pools, buses and emergency vehicles, while some cities also include motorcycles and alternative or hybrid fuel vehicles, regardless of the number of occupants. HOV lanes are able to significantly reduce journey times, and provide reliability and predictability for the user, particularly for bus schedules. Non-users also benefit from HOV lanes diverting traffic from the general-purpose lanes. The number of entry and exit points and their location in the HOV lane should take into account both flexibility in lane use and the possibility of producing bottlenecks. HOV lanes are developed by restricting certain roadway lanes, in or adjacent to the median.

Different types of lanes are identified according to the criteria used: temporary or permanent with concrete barriers; two-directional or reversible; and exclusive, concurrent or contraflow lanes working in peak periods. High-occupancy toll lanes are a class of HOV lanes where access is given to lower-occupancy vehicles for a toll.

Further reading

National Cooperative Highway Research Program (1998) *HOV Systems Manual*, Report 414, Washington, DC: National Research Council.

URBANO FRA PALEO

HIGH-RISE HOUSING ESTATES

All over the world massive numbers of inhabitants of cities and URBAN regions live in high-rise housing estates built during the second half of the last century. It is estimated that in Europe, excluding the former USSR, about 41 million people live in these estates, often in prefabricated dwellings. In Eastern Europe, these estates mainly consist of high-rise structures. In Western Europe, they often – but not always – consist of a greater mix of properties. But there also, high-rise housing dominates at least visually.

These estates were once seen as a complete break with the past. In contrast to many pre-war housing areas, the new estates were carefully planned. Whereas forms, architecture, land use and other features vary widely between and even within countries and urban areas, there were common social aims. These aims related to integrating people into modernity and into the social structures of industrialism, and were evident in Western Europe as well as in the state-bureaucracies of Eastern Europe. The main aim was to produce healthy mass housing in a 'sound' green environment, linked by roads and public transport to the CENTRAL CITIES, employment opportunities and all the commercial and cultural amenities of a fully employed society. Because the new dwellings were generally of a much higher quality than the dwellings in older areas, many households were eager to move to these areas. Also, the green environment and the quiet locations attracted many different kinds of households.

However, in many cases, these areas started to decline in different ways and much sooner than expected. Physical decline emerged, because the materials used were often not of a high quality, construction was sometimes poor, and repair and maintenance sometimes neglected. Social decline started because, for the 'winners' of the emerging post-industrial societies, more attractive housing opportunities became available elsewhere. In the West, the process of 'pulling out' by those who could afford to move, leaving dwellings vacant for those without alternatives, started in the mid-1980s or earlier and was associated with rising affluence and the heavy fiscal support for the home-owning middle classes. In Eastern Europe, a comparable process has commenced as these societies are experiencing a rigid social differentiation. As a consequence, the former 'decent and modern home' starts to lose value and appeal in comparison with re-emerging inner-city residential areas and new suburban environments. There is a clear danger that many of the large, post-war housing estates in Western and Eastern Europe will be the urban problem areas of the near future. Many of them have reached this questionable position already, housing large numbers of low-income households, unemployed people and households from ethnic minorities. These estates are increasingly associated with CRIME and social exclusion.

Further reading

Hall, P. (1997) 'Regeneration policies for peripheral housing estates: inward- and outward-looking approaches', *Urban Studies* 34(5): 873–90.
Power, A. (1997) *Estates on the Edge: The Social Consequences of Mass Housing in Europe*, London: MacMillan.

RONALD VAN KEMPEN

HINTERLAND

The term 'hinterland', which literally means back country (hinter = behind, land = land), was originally associated with the area of a port where materials for export and import are stored and shipped. The use of the word thereafter expanded to include any area under the influence of a particular human settlement. Thus, if an outlying area is strongly connected to a core settlement, for example by providing necessary energy and materials, the area is considered as hinterland.

The recognition of hinterland is important for the understanding of the nature of human settlement. Although a human COMMUNITY, such as a CITY, has a defined physical boundary, its existence is dependent upon the fundamental necessities provided by the areas outside of the boundary. Thus, an aggregated human settlement and its hinterland are often bound by their production–consumption relationship. Hinterland may also serve as an area for a core settlement's waste, whether that be waste material (i.e. in the form of pollution) or waste energy. The size and location of hinterland vary, depending on what it is contrasted to: it can be countryside as opposed to city, colonized country as opposed to imperial core regions, and an atmospheric layer as opposed to terrestrial human activities.

The relationship between settlement and hinterland has been characterized by the dominance of the former over the latter. Hinterland has long been regarded as the area of dependency in relation to the core settlement, with limited innovation and political competence, while the core settlement has been seen as the place of influence, controlling the development of the hinterland. The view towards hinterland is, however, changing, especially through the discussion of SUSTAINABLE URBAN DEVELOPMENT, which has evolved out of the reflection that URBAN demand is increasingly having strong impacts on the hinterland. It is widely

noted, for example, that North American residents require immense amounts of ecological resources, which are not necessarily coming from North America itself. The requirements of a big human community may thus have a significant impact on the resources originally identified with the different regions. The recognition of hinterland thus reminds us of the need to better understand the demands of a human settlement, and to minimize its impact on the accompanying areas.

The notion of a human community claiming physical areas outside its boundary is not new. Georg Borgstrom coined the term 'ghost acreage' in his 1965 book *The Hungry Planet* to argue that a country often requires extensive outside areas to support the home population. The idea is further strengthened by the concept of 'ecological footprint', which suggests that resources from outside spaces are needed to support a defined human population and its standard of living.

Further reading

Borgstrom, G. (1965) *The Hungry Planet: The Modern World at the Edge of Famine*, New York: Macmillan.
Wackernagel, M. and Rees, W. (1996) *Our Ecological Footprint: Reducing Human Impact on the Earth*, Gabriola Island: New Society Publishers.

AKI SUWA

HISTORIC CITY

A historic city represents a sector of the metropolitan area of great historic and symbolic value, originating in the pre-industrial era. Most of its important development occurred between the Middle Ages (ninth century) and the beginning of the transformations produced by the INDUSTRIAL REVOLUTION in the nineteenth and twentieth centuries. It is the URBAN area that was first constructed and inhabited in a city with important posterior growth. Although widely varying in origin, form and dimensions, the historic city makes up the nucleus of a modern-day city. Generally, it is identified as the space situated within city walls, although these may have been torn down in many places, especially during the nineteenth century. This was because of plans approved on interior reforms (London, Brussels, Paris, etc.) and urban expansion (Barcelona, Madrid, etc.) that were a consequence of demographic growth and the accumulation of wealth by the BOURGEOIS class.

The decline of the antique, pre-industrial city coincided with the technological advances associated with the Industrial Revolution, fundamentally with the improvements in transportation (railroads and street cars) and with the development of new hygienic and sanitation measures (networks supplying clean drinking water and sewage systems). These measures improved living conditions and reduced the high mortality rate in historic towns.

Traditionally, the historic city was synonymous with the antique and monumental areas of a metropolis. Its architectural richness converted it into the area most representative of a particular city, acting as a focal point for symbolic, artistic and economic values. This classic definition, however, has been modified over time. Because of pioneering work done in Italy, the habit of equating historic city with pre-industrial city was changed, and the definition of the term 'historic city' was altered to exclude those sectors of the metropolis which although situated within the old city walls did not contain characteristics pertaining to historic cities, and to include other areas that were developed in the nineteenth and twentieth centuries, yet contained sufficient elements to be categorized as part of the historic city. Within this new, broader definition, areas of great historic and architectural value which arose in the industrial or post-industrial era are included. Examples of note are the working-class SUBURBS built during the Industrial Revolution, certain neighbourhoods of the GARDEN CITIES, areas of social housing constructed during the early decades of the twentieth century, and even some of the characteristic 'ensanche', or area of urban expansion of Spanish cities in the second half of the nineteenth century.

By the year 2000, the historic city occupied a reduced part of the urbanized space because of expansion into peripheral areas. In spite of multiple transformations, many of them a result of speculation during the last 150 years (filling in of empty spaces within the old city walls, modification of the original layout through urban renewal projects, etc.), the historic city is easily differentiated from the contemporary city because of its irregular layout, narrow, winding streets, enclosed blocks and patios, etc.

Due to the fact that the historic city is both the oldest and the most complex and difficult area to put into order of any metropolis, it is the part where the greatest number of reforms have been

carried out. Literature on URBAN PLANNING often refers to the potential, the weaknesses and the threats to the integrity of the historic city. The difficulties encountered in adapting historic cities to twentieth-century lifestyles because of morphological, functional or traffic-related problems contributed to progressive physical and social decline. Early responses from the public and private sectors included drastic political measures for urban renewal that were incompatible with the conservation and recuperation of the original structures. This resulted in an increase in internal social SEGREGATION that contributed to the degradation of various sectors (an aging population, a decrease in the number of inhabitants, the creation of pockets of a marginal social class, physical deterioration of buildings, scarce economic dynamism, etc.) and to the transformation of these symbolic areas on both a social and aesthetic level. Later, after legal antecedents and proposals for integral rehabilitation were established in various European countries (the Housing Act of 1969 in Great Britain, active conservation and restoration in Bologna in 1969, the National Agency for the Rehabilitation of the Habitat in France in 1972), the objectives and methodologies for integral rehabilitation (physical and social) of historic cities spread throughout Europe to practically everywhere in the world. An essential part of the methodology for integral rehabilitation is the willingness to pass necessary urban legislation that includes general plans and specific tools for the recuperation of the historic city. Some examples are the plans for safekeeping, recuperation and interior reform that are in place in Portugal, Denmark, France, Italy, Belgium, Germany and Spain. Sensitivity towards the recuperation of historic patrimony and the rehabilitation of the historic city is a manifestation of the level of development and quality of life reached by a society; thus, the application of the politics of integral rehabilitation is common in almost all of the developed world.

Further reading

Council of European Municipalities and Regions (1994) *Guidelines for the Realization of Strategic Development Plans in Medium-sized Cities*, Lisbon: Oficina de Arquitectura, Lda.

Morris, A.E.J. (1974) *History of Urban Form: Before the Industrial Revolution*, London: George Godwin Ltd.

Rowntree, L.B. and Conkey, M.W. (1980) 'Symbolisms and the cultural landscape', *Annals of the Association of American Geographers* 70(4): 459–74.

Troitiño, M.Á. (1992) *Cascos antiguos y centros históricos: problemas, políticas y dinámicas urbanas*, Madrid: Ministerio de Obras Públicas y Transportes.

JESÚS M. GONZÁLEZ PÉREZ
RUBÉN C. LOIS GONZÁLEZ

HISTORIC DISTRICT

A historic district is a mechanism for designating as historically significant a distinct geographical portion of an URBAN or rural region. It is both an intellectual construct and a term recognized and defined by law. A district is first an acknowledgement that the historic character of an area is derived not just from the qualities of individual properties, but from the way these properties developed over time, and are related to each other formally, functionally, socially, and through the area's roads and other infrastructure, and natural conditions, such as climate and topography. In this sense, the district was the next step after the designation of individual sites in a long, continuing process within the historic preservation movement of identifying and interpreting ever larger scales of terrain in an attempt to understand a culture's or region's history and evolution. Secondly, the district is a way of protecting a 'SENSE OF PLACE', by developing legal mechanisms that regulate what can be done to both sites and their contexts. A historic district is often part of a larger urban setting, but it can also be part or all of a small town, or a rural area with historic agriculture-related properties, or even a physically disconnected series of related structures throughout a region.

Internationally, and in the history of charters and agreements that govern or suggest appropriate preservation practice, various terms are used to describe historic districts. These include 'conservation areas' (England), 'historic centres' and 'historic areas' (World Heritage Convention), 'historic places' (Canada, which includes districts but does not define them), 'old areas' (1975 Declaration of Amsterdam), and 'historic town or urban area' (1987 Washington Charter of the International Council of Monuments and Sites). In most countries, mechanisms to protect districts or historic areas lag considerably behind protections for individual properties.

While the effort that began in 1926 to preserve Colonial Williamsburg might be viewed as the forerunner of historic districts, the first official district in the United States was created in 1931 in Charleston, South Carolina. Next, the Vieux Carre Commission was created in 1937 to protect and preserve the French Quarter in New Orleans. In 1939, San Antonio, Texas, adopted an ordinance to protect La Villita, which was billed as the original Mexican village marketplace. In fact, it was an overcrowded and run-down residential neighbourhood that was cleared of its original inhabitants and turned into a tourist attraction, partially modelled after a similar effort undertaken at Olvera Street in Los Angeles a few years earlier. These early designations, coming during the Great Depression, were indeed aimed as much at tourism and economic development as at protection. Then, and too frequently since, internationally, the negative impact of designation on local inhabitants was either not a concern or an actual goal of the project. This helped to establish a pattern of GENTRIFICATION that has been associated with designation for many years. Over the last few decades, particularly as historic district designations have been increasingly applied to working-class residential and industrial areas, some communities have attempted to implement economic development strategies and other measures to help against involuntary dislocations. Often, these same measures are also intended to try to maintain the original 'gritty' character of the district against an influx of coffee shops, art galleries and other amenities that do not necessarily cater to the needs of the long-time inhabitants of the area. These anti-gentrification strategies have not been widely adopted, however, and their success is yet unclear, despite being called for as early as the Amsterdam Declaration at the 1975 Congress on the European Architectural Heritage.

Legal definitions of historic districts are developed at national, state and local levels. (International charters also recommend policies for districts, as seen above.) In the United States, the National Park Service defines a district as a significant concentration, linkage or continuity of sites, buildings, structures or objects united historically or aesthetically by plan or physical development. Through the National Register programme, the Park Service then lists four specific ways in which a district can become historically significant. A district is considered to be a single 'property' on the Register, so these criteria are also the ones applied to individual sites. In brief, they are (a) association with significant events, (b) association with significant people, (c) architecturally or aesthetically distinctive or significant, and (d) possessing information important in prehistory or history.

Besides significance, a district must also have integrity. Integrity is a term that describes the relative amount of historic fabric left in a site. Generally, in order for a district to maintain its historic designation in the United States, at least 40 per cent of its buildings must be considered 'contributing' structures, or individually eligible for the National Register. Contributing structures are those that help define the historic character of the district.

States and cities adopt their own definitions of historic districts to reflect the particular nature of the resources being considered, and to conform to state constitutions and local land use and other relevant laws. Thus, historic districts can be defined under ZONING, taxation and preservation ordinances. The district designation is the first step in enabling the government entity to regulate construction in the area, and to provide special economic incentives. Typically, a historic district ordinance will include the following:

1   a process for surveying or inventorying an area for the purpose of creating a district, along with ways to establish boundaries and develop thematic context statements;
2   a procedure for officially designating a district, with an appeals process;
3   a list of the types of activity to be regulated in the district (which can run the gamut from parking to signage to demolition and new construction);
4   a commission and/or design review board to evaluate proposed projects in the district; and
5   a process for one or more delays of demolition for contributing or eligible buildings.

In addition, many jurisdictions also provide property tax or income tax relief for property owners within a district, or for individual projects. There may also be grant or loan programmes available to support desirable projects; and fines or other disincentives for work that damages historic fabric or the area's character.

In the past few years, alternatives to historic district designation have arisen in response to

communities seeking the protections and economic incentives available under a preservation ordinance, but whose cultural resources do not meet the criteria established by the National Park Service for historic significance. One such example is the 'conservation districts' being established in San Antonio. These allow a neighbourhood to identify its own essential character-defining features, and work with the city on specialized regulation and review. While historic designations and regulations have been determined to be within the legitimate zoning powers of government by the US Supreme Court, the legality of conservation districts is yet to be tested.

Further reading

US National Park Service, National Register website: www.cr.nps.gov/places.htm

JEFFREY M. CHUSID

HISTORIC PRESERVATION

Historic preservation can be defined as a series of proceedings designed to maintain the oldest and most monumental elements of a city. Historic preservation is a philosophical concept that became popular in the twentieth century, which maintains that cities should be obligated to protect their patrimonial legacy. Cities are, by definition, products of centuries of development. Tourists and citizens are shown the origins and evolution of an area through the preservation of its oldest parts. It is generally thought that buildings and historic centres should be the reference points that create the public image of a metropolis. In a certain sense, the idea of the permanence of isolated MONUMENTS can be transferred to all of the older sectors of a city. Rome preserved its imperial past as an attraction and a sign of its identity; in Jerusalem, a recreation of its old town highlighting important religious symbols reaffirmed its role as a centre for the Jewish, Christian and Islamic faiths; in Seoul, the reconstruction of the Korean Imperial Palace is an exaltation to the history of the nation before the Japanese domination.

Over time, in carrying out the historic preservation of cities, an important shift of focus occurred. Traditional interest in preserving isolated monuments gave way to endeavours centred on preserving entire historic centres. Note that in the

Romantic tradition originating in France and England in the beginning and middle of the nineteenth century, the preservation of isolated monuments of Egypt and Greece was favoured. These ideas spread throughout Europe, where an impressive movement to restore cathedrals got underway, such as the one in France led by Viollet-le-Duc. Declarations were made to protect monuments and projects were undertaken to prevent deterioration in Italy, Germany, Spain and the Netherlands during much of the first half of the twentieth century. However, after the massive destruction of some European cities during the Second World War (Dresden, Coventry, Caen, Warsaw, Krakow), the idea to preserve larger historic areas came into being. In many instances, it was impractical to reconstruct the entire pre-existing metropolis; however, emblematic streets, plazas and points of special aesthetic interest were recuperated. The shift of focus from isolated monument to monument integrated in a harmonic and coherent setting was an important change. The reconstructed areas function as a sort of history lesson, preserving both monumental buildings erected by the rulers of the epoch manifesting their privileged status (cathedrals, palaces, commerce centres) as well as common buildings or structures where the majority of the population lived in houses or managed small businesses.

In fact, many cities (Florence, Venice, Rouen, Bath, Bruges, Santiago de Compostela, Santa Fe, Kyoto) have promoted visits to their historic centres, rather than to just one element within it, in their tourist campaigns. The URBAN image that they intend to portray is made up of many streets and plazas, not just focused on one monument in particular. As part of his leisure-time pursuits, the urban tourist can visit a historic centre, wander through its streets and enter its buildings to shop or eat. UNESCO has recognized this trend and generally gives the title of World Heritage City to entire historic urban areas, and only rarely just to isolated monuments within a city.

An interesting question arises in reference to the distinct readings of history that occur during the process of preserving historic urban areas. As with monuments, historic centres are interpreted based on certain momentous events and architectural highlights of their past, an interpretation which tends to devalue the importance of other periods of history. Thus, in Bruges, Reims and Santiago de Compostela, conservation efforts have focused on

the medieval epoch as a golden age, although the majority of their antique structures date from the sixteenth to the nineteenth centuries. In Paris and Vienna the aristocratic architecture from the nineteenth century is glorified, in Florence and Venice the Renaissance takes precedence, and in some cities in New England the colonial period of history is highlighted. However, the history of any given place is generally more complex than what is revealed through the efforts of historic preservation.

On occasion, historic preservation projects attempt to maintain urban centres without introducing any changes, while at other times history is reinterpreted from a present viewpoint. Historic preservation has been greatly influenced by artistic preservation. Thus, in numerous emblematic buildings, an attempt is made to recuperate not just the structure, but also the decoration and system of illumination. On the other hand, as an exact reconstruction of large urban areas would be impossible in practical terms, the past is often revealed through the preferences and choices made at the time of renovation. Sometimes, the decision is made to introduce even contemporary architecture into a historic centre, the underlying idea being that all epochs of history, including the present one, should leave their architectural legacy.

Further reading

Frampton, K. (1992) *Modern Architecture: A Critical History*, London: Thames & Hudson.

Merlin, P. and Choay, F. (eds) (1993) *Dictionnaire de l'urbanisme et de l'aménagement*, Paris: PUF.

Rossi, A. (1971) *L'architettura della città*, Milano: Francoangelli.

Zoido, F., de la Vega, S., Morales, G., Mas, R. and Lois, R. (2000) *Diccionario de geografía urbana, urbanismo y ordenación del territorio*, Barcelona: Ariel.

JESÚS M. GONZÁLEZ PÉREZ
RUBÉN C. LOIS GONZÁLEZ

HOME RULE

In the United States, under a constitutional doctrine known as Dillon's Rule, cities are legally seen as the administrative subunits or 'creatures' of the states, possessing only those powers that a state chooses to confer on them. A city charter, awarded by the state, defines the limits to a municipality's powers. *Home rule* refers to provisions that give cities and counties greater leeway to undertake various actions of their own without first having to obtain express state permission. Home rule cities and counties are generally free to enact laws of their own, just so long as they do not contradict existing state statutes.

The idea of home rule is to give municipalities greater control over their own affairs. Yet the extent of the powers devolved under home rule is still ultimately up to each state. According to Dillon's Rule, as a state grants, it may abridge or amend. Nearly all the American states permit some extent of local home rule, yet the degree of local autonomy varies greatly from state to state.

Despite home rule, state constitutional provisions and statutes continue to define just what a municipality may or may not tax and the maximum amounts that a locality may tax or borrow. Home rule has also been abridged when strong interest groups have pressed state legislatures to pre-empt or bar the passage of local gun control and tobacco regulation ordinances. In states like Michigan, state legislatures have also imposed governance changes on troubled local school districts.

Further reading

Advisory Commission on Intergovernmental Relations (ACIR) (1993) *State Laws Governing Local Government Structure and Administration*, Washington, DC: ACIR.

MYRON A. LEVINE

HOMELESSNESS

Homelessness has grown rapidly in advanced industrial nations over the past two decades. Definitions of homelessness vary, though all are based on the lack of decent, safe, permanent shelter. People living in emergency shelters or transitional housing, in places not designed for permanent human habitation (such as cars, tents or outbuildings), in addition to those living on the street or in public places (see also STREET PEOPLE), are defined as homeless. There is disagreement whether people living temporarily with friends or family should be counted as homeless, and over how to define the near-homeless.

The causes of homelessness are both structural and individual. The primary structural cause is the high cost of shelter. Some argue this results from higher building standards, or from speculative increases in prices. Some conservatives argue that high housing costs result from regulations (such as rent control) that increase the cost of housing and thus discourage development of affordable units. On the other side, social progressives argue that the high cost of housing results from the fact that it is treated as a commodity rather than as a basic right. A second cause is the loss of low-cost housing options, such as rooming houses or residential hotels, to housing abandonment and redevelopment driven by GENTRIFICATION. A third important cause has been the deinstitutionalization of people with mental illness or disabilities. Deinstitutionalization was designed to return patients to the community, but many communities were not prepared to meet their special needs and released patients soon became homeless.

Understanding the causes of homelessness is complicated by the diversity in the population, and thus the diversity of individual causes. Although single men have traditionally been the most visible, women and children (especially unaccompanied youth and single-parent households) make up a significant share of homeless people. Mental illness, including drug or alcohol abuse, is the primary cause for some. For others, family break-up, through domestic violence, youth runaways or divorce or separation, resulted in homelessness. Low wages and volatile employment result in homelessness for others. The lack of a social safety net means that minor crises soon escalate into loss of housing.

Counting the homeless and identifying their special needs raise difficult methodological issues, and most cities have only estimates of their homeless population. Estimates are complicated because the incidence of homelessness varies for individuals, from chronic protracted periods to short recurring or isolated periods. Most counts rely on service providers, although a few have attempted a census of homeless people. Developing permanent affordable housing, ensuring access to supportive services and protecting the rights of homeless people are difficult long-term policy issues that are often eclipsed by the need to provide shelter. Crafting effective solutions is difficult given the diversity of the population.

Further reading

Burt, M.B. (2001) 'Homeless families, singles and others: findings from the 1996 National Survey of Homeless Assistance Providers', *Housing Policy Debate* 12(4): 737–80.

Hopper, K. (1991) 'Homelessness old and new: the matter of definition', *Housing Policy Debate* 2(3): 757–813.

HEATHER MacDONALD

HOMEOWNERS' ASSOCIATION

Homeowners' associations in the United States are often seen as the modern-day incarnation of the Jeffersonian grass-roots ideal. Property owners in a neighbourhood form such an association in order to press for local policies that will protect their interests and the value of their property. Homeowners' associations are especially active in URBAN PLANNING, ZONING and land use, decisions that will affect the pace of growth, the quality of life, the level of taxation and the value of land in a community. Where families have bought homes on the basis of the reputation of the local school system, a homeowners' association may also be very involved in certain school matters. Increasingly, homeowners' associations are also concerned with the level of protective services provided by the public police and private security forces.

Homeowners' associations can be found in all parts of the United States. Such associations can be formed to represent the owners of town homes and the CONDOMINIUM units in high-rise buildings, not just the owners of single-family suburban homes. By the latter part of the twentieth century, homeowners' associations had increasingly taken to the idea of *common interest development*, where the homeowners' association itself is given responsibility for certain managerial services, for example the planting of trees on a development's common property, or the setting of a maintenance fee to be levied on all condominium owners in a development.

Homeowners' associations serve as a counterweight to the influence of corporate elites, ensuring that residential needs are not slighted in the pursuit of new economic development. Yet homeowners' associations often fall short of democratic ideals. Such residential organizations are often defensive and exclusive, seeking to protect the interests of established homeowners against newcomers,

outsiders and others who demand the heightened provision of public services. In cities like Detroit and Boston in the 1950s and 1960s, white home-owners' associations were formed to battle racial integration and to preserve property values. In more recent decades, residential associations have offered their members the self-service and pro-tective advantages of 'privatopias' and GATED COMMUNITIES.

As citizen organizations, homeowners' associa-tions are often plagued by the inefficiencies and ineffectiveness of part-time, amateur government. Few residential associations choose to assume the burden of paying for full-time professio-nals to administer the association's business. The association's elected office holders and a few citizen-volunteers often exert tremendous power in determining the association's day-to-day actions. But unpaid and untrained, these citizen-administrators often lack the time and skill neces-sary to assume important administrative functions. At times, a residential association will find it necessary to hire a professional manager to run key programmes and to direct, train and supervise the citizen-volunteers.

Further reading

Blakely, E. and Snyder, G.M. (1997) *Fortress America: Gated Communities in the United States*, Washington, DC: The Brookings Institution Press.
McKenzie, E. (1994) *Privatopia: Homeowner Associations and the Rise of Residential Private Government*, New Haven, CT: Yale University Press.

BERNARD H. ROSS
MYRON A. LEVINE

HOMOPHOBIA

Homophobia means literally fear (phobia) of sameness (homo). As a noun, 'homophobia' is used to characterize antagonistic behaviours and attitudes that target gay men, lesbians and, some-times, bisexuals. It can also connote such beha-viour towards transgendered individuals, although 'transphobia' has become a more accepted term in this regard. As an adjective, someone is homo-phobic when they exhibit these traits, namely fear, often expressed as hatred, of people who are attracted, or thought to be attracted, to people of the same sex. Some would argue that society in general is homophobic, given its emphasis on

what has been called compulsory heterosexuality, along with accompanying negative attitudes towards anyone who rejects these norms. This over-emphasis on heterosexuality is also called heterosexism.

Homophobia is manifested in cities as violence towards people and their homes, businesses and practices. It can be direct, such as physical or verbal assault, or it can be more subtly expressed in URBAN plans, policies, programmes and processes that discriminate against such things as spatial concentrations of gay men or lesbians, gay pride activities and proclamations, and gay bars.

While much homophobia is external, i.e. origi-nating from outside of a homosexual person, internalized homophobia is also prevalent. Inter-nalized homophobia comes from within a gay man, lesbian or bisexual and can be manifested as self-hatred or lack of acceptance regarding their, or their friends', sexuality. Internalized homophobia is generally thought to stem from many of the same forces as external homophobia: societal practices that frown upon, and actively discourage, homosexuality.

Further reading

Pharr, S. (1988) *Homophobia: A Weapon of Sexism*, Inverness, CA: Chardon Press.
Weinberg, G. (1973) *Society and the Healthy Homo-sexual*, New York: Anchor.

SUE HENDLER

HOUSING DISCRIMINATION

Housing discrimination is variously defined in the literature to mean the disparate treatment of a person on the housing market based on group characteristics or based on the place where a per-son lives. The latter is known as REDLINING in housing and credit markets.

Some definitions emphasize discrimination as intentional behaviour: to refuse to sell, to rent or to allocate any housing based on a person's group characteristics. The researchers, however, have widely varying views on the scope of group char-acteristics. Equal treatment of everyone is the starting point of human rights declarations, national constitutions and human rights legisla-tion. Yet many researchers still limit their list of group characteristics to race and/or ethnicity,

while others include religion, sex, national origin, disability and household type as elements defining group characteristics. Some researchers add 'income', or 'source of income', as a category on which discriminatory practices might be based. According to them, the percentage of income to be devoted to housing is misleading in individual cases; accordingly, a criterion based on income is not appropriate and discriminatory.

The researchers also have varying perspectives on the definition of intentional behaviour. To some, screening some prospective tenants more strictly than others and applying different application procedures to different groups of people fall within this category. In many countries, laws or regulations prohibit such practices. To others, intentional discrimination has a wider context, as the discriminatory behaviour of landlords or real estate agents may sometimes be reflected in market transactions, especially when persons who individually share some common characteristics can only get access to a certain type of housing at a higher cost. Some researchers also focus on discriminatory consequences of housing practices, which seem neutral on their own but disproportionately harm some persons who share some common characteristics.

Discrimination as intentional behaviour is sometimes called 'social discrimination'. Many researchers make a distinction between institutional and social/intentional discrimination. Institutional discrimination, which is sometimes called 'structural discrimination', results from regulation by legal instruments. Institutional/structural discrimination may occur when the actions of some people on the housing market are restricted by household-related regulations and/or dwelling-related regulations on irrelevant grounds. Institutional/structural discrimination results from circumstances independent of the intentional behaviours of landlords or estate agents.

An example may further clarify the differences between these two types of discrimination. Let us suppose that the eligibility requirement to take part in a housing development or a SQUATTER SETTLEMENT upgrading project is based on income level, assuming that it is neutral to other group characteristics such as ethnicity, race, age, household type or gender. If there are meaningful correlations between income levels and some of the other group characteristics, some groups will be subject to structural discrimination. Accordingly,

for example blacks, minority ethnic groups or female-headed households, whose current levels of income are generally lower, would have been discriminated.

There seems to be some agreement on the causes of housing discrimination. The researchers mostly mention that prejudice and prejudiced attitudes play an important role in shaping the intentional behaviours of many landlords or agents. There are others who claim that prejudice and prejudiced attitudes are also responsible for structural discrimination as human beings also determine the regulations. Prejudice is mostly perceived as a negative and unfavourable attitude towards the members of a particular group. In many cases, prejudice is based on stereotyped beliefs about certain groups, without any scientific testing to prove the validity of the assumptions.

Prejudice, like discrimination, has two sides. On the one side, there are individuals who are discriminated against regardless of whether they recognize certain behaviours or regulations as discriminatory practices. This is particularly true for minority ethnic groups, who lack the necessary language skills, information and understanding about the culture of the mainstream society. On the other side, there are other individuals who, rationalizing their behaviour, may not recognize or may not define their actions as discriminatory practices despite the fact that they consciously or unconsciously discriminate against some other individuals. There are, for example, landlords who discriminate against prospective tenants and base their rationalizing on the expected preferences of existing tenants. The evidence from the literature indicates that sometimes favouritism or rationalism shades into discrimination.

Statistics also play a role in the formation of discrimination, as many individuals tend to base their judgments on averages, disregarding the fact that each person has a specific character. For example, it may be demonstrated that those living in POVERTY neighbourhoods, on average, display deviant behaviours. Those who live in other types of NEIGHBOURHOODS may thus develop a prejudiced attitude towards all those living in poverty neighbourhoods, preparing the ground for intentional discrimination. 'Statistical discrimination' is a term used in the literature to describe such situations.

Housing discrimination has become a very popular theme among politicians and academics

during the last decades despite the fact that intentional and structural discrimination have always disadvantaged some groups within the borders of countries. In the city of Babylon, at least 5,000 years ago, the inner quarters were reserved for those in power. In many MEDIEVAL CITIES, the city centres were similarly inaccessible by the poor. Housing discrimination was particularly visible in many of the COLONIAL CITIES, and its imprint can still be seen in and around the HISTORIC CITY centres developed by the rulers. In much of colonial Africa and Asia, the Europeans often lived in a GARDEN CITY-type residential area surrounded by a 'cordon sanitare' (a GREEN BELT of open space) where no native was allowed to live. Also at that time, Africans experiencing slavery in the United States lived on plantations in one-room log cabins that mostly had one window and one door. The housing of URBAN slaves outside the master's lot was quite rare, and was prohibited by a municipal law.

Indeed, until the recent reforms, the apartheid system in South Africa denied, among other rights, the right of free movement on the basis of race, which kept the majority of the country's population in rural areas living in poverty. The end of the slave trade did not also mean in practice racial equality between blacks and whites. The differential treatment of individuals based on group characteristics still exists despite the Universal Declaration of Human Rights and the growing number of human rights charters and codes that condemn and try to combat discrimination. A substantial body of research has provided empirical evidence on housing discrimination, and illustrates various types of discriminatory practices. The research found private and public landlords who refused to rent their property to certain ethnic and racial groups. There were people who were provided with less information on housing options and/or others who were not shown available housing units. In the United States, there were some estate agents who believed that selling a house to blacks might lower the prices in the neighbourhood. There were also others who deliberately sold to blacks in order to create a chain reaction of sales by whites at knock-down prices.

In Europe, where both the scale and the nature of the problem is quite different, racial and ethnic minorities became visible only after the Second World War, when many ex-colonials settled in the

countries of ex-rulers. Also, those who were recruited as guest workers during the 1960s put their roots down in their new countries between the mid-1970s and early 1980s, and contributed significantly to the formation of ethnic communities.

At the present time, institutional discrimination in the housing market hardly exists throughout Europe. However, this is not to say that there are no European countries applying rules based on nationality. In Austria, some segments of rental housing are still closed to immigrants, and the eligibility criteria for a number of housing allowances are still based on nationality. In Spain, the researchers report that rents paid by immigrants are much higher than those paid by natives for comparable dwellings with respect to size and quality.

Research has shown that during the post-war immigration process in Britain, the black newcomers were forced to settle in declining housing within the city centres, and many of them experienced discrimination. Despite the introduction of anti-discrimination legislation in the 1960s, council housing remained closed to blacks for a long time because of residence qualifications and minimum waiting times on housing lists. An illegal dispersal policy was a barrier when blacks were eligible to live in council housing. Many cities attempted to disperse black groups from the inner-city areas to suburban locations. Indeed, the best-documented example of such a policy comes from Birmingham where the city council illegally limited the number of black households in residential areas between 1969 and 1975. The main rationale expressed was the prejudiced view that integration would be helped by dispersal.

A set ratio policy has long been illegal in Britain; however, a similar quota system is still applied in France. Housing commissions often apply selective criteria based on ethnic background and social position in an attempt to exclude foreigners, and to protect their neighbourhoods from the deterioration they believe the immigrants will cause. The most dilapidated stock of social housing or most remote estates are often reserved for immigrants. Although French housing law prohibits discrimination on the basis of nationality, research indicates that some municipalities have even tried to prevent immigrants from buying a house.

Immigrants also experienced both structural and intentional discrimination in the Netherlands

during the 1970s. In Rotterdam, the city council applied a set ratio policy in the residential areas until the issue was brought to court. In Amsterdam, immigrant access to a number of residential areas was closed following the complaints of existing tenants about the changing population structure of their neighbourhoods. The housing allocation system was changed in 1978, and since then no official barriers have existed preventing immigrants from gaining access to any kind of housing. The change in the housing allocation system has made a positive impact on the housing conditions of immigrants. However, there is some evidence indicating that the negative reactions of the existing Dutch tenants still prevent the immigrants from applying for housing in some residential areas. Sweden has never applied an official discrimination policy to limit the housing options of certain individuals based on their group characteristics. All types of tenures have always been theoretically accessible to immigrants. However, indirect forms of housing discrimination do exist in Sweden. There are municipalities which voted against refugee reception in local referendums, and there are Swedes moving out from residential areas where immigrants are concentrated, meaning that prejudice and prejudiced attitudes still have an impact on the housing choices of immigrants in both Sweden and the Netherlands.

Housing discrimination based on race or ethnicity still receives special attention in the European research community, as in the United States. However, intentional discrimination, as noted before, is not always based on only race and ethnicity. For example, it is mostly the single man who is discriminated against in Muslim societies, where discrimination research is scarce. Many landlords do not rent to single men because families do not like to live close to them. Similarly, single women are rarely accepted in neighbourhoods where they have no relatives. Indeed, prejudiced attitudes prevent many Muslim women from getting divorced, as after separation they would have experienced difficulties in gaining access to any kind of housing. Discrimination based on socio-economic status, although it is not well documented, is often experienced in many developing countries, where the poor are not allowed to enter many of the GATED COMMUNITIES of the rich, which may include PUBLIC SPACES and public roads for all.

Further research is expected to close the gaps in our knowledge on the complex issue of housing discrimination.

Further reading

Galster, G. (1992) 'Research on discrimination in housing and mortgage markets: assessments and future directions', *Housing Policy Debate* 3(2): 385–99.

Özüekren, S. and van Kempen, R. (eds) (1997) *Turks in European Cities: Housing and Urban Segregation*, Utrecht: ERCOMER.

Phillips, D. and Karn, V. (1991) 'Racial segregation in Britain: patterns, processes and policy approaches', in E.D. Huttman, W. Blauw and J. Saltman (eds) *Urban Housing Segregation of Minorities in Western Europe and the United States*, Durham, NC: Duke University Press, pp. 63–91.

Somerville, P. and Steele, A. (eds) (2001) *Race, Housing and Social Exclusion*, London: Jessica Kingsley Publishers.

A. SULE ÖZÜEKREN

HOUSING FOR SPECIFIC NEEDS

In many industrialized countries, the attention paid towards the elderly and people with disabilities has increased because their growing old has become a social and economic problem. Many surveys have recognized that, with the exception of the United States, many people are less attached to the idea of going into specific housing for the elderly and infirm. The majority of them would prefer to stay and live in their own domestic environment. From an economic point of view, this is a solution to be encouraged for the whole of society, guaranteeing at the same time other types of housing, with intensive care, such as nursing homes, for those periods in which the health status of the person requires specific medical treatments or rehabilitation interventions.

Many attempts have been made to reach a 'universal design' – to design solutions capable of creating accessible and usable environments, both inside and outside each building, for everybody, including the needs of the most frail. Starting from the space inside houses, the design process then involves the external areas of residential buildings, buildings open to the public, other types of building, URBAN areas, such as streets, squares, pathways, but also urban equipment; through such a process, the design is able to improve, if not guarantee, external and social contacts between the

inhabitants, who must be assisted in their daily lives.

Accessibility is a principal requirement in this regard, not only for new constructions but also for renewal interventions. Planning housing for the elderly does not necessarily mean designing specific environments, but incorporating certain exigencies into a more general design, in order to build for everybody. In addition, housing could help daily life activities if supported by particular equipment or aids or telecommunications services.

The first new idea in the 1980s was to build housing for the frail (particularly people with physical or cognitive disabilities, and older people with deteriorating health), rather than collective housing. New solutions tried to accommodate the needs of as many different categories of elderly people as possible – from those who required some support, to those who needed intensive care – which meant building not only houses but also the necessary related services. These included social and health services, some delivered directly at home (such as food or meals or nursing assistance), others included in the common rooms (recreational activities, laundry, a warden station, etc.).

The design can be more complex if aimed at people with disorientation problems. The principle of the late 1980s in the northern European countries was that everyone has the right to live independently, regardless of the type of cognitive or physical impairment they may be suffering from. This right must be combined with the support and care that every person might need, which means designing houses with assistance. A small flat, for a single person, with all the elements of independence (living room, kitchen, bedroom, bathroom, closet) and generally completed by a private outdoor space (balcony or garden) is therefore included in a group of similar flats within a larger complex with health assistance and social support. The type and amount of care provided is strictly related to each person's health status.

In other industrialized countries, the debate is oriented more towards the type of assistance, the right dimension, the type of management, etc. Thus, the design guidelines are for small units with single or double bedrooms (6–10) and related support spaces (kitchen, dining room, living and multipurpose room, assisted bathroom) located within a wider complex with other health and social environments. Some aspects of the complex are designed to better match the residents' needs, such as controlled exits so as not to lose a person, or specific outdoor spaces (e.g. Alzheimer's gardens) which improve the comfort and feeling of peace of the inhabitants.

Another design tendency, which came directly from user groups during the 1990s, is the construction of new cooperative housing in which pre-elderly people decide to live, moving from their existing homes, in order to face the third age all together, sharing not only common recreational spaces but also interests and hobbies.

Since the 1980s, new telecommunications systems have made it possible to support certain activities of daily life and improve safety. Telecare and teleaid, for example, were developed and implemented in different countries not only for emergency situations but also to reduce anxiety and complement the human care. The idea evolved in terms of other support systems that can improve the comfort or easy use of home appliances: first as prototypes, with the contribution of manufacturers, municipalities and user associations, then in small experimental home applications. The first level of the intelligent home is devoted to safety and security and the easy use of certain functions (switching lights on/off, opening/closing doors, windows, curtains and blinds, opening the entrance door, etc.). The second level is devoted more to health support and improving preventative rather than reactive care, with aids for memory, the monitoring of some health parameters, the control of wandering for people with disorientation problems, and the monitoring of daily lifestyles.

In trying to extract some technical specifications that could constitute guidelines for future interventions, the focus is on an appropriate design that can take into account the following requirements:

- sensibility, starting from inhabitants' specific needs;
- integration, related to both the specific NEIGHBOURHOOD and the social environment;
- innovation, utilizing the support of new communications and technological tools; and
- the right to choose among different types, from independent living to nursing homes.

Further reading

Canada Mortgage and Housing Corporation (CMHC) (1999) *Housing Options for People with Dementia*, Ottawa: CMHC.

Diaz Moore, K. (1999) *Towards a Language of Assisted Living*, Milwaukee, WI: School of Architecture and Urban Planning, University of Milwaukee.

Fisk, M.J. (2001) 'The implications of smart home technologies', in S.M. Peace and C. Holland (eds) *Inclusive Housing in an Aging Society*, Bristol: Policy Press, pp. 101–24.

The Finnish Environment (1999) *Housing of Older People in the EU Countries*, Helsinki: Ministry of the Environment.

ANNALISA MORINI

HOUSING MARKETS

The classic economic model of the market applies to the housing sector only with significant caveats. Housing is costly in comparison to household income, typically requiring long-term financing for its acquisition. It is bulky and durable. The supply of housing is relatively fixed (inelastic) in the short run. As a result, housing markets are dominated by the stock of existing units. It is generally not consumed, but rather used to generate a flow of services over an extended period of time. Indeed, owner occupied housing may appreciate in value over time, providing an important investment medium for many households. Government regulation of the construction and use of the housing stock, as well as of financial markets, also affects the housing market. Rental and sales housing are typically considered to be distinct markets. The information that is available to active participants in the housing market is limited and asymmetrical.

The housing market represents the interaction of supply and demand at a particular time and place to establish the price of housing. On the supply side, the housing market may be segmented along a number of dimensions. Because of the immobility of the housing stock, perhaps the most important of these is the location of the unit. Location, which uniquely differentiates each housing unit, is important because it determines accessibility to employment and public and private services. Housing units that are in close proximity to each other are often in the same market segment; housing is also differentiated by tenure, physical attributes (size, structure type, amenities) and price.

Housing markets are essentially local in nature, often coterminous with the local labour market. Within large metropolitan areas, distance may define sub-markets based on relative accessibility to employment, services and other amenities. Distinctions may also be made on the basis of the nature and condition of nearby properties. Location may also affect the quality of the public services that are available. In addition, the cost of these services, in terms of property taxes, will also have an effect on housing values.

The demand for housing is also segmented by the demographic and economic characteristics of households. For individual households, the choice of a particular unit often requires compromise among competing objectives. A preferred location in terms of accessibility may command a high price or provide limited space. Because the transaction costs of moving are often high, a household may continue to occupy housing that is not well suited to their needs. In this post-shelter society, owner occupants may be motivated in their housing choices primarily by investment considerations.

For most North American households the housing market works well, with the majority able to obtain adequate space and services while spending a reasonable portion of their income on shelter.

Further reading

Grigsby, W. (1963) *Housing Markets and Public Policy*, Philadelphia, PA: University of Pennsylvania Press.

Smith, W. (1970) *Housing: The Social and Economic Elements*, Berkeley, CA: University of California Press.

GARY SANDS

HOUSING SOCIETIES

Housing societies are organizations that are not part of the national, regional or local government and provide housing on a not-for-profit basis, predominantly to persons and families in the lower to middle income brackets. In the literature, the term sometimes has another meaning. Thus, in this entry, housing societies do not include organizations that provide COUNCIL HOUSING, such as those in the United Kingdom, or public housing, such as those in the United States.

In different countries, housing societies can have a quite different organizational structure and may vary in size greatly. In the Netherlands, for example, the housing associations are legally independent from the government and operate on

a non-profit basis. But activities they can or must pursue are regulated by law and the government has established a system of monitoring their activities. The associations are responsible for a large part of the SOCIAL HOUSING sector in that country, which constitutes roughly 36 per cent of the total stock of dwellings. The housing associations often own and manage more than 500 dwelling units, and some associations in the larger cities have more than 10,000 dwellings. In the United Kingdom, housing associations are also bound to government rules and are supervised by the Housing Corporation, which provides government-financed grants to the associations. But many of the associations are much smaller than in the Netherlands. They have on average some 275 dwellings, own some 3 per cent of the national stock and are of more recent vintage. In the United States, housing societies are predominantly small non-profits and COMMUNITY DEVELOPMENT CORPORATIONS and community based. They own or manage on average some 300 dwellings, but often much smaller numbers. In this country, the way in which these housing societies are organized and operate varies greatly from state to state and city to city. They derive their financial resources from a wide variety of funding agencies, including national, state and local grants, tax credits, and funding made available by philanthropic institutions. These examples illustrate that housing societies have a widely different meaning and function in different countries.

Both in the countries of north-western Europe and in the United States, the role of housing societies in providing affordable housing for lower- and middle-income households has increased over the past two decades. On both continents, central governments have reduced their spending on the construction and management of affordable housing and transferred responsibility to local governments and housing societies. The national budgets still available for housing are largely devoted to housing allowances for individuals and families, to help low-income households meet their housing bill, and to tax deductions on mortgages for homeowners. Housing societies are increasingly the main organizations engaging in the construction of new moderately priced dwellings, the rehabilitation of such housing and the management of rental stock for low-income households.

Further reading

Dieleman, F.M. (1999) 'The impact of housing policy changes on housing associations: experiences in the Netherlands', *Housing Studies* 14(2): 251–9.
Schill, M.H. (1994) 'The role of the nonprofit sector in low-income housing production: a comparative perspective', *Urban Affairs Quarterly* 30(1): 74–101.

FRANS M. DIELEMAN

HOUSING VOUCHERS

In the 1970s and 1980s, the US Department of Housing and Urban Development (HUD) shifted the focus of its housing policy from project-based assistance (i.e. assistance tied to units located in subsidized housing projects) to tenant-based subsidies in the form of vouchers provided to low-income families or individuals. Subsidized housing has generally been concentrated in URBAN areas marked by POVERTY, high unemployment rates and a wide range of social problems. By the late 1960s, the problems attending the concentration of low-income families were of increasing concern across the country. The shift in policy was designed to encourage the greater DISPERSION of subsidized housing residents, and to allow these families greater locational choice. Unlike project-based assistance, which is limited to use in specific units, a voucher allows a family to choose any housing that complies with the requirements of the programme, provided the owner of the unit is willing to participate in the programme.

Vouchers are provided to families under HUD's Section 8 housing choice voucher programme which is administered locally by public housing agencies (PHAs). A PHA determines a family's eligibility to participate in the programme based on total annual gross income and family size. In general, a family's income may not exceed 50 per cent of the median income for the county or metropolitan area in which the family chooses to live. Three-quarters of vouchers are reserved for extremely low-income families – those at or below 30 per cent of area median income.

Housing voucher recipients are required to pay 30 per cent of their monthly adjusted gross income for rent and utilities; the government subsidizes the balance of the costs up to a locally determined maximum, or *payment standard*. The local PHAs set the payment standards based on

fair market rents (FMRs), which are designated annually by HUD for housing markets throughout the country. FMRs reflect the rent and utilities charged in a particular housing market for a typical, non-luxury unit (adjusted by unit size). FMRs are usually set at 40 per cent of an area's median rent.

After receiving a voucher, a family usually has 120 days to find a unit. Once a participant finds a unit, the PHA inspects it to make sure it meets housing quality standards; reviews the lease before approving the unit for rental by the participant; and checks that the rent is reasonable. A voucher recipient may select a unit with a rent that is below or above the payment standard. If the unit costs more than the payment standard, the family is required to pay the difference, but the family may not pay more than 40 per cent of its income for housing costs. The PHA pays the rent subsidy directly to the landlord on behalf of the participating family; the family pays the difference between the actual rent charged by the landlord and the amount subsidized by the programme.

The housing choice voucher programme is designed to allow families to move without the loss of housing assistance. Under the programme, a family may choose a unit anywhere in the United States if they lived in the jurisdiction of the PHA issuing the voucher when they applied for assistance. *Portability* is the mechanism by which Section 8 recipients can move from one PHA's jurisdiction to that of another. It is an important tool in helping families move to neighbourhoods that offer better opportunities and improved living environments.

The potential significance of vouchers goes beyond their role in promoting decent, affordable housing. A growing body of social science research, such as the studies on the Gautreaux Assisted Housing Program and the Moving to Opportunity demonstration programme, suggests that housing mobility can benefit families by improving their access to educational, employment and other opportunities, and by allowing them to be near middle-income role models. However, the special demonstration programmes incorporate intensive counselling and set requirements that determine where voucher recipients can move (e.g. to areas with low levels of poverty or small proportions of racial minorities). Neither of these features is incorporated into the operation of the regular Section 8 programme.

Previous research on household mobility in the regular Section 8 programme shows that without intensive counselling, families receiving vouchers usually make short-distance moves in order to remain near relatives and friends and to ensure access to public transportation. As a result, the operation of the regular Section 8 housing voucher programme has had little impact on moving low-income renters into areas with a low incidence of poverty and a low proportion of racial minorities. Furthermore, the re-clustering of housing voucher families in particular nearby locations has led to concerns about increases in crime and declines in property values.

The challenge for policy-makers is to implement the regular Section 8 programme in a way that not only helps voucher recipients improve residential conditions and move towards self-sufficiency but also maximizes individual choice. Two HUD-funded studies of Section 8 vouchers by Varady and Walker (1998, 2000) show that it is possible to make such incremental improvements. A four-city comparative case study of 'vouchering-out', in which families living in severely distressed HUD-subsidized housing developments received Section 8 vouchers, showed that residents who moved only a short distance experienced positive outcomes, even though they remained close to their original neighbourhood. A case study of the operation of the regular Section 8 housing voucher programme in Oakland, Berkeley, and the newer suburban areas of Alameda County, California, showed that unlike voucher recipients in other US metropolitan areas, programme participants were able to exercise portability and to benefit from their move to the SUBURBS.

Further reading

Katz, B.J. and Turner, M.A. (2001) 'Who should run the housing voucher program? A reform proposal', *Housing Policy Debate* 12(2): 239–62.

US Department of Housing and Urban Development (2000) *Section 8 Tenant-Based Housing Assistance: A Look Back after 30 Years*, Washington, DC: HUD.

Varady, D.P. and Walker, C.C. (1998) *Case Studies of Vouchered-Out Properties*, Washington, DC: HUD.

—— (2000) *Case Study of Section 8 Rental Vouchers and Rental Certificates in Alameda County, California*, Washington, DC: HUD.

CAROLE C. WALKER
DAVID P. VARADY

HOWARD, EBENEZER

b. 29 January 1850, London, England;
d. 1 May 1928, Welwyn Garden City,
England

Key work

(1898) *To-morrow: A Peaceful Path to Real Reform*,
London: Swan Sonnenschein (revised and published
as *Garden Cities of Tomorrow* in 1902).

Ebenezer Howard, a clerical worker and social
reformer, has left the world a very important
legacy: the GARDEN CITY model. He was born in the
City of London, the financial core of the biggest
and liveliest town of the world at the time. Son of a
shopkeeper, he lived amid the crowded URBAN
environment of the growing industrial metropolis
up to the age of 7, when he was sent away to
school. This change allowed him to get acquainted
with the rural landscapes surrounding the schools'
locations. Back in London at 15, after completing
his education, Howard worked in several clerical
posts, where he became skilled in shorthand.

At the age of 21 he made a radical change,
emigrating to the United States with the aim of
farming. For a short time he established himself in
a small farm in Nebraska. Soon, he gave up,
making his way to Chicago, to continue his career
as an office worker. He arrived just after the great
fire of 1871, which had destroyed most of the
central business district, and witnessed the city's
regeneration and the rapid growth of its SUBURBS.
Again, in his American period, he was able to
experience both town and countryside life.

Back in England, Howard joined a firm of
official parliamentary reporters. By 1879 he had
engaged the English socialist movement in discussions on COMMUNITY organization along collective
socialist lines, criticizing the existing INDUSTRIAL
CITY and land distribution. Among his chief
influences, William MORRIS is worth mentioning.
The American novelist Edward Bellamy, who
published, in 1888, a futuristic utopian vision of
Boston in the year 2000, paved the way for
Howard's determination to create a plan for
bringing a better civilization into existence.

Combining the utopian dream of a new equitable society, the reformers' thoughts and his
practical mind, he was able to produce the garden
city model. His early experience, alternating
between life in the city and the countryside,
also played a role in the model, first published
in 1898.

The individual garden city, in Howard's model,
was merely part of a much larger system, which
proposed a cluster of cities around a CENTRAL
CITY, all interconnected and sharing leisure facilities and services. The creation of a community
blending the advantages of town and country and
providing local employment began in Letchworth,
planned by Raymond Unwin, in 1903. Howard
himself moved to the expanding city in 1905. Prior
to this, and shortly after Letchworth's construction
had begun, he acquired the land for the development of a much larger feature, a complete system
of garden cities, to be started with Welwyn Garden
City. He moved to the new town in 1921, remaining there until his death. Howard was knighted
in 1927.

Further reading

Choay, F. (1965) *L'Urbanisme, utopies et réalités: une
anthologie*, Paris: Éditions du Seuil.

IÁRA REGINA CASTELLO

HUMAN CAPITAL

The term 'human capital' generally refers to the
knowledge, skills and competencies embodied in
individuals that increase their productivity. There
is no general consensus, though, on this definition.
In the 1990s, the concept was extended to include
natural abilities, physical fitness and healthiness,
which are crucial for an individual's success in
acquiring knowledge and skills.

In 1961, Theodore Schultz formally introduced
the term 'human capital' in the *American Economic
Review*, although the concept was by no means
new. Adam Smith, back in 1776, had outlined the
idea that people, acquiring talents through education, study or apprenticeship, are akin to expensive
productive machines. As from the 1960s, the term
'human capital' became central to economic theory, in large part thanks to Gary Becker, who
published the 1964 book entitled *Human Capital*
that provided the first unifying framework on the
subject.

Investment in human capital is fundamental
for both individuals and society as a whole. Individuals who invest in human capital generally

accept lower earnings during their youth in anticipation of higher returns when they are older. Future earnings, indeed, are positively correlated with the stock of human capital accumulated, because presumably human capital leads to greater productivity. This means that their lifetime earning profile is steeper than that for people who decide not to invest and instead enter the LABOUR MARKET immediately. Human capital is a particularly valuable investment because, once acquired, it cannot be alienated from its owner, although, as with physical capital, human capital may depreciate in time. Individuals with a high level of human capital generally benefit from access to more remunerative and stimulating careers and are less likely to end up in dead-end jobs. From a social point of view, human capital fosters economic growth and is one component of a country's overall competitiveness. Furthermore, human capital produces positive external effects (also known as EXTERNALITIES), which improve the overall quality of life. It has been demonstrated that in an environment with a high level of human capital, the crime rate decreases and the productivity of physical capital and other workers increases (peer effect). The ensemble of all such positive externalities has been also referred to as SOCIAL CAPITAL.

There are three main ways of investing in human capital: formal education, on-the-job training and MIGRATION. Formal education is the most identifiable form of human capital. The longer individuals attend school, the higher is their human capital. The importance of on-the-job training has been often overlooked in the literature, even though it actually constitutes a high proportion of adult learning activities. Training can be either specific or general. The former represents those skills and knowledge which are tied to a particular firm and therefore cannot be used elsewhere, whereas the latter includes those which can be successfully employed in a wide range of jobs and firms (education can be considered as a particular form of general training). Firms are usually more inclined to incur the full costs of specific training, because of their ability to appropriate the benefits. The employer and employee, on the other hand, often share the cost of general training – normally in terms of lower wage levels – as the latter can easily take the acquired knowledge to a new job as a 'transferable skill'. This justifies the need for government to subsidize general training (and education). Migration is the third

way of investing in human capital. Sjaastad in 1962 was the first to acknowledge that migration can be considered an investment activity, which has costs and renders returns. By migrating, people acquire knowledge and experience, which increase their stock of human capital. It is also interesting to note that education appears to significantly increase an individual's tendency to migrate, suggesting that migration and education may be complementary rather than substitute forms of investing in human capital.

Human capital theory has been successfully employed in explaining several phenomena, such as the gender gap in earnings or wage differentials by age and occupation. As far as the gender gap is concerned, the fact that women on average earn less than men has been explained by observing that women have a different quantity and quality of human capital. Women generally invest less time in human capital formation, and even when they invest the same time, they tend to invest in a more specific human capital, which is likely to have higher non-market as opposed to market returns. Moreover, it has been noticed that the rate of utilization of human capital increases the value of the investment made and slows its depreciation. Women have on average a lower level of utilization, as they are often involved in other activities such as taking care of the house and raising children, and consequently they are likely to have a flatter lifetime earning profile.

One of the primary difficulties in dealing with human capital is identifying good indicators with which to measure it. The measurement of human capital should include both formal and informal investments. In practice, though, only formal education, simply quantified in terms of years of schooling, is traditionally used as a proxy for the human capital acquired, and this sets limits to the validity of the empirical results. Some critics have also observed how in reality it is impossible to separate the human capital investment from the consumption component of it. Individuals can decide to enter education or training not only for the potential of higher future earnings, but also partly for the simple enjoyment of being students or trainees.

Further reading

Becker, G. (1993) *Human Capital: A Theoretical and Empirical Analysis, with Special Reference to Education*, 3rd edn, Chicago: University of Chicago Press.

Laroche, M., Mérette, M. and Ruggeri, G.C. (1999) 'On the concept and dimensions of human capital in a knowledge-based economy context', *Canadian Public Policy – Analyse de Politiques* XXV(1): 87–100.

Schultz, T. (1961) 'Investment in human capital', *American Economic Review* 51(1): 1–17.

Sjaastad, L.A. (1962) 'The costs and returns of human migration', *Journal of Political Economy* 70: 80–93.

ALESSANDRA FAGGIAN

HUMAN ECOLOGY

Human ecology generally applies to studies of human communities using ecological principles; however, the exact focus and interpretation of human ecology varies from one discipline to the next. Even though the term 'human ecology' was popularized in the 1920s and 1930s, studies of this nature had been conducted since the early nineteenth century in France and England. Human ecology found its advocates among three sociologists: Robert PARK, Ernest Burgess and Roderick McKenzie, as well as a geographer named Harlan Barrows, all at the University of Chicago. However, Amos Hawley established a sound theoretical foundation for the study of human communities within human ecology with the publication of his 1950 book, *Human Ecology: A Theory of Community Structure.*

Intellectually, human ecology evolved from the convergence of research in plant and animal ecology, geography, and sociology. Social scientists familiar with biological studies of different species and some of the founding theories of ecology in studying the relationship of species with their environment eventually saw similarities between human communities and those of animals and plants. While early human ecologists, also known as classical human ecologists, theorized about the biotic and cultural spheres of communities, contemporary human ecology is more focused on environmental issues and the mutual relationship between human communities and their surrounding environments.

In addition to human ecology, URBAN ECOLOGY, SOCIAL ECOLOGY and cultural ecology are used to identify the applications of ecological principles to specific spheres of human communities. For example, whereas urban ecology is a body of knowledge about the URBAN milieu based on a human ecological framework (Wilson 1984), cultural ecology can be defined as a tool for studying the adaptive interaction between a culture's unique history and its environment (Hardesty 1986).

Though critical perspectives on human ecology which accuse the classical human ecologists of biological determinism abound, the importance of this intellectual endeavour has scarcely diminished. Human ecology has revived itself as an interdisciplinary field with a much wider interest in the environment than was initially envisioned by its founders.

Further reading

Hardesty, D.L. (1986) 'Rethinking cultural adaptation', *Professional Geographer* 38: 11–18.

Hawley, A.H. (1950) *Human Ecology: A Theory of Community Structure*, New York: Ronald Press Co.

Park, R.E. (1936) 'Human ecology', *American Journal of Sociology* XLII: 1–15.

Wilson, F. (1984) 'Urban ecology: urbanization and systems of cities', *Annual Review of Sociology* 10: 283–307.

ALI MODARRES

HUMANE CITY

Since the last quarter of the twentieth century, four schools of thought advocating for humane cities have emerged:

1   The NEW URBANISTS consider physical design to be the primary instrument for creating humane cities. They call for reorienting URBAN places away from the automobile and towards a more pedestrian-friendly scale.

2   The advocates of community-based development emphasize the empowerment of grassroots communities as the vehicle to more humane cities. John FRIEDMANN and others call for subordinating the needs of the economy ('economic space') to the needs and well-being of living communities ('life space').

3   SUSTAINABLE URBAN DEVELOPMENT advocates call for a balancing of economic, environmental and social equity concerns for the wellbeing of present and future generations. Amartya Sen's rights approach to sustainable development emphasizes the role of public

policy in protecting the poor and socially excluded, while fostering economic growth.

4   Dynamic systems theorists understand the city as a self-organizing organism. The humane city, in this view, is the result not of expert guidance or social advocacy, but rather of three basic iterative actions, or habits, that become embedded in the culture: the habit of listening to understand the 'other' before advocating a position; the habit of reflecting on, and revealing, one's own assumptions and values; and the habit of sensing together the emergent future of the whole organism.

Further reading

Atlee, T. (2003) *The Tao of Democracy*, Cranston, RI: Writer's Collective.
Short, J. (1989) *The Humane City*, New York: Blackwell.

PATRICIA A. WILSON

HUXTABLE, ADA LOUISE

b. 14 March 1921, New York, NY

Key works

(1976) *Will They Ever Finish Bruckner Boulevard?*, New York: Macmillan.
(1986) *Architecture, Anyone?*, New York: Random House.

The first architecture critic at a major US newspaper, Ada Louis Huxtable established architecture and URBAN DESIGN journalism in North America. Writing for a general audience, rather than for academics or professionals, she is credited with raising the public's awareness of the URBAN environment. Huxtable was educated at Hunter College and attended New York University. After working as an assistant curator of design at the Museum of Modern Art, Huxtable served as the architecture critic at *The New York Times* from 1963–82 and chronicled several periods of architectural style, most importantly MODERNISM and POSTMODERNISM, and changes in planning practice. In addition to architectural criticism, her writings documented both the failures of urban renewal (including the demolition of New York's Pennsylvania Station) and the

successes of HISTORIC PRESERVATION. Huxtable also celebrated in her writing the beauty of New York's skyscrapers and the advances in building technology that made them possible. Huxtable's books include *Will They Ever Finish Bruckner Boulevard?* (1976); *Kicked a Building Lately?* (1976); *Architecture, Anyone?* (1986); and *The Unreal America: Architecture and Illusion* (1997). In 1970, Huxtable received the first Pulitzer Prize for Distinguished Criticism. She has been a Fulbright and a Guggenheim fellow and was named a MacArthur fellow in 1981.

Further reading

Wodehouse, L. (1981) *Ada Louise Huxtable: An Annotated Bibliography*, New York: Garland.

HARRISON HIGGINS

HYGEIA

Hygeia provides an example of the purely theoretical utopian URBANISM postulated in the mid-nineteenth century by social reformers, as a weapon to combat the appalling conditions inflicted upon cities by the advances of the INDUSTRIAL REVOLUTION. Basic statements were published in England by Dr Benjamin Ward Richardson, whose ideals aimed at reaching perfect sanitary results or, at least, the co-existence of the lowest possible general mortality rate with the highest possible individual longevity.

Hygeia was conceived of as being large enough to accommodate 100,000 people homogeneously distributed on 4,000 acres, at an average density of 25 persons per acre. Richardson proposed radical changes in the usual construction standards. Roof-terraced houses were to be of stained brick. Chimneys would be connected to central shafts, channelling the unburned carbon to gas furnaces. Kitchens would be on the upper floor to facilitate brightness and ventilation, and to better distribute the hot water from the boiler. Small hospitals were advocated for every 5,000 people, to prevent the 'warehousing' of diseases attributed to the huge infirmaries of the time. The helpless, aged and mentally infirm were to be housed in modest-sized buildings. Railroads were to be underground, and no cellars or carpets of any kind were to be permitted in the houses. Public laundries, bathing

houses, gardens and areas for exercise would be generously provided.

Richardson's work gained widespread recognition and encouraged further endeavours to achieve healthier cities.

Further reading

Richardson, B.W. (1876) *Hygeia, A City of Health*, London: Macmillan.

LINEU CASTELLO

IMAGE OF THE CITY

Kevin LYNCH's *The Image of the City* was originally published in 1960. Its publication represented the end of an investigation, along with Gyorgy Kepes of MIT, on how people perceive cities and the aesthetics of the city.

Lynch builds his discussion on the visual image of the city on two notions. The first notion concerns 'legibility' – the ease with which the parts of the city are perceived and organized in a coherent structure. The second notion involves 'image-ability' – the capacity of a physical object to produce a remarkable image to the observer. Combined, the two notions imply an integrative condition through which the images become legible. In a sense, they become, for the observer, a structure.

Lynch's research centred on a research project conducted in three US cities – Boston, Jersey City and Los Angeles. He was attempting to integrate the perception of the URBAN environment and certain design principles that help to determine place legibility. The research was built on citizen interviews with citizens that used these three cities. He analysed legibility based on five elements: paths, nodes, edges, districts and landmarks.

According to Lynch, paths are elements along which an observer moves. The movement allows the observer to build a perception of the urban space. His research found that STREETS represent the predominant component in the description of URBAN FORM. The importance of streets varies according to an individual's knowledge of the city – the greater the familiarity with the area, the greater the differentiation of the space represented in the description of the visual form.

Edges represent lineal elements that define borders to the visual perception. They represent dividing lines between the various districts. They operate like elements that establish limits to the visual image.

Districts constitute wide and continuous surfaces on which the other elements – streets, nodes, landmarks and edges – compose the primary structure of the city. This group of elements allows an individual to form a homogeneous image of the area. Lynch also includes and acknowledges the importance of such variables as texture, space, forms, detail, symbol, type of buildings, habits, activities, densities, conservation state and topography in composing a homogeneous mental image of a district or NEIGHBOURHOOD.

A node represents convergences among different streets or concentrations of activities. They are identified as central places to the observer. A node could be a busy intersection where people come and go.

Lynch also acknowledges landmarks as elements that punctuate the visual structure of the city. They serve as key reference points to a city and act as prominent visual features to a city. They allow an individual an opportunity to see where they are in relation to a certain feature or physical object in a city.

The city, in all of its complexity, is an entity in movement. Lynch's approach goes beyond the description of the five aforementioned key elements. He continued to examine other issues pertaining to urban form, including the appreciation of various meanings that emerge from history.

The study of cities based on ENVIRONMENTAL PERCEPTION assumed a growing importance to the study of the modern city and to citizen participation in the decision processes of URBAN PLANNING.

Lynch continued his theme of the image of the city in several later works. In *What Time is this Place?* (1993 [1972]), he focused on the perception of time and of environmental transformations in the city. In *The Theory of Good City Form* (1981), he continued his theoretical proposition that the emergent sense of form, its capacity to produce an orientation, and the dimension of social control on the space define important qualities in the performance of the city.

Countless authors have contributed to the debate on the cognitive dimension of the city. Among the many names that can be mentioned are Jane JACOBS (1992), who denounces the sense loss in the modern city; Gordon CULLEN (1971), who focuses on the aesthetic dimension of the urban image; and Maurice Cerasi (1973), who discusses the complementarity between perception and experience and among topological and psychological spaces.

Further reading

Cerasi, M. (1973) *La Lettura dell'Ambiente*, Milan: CLUP.
Cullen, G. (1971) *Townscape*, London: Architectural Press.
Jacobs, J. (1992) *The Death and Life of Great American Cities*, New York: Vintage Books.
Lynch, K. (1960) *The Image of the City*, Cambridge, MA: MIT Press.
—— (1993 [1972]) *What Time is this Place?*, Cambridge, MA: MIT Press.

LEANDRO M.V. ANDRADE

IMMIGRATION

The decision to seek settlement in a country other than one's country of birth is one of the most important and complex events in a person's life. And yet thousands of people throughout the world embark on such a journey every day in pursuit of a future better than the one left behind.

A widely used theory to explain the origin of immigration is the 'push-pull' model. The model predicts the volume of immigration by analysing 'factors of expulsion' (economic and political hardships in the sending countries) and 'factors of attraction' (relative economic and political advantages in the receiving countries). Economic factors, in particular, play an important role. Empirical studies of wage differentials between sending and receiving countries, for example, show that the flow of migration is uniformly from poorer countries to wealthier ones.

Those who have looked at the behavioural aspects of immigration have found that only a small percentage of the sending country's population embark on the arduous and risky process of international migration – those whose aspirations have been raised by exposure to a better way of life either through education, through the popular media, through family members already living abroad or through a combination of the above. What seems to trigger the migration process is a large enough gap between migrants' aspirations and what is realizable in the country of origin.

Of course, individual decisions are embedded in social networks, societal structures and even global forces. At the macro level, structural imbalances in the world profoundly affect population movements around the world. The poorest countries of the world, for example, do not send large numbers of people to wealthy countries. Instead, studies show that international migration often starts in countries of intermediate levels of development, and among these countries some send more than others. Structural linkages between the sending and receiving countries are far more important in predicting the flow of people among countries, such as the history of colonization, the history of relations among nations for security (e.g. NATO), the presence of trade blocs and agreements (e.g. the North American Free Trade Agreement, the Organisation for Economic Co-operation and Development, the European Union), and the existence of guest-worker programmes. Following such historical threads, not surprisingly, larger numbers of people from Algiers have migrated to France, from India to Britain, from Turkey to Germany, and from Mexico to the United States.

Social networks are also capable of triggering international migration. Successful migrant experiences of 'pioneers' allow others in homelands to imagine futures abroad. As sociologists note, international migration is a complex process of network-creation and network-dependency.

Not all immigration takes place voluntarily. Thousands of people are forced out of their homelands and become refugees due to civil wars, famines, droughts or fear of persecution for their political beliefs.

Immigration was a central force in the URBANIZATION and INDUSTRIALIZATION process in the

United States following the Civil War, and it has always been fundamentally an URBAN phenomenon. Waves of people arrived from Europe at the end of the nineteenth century, mostly from Germany, England, Ireland and Scandinavia, and later from southern and eastern Europe, and settled in the urban areas of the Midwest and Northeast. Like the early settlers, newcomers tend to gravitate towards cities. In the late 1990s, the presence of immigrants was most pronounced in five major US immigration gateway regions: Los Angeles, New York, San Francisco, Miami and Chicago.

Immigration policy fundamentally shapes the make-up of the immigrant population. As in any country experiencing large flows of immigration, the question of who to admit into the United States has historically been controversial. For example, the Chinese Exclusion Act of 1882 (remaining in effect for the next sixty years) barred, for the first time, immigrants based on national origin. Barriers for migration dating from the 1920s were reduced after the passage of the 1965 Hart-Cellar Immigration Act, induced by the CIVIL RIGHTS MOVEMENT in the 1960s. The Act abolished European-oriented quotas on countries of origin, gave preference to family reunification and possession of scarce and wanted skills, and increased the numbers of immigrants to be admitted. As a consequence, the number of immigrants into the United States rose sharply, coming mainly from Asia and Latin America. At the same time, illegal immigration rose, especially from Mexico, by those who continued to use networks established during the Bracero programme of the Second World War era even though the programme had been abolished in 1964 – a programme that was set up to alleviate shortages in the US agricultural sector. The 1986 Immigration Reform and Control Act was passed giving special emphasis to controlling unauthorized flows.

In 2000, nearly 30 million immigrants (5 million estimated to be illegal) were living in the United States, about 10.4 per cent of the total population. A significant proportion of them came from Mexico (27.7 per cent), followed by China/Hong Kong/Taiwan (4.9 per cent).

Immigrants tend to cluster geographically. In the United States, Hispanics concentrate in the greater Los Angeles area, Asians in Los Angeles, New York and San Francisco, Cubans in Miami. Selectivity in migration coupled with legislation

allowing family unification has produced ethnic enclaves such as Chinatowns in New York and San Francisco, and Little Havana in Miami. In 2000, 69 per cent of all immigrants in the United States were family sponsored. Clustering is seen in Europe as well where 'foreigners' constitute a sizable share of the total population of major cities like Paris, London, Amsterdam, Brussels and Frankfurt.

While the influx of immigrants into cities has had beneficial impacts such as the revitalization of inner-city NEIGHBOURHOODS and the revival of local economies during the 1990s, scholars have also noted troubling bifurcation trends in employment opportunities for the newcomers. GLOBAL CITIES like New York, Los Angeles, London and San Francisco experienced ECONOMIC RESTRUCTURING resulting in highly educated immigrants working in the high end of the occupational spectrum and poorly educated and low-skilled immigrants working in low-paying jobs. Immigrants continue to arrive in cities where the NEW ECONOMY has fundamentally transformed the LABOUR MARKET and where prospects for the less-educated and low-skilled immigrants are uncertain.

Further reading

Light, I. and Gold, S.J. (2000) *Ethnic Economies*, San Diego, CA: Academic Press.

Massey, D.S., Arango, J., Hugo, G., Kouaouci, A., Pelligrino, A. and Taylor, J.E. (1998) *Worlds in Motion: Understanding International Migration at the end of the Millennium*, New York: Oxford University Press.

Portes, A. and Rumbaut, R.G. (1996) *Immigrant America: A Portrait*, Berkeley, CA: University of California Press.

Waldinger, R. (ed.) (2001) *Strangers at the Gates: New Immigrants in Urban America*, Berkeley, CA: University of California Press.

SEE ALSO: gentrification; global city; new economy

AYSE PAMUK

IMPERIAL CITIES

Imperialism is the policy of extending a nation's authority or rule over another either through territorial acquisition or by establishing political or economic hegemony. Imperial cities can be defined

as capitals of great empires and their conquered territories. Palatial buildings, grand avenues, magnificent squares and an architectural style that symbolizes the power of the empire characterize such cities. Babylon, a Mesopotamian city, assumed imperial dimensions during the fifth and sixth centuries BC. The Pharaohs of Egypt began to build the imperial cities of Memphis, Thebes and Tel-el-Amarna from the second millennium BC. Alexander the Great established the imperial city of Alexandria in Egypt in 331 BC to maintain his power throughout his huge Macedonian empire in the East. Under Roman imperialism (first century BC to fourth century AD), Rome became an imperial city. Emperor Constantine laid the foundations of the imperial Constantinople (Istanbul) in AD 330, with his decision to move the seat of the Roman Empire to this city.

The fall of the Roman Empire by the fifth century AD brought about the Dark Ages in Europe and the disintegration of European imperial cities. The rise of Islamic empires from the end of the eighth century brought about the rise of Islamic imperial cities. Baghdad and Cairo stood out as imperial centres of the Islamic empires in the tenth and thirteenth centuries respectively. In the early thirteenth century, the Venetians rebuilt Constantinople's imperial glory as the centre of their Latin empire. In fourteenth-century China, the Ming dynasty laid the foundations of the 'imperial forbidden city' in Beijing.

The next great wave of imperialism came about with Spain and Portugal's colonial expansion, starting in the early sixteenth century. In the same century, Lisbon became the imperial capital of Portugal, while their Indian possession, Goa, became the administrative centre of Portugal's empire in South Asia and the East. By 1763, Rio de Janeiro had become the capital of Portugal's most prized empire in Brazil. Mexico and Peru – the most favoured Spanish colonies in Latin America – saw the emergence of the imperial cities of Mexico City and Lima. Ironically, Mexico City was built on the finest imperial city of the New World, the Aztec capital of Tenochtitlán, after its fall in 1521. The Netherlands, another major imperial power of the era, had its share of imperial cities in Amsterdam and Batavia (Jakarta). France and Britain, which became two great imperial powers by the middle of the nineteenth century, had their own share of imperial cities at home (e.g. London and Paris) as well as in the colonies (e.g. Calcutta

and Rabat). Lesser imperial powers such as Tsarist Russia, Imperial Germany and Italy had their share of imperial cities in St Petersburg, Berlin, Rome and Tripoli (Libya). The early part of the twentieth century represented the height of imperial city-building. For example, the British built the new colonial capitals of New Delhi, Salisbury (Harare), Lusaka, Nairobi and Kampala. Colonies were seen as virgin territories where architects and planners could fulfil their imperial dreams.

Further reading

King, A.D. (1990) *Urbanism, Colonialism, and the World Economy: Cultural and Spatial Foundations of the World Urban System*, London: Routledge.
Morris, A.E.J. (1994) *History of Urban Form: Before the Industrial Revolution*, 3rd edn, Harlow: Longman.

SIDDHARTHA SEN

INCLUSIONARY HOUSING

Lack of affordable housing, together with racial and class SEGREGATION, are international problems. In the United States, a mechanism that addresses both problems is inclusionary housing (IH). Under an IH programme, low- and moderate-income housing units are included in an otherwise market-driven development. While voluntary IH programmes exist, the great majority of IH has been built as a result of local mandatory programmes requiring developers to include the affordable units in their developments. A major objective of IH is not only to increase the supply of affordable housing, but to foster greater social and economic integration. This is especially important in suburban communities where high housing costs – and in some cases racial DISCRIMINATION – have denied lower-income households the better jobs and educational opportunities found in newly developing areas. Mismatches between workers' earnings and housing prices have contributed to worsening traffic and air pollution problems. IH units are generally built concurrently with the market-rate units, sidestepping the NIMBY (NOT IN MY BACKYARD) problem, probably the most serious obstacle to the provision of affordable housing.

One of the earliest IH programmes, the 1972 Montgomery County, Maryland, Moderately Priced Dwelling Unit Program, is arguably the

most successful and largest local IH programme in the country, having produced more than 10,000 affordable housing units over a period of 25 years. At the state level, it is only in New Jersey and California that IH has had a lasting and significant impact, but for different reasons.

The origins of IH in New Jersey are tied to the use of EXCLUSIONARY ZONING in New Jersey municipalities to restrict development of multi-family housing in order to drive up the cost of housing and exclude lower-income households. The courts in New Jersey have led the attack against exclusionary zoning. With the landmark 1975 Mount Laurel decision (*Southern Burlington NAACP v. Township of Mount Laurel*), the New Jersey Supreme Court declared that zoning was being used unconstitutionally and that each municipality should provide their fair share of affordable housing. When the Mount Laurel township failed to provide affordable housing, the New Jersey Supreme Court decided in 1983, with the Mount Laurel II decision, that mandatory set-asides, i.e. IH, would be required and that courts would grant zoning relief and building permits to builders excluded from building low-income projects. Builders agreed to make 10 per cent of the units affordable to households earning under 50 per cent of the area median income, and 10 per cent to households earning between 50 per cent and 80 per cent.

The recalcitrant municipalities turned to the state legislature, resulting in the New Jersey Fair Housing Act of 1985. Under the Act, control over zoning was transferred from the judiciary to an administrative agency, the Council on Affordable Housing. This agency requires that each munici-pality develop a housing element of its master plan that determines its fair share obligation and iden-tifies the land appropriate for development.

In California, the most important factor explaining the widespread use of IH has been the housing affordability crisis, one of the worst in the nation. This crisis has led the state to mandate that all municipalities include a housing element in their GENERAL PLANS. The housing element, as in New Jersey, is supposed to indicate how a locality is going to meet its fair share of affordable hous-ing. The State of California Department of Housing and Community Development (HCD) reviews the housing element to ensure that it meets state requirements. HCD certification, however, does not guarantee that affordable units will be built. Nevertheless, incentives and disincentives linked to housing element compliance have prompted some localities to include IH as one of the mechanisms to produce affordable housing. In addition, a locality that fails to adopt a housing element that meets state requirements can be sub-ject to court order limiting the locality's approval powers. Even though litigation has been rare, research has shown that fear of litigation is an important motivator for the implementation of IH. Lacking a clear state mandate or a compelling court decision as in New Jersey, IH programmes in California are adopted locally. Their diversity, complexity and flexibility reflect the political and economic characteristics of each locality over time. A 1996 survey of IH in California found 75 pro-grammes in existence at the time, with significant variation in requirements and results.

IH programmes are difficult to implement, not only because of the opposition of the localities themselves, as in the case of New Jersey, but also because of the strong opposition of developers. They maintain that by developing housing afford-able to low-income people, they will either lose money or pass the added costs to the buyers of the market-rate units, making housing less afford-able for middle-income families. There is general agreement among researchers, however, that, in the long run, a third party – landowners – will pay for IH. When bidding for land after an IH pro-gramme is passed, builders would bid less because they would take into account the additional IH cost. Nevertheless, builders remain firmly con-vinced that IH is an additional regulatory and redistributional programme that runs against their financial interests and violates their ideological convictions. They often oppose IH with their considerable resources and strength and fre-quently defeat IH proposals. One way to attempt to appease the development industry is to provide cost-offsets in the form of financial assistance or regulatory relief, including state housing bonds and below-market-rate construction loans or den-sity increases and fast-track permit approval.

As the United States enters the twenty-first century, it is becoming abundantly clear that housing affordable to low-income households will become even more problematic and IH utilized more often as one of several programmes necessary to meet the needs of lower-income families. It is necessary to understand, however, that because of class and racial prejudices and insular local

controls, effective IH programmes will be enacted mainly as a result of the intervention of either a higher level of government or the courts.

Further reading

Calavita, N., Grimes, K. and Mallach, A. (1997) 'Inclusionary housing in California and New Jersey: a comparative analysis', *Housing Policy Debate* 8(1): 109–42.

Mallach, A. (1984) *Inclusionary Housing Programs: Policies and Practices*, New Brunswick, NJ: Center for Urban Policy Research, Rutgers University.

NICO CALAVITA

INCREMENTALISM

Incrementalism refers to the manner in which the majority of public policy development occurs in the United States. The basic premise of incrementalism is that government is better suited to address most issues in a stepwise, gradual fashion rather than holistically and comprehensively. Governmental structure and processes, specifically the separation of powers and checks and balances, impede cohesive policy-making.

Incrementalism, or successive limited comparisons, was advanced by Charles Lindblom as the major alternative decision-making model to the rational-comprehensive model. The rational-comprehensive model requires identifying and addressing the root causes or sources of public problems and considering all options for resolving them. Public policy-making is not amenable to this approach due to time constraints, inadequate information, insufficient resources and the cognitive limitations among policy-makers. Comprehensive and lengthy research is not politically expedient. Politicians are interested in results that correspond to elections and other political objectives. Another difficulty encountered with the rational-comprehensive model is the lack of agreement about the root causes of public problems. Policy-makers disagree on primary values and policy objectives. Consequently, they are more likely to agree on a partial solution to some aspect of a problem than to tackle a complicated situation in its entirety. For example, some researchers describe the multiplicity of municipalities in the typical metropolitan area as fragmented. However, because regional government is unobtainable in many places, regional special purpose districts are an incremental solution to the problem of multiple jurisdictions.

Agreement among diverse parties is easier to reach by employing a decision-making approach that entails making successive limited comparisons that differ only marginally from policies already in effect. The sequence of successive limited comparisons almost becomes an experiment to test what works best while not relying heavily on theory as a guide. It is theorized that policy-makers favour the incremental approach because it reduces the impact of error in the event that the wrong course is taken.

This approach does not lend itself well to policy matters, where more far-reaching action is appropriate. It is possible, therefore, that taking the incremental approach itself may pose a problem if the solution advanced does not adequately address the situation at hand. Some conditions might be addressed more effectively in a non-incremental manner, like the eradication of poverty, but minimal or moderate change maintains the status quo and existing power relationships and structures.

While incrementalism may be more politically palatable than comprehensive change, it may be less effective in solving problems. Michael Hayes (1992) discusses US involvement in the Vietnam War as a classic example of failed incremental policy-making. He describes a long series of relatively small missteps that went from limited military aid in the 1940s to large-scale war in the 1960s and challenges the notion that taking small steps allows for easier redirection or more immediate identification of undesirable consequences.

Non-incremental decision-making does occur such as with the US manned space exploration programme, which required non-incremental policy-making. Paul Schulman (1975) described the space programme as indivisible – not amenable to piecemeal or short-term approaches. The passage of welfare reform legislation in 1996 also occurred in a non-incremental fashion as it dramatically ended sixty years of federal government cash assistance to families under the Aid to Families with Dependent Children Program.

Further reading

Hayes, M. (1992) *Incrementalism and Public Policy*, New York: Longman.

Lindblom, C.E. (1959) 'The "science" of muddling through', *Public Administration Review* 19: 79–88.

Schulman, P.R. (1975) 'Nonincremental policy making', *American Political Science Review* 69: 1354–70.

MITTIE OLION CHANDLER

INDEX OF DISSIMILARITY

The index of dissimilarity measures the concentration of population in a geographical area (e.g. city, state or nation). Using the index of dissimilarity, for example, we could show the variation in population concentration between URBAN and rural areas, household income inequalities between certain races, levels of educational attainment by two distinctly different regions or ethnic groups, and racial SEGREGATION or concentration by specific spatial units. The index value of 100 per cent (or 1.0) indicates a complete dissimilarity (or segregation), while that of 0 per cent (or 0.0) denotes total integration. The Lorenz curve and the Gini index ratio (or Gini coefficient), used interchangeably, determine the index of dissimilarity.

The Lorenz curve is used to show a relationship between two observable variables (X_i and Y_i), often to indicate segregated (non-integrated) phenomena. To construct the Lorenz curve, each variable (X_i and Y_i) is categorized using a categorical or interval scale, and the value of each category is added together to produce a cumulative value of 1.0 or 100 per cent for each variable. Then, the cumulative percentages of each variable are plotted against each other, while a diagonal line of 45 degrees is drawn to illustrate the completely integrated situation.

Following the presentation of the Lorenz curve, the index of dissimilarity, Δ, can be calculated as the sum of the positive differences between the two variables, X_i and Y_i, using the same categories or intervals. For the index of dissimilarity, Δ, the differences between X_i and Y_i are the absolute value differences for the same category of each variable rather than the cumulative percentage differences of the two variables. The index of dissimilarity, Δ, is then:

$$\Delta = \tfrac{1}{2} \sum | X_i - Y_i |,$$

where n is the total number of categories for both variables, and X_i and Y_i are un-cumulated percentages of each category for two variables.

To plot the index value of Δ under the Lorenz curve, we first calculate the index value of Δ based on the above equation. Then, we draw a straight line between the corresponding values of X_i and Y_i, representing the index value of Δ.

Finally, the Gini concentration ratio (or Gini coefficient) is then the proportion of the total area under the diagonal, which can be shown as the area between the 45-degree line and the Lorenz curve. The Gini coefficient may be calculated using the following equation:

$$G_i = (\textstyle\sum X_i Y_{i+1}) - (\textstyle\sum X_{i+1} Y_i),$$

where X_i and Y_i are cumulative percentages for each variable observed (or independent variable) and n is the total number of categories or intervals.

Indeed, the index of dissimilarity is an important tool for policy-makers and politicians when making a wide variety of public funding decisions regarding racial integration for urban neighbourhoods, urban public school aids and other policy matters.

Further reading

Shryock, H.S. and Siegel, J.S. (eds) (1980, 1979, 1975) *The Methods and Materials of Demography*, Washington, DC: US Government Printing Office.

JOOCHUL KIM

INDUSTRIAL CITY

The transition from the MERCANTILE CITY to the industrial city was not a linear process. Different locations provided different answers to the innovations brought by INDUSTRIALIZATION, even though, generally speaking, all such locations grew very quickly. Size alone, however, cannot be used to distinguish the pre-industrial city from the industrial city.

Qualitative criteria should also be used to consider the pre-industrial town as a distinct type of settlement. Not only did industrialization revolutionize economies, it also affected the internal structure of many cities, so that by the end of the nineteenth century, the shape and functions of most cities appeared fundamentally changed, along with social relations. On the basis of observations made about Manchester, England, ENGELS

outlined not only a particular community, but what is considered to be the archetype of the industrial city.

The introduction of factories into what was previously a trading centre generated a concentration of production and commercial activities in the most central area, while the functional specialization of the different parts of the city became clear due to the separation of home and work. Furthermore, the new physical structure was increasingly being shaped by the competition for land, which had become costly merchandise. The results of the 'inevitable consequences of industrialization' were a structure of concentric areas around the commercial centre, the separation of the various functions and the SEGREGATION of a social class – the workers – from all the others.

While the typical industrial city in Europe was of average size, in the United States industrialization occurred mainly in large cities. However, such cities followed a similar pattern to their European counterparts, with a strong concentration of factories in the centre and a set of concentric and segregated quarters for the working, middle and upper classes.

The new social and economic institutions also brought about significant architectural transformations, with the use of new building materials and techniques and with the definition of new building typologies. The innovations, however, did not apply to the areas of the city that were not directly profitable, in particular the workers' dwellings. Everywhere, the growth of industrial cities was synonymous with URBAN SQUALOR, SLUMS and POVERTY, so much so that Lewis MUMFORD concludes that 'the factory and the slum are the two main elements in the new urban complex'.

Contemporary writers and reformers documented in great detail the crowding, the lack of the most basic sanitary facilities, and the high morbidity and infant mortality rates in an attempt to expose the conditions in which the working class were living. Many politicians, concerned that the new population could endanger the system's stability, tried to remedy the widespread insalubrious conditions by introducing regulations on the use of land. For these reasons, the birth of the industrial city is also considered as the time when URBAN PLANNING became a regulatory system.

Further reading

Briggs, A. (1963) *Victorian Cities*, London: Odhams Books.

Hall, P. (1988) 'The city of dreadful night', in *Cities of Tomorrow*, Oxford: Blackwell.

PAOLA SOMMA

INDUSTRIAL REVOLUTION

At the turn of the nineteenth century, many Western countries experienced a dramatic economic and social transformation. In the transition from a rural and agricultural economy to an URBAN and industrial one, several distinct stages can be defined as a phenomenon so accelerated and disruptive that it is commonly referred to as a revolution.

The Industrial Revolution began in Britain, and subsequently spread to Continental Europe and the United States, where industrialization started developing after the country's independence. The actual transition process from mercantilism to industrial capitalism accelerated only in the late 1830s and early 1840s. Not only was Britain the first country to be affected by this phenomenon, but the various phases and aspects of the development of large-scale industrial capitalism were expressed with a clarity and completeness that are not to be found elsewhere.

The Industrial Revolution was not an event but a process that developed over a considerable period of time, and its causes may be attributed to a number of simultaneous factors, namely an agrarian revolution, a rapid and significant increase in population, technological innovation in machinery, new forms of energy, and a significant accumulation of capital to be invested.

In the countryside, between the end of the seventeenth and the first half of the eighteenth centuries, the destruction of the open field system of cultivation with the enclosure of common and wastelands had resulted in an extension of cultivated land. The connection between the practice of enclosure and an increasing agricultural productivity was very close. Land privatization eased the introduction of new cropping techniques, rotation and soil improvement. Big land ownership destroyed rural medieval communities and transformed small farmers into renters. The landowning class became wealthy, but intensive agriculture

also meant a decline in the number of people engaged in agriculture, a rise in rents and the growth of pauperism.

In the same period, a significant decrease in the mortality rate, due partly to the consequences of the improved food production, generated an extraordinary increase in the population. This growth, together with the massive migration of surplus rural population to the emerging industrial cities, provided the industrial sector with the workforce it needed.

Beyond the increase in agricultural production and the substantial reserve of low-cost manpower, the switch to a production system focused on the factory was made possible by new mechanical discoveries and technological innovations in the textile, coal and iron sectors. Some inventions (the spinning jenny, the water frame, the fly shuttle) changed the features of cotton manufacturing. New power sources, together with the development of the steam engine, and the great upsurge in productivity gave rise to a new system of work organization. Lastly, a further growth of the factory system was due to the expansion of trade and the concentration of merchant capital. Thus, the Industrial Revolution was not the result of a single cause, but the cumulative effect of a number of revolutionary changes in agriculture, demography, technology and commerce.

At the same time, the expansion of capitalist production through the organized use of mechanical energy, beyond accelerating changes in all the above sectors, was accompanied by acute dislocation in the social and political structure, and by a radical change in the attitude of the state towards individual enterprise. As a consequence of the altered conditions in the accumulation of wealth, new social classes were created, i.e. the industrial capitalists and the unskilled factory workers.

The Industrial Revolution influenced the level of URBANIZATION, the urban hierarchy, and the size and structure of the cities. Urban growth fuelled by industrial requirements for labour, and by the pursuit of ECONOMIES OF SCALE, was extremely rapid and characterized by two phenomena, i.e. the prominence of certain new industrial centres, and the extension of existing cities, in general established market towns.

The consolidation of great urban centres providing access to water or railways was also favoured by the need to market finished goods. Transportation networks were organized to connect existing hubs and contributed to originating new ones, resulting in new relationships between factory locations and consumer markets.

The role of factories and railways in the location process is a general factor that assumed different forms in various contexts and historical moments. In Britain, for example, in the initial take-off stage, industrial investment focused on rural sites, but as it became clear that factories were to be located where coal and essential raw materials were available, or where unskilled labour was abundant, small and medium-sized towns experienced an unprecedented expansion. In America, by contrast, early factory industrialization took place in the country's largest cities, mainly in the East Coast harbours.

In the second stage, the situation reversed: in Britain development tended to be concentrated in urban centres, while in the United States factories began to leave cities and spread to the countryside. Furthermore, the construction of the railroad network in North America somehow preceded the location of industrial areas, and therefore expanded so that industrialists everywhere were able to take advantage of it.

Finally, it can be said that each country is unique in the way in which the development of a new system of production went together with the shift from an agrarian and mercantile economy to an industrial and capitalist one, because the different ways of accumulating capital echoed the previous social and economic conditions.

Despite all these differences, the Industrial Revolution had the effect of disrupting rural society all over the world, breaking down the urban/rural dichotomy and transforming both the urban and the rural landscape.

Further reading

Benevolo, L. (1967) *The Origins of Modern Town Planning*, Cambridge, MA: MIT Press.
Goheen, P.G. (1996) 'Industrialization and the growth of cities in nineteenth century America', in N.L. Shumsky (ed.) *American Cities*, New York and London: Garland.
Mumford, L. (1938) 'The insensate industrial town', in *The Culture of Cities*, New York: Harcourt & Brace.
Williamson, J.G. (1990) *Coping with City Growth during the British Industrial Revolution*, Cambridge: Cambridge University Press.

PAOLA SOMMA

INDUSTRIALIZATION

Industrialization marks the most fundamental economic, social, institutional, political, and environmental transformation of human life in recorded history. It is a process of transforming a country or region from its traditional agricultural production towards mass production based on modern technologies. Industrialization was originally a European phenomenon that made its decisive appearance in Britain in the second half of the eighteenth century with the INDUSTRIAL REVOLUTION. From there it spread throughout the world, with differences in intensity, timing and variety of technology. Following Britain, rapid industrialization took place in France, Germany and the United States (1870–1910); Sweden (1890–1920); Japan (1900–40); the Soviet Union (1900–60); Italy (1920–40); Canada (1915–50); and Australia (1920–40). In Latin America beginning in the 1930s, and in the rest of the developing world after the Second World War, political leaders favoured development strategies based on rapid industrialization. This was considered essential to post-independence nation-building, and ultimately to raising the living standards of the masses.

Almost every country that has experienced a rapid growth in productivity and improved living standards over the last two hundred years has done so by industrializing. Industrialization has entailed various changes in the spatial, economic and social structure of nations and cities. It has altered the modes of production and consumption of individuals and households; changed income distribution and increased inequalities; caused uneven development within and between countries, ethnic groups and cities; contributed to changes in household and family structure; encouraged mass migration from rural areas to the large cities and their SUBURBS, and accelerated URBANIZATION; changed the spatial pattern of the metropolis and cities; and, above all, it has altered the economic organization of society based on the social division of labour. Importantly, it has contributed to a sharp international division of countries along the lines of the level of industrialization (developed and developing countries), and interests (the G8); and has enhanced regional economic coalitions, such as those in the Middle East and Europe.

Industrialization is the main source of environmental degradation, including pollution, desertification and the catastrophic degradation of the planet's oceans, forests, atmosphere and biological diversity. As a response to the harmful environmental effects of industrialization, a worldwide environmental movement, including environmental activists and NGOs, has questioned, since the rise of environmental consciousness in the 1970s, the scope and ways of industrialization and its mode of production, and has raised the demand for sustainable development.

Industrialization is often understood as the principal agent in the making of modern society. Modernization theory, the subject of many critiques, holds that modernization is fundamentally achieved through industrialization, which is a largely technologically driven process. Therefore, most features of modern society can be traced to the influence of industrialization, which has its own dynamic, independent of social and political factors, and this 'logic of industrialism' will lead to a convergence between industrial societies irrespective of political ideology and social structure.

Further reading

Bensel, R.F. (2000) *The Political Economy of American Industrialization, 1877–1900*, Cambridge: Cambridge University Press.

Hewitt, T., Johnson, H. and Wield, D. (eds) (1992) *Industrialization and Development*, Oxford: Oxford University Press.

YOSEF JABAREEN

INFILL DEVELOPMENT

Infill development refers to the use of vacant or under-utilized land parcels and existing buildings within the CENTRAL CITY of a metropolitan area for the purposes of accommodating growth in lieu of the development of open space or farmland at the URBAN fringe. It is one tactic in a regional SMART GROWTH strategy. It typically involves one or more of three principal activities: the construction of new buildings on land that is currently underdeveloped or undeveloped; the rehabilitation of formerly unusable buildings; and/or the adaptive reuse of existing buildings.

Central cities usually have numerous small lots or larger parcels of land that are no longer used

at all, or that are only partially used. Some were created by unsuccessful urban renewal programmes. Others are properties that actually are, or are perceived to be, contaminated and are referred to as BROWNFIELDS. While much of this land is privately owned, a significant proportion is owned by the local government and is no longer on the tax rolls. Efforts to clean the brownfields and encourage new construction of housing and commercial and industrial uses on these sites constitute a common form of infill development.

When derelict buildings located in the central city are refurbished to make them usable again, another form of infill development has been achieved. Similarly, infill development may be said to occur when buildings that are no longer suitable for their intended use are changed to make them appropriate for a different use. For example, when a single-family house is rendered unusable by a change in surrounding land uses or increased traffic on the thoroughfare upon which it is located, it might be adapted for a use of a commercial nature.

Infill development can afford a variety of benefits to a metropolitan region that is successful in encouraging it. It is less costly to local government because it makes use of existing physical infrastructure rather than requiring the construction of new roadway, water and sewer extensions. It mitigates the loss of farmland and open space at the urban fringe by accommodating new growth in already developed areas. Infill development can provide new or rehabilitated housing in the central city, situated near cultural amenities and employment opportunities. This can, in turn, attract middle- and upper-income residents to the city, enhancing its tax base. By bringing people closer to jobs and sources of entertainment, it helps to limit automobile usage and its related energy and environmental costs. Furthermore, infill development can bring commercial and industrial jobs into the central city, benefiting current residents. It can also foster HISTORIC PRESERVATION, which helps to ensure cultural and architectural continuity. Such preservation has been found to be less expensive than building new structures.

Further reading

Bright, E.M. (2000) *Reviving America's Forgotten Neighborhoods: An Investigation of Inner City Revitalization Efforts*, New York: Garland.

Suchman, D.R. and Sowell, M.B. (1997) *Developing Infill Housing in Inner-City Neighborhoods: Opportunities and Strategies*, Washington, DC: Urban Land Institute.

THOMAS S. LYONS

INFORMAL SECTOR

Economic activities governed by private methods of regulation or informal rules that are outside of the government's legal framework. The seemingly unorganized and chaotic character of economic activities in the developing world generated interest among development economists who introduced the informal sector concept to describe activities that exist outside of government oversight. In the 1970s, International Labour Organization country studies – in particular the Kenya Report (1972) – made the formal and informal sector distinction to describe the URBAN economies of developing countries. According to this dualistic conceptualization, small-scale enterprises that rely on family labour and subsist without a guaranteed stream of income sustain the informal economy. Typical informal sector activities in the developing world involve petty commodity transactions on city streets (e.g. the sale of home-prepared food, clothing, watches, etc.). The activities of such small-scale enterprises are not licensed, regulated or protected by the government. Income generated in the informal economy is not reported to the government and therefore untaxed. In contrast to informal sector jobs, the government regulates formal sector employment, and professionals and civil servants who earn wages hold these jobs.

Much like in the LABOUR MARKETS, informality in the housing markets in the developing world is widespread. Many poor households can only afford housing built in the informal sector and live in informal housing settlements like the FAVELAS in Brazil and GECEKONDUS in Turkey. Not everyone living in informal housing settlements, however, holds an informal sector job, blurring the boundaries between the formal and informal housing sectors. There is empirical evidence in the literature suggesting that characterizing urban economies and housing settlements in dualistic terms (e.g. formal/informal) is too simplistic and can marginalize large groups of people in the informal economy whose contributions to productivity and wealth are otherwise significant. This strand of

analysis looks at the complex interactions between formal and informal sectors to explain disparities in income and in living conditions among groups.

The informal sector debate evolved in the late 1980s and 1990s with an emphasis on regulation. De Soto, in his widely read book *The Other Path* (1989), attributed the existence of the large informal sector in Lima, Peru, to the bureaucratic maze in housing construction. He argued that the multiple steps required for government approval forced developers to engage in informal and small-scale transactions. Others claimed that the legalist approach romanticized the self-employed and the micro-enterprises without paying adequate attention to inequalities. Disagreements remain in the literature.

Because the enforcement of laws and regulations is stricter in the West, the informal sector there is not as widespread as in the developing world. Nonetheless, informality in US labour markets (e.g. day labourers) and housing markets (e.g. the unauthorized conversion of garages into secondary units for rent) is known to exist.

Further reading

Portes, A., Castells, M. and Benton, L.A. (eds) (1989) *The Informal Economy: Studies in Advanced and Less Developed Countries*, Baltimore: Johns Hopkins University Press.

Rakowski, C.A. (1994) *Contrapunto: The Informal Sector Debate in Latin America*, Albany, NY: State University of New York Press.

SEE ALSO: favela; gecekondu

AYSE PAMUK

INFORMATIONAL CITY

The concept of the informational city originated in the 1980s as a response of URBAN theorists to the intensification of information society development and related urban processes. The most thorough early analysis of the informational city was presented by the urban sociologist Manuel CASTELLS in *The Informational City*, published in 1989.

The informational city is a macro-theoretical concept, grounded on an analysis of the relationship between the new information and communications technologies (ICTs) and the urban-regional processes, analysed in the broader context of the historical transformation of capitalism. Thus, this new city formation points to transformed spatial forms and processes as a manifestation of changes in technological and organizational development and the restructuring of capitalism. In this respect, it is fairly close to the concept of the TECHNO-CITY.

A precondition for understanding the essence of the informational city is to understand its broader context, the historical development of the socio-technical organization of advanced societies. Suffice it to say that after the historical states of development usually identified as the agrarian and industrial societies, a new societal formation started to emerge in the 1950s and radicalized in the 1980s and the following decade. It is commonly referred to as the information society, and its underlying logic is called the informational mode of development. Thus, in the sense that the information society or any similar conception depicts a new phase in societal development, the informational city reflects the formation of a new kind of urban setting.

To take the above-mentioned evolutionary view further, it is necessary to note that at the core of the informational society are specific technological arrangements used in production processes. The new technological paradigm emphasized knowledge creation and innovativeness. This new 'informationalism' started to emerge in the 1980s, and has had an enormous impact on society, modifying the material basis of social organization. Thus, just as the INDUSTRIAL CITY reflects the urban-regional aspect of the industrialization of the entire society, the same holds for the informational city. In brief, the term 'informational' indicates a specific form of social organization in which the creation, processing and transmission of information have become the fundamental sources of productivity and power.

In concrete terms, the analysis of the informational city concentrates on such topics as the formation of new industrial space, informational capitalism and space of flows, changes in capital–labour relationships, the dual city phenomenon, and the urban-regional aspects of globalization. This discussion reveals how techno-economic tendencies and social realities pose a challenge to city governments that need to design new urban policies to mediate the global networks of instrumental exchanges and local conditions. The informational city as a self-governing and democratic city government representing the locality needs to develop

e-government, e-democracy and the provision of e-services, and to utilize ICTs in urban-regional development processes, in order to meet the challenge of the information age.

Further reading

Castells, M. (1989) *The Informational City: Information Technology, Economic Restructuring, and the Urban-Regional Process*, Oxford: Blackwell.

ARI-VEIKKO ANTTIROIKO

INFRASTRUCTURE

While cities are and have always been centred on the people who inhabit and shape them, they are material artefacts. While buildings are what first meet the eye, infrastructure both seen and unseen gives cities their form and function. Contemporary societies and their economies could not exist without infrastructure. Infrastructure supports all economic activities and spatial development. Economic productivity rises and falls with investments and disinvestments in infrastructure. Infrastructure is the literal lifeblood of cities, providing water, energy, materials and treatment of wastes. Infrastructure provides a competitive advantage for those cities that have high-quality facilities, and high-quality environments made possible by infrastructure. Infrastructure systems exert a profound and pervasive influence on the shape and growth of cities, and vice versa, reflecting the reciprocal relation between URBAN places and infrastructure.

Infrastructure refers to built facilities and more generally networks of facilities – either above or below ground – that support health, safety and general welfare. Consider the origins of the term 'infrastructure'. Synonymous with the word 'base', it first referred to facilities built below the earth's surface: water, sewer, steam and drainage systems installed under streets. This original meaning has lost clarity as the complexity of our built world has increased over the years. Today, there are many meanings, and indeed, a multiplicity of distinct terms. Each country or jurisdiction has its own definitions. While not exactly synonyms, other terms used to convey infrastructure include 'public works', 'capital facilities', 'capital improvements', 'public facilities', 'utilities', 'public utilities', 'community facilities' and 'public development facilities'.

These words all constitute broad takes on physical capital assets that traditionally have included public and privately owned providers of facilities and systems such as utilities (gas and electricity, water supply and sewerage, waste collection and disposal, stormwater management); public works (roads and bridges, dams and canals, ports and airports, railways, transit and other transportation services); community facilities (schools, PARKS, recreation, hospitals, libraries, prisons, civic buildings); telecommunications (telephone, fax, INTERNET, radio, television, satellites, cable, broadband, multimedia); and knowledge networks (universities, research institutes, corporate research and development, government, philanthropic foundations, libraries, museums, archives).

Knowledge infrastructures, a new category, include human capital investments and research and development capital that support the creation, storage and distribution of knowledge and information. Another new category is that of virtual infrastructure, also referred to as wireless infrastructure or portable infrastructure. These refer to cell (mobile) telephones, notebook computers with wireless modems, and computer networking, among others. Is wireless infrastructure an oxymoron? In one sense, wireless is the ultimate infrastructure, as it pervades any other structure, including portable ones. In any event, wireless systems require plenty of tangible physical facilities, such as satellites, receiving and transmitting towers, switching systems, and more. Regardless of the conception of infrastructure, a key feature is its geographic organization in spatial networks. It is more useful, and appropriate to the etymology of the term, to think of infrastructure as a network rather than an individual project or facility. This framework applies to any scale of network, whether urban, regional, national or global. The interaction among infrastructure networks as they aggregate at territorial scales creates urban units, such as metropolitan regions.

Throughout history, until the land use planning and ZONING revolution of the late nineteenth and early twentieth centuries, city planning was largely infrastructure planning. That is, to shape the future of the city, and to determine the amount, location and type of urban growth, infrastructure was the primary tool to accomplish it. Before land use zoning, the survey and layout of a street

network, squares, plazas, parks and prominent civic buildings were the primary occupations of city planners, regardless of whether they were trained as architects, engineers, landscape architects, surveyors or colonial viceroys. A prime example of this approach was the Spanish LAWS OF THE INDIES of the sixteenth century, which specified street patterns, squares, urban infrastructure and their administration (along with numerous other requirements) for settlements in the Spanish New World.

This model was not limited to the Spanish New World. In early human settlements as disparate as Mesopotamia, Beijing and Ur, the construction of city walls, streets, aqueducts, temples and other civic monuments – all infrastructure – was what defined these settlements as cities, and distinguished them from the countryside, which had other forms of human occupation of land. Over history, other components of infrastructure were developed to further societal development, most of which were urban in nature. In Alexandria, Egypt, the great library and the port were critical to its growth. What helped to define Rome and the Roman Empire were aqueducts, fountains, baths, and roads such as the Appian Way. In the Rome of Pope Sixtus V (sixteenth century), new streets and monuments were used to re-plan the city. In the settlement of new territories, infrastructure played key roles. In the United States, for example, now classic city plans were completely infrastructure-based: Philadelphia (1682), Savannah (1733) and New York City (1811) to name but a few. In the Paris of the nineteenth century, HAUSSMANN's plan was principally the laying out of BOULEVARDS. In nineteenth-century Barcelona, the Catalan civil engineer CERDÀ founded modern city planning by conducting comprehensive surveys of the existing city and by designing multiple infrastructure systems, which were located and planned in accordance with his surveys of existing conditions.

These city planning pioneers sought to improve squalid and unhealthy urban conditions brought on by the INDUSTRIAL REVOLUTION by providing sanitary infrastructure such as sewage disposal, stormwater systems and modern transportation. They knew that to get a sustainable city, they had to build sustainable infrastructure. Then as today, sustainable infrastructure minimizes energy and material consumption, eliminates waste production, provides equitable access to all, and does

not cause harmful impacts through its location, construction and use.

In the United States, for example, most energy is consumed by infrastructure use, or by infrastructure systems, according to the Department of Energy. For example, in the year 2000, 27 per cent of all energy consumption in the United States was by the transportation sector and 38 per cent by residential and commercial structures. When the industrial sector is included, over 50 per cent of all energy was consumed by or in buildings and other structures. Thus, transportation and structures combine to consume over three quarters of America's energy. Similar statements can be made regarding water, wastewater, and materials, including wastes. In our physical environment, infrastructure accounts for significant energy and material consumption and waste production.

One criterion to increase infrastructure sustainability is to design and build it so that it is off-grid, in locally based, decentralized and discrete units; rather than inflexible, large-scale, capital-intensive monopolies, such as highways, other transportation facilities, water and sewer lines, and communications systems. For example, on-site rainwater harvesting – where feasible – which employs free energy (gravity) provides cheaper and purer water than a distant distributive system, with the latter's reliance on extensive collection, purification and distribution systems, each with their own energy, material and environmental costs. Centralized, distributive systems also have negative social and political costs, in that the poor and disenfranchised do not always have equitable access, and facility siting decisions are disproportionately made that have harmful public health impacts.

An integral part of infrastructure systems is overall management. Responsibilities for infrastructure management – planning, financing, designing, operating and repairing – are most often scattered among numerous entities, public and private, and various levels of government, and agencies within any level of government. This fragmentation makes infrastructure difficult to manage, and has led to significant problems. Solving the infrastructure management problem entails knowing how the institutions that build and operate infrastructure work.

There are four major pieces to infrastructure decision-making: planning, budgeting, financing and project management. Typically, one piece, budgeting, stands far above the rest in importance

in guiding decision-making. Moreover, the details of the budget process, and not broad social, environmental and economic goals, tend to guide infrastructure decision-making. Budget process details reward short-term budget balancing and political expediency instead of long-term capital assets management. In addition, typically capital budgeting, as well as its planning and financing, is conducted on a project-by-project basis, rather than by assessing the entire infrastructure network or system.

Financing is a politicized process that involves identifying the money to pay for infrastructure projects and programmes, and draws upon sources ranging from long-term bonds, user fees and dedicated taxes to federal grants and loans. In many public jurisdictions, infrastructure decision-making is finance-driven, and not driven by planning or strategy. This means that the amount of money available to pay for capital facilities determines, to a great degree, how infrastructure needs are assessed.

Ideally, planning precedes budgeting and financing. Planning can be done for five-, ten- and twenty-year time horizons, or longer. Project management, for the design and construction of individual projects, begins after the project has been planned and approved. In practice, planning and project management tend to be compartmentalized, for individual projects or networks, and not done in conjunction with other infrastructure systems, or an overall strategy for growth or improvement. This is due to operating agency procedures and governing legislation, which require planning to be done in a fragmented manner.

Typically, infrastructure planning and financing, while they are linked, often function in reverse. The availability of financial resources delimits what infrastructure is needed, rather than social demands specifying what facilities are needed, and then the funding necessary to build them. In this way, finance has constrained infrastructure development.

There is a wide array of methods used to finance infrastructure construction, operation and repair. The factors contributing to the selection of suitable financing options include, but are not limited to, market conditions, fiscal health, legislation, politics and the knowledge/negotiation skills of parties involved. Financing methods include bonds (two types: general obligation bonds to finance permanent infrastructure, and revenue bonds if the project will generate revenue from user fees), reserve funds (funds saved over time, and thus pooled in advance to pay for future construction), pay-as-you-go or user charges (financing from current revenues), grants from other levels of government, special districts (districts which assess special assessments or impact fees), joint development (public sector and the private sector developer co-finance), and lease/purchase.

Standard infrastructure provided today is limited by existing institutional practices. There are four main institutional limitations. One is infrastructure capacities that are inadequate to serve rapid growth. In this case, infrastructure cannot be provided at a sufficient pace or quantity to support new needs or lessen problems caused by existing infrastructure being overburdened by growth. Another limitation is aging and out-of-date infrastructure in need of repair or replacement. Thirdly, most infrastructure system operations squander resources and inequitably serve users. Finally, there is insufficient financial capacity to fund needs. In order to address these limitations, an institutional approach to remedy infrastructure shortcomings based on the life cycle of infrastructure planning is a preferred institutional approach. This approach posits the life cycle as:

- infrastructure needs assessment – all needs at any jurisdictional scale;
- life cycle costing;
- planning – comprehensive (all infrastructure systems link to the jurisdiction's growth strategy or master plan);
- budgeting – linking budgets to plans, both capital improvement plans and comprehensive development plans;
- financing – various methods that provide continuous, sufficient funding;
- designing – network scale, and the relation of the network to the urban scale;
- infrastructure impact analysis – comprehensive analysis of economic, social, environmental and fiscal impacts;
- construction;
- operation;
- repair and maintenance, rehabilitation, and replacement; and
- reassessment of infrastructure needs (repeat cycle).

Further reading

Batten, D. and Karlsson, C. (eds) (1996) *Infrastructure and the Complexity of Economic Development*, Berlin: Springer.

Graham, S. and Marvin, S. (2001) *Splintering Urbanism: Networked Infrastructures, Technological Mobilities, and the Urban Condition*, London: Routledge.

Hudson, W.R., Haas, R. and Uddin, W. (1997) *Infrastructure Management: Integrating Design, Construction, Maintenance, Rehabilitation and Renovation*, New York: McGraw-Hill.

Neuman, M. and Whittington, J. (2000) *Governing California's Future: Current Conditions in Infrastructure Planning, Budgeting, and Financing*, San Francisco: Public Policy Institute of California.

MICHAEL NEUMAN

INPUT–OUTPUT ANALYSIS

An analytical model used to examine the interrelationships between economic sectors in a nation, state or metropolitan area. It is widely employed in economic development to describe, assess and predict regional economic activity.

Leontif Wassily, an economist, developed input–output analysis in the 1930s. The analysis involves a matrix representation of the interrelationships among the sectors in a regional economy. It begins with the creation of a transactions table. This table displays sales and purchases in dollar amounts among the region's economic sectors in a given year. Transactions are considered either production or final demand. The former activities include sales and purchases among manufacturing, agriculture, service and other industries in the production sector. Households and imports also are shown as selling goods and services to the industrial sectors. For example, households sell labour to the production sectors, and firms outside the region provide raw materials and other inputs to producers. Final demand includes consumers of goods such as households, government, and exports.

The data in the transactions table can be manipulated to derive the direct requirements table. For a particular sector, for example service, the dollar amount of inputs from each sector to service is divided by the total dollar amount of purchases in the service sector and multiplied by 100. This calculation yields the percentage of inputs the service sector purchases from its own or another sector. Each cell of the table before multiplication by 100 represents the amount purchased from each sector for one dollar of output from the service sector; this value is known as a direct effect. The direct requirements table displays valuable information on the interrelationships among sectors and indicates the expected effects of changes in one industry sector on other sectors in the region. This table, however, shows only the effects of purchases in the initial transaction. Additional expenditures will occur as a purchase in one sector triggers additional purchases from and by other sectors; these expenditures are called indirect effects. The direct and indirect effects appear in a third table called the total requirements table. This table is computed using matrix algebra. The values in this table show the cumulative or total effect of one dollar of output in a given sector on that sector as well as the other sectors in the regional economy.

Input–output analysis is a useful tool for understanding the interrelationships among sectors in the regional economy. However, awareness of the model's assumptions, for example unlimited resource supplies and no ECONOMIES OF SCALE, is important in evaluating the results of the analysis. Data requirements also should be fully understood by potential users. While existing data are more readily available now than in the past, some information needs may require costly primary data collection. Given the data are available, the actual computations can be relatively easily achieved by using one of several computer applications available commercially.

Further reading

Miller, R. and Blair, P. (1985) *Input-Output Analysis: Foundations and Extensions*, Englewood Cliffs, NJ: Prentice Hall.

VICTORIA BASOLO

INSURANCE REDLINING

'Communities without insurance are communities without hope.' So concluded a US federal government advisory report in 1968. The reasoning for this observation is straightforward. No lender will provide a home or business loan if the home or business is not insured. So in the absence of insurance, homes cannot be purchased or maintained and businesses cannot be started or

grown. If insurance is available, but at exorbitant costs or on onerous terms, barriers to COMMUNITY development remain.

Problems of insurance availability or redlining persist in older URBAN communities around the world. While the relatively greater risk of loss in some NEIGHBOURHOODS accounts for part of the problem, unfair treatment of similar risks in such neighbourhoods continues and, consequently, undercuts redevelopment efforts. Outside the United States, many governments intervene by offering insurance products and services through the public sector. But in the United States, private insurance companies, nominally regulated by state insurance commissioners, dominate the market. Significant progress has been made in combating insurance redlining, but the stimulus has been non-profit fair housing groups utilizing civil rights laws rather than insurance regulators operating under the insurance laws of their states.

Among the specific practices that have been uncovered are the following:

1   refusing to provide insurance or charging higher costs for less coverage in older CENTRAL CITY neighbourhoods than in newer suburban neighbourhoods;
2   closing insurance agency offices in central cities and opening new offices in the suburban ring;
3   refusing to insure older or lower-valued homes, thus adversely impacting disproportionately minority central city neighbourhoods and favouring predominantly white suburban areas;
4   applying various standards inconsistently at the expense of city and in favour of suburban neighbourhoods; and
5   explicitly discriminating against racial minorities as in the case where a sales manager told an agent to 'quit writing all those blacks ... you got to sell solid-premium-paying white people'.

Since 1995, fair housing organizations around the country have filed lawsuits and administrative complaints resulting in favourable settlements with the largest insurers in the United States including State Farm, Allstate, Nationwide, American Family, Liberty Mutual, and others. In each case, the insurers have agreed to increase the marketing of their products in urban communities, in some cases targeting racial minorities in particular. They have eliminated the use of age or value of homes as a criterion for eligibility for insurance. In some cases, they have committed millions of dollars in reinvestment to support home purchase and repair in previously unreserved areas. And they have agreed to work with local community groups to monitor their progress.

A long-standing tradition of redlining appears to be turning into an emerging pattern of reinvestment. URBAN REVITALIZATION worldwide requires the active involvement of the insurance industry.

Further reading

Galster, G., Wissoker, D. and Zimmermann, W. (2001) 'Testing for discrimination in home insurance', *Urban Studies* 38(1): 141–56.
Squires, G. (ed.) (1997) *Insurance Redlining: Disinvestment, Reinvestment, and the Evolving Role of Financial Institutions*, Washington, DC: The Urban Institute Press.

SEE ALSO: discrimination; redlining; urban revitalization

GREGORY D. SQUIRES

INTELLIGENT TRANSPORT SYSTEMS

In general terms, 'intelligent transport systems' refers to those features of a transport system that improve its management and operation. It assumes the use of new technologies in the fields of electronics, computing and communication to improve efficiency, reduce costs, reduce environmental impacts and improve transport safety. However, usually it suggests that intelligent features provide the transport system with some kind of autonomous ability to control the processes of passenger and freight transport.

This intelligence can reveal itself in different components of the transport system and in many ways. Therefore, it is useful to make a distinction between the features that affect the organization of transport, i.e. transport services, and those qualities that influence the flows of vehicles on the INFRASTRUCTURE, i.e. traffic

services. Wide applications of information and communications technology (ICT), like electronic data interchange and global positioning systems (GPS), have accelerated the use of intelligence in transport systems in both the transport services and the traffic services domain over the last decade.

With regards to the efficiency of transport services, the emergence of more sophisticated and powerful computer systems and software, together with the possibilities offered by the INTERNET, have proved to be very valuable. Some examples of the resulting benefits include easier matching of supply and demand of transport capacity, better tracking and tracing possibilities for freight, better travel information for passengers, and improved coordination between different transport modalities. A new promising development is the emergence of transport agent technology. This technology is strongly based on autonomously controlled transport behaviour, i.e. decentralized and distributed decision entities, which are able to communicate and to negotiate with each other in a predefined way. It implies control of processes without intermediate human interference. The possible applications are wide and go far beyond the use for optimizing transport services. Agent technology in the field of transport is, however, still in its infancy, but much experimentation is underway. Implementation would mean a major step forward in the attempt to achieve more intelligent transport systems.

Probably more eye-catching are the past and current developments to handle traffic flows more efficiently by using intelligent tools. Signalling systems and ramp metering, made possible by ICT, have enabled road traffic management to become more dynamic and have therefore increased its effectiveness. In addition, GPS applications have enabled a fast and safe flow of transport vehicles of other modalities, mainly in sea shipping. Such applications are also increasingly being considered for inland shipping in order to improve transit times. Since the road infrastructure in particular is expected to become increasingly saturated, and since expansion of the network is unlikely to keep pace with the volume growth, there is a need to improve the utilization of existing infrastructure and to maintain the quality of traffic services. This issue will certainly be a seedbed for other intelligent solutions for traffic management in the near future. In this context, vehicle platooning is an emerging and promising phenomenon, but it still assumes the development of certain vehicle systems, such as collision avoidance and smart cruise control. The benefits are substantial: it has been shown that up to three times more vehicles moving in a platoon can pass through a given stretch of highway than can vehicles driven individually. In addition, dynamic trip booking systems, which for instance are widely applied in air transport, but not yet in road transport, could reduce waiting times and congestion.

Without any doubt the most conspicuous exponent of intelligent transport systems technology is automated vehicle guidance (AVG). In contrast to the developments mentioned above, AVG is focused on the behaviour of the individual vehicle and the role of the driver in particular. It assumes that driver tasks can be partly or even completely taken over by means of ICT. Generally, three task levels in AVG functionality are distinguished:

1 Informing and signalling tasks: vehicle control is still a task of the driver; information systems, for example navigation systems, lane departure warning systems, and obstacle or collision avoidance systems, are merely supporting this task.
2 Assisting and correcting tasks: supporting systems are integrated in the vehicle control task, but the driver can overrule these systems. An example is intelligent or adaptive cruise control, which will keep a desired driving speed while ensuring that safe progress is maintained.
3 Intervening tasks: the tasks of the driver disappear; vehicle control becomes autonomous. At this level, unmanned transport becomes an option.

Given the conditions that are required for operating unmanned transport, it is obviously more suited to use in public transport than in private transport. So-called automated people movers are already running successfully in several cities. These systems are able to cater very well to the needs of public transport in URBAN areas by offering transport services on short distances with a high frequency and high reliability. The automation reduces costs by saving on personnel costs,

which account for half or more of the total operational costs in public transport. In addition, it leads to an improvement in the quality of service. Thus, both aspects increase the attractiveness of public transport. The viability of these people movers is, however, largely determined by the fact that they operate on dedicated and segregated lanes in order to ensure safety. This condition is most easily met by rail-guided systems, but due to improvements in traffic control technology, the market for unmanned public transport is now gradually extending to road transport systems. Increasingly, automated guided vehicles or buses are used to carry passengers from airport terminals to parking areas or from public transport stations to business estates, but these systems are still closed system applications. Barriers still exist to expanding automated transport operations to situations where a mixture of manned and unmanned transport could be used. To solve this complexity is of course an even greater challenge for developing automated transport in the privately owned road transport sector. Vehicles that are equipped with AVG systems which inform, assist or correct will become more and more established in the short term, but despite significant research and development efforts on fully automated systems, implementation of such systems is still far off.

The great interest in the development of AVG systems is strongly inspired by several potential benefits, which range from operational cost advantages such as fuel consumption savings, to fewer incidents due to the elimination of human actions, to a better utilization of transport infrastructure. For unmanned transport systems in particular, the possible savings in personnel costs represent an additional benefit.

Further reading

Fabian, L.J. (1999) 'The exceptional service of driverless metros', *Journal of Advanced Transportation* 33(1): 5–16.

Hall, R.W. (1999) *Handbook of Transportation Science*, Boston, MA: Kluwer Academic.

Stough, R. (ed.) (2001) *Intelligent Transport Systems: Cases and Policies*, Cheltenham: Edward Elgar.

Tucker, P. (1998) *Intelligent Transport Systems: A Review of Technologies, Markets and Prospects*, London: Financial Times Automotive Publishing.

ROB KONINGS

INTERNAL STRUCTURE OF THE CITY

This notion is normally addressed from two alternative theoretical perspectives: on the one hand, as limited to a group of man-made physical features; and on the other, and more appropriately, within a framework of a comprehensive and integrated character, embodying a number of traits pertaining to a diverse set of thematic dimensions. As a consequence, the internal structure of the city would in the latter case be the outcome of the particular patterns of behaviour of the diverse traits taken into consideration, and their specific conditions of articulation.

The evolution in time of the behaviour of these traits involves each of them in specific ways, and as a side effect the conditions of their mutual relations usually change as well. These circumstances, involving both the alternatives considered, become highly significant in terms of the patterns of mutation of the internal structure of the city.

The conditions in which the above-mentioned significant traits become articulated are of a systemic nature, in the sense that they are universally related, and consequently, that changes in any one of them implicates the transformation of all the others, and that these relationships establish complex chains of a multiple causal character, subject to change in both their relative levels of precedence and the respective strength of each intervening member.

Furthermore, as has been mentioned, the prevalently large internal differences in the individual behaviour of most of the significant traits involved make it convenient to consider each of them in both aggregate and particular spatial settings, in which different measuring scales are normally applied.

Within the comprehensive perspective that we are considering, a number of thematic sectors and components pertaining to each of them become relevant, as follows:

- *Relative to social phenomena:* aggregate volume of population; birth, and infant and total death rates; gender, age, family composition and household structure differentiation; ethnic origin; educational level by age bracket; occupational structure; income distribution;

socio-economic stratification; patterns of social interaction.

- *Relative to economic phenomena:* sector composition of economic activity; economic base; LABOUR MARKET structure; market price and fiscal taxation of land and buildings; ownership and tenure condition of the fixed property stock; budgetary structure of the local government.
- *Relative to political and institutional organization:* local government structure; patterns of articulation between local, regional and federal governmental organizations; conditions of political representation; participation of the CIVIL SOCIETY in local government.
- *Relative to the natural support structure:* geomorphological features, including geological, altimetric and topographic characteristics; waterfronts, rivers and other surface and underground hydraulic resources; environmental quality of air, water and soil; thematic sectors, physical areas and time settings prone to natural or man-made disasters.
- *Relative to physical features:* land SUBDIVISION; water, rail, road and pedestrian mobility networks; utilities and their service areas; location patterns of singular activities, and service areas of social equipment; extensive specialized functional clusters, in which central (including central business district), institutional, residential, educational, medical, recreational, transportation, industrial and military activities, among others, as well as open areas, become predominant; land occupation and FLOOR AREA RATIOS; residential and other activities, day and night, and gross and net DENSITIES; building profiles; physical condition of the built stock; historic, stylistic and design qualities of the built environment.
- *Relative to functional phenomena:* communication demands; communication flows; passenger and goods transportation demands; car ownership; modal split of transportation; localized transportation flows of passengers and goods by road, rail, water and air.

The prevailing regional and global circumstances and the patterns of external articulation of the city, and the changing character of all of them, become crucial as regards its internal organization. Sometimes, initially, these impacts take place smoothly, or remain barely at a potential level;

eventually, under conditions of enhanced regional connectivity, they become effective to the degree of inducing significant changes in the former status of the internal structure of the city. In such situations, essential elements, as for example the characteristics of its economic base, the patterns in which the globally existing transportation demands and resources come into interaction, and the overall – state or federal – planning determinations becoming mandatory in the local URBAN circumstances, affect the performance of a particular set of urban functions and represent either their increased or diminished evolution, thus becoming crucial.

A rather generalized basic pattern of city structuring became characteristic in Western Europe and Anglo-Saxon North America following the strong wave of URBANIZATION associated with INDUSTRIALIZATION between the last quarter of the nineteenth century and the First World War. This consisted of compact, rather homogeneous urban developments centred upon public, mainly rail transportation, polarized upon rather extensive and diversified single main central areas in which production and consumption components became blended. Later on, increasingly heavy SUBURBANIZATION processes tied to conditions of deepened social differentiation and the generalization of private automobile transportation took place, and a parallel multiplication of scattered production and consumption sites gave rise to a very different type of overall urban organization, of a physically looser and socially highly stratified character.

The internal structure of the city is very highly significant in terms of the relative conditions of global urban efficacy, efficiency, equity and sustainability, to all of which it becomes causally related. As an outcome of their highly embracing and global character, the optimization of the above-mentioned set of notions is normally adopted as the target leading most processes of city planning. In instrumental terms, planning the city implies transforming within particular criteria and in specific respects and directions its internal structure.

Further reading

Castells, M. (1979) 'The urban structure', in *The Urban Question* (*La Question urbaine*), trans. Alan Sheridan, Cambridge, MA: MIT Press.

Stuart, C.F., Jr (1965) 'Toward a theory of urban growth and development', in *Urban Land Use Planning*, 2nd edn, Urbana: University of Illinois Press.

LUIS AINSTEIN

INTERNATIONAL EXPOSITIONS/ WORLD FAIRS

There is general agreement that the Great Exhibition of 1851 in London was the first true international exposition. Its success led to a succession of such expositions. Their scale and popularity grew over the nineteenth century: the London 1851 exposition attracted 6 million visitors, while the figure was 28 million for the 1889 Paris exposition. The first US international exposition took place in Philadelphia in 1876. It was succeeded by the much larger 1893 Chicago World's Columbian Exposition. The major international expositions of the twentieth century took place in St. Louis (1904), San Francisco (1915), Paris (1937), New York City (1939–40 and 1964–65), Brussels (1958), Montreal (1967), Osaka (1970) and Seville (1992). Some of these attracted over 50 million visitors. The first international exposition of the twenty-first century was in Hanover (2000).

International expositions invariably convey a message of optimism about the future and unwavering faith in progress. Emerging technologies were displayed with great fanfare – as in the case of electricity in Chicago (1893) and St. Louis (1904) and space exploration in Seattle (1962) and New York City (1964–65) – and visitors were introduced to new consumer products. Features of traditional fairs, such as amusement parks and a readiness to cater to visitors' taste for exoticism, carried through into international expositions.

The 1851 London exposition took place in the Crystal Palace, but subsequent ones occupied growing numbers of buildings on increasingly larger sites. The buildings of late nineteenth- and early twentieth-century expositions were extravagant expressions of the *beaux arts* style. By then, with their grand avenues, abundant water surfaces and fountains, and floodlit buildings and monuments, the sites of expositions had become as much an attraction as their exhibits. From the 1939–40 New York City exposition onwards, the previous stylistic coherence of buildings gave way to individual expression in the architecture of pavilions.

The tangible URBAN legacy of international expositions is limited to those components that have outlived the expositions in which they were featured: most famously, the Eiffel Tower, but also, for example, the Seattle Needle and monorail, Montreal's Habitat 67, and numerous museums and PARKS. International expositions have also impacted on URBAN PLANNING. The 1893 Chicago and the 1904 St. Louis expositions showcased CITY BEAUTIFUL planning principles. And later, the General Motors Futurama exhibit at the 1939–40 New York City exposition offered an idealized depiction of a 1960 car-oriented metropolis.

The golden age of international expositions may be over. Attendance at the 2000 Hanover exposition was disappointing. These expositions may be losing some of their relevance as means of communication improve and travelling increases. Moreover, DISNEY ENVIRONMENTS duplicate many of their features, and, probably most importantly, the ideology of progress, which pervades international expositions, does not sit well with prevailing postmodern thinking.

Further reading

Mattie, E. (1998) *World's Fairs*, New York: Princeton Architectural Press.
Rydell, R.W., Findling, J.E. and Pelle, K.D. (2000) *Fair America: World's Fairs in the United States*, Washington, DC: Smithsonian Institution Press.

PIERRE FILION

INTERNET

As the world's largest computer network, the internet was born in 1969 out of a cooperative research effort of the US federal government and Department of Defense. Known as the Advanced Research Project Agency Network (ARPANET), its original purpose was to support time-sharing applications among universities, and research and military supercomputer centres; however, ARPANET became the basis for future network interconnection. Soon referred to as the internet, it grew more rapidly after 1985 when the National Science Foundation accepted responsibility for funding additional computer network INFRASTRUCTURE in the United States, all of which would support the same open-architecture network environment and packet switching protocol (Transmission Control Protocol/Internet Protocol,

or TCP/IP). This move sponsored the integration of the internet with existing telephone, fibre optic and satellite systems and commercialization, making the internet what it is today.

No matter what the internet application, whether surfing the web, sending e-mail or streaming video, TCP/IP allows individual messages of data to travel network channels as manageable decomposed parts, or packets, only to be automatically reassembled into the original message at the destination. The internet has transcended the classic space and time constraint in synchronously linking distant places together. Thus, the power of this infrastructure has strong impacts on society. Businesses gain ECONOMIES OF SCALE through substitution for other locational constraints, leading to rapid globalization of markets. Individuals experience streams of data, voice and video from a multitude of sources at their fingertips, allowing them to enrol in online courses, maintain personal finances or participate in virtual communities of interest through online discussions. Government improves service delivery and information dissemination, which provides new electronic public spaces to enhance the interaction between government and its citizenry. Yet none of these internet effects are uniform across businesses, individuals, governments and cities.

With the rise of the internet, another categorization of the 'haves' and 'have-nots' arises. This digital divide exists because unequal access to the internet allows entities with access to take advantage of its power, flexibility and possibility. The case of cities in this internet age is a perfect example. Language labels favourable locations for high-growth business and information flows as 'GLOBAL CITIES', and these cities, like New York, London and Tokyo, are hosts to the international hubs, or Network Access Points, of the internet (Sassen 1991). As telecommunications networks supporting the internet have grown and dispersed, a network of networked cities enables another segment of cities to also compete for business and population (Townsend 2001). Washington DC, San Francisco and Seattle are now major internet hubs. However, these agglomeration trends indirectly affect the reputation and attractiveness of cities without internet hubs or, even more threatening, those that lack adequate internet access for business and citizens. Citizens, business and government must be able to leverage the internet to ensure benefits such as greater educational and

workforce training opportunities, private sector innovation and civic engagement. The internet is a factor that can accelerate the potential of cities with internet access and exacerbate the problems of cities without it. As an invisible infrastructure, the internet is a necessary condition for the survival of cities in this age.

Further reading

Sassen, S. (1991) *The Global City: New York, London, Tokyo*, Princeton, NJ: Princeton University Press.

Townsend, A. (2001) 'Network cities and the global structure of the internet', *American Behavioral Scientist* 44(10): 1697–716.

DARRENE L. HACKLER

ISHIKAWA, EIYO (HIDEAKI)

b. 7 September 1893, Obanazawa Village (now Obanazaw City), Japan; d. 1955

Hideaki Ishikawa majored in civil engineering at the Tokyo Imperial University and graduated in 1918. After graduation, he got a job in the architecture department of the American Trading Company and moved to the Nagoya Regional Commission for Town Planning in 1920 as an engineer. Although Ishikawa had desired to work in Tokyo, he devoted himself to town planning in Nagoya until he returned to Tokyo in 1933. Nagoya City conducted many land readjustment projects to produce subdivided URBAN land in providing public facilities. Eventually, Ishikawa was involved in various land readjustment projects, among which the Naka River canal construction project should not be forgotten as a good example of extensive land EXPROPRIATION for supplying planned development sites. During his assignment in Nagoya, he travelled in America and Europe in 1923 after his trip to China. Ishikawa met Raymond Unwin, the British government's chief engineer for town planning, and was impressed by Unwin's comments on his planning in Nagoya. Ishikawa was transferred from Nagoya to Tokyo in 1933 to be an engineer in the Tokyo Local City Planning Commission and served as acting head of the City Planning Bureau of the Tokyo metropolitan government until his retirement in 1951. His most important task in Tokyo was drafting the plan for reconstructing post-war Tokyo. GREEN BELT plans were designated as the

core of the Tokyo Reconstruction Plan in an attempt to create plazas and waterside PARKS, which Ishikawa considered very important. Ishikawa also incorporated hilltop views, which Unwin had thought important. The Tokyo Reconstruction Plan was implemented in a very limited way because of a restricted budget due to the introduction of the economic stabilization plan in 1949. Ishikawa quit the City Planning Bureau in 1951 to become a professor at Waseda University and at the same time Tokyo's first consultant. He joined the City Planning Institute of Japan in 1951. He died of acute yellow hepatic atrophy in 1955 at the age of 62.

TAKASHI ONISHI

J

JACKSON, JOHN BRINCKERHOFF

b. 25 September 1909, Dinard, France;
d. 31 August 1996, Santa Fe, New Mexico

Key works

(1972) *American Space*, New York: Norton.
(1994) *A Sense of Place, A Sense of Time*, New Haven,
CT: Yale University Press.

John Brinckerhoff Jackson is the pivotal figure in
cultural landscape studies. In the first issue of
Landscape, in 1951, a journal he founded and edited until 1968, Jackson named and defined the new
movement. Landscape was no longer to carry its
traditional meaning, derived from seventeenth-
century genre painting with views to distant natural
scenery or as background for human events. For
Jackson, landscape was the subject itself, composed
of the ordinary settings of everyday life in America –
houses, yards, commercial strips, and workplaces.
Landscape became the voice of the movement,
attracting leading scholars in several disciplines –
landscape architecture, geography, architecture,
journalism, and others. After 1968, Jackson began
a part-time teaching career in landscape architecture at Harvard and the University of California
at Berkeley, while continuing to write, compiling
several book-length volumes of essays.

Before turning his attention to cultural landscapes, Jackson's early life reveals three themes
that shaped his later work. His classical education
in European and American private schools and at
Harvard stimulated an early literary career. During the mid-1930s, Jackson published essays in
American literary magazines and a novel, with his
photograph appearing on the cover of a 1938
Saturday Review. He studied architecture briefly
at MIT and spent a year in formal drawing
instruction in Vienna, preparing him for his lifelong engagement in the visual landscape. He also
immersed himself in the land itself, as a New
Mexico cowboy and rancher in the 1930s, with
decades of motorcycling across America, and
during the Second World War as an army combat
intelligence officer in France. Reading local maps
and guides to help lead his units through enemy
territory convinced him that landscape evidence
could be read like a story, teaching about culture,
history and societal values.

Jackson's essays are like stories, composed
from the broad American landscape of cities, small
towns and rural countryside. The essays make
common things visible, expanding the meaning of
vernacular to encompass the everyday landscape of
trailer parks, flea markets and roadside buildings,
and enabling him to become an early critic of
modern architecture, planning and engineering.
Jackson was among the first to recognize the post-
war landscape of mobility: the commercial strip,
SUBURB, drive-in, and the truck's role in reshaping
the city. In later years, Jackson expanded his vivid
descriptions of the landscape to include the transformation of common places – the house, yard,
street and cemetery – from the medieval world to
the present.

Further reading

Horowitz, H.J. (ed.) (1997) *Landscape in Sight*, New
  Haven, CT: Yale University Press.
Wilson, C. and Groth, P. (eds) (2003) *Everyday America:
  Cultural Landscape Studies After J.B. Jackson*,
  Berkeley, CA: University of California Press.

RICHARD DAGENHART

JACOBS, ALLAN B.

b. 29 December 1928, Cleveland, Ohio

Key works

(1978) *Making City Planning Work*, Chicago: ASPO
  (trans. into Japanese, Minohara, Tokyo, 1998).
(1985) *Looking at Cities*, Cambridge, MA: Harvard
  University Press.
(1993) *Great Streets*, Cambridge, MA: MIT Press.

Allan B. Jacobs received his Bachelor of Archi-
tecture cum laude from Miami University of Ohio
in 1952, attended the Harvard Graduate School of
Design from 1952–53, and received his Master
of City Planning degree from the University of
Pennsylvania in 1954, where he studied under Lewis
MUMFORD. In 1954, he worked as a city planner in
Pittsburgh, Pennsylvania, until 1963, when he
received a Ford Foundation fellowship to work in
India for two years. In 1965, he became an associate
professor of city and REGIONAL PLANNING at the
University of Pennsylvania until 1967, when he
became the director of the San Francisco Planning
Department until 1975. From 1975 to the earlier
2000s, he was on the faculty of the Department of
City and Regional Planning at UC Berkeley, serving
as the department chair, 1977–81 and 1992–95, and
becoming an emeritus professor in 2002.

His *Making City Planning Work*, masterfully
presented in 1978, offers reflections on his experiences as the San Francisco planning director from
1967–75, providing invaluable guidance on how to
navigate the bureaucratic and political processes
that too often hamper the realization of desired
planning policies and outcomes.

As the planning director and through his writings,
Jacobs re-energized the integration of URBAN DESIGN
into local government planning. In particular, he
oversaw the development of a plan that demonstrated how to employ principles of design, scale,
livability and preservation in an effort to comprehensively build upon the needs of the city's individual
neighbourhoods. This plan included the initial work
on street livability by Donald APPLEYARD, also on
the faculty at UC Berkeley when Jacobs joined the
City and Regional Planning Department in 1975.

As colleagues, Appleyard and Jacobs collaborated on many fronts; among them was a seminal
article outlining principles for a more sensible
approach to future URBAN growth and development, 'Toward an urban design manifesto' (1987).

With the publication of *Great Streets* (1993),
Jacobs emerged as one of the most influential
urban design researchers on streets. By comprehensively capturing the elements that make certain
streets successful public spaces as well as effective
travel routes, Jacobs revolutionized the manner in
which academics and professionals viewed major
streets. This, along with other work with Elizabeth
MacDonald and Yodan Rofe, which included *The
Boulevard Book* in 2002, renewed interest in the use
of multiple roadway BOULEVARDS to create great
public places while, at the same time, serving
multiple modes, travelling at different speeds.

Throughout his career, Jacobs has pursued a
practical and grounded approach to planning by
encouraging colleagues and students to develop
powers of observation, a theme he explored in
Looking at Cities (1985). Jacobs fostered in his
students a high degree of creativity accompanied by
a pragmatic and professional foundation to help
them effectively address the needs of their clients, all
the while instilling a passion for city planning.

He has served in a professional capacity in
design and planning agencies and governments
worldwide, and has served as a juror for numerous
important urban design efforts. In 1999, MIT
honoured Jacobs with the Kevin LYNCH Award,
recognizing him for his 'rich blend of research,
teaching and practice, his humanity, and his
dedication to public purposes'.

Further reading

Appleyard, D. (1981) *Livable Streets*, Berkeley, CA:
  University of California Press.
Jacobs, A.B. and Appleyard, D. (1987) 'Toward an urban
  design manifesto', *Journal of the American Planning
  Association* 53(1): 112–20.
Jacobs, A.B., MacDonald, E. and Rofe, Y. (2002) *The
  Boulevard Book*, Cambridge, MA: MIT Press.

SEE ALSO: Appleyard, Donald; Lynch, Kevin;
Mumford, Lewis

BRUCE APPLEYARD

JACOBS (NÉE BUTZNER), JANE

b. 4 May 1916, Scranton, Pennsylvania

Key works

(1961) *The Death and Life of Great American Cities*, New
  York: Vintage Books.

(1969) *The Economy of Cities*, New York: Random House.

In the 1950s, Jane Jacobs wrote for *Architectural Forum*, of which she became an editor. In the 1960s, she fought to protect her West Greenwich Village NEIGHBOURHOOD from street widening, URBAN renewal and the proposed Lower Manhattan Expressway. Her objection to the Vietnam War caused her to leave the United States and move to Toronto in 1968. In Toronto, she maintained her activism, in particular by protesting against the Spadina Expressway, which would have segmented her new neighbourhood (the Annex). In the 1990s, she voiced her objection to the introduction of a tax formula detrimental to the inner city and to the amalgamation of Metro Toronto municipalities into a new City of Toronto.

Between 1961 and 1992, Jane Jacobs published five books. Her first, *The Death and Life of Great American Cities*, has had the most influence on URBAN PLANNING. It challenges all facets of the planning orthodoxy of the time: its more spectacular and controversial interventions such as urban renewal projects inspired by the LE CORBUSIER towers-in-the-park model, and routine planning practices such as those involving systematic land use separation. The book celebrates the diversity and complexity of old mixed-use neighbourhoods, while lamenting the monotony and sterility of modern planning. Jacobs contrasts the safety and stimulation of traditional neighbourhoods' 'intricate ballet of the street' with the desolation and susceptibility to crime of the abundant open space of urban renewal projects. The book objects to the considerable influence of experts and calls for the empowerment of citizens through small-scale governance. The disarmingly simple approach of the book is consistent with its author's distrust of expertise. Jacobs rests her arguments on her own direct observations of cities, many of which originate from her own street. The book was labelled as prophetic. Jacobs can indeed be credited for sounding the alarm about urban renewal when this enterprise was still young, and its worse shortcomings had yet to be experienced.

The impact of this book on urban planning was both immediate and lasting. Jacobs was instrumental in the renewed interest in traditional inner-city neighbourhoods, expressed in the halting of inner-city expressway developments and of conventional forms of urban renewal. This interest has also fuelled GENTRIFICATION and inspired NEW URBANISM.

Jane Jacobs's thinking has been remarkably consistent from the late 1950s onwards. In *The Economy of Cities*, she gives an economic treatment to the themes raised in *The Death and Life of Great American Cities* by linking a mixture of land uses and planning permissiveness with entrepreneurship, innovation and, ultimately, economic development. She revisits the themes of diversity, citizen empowerment and small-scale governance in her subsequent books.

Further reading

Allen, M. (ed.) (1997) *Ideas that Matter: The Worlds of Jane Jacobs*, Owen Sound, Ont.: The Ginger Press.

PIERRE FILION

K

KAHN, LOUIS ISADORE

b. 20 February 1901, Saaremaa,
Estonia; d. 17 March 1974, New York,
United States

Key works

(1951–53) Yale University Art Gallery, New Haven,
  Connecticut.
(1959–65) Salk Institute for Biological Studies, La Jolla,
  California.

American architect. Louis Kahn graduated from
the University of Pennsylvania in 1924, trained in
the *beaux arts* tradition by Professor Paul Cret.
He worked in the office of the city architect of
Philadelphia and was put in charge of the design
for the Philadelphia Sesquicentennial Exhibition
of 1926. In 1932–33, he was organizer of the
Architectural Research Group, focused on city
planning, SLUM REDEVELOPMENT and new concepts in housing. He became professor of architecture at Yale University in 1947, and in 1957 was
named professor at the University of Pennsylvania.

His monumental works are characterized by the
exploration of space and natural light using basic
geometric forms, with brick and concrete as
materials. Departing from the dominant international style, he remained close to the classical
forms. His view starts with making the building
dependent on its function and use, developing the
principle of servant and served spaces.

As a consultant architect for the Philadelphia
City PLANNING COMMISSION, he developed in the
1950s and 1960s several plans for the centre of
Philadelphia that were never executed. His plans
study the relationships between URBAN elements,
focusing on mobility and accessibility from a
different perspective, inserting the car in the urban
space as a key component, redesigning the traffic
pattern. Metaphorically, he conceived STREETS as
rivers and canals, integrating organically parking
buildings as harbours, and docks as accesses.

URBANO FRA PALEO

KELLEY, FLORENCE

b. 12 September 1859, Philadelphia,
Pennsylvania; d. 17 February 1932,
Philadelphia, Pennsylvania

Key work

(1905) *Some Ethical Gains Through Legislation*, New
  York: Macmillan (reprinted 1969 by Arno Press and
  The New York Times).

The eight-hour workday, minimum wages and
child labour laws that Americans now take for
granted were non-existent before progressive
reformers, Florence Kelley among them, waged
crusades against factory abuses in the INDUSTRIAL
CITY. One of a galaxy of prominent women reformers including settlement house leaders Jane
ADDAMS, Julia Lathrop and Lillian Wald, Kelley
pursued her work in their company. Living first at
Chicago's Hull-House and later at New York's
Henry Street Settlement, she observed bleak tenements where underpaid women sewed clothes and
children were crippled by work injuries. Kelley
incorporated her experiences into formal investigations that resulted in significant protective
labour legislation.

Kelley graduated from Cornell University in 1882, where she wrote a senior thesis on the legal status of children. Later, she studied law at the University of Zurich and embraced socialism. While in Zurich, Kelley translated Friedrich ENGELS's *The Condition of the Working Classes in England in 1844* for English publication. Her book and a brief marriage to a Russian socialist haunted Kelley politically throughout her life. In 1924, the Daughters of the American Revolution identified her as a dangerous communist and she was repeatedly attacked for 'nationalizing' children.

In 1892, Kelley arrived at Hull-House. She began collecting data on women who assembled garments at home (the sweating system) and on wage-earning children for the US Department of Labor. Based on her reports, Kelley was appointed chief inspector of factories for Illinois in 1893. Her frustration with the limits of labour legislation led her to pursue a law degree and eventually become a member of the Illinois bar.

In 1899, Kelley moved to New York City as secretary of the newly formed National Consumers League. From the Henry Street Settlement she continued her campaign to improve factory conditions by educating and mobilizing consumers. This work was quite successful, but Kelley recognized that higher pay was also necessary. She proceeded to lobby for a 'living wage' for women while unions were making the same argument for men.

Kelley remained active with the National Consumers League until her death in 1932. One of her accomplishments during that time was the federal Children's Bureau she established in 1912 with Lathrop and Wald. Kelley also organized New York's Committee on Congestion of Population in 1907, after which she and Mary Kingsbury SIMKHOVITCH sponsored an exhibit on the causes and consequences of CONGESTION and methods for alleviating it. That event was the catalyst for the first National Conference on City Planning in 1909. Kelley's career is a reminder of the importance of women's state and local leadership in shaping national action.

Further reading

Sklar, K.K. (1995) *Florence Kelley and the Nation's Work: The Rise of Women's Political Culture, 1830–1900*, New Haven, CT: Yale University Press.

DAPHNE SPAIN

KENT, THOMAS J. (JACK), JR

b. 30 January 1917, Oakland, California;
d. 26 April 1998, Berkeley, California

Key work

(1964) *The Urban General Plan*, San Francisco: Chandler.

Noted Californian URBAN and regional planner, Kent graduated in 1938 with a BA in architecture from the University of California, Berkeley, where in 1948 he founded the Department of City and Regional Planning, one of the most influential URBAN PLANNING programmes in the world. During 1938–39, he studied with Lewis MUMFORD in Europe and became acquainted with the Abercrombie Plan for London and its proponents. Returning to the United States in 1939, he co-founded Telesis, an influential organization that pioneered urban, regional and environmental planning in northern California. In the 1940s, he served as director of city planning for San Francisco, and influenced that city's development for half a century.

A principal author of the City of Berkeley's first master plan (1955), Kent served on the Berkeley City Council from 1957–65. In 1958, he co-founded a Bay Area citizens' REGIONAL PLANNING and conservation group now known as the Greenbelt Alliance, which became a model and inspiration for similar NGOs around the world. In 1961, he helped organize the Association of Bay Area Governments, a principal regional planning agency. His text, *The Urban General Plan*, remains influential, and was given National Landmark status by the American Institute of Certified Planners in 1990.

RICHARD W. LEE

KOOLHAAS, REM

b. 1944, Rotterdam, the Netherlands

Key works

(1987) Netherlands Dance Theatre, The Hague, the Netherlands.
(1996) Euralille's Master Plan, Lille, France.

Dutch architect, representative of deconstructivism and author of *Delirious New York: A Retroactive Manifesto for Manhattan* about modernist architecture, Rem Koolhaas graduated from the Architecture Association School in London. He later worked at Cornell University with Professor Ungers and at the Institute for Architecture and Urban Studies in New York.

Founder of the Office for Metropolitan Architecture (OMA) in London (1975), Koolhaas headed some housing projects before designing the Netherlands Dance Theatre in The Hague (1987), acknowledged as 'one of the nine greatest buildings of the twentieth century'. He was then commissioned to design the Kunsthal, Rotterdam (1992), a museum for temporary exhibitions, and the Educatorium, Utrecht (1997), a facility building for the University. His largest work is probably the Master Plan of Euralille (1996), which consists of the Lille Grand Palais and facilities for retail, culture and exhibitions.

After moving back to the United States, he designed the new Universal Studios Headquarters in Los Angeles as well as the new Campus Center of the Illinois Institute of Technology in Chicago.

Since 1995, he has been a professor at Harvard University, leading research tasks on worldwide URBAN environments.

Among his accolades are the American Institute of Architects Book Award (1997) and the Pritzker Architecture Prize (2000).

Further reading

Lucan, J. (1991) *OMA – Rem Koolhaas: Architecture 1970–1990*, New York: Princeton Architectural Press.

LUIGI BIOCCA

KRIER, LÉON

b. April 1946, Luxembourg, Luxembourg

Key works

(1992) *Léon Krier: Architecture and Urban Design 1967–1992*, London: Academy Editions.
(1998) *Architecture: Choice or Fate*, Windsor: Andreas Papadakis Publisher.

Léon Krier is a vociferous, traditional urbanist and architect. Younger brother of Rob KRIER, Léon is distinguished by his witty, polemical drawings and writing. He is perhaps best known for his master plan for Poundbury, in Dorset, England, Krier's first significant attempt to put his theory into practice. Krier developed the 1989 master plan for the project and oversees implementation in conjunction with its instigator, HRH The Prince of Wales. Krier has been a long-time personal adviser to the Prince on URBAN matters. In that capacity, Krier was a foundation member, the original theoretician, of the Urban Villages Group (now the Urban Network, part of the Prince's Foundation) in the United Kingdom (see URBAN VILLAGE). In addition, Krier is widely regarded as the intellectual 'godfather' of the Congress for the New Urbanism in the United States (see NEW URBANISM). Ultimately, Krier's influence via such groups has spread to other professional networks, as well as into European, American and Australasian institutional URBAN PLANNING policy.

Krier studied architecture for a short period at the University of Stuttgart. In the late 1960s and early 1970s, Krier worked with architect James Stirling in London. After that, reluctant to compromise his principles by being involved in building himself, Krier concentrated on drawing, writing and teaching. Krier's calling has been strongly directed by the adverse effects he perceives post-Second World War development has had on European towns. He is passionate in his critique of contemporary cities, and maintains a fervent commitment to defending and endorsing identifiable communities. In his work, Krier depicts a disintegration of the city over the last fifty years in terms of its patterns and public realm. This has resulted in built areas that, in his eyes, are 'anti-cities', based on over-expansion and segregated activities. In response, he battles for the 'reconstruction' of the genuinely urban city. The influence of Camillo SITTE is apparent in Krier's solutions for the city. He has taken a rationalist approach to seeking out the foundations of the city in terms of its enduring principles of design and form. His normative, organic paradigm for the city is a polypolis, based on the building block of the multifunctional urban quarter.

Since the 1990s, Krier has been more prolific in terms of bringing his ideas to concrete fruition, designing furniture (for Giorgetti, Italy), buildings (in Luxembourg, Portugal, Italy, the United States) and master plans (in the United Kingdom, Belgium), among other things.

Exhibitions of his work have been held throughout the world (in particular, personal exhibitions at the Museum of Modern Art, New York (1985), and in Bremen (1999)).

Krier has taught at the Architectural Association, the Royal College of Arts and the Prince of Wales's Institute of Architecture (all in London), and at Princeton University, the University of Virginia and Yale University (all in the United States).

Tributes to Krier include the Berlin Prize for Architecture (1977), the Jefferson Memorial Medal (1985), the Chicago American Institute of Architects Award (1987), the European Culture Prize (1995), the Silver Medal of the Académie Française (1997) and the inaugural Richard H. Driehaus Prize for Classical Architecture (2003).

Further reading

Thompson-Fawcett, M.M. (1998) 'Leon Krier and the organic revival within urban policy and practice', *Planning Perspectives* 13(2): 167–94.

SEE ALSO: polycentric city; public space; Sitte, Camillo; urbanism; vernacular architecture

MICHELLE THOMPSON-FAWCETT

KRIER, ROB

b. 10 June 1938, Grevenmacher, Luxembourg

Key works

(1979) *Urban Space*, New York: Rizzoli.
(1988) *Architectural Composition*, New York: Rizzoli.

Rob Krier is an architect, city planner, sculptor and author. He was educated at the Technical University of Munich (1959–64). From 1976–98, he held a professorship at the Institute for Architectural Design, Technical University of Vienna, and concurrently worked in architectural practice. For nearly twenty years, his office was in Vienna, and since 1993 has been Berlin-based, in partnership with Christoph Kohl.

Krier is a supporter of NEW URBANISM. He implements developments that reflect the form, street pattern, typology and architecture of traditional European cities. Krier's work pays respect to local history, emphasizing spatial context and

harmony with the existing built environment. Krier endeavours to enliven traditional approaches by using contemporary resources and innovative techniques. Krier is anti-MODERNISM. He advocates an architecture that is practical but not simply utilitarian. He sees the future of the city as lying in the downscaling of architecture, diminishing sensational elements, relying on variation from stylistic individuality and simplicity.

Krier's theoretical and practical contributions are evident in several key projects, including Ritterstrasse (1977–80) and Rauchstrasse (1980) in Berlin, Breitenfurterstrasse in Vienna (1981–87), and Kirchsteigfeld in Potsdam near Berlin (1992–97). Examples of Krier's master planning and architecture can also be found elsewhere in Europe, particularly the Netherlands. In terms of sculpting, Krier is notable for pieces in PUBLIC SPACES (e.g. in Barcelona, Berlin, Pforzheim, Potsdam, Luxembourg), reflecting his desire to reawaken an awareness of the public realm.

Further reading

Krier, R. (2003) *Town Spaces: Contemporary Interpretations in Traditional Urbanism, Krier Kohl Architects*, Berlin: Birkhäuser.

SEE ALSO: city typologies; Krier, Leon; sense of community; urban design; urbanism; urbanity

MICHELLE THOMPSON-FAWCETT

KROPOTKIN, PETER

b. 9 December 1842, Moscow, Russia;
d. 8 February 1921, Dmitrov, Russia
(buried in Moscow)

Key works

(1899) *Fields, Factories and Workshops: or Industry Combined with Agriculture and Brain Work with Manual Work*, London: Hutchinson.
(1902) *Mutual Aid: A Factor of Evolution*, London: Heinemann.
(1906) *The Conquest of Bread*, London: G.P. Putnam's Sons (first published as *La conquête du pain*, Paris: Stock, 1892).

Russian anarchist, geographer and prince, Peter Alekseyevich Kropotkin is internationally recognized both as an activist as well as a geographer.

After his secondary studies, he went as a military officer to Siberia, in 1862, took part in a geographical survey of Manchuria, in 1864–65, left the army in 1867 and returned to St Petersburg where he became a member of the Imperial Russian Geographical Society. In 1871–72, he travelled to Switzerland and adopted the anarchist ideals under the influence of the Paris Commune. He was arrested and imprisoned for anarchist views in 1874 in Russia. In 1876, he escaped from prison and went into exile in England, Switzerland and France and became a friend of Elisée Reclus. While in exile, he gave lectures and published widely on anarchism and geography. He returned to St Petersburg in the early days of the Russian revolution.

He left a remarkable intellectual legacy, being against all forms of massification. His texts provide a good example of the utopian URBANISM proposed at the end of the nineteenth century as a solution to the contradictions of the CAPITALIST CITY. He is clearly a 'de-urbanist' as he was against large URBAN agglomerations and in favour of an integration of town and country life, of agriculture and industry. But he refused to draw a definitive model of the future city form, as many radicals and utopians had done, because he thought this would be the work of a myriad of peoples in the decentralized society he advocated. He also influenced the concept of the GARDEN CITY proposed by Ebenezer HOWARD, which was, in a certain sense, a vision of anarchist cooperation.

Kropotkin's writings became influential through the first adherents of the radical geography approach, in the 1970s, especially *Fields, Factories and Workshops* (1899), in which he advocated the DECENTRALIZATION of industry combined with intensive agriculture and a combination of brain work and manual work, and *Mutual Aid: A Factor of Evolution* (1902), where he developed his system of economic cooperation that would make unnecessary any form of highly structured government. For Kropotkin, societies should be organized in free associations and geographically decentralized, in which the dichotomy between rural and urban areas would disappear, to be gradually replaced by a new and more egalitarian territory.

Further reading

Choay, F. (1965) *L'Urbanisme, utopies et réalités: une anthologie*, Paris: Éditions du Seuil.

Peet, R. (1998) *Modern Geographical Thought*, Oxford: Blackwell.

SEE ALSO: utopia

CARLOS NUNES SILVA

KRUMHOLZ, NORMAN

b. June 1927, Passaic, New Jersey

Key works

(1990) *Making Equity Planning Work*, Philadelphia, PA: Temple University Press (with J. Forester).
(1994) *Reinventing Cities: Equity Planners Tell Their Stories*, Philadelphia, PA: Temple University Press (with P. Clavel).

Norm Krumholz is professor of URBAN PLANNING at Cleveland State University, former director of the City of Cleveland Planning Commission, and former president of the American Planning Association and American Institute of Certified Planners. As planning director in Cleveland from 1969 to 1979, Krumholz and a dedicated core staff put forth the idea that planners should not act as politically detached, technical experts serving a vague unitary public interest. Instead, Krumholz argued, planners should play a politically active role in favour of the disadvantaged. In 1971, the group produced *Toward a Work Program for an Advocate Planning Agency*, which contained the now famous statement: 'In a context of limited resources, priority attention should be given to the task of promoting wider choices for those individuals and groups who have few, if any, choices.' Later, Krumholz called this planning approach EQUITY PLANNING.

Krumholz came into planning after an eight-year stint as a businessman. Inspired by the sharply diverging views on planning of Lewis MUMFORD and Robert MOSES that appeared in *The New Yorker*, he enrolled at Cornell University where he graduated in 1965 with an MA in city planning. On graduating, he accepted a job as assistant director in the Pittsburgh City Planning Department. In 1969, Mayor Carl Stokes and the Cleveland Planning Commission recruited him.

NICO CALAVITA

L

LABOUR MARKETS

There is widespread evidence that inner-city areas are characterized by an important concentration of labour market disadvantage. Two competing theories seek to explain the barriers to employment in URBAN locations. The prevailing skills mismatch perspective is essentially concerned with supply-side aspects. It emphasizes the gap between the education and skills held by CENTRAL CITY residents and those required by employers in a modern economy; it assumes that the local workforce can, but fails to, display a high level of occupational and spatial mobility, resulting in employment vacancies being filled by in-commuters. In contrast, proponents of a spatial mismatch point to local demand deficiencies, or the existence of a 'jobs gap'. They argue that SUBURBANIZATION and the deconcentration of jobs have caused high unemployment and under-employment in the metropolitan cores. In other words, the location of suitable jobs does not match the residential location of the unemployed who face spatial frictions when trying to access employment.

Local labour markets operate in ways that are structured by specific patterns of economic activity and socio-institutional determinants. These bear a strong influence on worker mobility and commuting patterns. Urban scholars have emphasized that the combined operation of the labour and housing markets establishes mutually reinforcing barriers to out-commuting and out-migration. The poor, unskilled and racial-ethnic minority populations remain trapped in deprived inner-city NEIGHBOURHOODS where they initially access low-cost housing. The low-wage and insecure service sector jobs found in these locations imply little

opportunity for progression on the occupational and pay ladder, which would in turn facilitate residential mobility. These factors are further compounded by inadequate or unaffordable transport, training and other public services, poor information about suitable vacancies outside the vicinity, often due to a lack of social networks, and employer DISCRIMINATION against inner-city workers. This residential SEGREGATION further constrains individuals' spatial horizons through its impact on social and cultural norms. Thus, both the financial and psychological costs of commuting to work may prove prohibitive to inner-city residents. For example, numerous studies have shown that large CONURBATIONS such as London or Los Angeles cannot be treated as single confined labour markets, but as arenas that generate a strong fragmentation of pay and employment conditions and the emergence of a 'contingent' workforce.

The focus on 'employability' found in decentralized WELFARE TO WORK programmes seeks to reconcile the skills and spatial mismatch perspectives through improved job search and commuting incentives, and more rarely, some consideration for demand-side initiatives. Several cities have established intermediate labour markets that offer subsidized training and employment opportunities to disadvantaged workers, and intermediary organizations that endeavour to change employers' perceptions and recruitment behaviour.

Nonetheless, a large body of empirical work in the Anglo-American literature has highlighted that the geography of labour market inequalities is proving resistant to predominantly supply-side oriented policy change, particularly in depressed metropolitan areas.

Further reading

Martin, R. and Morrison, P. (2003) *Geographies of Labour Market Inequality*, London: Routledge.

Simpson, W. (1992) *Urban Structure and the Labour Market: Worker Mobility, Commuting and Underemployment in Cities*, Oxford: Clarendon Press.

CORINNE NATIVEL

LAND BANKING

Land banking usually involves the acquisition of privately owned undeveloped land by a government entity. It originated in the 1920s and 1930s as a means of making low-priced land available for housing and ensuring orderly development. In most land banking programmes, the government acquires the land, retains total control of the type and timing of development on it, and leases development rights while maintaining ownership. These programmes can work only with widespread public support, clear legal backing and sufficient finances. However, they could stress public budgets, curtail expansion of property tax bases and lead to higher tax rates.

Successes in Europe and Canada encouraged use of land banking in THIRD WORLD CITIES in the 1960s and 1970s. Unfortunately, these programmes were not very successful because the administrative and financial burdens of land banking were beyond the institutional capacities of most municipalities. In the United States, because of a general aversion to government control of land, land banking has been seldom used. But some local governments in the United States have recognized its potential as a land conservation tool and experimented with it for open space protection and GREEN BELT creation, a few financing land acquisition using transfer of development rights.

More recently, the need for complying with environmental regulations has spurred the creation of mitigation land banks. These are managed by private companies, who, for example, buy undeveloped land, acquire permits to restore wetlands on them, agree to maintain these wetlands forever, and sell wetland mitigation credits to developers to facilitate development elsewhere.

JERRY ANTHONY

LAND SPECULATION

Land is a special commodity. Its supply is fixed and it is needed for all production. At the same time, land rent is an income for which no production is needed. So the price of land is not defined by the production costs but by the land use. These circumstances make the land market vulnerable to speculation. Land speculation can be defined as temporarily holding land out of use or withholding land from development in anticipation of some eventual gain. Possibilities to acquire these speculative gains occur when it is believed that future land development will be much more gainful than development for the current market.

Land speculation should not be mixed up with windfall profits made by original landowners, when they sell land for a more profitable use. Only when they intentionally withhold land from the market to increase their gain do we use the notion of speculation. Speculation is also involved when a site is sold to an agent (the land speculator) with knowledge of future rises in land rent, without sharing this information with the vendor. Land speculation is therefore always intended and never accidental.

In theory, each site can be used for any activity or use. However, the net return on land use will be greatest if it is used for the activity for which it is best suited. Because of changes in demand, the amount of land needed for each activity changes regularly and transactions in use will occur. When land is transferred to a better-suited use, its price rises. Speculation occurs when land prices start to rise beyond the site's current use value in anticipation of a future transaction in land use.

In general, these transactions are imperfectly predictable. So, knowledge and information on future use transaction before they become generally known are crucial to acquire speculative gains. Knowledge can consist of acquaintance with spatial processes. Information important to acquire a speculative gain consists of either information about demand for a specific site or information about changes in land use regulations.

Land speculation distorts the land market and shifts income unjustly to landowners. This shift of income is unjustified because it is unearned – no productive action was taken – or transactions in land use are the result of public decision-making. In the latter case, it is argued that it is unjustified to acquire private gains as a result of decisions

made in the public domain. Land speculation distorts the land market because it withholds land from the market, which may force land use to shift to less productive sites. Also, the process of speculation eventually can raise land rents beyond the marginal ability of its most productive use. To prevent these effects of land speculation, taxes on land values can be raised. However, the efficiency and/or effectiveness of such taxes to discourage land speculation is contested.

Further reading

Archer, R.W. (1973) 'Land speculation and scattered development: failures in the urban-fringe land market', *Urban Studies* 10(3): 367–72.

Foldvary, F.E. (1998) 'Market-hampering land speculation: fiscal and monetary origins and remedies', *American Journal of Economics and Sociology* 57(4): 615–37.

ERIK LOUW

LAND TRUSTS

A non-profit organization created to preserve land for some public purpose. Land trusts have been formed to protect agricultural land, open space, watersheds, cultural assets and other resources tied to real property. These organizations gained popularity in the United States in response to a reduction in federal land conservation efforts. In fact, land trusts exist in every state in the United States. They are also found in other countries including the United Kingdom, Australia and Canada.

A trust acquires land to protect it from development forces. The land may be donated or purchased with funds given to the organization by members and other supporters. These donations to the trust allow benefactors to contribute to the community, while earning a tax deduction. A trust may not have all the available funds on hand to purchase a piece of land. In this case, the trust may borrow funds from banks or individuals for the acquisition, and pledge future donations for repayment of the loan.

Some land trusts secure conservation easements to protect public resources. With the easement approach, the landowner retains title to the land but gives an interest in the land to the trust. The trust's interest typically involves a condition on the property's title that either prevents specific

development on the land or maintains the current use of the land. For example, a common easement secured by trusts is for the preservation of agricultural land. The landowner grants the easement to the trust, but continues to farm the land and profit from these activities. In addition, the owner can sell the land; however, the easement generally stays with the title, in perpetuity, and future owners must also keep the land in agriculture. The easement alternative has several advantages. First, it typically costs very little to acquire an easement. Secondly, the trust achieves conservation of the land, but without the responsibility of maintaining the property. Thirdly, easements often are much easier to acquire than outright title to the land. Conservation easements, however, can be problematic for land trusts, especially if they are challenged in court. Costs of a legal battle can be prohibitive to smaller trusts.

The land trust model has also been effectively applied to the preservation of affordable housing. This variant of the model is called a community land trust. The goal of the community land trust is to ensure that housing remains permanently affordable to lower-income persons. The trust may purchase land, existing housing, or develop new housing. Housing may be for rental or ownership by low-income persons. For the latter, the trust retains title to the land and the owner has title to the home. This arrangement gives the homeowner an opportunity to build limited equity through the home purchase, while the trust controls escalating costs from land speculation.

Further reading

Roakes, S.L. and Zwolinski, M. (1995) *The Land Trust as a Conservation Tool*, Chicago: Council of Planning Librarians.

Wright, J.B. (1992) 'Land trusts in the USA', *Land Use Policy* 9: 83–6.

VICTORIA BASOLO

LAND VALUE CAPTURE

Land value capture refers to the recapture by the public sector of increases in property values through the use of fiscal or regulatory policy tools. The increases in land value may occur due to market forces, government approvals, i.e. permissions for development, through government

provision of INFRASTRUCTURE, or a combination of these factors.

The concept of land value capture relates to the premise that land values in urbanized and urbanizing areas have or will increase above a base such as agricultural value. Consequently, there is some increase in land value that will accrue to the properties. The concept of land value capture relates to the attempt to capture these 'unearned increments' in land value that were identified by David Ricardo as economic rents. They are referred to as 'unearned increments' because they may not be the result of any productive activity by the property owner, but rather result from either public decisions or market forces responding to the public investment. There are several questions that need to be understood regarding this issue. Who created the increased land value? Who should capture it? If it is appropriate for the public sector to capture it, how can it be accomplished?

Increases in land value may occur due to investments or activities of the private sector; or investments or activities undertaken by the public sector. These are reflected by the market response of increased demand for the specific land that has been affected by the private or public sector activity. In addition, there may be simple market factors that lead to an increase in land values due to increased demand for land in a specific city or location.

On the private sector side, increases in land values may arise because the private sector, i.e. a land developer, engages in the planning process and in so doing either creates a planned community that makes the land/location a desirable place to locate, which leads to increased value; or obtains planning approvals/permission, which makes the land relatively more valuable because it is closer to being developed and has a higher value than land that has not obtained approvals. In this latter case, it could be argued that it is the public sector through its approval, or both sectors jointly, that has created the value.

In other cases, the public sector can clearly be identified as the actor who through its investment in infrastructure increases land value. This occurs through the provision of transportation infrastructure, sewer and water facilities, or other public facilities.

In addition to the question of who created the increased land value, the other side of the question is who should capture the land value increases. Where value has increased through public investments, decisions, or general market conditions, i.e. not through the investment or efforts of the landowner, it is often argued that the land value increase should be captured by the public sector. This was the argument of Henry GEORGE in the late nineteenth century. He suggested that the unearned increment or economic rent could be captured through the use of site value taxation. This early application demonstrates an example of a tax or fiscal policy that may be used to capture increased land values for the public sector. However, it is important to recognize that there are fiscal, tax and regulatory tools that may be used to capture increased land values for the public purse. In addition to the suggested use of site value taxation that has had a number of applications since the writings of George, there are a number of early applications that have taken place in England, some of which predate George's writings by hundreds of years. These early applications are consistent with what we refer to today as special assessments. The first documented case in England dates to 1250 in what is known as the Romney Marsh case, where a local ordinance allowed residents to be assessed for the repair of a sea wall. There were other documented examples for sewer projects in 1427, and road repairs in 1576. However, the first application of a special assessment based on changes in property value occurred in 1890 when London County Council applied special assessments to properties within a defined betterment area to help pay for the cost of road and bridge widenings. Throughout the twentieth century there were further studies and attempts at these approaches, with perhaps the best known being the Betterment Levy that was in place from 1967–71.

In more general applications today we see a number of fiscal and regulatory tools that are applied or considered by governments as policy tools that either intentionally or unintentionally have the impact of land value recapture. The fiscal tools that are applied include both taxes and fees. Examples of the former include the land value (site value) tax, property tax schemes that place higher values on land relative to improvements (structures), land speculation taxes, capital gains taxes and tax increment financing. Fees include the use of special assessment district fees, and impact fees. The regulatory instruments sometimes overlap in their terminology and applications as different

names may be used for similar applications in various jurisdictions. These instruments include bonus ZONING or density bonuses, inclusionary zoning, linkage fees, transfer of development rights schemes, EMINENT DOMAIN or EXPROPRIATION, LAND BANKING and leasing programmes, land readjustment programmes, and various applications of PUBLIC–PRIVATE PARTNERSHIPS. As each of these tools will have its own objectives, sometimes multiple in nature, and have various designs in different institutional settings, it is not possible to go beyond a simple listing of the tools. However, applications of land value capture clearly have a long history and still occupy a central place in both theory and practice in land policy today.

Further reading

Brown, H.J. (ed.) (1997) *Land Use and Taxation: Applying the Insights of Henry George*, Cambridge, MA: Lincoln Institute of Land Policy.
Hagman, D. and Misczynski, D. (eds) (1978) *Windfalls for Wipeouts: Land Value Capture and Compensation*, Chicago: Planners Press.
Smolka, M. and Amborski, D. (2000) 'Value capture for urban development: an inter-American comparison', Lincoln Institute of Land Policy Working Paper.

DAVID AMBORSKI

LATCH-KEY KIDS

(Also known as latch-key children.) 'Latch-key kids' is a vernacular term for children who are not supervised by adults, usually after school and until the parental working day ends. The term can refer to children as young as 5 years old who provide self-care or to older children who supervise their younger siblings. As women joined the formalized workforce in large numbers in the twentieth century and few childcare opportunities were made available, more children returned home to a house without adults after school. The term often has a negative connotation because of societal ambivalence about women in the formal workforce and problems (such as emotional damage or greater access to illicit items or activities) often associated with unsupervised children. Politically, the term has come to refer to children who are neglected. Research has not yet demonstrated a clear connection between children left unattended at home after school and neglect; however, some have raised concerns about the loneliness, boredom and

alienation which appear to arise as children and adolescents take on increasing responsibility for themselves outside of adult contact. The term is also associated with the decline of neighbourly qualities in an area; not only are the parents not around to supervise children, neither are the neighbours, themselves likely workforce participants, available to help keep an eye on NEIGHBOURHOOD kids.

Further reading

Hersch, P. (1998) *A Tribe Apart: A Journey into the Heart of American Adolescence*, New York: Ballantine.
Lamorey, S., Robinson, B.E., Rowland, B.H. and Coleman, M. (1999) *Latchkey Kids*, 2nd edn, Thousand Oaks, CA: Sage.

KIMBERLY KNOWLES-YÁNEZ

LAWS OF THE INDIES

Near the end of the sixteenth century, King Phillip of Spain set forth 148 ordinances for the organizing of settlements in New Spain. These ordinances would be used throughout what is now called South America, Central America, Mexico and the US American West. They were designed to guide the location of new settlements, manage city growth, protect public health, direct the design of major PUBLIC SPACES and buildings, and influence relations with native peoples. Classic Roman city design elements, and the French-style *bastide* town schemes, influenced site location, use of a grid system and the main plaza design.

While enacted on 13 July 1573, they represent the culmination of a longer history of planned Spanish settlement in the New World, starting with Santo Domingo in 1493, Panama City in 1513, and continuing northwards to San Antonio (United States). The Spanish crown wanted the settlements planned with uniformity. This made it easier for new residents to successfully occupy towns and provide a means by which the Church could spread its faith to the local Indian population, and assist in their religious conversion.

The ordinances provided instructions on choosing the town location such as selecting sites of moderate elevation allowing for north–south winds. Locations with good defence, land fertility, water, good roads, the possibility of passage by water and locally available building materials were specified. They mandated the orderly layout of

a town as it began to grow from the initial main plaza, church and housing and their replication in other town areas.

The main plaza was the town's central feature. It could be square or rectangular, with the rectangular dimension of length being one and one-half the width. The corners of the plaza were to point to the cardinal directions in order to lower the harshness of the winds. Secular, social, political and religious activities occurred in the plaza. It was the most important public space. Government and the main commercial buildings were placed there with portals on the plaza buildings to provide protection from rain and sun. The church was to be located adjacent to the plaza, and would be the dominant building in size and grandeur. A hospital was to be built near the main church. When the town was located on the sea coast, the main church had to be seen going out to sea.

Residential areas surrounded the main plaza. For the sake of civic beauty, the use of only one building style was recommended. Houses were arranged to provide a defensive barrier against invasion. Outside the central plaza area a common area of recreation and cattle grazing was to be located, followed by agricultural parcels. These were distributed by lottery. For defence reasons, local native peoples were not allowed to live inside the settlements.

Further reading

Crouch, D.P., Garr, D.J. and Mundigo, A.I. (1982) *Spanish City Planning in North America*, Cambridge, MA: MIT Press.

Violich, F. and Daughters, R. (1987) *Urban Planning for Latin America*, Boston, MA: Oelgeschlager, Gunn & Hain.

WILLIAM J. SIEMBIEDA

LE CORBUSIER (BYNAME OF CHARLES-ÉDOUARD JEANNERET)

b. 6 October 1887, La Chaux-de-Fonds, Switzerland; d. 27 August 1965, Cap-Martin, France

Key works

(1946 [1923]) *Towards a New Architecture*, London: Architectural Press.
(1957 [1942]) *La Charte d'Athènes*, Paris: Les Éditions de Minuit.

The professional life of Le Corbusier can be divided into two periods. The first runs from 1908 to the end of the First World War. This is when he developed his approach to architecture and URBAN PLANNING and became a foremost voice of the modern movement. He presented his views on architecture and cities in numerous books and in master plans for Paris, Algiers and Buenos Aires (none of which was implemented). His major architectural achievements over these years include the Villa Savoye in Poissy, France, the Palace of Centrosoyus in Moscow and the Ministry of Education and Public Health in Rio de Janeiro. However, this period was filled with frustration as many of his projects were ignored or turned down (most notoriously his plan for the League of Nations building in Geneva), in large part due to the still limited appeal of the modern movement.

Things changed after the Second World War when, thanks to the growing acceptance of MODERNISM, Le Corbusier was able to realize a number of major projects. He provided the original concept for the United Nations headquarters in New York City, but the two works that best reflect his thinking are Unité d'Habitation in Marseille and his contribution to Chandigarh, the new capital of the Punjab. In the post-Second World War decades his fame moved beyond architectural and planning circles as he became one of the leading intellectual figures of the time.

Le Corbusier took a clean-slate approach to cities. He perceived existing cities as chaotic, unhealthy and incompatible with the machine age. He proposed as an alternative a concept that would not only be adapted to the technology of the time, but make optimal use of this technology. Crowded cities would make way for towers surrounded by ample green space, thus providing residents with ample air, sun and contact with nature. Highways would assure rapid automobile accessibility. This vision inspired countless URBAN renewal projects and large suburban apartment developments across the world. But the latter part of the twentieth century was not kind to the Le Corbusier perspective on cities. Widespread disillusion with developments adhering to the towers-in-the-park model, in particular with their sterility and susceptibility to CRIME, and a renewed interest in the diversity, spontaneity and social interaction of traditional urban NEIGHBOURHOODS and their retail streets have turned planners and

public opinion alike against the Le Corbusier urban model. Yet the reputation of Le Corbusier as an architect endures. His ability to marry functionality and elegance, and to exploit the plasticity of reinforced concrete, as he did with brilliance in his Ronchamp chapel, are widely admired.

Further reading

Jencks, C. (1974) *Le Corbusier and the Tragic View of Architecture*, Cambridge, MA: Harvard University Press.
Serenyi, P. (ed.) (1975) *Le Corbusier in Perspective*, Englewood Cliffs, NJ: Prentice Hall.

PIERRE FILION

LEFEBVRE, HENRI

b. 16 June 1901, Hagetmau, France;
d. 29 June 1991, Pau, France

Key works

(1970) *La révolution urbaine*, Paris: Gallimard.
(1976 [1973]) *The Survival of Capitalism: Reproduction of the Relations of Production*, trans. Frank Bryant, New York: St. Martin's Press.
(1991 [1947]) *Critique of Everyday Life*, vol. 1, *Introduction*, trans. J. Moore, London: Verso.
(1991 [1974]) *The Production of Space*, trans. D. Nicholson-Smith, Oxford: Blackwell.
(2002 [1962]) *Critique of Everyday Life*, vol. 2, trans. J. Moore, London: Verso.

The French Marxist philosopher and sociologist Henri Lefebvre had a tumultuous and productive intellectual life. In addition to writing almost seventy books, he founded or took part in the founding of several intellectual and/or academic journals (e.g. *Philosophies*, *La Revue Marxiste*, *Arguments*, *Socialisme et Barbarie*, *Espaces et Sociétés*). In his early twenties, Lefebvre studied philosophy in Paris where he was attracted to Marxism. During that period, he got involved with the Surrealists who considered the possibility of creating a revolutionary movement that would have expanded beyond cultural concerns. He joined the French Communist Party in 1928, when communism was still regarded as a movement. He was suspended from the Party in 1957 – due to his open opposition to Stalinist dogmatism – but renewed with it at the end of the 1970s, when a growing number of French intellectuals of the left viewed this Party as a folk survivor. During the Second World War, Lefebvre was active in the Resistance. In the late 1950s and early 1960s, he shared with the Situationist International Movement a passion for invention, the shaping of new situations and the idea of revolution as a wild celebration. In 1961, Lefebvre was named professor of sociology at Strasbourg. He later accepted a position at Nanterre (an annex of the Sorbonne before it became Université Paris X-Nanterre in 1971) that he held until his retirement in 1973.

Lefebvre's contribution to Marxism has been considered both original and iconoclast. His vision was shaped by the Hegelian influence in Marx's work. He emphasized the notion of alienation in earlier works of Marx, bringing to the fore the importance of a humanist perspective and stressing the role of praxis. Lefebvre enhanced in different ways the knowledge of Marxism in France. In the 1970s, his analysis of URBAN spatiality made him famous among Anglo-American urbanists and geographers. Moving from rural sociology and the sociology of everyday life to URBAN SOCIOLOGY, Lefebvre related his understanding of urban issues to sweeping changes occurring within modern INDUSTRIAL CITIES and societies. He believed that these social changes could best be understood through urban culture and the mediating role played by space in reshaping social relations at the advanced stage of modernity. According to his ideas, space should not be considered as an empty receptacle but instead as a social practice informed by multidimensional processes including values, culture and power struggles. Looking at space from different angles, Lefebvre underscored the interplay of forms and forces in shaping the way the production of space is transformed by social practices. From his normative standpoint, the advent of urban society introduced new conflicts that challenged the traditional vision of centrality based on hierarchy and hence power relationships. In the late 1980s and early 1990s, some of Lefebvre's concerns about the production of space fuelled the geographical debate on the condition of POSTMODERNITY.

Further reading

Shield, R. (1999) *Lefebvre, Love and Struggle: Spatial Dialectics*, London: Routledge.

PIERRE HAMEL

L'ENFANT, PIERRE CHARLES

b. 2 August 1754, Paris, France;
d. 14 June 1833, Prince Georges County,
Maryland, United States

Key work

(1791) Plan for Washington DC.

French architect, military engineer and designer of
the original plan for Washington DC, L'Enfant
studied at the Académie Royale de Peinture et de
Sculpture (see ÉCOLE DES BEAUX-ARTS). In 1777, he
came to America to join the Continental army and
fight in the Revolutionary War. After retiring from
the military, he remodelled New York's old City
Hall. In 1791, George Washington appointed
L'Enfant to draw up plans for and supervise the
construction of the new capital city on the Potomac River.

Inspired by the baroque plan of Versailles,
L'Enfant rejected the simple traditional gridiron
pattern and adopted a grand proposal structured
around two perpendicular axes each with its own
visual terminus – the White House and the Capitol.
He also superimposed a network of diagonal avenues, circles, PARKS and public plazas onto an
irregular checkerboard plan.

In 1792, L'Enfant was dismissed because of
insubordination and his plan was forgotten. Having
failed to achieve recognition and proper remunera-
tion, he died penniless. His plan was rediscovered at
the end of the nineteenth century and revived by the
Senate Park Commission in 1901. A century later,
Washington DC reflected L'Enfant's plan.

Further reading

Caemmerer, H.P. (1970 [c.1950]) *The Life of Pierre
Charles L'Enfant*, New York: Da Capo Press.
Stephenson, R.W. (1993) '*A Plan Wholly New': Pierre
Charles L'Enfant's Plan of the City of Washington*,
Washington, DC: Government Printing Office.

CATHERINE C. GALLEY

LEWIS, OSCAR

b. 25 December 1914, New York City;
d. 16 December 1970, New York City

Key works

(1961) *The Children of Sanchez: Autobiography of a
Mexican Family*, New York: Random House.

(1966) *La Vida: A Puerto Rican Family in the Culture of
Poverty – San Juan and New York*, New York:
Random House.

Oscar Lewis was an American anthropologist,
most famous for his oral histories of poor SLUM
dwellers in Mexico, Puerto Rico, New York and
Havana, Cuba. Based on his research, Lewis
argued that the poor develop a 'culture of poverty'. Both his framing of his research topic
and his arguments about poverty generated
controversy.

Lewis's research methodology was ethnographic. During long periods spent in residence
with his research subjects, he would chronicle the
details of their daily lives and record oral histories
on his tape recorder. In a departure from the norm
in his field, his focus was on families rather than
communities. His first book to receive much public
attention was *Five Families*, published in 1959. In it,
he detailed the lives of five impoverished Mexican
families. The interviews with one family, re-edited,
became *The Children of Sanchez*, published in
1961. When published in Mexico, this book
angered nationalist intellectuals with its graphic
portrayal of Mexican poverty and was declared to
be 'slanderous and obscene' by the Mexican government, spurring an investigation. He subsequently published the account of another family as
*Pedro Martinez: A Mexican Peasant and His
Family* in 1964. His most controversial book was
La Vida, chronicling the life of a Puerto Rican
prostitute, living with her sixth husband, who was
raising her children in conditions unimaginable to
many middle-class American readers. Although
La Vida won a National Book Award, it also
offended many Puerto Rican leaders who feared
that Lewis's portrayal of this family would be
taken to be typical of Puerto Ricans.

Based on his research, Lewis argued that the
poor in capitalist societies adapt to their circumstances by developing a particular subculture that
they pass on to their children. He presented this
culture as both positive and negative, both maintaining their poverty status and providing them
with some rewards. Whether or not this subculture
is unchangeable has become the subject of intense
debate. Lewis himself offered evidence for its
malleability with his research in post-revolutionary
Cuba. He found evidence, reported posthumously in *Four Women – Living the Revolution*,
that the poor had improved their circumstances,

overcoming cultural practices developed prior to the revolution.

Lewis studied anthropology at the City College of New York (BSS 1936) and at Columbia University (Ph.D. 1940) with Ruth Benedict. He served on the faculties of Brooklyn College and Washington University before helping to establish the anthropology programme at the University of Illinois at Champagne-Urbana in 1948, where he taught until his death. Apart from his seminal work on URBAN slum dwellers, earlier work focused on the lives of peasants in northern India and Mexico.

ELIZABETH J. MUELLER

LIEBOW, ELLIOT

b. 4 January 1925, Washington, DC;
d. 4 September 1994, Silver Spring, Maryland

Key works

(1967) *Tally's Corner: A Study of Negro Streetcorner Men*, Boston, MA: Little, Brown.
(1993) *Tell Them Who I Am: The Story of a Shelter for Homeless Women*, New York: Penguin Books.

The son of Jewish immigrants from Latvia and Russia, Elliot Liebow received his Ph.D. from the Catholic University of America in 1966, then worked for twenty years at the National Institute for Mental Health.

Liebow is best known for his book *Tally's Corner: A Study of Negro Streetcorner Men* (1967), recognized as a classic ethnography of black street-corner society in Washington DC. It features the narratives of young unemployed and casually employed black men about work, their families and friends, and themselves.

In 1984, Liebow learned he had cancer, and decided to retire from federal government service. He volunteered at a soup kitchen and at emergency shelters for women. His ethnography, *Tell Them Who I Am: The Lives of Homeless Women* (1993), was based on ten years of participant observation in two women's shelters in the Washington DC area. Liebow testified to the courage and hard work of single homeless women living in shelters, as they faced life with the most limited resources and little family support. *Tell Them Who I Am* was written partly in collaboration with the homeless women and staff of the two shelters. They read and commented on the manuscript, and Liebow incorporated their sometimes humorous insights in footnotes throughout the book.

During his lifetime, Liebow received numerous awards in recognition of his scholarship, social activism and commitment to social justice.

RAE BRIDGMAN

LINEAR CITY

About a century ago, linear city models were presented as alternatives to the densely populated, concentric INDUSTRIAL CITY. The most famous model had been advocated by the Spanish urbanist Arturo SORIA Y MATA (1844–1920) in a series of articles published as early as 1882. In fact, he was the first URBAN planner designing an urban model based upon the integration of land use and INFRASTRUCTURE. Soria y Mata had strong misgivings about the often chaotic urban development in those days. The new nineteenth-century transport technology of tramlines could be turned into a liberating force: Soria y Mata advocated the idea that the pattern of urban extension has to be fully adjusted to the infrastructure necessary for efficient transport.

The 'Ciudad Lineal' takes the form of a city 400 metres wide, centred on a tramway and a thoroughfare running parallel. The ambitions of Soria y Mata were immense: he proposed a linear city running across Europe from Cadiz in Spain to St Petersburg in Russia. Only a few kilometres were ever built, just outside Madrid's perimeters, and these have now been completely swallowed up by the modern city. Although Soria y Mata advocated a new sort of land policy to make the Ciudad Lineal possible, from a social point of view his model was traditional because only dwellings for the better off would have immediate access to the central axis. Important from the perspective of urban models is the fact that the Ciudad Lineal is not the model of an alternative, linear city, but a model to extend existing cities. As such, the model has been influential since many regional plans, especially in Europe, have advocated some sort of linear extension of large cities, based upon infrastructure. The basic difference from the Ciudad Lineal concept in most cases is that an unbroken linear development was not advocated, but more

the model of beads along a string: smaller urban settlements grouped along an infrastructural line. The famous Copenhagen Finger Plan of 1947 is a clear example. The *unplanned* extension of cities focused on the road system, though many urban planners rejected that. Using the words of Lewis MUMFORD (*The Culture of Cities*, 1938), this would ultimately lead to the Townless Highway.

Today, the notion of linear urban development along infrastructure is taken up on a much wider scale. This has led to the corridor concept, which in most cases is used as an analytical concept, referring to a situation where urban growth seems to be the greatest in areas with immediate access to important infrastructure. This has led to the assumption that economic development in regions can be influenced by an improved provision of infrastructure. This idea has especially taken root in Europe, where many national governments (and the European Union) pursue the policy goal of balanced economic development.

Further reading

Bosma, K. and Hellinga, H. (eds) (1997) *Mastering the City: North-European City Planning 1900–2000*, Rotterdam/The Hague: NAi Publishers/EFL Publications, Part II, pp. 8–17.

Hall, P. (1996) *Cities of Tomorrow: An Intellectual History of Urban Planning and Design in the Twentieth Century*, updated edn, Oxford: Blackwell, pp. 112–13.

WIL ZONNEVELD

LINKAGE FEES

The costs associated with growth have traditionally been borne by the COMMUNITY at large and not by those who directly benefit from development. Beginning in the 1970s, however, INFRASTRUCTURE and public facilities costs were increasingly shifted to the private sector. In England, for example, planning authorities and developers routinely draw up legal agreements involving financial contributions from the developers to pay for needed infrastructure. In the United States, localities are charging developers development impact fees (DIFs) to fund increasingly larger shares of the capital facilities or infrastructure costs generated by new development. They represent the latest stage in the process of increasing regulation of land use, and of shifting development costs to developers. Legally, DIFs

are based on the 'rational nexus' test under which they are evaluated on the basis of whether the new development creates the need for new or expanded facilities, the fee charged is proportionate to the need created, and the fee is used to reasonably benefit the fee payer.

Linkage fees are a derivative of DIFs in the United States. They are exacted on developers by some cities and counties to pay for a vast array of facilities and services, including day care, public art, mass transit, downtown PARKS and public artworks.

But linkage fees are primarily used as a way to provide affordable housing. Housing linkage fees are based on the causal relationship – or 'nexus' – between the low-wage jobs generated by new commercial development – hotel, retail, office, etc. – and the ensuing need for housing affordable to those workers, housing that the market is generally unable to provide and the localities are unable or unwilling to subsidize directly. In the past, affordable housing production was mainly the responsibility of the federal government and the states, but their contributions are increasingly limited. Housing funds at the federal level, for example, have declined from a high of US$71 billion (in today's money) in 1978, to US$18.7 billion in 2000.

Linkage fees are one of a few tools that some localities in the United States are using to offset the federal government withdrawal from the housing arena at a time of increasing need. Similarly, in England, grants and subsidies from the central government for affordable housing production have been cut and growing emphasis has been placed on making private developers produce affordable housing.

The first cities to enact linkage fees were in California – Palo Alto in 1979 and San Francisco in 1981. While Palo Alto's programme applied to all commercial development in the city, in San Francisco it applied only to DOWNTOWN office construction, the result of the city's rise as a financial centre, and the concomitant boom in high-rise office building construction. While the need for housing the new workers skyrocketed, little new housing was being built; in fact, much housing was being destroyed to make room for office buildings or it was gentrified, particularly intensifying the plight of low-income residents. A strong anti-growth movement ensued that fought to restrict office development downtown and to respond to the housing crisis, and the

linkage programme was one of the measures adopted by the city in response.

The San Francisco programme targeted only office development, and only in the downtown area. A few other cities with strong downtown office markets followed suit, most notably Boston, Massachusetts; Hartford, Connecticut; and Seattle, Washington. This approach to linkage fees was limited to cities with a very active downtown development market, something rare in American cities in the 1980s. Linkage fees started to spread to other localities in the late 1980s and 1990s, most notably the City of San Diego, and the City and County of Sacramento in California. This second generation applied linkage fees to all commercial development.

Linkage fees vary among localities, but in no case do they contribute more than a fraction of the costs for affordable housing generated by different land uses, amounts established through 'nexus' studies performed by economic consultants. For example, in San Diego, the fees for office development are US$1.06 per square foot, hotel US$0.64, retail US$0.64, manufacturing US$0.64, warehouse US$0.27, and research and development US$0.80. For the last fiscal year, the San Diego fees generated US$4.5 million.

Setting fee levels is a political balancing act, weighed down at one end by the severity of the housing problem and at the other by a reluctance to discourage economic development. Office fees in Sacramento, for example, are lower than those in San Diego, while San Francisco's office linkage fees are nearly seven times higher, a fact that has hardly discouraged investors. The importance of establishing locally generated funds becomes even more evident if we remember that US$1 in local funds leverages US$5–10 from federal, state, foundation and private sources.

Opponents of linkage fees, mainly business and development interests, have steadfastly maintained that linkage fees are an unconstitutional taking of private property and discourage economic development. A 1991 decision of the US Court of Appeals for the Ninth Circuit upheld the constitutionality of Sacramento's linkage requirements, because they pay for a social cost that is reasonably related to the activity against which the fee is assessed. In 1992, the US Supreme Court declined to review this decision, thus upholding the constitutionality of linkage fees.

With respect to the economic viability of linkage fees, there is general agreement that they are successful only in places with a strong economy, where land values are high, and higher rents easily absorbed. Those are the same places that, in all likelihood, are suffering from a job–housing imbalance and a housing shortage. Not surprisingly, most cities that have adopted linkage fees are found in California, where high levels of economic growth are accompanied by a housing crisis of major proportions. Where such conditions exist, linkage fees can be very successful.

Further reading

Herrero, T. (1991) 'Housing linkage: will it play a role in the 1990s?', *Journal of Urban Affairs* 13(1): 1–20.
Nelson, A.C. (1988) *Development Impact Fees*, Chicago: Planners Press.

NICO CALAVITA

LIVABLE CITY

The 'livable' city refers to the city as experienced by its inhabitants. According to Lewis MUMFORD, 'livability' is not synonymous with 'standard of living'. In addition to providing clean water, clean air, adequate food and shelter, a 'livable' city must also generate a SENSE OF COMMUNITY and offer hospitable settings for all, especially young people, to develop social skills, a sense of autonomy and identity. It must reduce distrust by offering favourable environments for understanding and resolving social and cultural differences, and increase a sense of well-being through interpersonal experiences, beauty, festivity and conviviality.

Ten characteristics of a livable city facilitate achievement of these goals:

1   The essence of the city's livability is found in its public URBAN spaces, and the quality of social life they support. When public social life atrophies, depression, anomy, incivility, group conflict, even violence and tyranny may result. The public dialogue brings together different views and perspectives; the livable city must 'encourage the greatest possible number of meetings, encounters, challenges, between varied persons and groups'.

In a livable city, the public realm, as JANE JACOBS observed, is indispensable for social learning and socialization. Children

learn about caring, responsibility and trust by observing such behaviour.

A well-functioning public realm requires town squares that generate social life and NEIGHBOURHOOD places that foster COMMUNITY. These must be truly public, easily accessible and hospitable to all, multifunctional, used on a daily basis and as settings for community events, providing multiple reasons for people to talk, work together, coordinate activities and celebrate together.

2   The livable city is characterized by a compact, human-scale urban fabric of contiguous small-scale, mixed-use buildings forming blocks; these create continuous street walls and enclose public spaces.

3   The primary building block of the livable city worldwide is the 'shop/house', with shop, workshop or restaurant generating activity at street level, and residential dwelling above providing 'eyes on the street' (Jacobs). The close proximity of living, working, socializing, of the private and public realms, is what makes the public realm so hospitable, and the private dwelling so convenient.

4   A livable city is a just city: it does not segregate population groups but facilitates dialogue across ethnic and economic differences; GHETTOS and gated compounds are not appropriate. A fine, textured urban fabric provides flexibility for affordable and market-rate accommodation.

5   A livable city has a cellular structure: the neighbourhood is a microcosm of the city as a whole, containing diverse work opportunities, shopping, housing and all necessary infrastructure – schools, medical services, etc. – within a short radius. Only then is it possible to walk or bike to work, school or the shops, facilitating meetings and conversations, micro-social events that build community. This ideal is called in Europe the 'City of Short Distances'.

6   In a livable city, the transportation planner focuses on trips and accessibility, rather than vehicular movement; 'balanced transportation planning', prioritizing walking, biking and public transportation, aims to make all trips as pleasant, economical, safe, comfortable, simple and autonomous as possible, for children as well as for working adults.

Continuous city-wide networks of dedicated bicycle lanes and pedestrian routes extend accessibility, especially for children, younger people and the elderly. Traffic-calming techniques, common in European cities, are gradually appearing in North American cities.

To compete with the private automobile, public transportation must be more convenient, faster, less expensive and as comfortable as the car; with a compact urban fabric and sophisticated transit planning techniques, many European cities have achieved this goal.

7   Activities in the public realm such as farmers' markets, celebrations and festivals bring inhabitants together, not in their usual specialized roles, but as full human beings! Catalysts for social life, outdoor cafes and restaurants, ice-cream parlours, etc., encourage more time spent in public spaces, even into the night. These require traffic-free places suitable for conversation and children's play.

Community festivals are both an indicator of livability and a mechanism for developing community. The planning and practice, discussing themes, making costumes, creating beautiful works of art and music, and especially eating and drinking together in the public space at festivals bind a diverse community together.

8   The pleasure that inhabitants and visitors experience in a beautiful city is translated in the body into endorphins that increase mental and physical well-being. A beautiful city is aesthetic as a whole (identifiable boundary, centre and focal points); in the composition of buildings that complement each other; and in the design of individual buildings, places and works of art.

Accessible and playful fountains focus social life; public art enriches appreciation of the city. In a livable city, these works are not placed for corporate, but for community identity, representing the city and its traditions, cultural heritage, ethnic groups, crafts and industries.

9   Every livable city has an identity expressed in those architectural and spatial characteristics best loved by the city's inhabitants. These may consist of certain building materials and colours, a typical arrangement of scale and architectural forms, building lot size, roof lines, or the scale of public and semi-public spaces.

The DNA is codified in design guidelines; in order to fit into the context, new buildings must respect this 'genetic code', reflecting at least some existing patterns, or interpreting them in a contemporary idiom.

10 If we want our cities to be livable for all, then we must *first* make them livable for children. If our cities lack livability, children are the first to suffer. *Every aspect* of the city's urban design, built fabric, organization of streets or architectural forms impacts on children.

In a livable city, children feel at home; they can find their way around on their own, identify 'special places' and 'hold their city in the palm of their hand'. Children, after all, are the ones who will inherit the city, and become responsible for its future.

Creating a 'livable city' is a holistic and ecological task: like a living organism, the city's physical and social aspects are interdependent. No one element should be neglected or over-emphasized, for this disrupts the city's equilibrium and ability to heal itself through community participation. A livable city is a city in harmony with its citizens and with its natural environment.

The conception of 'livability' advanced here is illustrated in the work of the International Making Cities Livable (ICML) Conferences, and based on the ideas of such scholars and students of urban life as Hannah Arendt, Lewis Mumford, JANE JACOBS, Bernard Rudofsky, Wolf Von Eckardt, William H. WHYTE, Ben Thompson, Alexander Mitscherlich and Colin Ward. It is also compatible with the work of such organizations as Partners for Livable Places and Project for Public Spaces, and with some of the work of the New Urbanists.

Further reading

Crowhurst Lennard, S.H., von Ungern-Sternberg, S. and Lennard, H.L. (eds) (1997) *Making Cities Livable*, Carmel, CA: ICML Conferences.
Jacobs, J. (1961) *The Death and Life of Great American Cities*, New York: Random House.
Mumford, L. (1961) *The City in History*, New York: Harcourt Brace, p. 320.

SUZANNE H. CROWHURST LENNARD

LOCALLY UNWANTED LAND USE

Cities of all kinds face a large, distinct and fast-expanding number of development projects that are regionally or nationally needed or wanted, but objectionable to many people who live near them. Examples of such pariah land uses are low-income housing, halfway houses, hazardous waste facilities, power plants, airports, prisons and highways. Societies want and need them, but individuals – and often communities – do not want them close by. They are locally unwanted land uses, or LULUs.

LULUs strain the sense of equity. They gravitate to disadvantaged areas such as SLUMS, industrial NEIGHBOURHOODS and poor, minority, unincorporated or politically under-represented places that cannot fight them off and become worse places after they arrive. The United States typically deals with the difficulties LULUs present through devices such as land use planning, pollution controls, ENVIRONMENTAL IMPACT ASSESSMENTS, citizen participation, EMINENT DOMAIN and preservation areas. But new kinds of LULUs – for instance cellular (mobile) communication towers and mega-churches – continually appear, and since the late 1970s the American public has steadily become more sensitive to the harmful environmental and economic effects of LULUs and less willing to see them located or operated indiscriminately. The result, reflecting sentiments of 'NOT IN MY BACKYARD', has often been the blockage of LULUs.

Further reading

Popper, F. (1991) 'LULUs and their blockage: the nature of the problem, the outline of the solutions', in J. DiMento and L. Graymer (eds) *Confronting Regional Challenges: Approaches to LULUs, Growth, and Other Vexing Governance Problems*, Cambridge, MA: Lincoln Institute of Land Policy.

FRANK J. POPPER

LOS ANGELES SCHOOL OF URBAN STUDIES

The Los Angeles School redirects URBAN study away from notions of concentric zones and an ecological approach, used by the CHICAGO SCHOOL

during the 1920s, towards SOCIAL POLARIZATION and fragmentation, hybridity of culture, and auto-driven SPRAWL. As Chicago was seen as the new model urban place in the 1920s, Los Angeles is continuing to experience rapid growth in population and prominence, and serves as a new lens for examining urban places. The members of this school, most notably Edward Soja, Michael Dear, Allen Scott and Mike Davis, have been eager to promote Los Angeles as the capital of the twenty-first century.

Los Angeles, California, is used as the new paradigm for studying urban places because of its location in attracting global capital (both monetary and social), its quick and decisive reaction to industrial restructuring during the 1970s, and its social separation of two divergent classes of society. Los Angeles, unlike other cities such as Chicago and Pittsburgh, developed its industrial base not on Fordist production of automobiles or steel, but largely on flexible production sectors like financial and business, fashion and crafts, and motion pictures and music recording. This is not to say Los Angeles never had any steel or auto assembly plants, but they were always branch plants serving the West Coast market. Thus, when other cities, with heavy industrial bases, faced economic restructuring in the 1970s, they were not as successful in coping with losing growth in manufacturing sectors and diversifying their local economy. As this diversified economy developed, Los Angeles was able to attract financial capital from all over the globe, and its economy continued to grow, but in a different pattern – one not focused on a traditional central business district, but of sprawling growth centres throughout the region, creating a polycentric landscape. The automobile and freeway are the modes used to move people from place to place, and the population has continued to grow outward. With the shift from manufacturing to services, Los Angeles's demand for low-cost labour grew, and its employment base expanded with increasing migration streams from Latin America and Asia. These new immigrants became the supply of low-cost labour needed to clean offices, support restaurants, work in construction and provide sweatshop labour in the fashion industry. As the service sector continues to grow, so does the demand for migrant labour, creating a multi-ethnic, polyglot society with two distinct classes: the lower class, mainly composed of immigrant, low-skilled labour, and the upper tier of highly paid managers and professionals who live in fragmented and separate worlds within the same region. Residential enclaves have formed as traditional Mexican BARRIOS have expanded, disenfranchised African-Americans have remained in the South Central area, many Asian and Eastern European groups have settled in suburban locations called 'ethnoburbs', and the white population remains on the west side and outer suburbia. This leads to more social and political fragmentation.

With this being said, Los Angeles is believed to represent a new URBAN FORM. The analysis of this school of thought often centres around historical illustrations of Los Angeles as a place as well as scattered empirical vignettes used to highlight certain characteristics. A myriad of theoretical frameworks are applied to studying this urban place including Marxism, structuralism, libertarianism and POSTMODERNISM. The strength of these writings is their grounding in theory and Edward Soja's emphasis on asserting space into critical social theory. The writers have built upon the Chicago School work by adding ideas of the regulationists and critical social theorists in an attempt to better understand urban society in space. The study of class lends itself well to a Marxian or structural approach, while the fragmentation and multicultural nature of the place leads some authors to use postmodern theory to elucidate the many voices within the local society. The school's emphasis on social theory has redirected inquiry in URBAN STUDIES and geography from positivistic spatial science towards a broader, deeper examination of social society.

The Los Angeles School has not escaped without some criticism. First, the members often criticize the Chicago School for being overly deterministic, but the Los Angeles model is often applied with little consideration of places not fitting into the model. Secondly, the authors have often used the postmodern discourse to highlight the fragmentation and hybridity of Los Angeles, but fall into modernist thinking by proposing this model for urban development when postmodernism's goal is to reject grand narratives. This is an incongruous fit. Thirdly, the school has overemphasized pattern over process. Much of the writing stresses urban form and the look of the landscape over more broad structural influences shaping the urban process. Finally, some of the studies incorporated into the Los Angeles School

are quite clinical in their treatment of human subjects and their problems. The authors have been criticized for their method of looking out of their car window at the urban human condition without engaging with the fragmented society.

Further reading

Davis, M. (1990) *City of Quartz: Excavating the Future in Los Angeles*, London: Verso.

Dear, M. (ed.) (2002) *From Chicago to L.A.: Making Sense of Urban Theory*, Thousand Oaks, CA: Sage.

Scott, A. and Soja, E. (eds) (1996) *The City: Los Angeles and Urban Theory at the End of the Twentieth Century*, Berkeley, CA: University of California Press.

Soja, E. (1989) *Postmodern Geographies: The Reassertion of Space in Critical Social Theory*, London: Verso.

KEVIN ROMIG

LUNA PARKS

Luna parks are a particular type of leisure park and belong to the category of attraction parks. These can be traced back to the seventeenth and eighteenth centuries, often constructed at the whim of a rich landowner (Versailles and Prater in Vienna). Present-day attraction parks derive from the nineteenth century, when travelling funfairs settled on permanent sites (Tivoli, Copenhagen, 1843). They were then called amusement parks, located on the edge of the town. Attraction parks are distinguished from theme parks (see THEME PLACES) in that they usually do not reflect a common theme or set of themes. The key requirements for parks are location, accessibility and size. Today, luna parks are considered as small-scale attraction parks, easily accessed, potentially addressed to the permanent or temporary residential market, and located in the SUBURBS or even near the town centre. A basic characteristic is that they constitute temporary structures, thus referring to Virilio's 'nomadic existence'. Luna parks mainly offer classic funfair attractions (great wheel), newer features (electronic displays) and catering services. The oldest example in the United States is Coney Island (New York, 1903), founded by Frederic Thompson and Elmer Dundy. It was marketed as an 'electric city by the sea' and was an instant success. Another example is Cleveland's Luna Park (1905–29).

Further reading

Baud-Bovy, M. and Lawson, F. (1977–98) *Tourism and Recreation: Handbook of Planning and Design*, Oxford: Architectural Press.

Nasaw, D. (1993) *Going Out: The Rise and Fall of Public Amusements*, New York: Basic Books.

ALEX DEFFNER

LYNCH, KEVIN

b. 7 January 1918, Chicago, Illinois;
d. 25 April 1984, Martha's Vineyard, Massachusetts

Key works

(1960) *The Image of the City*, Cambridge, MA: MIT Press.

(1981) *A Theory of Good City Form*, Cambridge, MA: MIT Press.

Born into the bosom of a Catholic family of Irish origin, Kevin Lynch grew up in close proximity to Lake Michigan, studying at Francis Parker School, considered to be one of the first progressive schools in the United States. Being interested in architecture, Lynch entered Yale University, where he soon discovered Frank Lloyd WRIGHT's architecture, and in 1937, Lynch joined him at Taliesin, where he stayed for one and a half years.

In June of 1941, Lynch married Anne Borders, with whom he had four children. In 1944, during the Second World War, he was sent to the South Pacific. After returning to the United States, he rounded off his studies, obtaining a BA in URBAN PLANNING from MIT in 1947. In 1948, he was invited to join its Department of Urban Studies.

During the almost four following decades, Lynch built a rich legacy that continues today in the sense of a general theory of the urban environment. His interests were already apparent in his early texts, among which 'The theory of urban form' (1958, with Lloyd Rodwin) stands out. In this essay, the city is described through the complementarity of two systems – *flows* and *adapted spaces* – interpreted starting from a group of descriptive categories of URBAN FORM.

In 1960, Lynch published his more well-known work, *The Image of the City*, where he suggests

a means of interpreting the visual structure of the American city in terms of imageability and legibility, from the perspective of how the citizens perceive the environment in which they live.

An increased interest in these initial theoretical ideas characterizes his subsequent works. *Site Planning* (1962) offers a guide to URBAN DESIGN. *What Time is this Place?* (1972) discusses the perception of temporal flows and scales. *Managing the Sense of a Region* (1976) interprets the voracious process of metropolitan growth.

These efforts culminated in the publication of *A Theory of Good City Form* (1981), a wide-ranging and comprehensive discussion of urban theory, from the ancient city to the urban utopias of the twentieth century, which brings to light his general theory, founded in interpretative categories of the urban form.

Among his other numerous essays, two themes still deserve prominence: his reflections on UTOPIA vis-à-vis the possible city; and his critical attitude towards nuclear escalation. He also stood out as an environmental and city designer.

Kevin Lynch died suddenly on the morning of 25 April 1984. Posthumously, *Wasting Away* was published. On the day before his death, he was planning, on the telephone, a trip to China.

Further reading

Banerjee, T. and Southworth, M. (eds) (1996) *City Sense and City Design: Writings and Projects of Kevin Lynch*, New York: Oxford University Press.
Lynch, K. (1993) *What Time is this Place?*, Cambridge, MA: MIT Press.

LEANDRO M.V. ANDRADE

LYOTARD, JEAN-FRANÇOIS

b. 10 August 1924, Versaille, France;
d. 21 April 1998, Paris, France

Key works

(1984) *The Postmodern Condition: A Report on Knowledge* (*La condition postmoderne: rapport sur le saviour*), trans. G. Bennington and B. Massumi, Minneapolis: University of Minnesota Press.
(1988) *The Differend: Phrases in Dispute* (*Le Différend*), trans. G. Van Den Abbeele, Minneapolis: University of Minnesota Press.

French philosopher and author of 26 books and many articles, Lyotard is especially noted for his analysis of the impact of postmodernity on the human condition. A key figure in contemporary Continental philosophy, his writings covered a wide variety of topics including knowledge and communication; justice and freedom; the human body; inhumanity and technology; MODERNISM and POSTMODERNISM in art, literature and music; film; time and memory; culture and history; space, the CITY and landscape; the sublime; and the relation of aesthetics to politics.

Lyotard is best known for *The Postmodern Condition*, a report originally commissioned by the government of Quebec, where he examined knowledge, science and technology in advanced capitalist societies. Lyotard wrote that meta-narratives, the dominant orthodoxies that organized and defined the universal intellectual cultural objectives of modernity, such as beliefs in human progress or emancipation, are no longer considered credible justifications for evaluating truth or determining societal action. He argued that knowledge has been captured by the logic of market and technological efficiency. As a consequence, in place of modernist consensus and certainty, a new postmodern condition has emerged of chaos, unpredictability and dissensus.

Lyotard considered *The Differend* his most philosophical book and many critics consider it his most important work. In it, he sought to illustrate how the disadvantaged have their voices silenced by the dominant rules of a discourse, where politics is the threat of the 'differend' – the terror of having one's voice denied.

Lyotard spent ten years teaching philosophy in secondary schools (including a period in Constantine, Algeria, from 1950–52, where he became supportive of the struggle for Algerian independence), and over twenty years in teaching and research in higher education (Sorbonne, Nanterre, Centre National de la Recherche Scientifique, Vincennes). He spent twelve years doing theoretical and practical work devoted to the French group Socialisme ou Barbarie (Socialism or Barbarism), and a related newspaper, until becoming disenchanted with Marxism in 1966 and setting out to develop his own radical interpretation of philosophy and politics. Lyotard was active in the French student riots of May

1968. Lyotard was university professor emeritus at the University of Paris-VIII, and was for several years distinguished professor of French at the University of California, Irvine. He moved to Emory University in Atlanta in 1995, where he was professor of French and philosophy. He was also a visiting professor at Yale University, and other universities in the Americas and Europe.

Further reading

Malpas, S. (2002) *Jean-Francois Lyotard*, London and New York: Routledge.
Williams, J. (2000) *Lyotard and the Political*, London and New York: Routledge.

SEE ALSO: social justice

MICHAEL GUNDER

M

McHarg, Ian L.

b. 20 November 1920, Clydebank,
Scotland; d. 5 March 2001,
Philadelphia, Pennsylvania,
United States

Key work

(1969) *Design With Nature*, Garden City, NY: Natural
History Press.

Ian L. McHarg was one of the most influential
persons in the environmental movement, bringing
environmental concerns into broad public awareness and ecological planning methods into the
mainstream of landscape architecture, city planning and public policy. Growing up outside
Glasgow, Scotland, within sight of both the
Scottish Highlands and the gritty INDUSTRIAL CITY,
led to his life's work to reshape the relationship
between the city and the natural environment. For
more than six decades, beginning with his
apprenticeship to a landscape architect in Glasgow
at age 16, McHarg was a student, professor,
author, public figure and practising landscape
architect.

He began landscape architecture study at the
Glasgow School of Art and, after interruption by
the Second World War, continued at Harvard
University, where he received an undergraduate
degree followed by MAs in both landscape architecture and city planning. In 1954, McHarg
established the University of Pennsylvania's
Landscape Architecture Program, where he
developed his ecological planning methods and
continued teaching for more than four decades.
McHarg soon expanded his influence beyond

his academic setting through public lectures,
congressional testimony and television appear-
ances. In 1969, he wrote and produced the Public
Broadcasting System documentary *Multiply and
Subdue the Earth*, soon after publishing *Design
With Nature*, a book that is often compared with
Rachel Carson's *Silent Spring* for its broad envir-
onmental influence.

Design With Nature presented the rationale and
methodology for ecologically based planning,
using project examples drawn from his landscape
studios at the University of Pennsylvania and his
professional office, Wallace, McHarg, Roberts &
Todd in Philadelphia. Unlike many environmentalists, McHarg did not withdraw from the
city. His contribution was to establish ecological
analysis methods to guide city planning and
development decisions. Visually presenting ecological information in layers to establish areas of
environmental sensitivity helped to determine site
suitability for different land uses, densities and
design arrangements, anticipating the later development of GEOGRAPHICAL INFORMATION SYSTEMS.
The method was not an inventory but a combination with social values to reveal how nature and
cities could co-exist. The National Environmental
Policy Act of 1972, establishing the US Environmental Protection Agency and the requirements
and methods for environmental impact assessments, incorporated McHarg's ecological planning
method and his environmental terminology as
published in *Design With Nature*.

Among McHarg's many awards are the United
States National Medal of Art, presented by
President Bush in 1990, the Harvard Lifetime
Achievement Award, the Thomas Jefferson Foun-
dation Medal in Architecture, the Japan Prize in
City and Regional Planning, and a dozen other

international medals, awards and honorary degrees, in addition to several from the American Society of Landscape Architects.

Further reading

McHarg, I.L. (1996) *A Quest for Life: An Autobiography*, New York: John Wiley & Sons.

McHarg, I.L. and Steiner, F.R. (eds) (1998) *To Heal the Earth: Selected Writings of Ian L. McHarg*, Washington, DC: Island Press.

RICHARD DAGENHART

MacKAYE, BENTON

b. 6 March 1879, Stamford, Connecticut;
d. 11 December 1975, Shirley Center, Massachusetts

Key works

(1921) 'An Appalachian trail: a project in regional planning', *Journal of the American Institute of Architects* 9(10): 325–30.
(1928) *The New Exploration: A Philosophy of Regional Planning*, New York: Harcourt Brace.

Benton MacKaye's 1921 article, 'An Appalachian trail', triggered sixteen years of effort by many thousands of volunteers, organized through hundreds of local trail associations and COMMUNITY groups, to blaze and build a 2,140-mile trail along the crests of the Appalachian Mountains, covering fourteen states from central Maine to northern Georgia. The trail was mainly the work of dedicated volunteers who walked, built and negotiated with landowners, but several remote sections were finished by the US federal government, mainly through the work of the New Deal Civilian Conservation Corps. This was the miracle of MacKaye – how one individual, weak in oratory and with no wealth or religious or political movement, could motivate so many. The Appalachian Trail (AT) has encouraged outdoor recreation and nature conservation, and it has stimulated a worldwide trails and greenways movement.

Son of actor and theatre designer Steele MacKaye, Benton was brought up and lived most of his life in Shirley Center, Massachusetts, a classic New England village. In 1905, after completing a degree in forestry at Harvard, he went to work for the newly established US Forest Service directed by Gifford Pinchot. He shared Pinchot's conservationist ideas, he idealized rural living, and he was deeply concerned about the environmental problems and outward SPRAWL of large metropolitan areas. During and after the First World War, he advocated cooperative forestry as a means to resettle and employ large numbers of demobilized troops, simultaneously avoiding unemployment, misery and sprawl. His AT idea was embedded in a broader proposal for rural settlement, cooperative forestry and wilderness contemplation in the mountains, intended to limit the growth of the Atlantic Seaboard cities and to involve city people in summer camps, hiking and nature study. From its foundation in 1923 until it ceased to meet regularly in 1933, MacKaye was an enthusiastic member of the Regional Planning Association of America (RPAA), a group of intellectuals and activists meeting and working in the New York metropolitan region. RPAA member Lewis MUMFORD became a lifelong friend, and MacKaye did much of the work for the RPAA's REGIONAL PLANNING issue of the *Survey Graphic*, published in 1925. He gradually developed a theory of 'geotechnics' (regional planning) which advocated wilderness preservation, cooperative forestry and nature conservation along watershed divides, on steep slopes, in flood plains and in other ecologically sensitive areas. He proposed 'townless highways' (parkways through rural and wilderness areas) and 'highwayless towns' (GARDEN CITIES built in NEIGHBOURHOOD units with a system of paths, underpasses and overpasses so that residents could walk the whole community without contact with motor vehicles). In 1935, with Aldo Leopold, Robert Marshall and others, he founded the Wilderness Society.

Further reading

Anderson, L. (2002) *Benton MacKaye: Conservationist, Planner, and Creator of the Appalachian Trail*, Baltimore: Johns Hopkins University Press.

RAY BROMLEY

MAIN STREETS

Every town has one: main streets are a typical feature of any URBAN environment.

There are several ways to approach the study of a town's main street, and they vary according to the major area of interest of the numerous disciplines involved in the URBAN STUDIES field.

A city's main street may be analysed through the topic of its *economic* aspects, and it would be the dispute among land uses bidding for privileged locations offering best accessibility that would occupy the investigation priorities, but the implications of CAPITAL ACCUMULATION on a single strip, or, alternatively, the DISINVESTMENT risks the street is facing, would also be of interest.

Studies dedicated to the *historical* aspects of the city would invariably explore the HISTORIC DISTRICT in which the town was founded, often coincident with the location occupied by its main street, and often subject to HISTORIC PRESERVATION policies. The *sociological* approach would accompany any manifestation of CITIZENSHIP, and the occurrence of the social intercourses that might happen along the main street's crowded spaces; but would not forget to pay special attention to the STREET PEOPLE, often found on the main street's more neglected sites, or to the urban unrest, typically expressed on its SIDEWALKS. Those in the *administrative* orbit would probably prefer to examine where to establish the agencies of the CIVIL SOCIETY engaged in citizen participation and DIRECT DEMOCRACY alongside the street course, although they would not disregard to probe the LAND VALUE CAPTURE brought about by the CENTRAL CITY locational advantages earned by the street, as a means to warrant an equitable GOVERNANCE for the whole of the city. Centrality, main street shops, DOWNTOWN and the central business district would all draw the preferential attention of the *geographical* areas, in which the main street occupies a mandatory position, since it commands the structuring of the general layout of the city centre, and determines the basic configuration of the INTERNAL STRUCTURE OF THE CITY. Again in the spatial disciplines, *urban planners* would be worried about factors such as CONGESTION, urban renewal and its corresponding CAPACITY-BUILDING, or even the PEDESTRIAN MALL, features which are very frequently associated with the city's main streets. As for *architects*, it would be especially on the main street, of all the town's PLACES, that they would search for any signs of URBANITY, and would try to identify in the local built environment the presence of a likely SENSE OF PLACE, doubtless linked to elements of the CULTURAL HERITAGE, commonly portrayed in the main street's STREETSCAPE.

Notwithstanding the variety of paths available to study a town's main street, it was in the 1950s that awareness about the 'main street' as a concept of its own importance emerged in the urban studies field, eventually attaining the attention it encountered in the theoretical discussions of postmodern URBAN DESIGN, neo-traditional planning and META-URBANISM.

This interest was introduced by the inclusion of a fantasy version of a main street in the construction of Disneyland, the first of the DISNEY ENVIRONMENTS built by the Disney Corporation in the United States. Disney's strategy of portraying images of a small town's past is said to have had a substantial influence on many of the URBANISM experiences realized in the second half of the twentieth century. The setting of a paradoxical 'authentic' reproduction of a small town's main street encouraged the adoption of an architectural language which, emphasizing popular fantasy, allowed for a concrete mechanism to access the remote dream-world that populates people's imagination. Disney's designers were careful to thoroughly conceal all evidence of the working INFRASTRUCTURE, and masked all signs of physical obsolescence and social decay found in a real city. When it first appeared, in 1955, the resulting edited version of what would have been a real main street had been so conveniently filtered that it immediately assumed the role of symbolic bearer of urbanity messages. Perhaps an authentic main street would not have been as eloquent as their edited version, but Disney's main street was surely able to evince a substantial number of the emblematic urban-architectural values contained in the past.

In fact, the consequences of Disney's design are still active in the early twenty-first century. An unexpected revival of the interest in preserving historical sites, and the use of main streets as urban icons, are just a couple of the many cases that emerged from the original Disney scheme.

At the time of its opening, Disneyland's main street brought an unexpected boost to the preservation of historical buildings in several North American cities. Returning from Disneyland, and stimulated by the fascination brought about by the nostalgic images of the old main street, visitors longed to recover the romantic images of the past, lost in the main streets of their own cities. And an important debate was initiated. On the one hand, some preservationists alleged that the falsehood of imitations could never be a source of inspiration for the desire for the authentic. On the other hand, postmodern avant-garde urban designers did not

hesitate to engage in the practice of introducing fake architectural reproductions, using them as urban iconic elements, justifying this use as a strategy to ultimately achieve the preservation of the authentic pieces (see BARTHES).

Eventually, seen from a strict analytical point of view, it cannot be lessened that the main street concept has effectively opened up a path to reintroduce the important role performed by the memory of cities in city design, and has reinforced the importance of preserving the materialized mental representations of citizens' relationship with their urban spaces.

Further reading

Anderson, S. (ed.) (1986) *On Streets*, Cambridge, MA: MIT Press.

Castello, L. (1999) 'Understanding meta-urbanism: place making and marketing place', in *The Power of Imagination: 30th Conference of the Environmental Design Research Association*, Orlando, FL: EDRA, vol. 1, pp. 46–52.

Dunlop, B. (1996) *Building a Dream: The Art of Disney Architecture*, New York: Harry Abrams.

Rowe, C. and Koetter, F. (1978) *Collage City*, Cambridge, MA: MIT Press.

LINEU CASTELLO

MANUFACTURING CITY

In his *Economy and Society*, Max WEBER created a straightforward method for classifying cities. By examining who or what was driving their economies, Weber categorized cities as producer (manufacturing), MERCANTILE or consumer cities. In the manufacturing city, he stated that population growth and economic expansion were dependent upon (1) the successful export of locally manufactured goods to outside territories, and (2) revenues generated through the sale of these products, en masse, to indigenous entrepreneurs, workers and craftsman.

One of the best examples of a manufacturing city is Toyota City, the centre of automobile production in Japan. Home to seven Toyota Motor plants, just over half (50.16 per cent) of Toyota City's 186,000 workers were employed in industry in 1999. This represented the highest proportion of employment in manufacturing of any city in Japan. Conversely, only 14.32 per cent of the workers in the city were employed in trade in that year, and only 16.37 per cent in services.

While Toyota City's 1999 employment composition is representative of a manufacturing city, what really makes it a prime case study of this type is its historical development and the impact it has had on its surrounding region. Toyota was originally an agricultural village by the name of Koromo. It achieved town status in 1892. Then, in December 1935, the textiles firm Toyoda Automatic Loom Works purchased about 2 million square metres of undeveloped land in the town to build a factory for its new automobile division. Three years later, the company opened its Honsha (Main) Plant.

By the end of the Second World War, in addition to its textiles and aircraft operations, Toyota had ten plants in and near Koromo, including two passenger car plants, a truck and bus factory, a steel mill, a machine tools works, and an engine plant. However, it was its clustering of factories in Koromo which enabled the company to initiate its new 'just in time' production system. Between 1950 and 2000, as Toyota Motor's domestic production increased from 11,000 vehicles a year to more than 3.5 million, Koromo grew from a town of 32,000 people to a city of more than 350,000. In accomplishing this, Toyota Motor consolidated its production in the town, and encouraged affiliated industries and suppliers to relocate in Koromo. In appreciation for this, Koromo was renamed Toyota City, by municipal leaders, in 1959.

By the mid-1960s, a new regional core had developed around Toyota City. At the turn of the twenty-first century, as Toyota Motor transformed into the world's third largest auto producer, and Japan's most profitable company, its sub-region had grown to almost 2 million people. This area also had slightly more than 1 million in employment, including roughly 400,000 manufacturing workers.

Further reading

Jacobs, A.J. (2002) 'A Weberian look at Japanese cities', paper presented at the 72nd Annual Meeting of the Eastern Sociological Society, Boston, 7–10 March.

Weber, M. (1978) *Economy and Society*, Berkeley, CA: University of California Press.

A.J. JACOBS

MAQUILADORAS

Basically, maquiladoras are US-owned assembly plants, operating under special legal frameworks. Maquiladora development is predominantly, but

not exclusively, a Mexican phenomenon: it has also taken place in Central American, Caribbean and Asian countries.

Maquiladoras date back to 1964, when the Mexican government introduced the Programa de Industrializacion Fronteriza (BIP). The BIP aimed to attract investments by US manufacturing firms to the northern states of Mexico, which were then struggling with high unemployment rates and hence offered a large pool of cheap labour. The BIP offered the firms minimal company taxes, but did not allow them to locate outside a 10-mile zone along the border nor to sell their output on the Mexican home market. Until the late 1980s, the stereotypical maquiladora maintained no linkages with the local economy other than labour, employed lowly skilled young females, and was dedicated to standardized mass production. Due to an impressive growth, the sector was an important impetus to URBAN development in the BIP zone: MIGRATION and industrial growth turned many rural towns into rapidly booming INDUSTRIAL CITIES.

Throughout the 1990s, particularly after the North American Free Trade Agreement (1994), maquiladora regulations, introduced by the BIP but often adjusted, were abolished. Moreover, the sector has lost most of its stereotypical characteristics and has become a more integrated part of Mexican INDUSTRIALIZATION. Geographically, maquiladoras are no longer bound to the narrow border zone and can be found, mostly in small towns, down to the southern peninsula of Yucatan.

Further reading

Kopinak, K. (1997) *Desert Capitalism*, Montreal: Black Rose Books.

ARIE ROMEIN

MARSH, BENJAMIN CLARK

b. 22 March 1877, Eski Zaghra, Bulgaria; d. 31 December 1952, Winter Park, Florida, United States

Key works

(1909) *An Introduction to City Planning*, New York: private publication.

(1953) *Lobbyist for the People*, Washington, DC: Public Affairs Press.

Born in Bulgaria, the son of American missionaries, Benjamin Marsh became a leading US social reformer. After studies at Iowa College, Chicago and Pennsylvania, he worked with charitable institutions in Philadelphia. A sympathizer of the 'CONGESTION movement', led by Florence KELLEY, Mary SIMKHOVITCH, Lillian Wald and others in New York, he was appointed executive secretary of the non-governmental Committee on Congestion of Population. The Committee researched and advocated policies to reduce TENEMENT overcrowding, to improve housing, sanitation, recreation and services in SLUM neighbourhoods, to extend the subway system, and to create new communities both in the outer boroughs and outside New York City. Marsh travelled to see European housing and planning, wrote a pioneering textbook, and helped organize two US 'firsts' in 1909: the City Planning Exhibition held in New York City and the National Conference on City Planning and Congestion held in Washington DC. Despite his leading role, many early supporters of planning felt Marsh was too strident in condemning slumlords and speculators, and in advocating Henry GEORGE's 'single tax' – land value taxation. In 1912, Marsh left the United States to spend two years as a correspondent in the Balkan War. After his return, he devoted the rest of his life to national poverty and consumer lobbying based in Washington, working with the Farmers National Council and the People's Reconstruction League, and then with the People's Lobby.

RAY BROMLEY

MASS TRANSIT

Mass transit refers to high-capacity public transport systems in metropolitan areas. In its most general sense, the term applies to a spectrum of transport modes from heavy rail and metro systems to buses and people movers; in practice, it is often associated with the higher-capacity rail-based systems.

Mass transit developed originally in the nineteenth century, on two main fronts. On the city streets, there was the horse-drawn omnibus, and

later the horse-drawn streetcar or horsecar, and eventually the electric tram or streetcar. In parallel, railways developed suburban and COMMUTER services, and later underground railways and metro systems based on electric traction were created. Such metro systems proved successful and became commonplace in cities of over a million people on all continents.

These transit systems often originally grew incrementally, in haphazard fashion, constructed and operated by competing private companies, whose services did not always link up. Eventually, many of these systems were brought together under public control, where they could be planned as a whole system. Newer generation systems – for example the Mass Transit Railway (MTR) in Hong Kong and the Mass Rapid Transit (MRT) in Singapore – could be planned from the start as a whole, with the latest technology and rolling stock, and efficient networks and interchange stations.

The contemporary rationale for mass transit is that it is more efficient at moving large numbers of people, with potential benefits for sustainability, first directly since more people can be moved for a given amount of energy or emissions, and secondly in conjunction with more compact forms of city or TRANSIT-ORIENTED DEVELOPMENT, which may have more general URBAN sustainability benefits (e.g. reduced landtake and energy requirements).

There is effectively a hierarchy of urban transportation modes which together can form an integrated system to get the best out of all modes. At the upper end are high-speed, high-capacity and high-frequency modes such as heavy rail metro systems, serving high-density corridors, with relatively few stops. In the case of Paris, there are effectively two tiers of rapid transit, with the regional metro (RER – Réseau Express Régional) effectively forming a level 'above' that of the original metro. At the lower end of the hierarchy are local feeder services – such as buses and minibuses – serving less densely populated corridors, with more frequent stops and slower operating speeds.

Heavy rail-based transit systems are considered to occupy the top of the hierarchy due to their significantly higher capacities compared with other modes. This high capacity is achieved due to the higher speed and reliability of these systems, enabled by segregation, through the avoidance of delays due to CONGESTION. This gives such systems a crucial competitive advantage over street-running light rail transit or buses.

To achieve segregation typically implies elevated or underground rather than at-grade (ground level) construction. Underground is generally the most expensive option, but is potentially most flexible in route location, avoiding many surface constraints. Elevated systems may be unattractive due to visual intrusion. At-grade alignments may be suitable for out-of-town areas or SUBURBS, but may be impractical to retrofit into existing cities, unless using previous rights of way.

A variety of track, vehicle and propulsion technologies are available. Rail-based systems usually involve conventional electric propulsion and steel wheel on rail, but in principle could use monorail, rubber tyre, magnetic levitation or linear motor technologies. Modern systems may be automated with central control (driverless trains), and have platforms protected from tracks by screens and automatic doors.

Bus-based systems may also achieve relatively high speeds and capacities through use of articulated vehicles, segregated reservations or guided tracks. Notable examples include express busways in Curitiba, Brazil, and guided buses in Adelaide, Australia, and Essen, Germany.

Full metro systems generally have significantly higher costs than other transit modes. Such segregated rail-based systems have high capital costs, comprising the costs not only of rolling stock but also of dedicated INFRASTRUCTURE, stations, track, power and signalling systems, plus associated higher operating and personnel costs. To be economically viable, such systems need to be located in areas of high demand.

Mass transit systems accordingly support and are supported by cities of certain critical mass. These are typically cities with high central-area population density and high-density corridors linking with a strong central business district. These corridors are typically arranged in linear or radial formations. This is especially seen in the case of Hong Kong, where the MTR follows linear corridors of very high density along the coastal strips of the territory; here, a significant proportion of the population lives within walking distance of a station. Mass transit can also make good use of limited river or harbour crossings in cities divided by waterways – Hong Kong and New York City being prime examples. In general, mass transit has been a major influence on the development and ongoing viability of many such cities.

Mass transit has often been seen as a status symbol for a city – in growing cities, as a sign they have 'made it', or in declining ones, as a catalyst for URBAN REGENERATION. In North America, there has been something of a renaissance in mass transit in the post-war period, against a backdrop of general decline in use of transit, and decline in some cities and DOWNTOWNS. Sometimes, transit systems have been suspected of being built for political purposes – as a major tangible asset bequeathed by a given regime – when not necessarily justifiable on economic grounds. Nevertheless, these may yet perform an important social and economic function, and may be regarded as having become an essential component of large dense cities, where the travel demand could not be easily met by other means.

Overall, mass transit systems have become important public assets in many cities. The physical presence of their vehicles, the iconography of their maps, logos and posters, and their station areas acting as public spaces with their own social rituals have made them integral components of the identity and public life of their cities.

STEPHEN MARSHALL

MEDIEVAL CITIES

After the fall of Rome in the fifth century, URBAN civilization began to decline, reaching its lowest level in the eighth and ninth centuries when the Saracens, and the Vikings later on, made sea travel unsafe around the Mediterranean basin. According to PIRENNE, it is at this time that trade came to a halt, and with it urban civilization. Cities shrank sharply in size and population, and the middle class, the mainstay of an urban economy, disappeared.

At the end of the first millennium, population growth began to pick up and interacted with agricultural inventions to set the stage for a period of spiritual and economic renaissance. The West began its counter-offensive with the crusades and the maritime republics of Italy, starting with Venice and Amalfi, and later Pisa, Genoa and Barcelona in Spain, reclaimed the Mediterranean sea for Christendom, and in the process revived trade and, consequently, urban life.

Medieval cities began with a market outside the protective wall of a castle with the permission of the feudal landlord, who hoped to gain profits through the rental of the market stalls. In time, the burghers became independent of the land-based aristocracy to become free citizens. Medieval city-states were governed by the guilds of the merchants, artisans, money changers and manufacturers. The most powerful guild of thirteenth-century Florence was the Arte della Lana, makers of wool cloth. Socially and culturally, the greatest influence was that of the Church. Through the interpretation of the Scriptures, it established the standards of private and public morality, and by its law on usury, it conditioned the running of business.

Characteristics of the medieval city include:

1 A set of defensive walls that sharply separated town and country – no suburban SPRAWL here. As cities grew, the old circle of walls would be torn down and a larger set of walls constructed. The walls provided a sense of security and unity and their gates functioned as places of control and customs and tax collection.

2 Curvilinear streets, the natural response to the hilly terrain on which most medieval cities were built for defensive purposes.

3 A central core containing the religious and civic functions of the city. All lines converge at the centre, but indirectly, as in Siena, the quintessential medieval town, where all streets lead to the magnificent Piazza del Campo, without entering it directly. The effect is of surprise, complexity and variety, completely opposite to that of Baroque design, where straight lines converge directly on an end vista – a statue, obelisk or palace.

Many medieval towns expanded from earlier Roman settlements that retained in the centre, as in Florence, the original gridiron pattern. But most medieval towns and cities grew organically on hills and on the sides of mountains, slowly adapting to site conditions, local materials, climate and culture. City historians have labelled the medieval city as a model of city building in the cultural and vernacular mode, resulting from coherent and purposeful decisions made on the basis of a common culture.

With the birth of the Renaissance and the utilization of perspective in the fifteenth century, the organic medieval city was supplanted by the

RENAISSANCE CITY; as such, a city layout was no longer the result of multiple decisions made by individual citizens, but the outcome of a pre-conceived goal and design. The figure of the planner was reborn.

Further reading

Mumford, L. (1961) *The City in History*, New York: Harcourt, Brace & World.
Pirenne, H. (1925) *Medieval Cities: Their Origins and the Revival of Trade*, Princeton, NJ: Princeton University Press.

NICO CALAVITA

MEGACITY

Initially coined by urbanist Janice Perlman in the mid-1970s, the term refers to the phenomenon of very large URBAN agglomerations. The most common definition is based on population size. Depending on the source and date, megacities are those metropolises reaching a population of over 5, 8 or 10 million inhabitants (e.g. Asian Development Bank, United Nations). Some definitions stipulate a minimum population density of 2,000 persons per square kilometre; however, extensive CONURBATIONS with lower population densities such as the Ruhr District in Germany or the unified city-suburb of Toronto, Canada, are intermittently classed as megacities as well. A finite definition is futile considering the uncertainties of census data and differences in defining urban boundaries. What constitutes a giant city changes with context and time: ancient Rome with around 1 million residents would certainly have qualified as a megacity in its time.

Massive worldwide URBANIZATION in the twentieth century's latter decades led to a stark increase in the number of megacities. In 1950, only five cities (New York, London, Paris, Tokyo, Shanghai) had a population of 5 million or more. By 2000, nineteen (more likely twenty-four) cities were boasting more than 10 million, two-thirds of them located in developing countries and more than half in East and Southeast Asia.

Due to their historically unprecedented size, megacities have captured both public and professional interest. Aside from their size, there are remarkable differences among megacities in economic development, housing, SOCIAL POLARIZATION, INFRASTRUCTURE quality, crime rates, government efficiency, and growth rates. Many of the newest megacities in Africa and Asia, for example, are expanding at significant rates, whereas megacity growth in Europe and North and South America has slowed since the 1970s and 1980s. Accordingly, megacities in developing countries are surpassing the population figures of their industrial world counterparts in many instances. Still, they lack the global, economic and political influence that qualifies megacities such as New York, Tokyo, London or Paris as world or GLOBAL CITIES (see Saskia SASSEN; WORLD CITY HYPOTHESIS).

The breakdown of urban administration and services and mounting environmental problems such as air and water pollution, contamination of soils and insufficient provision of water and waste disposal were once seen as signalling the death of megacities. However, so far, megacities have not collapsed. The special treatment of megacities has indeed been criticized as smaller cities experience problems similar or worse, and air pollution is generally not associated with city size but the transport system, industrial production, topography and climate. Moreover, megacities tend to perform comparatively better in terms of healthcare and employment opportunities than smaller places.

Hence, only a few problems are specifically linked to megacity scale and dynamics. First, the complexity of urban GOVERNANCE and administration reaches new levels in megacities. Secondly, democracy and local participation are more difficult to implement and maintain. Thirdly, with a potentially weak administration and lack of proper planning, population concentrations in megacities are particularly vulnerable to supply crises (e.g. water, food, energy) and natural disasters. Finally, in countries where up to 40 per cent of the population is concentrated in one single urban agglomeration, social unrest can easily lead to political instability affecting the entire nation.

Further reading

Ezcurra, E. and Mazari-Hiriart, M. (1996) 'Are megacities viable? A cautionary tale from Mexcio City', *Environment* 38: 6–34.
Fuchs, R.J., Brennan, E., Chamie, J., Lo, F.-C. and Uitto, J.I. (eds) (1999) *Mega-city Growth and the Future*, Tokyo: United Nations University Press.

ANDREA I. FRANK

MEGALOPOLIS

Basically, megalopolis means a very large city. As such, the notion of megalopolis has often met resistance. The most well-known example of a *filippica* against metropolization on a large scale comes from Lewis MUMFORD, in particular the chapter 'Rise and fall of Megalopolis' in his *The Culture of Cities* published in 1938. Mumford depicts Megalopolis as the beginning of decline, as the overture for Tyrannopolis, characterized by a naked exploitation of colonies and HINTERLAND. About twenty years later, in a much less grim era, the geographer Jean GOTTMANN published his essay 'Megalopolis, or the urbanization of the Northeastern Seaboard'. Where Mumford reserved the term for the trend towards large cities, the megalopolis which Gottmann describes is the result of polynuclear URBAN growth, resulting in a continuous stretch of urban and suburban areas, the main axis of which is about 600 miles long, the Boston–Washington area ('BosWash'). The megalopolis concept became highly influential because it introduced a new, larger scale in thinking about urban patterns and urban growth. Since the concept indicates a blurring of the distinction between urban and rural areas, it gave an impetus to a vast array of studies on both sides of the Atlantic focusing on the question of how to delineate urban and metropolitan areas. Since the introduction of megalopolis, other concepts have been introduced addressing the super-regional level in URBANIZATION like 'urban field' or 'megacorridor', each focusing on different dimensions. DOXIADES went beyond megalopolis, presuming that by the middle of the twenty-first century a world city or ECUMENOPOLIS will be formed.

Further reading

Gottmann, J. (1957) 'Megalopolis, or the urbanization of the Northeastern Seaboard', *Economic Geography* 33: 189–200.
Mumford, L. (1938) *The Culture of Cities*, London: Secker & Warburg, pp. 283–92.

WIL ZONNEVELD

MELTING POT

One of the most widely used metaphors in the lexicon of American cultural discourse, the first usage of the term 'melting pot' is attributed to Israel Zangwill, who used it as the title of his play in 1908. Though contested, the concept refers to a specific view that American identity emerges from the cultural blending of its immigrant population. Despite its apparent naivety, the 'melting pot' ideology was in tune with the turn-of-the-nineteenth-century modernist theories of democracy and nation-building. Viewing America as an ideal nation of immigrants, the 'melting pot' concept assumes that in achieving a nation of equals, old allegiances and ethnic affiliations must be abandoned. However, it also assumes that in the construction of a new identity, all contributing cultures will be treated equally. Given that Israel Zangwill's play lacked any African-American, Native American or other non-European immigrant characters, the concept of equal contributions was immediately questionable. Furthermore, since American immigration is a continuous phenomenon, the identity of the nation must be ever evolving.

As an assimilation model, the melting pot enjoys continued usage in vernacular and political discourse; however, it has been supplanted by pluralism, multiculturalism and more inclusive models of assimilation in the academic debates on identity, adaptation and integration of immigrants into various political, social and economic spheres.

Further reading

Glazer, N. and Moynihan, D.P. (1970) *Beyond the Melting Pot – Revised 2nd Edition: The Negroes, Puerto Ricans, Jews, Italians, and Irish of New York City*, Cambridge, MA: MIT Press.

ALI MODARRES

MERCANTILE CITY

The prime catalyst in the mercantile city is commercial trade. According to Max WEBER, these commercial centres are dependent upon entrepreneurs who garner profits from the sale of domestic and foreign commodities, both locally and abroad. They also generate local revenues from the complementary economic opportunities that mercantilism provokes, such as the transport of goods to other locations. Therefore, a common thread among these trade centres is that they are located at major transloading points, where modes of transportation intersect. For example, major PORT

CITIES where shipping and rail yards meet are prototypical cases of mercantile cities.

Prime examples of mercantile cities historically include Venice, Chicago and Osaka. The latter two contain all the essential elements of a modern Weberian mercantile city. Both cities have long histories as major centres of international trade. Both have direct access to waterways and are close to international airports; their ports are also well connected to sub-regional centres via multi-lane expressways and rail. Among the three, Osaka is the least known in the West. Nonetheless, it has a long history as a city of commerce.

Originally settled off the coast of Osaka Bay, on the delta formed by the Yodo and Yamato Rivers, Osaka City first rose to prominence in the fourth century, as Japan's first imperial capital. However, it was not until the late sixteenth century that it really began to thrive, after the military leader, Hideyoshi Toyotomi, built a castle there. In response to this, scores of merchants and vassals from neighbouring areas took up residence in the city.

After Hideyoshi was defeated, Osaka remained a centre of wealth as the nation's commercial capital. In the seventeenth century, merchant families lined the streets of the city. In addition, ships with goods from all over Japan, and other parts of Asia, crowded its harbours. It was during this period that Osaka emerged as the principal market for the rice trade in western Japan, that nation's major food staple. City wholesalers, however, also traded hundreds of other items, including cotton cloth, lumber, paper, copper, tobacco, oils, seafood and spices. Commercial activity was so vibrant at this time that Osaka became known as the Venice of Japan, as well as *Nihon no daidokoro* (Japan's kitchen).

Osaka remained Japan's commercial capital until the early twentieth century, when Tokyo surpassed it. In 1999, its daytime population and number of jobs in retail and wholesale trade (860,000) trailed only Tokyo. Interestingly, in that year, Osaka's proportion of its private employment in trade was higher than that of Tokyo (37.7 per cent to 34.9 per cent respectively).

Overall, with 2.6 million residents in 2000, and 2.3 million in private employment, Osaka's legacy of mercantilism has made it the engine of economic growth in Japan's Kansai Region for over 400 years, a region of nearly 21 million people and 9 million jobs.

Further reading

McClain, J. and Wakita, O. (eds) (1999) *Osaka: The Merchants' Capital of Early Modern Japan*, Ithaca, NY: Cornell University Press.
Weber, M. (1978) *Economy and Society*, Berkeley, CA: University of California.

A.J. JACOBS

META-URBANISM

At the turn of the twenty-first century, daily life in the cities of the globalized society required the use of illusions as a way to evade the repetitiveness of the day-to-day routine. A trend in the area of urban-architectural projects, initiated in the late 1990s, responded to this by producing constructions filled with metaphors, sometimes merely symbols representing a reality placed only in people's imagination; in other words, by producing environments that might be seen as 'meta-real', i.e. which transcended the limits of local realities. As an outcome, customs more akin to a meta-reality could arise, since the expected structural coherence of a PLACE had been altered. In such circumstances, traditional URBANISM becomes converted into a sort of 'meta-urbanism', reasoning that led the present author to coin the term 'meta-urbanism' as a conceptual approach.

Meta-urbanism is a phenomenon found all over the world, mainly in the form of themed environments, such as theme parks, themed malls, the thematic re-urbanization of old historic central areas, DISNEY ENVIRONMENTS, THEME PLACES, the use of Disney's MAIN STREET concept in NEW URBANISM projects, and other manifestations involving the making of places filled with images that somehow evade the daily reality present in the cities – images that cannot be said to be part of the usual routine of the cities where they are located, either because they are representative of a past which has already disappeared, or because they represent a wholly imagined creation, exotic to the local environment.

The common point shared by the design of all these environments is the implicit objective to mask the crude reality lived by ordinary citizens in the cities – cities that in the second half of the twentieth century were either lacking minimal traces of quality in their physical environments, or presenting a permanent menace to their

citizens' quality of life, be it on the grounds of psychological stress or of physical violence.

Any interpretative appreciation of the phenomenon, though, should not preclude the contemplation of two aspects associated with its manifestation: the making of places anchored on a fantasized reality; and the corresponding marketing of such places. With that, two significant tendencies in postmodern urban analysis may be evinced: the important presence of the concept of place in the practice of URBAN DESIGN; and the also important strategy of place marketing in the activities of city management. These two factors, together with the imperative role played by the memory of cities in the context of citizenship, perform a triad that may bring useful methodological contributions to urban analysis and, hence, to the understanding of cities in the first decades of the 2000s.

Further reading

Castello, L. (1999) 'Understanding meta-urbanism: place making and marketing place', in *The Power of Imagination: 30th Conference of the Environmental Design Research Association*, Orlando, FL: EDRA, vol. 1, pp. 46–52.
—— (2000) 'Marketing, consumption, and the traditions of place. Marketing tradition: post-traditional places and meta-urbanism', in *Traditional Dwellings and Settlements Working Paper Series*, Berkeley, CA: University of California Press, vol. 124, pp. 1–21.

LINEU CASTELLO

METROPOLITAN STATISTICAL AREA

In general, the phrase 'metropolitan area' refers to a core city with a large population nucleus and the adjacent communities (or SUBURBS) that have a high degree of economic and social integration with that core city. Yet the exact statistical standards for just what constitutes a metropolitan area have changed over time as transportation and commuting patterns have reshaped metropolitan identities.

In the United States, the Office of Management and Budget defines the standards for what constitutes an official *metropolitan area*, standards that are then applied to the data gathered by the Census Bureau. The statistical criteria for a *standard metropolitan area* were defined in 1949, redefined as a *standard metropolitan statistical area*

in 1959, and redefined again as a *metropolitan statistical area* (MSA) in 1983.

The different definitions point to metropolitan areas of different sizes and characteristics. As of 30 June 1999, the United States had 258 MSAs, areas with a metropolitan population of at least 100,000 (75,000 in New England), with a core city of at least 50,000. The MSA definition includes relatively small as well as large metropolises. The designation *consolidated metropolitan statistical area* (CMSA) is reserved for the largest metropolitan areas, MSAs with a population of 1 million or more. There are eighteen CMSAs in the United States – and another one in Puerto Rico. The New York CMSA embraces New York City and neighbouring suburban counties in New Jersey and Connecticut as well as in New York State.

Further reading

US Census Bureau (2003) 'About metropolitan and micropolitan statistical areas', available online at www.census.gov/population/www/estimates/about metro.html

BERNARD H. ROSS
MYRON A. LEVINE

MEYERSON, MARTIN

b. 14 November 1922, New York City

Key work

(1955) *Politics, Planning, and the Public Interest: The Case of Public Housing in Chicago*, Glencoe, IL: Free Press (with E. Banfield).

Educated at Columbia (AB, 1942) and Harvard (MCP, 1949), Meyerson initially worked with the planning commissions of Philadelphia and Chicago, and taught on the planning programmes of both the University of Chicago (where he collaborated with Edward Banfield) and the University of Pennsylvania (adviser for Herbert Gans), before being tenured by Harvard in 1957. As the first director of the Harvard-MIT Joint Center for Urban Studies from 1959–63, Meyerson promulgated a vision of planning as an interdisciplinary endeavour that was moving from its utopian origins towards social scientific maturity.

Through his research, mentorship, essays and consulting, Meyerson exerted formative influence

on US post-war URBAN policy at the municipal and federal levels. Whether as coordinator with James Rouse of Eisenhower's American Council to Improve Our Neighborhoods, as initiator with Edward Logue of the Action for Boston Community Development pilot project during the Kennedy years, or as a senior member of Johnson's urban task force that formulated the MODEL CITIES programme, Meyerson embodied the optimism of city and regional planners prior to the urban crises of the mid-1960s.

When Meyerson left Harvard for a deanship at Berkeley in 1963, his career entered a new phase of academic administration that would lead him to the presidency of SUNY Buffalo (1966–70) and the University of Pennsylvania (1970–81).

CHRISTOPHER KLEMEK

MIES van der ROHE, LUDWIG

b. 27 March 1886, Aachen, Germany;
d. 17 August 1969, Chicago, Illinois,
United States

Key works

(1927) Weissenhof Appartment Building.
(1958) Seagram Building.

A pioneering modernist architect of the twentieth century, Mies van der Rohe began his architectural career in 1905, but it was his association with Peter Behrens (from 1908 to 1912) that had a profound influence on him as an architect. At his workshop, Mies met Walter GROPIUS and other up-and-coming German architects. Though influenced by Behrens, his style was based on advanced structural techniques, combined with simplicity and functionality of design. After designing the Monument to Karl Liebknecht and Rosa Luxemburg, Mies became the director of the Weissenhofsiedlung project in 1927, where he worked with LE CORBUSIER, Behrens, Gropius and many other architects. In 1929, he designed the German Pavilion at the Barcelona Exhibition. His reputation earned him the directorship of the Bauhaus after Gropius, in 1930. In 1937, four years after the Nazis closed down the Bauhaus, Mies left Germany for the United States. He taught and practised in Chicago for the remainder

of his years. Among his many projects in the United States, the Alumni Memorial Hall in Chicago (1946), Farnsworth House in Plano (1950), the Lake Shore Drive apartment blocks (1951) and the Seagram Building in New York (1958) are the most widely recognized. The latter was a joint project with Philip Johnson, which established Mies as one of the leading architects of skyscrapers and an inspiration for a generation of architects in the post-Second World War period.

ALI MODARRES

MIGRATION

Migration is a concept used in many different contexts. There is no universally accepted single definition; defining migration is largely a matter of convenience. In social sciences and politics, migration refers to people who for different reasons move from one place of living to another, for the purpose of establishing a new place of residence. There are four major forms: *invasion, conquest, colonization* and *immigration*. Sometimes, governments engage in *forced* population migration, a general term that refers to the movements of refugees and internally displaced people as well as people displaced by natural or environmental disasters, chemical or nuclear disasters, famine, or development projects.

People's decisions to migrate have both public and private consequences. This is one of the reasons why migration is a much-favoured topic of study in social sciences. Usually, migration is explained with a push and pull model. There are some factors in a person's, family's, ethnic group's or population segment's place of living – usually the native country – that are hard to stand, pushing inhabitants away. And there are some factors in another place or country that seem more attractive – subsequently people with more or less success move there. This process affects both the places left (out-migration) and the places people arrive at (in-migration). Migration behaviour is usually studied within the context of job search, and, in fact, for many people it is an integral element of upward mobility. The migration decision is often viewed as evolving in three stages. The first is the decision whether or not to move, perhaps in response to unemployment or persecution. The second stage is the gathering of information about

other places, including information about job and housing opportunities. In the last stage, a place is selected due to its relative attractiveness. Then the movement is planned, purposeful and deliberate. Migration in this meaning is one of the major forces transforming our current world. Migration shapes not only SEGREGATION and multicultural settings – mostly URBAN – but also ethnic minorities and diasporas around the world, as well as ethnically and socially fragmented societies. Global migration processes are likely to accelerate in the information age as individual entrepreneurs become increasingly able to set up their businesses wherever they want by using the web, cellular phones and satellite dishes.

Being a non-temporary and selective process, migration has important effects on both the supplying and the receiving regions. The most-discussed form today is, of course, IMMIGRATION. Immigration makes provision and survival possible for people in exposed situations. But it also creates social and economic problems, since it alters the demography, age composition, sex ratios and literacy rates of the affected areas. Since those who move away usually have more marketable job skills than those who remain behind, migration tends to lead to a concentration of poverty in the left area. Immigration is also a problem when newcomers have ethnic and/or socio-economic characteristics that differ from those of the resident population. But migration could also be positive to a country. One of the reasons why the American LABOUR MARKET is believed to function better than the European one is that American labour is highly mobile. Workers in areas with high unemployment rates are apt to move to areas where unemployment rates are low.

Migration concerns *asylum seekers*, *refugees* or *labour*, and there are different rights attached to these varying kinds of status. Asylum seekers are people who move across borders in search of protection, but who may not fulfil the strict criteria laid down by the 1951 Geneva Convention, defining refugees. Asylum seekers have applied for protection as refugees and await the determination of their status. 'Refugee' is the term used to describe a person who has already been granted protection. Asylum seekers can become refugees if the local immigration or refugee authority deems them as fitting the international definition of refugee. Many countries contribute to the international responsibility for displaced people

through the provision of safe havens and resettlement programmes.

When migration concerns labour, this can be as a result of nations and/or big companies needing comparatively unskilled labour or their search for advanced knowledge, the latter resulting in a more or less global category of cosmopolitan professionals. A more specific kind of migration is the 'brain drain', which is the result of high-school graduates never returning 'home', but leaving for more remunerative employment elsewhere. Many Third World countries lose their young intelligentsia when they never return after advanced studies. Another important form of urban migration is the movement of people from inner cities to SUBURBS, often discussed as SUBURBANIZATION, SPRAWL, urban flight or WHITE FLIGHT, when ethnic minorities are deliberately left behind. If the situation in the place migrants have left improves, or if the destination turns out not to be as attractive as the migrants had initially thought, there could be a *reversed migration*. Sometimes, this can be problematic. *Retirement* migrations, i.e. senior citizens rather freely picking a new and attractive place for living, are less problematic. Nevertheless, decisions on where to live have increasingly important economic implications. Sometimes, retirement migration is also a *return* migration, old people moving 'home' again.

Migration can be *domestic* or *international*. When migration is international, it is *legal* or *illegal*, as well as *voluntary* or *forced*. When domestic, it is often a movement of people from rural to urban areas. Places rapidly losing population have difficulties funding minimal levels of public services such as schools. They are also hit by falling property values. Local businesses lose revenue when people move away, and business owners may themselves be forced to shut down and move elsewhere. Loss of population is perhaps the single most widely recognized indicator that a COMMUNITY is in economic decline. Out-migrants also impose a negative externality on those they leave behind, since them leaving affects the SOCIAL CAPITAL that people in well-integrated places have in common. Receiving cities are often ill-equipped to deal with a rapid influx of people in terms of guarding against CONGESTION as well as when it comes to providing the newcomers with housing, basic sanitation, INFRASTRUCTURE and public services.

Since migration today all over the world is a regular and ongoing process, planners must acquire knowledge of population changes over time, and plan for the supply of public infrastructure and services. They have to plan for jobs, social services and adequate school facilities, so that there is neither an oversupply nor an undersupply of buildings, which is a most demanding challenge within a more or less unstable population. Since people make different kinds of social investments in their places of living, planners ought to also be aware of the formation of social capital, the glue that holds society together. Moving migrants often lose or severely reduce this capital, developed at some cost over time. This can mean one of two things: either they are less likely to move again, because they have become aware of the social costs, or they move more easily since they cannot anyway develop once again this feeling of being place-bound.

Further reading

Allen, J., Massey, D. and Pryke, M. (eds) (1999) *Unsettling Cities: Movement/Settlement*, London: Routledge.
Braziel, J.E. and Mannur, A. (2003) *Theorizing Diaspora*, London: Blackwell.
Castles, S. and Miller, M.J. (1998) *The Age of Migration: International Population Movements in the Modern World*, New York: Guilford Press.
Wrench, J., Rea, A. and Ouali, N. (eds) (1999) *Migrants, Ethnic Minorities and the Labour Market: Integration and Exclusion in Europe*, Basingstoke: Macmillan.

KARL-OLOV ARNSTBERG

MODEL CITIES

Model Cities was to a great extent the penultimate URBAN programme of President Lyndon Johnson's GREAT SOCIETY. Model Cities was a response to the growing awareness that the individual CATEGORICAL GRANT PROGRAMMES of Johnson's WAR ON POVERTY did not by themselves constitute an effective attack on the problems of the most distressed NEIGHBOURHOODS of the nation's largest cities. Model Cities represented a new approach, one that emphasized social programme as well as physical renewal. It sought to coordinate the actions of numerous government agencies in a multifaceted attack on the complex roots of urban poverty.

Johnson's earlier anti-poverty efforts had been impeded by agency parochialism. Each department and agency focused on its own housing, social welfare, education or urban renewal objectives. Each agency jealously guarded its domain and resources.

Model Cities grew out of the recognition that urban poverty could be fought effectively only through a comprehensive, coordinated multi-agency action. Formulated by a presidential advisory task force, the Demonstration Cities and Metropolitan Development Act was signed into law by President Johnson in 1966. It was initially labelled the Demonstration Cities programme as it was intended to show or demonstrate just what could be accomplished in riot-torn Detroit and a few other targeted cities if a critical mass of resources were to be concentrated in fighting the multitude of problems in distressed inner-city neighbourhoods. The hope was that a successful demonstration project would spur foundations and other public agencies to launch similar comprehensive anti-poverty efforts of their own. But the programme's name was soon changed to Model Cities as the continued outbreak of urban unrest and social protests throughout the late 1960s made any reference to 'demonstration' cities politically unpalatable.

Political forces, however, soon acted to undermine the logic of the Model Cities programme. Constituency-oriented Congress members were not likely to enact a programme that targeted considerable resources to only a small handful of cities. Legislators wanted to ensure that their states and districts would share in any such programme benefits. As a result, the Model Cities Task Force quickly broadened the concept to sixty-six cities across the nation. By the second year, the programme's benefits were spread even more thinly to neighbourhoods in a total of 140 Model Cities. To appease powerful legislative chairs and members, Model Cities were even set up in such unlikely states as Maine, Tennessee, Kentucky and Montana. Spread so thinly across the country, the programme's benefits were diluted. Model Cities no longer offered the resources for a sustained, concentrated, multifaceted attack on the poverty problem of any city.

The programme was further undermined as federal, state and local agencies continued to seek turf protectionism and resist the programme coordination efforts imposed by Model Cities. In

many cases, infighting among COMMUNITY groups, too, delayed and impaired Model Cities spending.

Yet Model Cities did have its successes. Despite the spread of resources, it did commit substantial aid to impoverished communities. The programme approach also emphasized social programmes and community rehabilitation and revitalization, not the clearance strategy of earlier urban renewal efforts that had focused only on the physical condition of SLUM areas. The programme also represented a mix of federal power and local control as the federal government worked directly with city hall and local communities. Indeed, the programme emphasized the importance of working with city hall to combat urban poverty; it strengthened the hand of city mayors whose position had been undercut by earlier federal efforts at community action. The states, in contrast, were virtually cut out of the programme as many state capitols at the time were viewed as anti-urban and obstructionist, were bypassed entirely. Despite the omission of a strong role for the states, the programme's emphasis on DECENTRALIZATION and local control to a great extent anticipated the NEW FEDERALISM direction of Republican policy that would soon follow. The Model Cities programme's requirements for citizen participation, although a retreat somewhat from the 'maximum feasible participation' of the poor requirements of the earlier community action programme, also helped to identify and train a new generation of inner-city and minority community leaders, a generation of local activists who would soon enter city hall.

With the election of Richard Nixon as president, Model Cities' goal of increased citizen participation was de-emphasized in favour of placing still further programme discretion in the hands of local elected officials. In 1974, the Model Cities programme came to an end as its social service grant money was merged into the new COMMUNITY DEVELOPMENT BLOCK GRANT programme along with urban renewal, urban PARKS and a number of smaller urban aid programmes. The new block grant allowed local elected officials even greater flexibility in their use of intergovernmental funds, undermining the strong anti-poverty emphasis, community empowerment and federal oversight features that had characterized the Model Cities programme.

Further reading

Frieden, B.J. and Kaplan, M. (1975) *The Politics of Neglect: Urban Aid from Model Cities to Revenue Sharing*, Cambridge, MA: MIT Press.

Kleinberg, B. (1995) *Urban America in Transformation: Perspectives on Urban Policy and Development*, Thousand Oaks, CA: Sage, ch. 6.

MYRON A. LEVINE

MODERNISM

Modes of artistic representation emergent in the late nineteenth and early twentieth centuries that sought to give formal expression to experiences of capitalist modernity. Modernist artists and writers sought to shatter the formal limitations of the prevailing artistic mode of expression, realism, and introduced a number of innovative techniques intended to capture the exhilarating, yet chaotic, modern URBAN experience.

The period broadly between 1850 and 1920 witnessed a series of profound economic, political, technological, cultural and geographical transformations that engendered a series of wide-ranging changes in the nature of everyday life in certain key sites. These changes included the restructuring of European capitalism, the rise of the industrial factory system and the implications of this in the massive growth of urban areas, new building technologies and architectural forms, such as the first skyscrapers, technological changes that gave rise to new forms of transport, communication and media, such as photography, film and advertising, and a heightened heterogeneity to social configurations such as the crowd and to urban social and cultural life more generally.

The imprint of modernity was uneven and its impacts were felt most acutely in the great cities of Europe and North America such as London, Paris, Berlin, St Petersburg and New York. As a consequence, much modernist art and literature was explicitly concerned with exploring and representing the development of a distinctly modern urban consciousness in these cities. The cities of modernity were spaces where few, if any, of the settled certainties of the pre-modern world remained. Commentators have argued that everyday experience in these cities was dualistic, being both exhilarating yet frightening, dynamic yet precarious, liberating yet disturbing. This led to a crisis of representation among artists and writers, as the

formal limitations of realism precluded its being able to give artistic form to this experience. As partly a response to, and partly a critique of, the changes capitalist modernity was undergoing, a number of avant-garde artistic movements emerged that embodied the major aesthetic concerns of modernism and which engaged directly with, and frequently took as their subject, the profound transformations of urban space, both physical and social. The most successful and developed of these was cubism, which, along with a number of related genres and techniques such as futurism, fauvism, montage and papier collé, radically altered the conventions and boundaries of artistic modes of expression across a range of media. Cubism's central techniques included the incorporation of juxtaposition whereby temporal and spatial orders were disrupted and replaced by montages of temporal fragments and multiple perspectives within the same frame.

Modernism produced many examples of avant-garde artists who, during the early twentieth century, took the modern urban experience as their subject. Works from this period include Robert Delauney's *The City of Paris* (1912), Ludwig Meidner's *I and the City* (1913) and *Berlin* (1913), George Grosz's *The Big City* (1916–17), and Fernand Leger's *The City* (1919). One of the most potent symbols of modernity was new technology. Numerous artists took this as their subject, including Delauney, whose *Eiffel Tower* (1910) viewed its subject from numerous perspectives at the same time, and Gino Severini, whose *Suburban Train Arriving in Paris* (1915) showed a train punching through the outskirts of the city scattering houses in its wake, a symbol of the disruptive impacts of new technologies on urban life. Examples from other media include Paul Citroën's photo-collage *Metropolis* (1923), films such as Walter Ruttman's *Berlin, Sinfonie Der Grosstadt* (1923) and Dziga Vertov's *Man with a Movie Camera* (1929), poems such as T.S. Elliot's *The Waste Land* (1922), and novels such as John Dos Passos's *Manhattan Transfer* (1929). Dos Passos's novel illustrates well the formal deconstruction that lay behind modernist modes of representation. The novel, written using a montage technique, charts the lives of a vast range of characters from a wide social spectrum in New York between 1892 and 1920, seeking to convey impressions of the city as a whole. Short episodic scenes rapidly follow one another with little to indicate transition from one to the other. The novel, covering three distinct time periods, is not presented in any logical chronological order. Finally, the pace of the narration varies wildly from scene to scene. Formally, the novel is more closely related to cinema than to writing. Using these techniques, the novel conveys the intensity and diversity of life in the city in ways impossible through a more realistic narrative.

Modernist representations of urban space have been criticized as being both highly gendered and sexualized, deriving from and celebrating the masculine subject position and erasing the presence of women or representing them as subservient to the male gaze in various ways. Modernism has also been identified as an architectural style based upon the writing and plans of architects such as LE CORBUSIER, which sought a form of purity by eschewing decoration and ornamentation. High modernism, as it was titled, became a dominant architectural style for commercial building and public housing projects during the 1950s and 1960s. However, it is difficult, if not impossible, to trace any connections between the philosophies and concerns that underpinned this and those underpinning the modernist artistic repertoire.

Commentators have recognized three distinct phases of aesthetic modernism. The emergence of modernism has been mapped broadly from 1850 to 1910 and was characterized by a crisis of representation among artists. The period roughly from 1910 to 1945 has been characterized as the period of 'heroic modernism' during which avant-garde artistic representations of modernity took shape; it was also a time of political engagement and of critiques of BOURGEOIS culture. The post-war period until around 1968 has been referred to as 'high modernism', during which the modernist aesthetic entered the mainstream and was seen in commercial architecture, film and advertising, and by which time few connections with the conditions within which aesthetic modernism was born remained.

Further reading

Berman, M. (1984) *All That is Solid Melts into Air: The Experience of Modernity*, London: Verso.

Brosseau, M. (1995) 'The city in textual form: *Manhattan Tranfer*'s New York', *Ecumene* 2(1): 89–114.

Harvey, D. (1989) *The Condition of Postmodernity: An Enquiry into the Origins of Cultural Change*, Oxford: Blackwell.

Lunn, E. (1985) *Marxism and Modernism*, London: Verso.

TIM HALL

MOLLENKOPF, JOHN

b. 16 March 1946, Bethesda, Maryland

Key works

(1975) 'The post-war politics of urban development', *Politics and Society* 5(3): 247–96.
(1983) *The Contested City*, Princeton, NJ: Princeton University Press.
(1991) *The Dual City: Restructuring New York*, New York: Russell Sage Foundation (ed. with M. CASTELLS).

John Mollenkopf has been a leading political scientist and is recognized for his analyses of US URBAN politics conducted in the latter part of the twentieth century. He has also been an active contributor to progressive debates and an expert observer frequently sought out by the media. He received his Ph.D. in political science from Harvard in 1974.

His initial work analysed the ways in which political entrepreneurs brought together new coalitions to bring about urban renewal in American cities in the 1950s and 1960s, and how their efforts produced intense COMMUNITY protest, ultimately yielding more NEIGHBOURHOOD-friendly coalitions. His seminal 1975 article, 'The post-war politics of urban development', advanced the concept of 'pro-growth coalition', which has become pervasive in the field. His subsequent book, *The Contested City*, has also become a classic.

In the early 1980s, he moved to New York City to become a division director of the City Planning Department, and subsequently joined the City University Graduate Center, becoming director of its Center for Urban Research. He was co-founder and chairman of the Research Committee on New York City of the Social Science Research Council, which produced four volumes that reinvigorated the study of New York and set a new research agenda for the 1990s. One volume, co-edited by him and Manuel Castells, *The Dual City*, has had a major influence on the debate about globalization and urban inequality relationships. He has also

published several important studies of politics and policy in New York City.

EMILIO HADDAD

MONO-CENTRIC CITY

The concept of a mono-centric city came into use to describe a city whose structure is dominated by a single centre. This model of URBAN structure became famous through an essay on the growth of the city written by CHICAGO SCHOOL member Ernest Burgess and published in 1925. Based upon the observation of a large number of American cities, Chicago in particular, Burgess imagined a city as a series of five concentric circles. The so-called central business district, or CBD, is the driving force behind the expansion of the city, a process in which Burgess was especially interested. According to Burgess, there is a clear tendency for each inner zone to extend its area by invading the next outer zone. Inspired by plant ecology, he called this process succession.

Burgess's theory became extensively amended and even rejected by other theories. One line of criticism was that Burgess's model is limited historically to a particular situation – Chicago – at a particular time – the industrial age – in a particular country – the United States. Above that, the model emphasizes clear-cut boundaries while any city is in fact a complex patchwork. Alternative models were introduced. One of the most important ones, the sector theory – or theory of axial development – was advanced in 1939 and is commonly associated with real estate economist Homer Hoyt. Hoyt advocated the idea that around the city centre, certain differences in land use arise, which continue in an outward direction as the city expands, following transportation corridors. The theory which distances itself most from Burgess's theory is the so-called multiple nuclei theory proposed by Chauncy HARRIS and Edward ULLMAN in 1945, which today lives on as the POLYCENTRIC CITY. This development in urban theory does not mean that the concept of the mono-centric city has become obsolete. For about half a century, Burgess's model was the starting point for many studies on land use within cities. Also, the concept is still employed in current language use to make a distinction between large, concentrated cities like London and urban regions characterized by a large

number of free-standing cities of various sizes at relatively small distances from each other, a pattern which is characteristic of large parts of Europe, for example.

Further reading

Burgess, E.W. (1925) 'The growth of the city: an introduction to a research project', in R.E. Park, E.W. Burgess and R.D. McKenzie (eds) *The City*, Chicago: University of Chicago Press, pp. 47–62.

WIL ZONNEVELD

MONUMENTS

A monument is a building or structure to which cultural, historical or artistic values are attributed. Its conservation, maintenance or rehabilitation is justified because of these values. Historically, the idea of a monument was associated with constructions dedicated to commemorate something such as an arch of triumph or the columns built by imperial Romans. This rigid definition has broadened over time, however, and has come to include the area immediately surrounding the structure as an integral part of the monument.

Within cities, monuments have become landmarks, reference points to identify or characterize a certain place. Monuments have also been given educational and political identity functions, to exalt the established regime for example, as well as commemorative and artistic values, such as in the case of the Statue of Liberty, the White House, or the Arc de Triomphe in Paris. There is a net relation between the expansion of nationalistic ideas and the construction of URBAN monuments at the end of the nineteenth century and up until the first third of the twentieth century. Representative examples of this abound in European cities, such as the altar to the fatherland built in Rome, the numerous statues dedicated to Joan of Arc in France, or a statue of Christopher Columbus in Spain from around the year 1892.

The concept of monument, or monumentality, has evolved through the ages. The deeper significance of the word is associated with the idea of remaining or remembering. Thus, some of the first monuments were dolmens or menhirs, megalithic constructions built for religious or funerary purposes. In early history, the idea of erecting immense structures that would dominate the landscape can be associated with the concept of monument. Examples are the pyramids of Egypt or the ziggurats of Mesopotamia. During the Renaissance, the idea of permanence through time was reaffirmed, and themes from the classical period were recuperated in the numerous palaces that proliferated in cities such as Florence, Venice and Rome. During the sixteenth and seventeenth centuries, a passion for collecting art objects was ignited, and monumental buildings were erected to house such collections, the forerunners of museums (Louvre, Hermitage, El Prado). The influence of Baroque tastes for monumentality in urban constructions can be seen in cities such as Paris, St Petersburg and Vienna, where the construction of palaces and great churches flourished. In the nineteenth century, the contemporary notion of a monument as a special structure that preserves the memory of the past came into fashion. Hence, the cities of Washington and Philadelphia were monumentalized, this concept was taken into consideration in the reconstruction of Paris, and in Berlin monuments recalled the power of the newly installed Prussians. This notion remained relevant to twentieth-century architecture, and can be evidenced in structures such as government buildings, sports stadiums, modern museums, commercial and office buildings, as well as in plazas and open spaces in urban centres.

Further reading

Ellin, N. (1999) *Postmodern Urbanism*, New York: Princeton Architectural Press.
Frampton, K. (1992) *Modern Architecture: A Critical History*, London: Thames & Hudson.

JESÚS M. GONZÁLEZ PÉREZ
RUBÉN C. LOIS GONZÁLEZ

MORRIS, WILLIAM

b. 24 March 1834, Walthamstow, England; d. 3 October 1896, Hammersmith, England

Key works

(1861) Established Morris, Marshall, Faulkner & Co.
(1890) *News from Nowhere: Or, An Epoch of Rest*, Boston, MA: Roberts Brothers.

Morris was a writer, designer, craftsman, typographer and socialist, who opposed modern forms of industrial production. He founded the handicraft firm of Morris, Marshall, Faulkner & Co., which specialized in stained glass, in an attempt to reconcile beauty with utility and to restore a direct relationship between producers and consumers. Morris was an authority on medieval craftsmanship and domestic design, who shunned the monotonous and cheaply built SUBURB. He became involved in politics in the mid-1870s, founding the Society for the Protection of Ancient Buildings, a pioneering conservationist pressure group. In 1883, Morris became a socialist and campaigned for the Democratic Federation. A year later, he became a founder member of the Socialist League. Between 1885 and 1895, Morris established and financed the newspaper of the Socialist League, *Commonweal*, in which his utopian fictions, *A Dream of John Ball* (1886–87) and *News from Nowhere* (1890), were serialized. In these fictions, his most famous literary works, Morris contrasted the negative social and environmental consequences of capitalism with his own socialist vision of an English Arcadia.

Further reading

Harvey, C. and Press, J. (1991) *William Morris: Design and Enterprise in Victorian England*, Manchester: Manchester University Press.
MacCarthy, F. (1994) *William Morris: A Life for Our Time*, London: Faber & Faber.

ADRIAN R. BAILEY

MOSES, ROBERT

b. 18 December 1888, New Haven, Connecticut; d. 29 July 1981, West Islip, New York

Key works

(1956) *Working for the People: Promise and Performance in Public Service*, New York: Harper & Brothers.
(1970) *Public Works: A Dangerous Trade*, New York: McGraw-Hill.

Robert Moses was a highly effective promoter and administrator of public works projects. From the mid-1920s until the mid-1960s, he oversaw much of the development of PARKS and parkways on Long Island, and parks, bridges, tunnels, expressways, URBAN renewal and housing projects in New York City. He sought to develop park and highway systems and modern housing for the New York metropolitan region, and he was also responsible for major hydro-power projects in upstate New York. Because many of his projects used EMINENT DOMAIN, evictions and demolition to clear areas for new construction, he has frequently been criticized as a destroyer of NEIGHBOURHOODS and COMMUNITIES. He relished publicity and his favourite response to critics was 'You can't make an omelette without breaking eggs.'

Born into a wealthy German-American Jewish family, Moses graduated from Yale in 1909, attended Oxford University and earned a Ph.D. in political science from Columbia University in 1914. In 1924, Governor Alfred E. Smith appointed him chairman of the State Council of Parks and president of the Long Island State Park Commission. In 1934, Mayor Fiorello La Guardia appointed him New York City park commissioner. In 1936, he became chairman of the Triborough Bridge Authority, and he made it his long-term centre of operations. He used the autonomy, revenues and legal powers of the public authority to acquire more official positions and to craft legislation to oversee the selection and execution of public works.

In the 1960s, Moses played the principal role in organizing the New York World's Fair of 1964–65, held in Flushing Meadow. Around that time, however, Governor Nelson A. Rockefeller sought to reduce his powers, ending his multiple state appointments in 1963 and forcing him to resign from the Triborough Bridge and Tunnel Authority in 1968. Moses dedicated the first part of his retirement to justifying his legacy with *Public Works: A Dangerous Trade*, but his image was forever changed by Robert Caro's monumental and highly praised biography *The Power Broker*, which depicted his life as a Shakespearean tragedy – an extremely talented and energetic man who becomes increasingly arrogant and ruthless in pushing his projects through. Caro crafted his book with prosecutorial zeal, and in the process he converted Moses into a symbol of all that was wrong with urban renewal and highway-building in the American city. Moses acquired notoriety, while his equally effective contemporary, Austin Tobin, who headed the Port of New York

Authority from 1942 until 1972, shunned publicity, lacked a zealous biographer, and is little known.

Further reading

Caro, R.A. (1974) *The Power Broker: Robert Moses and the Fall of New York*, New York: Alfred A. Knopf.
Rogers, C. (1952) *Robert Moses: Builder for Democracy*, New York: Henry Holt.
Schwartz, J. (1993) *The New York Approach: Robert Moses, Urban Liberals, and Redevelopment of the Inner City*, Columbus: Ohio State University Press.

RAY BROMLEY

MOYNIHAN, DANIEL PATRICK

b. 16 March 1927, Tulsa, Oklahoma;
d. 26 March 2003, Washington, DC

Key works

(1963) *Beyond the Melting Pot*, Cambridge, MA: MIT Press (with N. Glazer).
(1970) *Maximum Feasible Misunderstanding: Community Action in the War on Poverty*, New York: St. Martin's Press.

Daniel Patrick Moynihan proved to be one of the more remarkable and versatile US public figures of the last half of the twentieth century. A provocative scholar, he wrote numerous noteworthy books on a range of topics, from social policy to international affairs. Moynihan also held a number of prominent public positions. He was an adviser to both Democratic and Republican presidents alike, Ambassador to India (1973–75), US Representative to the United Nations (1975–76), and a four-term US Senator for the state of New York (1977–2001).

Born in Oklahoma and raised in New York City, Moynihan developed a strong interest in URBAN policy and policy analyses. One of his earlier works, *Beyond the Melting Pot* (co-authored with Nathan Glazer), examined the arrival and the acculturation of several ethnic groups in New York, calling into question the then prevalent 'MELTING POT' theory of assimilation. Moynihan and Glazer described the means by which various ethnic groups were able to maintain their distinct identities from one generation to the next. Moynihan served as director of the Harvard-MIT Joint Center for Urban Studies (1966).

Early in his career, while at the US Department of Labor during the Kennedy and Johnson administrations, Moynihan wrote a controversial monograph, *The Negro Family: The Case for National Action*. According to Moynihan, the disintegration of the Negro family had become so entrenched that it continued even during periods of national economic prosperity; outside intervention, Moynihan argued, was necessary to break the downward cycle. Moynihan's analysis became the centre of a national debate. President Richard Nixon named him as a special assistant and his advice was influential in the development of Republican welfare reform measures. More liberal groups and black power advocates criticized Moynihan's thesis for 'blaming the victim' by focusing too much attention on out-of-wedlock births and the breakdown of the black family, diverting attention away from DISCRIMINATION and other societal causes of the African-American poverty.

In *Maximum Feasible Misunderstanding*, Moynihan reflected on the controversy that surrounded Lyndon Johnson's WAR ON POVERTY, especially the infighting and conceptual confusion that plagued the administration's attempts to use political empowerment of the poor as part of an anti-poverty strategy. As his career continued, Moynihan argued for a sense of modesty in the development of social policy and welfare development, that it was not easy to alter human behaviour, and that there were no simple and workable answers for the poverty problem.

Moynihan received numerous honours and awards, including the Presidential Medal of Freedom.

BERNARD H. ROSS
MYRON A. LEVINE

MULTI-CENTRIC CITIES

One of the defining characteristics of metropolitan areas in advanced economies as well as in developing countries is that they are multi-centric or POLYCENTRIC. The majority of the jobs are often no longer found in the centre of the metropolitan area; the city is no longer a MONO-CENTRIC functional entity. Large concentrations of employment and services, sometimes referred to as edge cities, have developed far from the traditional DOWNTOWNS. In the very large metropolitan regions in

the United States, like Los Angeles and Dallas/ Fort Worth, less than 10 per cent of all jobs are located in the traditional core of the city. The majority of the daily COMMUTERS in these metropolitan areas travel from home to other employment centres other than the downtown.

A multi-centric URBAN landscape can be found not only in very large metropolitan regions, but also in smaller cities. In Europe, where more people live in cities of between 200,000 and 2 million inhabitants than in the United States, commercial and employment centres of considerable size have also developed outside the old historic cores. So the phenomenon of the multi-centric city can be observed at quite different geographical scales.

Multi-centric urban configurations have evolved in at least three different ways. First, there is the process of DECENTRALIZATION of the urban population and employment away from the city centre. Households move ever further away from the city centre in search of amenities such as more spacious housing, a healthier and safer environment, and better schools for their children. This decentralization is followed by new regional SHOPPING MALLS that wish to provide services to the dispersed population; and by firms seeking access to the suburban labour pool and sites where they can more easily expand than in locations closer to the city centre. This process is sometimes referred to as the centrifugal mode of the emergence of multi-centric cities. In this process of expansion of urban land use into the surrounding area, smaller centres that had been more or less self-sufficient in terms of employment and services now become incorporated in the larger metropolitan region (incorporation mode).

A third way in which multi-centric urban regions have evolved is through the process of fusion. Historically distinct and independent cities and towns of roughly similar size have become one more or less contiguous functional urban region as they grow in size; furthermore; commuter distances have expanded because such cities and towns are increasingly connected by rapid transport links. Notably, in Europe, many multi-centric urban regions have emerged in this way. Randstad Holland, the Rhine-Rhur metropolitan region and the Flemish Diamond are well-known examples. In the European Spatial Development Perspective, written by the EU member states, one of the key aims is to support the balanced development of such multi-centric urban systems throughout Europe.

Further reading

Champion, A.G. (2001) 'A changing demographic regime and evolving polycentric urban regions: consequences for the size, composition and distribution of city populations', *Urban Studies* 38(4): 657–77.

Clark, W.A.V. (2000) 'Monocentric to policentric: new urban forms and old paradigms', in G. Bridge and S. Watson (eds) *A Companion to the City*, Oxford: Blackwell.

FRANS M. DIELEMAN

MULTIFUNCTIONAL URBAN LAND USE

Land has different functions that often compete with each other. The more functions have to be fulfilled in the same area, the scarcer the land concerned is becoming. Especially in densely populated areas we witness many functional conflicts. For social and economic reasons this calls for a more efficient use of the available land (e.g. multi-level solutions and a better distribution of functions over time). Furthermore, in many countries there is emerging interest in spatial quality or spatial sustainability, which may be interpreted from an environmental angle as an increasing attention to quality of life in high-density areas.

Traditionally, spatial functions have been studied as alternative and mutually competing uses of scarce land (O'Sullivan 2000; Harvey 2000). In recent years, however, it is increasingly recognized that land may act as a critical medium for reconciling conflicting options. In other words, spatial functions are not necessarily conflicting, but may be complementary with respect to each other.

This development has prompted the birth of the concept of multifunctional land use. Through multifunctional land use, ECONOMIES OF SCALE and diversity may be achieved, which lead to a saving in the use of scarce land and which may improve environmental quality. Multifunctional land use can be seen as an empirical phenomenon emerging from economic forces, but it may also be conceived as a guideline for SPATIAL PLANNING initiatives. Hence, it can also be used as a planning concept in order to attain (URBAN) sustainability. The central purpose of multifunctional land use as a planning concept is to use scarce space as wisely as possible.

Multifunctional land use can be defined as follows: a land use pattern is said to become more multifunctional when, in the area considered, the number of functions, the degree of interweaving, or the spatial heterogeneity increases. An increased degree of multifunctionality may therefore result from the addition of functions to the area (multifunctionality by diversity), from an increase in dispersion of the number of functions over the area (multifunctionality by interweaving), or from an increase in the number of other functions touching a territory (multifunctionality by spatial heterogeneity).

Different forces influence the application of multifunctional land use on specific sites. These forces can be market driven, but may also be the result of government policy. Strong restricting and stimulating factors are to be found in the current spatial structure and current spatial developments. Except for technological developments, such restricting and stimulating factors can lead to a change in the degree of multifunctionality via the land market, i.e. price-making forces in the land market strongly influence the degree of multifunctionality.

Benefits associated with synergy in production and trade are not the only benefits that are related to multifunctional land use in urban development. By concentrating activities in a CITY, open areas suitable for nature development can be protected from urban SPRAWL. This effect might be intended or unintended by urban development and multifunctional land use.

Multifunctional land use is also an interesting concept in the field of urban redevelopment. Specific characteristics of urban redevelopment are fragmented ownership, the presence of existing INFRASTRUCTURE, businesses and activities, and stakeholder preferences. Spatial planning procedures regarding urban redevelopment are often not clear and there is also often a lack of a directing party. Multifunctional land use might be a useful instrument in urban redevelopment, not only to obtain higher densities, but also to cover the extra costs by means of capitalizing synergy effects that might arise as a result of the combination of economic activities. An example of multifunctional land use and urban redevelopment can be found in industry locations. Due to soil pollution, sanitization of the location is necessary, which leads to high costs for the redevelopment of the area. Also here, multifunctional land use might be a useful instrument to cover the extra costs by means of synergy effects.

Multifunctional land use tries to combine different land use functions on the same location. Concentration of activities in urban constellations can lead to advantages for each of the different land use functions concerned, but especially for the infrastructure function. In general, there exist strong interdependencies between land use patterns and transportation infrastructure, since locational decisions are, to a large extent, the result of the relative costs of travel to various activities. Next to this, the pattern of trips generated by these activities affects the costs of travel. Therefore, it can be said that the spatial organization of land use determines and, at the same time, is being determined by the design and characteristics of the transportation system. The presence of infrastructure in multifunctional urban land use projects provides certain flows of people that form a potential user group for other land use functions within the area. A focus on infrastructure in multifunctional urban land use is interesting because of economics of density in transport. These can be enhanced due to multifunctional land use. A sufficient population density and mixed-use city structures bring numerous positive effects by reducing transport demand, ensuring public transport is profitable, and enabling communities to share infrastructure costs (Procos 1976). Integrating the planning of transport, infrastructure, and urban and regional policies can help to reduce the need for travel and can decrease emissions, land use and resource consumption. This all may lead to a more sustainable use of land and transport. Planning should emphasize accessibility rather than mobility. A pattern of smaller urban areas is probably not suitable for attaining this goal, since they generate more traffic than compact centralized cities. Multifunctional land use could be used as an instrument to realize these goals.

Further reading

Harvey, J. (2000) *Urban Land Economics*, London: Macmillan.

O'Sullivan, A. (2000) *Urban Economics*, Boston, MA: Irwin/McGraw-Hill.

Procos, D. (1976) *Mixed Land Use: From Revival to Innovation*, Stroudsburg, PA: Hutchinson & Ross.

RON VREEKER
CAROLINE RODENBURG

MUMFORD, LEWIS

b. 19 October 1895, Flushing, New York;
d. 26 January 1990, Amenia, New York

Key works

(1938) *The Culture of Cities*, New York: Harcourt, Brace
& World.
(1961) *The City in History*, New York: Harcourt, Brace &
World.

When he appeared on its cover in 1937, the editors
of *Time* magazine called Lewis Mumford 'a new
type of public figure'. Others have called him
America's last great public intellectual. Throughout a career that saw the publication of some
twenty-five influential volumes, Mumford made
signal contributions to social philosophy, American literary and cultural history, the history of
technology, and, pre-eminently, the history of
cities and URBAN PLANNING.

In *The Culture of Cities* (1938) and *The City in
History* (1961), Mumford argued that the URBAN
experience is an integral component in the ongoing
evolution of the human personality. He insisted
that the economic functions of cities were secondary to their relationship to the natural environment
and to the spiritual values of human COMMUNITY.
Mumford applied these principles to his lively and
illuminating architectural criticism for *The New
Yorker* and his work with the REGIONAL PLANNING
Association of America.

From 1915 to the older man's death in 1932,
Mumford maintained an almost weekly correspondence with Patrick GEDDES, the Scots polymath and planning pioneer. Mumford addressed
Geddes as 'Master', named his own son Geddes,
and it was through Geddes that Mumford found

his true calling as a lifelong proponent of the
GARDEN CITY ideals of Ebenezer HOWARD.

Mumford's influence on urban planning can
hardly be overstated. His concept of the 'urban
drama' – the idea that the city is 'a theater of social
action' – resonates with a long line of architects,
planners and urban cultural analysts. His critique
of the sterility of much of modern planning practice and his denunciation of MEGALOPOLIS as 'anti-
city' inspired generations of NEIGHBOURHOOD
activists. And as a popular advocate of sound,
humane planning principles, Mumford invited the
democratic public into a discussion of what city life
could and should be.

The City in History is Mumford's masterpiece,
but the fullness of his intellectual curiosity is suggested by the range of his writings: *Sticks and
Stones* (1924) is a study of American architecture;
The Golden Day (1926) and *The Brown Decades*
(1931) examine nineteenth-century American literature; *Technics and Civilization* (1934), *Technics
and Human Development* (1967) and *The Pentagon
of Power* (1970) are trailblazing critical analyses of
the cultural effects of technology; *The Condition of
Man* (1944), *The Conduct of Life* (1951) and *The
Transformations of Man* (1956) are important
contributions to twentieth-century social philosophy; and *The Urban Prospect* (1968) is an outstanding collection of essays on city planning and
contemporary urban civilization.

Further reading

Hughes, T. and Hughes, A. (eds) (1990) *Lewis Mumford:
Public Intellectual*, New York: Oxford University
Press.
Miller, D. (1989) *Lewis Mumford: A Life*, New York:
Weidenfeld & Nicholson.
Novak, F. (ed.) (1995) *Lewis Mumford and Patrick
Geddes: The Correspondence*, London: Routledge.

FREDERIC STOUT

N

NASH, JOHN

b. 1752, London, England; d. 13 May
1835, Cowes, Isle of Wight, England

Key works

(1815–22) Brighton Pavilion.
(1817–28) Regent's Park.
(1827–33) Carlton House Terrace.

A contemporary of King George III, this British
architect made significant contributions to Victorian-
era architecture in nineteenth-century England.
Though his work spans both the Georgian and the
Victorian eras, his remarkable repertoire covers a
stylistic range that includes Gothic, neo-classical
and picturesque elements. John Nash was trained
briefly by Sir Robert Taylor, but before his career
could fully mature, he moved to Wales, because of
financial troubles. During this period, he partnered
with landscape architect Humphrey Repton. His
employment by the Prince Regent in 1811 enabled
Nash to embark upon a number of grand architectural projects. Among his most recognized contributions are Regent's Park (his first commission
from the Prince Regent), the Marble Arch (which
was designed for another project, but was later
moved to Hyde Park), the Brighton Pavilion
(with an exotic architectural style), Carlton House
Terrace, the Royal Mews, Haymarket Theatre, All
Soul's Church, St James's Park, the layout of
Trafalgar Square, and portions of Buckingham
Palace. Nash was dismissed before completing the
work he had planned for Buckingham Palace when
the former Prince Regent, by then George IV, died
in 1830. He returned to the castle he had built for
himself in Cowes in 1798, and died there in 1835.

Despite his expensive grand schemes, John
Nash contributed significantly to nineteenth-
century English architecture and his work inspired
many generations of architects.

ALI MODARRES

NECROPOLIS

This is an archaeological term of Greek origin,
meaning city of the dead. The term was borrowed
from the name of a SUBURB of ancient Alexandria
in Egypt, which was out of the peninsula and there
were many gardens, graves and places where the
dead were worshipped. Ancestor worship was
practised in Egypt from as early as the third millennium BC. It was later also adopted by the
Greeks and the Romans. The size of a tomb and its
degree of decoration in a necropolis were determined in part by the owner's wealth. Similarly,
a tomb's position relative to the centre of the
necropolis correlated with the owner's resources,
as the larger and more lavishly decorated tombs
were located there. The tomb façades in a necropolis generally faced eastwards towards the rising
sun, an arrangement that is traced back to Phar-
aonic times. In ancient Greece, the necropolis was
located outside the city walls along the roads
leading into the city and the graves were decorated
with little temples, statuettes, sculptures and ana-
glyphs with information about the social status of
the dead.

The most well-known necropolises are the
necropolises of Thebes and of Sakkara in ancient
Egypt, the necropolises of the royal graves in the
ACROPOLIS of Mycenae and of Ceramikos in

ancient Greece, and the Roman necropolis in Pompeii.

Further reading

Reeves, C.N. (ed.) (1992) *After Tutankhamun: Research and Excavation in the Royal Necropolis at Thebes*, Studies in Egyptology, New York: Columbia University Press.

KONSTANTINOS LALENIS

NEGROPONTE, NICHOLAS

b. 1 December 1943, New York City

Key works

Founder, MIT Media Laboratory.
(1995) *Being Digital*, New York: Vintage Books.

Nicholas Negroponte, Jerome B. Wiesner Professor of Media Technology, is the founder and the director of the Massachusetts Institute of Technology's uniquely innovative Media Laboratory. Negroponte studied at MIT, where he received a bachelor's and a master's degree in architecture. Already as a graduate student he specialized and became a pioneer in the new field of computer-aided design. He joined the Institute's faculty in 1966, and for several years thereafter spent his teaching time between MIT and visiting professorships at Michigan, Yale and the University of California at Berkeley. In 1968, he also founded MIT's pioneering Architecture Machine Group, a combination of lab and think-tank responsible for many radically new approaches to the human–computer interface. Negroponte was there at the birth of the INTERNET in the 1960s and the inception of multimedia in the 1970s. His Media Lab (www.media.mit.edu) is an interdisciplinary, multi-million-dollar research centre of unparalleled intellectual and technological resources where the focus is on the creative use of digital technologies to enhance the way people think, express and communicate ideas, and explore scientific frontiers. Negroponte has been very active in the private sector where he serves on the board of directors for Motorola, Inc. and he has provided start-up funds for more than twenty companies. Negroponte is also co-founder of and back-page columnist for *Wired* magazine. Out of this experience have come over two hundred published articles and three influential texts. He is the author of the 1995 best-seller *Being Digital*, which has been translated into more than forty languages.

TIGRAN HASIC

NEIGHBOURHOOD

The term 'neighbourhood' has no precise, agreed-upon definition. It is used to refer both to a physical place and to the group of people who occupy that place. In common parlance, it is often used as a synonym for 'COMMUNITY'. In fact, the terms are not entirely synonymous: while a neighbourhood may be a community, a community is not necessarily a neighbourhood. A neighbourhood is a geographically delineated subunit of a city whose residents (neighbours) share the circumstances that come with a common location. However, whether or not they interact in a way that makes them a community depends on a variety of factors.

This tension between neighbourhood and community has historical roots. As cities made the transition from commercial centres to centres of industrial production in the late nineteenth century, they became more functionally differentiated internally, and this reorganization also took on a geographic dimension. The implications of these changes for relationships among residents continue to be much debated. German sociologist Ferdinand Tönnies emphasized the shift away from ties based on the primary bonds of family (GEMEINSCHAFT) to those of the larger society (GESELLSCHAFT). This transition to URBANISM included a separation between home and work life and a much greater diversity of population living together in INDUSTRIAL CITIES than had occupied the smaller villages and towns of the previous era. While some argued that the movement of peasants to cities brought with it a transfer of village bonds to the URBAN neighbourhoods where they settled, others argued that the different internal logic of industrial cities undermined these bonds, leading to a decline in community.

Ecological models were the first and most influential explanations of the form taken by growing industrial cities, including the development of their neighbourhoods. This approach described city growth as a natural process of competition between social groups for resources, including land. Theorists described a variety of

particular city forms as resulting from this competition, with growth alternately pushing outward in concentric rings, along transportation corridors or clustering around various sub-centres. Generally, theorists argued that growth was pushing outward from the centre, with residential areas being located farther and farther from the centre. Neighbourhoods were differentiated, in this view, in a natural process through which immigrants sorted themselves into neighbourhoods in order to be close to jobs and to maintain their communal ties. The resulting neighbourhoods were viewed as internally homogeneous yet collectively quite diverse. Because there was little interaction between them, these neighbourhoods were thought to retain the basis for community: a common identity, cultural practices, and primary ties of friendship and kinship. Neighbourhoods bounded social interaction, thus reinforcing the overlap between residence and community.

Critics of the HUMAN ECOLOGY model lamented its use of natural metaphors for human processes, emphasized the decline in community that appeared to accompany much urban development, and noted how rare the tightly knit neighbourhoods predicted by ecological models appeared to be. Debate spurred alternative ways of viewing the basis for communal ties, with implications for how neighbourhoods would be defined. Gerald Suttles's 'community of limited liability' depicted residents as identifying with others on the basis of issues affecting their physical zone of the city, or to protect status or family needs, rather than community relationships. In this view, residents could belong to several communities at once, maintaining only partial allegiance to each. The scale of neighbourhoods, in this view, was not limited by the need to form intimate bonds; instead, neighbourhoods could be quite large. Others viewed communities more functionally, as social systems connecting neighbourhoods to the larger social system. Neighbourhoods, in this view, are not natural subunits of cities but instead are formed by larger political and economic forces. Community ties are influenced by forces external to the neighbourhood. Finally, social network analysts contend that the intimate ties that are the basis for community are alive and well, but not necessarily located in one's own neighbourhood. Community, in other words, does not require propinquity. Yet neighbourhoods do foster the casual bonds that are important to the formation of SOCIAL CAPITAL.

Empirically, researchers have found that neighbourhoods are identified differently by different groups, according to their relationship to the neighbourhood and to their social position within society. Those for whom the neighbourhood is the locus of social relationships will delimit a smaller area than others for whom it may be a market to be targeted or a zone for coordination of public services. Urban residents generally identify smaller neighbourhoods than do suburban residents. In addition, particular groups more commonly stress the social dimensions of neighbourhood, and base their neighbourhood definition on these factors. Such groups include African-Americans, the elderly, the unemployed, the unmarried, and long-term residents. Those less integrated into the larger society are thought to develop the strongest local ties. In contrast, those whose lives are more far flung – the young, the well educated and employed, and, generally, whites – are more likely to define neighbourhoods based primarily on physical rather than social characteristics.

By the late 1990s, an increasing number of public programmes and private initiatives, pursuing a wide variety of goals, were identifying neighbourhoods as the unit of action, building on lessons learned since the 1960s by COMMUNITY DEVELOPMENT CORPORATIONS and other neighbourhood-level organizations. Public programmes include city planning initiatives aimed at identifying local priorities for future public investments and foundation initiatives aimed at revitalizing low-income communities. The motivations for operating at the neighbourhood level include the smaller, more manageable scale, particularly for initiatives desiring to take a comprehensive approach in their programmes, the desire to target a particular population, or the belief that working at this scale makes engagement of residents more likely.

Further reading

Chaskin, R.J. (1995) *Defining Neighborhood: History, Theory, and Practice*, Chicago: The Chapin Hall Center for Children at the University of Chicago.

Jacobs, J. (1961) *The Death and Life of Great American Cities*, New York: Vintage.

Suttles, G.D. (1972) *The Social Construction of Community*, Chicago: University of Chicago Press.

SEE ALSO: Jacobs, Jane

ELIZABETH J. MUELLER

NEIGHBOURHOOD WATCH

Neighbourhood Watch schemes began in the United States in the late 1970s and have proved popular in many advanced capitalist societies as a COMMUNITY-based response to CRIME prevention and community safety. Schemes vary considerably in organization, scope and membership, but broadly speaking these schemes consist of community-based action (either directly from the community itself or through police or other agency sponsorship) through which the 'NEIGHBOURHOOD' (street, block or residential estate) comes together on a semi-formal basis with the objective of preventing and deterring crime in the immediate locality.

The fundamental focus of such schemes is the increased, collective vigilance of the immediate community in regard to suspicious acts, individuals and possible criminal acts and the consequent surveillance of people who might be taken to constitute a criminal threat to the community. This approach is often supplemented by other preventative activities such as property marking, home security checks and the raising of crime prevention awareness. Such focus highlights how such schemes rest on a conceptualization of crime as external to the community, primarily opportunistic and property oriented – which leaves considerable serious crime (e.g. domestic violence and abuse) unchallenged.

Proponents stress that Neighbourhood Watch schemes have positive direct and indirect effects on both crime and the communities that participate. Direct benefits include reduced opportunities for crime and, consequently, actual reductions in the number of crimes committed in the immediate locality. Indirect benefits include a reduced fear of crime, better relations with the police, an improved SENSE OF COMMUNITY, and other forms of 'civic renewal' that increase community integration and activity and stabilize social practices, norms and values. Critics maintain that Neighbourhood Watch extends state/official surveillance and control techniques right into the very local levels of community, acts to further exclude those at the margins of society (e.g. the homeless) and reinforces the siege mentality of some URBAN areas. Furthermore, evaluation of how effective such schemes actually are is limited and critics highlight crime displacement (rather than reduction) as the key effect of such schemes. Research also tends to demonstrate that schemes are most likely to develop and be well supported in predominantly middle-class, middle-aged, white and property-owning neighbourhoods where crime is relatively low. Conversely, the development of and participation in schemes in areas of higher crime and community fragmentation – arguably where the impact might be more beneficial – tends to be lower and schemes much less popular. However, such reservations do not appear to deter the further development of such schemes and the increased systematic and official support for these initiatives (in particular in the United States, Britain and Australia). Furthermore, the growing interest in more formal and proactive prevention measures, in the form of street wardens, estate rangers and other forms of citizen patrol, suggests that such approaches will remain politically and socially popular for the foreseeable future.

Further reading

Bayley, D. (1994) *Police for the Future*, New York: Oxford University Press.

RICHARD HUGGINS

NEW CHARTER OF ATHENS

The New Charter of Athens is one more expression of the reaction against the CIAM (Congrès Internationaux d'Architecture Moderne) discourse on URBANISM developed in Europe after the 1970s, with similarities to the NEW URBANISM movement, initially developed in North America.

The national town planning associations which were members of the European Council of Town Planners (ECTP) prepared, between 1995 and 1998, a charter of principles for the planning of cities in the twenty-first century, which was named, symptomatically, the New Charter of Athens (NCA), following the conference organized by the Greek Town Planners Association, in 1994, regarding the CHARTER OF ATHENS, during which the delegates, recognizing the negative effects of the CIAM Charter of Athens (1933) on URBAN structures, decided to prepare a new charter for better urban development.

The NCA puts citizens at the centre of urban policy, urging town planning to answer the social and cultural needs of present as well as future generations. Instead of a technocratic conception of planning, the NCA gives a prominent role to

citizens and city users and priority to their needs. This means that the city in the twenty-first century will be determined less by technocratic master plans than by processes of negotiation involving all members of the COMMUNITY, which calls for a different form of urban plan-making.

The NCA seeks to meet three key objectives: to define the present urban agenda; to define the role of town planning in that agenda; and to propose guiding principles to be followed by town planning professionals at all levels. For the first objective, the NCA proposal is based essentially on the EU urban policy documents on sustainable development, prepared during the 1990s (namely *Green Book on Urban Environment* (1990), *Europe 2000* (1991), *Europe 2000+* (1994), *European Sustainable Cities* (1996) and *European Spatial Development Framework* (1997)). In relation to the second objective, the NCA takes on board several principles adopted by town planning in the past, but still well accepted today, but also defines a set of new principles that should guide planning practice in the twenty-first century, among which we can mention the subsidiarity principle, a more fair distribution of urban resources, a larger proportion of green areas, protected areas and leisure spaces than in the past, and priority to BROWNFIELD redevelopment in relation to greenfield new development, containing suburban SPRAWL and therefore increasing urban compactness. For the third objective, the NCA adopted a set of ten recommendations, which reflect concerns very similar to those of the Charter of New Urbanism (1993), with the possibility of each national association introducing more specific guidelines to its members. The NCA is expected to be revised every four years.

Further reading

ECTP (1998) *New Charter of Athens*, London: ECTP.
Silva, C.N. (2002) 'A Carta do Novo Urbanismo e a Nova Carta de Atenas: a utopia urbana do Século XXI?' ('The Charter of New Urbanism and the New Charter of Athens: the urban utopia of the twenty-first century?'), *Cadernos Municipais* XVI(79): 35–47.

SEE ALSO: compact development; infill development; Smart Community; smart growth; sustainable urban development; transit-oriented development; urban growth boundary; walking city

CARLOS NUNES SILVA

NEW ECONOMY

The economic system that emerged in the 1990s in the United States, fuelled by four fundamental factors: (1) the information and communications technology revolution; (2) investments in research and development (R&D); (3) the globalization of economic activities; and (4) network-based growth.

The essence of the new economy is knowledge-based information. The capacity to create, process and commercialize information distinguishes 'new economy' industries from 'old economy' ones. The invention, rapid development and diffusion of information technologies between the early 1970s and the late 1990s transformed the US economy. Specific industries – information technology and finance – pushed the technology envelope by adopting and adapting available technology in new ways to expand their business capacity. The INTERNET industry, in particular, became the pulse of the new economy and saw rapid growth in late 1990s. The backbone of the industry comprises telecommunications firms, internet service providers, firms developing software for web-based transactions, web content providers, and companies that facilitate web-based transactions like Amazon.com. Job growth in high-tech industries – recognized as the new core of the US economy in the late 1990s – was unprecedented.

In addition to technological developments, the emergence of the knowledge-based economy is also attributed to significant investments in R&D. As Harvard University professors Michael Porter and Jeffrey Sachs note (2002), lowering costs, restructuring and raising quality are no longer sufficient for competitiveness in the new economy. To remain competitive, economic agents must not only create and process information efficiently with the latest information technology, but also invent and commercialize new products and processes rapidly. Cross-national studies show that the national capacity for innovation is highly correlated with technological sophistication and the size of the scientific and technical labour force. Porter and Sachs use international patenting – patents granted by the US Patent and Trademark Office to US and foreign inventors – as an indicator of a nation's capacity for innovation. International patenting has quadrupled in two decades, reaching 100,000 patents per year by the late 1990s – all located in the developed world, with the United States, Japan and Taiwan as the world's leading inventors.

Another distinguishing feature of the new economy is its organization on a global scale. The advent of the internet, for example, has transformed the way business is conducted in the financial sector. Deregulation of financial markets in most countries and the reduction of barriers for cross-national financial transactions have facilitated the globalization of financial markets. As a result, capital markets and institutions around the world are now interdependent, electronically connected, and transactions can take place instantaneously without recognizing national borders. Some scholars have even declared the end of the nation-state, noting that GLOBAL CITIES and global regions are more relevant for understanding the growth and decline of nations because they are the ultimate engines of economic growth and prosperity for nations. According to Manuel CASTELLS (2000), economic agents in the new economy have the institutional, organizational and technological capacity to conduct business transactions on a global scale. He attributes the collapse of the Soviet Union to its structural inability to adapt its industrial production processes in an increasingly informational and global world.

The new global economy is also networked. AnnaLee Saxenian has studied networks among highly skilled immigrants in California's Silicon Valley – home of the integrated circuit – and the extent to which networks create jobs and wealth in the new economy. She has found that one-third of Silicon Valley's engineers working in technology firms in the 1980s and 1990s were Indian and Chinese (Saxenian 1999). These immigrants have not only contributed to the growth of the technology sector, but have also established business linkages with select regions of India and China. The rapid expansion of transnational communities among the Indian, Chinese and Taiwanese in Silicon Valley underscores the global dimension of the new economy and its reliance on the process of 'brain circulation' rather than 'brain drain'.

The informational, R&D-based, global and networked nature of the new economy has unleashed innovative capacity most effectively in countries where physical INFRASTRUCTURE (e.g. ports, roads, and telecommunications infrastructure) and institutional and legal systems (e.g. protection of intellectual property) are well developed. Government involvement in inducing innovation (e.g. investment in HUMAN CAPITAL and physical infrastructure) is also key.

Despite predictions that the advancement in telecommunications technologies would create virtual communities, CYBERCITIES and decentralized human settlement patterns (e.g. EXURBIA), real cities still do matter. Internet firms (dot-coms), for example, flocked to San Francisco in the late 1990s. The sheer volume of venture capital and the influx of start-up technology firms, however, led to the GENTRIFICATION of NEIGHBOURHOODS. The ascendancy of the dot-com firms was particularly felt in one neighbourhood (South of Market) that was transformed from a primarily old industrial area into a 'multimedia gulch' or 'cyber-district'.

Clearly, not everyone benefits from the new economy, especially in global cities like New York, London and San Francisco where the headquarters of the new economy industries are located. Saskia SASSEN has noted the bifurcated earnings structure in global cities as a result of their particular industrial and occupational structure. Core functions of the new economy depend on a highly educated workforce capable of creating value and improving productivity in the new economy's knowledge-based industries. It also depends on a low-skilled workforce for low-paid service sector jobs. While the service industry where these two groups work continues to expand, the once well-paying manufacturing jobs in the 'old economy' have long since departed US cities for offshore locations or have completely vanished.

Further reading

Castells, M. (2000) *The Rise of the Network Society*, Oxford: Blackwell.

Porter, M.E. and Sachs, J.D. (2002) *The Global Competitiveness Report 2001–2002*, Oxford: Oxford University Press.

Saxenian, A. (1999) *Silicon Valley's New Immigrant Entrepreneurs*, San Francisco: Public Policy Institute of California.

Scott, A.J. (2002) *Global City-Regions: Trends, Theory, Policy*, Oxford: Oxford University Press.

SEE ALSO: economic restructuring; global cities; immigration; internet

AYSE PAMUK

NEW FEDERALISM

New Federalism seeks to reduce power in Washington and increase the power of states and localities. Begun during the administration of

Richard Nixon, the New Federalism was a reaction against the explosion in the number and scope of federal aid programmes that took place beginning with Lyndon Johnson's GREAT SOCIETY.

Advocates of the New Federalism argue that the power of the central government has grown in apparent violation of the US Constitution, where the Tenth Amendment provides that all powers not delegated to the national government 'are reserved to the States respectively, or to the people'. Advocates of the New Federalism further charge that national power is often dysfunctional, that federal grant programmes are too voluminous and complex to be understood and administered effectively. Federal rules and reporting requirements further deny sub-national officials much-needed programme flexibility.

When Nixon assumed the presidency in 1969, virtually all intergovernmental transfers took the form of CATEGORICAL GRANTS that allowed recipient governments very little discretion in the use of aid monies. Nixon promised states and localities greater freedom, a new American Revolution. He supplemented the existing categorical programmes with a new programme of *general revenue sharing* (enacted in 1972) that gave no-strings-attached money to every state and virtually every general-purpose local government across the United States. Revenue sharing was reduced over time and was finally terminated by Ronald Reagan in 1986 as part of his budget-cutting efforts.

Of greater long-term significance was Nixon's second alternative to categorical grants, his introduction of new *block grants*, where aid money was tied to a broad functional area but recipient governments were allowed great discretion in determining just how the monies would be used. The block grant concept became a permanent element in the intergovernmental aid system of the United States.

While Nixon had sought to give states and localities greater authority in the implementation of intergovernmental programmes, Ronald Reagan took the anti-Washington attitude of New Federalism one step further. Reagan cut back federal programme responsibility and spending; the responsibility of deciding on and funding domestic programmes would fall increasingly on the states and localities. Reagan terminated revenue sharing and Urban Development Action Grants; he reduced the funding of a great many other urban and social programmes. Only the

resistance of a Democratic Congress prevented Reagan from zeroing out the COMMUNITY DEVELOPMENT BLOCK GRANT programmes of the Economic Development Administration and the Small Business Administration. Declining cities had to cope with dramatic reductions in federal assistance. In just five years (1981–86), federal direct aid to Baltimore dropped from US$220 million to US$124 million. Detroit lost over half its direct federal aid. Faced with the Reagan aid reductions, many states and localities chose to curtail programmes rather than raise taxes. Reagan also pursued the consolidation of categorical grants into block grants, increasing the authority of the states over programmes that could not be eliminated.

There are criticisms of New Federalism. Calls for a New Federalism are often based on an exaggerated perception of federal power. While the number and funding of federal programmes have clearly grown over the years, it is still doubtful that the central government possesses the full extent of power alleged by its critics. The states and localities possess considerable discretion in the implementation of intergovernmental programmes that are largely administered by state and local officials. In some cases, the federal government has a difficult time in gaining state and local compliance with nationally mandated environmental standards and other programme objectives. The New Federalism gives states and localities greater freedom to pursue their own priorities in place of federally established programme goals.

Critics of the New Federalism further argue that a full reading of the Constitution shows that the federal government's responsibilities are not confined to a few enumerated powers. The Tenth Amendment is counterbalanced by an expansive clause that gives the federal government the right to make all laws 'necessary and proper' to the powers formally granted to it. In essence, the national government has a whole host of implied powers not explicitly written into the Constitution. The civil rights amendments give the national government further authority to undertake actions to protect the rights of citizens.

Federal aid has assisted sub-national problem-solving in a great many programme areas, from the building of new hospitals, sewers and highways to the provision of job training and childcare. Federal aid monies also stimulated local economies and provided for a level of equality of benefits among

programme recipients that could never have been realized by the states with their variations of wealth. Accompanying federal rules also mandated local citizen participation, giving previously ignored citizens new opportunities to be heard in the local arena. Federal action was also critical to civil rights gains, where anti-discrimination measures were often forced upon unwilling states and localities.

Despite the great many successes of intergovernmental programmes, the New Federalism philosophy, with its anti-Washington orientation, has proved to be a powerful and enduring idea. Even Democrats, who often seek national programmes for equity and racial fairness, have had to respond, although they have generally been less willing than Republicans to turn programme authority over unconditionally to the states. Bill Clinton's reinventing government initiatives sought to permit enhanced state and local programme innovation.

Advocates of New Federalism argue that government closest to the people governs best. While this is often undoubtedly true, at times New Federalism does not so much place power in the hands of the people as much as it reinforces the position of local elites, those who already hold the power in the state and local arena.

Further reading

Conlan, T.J. (1998) *From New Federalism to Devolution*, Washington, DC: The Brookings Institution Press.
Reagan, M.D. and Sanzone, J. (1981) *The New Federalism*, 2nd edn, New York: Oxford University Press.
Walker, D.B. (2000) *The Rebirth of Federalism: Slouching Toward Washington*, New York: Chatham House.
Wallin, B. (1998) *From Revenue Sharing to Deficit Sharing*, Washington, DC: Georgetown University Press.

MYRON A. LEVINE

NEW URBANISM

New urbanism, or neo-traditional town planning/design, is the most influential postmodern design paradigm for American SUBURBS. Developed by architects and URBAN designers since the early 1980s, it attempts to address the ills associated with urban SPRAWL and post-Second World War suburban development. Drawing from design principles of pre-Second World War urban fabric (especially US small towns of the early 1900s), the

paradigm aspires to promote the cherished American dream of a close-knit COMMUNITY with vitality, accessibility and efficiency. It also attempts to rectify racial and economic DISCRIMINATION associated with the suburbs. The proponents of new urbanism believe all this can be achieved through physical design that promotes pedestrian proximity (five- to ten-minute walks) to shops, jobs and community facilities; shared public cores; high-density housing; and mixed uses. The underlying assumption is that housing consumers will trade off larger suburban lots for pedestrian friendliness and proximity to NEIGHBOURHOOD amenities. The advocates of new urbanism founded the Congress for New Urbanism, in 1994. This is an international organization that is dedicated to replacing sprawl with neighbourhood-based alternatives.

The two most popular variations of the new urbanism theme are the traditional neighbourhood development or district (TND) and the transit-oriented development (TOD) or PEDESTRIAN POCKET (PP). Developed by the Miami-based architectural firm DPZ, of Andrés DUANY and Elizabeth PLATER-ZYBERK, TNDs are usually new suburban residential developments. The basic building block of the TND is the neighbourhood. Most TND homes are within a three-minute walk of neighbourhood PARKS, and within a five-minute walk of a central square or common with a meeting hall, childcare centre, bus stop and convenience store. A variety of housing types are included to accommodate different household types and incomes within each neighbourhood. Neighbourhoods are nested within larger units known as villages or towns. Another important characteristic of the TND is a grid-like street pattern with frequent connections and different types of streets such as BOULEVARDS, courts, roads, lanes and alleys. These streets have different specifications in terms of SIDEWALK widths, street planting, on-street parking, traffic speeds and pedestrian crossing times. TNDs are less regional in scope than TODs and rely on existing networks of roads and expressways for connections to major cities. The PP, developed by San Francisco-based architect Peter CALTHORPE, entails retrofitting of existing suburbs along with some new growth around public transportation hubs. Ideally, PPs should be located at strategic points along a regional transit system such as light rail, and linked to each other through such a transit system. PPs are

pedestrian-scaled mixed-used neighbourhoods with an average quarter of a mile walking distance to the transit stop. They are regional in scope and to a great extent are dependent upon investment in light rail and bus transit. Large-scale office development, stores and various types of residential units characterize such developments. TODs often include a layer of radial streets.

Seaside Florida, a sea resort, is the most well-known TND. Developed by architects Duany and Plater-Zyberk, the town follows the general TND design principles of fostering a strong SENSE OF COMMUNITY. Built on an 80-acre costal site, the town has a compact layout with an average density of about 12.5 units per gross acre. It is characterized by front porches, footpaths, street vistas, a quaint architectural style, and mixed-used development. To make the town a pedestrian-friendly place, walking times to one's daily needs are generally about five minutes. Seaside has been dismissed by critics as a resort town and hence not a true test of the TND. Furthermore, the housing prices also prohibit the inclusion of lower-income groups. Unlike Seaside, Kentland, Maryland, is the application of TND principles to a year-round, working community. It is located within the city of Gaithersburg, 23 miles north-west of Washington, DC. The master plan for the community was the result of a widely publicized design charette led by Duany and Plater-Zyberk. The plan has six neighbourhoods, each combining elements of residential, office, civic, cultural and retail use. In order to encourage diversity, in both age and income level, the community offers a range of housing types. But, contrary to expectations, the town remains unaffordable. Laguna West, in the Sacramento, California, area, is the first application of Calthorpe's TOD. As is common with TODs, the blocks in the town centre include residential as well as commercial uses. Light industrial and office workspace areas have been included along with community facilities and high- and medium-density housing.

Criticisms of new urbanism are wide and varied. Some critics argue that such suburban towns contribute to the further demise of the CENTRAL CITY by moving investment away from the city to the suburbs. They also argue that because of escalating housing prices in such developments, new urbanism promotes homogeneity and exclusion rather than diversity and inclusion. Empirical studies have indicated that such developments may actually promote the increased use of automobiles due to the gridiron pattern of streets. Studies also indicate that a large section of housing consumers are not willing to trade off large yards, privacy and access to automobiles for amenities associated with new urbanist communities. Despite criticisms, TNDs, TODs and their varied interpretations have been adopted throughout the United States as models for new and INFILL DEVELOPMENT. In fact, they are often seen as suitable tools for growth management, traffic control and sustainable development.

Further reading

Calthorpe, P. (1993) *The Next American Metropolis: Ecology, Community, and the American Dream*, New York: Princeton Architectural Press.

Duany, A., Plater-Zyberk, E. and Speck, J. (2000) *Suburban Nation: The Rise of Sprawl and Decline of the American Dream*, New York: North Point Press.

Katz, P. (ed.) (1994) *New Urbanism: Towards an Architecture of Community*, New York: McGraw-Hill.

Kelbaugh, D. (ed.) (1989) *The Pedestrian Pocket Book: A New Suburban Strategy*, New York: Princeton Architectural Press.

SIDDHARTHA SEN

NOISE

Noise is unwanted sound. Environmental, or URBAN, noise sources are transportation; stationary mechanical sources, ranging from large industrial activities to household air conditioners; or community-based, such as animals or amplification. Urban noise is generally not of an intensity that causes hearing loss but it interrupts sleep, disturbs communication and interferes with other human activities. Noise exposure also depreciates housing values. Annoyance from noise is widespread, but a connection between chronic noise annoyance and human health is, as yet, unproven. In industrialized cities, noise is the most ubiquitous pollution problem, with some 20 per cent of the dwellings subject to unacceptable levels of road traffic noise and a high proportion exposed to aircraft noise dependent on airport location.

Noise control is achieved at the source, in the propagation path, or at the receiver. Source quieting requires engineering design, driven by legislated limits in the case of motor vehicles and aircraft. Path control includes separation of source and receiver, and is amenable to land use planning

practice and the use of roadside noise barriers on no-access roadways. Receiver protection includes insulation of dwellings under flight paths. Models are available to predict levels from future stationary and transportation sources. Prevention of exposure through source and land use planning is more cost-effective than mitigating existing problems and economic instruments can be used such as higher landing charges for noisy aircraft. In many countries, management of urban noise suffers from a lack of political commitment.

Further reading

Berglund, B., Lindvall, J., Schwela, D.H. and Goh, K.J. (eds) (2000) *Guidelines for Community Noise*, Geneva: World Health Organization.

A. LEX BROWN

NOLEN, JOHN

b. 14 June 1869, Philadelphia, Pennsylvania; d. 18 February 1937, Cambridge, Massachusetts

Landscape architect, planning consultant, founding member of the American City Planning Institute and writer, Nolen operated a planning consultant practice from 1905 through to the mid-1930s and produced plans and studies for more than fifty cities and towns. Following graduation from the University of Pennsylvania in 1893 and ten years working in adult education, he pursued a master's degree in landscape architecture from Harvard University from 1903 to 1905. Nolen's Cambridge, Massachusetts, practice began with the planning of Myer Park, an affluent Charlotte, North Carolina, SUBURB, and launched a successful practice throughout the region, including city plans in Georgia, Virginia, Alabama, Tennessee and, in the 1920s, in Florida cities such as St. Petersburg and Venice. Plans for San Diego, California, Madison, Wisconsin, and Mariemont, Ohio, legitimized Nolen's national reputation.

A distinguishing trait of Nolen's work was its focus on smaller cities. His unique style incorporated elements of the CITY BEAUTIFUL MOVEMENT, such as a civic centre, and with the GARDEN CITY movement's emphasis on improved housing for the working class, preserving OPEN SPACE, and regulation of land uses to achieve a balanced COMMUNITY. Nolen was a tireless promoter of planning through his writings and extensive lecturing throughout the United States and abroad.

Further reading

Hancock, J.L. (1964) 'John Nolen and the American city planning movement: a history of culture change and community response, 1900–1940', unpublished Ph.D. thesis, University of Pennsylvania.
Silver, C. (1996) 'John Nolen: planner for the New South', *Journal of Planning Education and Research* 15: 101–10.

CHRIS SILVER

NON-PROFITS AND URBAN SERVICES

Non-profit organizations are distinguished by their private, voluntary character and charitable, COMMUNITY-benefit purposes. They are distinct from institutions of the state and market and are found throughout the world. Heavily concentrated in URBAN centres, non-profits form a core element of the city's social and institutional INFRASTRUCTURE. Non-profits are diverse enterprises, ranging from small NEIGHBOURHOOD soup kitchens to large professionalized hospitals and museums. Many non-profits provide critical 'street level' responses to the needs of the urban poor, operating emergency food and shelter programmes, and supplying job training, childcare, housing, healthcare, substance-abuse treatment and other social services. Other non-profits advance more generalized urban aims in arts and culture, the environment, and economic revitalization.

Non-profits are valued for their capacity to leverage private resources to meet community needs. They access private philanthropic funds through individual donations as well as corporate or foundation grants. And non-profits mobilize significant numbers of citizen volunteers, generating free labour supporting charitable efforts.

Historically separated, non-profits in the United States and elsewhere have become increasingly interlinked with government. Starting with national funding of non-profit urban service initiatives in the 1960s, the non-profitization of government services expanded steadily throughout the latter part of the twentieth century – fuelled by urban policies promoting PUBLIC–PRIVATE PARTNERSHIPS, and later by the PRIVATIZATION and 'reinventing government' movements of the 1980s and 1990s. These forces

catalysed new non-profit-based public service strategies that have been adopted into widespread practice: government contracting for non-profit delivery of public services, citizen purchase of non-profit services using government vouchers, and non-profit participation alongside government in partnerships and community coalitions. These trends have elevated non-profits to a more central, integrated role in urban administration and service delivery.

Authorities favour non-profits because they offer qualities distinct from government. In contrast to bureaucratized, rule-bound governmental institutions, non-profits can more rapidly and flexibly adapt to changing service conditions. They also offer unique strengths in understanding and serving highly heterogeneous city populations. Their community roots render non-profits capable of tailoring standardized programmes to better match the needs of diverse sub-groups, and more effective in reaching 'hard to reach' service populations, like the poor, and racial or ethnic minorities. Non-profits' charitable character also serves to notably reduce government's service delivery costs.

Ties to government remain controversial, however, because of potential impacts on other non-profit functions. Non-profits have historically assumed central roles in community-building efforts essential to the rejuvenation of urban SOCIAL CAPITAL. As non-profit attention gravitates towards services, questions emerge about whether those will atrophy. Further concerns surround impacts on advocacy functions. Non-profits have been the foremost players in advocating for governmental response to unmet social need. Closer financial ties hold the potential to stifle these efforts, silencing non-profits' important political voice. It remains to be seen whether non-profits can serve both as a force for change in government and as a service extension of it.

Nonetheless, expanded government reliance on non-profits has a global reach and signs of durability. It resonates with deepening citizen disenchantment with government's ability to solve social problems, and fits with prevailing turn-of-the-century political impulses to constrain or reduce the role of government.

Further reading

Boris, E.T. and Steurele, C.E. (1999) *Nonprofits and Government*, Washington, DC: The Urban Institute Press.

Salamon, L.M. (2002) *The State of Nonprofit America*, Washington, DC: The Brookings Institution Press.

Smith, S.R. and Lipsky, M. (1993) *Nonprofits for Hire*, Cambridge, MA: Harvard University Press.

DEBORAH A. AUGER

NON-SEXIST CITY

This term is best known from the title of Dolores HAYDEN's 1980 article 'What would a non-sexist city be like?' In the early 1980s, the term was sprinkled throughout the feminist URBAN STUDIES literature, although it has never been used widely. The critique of URBAN FORM and the vision of the future implicit in the term have, however, been integral to feminist scholarship in urban studies. In proposing a non-sexist city, Hayden and others necessarily identified how urban form is the product of, and continues to contribute to, GENDER DISCRIMINATION. One particularly problematic element of urban form, particularly acute in US cities, is the organization of residential space, which feminist scholars argue was informed by gender ideology which naturalized the roles of male breadwinner and female homemaker. ZONING practices that prohibit commercial services in many residential NEIGHBOURHOODS, gender norms that assign primary responsibility for unpaid family care to women, the private and individual nature of housework, and a material culture that calls for homes to be filled with items to be cleaned and maintained all contribute to the isolation of homemakers and compound the struggles of working mothers. Rather than supporting working women, the built environment presents the additional challenge of travelling between childcare, schools, worksites and residences.

To improve women's lives and empower women at home and in the workplace, there must be spatial as well as social changes. In the words of Hayden, 'a program to achieve economic and environmental justice for women requires, by definition, a solution which overcomes the traditional divisions between the household and the market economy, the private dwelling and the workplace' (1980: S176). In short, overcoming sexism requires a non-sexist city that serves employed women and their families and that provides support for women in a variety of situations. Hayden and others have described examples of housing and planning experiments aimed at

supporting working women. Many have also proposed changes in architecture and planning practices with the potential to improve the city for women. Many of these proposed changes, such as locating collective childcare services and food services in residential complexes, seek to remove the spatial constraints placed on working mothers.

Since the 1980s, feminists have identified additional sources of gender discrimination and proposed other modifications of the urban environment. One issue of particular importance has been public safety. Studies have shown the extent to which fear, especially fear of sexual assault, limits women's use of PUBLIC SPACE. Planners and activists have called for changes in the design, surveillance and lighting of dangerous public and quasi-public spaces. Others have sought to improve women's access to housing and social services, while others have looked to increase women's representation in planning and urban leadership. While the phrase 'non-sexist city' may have fallen out of use, women and men around the world continue to work towards its achievement.

Further reading

Greed, C.H. (1994) *Women and Planning: Creating Gendered Realities*, New York: Routledge.
Hayden, D. (1980) 'What would a non-sexist city be like?', *Signs: Journal of Women in Culture and Society* 5(Suppl. 3): S170–87.

DOREEN J. MATTINGLY

NOT IN MY BACKYARD (NIMBY)

Some facilities needed by society are difficult to site due to local opposition. These may be environmental (hazardous waste plants), or social (group homes). They must be somewhere, but each chosen site meets with local opposition. NIMBY characterizes opposition to LOCALLY UNWANTED LAND USES (LULUs) as selfish and parochial.

NIMBY opposition is sometimes selfish. Given a suitable site, we all ought to be willing to do our part, as we all contribute to problems. Over time, a COMMUNITY might recognize that a given land use is safe enough, and necessary for the common good. In some cases, through an open process and impact mitigation, NIMBY opposition has been overcome.

However, as ENVIRONMENTAL JUSTICE has shown, minority and lower-income communities have borne more than their share of burdens. While higher-income communities create more waste through consumption, they also have more capacity to resist. Whether due to racism or economics, distribution has not been equitable.

Not all NIMBY activism is parochial; some is NOPE (not on planet earth) activism. For example, opposition to nuclear plants and nuclear waste may be locally organized, but may oppose these land uses absolutely due to risk and waste. It may be hard to tell NIMBY from NOPE, as a siting decision may be the only forum for opposition. With more sustainable industry, there would be no need for many environmentally risky land uses.

Further reading

Rabe, B.G. (1994) *Beyond NIMBY*, Washington, DC: The Brookings Institution Press.

TARA LYNNE CLAPP

NUCLEAR FAMILY

The nuclear family is a term used in sociology and anthropology to describe a group of people united by the legal and biological ties of marriage and parenthood. In the United States and in Western literature, the term traditionally describes a group of people consisting of a man and a woman, legally related, and their legal and/or biological children. Some sociologists and anthropologists consider the nuclear family to be the most basic form of social organization.

Some authors have suggested that the traditional nuclear family is the norm for a household in most societies. Yet today we find that the core family roles of husband, wife, brother and sister, or father, mother, son and daughter, can be embodied by a variety of people both related and unrelated. The biological and legal relationships within a household do not necessarily conform to the traditional definition of nuclear family. Today, when we think of the relationships among people who share a common household and share the day-to-day elements of life, we recognize that there are many different arrangements of family. Because of the variation in the legal and biological ties at the closest core of a household, we no longer recognize the nuclear family as universal or basic.

In terms of the development of communities, policy-makers in many Western societies still

consider the traditionally defined nuclear family as identical to the typical household unit. Policy-makers, therefore, often make community planning and land use decisions with this in mind. The data, however, show a different picture. In the United States today, the traditional nuclear family is no longer the norm for households. In 1960, the traditionally defined nuclear family of a married couple with children under the age of 18 made up 44.3 per cent of all households. In 2000, these nuclear families composed only 23.5 per cent of all households. Still, much land use planning and the residential ZONING to implement it has been and remains based on what is today recognized as an inaccurate assumption of the nuclear family as the basic domestic unit.

Further reading

Otero, L.L. (1976) *Beyond the Nuclear Family Model: Cross Cultural Perspectives*, Beverly Hills, CA: Sage.

PATRICIA BARON POLLAK

O

OLMSTED, FREDERICK LAW, OLMSTED, FREDERICK LAW, JR AND OLMSTED, JOHN C.

Frederick Law Olmsted: b. 26 April 1822, Hartford, Connecticut; d. 23 August 1903, Belmont, Massachusetts

Frederick Law Olmsted Jr: b. 24 July 1870, Staten Island, New York; d. 25 December 1957, Malibu, California

John C. Olmsted: b. 1852, Geneva, Switzerland; d. 25 February 1920, Brookline, Massachusetts, United States

Frederick Law Olmsted headed the pre-eminent landscape architecture and planning consultancy of late nineteenth-century America, a practice carried on and expanded by his sons, Frederick Jr and John C., in the first half of the twentieth century under the name Olmsted Brothers. Olmsted Sr began his landscape architecture consulting practice in 1858 with architect Calvert Vaux when their Greensward plan was selected as the design for New York City's Central PARK. Olmsted had pursued several careers, including writer, journalist, farmer and general manager of a California gold mine, prior to joining with Vaux in 1865 to continue work on Central Park, but also to design Prospect Park in Brooklyn, a park system for Buffalo, New York, and the affluent Chicago railway SUBURB of Riverside, Illinois. Olmsted's design style was heavily influenced by the naturalistic tradition he had been exposed to by his father, John Olmsted. This was expressed in Central Park and in Brooklyn's Prospect Park through a 'Long Meadow' of undulating land that created

an appearance of limitless space within the otherwise narrow confines of a city park. Olmsted championed the URBAN park as a means to provide an antidote to the frenetic pace of city life, and incorporated into his designs rural elements such as walkways and horse trails, with separate grades for vehicular traffic. Park systems for Chicago and Boston introduced the concept of an integrated city-wide greenway system. Olmsted Sr planned the campus of Stanford University, the grounds of George W. Vanderbilt's palatial estate, Biltmore, in Asheville, North Carolina, and was the chief landscape consultant for the Chicago World's Columbian Exposition prior to his retirement in 1898.

Olmsted Jr, known to the family as Rick, apprenticed with his father on various job sites, in the 1880s and early 1890s, including thirteen months at Biltmore. He completed a bachelor's degree at Harvard University in 1894 – and was later named Charles Eliot Professor of Landscape Architecture – where he taught the first generation of professional planners, including students such as John NOLEN. He gained national recognition by filling in for his father on the Park Improvement Commission for the District of Columbia beginning in 1901, and by contributing to the famous McMillan Commission Plan for redesigning Washington according to a revised version of the original L'ENFANT plan. Unlike his father, Olmsted Jr's work was more centrally linked to city plans and advising planning bodies. He maintained his involvement in planning Washington DC as a member of the National Capital Park Planning Commission but he also prepared plans for cities such as Pittsburgh, New Haven, Rochester, Detroit and Newport. Like

Olmsted Sr, he planned new residential suburbs, such as Forest Hill Gardens in Nassau County (outside New York City) for the Russell Sage Foundation in 1909, the Palos Verdes Estates in California and an extension of the plan of Roland Park in Baltimore. Olmsted Jr was the first president of the American City Planning Institute, and a consultant to the US government through the Town Planning Division of the US Housing Corporation during the First World War, the Department of Commerce Advisory Committee on City Planning during the 1920s, and the National Park Service from the 1920s through to the mid-1950s.

John C. Olmsted was Frederick's nephew who he adopted as a son after marrying his brother's widow, Mary Cleveland Perkins Olmsted. Following graduation from the Sheffield School of Yale University in 1875, he joined his father's firm, securing an interest in 1878 and becoming a full partner in 1885. John was deeply involved in the planning of Boston's park system and was the most widely experienced of Frederick Sr's staff. Shy and quiet by nature, he was overshadowed by his younger brother, Frederick Jr, who was chosen to head Harvard University's new curriculum in landscape architecture. John's extensive park planning work also included designing systems in Buffalo and Rochester, New York, Hartford, Connecticut, Milwaukee, Chicago's South Park, Spokane and Seattle.

Further reading

Peterson, J.A. (1996) 'Frederick Law Olmsted, Sr. and Frederick Law Olmsted, Jr.: the visionary and the professional', in M.C. Sies and C. Silver (eds) *Planning the Twentieth-Century American City*, Baltimore: Johns Hopkins University Press, pp. 37–54.

Schuyler, D. (1984) *The New Urban Landscape: The Redefinition of City Form in Nineteenth-Century America*, Baltimore: Johns Hopkins University Press.

Zaitzevsky, C. (1982) *Frederick Law Olmsted and the Boston Park System*, Cambridge, MA: Harvard University Press.

CHRIS SILVER

OPEN SPACE

In URBAN PLANNING, this term refers to unbuilt areas not meant for building purposes which exist between the built-up areas, with the exception of transport INFRASTRUCTURE and public squares. The main components of open space are green areas (encompassing agricultural land, forests, meadows, pastures, PARKS, isolated green areas and URBAN greenery), as well as waters and other areas characterized by little or no transformation and which retain the properties of the landscape resulting from geographical location. The fundamental feature of the spatial setting of open space is the preservation of its continuity. In the concept of sustainable development, open spaces constitute an element of the development of space that is equivalent to the overbuilt areas and the ones meant for construction development. They require appropriate shaping and protection in order to preserve or improve the conditions of human life, which is of special importance within the intensively urbanized areas. The maintenance or preservation of the eco-efficiency of the open space system is influenced, in particular, by such characteristics as its adequate magnitude, preservation of spatial continuity, the natural resistance of the ecosystem, and proper use.

Open space systems are established for the purposes of the protection of their ecological functions and the specific character as well as beauty of the landscape, and the creation of leisure areas for the population. These objectives are being attained by safeguarding open spaces against uses that are incompatible with the assumed functions. Priority areas of open space protection include ecosystem and climate protection, air regeneration, the protection of landscape, waters, soils, etc.

The correctly shaped regional setting constitutes the spatially continuous system, which remains in contact with the open space in towns or permeates the inside of towns. The local open space system is characterized by contact with the open space located outside of the urbanized zone and its internal cohesion.

The creation of an open space system is particularly important in large urban agglomerations and in their peripheries, in view of the need to balance the existing and expected costs to the environment within the overbuilt space, which is associated with the preservation or improvement of the living standards of the present and future generations, especially in the strongly urbanized areas. The biggest threat to the preservation of open space is constituted by urban SPRAWL, and

the pressure from the construction sector related to the shortage of free construction space in many towns and the attractiveness of such areas in terms of investment. Spatial policy with respect to open spaces consists mainly in protecting them against overbuilding, with simultaneous maintenance or improvement of the natural and recreational functions.

BOŻENA DEGÓRSKA

P

PARK, ROBERT E.

b. 14 February 1864, Harveyville,
Pennsylvania; d. 7 February 1944,
Nashville, Tennessee

Key works

(1925) *The City*, Chicago: University of Chicago Press
(with E.W. Burgess).
(1928) 'The bases of race prejudice', *Annals of the
American Academy of Political and Social Science*
CXL.

As leader of the CHICAGO SCHOOL OF SOCIOLOGY,
Robert Park made significant contributions to the
study of URBAN communities, race relations and
the development of empirically grounded research
methods, most notably participant observation.
A former journalist, Park studied under SIMMEL in
Germany and William James in the United States,
retaining a fascination with the social and cultural
fabric of human society. In an autobiographical
note, he wrote that his interest lies in 'what is going
on rather than what, on the surface of things,
merely seems to be going on'.

For Park, the CITY is a product of human nature that may best be studied in terms of its physical
organization, its occupations and its culture. It has
a moral organization rooted in the customs and
habits of the people who live in it. It is these
human, historical and social attributes that turn a
geographical expression into a NEIGHBOURHOOD.
He viewed social, including racial, prejudice as a
key element in human relationships where individuals of the same race or class live separate and
segregated existences. A sense of neighbourliness,
and identification with a place often tend to fuse
with racial antagonism and/or class interests.

Applying the perspective of HUMAN ECOLOGY,
Park viewed the social world as composed of
competing groups and individuals in a constant
state of unstable equilibrium and readjustment.
MIGRATION and population movements disturb
the social order. The deterioration of local
attachments and the weakening capacity of primary groups like the family and COMMUNITY to
exert informal mechanisms of social control in the
urban environment were direct causes of CRIME,
juvenile delinquency and vice.

Re-education and reform required finding the
delinquent a social and physical environment that
would enable both free expression and the development of a healthy sociality. Repeatedly, Park
insisted that social relationships are formed by our
physical, psychological and biological make-up.
Anticipating Whyte's seminal ethnographic study,
Street Corner Society, he noted that it was
impossible to understand the gang, whether as a
menace or as an amusement, outside of its peculiar
habitat and social milieu.

First-generation immigrants may retain the
strict norms, values and mores of the old country
but they may break down in the second generation.
We are, he wrote, experiencing increasing indivi-
dualization and growing social atomization, with
fashion and public opinion replacing custom as the
main means of social control.

Concepts of competition, dominance and succession infused Park's sociological imagination.
The cultural community develops in a similar way
to the biotic, except 'the process is more complicated'. All big cities may exhibit similar patterns
(frames of reference) that can be described
abstractly – central business districts, LABOUR
MARKETS, GHETTOS, SUBURBS, satellite developments, light and heavy industry – but every new

specific enquiry should reaffirm or redefine, qualify or extend, the original hypothesis.

Social disorganization, a concept also discussed by Louis WIRTH, was sometimes presented as an inevitable product of twentieth-century life. Park used the automobile and the 'automobile bandit', the mass circulation newspaper, and the cinema as illustrative of the new demoralizing influences of contemporary civilization.

Further reading

Mathews, F.H. (1977) *Quest for an American Sociology: Robert E. Park and the Chicago School*, Toronto: University of Toronto Press.

Park, R.E. (1950) *Race and Culture*, New York: Free Press.

—— (1952) *Human Communities: The City and Human Ecology*, Glencoe, IL: Free Press.

Raushenbush, W. (1979) *Robert E. Park: Biography of a Sociologist*, Durham, NC: Duke University Press.

JOHN BLEWITT

PARKS

Preserving and assembling land where the public can gather, engage in physical activities, find solitude, and enjoy nature and the outdoors has been a major pursuit in the contemporary history of URBAN development.

There are many different types of parks, such as natural parks, recreational parks, sports parks, formal gardens (Boston Public Garden), cultural parks (Lowell National Historic Park, Massachusetts), playgrounds, pleasure parks (Tivoli Gardens, Copenhagen), conservation/ecological parks, etc. Parks can also be described by scale, such as vest-pocket parks, NEIGHBOURHOOD parks, regional parks and national parks. Many urban areas have extensive metropolitan parks systems.

Early opportunities for the creation of urban parks and greens in both Europe and the United States grew out of a practice dating from medieval Europe onwards, where common pasturelands were secured within the safe confines of villages and towns. The most famous US city park to evolve from this practice is the Boston Commons in Boston, Massachusetts (1634).

One of the earliest milestones in city park development was the creation in 1605 of Place Royale (renamed Place des Vosges) in Paris. Initially private, it was viewed across Europe as the prototype for the residential square and introduced the beneficial relationship between parks and urban residential real estate.

In the United States, early cities incorporated parks directly into their plans, most notably Philadelphia, Pennsylvania (1683 by Penn and Holm), and Savannah, Georgia (1733 by Oglethorpe), where park blocks were placed at regular intervals within the street grid.

In the mid-eighteenth century, the English landscape school emerged, creating a style that countered the formal gardens of the period. Its basic approach was to showcase the natural landscape with subtle man-made features. Influential designers of this period were William Kent (Rousham and Stowe) and Lancelot 'Capability' Brown (Blenheim). Building on this legacy was the author and landscape designer Humphrey Repton, who coined the term 'landscape gardener'.

At the beginning of the nineteenth century, royal gardens and hunting grounds in England and France provided opportunities for the creation of city parks. Two notable examples are Regent's Park (1812) and St James's Park (1823–29) in London, both designed by John NASH (1752–1835).

Galen Cranz has identified four eras in the contemporary history of US park development:

1   *Pleasure ground (1850–1900)*. By the mid-nineteenth century, parks were seen as antidotes for the many public ills associated with the INDUSTRIAL REVOLUTION. The most influential park project of this period was the 1858 Greensward plan for Central Park in New York City, by Frederick Law OLMSTED and Calvert Vaux. The plan built on the English landscape school's influence of presenting nature's beauty through the use of man-made features, such as constructing curvilinear paths and streets (as opposed to the gridiron pattern of the city) that would explore and frame peaceful, natural settings. Through many subsequent projects, Olmsted emerged as the father of landscape architecture, informing urban development practices of the time, as well as guiding the emerging CITY BEAUTIFUL MOVEMENT. During this period, George Edward Kessler, who created the influential parks plan for Kansas City in 1893, also did important work.

2 *Reform park (1900–30).* Around the turn of the twentieth century, the reform park style emerged in response to the relative lack of accessibility of pleasure ground parks to the rising number of new immigrants in the inner city. Having to work within existing city plans, these parks were smaller in scale. Playgrounds championed by Jane ADDAMS and others began to be included in park designs during this period (the Playground Association of America was established in 1906).

3 *The recreation facility (1930–65).* Strongly influenced by New York City Park Commissioner Robert MOSES, parks of this era were shaped by principles of functionality and programming, as well as by emerging standards, and were typified by facilities (recreation centres, play fields, etc.).

4 *Open space system (1965–present).* With the rise of the environmental movement, the preservation of OPEN SPACE within and around urban areas emerged, with an emphasis on keeping land in its natural state.

Further reading

Cranz, G. (1982) *The Politics of Park Design: A History of Urban Parks in America*, Cambridge, MA: MIT Press.

Schaffer, D. (1988) *Two Centuries of American Planning*, Baltimore: Johns Hopkins University Press.

BRUCE APPLEYARD

PEDESTRIAN MALL

The mall, now considered a uniquely North American contribution to the world of retailing, has its origins in market places in ancient cities in Europe and Asia. These market places, called the AGORAS in Greece, were also the first FESTIVAL MARKETPLACES. The INDUSTRIAL REVOLUTION resulted in very crowded and polluted cities in nineteenth-century Europe and North America. The GARDEN CITIES advocated by Frederick Law OLMSTED and Ebenezer HOWARD in the late 1800s appealed to the middle-class families that longed for the OPEN SPACES of the countryside. The STREETCAR SUBURBS that were built outside the city fulfilled this longing for green spaces. As these SUBURBS needed access to shopping facilities, malls became necessary. J.C. Nichols built one of the first such shopping centres in 1922. This mall – the Country Club Plaza near Kansas City, Missouri – was planned, owned and operated by a single entity. In 1935, Greenbelt, Maryland, was built as a planned COMMUNITY that included stores. Abraham Levitt's Levittown on Long Island built in 1946 was one of the first car-oriented suburbs on the edge of a large city. In 1954, Victor Gruen designed Northland, near Detroit, with over a hundred stores in, at that time, the world's largest shopping mall. At the same time, the number of cars on the road continued to increase and traffic engineers encouraged the construction of highway bypasses to allow traffic to flow smoothly. Routing highway traffic around the outskirts of town encouraged further growth in strip shopping centres and regional malls.

Pedestrian malls, as traffic-free zones, were first built in Essen, Germany, in 1926. After the Second World War, such malls continued to develop as traffic congestion grew in the relatively dense city centres. By 1966, there were over sixty such malls in Germany. To counter the growing isolation of cities due to SUBURBANIZATION in the United States during the late 1950s and 1960s, some cities banned cars from DOWNTOWN streets to create pedestrian malls. One of these attempts was in the city of Kalamazoo in 1958, also planned by the architectural firm of Victor Gruen. However, shoppers complained of the lack of convenient parking and of inner-city CRIME. In the late 1990s, the introduction of an access street through two blocks of the mall became the most controversial component of the revitalization plan for Kalamazoo. In 1999, after a long period of debate, two blocks of the mall were modified to accommodate car traffic. Thus, the first pedestrian mall in the United States is no longer a traffic-free zone. After the pedestrian mall of Kalamazoo was built, over two hundred pedestrian malls were constructed in the United States in the 1970s and 1980s. However, few of them are truly pedestrian now. Many downtown malls have since been rebuilt to allow automotive traffic. The recent interest in NEW URBANISM in the United States has spurred interest in the idea of mixed-use marketplaces. However, purely pedestrian malls are very rare.

Further reading

Rubenstein, H.M. (1992) *Pedestrian Malls, Streetscapes, and Urban Spaces*, New York: John Wiley & Sons.

Smiley, D.J. (ed.) (2002) *Sprawl and Public Space: Redressing the Mall*, Washington, DC: National Endowment for the Arts.

SUMEETA SRINIVASAN

PEDESTRIAN POCKET

A pedestrian pocket combines housing, offices and retail space within a quarter-mile walking radius of a transit system node. It provides a model for suburban development. The term 'pedestrian pocket' was coined and developed by architect-planner Peter CALTHORPE, in the mid-1980s, and introduced in a co-authored book, *The Pedestrian Pocket Book*, after a University of Washington design charrette in 1988. Calthorpe updated the conception to an average radius of 2,000 feet or a ten-minute walk, as TRANSIT-ORIENTED DEVELOPMENT in 1993. He and others have since applied its concepts to many development projects around the world, both as a strategy for controlling suburban growth and for guiding transit-oriented development infill in established URBAN areas.

The prescription is derived from the idea of a walkable NEIGHBOURHOOD node, and emphasizes a mix of uses balancing and clustering jobs, housing, shopping, recreation and childcare in proximity to a transit service. A system of safe walking paths connects the entire site. Pedestrian pockets are not meant as stand-alone developments but rather provide a long-range growth strategy of networked settlement patterns in a region that protects OPEN SPACE and agricultural lands.

The term is not a new concept. This is a settlement pattern that exists at many traditional transit station centres (rail, streetcar and bus stops) in the Netherlands, Germany, the United Kingdom and Sweden where joint land use/transport policies are increasingly applied. A related term, the 'pedestrian shed' ('ped shed'), is used in Australia for walkable streets and nearby destinations that support transit by providing easy access for riders, not cars. In many ways, the concept is consistent with the COMMUNITY planning principles of the GARDEN CITY and new towns.

Many others agree that development should be based on fairly small neighbourhood units, each combining homes with job opportunities and services, clustered along public transportation spines. The advocates of NEW URBANISM, particularly Calthorpe, have tried to extrapolate the concept into a model for sustainable metropolitan development as articulated in the Congress for New Urbanism charter. Such development nodes are now occurring along public transit routes and infill at higher densities within the existing urban areas. West Coast examples can be seen in the San Francisco Bay Area, Portland, and in Vancouver, Canada.

Further reading

Calthorpe, P. (1993) *The Next American Metropolis: Ecology, Community, and the American Dream*, New York: Princeton Architectural Press.
Kelbaugh, D. (ed.) (1989) *The Pedestrian Pocket Book*, New York: Princeton Architectural Press.

DAVID VAN VLIET

PEIRCE, NEAL R.

b. 5 January 1932, Philadelphia, Pennsylvania

Key works

(1983) *The Book of America: Inside Fifty States Today*, New York: W.W. Norton & Co. (with J. Hagstrom).
(1993) *Breakthroughs: Re-Creating the American City*, New Brunswick, NJ: Rutgers University Press (with R. Guskind).
(1994) *Citistates: How Urban America Can Prosper in a Competitive World*, Santa Ana, CA: Seven-Locks Press (with C. Johnson and J.S. Hall).
(1997) *Boundary Crossers: Community Leadership for a Global Age*, College Park, MD: Academy of Leadership Press (with C.W. Johnson).

Neal R. Peirce is widely regarded as America's premier journalist covering cities and URBAN regions. In stark contrast to most journalists, his career began in those much-coveted roles covering national issues, often for television. But Peirce concluded that America's most interesting scene was the state and local level. His *Washington Post* syndicated column, started in 1975, has appeared in 150 newspapers since. It remains the only national column dealing with wide-ranging local and regional issues, such as transportation, growth, environment, schools and the economy.

A prolific author, Peirce has written more than a dozen books covering the history of politics and civic life in every region of the United States, culminating with the internationally acclaimed

The Book of America: Inside Fifty States Today, which appeared in 1984.

In 1986, Peirce introduced what has become a new form of American journalism – an intensive examination, by outside journalists, of a region's strategic issues, reported in the form of a series of feature articles in the leading newspaper. Dubbed early on as 'Peirce Reports', such series have as of 2001 appeared in twenty-one American regions.

Though by the turn of the century the resurgence of metropolitan regions was well known, Peirce was among the earliest voices in explaining how an intensively international economy was reorganizing geography into what he called 'CITISTATES', explaining the argument in a 1993 book by that name (with Curtis Johnson and John Stuart Hall).

He is the founding chairman of a national network of journalists, analysts and consultants – called the Citistates Group – who focus on regions as the platform for competitiveness, livability, sustainability and equity. The ongoing work of the group can be found at www.citistates.com.

Cities, and the regions they compose, have rarely had more serious and steady study than the books, columns and special reports that Peirce has produced over the past generation, giving him a rare stature among American journalists. His work is often cited in scholarly journals and books and, because of the perspective he has accumulated, he remains in demand as a speaker in cities all over the world.

In addition, he has made sustained public service contributions at the highest international and national levels. Among his numerous international contributions, he has served as a faculty member of the Salzburg Seminar (1980, 1984, 1997), as moderator at the Great Cities of the World Conference, Boston (1980), and as a trustee of the German Marshall Fund of the United States (1988–97). He was a fellow of the Woodrow Wilson International Center for Scholars from 1971–74.

In the United States, he has served as a fellow of the National Academy of Public Administration; as a member of the Advisory Council, Trust for Public Land; and as a member of the boards of the Institute for Educational Leadership, Partners for Livable Communities, and the National Civic League.

JOHN S. HALL
CURTIS W. JOHNSON

PERFORMANCE MEASUREMENT

'Performance measurement' is the use of qualitative and quantitative indicators to demonstrate the performance and outcomes of organizations, programmes or activities. 'Performance measurement system' refers to an integrated system that links the development of performance measures across an organization in a way that includes a formal process for data collection and reporting. Performance measurement activities may stand alone, but they may also be (and frequently are) linked with other management processes, particularly strategic planning, quality management and budgeting.

The following definitions are commonly used to describe the different categories of measures that exist:

- *Input*. These measures reflect the resources that feed into the organization or activity.
- *Activity*. These measures reflect the level of activity or workload involved in regular ongoing operation.
- *Output*. These measures reflect the immediate product of the service or process reflected in the activity measures above.
- *Outcome*. These measures reflect the impacts or results of the outputs of the programmatic activity. They occur on an ongoing basis and may be identified as 'initial', 'intermediate' or 'long term' measures.
- *Efficiency*. These measures reflect the cost-efficiency of the outputs or outcomes of the programmatic activity, typically involving a ratio of inputs to outputs or outcomes.
- *Quality*. These measures reflect the qualitative aspects of programme activities, outputs and outcomes and are often measured using customer feedback and satisfaction.

In a comprehensive performance measurement system, measurements will be developed in each of these categories to provide an overall view of organizational activities, outputs and outcomes.

In the United States, requirements for the development and use of performance measures in government at all levels grew dramatically during the 1990s. At the federal level, in 1993, the Government Performance and Results Act (GPRA) was passed, requiring the establishment and use of performance measures in federal agencies. One of the stated purposes of the Act is to 'improve

congressional decision-making by providing more objective information on achieving statutory objectives, and on the relative effectiveness and efficiency of Federal programs and spending'. More specifically, as noted in Section 1115 of the Act, performance measures in the budget process should:

1 establish performance goals to define the level of performance to be achieved by a programme activity;
2 express such goals in an objective, quantifiable and measurable form unless authorized to be in an alternative form under subsection (b);
3 briefly describe the operational processes, skills and technology, and the human, capital, information or other resources required to meet the performance goals;
4 establish performance indicators to be used in measuring or assessing the relevant outputs, service levels and outcomes of each programme activity;
5 provide a basis for comparing actual programme results with the established performance goals; and
6 describe the means to be used to verify and validate measured values.

The tone set by the GPRA is reflected in many similar state and local Performance-Based Budgeting requirements. At the state level, Melkers and Willoughby (1998) found that all but three states (Arkansas, Massachusetts and New York) have legislative or administrative requirements for the development of performance measures and their use in the budget process, with most of these requirements established in the 1990s. A similar trend towards the use of performance measurement information is taking place in local governments, with many municipal and county governments requiring performance measurement with the intent to use it in the budgetary process. Internationally, Canada has had similar requirements to the GPRA at the national level since 1981, and other countries, such as Australia, have also made important strides in the development and use of performance measures.

Performance measurement systems are usually developed in conjunction with other management processes, such as strategic planning, quality management and especially budgeting processes.

First, performance measures are an important part of the development of a strategic plan. The strategic planning process relies on feedback loops as well as monitoring progress towards goals. Performance measures help to track this progress. Secondly, an important trend in both public and private organizations has been the emphasis on quality in terms of programme management and service delivery. To accomplish the goals of quality management means taking the client's perspective, requests and needs into account in programme planning and management. Developing client-oriented performance measures can be an important feedback to this process, addressing programmatic aspects of accuracy, timeliness, reliability and quality in the eyes of the beholder (client).

Thirdly, most performance measurement requirements point to the importance of integrating performance measures in the budget process. Using performance data in the budgetary process means integrating information about outcomes and impacts in decisions about the allocation of funds where the goal is to be able to allocate financial resources using performance information to make a more informed decision. Taking output, outcome and efficiency measures together gives policy-makers a full view of activities completed, the cost and value of the outputs and outcomes, and what has actually been accomplished with the actual expenditures.

Overall, performance measurement is an important activity that helps administrators to assess the outcomes of organizational and programmatic activities and track them over time. Specifically, performance data are useful for helping decision-makers do the following:

• understand the activities and objectives of funded programmes by viewing summary measures of performance;
• understand changes in performance over time;
• have more meaningful dialogues with public managers about agency activities, goals and performance;
• identify poorly performing and high-performing programmes and departments; and
• justify fiscal and other decisions using evidence rather than anecdotes or impressions.

Further reading

Hatry, H. (1999) *Performance Measurement: Getting Results*, Washington, DC: The Urban Institute Press.

Joyce, P.G. (1997) 'Using performance measures for budgeting: a new beat, or is it the same old tune?', *New Directions for Evaluation* 75: 45–62.

Melkers, J. and Willoughby, K. (1998) 'The state of the states: Performance-Based Budgeting in 47 out of 50', *Public Administration Review* 58(1): 66–73.

Tigue, P. and Strachota, D. (1994) *The Use of Performance Measures in City and County Budgets*, Chicago: Government Finance Officers Association.

JULIA ELLEN MELKERS

PERRY, CLARENCE ARTHUR

b. 1872, Truxton, New York;
d. 6 September 1944, New Rochelle, New York

Key works

(1929) 'The neighborhood unit', in *Regional Plan of New York and Its Environs, Regional Survey*, vol. VII, *Neighborhood and Community Planning*, New York: Regional Survey of New York and Its Environs.

(1939) *Housing for the Machine Age*, New York: Russell Sage Foundation.

Perry devised the NEIGHBOURHOOD unit plan, a residential COMMUNITY scheme disseminated through the Regional Plan of New York and Its Environs in 1929 that influenced planning in US cities and that was a notable feature of Sir Patrick ABERCROMBIE's famous Greater London Plan of 1944 and the first generation of British new towns after the Second World War. The neighbourhood unit plan was an effort to adapt city neighbourhoods to the automobile age by creating a residential design that required major vehicular traffic to use arterial roads and restricting internal traffic to local uses. Extensive use of cul-de-sacs supported this objective, as did the use of an internal system of pedestrian ways to access community institutions. Perry intended each neighbourhood to be only so large as to support a local elementary school, which would also serve as a community centre. Commercial activities would be restricted to edges of the neighbourhood through land use regulations. Moreover, the neighbourhood unit plan was intended to support a socially homogeneous residential environment through uniform housing construction.

Perry studied at Stanford University, received a BS degree from Cornell University in 1899, and later did graduate work at Teachers College, Columbia University. He joined the recreation department of the Russell Sage Foundation in 1909 after having been a high-school principal in Puerto Rico and working as a special agent for the US Immigration Commission. His work with the Foundation focused on advocacy for school playgrounds, which was an offshoot of the settlement house movement. He retired as associate director of recreation in 1937.

The GARDEN CITY movement influenced Perry in his approach to neighbourhood design, and in 1912 he moved his family into Forest Hill Gardens, a garden city-inspired community outside New York City. The first public presentation of the neighbourhood unit plan was at a joint meeting of the National Community Center Association and the American Sociological Association in 1923. Perry was also involved with the Regional Planning Association of America (RPAA) in the 1920s and his neighbourhood unit plan provided the basic footprint for Radburn, New Jersey, the partially built garden SUBURB of architect and RPAA stalwart Clarence STEIN begun in 1929. The neighbourhood unit plan was a model also used in the design of public housing in the 1930s and continued to be advocated in the post-Second World War era as the appropriate design for new URBAN neighbourhoods, although its application to suburban design was inconsistent with Perry's original concept of 'purposeful neighbourhood planning'.

Further reading

Dahir, J. (1947) *The Neighborhood Unit Plan*, New York: Russell Sage Foundation.

Gillette, H., Jr (1983) 'The evolution of neighborhood planning: from the progressive era to the 1949 Housing Act', *Journal of Urban History* 9(4): 421–44.

CHRIS SILVER

PIRENNE, HENRI

b. 22 December 1862, Verviers, Eastern Belgium; d. 24 October 1935, Ukkel, Belgium

Key work

(1925) *Medieval Cities: Their Origins and the Revival of Trade*, Princeton, NJ: Princeton University Press.

Pirenne was a Belgian historian, and a historian of Belgian history. But his lasting contribution to the

study of cities is a sweeping and controversial interpretation of the end of Roman civilization and the rebirth of MEDIEVAL CITIES. While the fall of the Roman Empire had been attributed to the invasion of Germanic tribes, Pirenne argued in his book *Mohammed and Charlemagne* that the invaders did not want to destroy the empire, but to share in its advances.

In *Medieval Cities*, Pirenne stressed that the Roman Empire's power and wealth had been based on the ease of movement that the Mediterranean sea, the *Mare Nostrum* (Our Sea) of the Romans, provided for political administration, military control and commerce. With the Arab invasion of the southern and western basin of the Mediterranean in the seventh and eighth centuries, and Viking raids later on, trade came to a halt, and with it URBAN civilization. The feudal civilization that emerged in its wake was a closed one, based on agricultural production for local consumption. Cities shrank sharply in size and population, and the middle class, which Pirenne saw as the mainstay of an urban economy, disappeared.

The beginning of a new millennium was characterized by a recrudescence of activity, vigour and spirit. Christendom began its counter-offensive with the crusades and the blossoming of the maritime republics of the Italian peninsula, making possible the renewal of trade and, with it, the growth of the middle class and the rebirth of urban civilization. Merchants settled under the walls of the fortified castles or in bishops' seats located along trade routes, incubating early commercial settlements that gradually blossomed into medieval cities.

NICO CALAVITA

PLACE

Place is a well-accepted theoretical construct of the area of spatial studies. The various disciplines involved in the area are used to approach the concept according to each individual rationale. In this way, one may say that there is a psychological interpretation of the concept, an architectural interpretation, and so on.

In the city studies area, the concept is frequently used to highlight environment – behaviour relationships, since the two major factors the concept deals with – people and space – are the central core of the studies. Notwithstanding that, the construct of place, though intensively used in the social sciences as a theoretical category, still demands further attention as far as environment–behaviour studies are concerned. Approximately at the end of the 1950s, and from the 1960s onwards, an encouraging approach was set out when joint efforts essayed by both architects and psychologists contributed to ascertain the scientific interest raised by the subject. This was the time when the initial intersection between research about the formation of places and exercises of design that intended to gather together the factors that make up a place encountered a receptive handling in the linked ventures developed by architects and psychologists. Important contributions were brought to the area by authors such as APPLEYARD, RAPOPORT and, especially, LYNCH, who advanced the idea of constructing what he called 'place utopias', where people and place would bond naturally, enjoying a pleasurable SENSE OF PLACE.

Places are regarded from several perspectives. They may be seen from the viewpoint of a behaviour setting, and, in this case, people's ENVIRONMENTAL BEHAVIOUR in them is what really matters in the analyses. Social representations of places are another way of approaching the subject, and there, the symbolic representation of society's territorial identity is what counts best. The appraisals of the physical-spatial images of the URBAN FORM people get affectionate with, occupy the efforts attempted by researchers of the ENVIRONMENTAL DESIGN area, and are usually captured by means of techniques such as COGNITIVE MAPPING. Trying to understand why people do so, however, may be best approached through ENVIRONMENTAL PERCEPTION research.

As a general rule, one can say that place, in urban-architectural theory, is a created environmental form, imbued with symbolic significance for its users. Environmental psychology moves in a similar direction, understanding place as the units where human experiences and physical form are fused together, creating a unitary context. Of course, there is no single factor that can explain all phenomenology involved in man–environment interactions. But if a place is imbued with symbolic attributes that evoke the role this place played in some of the most significant times the city has experienced, then the presence of memory in the structure of the place must be acknowledged, as an

additional attribute, since it is memory that can bring about the good (or bad) images a place evokes (see BOYER).

In the last decades of the twentieth century, though, subtle alterations in the original concept of place started to gain relevance. They probably came off as a consequence of the introduction of POST-MODERNISM into the practices of URBANISM. Their manifestation occurred chiefly in NEW URBANISM and META-URBANISM schemes. A growing disapproval of MODERNISM was popular at that time, soon to be joined by numerous criticisms of some unsuccessful attempts at designing places. This, and the relative success attained by the widespread proliferation of DISNEY ENVIRONMENTS – considered to be places designed as a whole – partially explains the changes introduced into the basic postulations of the concept. In short, a somehow particular interpretation of the construct's major delineations – which might be embodied under the single expression of 'place-making' – may have been responsible for the variances noticed in the construction of places; or, more correctly, of THEME PLACES. Born as a reaction to the rising criticisms that claimed that places were heading towards 'placelessness' or 'non-place places', the new interpretation of the concept favoured a type of design that came up so heavily compromised by the materialization of fantasy themes that, in the end, a disconcerting generalization of forms became the prevalent rule for places. This led some influential turn-of-the-century architects, like KOOLHAAS, to coin expressions such as 'generic cities', as an utter complaint about the sameness encountered in the GLOBAL CITIES found all over the world. In fact, this was only a typical spatial facet of the globalized market-economy society of the time – and the consumption patterns it entailed – responsible for producing URBAN landscapes marked by undifferentiated homogenizations, with consequent losses of the identity traces each urban environment should be able to individually demonstrate. Furthermore, if, on the one hand, some of these theme places were newly built constructions (usually comprising artificial replicas of old traditional places), most of them, on the other hand, were located in old historic places, where they could capitalize upon the accumulated perception people had already acquired about the place. In this way, the urban-architectural design of theme places also became responsible for inserting images of fantasy into the reality of real places, as espoused by meta-urbanism assumptions.

Another preoccupying interpretation that has been added to the concept's original theoretical framework relates to the need to incorporate well-planned marketing of the place into the process of place-making, as a strategic managerial procedure. Therefore, the making of places in the early twenty-first century has become a process integrating all dimensions involved in the conceptions of place, socio-psychological, marketing-managerial and urban-architectural.

Further reading

Banerjee, T. and Southworth, M. (eds) (1991) *City Sense and City Design: Writings and Projects of Kevin Lynch*, Cambridge, MA: MIT Press.

Bonnes, M. and Secchiaroli, G. (1995) *Environmental Psychology: A Psycho-Social Introduction*, London: Sage.

Canter, D. (1977) *The Psychology of Place*, London: Architectural Press.

Tuan, Y.F. (1977) *Space and Place: The Perspective of Experience*, Minneapolis: University of Minnesota Press.

LINEU CASTELLO

PLANNING COMMISSION

In the early twentieth century, US reformers and leaders of the planning movement viewed local governments as inefficient, ineffective and corrupt. In response, planning commissions were established in an effort to protect the planning function by placing it outside the influence of dishonest public officials, and to ensure that planning decisions were made based on objectivity and science as opposed to political power and influence. The first planning commissions in the United States were established in Hartford in 1907, in Milwaukee in 1908, and in Chicago, Detroit and Baltimore in 1909. In 1913, Massachusetts led the way for many states by making it mandatory for cities of more than 10,000 people to create official planning boards.

In 1926, the US Department of Commerce, under Secretary Herbert Hoover, published the Standard State Zoning Enabling Act, encouraging states to adopt laws that would allow local governments to form planning commissions responsible for overseeing the creation and enforcement of the ZONING codes.

In 1928, the US Department of Commerce followed with the publication of the Standard City Planning Enabling Act, which both promoted city planning and, at the same time, created several problems for it, including: (1) it confused the comprehensive plan with the zoning code, causing some communities to make zoning decisions without reference to long-range plans; (2) it recommended piecemeal adoption of the comprehensive plan, therefore undermining the pursuit of comprehensive planning; and finally (3) some believe that it may have weakened the emerging role of planning in local government through its recommendation that an appointed planning commission oversee planning operations rather than a directly elected legislative body.

During the 1930s and 1940s, planning commissions evolved as the primary clients of zoning codes and master plans, shifting influence away from the business and development interests. Additionally, they gained responsibility for the budgeting and planning of roads. Furthermore, by the 1950s they became integral factors in the many housing surveys, studies and plans that were undertaken in part due to the increased federal funds generated from Section 701 of the Housing Act of 1954.

In general, planning commissions are composed of appointed citizen planners, most of whom do not have formal training in planning. The responsibilities of a planning commission include:

- the creation of a locality's comprehensive plan and zoning code;
- the review of development proposals and zoning entitlements required by law;
- in some localities, the review of capital improvements programmes; and
- the hosting of a public hearing on any of the above items, when warranted.

In recent decades, the need for planning commissions has been questioned, as corruption has been largely eliminated from local government, and the extra layer of bureaucracy between the planning department staff and city council has been seen as unnecessary. Nevertheless, as a more apolitical body that is independent of the formal local government structure, a planning commission is likely better able to:

- oversee the development of the long-term vision for a COMMUNITY, and in turn be able to

make the tough, often politically controversial decisions necessary to realize it; and
- act as an arbiter for planning staff, applicants, citizens, and with other local governments.

Further reading

So, F. (ed.) (1988) *The Practice of Local Government Planning*, Washington, DC: International City/County Management Association.

Solnit, A. (1987) *The Job of the Planning Commissioner*, Chicago: Planners Press, American Planning Association.

SEE ALSO: general plan; zoning

BRUCE APPLEYARD

PLATER-ZYBERK, ELIZABETH

b. 20 December 1951, Bryn Mawr, Pennsylvania

Key works

(2001) *Suburban Nation: The Rise of Sprawl and the Decline of the American Dream*, New York: North Point Press (with A. Duany and J. Speck).

(2004) *Smart Growth Manual*, New York: McGraw-Hill Professional (with A. Duany and J. Speck).

Architect, town planner and teacher, married to well-known architect Andrés Duany, Plater-Zyberk is dean of the University of Miami's School of Architecture and co-founder and principal, since 1980, of the firm of Duany Plater-Zyberk & Company (DPZ). Educated in architecture at Princeton and Yale, Plater-Zyberk founded a master of architecture programme in SUBURB and town design at the University of Miami. She has been a visiting professor at many of the leading schools of architecture, a resident at the American Academy of Rome, and serves as a trustee of Princeton University.

Since the early 1980s, DPZ has created an impressive record of designing traditional new towns and retrofitting livable DOWNTOWNS in existing suburbs. DPZ's seminal success is demonstrated in their highly acclaimed COMMUNITY of Seaside, Florida, completed in the early 1980s. Thereafter, Plater-Zyberk began the codification of these principles through the 1991 Neighborhood Development Ordinance for Miami-Dade County,

Florida. Their pioneering work led to the creation of the URBAN DESIGN movement known as NEW URBANISM, which is a direct reflection of their concepts and principles. Both Plater-Zyberk and Duany are founding members of the Congress for New Urbanism.

DPZ's focus is on designing 'places'. They have completed more than 250 NEIGHBOURHOOD-based URBAN designs. DPZ also maintains an architectural practice, writes land use codes focused on SMART GROWTH and urban architectural regulations, and promotes the principles of new urbanism through an educational programme, including the highly acclaimed Seaside Institute.

Plater-Zyberk is the recipient of numerous awards including the Founders Award for Civil Leadership, the National Building Museum's Vincent J. Scully Prize, the Thomas Jefferson Memorial Medal of Excellence, and the American Institute of Architects' Silver Medal Award (shared with Duany) in recognition of an entire body of work.

DAVID C. PROSPERI

POLICE POWER

In all countries, governments have to have the power to regulate. In the United States, this power is called the police power and it is the inherent power of government to act to protect the public's health, safety, morals and welfare. It is not a power whose limits can be drawn with precision. Rather, the police power is defined in each jurisdiction by the legislative body, which determines the public purposes that need to be served by legislation. It is sometimes distinguished from the power of EMINENT DOMAIN, which is the power of government to purchase property.

The determination of the public interest is for the legislature. The power of the legislature to make that determination is quite broad. It is not for the courts to second-guess legislative conclusions. Over time, the courts have accepted the increased exercise of governmental power. For example, in the case of *Euclid v. Ambler Realty Company* (1926), the US Supreme Court decision upholding ZONING, the police power was viewed as intended to prevent nuisances. In *Berman v. Parker* (1954), the Court extended the power to include aesthetics, and in 1974 in *Village of Belle*

Terre v. Boraas, the Court upheld lifestyle regulation. Moreover, not only does the legislature decide which values it wishes to pursue, but it also determines the means used to achieve its goals: they too receive judicial deference.

In examining legislation, courts defer to legislative judgments that are reasonable. Furthermore, the burden of proof is on a plaintiff attacking governmental action to prove that it is unreasonable, not on government to prove reasonableness. This idea is that legislators are better able to determine public needs than courts and that courts intervene only when the action is unreasonable and in violation of the due process clause of the Fourteenth Amendment to the US Constitution. In certain other classes of cases, for example equal protection cases involving gender or racial DISCRIMINATION or denial of fundamental rights, the presumption of legislative appropriateness is forgone and the courts will use a more stringent standard of review and place the burden of proof on the government.

If the courts have upheld extensions of governmental power, they have also raised the bar for actions that go too far. This is done through the power of courts to declare that governmental regulation is a taking of property. The US Supreme Court has developed a complicated series of tests to decide such questions. But the result is that where a taking is alleged, government will have a higher burden of proof in many circumstances. In addition, the US Supreme Court has also ruled that where a zoning law or action results in a taking, not only must the law be revised but government will be liable for the economic losses suffered by the landowner as a result of an unconstitutional zoning action.

Further reading

Gitelman, M. and Wright, R. (2001) *Land Use in a Nutshell*, St. Paul, MN: West Publishing.
Mandelker, D. (1997) *Land Use Law*, Newark, NJ: Lexis.

KENNETH PEARLMAN

POLIS

Greek term for the CITY. Etymologically, 'polis' in Greek is the origin of terms like 'civilization' ('politismos'), 'politics' ('politiki'), 'civil rights' ('politika dikaiomata'), etc., since all these

presuppose a CIVIL SOCIETY which succeeded in forming a collective way of life, thus constituting the city.

In ancient Greece, it defined the administrative and religious city centre (polis – ACROPOLIS), as distinct from the rest of the city, which, as a whole, was called 'asty' (better translated as 'URBAN'). It also meant an independent political unit consisting of a city and the surrounding countryside, also seen in Italy and medieval Europe. Different states – and the same state at different times – had a variety of governments, ranging from absolute monarchy to pure democracy. Only citizens participated in the government of the city-state, and CITIZENSHIP was limited to those born of citizen parents. Participation by citizens in GOVERNANCE was considered an honour and a civil duty, but it was often limited by class distinctions. The first city-states were in Sumer, but they reached their peak in Greece, where there were several hundred in the fifth and fourth centuries BC. The first Italian city-states were Greek colonies. Later, Etruscan and native city-states emerged, including Rome. After the fall of the Roman Empire, many Italian cities (e.g. Florence, Genoa, Venice) were city-states until the nineteenth century, as were some German cities such as Bremen and Hamburg.

Further reading

Glotz, G. (1997) *The Greek City and Its Institutions*, London: Routledge.

KONSTANTINOS LALENIS

POLITICAL ECONOMY

The term 'political economy' connotes that the workings of the polity impact the economy and vice versa. It postulates an inextricable connection between political processes (variously understood as government, the public, or authoritative allocation) and economic processes (variously understood as material production, the private, or voluntary exchange). Political economy revives the eighteenth- and nineteenth-century classical tradition of studying political and economic phenomena as an inseparable whole, reuniting the disciplines of political science and economics severed from one another in the twentieth century. URBAN political economy emerged in the 1970s to analyse the modern metropolis.

Scholars work within one of three broad traditions, each rooted in the ideas of a seminal theorist. The Marxist tradition builds from Karl Marx's conceptions of the dynamics of capital accumulation, class struggle and historical materialism; the neo-classicist (or public choice) tradition from Adam Smith's conceptions of the dynamics of competition, market forces and individual self-interest; and the institutionalist tradition from Max WEBER's conceptions of authority relations, bureaucratic organizations and intentional action.

Marxian urban political economists view the city as a locus for class struggles in the workplace between owners and workers and in the COMMUNITY between those seeking to profit from urban land and those trying to live on it. The historical development of the URBAN FORM results from these struggles and the need to create conditions necessary for capitalist production and reproduction. The imperative to meet these needs shapes local politics in ways serving the interests of the capitalist class.

The basis for the neo-classicist tradition is that two factors of production – capital and labour – are highly mobile while the third – land (and a political jurisdiction (city) occupying it) – is not. The resultant competition among cities for capital and labour mirrors the competition among firms in a marketplace, and strongly biases local politics towards the interests of mobile businesses and higher-income citizens.

The first two traditions are commonly criticized for being overly economistic, determinist and structural: economistic because outcomes are explained by privileging economic processes, leaving little room for political ones; determinist because outcomes strictly flow from these economic processes, leaving little room for contingency; and structural because the context formed by economic processes is what determines outcomes, leaving little room for human choice (or agency).

The institutionalist tradition conceptualizes the relationship between the polity and economy to give a stronger role for the political, the contingent and the agent. URBAN REGIME THEORY – the dominant paradigm for the study of city politics – grows out of this tradition. Urban outcomes in capitalist democracies can be traced to the division of responsibilities between the institutions of the market and the state. This division estranges public power from economic activity so the local state must form organizational arrangements (regimes)

with local business interests to govern. Economic factors set the general context but contingent political processes and individual choices still matter fundamentally in shaping cities' social and physical environments.

Further reading

Peterson, P.E. (1981) *City Limits*, Chicago: University of Chicago Press.
Vogel, R.K. (1992) *Urban Political Economy*, Gainesville, FL: University Press of Florida.

DAVID L. IMBROSCIO

POLYCENTRIC CITY

In 1945, an article was published which for several years ranked among the most quoted articles in the literature on URBAN geography: 'The nature of cities', by Chauncy HARRIS and Edward ULLMAN. In this article, the authors unfold their 'multiple nuclei' theory as one of three models on the INTERNAL STRUCTURE OF THE CITY, the others being the concentric zone theory advocated by Ernest Burgess (see MONO-CENTRIC CITY) and the theory on axial development from Homer Hoyt. Harris and Ullman reject the central assumption of both models: that there is but one single urban core around which land use is arranged symmetrically in either concentric or radial patterns. Although, in the words of Harris and Ullman, the handicap of distance would favour as much concentration as possible in a small central core, it is highly unlikely that just one single city centre will develop. One of the reasons is very simple: such a concentration is in most cases physically impossible. Also, according to Harris and Ullman, there are so-called separating factors by which several distinctive nuclei arise. Sometimes, this is just a matter of history in cases where various nuclei have existed from the very origins of the city. Harris and Ullman name London as an example that is interesting because this city is often characterized as the quintessential mono-centric city. But London evolved around two centres, originally even separated by open country: the City, the centre of finance and commerce, and Westminster, the centre of political life. In other cases, several factors lead to the rise of separate nuclei and these factors all have to do with the location strategies of firms. For instance, certain activities require

highly localized specialized facilities or group together as the result of external economies.

The model of Harris and Ullman on urban structure bridges the gap between theories on the spatial organization of a single city and theories and models of systems of cities, like the CENTRAL PLACE THEORY. It is also a small step from the multi-centred city to the concept of a multi-centred city region: a region formed by a large city and numerous surrounding urban settlements that have expanded as the result of sub-urbanization. Often these settlements, originally being *sub*urbs, have acquired a certain independent status. The result is a polycentric city *region*. In various parts of the world, especially Europe (the Low Countries, for instance), an even more complex urban pattern has evolved, made up of numerous polycentric city regions. These could be referred to as polycentric urban regions. Actually, the present terminology of urban patterns and URBAN FORM is in a state of flux. The complexity of modern URBANIZATION patterns is difficult to express in language, in particular because many terms have already acquired a certain meaning in the past. The polycentricity concept itself is proliferating on all possible scales: from the level of an individual city to the level of an entire continent. The latter is especially the case in Europe: an EU policy document has the promotion of a more balanced and hence polycentric urban and economic development across the continent as one of its prime goals. The resulting confusion around polycentricity makes the concept almost useless, but alternatives are hard to find.

Further reading

European Commission (1999) *European Spatial Development Perspective: Towards Balanced and Sustainable Development of the Territory of the European Union*, Luxembourg: Office for Official Publications of the European Communities, pp. 19–21.
Harris, C.D. and Ullman, E.L. (1945) 'The nature of cities', *Annals of the American Academy of Political and Social Sciences* 242: 7–17.

WIL ZONNEVELD

POPULAR CULTURE AND CITIES

Popular culture is taken here to be any form of entertainment linked to the mass media, with the exception of film. This entry focuses primarily on television, radio, music and video games.

Television produced in the United States has many of the same prejudices as motion pictures, which is unsurprising as many production companies and studios overlap. Much of their output has a minimal URBAN consciousness (see CITIES AND FILM), which is consistent with media that are essentially escapist. This is revealed most obviously with soap operas, which are designed to be consumed anywhere. As Anger (1999) notes, it is hard to tell who the neighbours are, as American soaps are all about interiors. This stands in contrast to soaps produced in Europe and Australia, where class relations are less fluid and external shots of housing developments are important in establishing the literal and metaphorical location of different families; the most obvious example is the British soap *Coronation Street*, which has spanned decades and has always been based in the same working-class NEIGHBOURHOOD.

Even when US television shows are specifically located in a real place, there is minimal connection to reality. For example, at one time NBC had an entire slate of comedies set in New York (*Friends*, *Mad About You*, *Seinfeld*), a city with a rich ethnic demographic, yet all of the characters and virtually all of the extras were Anglos (a state of affairs that has been continually challenged by actors, unions and social activists). Despite the existence of some shows that have self-consciously tried to explore aspects of a city's social and political-economic uniqueness (*LA Law*, *Miami Vice*, *Chicago Hope*, and the *Drew Carey Show*, set in Cleveland), most network television seems determined to 'be of no place', as Meyrowitz (1985) puts it. An extreme example is *The Simpsons*, one of the longest-running comedies, which is self-consciously set in Springfield, an Everyman municipal name that is to be found in virtually all fifty states.

Since he developed his thesis, Meyrowitz's predictions (directed specifically towards the coverage of television news) have come about, and many US television stations now have minimal local content, while drawing extensively from network feeds for national and international coverage. Much of the same story is to be found with radio, which was once very localized in terms of ownership and content. Increasingly, radio stations have been bought up by conglomerates (many religious), and a relatively few corporations now own hundreds of stations across the United States. This diminishes the opportunity to reflect local tastes in music or to pass on information about local politics. Instead, most stations offer network news, standardized talk shows or sermons, and a recorded music playlist; when corporate executives pulled dozens of recordings after 11 September 2001, due to their perceived unpatriotic content, it had an impact on hundreds of stations and many millions of listeners. Exactly the same trend is witnessed in the newspaper market; most cities historically had at least one substantial newspaper that reflected local affairs, and while many of these publications still exist, they are increasingly the possessions of large media conglomerates that own television, radio and print outlets in multiple markets. Many stories are now nationally syndicated, and the muckraking journalism that once exposed corruption in city hall is now relatively unusual (a rare example is the Pulitzer prize-winning research by staff of the *Miami Herald* after Hurricane Andrew in 1992, which exposed the manner in which building contractors had systematically bribed the building inspectorate to pass shoddy houses, many of which did not survive the high winds; see GROWTH MACHINE). Now, only the very top of the urban hierarchy can boast top-line independent newspapers (New York, Chicago, Los Angeles).

The music industry stands in contrast to network television, insofar as it is much more explicitly linked to specific locales. Just as different performers have been connected to a simple national geopolitics – the British Invasion, of the 1960s, for example, brought rock groups to compete in the United States – the same is true of the music that emerges from different cities. The Mersey sound of that same decade competed with the Philly sound, the Motown sound, the Memphis sound, and so forth. In part, this was connected to specific recording studios (such as Chess, in Chicago), but over time, different musical 'scenes' emerged in a manner consistent with the cultural milieus discussed by Sir Peter HALL (see also CITIES AND THE ARTS). Southside Chicago musicians created an electric rhythm and blues that was discernibly different from the black music played in New Orleans or Houston in the 1940s. Jazz in Kansas City deviated from that created in New York in the 1950s, while the production of trippy, psychedelic long-form jams in Los Angeles and San Francisco was quite different from the harder and more avant-garde free-form music emerging from Warhol's New York at the end of the 1960s. These variations depended on a whole series of

contingencies; for instance, the size of a city dictated its potential audience and the number of performing venues it could support. The manner in which racial segregation was enforced in venues or on the radio could limit access to what was termed 'race music' (which was often more readily available in London and Paris than in the United States prior to the CIVIL RIGHTS MOVEMENT, when radio broadcasts began to cross communities for the first time). Similarly, the political climate of the 1970s dictated that punk in London was more about resistance ('Anarchy in the UK', 'God Save the Queen') than was the case in New York ('Blank Generation'). These free-standing urban cultural pockets have also maintained in subsequent years. While there were numerous cities with high unemployment in the 1990s, it was Seattle that generated the angular grunge style, precisely because it was so far removed from the established music business, as its biographers, like Charles Cross, argue. The rap industry was clearly divided into a West Coast and East Coast contingent throughout the decade, which was revealed as much more than a question of marketing strategies when rivalries turned into execution-style killings of artists like Tupac Shakur.

Do these cities in any way define the content of contemporary music? It is one of the great euphemisms of the industry that the designation 'urban' is not so much the antithesis of 'country' but that one is black and the other is white. And urban music has been very much about the so-called thug life – POVERTY, SEGREGATION, drug-dealing and death (e.g. 'Straight Outta Compton') – as much as it has been about the more generic American obsession with wine, women and songs about cars (something that it shares with country music, albeit in ways that neither side would probably care to acknowledge). The other clear fact about urban music is that it is consumed in both cities and SUBURBS, and hip hop is a cultural style that now effectively transcends its origins. It is also the case that its social concerns are relatively limited and effectively blunted by its relentless consumerism. For a more thoughtful critique of the urban condition, we would have to turn to bands such as Rage Against the Machine ('Battle of Los Angeles'), whose lyrical universe is defined by issues like globalization, poverty and exploitation, a concern which can also be found in Mexico (e.g. Molotov), the former Soviet Union, parts of Africa (many of which have developed their own versions

of hip hop, e.g. MC Solaar) and many European countries.

For a more sustained evocation of the urban condition, we would also need to turn to the loose collection of cultural critiques that come under the banner of cyberpunk. Frequently linked back to a seminal figure such as William Gibson and, later, Neal Stephenson, cyberpunk is a response to the challenges and opportunities implicit in computer technology and the INTERNET. With an often futuristic content, the genre is manifested in film, novels and computer games, and can also be seen as a dominant sensibility within movies (such as *The Matrix*), novels such as *Snowcrash*, and many Japanese animated presentations, or anime (such as the 2001 remake of *Metropolis*). One of the more compelling themes is the dichotomy between different spatial forms, especially the privatized spaces of the corporation and the interstitial PUBLIC SPACES that remain between them. Here is a street culture, where roam the willing and unwilling outcasts of a technologized society, and their lot is in contrast to what have been described as the luxury, security and grace of corporate enclaves. This focus on public and private space, how they are frequently conflated, how the former is monitored and the manner in which it is used and by whom, has numerous echoes within popular culture: skateboarding, for example, has become politicized because so much OPEN SPACE within cities has been appropriated by municipalities and by corporations so that the ways in which it gets used, and who has access to it, tie back into appearance and age.

There is a certain irony in the fact that the city, which has nourished the outlaw and the artist alike, has become a standardized dystopic cliché within so much popular culture. From *Blade Runner* on, it seems that virtually any evocation of the future URBAN FORM in movies, comic books or video games envisages a sharp differentiation between rich and poor, between private affluence and public squalor, all overlain with an apocalyptic sense of ecological catastrophe. Consequently, the events of 11 September 2001 have served only to pull popular culture and daily events closer, by emphasizing that fantastic images of mass destruction can turn into reality. This also adds to conjectures about different ways to use computers and gaming technologies to simulate new urban forms that reflect lower densities and less connectivity between vulnerable populations. The net effect may even be a new cycle

of popular culture that confronts the threats of a globalized world – while emphasizing the relative safety of the suburbs.

Further reading

Anger, D. (1999) *Other Worlds: Society Seen Through Soap Opera*, Peterborough, Ont.: Broadview.
Hall, P.G. (1999) *Cities in Civilization*, London: Phoenix Giant.
Meyrowitz, J. (1985) *No Sense of Place*, New York: Oxford University Press.
Moody, N. (1998) 'Social and temporal geographies of the near future', *Futures* 30(10): 1003–16.

ANDREW KIRBY

POPULISM

Populism is a political movement that has only partially developed as a consistent political ideology. It is more a political discourse articulating popular political identity in contrast to the politics of interest representation. Populism is a pattern of top-down mobilization that bypasses or subordinates institutional forms of political mediation. It can best be understood as a reaction to the political climate, a politics of resentment, a response to a generation of political experience. It aims to revise the prevailing methods of politics with paternalistic relationships between leaders and heterogeneous masses. Populism generally asserts the will of the 'people' against a mutually perceived social evil such as concentrated economic power, mundane power of monopoly, elites, privileges of the better off, establishment, political corruption, entrenched political class, traditional oligarchs, compromise, weakness of institutionalized channels of political representation, etc.

As a political movement it was historically developed in rural areas, but today it encompasses very diverse sections of the population, from economic interest groups, intellectuals and the middle classes to labourers and farmers, and can form cross-class alliances between urban labour, the middle sectors and domestic industrialists.

Populism was widely argued as a fixed stage in a sequential pattern of development; a transitional stage on the path between traditional and modern societies, a multi-class socio-political movement corresponding to the stage of import substitution and/or industrialization etc. However, it can be observed in the developed countries and at higher stages of development as well, with different discourses on anti-immigration, racism, power, inequalities, etc. Therefore, it is more reliable to accept populism as a spatio-temporal political discourse that can take various forms in different geographies and development stages. Its flexibility makes possible the application of populism to loosely connected empirical phenomena, from economic policies and development phases to political ideologies, movements, political parties, governments and social coalitions.

The method of populism is simple, direct and emotive. Spontaneity is essential. Populism stems from the word 'people', the common citizenry and their lifestyles, beliefs and values. It tries to reach large masses of poorly organized people and bring them into political action. By this token, populism enables individuals to view themselves as part of that large mass and act as such in politics. They see themselves as true democrats, voicing popular grievances and opinions systematically ignored by governing mainstream parties and the media. Populism searches for justice and extension of the rights and powers of the common people. These characteristics may keep democracy honest. It is radical and reactionary, but favours change if and only if it is to restore the past. It can be interpreted as an opportunity to bring structural dividing lines like class, gender and ethnicity together.

However, populism is generally associated with policies that are popular but not always economically rational. For example, economic populism is defined as an attempt to decrease social inequality through demand stimulation, but common consequences include inflation, public deficit, balance of payments problems and shortages. It may mean, therefore, that the subject of politics can be transformed into a mass without clearly defined interests, which can be mobilized against the general and long-term interests of the country. This possibility attributes a negative meaning to populism and makes it a frequently misused political term isolated from its positive attributes.

Further reading

Goodwyn, L. (1978) *Populist Moment: A Short History of the Agrarian Revolt in America*, Oxford: Oxford University Press.
Holmes, W. (ed.) (1994) *American Populism*, Boston, MA: DC Heath & Co.

MURAT CEMAL YALCINTAN

PORT CITY

Since the advent of trade among civilizations, human settlements have relied upon ports as a necessary condition for their existence. Even as we consider cities in antiquity for which proximity to a watercourse was a requirement for agriculture, the same watercourse made seasonal movement of agricultural products possible to their neighbouring settlements. Ancient cities such as Memphis, Thebes and Alexandria in Egypt, and Lagash, Ur or Babylon in Babylonia required proximity to water to thrive.

Most of the major cities of the past two thousand years have also required a port to flourish. Consider that the largest cities were consistently located at the water's edge, whether on a major river or the sea: Constantinople in Turkey, and Hangzhou, Beijing, Xian and Nanking in China are prominent among that number. In modern times, few of the major cities of the world have failed to be located on major rivers or the coastline. In 1996, the UN Centre for Human Settlements published its report, *An Urbanizing World: Global Report on Human Settlements 1996*, which noted the largest cities over the past millennium. For example, in 1975, the largest cities in the world were (in rank order) Tokyo, New York, Shanghai, Mexico City, São Paulo, Osaka, Buenos Aires, Los Angeles, Paris and Beijing. Of the ten, only Paris and Mexico City are landlocked. In 1900, the five biggest cities were (in rank order) London, New York, Paris, Berlin and Chicago. All but Paris lacked direct access to major waterways.

Considering that URBAN agglomerations develop largely because of their comparative advantage over other locations, a waterside location is a salutary attribute of that advantage. But port cities enjoy advantages greater than maritime commerce. Their suite of locational attributes includes the potential for fishing fleets, recreational use of marine or freshwater locations, irrigation for crops and a source of power from flowing water. In the modern era, urban settlements near large bodies of water have enabled industry to use the aqueous resource to cool power plants, as well as a host of other industrial uses.

In addition to their mercantile prospects (see MERCANTILE CITY), port cities tend to be cosmopolitan because they are ports of entry not only for goods but also for people. Certainly before the advent of air travel, international travel was dependent upon ports and they became MELTING POTS for civilization. Consider the influence of marine transportation on London, New York, Shanghai and, more recently, Los Angeles, Singapore and Vancouver. Each has experienced dramatic multi-ethnic and multicultural changes because of its role as a port of entry for immigrants who chose those cities largely because there was already someone there who shared a family tie or ethnic identity. Thus, the very nature of port cities permits civilizations to intermingle.

Further reading

UN Centre for Human Settlements (1996) *An Urbanizing World: Global Report on Human Settlements 1996*, London: Oxford University Press.

JAMES A. FAWCETT

PORTES, ALEJANDRO

b. 13 October 1944, Havana, Cuba

Key works

(1989) *The Informal Economy: Studies in Advanced and Less Developed Countries*, Baltimore: Johns Hopkins University Press (ed. with M. Castells and L. Benton).
(1993) *City on the Edge: The Transformation of Miami*, Berkeley, CA: University of California Press (with A. Stepick).

The three main axes of Alejandro Portes's work are URBANIZATION in Latin America, the informal economy, and immigrant communities in the United States. Portes was trained as a sociologist both in Latin America and in the United States, where he obtained a Ph.D. from the University of Wisconsin at Madison in 1970. His work on cities illustrates his Cuban origin and his immigrant experience. In the first stages of his work, Portes collaborated with John Walton to write, in 1976, *Current Perspectives in Latin American Urban Research* and *Urban Latin America: The Political Conditions from Above and Below*. Labour and class were elements he incorporated in his 1985 book co-authored with Robert Bach, *Latin Journey: Cuban and Mexican Immigrants in the United States*. This book also marks the beginning of his innovative research on global IMMIGRATION. In 1989, his ground-breaking edited volume on *The Informal Economy* prompted the study of the

INFORMAL SECTOR around the world. The theory and the cases presented show that the informal economy – the economy that is not regulated by the state – occurs across cultures and social classes in both 'developed' and 'developing' countries.

During the 1990s, Portes developed the role of SOCIAL CAPITAL, family and labour networks, entrepreneurship, and CULTURE in the formation of immigrant and ETHNIC ENCLAVES in the United States. He claimed that the assimilation and acculturation of new waves of immigrants varies not only according to their own characteristics and social capital, but also according to the recipient place. *City on the Edge* is the culmination of his identity and experience as an immigrant Cuban scholar. This book traces the impact of the different waves of Cuban immigration on the shaping of Miami as a GLOBAL CITY. In 1995, this book won awards for best book in urban and community sociology and in urban anthropology. Portes's interest in the impact of the global economy on Latin America's urbanization is reflected in his book *The Urban Caribbean*. This unique work compares urbanization in five countries of this understudied region. In 2001, Portes edited two volumes about the second generation of immigrants in US cities. Currently, Portes is a professor at Princeton University. He has authored more than two hundred journal articles and book chapters on URBAN SOCIOLOGY, URBAN STUDIES, global immigration and Latin American urbanization.

Further reading

Portes, A. and Rumbaut, R. (eds) (2001) *Legacies: The Story of the Immigrant Second Generation*, Berkeley, CA: University of California Press and Russell Sage Foundation.

Portes, A., Dore-Cabral, C. and Landolt, P. (eds) (1997) *The Urban Caribbean: Transitions to the New Global Economy*, Baltimore: Johns Hopkins University Press.

MARÍA TERESA VÁZQUEZ CASTILLO

POST-FORDIST SYSTEM OF PRODUCTION

It is widely accepted that there were some changes and transitions in the system of production in the 1970s, when Fordism had a structural crisis. Either these transitions signal a break off or a continuity.

A post-Fordist system of production can only be comprehended together with its Fordist precedent.

The logic of Fordism was based on mass production and consumption, which required a production process, labour, product, demand and consumer that were all standardized. After the Second World War, this standardization became a lifestyle that served the comprehensive regulations of the nation-state together with the Keynesian policies.

It is accepted that the transition from a Fordist to a post-Fordist system of production started with the deepening of the economic (oil) crisis that occurred in 1973. This crisis brought together transformations in the accumulation regime. Both mass production and consumption suffered from the crisis. Thus, supply and demand changed dramatically. On the other hand, there were innovations and changes in the lifestyles of people after the crisis. Innovations in technology introduced new methods of production for the supply side to meet the changes in demand. They re-established labour relations and the production system in social, economic and geographic terms.

The key concept of this new system of production was flexibility. In the new era, all factors of production, from management, facilities and product to labour and resources, had to be flexible for the supply side to meet the changing demands. The target had shifted from scale to special markets, from minimal cost to optimum cost for quality and speed. The huge factories producing standard products were divided into smaller units producing different products, externalized certain production stages, and employed part-time, temporary and contract personnel rather than permanent labour. By this token, the very strong labour unions started to dissolve as well. On the other hand, small enterprises re-established their significance and informal economic activities were re-discovered. These developments accelerated the decline of old MANUFACTURING CITIES with strong labour unions.

When compared to a Fordist production system, post-Fordism brings about localization instead of CENTRALIZATION, flexibility instead of rigidity, special production instead of mass production, small production units including houses instead of factories, product variety rather than standard products, part-time and temporary employment instead of permanent, lack of social security instead of strong social security through labour unions, strategic production planning

instead of comprehensive, and participation in decision-making rather than authoritarianism.

It is often argued that post-Fordist production systems are used as a means to legitimize a certain type of politics. Accordingly, flexibility is an ideological weapon brought forward in accordance with the economic logic of globalization. Although a post-Fordist system of production can be observed on a limited geographical scale, and thus has not yet become dominant, there is a huge effort on the part of transnational capital to extend its horizons. However, this is still a transition period that may have no end.

Further reading

Hall, S. and Jacques, M. (eds) (1990) *New Times: The Changing Face of Politics in the 1990s*, London: Lawrence & Wishart.

Harvey, D. (1990) *The Condition of Postmodernity: An Enquiry into the Origins of Cultural Change*, Oxford: Blackwell.

MURAT CEMAL YALCINTAN

POSTMODERNISM

Postmodernism is a concept used in virtually all disciplines, namely in architecture, URBANISM, the visual and plastic arts, literature, as well as in philosophy and the social sciences. It is not only difficult to define, due to the fluidity of its boundaries, but also difficult to locate temporally, because it is not clear exactly when postmodernism began, but certainly in different moments in each of the various fields. Nevertheless, in spite of its eclectic nature and semantic difficulties, it is possible to identify certain common ideas and procedures and, therefore, to talk about a postmodern discourse, embracing a whole range of subjects. And this will depend upon one's prior definition of MODERNISM, as this, as well as postmodernism, are polysemic terms applied either to define historical periods or to refer to different philosophies, social theories, architecture styles or developments in urbanism. For some, the prefix 'post' suggests that postmodernism emerged chronologically after modernism; for others, postmodernism is more a way of thinking than a period and therefore they can coexist temporally as different attitudes; from another point of view, such as the one expressed by David HARVEY or Fredric Jameson, it is no more than the adaptation to the reindustrialization of

URBAN and metropolitan areas in the framework of an emerging global economy of flexible accumulation; from yet another perspective, postmodernism should also be distinguished from postmodernity, as the latter refers to the new socioeconomic, political and cultural conditions in the developed world, associated with the rise of the post-industrial service-oriented global economy.

It seems to have been the French philosopher LYOTARD who introduced the term 'postmodernism' firmly into the philosophy discourse, and it is often thought to have been Charles Jencks who did the same in architecture, first in England, with his book *The Language of Post-Modern Architecture* (1977), a vision that from there passed to the United States and later to the rest of Europe, in the 1980s, although other authors had used the term before in the architecture context (e.g. Joseph Hudnut, in 1949, and Nikolaus Pevsner, in 1967). In his book, Jencks claims that the death of modernism occurred in 1972 when a group of buildings, built according to Congrès Internationaux d'Architecture Moderne (CIAM) principles, was imploded in Pruitt-Igoe (St. Louis, Missouri) after being considered unsuitable for living.

Postmodern philosophies in social science, contrary to what happened with structuralism and realism, do not take for granted the promises of modernism in terms of continuous progress based on reason and scientific knowledge. On the contrary, they aim to reflect the complexities of modern society. Indeed, a key facet of modernism is the fact that it aspires to set up a society in which reason and scientific knowledge will direct a continuous historical progress. These beliefs have been questioned in the past, among others by Nietzsche, but never as recently, especially after the late 1960s. In postmodern philosophy and social theory, in spite of the wide diversity of theories assembled under the name of postmodern (Rorty, Derrida, FOUCAULT, Deleuze, Lyotard, to mention just a few), reason is reinterpreted as a mode of social control, producing the truth and, through it, acceptable social practices. Postmodernism rejects not only metanarratives and the modern belief in progress based on reason, but also the utopian aim of imposing a normalizing order on society.

In architecture, postmodernism developed as a reaction against the academization of modern architecture and against the 'international style' (1920–70) and its ideology of regularity, rationalization, right angles, hygiene, purism,

standardization, etc. It was a reaction against the inhuman scale of modern architecture, seeking instead to reintroduce a SENSE OF COMMUNITY and PLACE, recognizing again the local or regional characteristics, therefore incorporating a variety of styles and influences. It was a reaction against stylistic uniformity, and against, for example, the idea of Adolph Loos that ornament was to be avoided. Important in this process were the books of American architect Robert VENTURI (*Complexity and Contradiction in Architecture* (1966), and, as co-author, *Learning from Las Vegas* (1972)), a manifesto of postmodern architecture, although without using the term, in which he argues in favour of vulgar architecture.

Postmodernism advocates everything that has been refused by the modernist (LE CORBUSIER, Walter GROPIUS, Loos, MIES VAN DER ROHE, among others): eclecticism, stylistic pluralism, tradition, vernacular styles, etc. But in spite of these key common elements, it is necessary to distinguish between the different trends within postmodern architecture's practices of authors such as Venturi, Charles Moore, Robert Stern, Léon KRIER, among many others.

Postmodern trends in contemporary urbanism follow closely those in architecture and consist explicitly in an effort to recreate a sense of continuity with the city's history, re-emphasizing aesthetics, aiming to improve the physical landscape by attacking the inhuman scale and sense of urban desolation in most twentieth-century cities. Postmodern urbanism is coherent with postmodern architecture because both mix eclectically elements and styles from the past and both combine traditional aesthetics with modern improvements in technology, without being revivalist. 'Postmodern urbanism' is therefore the term used to group those processes that aim to re-urbanize the contemporary city, in what some critics see as no more than an act of nostalgia, a shift towards concerns which are more related to symbolic issues than to social and political concerns. In contrast to CIAM's modern urbanism, based on a comprehensive and rational planning process, postmodern urbanism advocates traditional URBAN FORMS, social and cultural heterogeneity, the preservation of older residential housing and NEIGHBOUR-HOODS, commercial streets, as well as the construction of new buildings compatible with the city's historical background.

Following in the footsteps of early criticism of modernism by Jane JACOBS and Lewis MUMFORD, postmodern urbanism admits a diversity of life-styles and interests, whose combination produces a different city in each case. These eclectic reactions to modern urbanism, as defined by the CIAM movement, have so far been consolidated, in North America, around the movement initially termed 'neo-traditional architecture' and later named 'NEW URBANISM' and, in Europe, among other manifestations, through the movement inside the European Council of Town Planners that led to the adoption of the NEW CHARTER OF ATHENS, in 1998.

Further reading

Ellin, N. (1996) *Postmodern Urbanism*, Oxford: Blackwell.

Harvey, D. (1995) *The Condition of Postmodernity: An Enquiry into the Origins of Cultural Change*, Oxford: Blackwell.

Klot, H. (1984) *The History of Postmodern Architecture*, Cambridge, MA: MIT Press.

Woods, T. (1999) *Beginning Postmodernism*, Manchester: Manchester University Press.

SEE ALSO: modernism; post-Fordist system of production

CARLOS NUNES SILVA

POST-SOCIALIST CITY

The notion of the post-socialist city emerged as a result of the abandonment of the communist system in the 1990s in Central and Eastern Europe. The process of the transformation of the whole society and the cities there is still in the early stages. The existing structure, for example of large housing estates, will remain for decades. The speed of adaptation of the former socialist cities to the requirements of a market economy depends on the will of the citizens and the economic prosperity. In general, the prefix 'post' will be adequate for several decades to describe those cities transformed from a socialist past.

The first, most visible consequences of the transformation can be described as an increase in the socio-spatial differentiation, and in the substantial redesign of the technical INFRASTRUCTURE and the URBAN landscape. The most important political and economic processes having a direct impact on the transformation of urban space in

post-socialist cities concern first of all the return of the market mechanisms and particularly the importance of land rent. This has been followed by changes in the structure of land ownership, from not strictly defined or state ownership to local government and private ownership. As a result, the dominant rules of the spatial allocation of people and economic activities have become subject to market criteria (not ideological or political as under socialism). This has been followed by a radical increase in the intensity of land use together with the functional transformation particularly of the central parts of the city, from administrative and political functions to commercial ones. For example, the Warsaw Stock Exchange for one decade has been housed in the former headquarters of the Central Committee of the Communist Party. Also, the reassessment of urban space has caused the replacement of communist symbols by historic and national symbols. The former communist street names have been replaced by the pre-war names. The same has happened with the MONUMENTS.

The post-socialist city has become independent with the shift of control over space from central to local, mostly through the return of self-government and the formation of new local interest groups. The radical increase in the number of actors competing for particular locations in urban space has proved that local government was not always prepared to deal with such situations. The spontaneous development of the street market, the appropriation of PUBLIC SPACE and challenges to the planning regulations have become common problems.

Economic recession together with deindustrialization and a rapid increase in employment in the service sector have generated the formation of a new social structure. The post-socialist city has become dominated by the middle class (still under formation), POVERTY groups and the elite. The increasing social and wealth disparities have gradually separated the rich and the poor into segregated areas. For many citizens, the cities have become too expensive to live in unless, together with new, impoverished immigrants, they accept poorly paid jobs and poor housing in emerging SLUM areas.

Further reading

Enyedi, G. (1998) *Social Change and Urban Restructuring in Central Europe*, Budapest: Akademiai Kiado.

Weclawowicz, G. (1996) *Contemporary Poland: Space and Society*, London: UCL Press.

GRZEGORZ WECLAWOWICZ

POVERTY

The definition of poverty is politically contested. There are three main forms of definition (Pain *et al.* 2001). 'Absolute poverty' defines minimum requirements needed for physical survival, including food, water, shelter and healthcare. This standard can be applied to all societies and in that sense is objective. 'Relative poverty' is defined as what is required to participate in the 'normal' life of a particular society. This, however, has been criticized as merely expressing *inequality* rather than poverty as such. 'Hybrid' or 'multi-dimensional' definitions derive from the work of Amartya Sen. Sen has suggested that it is possible to consider certain features of human nature as being universal. These universal 'needs' and 'capabilities' exist in every society, but the means to achieve and express them differ from society to society. Poverty exists when such universal needs are not met. Criticisms of this approach have centred on the existence and measurement of universal needs. Despite these critiques, there is fairly broad agreement among academics and policy-makers on a hybrid approach, within which poverty has many dimensions beyond subsistence and includes minimum standards of well-being as perceived by the particular society.

Different explanations of poverty focus on different spatial scales. Individualistic explanations focus on supposedly 'genetic' characteristics such as intelligence or physical competence, or psychological traits such as fecklessness regarded as inherited or acquired. But these cannot explain the actual incidence of poverty. Other explanations argue that the way of life of poor NEIGHBOUR-HOODS, the 'culture of poverty', prevents poor people, individually or collectively, escaping from poverty, whether through low aspirations, disorganization, criminal activity, or poor education and health. Similar ideas are expressed in conservative discourses on the 'underclass', which see the poor as substantially responsible for their situation and evoke long-standing distinctions between 'deserving' and 'undeserving' poor.

Other explanations explain poverty in terms of economic and political, as well as cultural, processes, and in terms of processes at every scale including the individual and neighbourhood but stretching up to the global. One core, but discredited geographical theory links specific environments to poverty, namely environmental determinism. However, the most abstract, and largest scale of explanatory focus concerns debates around the role and properties of capitalism. The worldwide capitalist economy assigns a large set of workers to jobs which are poor in terms of wages, benefits, work processes and security. These affect the living standards of those workers after they retire, and (where they exist) state benefits to the unemployed since these are set below the low-wage level. Poor jobs are distributed unevenly geographically, being concentrated in the Third World and in certain cities and regions worldwide, though they are required also in prosperous and GLOBAL CITIES. Unemployment and underemployment are also unevenly distributed geographically; THIRD WORLD CITIES have especially high levels partly due to high migration from rural areas. Unemployment falls particularly on people from poor backgrounds since they lack job experience and personal contacts, and often lack currently and locally demanded skills. In cities with very high unemployment levels, particularly in the Third World, many of the poor have to survive through the informal economy in jobs which earn very little, are precarious, often hazardous and frequently criminal.

Poverty created by the waged economy is compounded by social reproduction. Capitalist housing provision for those on low or precarious incomes is of poor quality. Especially in cities of growing population it is often in short supply – in Third World cities grossly so (see HOMELESSNESS) – and exorbitant prices can be charged. The poor are more dependent on state-supplied services than the better off, and these are often systematically underfunded. The housing and land market confines the poor to particular areas of cities, whether in formal housing or shanty towns. These neighbourhoods usually have inferior state services, due to greater demands on them and politically determined underfunding; in the Third World, basic services such as water and sewerage are often absent. Poor neighbourhoods also have inferior or absent privately supplied services. Finally, the poor are subject to higher-than-average levels of crimes of violence and property (see CRIME).

These conditions of social reproduction within households and neighbourhoods then, in a vicious circle, further disadvantage the poor in competing for jobs. Malnutrition and ill health, poor education, poor clothing, inadequate transport, lack of facilities for caring for dependent children and adults, and the stigma attached to residents of poor neighbourhoods and to the homeless all present barriers to obtaining jobs. This is the kernel of truth in the 'culture of poverty' theory; but we now see that it is premised on processes at higher spatial scales.

Forms of social power in addition to class construct poverty. Women are more subject to poverty than men. They tend to be employed at lower wages, with less security and in worse conditions. In family businesses they often receive no money income. Their imputed responsibilities for work and caring within the home inhibit them from taking up (good) waged work. In many countries they do not inherit family property. Within households they are often allowed less food, space and access to education and healthcare. People of oppressed ethnicity and recent immigrants are also disproportionately represented among the poor through lack of relevant resources and contacts to obtain jobs, discrimination by employers, confinement to the worst housing, and racist violence. Undocumented migrants and refugees suffer these disadvantages exacerbated by legal restrictions. Older people suffer disproportionately from poverty because of receiving no (or low) wage income, inadequate or non-existent state pensions, and because their subsistence needs are often greater. Disabled people are overwhelmingly poor (on the 'hybrid' definition) due to discrimination in employment and society and because of their greater subsistence needs.

The generation of poverty is strongly conditioned by the major strands of national and city government policy: the regulation of industry and employment, taxation and state income transfers, funding and targeting of public services, and regulation of housing. Policies directed explicitly at URBAN poverty usually address symptoms or aspects rather than the set of interlocking processes considered above. Worse, these policies often blame the poor, discipline them (as in welfare-to-work programmes), or, as in some SLUM clearance and 'cleaner streets' policies, simply move them on. However, in many cities of the First and Third

Worlds, the poor have organized themselves to provide their own services and to place demands on businesses and governments to address their immediate needs.

Further reading

Merrifield, A. and Swyngedouw, E. (1996) *The Urbanisation of Injustice*, London: Lawrence & Wishart.

OECD (2001) *DAC Guidelines on Poverty Reduction*, available online at www.oecd.org/dataoecd/47/14/2672735.pdf

Pain, R., Barke, M., Gough, J., MacFarlene, R., Mowl, G. and Fuller, D. (2001) 'Geographies of poverty', in *Introducing Social Geographies*, London: Hodder Arnold, ch. 12.

Townsend, P. (1993) *The International Analysis of Poverty*, London: Harvester Wheatsheaf.

DUNCAN FULLER
JAMIE GOUGH
PHIL O'KEEFE

PRINCIPLES OF INTELLIGENT URBANISM

Principles of intelligent urbanism (PIU) are a set of ten axioms, laying down a value-based framework within which participatory planning can proceed. After review and amendment by stakeholders, PIU acts as a consensual charter around which constructive debate over actual decisions can be evaluated and confirmed. It also guides URBAN PLANNING procedures. PIU emerged from several decades of urban planning practice by Christopher Benninger in the Asian context (Benninger 2001). It was the basis for the new capital plan for Bhutan. The principles of intelligent urbanism are:

1 *A Balance with Nature* emphasizes the distinction between utilizing resources and exploiting them. It focuses on a threshold beyond which deforestation, soil erosion, aquifer depletion, silting and flooding reinforce one another in URBAN development, destroying life-support systems. The principle promotes environmental assessments to identify fragile zones, threatened natural systems and habitats that can be enhanced through conservation, density control, land use and OPEN SPACE planning.

2 *A Balance with Tradition* integrates plan interventions with existing cultural assets, respecting traditional practices and precedents of style.

3 *Appropriate Technology* promotes building materials, techniques, infrastructural systems and construction management consistent with people's capacities, geo-climatic conditions, local resources, and suitable capital investments. Accountability and transparency are enhanced by overlaying the physical spread of urban utilities and services upon electoral constituent areas, such that people's representatives are interlinked with technical systems.

4 *Conviviality* sponsors social interaction through public domains, in a hierarchy of places, devised for personal solace, companionship, romance, domesticity, neighbourliness, and COMMUNITY and civic life.

5 *Efficiency* promotes a balance between the consumption of resources like energy, time and finance, and planned achievements in comfort, safety, security, access, tenure and hygiene. It encourages optimum sharing of land, roads, facilities, services and infrastructural networks reducing per household costs, while increasing affordability and civic viability.

6 *Human Scale* encourages ground-level, pedestrian-oriented urban arrangements, based on anthropometric dimensions as opposed to machine scales. Walkable, mixed-use URBAN VILLAGES are encouraged, over single-functional blocks linked by motorways and surrounded by parking lots.

7 *Opportunity Matrix* enriches the city as a vehicle for personal, social and economic development, through access to a range of organizations, services and facilities, providing a variety of opportunities for education, recreation, employment, business, mobility, shelter, health, safety and basic needs.

8 *Regional Integration* envisions the city as an organic part of a larger environmental, socio-economic and cultural-geographic system, essential for its sustainability.

9 *Balanced Movement* promotes integrated transport systems comprising walkways, bus lanes, light rail corridors and automobile channels. The modal split nodes between these

become the public domains around which cluster high-density, urban hubs and pedestrian, mixed-use urban villages.

10 *Institutional Integrity* recognizes that good practices inherent in considered principles can only be realized through accountable, transparent, competent and participatory local GOVERNANCE, founded on appropriate databases, due entitlements, civic responsibilities and duties. PIU promotes a range of facilitative and promotive urban development management tools to achieve appropriate urban practices, systems and forms.

Interest in the concept of intelligent urbanism has spread to other contexts and its application is being widely discussed.

Further reading

Benninger, C. (2001) 'Principles of intelligent urbanism', *Ekistics* 69(412): 39–65.
—— (2002) 'Principles of intelligent urbanism', in *Thimphu Structure Plan*, Thimphu: Royal Government of Bhutan.
Williams, T. (2003) 'Smart advice for urban growth', *Regeneration & Renewal*, 6 June.

A. RAMPRASAD NAIDU

PRIVATIZATION

Privatization can be simply defined as any attempt to arrange for a greater private (and non-profit) sector role in the provision of services that were once provided by public sector agencies. More broadly, privatization refers to any attempt to reduce the government's role in service provision. Privatization reflects the viewpoint that private and non-profit service providers can often do a better job than government.

Privatization is based on the distinction that can be made between *providing* and *producing* public services. A government can act to ensure that a public service is provided without necessarily having a government agency itself produce the desired service. A public agency may produce the service (such as garbage collection by a municipal agency); alternatively, government can seek to utilize private sector firms (in this case, private garbage collection companies), non-profit agencies (including the involvement of church congregations in COMMUNITY clean-up efforts) and even self-provision (where, in a small community, each household is responsible for taking its garbage to the local dump). The government can choose which alternative mode – public or private production – permits the highest quality, most efficient service delivery.

There are various forms of privatization. Perhaps the most widely used variant of privatization is *contracting*, where private and non-profit firms competitively bid (in Britain, the process is known as *competitive tendering*) for the right to provide a service. Under *managed competition*, the salubrious effects of competitive bidding are maintained, but the workers in a municipal department are given a chance to keep their jobs as they are allowed to bid against private on non-profit firms for a public contract. Under *franchising*, a private firm is granted the exclusive right to provide a good, such as the concessions at an airport or a sports stadium.

There are numerous other vehicles for privatization. A government may offer *vouchers* that increase the ability of recipients to purchase services, such as schooling and housing, from private providers. A government may similarly offer *tax incentives* in an effort to induce the private provision of a service like child day care. Alternatively, a government may even arrange for the *private management of a public facility*, as was the case when the city of Philadelphia, at the beginning of the new millennium, arranged for a private educational firm to take over its troubled school system. The government may also encourage a community's *self-provision* of services, where volunteer efforts in such areas as PARKS maintenance and citizen safety patrols serve to reduce the need for municipal services. Under *load shedding*, the government simply divests itself from service responsibility by selling a facility (such as a water or electric utility) to the private sector. *Deregulation*, too, is a spur to privatization as it relaxes government rules and thereby allows the increased entry of private and non-profit firms into service areas that were once exclusively the domain of public agencies.

Privatization is the subject of a heated political debate. Proponents of contracting and other privatization efforts argue that:

- Contracting encourages competition and cost-efficiency. Contractors must be efficient in the

performance of their work; otherwise, they will lose the city's business to a competitor firm that submits a lower price.

- Contracting and privatization allow the government to circumvent the protections and rigidities of civil service personnel systems. Protected by civil service regulations, public sector workers are too often unconcerned with the performance of their jobs or with their responsiveness to clients. Workers in private businesses, in contrast, must perform effectively and respond to clients' wishes, or else they will lose their chances for promotions and perhaps even their jobs.
- Contracting permits certain economies, including ECONOMIES OF SCALE, available to the private firms that are not otherwise readily available to government. Private sector firms, for instance, usually enjoy greater flexibility than does government in assigning work tasks to employees.
- Contracting permits the introduction of successful private sector management systems into the production of public services.
- Contracting allows a city to draw on the expertise, experience, enthusiasm and commitment of the staffs of voluntary and faith-based organizations.
- Contracting can reduce large, initial capital outlays for INFRASTRUCTURE, equipment and training. Through contracting, for instance, a municipality can update its record-keeping without having to hire, train and tenure new permanent municipal workers or purchase new and expensive data-processing technology. Contracting allows the government the flexibility to expand services as needed without adding to the size of its permanent workforce.

Cities turn to contracting primarily in an effort to achieve cost savings. Cities have also resorted to contracting when they lack the necessary municipal staff and facilities to deliver services.

Opponents of privatization, however, reject what they see to be a crude attack on the public sector. They note that many public employees are highly qualified, energetic and dedicated public servants. They further charge that the privatization ideologues have unfairly contrasted the worst of the public sector with a highly idealized portrait of the way private business works. The private sector, too, is plagued with inefficiency and corruption. Opponents of privatization further argue that:

- Contracting is not always efficient. Contracting can cost more when all of the hidden costs are calculated. A city often incurs additional costs as a result of its need to monitor the contract, evaluate the work that is performed and deal with the service disruptions that result from contract turnover.
- Contracting savings are often illusory as firms may 'lowball' costs when initially bidding to win a service contract. In subsequent years, when the city is dependent on the contractor, the costs billed to the city rise dramatically in order to cover the true cost of producing the service.
- Collusion or 'bid-rigging' among contractors may also serve to reduce any real competition and cost savings in the bidding process.
- Contracting encourages contractors to cut corners and deliver inferior quality in an attempt to maximize profits.
- Contracting lengthens the accountability cycle, thereby making evaluation more difficult. Contracting reduces the local government's ability to make changes rapidly in response to client demands.
- Contracting can reduce the expertise of government employees. Should the contract be terminated, it would be difficult for local government to resume providing the service.
- Contracting with private firms may reduce the number of women and minorities in the workforce. The public sector often has the more successful record of affirmative-action hiring.
- Privatization is anti-union; unionized municipal workforces may be replaced by non-unionized firms. Conservative parties often support privatization efforts as a strategy to undercut what they see as wasteful public sector unionism.

A number of studies have compared the public and private agency service provision in an effort to determine if privatization offers any advantages. Scholars, for instance, have repeatedly found that private garbage collection offers similar or better service at a lower cost than do government agencies. Municipal collection in New York, for instance, costs more as municipal sanitation workers are unionized and receive higher pay and greater fringe benefits. Municipal unions also make it difficult for the city to

implement new productivity measures and cost-cutting techniques.

Yet privatization efforts do not always produce such dramatic gains in efficiency and cost savings in all service areas. Studies, for instance, have not found consistent cost savings when electric power utilities and water utilities are privatized. As utilities often constitute what economists call 'natural monopolies', there is no increased competition when a private monopoly replaces municipal service provision. Similarly, private provision does not always produce impressive gains in the area of social and health services, areas where business-like efficiency measures are not easily introduced. There are also questions as to the appropriateness of the privatization of schooling, where the teaching of a common history and other public values may be lost when parents and their children choose schools on the basis of their private interests.

Initially, it was more conservative forces, including the Republican Party in the United States, Margaret Thatcher's Conservative Party in Britain and the centre-right in France, that led the call for privatization in their efforts to limit government. In Russia and Eastern Europe, privatization was a means to develop economies free of direct government control. In many countries, public sector unions and political parties on the left strenuously opposed privatization efforts. But increasingly, national parties and local managers alike have come to view privatization in non-ideological terms, as a means of making scarce public resources go further. As a result, Tony Blair's 'New Labour' government in Britain has not sought to renationalize every industry that the Conservatives privatized; New Labour even continued the sale of certain public assets. Similarly, in France, the Socialists slowed, but did not stop, privatization.

Further reading

Green, J.D. (2002) *Cities and Privatization: Prospects for the New Century*, Upper Saddle River, NJ: Prentice Hall.

Hodge, G.A. (2000) *Privatization: An International Review of Performance*, Boulder, CO: Westview.

Savas, E.S. (2000) *Privatization and Public-Private Partnerships*, New York: Chatham House Publishers.

BERNARD H. ROSS
MYRON A. LEVINE

PROGRESSIVE PLANNERS

'Progressive planning' is a term that has been applied to many kinds of planning practice in the United States. This is partly because city planning in the United States was institutionalized in the progressive era of the late nineteenth and early twentieth centuries, when middle-class reformers tried to rid government of corruption. However, by the late twentieth century, professional planners in the United States had come to use the designation as an umbrella term for planning with an emphasis on promoting equality and social justice. Generally, progressive planners draw on citizen groups and social movements as sources of ideas, energy and support, and have a strong commitment towards involving the disempowered in planning. This differentiates them from the top-down social reform tradition that imposes reforms from above working through institutions, seeking, for example, to make government more rational and fair. However, progressive planners still acknowledge their own expertise, remaining progressive planners rather than COMMUNITY organizers.

The contemporary form of progressive planning emerged in the 1960s and 1970s when the CIVIL RIGHTS MOVEMENT as well as environmentalism, feminism and popular education inspired a number of planners. While traditional planning had seen itself as serving the public interest, planners in what was to become the progressive tradition came to criticize that position, pointing out that it was often only the interests of the most powerful that were served. In response, practitioners and theorists developed a number of new approaches to planning. ADVOCACY PLANNING came directly from the civil rights movement. In its earliest liberal form, articulated in 1965 by Paul DAVIDOFF, it aimed to give disempowered groups access to the process of planning, with planners acting as advocates in a way analogous to legal advocates in the civil rights movement. Later, advocates learned from critiques from POLITICAL ECONOMY perspectives in the 1960s and 1970s and became more sophisticated about the potential for such liberal advocacy to merely ameliorate the problems of disadvantaged groups. In the 1970s, a number of planners developed refined forms of advocacy with new names, such as EQUITY PLANNING focused on distributional issues, radical planning with roots in both socialist and anarchist

traditions, and feminist planning coming out of the women's movement (see GENDERED SPACES). While initially focused mainly on social and economic inequality, people attracted to these movements also became interested in environmental issues. By the 1990s, 'progressive planning' was probably the most widely used umbrella term for this cluster of approaches in the United States, although these forms of planning are often called advocacy planning because of their historical roots in this planning type.

Power is a central concern in contemporary progressive planning. While aware of political economy critiques of planning as serving the needs of business, property owners and the powerful state, progressive planners have been hopeful about the possibilities for progressive planning to empower people democratically and to promote SOCIAL JUSTICE. Combating racial and ethnic inequality has also been a core philosophy. From the role of planners in setting agendas and shaping attention to providing a setting for mutual learning and collaboration between planners and citizen groups, various types of progressive planners have also been intensely interested in the politics of information, knowledge and communication. This interest in the politics of communication and information has been in part an attempt to deal with planners' lack of political and economic power.

In the 1980s, Pierre Clavel analysed planning under a number of progressive governments in US cities. Hartford, Cleveland, Berkeley and Madison had such governments in the 1970s; Santa Monica, Santa Cruz, Burlington, Chicago and Berkeley in the 1980s. Outside the United States, Toronto City Council and the Greater London Council could be labelled as progressive at times in the late twentieth century. These governments obviously had a range of policies, not all of which were progressive, but they did support a range of non-profit and civic groups, provide housing for low-income groups, emphasize NEIGHBOURHOOD revitalization, promote participatory planning, and advocate for transportation designed to serve low-income residents. Progressive planning has also occurred under ordinary governments when individual planners have used their professional judgments to create progressive policies and projects. Others have acted as more radical 'guerrillas in the bureaucracy', the title of a book from the 1970s. In this situation, progressive planners have worked within governments to redefine the terms of the debate away from the definitions used by powerful government and corporate interests. They have also reallocated those few resources that planners have access to towards investments that have provided more equitable results. Many progressive planners have also chosen to work outside of government with non-profit groups, non-governmental organizations, universities, protest groups and citizen associations.

Progressive planners are represented in the profession in the United States by organizations such as Planners Network, founded by Chester Hartman in 1975, and the earlier Planners for Equal Opportunity. Outside the United States, planning that emphasizes values such as social justice and democratic participation exists in many countries under different names.

Further reading

Clavel, P. (1986) *The Progressive City: Planning and Participation 1969–1984*, New Brunswick, NJ: Rutgers University Press.

Fainstein, S. and Fainstein, S. (1996) 'City planning and political values: an updated view', in S. Campbell and S. Fainstein (eds) *Readings in Planning Theory*, Oxford: Blackwell.

Forester, J. (1989) *Planning in the Face of Power*, Berkeley, CA: University of California Press.

Friedmann, J. (1987) *Planning in the Public Domain: From Knowledge to Action*, Princeton, NJ: Princeton University Press.

ANN FORSYTH

PROPERTY RIGHTS

Property may be tangible, as an automobile or piece of land, or it may be intangible, such as a copyright or stock certificate. A right is a power, privilege or legally enforceable claim. Property rights are thus legally enforceable claims people have in things, tangible or intangible.

Property rights are fundamental to the operation of society. Economists, lawyers and political theorists have all opined on the nature of property rights.

John Locke was an early political theorist who espoused a natural rights view of property. In Locke's philosophy, the right to property is a natural right and precedes, and is not granted by, governmental power. An opposing view holds that property rights are granted by civil authority.

An economic approach to property rights examines the effects of the absence of property rights. The absence of property rights leads people to take as much of a resource, such as a forest, pasture or fishery, as they can for themselves. With no defined rights, no one has an incentive to manage the resource for maximum long-term returns, leading to the depletion or destruction of the resource. Property rights, it is argued, were developed contractually to maximize the value of resources to the COMMUNITY. Governments, which may have arisen by social contract, continued the development of property rights to maximize the welfare of their societies.

In the URBAN context, real estate is one of the most important types of property. Regardless of the debate over the origins of property rights, rights in realty now are typically defined by society or government. Tribal societies define rights and enforce them according to custom, usage and the economic needs of their communities. These systems of tribal property rights are known as 'customary tenure' and still exist today. In civil law countries, rights are defined exclusively by statute. In common law countries such as the United Kingdom, Canada, Australia and the United States, rights are defined both by judicial decisions (the 'common law') and by statutes. The various rights in land, whether established by customary tenure, civil law or the common law, form what is called the tenure structure. Tenure structure may be very simple, and include only such rights as the right to own the surface of the land, the right to use the land, the right of mortgage and the right of lease. Such a structure may be found in countries in transition from socialist economic systems. Tenure structure in common law countries, by contrast, may be complex and include the fee simple, the life estate, future interests, AIR RIGHTS, transferable development rights, time-share rights to resort properties, rights to minerals or to subsets of mineral deposits, as well as mortgage and easement rights. In common law countries, these various rights are sometimes collectively referred to as a 'bundle of sticks', meaning that each stick in the bundle is a separate legal right distinct from the others.

Ownership of property rights implies the right to exclusive possession of the property, the right to use the property, the right to exclude others from the property and the right to sell, lease, devise or otherwise transfer the property. Rights in real estate generally arise by express grant or conveyance. Most tenure systems also provide for acquisition of rights by prescription. This means that by occupying land for a certain period of time and by exhibiting other attributes of ownership, a person may acquire full title over the occupied property. A legal proceeding is usually required to confirm prescriptive acquisition.

Ownership of property rights necessitates a system of notice. Public registration or recording of property rights facilitates the market in those rights and aids in enforcement. Registration or recording is typically a government function, with property records accessible to the public to varying degrees, depending on the laws of the particular country.

Security of tenure is a precondition of economic development. Persons who have clear, enforceable rights in land are more likely to improve their properties and make them more productive. Defining, registering and enforcing land rights is one of the early components in donor strategies to assist developing and transition economies.

The tenure structure affects the spatial development of the city. This was seen most dramatically in the Soviet Union, where rights in land were largely owned by the state. State ownership of rights in real estate precluded development of a land market. With no land market, there were no price signals to guide the use and reuse of urban land. This led to large tracts of urban land being held by obsolete and unproductive industrial enterprises. This unproductive land was not recycled because there were no price incentives to direct reuse. New development thus took place on the urban periphery where transport, INFRASTRUCTURE and other costs were higher. The lack of private property rights led to inefficient cities and contributed to low economic growth.

Capitalist cities, by contrast, face a multiplicity of property rights. When regeneration or REDEVELOPMENT is required, resolution of competing property claims is difficult and increases the cost of the plan. Solutions generally require municipal leadership, PUBLIC–PRIVATE PARTNERSHIP and EXPROPRIATION.

In many developing countries, large segments of the population cannot afford to purchase property rights. This has a direct impact on cities, because these persons often simply occupy land at the periphery of cities and construct housing without the legal right to do so. Often, this housing

is constructed on environmentally sensitive lands, with tragic consequences when hillsides subside and floodplains are inundated. This occupation can in effect lead to title by prescription because governments often simply grant title to these occupiers. Title gives them security of tenure and the resulting incentive to invest in their properties and bring them into the legal economy.

Further reading

Anderson, T. and McChesney, F. (eds) (2003) *Property Rights*, Princeton, NJ: Princeton University Press.
Bertraud, A. and Renaud, B. (1995) *Cities Without Land Markets*, Washington, DC: The World Bank.
Fernandes, E. and Varley, A. (eds) (1998) *Illegal Cities*, London: Zed Books Ltd.

SEE ALSO: expropriation

R. JEROME ANDERSON

PRUITT-IGOE

Pruitt-Igoe was a public housing estate in St. Louis, Missouri, consisting of thirty-three apartment blocks, three of which were demolished on 15 July 1972. One year later, the remaining thirty blocks were also dynamited. The complex, on the south side of the city in an URBAN housing improvement area, was built in 1950–56, to an award-winning design by the architect Minoru Yamasaki. The buildings, each eleven storeys high, stood close to each other on a 57-acre site and accommodated almost exclusively African-Americans.

The project's recreational galleries and skip-stop elevators, once heralded as architectural innovations, became unsafe. Criminality in and around the complex was high and soon took on serious forms. At the beginning of the 1960s, some 10 per cent of the apartments stood empty; as that percentage increased, so did the vandalism. When the problems overwhelmed the St. Louis Public Housing Authority, demolition became the only solution. Demolishing a complex of 2,780 dwellings attracted worldwide attention and in the United States ended the belief that public housing was a successful strategy for solving the housing problems of people on low incomes.

The complex was a laboratory for social scientists. Many projects were conducted there, particularly in the 1960s. In 1963, a team of forty-five social workers was instigated. The University of St. Louis investigated ways in which the environment might be improved. A tenants' self-management project was initiated with Ford Foundation support. Plans were put forward to reduce the number of floors from eleven to three or four, but eventually continuing the management became impossible. Pruitt-Igoe has become an icon of failure in US public housing.

Further reading

Armstrong, H. (Helmuth, Yamasaki & Leinweber) (1951) 'Slum surgery in St. Louis: a new apartment type, patent pending', *Architectural Forum*, April: 128–36.
Meehan, E.J. (1974) *Public Housing Policy: Convention versus Reality*, New Brunswick, NJ: Center for Urban Policy Research, Rutgers University.

HUGO PRIEMUS

PSYCHOGEOGRAPHY

Psychogeography is a method of URBAN investigation created by members of the Situationist International (SI) in Paris of the 1950s and 1960s. The SI was a revolutionary movement influenced by Marxist and anarchist theory, but also by the attitudes and methods of Dadaists and Surrealists. They contended that the transformation of modern cities into centres of capitalism reduced life to mere production and consumption (see CAPITALIST CITY). People suffer extreme alienation, becoming mere spectators of life without any sense of being involved or interacting with one another; life is experienced as a kind of 'spectacle'.

The SI defined 'psychogeography' as 'the study of the precise effects of geographical setting, consciously managed or not, acting directly on the mood and behavior of the individual'. Proposed as a kind of action research, it was a means not only of *investigating* the effects of cities on their inhabitants, but also of *transforming* mental and physical space. In our everyday life, we tend to be unaware of the ways that the city can affect us. These effects need to be exposed and documented to form a basis for resistance and to formulate new ways of living in cities.

Situationists used one main method in their studies of urban geography, that of the 'dérive', defined as 'an experimental mode of behavior linked to the conditions of urban society: a technique for hastily passing through varied environments'. This passing through involved relinquishing our

habitual movements through the environment. By moving through the city with responsiveness to our surroundings, we notice aspects that are normally ignored. Thus, the frame of mind of the psychogeographer falls somewhere between concentration and unconsciousness: 'wanderings that express not subordination to randomness but a complete insubordination to habitual influences'.

As the Situationists carried out their research, wandering the streets of Paris in groups or alone, they would find themselves drawn to some places, repelled by others and even excluded altogether. Insofar as cultural and social forces shape these inclinations, a detailed mapping of these inclinations may reveal such forces at work in the city. The series of psychogeographical maps of Paris produced in the late 1950s were constructed by another SI method, the process of 'détournement', where fragments of existing works (in this case published maps) are taken and rearranged or juxtaposed to produce new meanings. Thus, whereas conventional maps convey an abstract, geometric kind of 'truth' about the urban environment, once they are cut up and rearranged, on the basis of findings from psychogeographical studies, the resulting maps convey an alternative experiential or existential 'truth'. They show experience of space as fragmented and discontinuous; areas *experienced* as distinct are literally pulled apart on the map. The maps simultaneously deconstruct cartography (literally and figuratively) and give graphic expression to psychogeographical surveys.

Further reading

Bonnett, A. (forthcoming) *Psychogeography: The Revolution of Everyday Space and the Escape from the Avant-garde*, London: Continuum Books.
Sandler, S. (1998) *The Situationist City*, Cambridge, MA: MIT Press.

SEE ALSO: Benjamin, Walter; de Certeau, Michel; Lefebvre, Henri

SIMON UNGAR

PUBLIC ART

Public art is commonly defined as including any visual arts practice commissioned for sites of open public access. Traditionally, this took the form of formal statuary and MONUMENTS, though more recently the scope of public art has multiplied as artists explore new media, including light sculpture, installations, hoardings and performance art. This means that traditions of formal, realist and heroic public art have been largely superseded by conceptual, figurative and abstract forms of expression, to the extent that it is sometimes difficult to distinguish between commissioned art and unlicensed works (e.g. GRAFFITI or murals).

It is being 'on display' in a public setting rather than in the private space of the gallery that lends public art its evidential status and power, making it a significant locus in civic rituals. This is particularly the case when public art commemorates an event regarded as significant in the history of a particular city, contributing to what Iris Marion Young refers to as 'collected' rather than 'collective' memory. As a repository of memories and stories, public art thus plays an important role in the making of civic identities. Furthermore, when a COMMUNITY lives with public art for a long period, it may take on new significance as an anchor point in people's COGNITIVE MAPPING of the city. This implies that the ubiquity of public art in Western cities does not in any way undermine its importance in the making of identities. Indeed, it is the very familiarity of public art that renders it significant as a reminder of the 'imagined community' of the city.

Advocates of public art thus suggest that public art offers a means of humanizing the built environment, turning abstract URBAN spaces into meaningful PLACES. Against this, many urban researchers have adopted a critical perspective, suggesting that public art generates ideological effects insofar as it mobilizes meaning in the built environment to sustain relations of domination. The power of public art has certainly not been lost on those in positions of authority, and much research has focused on the role of public art in legitimating specific socio-spatial relations. For example, Sharon Zukin has theorized public art as 'cultural capital', suggesting that public artworks project an official symbolism that represents the city in the image of its cultural elite. Of course, this symbolism is constantly reinvented as different elites seek to write their control into the urban landscape. However, some recent interpretations of public art have stressed that the meaning of public art is not fixed, and that attempts to produce a single official narrative or civic identity are

inevitably contested by popular memories and myths that are fluid and plural. Open to oppositional readings, the intended meaning of public art may therefore be incorporated into an alternative set of place myths that conflict with official narratives, making public art a locus for both power and resistance.

Further reading

Miles, M. (1998) *Art, Space and the City*, London: Routledge.

PHIL HUBBARD

PUBLIC HEALTH SURVEILLANCE

Increases in mass population movements, growth in international travel and trade, and transportation of live animals and animal products have provided opportunities for diseases to spread across international boundaries. An outbreak anywhere in the world must now be treated as a threat to all countries. The global reach of the AIDS pandemic and the appearance of West Nile virus in the United States demonstrate this point. In 1985, the Asian tiger mosquito slipped into the United States inside a shipment of tyres from Asia. Within two years, the mosquitoes, capable of transmitting yellow fever and dengue fever, had established themselves in seventeen states. The global nature of emerging/re-emerging disease and the threat of bioterrorism require international cooperation in disease surveillance.

Surveillance is defined as the ongoing systematic collection, analysis, interpretation and dissemination of descriptive information for monitoring health problems. Surveillance systems are generally called upon to provide information regarding when and where health problems are occurring and who is affected. Basic surveillance functions include case detection, confirmation, reporting, analysis, investigation, response, feedback and monitoring.

Reporting systems are the intelligence network that underpins disease surveillance, control and prevention. Without this framework in place it is impossible to track where diseases are occurring, measure progress in control activities, monitor anti-microbial drug resistance, or provide an early warning system for outbreaks and the emergence of new diseases. Surveillance data are also needed to assess where resources should go for maximum cost-effectiveness.

The surveillance of communicable diseases is concerned with the dynamics of the spread of the disease not only within a country but also from one country to another. In addition to providing necessary information for monitoring diseases and evaluating control measures, global surveillance serves as an early warning system for epidemics and provides the rationale for public health intervention. Early detection of communicable diseases and immediate public health intervention can limit infectious disease morbidity and mortality and minimize negative effects on international travel and trade.

At the global level, surveillance functions are carried out through a framework that links elements of national healthcare systems with various entities, including media channels, non-governmental organizations active in health, and laboratories and other institutions participating in networks focusing on particular diseases and/or regions. The Centers for Disease Control and Prevention is the single largest source of expertise and resources available to the international surveillance and response system. The World Health Organization (WHO), the lead agency in international health with over 190 member states, is in the unique position to coordinate infectious disease surveillance and response at the global level partly through its role in strengthening international surveillance capacity. A global alert system has been developed by WHO which prioritizes surveillance for diseases such as influenza (Flu-Net), rabies (RABNET), AIDS and Creutzfeldt-Jakob disease. Foreign assistance agencies such as the World Bank, as well as private foundations, are important sources of support for strengthening surveillance operations, particularly those taking place in developing countries. The US Department of Defense also contributes to global surveillance.

Further reading

Heymann, D.L. and Rodier, G.R. (1998) 'Global surveillance of communicable diseases', *Emerging Infectious Diseases* 4(3): 62–5.
Teutsch, S.M. and Churchill, R.L. (eds) (2000) *Principles and Practice of Public Health Surveillance*, 2nd edn, Oxford: Oxford University Press.

LOUISE S. GRESHAM

PUBLIC SPACE

Conventionally defined as those spaces that can be freely (and legally) accessed by all citizens, the public spaces of the city consist not only of STREETS, plazas and squares, but also the internal spaces of public buildings such as libraries or town halls. Located between the private spaces of the home and the workplace, such spaces have often been valorized as democratic spaces of congregation and political participation, where marginalized groups can vocalize their rights. The use of such spaces for political protests, parades and pickets emphasizes their political role: historically, it is apparent that revolutionary movements often take shape on the streets, and that public space may become a battleground between the state and law and those seeking to impose a new social order. Examples here are legion, from the French revolutions of the nineteenth century played out on the streets of Paris to the more recent ritualized confrontations between anti-globalization protestors and riot police in Seattle.

Surprisingly, writing on cities has often ignored the seemingly mundane rhythms and rituals of public space in favour of broad analyses of urban POLITICAL ECONOMY. Notable exceptions include Walter BENJAMIN's unfinished Arcades project, which sought to theorize the life of modern cities by examining the minutiae of street life in Paris. Tacking constantly between detailed observations of public space and analysis of socio-economic processes, it was some fifty years after his death before Benjamin's writing began to inspire more inventive ways of 'writing the streets', from the PSYCHOGEOGRAPHIES of Iain Sinclair to the lucid prose of Mike Davis. In addition, there continues to be a vital tradition in environmental psychology and URBAN PLANNING exploring how the design of public space encourages social interaction (or conversely, how design interventions may discourage public interaction and socializing).

Currently, a major preoccupation among URBAN theorists is the putative 'death' of public space, as manifest in the ongoing PRIVATIZATION of urban space. This involves the increasing replacement of shopping streets by private PEDESTRIAN MALLS, the heightened surveillance of public space by CCTV and the emergence of GATED COMMUNITIES. Additionally, it is apparent that there is increased intolerance of 'Other' groups in the public spaces of Western cities, with policies targeting prostitutes, the homeless and teenagers all raising concerns about the inequality of access to the public realm. However, several commentators have conversely asserted that such exclusions are nothing new, and that public space has always been predicated on a narrow notion of CITIZENSHIP that has favoured particular social groups. Either way, studying the social relations played out in public space begins to reveal the power relations that are characteristic of contemporary URBANISM.

Further reading

Lofland, L. (1973) *A World of Strangers: Order and Action in Urban Public Space*, New York: Basic Books.
Sennett, R. (2001) 'Reflections on the public realm', in G. Bridge and S. Watson (eds) *A Companion to the City*, Oxford: Blackwell.

PHIL HUBBARD

PUBLIC–PRIVATE PARTNERSHIPS

Public–private partnerships involve nothing grander than an arrangement between a unit of government and a business that brings better services (e.g. healthcare, education) or improves the city's capacity to operate effectively (e.g. make more housing, build bridges). Common to all such arrangements is the idea that whatever is produced benefits many persons or perhaps the city as a whole and thus can be viewed as serving a broader public good. Though sometimes criticized as way too cosy an arrangement and all too profitable for the businesses in question, whatever is accomplished presumably could not have been undertaken alone by either the business or the government, much less carried off successfully.

Contemporary observers sometimes portray these partnerships as a new and radical departure from the past when businessmen kept to business and government leaders took care of matters that fell within the public domain, which included watching what the businessmen did. The wall between business and government, however, was never very tall or especially difficult to get around. Indeed, had that wall been tougher to scale it is unlikely that cities as we know them today would have been built. Part of the reason why these walls were so easily circumvented, of course, was that commercial and government leaders in early

American cities were often the same persons. They had no difficulty seeing how the public might be served by new business ventures, the jobs they might yield, the taxes they produced and the inflated land values that came with improvements to the town.

The history of such partnerships in the United States extends all the way back to deals struck between businessmen who underwrote the first expenses of colonial settlers and governments that provided groups with royal lands from which all might someday profit. Much more familiar to us today, however, are the kinds of arrangements already alluded to that often brought business-men and government leaders together: dredging harbours and building docks, paving roads, add-ing streetlights, improving the schools, and so forth. The idea behind all such ventures is that of shared risk.

What contemporary critics of public–private partnerships argue, of course, is that these deals turn out a whole lot better for the businessmen than they do for the public. The risk is more on the public side of the equation, in terms of increased taxes and bond indebtedness, than it is on the private side or for the elected officials who push big public projects and have been known to benefit either politically or financially from these 'part-nerships'. That businessmen usually have a great deal invested in whatever venture is undertaken and politicians can come out of these deals looking stupid, or worse, is either overlooked or considered not very important. Much more relevant are the immediate risks and decidedly long-term and rather diffuse benefits that would flow to indivi-dual citizens, even if the partnership accomplishes what it sets out to do.

Notwithstanding the long and often storied history of public–private partnerships in American cities, there certainly have been many instances when the public had to bail out ill-conceived or poorly executed projects. Also contributing to this perception is the fact that many projects under-taken by such partnerships in the last fifty years or so have grown increasingly more expensive and popular, as public officials have sought new ways to offload the provision of services to private vendors.

The character and substance of the current debate over these partnerships was probably set in the years immediately following the conclusion of the Second World War. That was when the US government started underwriting much of the upfront expense associated with massive rebuilding campaigns in many older and rundown inner-city commercial and residential districts. URBAN renewal projects did not exactly break the bank; but they did draw immediate and serious attention to the kinds of problems that critics of public–private partnerships talk about.

Federal money, lots of it, was spent buying up property, clearing land and providing for INFRA-STRUCTURE improvements in the hope that private developers would build something better suited to the demands of a modern, post-industrial city and economy. The money came with many federal strings – programmes, regulations and bureau-crats – and little apparent accountability. Many low-income and minority households were dis-placed and the land often stayed empty for a number of years before any businesses stepped forward to build on the vacant property. The result in many cities was both hugely expensive and unsightly.

Subsequent initiatives to rebuild cities were still dependent on private initiatives and artfully craf-ted partnerships with government agencies. How-ever, a lot more emphasis was put on rehabilitating standing buildings and lining up developers to work on the property and businesses to fill it afterwards before the project went forward. The results, even when large corporations were in-volved in the reclamation process, were decidedly better. Companies had better-looking and safer NEIGHBOURHOODS to be located in and residents had rehabilitated apartments or houses to rent and buy. In a few notable and studied cases, the resi-dential population of neighbourhoods reclaimed this way turned out to be more racially and eco-nomically mixed *after* they were redeveloped than they had been before.

The lesson to be drawn from this brief history lesson is not that public–private partnerships are inherently good or bad, only that they have been an important feature of city-building for a long time and are not going away. More importantly, perhaps, these partnerships speak to an even older problem in cities, one that the ancient Greeks worried about a great deal. That was how to bal-ance the rights assigned to each individual with his larger public obligations. In the case of American cities, one finds a rich (and understandably chequered) history in the use of such partnerships but an even more important contribution to our discourse over the rights and obligations of

wealthier persons and larger and more powerful corporations to the commonweal.

Further reading

Clavel, P. (1986) *The Progressive City*, New Brunswick, NJ: Rutgers University Press.

Monti, D. (1990) *Race, Development, and the New Company Town*, Albany, NY: State University of New York Press.

Tabb, W.K. and Sawers, L. (eds) (1984) *Marxism and the Metropolis*, New York: Oxford University Press.

DANIEL J. MONTI Jr

Q

QUARTERED CITY

The metaphor of the quartered city, introduced by Peter Marcuse (1989), apparently serves a dual purpose. One is to establish the power of a metaphor in the analysis of URBAN development by exploiting its inherent ambiguity. The other is to underline the increasing social and geographical complexity of cities in the post-Fordist era. The method of representing social phenomena through metaphor has gained widespread currency in social science writing that is strong on interpretation rather than rigorous description and analysis. The way Marcuse applies this method has helped erode the concept of the dual city, which had long been used to highlight conflicting trends in urban society. The opposition inherent in the dual city concept is expressed in sources as disparate as a novel on social conditions by Charles Dickens, *A Tale of Two Cities* (1960 [1859]), and an early social survey, *How the Other Half Lives*, by Jacob RIIS (1997 [1890]), both of which had a profound effect on the social reform movement of the time. The dual city concept has also been used to describe the effects of colonialism and the representation of the COLONIAL CITY where Western administrations constructed their own compounds and settlements alongside the indigenous centres. And lately, the concept has been applied to express opposing tendencies in urban societies as they adopt the contemporary communications technology associated with economic restructuring, to the benefit of the well educated and to the detriment of those whose skills have been made redundant (CASTELLS 1989).

By introducing an alternative concept – the *quartered* city – Marcuse (1989) rejects the dual city metaphor on the grounds that most people in cities are neither very rich nor very poor but somewhere in between. He distinguishes new social categories, which make up an increasingly fragmented society, and places them in specific geographic and symbolic locations. These locations – quarters of the city – are separated by walls of various kinds (Marcuse 1995): for example, the intimidating signs of status; the fences, security guards, and rules and regulations of 'GATED COMMUNITIES'; and distance as an acknowledgement that geographic remoteness and lack of access can constitute a significant barrier. The quarter where he locates the powerful members of the business elite is the socially impenetrable 'CITADEL'; the new professional class is placed in the gentrified quarters of the city; and the partly racially and/or ethnically defined group of 'losers' is situated in the 'GHETTO'. At the same time, the middle and working classes are located in tenement quarters and bland urban NEIGHBOURHOODS but also in socially and economically circumscribed suburban settlements, contingent upon the preferred lifestyles of the residents, the constraints of their employment situation, and the transportation landscape that determines the pattern of accessibility.

The quartered city is thus a multifarious representation – metaphorical and real – of the socio-economically and geographically fragmented state of contemporary URBAN society under conditions of neo-liberal regulation by market forces.

Further reading

Castells, M. (1989) *The Informational City: Information Technology, Economic Restructuring, and the Urban-Regional Process*, Oxford: Blackwell.

Dickens, C. (1960 [1859]) *A Tale of Two Cities*, Signet Classic, New York: New American Library.

Marcuse, P. (1989) 'Dual city: a muddy metaphor for a quartered city', *International Journal of Urban and Regional Research* 13(14): 697–708.

—— (1995) 'Not chaos but walls: post-modernism and the partitioned city', in S. Watson and K. Gibson (eds) *Postmodern Cities and Spaces*, Oxford: Blackwell, pp. 243–53.

Riis, J.A. (1997 [1890]) *How the Other Half Lives: Studies Among the Tenements of New York*, Penguin Classics, New York: Penguin Putnam.

JAN VAN WEESEP

R

RADIANT CITY

The radiant city constitutes one of the most influential and controversial URBAN doctrines of European MODERNISM. The term was introduced by LE CORBUSIER in the early 1930s and found a systematic exposition in his book of the same title. Commonly regarded as the culmination of Le Corbusier's urban thinking, *La Ville Radieuse* (1935) put forward a radical vision for a machine-age society, in which the resources liberated by modern techniques were harnessed towards a new realm of human freedom.

The radiant city was conceived of as a universally applicable model. Its principles were formalized in a ZONING plan based on the rigorous distribution of functions across a territory. A cluster of cruciform office towers, housed in skyscrapers built on pilotis, was situated at one extreme of the plan, while the industrial area lay at the opposite end. The central sector was occupied by public facilities, flanked on both sides by ribbons of high-rise apartment blocks, also raised above the ground. These housing units, designed as 'recessed' buildings, were immersed in landscaped PARKS so as to allow residents direct access to leisure and sporting activities outside their homes. In response to the perceived individualism of the GARDEN CITY, Le Corbusier proposed a model of urban concentration in which the ground surface was put to maximum collective use. The residential quarter (also called 'green city') was based on an egalitarian provision of domestic space for all inhabitants and on the cooperative sharing of communal services.

This generic scheme parcelled out activities in parallel bands following an expandable linear grid.

A multi-level traffic system ensured high-speed circulation for subway trains and automobiles, which were kept strictly separated from pedestrian movement. By declaring the demise of corridor streets and central COURTYARDS, Le Corbusier envisioned a sanitized environment in which the basic human needs – which he identified as sun, space and greenery – would be ultimately satisfied.

The social UTOPIA embodied by this theoretical model is inseparable from its underlying political vision. Le Corbusier's attempt to achieve a synthesis between individual freedom and authoritarian rule reflected his adhesion to the hierarchical principles of French syndicalism. Hence, while the radiant city was planned as a setting for a classless society, its implementation required a strong political authority capable of executing the plan from above.

Le Corbusier proposed a series of projects for specific cities (most notably Algiers, Antwerp, Barcelona, Nemours and Stockholm), with a view to persuading political leaders and the general public alike of the viability of his ideas. Although none of these schemes was ever executed, the radiant city model exerted a significant impact on URBAN PLANNING worldwide, especially in the post-war period. It equally attracted wide criticism, particularly for its allegiance to a totalitarian notion of URBANISM threatening to erase the historical stratification of places and the vital role of urban diversity.

Further reading

Fishman, R. (1977) *Urban Utopias in the Twentieth Century*, New York: Basic Books.

Le Corbusier (1967) *The Radiant City* (*La Ville Radieuse*), trans. P. Knight, E. Levieux and D. Coltman, London: Faber & Faber.

DAVIDE DERIU

RAPOPORT, AMOS

b. 28 March 1929, Warsaw, Poland

Key works

(1969) *House, Form and Culture*, Englewood Cliffs, NJ: Prentice Hall.
(1977) *Human Aspects of Urban Form: Towards a Man-Environment Approach to Urban Form and Design*, Oxford: Pergamon Press.

Leading researcher in anthropology, environmental cognition and behavioural studies as applied to architecture, planning and URBAN DESIGN, Rapoport's work is widely published and influential in many countries. He was one of the founders and is an active member of the Environmental Design Research Association, has held honourable and visiting positions in many universities around the globe, and is professor emeritus at the University of Wisconsin-Milwaukee.

Rapoport argues strongly for a design process that is based on rigorous investigation, and believes that advancements in the design disciplines can only happen through the development of theory and applied research. His first major book, *House, Form and Culture*, pioneered the study of meaning and cultural issues in design, showing that housing is to be understood as an expression of the individual as part of a larger social context. His views were to be further developed in *Human Aspects of Urban Form*, a seminal book that marked generations of designers in the late 1970s who were turning away from modernist tenets towards a more humanistic and contextual approach. His current research interests include vernacular design, traditional environments, and the use of history and precedents.

Further reading

Rapoport, A. (1990) *History and Precedent in Environmental Design*, New York: Plenum Press.
—— (2000) 'Science, explanatory theory and environment-behavior studies', in S. Wapner, J. Demick, T. Yamamoto and H. Minami (eds) *Theoretical Perspectives in Environment-Behavior Research*, New York: Plenum Press.

VICENTE DEL RIO

REDEVELOPMENT

Redevelopment represents a process of land development used to revitalize the physical, economic and social fabric of URBAN space. Redevelopment efforts generally target areas in cities where one or more of the following conditions are present: (1) land is underutilized; (2) the built environment (buildings and INFRASTRUCTURE) is deteriorating; and (3) economic opportunities are limited. Redevelopment occurs in cities throughout the world and reflects the goals of physical and economic revitalization. The policies, laws and priorities of redevelopment, however, vary among countries, and among states or provinces within countries. The United States has a long and evolved tradition of redevelopment and, therefore, serves as a good example of the process.

Today, local governments in the United States lead redevelopment efforts. However, the federal government motivated much of the redevelopment activity in cities from the 1930s to the 1960s. The roots of redevelopment can be traced back to the New Deal and the federal government's programmes aimed at improving housing conditions and eliminating SLUMS. In the 1940s, a few states followed the federal lead by adopting redevelopment legislation and working with private redevelopment agencies to develop and rehabilitate housing.

The federal government initiated large-scale, urban redevelopment with the passage of the Housing Act of 1949. This Act began an era of urban renewal that would substantially change some urban areas and create an administrative infrastructure for the redevelopment process. Three Housing Acts, passed in 1954, 1956 and 1959, addressed some aspect of the urban renewal process including rehabilitation of slum areas, relocation payments for displaced individuals and businesses, and BLIGHT analysis and planning requirements for localities. Also, this legislation created mortgage insurance programmes through the Federal Housing Administration that assisted with relocation of displaced residents as well as with rehabilitation and new construction in blighted areas.

The early 1960s witnessed the passage of additional federal legislation in support of urban renewal. Congress authorized increased funding for urban renewal and made adjustments to existing policies to facilitate renewal efforts.

The goals of urban renewal, the eradication of blight and slums in cities, resulted in the displacement of many persons and the obliteration of close-knit NEIGHBOURHOODS in some areas. For this reason, the federally funded renewal era has been criticized by many urban scholars. While renewal may be recalled as a destructive period in urban history, it provided the basis for current redevelopment efforts in cities.

In the current era, redevelopment typically occurs in older sections of a city. Since these areas reflect the needs, constraints and land use regulations, or lack of them, at the time of initial development, older urban areas may become functionally obsolete over time; parcels of land may be oddly shaped or too small under the current ZONING to develop effectively. These areas may be aesthetically unattractive with eclectic development styles and patterns involving original uses and structures alongside newer developments. The combination of land underutilization and irregular development can result in a blighted area.

Cities use the redevelopment process to revitalize blighted areas. A typical redevelopment plan aims to bring new development to these areas. To do so, the city often must acquire private property, demolish existing structures, relocate residents and assemble land suitable for new development projects. The process of land assemblage begins with the city identifying property owners in a redevelopment area and attempting to purchase the requisite properties at market value. In some cases, this process proceeds without complications. However, in other cases, property owners may wish to retain their land or may reject the price offered by the city for their land. For this reason, redevelopment projects often involve the use of EMINENT DOMAIN. Eminent domain, a power given to cities through state enabling legislation, allows cities to acquire property from a private owner by paying just compensation or fair market value for the property. The use of eminent domain, however, can result in political costs to a city's elected officials. Therefore, some cities will negotiate with a recalcitrant property owner in order to avoid a public confrontation.

Redevelopment plans may face several other implementation obstacles. First, relocation of businesses and residents can be problematic. Resistance to displacement and relocation may be significant in some neighbourhoods, although generous relocation benefits typically are paid to displaced parties.

Secondly, the demolition of older buildings may be challenged by citizens concerned with HISTORIC PRESERVATION. Redevelopment plans may integrate historic buildings or their façades into new development projects to circumvent this type of challenge. Finally, environmental problems may prohibit redevelopment from proceeding in some areas. Redevelopment project areas may contain BROWNFIELDS, sites with environmental contamination, that are very costly to clean up and prepare for new development.

Cities can experience significant costs when undertaking redevelopment projects. Federal dollars for major redevelopment activities in the United States disappeared decades ago, leaving cities in need of other sources of funding. Some states allow for a portion of the local property tax to be redirected to redevelopment activities. TAX INCREMENT FINANCING can provide a stream of funds to redevelopment agencies that can be spent directly on projects or used to pay back bonds issued by the redevelopment agency.

The emphasis of redevelopment shifted during the twentieth century in the United States. Current efforts in cities emphasize local economic development, unlike early federal efforts with a focus on housing conditions. Urban fiscal stress and a recession in the 1980s prompted cities to identify new approaches to raising revenues and boosting the local economy. For some cities, redevelopment provided a powerful mechanism for increasing property taxes, sales taxes and employment rates through development of underutilized land and by providing attractive incentives such as a fully assembled site to businesses willing to locate in their communities.

The specifics of redevelopment vary across the United States and reflect differences in state legislation. Moreover, redevelopment can serve many goals depending on a city's priorities. However, the fundamental concept of redevelopment, the physical and economic improvement of a previously developed, but currently blighted, urban area, is reflected in all redevelopment efforts in US cities as well as urban areas throughout the world.

Further reading

Gotham, K.F. (ed.) (2001) *Critical Perspectives on Urban Redevelopment*, Amsterdam, NY: JAI Press.

VICTORIA BASOLO

RED-LIGHT DISTRICTS

While prostitution has never been a solely URBAN phenomenon, it is in towns and cities that it has generally been most visible. This visibility is particularly obvious when sex work has been concentrated in red-light districts, areas mythologized as spaces of sexual opportunity, deviation and experimentation. In most cases, these areas are particularly associated with female street prostitution, though in some instances, they are also characterized by an agglomeration of 'adult-oriented' businesses, sex clubs, theatres and peep shows. In some cities, these areas may coincide with spaces of male prostitution and gay venues, though the visibility of these in the landscape is often less pronounced.

The concentration of 'vice' and prostitution in specific areas has long fascinated urban sociologists, with the pioneering work of the CHICAGO SCHOOL OF SOCIOLOGY including several detailed ethnographies of the lifestyles of those occupying these areas of 'immorality' and deviance. More recent work has suggested that these areas cannot be understood merely as the outcome of supply and demand economics, but as the outcome of a complex interaction of moral codes, legal strictures and policing practices that encourage the containment of vice in inner-city areas away from whiter, wealthier suburban populations (who, ironically, are the principal clients of sex workers). However, recent efforts to clean up long-established vice areas, together with the tendency for clients to contact sex workers via the INTERNET and mobile phone, mean that red-light districts are becoming less numerous in Western cities.

Further reading

Hubbard, P.J. (1999) *Sex and the City: Prostitution in the Urban West*, London: Routledge.

PHIL HUBBARD

REDLINING

Redlining is the explicit spatial definition of areas or NEIGHBOURHOODS where credit, investment or other services are less available than in other neighbourhoods. Banks may discriminate against applications for credit in a redlined neighbourhood, regardless of the characteristics of the applicant. Loans may be denied at a higher rate in redlined neighbourhoods than in comparable neighbourhoods, or they may be available only at a higher price. In the United States, redlining is closely associated with race. Neighbourhoods with high proportions of minority residents are more likely to be redlined than other neighbourhoods with similar household incomes, housing age and type, and other determinants of risk, but different racial composition. Redlining is closely related to DISINVESTMENT.

The term originated in the United States, reputedly in the practices of the Federal Housing Administration (FHA), a federal agency established to provide mortgage insurance and expand home ownership. The FHA followed an explicit practice of identifying neighbourhoods that were 'good' and 'bad' business risks; neighbourhoods with a red line around them were ones with high proportions of minority residents which, according to the FHA, would result in declining property values and were thus bad risks. The conventional finance industry followed the lead of the FHA, and remained reluctant to make loans or insure property in some areas long after the FHA had explicitly changed its racially based policies.

The extent of bank redlining in the United States became clear when data were first collected on the location of loans originated, under the Home Mortgage Disclosure Act (HMDA) of 1976. The evidence of bank neglect of some neighbourhoods led to new legislation the following year, the Community Reinvestment Act (CRA). The CRA established an examination procedure for federally insured banks, to ensure that they were satisfying their charter obligations to meet the legitimate credit needs of the communities in which they did business. Banks that were systematically redlining neighbourhoods could be denied permission to merge or expand. For some time, banks were able to continue redlining because the data were limited only to loans they originated, and they could claim there was little demand for loans in other locations. However, extensions to the HMDA in 1989 included data on all loans applied for in each census tract, as well as the race, gender and income of the applicant. The new data revealed that mortgage redlining was widespread in many cities, and provided the basis for much stronger enforcement of CRA requirements.

COMMUNITY advocates have also examined whether other services, such as home insurance, show evidence of redlining. Advocates have used the concept of redlining to argue for its opposite – affirmatively targeting resources to neglected neighbourhoods, or 'greenlining'. Another variant on the concept is 'brownlining' – the difficulty that prospective developers of BROWNFIELD sites face in assembling financing for the project.

Further reading

Squires, G.D. (ed.) (1992) *From Redlining to Reinvestment: Community Responses to Urban Disinvestment*, Philadelphia, PA: Temple University Press (see esp. ch. 1).

Yinger, J. (1995) *Closed Doors, Opportunities Lost: The Continuing Costs of Housing Discrimination*, New York: Russell Sage Foundation.

HEATHER MACDONALD

REGIONAL PLANNING

Regional planning basically covers an area broader than a city, but smaller than a country as a whole. As Peter HALL (2002) has noted, there are two types of regional planning: national/regional planning and regional/local planning. According to his definition, the large-scale, economic type of planning is called national/regional planning, because it relates the development of each region to the progress of the national economy. And the small-scale, physical type of planning can be called regional/local planning, because it attempts to relate the development of a whole URBAN region to each local part of it. National/regional planning is also called top-down planning because higher-level governments, such as central or regional governments, are responsible, whereas regional/local planning is called bottom-up planning because city governments or combined organizations of local governments are in charge.

Since physical planning is generally considered to be the concern of local governments, national/regional planning has been losing its essential role in planning. Today, the Netherlands and Japan are exceptional cases where national planning is periodically made to reflect the direction of nationwide development. On the other hand, regional/local planning is considered necessary because many

urban activities are carried out beyond municipal jurisdictions. The plans made beyond municipal administrative boundaries to cover substantial living areas are called regional plans. Regional planning can be comprehensive by covering various subjects, but it more often specifies a particular subject, which requires region-wide consideration. The following are typical subjects of regional planning:

- *Transportation and telecommunications.* People, goods and information travel beyond municipal jurisdictions. The area in which daily journeys to work, school and the shops take place is important as a unit of regional transportation planning. This is especially true for transportation-related organizations, such as public works departments, railway companies, bus companies and taxi companies. Today, a rapid broadband telecommunications network has become a more important subject for regional planning, as telecommunications begin to complement transportation or even replace it. Private companies play major roles in telecommunications fields, but local governments may work in order to coordinate the various demands of users with the services supplied by telecommunications-related companies.
- *Water supply, sewers and sewage treatment, and solid waste disposal.* A water supply system should be designed with topography and hydrology in mind. When the populations of each municipality are small, a multi-municipal system becomes more efficient and economical, for example for water supply or sewage treatment. This is especially true in a recycling system, in which used materials are reused, reproduced and recycled. The locations of the plants and the areas to be covered must be considered in order to make the entire system efficient and effective.
- *Pollution prevention.* Air pollutants are disseminated over a wide area from the place of origin. Therefore, air quality management must be conducted in a region where the causes and effects are monitored and managed.
- *Disaster prevention.* Some natural disasters cause a lot of damage in a city. Serious damage, however, tends to occur in rather narrow areas, while light damage can be spread out over a large area. Therefore, regional planning can be effective in promoting mutual support through

agreements whereby lightly damaged municipalities help heavily damaged ones by dispatching personnel or sending goods when a serious disaster occurs.

- PARKS, *outdoor recreation and* OPEN SPACE. Because population densities and land values in the core of a region are vastly greater than on the periphery, it makes sense for outlying areas to provide more of their share of open space and parkland. This arrangement does not obviate the need for such efforts at the municipal level, but it does make a case for multi-jurisdictional efforts as well.

- *Economic development.* Business activities are conducted beyond municipal boundaries, and attract employees living in areas where people can commute to the central business district or other workplaces. And their markets also develop towards regional, national or even international areas. From a regional strategy point of view, if the region unites for this purpose, it can sell itself as a single entity to the rest of the world rather than expend funds on competition between municipalities that are all part of the same regional LABOUR MARKET. Thus, some zero-sum game activity can be avoided. A regional rather than a municipal approach may also achieve marketing ECONOMIES OF SCALE, for example in having a single development office represent the entire region. For better cooperation among the local municipalities in the region, however, the local taxation system needs to be arranged on a cooperative basis to realize the economic development of the region as a whole.

- *Housing.* Housing and land use policies in one municipality affect housing prices, rents and vacancy rates in the entire region. Employment growth in one municipality affects housing demand in many other municipalities. Thus, one can argue that housing is a regional as well as a local issue. In some countries, population growth is still regarded as an important target of municipal policies. Thus, cooperation among the municipalities to create a regionally harmonized housing policy is indispensable.

- *NIMBY and NIMTO facilities.* 'NOT IN MY BACK-YARD' and 'not in my term of office' syndromes are worldwide tendencies that local people are likely to adopt. Local people are often opposed to the construction of certain types of facilities in their vicinities, although they well understand their necessity. Some officials may be inclined to put off their decision-making regarding such facilities, or some of the local people may be obviously opposed to them. Eventually, necessary facilities or policies are not implemented in both cases. Taking care of the minority's opinion results in neglecting the majority's interests. Regional planning can be useful in finding better solutions in those cases since the choices on the location of such disputed facilities will increase when the sites are selected from within a broader area. Furthermore, each local municipality may accept a smaller number of such facilities instead of asking other local municipalities to accept the others through regional cooperation.

Further reading

Hall, P. (2002) *Urban and Regional Planning*, London: Routledge.

TAKASHI ONISHI

RENAISSANCE CITY

A Renaissance city was designed under principles of order and unity generated by a renewed interest in the classic ideals of Greece and Rome. This movement was started in Italy at the beginning of the fifteenth century, and then extended to other countries during the eighteenth century. It involves a conscious arrangement of buildings into a predetermined, often geometric form. It represents one aspect of a complex process of cultural, social and religious transformation. This movement is easily followed in art and landscape design. However, some city historians believe that the Renaissance city does not exist.

It is clear that there was a desire to create new cities or to improve existing ones by using the new principles. At the time of the Renaissance, many of the cities were fully developed and there was little opportunity to build new cities. However, designers were inspired by the recently discovered writings of the Roman architect Vitruvius (first century AD). Several designers proposed models of ideal cities using geometrical and symmetrical forms that also accommodated the need for elaborate defensive systems. For example, Antonio

Averlino (1404–72), known as Filarete, proposed Sforzinda (1457–64), a city shaped as a geometric star. One of the few cities constructed under these classical ideas is Palma Nova (1593), a military garrison in northern Italy. It is attributed to Vincenzo Scamozzi (1522–1616). It has a hexagonal central square, radial streets (only some of which have access to the outside) and a complex defensive system. If there was an opportunity to build a new area, it was developed into a unified grid design, particularly in residential districts.

The straight, Renaissance street, designed as a whole project, is a clear departure from the narrow, organic medieval street. The earliest example is the Via Nuova in Genoa (1470). As cities grew in population, it became difficult to accommodate larger populations within the existing medieval fabric. Streets were widened to facilitate circulation, link important structures and connect regional routes.

The Renaissance public square is illustrated by the Piazza de la Annunziata I in Florence (1419–1516). It was created over time by three designers who followed similar organizing concepts and produced a unified design (see also BAROQUE CITY). In addition to this square, the city of Florence, as the cradle of the Renaissance, received many other improvements such as an updated pedestrian system and the treatment of façades to bring unity and order to the city. Spain, who had few orderly cities before the Conquest, brought a desire for orderly plans to the New World which reflected Renaissance ideals (see LAW OF THE INDIES). It also adopted these ideals in its geometric and unified *plaza mayores*, or central squares, which were carved within the medieval cities.

Further reading

Argan, G. (1969) *The Renaissance City*, New York: George Braziller.
Morris, A.E.J. (1994) *History of Urban Form*, 3rd edn, New York: John Wiley & Sons, pp. 157–90.

ANA MARIA CABANILLAS WHITAKER

RENT CONTROL

Rent control has been defined as a policy designed to protect tenants from the high market rents which otherwise would result from a shortage in the supply of rented housing (Robinson 1979: 76). A better, more general definition would be a policy designed to determine rents and/or rent development. The concept also includes policies in which controlled rents differ little from market rents.

Rent control, and rent freezing in particular, is probably the earliest consciously applied instrument of housing costs policy in America and Western Europe. Certainly in the Middle Ages, and possibly even in classical Rome, rent control was imposed. Rent freezing is like an emergency brake applied by a government finding itself in dire straits. The First and Second World Wars precipitated such circumstances. Many dwellings were destroyed, the building industry lay dormant for years, and in a very short time the housing shortage became immense. Concern was expressed that tenants should not be obliged to suffer from landlords charging excessive rent increases.

One argument for rent control was the desire to prevent landlords from making excessive profits. The major political support for rent control probably rests on fundamentally ambiguous views towards profit, privately owned wealth and the 'social good'. Many consider that rent *ought* to be related to costs. This assertion is also a normative statement, going back to the medieval principle of *iustum pretium* (Thomas Aquinas).

A second motive for the imposition of rent control lies outside housing. There was apprehension concerning the upward pressure on labour costs resulting from scarcity prices in the rental sector (less concern was expressed over the owner-occupied sector). The resurgence of industry and strengthening of the international competitive position benefited from low rent levels. In many Western European countries, the relationship between rent control and wages policy played an important part, especially after 1945.

A third motive, one that could well be cited as the cause of the rebirth of rent control in many countries in the 1970s, is the endeavour to curb inflation. Low rents made it easier to keep wages low, and low wages meant low prices.

Rent freezing was only acceptable as a *temporary* measure. The liberalization of rents remained the intention. The following motives for rent control in Western Europe can be listed, many of which have been used to justify not abolishing the instrument rather than as arguments supporting its introduction:

1 Stimulation of general housing demand, thereby raising the consumption of housing

services ('the general housing consumption goal').

2   Keeping rents low for newly built housing to stabilize the costs and volume of the construction market ('the construction market stabilization goal').

3   Prevention of a potentially enormous siphoning off of the income and capital of tenants to house owners ('the general income-distribution goal') and the promotion of a fairer distribution of real income between tenants ('the inter-tenant income-distribution goal').

4   Enabling low-income households (in particular those with children) to compete on the housing market with other categories and acquire a larger share of the housing consumption than would have been the case without rent control ('the housing consumption distribution goal').

5   Directing the total demand towards large, well-appointed dwellings ('the housing-demand composition goal').

6   Restraining cost inflation ('the general anti-inflation goal').

It may be noted that objectives 1, 2, 4 and 5 are housing objectives, while objectives 3 and 6 lie largely outside housing.

Some see rent control as a point of action around which tenants can organize themselves. The value of rent control, apart from the progressive income transfers it effectuates, is that it is an immediate gut issue around which people can organize. As they work to improve their own housing conditions, their consciousness is raised about the working of the housing system.

Rent control can refer to (a) rent freezing, so that housing circumstances are 'fossilized'; or (b) rent increases set by the government, so that rent levels (probably) remain lower than the market rent.

In New York, the term 'rent stabilization' was reserved for (b) (because it was introduced in 1969 with the Rent Stabilization Law). In Great Britain, this type of rent control was referred to as 'rent regulation', aimed at the attainment of fair rents. Rent stabilization and rent regulation are characterized by more flexible programmes than the more orthodox forms of rent control. Making general comments about rent control policy is difficult, since the programmes differ so widely.

Public authorities defend rent control, referring to social advantages and positive impacts,

although in practice negative effects appear to be numerous. On the demand side of the housing market, processes operate redistributing prosperity between tenants, between tenants and landlords, and between tenants and owner-occupiers; not all of these outcomes are either justified or intended. Demand for housing is stimulated by rent control, which encourages uneconomic use of the housing stock. Purchasing power outside housing is increased. Rent control can reduce mobility on the housing and LABOUR MARKETS, so that tenants in the regulated sector tend to hang on to their apartments.

On the supply side of the housing market, a capricious distribution of prosperity between tenants and landlords is created. New building initiatives are discouraged, so that housing shortages continue. Housing improvement is neglected, and poor maintenance generates DISINVESTMENTS. Eventually, a flight of capital can take place, including REDLINING, abandoned properties and accelerated demolition. In the course of time, many private landlords withdraw, bringing about a negative selection. The remaining landlords are invariably guilty of tenant selection and DISCRIMINATION. The reaction to this state of affairs is often a call for municipalization. Black markets are created, key money and excessive transfer costs demanded. A switch to the free rent sectors takes place (rented furnished accommodation, for example) and from rented dwellings to home ownership.

Marks (1984) argues that rent control is rarely straightforward. It is virtually impossible for a government to control and regulate the entire supply of a commodity. Once a shortage appears, alternative markets and black markets arise. A government may criminalize these markets and prosecute suppliers in a draconian fashion. Or, a government may tolerate these markets to relieve shortages. Governments may deliberately leave a portion of the market untouched by regulation to serve as a safety value for excess demand. This unregulated portion of an otherwise regulated market becomes the 'shadow market'. Here prices will be driven above their normal market value.

Long-term, substantial rent control leads to a distorted rent pattern with little correlation between rent and quality, poor use of the rental housing stock, and persistent shortages – reasons enough for the continuation of rent control and the development of policy satellites: object subsidies, subject subsidies and the distribution of

housing accommodation. Over the whole line, prosperity falls, municipal finances are eroded and bureaucracy flourishes.

Most empirical information on the consequences of rent control refers to the reaction of private landlords, particularly in the United States. Researchers have not properly established the relationship between rent control and non-profit rental housing. Where government rent regulation seeks to imitate ideal markets and social housing providers dominate – as is the case in the Netherlands – the impact of rent policy may be more positive than has been observed in American housing markets.

Further reading

Block, W. and Olsen, E. (eds) (1981) *Rent Control: Myths and Realities*, Vancouver: Fraser Institute.

Brenner, J.F. and Franklin, H.M. (1977) *Rent Control in North America and Four European Countries*, Washington, DC: Potomac Institute.

Marks, D. (1984) 'The effects of partial-coverage rent control on the price and quantity of rental housing', *Journal of Urban Economics* 16: 360–9.

Nevitt, A.A. (1970) 'The nature of rent-controlling legislation in the United Kingdom', *Environment and Planning* 2: 127–36.

Robinson, R. (1979) *Housing Economics and Public Policy*, London: Macmillan.

HUGO PRIEMUS

RESTORATIVE JUSTICE

CRIME is taking a heavy toll on contemporary cities. A major challenge is the development of effective responses to crime and the building of safe communities.

The dominant response is a retributive criminal justice system, whereby crime is reduced to legal questions of guilt and punishment. 'Get tough' policies and the use of mass imprisonment in the United States epitomize this approach. The United States, in its war on crime, incarcerates over 2 million people; half of those are non-violent offenders. Most individuals who are processed by the criminal justice system are poor and/or an ethnic minority. For example, African-Americans constitute 13 per cent of the general population, yet they make up nearly 50 per cent of prison inmates.

The current justice system does not resolve crime and does not address the needs of victims and communities. A slow and inefficient adversarial battle takes place between the offender and the state. Large numbers of offenders are removed from city NEIGHBOURHOODS, only to return with a stigma at a future date. Offenders and communities are not provided with sufficient resources to overcome the root causes of crime.

Money to cover the spiralling costs for the war on crime is taken from education, social services and healthcare budgets. Communities and victims are disempowered, while fear of crime and crime rates remain high. Many URBAN neighbourhoods as well as the overburdened justice system are in a state of crisis.

Restorative justice offers an alternative paradigm for responding to crime and achieving justice. Restorative justice begins with the understanding that crime is not simply lawbreaking. Crime is an offence against a real victim rather than an offence against the state. Crime harms victims and communities; it is a violation of people and relationships. Violations create an obligation to make things right.

A restorative process creates an opportunity for victims, offenders and communities of interest to come together to discuss the wrongdoing and develop a plan of action for resolution. Through this process, offenders are held accountable and take action to repair the harm. The various dimensions of harm to victims and communities are addressed, such as financial loss, physical injury, emotional suffering, property damage, broken relationships, etc. Attention is also given to uncovering root causes of criminal behaviour and building offender competencies in areas such as education or employment.

Rather than permanently stigmatizing and ostracizing offenders, offenders who take responsibility are positively reintegrated into their communities. Note that reintegration can also apply to victims since they may feel disconnected or even stigmatized as well. Victims and offenders remain valuable members of our city neighbourhoods.

Restorative justice takes the need for safety of city residents seriously. To reduce crime and recidivism, offenders are assessed and responded to according to their individual situation, including their risk to others. Offenders are monitored in terms of fulfilment of any agreements. While the use of incarceration should be limited, certain offenders require secure custody when they pose too great a threat to others. Restorative values can,

nonetheless, be applied in any setting, including prison. Public safety, especially in the long term, is ultimately achieved by building the capacity of communities to control and prevent crime.

Crime is a COMMUNITY problem. Restorative justice strengthens communities by advocating problem-solving between victims, offenders, families, friends, community members, non-governmental organizations and government officials. Justice requires a collaborative effort.

Transformation in the roles and responsibilities of individuals, communities and government is another core feature of restorative justice. Authority is decentralized and partnerships are created to solve crime problems and to achieve effective crime prevention. Government is responsible for preserving a just order and the community for establishing just peace, while individuals are collectively responsible for creating cooperative community relations. There is a commitment to a broader transformation of cities to overcome inequities and increase the overall quality of life for all (see SOCIAL JUSTICE).

Restorative justice is thus not a programme but rather a different way to understand and approach crime. Since the early 1990s, there has been a growing global movement towards restorative justice. Various programmes have been implemented in numerous cities around the world, such as family group conferencing, victim–offender mediation, circle sentencing and community reparative boards. A board of community volunteers in Vermont, United States, meets with a woman who admits she was drinking and driving. A police officer in Australia facilitates a conference with a youth who stole a car; his mother and the victim participate. In Canada, a judge sits in a circle with a large representation of the community to discuss what to do about a man convicted of assault.

A particular project can be deemed restorative based on its commitment to the principles of the paradigm. Following the Tenth United Nations Congress on the Prevention of Crime and the Treatment of Offenders in Vienna in 2000, there has been further international discussion and endorsement of basic principles on restorative justice.

The restorative paradigm and its practical implementation remain in the infancy of development. As shown in this brief overview, restorative justice is a process as well as an outcome. Restorative justice addresses the injustices of crime. Victims, offenders and communities are given the opportunity to come together as respected participants in the process. Justice is about accountability, reparation and healing. Restorative justice provides a sound path towards healthier and safer cities.

Further reading

Bazemore, G. and Schiff, M. (eds) (2001) *Restorative Community Justice: Repairing Harm and Transforming Communities*, Cincinnati, OH: Anderson Publishing.

Braithwaite, J. (1989) *Crime, Shame and Reintegration*, Cambridge: Cambridge University Press.

Galaway, B. and Hudson, J. (eds) (1996) *Restorative Justice: International Perspectives*, Monsey, NY: Criminal Justice Press.

Van Ness, D. and Heetderks Strong, K. (1997) *Restoring Justice*, Cincinnati, OH: Anderson Publishing.

EVELYN ZELLERER

RIBBON DEVELOPMENT

No better metaphor could be used to describe the long and narrow lines of one-plot-deep development fronting onto roads radiating from the outer boundaries of cities into the surrounding countryside. In many countries, the resulting ribbons of development are clearly visible on land use maps and aerial photographs, giving cities and the countryside a particular character. Most towns start out as linear developments along a short length of road, but ribbon development refers to the unchecked linear growth of roadside development largely unrelated to any centre.

The attractiveness of ribbon development for developers is that it can save the costs incurred in providing land, roads and INFRASTRUCTURE in estate layouts. For homeowners and businesses it provides direct access to the main highway and in some countries it may provide access to both city and agricultural employment for different family members. The general idea has been taken forward as a positive approach to planning the 'LINEAR CITY'. One example is the 1910 linear SUBURB proposal of Soria y Mata in Madrid. But in most developed countries, ribbon development is used in a pejorative sense. It is generally thought of as a particular facet of the general problem of SUBURBANIZATION and urban SPRAWL. It damages the landscape value of the city's HINTERLAND, it exacerbates road CONGESTION and creates danger for road users by mixing local and through traffic,

and it will often be poor access to infrastructure and services.

Ribbon development became a feature of the massive URBAN expansion during the early part of the twentieth century in developed countries and was encouraged by the exponential growth in car ownership which reduced the value of CENTRAL CITY locations. It later became a main feature of the urban pattern in developing countries. Opposition to this form of development (as part of urban sprawl generally) was a major factor in the introduction of city planning. In the United Kingdom, early action was taken through the Restriction of Ribbon Development Act 1935. This gave powers to the highway authority (rather than the planning authority) to prevent development along the main highways. The controls have continued in different forms with the highway authority still able to prevent development where access to a main road is required. The results were dramatic and examples of ribbon development in most of the United Kingdom today date from the 1930s.

This is in marked contrast to experience in other countries. In Europe, for example, Belgium has extensive ribbon development which was only brought under control towards the end of the twentieth century. In the United States, attitudes to ribbon development are generally more relaxed, with ribbon or strip development being the 'natural' form of many cities in the Midwest. Nevertheless, some states have successfully restricted this form of development through urban limit lines and other means.

VINCENT NADIN

RIIS, JACOB

b. 3 May 1849, Ribe, Denmark;
d. 26 May 1914, Barrie, Massachusetts, United States

Key works

(1890) *How the Other Half Lives: Studies Among the Tenements of New York*, New York: Charles Scribner's Sons.
(1902) *The Battle with the Slum*, New York: Macmillan & Co.

Danish-born journalist, photographer and activist, Jacob Augustus Riis contributed significantly to the cause of URBAN reform in America at the turn of the twentieth century. An immigrant himself who experienced the hardship of homelessness in New York, Riis engaged in a lifelong battle for SOCIAL JUSTICE fuelled by deep-rooted Protestant ethics. During his career as a police reporter, he pioneered a 'muckraking' style of investigation aimed at exposing the social and political shortcomings that tacitly allowed the spread of POVERTY, unhealthiness and squalor in the SLUMS.

Riis's published work owed its vast success to the compelling evidence presented through texts, statistics and pictures. With the help of newly developed flash-lighting, in the late 1880s Riis embraced photography to depict the darkest corners of the city. This unprecedented use of snapshots, which earned him the reputation of America's first journalist-photographer, exerted an enormous influence upon American social reportage. However, the picturesque taste underlying Riis's illustrations, often coloured by racial and ethnic stereotypes, has also been criticized for turning the slums into the stage of an urban spectacle.

His first book, *How the Other Half Lives* (1890), contained a vivid portrayal of New York's urban poor. By describing and contrasting the characteristics of different immigrant communities, Riis mapped a social reality that had largely gone unreported. His reportage lent moral support to the reformist movement that demanded, alongside changes in sanitary law, the urgent remodelling of existing TENEMENTS and the construction of new model housing.

Riis's first book contained most of the themes he was to investigate in subsequent years. Its sequel, *The Children of the Poor* (1892), documented in particular the problems that beset the child population of the East Side tenements, and proved instrumental to the enactment of child labour regulations. Riis continued to deal with the plight of the urban dispossessed in further writings, which consistently advocated immediate practical measures. His work, which stirred American public opinion towards a greater awareness of slum conditions, contributed to the demolition of insalubrious tenements and the provision of such facilities as playgrounds and neighbourhood PARKS.

Riis lectured publicly to a broad range of religious, political and scientific institutions. His speeches, usually illustrated with 'magic lantern' slides, became the central focus of Riis's civic mission after he gave up journalism in 1901 (the year in which his autobiography, *The Making of an*

American, was published). Among Riis's admirers was Theodore Roosevelt, who collaborated with him as a police commissioner in several investigations, and for whom Riis wrote a celebratory biography on the occasion of the 1904 presidential elections.

Further reading

Fried, L.F. and Fierst, J. (1977) *Jacob A. Riis: A Reference Guide*, Boston, MA: G.K. Hall & Co.

Lane, J.B. (1974) *Jacob A. Riis and the American City*, Port Washington, NY: Kennikat Press.

DAVIDE DERIU

RIOTS

Riots have been associated with everything from NEIGHBOURHOOD brawls, rowdy street gatherings, crowds, and panics to violent protests over food prices, communal attacks against immigrants or deviant persons, and better-organized fights between political parties or workers and their bosses' hired goons. Widely disparaged for their destructive power and thought to be an ineffective way for groups to register their grievances, rioting has often been lumped with many other unattractive aspects of life in modern URBAN settings.

Many nineteenth-century theorists and public officials made an explicit connection between cities and various expressions of destructive behaviour by individuals and groups. There was something about cities that either made otherwise rational persons behave in violent and seemingly irrational ways (i.e. so-called 'contagion' theories of crowd behaviour) or brought people together who were disposed to acting out violently without much provocation (i.e. so-called 'convergence' theories). Laid out systematically in the work of 'crowd psychologists' like Gustav LeBon, the idea was that crowd violence or rioting was a spontaneous outburst that violated conventional standards of right thinking and customary ways of acting.

In truth, riots actually have a long and storied history in a variety of communal settings. Whether to express their displeasure with changes overtaking them or to show their support for traditional ways of doing things, people have long used relatively brief but violent attacks against persons and property to make a point about things they do and do not like about the way their world works.

More often than not, the message conveyed by this kind of violence was taken seriously. No doubt that was part of the reason why established leaders usually portrayed riots and other challenges to their authority in such unflattering ways.

There have been moments, of course, when crowd behaviour was not viewed so negatively. Riots and violent street protests against the British prior to the American Revolution, to cite one particularly obvious and patriotic example, were deemed necessary and effective then and are seen in no less favourable a light today. Yet these acts stand out as notable exceptions in an otherwise long and sometimes quite nasty history of conflicts between people belonging to different races or ethnic groups or between citizens and representatives of public officials like the police. Looked at in this way, civil disorders occur outside of conventional political and social routines and should be treated as aberrations of an otherwise stable, if fluid, social system.

Attempts to 'rehabilitate' acts of civil unrest in the eyes of the public were made by many social scientists and some public figures during the 1960s and early 1970s when protests against racial SEGREGATION generated a lot of sympathy for civil rights activists. Some writers thought that civil unrest, and even the GHETTO riots of those years, could be viewed as legitimate expressions of protest against unjust rules and customs. Violence and disorder in this case ushered in new and decidedly better ways for people to act (i.e. a so-called 'emergent norm' theory of collective behaviour).

The problem with all these explanations, of course, is that they missed the most important feature of rioting and other forms of civil unrest. Most of the time the acts did not permanently upset prevailing routines and institutions. Nor did the persons who initiated them have much of a plan to make over the world with their disruptive and sometimes violent actions. Viewed in a broader historical context, intermittent outbursts of crowd violence were far more likely to re-enforce prevailing standards for right conduct and existing institutional routines than they were to change them.

Some American social historians would be sympathetic with this point of view, noting that our legacy of collective violence is both closely tied to routine political life and more conservative than its European counterpart. Still, they too have found it difficult to relate the late twentieth-century

rioting that took place in many inner cities to this conservative tradition.

There are several ways to make this connection clearer. First, we would note that it has only been in the last eighty years or so that persons from working lower classes gained anything like a monopoly over the use and timing of rioting and experimented with it as a means of helping themselves acquire new economic and political resources. Even then, however, their behaviour has reflected many lessons passed down from a time when civil disorders were led by persons with more prestige and clout or had the tacit support of such persons. Specifically, violence was much more likely to be directed at 'outsiders' and take place inside the confines of their own neighbourhoods than it was against really wealthy and powerful persons or the symbols of their wealth and power wherever they could be found.

Secondly, most violence continues to be launched from a 'communal' base rather than an 'associational' one. If not unorganized, civil unrest remains outside the normal repertoire of special interest groups that are better organized like labour unions and political parties. Almost completely absent as well from rioting and virtually all of the well-organized, armed and funded street gangs of the late twentieth century is any hint of a political agenda, radical or otherwise.

Thirdly, and finally, even if rioting has lost most of its moral force in urban politics, its strategic importance has not diminished. What little we know about the use of violence in relation to other means of raising issues or redressing grievances indicates that violence probably speeds up the introduction of new reforms and promotes the introduction of a new group to the polity. It is more likely to be used as a last resort than a frivolous gesture, demonstrating that the conflict being waged has reached some symbolic crest. This affords antagonists with an opportunity to reassess their positions and to begin the delicate process of making conciliatory gestures to one another. More organized and politically sophisticated types of violence, like terrorism, typically do not.

Further reading

Brown, R.M. (1975) *Strain of Violence: Historical Studies of American Violence and Vigilantism*, New York: Oxford University Press.

Lipsky, M. and Olson, D.J. (1977) *Commission Politics: The Processing of Racial Crisis in America*, New Brunswick, NJ: Transaction Books.
Richards, L.L. (1970) *'Gentlemen of Property and Standing': Anti-Abolition Mobs in Jacksonian America*, New York: Oxford University Press.
Tager, J. (2001) *Boston Riots: Three Centuries of Social Violence*, Boston, MA: Northeastern University Press.

DANIEL J. MONTI Jr

ROBINSON, CHARLES MULFORD

b. 30 April 1869, Ramapo, New York; d. 30 December 1917, Albany, New York

Key works

(1901) *The Improvement of Towns and Cities*, New York: G.P. Putnam's Sons.
(1903) *Modern Civic Art*, New York: G.P. Putnam's Sons.

Robinson advocated aesthetic URBAN improvement. A former Rochester journalist, he became the first professor of civic design in the United States at the University of Illinois in 1913. Travelling to Europe from the 1890s, he reported on civic betterment for progressive journals like *Atlantic Monthly*. His first book became a touchstone for improver groups unifying their work under the banner of 'the CITY BEAUTIFUL'. His second book fashioned a broader planning perspective. His last book moved towards pragmatic concerns: *The Width and Arrangement of Streets* (1911). Robinson had the greatest influence as a missionary for urban beautification. He generally eschewed monumentalism in favour of more realizable efforts adapted to local conditions within the framework of a harmonious general plan. Little things mattered: trim SIDEWALKS, decorative lighting, pocket PARKS, tree-lined streets. As a self-styled 'civic adviser', he prepared plans for over twenty mostly small to mid-sized American cities. Brief and boosterish, these relied on words not images to stir the public imagination. Robinson's significance in planning history rests on his internationally informed publicity for civic art, his networking of populist groups, and his pioneering roles as planning practitioner and educator.

Further reading

Peterson, J. (2003) *The Birth of City Planning in the United States, 1840–1917*, Baltimore: Johns Hopkins University Press.

ROBERT FREESTONE

ROSSI, ALDO

b. 3 May 1931, Milan, Italy;
d. 4 September 1997, Milan, Italy

Key works

(1987) Hotel Il Palazzo, Fukuoka, Japan.
(1992) Bonnefanten Museum, Maastricht, the Netherlands.

Italian architect, artist, teacher and theorist, Aldo Rossi achieved distinction all over the world. He was one of the leading exponents of the post-modern movement. He graduated in architecture at the Milan Polytechnic in 1959 and joined the Italian magazine *Casabella-Continuità*, serving later as its editor until 1964. From 1965, he taught at the universities of Milan, Zurich, New York and Venice.

In 1966, Rossi published *L'architettura della città* (*The Architecture of the City*), an important treatise about concepts of URBAN morphology, establishing himself as one of the greatest architectural theorists. This book was translated into several languages and became a reference point for international architectural culture.

Aldo Rossi imagined the city as an organism consisting of many assembled parts; these parts, broken down in simple forms as cubes, cylinders, prisms, columns or cones, are reassembled to form the architectural thing. Rossi's projects are founded on a repetitive use of simple forms and we can always see these architectural elements differently combined in his works.

Rossi's works are closely related to his writings. In his book *A Scientific Autobiography* (1981), based on notebooks written since 1971, he describes how his personal history influences and mixes with his architectural projects. In 1971, he had an automobile accident that brought about a change in his life, ending his youth. While he was in the hospital, as a patient, from meditations on death and skeletons sprang the project for the San Cataldo Cemetery in Modena. In this work, Aldo Rossi wanted to define the ideal city of the dead.

Among Rossi's first projects we find the war monument in Segrate (1965); the Gallaratese residential complex in the outskirts of Milan (1969); the student housing in Chieti (1973); and the Theatre of the World in Venice (1979), a temporary floating theatre built for the Venice Biennale. But Rossi built all over the world: the Pocono Pines House in Pennsylvania represents one of the first buildings he completed in the United States; in 1987, the Toronto Lighthouse Theatre on the banks of Lake Ontario, Canada, was realized; and in the same year the Hotel Il Palazzo was built in Fukuoka, Japan.

He participated in the Internationale Bauausstellung in 1987 with a project in Berlin, winning a special first prize. Many important projects were carried out in Italy in 1988. One year later, Rossi won the international competition for the design of the Deutschen Historischen Museum in Berlin, and in 1990 in Venice he was awarded the Pritzker International Prize, the highest honour for an architect. Among his last projects were the Bonnefanten Museum in Maastricht, the Netherlands (1992), and the Quartier Schutzenstrasse in Berlin, Germany (1996).

Further reading

Ferlenga, A. (2002) *Aldo Rossi: The Life and Works of an Architect*, London: Konemann.
Rossi, A. (1984) *The Architecture of the City*, trans. D. Ghirardo, Cambridge, MA: MIT Press.

RITA POMPOSINI

ROUSE, JAMES

b. 26 April 1914, Easton, Maryland;
d. 9 April 1996, Columbia, Maryland

Key works

(1967) Columbia, Maryland.
(1978) Faneuil Hall Marketplace, Boston.
(1980) Harborplace, Baltimore.
Founder, Enterprise Foundation.

American entrepreneur, developer and philanthropist, James Rouse pioneered in developing both new forms of URBAN development and fresh approaches to urban problems. In 1995, he received the US Presidential Medal of Freedom, the nation's highest civilian award, for his lifetime achievements.

After working his way through law school, Rouse partnered with a colleague to form a mortgage banking company. By 1954, he had transformed the firm into The Rouse Company, a real estate development firm that opened one of the first enclosed SHOPPING MALLS, Harundale Mall, in 1958 – the first of many commercial projects that established Rouse as one of the nation's leading developers. However, serious decay in the DOWNTOWN core of Baltimore and cities like it drew Rouse's attention back to the city. His pioneering Faneuil Hall Marketplace, opened in Boston in 1978, and Baltimore's Harbortown, opened in 1980, created the genre of the FESTIVAL MARKETPLACE: a novel approach to REDEVELOPMENT intended to create a new urban image and anchor broader downtown revitalization.

Despite the success of The Rouse Company, Rouse was dismayed by escalating urban decay and suburban SPRAWL, both of which he viewed as destructive of the human spirit. His civic efforts starting in the early 1950s to promote comprehensive, context-sensitive NEIGHBOURHOOD redevelopment in Baltimore had failed, so Rouse took a new tack. In 1963, seven years before federal legislation sought to stimulate the development of new towns, The Rouse Company began buying land in the Baltimore–Washington corridor to build Columbia, Maryland. Opened in 1967 and built out over the next three decades, Columbia extended the GARDEN CITY tradition that preserved green space and strategically located commercial centres to encourage sociability. Rouse's cherished goal of economic integration proved elusive despite inclusion of more multi-family housing than other new towns of the period. However, his insistence that racial DISCRIMINATION be banned from Columbia had a striking impact, and Columbia remains distinctive for its stable racial balance. Never incorporated, it also served as a model of the HOMEOWNERS'-ASSOCIATION-as-local-government form later used by GATED COMMUNITIES to achieve very different goals. Columbia is widely regarded as one of the most successful of the nation's new towns.

In his later years, Rouse refocused in earnest on urban POVERTY. In 1972, members of his church sought his help in rehabilitating two decrepit but occupied apartment buildings in Washington DC. The project's success led to the formation of Jubilee Housing, a COMMUNITY DEVELOPMENT CORPORATION committed to providing affordable housing and social services to poor households. Inspired by this model, in 1982 Rouse founded the Enterprise Foundation, a national financial and technical assistance intermediary that helped to support the dramatic expansion of the COMMUNITY development movement over the next two decades. Rouse described the foundation's work as the most important of his life.

Further reading

Gillette, H., Jr (1999) 'Assessing James Rouse's role in American city planning', *Journal of the American Planning Association* 65(2): 150–67.
Sachs, A. (1996) 'Marking the milestones of James Rouse's career', *Columbia Flier*, 11 April, pp. 40–3.

AVIS C. VIDAL

ROWE, COLIN FREDRICK

b. 27 March 1920, Rotherham, Yorkshire, England; d. 5 November 1999, Arlington, Virginia, United States

Key works

(1976) *The Mathematics of the Ideal Villa and Other Essays*, Cambridge, MA: MIT Press.
(1978) *Collage City*, Cambridge, MA: MIT Press (with F. Koetter).

Architecture historian and teacher Colin Rowe studied at the Liverpool School of Architecture and, under Rudolph Wittkower, at the Warburg Institute in London. Rowe taught at Yale University, the University of Texas in Austin, and at Cambridge University before joining the architecture faculty at Cornell University in 1962. He remained at Cornell for 28 years. His scholarship and teaching affected both architectural history and design education. Many of Rowe's students became internationally prominent architects, extending his influence throughout the architecture and planning professions.

Rowe was associated with two ideas central to late twentieth-century architecture and URBAN DESIGN theory: formalism and contextualism. Beginning in the 1940s, Rowe championed understanding architecture through the analysis of its visual form rather than through historic, social or political factors. His arguments influenced a 1960s group of architects, dubbed the 'Whites', that

included Richard Meier and Charles Gwathmey. Rowe later became interested in the URBAN context's relationship to building design, an issue he addressed in *Collage City*, written with Fred Koetter. Rowe's and Koetter's call for a retreat from MODERNISM's architecture and planning ideals and their positive evaluation of traditional city elements fundamentally influenced the NEW URBANISM movement.

HARRISON HIGGINS

RYKWERT, JOSEPH

b. 5 April 1926, Warsaw, Poland

Key works

(1963) *The Idea of a Town: The Anthropology of Urban Form in Rome, Italy, and the Ancient World*, Princeton, NJ: Princeton University Press.
(1972) *On Adam's House in Paradise: The Idea of the Primitive Hut in Architectural History*, New York: Museum of Modern Art.

Joseph Rykwert has taught the history and theory of architecture at several institutions in Europe and North America. His longest academic association has been with the University of Pennsylvania, where he was named professor of architecture in 1988 and where he is currently professor emeritus. Rykwert has been recognized for the range of his research, which includes the history of gardens, the architecture and URBANISM of CLASSICISM in Italy and England, and modernist architecture and urbanism. Rykwert's writing has consistently focused on relating the city's physical and cultural aspects. A critic of rationalist, economic and organic interpretations of building and URBAN FORM, Rykwert has argued for architecture's and URBAN DESIGN's engagement with narrative understandings of place observed in poetry and myth. His publications include *The Idea of a Town* (1963); *On Adam's House in Paradise* (1972); *The Dancing Column* (1996); and *The Seduction of Place* (2000). With Robert Tavernor and Neil Leach, he translated ALBERTI's architecture treatise, *On the Art of Building in Ten Books* (1989).

Further reading

Dodds, G. and Tavernor, R. (eds) (2002) *Body and Building: Essays on the Changing Relation of Body and Architecture*, Cambridge, MA: MIT Press.

HARRISON HIGGINS

S

SAFE CITY

A safe city regards safety as a primary consideration in all of its practices. In a safe city, one thinks about such things as the layout of a NEIGHBOURHOOD, the design of a bus shelter and the nature of a needle exchange programme with regard to the safety of all residents but especially vulnerable groups. These groups include women, people of different races and ethnicities, gays, lesbians and transgendered people, elderly people, and people with disabilities, among others.

Safety is multifaceted in that it can refer to falling victim to CRIME or to physical hazards such as snow and ice on SIDEWALKS. Safety is also a perception. One does not have to be physically harmed in order to live in fear. Thus, a safe city approach includes reference not just to crime statistics but also to the fears that people have. Avoiding taking public transportation because one is afraid of being stalked is as important in this regard as a reported assault in a subway station.

Safety in cities is often assessed with safety audits that focus on lighting, sightlines, signage, vegetation, proper maintenance and the accompanying perceptions of auditors. This is also a useful way of involving representatives of vulnerable groups in that they are usually invited to help conduct the audits. The results of a safety audit can be addressed using Crime Prevention Through Environmental Design, self-defence training, litter collection and other initiatives aimed at reducing crime and hazards, as well as the risks and perceptions thereof.

Further reading

Wekerle, G. and Whitzman, C. (1995) *Safe Cities: Guidelines for Planning, Design, and Management*, New York: Van Nostrand Reinhold.

SUE HENDLER

SANDERCOCK, LEONIE

b. 1949, Adelaide, Australia

Key works

(1998) *Towards Cosmopolis: Planning for Multicultural Cities*, Chichester: John Wiley & Sons.
(ed.) (1998) *Making the Invisible Visible: Insurgent Planning Histories*, Berkeley, CA: University of California Press.

A significant international academic whose work spans the interdisciplinary fields of URBAN STUDIES, URBAN policy and planning, and is characterized by a sustained concern to elucidate issues of difference, SOCIAL JUSTICE and possibility. In 1970, Sandercock completed an honours degree in arts at the University of Adelaide, before undertaking her doctorate in the Urban Research Unit at the Australian National University in Canberra. Her first book, *Cities for Sale: Property, Politics and Urban Planning in Australia* (1975), was drawn from that work (see *Public Participation in Planning*, from 1975, and *The Land Racket*, from 1979). Between 1981 and 1986, she was foundation professor of urban studies at Macquarie University in Sydney, publishing – among other pieces – *Urban*

Political Economy: The Australian Case (1983), with Mike Berry.

During the period to 1996, Sandercock then pursued screenwriting in Hollywood (she holds an MA in fine art in screenwriting from the University of California in Los Angeles); one of her screenplays was produced as Movie of the Week and there are nine others to her credit. In this time she was also professor in URBAN PLANNING also at the University of California, Los Angeles. Returning to Australia in 1996, Sandercock was appointed professor and chair of the Department of Landscape, Environment and Planning at the Royal Melbourne Institute of Technology, moving to the Faculty of Architecture, Building and Planning at the University of Melbourne in the late 1990s. In July 2001, Sandercock was appointed professor in the University of British Columbia's School of Community and REGIONAL PLANNING, where she teaches and undertakes research on planning theory and history, and social policy.

Sandercock's interests include the multicultural; GLOBAL CITY regions; issues of GOVERNANCE in 'CITIES OF DIFFERENCE'; participation, democracy, information and communication; fear, the other and the city; new models of planning; storytelling in planning theory and practice; and urban governance. Each strand of her work advances alternatives to modernist planning theory (see MODERNISM) and demonstrates her capacity to synthesize across the borderlands that comprise planning practice and real experiences of the city. These novel and productive associations – drawn from POSTMODERNISM, post-colonialism and feminism – also inform her agenda for planning education in the twenty-first century, explained in an appendix to *Towards Cosmopolis*. Drawing on John FRIEDMANN's ideas about the core curriculum in planning and its relationship to the 'life space' which is inhabited by all, Sandercock proposes six macro-social processes that are central to planners' concerns: URBANIZATION; regional and inter-regional economic growth and change; city-building; cultural differentiation and change; the transformation of nature; and urban politics and empowerment. Given these, she has advanced a shift in emphasis in planning education from methods and skills to literacies – technical, analytical, multicultural, ecological, design and ethical.

ELAINE STRATFORD

SASSEN, SASKIA

b. 5 November 1949, The Hague, the Netherlands

Key works

(1991) *The Global City: New York, London and Tokyo*, Princeton: Princeton University Press (new edn 2001).
(1998) *Globalization and Its Discontents: Selected Essays 1984–1998*, New York: The New Press.

Saskia Sassen is currently the Ralph Lewis Professor of Sociology at the University of Chicago. She is the author of eight books and many articles.

Sassen is an internationally well-known scholar who has published widely on the processes of globalization and their impacts on the city and the state. Her studies focus on power and the materiality of power on place. Sassen argues that the city of today emerged as a strategic site for the new type of operations introduced with the globalization of the economy. The city becomes a nexus, where new claims materialize and assume concrete forms. These formations bring about a new geography of politics by linking cities and altering the way politics used to be. This type of politics is able to contain non-formal political actors, too.

Sassen argues that the global economy materialized on a network of cities constituting a new economic geography of centrality. Although this centrality continues to reproduce the existing inequalities, it also brings about new dynamics of economic growth. Centrality occurs in the major international financial and business centres, starting with New York, London and Tokyo, and including many cities of both the developed and the developing world. These are the GLOBAL CITIES that concentrate a disproportionate share of global corporate power and are the key sites for its overvalorization. Global cities are the hubs of international financial and business communities with the command functions they contain. They are the post-industrial production sites for the leading industries of the global era – finance and specialized services – and are the national and/or transnational marketplaces for these industries' products.

Sassen argues that global cities contain marginality together with centrality because they concentrate the disadvantaged as well and are key sites for their devalorization. Labour has a distinctive place in this marginality. On the one hand, there

grows a highly sophisticated professional class with very high wages serving in the finance sector and producer services. On the other hand, there is a faster growth in the low-wage LABOUR MARKET. This demand of the global economy is generally met by immigrants and women with their cheap labour force by transforming the labour market relations dramatically. When combined with other factors, the inclusion of immigrants and women in the labour market increases informality in the URBAN economy. Sassen sees informalization as the low-cost equivalent of deregulation at the top of the system, a way of producing and distributing goods and services at a lower cost and with greater flexibility.

In her work *Guest and Aliens* (1999), Sassen examines IMMIGRATION patterns in Europe in the last three centuries. She presents how the historical patterns of immigration can illuminate contemporary policy-making processes. By doing this, she suggests that policy-makers will have the opportunity to widen their options in dealing with the problem rather than merely closing up all the gates to the migrants as a response to the growing xenophobia and/or racism.

MURAT CEMAL YALCINTAN

SATELLITE TOWNS

Satellite towns possess a certain functional autonomy (public administration, commerce, industry, etc.), but depend on a nearby major city for development of economic activities. They are influenced by the lure of employment opportunities and diverse leisure-time pursuits. The establishment of a satellite town is linked to the construction of a dense communication network of highways and trains that serve as a connection to the larger metropolis. Its origins can be found in Western Europe in the 1940s within the broader scope of URBAN PLANNING and regional proposals set down by social democratic governments with the aim of curbing the excessive growth of some URBAN centres, as had been detected by the Barlow Commission for London in 1940. These towns either develop in areas with a pre-existing population or, on occasion, are planned and created in vacant areas. The terms 'new town' and 'satellite town' are frequently used interchangeably,

given the coincidences in their origins and characteristics.

The first examples can be found in Great Britain, where the so-called new towns were planned, influenced to a large degree by the precepts of Ebenezer HOWARD and some of the principal ideas of the GARDEN CITY. The Greater London Plan of 1944 established maximum limits for population density (250 inhabitants per hectare) within the County of London. As a consequence, it was necessary to relocate more than a million people that were residing within the inner city of London to ten new towns, which were to be located outside a permanent GREEN BELT that surrounded the constructed area. When the Labour Party came to power, new and important urban legislation was passed: the New Town Act (1946), and the Town and Country Planning Act (1947). The latter established the independent and autonomous character of these new towns; more than mere satellite towns, these new urban centres were located between 40 and 50 kilometres from London and, following Howard's approach, reduced population density limits were set for each one (between 20,000 and 60,000 inhabitants). Between 1946 and 1951, the government approved the creation of fourteen satellite towns, and seven more were created in the following years. With time, British satellite towns evolved, and were classified in three distinct generations. Of the fourteen new towns of the first generation (1946–51), eight were created with the aim of diverting industry and population from London. In general, they have a low demographic density and are composed of three distinct zones: the urban centre, the industrial sector and the residential area, which, in turn, is subdivided into NEIGHBOURHOOD units characterized by low density (65 inhabitants per hectare) and a maximum total population of less than 10,000 inhabitants. Of this group, the pioneer towns of Stevenage (1946) and Harlow (1947) could be mentioned. The second generation (1951–60) opted for a higher density and a conceptual reorientation of the principle of a compact town. The urban centre was strengthened, yet understood as a bi-centric nucleus, which should both concentrate the principal collective activities and be accessible to the residential area. A maximum limit of ten minutes was set for the amount of time it should take to walk from the urban centre to the residential area. The towns of Hook and New Cumbernauld are the only examples of

this generation. Important methodological changes were introduced in the plans for satellite towns of the third generation: the models for URBAN DESIGN were diversified, including a system for adding self-sufficient neighbourhoods to the model of the compact town; the size of the population was increased; occasionally pre-existing communities were included; an increase in private initiative was projected, etc. Towns of note from this phase are Runcorn (1964–65) and Milton Keynes (1968–71).

This idea of REGIONAL PLANNING which originated in Britain spread to numerous other nations within a few years. The Scandinavians adopted much of the ideological, technical and formal content of the system of territorial re-equilibrium, as in the case of the well-known general plan of Stockholm (1945–46) directed by Steven Markelius. His project was based on the complementary ideas of a high-density urban centre and the construction of a crown of satellite towns around the capital, namely Vällingby (1950–55), Farsta (1952–59), Skärholmen (1966), Tenska-Rinkeby (1975) and Norra Järvafältet (1977). According to his plan, each one of these towns, which were never conceived of as entirely self-sufficient, should have a population of up to 50,000 inhabitants, and be composed of three or four neighbourhoods grouped around a metro station. Thirty minutes was allotted for commuting to the centre of Stockholm. In France, the construction of 'villes nouvelles' began in the mid-1960s. The regional plan for Paris of 1965 proposed the decongestion of the capital through the creation of five new towns that were to be distributed along the principal communication arteries. Unlike in previous examples, these satellite towns were conceived of as important population centres (500,000 inhabitants), and were a monument to URBANISM. Equally important were the attempts to reduce urban CONGESTION in Finland (Tapiola, the first satellite town of Helsinki), in Belgium (Charleroi), and in the USSR. In a radius of 70 kilometres around St Petersburg, almost forty satellite towns or 'sputniks' were constructed. In the United States, as a consequence of the regional plan 'Year 2000' for Washington DC (1961), two satellite towns were created in the HINTERLAND of the national capital: Reston and Columbia.

Within the new towns, other towns with a lesser degree of self-sufficiency and less closely linked to the politics of DECENTRALIZATION were sometimes incorporated: the so-called bedroom communities. Physically, these communities are dominated by high-rise structures and monotonous architectural prototypes of poor-quality construction and socially they are characterized by a homogeneous population of blue-collar young married couples and have high population densities. The initial lack of INFRASTRUCTURE in these communities was gradually corrected over time; however, due to the low level of endogenous employment, workers were obliged to commute to nearby urban or industrial centres. By the year 2000, bedroom communities were present around all of the major cities of the world.

Further reading

Galantay, E.Y. (1975) *New Towns: Antiquity to the Present*, New York: George Braziller, Inc.

Gravagnuolo, B. (1991) *La obra progettazione urbana in Europa, 1750–1960*, Rome: Gius. Laterza & Figli.

Grupo ADUAR (2000) *Diccionario de geografía urbana, urbanismo y ordenación del territorio*, Barcelona: Ariel.

Rodwin, L. (1956) *The British New Towns Policy*, Cambridge, MA: Harvard University Press.

JESÚS M. GONZÁLEZ PÉREZ
RUBÉN C. LOIS GONZÁLEZ

SCIENCE PARKS

The term 'science park' has been coined to describe the planned development of high-quality buildings and business services in a bucolic setting, serving high-technology research and production and generally located in proximity to a leading university. While it is often used interchangeably with 'technology park', 'research park', 'innovation centre' and the like, the differences between them are practically indistinguishable. Both the public and the private sectors are engaged in science park development, from the stage of inception through to day-to-day management. Public sector involvement is often based on a view of the science park as an instrument for improving the local economy through promoting employment, increasing the local tax base and encouraging local linkages between on- and off-park firms. The science park is also considered an important link in facilitating local technology transfer between a university and high-technology companies.

Additionally, the science park is often perceived as contributing to national economic development, stimulating the formation of new high-technology firms, attracting foreign investment and promoting exports. Private sector involvement is generally limited to those science park projects anticipated to develop into high-yield, prestigious real estate developments, invariably in favourable locations.

The archetypal and most successful science parks such as the Stanford Research Park (United States) or the Cambridge Science Park (United Kingdom) have defined the parameters of the science park 'model'. These include planned, high-quality, low-density buildings, greenfield, outer-city locations, incubator premises for spin-offs, and on-site business services (marketing, venture capital, patenting and legal) available for tenants. While Stanford Park is a creation of the 1950s, an organized science park movement with national structures, political lobbies and membership criteria only emerged as a worldwide phenomenon in the 1980s. It has subsequently spawned many permutations of the original science park concept in order to meet the contingencies of different national circumstances. The contemporary science park landscape now includes inner-city science parks, converted industrial manufacturing premises that house science parks, parks unconnected to universities and parks with only a tenuous connection to leading-edge technology.

Despite high public expectations of local technological development and growth, empirical evidence shows the science park reality to be rather more mundane. Research from a variety of national contexts has demonstrated that science park firms do not have stronger linkages with local universities than off-park firms. Neither are they any more innovative or likely to exhibit high rates of new firm formation or spin-off. In some instances, they operate as 'enclaves', strongly linked globally but weakly connected locally. The continuing popularity of the science park concept and the willingness of high-tech firms to pay prime rents for a successful science park location suggest other forces at work. The utility of a science park address may be more related to the prestige and symbolic value it affords than to the material, economic and technological advantages to be realized from the park itself.

Further reading

Massey, D., Quintas, P. and Wield, D. (1992) *High Tech Fantasies: Science Parks in Society, Science and Space*, London: Routledge.

SEE ALSO: technopoles

DANIEL FELSENSTEIN

SCOTT BROWN, DENISE
b. 1931, Nkana, Zambia

Key work

(1972, 1977) *Learning from Las Vegas*, Cambridge, MA: MIT Press (with R. Venturi and S. Izenour).

Architect, designer, urban planner and postmodern theorist, married to architect Robert VENTURI and partner in the firm Venturi, Scott Brown and Associates. Scott Brown was born in Zambia, raised in South Africa, and educated in London and at the University of Pennsylvania. At Yale in the late 1960s, she developed courses that encouraged architects to study problems in the built environment employing both traditional empirical methods of social science but also media studies and pop culture. The book *Learning from Las Vegas* suggests that the study of parking lots and isolated commercial centres and other forms of 'low art' offer valuable lessons for design.

The holder of ten honorary doctorates and numerous awards, Scott Brown's project work through the last quarter of the twentieth century includes both civic and institutional projects steeped in popular symbolism. She has pioneered preservation planning as a model for historical locations and DOWNTOWNS in Tennessee, Texas, and Miami Beach, Florida. Major civic and institutional projects include the Denver City Center Cultural Complex, the Smithsonian Institution's Museum of the American Indian, and the Departement de la Haute-Garonne provincial capitol building in Toulouse, France. Finally, Scott Brown has also been involved in campus planning with clients including the Universities of Kentucky, Michigan and Pennsylvania. Her most recent efforts are in life science and medical complexes.

DAVID C. PROSPERI

SCULLY, VINCENT J.

b. 21 August 1920, New Haven,
Connecticut

Key works

(1977) *Shingle Style and the Stick Styles: Architecture,
Theory, and Design from Richardson to the Origins of
Wright*, New Haven, CT: Yale University Press.
(1991) *Architecture: The Natural and the Manmade*,
New York: St. Martin's Press.

Professor, historian and critic, Scully is currently
the Sterling Professor Emeritus of the History of
Art at Yale University (where he has taught for
over fifty years) and the Distinguished Visiting
Professor in Architecture at the University of
Miami. Among his former students are the
MONUMENT designer Maya Lin and architect/
educator Elizabeth PLATER-ZYBERK.

Throughout his career, Scully has argued that
modernist planning popular in the United States in
the 1950s – in the form of highways that erase
NEIGHBOURHOODS, superblocks that contain either
commercial skyscrapers or housing, and other
forms of building based on high modernism's
instructions for living – is incompatible with
communal values. Moreover, this mode of planning and building so dominant throughout the last
half of the twentieth century is still inhibiting more
recent planners and architects in their efforts to
build communities. In this regard, Scully is often
viewed as one of the intellectual forefathers of the
NEW URBANISM movement.

Scully's unique approach to scholarship and
criticism in architecture, city planning and URBAN
DESIGN has earned him numerous awards, including becoming the first winner of a prize offered
by the National Building Museum in Washington
DC named in his honour.

DAVID C. PROSPERI

SEGREGATION

Segregation is the spatial separation of various
groups across different places. The groups in
question may be defined by any readily identifiable
trait, such as gender or age. Most often in URBAN

contexts groups have been defined in terms of
race, ethnicity, religion or income. The places in
question may be defined in a variety of geographic
terms. NEIGHBOURHOODS, school buildings, political jurisdictions or departments within a workplace are often used.

Implicit in the places in question are two levels
of geographic scale: unit and extent of observation.
Both will affect the measured degree of segregation. Unit of observation scale is the size of the
area in which groups are tallied. Larger units of
observation typically manifest less segregation,
because they will have a tendency to include a
wider variety of groups within them than a smaller
entity. For example, racial segregation measured
across city blocks will typically register higher than
if measured across multi-block territories called
census tracts, and much higher than if measured
across even larger units of observation like municipalities. Extent of observation scale is the number of areas that are employed as units of
observation in the estimation of segregation. One
might use all the neighbourhoods in the CENTRAL
CITY municipality, or all within the larger metropolitan area. There is no predictable relationship
between extent of observation scale and measured
segregation, however, because extending the
boundary of investigation may take in areas that
differ in their segregation from the original, smaller
territory.

Two aspects of segregation are typically
analysed: evenness and exposure. Evenness refers
to the similarities of the spatial distributions of
two or more groups across the places that constitute the units of observation. Exposure is the
extent of potential contact between alternative
groups within the given unit of observation.
Segregation is associated with lower evenness or
exposure.

Segregation is the norm for contemporary
cities, though it differs greatly in nature, degree
and causes depending on national context. Since
1970 in the United States, for example, neighbourhood segregation of blacks and whites has
steadily, if modestly, fallen, though they remain
the most segregated groups. Hispanics are the next
most residentially segregated from (non-Hispanic)
whites, followed by Asians. Hispanics' and Asians'
residential segregation changed little during the
1990s. After declines in school segregation among
blacks, Hispanics and whites from 1970 to 1990,

the trend has since reversed. Residential segregation of economic groups has increased since 1970 in the United States.

One cannot identify precisely the causes of segregation without first detailing the group and venue in question. The segregation of income classes across neighbourhoods is explained by a different set of factors than, for example, the segregation of races across schools. Nevertheless, certain forces are common across most types of segregation relevant to the city: preferences, market constraints, private DISCRIMINATION and public discrimination.

Segregation can be caused by preferences for homogeneous places where the individual's own group predominates. These preferences for self-segregation may involve perceptions that the own group possesses many positive attributes, institutional similarities, or historical, linguistic and cultural commonalties. Alternatively, self-segregation may be generated by an aversion to other groups who are perceived as having undesirable personal qualities, conveying lower status, or bringing about undesirable social processes or market reactions.

Three primary sorts of market factors can cause segregation as well: differences in economic means, proximity and information. When ability to pay governs access to such places as residences, private schools or clubs, intergroup differences in financial resources will translate into segregation. Two groups may not be proportionally represented across neighbourhoods differing in their affordability, for example, because they are not equally well off financially. Occupancy of a place is generally associated with proximity because of the time and out-of-pocket expenses associated with travel. Groups may thus frequent quite different sets of restaurants, workplaces or schools if they happen to live large distances apart. Choices of places are also influenced by the amount and accuracy of information available to the decision-makers. One group with little or unflattering information about a COMMUNITY will have fewer tendencies to move there than groups possessing convincing, affirming information about its attributes.

Private discrimination may cause segregation by erecting barriers to the choice of particular places by particular groups, producing involuntary segregation. This discrimination can take many forms, some of which are visible to those being disfavoured, others of which are not. For instance, institutions can create formal rules forbidding the entrance of certain groups, such as restrictive covenants attached to real estate deeds or group prohibitions in club or school admission policies. In other cases, discrimination can involve limited or false information provided by agents of real estate concerns or other institutions, such as steering prospective homebuyers to vacancies in neighbourhoods where their group is already over-represented. In yet other situations, the discrimination can be perpetrated by the current occupants of the space who, when confronted with a new occupant representing an undesired group, engage in acts of intimidation or harassment. Such acts of HOUSING DISCRIMINATION are rendered illegal only if the nation in question legislatively grants a group protected status. In the United States, for example, discrimination on the basis of race, gender or religion is prohibited, but not discrimination based on ability to pay.

Public policies and practices can also cause segregation through many analogous ways as private discrimination. Past US regulations have, for example, denied low-interest, federally insured mortgages to all but segregated, predominantly white-occupied neighbourhoods. Lower levels of government can promulgate laws that are explicitly segregationist, as in the case of states in the American South in the first sixty years of the twentieth century. Local governments can segregate through their regulatory functions. For example, many school boards in the past delineated the boundaries of elementary school districts to produce homogeneous student bodies.

Further reading

Kaplan, D.H. and Holloway, S.R. (1998) *Segregation in Cities*, Washington, DC: American Association of Geographers.

Massey, D.A. and Denton, N.A. (1988) 'The dimensions of residential segregation', *Social Forces* 67(2): 281–313.

——(1993) *American Apartheid*, Cambridge, MA: Harvard University Press.

SEE ALSO: desegregation; ethnic enclaves; exclusionary zoning; index of dissimilarity; redlining

GEORGE C. GALSTER

SENSE OF COMMUNITY

There is no clear-cut definition for COMMUNITY. As a concept, it evokes the bucolic imagery of small-town life. Indeed, nearly three out of every four Americans define community from a rural, small-town perspective. Surveys suggest those migrating newcomers from the metropolitan areas are relocating to small towns in the belief that they provide a sense of belonging not afforded in contemporary URBAN areas.

To be socially connected, the late Kevin LYNCH claimed a city dweller needed a mental map that gives meaning to symbolic landmarks, activity nodes and boundaries. However, as the NEW URBANISM asserts, modern telecommunications, the automobile and unbridled SPRAWL are the chief culprits for the destruction of such symbolic mental maps. The commercial MAIN STREET, the friendly city hall and the parish church, as social activity nodes, have all but disappeared. The modern community has been redefined. Community is fast becoming to mean the place where some person works or where he/she plays rather than where he/she resides. Known as the 'community of interest' rather than the 'community of place', for many social scientists this is troublesome because without spatial boundaries a citizen is more likely to ignore his/her local attachments, making the suburban town or urban NEIGHBOURHOOD nothing more than a collection of non-connected dwelling units.

J.B. JACKSON believed that Main Street, Elm Street and Courthouse Square were critical linkages for defining a sense of community. The spatial arrangement of these sacred symbols helped forge the formal and informal networks of social interaction; both networks were needed to establish civic loyalty and civic engagement. These spatial mental maps that have historic and symbolic meaning helped identify not only a place where a citizen belongs but also his/her civic institutions. The CITY BEAUTIFUL MOVEMENT is one such manifestation of this emotional tie between civic institutions and social connection. Without collective spatial reference points, the social fabric of civic engagement is in jeopardy. Indeed, when America's founding fathers were grappling with GOVERNANCE, they stressed the importance of loyalty to place. For this reason, the Land Ordinance of 1785 dictated that every surveyed township would have land allotted for civic institutions, fostering a rededication to civic virtue with each succeeding generation. A sense of community can only exist when it is a place where people live their lives and fulfil their collective hopes and dreams within the context of their acceptance of the civic obligations owed to that community.

Further reading

Freie, J.F. (1998) *Counterfeit Community*, Landham, MD: Rowman & Littlefield.

Mattson, G.A. (1997) 'Small community governance: some impediments to policy making', in J. Gargan (ed.) *Handbook of Local Government Administration*, New York: Marcel-Dekker.

SEE ALSO: City Beautiful movement; gated communities; homeowners' association; Jackson, John Brinckerhoff; Lynch, Kevin; new urbanism; sense of place

GARY A. MATTSON

SENSE OF PLACE

Sense of place is a multidimensional, complex construct used to characterize the relationship between people and spatial settings. Sense of place can be established at different geographic scales: the NEIGHBOURHOOD, city or country. The term unites two related meanings of sense: (1) understanding and order, as in 'making sense', and (2) feeling and sensation, such as smell, taste and sight. As a concept, sense of place acknowledges that places are not merely points in space, but locations that assume the meanings that people assign to them based on experiences, memories and feelings. Having a sense of place, thus, means to know a place, not necessarily just its spatial structure, but also its functions, users, ownership, and so forth.

The notion of sense of place is related to the Roman concept of *genius loci*, spirit of place. However, while in Roman times each place was thought to 'own' a special character (*genius loci*), waiting to be revealed, modern thought conceives sense of place as a social construct dependent on human interpretation of the place setting. Hence, for the same location, sense of place will vary from person to person and over time.

Scientifically, sense of place serves as a meta-idea or attitude that subsumes related concepts

such as place attachment, identity formation, dependence, security and belonging. People have various means to gain place knowledge. Social networks, relationships, and length and status of residency are key factors influencing a person's development of sense of place. Short-term, transient residents, therefore, rarely develop feelings of attachment and belonging to a place akin to those felt by long-term or native residents. Yet travellers and immigrants often derive some sense of place of their destination prior to arrival from secondary information. In addition, having a (strong) sense of a place does not inevitably imply a feeling of belonging and security inasmuch as knowing a place can also entail the feeling of being 'out of place' or excluded.

Studies in ENVIRONMENTAL BEHAVIOUR and psychology have developed quantitative measures of sense of place that equate the level of a person's cognition and emotional attachment to a place with different levels of sense of place. A sense of place increases with a person's knowledge and place attachment. Levels of sense of place appear positively correlated with people's willingness in terms of personal intervention or sacrifice to maintain, protect or preserve a place's condition. Thus, enhancing people's sense of place could become an important policy to foster environmental stewardship and protection.

Architecture and URBAN DESIGN can contribute significantly to the development of a recognizable character of the built environment, which in turn can invoke a sense of place. While ethnic communities tend to employ architectural elements, materials, and ornaments reminiscent of their country of origin to create a sense of home away from home, the NEW URBANISM movement (see also Andrés DUANY; Elizabeth PLATER-ZYBERK; Rob KRIER; Léon KRIER) aims to recreate the look and feel of traditional neighbourhoods to instil a sense of place, security and COMMUNITY in new URBAN developments.

Further reading

Jorgensen, B.S. and Stedman, R.C. (2001) 'Sense of Place as an attitude: lakeshore owners attitudes toward their properties', *Journal of Environmental Psychology* 21: 233–48.
Steele, F. (1981) *The Sense of Place*, Boston, MA: CBI Publishing.

ANDREA I. FRANK

SEXUALLY TRANSMITTED DISEASES

Sexually transmitted diseases (STDs) are a group of communicable diseases that are transferred predominantly by sexual contact. Curable STDs are *Neisseria gonorrhoeae*, *Chlamydia trachomatis*, *Treponema pallidum* (syphilis), and *Trichomonas vaginalis* infections (trichomoniasis), while non-curable STDs are mostly of viral origin such as Human immunodeficiency virus (HIV), herpes simplex virus (HSV), human papilloma virus (HPV) and hepatitis B virus. More than thirty bacterial, viral and parasitic diseases have been identified that can be transmitted by the sexual route.

Sexually transmitted infections are a major cause of acute illness, infertility, long-term disability and death, with severe health consequences for millions of men, women and infants. The global problem of STDs is influenced by a number of factors including age group, social status, sexual orientation, sexual behaviour patterns, therapy guidelines and susceptibilities to therapy, absence or presence of prevention and control programmes, and reliability and registry of data. Low socio-economic status has been associated with increased STD rates in the United States and the developing countries.

The epidemiologic trends of STDs are strikingly different in various parts of the world. In Northern and Western Europe, there has been a spectacular decline in the incidence of STDs, particularly gonorrhoea and syphilis. The situation in North America is far more complex, with geographic areas and large population groups having low levels of STDs and others continuing to experience an epidemic of STDs, particularly inner-city minority populations in the United States.

In most developing countries, STDs exhibit a higher incidence and prevalence, an alarming rate of anti-microbial resistance, and a higher rate of serious complications and interactions with HIV infection. In its World Development Report for 1993, the World Bank estimated that for adults between 15 and 44 years of age in the developing world, STDs excluding HIV infection were the second cause of healthy life lost in women, after maternal morbidity and mortality. In 1995, the total number of new cases of the four major curable STDs in adults between the ages of 15 and 49 was estimated by the World Health Organization

to be just over 330 million. South and Southeast Asia accounted for the largest number of new cases, followed by sub-Saharan Africa.

At the end of 1999, an estimated 34.3 million people around the world were living with HIV/AIDS, 95 per cent in the developing world. Sub-Saharan Africa is most at risk: of 5.4 million people with HIV in 1999, 4 million were living in this region, an area that held only one-tenth of the world's population.

Reduction of the prevalence and burden of STDs requires a concerted effort by national public health services, international and bilateral agencies, non-governmental organizations, private doctors, and research institutions. A comprehensive and multifaceted strategy is required to address a problem that originates in the complex web of social and biological systems. Prevention programmes need to incorporate health promotion, improved access to STD services, improved access to technologies, reduction of the stigma associated with STDs, and improved surveillance.

Further reading

Agacfidan, A. and Kohl, P. (1999) 'Sexually transmitted diseases (STDs) in the world', *FEMS Immunology and Medical Microbiology* 24: 431–5.
Gerbase, A.C., Rowley, J.T., Heymann, D.H.L., Berkley, S.F.B. and Piot, P. (1998) 'Global prevalence and incidence estimates of selected curable sexually transmitted diseases', *Sexually Transmitted Infections* 74(Suppl. 1): S12–16.

LOUISE S. GRESHAM

SHEVKY, ESHREF

b. February 1893, Istanbul, Turkey;
d. 13 October 1969, Altadena, California, United States

Key works

(1949) *The Social Areas of Los Angeles: Analysis and Typology*, Westport, CT: Greenwood Publishing Group (with M. Williams).
(1985 [1955]) *Social Area Analysis: Theory, Illustrative Application, and Computational Procedures*, Palo Alto, CA: Stanford University Press (with W. Bell).

Shevky was born in Istanbul to a family with a distinguished legacy in public service. As a young scholar, his work reflected the French sociological school of LaPlay that stressed area studies. Emigrating to the United States, Shevky was awarded a Ph.D. in experimental medicine from Stanford University. From 1935 to 1950, Shevky worked outside the academy for, among others, the US Soil Conservation Service and the Haynes Foundation. In 1945, he was the only non-university member of the Pacific Coast Committee on Community Studies of the Social Science Research Council. While working at the Haynes Foundation in Los Angeles, Shevky and Marilyn Williams developed the classic formulation of SOCIAL AREA ANALYSIS. Subsequently titled 'HUMAN ECOLOGY', Shevky's method has spawned a research tradition in URBAN SOCIOLOGY that is practised globally. Joining the faculty at UCLA in 1950, Shevky was instrumental in organizing the Eastern Area Studies Program, which focused on applying modern research skills to contemporary Middle Eastern URBAN problems.

DAVID C. PROSPERI

SHOPPING MALLS

The shopping mall houses a variety of commercial facilities under one roof in order to serve a large territory. Its size and composition depend on the dimension of the trade area, from a single SUBURB to a whole region. The various stores and services are planned to be complementary; they are organized spatially so as to influence the shopping behaviour of customers.

Between the two World Wars, 'shopping villages' and 'business centres' appeared in new SUB-DIVISIONS on the outskirts of US cities. The Country Club Plaza (Kansas City, 1922–24) is often seen as the first modern regional shopping centre because of the developer's attention to store composition and access by car.

Rapid DECENTRALIZATION of people and activities in the post-war years prompted the multiplication of suburban shopping malls (and vice versa). With this growth in numbers came an increase in standardization; the shopping mall became a specific building type, the subject of much research on the part of developers, architects and consultants (for instance Victor GRUEN). Shoppers World (Framingham, Massachusetts, 1951) and the Northland Regional Shopping Center (Detroit, 1952) are early examples of this

new type, with their PEDESTRIAN MALL (still uncovered at the time) as a safe spatial enclave. A covered mall appeared for the first time in the Southdale Shopping Center (Minneapolis, 1955), transforming the shopping mall into a single, autonomous building.

Thus, the 1950s saw the crystallization of spatial patterns that characterize the modern shopping centre. The building is located in the middle of the plot, leaving large spaces for parking along the perimeter. Inside, rows of shops define the pedestrian mall itself, which is 'anchored' by a department store at one end or, preferably, at both ends ('dumbbell' plan). The 'magnet stores' attract customers to the mall, while smaller shops along the mall benefit more from impulse buying. This spatial pattern has become more complex over time with the addition of malls and anchors.

In the 1970s, the shopping mall started to integrate other activities besides shopping, mostly in the realms of entertainment and services. With its office building, hotel and ice rink, The Galleria (Houston, 1971) marks the beginning of this change. The trend towards large-scale, multifunctional complexes reached its apex with the West Edmonton Mall (Edmonton, Canada, 1981) and the Mall of America (Bloomington, Minnesota, 1992), where the interaction of shopping and entertainment activities is maximized. Thanks to their accessibility (by car), their mix of uses and their controlled climate, malls increasingly serve as COMMUNITY spaces, even though they remain formally private.

The shopping mall was born in the American suburb, home of the car. From there, it has been disseminated to other environments, to DOWN-TOWN areas serviced by public transportation, to airports and train stations, and so on on all continents. In these new settings, the close relationship between commerce and mobility has remained a fact, as has people's desire for safe, air-conditioned spaces and the ability of developers to use excitement for economic profit.

Further reading

Gruen, V. and Smith, L. (1952) 'Shopping centers: a new building type', *Progressive Architecture* 33(6): 67–94.
Maitland, B. (1985) *Shopping Malls: Planning and Design*, London: Construction Press.

GIANPIERO MORETTI

SIDEWALK

A sidewalk is a path, often constructed of concrete or cement, though occasionally asphalt, that is parallel to a road, and is designed for pedestrians. The British and Australians prefer using 'pavement' to mean the same.

Ancient ruins of Pompeii testify to the fact that separate pathways have existed for a long time to offer comfortable passage for pedestrians. Not only was the pedestrians' safe passage important but their comfort and convenience was also accommodated. Stepping stones were placed on the streets so that the pedestrians could avoid wading through the poorly drained streets.

Professionals interpret sidewalk designs differently. To an URBAN designer, a well-designed sidewalk (safe, comfortable, attractive and user-friendly) offers a diverse array of activities that enrich the streetscape. Thus, sidewalk cafes, amateur music bands, and vendors are all integral parts of a lively sidewalk. To engineers, however, the throughput of a sidewalk is very important; in particular, the ability to move pedestrians at high speeds with adequate space per person is paramount. Certain standards (levels of service) were developed and sidewalks are graded based on the level of CONGESTION and lack of space per pedestrian. Thus, high levels of congestion would be given a poorer grade.

The effort and care taken in designing sidewalks can reflect on the attitude of a city or COMMUNITY towards acceptance of pedestrians and associated activities. A narrow sidewalk that barely allows one adult to walk on it is an example of a lack of consideration for the needs of pedestrians. A wide sidewalk with landscaping and appropriate street furniture (benches, lights, kiosks) that encourage pedestrians to walk or linger shows a stronger commitment to pedestrians.

Further reading

Fruin, J.J. (1971) *Pedestrian: Planning and Design*, New York: Metropolitan Association of Urban Designers and Environmental Planners, Inc.
Untermann, R.K. (1984) *Accommodating Pedestrians*, New York: Van Nostrand Reinhold.

SHEILA SARKAR

SIMKHOVITCH, MARY KINGSBURY

b. 8 September 1867, Chestnut Hill,
Massachusetts; d. 15 November 1951,
New York City

Key works

(1926) *The Settlement Primer: A Handbook for Neighborhood Workers*, New York: National Federation of Settlements.
(1938) *Neighborhood: My Story of Greenwich House*, New York: W.W. Norton & Co.
(1942) *Quicksand: The Way of Life in the Slums*, Evanston, IL: Row, Peterson & Co. (with E. Ogg).

Mary Melinda Kingsbury was born to a middle-class family and brought up in the SUBURBS of Boston. She got her BA in economics at Boston University and then went on to study economic history and political science at Radcliffe College. While at Radcliffe, she joined the Church of Carpenter, led by the Reverend William Bliss, a socialist reformer who encouraged COMMUNITY service in SLUM neighbourhoods. She often visited Denison House, a pioneering Boston settlement house, and she played organ and led a girls' club at St. Augustine's Chapel, a local black congregation. She visited black and immigrant families in Boston's TENEMENT slums, observed and documented their POVERTY, and became aware of the power and wealth of the city's slumlords. In the mid-1890s, she spent a year studying in Berlin and then took up residence in New York City, where she worked at the College Settlement on the Lower East Side and then as head resident of the Friendly Aid House, a Unitarian philanthropic institution. In 1899, she married Vladimir Simkhovitch, a Russian immigrant, and she adopted his last name.

Mary Simkhovitch associated with Jacob RIIS, Felix Adler and other social reformers who were very critical of charitable programmes for the poor, seeing them as treating symptoms of poverty but not the underlying causes. Simkhovitch was one of the founders of the New York Association of Neighborhood Workers in 1900, and then the Cooperative Social Settlement Society in 1901. In 1902, she founded Greenwich House, a new settlement house in Greenwich Village, then a poor, densely populated area with a predominance of immigrants. Under Simkhovitch's directorship, Greenwich House helped establish the Greenwich Village Improvement Society in 1902, New York's first NEIGHBOURHOOD association, and it published the country's first *Tenants' Manual* in 1903, authored by Emily Dinwiddie. By 1908, it had an infant care centre, a home visitors programme and a visiting nurse programme, and it organized cooking and diet instruction, street pageants, plays and concerts. Greenwich House was a national pioneer in nursery schooling, after-school programmes, and arts programmes, attracting the support of such eminent figures as Eleanor Roosevelt, Amelia Earhart and Gertrude Payne Whitney. In the 1930s, Simkhovitch broadened her campaigning to embrace national housing reform, advocating public housing for working people. She led the Public Housing Conference from 1932 to 1943, she was appointed vice-chairman of the newly created New York City Housing Authority (NYCHA) in 1934, and she worked with New York Senator Robert Wagner to develop the National Housing Act of 1937. She continued as director of Greenwich House until 1946 and as vice-chairman of the NYCHA until 1947.

RAY BROMLEY

SIMMEL, GEORG

b. 1 March 1858, Berlin, Germany;
d. 26 September 1918, Strasbourg, France

Key work

(1950) *The Sociology of Georg Simmel*, trans., ed. and with an introduction by K.H. Wolff, New York: Free Press.

Born into a Berlin Jewish family converted to Christianity, Georg Simmel studied history and philosophy and received his doctorate in philosophy in 1881 from the University of Berlin, where he became a lecturer in 1885. In 1914, he obtained a full professorship at the University of Strasbourg, where he remained until his death in 1918. He was active in different intellectual circles in Berlin and established, with Max WEBER and Ferdinand Tönnies, the German Society for Sociology. Considered by many as one of the founding fathers of sociology, he did not limit himself to disciplinary boundaries; rather, he combined different approaches and topics

(history, philosophy, sociology, aesthetics and ethics). Simmel developed an original line of thought on society, stressing the ambivalence at work in modern culture and history. This led him to consider the interactions between individuals through what he called the process of 'sociation', meaning reciprocal relationships established by individuals in their interaction with others. Using this perspective, he produced a provocative analysis of the modern metropolis ('The metropolis and mental life', 1902) focusing on the new challenges posed to individuals faced by uncertainty, particularly the issue raised by the need to express their individuality. According to this idea, URBANITY has a different form and meaning in small cities in comparison to larger ones. His influence has been substantial, especially in the field of URBAN SOCIOLOGY where researchers like Robert PARK, Ernest W. Burgess and Louis WIRTH of the CHICAGO SCHOOL OF SOCIOLOGY carried on with some of his insights concerning the production of space.

Further reading

Leck, R.M. (2000) *Georg Simmel and Avant-Garde Sociology: The Birth of Modernity, 1880–1920*, Amherst, NY: Humanity Books.

PIERRE HAMEL

SITE VALUE TAXATION

Site value taxation, or land value taxation as it is sometimes called, refers to an alternative system of taxing property. Property owners are generally taxed on the combined value of their land and improvements such as buildings and landscaping. In a pure site value taxation system, governments charge a tax to property owners based exclusively on the value of their land. This is also referred to as a single tax on land. The more common form of site value taxation is one wherein land is taxed at a differential and higher rate than the real estate improvements on the land.

The earliest idea of site value taxation is attributed to the noted nineteenth-century political thinker Henry GEORGE. In observing the simultaneous rise in economic prosperity and POVERTY in the United States during the nineteenth century, George concluded that the poverty persisted in part because owners of properties accumulated wealth simply through land ownership without participating in productive processes that contributed to the generation of wealth. George's solution to this inequity – site value taxation – was based on the premise that the land accrues value due to the presence of natural amenities and/or the provision of INFRASTRUCTURE by the society at large and that this value should be recuperated through a tax on land.

Various forms of site value taxation exist in several countries including Australia, Canada, Denmark, New Zealand and the United States. In the United States, several states have used the HOME RULE provision to adopt legislation that enables local governments to implement some form of site value taxation. States that have adopted enabling legislation for site value taxation include Maryland, New York, Pennsylvania and Washington DC. One of the earliest site value taxation systems in the United States was put in place in 1913 in the cities of Pittsburgh and Scranton in Pennsylvania, wherein built structures were taxed at a higher rate than the land.

Proponents of site value taxation argue that reducing or eliminating the tax on real estate improvements lowers the tax burden on productive labour and capital and simultaneously recuperates publicly created value. Site value taxation encourages owners to make efficient use of their land by making significant improvements to it. This feature makes site value taxation a popular proposal for older INDUSTRIAL CITIES that have experienced significant DISINVESTMENT by owners who allow their properties to fall into a state of disrepair. Efficient use of land due to site value taxation can also encourage development to occur at higher densities, thereby offering a disincentive for SPRAWL. The future popularity and use of site value taxation depends on a robust empirical evaluation of its impact on the fiscal and economic health of regions.

Further reading

George, H. (1912) *Progress and Poverty*, New York: Doubleday, Page & Co.
Netzer, D. (ed.) (1998) *Land Value Taxation: Can It and Will It Work Today?*, Cambridge, MA: Lincoln Institute of Land Policy.

SAMINA RAJA

SITTE, CAMILLO

b. 17 April 1843, Vienna, Austria;
d. 16 November 1903, Vienna, Austria

Key work

(1965 [1889]) *City Planning According to Artistic Principles*, trans. G.R. Collins and C.C. Collins, New York:
Random House.

An architect and urban planner, Camillo Sitte
lived through a turbulent historical period in
Austria. Having received artisan training from his
father, Sitte was educated and influenced by
Rudolf von Eitelberger and Heinrich von Ferstel,
who were champions of the applied arts. It was on
the recommendation of Eitelberger that Sitte
became the head of the new State Trade School in
Salzburg in 1875, but Sitte returned to Vienna in
1883 to establish a similar school there. As a supporter of nationalist concepts and an advocate of
arts and crafts, Sitte emphasized history and the
importance of landmarks and a medieval aesthetic
sensibility in design. He was in many ways similar
to his inspiration, Richard Wagner, in elevating
the importance of history, tradition and artistic
aesthetics to the centre of planning and design. He
argued profusely against the straight-line BOULE-
VARDS of his time, opting instead for the curving
and irregular streets of the medieval period. This
nostalgic return to the past, which failed to
accommodate the emerging technological innovations (and their required spatial logic), stood in
opposition to the views of some of his contemporaries, including modernists such as Otto
Wagner. Sitte's revival of picturesque aesthetics
was unacceptable to the modernist planners of the
post-*Ringstrasse* period, in Vienna and elsewhere,
at the turn of the twentieth century. While he was
later admired by Raymond Unwin and Lewis
MUMFORD, many modern planners, such as LE
CORBUSIER and Walter GROPIUS, emulated Otto
Wagner in their functionalist interpretation of the
city and its need to accommodate automobiles and
technology. For Sitte, planning became a tool to
generate CULTURE, to remind citizens of a lost past
through monumental spaces, such as picturesque
squares. However, his grand vision of cities,
though important, took a back seat to Otto
Wagner's emphasis on function, mobility and

advocacy for rational URBAN DESIGN that adapted
to modern life.

Further reading

Schorske, C.E. (1981) *Fin-de-siècle Vienna: Politics and
Culture*, New York: Random House.

ALI MODARRES

SKID ROW

Skid row refers to the section of the city, according
to the stereotype, populated by individuals, typically men, on the down-and-out: hobos, tramps,
vagrants, bums, alcoholics and the homeless.
Downtrodden skid row tracts are often characterized by poorly maintained single room occupancy
(SRO) hotels that are let by the night or the week.
Historically, skid row is also the site of dormitory-
style shelters, where a person of very limited means
can secure a bed in a room lined with beds; and
'flophouses', where the poorest of the poor have
had to sleep on the floor of a barren room. This
supply of cheap, albeit unsatisfactory, housing
offered persons of marginal or intermittent income
a last opportunity for shelter, an alternative to
living on the streets. Skid row also denotes that
part of the city where church-based and evangelical
missions have offered meals and a cot for the night
to those persons most in need.

   The skid rows of American cities are well
known and historically have included the Bowery
in New York, West Madison Street in Chicago, the
DOWNTOWN along Fifth Street (often referred to as
the 'nickel') in Los Angeles, the Tenderloin in San
Francisco, and Larimer Street in Denver. But the
size of skid row areas and the supply of cheap,
affordable SRO housing that they offered have
been shrinking over the years. In the 1950s and
1960s, URBAN renewal and SLUM clearance programmes bulldozed many skid row areas, greatly
reducing the number of SRO housing units in the
process. In more recent years, many of the
remaining SROs and missions have fallen victim to
the forces of downtown REDEVELOPMENT and
URBAN REVITALIZATION, which, in numerous cases,
has resulted in the GENTRIFICATION of these previously undesirable areas. In Chicago, four luxury
high-rise Presidential Towers were built on the site
of a skid row renewal area, catalysing the extension
of the city's downtown (the 'Loop', as it is referred

to in Chicago) to the west, transforming the entire nature of West Madison Street. The redevelopment of the area and the concomitant loss of SRO housing were further abetted by the nearby construction of a new CONVENTION CENTRE and sports arena for the Chicago Bulls basketball team. In San Francisco, the Tenderloin area has similarly been threatened by redevelopment pressures (generated in part by the construction of the George Moscone convention centre and a new modern art museum) that have resulted in the expansion of the city's traditional downtown to the South of Market Street area.

Skid row is often stereotypically portrayed as pathological; life on skid row is often seen to be the result of alcoholism, drug abuse, laziness and individual failure. In such cases, counselling and social services are offered to help a person overcome his substance abuse patterns and other bad habits and put him back on his feet.

Yet skid row was never entirely the result of individual failure. Despite stereotypical perceptions, most of the residents on skid row were not vagrants or transients. The residents of skid row worked, but they had only limited skills and were quite marginal in the LABOUR MARKET. These manual workers often worked intermittently or only part-time and received very low wages. Skid row was also the site of exploitative day labour practices, where unskilled workers were recruited on the spot for temporary, ill-paid work. Skid row was produced by larger structural forces, not just by individual failure.

Skid row was also quite functional. It was initially a place of working men's housing, where railroad workers and other mobile labourers and economically marginal workers, especially seasonal workers, could find a place to stay, where shops and support services were available nearby. Boarding houses, SROs and local bars provided single adults, especially the elderly, with opportunities for friendships and for socializing. Support services – restaurants (especially necessary for residents lacking kitchen facilities), shops, employment agencies, downtown jobs, and missions – were all available nearby. These services and opportunities would disappear amid post-industrialism, with the gentrification and transformation of the NEIGHBOURHOODS and the decline of manual labour jobs in the CENTRAL CITY.

The disappearance of SRO housing has been an important factor in the rise of HOMELESSNESS.

Deindustrialization changed the nature of skid row, marginalizing its people even further; the shift to a service economy has meant the loss of decent wages earned by seasonal and part-time manual labourers (again, typically men). These workers simply lacked the means to buy housing. A changed economy also altered the social composition of skid row, as untrained younger workers, too, began to find themselves increasingly on the edges of a skills-based service economy.

American cities have at times sought measures to deal with the decline of skid row and the problems posed by the disappearance of their stock of SRO housing. At times, they have enacted ordinances limiting SRO destruction and conversion. Some cities have sought to build higher-quality SROs and other forms of assisted housing that would provide those on the margins of American life with a more livable alternative – an SRO that affords its residents privacy, dignity and skills training – to mass shelterization or life on the streets. Limited federal monies have been made available for such efforts. But all of these programmes have been sharply constrained by a shortage of resources and by a city's competitive interest in upgrading neighbourhoods in the pursuit of new economic development, an interest that is magnified by the pressures of a city's GROWTH MACHINE.

Further reading

Bogue, D. (1963) *Skid Row in American Cities*, Chicago: University of Chicago Press.

Hoch, C. and Slayton, R.A. (1989) *New Homeless and Old: Community and the Skid Row Hotel*, Philadelphia, PA: Temple University Press.

Timmer, D.A., Eitzen, D.S. and Talley, K.D. (1994) *Paths to Homelessness: Extreme Poverty and the Urban Housing Crisis*, Boulder, CO: Westview.

MYRON A. LEVINE

SLUM

Parts of cities where the housing quality is bad and living conditions are poor are called slums. They are known by various names: 'FAVELAS' in Latin America, 'kampungs' in Southeast Asia, 'shanty towns' in Africa and 'jhuggi-jhompris' or 'BUSTEES' in South Asia. Worldwide, over 400 million people live in slums. About one-third or more of the population in the larger THIRD WORLD CITIES live in

slums – in Mumbai, India, slums such as Dharavi house about 4 million residents, while in Rio de Janeiro, Brazil, over 1.2 million of the city's 6 million residents live in slums a short distance away from the famous Copacabana beach.

While the majority of slums are now found in poor countries, the earliest slums in recent history were formed in industrializing cities of Western Europe in the eighteenth century and later in the Americas. Their genesis was catalysed by the INDUSTRIAL REVOLUTION and IMMIGRATION – from rural areas to cities and from Europe and Asia to the Americas. The new immigrants needed cheap housing within walking distance of their job sites. As demand for housing increased, landlords bought single-family houses, tore them down, and erected 2–6-storey buildings with small rooms to rent. Many rental units did not have windows – thus, these units did not get any natural light or fresh air. Families shared a bathroom located in an outhouse on the ground floor. The appalling conditions in London and New York slums inspired many creative works, newspaper articles, political debates and policy responses. In Britain, the earliest interventions involved demolition facilitated by slum clearance laws – often with much forethought and consideration, especially after Sir Patrick GEDDES's notion of 'conservative surgery' (i.e. selective demolition of slums) became popular in the late nineteenth century. In the United States, early attempts were focused on checking the proliferation of slums using laws such as New York's Tenement Housing Act of 1906 and from the 1920s onwards through ZONING regulations. However, the 1949 federal Housing Act created an URBAN renewal (or slum clearance) programme that resulted in extensive demolition of slums in US cities, displacing thousands of families and disrupting communities. Much of this happened without good planning and several, allegedly, to facilitate new high-end commercial and office space construction.

In spite of slum-prevention efforts, they continue to exist in developed countries. In the United States, urban SPRAWL provided opportunities for richer inner-city residents to move from crowded DOWNTOWN areas to larger houses in less dense NEIGHBOURHOODS and newer houses. Facilitated by a variety of factors including federal government policies and NIMBY (NOT IN MY BACKYARD) attitudes towards housing for urban poor, poor families that could not afford to move out were stuck in areas that soon became concentrated pockets of POVERTY in CENTRAL CITY slums. In the recent past, federal, state and local policy efforts have been initiated to improve slum areas. The US Department of Housing and Urban Development allocates a portion of its annual budget for programmes focused on slums. Improvement efforts led by residents of such areas have also gained momentum and considerable success, none more than the Dudley Street Neighborhood Initiative that helped reverse the decline of an old Boston, Massachusetts, neighbourhood.

In developing countries, slums first sprang up near employment centres. Several features characterize Third World slums. Houses in slums are old and dilapidated (or built of materials such as cardboard and sheet metal) and may not provide adequate protection from the elements and may pose health hazards for the occupants. Most slums do not have access to potable or running water, water-borne or other sanitation systems, garbage collection, or stormwater drainage facilities. In turn, this results in a high incidence of easily communicable diseases, causing high infant mortality, high levels of personal and family stress, and low labour productivity. Motorable road access – even for emergency vehicles – is often non-existent, as are developed pedestrian pathways. Residents of these areas do not have easy access to schools, primary healthcare facilities, or spaces for active and passive recreation and socializing. CRIME rates within cities are often highest in slums.

With the advent of URBAN PLANNING in Third World cities, municipalities focused on creating new, better-quality areas elsewhere, ignoring the slums. Faced with severe financial constraints, many local governments did not have enough funds to plan new INFRASTRUCTURE as well as improve slums. When conditions in slums degenerated to such levels that public and political opinion demanded action, many cities demolished slums. Such demolitions cleared urban eyesores from some parts of cities; soon, however, new slums would form elsewhere. Moreover, demolitions displace families, disrupt social networks, and restrict the availability and price of low-skilled labour essential for the labour-intensive economy of Third World cities. Relocating displaced families is often hampered by the scarcity of cheap land and governments' budgetary constraints.

Of late, demolition is no longer seen as a desirable slum intervention policy in Third World cities – instead, slum improvement is. This involves building infrastructure such as new motorable roads and pedestrian pathways, laying down water supply and sanitation pipes, and electricity lines. Often, these efforts involve contributions from slum dwellers, either in cash or through free labour. Many private voluntary agencies with aid from local, state, national and international sources such as the World Bank, Oxfam and the United Nations Children's Education Fund are active in Third World slums to improve access to potable water and sanitation facilities, provide educational opportunities for children and adults living in slums, and provide childcare and job-skills training for young adults. For such efforts to be successful, several factors must be present. First, the efforts must respond to the felt needs of slum dwellers. Secondly, government entities, voluntary agencies and the residents should be highly committed to make the efforts succeed. And thirdly, the interventions must be designed to be sustainable. In spite of these odds, improvement efforts often succeed. Indonesia's Kampong improvement programmes are a widely acclaimed success. In New Delhi, India, Asha – a private, non-profit agency – provides free primary healthcare to over 160,000 slum residents, improving health and school attendance among children. In Rio de Janeiro, the Edisca School of Dance for slum children has offered a way out of poverty for many. Yet much remains to be done. Recognizing this, the United Nations adopted the Cities Without Slums Action Plan at its Millennium Summit in 2001 – a plan that, if successful, will improve the lives of over 100 million slum dwellers by 2020.

Further reading

Johns Hopkins University (2002) 'Making urban areas work', *Population Reports* 30(4): 14–21.

Kirby, D.A. (1979) *Slum Housing and Residential Renewal: The Case in Urban Britain*, Longman: London and New York.

United Nations Centre for Human Settlements (Habitat) (1982) *Survey of Slums and Squatter Settlements*, Dublin: Tycooly International.

World Bank (1994) *World Development Report: Infrastructure for Development*, New York: Oxford University Press, for the World Bank.

JERRY ANTHONY

SMART COMMUNITY

The Smart Community, as a concept, is rooted in a number of economic, cultural and technical trends, which converged in the mid-1990s and provided communities with both the incentive and the means to become 'Smart'.

During the late 1970s and early 1980s, many communities in Western Europe, North America and Asia began experiencing challenges to their traditional economic development models. Expanding populations, the growth of SUBURBANIZATION and traffic CONGESTION caused by COMMUTERS placed an increasing demand on their tax bases. In addition, the trend towards the globalization of commerce resulted in a major shift of traditional manufacturing capacity away from developed countries to lower-wage regions of the world. The loss of highly paid manufacturing jobs and industrial facilities increased local unemployment while reducing tax revenues in the communities concerned.

On the technical front, the INTERNET was maturing from a strategic communications tool, originally created for the US military, to embrace educational and research institutions worldwide. By 1993, it had evolved into the web, giving the commercial sector and the general public access to a global information and communications resource. The expansion of the internet paralleled the exponential growth of the telecommunications industry. Hardware and software were developed that allowed vast amounts of information to be transmitted quickly and cheaply on a global scale while the convergence of information and communications technologies (ICTs) blurred the lines between computers, telephone, radio, television and the print media.

Communities which had suffered the most from deindustrialization, as well as those in which the telecommunications industry was already significant, such as Silicon Valley in California, started looking at the potential of ICT as a tool for URBAN renewal and enhanced economic growth. These communities developed a vision of how ICT could link local economies with the global marketplace while helping municipal governments provide their citizens with innovative services at reduced cost. The concept of the Smart Community was born.

There have been numerous attempts to define the term 'Smart Community'. One definition was

coined by a blue-ribbon panel of experts, created by the Canadian government in 1998, to provide advice on a potential national Smart Communities programme. This panel defined a Smart Community as 'a community, ranging from a neighborhood to a nation-wide community of common or shared interests, whose members, organizations, and governing institutions are working in partnership to use information and communication technologies to transform their circumstances in significant ways'.

In practical terms, Smart Communities deploy a range of ICT-based (Smart) services to enhance local economic, social and cultural development. A typical Smart Community might deploy an integrated suite of services, available through a single community portal, in areas such as TELECOMMUTING, electronic commerce, tourism, tele-health, distance education, social services, law and public safety, transportation, access to government, and resource/environmental management.

Further reading

International Center for Communications, San Diego State University (1997) *Building Smart Communities – Smart Communities Implementation Guide*, California Department of Transportation (CALTRANS).

Report of the Panel on Smart Communities (1998) *Smart Communities*, Ottawa: Communications Branch, Industry Canada.

ELINOR BRADLEY
STEPHEN GURMAN

SMART GROWTH

For years, American planners' nemesis was SPRAWL – the leap-frogging, low-density, automobile-dependent, single-use pattern of post-Second World War development. And for years nobody seemed to listen. Development continued, and still continues, in a sprawling fashion, consuming ever-increasing quantities of open land, even in areas with little or no overall growth. But during the 1990s, a revolution of sorts began to take place, and the American Planning Association was joined by an unlikely coalition of business, environmentalists, labour unions, URBAN minorities, fiscal conservatives, farmers, historic preservationists, government officials and, reluctantly, even developers. All these groups are trying to work together, but they come to the table with very different goals.

Nevertheless, they all agree, with various degrees of enthusiasm, that sprawl is unsustainable. It continues to pave over prime farmland, forestland, pasture and range; requires huge capital INFRASTRUCTURE and public service costs; increases the use of motor vehicles with their concomitant emission of air pollutants and the greenhouse gases that contribute to global climate change; and finally, it sharpens the SEGREGATION of metropolitan areas into rich suburban enclaves and poorer urban communities.

To counter these trends, smart growth proposes:

1   Limiting outward development and fostering COMPACT DEVELOPMENT. The most important tool to limit or shape the geographic extension of growth is URBAN GROWTH BOUNDARIES (UGBs) that separate developable land from rural land. UGBs can be enacted by individual localities or mandated by state legislation, as in the case of Oregon and Washington. Sprawl limitation can be fostered through incentives and disincentives as well. Maryland and other states have designated priority growth areas for funding, while sharply limiting subsidies for new roads, schools or sewers in rural areas. Limitation of outward growth is also shaped by natural resource preservation. Acquisition of land is being combined with regulatory measures such as habitat conservation plans in California, to preserve land at the metropolitan fringe. Voters throughout the country have approved measures to acquire land for resource protection, PARKS, trails and OPEN SPACE in general.

2   Revitalization of existing communities. It is assumed that existing communities enjoy excess public facility capacities that can be utilized through infill of vacant sites and the REDEVELOPMENT and densification of underused parcels. This maximization of existing resources is to be accompanied by an effort to focus governmental, private and non-profit resources in the revitalization efforts of neglected communities.

3   NEW URBANISM. With new urbanism, development is designed to expand opportunities for social interaction and cultural exchange, thus fostering a SENSE OF COMMUNITY. This is achieved through the creation of walkable

and interconnected streets, higher housing densities and a variety of housing types, NEIGHBOURHOOD/village/town centres with public buildings, plazas, mixed uses, and transit stations.

4   Reorientation of transportation. Smart growth calls for a revolution in transportation planning with an emphasis on modes of transport besides the automobile. This implies new approaches at all levels of government and the design professions, especially highway engineers. This trend was supported by the federal government during the 1990s with legislation that not only required that social and economic issues be considered when planning transportation facilities, but also dedicated unprecedented funding to mass transit.

5   A regional approach. The region, not cities, states or nations, is emerging as the dominant economic unit of the twenty-first century. Related industries form clusters at the regional level, establishing networks that tie the region together – hence the need to plan regionally for transportation, public facilities, open space, land use, and social and cultural networks. To be successful, smart growth must go beyond piecemeal strategies and focus on the regional interconnections between older neighbourhoods and newer SUBURBS, land use and transportation, tax-base inequities and the social and economic health of communities.

The smart growth agenda is comprehensive and ambitious. The problem is implementation. Control of outward movement means limiting availability of single-family homes and reliance on the automobile, the mainstay of the traditional American lifestyle. Increasing densities in existing neighbourhoods clashes with NIMBYism, as residents fear a decreased quality of life in their immediate area while being oblivious to the benefits to the greater good. To make matters worse, claims about the availability of public infrastructure in older neighbourhoods are often unwarranted. Los Angeles has a public facilities deficit in its urbanized communities of at least US$5 billion and San Diego a similar deficit of US$2.5 billion. It is unlikely that those cities will raise taxes to make up the deficits, or that the residents will accept additional densities without

additional amenities, let alone without the parks and libraries, SIDEWALKS, schools and other public facilities to serve the existing population. On the other hand, if revitalization efforts are successful, the door would open to the gradual GENTRIFICATION of inner-city neighbourhoods. If that happens, smart growth and new urbanism will engender what urban renewal did fifty years ago, without the demolitions. Finally, REGIONAL PLANNING is rarely practised in the United States, because land use planning powers are exercised at the local level. Many localities in the faster-growing regions are engaging in ballot-box planning to slow down residential growth. It is unlikely that they would accept the increasing densities called for with the TRANSIT-ORIENTED DEVELOPMENT proposed by smart growth proponents.

The greatest challenge facing smart growth is to ensure equitable development in the region ('COMMUNITY-based regionalism'), to address the problems of decline in the centre and sprawl at the periphery. To succeed, two essential but politically contentious measures need to be pushed through. First, the tax system must be restructured to distribute tax revenues more equitably at the regional level. Currently, wealthy suburban communities, through sales and property taxes, generate a stream of revenue that allows them to attract even more revenue-generating activities. Conversely, poor communities, usually the inner city and first-ring suburbs, fall in a downward spiral that eliminates any possibility for them to share in the benefits of the growth of the region.

Secondly, affordable housing must be distributed fairly in the region, close to the jobs and other opportunities found in the suburban communities, to achieve what the urban economist and planner Anthony Downs in 1973 called 'dispersed economic integration'. Without such measures – and regional government to implement them – smart growth will end up implementing only those measures that favour the more affluent in the region, especially the preservation of natural resources and open space close to where they live.

Finally, the loose alliance of smart growth proponents might come unglued on the issue of growth. Environmental organizations have agreed to growth accommodation on the premise that growth would be pushed into existing cities. They fear that the developers' agenda is not only to intensify development in existing neighbourhoods

but at the same time to continue to build at the metropolitan fringe. Smart growth for the developers might be reduced to applying new urbanist principles to the design of new development.

It is unlikely that the smart growth programme will be implemented in its entirety. It remains to be seen what elements will survive and what their impact will be. One thing is certain, though. It has been a long time since urban and regional planning and community design have been the subject of such intensive debate.

Further reading

Burchell, R.W., Listokin, D. and Galley, C.C. (2000) 'Smart growth: more than a ghost of urban policy past, less than a bold new horizon', *Housing Policy Debate* 11(4): 821–80.

Calthorpe, P. and Fulton, W. (2001) *The Regional City*, Washington, DC: Island Press.

Orfield, M. (2002) *American Metropolitics: The New Suburban Reality*, Washington, DC: The Brookings Institution Press.

O'Toole, R. (2001) *The Vanishing Automobile and Other Urban Myths: How Smart Growth Will Harm American Cities*, Bandon, OR: Thoreau Institute.

NICO CALAVITA

SOCIAL AREA ANALYSIS

Social area analysis is a technique devised by Eshref SHEVKY and Wendell BELL in the 1950s leading to a classification of residential (mainly URBAN) units in respect to features of their social structure. It consisted of calculating and comparing standardized indexes for each unit, related to a small number of variables chosen as measurable expressions of three pre-defined axes ('social rank', 'URBANIZATION' and 'SEGREGATION', subsequently renamed 'economic status', 'family status' and 'ethnic status' respectively). The combination of indexes on the three axes served as classifiers for each unit.

These three axes were meant to adequately summarize the socio-spatial dynamic, conceived as a linear process of economic/technologic development under the form of increasing *societal scale* (complexity) produced by INDUSTRIALIZATION and urbanization. The axes were considered a priori independent, and an empirical validation of this hypothesis was sought in the clear differentiation (reduced correlation) between the spatial division patterns along each axis. This validation was implicitly considered confirmatory for the technique as a whole.

Social area analysis has been applied in several contexts with controversial results. Its basic assumptions were considered more or less confirmed in US cities, but not in Rome or Newcastle, where substantial correlation appeared between variables belonging to different axes.

Criticism was focused on the technique's inability to account for all possible URBAN situations, and identified the pre-definition of the content and the number of axes shaping the analysis as its major limitation. A refinement of the statistical procedure was introduced with multivariate (factor analysis) models, where the number and the content of the basic axes, summarizing the information contained in a much larger number of initial variables, became the outcome rather than the starting point of the analysis. The abundant work on cities using such models has taken the name of *factorial ecology*.

The most important problem with social area analysis, however, was not of a technical/statistical nature. It was its lack of serious theoretical foundation, substituted by an intuitive and rather simplistic conception of the changing social structures that made it highly dependent on the features of the early post-war urbanization in the United States (rapid SUBURBANIZATION, proliferation of specific types of household and housing, and highly segregated minorities). These historical and geographical contingencies became general principles in the technique's constitutive logic.

Social area analysis has been part of the turn from naturalist-ecological explanatory models and methods to positivist ones. It was a step further from Hoyt's sectoral model of the city's social structure and his critique of the ecological foundation of the Burgess zonal model; an important link in the chain that has led from the HUMAN ECOLOGY of cities to their factorial ecology.

With the decline of positivism, the interest in social area analysis and factorial ecology was reduced dramatically. However, parts or variants of these techniques were used subsequently in productive ways within more theoretically informed work.

Further reading

Johnston, R.J. (1971) *Urban Residential Patterns: An Introductory Review*, New York: Praeger.

Shevky, E. and Bell, W. (1955) *Social Area Analysis: Theory, Illustrative Application, and Computational Procedures*, Palo Alto, CA: Stanford University Press.

Timms, D. (1971) *The Urban Mosaic: Towards a Theory of Residential Differentiation*, Cambridge: Cambridge University Press.

THOMAS MALOUTAS

SOCIAL CAPITAL

A building requires continued capital investment in its physical INFRASTRUCTURE or support framework. Similarly, capital investment in the physical infrastructure of a city – in its airport, roads and port facilities – is essential for URBAN growth and expansion. The concept of social capital extends the same logic further to social policy and the health of communities. According to political scientist Robert Putnam, *social capital* refers to the networks, norms and sense of trust that enable persons to work effectively together. The norms and networks that enable reciprocity and successful collaboration are largely developed outside of government in labour unions, churches and a variety of COMMUNITY organizations.

A myriad of NEIGHBOURHOOD, civic, philanthropic, faith-based, ethnic, and even recreational and membership benefit organizations characterize healthy communities. Membership in such groups generates the social capital of a community, a sense of shared fate, empathy, care and responsibility for others, and the beginning of networks that can be formed for effective community problem-solving. Through active engagement in community groups, a person also builds the skills and habits of effective democratic participation.

Putnam developed his thesis by observing the performance of the new regional governments that Italy created in the 1970s. Some of the new local governments performed excellently; others were dismal failures. According to Putnam, the variance was the result of differences in regional traditions. Some regions had a strong tradition of civic engagement; in other regions, the CULTURE emphasized private gain and patronage and residents deferred to the dominance traditionally accorded local BOSSES.

Putnam further argued that the late twentieth-century United States was suffering from a decline in social capital, signified by the decline of bowling leagues. Putnam's thesis proved very popular but was very controversial. Critics such as Everett Carll Ladd charged that Putnam had overstated the decline of communal life in America. They countered that Americans were not living isolated lives but enjoyed a rich associational life where new fraternal, social and voluntary organizations (including women's groups, environmental clubs and even youth soccer leagues) emerged in the place of unions and other more traditional organizations.

Government actions can affect a city's supply of social capital. In the United States, urban renewal programmes that replaced SLUMS with high-rise public housing unfortunately and inadvertently destroyed the dense social networks that had characterized the old neighbourhoods; the sense of neighbourhood and feelings of mutual dependence and responsibility were not re-established in the high-rises.

A healthy community requires continued investment in its social capital and what the National Civic League has termed its 'civic infrastructure'. Advocates of social capital argue for public policies that promote PUBLIC–PRIVATE PARTNERSHIPS, that extend the reach of community and civic organizations, and that foster new opportunities for intergroup collaboration and cooperation.

Further reading

Putnam, R.D. (2000) *Bowling Alone: The Collapse and Revival of American Community*, New York: Simon & Schuster.

Putnam, R.D., Leonardi, R. and Nanetti, R.Y. (1994) *Making Democracy Work: Civic Traditions in Modern Italy*, Princeton, NJ: Princeton University Press.

MYRON A. LEVINE

SOCIAL ECOLOGY

Social ecology represents a general framework for understanding the interrelations between individual and collective behaviour and the physical, institutional and socio-cultural contexts in which behaviour occurs. The concept has been employed to study a diverse array of social problems and policies within the behavioural and social sciences. In URBAN STUDIES, for example, social ecology is often used to reflect the dynamic interplay between characteristics of people and places to understand the spatial distribution of

populations and social behaviour. Accordingly, the traditional units of analysis have been characteristics of the NEIGHBOURHOOD, city, region or nation as they interact with demographic and social variables (e.g. race, family status, POVERTY).

The field of ecology has its roots in biology, which developed the concept to explore the inter-relations between animal and plant life and their surrounding habitat. Sociologists, particularly those of the CHICAGO SCHOOL, applied the biologists' concepts to problems of the human COMMUNITY. HUMAN ECOLOGY emphasizes biological, economic and geographic factors as they relate to individual and population behaviour and well-being; however, the field has been evolving to include the role of CULTURE, architecture and social dynamics as well. Social ecology represents the next step in this evolution, placing greater emphasis on socio-cultural and institutional contexts than did earlier conceptions of human ecology.

A social-ecological perspective encompasses four core assumptions concerning the nature of environment–behaviour transactions. First, the perspective focuses on the interactions among attributes of individuals and groups (e.g. attitudes, racial composition) and facets of the environmental setting (e.g. density). The second assumption concerns the multidimensionality of environments. For example, environments can be described in terms of their subjective or objective qualities, their proximity to individuals and groups, and composite variables such as person–environment fit or social climate. In addition to describing features of the environment, a social-ecological perspective may examine behaviour from multiple levels of analysis, focusing on individuals, groups, organizations or populations. Fourthly, social ecology contains concepts from systems theory such as interdependence, homeostasis, negative feedback and deviation amplification as a means to capture the dynamic nature of environment–behaviour relationships in which the environment shapes individual and collective behaviour and individual and collective actions in turn shape the environment. Finally, environments are characterized as a set of nested contexts in which local, proximate settings are embedded in more distal ones; a social-ecological framework considers the interdependencies among these layers.

While social ecology affords a more interdisciplinary and holistic perspective, one limitation is that analyses can be overly broad with respect to illuminating the most critical factors at each level of analysis. Identifying high-impact leverage points and focusing on the most problematic phenomena, particularly among the most affected populations, are two ways that social-ecological analyses may yield keener, more instructive insights.

Further reading

Park, R. and Burgess, E. (eds) (1925) *The City*, Chicago: University of Chicago Press.
Stokols, D. (1992) 'Establishing and maintaining healthy environments: toward a social ecology of health promotion', *American Psychologist* 47(1): 6–22.

JENNIFER E. GRESS

SOCIAL HOUSING

Social housing refers to housing for lower-income households that is made more affordable through public subsidies. Subsidies typically reduce capital or financing costs, but may be used to reduce operating costs as well. Usually, such housing is offered for rent. Social housing is generally rationed by some form of means testing or administrative measures of housing need.

In North America, social housing policies of the national governments have been shaped by the widely held view that primary responsibility for the provision of housing belongs to the private sector. In both Canada and the United States, the term 'public housing' is used to refer to low-income housing units in developments that are owned and operated by a local government agency. Public housing represents less than 2 per cent of all rental housing in both the United States and Canada.

The United States Public Housing programme began during the Great Depression of the 1930s. It was envisioned as a means of helping the 'deserving poor', those who were temporarily disadvantaged but who were expected to move up and out as the economy recovered. The supply and rental rates of public (rental) housing were limited so that it would not compete with the private housing market. Income restrictions were placed on occupants to force them back into the private housing market as soon as their economic circumstances had improved.

Some 1.3 million public housing units had been built by the mid-1990s. New York City has built over 150,000 units, more than 10 per cent of the national stock of public housing. Chicago and Philadelphia, along with the Commonwealth of Puerto Rico, also have large stocks of public housing. In Canada, a total of some 205,000 public housing units were built between 1949 and 1984, 55 per cent of them in Ontario, most often in large cities.

The term 'public housing' is not synonymous with social housing, however, since both governments provide housing assistance to low-income households that reside in privately owned housing as well. The US and Canadian governments in the early 1960s introduced a number of new subsidy programmes that were aimed at low- and moderate-income households. These new developments were owned and managed privately, usually by non-profit organizations. By the mid-1990s, the number of privately owned social housing units produced exceeded 1.8 million units. For a number of reasons (newer units with more amenities, higher income limits, better management), these privately owned social housing units were more attractive to prospective tenants than the older public housing projects. The Canadian government also shifted to private sector ownership of social housing, with about 250,000 non-profit and cooperative units built since the early 1980s.

US housing policy has shifted even further towards reliance on the private sector through the increased use of housing vouchers. This programme allows participating low-income households to rent privately owned units, with part of the monthly rent paid by the federal government. More than 1.4 million, primarily single-parent, households participated in this programme in 1997.

In general, Western European countries have a much higher proportion of social housing than is the case in North America, with most EU members having at least 10 per cent of their housing in the social sector. The tenure share of social housing is highest in the Netherlands (36 per cent), Sweden (31 per cent), United Kingdom (26 per cent), Austria (21 per cent) and Germany (20 per cent). Belgium and the Mediterranean countries (Portugal, Spain and Greece) have relatively low proportions of social housing.

The providers of social housing vary from country to country. Local authorities typically have been the major providers in the United Kingdom and Ireland, although not-for-profit housing associations have become more prominent in recent decades. In most other EU countries, housing associations are the dominant model.

The differences in the proportion of social housing have little to do with overall housing quality. Among EU member countries with approximately equivalent GDP per capita, there is little difference in housing quality indicators such as space or crowding, regardless of the proportion of the stock in the social housing sector.

Along with the shift to private sector ownership of social housing, there has also been a shift to a decline in the tenure share of social housing in most EU countries. In part, this has been the result of the rise in owner occupied housing as the public investment in social housing has declined. In the United Kingdom, the sale of council housing to sitting tenants has also contributed to this shift.

Over the past two decades, there has been a significant shift from social housing development to providing housing aid to households. In 1984, for example, Germany, the United Kingdom and France each devoted more than half of their housing assistance expenditures to housing unit subsidies. By 1998, the proportion had declined by more than half in the United Kingdom and France, and by almost 20 percentage points in Germany.

In the prosperous countries of North America and Western Europe, social housing has undergone a significant transformation since 1980. Production levels of new social housing units have declined, often accompanied by a shift to person-based subsidies. At the same time, housing associations and other not-for-profit organizations have increasingly been favoured over municipal ownership. In the United Kingdom, the privatization of much of the existing stock of council housing has brought about a significant shift in tenure share. These changes are consistent with the increasing predominance of market-driven economies in post-shelter societies.

Further reading

Abrams, C. (1964) *Man's Struggle for Shelter in an Urbanizing World*, Cambridge, MA: MIT Press.
Maclennan, D., Stephens, M. and Kemp, P. (1997) *Housing Policy in the EU Member States*, Luxembourg: European Parliament Directorate General for Research.

Vale, L. (2000) *From the Puritans to the Projects*, Cambridge, MA: Harvard University Press.

Varady, D., Preiser, W. and Russell, F. (eds) (1998) *New Directions in Urban Public Housing*, New Brunswick, NJ: Center for Urban Policy Research, Rutgers University.

GARY SANDS

SOCIAL INEQUALITY

Social inequality refers to the disproportionate distribution of resources and/or rewards among different individuals, social groups and/or segments of society. Social inequality usually implies the lack of equality of outcome, but may alternatively be conceptualized in terms of the lack of equality of access to opportunity. The existence of a high degree of social inequality is usually considered morally unacceptable, particularly by those who understand it in terms of a lack of equality of outcome. However, those who prefer to view social inequality in terms of access to opportunities will often tolerate unequal outcomes as long as there is relative equality of opportunity. Therefore, differences in approaches to social inequality are tied closely to differing notions of SOCIAL JUSTICE.

Social inequality is related to the concept of social stratification, in which society is hierarchically divided into sub-groups, based on class, race, gender, religion and/or political power. A highly stratified society is one in which there is minimal intra- and inter-generational social mobility. Disagreements exist between neo-liberal, functionalist, Weberian, Marxist and post-structuralist theorists concerning how social stratification and inequality should be understood, and the degree to which social stratification and/or social inequality is a necessary, desirable or unavoidable feature of modern industrial society. Nonetheless, most agree that social inequality and stratification are present in some form, albeit to different degrees, in all cities and nations.

Although a certain amount of social inequality may be voluntary and unstructured (for example, some individuals or social groups may voluntarily decide to work fewer hours, etc.), it is generally agreed that larger social, economic, cultural and political institutions are important factors producing and structuring the articulation of social inequality. Social inequality may be the direct or residual result of the structure of the labour,

housing and property markets; of systemic GENDER DISCRIMINATION, racism, and DISCRIMINATION based on religion or ethnicity; of differences in family structure or of state policy towards different social groups (based on age, family status, etc.); or of a caste system. A prominent perspective sees social inequality within and among cities (and nations) as the result of UNEVEN ECONOMIC DEVELOPMENT, rooted in the contradictions of CAPITAL ACCUMULATION.

Social inequality is most commonly measured in terms of individual, family or household income level, and the most common tool used to measure income inequality is the INDEX OF DISSIMILARITY (with the individual, family or household as the unit of analysis). However, studies of social inequality may also examine, among other things, occupational or educational attainment; underemployment and unemployment; mobility within and between social strata; differences in health and mortality rates; rates of CRIME, victimization and incarceration; and access to public benefits and services, or to private services.

Social inequality may be geographically articulated in different forms in the social and physical space of cities, regions and nations. Within URBAN areas, social inequality is typically expressed spatially as the uneven distribution or SEGREGATION across NEIGHBOURHOODS of people sharing certain attributes or social categories, such as income, race, religion and gender. Again, among the range of statistical tools employed to measure segregation and spatial forms of inequality, the most common is the index of dissimilarity, in this case with the neighbourhood or other spatial unit as the unit of analysis. DIVIDED CITIES are said to be the result of a high degree of both social and spatial inequality within urban areas (see also QUARTERED CITY).

Explicitly spatial processes may operate independently from (aspatial) social, economic and/or cultural processes in influencing the articulation of social stratification and social segregation. In the United States, for instance, not only has racial discrimination in hiring practices reduced blacks' job opportunities, but POVERTY, coupled with HOUSING DISCRIMINATION, has worked to spatially segregate blacks into urban neighbourhoods characterized by low accessibility to places of potential work, a situation known as the 'SPATIAL MISMATCH' between the location of labour and employment. Both these spatial and aspatial

processes operate somewhat independently, but nonetheless forcefully, to reduce employment opportunities for many blacks living in urban areas. It should be noted that not all instances of spatial concentration are also cases of socio-spatial inequality. Many ethnic groups, for instance, voluntarily cluster in particular neighbourhoods or sections of town in order to be able to draw on COMMUNITY resources that improve their standard of living. The usual test in such cases is the degree to which such social segregation is forced or compelled. Social geographers in particular have been at the forefront of studies examining the socio-spatial impact of such phenomena as ECO-NOMIC RESTRUCTURING, MIGRATION and IMMI-GRATION, GENTRIFICATION, GATED COMMUNITIES, and REDLINING within urban regions.

In both its aspatial and spatial manifestations, social inequality is closely related to the concept of SOCIAL POLARIZATION. Although occasionally considered analytically distinct, social polarization, at least when defined as a frequency distribution with an 'hourglass' shape characterized by lower shares of the population in the middle of the distribution, is but one form of social inequality. Other common forms of social inequality include the pyramid-shaped distribution in which one finds a greater share of the population (persons, households, neighbourhoods) in each subsequent segment lower down the ladder, the reversed pyramid-shaped distribution in which greater shares of the population are found in segments higher up the ladder, as well as the situation in which relatively equal shares of the population are found across different income (or social) categories.

Further reading

Bourne, L.S. (1997) 'Social inequalities, polarization and the redistribution of income within cities: a Canadian example', in B.A. Badcock and M.H. Browett (eds) *Developing Small Area Indicators for Policy Research in Australia*, Adelaide: University of Adelaide Monograph Series 2, pp. 21–44.
Goldsmith, W.W. and Blakely, E.J. (1992) *Separate Societies: Poverty and Inequality in U.S. Cities*, Philadelphia, PA: Temple University Press.
Hamnett, C. and Cross, D. (1998) 'Social polarization and inequality in London: the earnings evidence 1979–1995', *Environment and Planning C: Government and Policy* 16: 659–80.
Sen, A.K. (1992) *Inequality Reexamined*, Cambridge, MA: Harvard University Press.

R. ALAN WALKS

SOCIAL JUSTICE

The concept of social justice is elusive and it is unlikely that a generally accepted definition of social welfare can be formulated for contemporary URBAN systems. It is nonetheless a consuming topic in URBAN STUDIES. At its most base, justice is a set of principles for resolving conflicting claims. Social justice is a particular application of these principles when conflict arises out of the need for social cooperation when individuals and/or groups collide in getting needs met or attaining advancement.

Contemporary academic focus on 'social justice and the city' began with the publication of David HARVEY's (1972) eponymous volume on the topic. This complex, controversial text shattered traditional boundaries and categories of how scholars approached urban social issues. The book is divided between what Harvey calls liberal transformations and social formulations. The former are the staples of geography and urban studies, focusing on spatial differentiation, and the relations between people's activities in spaces and the distribution of opportunities. Social justice, in this context, is about the division of benefits and the allocation of burdens to specific populations in the urban arena. Harvey's notion of social formulations was more radical, focusing on the interdependencies between spatial phenomena and modes of production. Social justice is broadened to include not only the division of benefits but also the allocation of burdens arising out of the socio-spatial processes of undertaking joint labour. The book had an immediate and spectacular effect on urban studies, virtually redefining the objects of urban analysis. Harvey argued that all social processes are inherently spatial and, concomitantly, all spatial forms incorporate social processes. The dovetailing of spatial concerns with post-Marxist theory on social relations opened a wealth of concern about the precise nature of social justice, spatial theory and the effects of URBANISM.

Of course, concern about social justice predates Harvey's opus. For much of the twentieth century, academic research focused upon who lived where, tensions over who had what, and to what parts of the city they had access. In the 1970s, sociologist Gerald Suttles argued that the effects of market, political and legal forces on specific groups actually differentiate urban communities. Boundaries between, and the central identity of, NEIGHBOUR-HOODS and functional areas are provided by

the relationships between patterns of social interactions and the settings in which they occur. Ethnic and racial groups, occupational status and people with similar lifestyles, he argued, provide the basis for urban differentiation out of which social justice arises. In the 1980s, Peter Saunders took this point further, arguing that the problem of space should be severed from concern with specific social processes, although he conceded that all social processes occur in a spatial and temporal context. Heavily influenced by the Chicago School of Urban Ecology, these perspectives viewed space simply as a container of people's activities. During the 1970s and 1980s, urban researchers posed sophisticated quantitative and conceptual models that sought to understand spatial differentiation but the application of urban social justice always eluded these beguiling generalizations.

With the English translation of Henri LEFEBVRE's *The Production of Space* (1991), attention was refocused on the importance of understanding the relations between theoretical abstractions and concrete social outcomes. Lefebvre articulated a commitment to the geographical and the local as key issues in realizing social justice. Pulling from this work, Ed Soja notes that a growing COMMUNITY of scholars and citizens is beginning to think about the *spatiality* of urban life in much the same way that past wisdom persistently focused on its intrinsically revealing historical and social qualities. With its focus on the ways different groups produce space rather than on how they act out in space, the 'spatial turn' in scholarship and popular culture suggests a way of thinking about social justice that was undertheorized in urban studies prior to the 1990s.

In the last decade, new expressions of social and spatial justice arose through identity politics. Empirical studies focusing on everyday lives, articulated especially but not exclusively by feminist scholars, suggest new ways of understanding the varied and multi-layered contexts of day-to-day urban experience. As a consequence, research on difference and diversity is perhaps the most enduring hallmark of contemporary social justice studies. Much of this change resonates with an intensification of political struggles by women, minorities and other oppressed groups. Feminists, students of queer theory, anti-racist scholars, and scholars concerned with ability and disability raised academic consciousness about the ways in which difference promotes or enervates social justice.

Of particular influence in this new contextualized thinking about social justice is Iris Marion Young's (1990) elaboration of *Justice and the Politics of Difference*. Her enduring point is that although contemporary society appears committed to equality, at the ideological level injustices towards those categorized as 'other' are veiled in everyday habits and cultural meanings. The geographic form of Young's social justice challenges the liberal democratic model of small, decentralized, autonomous communities for urban decision-making such as HOMEOWNERS' ASSOCIATIONS and community interest developments. Instead, she wants to promote social justice through the promiscuous mingling of different peoples that Lewis MUMFORD long ago identified as the essence of URBANITY. Her solution is to highlight the 'inexhaustible' experience of difference in cities by contriving larger-scale regional authorities that can ensure power is not based solely upon local sovereignty. By so doing, Young is promoting space and scale as deliberate political acts; hers is a space where groups and classes can constitute themselves and recognize one another. Although Young's solution may appear utopian, it is important for urban studies precisely because it points to the problematizing of sites – and spaces – within the city as social practices as well as political acts. If social justice is possible, it must begin by intervening in the contradictions and tensions of existing society.

Further reading

Harvey, D. (1972) *Social Justice and the City*, Oxford: Basil Blackwell.
—— (2000) *Justice, Nature and the Geography of Difference*, Oxford: Basil Blackwell.
Lefebvre, H. (1991) *The Production of Space*, Oxford: Basil Blackwell.
Young, I.M. (1990) *Justice and the Politics of Difference*, Princeton, NJ: Princeton University Press.

STUART C. AITKEN

SOCIAL POLARIZATION

Social polarization is a state and/or a tendency denoting the growth of groups at the extremities of the social hierarchy and the parallel shrinking of groups around its middle. A polarized social structure would resemble an hourglass, whereas a

non-polarized one would be expanded in the middle.

Social polarization is related to images of social dichotomy and to impressionistic concepts such as 'two-thirds society' and 'divided' or 'dual' cities. It is frequently used inappropriately, however, interchangeably with increasing SOCIAL INEQUALITY (which refers to the distance between social extremes and not to the shape of individuals' or households' distribution along this distance) or SEGREGATION (which is a spatial expression of inequality not necessarily related to polarization).

Immigrants and minorities are an important part of the polarization image since they are systematically participating and largely overrepresented in the growth of the lower social pole. In post-industrial societies, members of the dominant groups in multi-ethnic and multi-racial countries as well as natives in countries with reduced diversification along these lines have presented an increased social mobility in the post-war decades. The emptied lower echelons of the social scale have been massively invested by minorities and especially by new immigrants that often lack basic citizenship rights and are relegated to the status of pariahs.

The social polarization thesis was first developed in the early 1980s, and is subsequent to the mutually competing (theoretically and politically) theses of *professionalization* and *proletarization*, which maintained the unilateral growth of the upper/lower pole of the social structure respectively.

Saskia SASSEN gave theoretical substance to the polarization thesis in the early 1990s, relating it to the emergence of new social processes in the GLOBAL CITY. Globalization and ECONOMIC RESTRUCTURING have immensely upgraded, according to this author, the role of a few leading cities of the advanced capitalist world (like New York, Tokyo and London) in the management of the world economy. The geographic dispersion of the corporate structure and activity around the world has reinforced the strategic role of such cities, not necessarily as the locus of corporate headquarters, but mainly as spaces of concentration and synergy between the high-end services to multinational corporations.

The proliferation of these services, ranging from stock markets and firms of legal advisers to computer analysts and accountancy services, has an important impact on the LABOUR MARKET, since they primarily attract highly skilled and remunerated professionals. This growth in the upper pole of the occupational structure produces a new kind of demand, with consumption models being increasingly distantiated from the mass consumption products and services of Fordism. In a deregulated labour market, the demand for more personalized products and services tends to reinforce informal production and the black economy, leading to the growth of the lower occupational echelons as well. Moreover, polarization is intensified by the important reduction in the semi-skilled industrial jobs that occupied the middle and formed the backbone of the occupational structure in the context of developed Fordism, both in terms of status and revenue.

Economic restructuring has therefore a socially polarizing effect which is specific to global cities because of their strategic role in the globalizing economy and particularly because of the impact of this role on their labour markets. The polarization thesis is closely related to the claim that national states lose a great deal of their regulating power in a globalizing economy and what happens to the social structures of global cities is much more the effect of international economic structures and processes (related to the accelerated and unobstructed movement of capital and labour) than that of national policies of social regulation.

Sassen's polarization thesis has been criticized on several grounds, but foremost for being an oversimplified version of the changes in the social structures of the leading world cities, and at the same time for being too dependent on the US situation. Chris Hamnett and Edmond Preteceille have shown that in London and Paris the occupational structure is tending towards professionalization rather than polarization. Evidence on US cities has not been interpreted unanimously as polarization either. Peter Marcuse writes about a 'quartered' instead of a dual New York, while John MOLLENKOPF and Manuel CASTELLS are not comfortable with the reduction of the complexity of New York's social structure to the dichotomy between the two extremes of the scale of income distribution.

The comparison between European and US global cities has brought to the fore the question of different welfare arrangements and their impact on social polarization. Costa Esping-Andersen's work

on welfare regimes has served as an interpretative basis for the different impacts of polarization tendencies on the social structures in different welfare contexts. Liberal types of welfare states, like the United States, tend to leave a much wider margin to market mechanisms in exercising economic coercion to accept low-paid and precarious jobs, whereas the social-democrat welfare model in Scandinavian countries tends to protect citizens from such coercion, and to reduce growth at the lower end of the social structure. Moreover, housing policies with an ecumenical approach in the latter have reduced to a certain extent segregation tendencies, which in the liberal context tend to appear as a corollary to polarization. Such differences substantiate the claim that politics and policies still matter and that social polarization is not the inevitable product of economic restructuring and globalization.

Further reading

Hamnett, C. (1994) 'Social polarisation in global cities: theory and evidence', *Urban Studies* 31: 401–24.
—— (1996) 'Social polarisation, economic restructuring and welfare state regimes', *Urban Studies* 33: 1407–30.
Mollenkopf, J.H. and Castells, M. (eds) (1991) *Dual City: Restructuring New York*, New York: Russel Sage Foundation.
Sassen, S. (1991) *The Global City: New York, London, Tokyo*, 2nd edn, Princeton, NJ: Princeton University Press.

THOMAS MALOUTAS

SOCIALIST CITY

The concept of the socialist city has its roots in the attitude that cities are a product of socio-economic systems. Moscow was the first large city reshaped and planned according to the socialist ideology. It became an important model for the other cities in the expanding twentieth-century socialist world. The first landscaped characteristic identified as socialist was the Stalinesque skyscraper, and domination of the 'symmetrical patterns'. The formation of a universal socialist URBAN pattern, however, was constrained by the historically short period of the ideological impacts. The resisting factors were also the different levels of URBANIZATION, history and CULTURE at the starting point of socialism in the particular country.

Under socialism, the main process in the formation of the socialist society was forced INDUSTRIALIZATION that had a decisive impact on the formation of a new city and the transformation of an existing city to meet the requirements of the new ideology. The most significant features of the type of socialist city developed under communist rules are as follows.

In general, URBAN DESIGN and town planning under socialism followed 'Western' patterns. The absence of private property and land value, particularly in central locations, resulted in the existence of empty spaces. Uniformity of architecture and urban landscape created a higher proportion of wasteland, and led to the deterioration of old quarters of cities. Physical planning aimed for the permanent redistribution or elimination from city space of any non-communist symbols. Ignorance of the environmental problems caused by industry and urban development created ecological catastrophes in some industrial regions. Architectural design and physical spatial structure were not prepared for the automobile.

The city was dependent on the central government for its finances and was 'organizationally divided'. The centralized authoritarian system had split off different decisions concerning the city, which came from different government departments and, at the smaller scale, from the authorities of particular cities. Municipal offices became units subordinated to the state administration.

The objective of the ruling communists, which was imposed on planners and builders, was to quickly provide housing for the labour force. The construction of large and homogeneous housing estates became the norm. Adequate service facilities, due to the constant investment shortages, usually lagged behind the development.

Inhabitants of the socialist towns and cities consisted mainly of the working class – the proletariat. Authorities controlled the inflow of people and recruited the labour force from those social categories acceptable from an ideological point of view. The egalitarian principle resulted in a relatively low level of wealth differentiation. The central allocation of inhabitants in relation to the localization of dwellings created a social mix that, together with the organization of the social life of urban dwellers around the place of work, diminished the chances of creating local communities.

Further reading

French, R.A. and Hamilton, F.E.I. (1979) *The Socialist City: Spatial Structure and Urban Policy*, Chichester: John Wiley & Sons.

Smith, D. (1989) *Urban Inequality under Socialism: Case Studies from Eastern Europe and the Soviet Union*, Cambridge: Cambridge University Press.

GRZEGORZ WECLAWOWICZ

SOLERI, PAOLO

b. 21 June 1919, Turin, Italy

Key works

Arcosanti Project, Scottsdale, Arizona.
(1970) *The Bridge Between Matter and Spirit is Matter Becoming Spirit: The Arcology of Paolo Soleri*, New York: Anchor Press/Doubleday.
(1983) *Arcosanti: An Urban Laboratory?*, Arizona: The Cosanti Press.

Italian architect, constructing an ecological city with his sympathizers in the Arizona Desert in the United States, Paolo Soleri was granted his Ph.D. in architecture from the Torino Polytechnico in 1946. He spent a year and a half in fellowship with Frank Lloyd WRIGHT at Taliesin from 1947, after his exile in France for staying out of Fascism. Though he left for Italy because of a difference in views from Wright, he returned to and settled in Arizona in 1956.

Soleri's concept of 'Arcology' is the synthesis of architecture and ecology as the philosophy of democratic society, inspired by the cosmogony of French philosopher and palaeontologist Pierre Teihard de Chardin. Arcology is a gigantic URBAN structure using an analogy of living things with complexity and miniaturization, which aims at both maximum interaction and accessibility in a high density inside and minimum use of energy and land, reducing waste and environmental pollution outside.

Soreli's major project is Arcosanti, a town for 5,000 people under construction in the desert since 1970, which is a version of 'Mesa City', a self-sufficient city model with a high-degree water system based on Arcology for 2 million people in 55,000 acres of land.

TOMOKO KURODA

SORIA Y MATA, ARTURO

b. 15 December 1844, Madrid, Spain;
d. 6 November 1920, Madrid, Spain

Spanish road engineer, politician, philosopher and journalist, Soria y Mata is widely known for the LINEAR CITY concept. Interested in new communications and transportation systems, as well as in traffic problems, he invented telegraphic apparatus and worked to have Madrid equipped with telephone networks, a subway system and tramway lines, achieving the city's first tramline in 1875. The belief that city form should be derived from the necessities of locomotion led him to a planning theory based on rapid URBAN transit by rail, further developed as the linear city concept in a series of articles for the Madrid newspaper *El Progreso* from March 1882 on.

The structure of the linear city would be composed of 'a single street of 500 meters' width and of the length that may be necessary, whose extremities could be Cadiz and St. Petersburg, or Peking and Brussels'. A 40-metre-wide central spine would contain rail lines, a BOULEVARD, the main INFRA-STRUCTURE, gardens, public services and equipment. Along two parallel strips on either side, housing, work places and recreation facilities would be interconnected by secondary streets, perpendicular to the main strip. Housing would consist of single-family units in 400-square-metre lots, on a 20 per cent ground occupation. The linear city was meant to 'ruralize the city and urbanize the countryside', and to be universally applicable as a ring around existing cities, as a strip connecting two cities, or as an entirely new linear town across an unurbanized region.

Soria defended his ideas in lectures, booklets, and in his magazine *La Ciudad Lineal* (*The Linear City*), which ran for nearly three decades. His Compañia Madrileña de Urbanización (Madrid's Urbanization Company) was established in 1892 to apply the concept to the outskirts of Madrid, as a 30-mile ring between the villages of Fuencarral and Pozuelo de Alarcón. After the crucial approval for a loop tramway line, construction began in 1894 but only a minor section was completed by the 1910s, due to a lack of capital to purchase the necessary land, and difficulties in adapting the concept to existing property boundaries.

Despite the practical failure and the model's diagrammatic simplicity, the linear structure,

affording unlimited possibilities for urban extension, proximity to the countryside, and an open and functional pattern based on traffic flow, was highly influential in twentieth-century URBANISM. Soviet architects in the 1920s, such as Milyutin in several linear proposals, and Leonidov in the Magnitogorsk new town project (1930), considered it an ideal tool for production; LE CORBUSIER used the concept in his proposal for the development of Zlin in Czechoslovakia (1935), and as one of the three ideal human settlements in his thesis *Les trois établissements humains* (1945), without crediting Soria.

Further reading

Collins, G. and Flores, C. (eds) (1968) *Arturo Soria y la Ciudad Lineal* (*Arturo Soria and the Linear City*), Madrid: Revista de Occidente.

SILVIO ABREU

SORKIN, MICHAEL

b. 2 August 1948, Washington, DC

Key works

(ed.) (1991) *Variations on a Theme Park: Scenes from the New American City*, New York: Farrar, Straus & Giroux.
(2001) *Some Assembly Required*, Minneapolis, MN: University of Minnesota Press.

Architect, professor, critic and URBAN designer, Sorkin is one of the most provocative and polemical (yet humorous) voices in contemporary CULTURE and in the design of urban places at the turn of the twenty-first century. He was educated at the Universities of Chicago, Columbia and Harvard and at MIT. In the 1980s, Sorkin was the sole architectural critic for the *Village Voice* in New York City. He has held visiting professorships and named chair positions at over twenty-five schools of architecture. In the 1990s, Sorkin was director of and professor of URBANISM at the Institute of Urbanism at the Academy of Fine Arts in Vienna. Currently, he is professor of architecture at CUNY and principal of the Michael Sorkin Studio, New York City.

Highly respected and internationally known for both his theoretical and his experimental work, Sorkin's main premise is the reinvention of architecture based on the idea that the contemporary city is both the primary source of architecture's social meaning as well as its main imaginative challenge. The Michael Sorkin Studio is known for its imaginative drawings, models, exhibitions, writings and installations.

Further reading

Sorkin, M. and Zukin, S. (eds) (2002) *After the World Trade Center: Rethinking New York City*, New York: Routledge.

DAVID C. PROSPERI

SPATIAL MISMATCH

Since its conceptualization in the late 1960s, the spatial mismatch hypothesis has been widely cited to explain the economic problems encountered by inner-city minorities. Scholars have noted an increasing geographic separation between employment opportunities and African-Americans, who have remained trapped in urban GHETTOS by HOUSING DISCRIMINATION while jobs have decentralized into the SUBURBS. Although the central business district remains the largest job centre in most metropolitan areas, relatively few inner-city residents possess the skills for many of the remaining jobs. Moreover, the available positions are low-skill jobs subjected to intense competition. Geographic distance limits access to suburban jobs by hindering job search and imposing long and costly commutes. The scarcity of nearby employment and the physical barriers to faraway jobs lead to high unemployment and depressed incomes. Spatial mismatch has also been tied to the development of an URBAN UNDERCLASS. African-American communities have experienced class SEGREGATION driven by a middle-class exodus from the established ghettos. This out-migration has weakened COMMUNITY institutions and social networks, created a paucity of positive role models and undermined local businesses. The result is the formation of underclass NEIGHBOURHOODS, characterized by extensive joblessness and social alienation.

The economic problems associated with physical distance are aggravated by three related place and space phenomena. The first is a racial disparity in transportation, which determines the relative ability to overcome distance. The URBAN structure

is predicated on the automobile, but inner-city minorities are disproportionately dependent on a public transit system that tends to increase the burden of commuting. The low car ownership rate in the urban core is due to both low income and higher operating costs, particularly higher insurance premiums. Transportation mismatch magnifies the negative impacts of geographic distances. The second phenomenon is weak informational networks. Residential segregation diminishes informal interactions with individuals, organizations and employers outside minority neighbourhoods. Weak external linkages create an additional barrier to economic opportunities. Finally, employers are reluctant to hire people from inner-city neighbourhoods. Firms often avoid recruiting in these areas, and applicants are at times stigmatized by stereotypes ascribed to their neighbourhoods.

Spatial mismatch is linked to public policy. Urban policies have contributed to the isolation of minorities. Transportation funding for highways and tax incentives for home ownership have facilitated SUBURBANIZATION, while housing assistance programmes have confined most subsidized units to the urban core. Ghettos persist in part because fighting housing discrimination has not been a national priority. Spatial mismatch, in turn, impedes the effectiveness of public policies. For example, geographic barriers to employment opportunities have hindered the implementation of the 1996 welfare reform. Many recipients do not have adequate transportation to find and hold jobs far away from their homes. Spatial mismatch is also tied to the failure of the urban education system. The same urban structure that isolates minorities from economic opportunities also concentrates minority students in low-performing, underfunded, inner-city public schools. Inadequate education will handicap them as adults in the LABOUR MARKET, thus reproducing economic inequality across generations.

Empirical research generally supports the spatial mismatch hypothesis. The evidence regarding the relevant urban features is consistent and compelling. Despite some progress over the last couple of decades, racial segregation remains a prevalent and unfortunate reality. The DECEN-TRALIZATION of employment has not been well documented, and there is little sign of a reversal. Research on the independent impact of spatial mismatch on employment outcomes is supportive but not conclusive. Labour-market studies find a correlation between low job access and adult joblessness, but the correlation does not necessarily prove that spatial mismatch causes the negative outcome. The statistical relationship may be associated with a reverse causality, i.e. those with weak labour-market attachment choose to live in job-poor neighbourhoods. Studies showing an adverse impact on youth employment are less likely to be biased by this reverse causality, thus lending credence to the hypothesis. Additional evidence comes from research on inner-city residents relocated to the suburbs. Chicago's Gautreaux programme, which moved several thousand black families out of the inner city during the 1970s and 1980s, produced positive outcomes, with the major benefits going to children. Evaluations of the US Department of Housing and Urban Development's Moving to Opportunities demonstration programme indicate that some results can be replicated in other metropolitan areas. The impact of relocation on employment, however, appears to be minimal, at least for the short run. This implies that greater proximity to employment opportunities may not be sufficient to overcome other disadvantages. Finally, spatial mismatch appears to be marginally applicable to poor Latino and Asian neighbourhoods. These areas have a visible economic base, and a large majority of the residents are able to secure employment outside their neighbourhoods.

Despite its limitations, spatial mismatch is still useful to understanding an important dimension of the economic problems of the inner city. Put in proper context, the hypothesis provides insights into how the existing urban structure reinforces and accentuates racial disparities. Eliminating the problems requires policies and programmes to promote residential integration, improve public transit and access to automobile ownership, and bring jobs back to the urban core.

Further reading

Kain, J. (1968) 'Housing segregation, Negro employment, and metropolitan decentralization', *Quarterly Journal of Economics* 82: 175–97.

Rubinowitz, L. and Rosenbaum, J. (2000) *Crossing the Class and Color Lines: From Public Housing to White Suburbia*, Chicago: University of Chicago Press.

Taylor, B. and Ong, P. (1995) 'Spatial mismatch or automobile mismatch? An examination of race, residence, and commuting in the U.S. metropolitan areas', *Urban Studies* 32(9): 1453–74.

Wilson, W.J. (1996) *The Truly Disadvantaged: The Inner City, the Underclass, and Public Policy*, Chicago: University of Chicago Press.

PAUL ONG

SPATIAL PLANNING

Numerous authors have used the term 'spatial planning' since the 1950s with different purposes in mind. The term is also the literal translation of the name of the planning system in some other countries; for example in Germany the system is known as *Raumplannung*, which translates as 'spatial planning'. Until the 1990s, the term 'spatial' was used primarily to refer to the way that planning should deal with more than simply ZONING, land use planning, or the design of the physical form of cities or regions. Its use generally suggested that planning should also address the more complex issues of the spatial relationship of activities such as employment, homes and leisure uses.

During the 1990s, the term 'spatial planning' began to be used more widely in Europe, signalling a more significant change in the understanding of the role of planning. This change has been heavily influenced by thinking about planning at the European level and, in particular, publication of the European Spatial Development Perspective in 1999. It was also influenced by the commitment of the European Union and national and regional governments to promote more SUSTAINABLE URBAN DEVELOPMENT. Subsequently, the spatial approach has been a major feature in the reform of national planning systems at the turn of the twenty-first century.

Spatial planning seeks to influence *spatial development* in the broadest sense and particularly the obvious contradictions in public policy that contribute to unsustainable development. Spatial development refers to the physical distribution of built and natural features and human activities across a territory (city or region), and the distribution of economic and social activities in terms of their quality, for example disparities in access to opportunities from one NEIGHBOURHOOD to another. So spatial planning incorporates aspects of regional and URBAN policy that deal with issues such as social disadvantage and economic competitiveness.

If planning is to address this wider range of issues, then it needs to have some influence over policies and action in other sectors such as health, education and economic development. This is because all policy sectors will have an impact on spatial development. For example, if a health authority decides to centralize its facilities in a smaller number of larger hospitals, there are implications for the distribution of access to those facilities. In Europe, this spatial development dimension of other sector activities is known as *spatial policy*. Most zoning or land use planning systems have relatively little influence over important spatial policy decisions. Thus, the spatial planning approach concentrates on establishing better coordination among various sectoral policies, both *horizontally* across different sectors, and *vertically* among different levels of jurisdiction, for example transnational, federal, state and city.

Further reading

Cullingworth, B. and Nadin, V. (2001) *Town and Country Planning in the UK*, London: Routledge, ch. 4.
Faludi, A. (ed.) (2002) *European Spatial Planning: Lessons for North America*, Cambridge, MA: Lincoln Institute of Land Policy.

VINCENT NADIN

SPRAWL

Sprawl is the undue spread of an URBAN settlement. In that sense it was seen in every ancient town in which population pressure resulted in building outside the usual fortified walls, and the common solution was to build new walls further out, as soon as means permitted. In the post-industrial era, sprawl has entailed no clear-cut disadvantage, such as the loss of protection, nor resulted in a uniquely identifiable cost such as investment in new fortifications. The disadvantages and costs have been incremental – increased travel time, transport costs, pollution, destruction of countryside, and so on.

Sprawl is especially a characteristic of US cities, and in that country it was in effect promoted by Frank Lloyd WRIGHT in his utopian Broadacres, or Broadacre City, of 1935. At this time, British planning ideology, because it was concerned largely with the 'CONGESTION' of existing settlements, tended somewhat in the same

direction. Wright's practical influence in the United States may not have been great, but sprawl more directly resulted from the triumph of the petroleum suppliers and the automobile, at the expense of public transport.

The pejorative meaning of sprawl postdates the Second World War, and even then was at first less to do with the total spread, or the wastage of land, than with the lack of coordination, economic inefficiency and aesthetic undesirability of this type of development. Sir Patrick ABERCOMBIE said in 1948 that on the outskirts of any city, one would find 'a lot of loose, straggling building'. The problem was not the low density of the housing development itself, but the fact that the houses were interspersed among vast areas of undeveloped land.

The planning of new or substantially new cities continued to permit or even encourage such sprawl. C.A. DOXIADES' first plan for Riyadh, Saudi Arabia, in 1971, was a rectangular grid along a main spine, resembling his earlier designs for cities like Karachi, and it did not provide for the massive population growth which eventuated. In 1989, the National government imposed the Urban Limits policy, which was supposed to allow for fifteen years' development, but provided enough land for twenty-five years, and hence further encouraged the sprawl. Land was at first granted for nothing, there was no control over SUBDIVISION, there was strip development on all main roads, and public transport was (and is) nugatory. Between 1986 and 1996, the road length doubled, and the average vehicle trip increased from 10 to 14 kilometres.

In response to such conditions, the current meaning of sprawl has been refined, mainly since the 1970s, as the environmental movement – and actual petrol crises – have increased public awareness of fuel shortages, pollution and the destruction of natural countryside. It is now more generally accepted that a medium to high DENSITY is a desirable characteristic of all development, as well as a sharp cut-off between the new development and the surrounding countryside. Laissez-faire planning policies in the Western world have meant that these goals are rarely achieved, but there are cities, such as Beijing, in which they have been put into effect, with the result that sprawl has been largely eliminated.

MILES LEWIS

SQUATTER SETTLEMENTS

Squatter settlements are the autonomous, self-help housing developments created by low- and very low-income individuals and families around the world. They are prominent in developing and least developed nations where the formal sector cannot meet the ever-growing demand to house these households. The term 'squatter' is derived from the central characteristic of individuals and groups *sitting* or *squatting* upon private property that is not legally theirs. They are known throughout the world by other local names. For example, they are known as 'FAVELAS' in Brazil, 'barriadas' in Peru and Central America, 'invaciones' and 'BARRIOS piratas' in Colombia, 'BUSTEES' or 'jhuggis' in India, 'barong-barong' in the Philippines, 'GECE-KONDU' in Turkey, and 'kampung' in Indonesia.

These settlements are also referred to in the housing literature as 'spontaneous settlements'. This term is associated with perceptions about the way these settlements appear to start and evolve, out of nowhere, unplanned and not adhering to legal regulations (e.g. land use and building codes). They are also identified as SLUMS, due to the overcrowded and very poor physical conditions. However, a central difference is that traditionally, slums originate from within the formal housing sector, and are created through a process of deterioration due to neglect.

Actual understanding

From an actual perspective, some of the central characteristics describing squatter settlements include:

- informal sector development outside of government/private sector delivery programmes;
- acquired through invasions of private property without landowner's consent;
- illegally developed on vacant or underutilized land, often without legal property title;
- low security of land tenure;
- unregulated land use and construction practices;
- low initial quality with evolving capabilities (quantity as well as quality);
- initiated through well-planned invasion phase;
- incremental growth and evolution of both housing units and COMMUNITY realized through self-help labour by individuals or groups; and
- bottom-up housing approach.

Other descriptive aspects are demonstrated by squatter settlements over time, such as heterogeneity of physical quality, with improvement over the years in some cases.

Characteristics such as development opportunity and security of housing tenure (i.e. permanence through ownership) are inversely associated with the characteristics of the land selected as a site for development (slope, access, views, size). For example, combinations of these characteristics may increase the market value of a property; thus, either the expected short- or long-term security of tenure decreases through quick government response, or stringent legal enforcement leads to eventual evictions from the land. On the other hand, with sites on private or public lands where combinations of these characteristics result in underutilization or vacancy due to low development value, the likelihood of legal enforcement and eviction is considerably less, and legal systems in different countries are found to be more accommodating on behalf of squatters. Therefore, squatter settlements tend to spring up around sites with combinations of very undesirable physical characteristics, often exposing them to natural hazards, for example mud slides or flooding.

Ongoing processes and trends responsible for the need for squatter settlements as expanding alternatives to the provision of housing for the poor and very poor principally include:

- Population growth – The rapid rate of growth represents the central challenge to the carrying capacity of natural and created systems, and therefore to equitable and sustainable access to resources such as adequate land or water. This exponential growth rate is best observed over the past 200 years. In that period, we have seen the rate of growth increase from close to 1 billion in 1800, to 1.7 billion in 1900, to the 4.3 billion increase of the last century, resulting in approximately 6 billion people living on earth at the end of the twentieth century (United Nations Centre for Human Settlements 2001).
- URBANIZATION – The population growth is mostly concentrated in cities. Approximately half of the planet's 6 billion people were living in cities at the beginning of 2001. This reflects an urbanization trend last century from less than 0.3 billion in 1900 to around 3 billion in 2000 (*ibid.*). Unfortunately, urbanization is not evenly distributed. Annual rates in industrialized nations are expected to decline from 1.1 per cent to 0.6 per cent between 1995 and 2015. Projections point to an annual urbanization rate of as much as 4.6 per cent for the least developed countries, making their cities the challenge settings for the effective performance of their governments' and international agencies' programmes to provide adequate housing.
- Globalization – This is a new trend evolving from the economic sector. Its inevitable outcomes appear to increase the 'probability of land speculation, and therefore, of increased land prices' (*ibid.*). Such a trend, without institutional interventions, most likely will increase difficulties for the urban poor.
- Government and formal sector limitations – In most industrialized nations, the formal sector has been able to provide accessible, adequate, available and affordable housing to the majority of the population. Unfortunately, this has not been the case to the greatest extent in least developed and developing economies. Their inability is compounded by the observation that it is precisely in these countries where population and urbanization trends are expected to be greatest.

Therefore, in such a context, even squatter settlements as solutions to the housing problems of these countries face a bleak future when looking at how to meet their needs quantitatively, and an even bleaker one when thinking about how quality could be improved, or even maintained. As to an understanding of the actual extent of the need, accurate figures of the population living in such settlements around the world are really not available. However, it is estimated that squatter communities in many of these countries presently house from 30 to 60 per cent of their urban populations. Therefore, using these estimates and the projections that by 2015, 49 per cent of the total world population will live in the urban centres of all the least developed and developing countries, it is expected that by 2015, between 0.9 billion and 1.4 billion people will be living in such settlements (*ibid.*).

Typological understanding

Based on these trends, it becomes imperative to understand *how* these residential environments come into existence. Three distinct types of approaches to their development are prominent in Latin America: *spontaneous invasions* by squatters, *pirate invasions* or illegal developments by developers, and as *government solutions* through the development of subsidized and planned communities. This typology, with some differences in 'labels', not situations, is also representative of approaches found elsewhere in the world.

The spontaneous invasion represents the squatting of literally hundreds of families on land that is either publicly or privately owned. The acquisition of property is through the illegal process of squatting. Much has been written about the squatter settlement in terms of the conditions of these settings, their physical characteristics, and the social and behavioural processes and their outcomes (Turner 1977; Aldrich and Sandhu 1995). 'Spontaneous invasions' is definitely a misnomer. There is nothing spontaneous about them. On the contrary, they are outcomes of some of the best-organized activities by groups of families in developing or least developed countries. Examples of these settlements abound, particularly in the peripheries of most, if not all, of their cities.

The pirate invasions are principally illegally developed outside the laws of municipalities and formal sector norms. They result from a so-called 'SUBDIVISION process' of private properties. They usually take place in areas which are neither planned for development by government agencies nor served by public INFRASTRUCTURE utilities such as water, sewerage or electricity. Thus, from the start, their development is illegal. The major difference between these pirate invasions and squatter settlements is that the residential plot is bought by the individual from the 'pirate developer', who may or may not be the legal owner of the land, at costs characterized by exorbitant interest rates. However, their physical evolution in terms of dwellings and infrastructure improvements follows closely the development of the squatter settlement through the initiative and self-help practices of the residents themselves.

Governments also create residential solutions for these households. These efforts follow financial, legal and development regulations associated with formal sector housing. They are created and delivered either through 'site and services' or as 'turnkey' types of projects. Their production maximization of quantity often takes precedence over any standards sensitive to most behavioural requirements, thus leading to their failures identified in the literature. However, there are many examples of this type of development with sensitivity to the residents' behaviour, needs and economic means, such as the bamboo projects in Manizales, Colombia, or in various regions of Costa Rica.

While squatting and pirate developments represent the most common alternatives, most of the three types of residential environments come about in great part through self-help practices. These practices depend on individual and group participation and are fundamental to the improvement of the settlements' physical quality. This is noticeable over time in spontaneous and pirate settlements. For example, in cases such as 'Las Cruces' in South Bogota, the physical condition of these environments improves from cardboard shacks at the time of an invasion, to concrete block structures with public infrastructure and utilities servicing the community. In contrast, over a similar period of time (twenty years), the condition of some major US public housing projects has decreased from excellent after construction, to the point of complete demolition, as has been the case with projects such as PRUITT-IGOE in St. Louis, Missouri.

A conceptual understanding

In addition to the understandings from the definitional and actual perspectives above, squatter settlements can also be understood conceptually in terms of four basic components: the *individual*, the *group*, the *natural systems* and the *created urban systems*. That is, an environment where the resident, as the individual and as the group, constitutes the basic *non-physical social part*; and where the natural systems (topography, geology, hydrology, vegetation, etc.), with the created systems (paths, buildings, land uses, etc.) superimposed over the natural ones, compose the *physical part* of the settlement.

Using this conceptualization, these settlements are additionally defined as the totality of relationships which exist both between components of each part, as well as between the non-physical social and physical parts. It is these relationships

which allow us to understand the more complex, comprehensive and meaningful aspects of a settlement's substantive nature and processes when analysing it at any one point along its development. For example, we can begin to understand how the interactions between the components and parts can represent the settlement's dynamic capacity to include change and evolve over time; or how at the inception stage a resident as a group makes it possible to hold a claim in the invasion of a property; or how an individual ascribes meaning to his/her squatter housing; or how the fit between housing and behaviour is optimized through physical reorganization, residential relocation, or behavioural relocation.

Further reading

Aldrich, B.C. and Sandhu, R.S. (1995) *Housing the Urban Poor: Policy and Practice in Developing Countries*, London: Zed Books Ltd.
Arias, E.G. (ed.) (1993) *The Meaning and Use of Housing: International Perspectives, Approaches and their Applications*, Ethnoscapes Series, Aldershot: Avebury.
Turner, J.F.C. (1977) *Housing by People: Towards Autonomy in Building Environments*, New York: Pantheon Books.
United Nations Centre for Human Settlements (Habitat) (2001) *Cities in a Globalizing World: Global Report on Human Settlements 2001*, London: Earthscan Publications Ltd.

ERNESTO ARIAS

STADIA

Stadia are typical modern constructions, in a world where competition sport has become a universal value, and are the successors to the ancient arenas. The history of stadia, starting in Olympia (Greece), reflects the history of modern societies: Panathinaikon Stadion in Athens (revival of the Olympic Games in 1896), Wembley in London (1924, temple of football), Berlin (1936 Nazi Olympics). In the 1960s, the construction of stadia focused on the comfort of spectators. However, several football tragedies in the 1980s (the worst occurred in Moscow in 1982) have led to new security rules. Contemporary stadia are usually located at the edge of the city. Stadia for URBAN football have a moderate land usage, are relatively expensive for participants, and have high costs supported by local authorities. Stadium construction is considered a factor of URBAN REGENERATION,

urban development and, consequently, urban TOURISM. This was initiated in the United States in the 1950s, where the existence of a professional team constitutes an intrinsic characteristic of urban CULTURE. In the 1990s, inter-urban competition for franchises became increasingly intense. Britain had shown, during the 1980s and the 1990s, an unprecedented level of stadium development. Stadium construction is also considered an important design challenge; in the 1990s in particular, various polyvalent (Stade de France) or innovative (Sapporo Dome) stadia were built.

Further reading

Bale, J. (1994) *Landscapes of Modern Sport*, Leicester: Leicester University Press.
John, G. and Sheard, R. (1994, 2000) *Stadia: A Design and Development Guide*, Oxford: Architectural Press.

ALEX DEFFNER

STEIN, CLARENCE S.

b. 19 June 1882, Rochester, New York;
d. 7 February 1975, New York, New York

Key works

(1924–28) Sunnyside Gardens.
(1928–29) Radburn.
(1951) *Toward New Towns for America*, Cambridge, MA: MIT Press.

After attending Columbia University in New York and the ÉCOLE DES BEAUX-ARTS in Paris, Clarence S. Stein began his career as an architect. He apprenticed with the classicist architect Bertrand Grosvenor Goodhue in New York where his projects included the San Diego World's Fair (1915) and the copper-mining town of Tyrone, New Mexico (1916), leading him towards a lifelong involvement with housing reform. After leaving Goodhue's office, Stein organized the Regional Plan Association of America (RPAA) in 1923 with Lewis MUMFORD, Benton MACKAYE, Henry Wright and others, after which he served as its president (1925–48).

Although Lewis Mumford introduced Ebenezer HOWARD'S GARDEN CITY ideas into America, Clarence Stein adopted garden city design ideas from Raymond Unwin, the British town planner and housing reformer. Travelling to England in 1924, Stein and Wright met Howard, but more importantly visited Unwin and his two influential

projects: Letchworth Garden City (1903–14) and Hampstead Garden Suburb (1905–12). Upon returning to New York, Stein, with Wright and colleagues from the RPAA, set out to build the garden city in America.

Their first project, Sunnyside Gardens in Queens (1924–28), brought Unwin's COURTYARD housing to America. In the widely publicized project, housing units were turned away from the street to face communal recreation space in the centre of the block, rejecting traditional block SUBDIVISIONS with front yards on the street and private backyards. Stein's later projects, such as Phipps Garden Apartments (1932) and Hillside Homes (1935), provided models for higher-density courtyard housing as alternatives to traditional TENEMENTS and later, with Stein as a consultant, formed the basis for the US Housing Authority's public housing design standards.

Stein's Radburn, New Jersey (1928–29), adapted Unwin's superblocks and culs-de-sac at Hampstead into an automobile-oriented SUBURB, while adding Clarence PERRY's NEIGHBOURHOOD unit, the first hierarchical-arrangement streets, and total separation of automobiles from pedestrians. Radburn provided the model for the Roosevelt administration's GREEN BELT towns, several post-Second World War British new towns, and later projects in the United States, including Columbia, Maryland, and Reston, Virginia. The Federal Housing Administration adopted Radburn's superblocks, culs-de-sac and street arrangements into design standards for its loan guarantee programmes for suburban housing, again with Stein and others from the RPAA as consultants.

Clarence S. Stein's legacy of housing innovation was recognized with the Gold Medal of the American Institute of Architects (1955) and the Distinguished Service Award from the American Institute of Planners (1958). However, that legacy has been diminished by the bureaucratization of Sunnyside and Radburn into simplistic design standards that shaped much of the negative legacies of public housing and the post-Second World War American suburb.

Further reading

Parsons, K. (ed.) (1998) *The Writings of Clarence S. Stein: Architect of the Planned Community*, Baltimore: Johns Hopkins University Press.

RICHARD DAGENHART

STRATEGIC PLANNING

The word 'strategy' derives from the Greek *strategos*, which refers to military leadership or generalship. It entered English vocabulary in the late seventeenth century. Originally, strategy meant the overall view of what must be done to win the war, compared to tactics, which were about immediate actions needed to beat the enemy in a battle. It was after the Second World War that the idea of strategic planning became a new fashionable planning doctrine. In a general sense, strategic planning is a disciplined effort to produce fundamental decisions and actions that make it possible to achieve an organization's long-term goals.

Strategic planning originated around the early 1960s in the business sector in the United States. It was developed as a response to a rapidly changing environment and increased competition, thus providing corporations with new tools for taking care of their planning functions. From the late 1970s, it gradually came to be applied in various private and public organizations all over the developed world. Later, the resources devoted to planning in most large firms steadily declined throughout the 1980s as the expected benefits of rationalistic strategic planning often failed to materialize.

The idea of public planning long remained separate from corporate strategic planning. Professionals of URBAN PLANNING, for example, were decidedly reserved as to the applicability of business-originated models to an URBAN context. After the early 1980s, it gradually gained recognition for its applicability to public planning. And what is worth stressing, strategic planning was not, after all, so significantly different from the sophisticated comprehensive planning practised in cities. It also became clear that because of fundamental changes within governments, they needed to anticipate rather than merely react to those changes, and strategic planning was essential to that process.

A classic study behind this new planning doctrine was Igor Ansoff's *Corporate Strategy* (1965). Since then, various approaches have emerged. During the 1980s, strategic planning was largely replaced by a broader concept of strategic management. In this sense, strategic planning can be seen to represent the early 'planning-oriented' phase of strategic thinking. Present-day strategic planning is a decentralized, process-oriented,

goal-oriented and focused approach to an organization's integrated planning function. Thus, its core idea is still highly relevant: understanding what is happening in the external environment is a precondition for the attainment of an organization's goals.

The basic elements of the strategic planning process are the following: scanning the environment, identifying strategic issues, setting mission statements or broad goals, undertaking external and internal analyses (e.g. SWOT ANALYSIS), designing strategies and implementation plans to carry out actions, and lastly monitoring and assessing performance. The key outcome of this process is a strategic plan. Yet the process itself is valuable for the organization, for it serves as a comprehensive and well-structured learning process.

Further reading

Bryson, J.M. (1995) *Strategic Planning for Public and Nonprofit Organizations*, rev. edn, San Francisco: Jossey-Bass.

Mintzberg, H. (1994) *The Rise and Fall of Strategic Planning*, New York: Free Press.

ARI-VEIKKO ANTTIROIKO

STREET PEOPLE

Street people are part of the homeless population living in URBAN areas. HOMELESSNESS incorporates a much broader group than this most visible population living on the streets. The lines of demarcation between other homeless populations, for example refugees, those in shelters, migrant workers, squatters, etc., and street people are not entirely distinct. The delineation of street people is primarily determined by residential arrangement and their location in the urban setting.

Street people have no permanently fixed place to stay the night that is intended as a residence. They may stay in places ranging from PARKS to SIDEWALKS to shanties. Many develop informal residences under overpasses or bridges, in tunnels, in alleys, or in shanty towns or FAVELAS. Usually, there is no legal claim to the space and often no way to secure possessions at the informal residence. Having no jurisdictional claim to the residential space, street people are in the insecure position of facing a further displacement from

their residential area by reclamation by the titled owner of the property, a municipal decree, or more violent removal by other private citizens.

Because the life of street people primarily transpires in PUBLIC SPACE, they face great exposure to CRIME, the elements and disease, and thus they have higher morbidity rates than in the general population. Street people can informally carve out private space with a shanty, but it is a fragile privacy, easily eroded. Because the lives of street people take place in public, they must identify places for bathing, eating, retrieving potable water, and for personal relief or rest.

While in parts of the world street people include a large population with mental health or substance abuse problems, economic factors overwhelmingly create the conditions in which people end up living on the streets. Street people are among the poorest of the urban poor. Underemployment and unemployment, rural migrations to urban areas, and insufficient housing for the poor contribute to the urban POVERTY of street people. Many wind up in cities following large-scale displacement from wars or natural disasters, thus flooding the urban job and housing markets.

Street people often do not assimilate into the formal economy; rather, their subsistence comes from informal means such as begging, peddling, minor services like cleaning cars, sex work, non-violent crime like theft, or foraging. Incomes vary widely, including those with jobs in the formal economy but living in areas with inadequate affordable housing. In some cities, charities provide access to basic resources including food, medicine and clothing.

The experiences of life on the streets vary based on a combination of social, geopolitical, economic and individual factors. Many street people form communities, ranging from a group of friends to extensive shanty towns, yet others experience great amounts of social disaffiliation and isolation.

Further reading

Huth, M.J. and Wright, T. (eds) (1997) *International Critical Perspectives on Homelessness*, Westport, CT: Praeger.

Snow, D.A. and Anderson, L. (1993) *Down on Their Luck: A Study of Homeless Street People*, Berkeley, CA: University of California Press.

PHILIP RODNEY WEBB

STREETCAR SUBURBS

'Streetcar suburbs' refers to those SUBURBS in the United States that developed in the years before the automobile, when the introduction of the electric trolley or streetcar allowed the nation's burgeoning middle class to move beyond the CENTRAL CITY's borders. Typically, this initial wave of American SUBURBANIZATION developed in a finger-like or spoke pattern: dense settlements followed the streetcar tracks, but the areas between the streetcar lines remained largely rural. It took the development of the automobile to fill in previously underdeveloped areas between the streetcar tracks and to allow Americans to move beyond the streetcar suburbs.

The streetcar suburbs mark a heightening of class segmentation in URBAN America, as those who could afford the costs of new housing and the streetcar commute could now leave the city. At first, cities used their ANNEXATION powers to adjoin the neighbouring areas of growth to the city, in effect extending the city's boundaries outwards with the growth of the suburban population. But a landmark event occurred in 1873, when the residents in the growing streetcar suburb of Brookline, surrounded on three sides by the city of Boston, sought refuge from the industrial and 'corrupting' central city and refused to be incorporated into it. Brookline's resistance to annexation marked the beginning of a new wave where suburbs assertively guarded their political autonomy, their separation from the central city.

Further reading

Warner, S.B., Jr (1978) *Streetcar Suburbs: The Process of Growth in Boston, 1870–1900*, 2nd edn, Cambridge, MA: Harvard University.

MYRON A. LEVINE

STREETS

In the earliest days of URBAN development, streets tended to be named for some prominent land use or landmark, like 'Church', 'Market', 'Monument', 'Canal', 'Wall', 'Court', 'Dock', etc. Often, major streets would be named for symbols of power and authority, like 'State' or 'King' or 'Queen'; after the American Revolution, the names of heroes and leaders like Washington and Jefferson also served this role. There were important exceptions (the planners of New York's numbered streets and Philadelphia's tree names were ahead of their time), but in most American cities that was the pattern until the middle of the nineteenth century.

The street has been the most ubiquitous land use in the land use fabric. Unfortunately, streets have been considered more as facilitators of traffic to different types of land use than as a land use. When designed well, streets can be the most attractive PUBLIC SPACES that offer a COMMUNITY the opportunity to address their needs for infusing harmony and safety within their NEIGHBOURHOODS.

The streets that linger in the minds of users are often associated with the presence of people, music, unique landscaping, attractive architecture and low vehicular speeds. A successful street encourages the feeling of belonging, inspires residents or stakeholders to keep the street and the adjacent SIDEWALKS clean and well landscaped, encourages residents to have inviting front yards or stoops, and emanates an overall feeling of hospitality.

Further reading

Appleyard, D. (1981) *Livable Streets*, Berkeley, CA: University of California Press.
Vernez-Moudon, A. (ed.) (1989) *Public Streets for Public Use*, New York: Columbia University Press.

SHEILA SARKAR

STREETSCAPES

Streetscape can be taken to mean the physical elements that make up the public realm, including relationships between the buildings, floorscape, other landscaping (hard and soft) and functions that make up the STREET and the resulting space so delimited.

Being primarily a visual concept, streetscape relates to one – albeit important – part of the URBAN DESIGN agenda. Nevertheless, it will always be a mistake to look at the visual qualities of streets in isolation from the functions they accommodate and which in large part also determine the appearance of streets.

New streets are designed in their entirety and offer a one-off chance to consider the total design of the streetscape. This includes the street layout

and interconnections, the three-dimensional street form, architectural design, relationships of uses to the street, hard and soft landscaping, road design and specification, and lighting and signage. It includes the total streetscape environment. For this reason, ensuring that streetscape quality is considered in all new developments is of vital importance.

Streets start to adapt and change almost immediately after being created. Therefore, the quality of the streetscape does not rely solely on the initial design of streets, but also on the subsequent and ongoing changes. Changes provide more limited opportunities to influence overall streetscape quality but are nevertheless vitally important. Their importance reflects the fact that established streets account for most of the URBAN environment. Even high-quality streetscapes can be easily undermined by poor-quality and uncoordinated incremental changes over time.

The processes that determine the overall quality of established streets concern management and maintenance as much as design and development. Nevertheless, within the three-dimensional envelope provided by the street walls, ground surface, and already established connections with surrounding streets, a considerable range of variables remain to be influenced: the road and footpath design, the hard and soft landscaping, the lighting and signage, and changes to surrounding uses and façades.

Both new and established streets are frequently criticized for the quality of their streetscape. The problems relate to different processes inherent in the production of streetscapes, with the emphasis varying between new and established streetscapes.

New streets are criticized on several grounds:

- for being determined excessively by highways standards – road widths, specifications, safety standards, corner radii, traffic calming, signage, barriers, etc.;
- for being determined excessively by planning standards – space between buildings, parking standards, ZONING, etc.;
- for a lack of visual interest – both in townscape quality and architectural form and detail, and in the variety and mix of uses.

Depending on when and where they were created, established streets are sometimes criticized on the same grounds. They are also criticized on other grounds:

- for being dominated by traffic (and unfriendly to pedestrians) – noisy, smelly, unsafe, inaccessible and visually chaotic;
- for being poorly managed – poorly lit, dominated by parking, signage and uncoordinated street furniture, and suffering from layer upon layer of minor development activity (inappropriate shopfronts, shutters, windows and doors, cladding, boundary treatments, off-street parking, etc.);
- for being poorly maintained – poor quality materials, GRAFFITI, rubbish, damaged and poorly repaired paving, and uncoordinated or sometimes over-elaborate hard landscaping.

The streetscape quality of new streets is determined by the relationship between design and development processes, and the extent to which design is understood and prioritized by the public sector through the planning and highways adoption process, and by the private sector through development activity. For existing streets, a more complex series of processes and interactions is apparent. Incremental changes are made by a broad range of local government activities, and by the incremental changes made by property owners and small-scale development activities over time. The challenge is to coordinate the changes so that the total adds up to more than the sum of the parts.

Further reading

Carmona, M. (2001) 'Streetscape: a plea for quality', *Town and Country Planning* 70(12): 338–41.

Davies, C. (1997) *Improving Design in the High Street*, London: Royal Fine Art Commission.

MATTHEW CARMONA

SUBDIVISION

A subdivision is a simple real estate transaction between a seller, who is subdividing a parcel of land, and the buyer of the resulting parcel. Subdivisions may be simple, involving only a single seller and buyer, or complex, involving large tracts of land divided into many smaller parcels. Repeated again and again, these transactions create the

form of towns, cities and SUBURBS. The complex arrangement of streets and blocks and squares in Savannah, the simple and uniform grid of Manhattan, and the curvilinear streets and culs-de-sac of contemporary GATED COMMUNITIES are the results of subdividing. Subdivisions have created the form of URBAN America.

The earliest American land subdivisions accompanied colonial settlement. City plats – simple maps recording subdivided parcels of land – set out frameworks for private ownership, common property and street rights-of-way. Some of these cities were unique, such as Philadelphia or Savannah. More often, the plats continued legal traditions of colonial powers. In the Southwest, the Spanish LAWS OF THE INDIES created towns with a central plaza, a block for a church, and orthogonal arrangements of small lots and blocks, such as the original plat of San Antonio. In English tidewater colonies, settlements often followed a simple grid of blocks and lots, contrasting with New England towns of the same era, such as Boston, with ad hoc arrangements following English common law. French colonial towns, including New Orleans, followed forms similar to French medieval bastides, with their plans of streets and blocks, a central market square, a cathedral site and enclosing defensive walls. Colonial subdivision plats followed simple orthogonal arrangements of lots, blocks and streets but produced an infinite variety of cities across America.

Westward expansion marked a second period of land subdivision. The Northwest Land Ordinance, proposed by Thomas Jefferson and passed by Congress in 1785, created a national subdivision plat to guide the distribution of public lands in the thirty states west of the thirteen colonies. A checkerboard of 1-square-mile sections, arranged in uniform 36-square-mile townships, stretched from Ohio to California. These sections, when transferred to private ownership by land grants or auctions, were easily subdivided into smaller and smaller parcels. The 640-acre section became the framework for the 160-acre quarter-section homestead, the 40-acre family farm, and town plats with further equal subdivisions into common 300- and 600-foot blocks. The Northwest Land Ordinance created the forms of nineteenth-century American towns and cities, just as the traditions of Spain, France and England shaped colonial America. The simple national grid and its

subdivisions are revealed in the still-visible structure of the rural and small-town West, as well as the simple lot, block and street dimensions of Chicago, San Francisco and hundreds of other nineteenth-century towns and cities.

At the end of the century, America's growing middle class found single-family home ownership affordable for the first time, creating a first wave of suburban development. The residential subdivision – and the subdivision plat – became the standard unit of urban growth, replacing the town plats of colonial settlement and westward expansion. The subdivision also became the foundation of the emerging real estate industry, composed of small-scale developers, homebuilders, real estate brokers and mortgage lenders. The subdivision plat made urban development simple and efficient. A developer purchased land, prepared a subdivision plat divided into streets and blocks with sequentially numbered lots, and filed the plat with the municipal clerk. At first, subdivision plats simply extended existing streets and blocks, following the routes of new electric streetcars. After the turn of the century, automobiles fuelled urban expansion and land speculation, resulting in subdivisions with unbuildable lots, no utilities and unpaved and disconnected streets.

Controlling subdivision platting became a critical problem for local governments. Early proposals for subdivision reforms came from Charles Mulford ROBINSON, an advocate for the CITY BEAUTIFUL MOVEMENT. His books, *The Width and Arrangement of Streets* (1911) and *City Planning* (1916), set out the main points of subdivision reform that guided practices throughout the twentieth century. Robinson argued for the functional classification of streets, with radial and circumferential arterials to carry high-volume traffic and small and less expensive local streets to support middle-class and working-class domestic life. He also argued against traditional platting in gridiron patterns, favouring street and block arrangements responding to topography and reflecting the informal street arrangements and cul-de-sac ideas of Raymond Unwin and the English GARDEN CITY. Equally important was Robinson's advocacy of the NEIGHBOURHOOD as the fundamental unit of urban growth, instead of incremental residential subdivisions. Robinson's proposals preceded later influential projects, including Clarence PERRY's neighbourhood unit

(1929) and Clarence STEIN's Radburn, New Jersey (1929), which incorporated the neighbourhood unit along with superblocks, hierarchical street networks and culs-de-sac.

The US Department of Commerce's Standard City Planning Enabling Act (1928) provided model legislation authorizing use of POLICE POWERS to regulate subdivisions. The Federal Housing Administration (FHA), created by the Housing Act of 1934, acted to implement subdivision reforms. Although intended only as a guide for mortgage insurance underwriting, the FHA's standards, published in a series of technical bulletins during the 1930s, resulted in a national template for subdivision design as they were adopted into subdivision regulations in hundreds of cities across America. After 1945, subdivision reform transformed city-building. Instead of the orthogonal arrangement of interconnected lots and streets and blocks, American cities were designed with superblocks, bounded by wide arterial roads, incorporating hierarchical street networks, with inward-oriented arrangements of curvilinear streets and culs-de-sac. The contemporary urban structure of the highway and the gated community is an outgrowth of subdivision reform begun early in the twentieth century. The process of land subdivision remains the primary influence on the form of urban America.

Further reading

Reps, J. (1965) *The Making of Urban America*, Princeton, NJ: Princeton University Press.

Robinson, C.M. (1915) *City Planning*, New York: G.P. Putnam's Sons, The Knickerbocker Press.

RICHARD DAGENHART

SUBURB

The term generally refers to an outer district lying within commutable distance of a CITY or URBAN area. In medieval times, defensive walls often demarcated the boundary between a suburb and a city. In modern times, however, the boundaries between the city and suburb are more difficult to identify. Modern suburbs might have their own political jurisdiction, especially in the United States, but this is not always the case, especially in the United Kingdom where suburbs are located within the administrative boundaries of cities.

Suburbs might be associated with a particular social function, as a residential space of consumption and family reproduction, but again, these functions are shared by cities and do not constitute a comprehensive definition. Single definitions of the suburb may be impossible to articulate because the concept is used to describe a variety of morphological forms, landscapes, communities and political spaces. Consequently, the suburb is best understood through the historical and cultural analysis of URBANIZATION.

During the early nineteenth century, the upper middle classes, consisting of wealthy merchants, bankers and professionals, sought to emulate the rural estates of the aristocracy by building large suburban villas to escape the increasingly squalid conditions of the INDUSTRIAL CITY. In the industrializing economies of America and England, therefore, the historical enlargement of cities was marked by a change, with distance from the city centre, from continuous building façades to more open landscapes. These open suburban landscapes were typically dominated by spacious terraced, semi-detached and detached dwellings, which were separated from the road and each other by private gardens; for example, Brooklyn Heights in New York and the Eyre Estate in north-west London. The development of these exclusive residential suburbs was entwined with the BOURGEOIS ideal of the family, which was based upon companionate marriage, a sexual division of domestic labour and the separation of public and private space. This arrangement effectively restricted women's access to paid employment and services, while providing a male retreat from the 'evils' of the city. The suburb also provided the middle classes with opportunities to engage in forms of respectable consumption, including the maintenance of gardens and 'moral' forms of recreation. Consequently, the suburb was imagined and often experienced as an ideal environment, which incorporated the best aspects of town and country life. The subsequent construction of suburbs in England and America, therefore, was as much a moral and cultural phenomenon as a morphological one.

Suburban expansion became possible with technological advances in transportation. Steam ferries and railways provided affordable transport for the middle classes who took up residence in rapidly expanding suburbs during the mid-nineteenth century. The form of these growing suburbs mimicked the open landscapes created by

the upper middle classes in the early part of the century and were dominated by single-family homes. During the late nineteenth century, the development of electric trams facilitated the construction of working-class suburbs, which were built to the minimum standards required by environmental health legislation. In the United Kingdom, this provoked a response from the GARDEN CITY movement and its supporters, which was a direct attempt to create good-quality housing and communities for labour artisans and the working classes. Inspired by the vision of Ebenezer HOWARD, garden cities were perceived as a means of incorporating the best features of town and country. In effect, however, garden cities became no more than garden suburbs, with the majority of residents commuting to work in neighbouring cities.

With their emphasis on low-density single-family houses, the morphology and architecture of the garden suburb greatly influenced twentieth-century municipal and private housing provision. The inter-war period was a time of rapid suburban expansion, which was made possible by the escalating demand for housing, declining costs of housing construction, government building programmes and the mass production of the automobile. In North America, Europe, Australasia and parts of southern Africa and Latin America, the state paid for the construction of roads, which increased the distance over which workers could commute and provided the automobile with a competitive advantage over public transport. The recession of the 1930s halted suburban growth in the United States, but signalled an unrivalled period of expansion in the United Kingdom, which gave rise to fears about unplanned suburban SPRAWL. Due to the relatively dense population of the United Kingdom, Town and Country Planning Acts were implemented to control RIBBON DEVELOPMENT and to contain suburban expansion. After the Second World War, suburban expansion was stimulated by the construction of motorways and interstate highways, which provided the opportunity for workers to commute over even greater distances. In America, this post-war expansion had a racial dimension, which has been described as the middle-class 'WHITE FLIGHT' from the city. Such labels, however, underplay the heterogeneity of post-war suburbs, which contain a diversity of housing forms, age groups, household types, classes and racial and ethnic groups. Of particular interest is the contemporary development of the

TECHNOBURB, neo-traditional planning, GATED COMMUNITIES and edge cities.

Since their inception, academics, architects, planners and artists have invariably criticized modern suburbs. Uniform suburban landscapes are considered to be monotonous and uninspiring. The suburban separation of public and private life has been blamed for the production of dysfunctional social and physical environments. Suburbs are also considered to be a structural component of gender, racial and class inequalities. These criticisms withstanding, there is still relatively little research on how residents actually construct and experience suburban life within the physical restrictions imposed by developers and planners. The available evidence suggests that the suburb continues to be a popular living environment, which, through the global translocation of planning knowledges and cultures, is being increasingly adopted outside the Anglo-American context in the cities of the developing world.

Further reading

Fishman, R. (1987) *Bourgeois Utopias: The Rise and Fall of Suburbia*, New York: Basic Books.

Jackson, K.T. (1985) *Crabgrass Frontier: The Suburbanization of the United States*, New York: Oxford University Press.

Silverstone, R. (1997) *Visions of Suburbanization*, London: Routledge.

Whitehand, J.W.R. and Carr, C.M.H. (2001) *Twentieth-Century Suburbs: A Morphological Approach*, London: Routledge.

ADRIAN R. BAILEY

SUBURBANIZATION

Suburbanization is the movement of households and businesses out of city centres and the consequent growth of low-density, peripheral URBAN areas. Though it has existed for centuries, the process was greatly accelerated in the nineteenth and twentieth centuries, and the modern SUBURB may be seen historically as a product of INDUSTRIALIZATION. Today, urban DECENTRALIZATION also takes the form of exurbs and edge cities.

Suburbanization is driven by a variety of factors. 'Push' factors contribute to people's desire to move away from the city; 'pull' factors are complementary and reinforce people's attraction

to the suburbs; 'enabling' factors account for their ability to translate desire into action.

Chief among the phenomena that motivate departures to the suburbs is the CONGESTION of the city and its more taxing environment, both in terms of demands placed on mind and body and in terms of fiscal imposition. The concentration of people and businesses also causes land values to climb in central areas; rising values, in turn, prompt an even more intensive use of land and foster the replacement of residential functions by commercial ones. Suburbs, on the other hand, though less well provided with public services and facilities, are greener, more sparsely populated, and less affected by pollution and CRIME. They also feature lower land values and lower levels of taxation; in many cases, they afford greater political autonomy and better citizen control over local government.

Suburbanization turns the countryside into a variety of urban settings, each with its distinctive advantages. In the form of a well-planned, socially segregated residential COMMUNITY, suburbia enables families to maintain their social, cultural and financial capital. In the form of a SUBDIVISION of cheap homes, it gives workers and employees access to home ownership. In the form of an industrial or commercial settlement, it reduces companies' costs of doing business and it permits activities (from polluting the air to gambling) that are more strongly regulated in the city.

Suburbanization is not only a response to objective conditions but also an effect of preferences and prejudices. In the Anglo-Saxon countries, feelings of dread for the INDUSTRIAL CITY and feelings of attraction to the suburbs have been particularly strong. In eighteenth- and nineteenth-century England, the early development of industrialization, the prevalence of laissez-faire attitudes, the rise of a moral-religious cult of the nuclear family within the bourgeoisie, and the high status of the landed gentry all helped to turn the country estate and its historical child, the suburban villa, into a residential ideal. (In post-Second World War America, racial prejudice also contributed to the phenomenon of 'WHITE FLIGHT' into the suburbs.) In other countries, such as France and Austria, where industrialization proceeded more slowly, where strong monarchies had concentrated status and wealth in capital cities, and where love of privacy and of nature were less intense, the upper classes remained more attached to city life and used the power of the state

to clear, clean and beautify city centres. Suburbanization became in large part a movement of industries and of workers into the periphery. This historical development gave the French suburb a lesser status than its British or American counterpart. Despite these differences, suburbanization has become a near-universal phenomenon in developed and in industrializing countries.

What made suburbanization at all possible was first and foremost a number of economic and technological changes. The wealth generated by industrial development afforded massive investments in the built environment. The industrial mode of production also enabled the industrialist to live at a distance from his plant, keeping his family safe from the dangers of the city and leaving the organization and monitoring of daily operations to a cadre of professional managers. His suburban home benefited from recent innovations in the delivery of water, gas and electricity and the removal of human waste.

More recently, the adoption of assembly-line and 'just in time' systems of production has driven business owners to look for cheaper land on the periphery in order to accommodate larger (one-storey) plants and trucking facilities. In the retail sector, the diffusion of new building types such as the SHOPPING MALL, the big-box store and the power centre has also put a premium on peripheral land located near a highway, preferably near a highway interchange.

Transportation technology has no doubt been the most important enabling factor in suburbanization. Fast buggies and steam locomotives helped early suburbanites bridge long distances between home and city. Canals and rail lines gave corporations access to power and raw materials and became axes of industrial development. Trolleys and suburban trains made it possible for large numbers of managers, employees and even workers to live beyond walking range of their place of employment. Cars and trucks, moving over a rapidly growing highway network, made vast expanses of cheap land accessible to households and businesses. Aeroplanes, more recently, have allowed for the development of business hubs around airports. Innovations in communications technology, too, have greatly facilitated suburbanization. The telegraph, the telephone, the fax and the computer have decreased the need for physical proximity and have steadily lessened the dependence of the suburb on the CENTRAL CITY.

Suburbanization, like URBANIZATION in general, is nourished by population growth, economic development and technological innovation, but not determined by them. Dissatisfaction with urban conditions and competition in the urban land market tend to drive households and businesses to the periphery. Who actually settles in the suburbs depends on financial and technical constraints but also on personal, corporate and public priorities. What form the suburbs actually take depends on geographic conditions but also on individual and collective interests. Suburbanization is but one way in which people appropriate space to meet their needs, using the unequal means at their disposal to get closer to what they like and further from what they dislike.

Further reading

Fishman, R. (1987) *Bourgeois Utopias: The Rise and Fall of Suburbia*, New York: Basic Books.
Jackson, K.T. (1985) *Crabgrass Frontier: The Suburbanization of the United States*, New York: Oxford University Press.
Mumford, L. (1961) *The City in History: Its Origins, Its Transformations, and Its Prospects*, New York: Harcourt Brace Jovanovich.
Walker, R.A. (1978) 'The transformation of urban structure in the nineteenth century and the beginnings of suburbanization', in K.R. Cox (ed.) *Urbanization and Conflict in Market Societies*, Chicago: Maaroufa Press.

RAPHAËL FISCHLER

SUNBELT

'Sunbelt' commonly refers to the southern third of the United States, a region where the rate of growth and URBANIZATION exceeded that of the nation as a whole from the 1940s to at least the 1990s. The most common definition draws a line along the northern borders of North Carolina, Tennessee, Arkansas, Oklahoma, New Mexico and Arizona and extends the line westward to include southern Nevada and California. Some critics have argued that the entire Pacific coast, much of the mountain West, and Virginia have shared the same patterns and sources of growth.

As early as the 1940s, military planners used 'sunbelt' as a shorthand for sections of the country with mild climate for outdoor training and clear skies for aviation. Its popularity, however, dates from Kevin Phillips, *The Emerging Republican Majority* (1969). A political journalist, Phillips characterized the sunbelt as a band of rapidly growing states across the southern United States where conservative Republicans were replacing Democrats as the majority party. The concept then attracted widespread attention as a way to understand the shifting regional balance in the United States, particularly as it offered a way to recharacterize the South as future-oriented rather than held back by its past.

The sunbelt economy has been distinguished by a heavy reliance on (1) federal spending for defence production and military bases; (2) tourism, entertainment and retirement spending; (3) sunrise industries such as electronics, software and aerospace, all of which have been closely tied to defence needs; and (4) transportation, financial, trade and service activities for rapidly growing regional populations.

The role of disproportionate federal spending and investment in fuelling the sunbelt made its emergence a national policy issue in the late 1970s and early 1980s. Observers contrasted the sunbelt with the economically mature and slowly growing FROSTBELT, or old industrial heartland of the Northeast and Midwest, and frostbelt leaders organized to fight for changes in the allocation of federal spending. The sunbelt also contrasts with the rustbelt of old MANUFACTURING CITIES with declining or outmoded industries.

Criticism of the term has focused on two issues. One critique is that the idea of a single sunbelt elides substantial differences in the political, economic and demographic character of the South Atlantic states and the Southwest/Pacific states. The other is that promotional use of 'sunbelt' has ignored the continued existence of substantial pockets of POVERTY in the rural South and Southwest; such internal differences have been characterized as 'shadows on the sunbelt'. This latter objection points up the extent to which the sunbelt has been the product of rapid urbanization and the rise to national or international importance of metropolitan areas such as Atlanta, Charlotte, Miami, Houston, Dallas-Fort Worth, Phoenix, Las Vegas, San Diego and greater Los Angeles.

Despite criticism, the 2000 Census showed that the region remained the fastest-growing part of the United States during the 1990s. Seventeen of the fastest-growing metropolitan areas were in sunbelt states, including Las Vegas, Austin, Phoenix, Atlanta and Orlando.

Further reading

Abbott, C. (1987) *The New Urban America: Growth and Politics in Sunbelt Cities*, Chapel Hill: University of North Carolina Press.

Bernard, R. and Rice, B. (eds) (1983) *Sunbelt Cities: Politics and Growth since World War II*, Austin: University of Texas Press.

Mohl, R. (ed.) (1990) *Searching for the Sunbelt: Historical Perspectives on a Region*, Knoxville: University of Tennessee Press.

Sale, K. (1975) *Power Shift: The Rise of the Southern Rim and its Challenge to the Eastern Establishment*, New York: Random House.

CARL ABBOTT

SURVEILLANCE

'Fortified cities', 'carceral places' and 'interdictory spaces' are terms used to convey the level to which URBAN space is technologically and physically under surveillance. Satellite remote images, global positioning systems and geographic information science enable a degree of monitoring and observation that was scarcely believable a decade ago. Electronic access, gates, speed bumps and security guards increasingly control private residential spaces. There is growing use of, and widespread popular support for, CLOSED CIRCUIT TELEVISION surveillance of city STREETS, PARKS, and pedestrian districts in Britain. Devices as basic as pagers and cell phones enable employers to keep track of their employees. New forms of surveillance and control extend from garbage collecting, to HOMELESSNESS intervention, to policing itself. In the United States, technological surveillance of city streets extends to the use of sensitive sound-pinpointing devices that guide police officers to the scene of gunshots or other disturbances (Graham and Marvin 1996). Police increasingly use sophisticated computer-based 'profiling' techniques to detain or remove potentially 'dangerous' persons from parks and city streets. In some US states, and particularly in the case of convicted child-molesters who have served sentences, these profiles are made public on the INTERNET. While such technological innovations in the field of surveillance are not complete, some critics argue that their panoptic reach, coupled with the increased privatization of traditional publicly supported functions, add up to a vision of cities as perfectly knowable, perfectly planned and perfectly controlled.

Further reading

Graham, S. and Marvin, S. (1996) *Telecommunications and the City: Electronic Spaces, Urban Places*, New York and London: Routledge.

STUART C. AITKEN

SUSCEPTIBILITY-TO-CHANGE ANALYSIS

Susceptibility-to-change analysis (STC) is a simulation approach utilized in physical planning and URBAN DESIGN to analyse and evaluate the permanence of physical systems in a city. It proactively manages development. It searches for and identifies alternative locations to house future growth actions. It does so by identifying and assessing locations within the existing city based on condition and performance of the physical elements on them that:

- *are susceptible to change*, and therefore could be *suitable for redevelopment*; or
- *are not susceptible to change*, and *should remain* as functioning parts in the city.

To this end, the STC simulation assesses elements of the built system found in a specific location (buildings, STREETS or pedestrian paths) as to the *value* of the activity they house or support (economic, historic), or their *physical condition* (deteriorated or obsolete vs. well maintained).

Justifying the concept

Its value to planning and urban design is as a proactive growth management tool by maintaining an up-to-date evaluative information base on the built inventory, and on potential areas for locating future development. It is an approach to recycle the city and possibly address urban SPRAWL by identifying alternative sites for new development actions within existing URBAN areas rather than at their peripheries.

The STC becomes inclusive in its evaluative and locational performance when integrated with participatory planning and design approaches. It provides the necessary flexibility in the resolution of the 'wicked problems' inherent to planning and design (Rittel and Webber 1973). The integration of notions such as the CRITICAL COALITION ensures that the STC outcomes will reflect more accurately

the values, attitudes and needs of a COMMUNITY than when it is carried out only by experts.

Simulation process

The STC comprises one descriptive and two evaluative phases. All descriptions and evaluations in these phases are carried out at a selected 'unit of analysis' (e.g. an individual property, city blocks, city districts, or even areas within regions). Its identification is related to the desired resolution and accuracy of the results for a given application. Such a unit is the smallest spatial entity. Once selected, it can be aggregated into larger units, but not subdivided. It can also be easily integrated with most GEOGRAPHIC INFORMATION SYSTEMS (GIS), since the units and their descriptors are similar in concept to the graphic objects and attributes in a GIS.

The base description

In the first phase, an inventory describing systems and their respective elements is carried out for the study area. It identifies systems as land use (residential, production or service activity), movement (private and public modes, or paths), utilities (water, waste or energy), and services (facilities and service radii). It includes objective descriptions such as their 'past evolution', 'present state' or the nature of 'proposed improvements' to these systems, for example their present operating capacity or future plans for expansion.

The structural givens

The second phase focuses on the past and present state of a setting under analysis through an evaluation of existing systems and elements. They are assessed either as *structural givens*, which for the purpose of the simulation are permanent in the physical structure of the urban setting, or as elements *susceptible to change*. The latter indicate those locations suitable for future redevelopment. These identifications represent the 'informed' perceptions of participants, based on their assessments of the economic, cultural or policy functions and value these elements and systems presently perform. For example, the *economic given* is a designation applied to elements and systems in each unit of analysis after a cost-analysis of maintenance vs. replacement is carried out to make the assessment. Locations evaluated as *policy*

givens are associated with systems designated as permanent or unchangeable for any proposed development through regulatory policies such as 'special districts'. Likewise, *social givens* are those properties (public and private) which are deemed as possessing attributes critical to establishing the cultural continuity and identity of the urban setting, for example historic buildings and districts. The resulting characterization for the unit of analysis ranges on a scale from *least* to *most* susceptible to change, where the structural givens are evaluated around the 'least' end, while those units assessed around the 'most' end of the scale are considered as locations suitable for future redevelopment actions.

The probabilities

The third phase focuses on the evaluation of the *probabilities* of future plans. The participants designate through discussions the anticipated completion and utilization of proposed developments identified in the inventory phase. This identification is carried out as a set of probable future categories (P_1, P_2, P_3, ..., P_n); for example, a probability P_1 may include projects actually in progress which may be expected to be structural givens within a period of time, say two years (t_1), while P_3 may be developments currently under study with an anticipated completion/utilization period of seven years or more (t_3). The summation of the various combinations for the development probabilities identifies in each unit of analysis the type, number and location of all probable developments anticipated to be completed and utilized in times t_1, t_2, t_3, and so on. Thus, probabilities identify the relative permanence based on expected futures. Utilizing a similar scale as for the 'givens', the characterization of the unit of analysis ranges from 'least' to 'most' susceptible to change, where those units having probabilities of two years or less (P_1) for their implementation are assessed as least susceptible to change, while those with seven or more years (P_3) before implementation are evaluated as most susceptible to change.

Following this process, the overall susceptibility to change of the study area may be found in terms of the different perceptions and values of the groups identified. Once the givens and probabilities are completed, not only elements and systems useful and meaningful in a city are identified, but also an inventory of locational

alternatives suitable for redevelopment is attained for districts of the city.

Evolution and applications

The STC, as described, is a recent approach which can be used to guide development by supporting locational decisions in physical planning as well as in urban design or landscape architecture applications. It has evolved from the professional applications of the original two-dimensional mapping technique utilized professionally in the redevelopment of central districts of various cities in the United States (Wallace *et al.* 1979), to the identification of suitable locations for public housing for the elderly (Anselin and Arias 1983), to three-dimensional participatory games for NEIGHBOURHOOD redevelopment (Arias 1996), to interactive computer-based simulation and gaming formats developed at the Urban Simulations and Information Systems Lab of the College of Architecture and Planning and the Center for LifeLong Learning and Design at the University of Colorado (see ENVISIONMENT AND DISCOVERY COLLABORATORY).

Further reading

Anselin, L. and Arias, E.G. (1983) 'A multicriteria framework as a decision-support system for urban growth management applications: central city redevelopment', *European Journal of Operational Research* 14: 300–9.

Arias, E.G. (1996) 'Bottom-up neighborhood revitalization: participatory decision support approaches and tools', *Urban Studies Journal – Special Issue on Housing Markets, Neighborhood Dynamics and Societal Goals* 33(10): 1831–48.

Rittel, H. and Webber, M. (1973) 'Dilemmas in a general theory of planning', *Policy Sciences* 4: 155–69.

Wallace, D.A., McHarg, I., Roberts, W. and Todd, T. (1979) 'Downtown Detroit redevelopment plan: 1979', City of Detroit, MI: WMRT technical report, Philadelphia, PA.

ERNESTO G. ARIAS

SUSTAINABLE URBAN DEVELOPMENT

Sustainable urban development is not new. Cities have historically been built to symbolize their civilizations. Ancient Alexandria or Athens are enduring examples. However, sustainable urban development involves more than timeless architecture. A city involves all dimensions of human life. For example, classical Greek cities built on hillsides allowed airflow and gravity to aid public sanitation. Their limited size avoided depleting the HINTERLAND. Socially, they emphasized the POLIS to give solidarity to the life of the city. Finally, their built environment created an indelible SENSE OF PLACE.

Unlike the Greek polis, however, the modern city ignores ecological limits. Considerable evidence suggests that the modern city cannot sustain its current development patterns. Besides its economic, social and aesthetic functions, a city is also a living organism whose environmental impacts transcend its physical boundaries. For example, scientists estimate that the sprawling Phoenix, Arizona city-region hourly consumes an acre of fragile desert ecosystem. City dwellers deplete many natural resources, often exceeding their carrying capacity. Cities create entirely new habitats for wildlife such as wastewater treatment plants, divert rivers from their ancient courses, and endlessly manipulate other species sharing the same environment.

Photosynthesis powers most biological systems; external sources power the city. A typical city daily consumes much more energy per unit of land than the same amount of undeveloped land. Traffic generated by our sprawling metropolitan regions causes air pollution, which then enters the local water system and causes water pollution. The MEGALOPOLIS also consumes vast amounts of OPEN SPACE and farmland, depleting vital vegetative cover. This process further weakens the city's ability to clean its own pollution.

The environmental impacts of modern cities reach beyond their surrounding regions. The size, scale and connections of the modern metropolis exert global impacts. The ecological footprint is one measure of these impacts. The ecological footprint of a city is defined as the total amount of productive land required to sustain its current activities and remove its waste products. The ecological footprints of cities like New York and Tokyo are hundreds of times their actual size and have also been implicated in the problems of acid rain, ozone depletion and global warming. The Secretary-General of the United Nations concluded that the world's battle for sustainability will be won or lost in the cities of the world.

Sustainable urban development therefore requires a deeper understanding of HUMAN ECOLOGY. Our understanding of cities as biological

organisms lags far behind the challenges. To foster such understanding, the National Science Foundation has added two URBAN sites, Baltimore and Phoenix, to its Long-Term Ecological Research programme. GEOGRAPHIC INFORMATION SYSTEMS allow far more precise mapping of environmental problems. Along with research on the ecological footprints of cities, sustainability indicators are also being developed and measured. Sustainable urban development, while not a new concept, reflects growing scientific concerns and research about the impacts of URBANIZATION.

Several key indicators of sustainable urban development have clearly emerged. First, GROWTH MANAGEMENT minimizes use of non-renewable natural resources. It protects critical regional areas such as farmland or wetlands, often acquiring and integrating them into a system of protected public lands (see also LAND TRUSTS). GREEN BELTS such as Minneapolis's chain of lakes, Portland, Oregon's celebrated URBAN GROWTH BOUNDARY, and Boulder, Colorado's 20,000 acres of surrounding open space provide useful examples. Sustainable urban development also promotes COMPACT DEVELOPMENT patterns, as found in Curitiba, Brazil, which then enhances resource conservation through increased MASS TRANSIT.

Economic development, the foundation for human settlements, seldom acknowledges ecological limits in either capitalist or socialist systems. However, the ecological footprint demonstrates the need for ECONOMIC RESTRUCTURING aligned with the natural world. Sustainable urban development therefore needs an ecology of commerce. Such an economic system would move beyond resource conservation to promote ADAPTIVE REUSE of existing natural and built resources, emphasize renewable resources, and restore environmentally degraded areas such as BROWNFIELDS. Chattanooga, Tennessee is committed to ecological commerce. It has created lucrative new industries such as electric vehicle production, ecotourism (see TOURISM) and an international consulting role sharing its special expertise.

SOCIAL JUSTICE is another key element of sustainable urban development. In fact, the term 'sustainable development' grew out of the United Nations' concern about the increasing disparity between the developed and developing nations. Sustainable urban development creates spatial arrangements and living environments that minimize environmental hazards to citizens, provide access to the resources and opportunities of the city, and enable CITIZEN PARTICIPATION. With extensive and ongoing regional visioning efforts, pollution clean-ups in poor NEIGHBOURHOODS, and an affordable housing initiative, Chattanooga once again shows how to address this key indicator.

Like the cities of classical antiquity, sustainable urban development seeks to create a strong sense of place. Meeting basic human needs for human scale, significant form and aesthetic satisfaction appears necessary to sustain a city. The NEW URBANISM movement reflects these elusive but enduring concerns. New urbanism promotes smaller lot sizes, attention to building form and street patterns, with clearly identified civic spaces to anchor human activities.

Ultimately, these four key indicators of sustainable urban development do not exist in isolation. URBAN ECOLOGY, the foundation for sustainable urban development, insists that we consider all elements of urban life in relation to one another. Sustainable urban development, then, is not an end-state but rather a process by which all elements of urban life are balanced to foster the greatest human health and happiness, like the golden mean of classical antiquity (see HUMANE CITY). It is not a new concept, but it is a new vision of the Ideal City.

Further reading

Hawken, P. (1993) *The Ecology of Commerce: A Declaration of Sustainability*, New York: HarperCollins.

Katz, P. (1994) *The New Urbanism*, New York: McGraw-Hill.

McHarg, I. (1971) *Design with Nature*, Garden City, NY: Doubleday.

World Commission on Environment and Development (1987) *Our Common Future*, New York: Oxford University Press.

W. ARTHUR MEHRHOFF

SUTCLIFFE, ANTHONY

b. 28 September 1942, Northampton, England

Key works

(ed.) (1984) *Metropolis 1890–1940*, London: Mansell; Chicago: University of Chicago Press.

(1993) *Paris: An Architectural History*, New Haven, CT: Yale University Press.

English historian Anthony Sutcliffe graduated from Merton College, Oxford (1963), and with a doctorate from the University of Paris (Sorbonne, 1966). After several academic posts held since 1966 in the United Kingdom and North America, Sutcliffe, since 1988, has been faculty member of the Department of Economic and Social History, University of Leicester. He became emeritus professor at Leicester in 1998. The same year, he was appointed special professor of the Department of History, University of Nottingham.

One of Sutcliffe's major contributions has been in the fields of URBAN and planning history, which he pioneered in Britain with Gordon Cherry from the early 1970s, through the creation of the Planning History Group (PHG). In 1974, he organized with Cherry a landmark meeting on the subject, at the Centre for Urban and Regional Studies, Birmingham University. Another meeting of the PHG in London in 1977 was the basis for the creation of the International Planning History Society (IPHS). Sutcliffe's fundamental work in this field also included a book series in planning history in the early 1980s, the promotion of the IPHS bulletin, and the publication from 1986 of the journal *Planning Perspectives*. While establishing the international network of urban and planning history as an epistemic field, Sutcliffe's research work was focused on the emergence of planning as an international movement, an administrative practice and an academic discipline. His major publication in this respect was *Toward the Planned City: Germany, Britain, the United States and France, 1780–1914* (1981).

Understanding urban history is a variant of social history with the emphasis on the physical form; Sutcliffe set the city and its planning against its social and cultural background. Related to his interest in the study of the metropolises of the industrial era – epitomized by Victorian London and Second-Empire Paris – the 'giant city' became a related focus of Sutcliffe's publications, in terms of both its intellectual representation and its technical components. A major result of this research was the edited collection *Metropolis 1890–1940* (1984).

Sutcliffe's work on metropolitan CULTURE has made him an internationally respected expert on urban cultural history, one of whose features is the exploration of artistic representation for reconstructing the cultural climate in which urban and planning changes take place. Among other artistic discourses that Sutcliffe has explored in order to contextualize the rise of the industrial metropolis and institutional planning between the late nineteenth century and the beginning of the twentieth, he has worked and taught on how the cinema mirrored that crucial period of Western URBANIZATION.

Having studied the Parisian urban fabric in his doctoral thesis published in 1970, Sutcliffe picked up the subject in *Paris: An Architectural History* (1993), which was awarded a major prize by the American Association of Publishers. He has also worked on the architectural history of London.

Further reading

Stave, B.M. (1981) 'A conversation with Anthony Sutcliffe', *Journal of Urban History* 7(3): 335–79.

Sutcliffe, A. (1981) *Toward the Planned City: Germany, Britain, the United States and France, 1780–1914*, Oxford: Basil Blackwell; New York: St. Martin's Press.

ARTURO ALMANDOZ

SWOT ANALYSIS

SWOT (Strengths, Weaknesses, Opportunities and Threats) analysis provides a simple but useful tool for evaluation of the strategic position of a city or organization. It is a strategic management tool designed for use in the preliminary stages of decision-making processes and is usually associated with STRATEGIC PLANNING. Like other management tools originally intended for business, it can be tailored for application in public policies, including URBAN and regional policy.

The use of SWOT analysis as a tool in strategic management reflects the increasing importance of external ('environment') impacts on organizations or territories and the need to react accordingly and the conviction that the most important factor for success is the adaptability to new circumstances. SWOT analysis allows an understanding of internal factors (strengths and weaknesses) and external factors (opportunities and threats) that structure any vision of the city's future.

The internal analysis examines the city's or organization's capabilities, through the identification of its strengths and weaknesses. This

component includes questions about the advantages of the city, about those things that the city does well and about what other people see as the city's strengths (e.g. key skills, productivity, costs, innovation capacity, adequate public funds, etc.). For weaknesses, the inquiry should try to identify what could be improved, what sort of things the city does badly, what should be avoided, what other people perceive as weaknesses and the sort of activities that the city's competitors do better (e.g. the opposite of strengths, such as obsolete INFRA-STRUCTURES, low productivity, high costs, low innovation capacities, lack of public finance resources, etc.).

The external analysis examines features that create opportunities (anything with the potential to increase the city's or organization's strengths) and those that create obstacles or threats to performance (anything with the potential to hurt a city's or organization's capacities). Indeed, national and regional influences are important when deciding what new initiatives need to be added or which existing ones need to be modified or eventually abandoned. For the identification of opportunities, questions shall be directed to the identification of good opportunities facing the city (e.g. changes in markets, diversification of activities, changes in lifestyles, in technology, in government policies, in demographic patterns, etc.). For the identification

of threats, questions shall be directed to the identification of obstacles that the city faces (e.g. slow growth, vulnerability to recession, adverse demographic shifts, etc.) and what the other competing cities are doing, among other similar questions.

The requirements for undertaking a SWOT analysis are simple and easy to implement and can be carried out through surveys, interviews with stakeholders, focus groups, etc. The main challenge is probably to make the correct judgment of strengths and weaknesses as well as of opportunities and threats.

In sum, the SWOT framework helps to focus urban policy activities into sectors with the greatest opportunities and where the city is strong, minimizing weaknesses and counteracting threats.

Further reading

Ivancevich, J., Lorenzi, P., Skinner, S. and Crosby, P. (1994) *Management Quality and Competitiveness*, Boston, MA: Irwin.
Stacey, R. (1994) *Strategic Management and Organisational Dynamics*, London: Pitman Publishing.

SEE ALSO: strategic planning

CARLOS NUNES SILVA

T

TAKAYAMA, EIKA

b. 1910, Takanawa, Tokyo, Japan

Key works

(1963–80) Tsukuba Science City.
(1964) Olympic Annex, Tokyo.

Eika Takayama started his career as a city planning researcher when he was appointed as a research associate in the Department of Architecture, Tokyo Imperial University, right after he graduated in 1934. After that, he became an associate professor in 1938 and a professor in 1949 when he was awarded an engineering doctorate for a dissertation entitled 'A study on density in city planning'. He founded the Department of Urban Engineering at the University of Tokyo in 1962, the first academic programme in city planning in Japan. He served as a professor until his retirement from the university in 1970. Besides a career as an academic, he made use of research results in the practice of city planning. He was heavily involved in most large-scale URBAN development projects in Japan in the 1970s and 1980s when urban development was a boom in this country due to strong URBANIZATION pressures, as either project adviser or chief planner. Among such projects, the development of Kozoji New Town in the suburbs of Nagoya, the facility planning for the Tokyo Olympics, the new village development on the Hachiro Lagoon reclamation area in Akita Prefecture, the master planning for the Tsukuba Science City in Ibaragi Prefecture and the site plan for the Okinawa Marine Exposition in Okinawa Prefecture were the development or plan-making projects in which Takayama played a leading role.

He was awarded the Ishikawa Prize of the City Planning Institute of Japan and the Grand Prize of the Japanese Institute of Architecture. He devoted himself to the foundation of various new organizations, which enabled urban planners or researchers in city planning to work together with related experts, such as the Japanese Centre for Area Development, the Urban Disaster Prevention Institute, the Urban Redevelopment Coordinators' Association and the Mori Memorial Foundation. Although he worked for central and local governments in most cases as an adviser on their projects, he maintained his point of view that the way to think about how cities should be is as one of the scholars or even one of the citizens. Those who know him well remember that he often said there would be tremendous things to do in Japanese cities while looking around high-density, built-up areas in Tokyo. He suggested that Japanese city planners had to restructure Tokyo for realizing more beautiful and safer cities.

TAKASHI ONISHI

TANGE, KENZO

b. 4 September 1913, Imabari, Ehime Prefecture, Japan

Key works

(1949–55) Hiroshima Peace Centre.
(1957) Tokyo Metropolitan Government Offices.
(1964) St Mary's Cathedral, Tokyo.
(1991) Tokyo Metropolitan Government Offices.
(1992) United Nations University, Tokyo.
(1994) Shinjuku Park Tower Building, Tokyo.
(1994) Singapura Indoor Stadium, Singapore.

Kenzo Tange is Japan's best-known and most influential architect. His long career spanned the entire second half of the twentieth century and still continues, producing numerous important and distinctive buildings in Tokyo, other Japanese cities and cities in various countries around the world, as well as ambitious physical plans for Tokyo and its environs.

Many of his buildings blend themes from modern architecture with those of traditional Japanese CULTURE. Tange is also credited with architectural innovations such the 'shell structure' and various buildings with OPEN SPACES that can be enclosed later to meet expansion needs.

Tange was educated at Tokyo University, from which he graduated with honours in 1938 from the Department of Architecture. From 1942 to 1945, he was a graduate student at the same university and, in 1949, became a professor there. He was influenced during these formative years by the modern architecture and URBAN visions of LE CORBUSIER, Walter GROPIUS and Sigfried Giedion. At Tokyo University he organized the Tange Laboratory, where his students included Maki Fumihiko, Isozaki Arata and Kurokawa Kishō.

Tange's first completed building was the Peace Centre (1949–55), developed at the site of the epicentre of the Hiroshima atomic bomb. Among his notable buildings in Tokyo are the Sōgetsu Art Centre (1955–57), Tokyo Metropolitan Government Offices (1957), St Mary's Cathedral (1964), the United Nations University (1992) and the Shinjuku Park Tower Building (1994). His Yoyogi National Gymnasium building, completed in 1964 for the Tokyo Olympics, is especially highly regarded, particularly for its striking suspended roof. Perhaps his crowning achievement is the massive, new Tokyo Metropolitan Government Offices, completed in 1991 to replace the structure he designed in 1957. The new 'City Hall' is a complex of three interconnected buildings, including the two tallest in Tokyo, and incorporates a mix of features taken from major landmarks around the world and the early history of Tokyo. Some critics regard it as an extravagant monument to Tange himself and his political associates, particularly multi-term Tokyo governor Suzuki Shunichi. Tange is also known for URBAN PLANNING proposals for Tokyo, including a REDEVELOPMENT plan for the Tsukiji district (1961–64), his 'Plan for Tokyo 1960' that called for extending the city to man-made islands in Tokyo Bay, and more recent proposals for a futuristic new city centre on reclaimed land. Among Tange's many projects outside of Tokyo is the reconstruction of the city of Skopje, Yugoslavia, after destruction by an earthquake (1965), and the expansion of the Minneapolis Art Museum (1975), his only American project. His major awards include Order of Culture, Japan (1987); Commander dans l'Ordre des Arts et Lettres, France (1984); Grand Prize, Architectural Institute of Japan; and the Pritzker Architectural Prize (1987).

Further reading

Boyd, R. (1980) *Kenzo Tange*, New York: George Braziller.
White, G. (1990) *Kenzo Tange: A Selected Bibliography*, Monticello, IL: Vance Bibliographies.

ROMAN CYBRIWSKY

TAX ABATEMENTS

With the advent of improved rail and roadway networks, civic boosters have realized that cost factors other than transportation have become equally important in the employment opportunities game. Consequently, policy efforts to retain and attract businesses to their COMMUNITY are more complex. The tax abatement tool has been such a complex policy mechanism used by state and local governments to give some communities the competitive edge in the economic development process.

Viewed as the flip side of tax increment financing, tax abatements provide tax relief, usually in the form of property tax concessions to assure the investment of new businesses or the retention of existing ones. Tax concessions reduce the cost of doing business within the region, allowing a firm to improve its competitiveness in the regional or national marketplace. The most common practice is to freeze the property tax assessments on particular parcels of land within a designated district, making that district attractive to new investment or reinvestment by the private sector. As a case in all tax subsidy programmes, the cost-effectiveness is in dispute. In theory, such tax incentives are necessary so that the local labour pool has improved opportunities. Over time, the tax concessions are offset by additional wage and sales taxes derived from the improved incomes of residents.

Critics are not so sure. First, the American LABOUR MARKET is quite fluid, making it likely that job seekers may not be local residents but COMMUTERS from adjoining communities. Secondly, tax abatements do impose opportunity costs. That is, tax dollars are forgone for other uses such as a job-training programme, which in the long run is a better predictor of improved job opportunities. Furthermore, state tax abatement programmes tend to merely spatially reshuffle employment opportunities within a state, negating any possible region-wide expansion of the employment pool. In reality, locally elected officials are rarely concerned about the economic plight of their regional neighbours. Besides, tax abatements are easier to implement than comprehensive job-training programmes. Thus, in the United States it is a popular policy strategy, with at least three-dozen states having such authorizing legislation. Nevertheless, studies indicate that taxation alone does not determine business locations, but businesses do consider the relationship between taxes they are paying and the public services they receive in turn. If state and local governments begin with the assumption that a tax concession is a significant factor in business location decision-making, then tax concessions ought to yield additional business relocations. Taking it to its extreme, then, business taxes ought to be zero. Yet, to do so, a community will place the tax burden inequitably on the shoulders of the homeowner. Tax abatement appears to be an adequate compromise.

Further reading

Fisher, P.S. and Peters, A.H. (1998) *Industrial Incentives*, Kalamazoo, MI: W.E. Upjohn Institute.

Lebedur, L. and Hamilton, W. (1986) 'The failure of tax concessions as economic development incentives', in S. Gold (ed.) *Reforming State Tax Systems*, Denver, CO: National Conference of State Legislatures.

SEE ALSO: tax increment financing; urban revitalization

GARY A. MATTSON

TAX INCREMENT FINANCING

Tax increment financing (TIF) has become a popular creative financial planning tool. It is quite attractive because it does not impose a new tax burden on the homeowner. The original intent of a TIF programme is to stimulate private investment within a blighted area that has been designated, through public hearings, to be in need of economic revitalization. A typical TIF programme will finance a variety of INFRASTRUCTURE and site improvements that will attract new business and will hopefully raise property values within the TIF district.

The implementation scheme is for the city to map a designated TIF district, representing the area in which the city wishes to stimulate investment. Next, the city will likely borrow money by floating industrial revenue bonds. Industrial revenue bonds are of little risk to the city because the city does not pledge its full faith and credit, nor is there a moral obligation for the homeowner to repay debt holders in the event of a default. At this point, the city will establish a REDEVELOPMENT agency to oversee a TIF enterprise fund account for the district. Any new tax revenues beyond the baseline, which is defined as the increase difference in property tax revenues collected from the district (current year revenues minus revenues collected in the year of the district's designation), are placed within a TIF enterprise fund account. Such tax receipts are then used to pay off the district's bonds within the certain time frame or to make further improvements within the TIF district itself.

There are certain political advantages to this revitalization strategy. First, TIF is used to improve the economic viability of a designated blighted area. Secondly, most of the improvements are nuts-and-bolts activities that have distributive overtones. Furthermore, voter approval is not required for the issuance of the bonds, as the citizen-taxpayer is not legally obligated for the bonds. Critics note that TIF projects can divert scarce tax revenues from equally important social programmes. Moreover, smaller cities are less likely to gain an economic advantage. Furthermore, there are few guarantees that the districts' property values will improve and be sufficient to pay off the TIF bonds. Finally, it is likely that school and county districts will not benefit from the additional tax revenues generated, making it particularly irritating to both governmental entities.

Further reading

Klemanski, J.S. (1989) 'Tax increment financing: public funding for private economic development projects', *Policy Studies Journal* 17(3): 656–71.

Stinson, T.F. (1992) 'Subsidizing local economic development through tax increment financing', *Policy Studies Journal* 20(2): 241–8.

SEE ALSO: community-based financing; public–private partnerships; urban revitalization

GARY A. MATTSON

TECHNOBURBS

Technoburbs reflect a new economic dynamism and the changing nature of suburbia. No more can SUBURBS be viewed solely as bedroom or dormitory communities.

Strictly speaking, the term 'technoburb' refers to that growing slice of suburbia centred around high-tech industrial parks and their nearby retail malls, cultural centres, and entertainment and educational complexes. Technoburbs are a variant of the edge city concept where shopping and entertainment galleries, office and research and development complexes, and high-quality residences are all increasingly found in suburban communities.

In the United States, technoburbs are found quite widely, including the Route 128 corridor outside of Boston; the Silicon Valley region between San Francisco and San Jose; the 'Silicon Prairie' between Dallas and Fort Worth; the greater Schaumburg area by O'Hare Airport north-west of Chicago; Orange County outside of Los Angeles; White Plains, New York, and a number of New Jersey towns outside of New York City; the Research Triangle in North Carolina; the booming area near Houston's international airport; and Auburn Hills, Michigan, the site of Chrysler's headquarters, more than twenty miles away from Detroit. Technoburbs are not confined to the United States. France, for instance, has pursued a TECHNOPOLE or GROWTH POLE strategy, developing concentrations of technology-oriented areas outside of major French cities, including Roissy and Massy just outside of Paris.

Breakthroughs in the field of telecommunications have allowed high-tech and white-collar offices to decentralize. With fax technology and other advances in communications, computerization and transportation, even those industries that once required a CENTRAL CITY setting for face-to-face contact can now conduct their business from sites on the suburban rim. The result is a weakened URBAN core where technoburbs, with their high-end job, retailing and entertainment facilities, become attractive sites for globally oriented, foreign-owned firms.

Technoburban development results in increased SPRAWL, highway CONGESTION, green space exhaustion and difficulties in planning mass transportation. Pedestrians have a near-impossible time navigating the new suburban areas, where shopping galleries, entertainment, restaurant complexes and corporate office parks are separated by large parking lots and highway cloverleafs. The central city poor also face great difficulty in commuting to the growing number of jobs in suburbia. In some cases, firms located in the technoburbs have had to provide minivans in order to recruit lower-income manual and service workers who reside in the old central city and inner-ring suburbs.

Life in technology-oriented suburbs can compound problems of citizen detachment and a sense of anti-urbanism. With jobs, residences and entertainment all found in technoburbia, residents do not feel a responsibility to confront the problems of the urban core. Residents come to see technoburbs in terms of constituting fully fledged decentralized cities, not as suburbs. As Joel Fishman (1987) observes, the growth of technoburbs renders the central city and older suburbs obsolete. The result is DIVIDED CITIES or, more accurately, an even more greatly divided and segmented metropolis.

Further reading

Fishman, R. (1987) *Bourgeois Utopias: The Rise and Fall of Suburbia*, New York: Basic Books.
Garreau, J. (1991) *Edge City: Life on the New Frontier*, New York: Doubleday.

MYRON A. LEVINE

TECHNO-CITY

'Techno-city' refers theoretically to the URBAN area in which most of the economic, political and socio-cultural processes are mediated and assisted by new technologies. This term is commonly used by critical urban theorists.

In the discussion of the impact of technology on societies and cities, three different normative

stances can be identified. Technophilic viewpoints are typically presented by the representatives of a business COMMUNITY and technocrats in governments throughout the world. The opposite view, in which the risks and negative impacts are put on the agenda, is typical of critical intellectuals (e.g. Theodore Roszak, or Neil Postman). Somewhere in the middle are those who apply technographic and dialectical approaches (e.g. Douglas Kellner, or Manuel CASTELLS).

The techno-city is an urban formation with technologically mediated economic, power and socio-cultural processes. In this sense, it is close to the concept of the INFORMATIONAL CITY. In the analyses of the techno-city, a critical connection between technology, economy and urban life is central. They point to an urban setting in a techno-capitalist society in which technical and scientific knowledge and information and communications technologies play a role in the processes of production and reproduction. In the critical analyses of techno-cities, attention is paid to such themes as INFRASTRUCTURE, the transformation of urban space, the dual city phenomenon, the DIGITAL DIVIDE, and corporate power.

Further reading

Downey, J. and McGuigan, J. (eds) (1999) *Technocities*, London: Sage.
Graham, S. and Marvin, S. (1996) *Telecommunications and the City: Electronic Spaces, Urban Places*, London: Routledge.

ARI-VEIKKO ANTTIROIKO

TECHNOPOLES

A technopole is a geographic cluster of related technology-based activities. The components of this complex typically comprise local firms, universities, financial institutions and public research organizations. Akin to the cities of the INDUSTRIAL REVOLUTION, technopoles act as engines of regional and national growth for the high-technology economy. They differ from SCIENCE PARKS in terms of both scale and scope. While a particularly successful science park may act as a nucleus for a technopole (as does Stanford Research Park for Silicon Valley), generally a science park is but one component of the technopole ensemble. A technopole can take the form of a localized cluster of innovative activity (e.g. Silicon Valley), government-planned science cities (the Japanese Technopolis programme), high-technology districts within established metropolitan areas (Southern California, Munich), new growth nodes (e.g. Bangalore, India) and occasionally, individual science parks that have generated a regional growth dynamic (Cambridge Science Park, Hsinchu Science Park in Taiwan). The scope of the technopole goes beyond providing conducive premises for innovative activity. The hallmark for a successful technopole is effective regional development and not just aggregate regional growth. A technopole will allow for upgrading regional human capital, large-scale INFRASTRUCTURE investment and region-wide 'seeding' of innovation and know-how.

The processes that feed the growth dynamic of the technopole are similar to those accounting for localization ECONOMIES OF SCALE: LABOUR MARKET pooling, easily accessible intermediate inputs and the existence of technological spillovers. These factors are complemented by local SOCIAL CAPITAL. This describes the local norms, business practices, entrepreneurial cultures and voluntary associations that act as the oil that greases the wheels of the more formal economic processes. For example, the success of Silicon Valley as a technopole is grounded in the informal interactions between entrepreneurs, venture capitalists, law firms, universities and government agencies. These give way to a unique regional way of doing business grounded in performance, competition and collaboration.

The ability to assess the success of technopoles is undermined by the diverse array of initiatives that lay claim to the title. However, the technopole experience so far seems to coalesce around a few common lessons. First, technopoles would seem to take a long time to incubate. Regional development effects are only realized after many years of investment and the 'carrying costs' of an unsuccessful technopole are high. Secondly, most successful technopoles have benefited at some time from public investment in infrastructure, universities or even by government procurement and contracting. Finally, successful technopoles are those that can constantly 'reinvent' themselves. The boom and slump nature of technology-based development means that technopoles need to change their main product base over time. Silicon Valley, for example, developed from a defence

platform in the 1950s and 1960s through integrated circuits in the 1970s and 1980s to the INTERNET and networking base of the 1990s and 2000s.

Further reading

Castells, M. and Hall, P. (1994) *Technopoles of the World: The Making of 21st Century Industrial Complexes*, London: Routledge.

SEE ALSO: science parks; technoburbs; techno-city

DANIEL FELSENSTEIN

TELECOMMUTING

Telecommuting is a subset of telework. Telework is any form of substitution of information and communications technologies (ICTs) for work-related travel. Telecommuting, a term primarily used in the United States, is the partial or total substitution of ICTs for the commute to work. Telecommuting came into prominence in the 1970s to describe work-related substitutions of tele-communications and related information technol-ogies for travel. Telecommuting, by allowing the efficient use of URBAN space (i.e. live/work) and the reduction in the consumption of material and energy resources (i.e. less travel for work), fits into the framework for urban sustainability. While telecommuting was seen as a work option that reduces dependence on transportation, it is of primary interest now to both the private and the public sector because it produces a mobile, flexible labour force and reduces overhead costs.

Telecommuters are primarily highly skilled upper-middle-class professionals. They work away from the employer's office or production facility, often in employees' own homes, usually on a part-time basis (i.e. part-time at home and part-time in an office), using telecommunications and infor-mation technologies to communicate with their offices. Their main reason for opting for such a work situation is work flexibility and they appreciate the control over their time that working at home affords them. Telecommuters are an esti-mated 8.8 per cent of total workers in the United States while teleworkers constitute 16.7 per cent of the total worker population (Helling 2000) and include independent contractors and self-employed entrepreneurs.

First seen as an energy saviour in the 1970s, telecommuting was also linked with a recognized need to humanize corporations and the freedom to choose a work style for greater personal fulfilment. With the widespread introduction of computers in the 1980s, the arguments supporting telecommut-ing became more utilitarian. Corporations, recog-nizing that to survive they had to be leaner, were looking for a way to reduce overhead costs and increase organizational adaptability and saw a flexible workforce, based not in a corporate head-quarters, but at home, or in some other alternative worksite, as one solution.

The 1990s and the early 2000s have seen a sig-nificant increase in job opportunities that are conducted entirely online, from website designers, computer graphics artists, systems analysts and programmers to online stock traders. Often, these employment opportunities are not based on full-time employment as telecommuters but as con-tract workers. This is resulting in far more worker mobility and companies who are growing and shrinking as the economy demands.

The integration of the computer, telephone and television, which has allowed opportunities for tel-ecommuting, has made possible in the urban sphere the dispersion of services across a wide geographical area, and the concentration of specialized services and like-minded people in particular areas. In addition, telecommuting is intensifying the expan-sion of the private domestic sphere and the blurring of the boundaries between home and workspa-ces and the private and public spheres. There is inconclusive evidence that telecommuting allows opportunities for greater COMMUNITY involvement in residential NEIGHBOURHOODS.

Further reading

Gurstein, P. (2001) *Wired to the World, Chained to the Home: Telework in Daily Life*, Vancouver, BC: Uni-versity of British Columbia Press.
Helling, A. (2000) 'A framework for understanding tele-work', in *Telework and the New Workplace of the 21st Century*, Washington, DC: US Department of Labor.

PENNY GURSTEIN

TENEMENTS

The word 'tenement' can refer to a dwelling unit within a larger building (English usage) or to the building as a whole (Scottish usage). The

expression 'tenement house' was coined in the United States to designate a building containing cheap rental accommodations. In the hierarchy of housing types, the tenement occupies a place between lodging house and apartment building. It is distinguished from the former by the longer terms of rentals and the larger size of domestic units and from the latter by the lower quality of accommodations and the lower socio-economic status of the tenants.

Tenements emerged in INDUSTRIAL CITIES under the pressure of massive rural and/or foreign IMMIGRATION. In their first incarnation, they were single-family dwellings subdivided ('tenemented') to accommodate several households. Thus, during the first decades of the nineteenth century, large numbers of previously fashionable homes underwent a dramatic FILTERING PROCESS and saw the number of their residents expand exponentially in the process. Tenements were soon also built on open land or in replacement of smaller structures, specifically for the purpose of maximizing the number of tenants on a given parcel. The first characteristic of the American tenement house, of the German *Mietskaserne* (rental barracks) and of similar structures, then, is the intense crowding of their 'inmates' in a limited amount of space, the large number of people per room and the high population DENSITY on the land. Although tenements can in principle be of good quality, other key features are often the deficiency of internal and external INFRASTRUCTURE, the paucity of sanitary facilities, the low grade of building materials, the lack of maintenance, the large size of the building relative to its lot and, hence, the lack of light and air accessible to each unit. An agglomeration of such substandard buildings is often identified as a SLUM or rookery.

It is in New York City that the tenement took on its most extreme forms. Starting in the 1850s, the city saw many of its very deep lots being nearly completely covered by tall 'railroad tenements', buildings with two long strings of rooms between the street and a small backyard. On each floor, of which there could be up to six, only four rooms had direct access to air and natural light, while another twelve or so did not. Following a competition organized in 1879 by the editors of *The Plumber and Sanitary Engineer* for the best design of a tenement house at once salubrious and economical, the 'dumb-bell' plan, with its narrower middle part, became widely applied. Its

characteristic plan-shape was the result of the introduction of two light shafts along the long sides of the building, thin wells that supplied hardly any light or air to the middle rooms but brought them noises from adjoining units and smells from garbage thrown to the bottom.

The typical image of life in a tenement house in the nineteenth century, as documented by Jacob RIIS, among others, was that of several families sharing a water pump and wooden privies in the yard (or a sink and a toilet on their floor), each occupying one, two or three small rooms, all paying a large portion of their meagre income to an absentee landlord as careless as greedy (though often of modest income himself), doubling up with lodgers who augment their income but reduce their privacy, plying mind-numbing or dangerous trades at home, seeing children play among piles of refuse or in pools of stagnant water, and finding in their NEIGHBOURHOOD little solace from toil and misery other than the gloomy street and the smoky saloon. It is against all these problems, in particular the CONGESTION of population, the lack of light and air, and the health hazards to which residents were exposed under such living conditions, that tenement reformers struggled in the later decades of the century.

Reform efforts in Europe and North America took two general forms. Progressive businessmen, union leaders and their professional allies, organized in some sort of HOUSING SOCIETY, attempted to make better housing accessible to workers. For many, the right alternative was the cottage on cheap suburban land, rendered accessible by modern transportation networks. For others, it was simply an improved type of tenement building in a central location. The construction of model tenements, in the third and especially the fourth quarter of the nineteenth century, was meant to demonstrate the economic feasibility of housing the 'deserving poor' in decent yet affordable units. Larger in scale than regular tenement projects (so as to offer shallow building volumes and ample OPEN SPACE), offering higher-quality units and better facilities and managed by a professional staff, these projects did offer their residents much improved living conditions (as did public housing projects, later on). But they remained few in number and modest in their effects on the housing market.

Much more important, at least in terms of impact, was the second form of housing reform,

the legal imposition of stricter building standards. Epitomized in the United Kingdom by the local measure adopted under the Public Health Act of 1875 and in the United States by the legislation designed by Lawrence VEILLER at the turn of the twentieth century, restrictive legislation used the POLICE POWER of the state to prevent the worst in tenement construction and applied increasingly strict requirements to raise housing standards for the working class. Although building by-laws indeed checked the proliferation of unsanitary housing and generally helped to improve the quality of new housing, they also added to the cost of construction and rendered new residential units less affordable. Economic necessity has always forced people to share accommodations originally developed for fewer residents or to occupy sub-standard units. Despite the enactment of housing codes and related efforts at code enforcement, tenements, a hallmark of the industrial city, remain present in the post-industrial one.

Further reading

Burnett, J. (1980) *A Social History of Housing, 1815–1970*, London: Methuen.
DeForest, R. and Veiller, L. (1903) *The Tenement House Problem*, 2 vols, New York: The Macmillan Company.
Plunz, R. (1990) *A History of Housing in New York City*, New York: Columbia University Press.
Sutcliffe, A. (1981) *Toward the Planned City: Germany, Britain, the United States and France, 1780–1914*, New York: St. Martin's Press.

RAPHAËL FISCHLER

THEME PLACES

One of the common characteristics of URBAN PLANNING, CULTURE and TOURISM is the importance of inventing new spaces. Typical examples are theme places, or theme environments, which have grown since the 1960s, especially after the expansion of POSTMODERNISM – that focuses on symbolic representations fostered by commercialization aiming to promote mass consumption. Examples of theme places are large SHOPPING MALLS (West Edmonton Mall), restaurants, casinos (Las Vegas) and PARKS. Examples of chains of theme restaurants are Hard Rock Cafe and Planet Hollywood.

Theme parks constitute the most popular theme places. These family entertainment parks are designed around landscapes, settings, rides, performances and exhibitions. Their main peculiar characteristics are self-containment and reflection of a common theme or set of themes. The origin of theme parks lies in circuses, festivals and fairs, and specifically in entertainment parks, expositions and city plans. Probably, the first entertainment park, which opened in 1583, was Bakken, near Copenhagen.

There are two main reasons for the appeal of theme parks: they offer a safe, controlled and clean recreational environment, and they present an easy view of very different geographical environments that can be readily labelled and consumed by visitors. Theme parks include many of the elements of LUNA PARKS, or attraction parks, but present a more specific image, based on children's fairy tales and stories (DISNEY ENVIRONMENTS), miniature towns, future science, the Wild West, rural worlds, and TV series.

The prerequisite of a large, and relatively inexpensive, piece of land does not facilitate the location of theme parks in cities. Theme parks are typically sited within a one-hour 'drive time' from a city or cluster of cities. In reference to existing tourist development, theme parks tend to locate in two general types of areas: established tourism areas (Florida) and greenfield sites in places where tourism is absent (Disneyland in Anaheim). The majority of the most popular European theme parks are located in the western maritime climatic zone of northern Europe.

Theme parks constitute examples of globalization with the Pacific Rim being the area of fastest growth. The most characteristic peculiarity of European theme parks is the cultural theme park. The growth of attendance at theme parks is impressive: in 1994, all US parks received 94 million visitors out of the total figure of 206.6 million visitors to parks. The majority of users are day visitors.

The central locus of the materialization of theming has been the city, and this process has a twofold dimension: the consideration of the city as a theme park and of the theme park as a city. As far as URBAN development is concerned, theme parks have affected (in connection with SUB-URBANIZATION tendencies) habitation, transportation and economic development. Theme parks can also have more intangible kinds of impacts on

cities through their capacity to provide working models of city-like environments and public places and spaces.

Further reading

Cerver, F.A. (1997) *Theme and Amusement Parks*, New York: Hearst Books International.
Gottdiener, M. (1997, 2001) *The Theming of America*, Boulder, CO: The Perseus Books Group.

ALEX DEFFNER

THEORY OF DEMOGRAPHIC TRANSITION

The theory of demographic transition dates back to the second quarter of the twentieth century. Taking Western Europe as the example, it describes a demographic progression through different phases of fertility–mortality relationships. This development was conditioned by the region's transformations from a traditional to a modern URBAN-industrial society. Abstracting from considerable regional differences, both birth and death rates were high and populations increased at most slightly until the early nineteenth century. Demographic growth in this 'traditional phase' of the transition was mainly determined by mortality, given populations' vulnerability to natural disasters like epidemics and famines. In the second phase, mortality rates dropped steadily as the economy expanded and hygiene and health conditions improved. Fertility rates remained high and populations sharply increased in numbers. It was only several decades later that fertility also started to drop and the transition entered its third phase, characterized by slowly declining demographic growth. The transition ended in the mid-twentieth century with the fourth, 'modern phase', when both birth and death rates stabilized at low levels and natural population growth returned to very modest levels. Some demographers distinguish a fifth stage of diminishing population size.

The early contributions to the theory were descriptions rather than explanations of observed demographic trends. Since the 1950s, growing attention has been paid to the question what causes the decline of fertility. From a policy point of view, more insight in this process has been considered relevant during most of the post-war era given the alarming demographic growth of Third World regions. Generally speaking, declining fertility is a matter of interplay of socio-economic, cultural and environmental processes, due to the modernization of societies at large (such as urban industrial development, secularization and mass schooling), and behavioural adjustments like the increase of the average marital age and a growing use of contraceptives. However, due to its complexity, historic nature, and huge variations in space and time, many questions on fertility decline still remain unanswered.

Kingsley Davis's contribution to our understanding of the declining fertility is the elaboration of the theory of multi-phasic response (1963). This theory starts from the notion of growing disadvantages of the sustained high rates of natural demographic growth that occurred during the third phase of the transition. In the Western societies of the late nineteenth century, where labour increasingly proletarianized, more children meant 'more mouths to feed' and contributed to household POVERTY rather than to production. Regarding the central question of how populations responded to this, the theory mentions methods that lower fertility, such as increasing celibacy, delay of marriage, contraception and induced abortion, but also out-migration. Out-migration could be emigration (the Irish), but was most often migration from peasant communities to rapidly growing INDUSTRIAL CITIES. Clearly, the relative importance of the various responses varied between societies and depended on specific conditions.

Further reading

Andorka, R. (1978) *Determinants of Fertility in Advanced Societies*, London: Methuen.
Davis, K. (1963) 'The theory of change and response in modern demographic history', *Population Index* 29(4): 345–65.
Jones, H. (1990) *Population Geography*, 2nd edn, London: Paul Chapman Publishing.

ARIE ROMEIN

THIRD WORLD CITIES

'Third World cities' typically refers to cities in the developing world. Although the term 'Third World' is not universally accepted, there is general consensus that it was coined in the mid-twentieth century to differentiate between the capitalist

industrialized nations led by the United States and Western Europe (the First World, dominated by the market) and the former communist countries led by the Soviet Union (the Second World, dominated by the apparatus of the state). Emerging from colonization, Third World nations adopted different policies and ideologies to manage their own affairs and have evolved in different ways. Therefore, they cannot be characterized as a homogeneous bloc.

Third World countries and cities are in a state of transformation. A few (e.g. the city-state of Singapore) have transformed, or are in the process of transforming themselves into First World countries through successful investments and experiments in industrialization, education, or by the strategic use of available natural resources to gain geopolitical power. However, the majority of Third World cities found in Central and South America, the continent of Africa, the Indian sub-continent, China, and Southeast Asia are poor.

Population growth and overcrowding are some of the most significant problems affecting Third World cities. At present, much of the world's population lives in the Third World and it is estimated that approximately 60 per cent of the Third World's population will live in cities by 2025. Third World cities are vastly under-prepared for this population explosion. The influx of migrants into Third World cities has created chronic housing shortages, resulting in the emergence of SQUATTER SETTLEMENTS (see also FAVELA, SLUM). Unsanitary conditions in these informal housing settlements and the lack of access to clean drinking water have direct implications for the spread of infectious diseases in these cities. The majority of citizens in Third World cities live in crumbling INFRA-STRUCTURE and lack access to basic human services and well-paying jobs. Women and children bear a disproportionate share of the POVERTY burden in these cities. Unsustainable use of natural resources and the absence of strong environmental safeguards have created additional environmental problems such as air pollution.

Third World cities are also characterized by vibrant civic networks where citizens develop creative solutions to solve complex problems. Supported by non-governmental organizations, civic engagement and citizen participation are helping to ameliorate some of the more pressing URBAN problems. International agencies such as the International Bank of Reconstruction and

Development and associated agencies (the World Bank Group) and the International Monetary Fund significantly influence the GOVERNANCE and managerial policies in many Third World cities. The availability of educated cheap labour and the gradual diffusion of advanced information and telecommunications technologies have created new work opportunities for a privileged few. These developments have spurred private sector investment in some Third World cities. Despite these efforts, there is a growing and persistent disparity between the rich and poor in Third World cities.

Further reading

The Development Gateway: www.developmentgate way.org
The World Bank Group: www.worldbank.org

LAXMI RAMASUBRAMANIAN

TIEBOUT, CHARLES MILLS

b. 12 October 1924, Norwalk, Connecticut; d. 16 January 1968, Seattle, Washington

Key works

(1956) 'A pure theory of local expenditures', *Journal of Political Economy* 64: 416–24.
(1962) 'The community economic base study', Supplementary Paper No. 16, Committee for Economic Development.

Charles Tiebout's undisputable recognition in the arena of local government and fiscal federalism goes back to his famous and widely cited paper: 'A pure theory of local expenditures'. Tiebout's much quoted 'vote with their feet' began as a light-hearted comment during a seminar in graduate school. Eventually, he developed this into the TIEBOUT HYPOTHESIS which recognized that individual households show a clear preference for public goods provided by local governments within a geographically distinct jurisdiction. By choosing different local communities, each with its preferred mix of taxation and public services, households 'vote with their feet'.

Most of Tiebout's research, however, focused on regional economic development and the effects of export growth on communities. Among others, Tiebout established a proven relationship between

the economic base, so popular in planning, and the Keynesian multiplier framework.

Charles Tiebout received a Ph.D. from Michigan in 1957. Tiebout held appointments at Northwestern University (1954–58), the University of California at Los Angeles (1958–62) and the University of Washington at Seattle (1962–68) until his sudden, unexpected death in 1968.

Further reading

Fischel, W.A. (2001) 'Municipal corporations, home-owners, and the benefit view of the property tax', in W.E. Oates (ed.) *Property Taxation and Local Public Finance*, Cambridge, MA: Lincoln Institute of Land Policy.

RAINER VOM HOFE

TIEBOUT HYPOTHESIS

The Tiebout hypothesis, expounded by Charles TIEBOUT in a highly regarded paper, explains how private markets can efficiently provide public goods and why the range of public services varies between jurisdictions within a metropolitan area. The provision of public goods in the city (street lighting, PARKS, roads) is problematic. The price mechanism cannot discern between large and small consumers and *free-riders* will typically consume more public goods than they pay for. The Tiebout model shows that in the case of *local* public goods (i.e. goods provided over limited spatial areas such as municipalities) and competing jurisdictions, a market for public goods will develop. Consumers will 'vote with their feet', choosing one jurisdiction over another based on the combination of public goods offered and local tax rates.

Critiques of the Tiebout model point to its US-centric view of local public finance and its emphasis on suburban development to the exclusion of processes affecting the inner city. The model is also said to ignore the spatial spillovers inherent in certain public goods such as roads and transportation systems and to focus on the property tax in creating fiscal surplus, overlooking the existence of commercial and industrial taxes. For all these limitations, the Tiebout mechanism has been particularly influential in the study of URBAN public finance, providing a simple and plausible explanation of the variation in quality and quantity of urban services in a metropolitan area.

Further reading

Tiebout, C.M. (1956) 'A pure theory of local expenditures', *Journal of Political Economy* 64: 416–24.

SEE ALSO: Tiebout, Charles Mills

DANIEL FELSENSTEIN

TOFFLER, ALVIN

b. 4 October 1928

Key works

(1970) *Future Shock*, New York: Bantam Books.
(1980) *The Third Wave*, London: William Collins.
(1991) *Powershift*, New York: Bantam Books (with H. Toffler).
(1993) *War and Anti-War*, Boston, MA: Little, Brown (with H. Toffler).

Alvin Toffler is regarded as one of the world's outstanding futurists. He has published a number of best-sellers starting in 1970 with *Future Shock*. This book emphasizes the personal and social costs of change, analyses the premature arrival of the future and deals with the processes of change.

In 1980 *The Third Wave* followed, dealing with the direction of change: the revolutionary transformation from an industrial economy to a knowledge society, characterized by a shift from mass production in industrial plants to the production of individual tailor-made goods and services in the office and – more and more – in houses. The prosumer will increasingly replace the consumer. Toffler: 'We are about to revolutionize our homes.'

Powershift followed in 1991, in which the Tofflers describe the power struggles to come – in business, in politics and in global affairs – as a new 'system of wealth creation' clashes with today's vested power blocs. The Tofflers propose that power is shifting from its old basis in violence and wealth towards a new dependence on knowledge, and 'knowledge about knowledge' in particular. This shift is about to set off upheavals in existing power relationships as individuals, companies and countries scramble for control of the new knowledge resources.

In *War and Anti-War* (1993) the Tofflers sketch the emerging economy of the twenty-first century again, this time presenting a new theory of war and

revealing how changes in today's military parallel clearly the changes now taking place in business.

HUGO PRIEMUS

TOURISM

URBAN tourism has increased dramatically during the past several decades. Both domestic and international tourism are important components of cities' economies. As manufacturing has become less significant as a source of jobs for urban residents and of tax revenue for city governments, the service sector of the economy has become more vital. Tourism is a growing component of the service sector within cities.

One can examine urban tourism from the perspective of both the tourist and the city. Tourists are attracted to cities for a variety of reasons, including historic heritage, amusement parks, climate, and proximity to recreational activities such as boating or skiing. Cities also actively pursue tourists. This often includes significant government financing for projects such as museums. Examples of other efforts are the creation of entertainment districts and HISTORIC PRESERVATION areas. Also, cities finance CONVENTION CENTRES in order to attract business travellers.

In addition to the tourist and the city, an industry has evolved to promote and profit from tourism. Among the key players are cruise ships, rental car agencies, hotel and motel developers, and airlines. Although local entrepreneurs often play a key role in promoting tourism, these national and international businesses are increasingly important.

The major businesses in the tourist industry and the city governments trying to attract tourism emphasize the benefits that result from increasing numbers of visitors. These include jobs for local residents, entrepreneurial opportunities for business people to cater to visitors, more revenues for existing businesses such as hotels and restaurants, a more vibrant city, and greater tax revenues for cities from the new businesses and tourists.

Critics, however, point to potential downsides of urban tourism. Among these are environmental degradation, the prevalence of primarily low-paying jobs generated by tourism, and the displacement of residents for tourist-related projects.

City expenditures for tourism can also result in decreasing services for residents if the tourist projects fail to generate significant revenues. And even if a particular project section of the city attracts tourists, it is often no more than a tourist bubble, which produces few benefits for most of the city's residents. Also, in the eyes of some critics, even successful tourist venues contribute to a loss of authenticity in the COMMUNITY as both hosts and guests focus on the tourist venues and ignore the city's cultural and historical roots.

It is not possible to generalize about the benefits and costs of tourism. Among the factors that influence the distribution of costs and benefits are the levels of unionization among tourist industry workers, whether or not a city has a living wage ordinance, and what type of environmental protection measures the city adopts. Community participation in the planning of tourism can contribute to wider community benefits.

Further reading

Judd, D.R. and Fainstein, S.S. (eds) (1999) *The Tourist City*, New Haven, CT: Yale University Press.
Rothman, H. (1998) *Devil's Bargains: Tourism in the Twentieth Century American West*, Lawrence, KS: University Press of Kansas.

ROBERT KERSTEIN

TRAFFIC CALMING

Traffic calming has become a household word, a tool to combat speeding and other unsafe behaviours of drivers in NEIGHBOURHOODS. Although originating from the German word *Verkehrsberuhigung* which means to quieten, pacify, soothe, comfort, mitigate traffic, traffic calming devices were in the STREETS of Pompeii during Roman times. The term 'traffic calming' began in the 1970s with the research project *Verkehrsberuhigung in Wohngebieten* (Traffic Calming in Residential Areas), in residential areas in North Rhine-Westphalia, Germany. The success in reducing collisions demonstrated its value. It soon became popular. Over the years, researchers have developed their own interpretations of traffic calming, involving a range of design measures aimed at reducing speeds and making clear through design that priority is no longer given to motor vehicle traffic.

In the Dutch Woonerf neighbourhood, vehicle speeds were lowered to walking speed. These designs rethink the role of streets and appreciate them as PUBLIC SPACES to be enjoyed. The concept has been emulated in other European countries, Japan, and in the neo-traditional towns of URBAN designers in the United States.

The most widely used and despised traffic-calming measure is the speed bump. For streets, bumps have been modified to relatively low and wide humps. Different variations of speed humps have been engineered to accommodate emergency vehicles and speeds varying from 25 kph (15 mph) to 40 kph (25 mph). Other traffic-calming measures that include landscaping and enhance the street are more widely accepted but are also more expensive. Street segments may include traffic-calming additions such as chicanes (which narrow the street width and add mild curvature) or islands or medians to reduce the width of the street. In intersections, traffic circles or roundabouts may be used to reduce speeding and unsafe turning movements and protect pedestrians and vehicles.

Historically, traffic calming has been used primarily in residential areas where support is strong and often passionate. The opportunity to explore different ideas in street design was easily accepted in residential precincts. Their effectiveness in reducing crashes and injuries led to wider use of these devices in local and arterial streets which serve mixed retail and residential areas, schools, and PARKS. The promotion of traffic calming resulted in wider acceptance of alternative transport using the 'green modes', i.e. walking and cycling. Towns and neighbourhoods have been designed to integrate traffic calming and embrace a balanced transportation system that enables the residents to choose modes. A case in point is Houten in the Netherlands, a town that offers a bicycle network giving priority to bicycles over motor vehicle traffic.

Traffic calming has evolved to shape a transportation system that enables residents to choose 'green modes', explore their public spaces and reduce the incidence of negative interaction with fast-moving vehicles. The success of traffic calming lies in innovative policies that embrace a more environmentally sensitive land use and transportation system, enabling planners and urban designers to explore designs for streets as public places.

Further reading

Hass-Klau, C., Nold, I., Bocker, G. and Crampton, G. (1992) *Civilised Streets: A Guide to Traffic Calming*, Brighton: Environmental & Transport Planning.

SHEILA SARKAR

TRANSFER OF DEVELOPMENT RIGHTS

The underlying legal concept of a transfer of development rights (TDR) programme is the notion that all land has a bundle of property rights. Each component property right may be severed from the rest and transferred to someone else, leaving the owner with all other rights of ownership. One of these component rights is the right to improve and/or develop one's own property. As a modern legal invention, a development right is associated with the common law concept of a property easement. An easement allows the easement holder a legal right to make limited use of another's property. Each easement requires some sort of purchase from the property owner. As a leasehold, an easement is expressed in a formal document that sets forth the terms of agreement, its duration and provisions for termination. An easement cannot be revoked by a subsequent buyer of the property, unless the easement holder is willing to release it.

The impetus for the adoption of a TDR programme is the potential development threat imposed upon a historic property, prime farmlands or an environmentally sensitive area due to URBANIZATION. Normally tied to ZONING bonus densities, a TDR was first used as a means of preserving historic buildings within a central business district by allowing the developer to purchase the zoning density rights from the property owner and then transferring those rights to another site within the city. Presently, it is widely used to protect either the natural landscape or prime farmland on the URBAN fringe within established URBAN GROWTH BOUNDARY districts.

A TDR programme is a simple idea that imposes a complex method of implementation. First, the local government must update its comprehensive plan and official zoning map. Secondly, it must establish a credit formula so that participating landowners can sell their zoning density credits to a potential developer. Next, there must

be a designated district that has been deemed eligible to obtain these additional DENSITY BONUSES. Once these three conditions are met, then the programme can be implemented. The sale and transfer of development rights become a market transaction. The original landowner obtains compensation for the sale of development restrictions directly from the purchaser. As a land preservation tool, it is a compromise between acquisition and regulation.

The TDR programme has its limitations. To work, it requires a very active real estate market. Moreover, it requires an extensive land use inventory analysis on a regional level. If the real estate market is active, however, there is no guarantee of success. A developer can easily leapfrog to an adjoining COMMUNITY without a TDR programme. Overall, a TDR programme is least attractive to those people who own prime development properties, and most attractive to those residents with low development prospects.

Further reading

Gottsegon, A. and Gallagher, C. (1992) *Planning for Transfer Development Rights*, Morristown, NJ: The New Jersey Conservation Foundation.

SEE ALSO: density bonus; land banking; urban growth boundary

GARY A. MATTSON

TRANSIT-ORIENTED DEVELOPMENT

Transit-oriented development is a form of URBAN development that promotes a symbiotic relationship between dense, compact URBAN FORM and public transport use. Transit-oriented development usually refers to contemporary developments consciously devised in relation to transit, often with a neo-traditional style of buildings arranged in grids of STREETS. The case for transit-oriented development (also known as TOD) was advanced by Peter CALTHORPE in *The Next American Metropolis*. Transit-oriented development has become a key component of NEW URBANISM, where it may form part of a compact, sustainability-oriented urban and regional package.

Urban forms based on transit access have been advocated and constructed for many years. These include plans for LINEAR CITIES, such as Arturo SORIA Y MATA's nineteenth-century 'Ciudad Lineal', and futuristic urban forms based on 'pods' of development strung along monorails. Historical precedents for transit-oriented development can also be found in traditional railway and STREETCAR SUBURBS.

Before widespread automobile accessibility, dense compact urban forms were naturally convenient for the WALKING CITY. Settlements naturally became transit-oriented following the introduction of various forms of MASS TRANSIT – whether railways, buses, streetcars or underground systems. Transit allowed the overall scale of cities to expand – often creating radial forms of development at the macro scale – while retaining the compact character at the local scale (e.g. streets of TENEMENTS and COURTYARDS). The advent of mass automobile use precipitated more dispersed urban forms; new development no longer needed to be particularly close to existing urban facilities, or tied to transit access.

Now, transit is a favoured form of transportation for reasons of operational efficiency (the ability to move large numbers of people), sustainability (through energy efficiency and reduced emissions) and social policy (providing a service for those without access to private transport).

In general, the attractiveness of transit may be influenced as much by the quality and convenience of access as by the quality, frequency and reliability of the transit services themselves. This means that transit-oriented design is concerned not only with issues of transport technology and operations, but with complementary modes of access, such as walking and cycling, and the consideration of a variety of URBAN DESIGN issues such as layout, enclosure, shelter, visual interest and personal security – a range of physical and psychological factors that influence the desire to walk or cycle in the first place.

There are several variants of transit-oriented concepts, including 'transit corridor districts', 'transit villages' and PEDESTRIAN POCKETS in the United States, sustainable urban NEIGHBOUR-HOODS and 'transport development areas' in the United Kingdom, and 'ped sheds' (the 'watersheds' of pedestrian flow focusing on public transport stops) in Australia. While optimal layout features will depend on local site conditions, and the type of transit concerned, the different variants of transit-oriented development tend to have an identifiable set of common features.

Each transit stop tends to have an associated catchment area, within which the activities served by the stop are located. The size of this catchment area is usually determined by a desirable maximum walking distance, normally based on about half a kilometre or a 5-minute walk, but may vary with the type of transit or travel purpose involved. The shape of the catchment area is often equated, at least theoretically, with a circular locus around the transit stop. The loci for adjacent stops may overlap, forming a corridor, or there may be clear green space between successive loci, giving rise to a characteristic 'beads on a string' pattern. Streets may be laid out to radiate from the transit stops, although conventional grid patterns are also suitable. Small block sizes are favoured, to encourage permeability and directness of route, and diversity of uses in proximity to the station. Transit-oriented plans typically allocate higher densities and intensities of use nearer to the transit stops. The location of retail or COMMUNITY services there helps to create mixed-use focal points.

Principles of transit orientation have been applied to new developments and planned towns, as well as being retrofitted to existing urban corridors. Building transit accessibility into plans for new development may be regarded as common sense in contexts where there is a tradition for compact development and transit use. Not surprisingly, transit-oriented development has been successful where demand for that kind of development and demand for transit coincide. However, these conditions cannot be taken for granted, and some caveats may be noted.

First, there is a broad question regarding the extent to which environment influences behaviour, and more specifically, whether the form of urban development can actually encourage transit use. Development that replicates traditional urban forms will not necessarily give rise to traditional patterns of travel, or transit use. Indeed, some would-be transit-oriented developments may fail to generate demand for transit, or even be provided with a transit service at all.

Conversely, the provision of new transit services or stations may not necessarily stimulate urban development. The existence of a transit stop is not sufficient in itself to establish a successful transit-centred development, or to support local retail or communal facilities. In the United States, some transit-centred developments have failed to

flourish in the absence of a robust local economy, political will and consumer demand.

There is also a wider question about the urban context into which a transit-oriented development is inserted. While any particular development may have all the right design features, if it is set in the context of a dispersed, car-oriented urban area, it may fail to provide a competitive package: both the origin and the destination have to be transit-accessible for transit to become an attractive proposition.

As more dedicated transit-oriented developments are built out and evaluated, in a variety of contexts, lessons can be learned to maximize the chance of future success.

Further reading

Calthorpe, P. (1993) *The Next American Metropolis: Ecology, Community, and the American Dream*, New York: Princeton Architectural Press.

Cervero, R. (1998) *The Transit Metropolis: A Global Inquiry*, Washington, DC: Island Press.

STEPHEN MARSHALL

TRANSPORT AND THE ENVIRONMENT

For people, access to safe, convenient, reliable and affordable transport, or conversely non-access, can define the quality of life in URBAN areas. For industry and services, transport is an essential support function providing the lubricant for most consumption and production activities. Access requires transport facilities to be in close proximity to where people live, work and spend their leisure time, and to the sources and destinations of goods and materials. Transport in urban areas includes movement of people and goods by motorized means, by private modes, mainly the private car, or by public modes of bus, rail or water transport, and by non-motorized means including cycling and walking. Almost all freight in cities is carried by road because of the diversity of this transport task, the short distances and the time-sensitive nature of many deliveries.

The transport task in all cities is growing and changing in nature. The growth is a function of increase in population, increased vehicle ownership related to economic growth and affordability of private vehicles, increased distances travelled

per capita and per vehicle, and global changes in the way that goods are sourced, manufactured and distributed. Demographic changes, including smaller households and the dispersal of dwellings and employment through metropolitan areas and CONURBATIONS, play a major role in increasing both the vehicle kilometres travelled (VKT) and the difficulty of servicing this travel by non-private transport modes. In most cities, the proportion of trips that are circumferential rather than radial is also increasing. This is compounded by changes in the workforce including increased participation by women, and a concomitant preference for trip chaining – households incorporating a number of stops for different purposes into one trip – which makes transit use difficult. Occupancy rates of the private motor vehicle also remain extremely low. All these URBAN FORM and societal changes have important consequences on efforts to manage the environmental effects of transport.

With the exception of walking and cycling, all transport activities place heavy demands on the environment. First, the provision of transport systems can displace existing populations, divide communities through linear structures, and take land for INFRASTRUCTURE, including OPEN SPACE such as creeks and parklands that may appear to offer the lines of least resistance for new construction. Secondly, there are the energy and resources required in both the ongoing operations – energy in the form of petroleum, liquefied gas or electricity – and the resources and embedded energy in individual vehicles and infrastructure. The third environmental dimension includes noise, air and water pollutants. Toxic air pollutants such as carbon monoxide, lead and particulate matters expose dwellings or pedestrians near transportation routes and transport also contributes to the overall pollutant load in regional air sheds and catchments. A significant proportion of the population is exposed to environmental noise from transport. Fourthly, closely related to energy consumption, are greenhouse gas emissions. Of increasing concern too are the SOCIAL JUSTICE issues of transport in cities, in terms of both mobility and access of different social groups, and unequal burden of transport pollution load, particularly on the poorer and disadvantaged.

Motor vehicles contribute significantly to urban air pollution. For example, in Australian cities, motor vehicles contribute 79–88 per cent of carbon monoxide, 50–80 per cent of nitrogen oxides, some 50 per cent of hydrocarbons, and some 20 per cent of particulates. The hydrocarbons and nitrogen oxides are the primary ingredients in the production of photochemical smog and the particulates are a major consideration in health. A response has been legislation requiring cleaner fuels and emission reduction for individual vehicles, leading to catalytic systems for automobiles. As a result, in countries where controls have been effective, there has been significant reduction in emissions of local and regional area pollutants, but populations are still exposed to levels that pose health risks. There have been declines in the levels of carbon monoxide and hydrocarbons, and lead emissions have been almost eliminated. There have also been reductions in particulate emissions. Only emissions of nitrogen oxides remain above 1970 levels. The effect of these controls would have been greater except for the growth in VKT in cities and the increasing average age of private vehicles. Benefits are attenuated by the slow penetration rate of new technologies into the vehicle fleet.

Improved technologies have also had an effect on energy consumption, with energy use of automobiles per passenger kilometre falling slightly over the past decades. However, this has been nullified by the growth in travel. In the United States, for example, transportation energy use has grown at 1.5 per cent per year for the past two decades, and the transportation sector accounts for a quarter of national energy consumption, nearly all petroleum. While the figures will vary between countries, some 60 per cent of all passenger vehicle travel in Australia, some 20 per cent of articulated truck travel, is within metropolitan areas. Growth in energy consumption is causing a corresponding increase in greenhouse gas emissions, and in the United States the transportation sector now surpasses the industrial sector's carbon dioxide emissions.

Beyond the control of emissions, approaches to the management of environmental effects include demands for city spatial planning and transport planning to become better coordinated. Shifting the modal split from private vehicles to public transport systems would have major benefits, but while there are major attempts to do this, in most cities there is still a greater investment in roads than any other public transport infrastructure. In urban areas, passenger transport continues to shift towards cars for many of the societal and urban form reasons described above and this growth

remains linked to economic growth. Decoupling growth in VKT, and consequent environmental effects, from economic growth is recognized as a major environmental issue in Europe.

Air transport is also a major component of many urban environments as airports need to be located near to their client populations – either in cities, on their fringes or connected by fast surface transport routes. The largest environmental problem related to airports is that of exposure to noise from flight paths. Even where airports are located distant from cities initially, they are major attractors of commercial, industrial and residential development and, in the absence of extremely strong land use controls, get caught up in overall metropolitan expansion.

A. LEX BROWN

TUGWELL, REXFORD G.

b. 10 July 1891, Sinclairsville, New York;
d. 21 July 1979, Santa Barbara, California

Key works

(1933) *The Industrial Discipline and the Governmental Arts*, New York: Columbia University Press.
(1946) *The Stricken Land: The Story of Puerto Rico*, Garden City, NY: Doubleday.

Rexford Tugwell got his Ph.D. in economics from the University of Pennsylvania in 1922 and taught economics at Columbia University from 1920 until 1932. He advocated national economic planning, scientific management of industry, and close coordination of government and corporations. In 1927 he accompanied a trade union delegation to Russia and co-edited their report as a scholarly book describing the Soviet system.

In 1932 Tugwell joined the small group of advisers to presidential candidate Franklin Delano Roosevelt known as 'the Brains Trust' and helped devise the New Deal strategy which Roosevelt implemented when he came to power in 1933. Tugwell was appointed Assistant Secretary and then Under-Secretary of Agriculture, and in 1935 he became director of the new Resettlement Administration (RA), created to promote rural rehabilitation and rural and URBAN resettlement. Rural rehabilitation provided loans and training to farmers and retired over 9 million acres of marginal and disaster-prone land from cultivation. Rural resettlement developed 'subsistence homesteads', agricultural communities for displaced rural people. Urban resettlement developed 'GREEN BELT towns', new residential communities for former city SLUM dwellers. Over 100 subsistence homesteads were established, but only three green belt towns were built, the largest being Greenbelt Maryland with 7,000 people. The RA emphasized cooperatives and COMMUNITY development, and Tugwell was widely criticized as 'Red Rex', supposedly the most left-wing of Roosevelt's policymakers. He resigned from the federal government in 1936 and the RA was merged into the Department of Agriculture. The green belt towns were publicized as model communities and helped generate enthusiasm for new town building after the Second World War.

Tugwell briefly served as vice-president of the American Molasses Corporation and then as chairman of the recently established New York City Planning Commission. He developed a vision of planning as 'the fourth power', a technical arm of government equal in significance to the executive, legislative and judicial functions. He sought approval for a Master Plan of Land Use for New York City, but many groups opposed his vision and Robert MOSES was a prominent critic. Tugwell resigned his chairmanship in 1940 and soon afterwards Roosevelt appointed him governor of Puerto Rico. He served five years as governor, instituting a planning system and developing many of the ideas that were subsequently extended by Luis Muñoz Marin, governor from 1949 to 1964, in the influential economic development strategy called Operation Bootstrap. In 1946 Tugwell joined the political science faculty of the University of Chicago, where he developed a graduate planning programme. From 1964 onwards he worked at the Center for the Study of Democratic Institutions in Santa Barbara, California, developing proposals to reformulate America's system of government.

Further reading

Chase, S., Dunn, R. and Tugwell, R. (eds) (1928) *Soviet Russia in the Second Decade*, New York: John Day.
Sternsher, B. (1964) *Rexford Tugwell and the New Deal*, New Brunswick, NJ: Rutgers University Press.

RAY BROMLEY

TYRWHITT, MARY JAQUELINE

b. 24 May 1905, Pretoria, South Africa;
d. 20 February 1983, Sparoza, Greece

Key works

Correspondence course for Allied soldiers.
Founding editor, *Ekistics*.

Jaqueline Tyrwhitt, a town planner, editor and educator, helped shape modern URBANISM from the 1940s to the 1970s. An influential node for the global exchange of ideas, information and models for human settlements, she was instrumental in sustaining the creative East–West dialogue that had given rise to pre-war MODERNISM, thus facilitating the further development and GLOCALIZATION of this collective civic ideal in the new field of URBAN DESIGN; through the work of the UN; and in the EKISTICS movement.

Tyrwhitt's contributions include her skill in developing methods for the application of the ideas of Patrick GEDDES, as well as in publicizing them. In London during the Second World War under the auspices of the Association for Planning and Regional Reconstruction, Tyrwhitt ran a correspondence course employing Geddes's approach, to prepare soldiers in the Allied forces for the task of post-war reconstruction. Her instructional materials included the first descriptions, in English, of techniques for analysis of thematic map overlays – a pioneering GEOGRAPHIC INFORMATION SYSTEM (GIS). Tyrwhitt explained Geddes's relevance to modern planners with her compilation, *Patrick Geddes in India* (1947), at the Vancouver Conference on Human Settlements (Habitat, 1976), where she played a prominent role, and through her consulting work in Asia.

Tyrwhitt continued to perpetuate Geddes's legacy from her position at the centre of the post-war activities of Les Congrès Internationaux d'Architecture Moderne (CIAM), as acting secretary from 1951. When José Luis Sert became dean of the Harvard Graduate School of Design, Tyrwhitt was among the first people he hired, to introduce modernist thought into the curriculum. As assistant professor of city planning, in 1956, Tyrwhitt organized the lecture series 'Ten Discussions on the Shape of Our Cities', which paved the way for Harvard to launch the first urban design programme in the United States, in 1959. As associate professor of urban design from 1958–69, Tyrwhitt brought her synoptic perspective to bear in training a generation of professionals, many from developing nations, who saw urban design as a field for the socially committed.

In 1969 Tyrwhitt retired from Harvard to work full time with Constantinos DOXIADIS at his Athens Centre for Ekistics, helping establish the new science of human settlements through a programme of research, education and international cooperation. Tyrwhitt helped Doxiadis start *Ekistics*, the journal, in 1954. Under her editorial direction, *Ekistics* highlighted non-Western approaches to planning and urbanism, particularly from Japan, which directly influenced theorization in the field in the 1960s and 1970s. She also provided the organizational genius for the Delos Symposia, one of the most influential professional forums of its era (1963–72). The Delos helped build world support for the UN's conference on the environment and human settlements, which in turn forged general agreement on the concept of sustainable COMMUNITY development.

Further reading

Tyrwhitt, J. (ed.) (1947) *Patrick Geddes in India*, London: L. Humphries.

ELLEN SHOSHKES

U

ULLMAN, EDWARD

b. 1912, Chicago; d. 1976, Seattle

Key works

(1945) 'The nature of cities', *Annals of the American Academy of Political and Social Sciences* 242: 7–17 (with C.D. Harris).
(1953) 'Human geography and area research', *Annals of the Association of American Geographers* 43: 54–66.
(1954) 'Geography as a spatial interaction', *Interregional Linkages*, Proceedings of the Western Committee on Regional Economic Analysis, Social Science Research Council, Berkeley, CA, pp. 1–12.
(1957) *American Commodity Flow: A Geographical Interpretation of Rail and Water Traffic Based on Principles of Spatial Interaction*, Seattle: University of Washington Press.

Along with Chauncy HARRIS, Ullman's name has become synonymous with the 'multiple nuclei' model and the famous 'The nature of cities' article, which they wrote together. As an URBAN geographer, transportation researcher and regional development specialist, Ullman became the champion of applied geography. In the last three decades of his life, after the publication of that landmark article, he dedicated his career to advancing the cause of 'geography as spatial interaction' and its associated quantitative techniques. He was especially instrumental in advancing knowledge of the urban spatial structure, and as early as 1941, while working on his Ph.D., he began to publish on CHRISTALLER'S CENTRAL PLACE THEORY ('A theory of location for cities', *American Journal of Sociology* 46: 853–64).

Ullman received his BA (1934) and Ph.D. (1942) from the University of Chicago, and his MA from Harvard (1935). He served within various government agencies during the 1940s, especially during the Second World War, when he became a member and later the director of the Joint Intelligence Study Publishing Board (1943–46). From 1946 to 1951, he became assistant professor and later associate professor of REGIONAL PLANNING at Harvard. He joined the University of Washington in 1951 and stayed there until his death in 1976. During his career, Ullman worked on various consulting and planning projects, based on which he published numerous documents and manuscripts. As his friend Harris (1977) observed, Ullman may have taken on these responsibilities in part to demonstrate the value of geography. In fact, Ullman was active in many scholarly associations, including the Association of American Geographers, from which he received the title of Distinguished Geographer in 1972, and the Regional Science Association, for which he served as president in 1961. His papers appear in sociology, political science, geography, and regional science journals and proceedings. However, Ullman's interdisciplinary research and publication always included spatial reasoning and never fell too far from the geographic tree.

Further reading

Eyre, J.D. (ed.) (1977) *A Man for all Regions: The Contributions of Edward L. Ullman to Geography: Papers of the Fourth Carolina Geographical Symposium*, Chapel Hill: Department of Geography, University of North Carolina at Chapel Hill.
Harris, C. (1977) 'Edward Louis Ullman, 1912–1976', *Annals of the Association of American Geographers* 67(4): 595–600.

ALI MODARRES

UNEVEN ECONOMIC DEVELOPMENT

Diversity (anisotropism) is an inherent condition to territory and can be used by local development strategies, but differences can be increased during the economic development process. After a political decision there are winners and losers. It is known that economic development is unbalanced on a territorial scale and does not always mean a better level of welfare in the population. These social and territorial inequalities have their origin in the capitalist production system. The main challenge for economic development in the twenty-first century is to learn how to manage space in a better way.

The current globalization process needs a new vision between scales in order to explain the economic change process and competitiveness between territories. In the past, Fordism represented the search for ECONOMIES OF SCALE in production, REGIONAL PLANNING in the regional space, and architectonic functionalism (LE CORBUSIER) and metropolization in the URBAN space. But the economic crisis and diseconomies of agglomeration favoured the extension of a new, more flexible POST-FORDIST SYSTEM OF PRODUCTION. The new nature of economic, social and cultural phenomena has promoted a new relationship between the global and local scales (see GLOCALIZATION). Local initiatives are now the foundation of regional development policies. This is partly due to the important changes that occurred after the beginning of the 1980s, influencing the nation-state and its public policies.

Following the two great recessions (1973 and 1981), the patterning of uneven economic development between sectors and geographical regions has shifted and uneven spatial economic development is mirrored by disparity in the quality of work. The radical transformation of production and employment geography has questioned the traditional theories about industrial localization and regional inequalities. The global enterprise uses territorial hierarchical structuring strategies in order to reduce incertitude and increase its control capacity. This way, one can expect more benefits although it means the production of territorial inequalities.

In this context, regional development policies try to reduce uncertainty for the enterprises through regional supply policies in order to improve working conditions of the local companies: for example INFRASTRUCTURES, taxation incentives, technology, information and qualified workers access. The last three are examples of intangible factors like regulation systems, SOCIAL CAPITAL (through GOVERNANCE and PUBLIC-PRIVATE PARTNERSHIPS) and its result, confidence, that reduce transaction costs and allow a good use of the endogenous potential (including natural and CULTURAL HERITAGE). The intangible capital is more and more decisive for competitive success. Every time locality is more important, because it can present better supply strategies in a new context of competitiveness between territories and cities inside the international urban system.

If neo-positivists paid little attention to the feeling of identity, CULTURE becomes in POST-MODERNISM the referent that structures COMMUNITY, consciousness and behaviour.

Overcoming structuralism and elaborating on the concepts of 'agency' and 'structure' (see GIDDENS and POLITICAL ECONOMY), individuals socialize within the framework of a particular culture (social system) that interacts with supraregional cultural developments. Cultural consciousness (regional identity), which may or may not have political expression, completes and rounds the social interpretation of the region derived from economic aspects.

Further reading

Olson, M. (1996) 'Big bills left on the sidewalk: why some nations are rich and others poor', *Journal of Economic Perspectives* 10(2): 3–24.

Smith, N. (1990) *Uneven Development: Nature, Capital and the Production of Space*, Oxford: Basil Blackwell.

JOAQUÍN FARINÓS DASÍ

URBAN

Scholars studying cities must wrestle with the difficult definitional issue of how to differentiate geographical areas that are sufficiently city-like that they can be considered *urban* from geographical areas that do not qualify as urban. Precisely defining urban proves difficult because human settlements are so varied.

The word 'urban' comes from the Latin word 'urbs'. This was the Roman political/legal term for the *physical aspect* of a new settlement as opposed to the Roman word 'civitas' that referred to the

new social organization that came into being when a new urbs was created and from which our word 'city' is derived. The Romans did not concern themselves with technical measurement and definitional issues of what exactly was and was not urban the way the US Census Bureau does. Roman urbs tended to be reasonably dense, but varied enormously in size. We would consider some of the areas the Romans called urbs urban; others not.

Political/legal definitions of what is and is not urban do not work very well for modern cities in the United States (and other countries). The incorporated city of Juneau, Alaska, for example, includes a large area of empty surrounding land. While the built-up area of Juneau is quite dense, a scholar analysing population density using Alaska's *political* definition of Juneau would conclude that it was a very low-density city if she divided the total population by the total area of the city. Many very low-density areas that are not very urban have been incorporated under widely different and historically permissive state laws regarding the incorporation of cities. On the other hand, some areas that are quite populous and dense choose never to incorporate as separate cities.

The US Census Bureau has developed the most precise and most commonly used definition of what is urban. Most urban planners and policy analysts use Census definitions for land use and programme planning, though not everyone agrees with the Census definitions.

In the 2000 Census, the US Census Bureau used two main concepts to define urban territory – *urbanized areas* (UAs) and *urban clusters* (UCs). The Census classified as urban *all* territory, population and housing units located within either a UA or a UC. The Census also uses the term 'place' and classifies some additional places as urban. The Census distinguishes two kinds of places – *incorporated places* (defined by the legal criteria of the state in which they are incorporated) and *Census-defined places*, which the Census considers places based on criteria of population size and density regardless of whether or not they are incorporated. Outside of UAs and UCs, any incorporated place or Census-designated place with at least 2,500 inhabitants is defined as an *urban place*.

The Census drew UA and UC boundaries to encompass as precisely as possible all densely settled territory in the United States. Urban territory defined this way generally consists of (a) a cluster of one or more block groups or census blocks each of which has a population density of at least 1,000 people per square mile; (b) surrounding block groups and census blocks each of which has a population density of at least 500 people per square mile; and (c) less densely settled blocks that form enclaves or indentations, or are used to connect discontiguous areas with qualifying densities.

A UA consists of densely settled territory that contains 50,000 or more people. At least 35,000 people in a UA must live in an area that is not part of a military reservation. The US Census Bureau delineates UAs to provide a better separation of urban and rural territory, population, and housing in the vicinity of large places.

A UC consists of densely settled territory that has at least 2,500 people, but fewer than 50,000 people. A UC can have 50,000 or more people if fewer than 35,000 people live in an area that is not part of a military reservation.

Scholars sometimes use a three-way classification system – classifying territory between urban, *suburban* and rural. Conceptually, this makes sense because in the United States and most of the rest of the world, denser CENTRAL CITY areas of intense economic activity are surrounded by rings of land that is mostly developed (no longer rural), but at lower densities and proportionally more residential than in the central city itself. Distinguishing suburban land from urban land at one end of the spectrum and from rural land at the other end proves very hard in practice. New York, Los Angeles, London, Tokyo, São Paulo and many other cities have suburban territory outside the central city area that has more population size, density and economic activity than many cities. On the other hand, at their fringes are low-density residential areas that blend into farmland.

While the Census Bureau has defined metropolitan areas, rural areas and central cities, it has chosen not to use the word 'SUBURB' in its system of classifying geographical areas. It is possible to identify areas that are within Census-defined METROPOLITAN STATISTICAL AREAS (MSAs) – hence not rural – but outside of Census-defined central cities – and hence not central city areas. For analytic purposes, this in-between area qualifies as a rough proxy for suburbs. Careful analysts choose to stick to the precise, if awkward, Census approach of distinguishing between central cities

and land within MSAs that is not in central cities. The Rutgers University Center for Urban Policy Research uses this approach in a database they developed for the US Department of Housing and Urban Development (HUD) from which HUD's *State of the Cities* reports were developed during the Clinton administration.

At the international level, UN-Habitat, an arm of the United Nations Centre for Human Settlements, is mandated to provide information on urban conditions. UN-Habitat's *Global Report on Human Settlements* summarizes data UN-Habitat and the UN statistical division collect on population, housing, social and economic conditions in 315 cities in 91 countries. The World Bank's data on world development indicators are another authoritative source helpful in understanding URBANIZATION at the world scale.

Further reading

Center for Urban Policy Research (CUPR) (2000) *State of the Nation's Cities*, New Brunswick, NJ: CUPR.
US Department of Commerce, Bureau of the Census (2000) *Geographic Area Reference Manual*, Washington, DC: US Government Printing Office.
—— (2000) *Geographic Terms and Concepts*, Washington, DC: US Government Printing Office.
World Bank (2002) *World Development Indicators*, 6th edn, Washington, DC: World Bank.

SEE ALSO: city; urbanization

RICHARD LEGATES

URBAN AGRICULTURE

The practice of urban agriculture has long contributed to satisfying a basic need of all cities, their food security. In many cities in the less developed world, URBAN farming is an important means of increasing household food security; indeed, the growing incidence of POVERTY in Third World urban households has given added emphasis to the importance of self-production of staple food needs. The development of urban agriculture in contemporary THIRD WORLD CITIES is matched by the encouragement given to urban farming in cities in the First World, and particularly in Europe, in the nineteenth and much of the twentieth centuries. Its encouragement was seen as a means of supplementing the food needs of the urban poor. Nor has urban farming disappeared from European cities,

although with increasing affluence, its advocacy is based on recreational and ecological grounds more than it is on food security.

It is in the cities of sub-Saharan Africa and those in the poorer Asian and Latin American countries that urban agriculture has become the most significant, and a visible component of the landscape. Pioneering research in the early 1980s by the Mazingiri Institute in Nairobi established its importance to urban food production and the contribution it could make to meeting nutrient and vitamin requirements for healthier diets. NGO activity has become important in fostering the spread of best practice in urban farming, while lobbying urban governments to ensure that town planning gives adequate recognition to urban farming within land use ZONING proposals. Unsympathetic attitudes towards urban farming by city officials, in part because of perceived environmental and health risks of farming within urban areas, combined with the effects of urban SPRAWL mean that securing adequate space for urban agriculture is often a source of local conflict.

Further reading

Freeman, D.B. (1991) *A City of Farmers: Informal Urban Agriculture in the Open Spaces of Nairobi, Kenya*, Toronto: McGill-Queen's University Press.
Urban Agriculture Magazine, available online at www.ruaf.org

RONAN PADDISON

URBAN CRISIS

An urban crisis exists when URBAN problems seem insurmountable and the CITY's fate at a critical juncture. At such times, governments seem powerless and a sense of dread permeates the city. Yet a crisis is always a matter of perspective, always a representation. Cities never lack 'problems', but only intermittently do commentators declare them to be 'in crisis'.

URBAN DECLINE has a similar quality. Cities have struggled with endemic problems – CONGESTION, SLUMS, social unrest, municipal mismanagement and epidemics – since their inception. Not until the last half of the twentieth century, however, did commentators portray the city as in decline and only in the 1960s and 1970s as facing a crisis. Even then, these claims were almost wholly confined to the United States. At other times

and in other places, the problems of the city have been considered insignificant, temporary or manageable.

In the United States, urban problems were first discovered in the INDUSTRIAL CITY of the late nineteenth century. The industrial city spawned a middle class that gave rise to cadres of social reformers. These men and women documented social ills and physical decay and then acted to improve the lives of city dwellers. Progressive reformers of the early twentieth century recognized the 'shame of the cities' (to borrow a 1904 phrase from the journalist Lincoln Steffens). Whatever ills that cities suffered, though, would be cured through philanthropic efforts and government regulations.

In the post-Second World War era, the urban crisis appeared twice, once as a racial crisis and later as a fiscal crisis. The onset of these crises was gradual. The sense that urban problems could be solved began to wane in the 1950s. Then, massive SUBURBANIZATION indicated that the middle class was rejecting the urban way of life. In response, a few commentators began to talk and write about the 'crisis of the cities'.

The belief in an urban crisis became widespread in the 1960s. Then, black MIGRATION from the South and WHITE FLIGHT to the SUBURBS came together in a particularly powerful way. Blacks had entered the industrial cities as the number of jobs was declining. Once there, they confronted DISCRIMINATION that confined them to racially segregated NEIGHBOURHOODS and pushed them into old and inadequate (often public) housing. The black GHETTO was a place of high unemployment, concentrated POVERTY, and CRIME. Antagonistic relations between the black COMMUNITY and the police, a CIVIL RIGHTS MOVEMENT intent on redressing the injustices of the inner city, and a high level of frustration among blacks combined to spark violence.

During the 1960s, RIOTS were frequent occurrences in the older, CENTRAL CITIES. Typically triggered by a police incident, blacks rampaged, often for days. Rioters set fires, exchanged gunfire with the police, and looted stores. A number of rioters lost their lives. Property damage was often extensive. Racial unrest seemed to have no solution. Fear spread throughout the metropolitan area, and white households continued to flee. The urban crisis was at its peak, and it was a racial crisis.

Racial disturbances ebbed by the early 1970s only to be replaced by near-bankruptcy on the part of many central cities. Urban conditions were worse than they had been ten years earlier; the riots had exacerbated white flight, owners were closing stores and factories, and the DOWNTOWNS were becoming desolate. When manufacturing collapsed, the industrial cities went into decline. This decline became a crisis when the country was struck by rapidly rising energy costs and a severe recession.

City governments experienced a near-fatal financial blow. Tax-paying households were leaving, businesses were closing, and property values were falling. Tax revenues plummeted. Yet the demand for public education, social services, road maintenance and police protection continued to grow. The result was a fiscal crisis. Numerous cities – New York City being the most prominent – approached bankruptcy. Here was another intractable problem causing hardship and repelling households and investors. Pessimism reigned.

The crisis soon ended. In the late 1970s, commentators discovered GENTRIFICATION and, a few years later, developers embarked on new office and retail development that reinvigorated the once-declining downtowns. An URBAN RENAISSANCE was underway. Since then, the problems of the city have seemed manageable and less threatening. And, while one commentator or another has deployed the 'crisis' claim hoping to draw attention to a specific issue – housing, education, GRAFFITI, crime, the environment, suburban decline – the urban crisis has passed. Even urban decline no longer dominates academic and popular commentary.

Industrial cities in other countries did not experience the racial violence or fiscal problems that occurred in cities in the United States, though many (particularly in the United Kingdom) did suffer from the collapse of manufacturing. It was mainly in the United States, where anti-URBANISM is a social fact and suburbanization an affordable alternative, that the urban crisis became the prevailing assessment of cities.

The only other significant use of the claim involves the accelerated URBANIZATION of the late twentieth century that gave rise to MEGACITIES. Cities such as Lagos, Bombay, Mexico City, São Paulo and Shanghai have added millions of people in only a few years. In the face of this population explosion, large cities in developing countries

struggle to provide housing and municipal services, protect the environment, and manage the local economy. For many contemporary commentators, the urban crisis has returned – this time as a crisis of sustainability.

Further reading

Banfield, E.C. (1968) *The Unheavenly City: The Nature and Future of Our Urban Crisis*, Boston, MA: Little, Brown.
Beauregard, R.A. (2003) *Voices of Decline: The Postwar Fate of U.S. Cities*, 2nd edn, New York: Routledge.

ROBERT A. BEAUREGARD

URBAN DECLINE

Cities have come and gone, with former ancient, mercantile or imperial glories reduced to archaeological remains or paintings of civic notaries. Although definitions of the CITY and of the URBAN vary, most commentators agree that key elements include population density, complex networks of social interaction and significant levels of economic activity. Urban decline occurs when people move from urban NEIGHBOURHOODS, businesses cease to operate and local authorities lose the capacity or political will to maintain them. Urban decline is closely related to industrial and economic degeneration. When the major heavy extractive and manufacturing industries developed in nineteenth-century Europe and the United States, towns and cities developed too. Domestic residences, communication routes, neighbourhoods, and places to worship and purchase goods and services followed. When these industries disappeared in the second half of the twentieth century, a great deal of full-time, usually male, employment disappeared in the former thriving mining communities and steel, motor manufacturing and shipbuilding towns. Although organized labour in places fought hard against this, a combination of emerging post-Fordist production processes, globalization and the political dominance of free market economics, with Southern countries offering cheaper labour for 'footloose' capital, led the Northern economies to restructure themselves to exploit their comparative advantage in new information and communications technologies, and financial and consumer services.

With the loss of their economic life-force, many towns in the 'rustbelt' and inner-city areas suffered a moral, spiritual and physical decline. The traditional skills needed for manufacturing and other jobs were no longer needed and many of those who were able to retrain left the declining districts to search for work elsewhere. The demographic shift was compounded by the growing belief that the urban increasingly meant a lower quality of life characterized by dirt, CRIME, social disorder, bad housing, unemployment, physical degeneration, HOMELESSNESS, poor public services and all too frequently ethnic conflict. Media coverage of disturbances in Los Angeles in the United States and Bradford, Burnley and Oldham in England in the 1990s reinforced this popular feeling. Urban decline and deprivation frequently fuelled social and political frustrations leading to serious civil disturbances in many cities with racial and ethnic disadvantage at their core. Many areas suffered from multiple deprivation and political alienation, soon to become known as 'social exclusion'. This term was first developed in France in the 1970s and then adopted by Third Way politicians in Britain and the United States in the 1990s. Although attempts to remedy this became key aspects of national and federal policy, many urban areas remained severely disadvantaged as social, physical and economic investment failed to materialize.

Urban decline is not simply due to a change in economic circumstances alone. One key factor facilitating decline has been identified by sociologists such as Amitai Etzioni, urbanists such as Jane JACOBS and political scientists such as Robert Putnam as the loss of a SENSE OF COMMUNITY or those relationships of trust, mutual reciprocity and obligation known as SOCIAL CAPITAL. When people cease to look out or care for each other, they cease to care for the neighbourhood they live in. When particularly young people sense their future holds very little or they experience systemic prejudice and DISCRIMINATION, drug-related crimes, vandalism and anti-social behaviour increase. The 'broken window thesis' first developed in the 1980s suggested that a sign that people no longer care is when a window in an abandoned residence or office block gets broken and remains broken. One act of vandalism elicits similar acts and a spiral of decline ensues leading to more residents and businesses seeking peace and economic security elsewhere.

Apart from these push factors, there are also several pull factors influencing urban decline. The

belief that a good quality of life can only be achieved in a sub-urban environment is one. The sprawling SUBURBS of many cities in the United States, of Sydney in Australia and of Auckland in New Zealand, together with the nostalgic association of the rural with the healthy, have nurtured a growing anti-urban sentiment. The suburban detached or semi-detached owner occupied home with garden is a powerful ideal. Property developers and mortgage companies invest in greenfield and out-of-town retail developments while the poorer urban areas are redlined. The cultural justification for SUBURBANIZATION has its roots in the work of architectural luminaries like Frank Lloyd WRIGHT and visionary planners associated with the GARDEN CITY movement of Ebenezer HOWARD. Suburbanization has a corollary in the growth of GATED COMMUNITIES where the rich exclude themselves from the perceived dangers of the contemporary urban environment. Suburbanization also means increased car use creating demand for more highways that cut into the countryside and frequently physically dissect neighbourhoods, towns and cities. The car is the epitome of the privatized, unsustainable and self-regarding consumer culture that prizes mobility above accessibility and ecological integrity.

Urban decline may also be seen as a consequence of MODERNISM. Urban development and the thrill of URBANISM have been closely associated with the skyscraper. No emerging city or city economy is complete without the biggest or the highest as the history of Chicago, New York, Shanghai, Kuala Lumpur, São Paulo and even London testifies. The high-rise apartment blocks built in the immediate post-Second World War period were conceived as an efficient and economical way of housing the urban poor, clearing away SLUMS and their diseases, and accommodating the increasing demand for homes from overseas and internal migrants. Many of these new estates or projects became veritable ethnic GHETTOS reproducing and amplifying the social ills they were designed to eradicate. Many were inadequately built and so poorly maintained that damp, condensation, noise, broken-down elevators and other services nurtured social disorganization, crime and disorder. Demolition, most famously with the PRUITT-IGOE buildings in St. Louis, became the only solution and many new estates that have replaced them owe much more to the insights of Jacobs than the vision of LE CORBUSIER.

People of every social class and ethnic background need more than machines to live in.

Further reading

Jacobs, J. (1962) *The Death and Life of Great American Cities*, London: Jonathan Cape.
Pile, S., Brook, C. and Mooney, G. (1999) *Unruly Cities*, London: Routledge.
Sugue, T.J. (1996) *The Origins of the Urban Crisis: Race and Inequality in Detroit*, Princeton, NJ: Princeton University Press.

JOHN BLEWITT

URBAN DESIGN

Urban design (UD) is a professional specialization of architecture, planning and landscape architecture, yet it is taught as a field of studies and not as a specific professional discipline evolving from practice. Academically, it borrows substantive and procedural knowledge from public administration, sociology, law, URBAN geography, urban economics and other related disciplines from the social and behavioural sciences, as well as from the natural sciences (Lang 1994).

In its applied form, UD addresses the scale between architecture and city planning. To this end, its nature can be viewed in terms of settings, processes and outcomes as follows:

Settings

- emphasis on political and administrative dimensions with different levels and types of decision-making;
- recognition of a pluralistic client whose roles of implementation, paying and using are often separated into different entities, e.g. government and private sector institution bureaucrats who help implement, taxpayers who pay, and the public citizen who uses;
- awareness of the importance of the built environment, recognizing its spatial attributes and limitations and focusing on either districts as the intermediate scale of aggregation between buildings and cities, or selected systems across districts;
- understanding of the relationships of physical systems, e.g. transportation, OPEN SPACE, land use, to non-physical systems, e.g. social, economic or administrative systems;

- inclusion of multiple designers and other experts and therefore interdisciplinary interactions.

Processes

- understanding that processes in the activity are characterized by change, ill-structured and ill-defined problems, and multiple stakeholders with multiple objectives and competing agendas;
- seeking the organization of complexity through processes aimed at incremental growth over long time spans;
- enabling change through frameworks that guide design activities but minimize constraints to the creativity of actual interventions;
- recognition that process is as important as outcomes or products;
- inclusion of methods developed to address the client's separation of roles and functions (implementing client, paying client, user client).

Outcomes

- civic dimension where the central focus of the intervention is the public environment;
- concern of the interventions is at larger scales of aggregation than just one building, thus focusing on systems and their relationships, e.g. buildings, open space, movement, INFRASTRUCTURE, etc.;
- longer time frame and indefinite products (guidelines) as opposed to shorter terms of architecture and landscape architecture's finite and specific outcomes (building);
- unlike architecture, outcomes are not necessarily the products of great urban designers;
- scales of aggregation and longer time frames lead to a greater need for accountability as well as a greater anonymity of the practitioner than in other design professions.

From a historical perspective, UD appeared academically during the late 1960s, evolving from the city design and civic/COMMUNITY design efforts and academic courses of the 1950s. Its introduction as presently practised in the United States took place through various academic and professional developments during the 1950s and early 1960s, for example the first urban design course and the urban design conferences by Wilhelm von Moltke

at Harvard's School of Design; the first degree-granting programme at the University of Pennsylvania; and the introduction of urban design as a category in the annual awards programme of the professional magazine *Progressive Architecture*.

Educationally in many parts of the world, the UD activity as characterized above has emerged from the widening gap at the nexus of architecture and planning. In reaction to the Congrès Internationaux d'Architecture Moderne, architecture shifted its focus from urban design towards the design of single buildings. On the other hand, the trend in city planning shifted its focus from urban design to urban research concerns and approaches. These shifts have led by default to the gap's emerging concern with community design, prevalent to a large extent in the North American and European contexts. The antecedents of community design as practised by most urban designers today may be found in (1) UD practices of the 1950s which provided the basis for urban development design; (2) the ADVOCACY PLANNING movements of the 1960s; (3) the ecology movement resulting in urban conservation and environmental quality design; and (4) the trend towards historicism during the 1970s which led to HISTORIC PRESERVATION. In the US context, this widening gap was perceived during the late 1960s and early 1970s to be created by the focus on policy planning in city planning curricula and the continuing emphasis on the design of individual buildings in architecture (Ferebee 1982).

Teachers, thinkers and practitioners have consistently viewed UD over the years very differently both conceptually and practically. This lack of consensus about the activity was the case back at the 1981 Urban Design Retreat (*ibid.*), and continues to be the case today. Recently, a group of UD practitioners who attended an international meeting were asked: What is UD? What does it contribute? And who practises UD? Exemplifying this definitional diversity, their responses to the three questions varied broadly (Van Alen Report 2002).

This wide spectrum of definitions and interpretations of the activity persists in UD education around the world today. For example, in some European and Latin American countries, UD continues to be part of the architectural curriculum, in contrast to some North American programmes that have seen it as part of architecture,

landscape architecture, city/URBAN PLANNING or even public administration. For example, of the 120 members of the Association of Collegiate Schools of Planning, 85 per cent have faculties with UD as their specialty with curricula that include the categories of design studios, policy/process, history/theory, and environmental psychology/behaviour; as well as specialty courses that include various critical aspects of UD such as real estate development, historic preservation, GEOGRAPHICAL INFORMATION SYSTEMS/computer-aided design approaches to URBAN FORM, and NEW URBANISM.

Thus, this ongoing diversity of views is in a way responsible for the continuing discussions on fundamental educational and professional aspects of programmes, such as whether to place emphasis on the traditional architecture vs. the landscape curriculum; the design studio vs. the case study approach; the policy vs. the design approach; systems vs. non-systems thinking; or the theoretical vs. the practical understanding. A positive outcome from these discussions, at least in the United States, is that UD education is constantly being assessed and evolved to better reflect the changing contexts of practice.

Further reading

Bacon, E. (1974) *Design of Cities*, New York: Penguin Books.
Ferebee, A. (ed.) (1982) *Education for Urban Design*, Purchase, NY: Urban Design Institute.
Lang, J.T. (1994) *Urban Design: The American Experience*, New York: Van Nostrand Reinhold.
Van Alen Report (2002) *Urban Design Now* (April).

ERNESTO G. ARIAS

URBAN ECOLOGY

Ecology – from the Greek word 'oikos', meaning 'home' or 'house' – is the study of the relationship of living organisms with each other and with their physical and biological environment.

Urban ecology studies the interactions between biological communities and the URBAN environment; its goal is to achieve a balance between human CULTURE and the natural environment.

The science of urban ecology deals with the structure and function of urban areas as living spaces, ecosystems and landscapes, binding together both pure and applied research. When considering URBAN FORM, it is not hard to think only about its external characteristics: a city, in fact, to be a physical place, is further a net of relationships.

Cities are delicate organisms. Their health and the health and welfare of their citizens vitally depend on the function and fitness of numerous symbiotic systems that control the quality and quantity of energy, material and human capital throughout the city and beyond. Although these systems clearly function at the intersection of the built, natural and human environments, many were designed, managed and optimized as independent components. As a result, conflicts between systems now threaten the entire urban organism's health. Clearly, many of the problems afflicting urban centres cannot be addressed by any single discipline; therefore, the principal aim of urban ecology is, through the participation of many relevant disciplines, to discover how systems are linked and how they can thus contribute to management concepts.

Urban ecology can only be conducted in an interdisciplinary manner.

Further reading

Breuste, J., Feldmann, H. and Uhlmann, O. (eds) (1998) *Urban Ecology*, Berlin: Springer-Verlag.

ANTONELLA VALITUTTI

URBAN FORESTRY

Urban forestry involves both planning and management, including the programming of care and maintenance operations of the URBAN forest. Strategic or master plans establish the overall long-range goals, while management plans are particularly focused on field operations, identifying and prioritizing site-specific tree planting, maintenance and removal activities within a certain time frame.

The urban forest is the system formed by trees and other associated vegetation – shrubs, flowers and grass – beyond just canopy cover, forming green areas of various extent and size located within and around a COMMUNITY – city, village or development. This network of fragmented green spaces includes PARKS, tree-lined STREETS, areas classified as urban where development is expected in the mid- or long term, vacant lots, schoolyards,

cemeteries, lakefronts and river banks, roadways, and utility rights-of-way. Managing it as a system and not as a sum of single elements is the result of understanding its complexity. Although most of the landownership is public, some of these spaces are owned by communities and utility companies. The limits of the urban forest are ill-defined since development is always extending beyond the limits of the town.

Urban forestry looks for the environmental, economic and social benefits of the green areas. As an environmental asset, the forest cleans out the air, attenuates the heat island effect, and reduces soil loss and stormwater run-off, improving quality downstream. Among the economic benefits, it increases property values, attracting residents and businesses. Both factors improve public health and quality of life.

Further reading

Miller, R.M. (1996) *Urban Forestry: Planning and Managing Urban Greenspaces*, 2nd edn, Englewood Cliffs, NJ: Prentice Hall.

URBANO FRA PALEO

URBAN FORM

Urban form refers to the physical form of URBAN areas in three dimensions at a variety of scales. While urban form in principle refers to overall physical form, it may often be represented by specific properties which can be quantified, such as DENSITY.

Urban form is a broad term that may be distinguished from a series of other urban- and form-related terms. *Urban structure* may be equated with the two-dimensional organization of the ground plan of an urban area, such as the STREET pattern or the structure of land parcels. In this sense, it can be regarded as a specialized aspect of urban form; however, urban structure can also have socio-economic interpretations which have no direct associations with physical form. *Settlement form* is more specific than urban form in implying the form of discrete units such as cities, towns and villages. In contrast, urban form could apply to any portion of URBANITY, whether constituting part of a city, town or other urban accretion. *Development pattern* implies the layout of an urban area in deliberate formations, and might also

connote the chronological development of a settlement, as where an original core gains a gridded extension and then a suburban fringe. *Built form* has the connotation of the construction of a pre-conceived artefact rather than an emergent accretion of independently assembled parts. Built form typically implies urban form in three dimensions, at the scale of individual buildings. Meanwhile, *urban fabric* has the connotation of being a continuous surface, often a pre-existing form that may be 'torn' by new interventions (e.g. urban highways), or 'repaired' again (e.g. by sensitive INFILL DEVELOPMENT). Urban form is perhaps the most all-encompassing of these terms, that can imply either design or emergence of form, in two or three dimensions, from the scale of COURTYARDS to CONURBATIONS.

Although urban form includes all three dimensions in principle, at the widest scale, an urban area approximates to a two-dimensional surface, akin to an image on a map or the view obtained from satellite photography. From this point of view, urban form may refer to the overall size or shape of the urban area (e.g. a linear or star-shaped form), or its degree of articulation into discrete settlement units.

Zooming in to the scale of resolution at which we observe an urban area from an aeroplane, the heterogeneity of constituent patterns of development becomes clearer, and overall form becomes articulated into distinct corridors, SUBDIVISIONS or discrete clusters or 'pods' of development. Here, the compact or sprawling character of an urban area could be interpreted. The urban grain and texture become discernible, and it becomes possible to interpret irregular or 'organic' versus orderly or 'geometric' forms.

If we imagine our aeroplane view when close to landing, the three dimensional resolution of urban form becomes more pronounced. Distinctions between high-rise and low-rise development become apparent, and settlement types such as hill towns and WALLED CITIES would be recognizable. Our perspective of urban form zooms in further, from the general arrangement of buildings – in blocks, terraces, courtyards or superblocks – to the sizes, shapes and styles of individual buildings. At this scale – ultimately, the view from the ground – streets themselves become three-dimensional spaces, channels lined by façades.

Some urban form descriptors imply packages of features or associations. TRANSIT-ORIENTED

DEVELOPMENT implies a mixture of land uses in a fine-grained street grid arranged along a superstructure of transport nodes. SPRAWL has connotations of unplanned growth, as well as low-density extension over a large area. Different associations may attach to the same basic form. Whereas the LINEAR CITY label is often positively identified with efficient and transit-friendly planning, RIBBON DEVELOPMENT – also a linear form – is often associated with inefficiency and sprawl.

Some urban form packages imply functional correspondences. Decentralized concentration implies functions spread over an urban region in relatively dense nodes or clusters. Similarly, the distinctions between MONO-CENTRIC and MULTI-CENTRIC or POLYCENTRIC CITIES, or between CENTRAL CITIES, SATELLITE TOWNS and URBAN VILLAGES, are more than matters of physical arrangement, but imply functional relationships (see CITY TYPOLOGIES).

City planners have traditionally been concerned with proposing ideal or optimal urban forms (see LYNCH's *Good City Form*). This typically involves envisaging certain combinations of size, density, structure and built form, married explicitly or implicitly to various urban functions.

Much of twentieth-century city planning was concerned with reducing the overcrowding of the nineteenth-century INDUSTRIAL CITY, in the quest for better living and housing standards. This implied DISPERSION and reducing density, and often entailed the provision of gardens (see GARDEN CITIES) or landscaped OPEN SPACES. However, the late twentieth-century rise to prominence of the environmental movement and concern for sustainable development led to renewed favour for compact cities whose dense form would minimize landtake and reduce travel distances.

In arguments about optimal urban forms, density is often a contested issue, not least since it may be specified in different ways – for example residential or employment density; net or gross density. High rise and high density, though often associated in practice, are in principle independent properties. In some circumstances, planners have pointed out the low densities achievable with high-rise buildings, while others have drawn attention to the high densities achievable with low-rise buildings.

The diversity of ways of characterizing urban form means that care is required with interpretation and comparison. The scale at which an urban area is regarded can affect the interpretation of its shape and structure. Different observers might look at the same city and see different things – for example a monolithic bounded area or a polycentric network structure. Conversely, several quite distinct structural forms could be given the same shape label – a spine-and-pod system and an elongated street grid could both be described as linear corridors. In comparing forms, therefore, it is necessary to be clear about both which aspect (e.g. shape, structure or texture) is being considered, at which scale of resolution.

Further reading

Lynch, K. (1981) *Good City Form*, Cambridge, MA: MIT Press.

STEPHEN MARSHALL

URBAN GROWTH BOUNDARY

One type of GROWTH MANAGEMENT programme that seeks to guide or shape growth without restricting its overall rate. It seeks a more compact and contiguous development pattern that is economically efficient from government's view and is protective of OPEN SPACE, agricultural land and environmentally sensitive areas. A line is drawn on a map showing the outermost intended limit of URBAN development within the planning time frame. The intent is to accommodate projected growth over, for instance, a twenty-year time period but within a more contained area of space than might otherwise happen without the planning boundary.

Urban growth boundaries (UGBs) are increasingly discussed as part of SMART GROWTH strategies in the United States because they seek a balance between development and conservation and are a tool that confronts, and tries to combat, the issue of urban SPRAWL. Other terms used to describe similar goals include 'urban service limits', 'urban limit lines', 'development policy areas', 'urbanizing tiers' and 'designated growth areas'. Programmes that use UGBs encourage growth within boundaries through regulatory incentives, increased planned densities of growth, and public financing of INFRASTRUCTURE, and discourage growth outside through restrictive regulations, limited public funding of infrastructure, and public acquisition of open space.

Examples of UGB use are Portland, Oregon; San Diego, California; Sarasota County, Florida; Montgomery County, Maryland; and Minneapolis/ St. Paul, Minnesota. In Portland, the UGB was drawn in the 1970s to accommodate an expected twenty years of growth. San Diego uses three growth management tiers – urbanized area (UA), planned urbanizing area (PUA) and future urbanizing area (FUA). In the UA, INFILL DEVELOPMENT of existing built areas is encouraged through full public provision of infrastructure. In the PUA, growth is funded through impact fees that pay for that portion of COMMUNITY infrastructure needed by the new project. In the FUA, ZONING and property taxation facilitates continued agricultural use.

Some state governments (such as Oregon and Washington) require metropolitan urban growth boundaries for many of their urban regions. In Oregon, the state's URBANIZATION goal is 'to provide for an orderly and efficient transition of rural and urban land use'. The state requires that 'urban growth boundaries shall be established to identify and separate urbanizable land from rural land'. Studies of the Oregon UGB programme have shown increased densities of development within UGBs and a more contiguous form of urbanization; however, they have also found some leakage of growth outside the boundary. A few state governments (such as Maryland and New Jersey) use a variation of UGBs and attempt to channel state spending on new infrastructure to areas inside of existing urbanized or serviced areas.

According to Porter (1997), growth boundaries should build on and link to comprehensive land use planning policies, zoning requirements and CAPITAL FACILITY PLANNING. They should be based on realistic projections of growth and include procedures for periodic review and boundary adjustment. In some cases, a 'market factor' or 'excess supply' is added to the projected needed amount of land to assure that the regulation is not overly interfering with the private development market. GEOGRAPHIC INFORMATION SYSTEMS play a critical supportive role in UGB systems because they are able to continuously monitor available land supply within the boundary relative to forecasted demand and to track the specific characteristics of that land supply, including land ownership and geographic and site constraints.

Potential downsides of UGBs include an arbitrary setting of the boundary that does not take into account actual characteristics of land supply within its borders, inflation of housing prices if the boundary is drawn too far inward, and difficulty of administration especially if the URB contains within it multiple local governments. Benefits from a well-designed UGB programme – in the form of environmental, fiscal and community character – are considerable. The UGB strategy, because it accepts and accommodates growth at the same time as it seeks to shape it, is one of the more comprehensive and integrated forms of growth management and likely has greater support from developers than policies perceived as more restrictive of the development process.

As part of an urban containment policy, there is often an attempt to establish a GREEN BELT of protected lands around the projected urbanization limit. Public purchase of land in these areas absolutely forecloses urban sprawl there. In the United Kingdom, green belts have been used more extensively and effectively than in the United States to contain urban growth. This is due in part to the British 1947 Town and Country Planning Act and its emphasis on urban containment and countryside protection.

Further reading

Easley, V.G. (1992) *Staying Inside the Lines: Urban Growth Boundaries*, Planning Advisory Service Report 440, Chicago: American Planning Association.

Freilich, R.H. (1999) *From Sprawl to Smart Growth: Successful Legal, Planning, and Environmental Systems*, Chicago: American Bar Association, Section on State and Local Government Law.

Knaap, G. and Nelson, A. (1992) *The Regulated Landscape: Lessons on State Land Use Planning from Oregon*, Cambridge, MA: Lincoln Institute of Land Policy.

Porter, D.R. (1997) *Managing Growth in America's Communities*, Washington, DC: Island Press.

SCOTT A. BOLLENS

URBAN HOMESTEADING

Urban homesteading is a concept rooted in the settlement of the western United States during the nineteenth century, when the federal government offered land as an incentive for people to develop unchartered land. The modern notion of URBAN pioneers salvaging deteriorated housing gained popularity during the late 1970s as a solution to alarming rates of housing abandonment. Urban

homesteading occurred primarily in CENTRAL CITIES where vacant housing dotted NEIGHBOUR-HOODS due to massive SUBURBANIZATION and the consequences of public policies. Some houses went into default, were ultimately abandoned by their owners, and became part of the federal government's inventory as the insurer of the mortgages. Municipal governments and private owners, however, held the majority of abandoned properties in most cities. Urban homesteading programmes transferred vacant properties to prospective owners for little or no financial consideration.

The federal government, states and some city governments passed urban homesteading programme legislation. Several cities operated urban homesteading programmes before the federal government authorized one; others formed local programmes after the federal programme was established. Cities established unique guidelines for their programmes, while federal programme participants were bound by programme policies.

The federal programme was promulgated as Section 810 of the Housing and Community Development Act of 1974. States and local units of government were allowed to acquire vacant single-family homes that had reverted to the US Department of Housing and Urban Development (HUD), the Veteran's Administration and the Farmers Home Administration for use in urban homesteading programmes approved by HUD. States and localities designated Local Urban Homesteading Agencies (LUHAs) to operate the programme. LUHAs were states, local governments, other public agencies or non-profit agencies. The federal government stipulated that houses had to be transferred to homesteaders in unrepaired condition and without substantial consideration – usually one dollar.

Homesteaders were required to complete all repairs within a set period of time, usually up to three years. After residing in the home for a required period of time, usually three to five years, the homesteader would acquire fee simple title to the property. Section 810 funding did not cover administration or rehabilitation costs.

Urban homesteading is also associated with the concept of sweat equity where participants' labour replaces or augments financial investment. Most homesteaders, however, used either private financing, conventional loans from financial institutions, the HUD Section 312 low-interest loan programme or COMMUNITY DEVELOPMENT BLOCK GRANT funds to make repairs.

The impact of the federal urban homesteading programme was minimal when it ended in 1992. Although HUD held most of the federal foreclosed properties, only 3 per cent of HUD properties were disposed of through this programme. Homestead units constituted barely 2 per cent of the structures in targeted neighbourhoods. Evaluations of urban homesteading programmes, in general, concluded that they contributed to the improvement of nearby properties and neighbourhood stabilization.

Urban homesteading has been attempted using multi-family structures, primarily as a response to DISINVESTMENT by landlords. The Urban Homesteading Assistance Board, formed in New York in 1973, incorporated a self-help approach to help tenants in city-owned properties to become owners of their homes.

The term 'urban homesteading' is now used generically to describe initiatives that allow lower-income persons to secure ownership of properties using self-help strategies.

Further reading

Chandler, M.O. (1988) *Urban Homesteading: Programs and Policies*, Westport, CT: Greenwood.
—— (1991) 'The evolution of urban homesteading: planning for lower-income participation', *Journal of Planning Education and Research* 10(2): 40–54.

MITTIE OLION CHANDLER

URBAN IDENTITY

There are two ways in which place is related to identity. The first is *place identification*. This refers to a person's expressed identification with a place, which becomes part of social identity. The second way in which place has been related to identity is through *place identity*, as a specific aspect of identity comparable to social identity. One important mechanism which supports place identity is *place attachment*. Place attachment refers to an emotional bonding between the individual and his or her life space, which could be the home, the NEIGHBOURHOOD, or places and spaces on a larger scale.

Appropriation can be defined as a particular affective relation to an object that may then become part of the identity of the individual.

Appropriation of one's living space is a condition of feeling 'safe' and 'at home': it is essential for the construction of spatial identity. It is important for the individual to be able to organize and personalize space. This may be crucial not only in one's home, but also in the work environment or any other place in which the individual makes a temporal investment. Steady or transitionally occupied places produce place attachment and are often accompanied by ties to personal objects like furniture, pictures and souvenirs, which mark the appropriation of places.

Urban identity is essentially acquired through various territorially bonded *social networks*. Appropriation operates not only at the level of the dwelling, but extends to places like the STREET, the district, the town or even the country, and is accompanied by social networking (family, friends, neighbours, communities). The feeling of being at home in one's neighbourhood is linked to the frequency of encounters, the nature of local relationships, and the satisfaction these provide. It implies social integration extended to local service providers like physicians, shopkeepers and others, and constitutes the framework for the different individual networks (workplace, leisure, school, etc.). Furthermore, the social relations that a place signifies may be more important than the place itself for feelings of attachment. Taking root corresponds to desires for stability and permanency in one's linkage to a certain place and one's involvement in it over the long run. In traditional URBAN structures, the residential environment includes the district in its socio-spatial aspects, delimited by architectural, social and administrative boundaries. Mono-functionalism of city structures has extended the relation to the city as the place of daily life beyond the traditional local district.

Increased residential mobility of the society has provoked a shift from place investment in one's housing to the furniture and other 'belongings' which contribute to the individual's identity. Moreover, *settlement identity*, referring to individual preferences for certain types of habitat, allows a residentially mobile individual to conserve coherence and identity, spanning across various residences.

Further reading

Lally, M. (1992) 'Urban related identity: theory, measurement, and empirical findings', *Journal of Environmental Psychology* 12: 285–303.

Sadalla, K.E. and Stea, D. (1978) 'Approaches to a psychology of urban life', *Environment and Behavior* 10(2): 139–46.

GABRIEL MOSER

URBAN INDICES

Urban indices may be defined as planning standards incorporated into local plans in order to exercise the control of land use and spatial occupation in URBAN areas. In their simplest form, these standards express themselves as mathematical expressions of a desired relationship between socio-economic measures and the space in which they occur. Dealing with highly dynamic processes, determined by development patterns, lifestyle and social demands, these standards are variable in both space and time. Traditionally, they have been divided into two main groups: those that relate spatial and socio-economic values, and the ones that indicate the relationship between two kinds of space. Examples of the first group are the residential DENSITY, i.e. persons per unit area of land, and the OPEN SPACE area per capita. In the second group, one can point out the plot ratio, meaning the relationship between a building and its site.

Planning standards have long been used both in the preparation of plans, where they help to evaluate urban conditions and to access acceptable quantities of various uses of land, and in the function of development control, exercising the continuous regulation of changes in the urban fabric. In this way, they contribute to the design of the preferred physical pattern and, due to their restrictive power, to determining the future characteristics of the city. They were very popular worldwide throughout the twentieth century, used as effective planning tools to guide urban growth, to control the fast city SPRAWL brought on by the urban revolution that followed INDUSTRIALIZATION, and to give shape to the URBAN FORM.

Urban indices were introduced as efficient planning tools by British planning law. Translated into laws and norms, they were, supposedly, capable of providing an adequate spatial organization to urban activities, while assuming desirable levels of environmental quality. However, the way in which such planning techniques have been used for development control has been, in most cases, unduly rigid. Up to the 1960s, they were conceived

and applied without any flexibility, aiming at achieving an ideally organized urban landscape, based on past experiences and combining concepts derived from the CITY BEAUTIFUL and hygienist movements. Moreover, local circumstances were not accounted for. In this approach, cities planned in accordance with these physical criteria might well be acknowledged as architectural compositions. Soon it became quite clear that standards relating to physical conditions alone would not be enough; some type of socio-economic indicator should be incorporated, both in the analytical process and in the development control mechanisms.

Elsewhere in the world, planning practice did not use urban indices with the same intensity as in the British experience. In North America, ZONING was, undoubtedly, the chief mechanism, concerning land use planning and land occupation control. In France, the planning process introduced a different concept, that of the priority occupation area, which, in its basic form, dealt with the same issues as zoning. In Latin America, where the British town planning tradition was prevalent in many countries, development control started considerably late, mainly as an answer to the late URBANIZATION process. Several attempts were made to control the explosive urban growth through the application of indices such as a maximum residential density; a determined plot ratio and floor space proportion; and a minimum open space area per capita. The results were not very encouraging, though. Different social conditions and cultural backgrounds allied to a lower development level and to an accelerated urbanization rate are said to be the main causes of the intended control failure.

In time, the introduction of new planning concepts, such as comprehensive planning, ADVOCACY PLANNING and COMMUNITY participation planning, lessened the widespread use of standards and urban indices in the town planning process. Nevertheless, they were still considered as effective control techniques in the implementation of particular URBAN DESIGN strategies. General rules do not apply everywhere. A piecemeal approach is needed, so indices should be formulated accordingly.

Towards the end of the last century, in the late 1980s, North American planners started to shape a new theoretical approach concerning planning studies and town planning. The movement began as a reaction to the urban sprawl brought about by the suburban planning practised in the United States over the last fifty years, and eventually evolved into the NEW URBANISM. It deals, primarily, with the creation and revitalization of communities, emphasizing the importance of community life. In order to accomplish this goal, it proposes a limited-size NEIGHBOURHOOD integrating different types of land uses, and well provided with accessibility through walking BOULEVARDS and PUBLIC SPACES. Density is to be left on the low side, allowing for a residential mix that includes detached and semi-detached houses. The neighbourhood design is clearly inspired by principles advanced by the GARDEN CITY and City Beautiful movements. To guarantee the application of these guidelines, neighbourhood plans need to include development control tools, which regulate the implementation process. This has, obviously, led to a revival of the use of urban indices, in a revised and even stricter version, as control tools regarding the neo-traditional planning of neighbourhoods, and therefore of all urban development.

Further reading

CEPAM (Centro de Estudos e Pesquisas em Administração Municipal) (1971) *Índices Urbanísticos* (*Urban Indices*), São Paulo: Secretaria de Estado dos Negócios do Interior, Governo do Estado de São Paulo.
Fulton, W. (1996) *The New Urbanism: Hope or Hype for American Communities?*, Cambridge, MA: Lincoln Institute of Land Policy.
McLoughlin, J.B. (1973) *Control and Urban Planning*, London: Faber & Faber.
Roberts, M. (1974) *An Introduction to Town Planning Techniques*, London: Hutchinson Educational.

IÁRA REGINA CASTELLO

URBAN PLANNING

Urban planning refers to the process of envisioning alternative futures for an URBAN area, setting goals and objectives, and formulating implementing strategies to reach the alternative future. Urban planning is the *process* that results in urban plans.

Urban planning is inherently political – local governments are involved in formulating and carrying out urban plans and urban planning requires political skills in negotiating among competing interest groups such as developers, environmentalists, historic preservationists, OPEN SPACE

advocates, NEIGHBOURHOOD activists, and advocates for low-income and minority groups.

Urban planning is distinguishable from *rural planning* (which deals with areas that are not yet urban) and REGIONAL PLANNING (which deals with large areas that may include both urban and rural areas). In the United Kingdom, city planning is called town planning and regional planning is called country planning.

The term 'urban planning' is sometimes used restrictively to refer only to *land use planning* – physical planning focusing on what uses go where in an urban area. While urban planning involves designing the built environment, much more than the physical arrangement of parts of the built environment is at stake. More frequently, the term 'urban planning' encompasses environmental, transportation and housing planning. Sometimes it includes economic development planning, social planning, INFRASTRUCTURE planning, open space planning or other specialized planning.

Historically, monarchs, military commanders, engineers and architects engaged in urban planning from very early times. Ur in Mesopotamia, Mohenjo-Daro and Harrapa in Pakistan, Periclean Athens, Rome and the Roman Empire in classical times, medieval bastide new towns, Florence in the age of the Medicis, the Aztec city of Tenochtitlan, and the African city of Zimbabwe are all impressive examples of urban planning during the last six millennia.

Precursors of modern urban planning arose in the second half of the nineteenth century in response to the problems of nineteenth-century INDUSTRIAL CITIES. Appalled at epidemics, air and water pollution, squalid SLUM conditions, and SOCIAL INEQUALITY, communists, socialists, anarchists and other ideologues; landscape architects, engineers, doctors, settlement house workers and other professionals; and progressive politicians and businessmen called for urban planning to make urban areas healthier, more just and more pleasant places to live.

The year 1909 is a watershed in the history of urban planning. In that year, England passed its first Town Planning Act and the first curriculum in town planning in the world was established at the University of Liverpool. The first course in city planning in America, taught by Thomas ADAMS at the Massachusetts Institute of Technology, and the first national planning conference in the United States also occurred in 1909. Forty people attended the charter meeting of what was to become the American Planning Association.

Since 1909, more and more trained professional urban planners have pursued urban planning throughout the world. Courses and programmes granting academic urban planning degrees have proliferated. Cities hire planners as part of their regular staffs. National professional associations, academic and practitioner publications, and the other trappings of a planning profession have grown steadily. Today, urban planning is an established feature of cities everywhere in the world.

In the United States, local governments (cities and counties) have the ultimate say in local urban planning. City councils and county boards of supervisors vote on urban plans and usually hire the planning director. Most decisions of the city or county PLANNING COMMISSION may be appealed to them. City governments typically have a five- to seven-person *planning commission* that sets policy. A *planning director* reports to the commission and oversees a *planning department*. Professionally trained urban planners staff planning departments. Increasingly, planning staffs have received academic education in city planning usually at the master's level. Planning departments also employ architects, geographers, computer technicians, economists and a range of other professionals. Urban planners who work for cities are generally civil servants. Most urban planners are part of the civil service system – hired through a competitive exam, promoted on the basis of merit, and protected from being laid off as administrations change.

Planning departments engage in both *advance planning* (looking ahead to the long term and formulating comprehensive land use plans) and *current planning* (evaluating development proposals against plans and land use regulations and approving or disapproving specific development projects). Most planning departments administer SUBDIVISION, ZONING and other land use regulations and sometimes local building and housing codes as part of the urban planning process.

Large urban planning offices may have dozens of urban planners working on land use planning, environmental analysis, transportation planning, and perhaps also URBAN DESIGN, housing, HISTORIC PRESERVATION, PARK and open space planning, economic development, and a variety of related specialties. Small urban planning departments may have only a planning director.

Land use planners prepare base maps of their COMMUNITY and undertake studies of the economic and demographic trends that are shaping it. They study what land is available and estimate how much should be allocated to residential, commercial, industrial, open space, institutional and other uses. They specify appropriate densities for different parts of the community. Environmental planners conduct environmental assessments to inform urban planning. They look at how development will impact the natural environment and identify ways to mitigate negative environmental impacts. Transportation planners study trip generation, commuting patterns and needs for transportation infrastructure. They often have strong quantitative skills and model alternative transportation futures. Housing planners analyse local housing markets, housing needs of different segments of the population, and how to deploy government and non-profit housing programmes to meet housing needs.

The nature of urban plans varies widely depending on the context. In fast-growing areas like the San Francisco Bay Area, GROWTH MANAGEMENT is essential. In declining urban areas like Flint, Michigan, the emphasis is on managing decline.

Central to urban planning is preparing a *comprehensive land use plan*, variously referred to as a general or master plan. For almost a century, local governments have been developing comprehensive land use plans as the key document guiding the future physical development of the community. Comprehensive land use plans vary widely in sophistication and content, from highly technical multi-volume plans for large cities that have well-developed planning to small and rudimentary documents in smaller cities with less planning capacity. They almost always contain both maps and text illustrating the projected future of the community. They may include visions, goals, objectives, policies, standards and implementation strategies. Descriptive material on the physical features of the community and its changing demography, housing stock, environment and economics may be in an accompanying volume or incorporated into the plan itself. Comprehensive land use plans generally address land use, transportation, housing and open space, and may address urban design, infrastructure, noise, safety, recreation, public facilities, economic development, arts, CULTURE and many other aspects of the

community. Some contain sub-plans for areas within the city.

University of North Carolina planning professors Edward J. Kaiser and David R. Godschalk call land use planning 'a stalwart family tree' which, after a century of evolution, shows no signs of disappearing. They trace the roots of the tree to early theorists and practitioners such as Edward Bassett, T.J. KENT Jr and Stuart Chapin and influential early plans such as the 1909 BURNHAM Plan for Chicago. They trace the growth of land use planning's *trunk* from early GENERAL PLANS to modern hybrid plans that deal with design, policy and management. Kaiser and Godschalk identify *branches* of planning such as verbal policy plans, land classification plans and development management plans emerging from the trunk.

Urban planning is an open and often contentious process. A citizen's advisory committee may be involved and drafts of the plan may be discussed at neighbourhood meetings. There will always be public hearings with an opportunity for input before the plans are finalized.

Comprehensive land use plans 'sit atop the pyramid' of local land use regulation in the words of Harvard law professor Charles Haar as an 'impermanent constitution' for the community. Zoning, subdivision ordinances and other land use regulations form the base of the pyramid.

In the United States, the American Planning Association is the professional association for urban planners. Its counterpart in the United Kingdom is the Town Planning Institute. In North America, the most common academic degree for urban planners is a master's degree in city planning (MCP) or in city and regional planning (MCRP). The Planning Accreditation Board accredits over a hundred master of city and regional planning programmes in North America and a small number of undergraduate planning programmes. Professors teaching city and regional planning in the United States are now organized into the Association of Collegiate Schools of Planning. A planning accreditation board accredits professional urban planning programmes. In Europe, the Association of European Schools of Planning, and in Asia, the Asian Planning Association are organizations of professors teaching urban planning.

Urban planning is informed by a body of knowledge called *planning theory* which seeks to describe how planning occurs, and how it *should* occur. At first, urban planning was closely allied to

architecture and design. Early planning theorists at the University of Liverpool, MIT and elsewhere defined urban planning as physical site planning writ large. They encouraged their students to produce static, end-state physical plans. Somewhat later, planning theorists developed and elaborated what is called the rational planning model. They argued that urban planning should proceed rationally and predictably as technically expert planners formulated goals, selected the best course of action to achieve the goals, proposed implementation strategies, and incorporated learning from feedback into the process. During the last four decades of the twentieth century, many competing paradigms shaped urban planning theory: systems theorists argued that planning should consist of applying mathematical models to quantitative measures of urban systems; Marxists saw planning as an aspect of class struggle and urged planners to examine the class basis of society and reshape it; advocacy planners felt that urban planning should not be the sole province of local government, but that planners should develop multiple plans and advocate for clients with different points of view.

Urban planning methodology is evolving rapidly – particularly with the growth of quantitative methods and the possibilities information technology creates for planning. Most urban plans are data driven – based on data about land use, population, economics and transportation systems. Urban planners use statistical analysis, econometric modelling, cost-benefit analysis and other techniques to quantify and forecast phenomena the plan seeks to accommodate and shape. Census and other secondary data are fundamental. Increasingly, urban planning involves spatial analysis using GEOGRAPHIC INFORMATION SYSTEMS (GIS) software. Qualitative methods are also used in urban planning. Observation, case studies, depth interviews, use of unobtrusive measures, simulations and focus groups can all contribute to planning.

Further reading

International City Management Association (ICMA) (1999) *The Practice of Local Government Planning*, 3rd edn, Washington, DC: ICMA.

Kaiser, E.J. and Godschalk, D.R. (1995) 'Twentieth century land use planning: a stalwart family tree', *American Planning Association Journal* 61(3): 365–85.

—— (1995) *Urban Land Use Planning*, 4th edn, Chicago: University of Illinois Press.

Stein, J. (1995) *Classic Readings in Urban Planning*, New York: McGraw-Hill.

SEE ALSO: urban studies

RICHARD LEGATES

URBAN PRIMACY

Urban primacy is a defining feature of world population distribution wherein historical and economic forces have created cities far larger than others in the same country. In ancient times, religious and administrative elites fostered city growth by controlling the resources of surrounding agrarian areas. In early modern times, the development of nation-states was central to the expansion of national capitals as centres of diplomacy and finance. London, for example, shared these characteristics but also prospered after 1500 as a result of international commercial activities such as piracy and the slave trade. Paris primarily benefited from efforts to centralize the authority of the French monarchy, though it, too, grew as a result of similar overseas incursions. Other European capitals such as Amsterdam, Vienna and Madrid also achieved national pre-eminence with the acquisition of colonial empires accumulated during the sixteenth and seventeenth centuries. Their regions still contain a significant proportion of national population.

By the end of the eighteenth century, INDUSTRIALIZATION began to eclipse mercantile trade as the leading engine of economic growth. The factory system and its large-scale organization of production led to extraordinary URBAN expansion in new locations such as Manchester and Liverpool and then subsequently across Europe and the eastern United States. The rapid progress of technological change continued to create very large urban centres such as Los Angeles, with its early dependence on oil and the film industry, or Denver and Chicago as transportation-based hubs at the edge of the American prairie.

However, urban primacy is most dramatically encountered in less developed countries (LDCs). Hence, Bangkok has more than fifty-five times more population than does Khon Khaen, Thailand's next largest city, and Buenos Aires has ten times the population of Cordoba, second in size in Argentina. A number of forces are

responsible for the extraordinary growth of giant cities in Asia, Africa and Latin America. First, urban centres in LDCs were initially created in response to raw materials exportation such as coffee, cotton, spices, jute and grains. It therefore follows that almost all major cities in Africa and Southeast Asia are seaports. Secondly, these ports have enjoyed the cumulative benefits of investments in INFRASTRUCTURE, finance and communications. Because of these locational imperatives, a political dimension was inevitably superimposed, invariably resulting in status as a colonial capital. After independence, that function accelerated with the growth of national bureaucracies and multinational commercial ventures which, in turn, required additional infrastructure and commercial and diplomatic facilities.

Urban primacy also occurs in inland locations. For example, the Spanish LAWS OF THE INDIES legislation prohibited the founding of major towns in port locations for reasons of security and settler industriousness. Ports were open to attacks from pirates or foreign navies and offered too many distractions for colonists. Even so, inland capitals in Spanish America maintained control over their ports, further aggrandizing primate city interests. An example would be the relationship between Lima and Callao in Peru or Quito and Guayaquil in Ecuador.

In Africa, capital cities are often the only urban centre of any importance. In twelve nations they have more than ten times the population of the next largest city. In all but five countries the capital is the largest city and is often the only urban centre of any importance, a factor that accelerates urban primacy. Since capitals are the fastest-growing city, attracting manufacturing, warehousing and a plethora of other economic, political and cultural functions, rural–urban migration is an unrelenting factor in unchecked primate city growth. In addition, African capitals are marginally located, often in a coastal site convenient to the former colonial metropole. While these cities at one time may have functioned effectively in a colonial role, they are poorly situated from the standpoint of a national capital as they are far from the nation's geographical centre. This factor introduces cross-border migration as another compelling factor in primate city population growth.

The implications of urban primacy in LDCs are immense. While very large cities were most likely to be found in India, China, Europe and North America prior to the Second World War, in the early part of the twenty-first century about two-thirds of the 35 or 40 largest world cities are situated in LDCs. Current trends suggest that cities with more than 4 million inhabitants will double between 1985 and 2010 and will number more than 130 by 2025; all but fifteen of these cities will be located in LDCs. Urban services will be impossible to manage. The absence of sanitation and potable water will add to the already staggering impact of the AIDS epidemic. To this, one can add the virtual absence of healthcare facilities. Furthermore, populations will have to find or create housing units in dense, centrally located areas or on the urban periphery, far from the jobs they seek. In close-in NEIGHBOURHOODS, natural hazards are part of the trade-off, with environmentally dangerous sites (such as riverbanks) posing risks from flooding and landslides. As additional settlers arrive in the primate city, they will be difficult to absorb into the LABOUR MARKET, creating the potential for high rates of CRIME. The transportation system will overload and gridlock will characterize urban STREETS for nearly twenty hours a day. Funds to grapple with these problems are virtually non-existent and quality of life is now little more than a textbook phrase. And yet, people will continue to seek a better life due to rural POVERTY, lack of access to land ownership and inferior levels of public services. Despite the challenges that await, migrants will continue to view primate cities as unrivalled gateways to opportunity.

Further reading

Dogan, M. and Kasarda, J.D. (1987) *The Metropolis Era: A World of Giant Cities*, Newbury Park, London and New Delhi: Sage.

Gilbert, A. and Gugler, J. (1992) *Cities, Poverty and Development*, 2nd edn, New York: Oxford University Press.

Griffiths, I.L. (1995) *The African Inheritance*, London and New York: Routledge.

DANIEL GARR

URBAN REGENERATION

Urban renewal or regeneration is nothing new. The building of new roads and paved sidewalks, the creation of public health schemes and sewage systems helped revitalize and clean up some big

European cities like London, Paris and Berlin in the nineteenth century. Much decrepit SLUM housing of the URBAN poor was cleared away in the twentieth century in favour of new modernist apartment blocks (which frequently became slums themselves) or new SUBURBS or 'new towns' modelled on the greener ideas of the GARDEN CITY urbanists or the new urbanists of the later twentieth century. With the more recent experience of URBAN DECLINE in Europe and North America, urban regeneration has become a political and social priority of local and national governments. It has also created a new class of urban regeneration professionals.

Regeneration involves many interconnected elements which if tackled separately have less impact than if addressed in a holistic manner. A primary purpose of regeneration is to restore economic viability to a given area often by attracting external private and public investment and by encouraging business start-ups and survival. All businesses require markets to sell their goods and services. These may be local with the corner store or more widespread with a manufacturing or service industry. They require capital to develop and innovate and a skilled and effective workforce. If this workforce is to be recruited locally, the educational INFRASTRUCTURE must be able to meet this need. One approach to business development has involved the state and the private banks offering credit deals, exemption from local business taxes (at least in the short term), other forms of tax incentives and a relaxation of planning laws and regulations. Particularly in the free market 1980s, some politicians heralded enterprise zones as the way forward although the vagaries of the market – 'market failures' – did not always satisfactorily deal with the entrenched social problems of POVERTY, civic alienation and social disorganization.

The approach to ZONING was extended in the 1990s. The Clinton administration created the federal EMPOWERMENT ZONE programme in 1994. Various Urban Development Corporations have significantly improved many inner-city NEIGHBOURHOODS like the South Bronx in New York. However, the problem of relative poverty persists. In Britain, the Labour government's National Strategy for Neighbourhood Renewal has involved the creation of Education Action Zones, Health Action Zones and Local Strategic Partnerships with a strong emphasis on the need for grass-roots

participation, COMMUNITY planning and local people's ownership of change. This bottom-up approach has helped foster the growth of social and community-based economic enterprises which, although not sufficient themselves for complete urban revitalization, are able to address the issue of reskilling the local labour force by helping to create intermediate LABOUR MARKETS and a modest entrepreneurial culture. The practical realization of community empowerment through community action, organizing, leadership and CAPACITY-BUILDING has led to remarkable changes. Community-led regeneration helps build SOCIAL CAPITAL and can help avoid the emergence of local mistrust and suspicion that may accompany top-down approaches. The Dudley Street Neighborhood Initiative in Boston described by Medoff and Sklar in their *Streets of Hope* (1994) is one example of community-based action. In Bradford, in the north of England, local people administered the state grant known as the Single Regeneration Budget for the physical, economic and social renewal of the working-class housing estates (Royds), setting up their own community association and consultancy firm to generate income after the grant monies expired and so sustain the renewal process and ensure they, rather than government, controlled their future.

Multi-agency partnership working has been a key feature of regeneration and neighbourhood renewal initiatives in the 1990s and early twenty-first century. The aim to create 'joined up' thinking and action may occasionally lead to interagency or institutional tension but government sees this as the consequence of developing a more holistic worldview and a new mode of urban management and GOVERNANCE. Multi-agency partnership working complements the existing electoral system by encouraging greater participation and involvement in civic affairs. Health, housing, education, local government, private business, education and voluntary sector bodies are all represented on the partnership bodies in the United Kingdom. Each partnership applies a multifaceted approach to a given problem or issue. Dealing with CRIME is no exception.

Reducing crime and disorder – 'reclaiming the streets' – and improving the physical infrastructure of an urban area – 'fixing broken windows' – have consequences far beyond the immediate and obvious. Businesses leave areas if they are regularly robbed. Shoppers avoid places if they feel unsafe.

Effective policing and community cooperation in crime-reduction activities requires much more than questionable zero-tolerance policies as many deprived urban areas have experienced a breakdown in trust and police–community relations. This has been the case particularly in areas with significant black and ethnic minority populations. Civil disorder and high crime rates lead to what Skogan (1992) has termed 'a spiral of decay' and combating these issues requires both institutional partnerships and action on the ground by the communities themselves. Vandalism, vagrancy, prostitution and anti-social behaviour are important issues but these are often symptoms of deeper social problems and purely punitive approaches easily compound social injustices rather than relieve them. Wielding a baton or locking someone up does not achieve social cohesion and inter-ethnic trust. The authors of the influential book *Comeback Cities* (2000) rhetorically ask the question: can it just be a coincidence that the two cities that did the best job rejuvenating neighbourhoods also lowered crime the most?

New approaches to business thinking and activity have led to a variety of financial, economic and social dividends. Investments in housing and increased home ownership and residential stability offer increased business opportunities. Although individual households in poorer neighbourhoods may not have as high a level of disposable income as those living in richer areas, their population density is often greater. If neighbourhoods are also safe, new retail outlets and new customers emerge. Some companies including discount stores like Payless, supermarkets like Tesco's and fast food chains like McDonald's have combined a sense of social responsibility with a keen eye for the bottom line. A new Tesco supermarket in Leeds (northern England) was sited in a deprived urban neighbourhood, hired and trained local people, and prospered. Other businesses may emerge to meet the needs of a particular ethnic community. The return of retail businesses to the inner urban areas also facilitates the development of social interaction, reduces the need for travel to out-of-town shopping centres and so lessens the reliance on the private car, which many poor households may not own, and a public transport facility that is irregular and unreliable. The growth of local businesses meeting the needs of local people locally is also an important aspect of sustainable

urban regeneration. Just as the construction of out-of-town SHOPPING MALLS may adversely affect the economic vitality of DOWNTOWN areas, a re-channelled consumer culture may revitalize an urban area.

Environmental and design issues are important factors in urban regeneration. The architect Richard Rogers led Britain's Urban Task Force in the late 1990s, publishing its *Towards an Urban Renaissance* in 1999. The report argued for compact urban developments rooted in a commitment to excellent and sustainable design principles, the creation of integrated urban transport systems prioritizing the needs of cyclists, pedestrians and public transport passengers, streamlined planning processes, the location of as much development as possible on BROWNFIELD sites and the reuse of existing buildings wherever possible, the establishment of public–private investment funds and urban regeneration companies, mixed-income housing developments, and much more. Although criticized for being design led and insufficiently attentive to practical social and political problems, the report offers an inspiring and radical vision of an urban regeneration practice. Rogers's Millennium Village in Greenwich, south London, built adjacent to the infamous Millennium Dome, is an attempt to realize these principles.

Regeneration can be boosted by prestigious cultural or arts projects, which have many economic and social benefits. The hosting of international sporting events like the Olympic Games at Homebush in Sydney in 2000 or the British Commonwealth Games in Manchester in 2002 were linked to major renewal initiatives involving significant physical regeneration, the decontamination of polluted land and the construction of new STADIA. Major sporting events are closely connected to the leisure and tourism industries that provide local people with incomes and jobs. The development of 'cultural industries' may also have similar effects. When Glasgow won its bid to become European City of Culture in 1990, a range of regeneration multipliers were put into effect. Gateshead in the north of England has experienced significant arts-led regeneration following the commissioning of an iconic piece of PUBLIC ART known as the Angel of the North. The same may be said for Barcelona (the Olympic host in 1992), whose own regeneration in the 1980s and 1990s centred on global sport, high art and culture, symbolized by Frank Gehry's postmodern

Guggenheim Museum in Bilbao, and advanced industrial production. Its regional Catalan, as opposed to Spanish, identity is a factor too.

Other urban areas have capitalized on conserving their architectural heritage or focusing on one particular characteristic as their Unique Selling Point. Stratford-upon-Avon in England is 'Shakespeare country', Trier in Germany expresses a time of bishops and princes, other towns may become associated with theatre, film or literary festivals. Old industrial buildings, docks and warehouses may be turned into highly sought-after and, if on a riverside, extremely expensive developments with a cafe culture, vibrant nightlife, casinos and the obligatory marina. Hannigan (1999) has written of the 'fantasy city' where entertainment, '*eat*ertainment' and '*edu*tainment' are aspects of a combined Disneyfication and McDonaldization of urban economic and social relationships. All this is an important aspect of how a revitalized urban centre may redevelop its social and physical infrastructure and consequently market itself. The absurdly expensive Millennium Dome theme park in south London was justified for its regenerative effect on a relatively deprived urban area. A district, a whole town or even a city (Las Vegas?) can become thematized.

Whether economic (rent gap theory) or consumer factors predominate urban regeneration is frequently associated with demographic change. Young professional singles or couples may enjoy the lifestyle offered by a vibrant inner-city locality and may wish to live and work in that area rather than travel hours to a suburban home. Loft living has been key to development in a number of cities, most notably New York, and the GENTRIFICATION of former working-class localities and working-class housing is a phenomenon common to many cities in Europe and North America. Some urban commentators argue that gentrification creates vacancies in other parts of a city allowing the displaced residents to find alternative accommodation. There is also the expectation that many people rationally desire to ascend the housing ladder similar to moving up the career ladder to maximize their status and income. For this to occur, apart from people acting rationally, a degree of job security and a degree of stability in the property market are needed. Smith argues in *The New Urban Frontier* (1996) that gentrification is a key element in a new urban revanchism

embodying 'a revengeful and reactionary viciousness' against working-class and ethnic populations who have stolen the city from the white upper and middle classes.

The rapid development of information and communications technologies and post-Fordist industrial structures and production processes have reconfigured urban space and regional and urban development. The sociologist CASTELLS has characterized the contemporary (First) world as a network society and the successful modern city as one centred on information processing, new INTERNET technologies and the clustering of industrial innovations. Hi-tech industrial developments in California's SUNBELT and the increasing dominance of a number of key global or world-class cities may lead to two- or three-speed urban economies in many areas with the need for urban renewal and regeneration to be an ongoing process.

Further reading

Grogan, P.S. and Proscio, T. (2000) *Comeback Cities: A Blueprint for Urban Neighborhood Revival*, Boulder, CO: Westview.

Hannigan, S. (1999) *Fantasy City: Pleasure and Profit in the Postmodern Metropolis*, New York: Routledge.

Keating, D.W. and Krumholz, N. (eds) (1999) *Rebuilding Urban Neighborhoods*, London: Sage.

Medoff, P. and Sklar, H. (1994) *Streets of Hope: The Fall and Rise of an Urban Neighborhood*, Boston, MA: South End Press.

Skogan, N. (1992) *Disorder and Decline: Crime and the Spiral of Decay in American Neighborhoods*, Berkeley, CA: University of California Press.

Smith, N. (1996) *The New Urban Frontier: Gentrification and the Revanchist City*, London: Routledge.

JOHN BLEWITT

URBAN REGIME THEORY

An urban regime might be defined as the informal arrangements by which government and private actors cooperate in order to make and carry out governing decisions and achieve a lasting influence in key policy areas. Developed by Stephen Elkin and Clarence Stone in the 1980s, urban regime theory provides a new method to analyse what actors have power in URBAN politics. This theory serves as a paradigm shift in urban politics. It differs from both pluralism and Paul Peterson's

model in *City Limits*, both earlier explanations for who has power in cities.

According to pluralism, power is spread widely between various actors in the urban political arena. Best articulated at the urban level by Robert DAHL in his 1963 book, *Who Governs?*, this theory finds that a specialization of influence exists in which different groups and leaders dominate different issue areas. Power is not concentrated in one group or leader.

In his influential book written in 1981, *City Limits*, Paul Peterson offers a model that opposes pluralism in explaining what actors have power in urban politics. According to Peterson, pluralists understate the degree of power that the business COMMUNITY imposes on local decision-making. He uses the concept of mobility of capital to show the power that business possesses in the urban arena, because business owners may threaten to relocate to another city or state. Cities cannot afford to risk this loss to their job and tax bases according to Peterson. Municipal officials must anticipate what businesses want, with the city as a whole acting as if it had a unitary interest in economic development. Politics becomes secondary, and policy drives politics as a result.

Urban regime theory might be considered to fall between pluralism and Peterson's model. Unlike pluralists, urban regime theorists do not find power to be distributed evenly or widely between different actors. Instead, urban regime theory finds that local officials usually favour upper-strata interests such as businesses over the long term.

However, urban regime theorists differ from Peterson, because they do not believe that businesses are most powerful in all cases or that a unitary interest in development will always emerge. A regime oriented to business is not always guaranteed to form. In other words, politics matters. Politics drives policy.

Also, they differ with Peterson in explaining why businesses are usually favoured. Regime theorists do not believe mobility of capital fully explains the power of business. Rather, they find that businesses may hold a privileged position and control a disproportionate share of the resources that city officials value. However, it is not completely accurate to state that local officials desire to serve business and land interests. Their major concern is to induce economic performance, not serve these interests. Although city officials may

resist the wishes of business leaders, they usually come to realize the importance of forming a viable alliance with key business interests due to the additional revenue and enhanced economic performance that such an alliance might bring to the city. Such an alliance is natural and equals the status quo. Those such as community organizations who hold a different conception must fight an uphill and often losing battle. Incentives for local officials to take a broader view that includes community organizations and other groups are largely absent, because an alliance with these groups often does not enhance a city's economic performance.

As a result, of the three urban regime types elaborated by Stone, corporate regimes may be the norm in most large cities. In a corporate regime, business interests exercise a major role in guiding a city's development policy, and the regime's central concern usually involves promoting development interests. Regarding the use of public resources, a corporate regime's prevailing coalition would likely choose to subsidize investment through measures such as TAX ABATEMENT for business development. Corporate regimes also take the position that investment is most readily promoted by pushing development costs towards the public sector in areas such as INFRASTRUCTURE. Finally, this regime type seeks development benefits to its own business partners rather than collective benefits to the community. In other words, it views large profits to private actors as a justifiable reward for their investment and a low tax rate as necessary mechanisms to maximize private choice.

In the second type of urban regime, a progressive regime, community organizations and groups representing persons of diverse race and income levels play a major role in guiding development policy. A progressive coalition usually chooses to use public resources to enhance equality and believes that equality can be promoted without hindering economic productivity. In addition, progressive regimes want to push development costs towards the private sector, since it is the one that stands to make profits. In terms of benefits, progressives attempt to use development in order to gain collective benefits. As an example, the regime may seek a linkage policy that requires corporate developers to make monetary contributions to a city's low-income housing. Finally, progressives may also seek communal amenities such as HISTORIC PRESERVATION, enhanced PARKS

and recreational areas, environmental quality, and planned growth.

In Stone's third urban regime type known as a caretaker regime, small property holders such as homeowners and small businesses exercise a key role in guiding development policy. The caretaker regime would like to rely on the free market rather than the use of public resources, because it views all subsidies as an unfair redistribution of what should be privately held funds. In addition, caretaker coalitions try to minimize development costs and may seek a no-growth policy when confronting costs caused by private actions. This regime holds these views because its coalition is mainly interested in low taxes and a non-interventionist government.

Further reading

Dahl, R.A. (1963) *Who Governs? Democracy and Power in the American City*, New Haven, CT: Yale University Press.

Elkin, S.L. (1987) *City and Regime in the American Republic*, Chicago: University of Chicago Press.

Peterson, P.E. (1981) *City Limits*, Chicago: University of Chicago Press.

Stone, C.N. (1989) *Regime Politics: Governing Atlanta, 1946–1988*, Lawrence, KS: University Press of Kansas.

SUSAN E. BAER

URBAN RENAISSANCE

Urban renaissance represents an attempt to deal with the negative impact of three major factors driving change in URBAN areas:

1   The technical revolution – centred on information technology and exchange.
2   The ecological threat – based on greater understanding of the implications of our rapid consumption of natural resources and the importance of sustainable development.
3   The social transformation – flowing from increased life expectancy and new lifestyle choices.

The result has contributed to the massive peripheral expansion of most Western towns and cities driven along in the latter part of the twentieth century by:

• increased car ownership and mobility;
• deregulation of planning processes;
• increased owner occupation;
• industrial decline and development of the service-based economy;
• single-use low-density ZONING; and
• a lack of investment in public transport and in urban environments.

Increasingly, the suburban housing environments that have spread out from established urban areas have in recent times been joined by out-of-town business parks, retail centres and leisure facilities, further weakening the connection between homes and urban centres and increasing reliance on the private car. These factors have resulted in profound changes in lifestyles which are both reflected in, and influenced by, the outward spread of cities; in the legacy of empty buildings and sites left behind in the urban landscape of many towns and cities; and in the uniformity of the new urban landscape that results.

The idea of an urban renaissance is a complex one and one that defies simple definition, but one that attempts to address the urban malaise that results from these trends. It has, for example, been simplistically likened to spreading the 'cafe culture' found in many vibrant European cities. Thus, comparisons tend to be made with large cities like Barcelona or Amsterdam. Conversely, the concept has on occasions been simplistically regarded as interchangeable with the need to develop more housing on BROWNFIELD sites at higher densities. Although these factors make a desirable contribution to delivering an urban renaissance, they are unlikely by themselves to lead to long-term sustainable change. The concept has even been used to mean reviving the kind of urban places associated with model Georgian towns like Brighton and Tunbridge Wells in the United Kingdom.

Clearly, the concept moves decisively beyond the post-war concept of urban renewal and the more recent concept of URBAN REGENERATION. Thus, renewal was primarily concerned with public sector-driven, large-scale and often highly zoned physical change of inner-city areas; regeneration with economic growth by using public funds to lever in subsequently largely undirected market investment. Urban renaissance on the other hand seeks partnerships to drive forward public and private goals and investment on a number of policy fronts, directed towards the reinvigoration

of urban areas. In broad terms, the concept encompasses:

- a change in attitudes to urban environments and urban living – from negative to positive;
- a reinvestment in urban environments in their economic, social and environmental INFRA-STRUCTURE; and
- to deliver this, positive and integrated leadership and management.

Urban renaissance therefore represents more a set of 'processes' of change and adaptation than a set of 'products' or clearly defined solutions to the way towns and cities work. In this context, the concept goes beyond physical environmental solutions and objectives to encompass concerns for social exclusion, wealth creation, sustainable development, urban GOVERNANCE, health and welfare, CRIME prevention, educational opportunity, freedom of movement, as well as environmental quality and good design. Indeed, architect Richard Rogers in his introduction to *Towards an Urban Renaissance* argued that although regeneration needs to be design led, to be sustainable it has to be placed within its economic and social context. The Urban Task Force argued that change in the form of an urban renaissance should be founded on the joint principles of:

- design excellence;
- economic strength;
- environmental responsibility;
- social well-being; and
- good governance.

These areas of concern will also involve more than just the land use planning system in delivering their combined potential; they equally affect transport planning, housing policies, area management and maintenance, health, social and education services, police forces, and local and national fiscal frameworks. Inevitably, therefore, the delivery of long-term urban renaissance will also require some degree of holistic 'joined up' thinking and the coming together of stakeholders – public, private and COMMUNITY – in formal or informal partnerships.

Further reading

Carmona, M. (2001) 'Implementing urban renaissance – problems, possibilities and plans in South East England', *Progress in Planning* 56(4).

Urban Task Force (1999) *Towards an Urban Renaissance*, London: E&FN Spon.

MATTHEW CARMONA

URBAN RESTRUCTURING

Urban restructuring refers to major changes in the spatial structure and organization of cities, and usually to the underlying restructuring of the economy, society and politics of cities. Soja (1989) describes restructuring as a 'brake' or 'break' in trends – a shift towards a different social, economic and political order. It is structural as opposed to piecemeal change, but not complete revolution or transformation.

Literature on urban restructuring emerged in the 1980s as a response to the major changes occurring in cities following the political and economic crises of the 1970s. Previous URBAN theories were criticized for their inability to capture the dramatic changes occurring in the late twentieth-century city. The emerging literature was initially located within Marxian theory, but later work embraced POSTMODERNISM. Influenced by HARVEY's work on the historical geography and spatiality of capitalist development, it reflected a recognition of the restless nature of capitalist URBANIZATION. His ideas of the 'spatial fix', and his theories of geographically uneven development, were important precursors to theories of urban restructuring. Harvey's own work (1989) links epochs of development to the changing nature of capitalist urbanization, although he does not use the term 'urban restructuring'. Compared to the earlier work of Harvey and CASTELLS, the restructuring literature places much greater attention on production, and the specific dynamics of ECONOMIC RESTRUCTURING. The links between global economic restructuring and urban restructuring in the late twentieth century has been a dominant theme. Indeed, Soja argues that the term 'urban restructuring' is synonymous with the dynamics of capitalist urbanization in this period.

Earlier work in the 1980s drew on theories of the new international division of labour as low-waged industries decentralized to peripheral areas and countries. Authors such as SASSEN examined the impact of job loss within industry, the growth of services, the informalization of work, and the growth of sweatshop and subcontracted production

on cities in developed countries. Some cities or spaces within them declined, while other areas grew, and new economic spaces emerged. Several governments (often in alliance with local capital) responded to economic restructuring by redeveloping cities to reposition and re-image them as consumer and cultural centres, for example through waterfront, shopping, upmarket residential and leisure developments, usually in areas of economic decline. Restructuring also affected households and communities within cities, and reshaped urban politics. Urban space was thus restructured through the interplay of global capital, the state and activities of urban residents.

Later research linked urban restructuring to broader changes in the regime of accumulation – to changing patterns of production, technologies, labour processes, and shifts in associated patterns of consumption, CULTURE, ideology and state. Soja (1989) linked the evolution of URBAN FORM to rounds of restructuring, focusing particularly on the shift to post-Fordism following the 1970s crisis. The introduction of flexible forms of production, globalization, the rise of the service and informational economy, the replacement of the welfare state with entrepreneurial GOVERNANCE and neoliberal policies, along with growing SOCIAL INEQUALITY, shaped what Soja (1995) termed 'postmodern urbanization' – not a complete break with modernism, but a series of restructurings – as exemplified in Los Angeles. Globalization underpinned the growth of this large city-region as a multicultural world city. Industrial restructuring led to a new industrial geography with the rise of new TECHNOPOLES, and patterns of deindustrialization and reindustrialization. The urban form was restructured towards polycentric development, increasing urban SPRAWL and the growth of edge cities, but with patterns of both DECENTRALIZATION and recentralization. New patterns of SOCIAL POLARIZATION and SEGREGATION mirrored growing income inequalities. The city became more ungovernable: a carceral city, divided behind walls, razor wire, surveillance systems. Hyper-reality became pervasive.

While Los Angeles was seen as the paradigm case, and much of the subsequent literature focused on major world cities in the West, in the 1990s, several studies considered the influence of globalization on urban restructuring in large, rapidly growing cities in Asia, such as Tokyo, Jakarta and Shanghai. The REDEVELOPMENT of Shanghai to reposition it as a financial and trade centre post-socialism received considerable attention. Other studies in the 1990s included Brazilian, Australian, South African and European cities.

Perspectives on the forces driving urban restructuring vary, mirroring broader debates on economic restructuring, and the structure–agency debate. Although urban restructuring might be viewed as a simple reflection of global economic restructuring, several authors insist that economic, social and political dynamics and histories particular to specific nations and cities shape processes of urban restructuring in critical ways. Feminist authors have also shown how women's ways of coping, priorities and life strategies contribute to the constitution of places and the transformation of urban space. While there are commonalities in patterns of urban restructuring, there are also important variations, and outcomes are not predetermined.

A smaller body of literature in the 1990s considered urban restructuring as a planning goal in its own right, generally linked to transformatory ideals, such as Hahn and Simonis's (1991) ecological urban restructuring, policies to restructure South African cities post-apartheid, and restructuring to achieve COMPACT DEVELOPMENT. The use of planning to achieve urban restructuring is a long-standing theme in urban literature, but several earlier initiatives, such as urban redevelopment discussed by Fainstein et al. (1983) and others, tended to restructure for capital. Much research on post-Fordist urban restructuring has reached similar conclusions. Whether these idealistic approaches can achieve their aims in the context of contrary restructuring trends remains an important debate.

Further reading

Fainstein, S. (1996) 'The changing world economy and urban restructuring', in S. Fainstein and S. Campbell (eds) *Readings in Urban Theory*, Oxford: Blackwell.

Fainstein, S., Fainstein, N., Hill, R.C., Judd, D. and Smith, M.P. (1983) *Restructuring the City: The Political Economy of Urban Redevelopment*, New York: Longman.

Feagin, J.R. and Smith, M.P. (1987) 'Cities and the new international division of labour: an overview', in M.P. Smith and J.R. Feagin (eds) *The Capitalist City: Global Restructuring and Community Politics*, Oxford: Blackwell.

Hahn, E. and Simonis, U. (1991) 'Ecological urban restructuring', *Ekistics* 58(348/349): 199–209.

Harvey, D. (1989) *The Urban Experience*, Baltimore: Johns Hopkins University Press.

Soja, E. (1989) *Postmodern Geographies: The Reassertion of Space in Critical Social Theory*, London: Verso, ch. 7.

—— (1995) 'Postmodern urbanization: the six restructurings of Los Angeles', in S. Watson and K. Gibson (eds) *Postmodern Cities and Spaces*, Oxford: Blackwell.

SEE ALSO: economic restructuring

ALISON TODES

URBAN REVITALIZATION

Urban revitalization represents a process aimed at overcoming problems stemming from deterioration and crisis in cities and URBAN areas. The term 'revitalization', defined broadly, can include operations on physical structures, as well as interventions in economic activities or social conflicts. Generally, revitalization efforts are directed at cities that have experienced a decrease in population, that present high levels of unemployment or delinquency, that have been affected by the closing of numerous businesses or that exhibit physical degradation in some of its most characteristic sectors. Urban revitalization programmes have been initiated in areas that have experienced a severe economic crisis, such as Glasgow or Liverpool, areas that have lost competitiveness, as in Dortmund or Liège, and where serious problems of deterioration are evidenced in historic structures, such as in Venice, Lisbon or Salvador de Bahia. Thus, various types of urban revitalization can be specified: those that include a physical recuperation of buildings or zones threatened by degradation, those carried out in urban areas that have experienced economic crisis due to failing industries such as mines or shipbuilding, and those that promote sustained development in social terms (elimination of pockets of POVERTY or marginality) or on an environmental level (reduction of high levels of pollution, responsible management of solid wastes, etc.).

Urban rehabilitation is the name given to a series of physical acts of revitalization in a city. This consists of improving the conditions of existing structures through their substitution or modernization. Rehabilitation can refer to improvement in the diverse elements common to many structures (façades, cornices, SIDEWALKS,

etc.), or it can include functional spaces (housing or commercial space). Either of these can be carried out independently of the other. Although this process is traditionally undertaken in constructed areas, or in the surrounding OPEN SPACES, the concept of integral rehabilitation appeals to modern sensitivities. This implies not only the generic recuperation of a building and its surrounding area, but also the recuperation of its resident population and the traditional activities that were carried out there. These ideas are present in progressive politics for the safekeeping of a city's heritage. Examples of physical rehabilitation can be found in numerous cities around the world, from Berlin to Dubrovnik, Quebec, San Juan de Puerto Rico or Marrakech. It is more difficult, however, to cite examples of integral rehabilitation, where physical reforms were accompanied by the residential reoccupation of an abandoned area such as in the DOCKLANDS of Glasgow or the Olympic village of Barcelona.

Related to this broader sense of the word 'rehabilitation' are other more precise terms: 'restructuring' and 'restoration'. Restructuring is understood as the projects undertaken in a given building that affect its structural elements, causing modifications in its internal morphology. Partial restructuring occurs when the operation does not result in the complete disappearance of the building, and total restructuring refers to the elimination of all interior elements. One or the other of these formulas has been followed in many cities where the historic centre was badly degraded. Occasionally, the process of GENTRIFICATION in old NEIGHBOURHOODS has caused the original SUBDIVISIONS (vertical and horizontal) to disappear, making room for the construction of more ample housing, as witnessed in numerous population centres in Northern and Western Europe. On other occasions, the opposite has occurred: a building that was once a one-family dwelling becomes subdivided into various apartments. This is frequently seen in Latin American cities, and has the opposite effect of that which is intended by urban revitalization; namely the degradation of living conditions. The term 'restoration' refers to the restitution of an entire building, or a portion thereof, to its original state and conditions. The materials used should be in keeping with those employed at the time the structure was first erected. Traditionally, only isolated urban elements, such as MONUMENTS, were restored. However,

through time, restoration projects extended to entire neighbourhoods within a city, where façades were painted in their original style, indigenous wood was used for doors and windows, traditional construction techniques were employed, etc.

Although on numerous occasions urban revitalization is a process that is initiated with a physical rehabilitation project, at other times it has a broader scope. Thus, in some older cities greatly affected by the closure of factories or other businesses that generated mass employment, plans for integral rehabilitation have been developed. Where industrial spaces, often located along waterfronts, have declined to a state of considerable degradation, public administrations and some relevant citizens groups have intervened and attempted to renovate physically and re-evaluate the area, and at the same time have prompted new economic activities to substitute those that became obsolete. In Great Britain or the Ruhr basin in Germany, as well as in the north-east or Great Lakes region of the United States, numerous examples of this process can be found. Normally, the old factories or abandoned docks are converted into installations for commercial or leisure activity, or are used by businesses whose activity is far different from the earlier industries. Many riverfront or coastal properties as well as peripheral urban areas have been renovated. Often, the new uses of the property recuperate the previous levels of employment and productivity. A positive example of this is Liverpool, although in Dortmund the recuperation has been much more difficult. Nevertheless, in all cases, the interventions have been well accepted by citizens, and have generated a favourable image of the city.

In operations aimed at overcoming urban decadence there is generally a relationship between physical rehabilitation and economic dynamism. There is an attempt made to couple the high-level investment designated for the physical recuperation of the sector with incentives aimed at developing new businesses in order to reactivate the area in question. Several successful examples are found in London, the dock area of San Francisco or in the previously mentioned case of Liverpool. Yet often the expected economic recovery does not accompany the work of the architects and urban planners, and the city does not recuperate its original dynamism, despite undoubted improvements in quality and public image. Contrary to what was once thought, the form of an urban area does not determine the living conditions of the population,

although it is a variable to be taken into consideration in the processes of urban revitalization.

It can be said that plans for urban revitalization always introduce substantial changes in the various components of a city. On the one hand, the process of rehabilitation produces the rejuvenation of older, degraded sectors, and on the other, urban planners manage to convert decadent cities (those that never modernized or lost the activities that sustained them) into settings for new architectural design. Simultaneously, efforts are made to initiate the development of certain types of businesses or special service sectors. However, uneven results are generally obtained from this attempt. Success depends upon the local authority's capacity for negotiation, the location of the city in question and the qualification of the entrepreneurs involved.

As previously mentioned, the problems of crisis in numerous cities are aggravated by social conflicts or citizen unrest. In many cities in Latin America or southern Asia, urban revitalization projects include the development of a series of basic social services and the improvement of living conditions. In poor neighbourhoods, it is necessary to enforce obligatory education, put sanitation programmes into action and offer employment opportunities. These actions, along with improving construction materials, and providing electricity, centralized water distribution or waste treatment, constitute the foundations for recuperating cities where massive IMMIGRATION has accompanied acute urban degradation for decades. Projects of this sort have been carried out in various cities in Brazil, as well as in Lima, Peru, exemplified by the work of J. Turner.

Likewise, in the developed world, economic incentives and rehabilitation of certain urban sectors may be supplemented by active social politics. One school of thought maintains that urban revitalization should be accompanied by a return to more open social relationships among members of a COMMUNITY, respect for the diverse character of different neighbourhoods within the city (avoiding gentrification or the concentration of certain groups of the population with problems), and a fusion of residential zones with areas of small businesses. The quality of life in any given area and the dynamics of urban revitalization can be directly influenced by reducing the differences in levels of income, reversing the process of residential SEGREGATION, and limiting sectors inhabited by only one class (social and economic) of citizen.

In some concrete examples, any attempt at revitalization must first overcome social and political obstacles. Thus, in cities in Northern Ireland, the suspension of violence between Catholics and Protestants is a precondition to any plan for urban recuperation. The objectives established by these plans should be the disappearance of internal barriers within the city and residential integration, and they should take advantage of the economic progress that follows a greater sense of citizen security. In South Africa, the disappearance of the apartheid regime ushered in the development of a new type of multiracial city that has served to revitalize numerous urban spaces, although it has created some other problems, such as an increase in insecurity. Some of the old townships have acquired a new central status, as in Soweto, and at the same time a new class of elite blacks are demanding urban housing in the cities previously inhabited only by whites, which has stimulated the real estate market and served to diversify economic activity.

By the year 2000, the term 'revitalization' broadened, and acquired environmental connotations. After the fall of pro-Soviet regimes in Eastern and Central Europe, and the crisis in older industrial regions around the world, water and air pollution and the accumulation of massive amounts of garbage became primary factors in the process of urban deterioration. Thus, in addition to politics of rehabilitation and economic promotion, programmes were begun that promoted HEALTHY, clean or simply LIVABLE CITIES. This includes cleaning up the city, along with the other aspects of renewal. The vitality of an urban area depends on its environmental sustainability. The reduction of traffic, recycling of wastes, incentives for the use of electrical vehicles, transfer of industries to alternative locations and reduction of CONGESTION in some sectors are universally accepted measures to improve the environmental quality of cities so that they are better prepared to face an uncertain future.

Further reading

Council of European Municipalities and Regions (1994) *Guidelines for the Realization of Strategic Development Plans in Medium-sized Cities*, Lisbon: Oficina de Arquitectura, Lda.

Ellin, N. (1999) *Postmodern Urbanism*, New York: Princeton Architectural Press.

Merlin, P. and Choay, F. (eds) (1993) *Dictionnaire de l'urbanisme et de l'aménagement*, Paris: PUF.

Zoido, F., de la Vega, S., Morales, G., Mas, R. and Lois, R. (2000) *Diccionario de geografía urbana, urbanismo y ordenación del territorio*, Barcelona: Ariel.

JESÚS M. GONZÁLEZ PÉREZ AND RUBÉN C. LOIS GONZÁLEZ

URBAN SOCIOLOGY

Urban sociology, a sub-discipline within the field of sociology, scientifically examines topics unique to URBAN environments. Urban sociology is one of the oldest sub-disciplines in the field of sociology, its early works dating back to the mid-nineteenth century. Inspired by the development of small villages into bustling and chaotic cities during the INDUSTRIAL REVOLUTION in Europe, sociologists such as Emile Durkheim, Karl Marx, Friedrich ENGELS, Georg SIMMEL, Ferdinand Tönnies and Max WEBER were among the first sociologists whose work examined the changing social world found in cities. Their lasting contributions to the field are theoretical formulations of the social structure and social organization of cities.

Although these early works by European sociologists are considered crucial to our understanding of urban sociology today, in the United States, the academic discipline developed its identity at the University of Chicago in the early twentieth century. The scholars of this CHICAGO SCHOOL, as it was known, believed that an understanding of cities involves systematic exploration and observation of human behaviour. Unlike the European works that provided abstract theorizing, the American tradition involved first-hand fieldwork of city life.

In their observations of urban life, these early urban sociologists found that there were radical differences between the demographic, geographic, social, economic and political nature of urban environments and their rural counterparts. The early works in urban sociology brought to light an awareness that the city was an interesting and complex geographic entity to study and explore. Furthermore, these new cities were believed to be changing the way humans interact with one another and with their environment.

The human ecological perspective pioneered by Robert PARK and Ernest Burgess at the Chicago School greatly influenced twentieth-century urban sociology. This school of thought promotes the

idea that people sift and sort themselves into homogeneous areas throughout the city through a benign process of competition. Human ecologists believe that there is a need for balance and cooperation in geographic spaces such as cities and therefore the organization of people in cities seeks a state of equilibrium. Although the human ecological perspective continues to influence urban sociology, it began being criticized in the 1970s for being overly reliant on macro-structural factors and ignoring individual choice among other factors in explaining the organization of cities' social world.

In general, any urban phenomenon that exists in the social world can be considered part of urban sociology, which contributes to the interdisciplinary nature of the discipline. Studies in other fields such as anthropology, economics, geography, history, political science, psychology and URBAN PLANNING are often linked with urban sociology to help advance our understanding of urban environments. Although urban sociology is a broad field, there are a number of topics, such as spatial structure, urban lifestyles and social organization, that have dominated the area of study.

Further reading

Park, R.E. (1925) *The City*, Chicago: University of Chicago Press.
Wirth, L. (1938) 'Urbanism as a way of life', *American Journal of Sociology* 44(1): 1–24.

MAI NGUYEN

URBAN SQUALOR

As the 'dark side' of the CITY, urban squalor is the most obvious visible reminder of the problems, particularly the inequalities, that characterize URBAN life. Commonly identified with the SLUM or the GHETTO (or in the parlance of Victorian England 'the rookery'), the squalor of the city becomes defined in physical and social terms – NEIGHBOURHOODS characterized by poor physical environments and by their (apparently) socially dysfunctional nature. As a value-laden term, however, it is important to identify who is defining what is to be considered urban squalor, and how; like its antonym, what constitutes squalor is at least partly in the eye of the beholder. If the 'dominant' view has been to link squalor with particular parts of the city, the slum, alternative interpretations may highlight other features of the city – dirt, litter, pollution, disorder – problems that might be worse in certain neighbourhoods, but are more endemic to the nature of the city. Defining what constitutes urban squalor varies too across space-time; our understanding of it needs to be sensitive to its relativism.

What, then, was a defining feature of the squalor of cities in earlier historical periods, their smell, arising from the lack of proper sewage and cleansing systems, has been all but eliminated as a problem of the contemporary city in the advanced economies. Nor was such an absence only an assault on the nose; though not fully recognized until the nineteenth century, the absence of effective sewage disposal was life threatening. The rapid growth of cities in the early industrial period exacerbated the sanitary problems of the city (as has been the case more recently in many THIRD WORLD CITIES). In the first industrial nations of Europe – Britain, for example – the sanitizing of the city became a crusade against urban squalor that, particularly through the cholera and typhus epidemics that periodically raged through the city, threatened the BOURGEOIS as well, if not as much, as the poor.

Urban growth was to give rise to new forms of squalor. Urban INDUSTRIALIZATION was accompanied by the rise of atmospheric and other forms of pollution. Increased consumption of domestic coal fostered smoke fog, while industrial effluents polluted river courses. Both were identified as health as well as economic costs of urban growth in the nineteenth century, yet their resolution has been relatively intractable. In spite of regular smoke fog crises in cities such as London, culminating in the severe smog of December 1952, in which no fewer than 4,000 died, it was not until the passing of the Clean Air Act of 1956 that the demands of the National Smoke Abatement Institution, established in the 1880s, were to be met. Cleaning the city, its water courses as well as its air, met powerful opposition from capital interests.

Other aspects of city life were labelled as symptomatic of its more squalid side, a threat to the social order and at variance with the dominant (elite's) view of the 'modern' city. The vagrant, often coupled with street begging, prostitutes, informal street traders, each by no means new to the city, became more visible partly because of their greater number in the expanding cities of the

last two centuries. Such activities needed at least to be regulated, and if not eliminated at least spatially marginalized so that they become less visible. Regulating cities in these ways in the developed nations has its counterpart in the regular clearance of street hawkers in present-day Third World cities, the poor 'invading' the modern. What constitutes the squalid is as so often a question of appearances, its treatment subject to moralizing codes.

Nowhere is this truer than in the dominant perception of what constitutes urban squalor, the 'slum'. The squalor of the archetypal slum is reflected by the dirt and the decay of the physical environment, which in turn are read as the product partly of the awesome population densities, the POVERTY and social disintegration. Historically, the slum played a crucial role as reception centre for newly arrived immigrants to the city, as in the New York TENEMENT districts which the (Tenement House) Commissioners described as 'places in which thousands of people are living in the smallest place in which it is possible for human beings to exist – crowded together in dark, ill-ventilated rooms, in many of which the sunlight never enters and in most of which fresh air is unknown'. It was a description that could have been applied equally to the notorious slums of the East End of London, or of Glasgow of the nineteenth and early twentieth centuries. In Europe, in particular, the solution to such squalor was the wholesale demolition of the slums; in Glasgow, this began with City Improvement housing in the very worst areas, continuing through to the post-Second World War clearance of infamous slums such as the Gorbals.

Demolition, the extirpation of urban squalor, has continued to be practised, as of high-rise social housing in many British cities built in the 1950s and 1960s that had become difficult to let. As in the notorious PRUITT-IGOE estate, the housing becomes entrapped within a vicious downwards spiral, in which in the case of the St. Louis neighbourhood the social problems of racial DISCRIMINATION, CRIME and drug-related problems had become the modern-day equivalent of the squalor of the nineteenth-century city. Squalor in this sense becomes associated with the 'underclass' and the 'socially excluded'. Nor should we imagine such ideas are new; notions of the underclass or of the 'cycle of poverty' are not dissimilar to the concept of the 'residuum' of Victorian Britain. Yet demolition may not necessarily be the method by which

to deal with the problems of (perceived) urban squalor. Though by no means stopped, the bulldozing of squatter settlements in Third World cities is less commonplace than was the case before research showed that while physically conditions were often squalid, socially they were far from dysfunctional. Once endorsed by the WORLD BANK, shifting the boundaries of what constituted urban squalor became a means of reducing the housing problems of many Third World cities by providing assistance to shanty-town dwellers to upgrade their environment.

Further reading

Driver, F. (1988) 'Moral geographies: social science and the urban environment in mid-nineteenth century England', *Transactions of the Institute of British Geographers* 13: 275–87.

RONAN PADDISON

URBAN STRESS

Three aspects characterize URBAN environments: ambient physical conditions (noise, pollution), social conditions (DENSITY) and overstimulation (exposure to a high number and variety of visual and other stimulations). Cities are more noisy, more dense and more polluted.

Noise represents the most spectacular stress to which city dwellers are daily exposed. Noise levels augment with the size of the agglomeration. It is the form of stress most often mentioned and which leads to the most frequent complaints. One-quarter of urbanites are daily exposed to at least one loud noise condition at work, at home or in transportation. Exposure to *pollution* is more frequent in big cities, due to traffic, resulting for instance in higher numbers of respiratory problems for children and the elderly. Cities often accumulate different forms of *density* due to population concentrations. City dwellers may be exposed to social density outdoors (high number of individuals on the same spot) and to spatial density in their habitat (little space per person), and overall may face a struggle for resources due to a low ratio between population and amount of available resources, i.e. a large number of people in relation to the available transport facilities, shops and services.

The multiplication of visual and auditory stimuli to which city dwellers are exposed constitutes

informational overload. Such environmental over-stimulation produces increased fatigue. *Urban mobility* is more constrained than in small towns: commuting daily from residence to the workplace, facing frequent conditions of gridlock or crowded mass transit, is stressful. Available leisure time is compressed due to more time spent commuting, and there is less time available for social interaction. *Bureaucratic formalism* is also more frequent in big cities than in small towns. Due to these various factors, urbanites more often experience pressure, compared to the inhabitants of small towns.

The level of *criminality* is higher in big cities. Residents of big cities have a greater probability of being exposed to violence and CRIME. This is commonly explained by the numerous opportunities the city offers to delinquents, along with a low probability of being recognized. Fear of crime (which is not necessarily correlated with objective crime rates) restricts people's behaviour by making them feel vulnerable. It is exacerbated by an environment which appears poorly maintained, characterized by littering, VANDALISM and GRAFFITI, uncleanness, lack of illumination, and deviant behaviours in the NEIGHBOURHOODS.

Overall, city life appears constraining and demanding. These various aspects of urban environments lead city dwellers to engage in protective adaptation processes, creating specific behavioural norms (see URBANITY). However, despite being generally more exposed to environmental stressors, inhabitants of big cities, with the exception of newcomers to the city, do not mention these factors as more frequently annoying them, compared to those living in small towns or even people in rural habitats.

Further reading

Evans, G.W. and Cohen, S. (1987) 'Environmental stress', in D. Stokols and I. Altman (eds) *Handbook of Environmental Psychology*, New York: Wiley-Interscience, vol. 1, pp. 571–610.

Moser, G. (1992) *Les stress urbains* (*Urban Stress*), Paris: Armand Colin.

GABRIEL MOSER

URBAN STUDIES

Interdisciplinary field concerned with understanding the multifaceted nature of cities. The complexity of contemporary URBAN environments and urban societies requires an interdisciplinary approach in conceptualizing the city. Not surprisingly, the urban studies field has attracted scholars from a wide range of disciplines devoted to exploring questions about cities. The core theoretical and methodological concerns of the urban studies field come from the social science disciplines of history, economics, sociology, geography, political science, anthropology, and the professional fields of URBAN PLANNING, architecture, landscape architecture, and URBAN DESIGN. Richard LeGates and Frederick Stout have organized key writings in the field in their anthology – *The City Reader* (2000). Scott Campbell and Susan Fainstein's anthology – *Readings in Urban Theory* (2001) – includes other writings by key scholars in the field. Ronan Paddison, the managing editor of the journal *Urban Studies*, has brought together syntheses of urban research produced in different disciplines in his edited book – *Handbook of Urban Studies* (2001).

Urban economists, urban historians, urban sociologists, urban geographers, urban anthropologists and urban planners have produced much of the urban studies scholarship not only by examining cities from the traditional lens of their respective disciplines, but also by pushing traditional boundaries of the disciplines. For example, urban economists' explicit focus on cities and integration of 'space' into the traditionally 'aspatial' discipline of economics have opened up new avenues of research such as urban land and housing economics. Urban economists like William C. Wheaton have studied the spatial organization of US cities. Economic sociologists like Alejandro PORTES have studied IMMIGRATION as the outcome of socially oriented economic action, highlighting the role of social networks and SOCIAL CAPITAL in international migration.

Urban studies, as an academic field, emerged in the early twentieth century in the United States and England simultaneously following the rapid INDUSTRIALIZATION and URBANIZATION of the late nineteenth century. Sir Peter HALL notes that a degree programme in town and country planning was established at the University of Liverpool in 1909. During the same year, the Massachusetts Institute of Technology began offering courses about cities for the first time in the United States. San Francisco State University urban studies professor Richard LeGates notes that the field grew slowly through the end of the 1950s followed

by a significant expansion in the 1960s in the United States. The growth in the field is attributable to an increased public policy interest in solving a wide range of urban problems including POVERTY, NEIGHBOURHOOD decline, residential SEGREGATION by race, and the practice of RED-LINING of low-income minority neighbourhoods by commercial banks in mortgage lending.

One of the earliest contributions to the field of urban studies was the research carried out by the CHICAGO SCHOOL OF SOCIOLOGY in the 1920s. Robert PARK, Ernest Burgess and others in this school developed a theory about the relationship between the physical and social organization of cities, now known as the HUMAN ECOLOGY approach. The field has evolved and developed in the past eighty years with efforts to understand and solve urban problems.

Urban studies programmes are housed in different academic units within universities – social and behavioural sciences, liberal arts, humanities, and architecture – and are predominantly offered at the undergraduate level. A handful of universities in the United States offer graduate degrees in urban studies including Temple, Michigan State, Portland State and the University of New Orleans.

The Urban Affairs Association (UAA) and the Association of Collegiate Schools of Planning (ACSP) are two of the main professional organizations for urban scholars, researchers and practitioners. Founded in 1969, the UAA disseminates research findings about URBANISM and urbanization through annual conferences and its quarterly publication – *Journal of Urban Affairs*. The ACSP brings together faculty teaching in city and regional planning programmes and publishes articles reporting planning research in the *Journal of Planning Education and Research*. Urban studies scholarship is also published in other peer-reviewed journals such as *Urban Studies, International Journal of Urban and Regional Research* and *Journal of American Planning Association*.

Urban studies scholars use a range of research methods including survey research, qualitative field research, quantitative data analysis, and applied statistics. Survey research and inferential statistics are core urban studies methodologies, especially among scholars closely aligned with social science disciplines like sociology and political science, and professional fields like city and regional planning. Spatial analysis and data visualization with GEOGRAPHIC INFORMATION SYSTEMS (GIS) have also entered urban studies curricula. The availability of uniform and high-quality secondary data down to the block level in the United States from the US Census Bureau coupled with advances in computing and telecommunications technologies has accelerated the quantitative and spatial analysis of demographic, social and economic characteristics of US cities.

Among qualitative research methods, participatory action research and ethnography are commonly used to gain an in-depth understanding about the life of urban residents. The work of sociologist Herbert GANS – *The Urban Villagers* (1962) – describing the life of Italian immigrants in Boston's West End neighbourhood is a classic example.

Like their counterparts in the developed world, urban researchers in the developing world carry out qualitative and quantitative analyses as data permit. Because high-quality secondary data as from the US Census Bureau is not readily available in much of the developing world, most researchers undertake limited-scope original surveys. Only through such original field research, for example, researchers have been able to document the phenomenon of the INFORMAL SECTOR and the development of informal housing settlements in major cities of the developing world. Much of this research is reported in journals specializing in human settlement issues in the developing world such as *International Journal of Urban and Regional Research, Habitat International* and *Third World Planning Review*.

The world is fast urbanizing. The United Nations reports that by 2020, more than half of the population in the developing world will live in cities as rapid urbanization processes continue. By 2015, nearly half of the world's 358 cities with more than 1 million people will be in Asia. Urban studies scholars have produced a body of research on 'MEGACITIES'. As the world's population increasingly becomes urbanized, the field of urban studies is seeing an increase in research on urban and environmental management issues in the developing world and on GLOBAL CITIES in the context of globalization and transnationalism.

An important emerging concern of urban studies scholars is an understanding of how globalization of economic activities plays out at the local level. University of California, Davis professor Michael Peter Smith uses the metaphor 'transnational urbanism' to describe the criss-crossing

transnational circuits and their cultural affect on cities. Others use different terms like 'GLOCALIZATION'. Postmodern geographer Edward Soja has explored postmodern geographies especially with reference to Los Angeles as a real and imagined place. Soja, his colleagues at the University of California, Los Angeles, and others like Mike Davis at the University of Southern California have developed a body of work focused on the transformation of Los Angeles in the context of POSTMODERNISM, also known as the LOS ANGELES SCHOOL OF URBAN STUDIES.

The concerns of urban studies scholars in the world are varied and influenced by current thinking in different disciplines. University of Toronto political science professor Richard Stren has found that in the 1990s, Asian researchers were primarily concerned about the finance of urban INFRASTRUCTURE (influence of economics), African researchers were focused on the management of urban services (influence of geography), and Latin American researchers were most concerned about social policy (influence of sociology). Broadly, he has found distinct themes being studied in the developing world: in the 1960s, modernization; in the 1970s, underdevelopment and dependency; in the 1980s, urban social movements; and in the 1990s, urban management and GOVERNANCE issues at the local level.

Urban studies researchers have also increasingly become part of the 'policy community' and have joined efforts to solve urban problems with policy-makers and urban planners. The concerns of the professional field of policy analysis (since the late 1960s) as well as the professional field of urban planning (since 1909) overlap with the field of urban studies. All three fields explicitly attempt to link knowledge and practice. Urban studies scholars integrate research findings from policy analysis and urban planning into efforts to solve public policy and urban planning problems. The problem-solving strand of the urban studies field emphasizes forging sustainable links with urban researchers and practitioners. This effort requires steady funding for basic and applied social science research. In the international realm, the World Bank has become a leading sponsor of policy-relevant urban research in the developing world. Key sponsors of basic social science research on cities in the United States are the National Science Foundation, the Social Science Research Council, and the Ford Foundation.

Another key concern of the field of urban studies at the start of the twenty-first century is the role of advanced information and tele-communications technologies in shaping different settlement patterns. Despite the fact that technology enables firms to locate anywhere on earth, much of the core decision-making functions of the technology-driven NEW ECONOMY industries remain grounded in select cities where face-to-face contact remains critical.

Further reading

Hall, P. (1996) *Cities of Tomorrow: An Intellectual History of Urban Planning and Design in the Twentieth Century*, 2nd edn, Oxford: Blackwell.

LeGates, R. and Stout, F. (eds) (2000) *The City Reader*, 2nd edn, London: Routledge.

Lloyd, R. and Sanyal, B. (eds) (2000) *The Profession of City Planning: Changes, Images, and Challenges: 1950–2000*, New Brunswick, NJ: Center for Urban Policy Research.

Stren, R. with Bell, J.K. (eds) (1995) *Urban Research in the Developing World*, vol. 4, *Perspectives on the City*, Toronto: Centre for Urban and Community Studies, University of Toronto.

SEE ALSO: urban planning; urban sociology; urbanism; urbanization

AYSE PAMUK

URBAN UNDERCLASS

The term 'urban underclass' refers to a relatively permanent substratum of the population marginal to the economic mainstream. Conceptualized and popularized in the early 1980s, the term suggested the emergence of a new phenomenon in the class structure of advanced industrial societies. Both journalists and scholars identified a social group that seemingly remained chronically and deeply impoverished irrespective of broader macroeconomic conditions. The term implies not only a deep and incessant condition of POVERTY but a behavioural dimension as well. CRIME, teenage pregnancy, female-headed families, welfare dependency, weak labour force attachment, violence, low educational attainment, and drug addiction are all behavioural correlates of underclass existence. Such non-mainstream behaviours are variously labelled pathological, socially dysfunctional, undesirable or destructive.

Although the rudiments of an urban underclass exist in Western European societies among immigrant populations from the developing world, the concept is largely an American phenomenon.

In the United States, the concept has both a racial and a spatial dimension. It is disproportionately African-American (and to a lesser extent Latino). Its locus is the inner city. It is also associated with conditions of *concentrated* poverty. The underclass population resides predominately in NEIGHBOURHOODS where nearly half of all residents are poor.

The overall size of the urban underclass in the United States is not large. It constitutes just a fraction of those persons officially counted as poor. Estimates of underclass size hover well under 1 per cent of the overall US population and under 10 per cent of the US poverty population. Nevertheless, because of the severity of the experienced economic depravation, the intractability of the social condition, and the broader impact on American life, the presence and durability of the urban underclass stands as America's deepest URBAN problem.

The concept itself – underclass – has become highly contentious. Some scholars find it to be a pejorative term that stigmatizes the urban poor as undeserving of concern and assistance and responsible for their own poverty. The GHETTO poor, the estranged poor and the lower-lower class have been suggested as conceptual alternatives.

The most seminal study of the urban underclass is William J. WILSON's *The Truly Disadvantaged*, published in 1987. This book drew a significant amount of attention – both popular and academic – to the problem of concentrated urban poverty in the United States, while sparking an explosion of additional underclass research. Wilson put forward what many observers deemed to be an innovative, provocative and cogent explanation for the emergence of the urban underclass. His explanation diverged from more familiar rivals rooted in either the perverse incentives of social programmes that encourage welfare dependency, the effects of racial DISCRIMINATION in employment and housing, the inadequacy of the American welfare state compared to the European experience, or the internalization of dysfunctional values and attitudes – a culture of poverty – pervading underclass neighbourhoods.

In contrast, Wilson emphasized structural change and socio-spatial factors. In his view, the American urban underclass developed as a result of the structural transformation of city economies from centres of goods production and distribution to centres of information processing and higher-level service provision. This ECONOMIC RESTRUCTURING led to a decline in employment opportunities in manufacturing and other blue-collar industries, which required little formal training, and a growth in employment opportunities in the advanced service sector, which required high-level skills and education. The resulting disparity between the low skill level of ghetto residents and the educational and training requirements necessary to obtain quality city jobs (the jobs–skills mismatch) causes chronic and enduring unemployment and underemployment. The effect of widespread joblessness is exacerbated by the social and spatial isolation of the inner-city poor that resulted from the MIGRATION patterns of whites and, after the CIVIL RIGHTS MOVEMENT of the 1960s, middle-class African-Americans to SUBURBS. This selective out-migration leads to a concentration of the poor in ghettos, as well as the subsequent decline of the viability of basic COMMUNITY institutions (churches, schools, etc.) in the inner city and the absence of stably employed role models. This estrangement from mainstream society and associated values leads to what Wilson calls the tangle of pathological behaviours of the urban underclass.

Wilson's explanation attracted a plethora of criticism. Critics challenged his view on a variety of fronts. A prominent critique charged that he imparted too little causality to race by largely dismissing contemporary, rather than merely historical, racial discrimination and residential SEGREGATION. Other voices criticized Wilson for underestimating the deep embeddedness of underclass behaviour, contending that an entrenched dysfunctional subculture suffuses the fabric of ghetto life and is unlikely to be modified (in the near term) by improved economic conditions. Another objection was that he overestimated the role played by the decline of urban manufacturing because this sector never provided many jobs for African-Americans in the first place. Feminists challenged Wilson's emphasis on male unemployment and low marriage rates as the causes of a phenomenon, underclass poverty, disproportionately experienced by women and children. Wilson was also criticized for attributing salient transformations in urban life to the workings of

blind economic (structural) forces, while ignoring the impact of conscious and intentional actions taken by decision-makers (i.e. human agency) that also contributed much to the making of the urban underclass.

An important element of all serious research on the urban underclass, including Wilson's work, is how to address the problem via public policy. The 1990s witnessed increased enthusiasm for efforts to reduce concentrated urban poverty by dispersing the inner-city poor to surrounding suburbs where their employment prospects, educational opportunities and overall quality of life supposedly would be enhanced. This enthusiasm rekindled a spirited policy debate over whether the plight of the urban underclass can be best ameliorated by these measures or by efforts to improve the social and economic conditions of the inner-city neighbourhoods where the urban underclass now resides.

Further reading

Jargowsky, P.A. (1997) *Poverty and Place: Ghettos, Barrios, and the American City*, New York: Russell Sage Foundation.

Murray, C. (1984) *Losing Ground: American Social Policy, 1950–1980*, New York: Basic Books.

Wilson, W.J. (1987) *The Truly Disadvantaged: The Inner City, the Underclass, and Public Policy*, Chicago: University of Chicago Press.

—— (1996) *When Work Disappears: The World of the New Urban Poor*, New York: Alfred A. Knopf.

DAVID L. IMBROSCIO

URBAN VILLAGE

As URBAN REGENERATION became an agenda buzzword for professionals, decision-makers and developers worldwide in the late 1990s, the urban village concept for creating mixed-use URBAN developments on a sustainable scale became an important form of development. The idea behind its design philosophy is to identify and deliver solutions that work in a twofold manner: in defining sustainable and identifiable URBAN FORM and also supporting and establishing a real sense of vibrant COMMUNITY. The term first emerged in the 1960s in the work of Herbert GANS, an American sociologist and educator. He made an important analysis of second-generation Italian-Americans in his famous work *The Urban Villagers* (1962). His study illustrated clearly the traditional notion of NEIGHBOURHOOD and family defining a social group. All of his 'urban villagers' live in a certain neighbourhood and share a great deal of time together. Even though the term might seem an oxymoron at first sight, it sums up the human desire for both autonomy and community.

The contemporary urban village ideas are closely related to NEW URBANISM and SMART GROWTH ideas initiated in the United States. In the United Kingdom, this movement was manifested through the pioneering work of Leon KRIER (initiated by HRH The Prince of Wales) and found its materialization in the Urban Villages Forum. The work of Jane JACOBS and Christopher ALEXANDER provided the background for these ideas in many respects. The core of the idea lies in the sustainable human-scale mixed-use development matrix suitable for both creating new neighbourhoods and repairing the fabric of existing ones. As the idea is characterized by economic, environmental and social sustainability, the areas should be more attractive for living, working, playing and investing, in a way that would combat dispersal, exclusion and alienation, which have undermined the viability, vitality and livability of towns and cities.

In these concepts and solutions, the neighbourhood centre becomes both a civic focus and an informal place of gathering for the people in the community. The long-term functional viability of an urban village is secured by the inclusion of some critical components: a variety of uses, such as shopping, leisure and community facilities, should exist alongside housing units, which should offer a choice of tenures, both residential and commercial. Even though places should be compact, in the sense of creating a recognizable neighbourhood, such density of development should be made to encourage all of the above non-housing activities. Walkability, discouragement of excessive car use and a strong SENSE OF PLACE are pertinent. Strong community participation and a high level of involvement by all the local residents in the planning and management of their neighbourhoods become sine qua non for the long-term sustainability of urban villages. At the core of this idea lies a commitment to the concepts of strong citizen participation, affordable housing, and social and economic diversity.

Further reading

Aldous, T. (1992) *Urban Villages*, London: Urban Villages Forum.

Sucher, D. (1995) *City Comforts: How to Build an Urban Village*, Seattle: City Comforts Press.

TIGRAN HASIC

URBANISM

The term 'urbanism' originated in the late nineteenth century with the Spanish engineer-architect Ildefons CERDA. Cerda's intent was to create an autonomous activity focused on the spatial organization of the city, an activity that treats the physical city as a totality.

Widely embraced in Europe, this meaning of urbanism – as URBAN FORM – has been less prevalent in the United States, with the exception of early city planners such as Daniel BURNHAM. Specifically in the twentieth century, urbanism has more often been used in the United States to express a way of life, a notion taken from a famous essay in 1938 by the URBAN sociologist Louis WIRTH. A third meaning subsequently appeared which cast urbanism as a critical practice; that is, as an intellectual engagement with the city.

Architects, urban designers and city planners are most likely to define urbanism in terms of the city's morphology. Their concern is the spatial arrangement of commercial and civic buildings, OPEN SPACES, transportation nodes and corridors, and homes. Looking inward, the issue is functionality. What specific juxtaposition of built forms enables the city to function efficiently and effectively? Is the quality of life enhanced? Viewing the city as an ensemble, the issue is aesthetic coherence. To what extent does the city express society's values?

A long history of city planning in Europe created a legacy of large-scale urban interventions. Examples include Baron Georges-Eugène HAUSSMANN's rebuilding of Paris in the mid-nineteenth century as well as utopian proposals for addressing the ills of the INDUSTRIAL CITY developed by Tony GARNIER and Ebenezer HOWARD among others. Towards the middle of the twentieth century, LE CORBUSIER with his 'RADIANT CITY' and the CHARTER OF ATHENS of the Congrès Internationaux d'Architecture Moderne (CIAM) established a high modernist version of the ideal city. High modernists celebrated movement, technology and the aesthetics of large urban assemblages. Skyscrapers became particularly iconic. Functions were kept separate for maximum efficiency of land use and orderliness was paramount.

Even as Western governments embraced high modernist urbanism in post-Second World War reconstruction, another type of urbanism was emerging. In the United States, SUBURBANIZATION and its attendant SPRAWL redefined the physical city at the metropolitan scale. The city centre became less important than the periphery as the automobile supplanted MASS TRANSIT and as edge cities became established. The modernist separation of functions, however, was maintained.

The late twentieth century witnessed another American contribution to urbanism – NEW URBANISM. New urbanism is meant to be an antidote to sprawl and the loss of COMMUNITY brought about by suburbanization. It attempts to reintegrate land uses and recapture the social qualities of small towns. New urbanists call for increased suburban densities, mixed land uses, and a shift of emphasis to the pedestrian and away from the automobile.

The second use of urbanism, mainly confined to the United States, is as a way of life; that is, as the social relations, institutions, attitudes and ideas that appear specifically in cities. Numerous urban scholars have argued that life in cities is different from life outside of them. Louis Wirth, for example, pointed to the lower proportion of primary relations in cities and higher proportion of impersonal interactions. He traced this shift to cities' greater size, DENSITY and heterogeneity when compared with their rural counterparts.

The notion of urbanism as a way of life received powerful expression in Jane JACOBS's 'SIDEWALK' urbanism. In *The Death and Life of Great American Cities* (1961), Jacobs vilified high modernist urbanism. Instead, she celebrated diverse NEIGHBOURHOODS and vibrant street life where tolerance, community and creativity could flourish. Her emphasis was on everyday life, a theme also central to the writings of Henri LEFEBVRE. In Jacobs's writings, physical form and a way of life are tightly bound, thereby conflating these two meanings of urbanism.

An understanding of urbanism is central to the American anti-urbanism that was particularly virulent in the early twentieth century. Anti-urbanism was then and continues to be a rejection of the way people live in the city. It condemns the city because of the vices – smoking, gambling, prostitution, drinking – that it seems to breed.

The city compels innocent people to sin, children to beg on the streets, and politicians and businessmen to solicit graft. The industrial city was dirty and chaotic; more importantly, it was immoral.

When anti-urbanism reigned in the 1920s, urban elites retaliated with a snobbish URBANITY. Urbanity is a way of life once characteristic of the upper class. To be urbane is to be sophisticated and discriminating. Leisure, entertainment, conspicuous consumption, and social engagements are its hallmarks. Urbanity is life immersed in the city's public culture.

A third meaning of urbanism is as a critical practice. This meaning is less common and confined mainly to the fields of architecture and URBAN DESIGN. The emphasis is on the perspective that one brings to intervention. It encompasses assumptions about human behaviour, the nature of social institutions, and the role of perception in shaping the experience of the built environment. In Françoise CHOAY's writings, this urbanism is presented as a means of uncovering and expressing political values; that is, as an ideological practice.

Of course, the three urbanisms are interrelated. The form of the city influences, even if it does not wholly determine, how people live there. And when urban designers, architects and city planners address the city's development, they engage in urbanism as a practice. As the city is used, so is it produced; its form and its way of life inextricably intertwined.

Further reading

Choay, F. (1988) 'Urbanism', in P. Merlin and F. Choay (eds) *Dictionnaire de l'urbanisme et de l'aménagement*, Paris: PUF.

Ellin, N. (1999) *Postmodern Urbanism*, New York: Princeton Architectural Press.

ROBERT A. BEAUREGARD

URBANITY

Traditionally, cities have been a pole of attraction, since they concentrate work opportunities, CULTURE, entertainment and recreation. Even if the URBAN environment represents for some of its inhabitants an ideal level of stimulation, an optimal context for their activities and numerous possibilities due to its diversity of places and opportunities for interpersonal contacts, daily conditions of life have progressively become more and more difficult. Already at the beginning of the last century, SIMMEL pointed to urbanites' social withdrawal and egoistic behaviour, as well as detachment and disinterest towards others.

Stressful urban conditions (see URBAN STRESS) constrain individuals to engage in particular *adaptation* processes. Adaptation is a dynamic process of behavioural and/or cognitive change in order to achieve congruence between the individual and the environment, resulting in specific urban behavioural norms. Milgram distinguished between three *urban adaptation behaviours*: coping by choosing priorities; self-protection by erecting psychological barriers around oneself; and creating rules and institutions. Urban behavioural particularities consist of the following: segmented and functional way of interacting with one another; anonymity and lack of involvement; the consequences on cognitive processes (the impossibility of identifying most of the people seen during the day, selection of stimuli, indifference towards deviant and bizarre behaviours, restriction and selectivity of responses to others' demands); and finally increased competition for services (subway, taxis, lining up for cinemas, etc.). Environmental constraints like gridlock, overcrowding and the fear of CRIME enhance behaviours aimed at gaining control and having freedom of action. The pace of life is more hectic. A certain number of activities are performed more rapidly in big cities than elsewhere. People walk faster in metropolises than in small towns. Fear of crime provokes behavioural restrictions especially among elderly people. The constant expression of such adaptation behaviour in urban settings points to the fact that it represents a normative behaviour.

Overall, urban behaviours are characterized by *individualism* and an *indifference towards others*. Urbanites pay less attention to others, are less friendly and less helpful. Typical reactions to stress are withdrawal into oneself and less responsiveness to the diverse solicitations of others. Numerous research studies undertaken several years ago consistently demonstrate that the conditions of urban life reduce the attention given to others and diminish one's behavioural availability to help others. *Helping behaviour* implies a deliberate interaction with a particular, identifiable individual. It occurs significantly less frequently in urban environments. Confronted by the overstimulation characteristic of urban environments, people filter

the inputs and focalize their attention on the most important solicitations, neglecting peripheral stimuli. Such a reaction allows urbanites to avoid unwanted interactions. Helping concerns two different behaviours: either responding to an explicit demand for help, or deciding to react to an implicit demand. Both are less frequent in urban settings. Urbanites, in concentrating on their own needs, not only deliberately avoid contact with others, they are also less attentive to what is going on around them.

In terms of *interpersonal relationships*, individuals react by establishing interaction priorities, excluding a certain number of contacts. They tend to isolate themselves and be suspicious of others. In urban settings, interpersonal relationships are governed by rigorous rules which enable individuals to preserve, in spite of high DENSITY situations, the minimum of privacy needed to protect themselves from intrusion by others. Encounters with friends are more often programmed in advance, while informal encounters are rare and limited to the home setting or the workplace.

The dimensions of interpersonal exchanges in the urban environment tend not only to make social contacts more superficial and eliminate those interactions which bring no personal benefit to the individual, but also to lead to the neglect of expressions of *politeness*. In big cities, individuals less often hold open the door for others or say 'hello' when they enter a shop or a public transportation utility. While, within our society, helping behaviours reflect the rule of morality (it is 'good' to help one's neighbour), behaviours reflecting *civility* represent a 'cooler' type of sociability, to the extent that they are an expression of codified and formal relations, which provide the individual with a minimum guaranteed autonomy. 'Civility' refers to tacit rules governing social behaviours regulating interaction. These rules embrace shared conventions concerning what it is appropriate to do in the normal activities of everyday life, and which behaviours, if not performed, lead to negative social sanctions. Civility is a disinterested act, concerning relations with people who are unknown to the actor. It involves a common code of conduct which is indispensable for maintaining the social tissue, based on respect for the other, attention to others, but also a certain modesty and self-effacement.

PUBLIC SPACES are locations for expressing COMMUNITY values, rules and codes which are indispensable to individuals to maintain their social distance, either to protect their personal space, or, on the contrary, to get closer to others. Social densities such as can be observed in large cities create both physical proximity and social distance. Thus, the regulation of social life in a densely populated society is based on adhering to routines which can be anticipated and predicted. Civil behaviours are indispensable for social interaction in the urban setting, all the more so since the ambient conditions are constrained by exposure to numerous stressful situations. On the other hand, incivilities can be considered as the negation of the processes of the civilization of morals or standards of behaviour, as breaches of the social mores.

Further reading

Korte, C. (1980) 'Urban-nonurban differences in social behavior and social psychological models of urban impact', *Journal of Social Issues* 36: 29–51.

Krupat, E. (1985) *People in Cities: The Urban Environment and its Effects*, New York: Cambridge University Press.

Milgram, S. (1970) 'The experience of living in cities: a psychological analysis', in F.F. Korten, S.W. Cook and J.I. Lacey (eds) *Psychology and the Problems of Society*, Washington, DC: American Psychological Association.

Simmel, G. (1950) 'The metropolis and mental life', in *The Sociology of Georg Simmel*, trans. and ed. K.H. Wolff, Glencoe, IL: Free Press.

GABRIEL MOSER

URBANIZATION

Urbanization is the *process of becoming urban*. Scholars have identified recurring patterns in the way in which raw land and scattered development becomes URBAN. The US Bureau of the Census and UN-Habitat have developed concepts to precisely define and measure urbanization and collected data to chart the course of urbanization in the United States and throughout the world.

Urbanization of the human population accelerated rapidly beginning in the middle of the eighteenth century. Today, urbanization has tapered off in the most economically advanced countries, but is proceeding very rapidly in developing countries.

Intuitively, *urban* places are places that have enough *total population*, sufficient *population density*, and sufficient *human activity* to be

distinguished from places which have few people, no or low density development, and little human activity – raw land, sparsely populated rural areas, and small towns.

The US Census supplements this intuitive definition of urban places with a precise technical definition of urbanization. Under the Census definition, urban territory consists of *urbanized areas* of densely settled territory containing 50,000 or more people, *urban clusters* generally consisting of densely settled territory with at least 2,500 people, but fewer than 50,000 people, and some less densely settled blocks that form enclaves or indentations, or are used to connect discontiguous areas with qualifying ones. The 2000 Census uses these concepts and the actual census data collected to classify all land in the United States as urban or not urban. Other countries' definitions of what is and is not urban and the quality of their census data vary.

Greek urbanist Constantinos DOXIADES notes that there are a small number of recurring patterns by which land everywhere in the world and through human history urbanizes. Often, undeveloped land is *annexed* to existing urban land (usually within an incorporated city), the land is subdivided, INFRASTRUCTURE is installed, and as houses and other buildings are built on the land, it qualifies as urban. A second common process of urbanization that Doxiades discusses is when enough scattered individual developments have occurred that previously rural land qualifies as urbanized. A third process is the expansion of small towns and infill of land between them so that places which were previously too small to qualify as urban merge in a single urban area. Occasionally, previously raw or scarcely developed land is urbanized by design all at one time. Greek colonies that spun off from existing polei in the fifth century BC, English and French bastide towns built in southern France in the thirteenth century, the early twentieth-century English GARDEN CITIES of Letchworth and Welwyn, and the new national capital of Brasilia, built in the second half of the twentieth century, are examples.

Historical urban demographer Kingsley Davis distinguishes urbanization from *population growth*. He focuses on the *proportion* of the population in a society which lives in urban areas. According to Davis's definition, in a highly urbanized country like the Netherlands, almost the entire population lives in urban areas, the proportion of the population that is urban approaches 100 per cent, and the society can be characterized as fully or nearly fully urbanized. In a country like Upper Volta, only a very small proportion of the total population live in urban areas and the society can be characterized as barely urbanized at all.

In Europe, urbanization proceeded slowly and erratically – with a major de-urbanization after the collapse of the Roman Empire. Until the INDUSTRIAL REVOLUTION, most of Europe's population was rural. Davis argues that the overall *rate of urbanization* in Europe proceeded at a snail's pace from the late Middle Ages until the beginning of the Industrial Revolution. Beginning about 1750, the rate of urbanization in England, then other European countries and North America accelerated dramatically. By the latter part of the twentieth century, the developed countries of Europe and North America were nearing full urbanization and the rate of urbanization declined. Sir Peter HALL and other scholars argue that the most advanced industrialized countries may now be *de-urbanizing* as more people choose to live in small towns and rural areas within commuting distance of urban areas. A current debate is whether the ability of information technology to connect every place on the globe will make urban areas less important and lead to a slowing or reversal of urbanization. Research by Saskia SASSEN and others has found that the opposite is occurring. Information technology appears to be increasing the size and importance of large cities.

In the United States, decennial censuses have reported a consistent process of urbanization. (Only the census of 1810 reported a slight drop in the proportion of the urban population.) The US Department of Housing and Urban Development has contracted with the Rutgers Center for Urban Policy Research to systematize data on US cities. Their database and the periodic *State of the Nation's Cities* reports provide detailed data on urbanization in the United States.

Urbanization in the developing world today is occurring rapidly and on a huge scale. Since urbanization in developing countries is almost everywhere accompanied by simultaneous rapid population growth, not only is the proportion of the population of most developing countries that is urban increasing rapidly, but the *absolute size* of the cities and conurbations that are emerging is much larger than ever before. MEGACITIES of 10 million and more inhabitants are growing rapidly. The United Nations has established

a research institute in Nairobi named UN-Habitat to monitor and report on world urbanization. Their periodic *Global Report on Human Settlements* provides data on population, housing, social and economic conditions in 315 cities in 91 countries.

Doxiades has explored the total urbanization of planet earth. He concluded that if existing trends continue, within a few centuries the entire habitable area of the earth will be completely urbanized. Doxiades termed the resulting world city – the interconnected, urbanized, society of all mankind – ECUMENOPOLIS. As urbanization proceeds, the search for forms of urban development that will provide for a sustainable and livable ecumenopolis is one of mankind's greatest challenges.

Further reading

Davis, K. (1965) 'The urbanization of the human population', *Scientific American* (September).

Doxiades, C. (1975) *Anthropopolis: City for Human Development*, New York: W.W. Norton & Co.

Palen, J.J. (1996) *The Urban World*, 5th edn, New York: McGraw-Hill.

United Nations Centre for Human Settlements (Habitat) (2001) *Cities in a Globalizing World: Global Report on Human Settlements 2001*, London: Earthscan Publications Ltd.

SEE ALSO: city; urban

RICHARD LEGATES

URBAN–RURAL TENSION

In popular understanding, as well as among some academics, URBAN and rural are to be interpreted as antonyms: the urban and its meaning is definable in terms of that which (self-evidently) it is not, the rural. By their nature, the importance of primary sector activities, particularly agriculture, the density of population and their 'cultures', rural areas are the opposite of what the urban represents. Little wonder, then, that urban and rural interests vary and that urban–rural conflict is commonplace. However intuitively attractive such an argument, it obscures the entangled relationships between urban and rural areas and populations. In reality, as the social and cultural historian Raymond Williams has suggested, town and country are better understood in terms of their interdependency. Yet interdependency does not preclude the possibility of urban–rural tensions, which, if to varying degrees, characterize all societies.

Both perspectives, oppositional and interdependent, are of value in understanding the tensions that commonly underpin the urban and the rural. Where they are read as opposites, reflecting differences in activities, values and interests between urban and rural dwellers, the potential for conflict is the more obvious. But even where their relationship is read as more interdependent, the perceived domination of rural interests by the towns (or, less commonly, vice versa), notwithstanding their complementarity, is an underlying tension within many societies. Equally, in the urbanized and geographically mobile and affluent societies of the advanced economies, new forms of urban–rural tension are emergent as the 'functions' of the rural for the city dwellers (as 'wilderness', as production zone, and as residential haven from city stress) are (re)imagined and become the source of contestation.

Typically, urban–rural tensions arise in several ways. The encroachment of the city into the countryside through urban SPRAWL and the consequent loss of agricultural land is commonplace. But this steady growth, linked to the progressive URBANIZATION of society, is the harbinger of conflict in less obvious ways, perceived as threatening rural cultures, besides raising the spectre of political domination, if not actual annexation, by the city. Conflicts arise too because of the different values and interests between rural groups and those located in the city, tensions that may be played out in national political arenas. Increasingly in postmodern (and hence urbanized) societies, these conflicts arise from the differing perceptions of what rural life and areas should be. To understand these tensions it is useful to distinguish broadly between urban and rural relations in societies undergoing modernization (and hence the development of urbanization) and those in late (or post-) modernity. In the former, the physical, if less so the economic, separation of city and countryside is more distinct, boundaries which become more blurred with the trends towards DECENTRALIZATION and counter-urbanization characteristic of late modernity. Such a distinction is not to suggest that we should consider urban–rural relations as developing in (two) stages, where in reality historically rooted conflicts become superimposed by more nuanced tensions created by the spatial fluidities of late modernity.

In societies undergoing modernization, the development of cities alters the relationships between town and country. The development of the INDUSTRIAL REVOLUTION heralded the development of urban-industrial societies, shifting the economic base from the rural areas to the cities. Their development altered the demographic balance; in the most advanced economies, urbanization led to the residualized rural population (accounting for less than a quarter of the total population, and in which those engaged in agricultural and other rural activities accounted for significantly lower proportions of the workforce).

The development of large cities and of urbanized societies, unprecedented historically, emphasized the differences between the town and country. Urban and rural interests were separate, and hence they needed different solutions. In Britain, the system of local government was restructured successively in the Victorian period, boundaries being drawn between town and country. The cities had particular problems – of circulation, disease, overcrowding and POVERTY – which required separate solutions. The development of the new urban-industrial elite challenged the interests of old landed elites, for free trade against protectionism (as in Britain), conflicts played out in the national political arena between political parties vying for control of the state. In some societies – Australia, for example – political parties were created exclusively for rural areas. Nor, given that progressively urbanization tipped the demographic balance in favour of the cities, were rural areas to be inevitably outvoted; in Australian states, notably Queensland and Western Australia, a complex system of weighting, in which the value of the rural vote was a multiple of the metropolitan vote, ensured that rural interests were represented in state legislatures.

In developing economies undergoing modernization, urban and rural conflicts reflect opposing interests. Typically, such conflicts have been dominated by urban interests, in spite of the fact that, as in much of Africa, urbanization, though developing rapidly, accounts for considerably less than half, and usually a third, of the population. Theorists of the 'urban bias' show how the terms of trade are skewed to favour the cities; state spending on urban projects, the maintenance of urban food subsidies at the expense of the rural producers, and the encouragement of cash-crop farming have been commonplace policies favouring urban elites. The weakening of state controls, part of the conditionalities imposed by structural adjustment packages, may have diluted the force of the urban bias, but it has led in turn to new forms of conflict between urban and rural areas. Peri-urban areas in African and other cities have become the sites for enhanced tension as the demands for urban expansion conflict with the need to retain agricultural land, not least for the contribution it makes to feeding the growing city.

In the advanced economies, such conflicts have become overlain by new tensions arising from the spread of urban populations into the countryside and of the increasing importance of consumption. Ironically, the spread of metropolitan populations to the countryside may face opposition not only from agricultural interests, but also increasingly, as a result of the spread of NIMBYism ('NOT IN MY BACKYARD'), from affluent commuters already resident in rural areas. Functional interdependency between the city and countryside has led to the demand for local governments based on the city-region, read often as threatening to rural interests. The perceived domination of urban over rural interests can become rekindled and the source of national political controversy, as in Britain at the turn of the millennium. Conflicts (such as renewed attempts to ban fox-hunting) have become portrayed as tensions between urban and rural dwellers; while an oversimplification, not least because so much of the rural population is in fact urban, the issue along with others such as access to the countryside by city dwellers have renewed the perception of urban and rural as opposites, in turn rekindling tensions.

Further reading

Barnett, A. and Scruton, R. (eds) (1998) *Town and Country*, London: Jonathan Cape.

RONAN PADDISON

UTOPIA

In 1516, at the time of the great discoveries undertaken by the Iberian navigators, Thomas More's (1478–1535) *Utopia* was published. Divided into two tomes, Book One registers Raphael Hythlodaeus's speech, tracing, in an accusatory tone, a social and economic picture of England at the beginning of the sixteenth century. To the

reader, it fits to await the description of the island of Utopia, referred to by Raphael as *the best of the republics*, revealing its Platonic and Aristotelian origins, and that is the detailed object of Book Two.

The second part, dedicated to a description of the way of life in Utopia, tells first of all of the spatial organization. For More, geometry and hygiene are equivalent to discipline. The territory, the city and the house define a structure that is based on territoriality and the presence of institutions that regulate relationships at all levels, from the daily life of the citizens to the organization of the state and of the nation. The utopian programme configures a system of rules that operate on this imaginary society, through the statement of conduct suitable for work, for family life and in the collective, defining the system of faiths, subordinate to the hegemonic plan of the state, among other significant aspects. Knowledge is a distinguished good; the sense of freedom never appears explicitly restricted, but managed for an ordered state.

Colin ROWE interprets More's Utopia as a heuristic device. It is a logical consequence of what is defined, in a causal relationship, in the plan of that hygienized and disciplined environment.

Others such as CHOAY and LYNCH differed with More's views on Utopia.

More's Utopia introduced a tradition of thinking of the city as the *locus* of a practice (social) connected to a place (spatial), a condition that can be seen, in different fields, in varied works and experimentations: in Owen's New Harmony (1825), in Fourier's phalanstery (around 1829) and in HOWARD's GARDEN CITY; in HYGEIA, the sanitarian utopia of Richardson (1876); in the thought of LE CORBUSIER, especially through *Urbanisme* (1924, *The City of Tomorrow*) and the CHARTER OF ATHENS; or still, in the field of social psychology, in *Walden II* (1948), the behaviourist utopia of B.F. Skinner, among countless other examples.

Further reading

Choay, F. (1997) *The Rule and the Model: On the Theory of Architecture and Urbanism*, Cambridge, MA: MIT Press.

Hexter, J.H. and Surtz, E.L. (eds) (1965) *The Yale Edition of the Complete Works of St. Thomas*, vol. 4, *Utopia*, New Haven, CT: Yale University Press.

Rowe, C. and Koetter, F. (1984) *Collage City*, Cambridge, MA: MIT Press.

LEANDRO M.V. ANDRADE

V

VAN EESTEREN, CORNELIS

b. 1897, Kinderdijk, the Netherlands;
d. 21 February 1988, Amsterdam,
the Netherlands

Key work

1935 Amsterdam Expansion Plan.

Dutch architect and planner Cornelis van Eesteren is known widely for his chairmanship of the Congrès Internationaux d'Architecture Moderne (CIAM) (1930–47) and his 1935 Amsterdam Expansion Plan (1929–35). Less known are contributions to the De Stijl movement, with its founder Theo van Doesburg, the artist Piet Mondrian, and others. It is van Eesteren's association with De Stijl that helps explain city planning's shift from classical URBAN forms to modernist, abstract compositions.

Trained in architecture at the Academy of Visual Arts and Technology in Rotterdam (1914–17), followed by studies in Amsterdam, van Eesteren apprenticed in The Hague before opening his own office. His early architecture and urban projects followed the traditional forms set out by H.H.P. Berlage in Amsterdam South (1914–15).

Winning the 1921 Rome Prize shaped van Eesteren's lifelong career in planning, enabling him to travel widely to European cities and meet leading figures in town planning and the emerging artistic and architectural avant-garde. These included Fritz Schumacher in Cologne, then the most progressive town planner in Europe, van Doesburg at the Bauhaus in Weimar, and many others. Afterwards, van Eesteren resided in Paris where he studied URBAN PLANNING, while also collaborating with van Doesburg for the famous De Stijl exhibition at the 1923 Solon d'Autonme.

During the next ten years, van Eesteren developed his idea of the Functional City, while entering design competitions – especially his prize-winning Unter den Linden project in Berlin (1925) – and with his European lecture tours. His Functional City had three key parts. First was a focus on scientific analyses and forecasts of land uses rather than URBAN FORM, radically breaking from classical planning traditions and his contemporaries, particularly LE CORBUSIER. Second was rejection of the GARDEN CITY as a futile attempt to recreate medieval villages and, instead, arguing for planning the metropolis as a whole with INFRASTRUCTURE, especially transportation, as a systemic framework for urban elements. Third, and most important for the modernist city, was substituting radically new compositional methods in place of traditional urban forms. These compositional ideas, derived from his collaboration with the De Stijl movement and elaborated in his competition entries, replaced traditional perimeter blocks and picturesque views with modernist forms, featuring asymmetrical arrangements of geometrically simple buildings and spaces on open blocks, turning the classical city inside out.

van Eesteren's Functional City shaped his planning and construction supervision of the 1935 Amsterdam Expansion Plan (1929–35), the most important built example of modernist city planning. The Functional City was also the theme of the 1932 CIAM meeting, which Van Eesteren organized, and it became the initial framework for the CHARTER OF ATHENS, published only long afterwards with Le Corbusier's revisions and additions. The Amsterdam Plan and the Charter of Athens were to become the primary influences for

urban planning and urban projects worldwide until the 1970s.

Further reading

van Eesteren, C. (1997) *The Idea of the Functional City*, Rotterdam: NAI Publishers.

RICHARD DAGENHART

VANDALISM

Vandalism is a familiar figure in most cities. It is a hodgepodge concept that covers extremely different behaviours. Considering vandalistic behaviour as resulting from a particular relation to the environment raises the issue of knowing why some objects are damaged and others are not. Looking at the actor's intention to destroy allows distinguishing between casual and purposeful destruction. Thus, *vandalism is an intentional act aimed at damaging or destroying an object that is another's property*. Such a definition permits to distinguish hostile behaviour, the aim of which is to damage or destroy the object, instrumental behaviour, which consists of damage or destruction caused to an object as a means to achieve other goals (appropriation of another's property, sabotage), and behaviour motivated by a desire to express oneself through the degradation of objects as well as play vandalism (breaking window panes). Obviously, slashing seats on public transport is vandalism, but GRAFFITI added to advertising posters or walls in the street is mainly expressive.

Environments are not all vandalized in the same way. What must thus be searched for is what becomes a target for vandalism in the environment, why it so becomes, and what mechanisms rule the choice of the object. Social norms determine our behaviour towards objects of our surroundings. For every environmental site, certain behaviours are accepted, others tolerated, others judged unacceptable (to trample a cigarette butt in the street is accepted, tolerated in a public place, but is not accepted, and thus an act of vandalism, on the carpeted floor of an office). *The relation between the users and the environment explains that some environments are vandalized and others preserved*. This point of view has led to evidence of certain mechanisms concerning the role assigned to the object by the individual. Consequently, if the object as such is aimed at, we can speak of 'targeted vandalism'; if the object's selection results only from a concurrence of fortuitous circumstances, we then speak of 'untargeted vandalism'. Targeted vandalism may be explained by bad environmental insertion of an object, environmental inadequacy, and induced neglect. The idea that novelty attracts vandalism was demonstrated, for instance, about playgrounds as well as about URBAN furniture; it can be explained by ill-suited environmental insertion of the objects concerned (selection of location, for example). In the same way, if an environmental object prevents the individual from reaching fixed goals, aggressive behaviour towards the object may result. Finally, vandalism increases rapidly where environments seem to be neglected. Damage is the result of the accumulation of micro-behaviours expressing an attitude of neglect to the object (kicking a door to open it, for example) or its deviated use. Untargeted vandalism may reflect a bad social climate, and the selection of the target is then explained by how fragile and accessible the vandalized object is.

Reference to the intention as well as to the target of vandalism may usefully guide prevention and intervention.

Further reading

Lévy-Leboyer, C. (ed.) (1984) *Vandalism, Behaviour and Motivations*, Amsterdam: North Holland.

GABRIEL MOSER

VEILLER, LAWRENCE

b. 7 January 1872, Elizabeth, New Jersey; d. 30 August 1959, New York City

Lawrence Veiller (originally pronounced 'Vay-ay', now 'Veyler') was the premier housing reformer of his time and a major figure in the GOOD GOVERNMENT and URBAN PLANNING movements of the early twentieth century. During the 1890s, he became intimately acquainted with the housing conditions of New York's poor and with the public regulation of housing construction. He worked tirelessly to improve both over the next decades. He headed the Charity Organization Society's Tenement House Committee, set up in 1898 at his urging, acted as secretary of the New York State Tenement House Commission of 1900, organized

and ran New York City's Tenement House Department in 1902–04, and founded the National Housing Association in 1911 and directed it until 1936. He also participated actively in the institutionalization of ZONING, in the fight against tuberculosis and in court reform.

Veiller single-handedly shaped housing reform legislation in the United States, authoring the New York State Tenement House Law of 1901 and publishing three key texts: *A Model Tenement House Law* (1910), *Housing Reform: A Hand-Book for Practical Use in American Cities* (1910) and *A Model Housing Law* (1914, revised 1920). A skilled and committed reformer, he was conservative in his social outlook and politics and objected strenuously to the adoption of European-style policies of SOCIAL HOUSING and public subsidies. He believed that housing and social conditions could be improved by 'negative' rather than 'positive' legislation, i.e. by the adoption and strict enforcement of minimum standards, as well as by mass SUBURBANIZATION.

Further reading

Lubove, R. (1962) *The Progressives and the Slums: Tenement Reform in New York City, 1890–1917*, Pittsburgh: University of Pittsburgh Press.
Veiller, L. (1914) 'Housing reform through legislation', in C. Aronovici (ed.) *Housing and Town Planning*, Philadelphia: American Academy of Political and Social Science.

RAPHAËL FISCHLER

VENTURI, ROBERT

b. 25 June 1925, Philadelphia, Pennsylvania

Key works

(1966) *Complexity and Contradiction in Architecture*, New York: Museum of Modern Art.
(1977) *Learning from Las Vegas*, rev. edn, Cambridge, MA: MIT Press (with D.S. Brown and S. Izenour).

Robert Venturi received a Bachelor of Arts in architecture in 1947 and a Master of Fine Arts in 1950 from Princeton University. He later received a Rome Prize Fellowship, allowing him to study at the American Academy in Rome from 1954 to 1956. Influenced by early masters such as Michelangelo and Palladio, and modern masters including LE CORBUSIER, Alvar Aalto, Louis KAHN

and Eero Sarrinen, Venturi began his own architectural design firm in the late 1950s. He later formed a partnership with architect John Rauch, and in 1967 expanded that partnership to include his present wife, architect and theorist Denise SCOTT BROWN.

Venturi has been categorized as exhibiting the tenets of the postmodern period. Venturi himself maintains that he has never been a postmodernist. His academic affiliations include participating in lectures, conferences, juries and panels throughout North America, Europe, Asia and Africa. Venturi has published numerous works relating to design and architectural theory, including more recently *Iconography and Electronics Upon a Generic Architecture: A View from the Drafting Room* (1996). Along with partner Denise Scott Brown, Robert Venturi is considered a master in the field of architectural theory and design.

Further reading

www.archpedia.com/Architects/Robert-Venturi.html
www.vsba.com

WENDY TINSLEY
ROGER W. CAVES

VERNACULAR ARCHITECTURE

Vernacular architecture usually describes buildings created by individuals or groups not trained in formal (Western) design principles. Vernacular architecture represents the majority of buildings and settlements created in pre-industrial societies and includes a very wide range of buildings, building traditions, and methods of construction. Terms such as 'unselfconscious', 'common', 'folk' and 'indigenous' are often used interchangeably to describe vernacular architecture, but each of these terms does not individually capture its essential qualities. Some of the more pejorative connotations surrounding vernacular architecture (e.g. provincial, primitive) appear to have been eroded over time. However, the tendency to describe vernacular architecture in sole contrast to the 'other architecture' (high-style design) still endures.

Examples of vernacular architecture abound in almost every continent. Vernacular architecture evolves over time through a process of trial and error. These buildings and settlements can be recognized by their integration into the

socio-cultural landscape within which they are set. Therefore, scholars argue that it is impossible to understand vernacular architecture in terms of its form qualities alone. Vernacular buildings hold great perceptual and associational meaning for their users. For example, religious or sacred buildings in traditional cultures are organized to conform to divine cosmologies that structure space and to rigorous guidelines that direct building construction. The organization of these spaces is understandable only in terms of the underlying sacred meanings. It is imperative that vernacular architecture study occurs within the cultural context within which it was created.

Vernacular architecture has responded to changes in availability of building materials and construction technologies, and socio-cultural transformations. For example, in the United States, in the last fifty years, demographic shifts, SUBURBANIZATION and the advent of the automobile have collectively contributed to the development of new utilitarian buildings that cannot be considered 'architecture' in an elite sense. These include new building types such as gas stations, supermarkets, fast food outlets and even mass-produced housing. These building types are now distinctive and enduring elements of the American landscape. Some architects argue that it is important to understand these emergent architectural forms and the meanings they hold for present-day society.

To capture this complexity, scholars have introduced an intermediate category called 'popular architecture' that is primarily URBAN and post-industrial. Popular architecture mediates between high-style architecture and vernacular architecture. Vernacular architecture studies is now a multidisciplinary field that integrates concepts and methodologies from disciplines such as anthropology, archaeology, geography and history in addition to the traditional fields such as architecture, landscape architecture and art history.

Further reading

Oliver, P. (ed.) (1997) *Encyclopedia of Vernacular Architecture of the World*, Cambridge: Cambridge University Press.
Upton, D. and Vlach, J.M. (eds) (1986) *Common Places: Readings in American Vernacular Architecture*, Athens, GA: University of Georgia Press.

LAXMI RAMASUBRAMANIAN

VICTORY GARDEN

The term 'Victory Garden' was coined during the First World War, giving an old practice a new name; city dwellers have long used backyards or vacant lots to grow fruits and vegetables. But when the disruption of trade and the shift of manufacturing to support the war effort created food shortages in Britain and Canada, and later in the United States and Europe, governments encouraged people to plant kitchen gardens – not only to supplement their rations but also to boost morale, hence 'Victory' Garden. The policy was revived during the Second World War for much the same reasons.

The practice still persists, though for different reasons. In the nineteenth century, COMMUNITY gardens and private allotments had been established to counteract revolutionary tendencies. Today, they are maintained as domestic urban OPEN SPACES (Woolley 2003) – green oases, either on marginal strips of land (e.g. along railroad tracks) or in large complexes. Many gardening associations allow cottages for overnight stays and organize social activities. With the advent of the ecological movement and rising demand for organic food, there was renewed interest in URBAN farming.

This 'marginal' activity came under pressure of being displaced by developers seeking control over what had over time become prime locations. In response, gardening associations from New York to Berlin and Tokyo started to fight back – and with some success. Whereas community gardens were once recognized as an economic necessity, these small-scale projects are now seen as contributing to the quality of urban life and the greening of many inner-city areas.

Further reading

Buswell, S. (1980) 'The garden warriors of 1942', *Urban Agriculture Notes*, City Farmer, Canada's Office of Urban Agriculture, available online at www.cityfarmer.org/victgarA57.html
Woolley, H. (2003) *Urban Open Spaces*, London: Spon. www.cityfarmer.org
www.victorygardens.net

JAN VAN WEESEP

VIEW CORRIDORS

View corridors provide scenic vistas of important landmarks, usually following the route of a path, SIDEWALK, STREET, highway, river or other linear

element. They play an important role in URBAN landscapes, providing static and dynamic visual experiences for observers. View corridors, which often use the termini or intersections of important streets as focal points, were used frequently during the CITY BEAUTIFUL era of urban reconstruction (1890s–1915). Garnier's Opera House (1875) in Paris is a splendid example of the placement of an important building at the terminus of a view corridor, in this case the Avenue de l'Opera. In Washington DC, L'ENFANT's plan (1791) featured two great axes aligned with the Capitol and the White House from which diagonal avenues radiate. This assemblage of streets and avenues provides abundant opportunities to highlight buildings, MONUMENTS and fountains at the termini of view corridors. In other cities, view corridors help draw distant parts of a city together using varied visual experiences to accentuate important locations. View corridors are not limited to cities; manmade and natural view corridors in suburban and rural places enhance the natural environment by providing vistas of mountains, rivers and other natural landmarks. The preservation of view corridors in urban and rural landscapes has become increasingly important. Protection of view corridors can be achieved using design controls such as ZONING requirements and OPEN SPACE preservation in addition to building height maximums, setback restrictions and bulk requirements.

Further reading

Wilson, W.H. (1989) *The City Beautiful Movement*, Baltimore: Johns Hopkins University Press.

SEE ALSO: City Beautiful movement; urban design

DANIEL BALDWIN HESS

VISUAL PREFERENCES

The use of visual preferences in URBAN PLANNING has gained increasing popularity since the 1990s, in particular through the use of Visual Preference Surveys™ as designed and used by A. Nelessen Associates of New Jersey. During a COMMUNITY 'visioning' exercise, participants are shown dozens of photographs or slides depicting landscapes. These images are then rated on a scale in response to specific questions, and then analysed to

determine which design elements are associated with positive and negative ratings. These results are compared to a questionnaire and then analysed in light of the proposed development or ZONING regulation. Images representing current conditions are often augmented with realistic computer-generated simulations demonstrating different development options. Surveys frequently reveal a notable preference for neo-traditional planning elements over standard SUBDIVISION designs.

Despite the popularity of visual preference tools in both empirical research and urban planning, when it comes to their interpretation, some scholars advise caution. Because of the difficulty and expense of transporting participants to an environment, preferences are most often selected from photographs of the landscapes in question. Yet humans experience environments by actually being physically present within them; photographs are only surrogates for the actual environment and so require interpretation based on the individual's memories of such environments. Participants' interpretation of the images may therefore be influenced by any number of factors, not the least of which is that all senses save sight are unaffected by the content of the image. Photography is, of course, an art form with its own aesthetic principles – such as lighting and composition – to which participants are also likely to respond.

Besides their application in urban planning, there is a long tradition of the use of visual preferences in the discipline of ENVIRONMENTAL BEHAVIOUR research, most notably in landscape assessment and environmental simulation studies. Such research has been conducted under the influence of several paradigms, including the psychophysical paradigm, which assumes that the represented landscape acts as a stimulus to which the passive observer responds; the cognitive paradigm, which is more concerned with learning what meanings the active observers construct regarding the landscape, rather than what is valued in the landscape; and the experiential paradigm, which focuses on the experience of the interaction between the individual and the landscape. The Visual Preference Survey™ falls into the psychophysical tradition, in that it attempts to determine which features of the depicted landscapes will be consistent with general perceptions of aesthetic beauty, while eschewing the added dimension of the meanings and values behind the preferences. As such, while tools measuring visual preferences

are justifiably popular for their usefulness in urban planning, their limitations should also be recognized.

Further reading

Scott, M.J. and Canter, D.V. (1997) 'Picture or place? A multiple sorting of landscape', *Journal of Environmental Psychology* 17: 263–81.

Taylor, J., Zube, E.H. and Sell, J.L. (1987) 'Landscape assessment and perception research methods', in R. Bechtel, R. Marans and W. Michelson (eds) *Methods in Environmental and Behavioral Research*, New York: Van Nostrand Reinhold, pp. 361–93.

MICHAEL DUDLEY

W

WADE, RICHARD C.

b. 14 July 1922, Des Moines, Iowa

Key works

(1959) *The Urban Frontier: The Rise of Western Cities, 1790–1830*, Cambridge, MA: Harvard University Press.
(1964) *Slavery in the Cities: The South, 1820–1860*, New York: Oxford University Press.
(1969) *Chicago: Growth of a Metropolis*, Chicago: University of Chicago Press (with H. Mayer).

Richard C. Wade has been one of the key figures in the emergence of URBAN history as a recognized specialization within American scholarship. He took an MA in history at the University of Rochester and a Ph.D. at Harvard, where he studied with Arthur Schlesinger, a leading social historian who had pioneered American urban history with *The Rise of the City*. Wade went on to teach at Washington University, the University of Chicago, and the Graduate Center of the City University of New York.

Wade's doctoral dissertation, published as *The Urban Frontier*, was a path-breaking book that inverted the standard narrative of North American settlement associated with Frederick Jackson Turner. In Wade's version, cities were the 'spearheads of the frontier', the economic and political centres from which settlement and investment spread into newly developing territories. By calling attention to the importance of urban growth in the first years of national expansion, the book forced a rethinking of the relationships between agrarian and commercial systems. *The Urban Frontier* was also noteworthy for setting a standard of energetic comparative research on urban development.

The same energy characterized Wade's second book, *Slavery in the Cities*. Here too he undermined

a common understanding – that of slavery as a purely agrarian labour system. The book emphasized the importance of the South's commercial cities and applied the social theory of Robert PARK and Louis WIRTH to argue that the conditions of urban life were incompatible with the rigidities of slavery. In Wade's view, the realities and importance of urban life in the ante-bellum era were undermining the South's peculiar institution and connecting the region more closely to the North.

Wade extended his influence on the field of urban history in several additional ways. He worked with several cohorts of graduate students who helped to establish urban history at major universities throughout the country. He edited the first, and very influential, series of books on urban history in the 1960s and 1970s. His co-authored pictorial history of Chicago led the way in the use of photographs as historical evidence. And he has offered a number of programmatic statements of the importance of cities and city life to understanding the major contours of US history. In these essays, he argued convincingly for the centrality of urban growth to the shaping of American society and American political choices.

Wade has long been active in Democratic Party politics, both in Chicago and in New York. He played an important role in helping to shape the campaign strategy of fellow historian George McGovern in 1972.

CARL ABBOTT

WALKING CITY

A term describing cities where the majority of people walked to get from one place to another before improvements in transportation technologies made

it possible and economically feasible to travel by train, steamboat, cable car, electric streetcar and automobile. The increase in URBAN population that paralleled the rise in new transportation technologies spread cities out so that eventually a significant percentage of the cities' residents could (and needed to) use means other than their own feet to move about. All cities in the world before 1800 were walking cities. Well into the nineteenth century almost every city everywhere in the world could be traversed by foot in less than half an hour.

Until about 1830, horses provided the only practical alternative to walking for all but a very few city dwellers. Horses were too expensive for the majority of the population to use. Steamboats and railroad trains were primitive, expensive and scarce – not sufficient to create a commuting workforce or to push the boundaries of cities very far from the centre. Horse-drawn streetcars existed only in the largest cities. The transportation technologies that would be invented in the last decades of the nineteenth century – the cable car, electric streetcar and automobile – did not exist.

Cities had to be built compactly enough to work for a population that walked. Walking cities were rarely created by design. They evolved as walking cities because that is the way they had to be.

Walking cities present an apparent paradox. At a time when the world's population was only a fraction of its present size and when only a small part of the population lived in cities, land was much more abundant and far less expensive than it is today. One would expect that cities would take advantage of this situation and spread out. Nonetheless, because they were built to accommodate people who moved about by foot, all cities, everywhere in the world, were quite dense. Even in North America, COLONIAL CITIES were all walking cities – specks in the wilderness surrounded by inexpensive land, but built at higher densities than most modern SUBURBS and many modern cities.

Beginning in the first decades of the nineteenth century, the walking city began to change. Some affluent COMMUTERS commuted into New York City by steamboat. Risk-averse people tolerated smoke and cinders and the possibility of exploding boilers and took the train. By mid-century, the INDUSTRIAL REVOLUTION pushed up urban populations and made cities more polluted, congested and dangerous at the same time that the number of steamboats soared and trains stopped blowing up.

More people moved further out from city centres. By the 1880s, electric streetcars definitively ended the era of walking cities. The movement from city to metropolis was underway.

The term 'walking city' does not refer to portions of cities closed to traffic and turned back to pedestrian use or to pedestrian-friendly NEW URBANISM cities. It refers to a distinct historical city form, dictated by limitations in transportation technology.

Further reading

Nash, G. (1991) *Forging Freedom: The Formation of Philadelphia's Black Community, 1720–1840*, Cambridge, MA: Harvard University Press.
Warner, S.B., Jr (1987) *Private City: Philadelphia in Three Periods of Its Growth*, 2nd edn, Philadelphia, PA: University of Pennsylvania Press.

RICHARD LEGATES

WALLED CITIES

A walled city is a type of FORTIFIED CITY surrounded by walls, which provide protection, define the city boundaries, control access, provide a customs barrier to collect fees and/or symbolize power. The walls can be simple palisades or earthworks to extensive military fortifications with towers and bastions. They all include gates to provide access to the city.

Since walls were built to be as strong as the technology of the time permitted, they are often the last evidence of a city's existence and many are still standing, providing examples from the different historical periods. The Sumerian CITY OF UR has a brick wall. In early Greece, the high CITADEL of Mycenae was surrounded by a wall with a space for the palace and the royal tombs. The wall had two gates, one being Lion Gate. They were approached by a walled ramp controlled by a high tower. In classical Greece, walls were combined with natural barriers to protect settlements, but the structures were often primitive. The Romans used walls for protection (an existing example may be found in Pompeii, Italy) and also to define the city territory. During their colonization of Europe, they used walls first to define the rectangular boundaries of their planned military camps (castra) which, in turn, served as models for their newly founded towns. During the medieval period, constant conflict required walls particularly in the

burgs or fortified cities. One exception was England, which achieved unity earlier and was protected by water. However, some cities such as Chester retained their Roman walls, and walls were built around frontier towns such as Conwy and Caernarfon in Wales.

From the fourteenth century, double walls were used to increase protection throughout Europe as illustrated by one of the most famous walled cities in existence, the old 'cité' of Carcassonne, in southern France. This city, whose restoration in the nineteenth century by Viollet-le-Duc was highly contested, illustrates a complex defence system with two rows of ramparts, and towers. The walls enclosed a space occupied by the palace, the church and residential buildings still in use today. Some of the new, planned, 'bastide' medieval towns in France were also surrounded by walls with gates. In Spain, the city of Avila still has its extensive one-wall system (started in 1090 over Roman walls), with eighty-eight towers and twelve gates. The church was built into the city walls.

Some cities had a series of walls to accommodate growth. Florence had two medieval walls surrounding the fortified Roman nucleus. Chinese cities used walls not only on the outer perimeter of their cities but also to control access to different city areas and even to city blocks. When the walls became unnecessary for defence, they were often demolished and the space used to develop BOULE-VARDS, such as in Paris and Vienna.

Further reading

Kostof, S. (1991) 'The walled edge', in *The City Assembled*, Boston, MA: Little, Brown, pp. 26–34.
Morris, A.E.J. (1994) *History of Urban Form*, 3rd edn, New York: John Wiley & Sons.

ANA MARIA CABANILLAS WHITAKER

WAR

The most evident impact of war on cities is the destruction of extensive areas of built environment and their reconstruction. URBAN history features many examples of cities built by conquerors, who generally cancelled the traces of the past in order to enhance their own dominion over the occupied country, but the best documented post-war reconstruction projects are those carried out inside and by the country concerned.

Reconstruction processes are often the occasion to experiment with urban organization models on a large scale and in short periods of time. This might explain why architects, fascinated by a tabula rasa – the opportunity to work on a clean sheet, which presupposes the availability of an extensive area without buildings (or inhabitants) – often look on the destruction wrought by war as an 'opportunity' or a 'challenge'.

In the years between the two world wars, some of the town planning theories illustrated in famous manuals were based on the assumption that there are urban typologies more apt than others to resist aerial bombings. LE CORBUSIER, for instance, believed that excessive urban DENSITY was the greatest danger and tried to raise consensus on his idea of Ville Radieuse, made up of a series of high-rise buildings in between large unbuilt spaces, claiming its minor vulnerability.

In 1945 there were innumerable examples of European cities laid waste. The reconstruction projects exemplify two substantially different attitudes. The first, in an attempt to reaffirm continuity with the past, tries to restore the situation to what it was before, to rebuild or reproduce 'as it was, where it was', at least in those areas with a strong emotional and symbolic significance. Warsaw and Frankfurt are the best examples of this approach. In both cases, the goal of restoring the image of the old historical city centre has been sought by employing old materials and building fragments, imitating, with the help of archival documents, the pre-existing outline of façades and skyline. Even though this procedure is confined to a small portion of the city, it bears a strong emotional impact and is generally widely agreed upon by the population.

The second attitude, adopted by Rotterdam, for example, where the bombardment had caused the wholesale destruction of the DOWNTOWN area, is more complex. Here the effort is to combine reconstruction with the introduction of innovative alterations. Starting from the desire not only to see the city rise again but also to use its reconstruction as a means to achieve improved living conditions, architects and planners, central government and local authorities joined forces to devise plans that paid attention not only to building volumes and physical INFRASTRUCTURES but also to social rules and models.

The reconstruction project goes beyond the idea of a simple restoration. The layout of each unit is

part of a general plan, which attempts to experiment with new building typologies and models of urban organization, like the creation of large pedestrian zones in the commercial district, a wide network of public transport, and most of all a new relationship between owners' rights and the collective interest. The government expropriated all the land by compulsory purchase and the reconstruction was used as the occasion to foster the idea that large-scale decisions justified in the name of the public good should prevail over the individual faculty of choice, in a word 'to use the damage done by bombs to good purpose'.

A similar attitude, which considers the reconstruction as an opportunity to speed up the practical application of previously elaborated theories, which had seemed of difficult realization up till then, is also present in the English experience. Immediately after the first bombings of London in 1940, the reaction of many technicians and politicians was that of being faced not only with the necessity of reconstructing the destroyed cities, but also with the chance of creating a totally new and better Britain.

Even before the outbreak of war, some official reports had shown the negative effects of the excessive concentration of housing and workplaces. On this basis, the commission was given for further research in order to elaborate indications for future developments, with particular reference to the localization of industries, to the creation of a central planning authority and to the need to control land prices.

With regard to the London region, ABERCROMBIE captured the dominant planning themes and wove them into a coherent strategy. The main criteria were those of limiting the expansion of London by creating a GREEN BELT and guiding future development into a series of new towns.

On the one hand, the reconstruction proceeded in accordance with the ideology of the decentralized city; on the other, the extent of warfare destruction contributed to strengthening the idea that more planning was indispensable.

In the same period, DECENTRALIZATION, of both residences and industries, was implemented also in the United States. American cities had not suffered during the conflict, but DISPERSION was systematically presented as a method of reducing hypothetical attacks. War had thus an indirect, nonetheless strong and lasting, influence on territorial organization. The military took over the URBAN PLANNING models and the urban planning projects were reinforced by defence requirements.

The last five or six decades have seen an almost endless series of armed conflicts that have had serious consequences for many cities, especially in the developing countries. Here war, more than just destruction, can determine heavy and rapid movements of citizens towards cities and alter the relationship between the number of inhabitants, houses and services. Evacuations to avoid forthcoming bombings, or migratory fluxes at the end of battles, especially when a crisis of agricultural production is connected to the general disruption of economic activities, have the result of worsening the already low quality of life.

Further reading

Cherry, G.E. (1988) 'Britain at war: plans for reconstruction', in *Cities and Plans*, London and New York: Edward Arnold.

Hasegawa, J. (1992) *Replanning the Blitzed City Centre*, Buckingham: Open University Press.

Thrift, N. and Forbes, D. (1986) *The Price of War: Urbanization in Vietnam, 1954–1985*, London: Allen & Unwin.

PAOLA SOMMA

WAR ON POVERTY

The War on Poverty refers to a group of programmes created as part of President Lyndon Johnson's GREAT SOCIETY agenda. In his first state of the union address in 1964, President Lyndon Johnson called for a 'war' on POVERTY. Johnson revised and expanded proposals developed to address the problem of poverty by advisers to President Kennedy shortly before his assassination in 1963, incorporating them into the Economic Opportunity Act of 1964. The forty programmes established by the Act were collectively aimed at eliminating poverty by improving living conditions for residents of low-income NEIGHBOUR-HOODS and by helping the poor access economic opportunities long denied them. A new agency was created to administer the programmes, the Office of Economic Opportunity (OEO.) The Act also introduced the idea that the poor themselves should participate in designing social programmes, by mandating that the new COMMUNITY agencies created to receive and administer funds be developed and administered with 'maximum

feasible participation' of residents of low-income communities.

OEO programmes included VISTA (Volunteers In Service To America); the Job Corps; the Neighborhood Youth Corps; Head Start; Adult Basic Education; Family Planning; Community Health Centers; Congregate Meal Preparation; Economic Development; Foster Grandparents; Legal Services; Neighborhood Centers; Summer Youth Programs; and Senior Centers. All but the Neighborhood Youth Corps were still operating in the late 1990s. Over a thousand Community Action Agencies (CAAs) were created at the local level to implement anti-poverty programmes. CAAs varied greatly, from non-profit groups to city agencies and community-controlled groups. Such centres helped train a new generation of community activists and leaders, many of whom became federal anti-poverty workers. Most continue to operate, although their focus has shifted away from advocacy and citizen activism to programme administration. CAAs were initially funded directly by the federal government. By 1967, in response to claims by many Democratic mayors that the federal government was funding their opponents, funds began to be passed through the hands of state governors before going to community agencies.

The designers of the War on Poverty drew on lessons from community-based pilot programmes sponsored by the Ford Foundation, the National Institute of Mental Health and the Kennedy administration's President's Committee on Juvenile Delinquency. These efforts attempted to make changes at the city and community level in order to prevent problems like juvenile delinquency and the decline of inner-city neighbourhoods. Those involved in these efforts concluded that these problems were inextricably rooted in the underlying issue of poverty and that this was beyond the capacity of local government to address. Conceptually, attention was focused on poverty at the community – rather than the individual – level. Community action was originally conceived, in this framework, as an effort to help increase the capacity of the community to reform the systems blocking access to opportunities. Powerlessness was defined as part of the problem of poverty.

While the OEO programmes are most closely identified with Johnson's War on Poverty, other programmes, not officially labelled as part of the War on Poverty, were also an important part of anti-poverty efforts. In fact, the OEO programmes were dwarfed by both the scale of spending and the scale of impacts fostered by these less targeted Great Society programmes. At their peak, OEO programmes constituted only 6 per cent of federal social welfare spending. The Johnson administration was arguably more effective in addressing poverty through larger, less targeted programmes. For example, civil rights legislation ending the legal basis for racial DISCRIMINATION in access to the vote, to employment, to use of public facilities (broadly defined) and to housing all set the preconditions for the minority poor to move out of poverty. Expansion of access to healthcare for the elderly and the poor, through the Medicare and Medicaid programmes, helped improve the health status of both groups. A cabinet-level Department of Housing and Urban Development was established, administering a new programme of rent subsidies. The minimum wage was increased, pushing full-time workers above the poverty threshold for a family of three. Finally, Johnson's economic policy, centred on a US$10 billion tax cut, spurred a period of economic expansion that arguably helped reduce poverty as well.

Perhaps the most important change spurred by the War on Poverty was the definition of poverty as a national issue and responsibility for it as a federal responsibility. Supporters point out that the national poverty rate declined from 22.2 per cent in 1963, when President Johnson took office, to 12.6 per cent in 1970, when the anti-poverty programmes had arguably taken effect. Health indicators showed considerable improvements: infant mortality fell by one-third in the decade after 1965, due to expansion of federal medical and nutritional programmes. The percentage of the poor who had never been seen by a physician fell from 20 per cent to 8 per cent. The proportion of families living in housing lacking indoor plumbing fell from 20 per cent in 1960 to 11 per cent in 1970.

Despite these indicators of success, the legacy of the War on Poverty is clouded by the controversy provoked by the community action programme, particularly the requirement for the 'maximum feasible participation' of the poor. By 1980, the image of these programmes was so negative that President Ronald Reagan could claim that 'we waged a war on poverty and poverty won'. Liberal critics note that despite the emphasis on blocked opportunities as the root cause of poverty, Johnson's anti-poverty programmes did not

attempt to address the lack of employment opportunities directly, by creating jobs.

Further reading

Katz, M.B. (1986) 'The War on Poverty and the expansion of social welfare', in *In the Shadow of the Poorhouse: A Social History of Welfare in America*, New York: Basic Books, pp. 251–73.

Lemann, N. (1991) 'Washington', in *The Promised Land: The Great Black Migration and How It Changed America*, New York: Knopf, pp. 110–221.

Marris, P. and Rein, M. (1967) *Dilemmas of Social Reform: Poverty and Community Action in the United States*, New York: Atherton Press.

SEE ALSO: advocacy planning; civil rights movement; Moynihan, Daniel Patrick; poverty

ELIZABETH J. MUELLER

WARNER, SAM BASS, JR

b. 6 April 1928, Boston, Massachusetts

Key works

(1962) *Streetcar Suburbs*, Cambridge, MA: Harvard University Press.
(1969) *The Private City*, Philadelphia, PA: University of Pennsylvania Press.
(1972) *The Urban Wilderness*, New York: Harper & Row.

Sam Bass Warner Jr has had an enduring influence on the twentieth-century study of cities. An early 1990s survey about influential URBAN historians confirmed this, ranking Warner at the very top. First attracted to the law and to newspaper publishing, Warner became curious about people's nineteenth-century residential choices; the result was his Harvard Ph.D. thesis, *Streetcar Suburbs* (1962). As one of the first to explore the relationship between a CITY and its SUBURBS, and to closely examine the historic suburban experience, Warner helped to create the new urban history in the 1950s and 1960s.

Warner began his academic career at Washington University in St Louis, and then moved to the University of Michigan. He went on to hold endowed chairs at Boston University and at Brandeis University, in the region where he was raised and still lives. Following his 'official' retirement, Warner now volunteers his time, working with students and faculty at MIT.

Warner exhibited a multifaceted interest in cities throughout his long career. His first book established a new standard for urban scholarship. From this base, he interweaved historical, architectural, environmental and policy concerns through his subsequent dozen books and nearly thirty articles and book chapters. Warner's best-known works, in addition to his first, are *The Private City* (1969) and *The Urban Wilderness* (1972). These established what might be termed the 'Warner way' in urban history – examining the interactions and impacts of power, policy, choices and actions (public, private and corporate).

The Way We Really Live (1976), a series of Boston Public Library lectures, elegantly highlighted Warner's primary concern that cities first be healthy places for human beings to live and work. Here he first explored more contemporary themes – the actualities of local urban life, COMMUNITY gardens, and healthcare settings. In the 1980s, Warner laboured to extend the concerns of urbanists beyond their traditional axis of power and policy, and continued to balance between 'traditional' history and contemporary concerns. *The Province of Reason* (1984) presented an evocative set of collective biographies, and simultaneously questioned the lingering effects of past choices on the present.

Warner's strong environmental interests and his sensitivity to the lives of ordinary people burst forth in *To Dwell is to Garden: A History of Boston's Community Gardens* (1987) and in his co-authored *Restorative Gardens: The Healing Landscape* (1998). Warner's interests are still expanding, now encompassing a metropolis, and navigating a dual interest in the contemporary lifeways of people and place. His recent *Greater Boston: Adapting Regional Traditions to the Present* (2001) uses words and drawings to illuminate his multi-state region. His current project, 'The Inquiring Citizen's Guide to Urban Environment', will illuminate urban uses of land, air and water.

Sam Bass Warner is also a concerned citizen. A sensibility about the way that the world should be has infused his writing for decades. Those who read Warner's work may suspect that a person of great curiosity and generous spirit is behind the words. This is true.

JUDITH A. MARTIN

WATERFRONT DEVELOPMENT

In the early 1970s, the changes in maritime technologies and the relocation of industrial and shipyard activities resulted in the abandonment of large stretches of land in potentially valued central areas. In many cases, the relocation of the port functions led to economic decline, social distress and physical decay. Reclaiming and transforming these areas for residential, commercial and recreational purposes quickly became an increasingly widespread phenomenon in geographical terms, and very important in terms of mobilized resources, so much so that Peter HALL defined it as 'the major urban event in urban planning in the 1980s'.

The implemented actions share some common characteristics. As far as design criteria are concerned, emphasis was put on the need to reassess the link between the city and its port. It is argued that in the past, waterfronts were the core of the main economic activities and part of the MERCANTILE CITY tissue. However, this strong unity was broken in the industrial period. Now, the new waterfront should be visually and physically accessible to the public and attractive to visitors, residents and investors. Moreover, the waterfront development initiatives were associated with the introduction of significant changes in the decision-making processes and in the institutional and financial set-up. New patterns of role division were experimented with between public and private capital. In general, these operations were made possible by substantial public investments and by the transfer of land from public organizations, while the big property companies dealt with the implementation and commercialization of the projects.

The early American experiences, such as Baltimore and Boston, are emblematic examples of the mixture of public and private speculation and the role of this new type of mega-developer, in particular ROUSE who started the FESTIVAL MARKETPLACES era.

In Europe, the waterfront development experiences are more varied. In some instances, abandoned areas were transformed into attractive residential areas but the social impact was neglected, while in others, attempts were made to associate research on economic development with environmental and social issues.

The most ambitious and controversial project is that of the London Docklands, which raised questions on equity and doubts about the economic viability of the scheme. A more balanced approach was found in Barcelona and Rotterdam, where the scheme was not limited to the opportunities of immediate financial return for investors, but focused on the social issues of the whole metropolitan area, by creating jobs and improving the conditions of some decayed residential areas without expelling the residents.

Among the many scholars who have studied the waterfront development phenomenon, there is no unanimity on whether the waterfront is a special place where innovative ideas are experimented with or just one of the many places where the recurring phases of DISINVESTMENT and reinvestment are exercised in a new form.

Further reading

Hall, P. (1991) *Waterfronts: A New Urban Frontier*, Working Paper 538, University of California Institute of Urban and Regional Development, Berkeley, CA.

Hoyle, B.S., Pinder, D.A. and Husain, M.S. (eds) (1988) *Revitalizing the Waterfront: International Dimensions of Dockland Redevelopment*, London: Belhaven Press.

PAOLA SOMMA

WEBBER, MELVIN M.

b. 6 May 1920, Hartford, Connecticut

Key works

(1964) 'The urban place and the nonplace urban realm', in M. Webber (ed.) *Explorations into Urban Structure*, Philadelphia, PA: University of Pennsylvania Press.

(1968–69) 'Planning in an environment of change' (2 parts), *Town Planning Review* 39(3): 179–95 and 39(4): 277–95.

Melvin ('Mel') Webber is professor emeritus at University of California, Berkeley, past director of the university's Transportation Center, the author of classic theoretical papers and of major consulting reports, and an active contributor to debates on transportation policy, regional development and planning theory. His education in economics, sociology and city and REGIONAL PLANNING gave him a broad view of transportation issues and of URBAN problems in general. A deep-seated concern with equity and freedom permeates his writings. Yet he recognizes the inherent

complexity and amorphous character of policy problems and, hence, their ultimate insolubility.

Webber's most important contributions were prescient articles of the 1960s on the impacts of technological and social change on cities and on planning. New means of communication and new institutions allow, he saw, for the maintenance of individual and communal bonds irrespective of spatial location and for the emergence of dispersed metropolitan areas (or, in his felicitous phrases, for 'community without propinquity' and the rise of a 'nonplace urban realm'). At the same time, he noted, the increasing diversity of needs and wants in urban populations makes the definition of a public interest impossible and forces planners to trade their identity of technical problem-solvers for that of enablers of public debate and organizers of market-like processes of interaction.

RAPHAËL FISCHLER

WEBER, MAX

b. 21 April 1864, Erfurt, Thuringia, Germany; d. 14 June 1920, Munich, Germany

Key works

(1930 [1904]) *The Protestant Ethic and the Spirit of Capitalism*, trans. T. Parson, New York: Charles Scribner's Sons.
(1958 [1921]) *The City*, trans. D. Martindale and G. Neuwirth, Glencoe, IL: Free Press.

Max Weber, a German-born sociologist, is one of the most important theorists on the development of modern Western society. In *The City* (1921), Weber found the rise of the modern, Western city in the Middle Ages, where the growth of guilds and the bourgeoisie undermined traditional patterns of authority and gave rise to a new emphasis on self-governance, written rules, and citizenship rights. In *The Protestant Ethic and the Spirit of Capitalism*, Weber identified an affinity between the virtues espoused by Puritanism – hard work, planning, frugality, self-restraint and the deferment of personal pleasure in favour of capital reinvestment – and the requirements of economies beginning INDUSTRIALIZATION.

Possibly even more influential for the GOVERNANCE of cities was Weber's path-breaking work on bureaucracy. Weber saw bureaucracy as a hallmark of modern society, an orderly and efficient alternative to pre-modern societies with their irrational traditions and unreliable charismatic leaders. Weber's 'ideal type' bureaucracy has the following characteristics:

1 *Permanence:* Official business is conducted on a continual basis. Citizens can rely on bureaucracy; service is not disrupted simply because a person leaves office.
2 *Specialization:* Each member of the organization is assigned a specific area of work and becomes an expert in that specialized area of performance.
3 *Hierarchy:* A chain of command assures a uniformity of performance; each official knows to whom to report.
4 *The impersonal application of written rules and procedures:* The agency's rules, not the personal or political opinions of the bureaucrat, determine performance.
5 *Full-time, salaried officials with a careerist orientation:* Officials who depend on the organization for their financial well-being will tend to follow the rules of the organization. Part-time, volunteer or temporary workers, in contrast, will likely be more willing to substitute their own views for the policies of the organization.

While Weber argued for the superiority of bureaucratic organization, he also observed its concomitant dangers of secrecy and the inability of outsiders to effectively challenge decisions of bureaucratic specialists.

During the late nineteenth and early twentieth centuries, the GOOD GOVERNMENT MOVEMENT in the United States and progressive reformers worldwide turned to Weberian bureaucracy as a normative model of fair and efficient government. Service delivery based on written rules, norms of neutrality, and expertise was seen as preferable to the favouritism, corruption and inefficiency of local parties and elites.

Yet the new URBAN service bureaucracies also brought new problems as the large organizations seldom functioned as smoothly as Weber's 'ideal type'. Dissatisfaction with inefficient and unaccountable bureaucracy eventually led to demands for a new generation of reforms – including DECENTRALIZATION, citizen participation, COMMUNITY control, self-service provision, merit pay,

PERFORMANCE MEASUREMENT, PRIVATIZATION and other innovations designed to improve the performance of bureaucratic organizations or else to bypass them entirely.

Further reading

Fry, B.R. (1989) *Mastering Public Administration: From Max Weber to Dwight Waldo*, Chatham, NJ: Chatham House, pp. 15–46.

Gerth, H. and Mills, C.W. (1946) *From Max Weber: Essays in Sociology*, New York: Oxford University Press.

BERNARD H. ROSS
MYRON A. LEVINE

WELFARE TO WORK

In the United States, from 1935 until 1996, the federal government, in partnership with the states, provided a safety net to eligible poor families with children through the Aid to Families with Dependent Children (AFDC) programme. Though not an entitlement, AFDC did ensure a steady source of cash assistance and other benefits as long as applicants, mainly single mothers and their children, met eligibility criteria.

By the early 1990s, some policy-makers became concerned that the programme discouraged work by enabling recipients to survive on welfare payments. Although at least half of all AFDC recipients remained on welfare only temporarily, and others cycled back and forth between welfare and work, it was still felt that a significant number of welfare recipients were caught in a 'cycle of dependency'.

Vowing to 'end welfare as we have known it', President Bill Clinton in 1996 signed legislation which eliminated the sixty-one-year-old AFDC programme, and replaced it with the Temporary Assistance for Needy Families (TANF) programme. Like AFDC, TANF provides cash income to poor families, but under strict time limits. Unlike AFDC, TANF emphasizes work; families must quickly find employment as soon as they are determined to be ready to hold a job. If they do not, TANF requires states to sanction them, either by reducing the family's assistance or by terminating it altogether. Unlike AFDC, there is no longer one national model for providing welfare assistance to families with children. Instead, the states were given a lump sum of money

and the flexibility to design and implement their own assistance programmes as long as they comply with very broad federal guidelines.

It is still too early to know the impact of the new welfare laws on poor families and their children. We do know that significant numbers of families left welfare between 1996 and 2001, a period of economic boom. We do not know what proportion of those families found jobs, or sought other sources of income. Those who found jobs have been among the better educated, with work experience, and often with older, and fewer, children. It is much harder to find jobs for those with limited skills and work experience, and younger children. Even among those who leave welfare for work, a significant number are unable to remain off welfare for long, since the jobs they find are often short term, unreliable and poorly paid.

Going from welfare to low-wage employment will not move families out of POVERTY. What is needed are opportunities to gain higher-level training and education, more supportive services such as childcare, medical benefits and transportation, and better-paying, full-time jobs.

Further reading

Edin, K. and Lein, L. (1997) *Making Ends Meet: How Single Mothers Survive Welfare and Low-wage Work*, New York: Russell Sage Foundation.

Mink, G. (1998) *Welfare's End*, Ithaca, NY: Cornell University Press.

ALMA H. YOUNG

WHITE FLIGHT

Metropolitan areas have continued to grow outward from older cores since the advent of motorized train and automobile transportation in the late nineteenth century. As a result, spatial separation between both ethnic groups and socioeconomic classes has continued or increased in many multi-ethnic URBAN areas. While there may be some debate whether such self-selected SEGREGATION is specifically ethnicity-based or particularly damaging, it is clear that inasmuch as income and wealth are correlated with white ethnicity, the flight of such households away from metropolitan cores may bring a concentration of POVERTY and concomitant social problems in older urban cores, and the speeding away of resources – particularly those which contribute to shared goods such as

public schools, INFRASTRUCTURE and services – towards areas of relatively homogeneous and better-off NEIGHBOURHOODS. The material results of spatially and socially concentrating lower-income households without a critical mass of resources to counter these problems are the central concerns around white flight.

The implication is that middle-class white households – in an effort to ensure the highest-value return on their tax monies and other resource investments in schools, services and residential property values – self-select into neighbourhoods of similar households, resulting in the concentration of both 'choice' white households and 'constrained' non-white households. As a result, white households are able to reap the benefits of their contributions to shared public goods – particularly schools – while non-white households are left both without resources to improve their socio-economic situation and without public services and infrastructure which might aid in their efforts.

D. GREGG DOYLE

WHYTE, WILLIAM H.

b. 1 October 1917, West Chester, Pennsylvania; d. 12 January 1999, Manhattan, New York

Key works

(1980) *The Social Life of Small Urban Spaces*, Washington, DC: The Conservation Foundation.
(1988) *The City: Rediscovery of the Center*, New York: Doubleday.

American writer and public intellectual William H. Whyte identified the elements that create vibrant PUBLIC SPACES within the city. In the 1970s, Whyte filmed a variety of URBAN plazas in New York City. By systematically observing behaviour, Whyte identified the physical elements of successful spaces and proposed guidelines to enhance users' physical and psychological comfort and to encourage casual interaction.

Whyte observed that plazas with people attracted more people. To encourage people to linger in a space, seating was necessary. Whyte recommended that urban plazas should have a minimum of 1 linear foot of seating space for every 30 square feet of plaza space. The optimum seating height was 18 inches and Whyte promoted the use of movable chairs because of their flexible placement. If ledges were intended to provide plaza seating on both sides, they should have a minimum depth of 30–36 inches. Microclimate was important: shade and sun opportunities within the same space extended the season for outdoor sitting. Food attracted people and plazas with small kiosks or mobile food carts were more active. After much observation, Whyte concluded that CONGESTION was not a problem as plaza users were adept at self-regulating DENSITY. He wrote that concentration was the genius of the city and most spaces failed because of a lack of congestion.

Some of Whyte's guidelines considered the psychological comfort of public urban spaces. People were attracted to the edge of water and many vibrant public spaces used falling water to mask street noise. Clear sightlines between the plaza and the street were important for safety and Whyte advised against raising or lowering the grade of urban plazas more than 3 feet from street level. Urban 'undesirables' were less threatening in busy urban plazas and therefore by increasing the number of people using the space the perception of safety was enhanced. Whyte observed that many successful urban public spaces had a 'mayor' or guardian. The mayor's regular presence deterred destructive or illegal activities and the mayor often served as an unofficial ambassador for the area.

Whyte believed the simplicity of modern architecture and the increasing dominance of the vehicle had reduced the desirable pedestrian life in many urban centres. The creation of interesting and active public spaces was an effort to return people to the city centre and to encourage the unplanned, informal social encounters that enrich urban life. Street performers added entertainment and provided observers with a topic for casual conversation. Sculpture served as an engaging visual landmark. Changing displays in retail windows and SIDEWALK vendors injected other forms of visual interest into public spaces. Whyte's observations recognized that dynamic public gathering spaces were an asset of the city and his guidelines, focused on maximizing the popularity of these spaces, have helped to humanize URBAN DESIGN.

LARISSA LARSEN

WILLMOTT, PETER

b. 18 September 1923, Oxford, England;
d. 8 April 2000, London, England

Key works

(1957) *Family and Kinship in East London*, London: Routledge & Kegan Paul (with M. Young).
(1963) *Evolution of a Community*, London: Routledge & Kegan Paul.

Peter Willmott was co-founder with Michael Young of the Institute of Community Studies in 1954, whose studies of family life and housing influenced both social policy and the development of applied social research in Britain after the Second World War. Their first book, *Family and Kinship in East London* (1957), examined the effect of SLUM clearance on extended family relationships, and became a classic of British sociology. Later, in *Evolution of a Community* (1963), Willmott showed how kinship networks eventually re-established themselves in a suburban rehousing estate. The studies challenged the validity of housing policies which forced generations apart, isolating both old people and young families. His work includes studies of housing, kinship, social networks and urban POVERTY, in both Britain and France, and a forthcoming history of the English family.

Willmott left home at the age of 16. During the war, after being briefly drafted into the coal mines, he worked for the Society of Friends. In 1948 he married Phyllis Mary Noble, who became a frequent collaborator in his research. In 1949 he joined the Labour Party's research unit, where he met Michael Young. He left the Institute of Community Studies in 1978 to direct the Centre for Environmental Studies. From 1981–83 he headed the Central Policy Unit of the Greater London Council, and from 1983–97 was a senior fellow at the Policy Studies Institute.

Further reading

Krausz, E. (1969) *Sociology in Britain: A Survey of Research*, London: B.T. Batsford.
Platt, J. (1971) *Social Research in Bethnal Green*, London: Macmillan.

PETER MARRIS

WILSON, WILLIAM JULIUS

b. 20 December 1935, Derry Township, Pennsylvania

Key works

(1987) *The Truly Disadvantaged: The Inner City, the Underclass, and Public Policy*, Chicago: University of Chicago Press.
(1996) *When Work Disappears: The World of the New Urban Poor*, New York: Alfred A. Knopf.

American sociologist William Julius Wilson identified the importance of NEIGHBOURHOOD effects and demonstrated how limited employment opportunities and weakened institutional resources exacerbated POVERTY within American inner-city neighbourhoods. In 1959, one-third of the poor within the United States lived in large cities. By 1991, this percentage had increased to nearly one-half and inner-city African-American neighbourhoods witnessed a disproportionate increase in poverty. Wilson's work addressed this changing pattern and he introduced the concepts of social isolation and concentration effects and recognized the importance of institutions as social buffers.

Wilson's book *The Truly Disadvantaged* identified how changes in the post-industrial economy and URBAN FORM concentrated poverty within inner-city neighbourhoods. In the post-industrial economy, the number of manufacturing jobs declined while the number of lower-paying service jobs increased. During this same period, many of the largest employers relocated from the city's core to the suburban fringe. This geographical shift separated the URBAN labour pool from the sites of suburban employment. Wilson believed that the concentration of poverty had also weakened the traditional family structure. High rates of unemployment among young black males reduced their desirability as marriage partners and subsequently, many young black women remained unmarried and raised their children alone. Wilson used the term 'neighbourhood sorting' to describe the exodus of successful families from impoverished inner-city neighbourhoods. The loss of successful families weakened the neighbourhood's social institutions, such as schools and churches, and consequently these weakened institutions were less able to buffer the remaining residents from the impacts of poverty.

Wilson's identification of how structural eco-nomic changes impacted the residents of poor inner-city neighbourhoods challenged the prevail-ing cultural explanation of poverty. The cultural explanation hypothesized that members of poor neighbourhoods possessed different values and behaviours and these differences produced an 'URBAN UNDERCLASS'. Wilson argued that social isolation worsened the impacts of poverty and the disappearance of work from the neighbourhood negatively eroded the neighbourhood's potential for social organization and reduced the presence of positive adult role models.

Wilson's explanation of the social and economic problems facing many inner-city residents empha-sized the importance of class over race. Massey and Denton's work (1993) challenged this assumption and their research demonstrated that racial SEGRE-GATION disproportionately constrained the resi-dential options of black residents. Jargowsky (1997) strengthened Wilson's structural explanation of poverty by demonstrating that increases in the metropolitan area's economy reduced the poverty concentrations within inner-city neighbourhoods.

Wilson's identification of neighbourhood effects has initiated new research investigating the importance of context and how social networks can mediate outside threats.

Further reading

Jargowsky, P.A. (1997) *Poverty and Place: Ghettos, Barrios, and the American City*, New York: Russell Sage Foundation.
Massey, D.S. and Denton, N.A. (1993) *American Apartheid: Segregation and the Making of the Under-class*, Cambridge, MA: Harvard University Press.

LARISSA LARSEN

WIRTH, LOUIS

b. 28 August 1897, Gemunden, Germany; d. 3 May 1952, Buffalo, New York, United States

Key works

(1938) 'Urbanism as a way of life', *American Journal of Sociology* XLIV.
(1956) *The Ghetto*, Chicago: University of Chicago Press.

Louis Wirth was a key member of the CHICAGO SCHOOL OF SOCIOLOGY. Emigrating from Germany at the beginning of the First World War, Wirth's academic, COMMUNITY and political interests were wide-ranging. His doctoral dissertation completed in 1925 was later published as *The Ghetto*. It is an important work in historical sociology and Judaic studies. His lectures and publications on minority groups, URBANISM, planning, HUMAN ECOLOGY, intellectual life and social organization remained significant areas of concern until his premature death from coronary thrombosis in the early 1950s.

Wirth's essay of 1938, 'Urbanism as a way of life', is seminal. Using WEBER's notion of the ideal type, Wirth saw the URBAN and the rural as con-stituting two distinct types of community at opposite ends of a continuum. For Wirth, the urban was more than a physical space because its technologies, modes of communication, cultural attractions, economic opportunities together with the social relationships the city nurtured exerted a pull on the rural populations offering an identifi-able and distinctive way of life.

His definition of the city as a 'relatively large, dense and permanent settlement of socially het-erogeneous individuals' forms the basis of his theory of urbanism. As a product and consequence of history, Wirth viewed the city as 'a MELTING POT of races, cities and cultures' with possibilities for conflict and consensus, segmentation and cohe-sion, values and ideologies, social disorganization or organization. The relationship between size of the urban population, DENSITY of settlement and heterogeneity of the inhabitants was understood primarily in terms of the possibilities afforded for group life and the social interaction among a variety of urban 'personality types'. He said the most important thing to know about a society was what it took for granted and what it believed to be sacred and inviolable.

Wirth outlined three perspectives for research-ers in URBAN SOCIOLOGY. First, borrowing from biology, he suggested that the concept of ecology should influence our understanding of the urban mode of life, for the pattern of land use, land values, housing, transport, communication and public utilities were not isolated or unrelated phe-nomena. Social activity could not be properly separated from these factors. Secondly, although it might be easier to remain anonymous in the city, individuals sharing similar interests of necessity come together in organized groups to realize their interests and goals. Urban life entails a high but

potentially fragile, complicated and possibly volatile degree of interdependence. Thirdly, collective behaviour, understood as the purposive activities of voluntary and other groups, enables the urban individual to acquire status, his 'urban personality' and life career.

Although cognizant of Marxism, class did not figure prominently in Wirth's social analyses. Writing when the mass media was becoming increasingly powerful and susceptible to the manipulation of big corporations, politicians and totalitarian demagogues, Wirth became sensitive to the lessening importance 'of the territorial unit as a basis of social solidarity'. However, the recent interest in the concepts of CIVIL SOCIETY and SOCIAL CAPITAL renders Wirth's views on urban associational life continually relevant.

Further reading

Reiss, A.J., Jr (ed.) (1964) *Louis Wirth: On Cities and Social Life*, Chicago: University of Chicago Press.
Salerno, R.A. (1987) *Louis Wirth: A Bio-Bibliography* (Bio-Bibliographies in Sociology, No. 1), Westport, CT: Greenwood Publishing.

JOHN BLEWITT

WOODS, ROBERT ARCHEY

b. 9 December 1865, Pittsburgh, Pennsylvania; d. 18 February 1925, Boston, Massachusetts

Key works

(1897) *English Social Movements*, New York: Charles Scribner's Sons.
(1911) *Handbook of Settlements*, New York: Russell Sage Foundation (ed. with A. Kennedy).

Robert Woods was the most highly visible man in a field – the settlement house movement – dominated by women like Jane ADDAMS, Vida Scudder (founder of the College Settlements Association) and Mary Kingsbury SIMKHOVITCH. A deeply religious man, Woods attended Andover Seminary after graduating from Amherst College. While at Andover he was encouraged to travel to London to learn how to address the problems of CONGESTION and crowding in the INDUSTRIAL CITY. Woods lived at Toynbee Hall, the first settlement house, for six

months during 1890. There he joined debates on the merits of philosopher John Ruskin's theories of POLITICAL ECONOMY, housing reformer Octavia Hill's experiments with model TENEMENTS, and social reformer Charles BOOTH's empirical study of the London poor.

Woods was a prolific author and tireless public lecturer. He took every opportunity to promote the settlement house philosophy that the educated classes owed something to the working masses. Woods believed settlement workers should live in the poorest NEIGHBOURHOODS to experience their multiple deprivations in order to redress social inequalities and promote CITIZENSHIP. Woods found a willing audience in American cities, where rapid INDUSTRIALIZATION and overwhelming IMMIGRATION resulted in highly concentrated POVERTY. Jane Addams, for example, invited him to speak at Hull-House after reading his *English Social Movements*.

When Woods returned to the United States, he became head resident of the newly established Andover House in Boston's South End, renamed South End House in 1895 to better identify with the neighbourhood. He was among the last of the original six Americans who visited Toynbee Hall to establish a settlement house. Woods served at South End House until his death in 1925.

Residents of South End House used Charles Booth's study as a model to investigate district housing and organize voters to elect aldermen at large. They also served on various city commissions dealing with education, housing, sanitation, child labour laws and public health. One of their goals was to ensure that the district of Irish immigrants received its fair share of municipal resources.

Woods never worked alone. Leaders of the settlement house movement were in constant contact. After Addams, Scudder and Woods met for the first time in 1892 at a conference of settlement workers sponsored by Felix Adler's Ethical Culture Society, they maintained strong personal and professional ties. Woods and Addams became the leading public intellectuals of the settlement house movement; Adler, Simkhovitch and Jacob RIIS eventually established Greenwich House in New York City. Woods was both reformer and incipient URBAN planner, collecting data on living conditions and lobbying constantly for neighbourhood improvement.

Further reading

Woods, R.A. (1898) *The City Wilderness: A Settlement Study by Residents and Associates of South End House*, Boston, MA: Houghton Mifflin.

DAPHNE SPAIN

WORLD BANK

The World Bank has been directly engaged in URBAN affairs for over three decades. Over this time, urban activities have evolved from projects focusing on specific material issues, such as housing, to programmes on structural issues, such as GOVERNANCE. The World Bank's initial urban activities (as outlined in *Urbanization World Bank Sector Working Paper* (1972)) focused on pump-priming policies designed to facilitate self-help and provide exemplars for later self-sustaining policies. Starting with two projects a year, the Bank involved itself in housing (sites and services and SLUM upgrading), urban transport, integrated urban development and regional development. Between 1972 and 1981, sixty-two urban projects in thirty-six countries were sponsored by the Bank at a cost of US$2 billion.

Many of these 'demonstration' projects failed to adequately keep pace with demand, however, and were rarely adopted by urban elites beyond one-off showcases. In its ten-year review of urban policy in 1981, *Learning by Doing*, the Bank noted its concern with wider economic and political processes, which were seen as impediments to change. Urban initiatives became increasingly reflective of the Bank's reform agenda and promotion of neo-liberalism. Throughout the 1980s, a shift away from single projects and towards multiple-sector reform was apparent. The New Urban Management Programme (NUMP) in the 1990s, a collaboration with the United Nations Development Programme and the United Nations Centre for Human Settlements, further represented this shift to market approaches, and away from projects led by government and Bank lending. A key objective of the NUMP was on realizing urban productivity and cities as engines of growth.

By the 1990s, new concerns (such as the environment) appeared. In its *Urban Policy and Economic Development: An Agenda for the 1990s* (published 1991), the focus was on increasing the productivity of the urban economy; the productivity of the urban poor themselves (through increasing access to goods and services); the urban environment; and a greater understanding of the city through increased research. The agenda sought to build on the capacity of urban managers to be more responsive, but was criticized for moving too far towards policy concerns and away from directly helping the poor. The onus was clearly away from the state (the 'enabler') and on CIVIL SOCIETY and the private sector.

The promise of cities weighed against their failures has been a theme of early twenty-first-century thought. *Cities in Transition* (2000) sees cities at the 'frontline' of development: they are simultaneously centres of great POVERTY and of productivity and globalization. The report argues that cities must be livable, competitive, well governed and managed, and bankable (financially sound), but is concerned about the sustainability of cities, CULTURE, disaster prevention and mitigation, and equity. Neither the state nor the market is now seen as capable in isolation of turning cities around, however. Civil society is now expected to take a greater and proactive role in urban affairs.

These trends are also evident in the World Bank's chapter on cities ('Getting the best from cities') in the 2003 *World Development Report* which views city lights as both 'beacons of hope and warning flares'. For cities to be sustainable and positive, they need appropriate institutions; greater access of their citizens to ideas, knowledge and technology; employment creation (increasing productivity); and to maintain urban environments in the face of sustained demographic pressures. There is a more political tone in terms of creating inclusive cities through building constituencies to balance the power of urban elites and a warning on the dangers of divergent interests and urban unrest.

The World Bank's urban budget has never been large – it has typically averaged between 2 and 8 per cent of its total lending portfolio. However, it has been significant in terms of agenda-setting. There has been both change and continuity in the Bank's focus since 1972. It has always viewed cities as places of potential productivity, but also of unrest. The World Bank's priorities have never strayed far from the market, though its concerns have not been solely on economic policy. Today, it grapples with how to balance the material and

visible world of projects that can be seen with that of the political and economic context of urban development that is so critical in making cities functional and sustainable for their growing populations.

Further reading

World Bank (2003) *World Development Report 2003: Sustainable Development in a Dynamic World: Transforming Institutions, Growth, and Quality of Life*, Washington, DC: World Bank.

SEE ALSO: Third World cities

DONOVAN STOREY

WORLD CITY HYPOTHESIS

Globalization forms a BORDERLESS SOCIETY. But various economic functions are not just dispersed freely, and capital and industries tend to converge geographically on a specific city because of its relationship with global economic functions. During the process, head offices of multinational companies, functions of business services provided to such offices, international financial institutions, other international organizations and information-telecommunications functions will integrate, building up a geographical centre for communications and information processing. And they become the reference points for financial and cultural flow supporting global order. The world city hypothesis refers to the spatial organization of the new international division of labour, which was pointed out by John FRIEDMANN and others in the 1980s and presented as seven propositions.

Furthermore, Friedmann, after reviewing the development of the studies since, summarized the paradigm of a world city. According to the paradigm, a world city functions as a node for the global economic system, but most parts of the world are practically excluded from this capital space. A world city is a large-scale URBAN area that realizes high-level reciprocal socio-economic activities and has a hierarchical structure according to economic power. Friedmann also stated that the administrative centre of a world-city hierarchical structure is a special social class of ultra-national capitalism and a paradigm for study rather than a hypothesis.

Further reading

Friedmann, J. (1986) 'The world city hypothesis', *Development and Change* 17: 69–84.
Knox, P. and Taylor, P. (1995) *World Cities in a World-System*, Cambridge and New York: Cambridge University Press.

YUICHI TAKEUCHI

WREN, SIR CHRISTOPHER

b. 20 October 1632, East Knoyle, Wiltshire, England; d. 25 February 1723, London, England

Key works

(1664–68) Sheldonian Theatre, Oxford.
(1675–76) Royal Observatory, Greenwich.
(1675–1710) St Paul's Cathedral, London.
(1682–92) Royal Hospital Chelsea.
(1689–1700) Hampton Court Palace, London.
(1696–1712) Greenwich Naval Hospital.

Although we might think of him as, first and foremost, an architect, Christopher Wren was also a brilliant mathematician, astronomer and scientist. Wren entered Wadham College, Oxford, in 1649, graduating with BA and MA degrees, and was elected a fellow of All Souls, Oxford. At Oxford, Wren carried out many scientific experiments, working on anatomy and making drawings of the human brain. During his twenties, he devised a blood transfusion method; an instrument to measure angles; instruments for surveying; machines to lift water; ways to find longitude and distance at sea; military devices for defending cities; and means for fortifying ports. He mapped the moon and stars and investigated the rings of Saturn and provided ground-breaking work that would later assist Newton, Halle and Boyle in their work on meteorology and muscular action. At 25 years of age, Wren became professor of astronomy at Gresham College, London. Wren was part of a scientific discussion group at Gresham College that, in 1660, formally became the Royal Society. Wren became Savilian Professor of Astronomy at Oxford in 1661 and it was after this appointment that he made important contributions to mathematics. Wren's interest in architecture had arisen at Oxford. In 1663, he designed the chapel at Pembroke College, Cambridge, and submitted a model of a design for the Sheldonian

Theatre, Oxford. In 1668, building work began on Wren's designs for Emmanuel College Chapel, Cambridge, and the Garden Quadrangle, Trinity College, Oxford. Wren's greatest opportunity in architecture came with the rebuilding that followed the fire of London of 1666 when, as Commissioner for Rebuilding the City of London, he surveyed the area destroyed by the fire and replanned the entire city, supervising the rebuilding of fifty-one churches. In 1669, Wren was appointed Surveyor of the King's Works and is perhaps best known today as the architect of St Paul's Cathedral. Although his first and second designs for the new cathedral were rejected, both by London City Council and by the clergy, Wren's third design, based on a Latin cross with a large dome, formed the basis for the cathedral we see today. In 1675, he received a commission from King Charles II to build a royal observatory for John Flamsteed, the Astronomer Royal. His other architectural works include the Royal Hospital Chelsea, Hampton Court Palace, the Palace of Whitehall, and the Greenwich Naval Hospital. Dying at the age of 90, he was buried at St Paul's Cathedral, completed twelve years previously.

Further reading

Jardine, L. (2002) *On a Grander Scale: The Outstanding Career of Sir Christopher Wren*, London: Harper-Collins.
Summerson, J. (1953) *Sir Christopher Wren*, London: Collins.
Tinniswood, A. (2001) *His Invention So Fertile: A Life of Christopher Wren*, London: Joanthan Cape.

MARK TEWDWR-JONES

WRIGHT, FRANK LLOYD

b. 8 June 1867, Richland Center, Wisconsin; d. 9 April 1959, Taliesin West, Arizona

Key works

(1934–36) Fallingwater, Pennsylvania.
(1956–59) Guggenheim Museum, New York.

American self-taught architect and one of the greatest masters of modernist architecture, Frank Lloyd Wright started his career in 1887 with Louis Sullivan of the School of Chicago, developing a new concept of interior space in architecture and proposing a new design based on overlapping and interpenetrating functional areas. Considered as the founder of organic architecture, Wright played a key role in the architectural movements of the twentieth century, influencing three generations of architects worldwide through his several works.

In 1893, he started his own design office and created an independent designing method, evolving from its vital nature and bringing out an instinctive style, with no intrusiveness of abstract rules, boxes, functions and schemes. He was a valuable contributor to the generation that followed the first modern architects and, as a unique figure, was difficult to classify in terms of any one architecture movement of the modernist age. He stood out from the influences of the European masters.

His career had two relevant stages. In the first, until 1910, he experimented with the concept of the prairie house, a long, low building with hovering planes and horizontal emphasis. The most relevant works of this period are the Robie House, the Moore House and the Unity Church.

Wright summarized the key points of the prairie house as follows: reduced inside walls; a well-lit house; a house in harmony with the outdoors, emphasizing the foundations as a connection to the ground; larger open functional areas rather than rooms enclosed as boxes; use of one predominant building material; a straight and unadorned style; and systems, pipes and furniture integrated into the building process.

After a short stay in Europe and a journey in Japan, he was commissioned to design the Imperial Hotel in Tokyo (1922). Wright returned to the United States and gradually reduced his activity, before opening a new, successful stage of his career in the 1930s, when he designed multiple works, including the pureness of Fallingwater (1936).

Other relevant works of this period are the Johnson Wax Building (1939) and Taliesin West School (1938). As an author, he published *The Disappearing City* (1932), an essay on his own ideal city with large residential and green areas separated from the working buildings.

As Wright's design concept developed fully through the works of his mature age, he was able to keep a unique style as well as a self-renovated mind in approaching design solutions. However, his last works reflect less spontaneous, but more

intellectual issues of further potential design expressions.

Further reading

Heinz, T.A. (2000) *The Vision of Frank Lloyd Wright*, Edison, NJ: Chartwell Books.

Rattenbury, J. (2000) *A Living Architecture: Frank Lloyd Wright and Taliesin Architects*, San Francisco: Pomegranate Communications.

LUIGI BIOCCA

Y

YOUNG, MICHAEL DUNLOP

b. 9 August 1915, Manchester, England;
d. 14 January 2002, London, England

Key works

(1957) *Family and Kinship in East London*, London:
Routledge & Kegan Paul (with P. Wilmott).
(1958) *The Rise of the Meritocracy, 1870–2033: An Essay
on Education and Equality*, London: Thames &
Hudson.

Michael Young was an urbanist of many
different dimensions: academic researcher, pole-
micist, institution-builder. From a career in politics, at the relatively late age of 39, he took a
Ph.D. at the London School of Economics. The
Institute of Community Studies, which he founded in 1954 and for which he was best of all
known, was apparently started to house his
Ph.D., which grew into the book he wrote with
Peter Willmott, *Family and Kinship in East London*
(1957). It argued a political point: London's
working-class East Enders lived in close relationships within extended families, and if planners
tried to move them out to satellite estates, these
close relationships would be lost. This thesis was
controversial, as it was meant to be. The book
became a sensation and an academic best-seller.

Young's collaboration with Peter Willmott
resulted in two further major books, *Family and
Class in a London Suburb* (1962) and *The Symmetrical Family* (1973). They deftly combined statistical analysis and anecdotal reinforcement derived
from interviews, in the best tradition of British
empirical social science. This was an unusually
productive relationship, and when Peter Willmott
left the Institute, Young never again produced
academic work of such quality.

Instead, he devoted even more of his time to
institution-building. In the 1950s, he had taken a
leading role in the creation of the Consumers'
Association, and in the 1960s the Open University;
now they were followed by a flood of others. At the
age of 80, he was commemorated by a *Festschrift*
from his many ex-colleagues and admirers, titled
Young at Eighty.

Further reading

Briggs, A. (2001) *Michael Young: Social Entrepreneur*,
Basingstoke: Palgrave.
Dench, G., Flower, T. and Gavron, K. (1995) *Young at
Eighty: The Prolific Public Life of Michael Young*,
Manchester: Carcanet Press.

SIR PETER HALL

Z

ZERO TOLERANCE

Zero tolerance (or selective *in*tolerance as it is sometimes called) is an approach to CRIME control whereby all offences, criminal action and activity, and, for some, anti-social behaviour are seen and treated as unacceptable and to be tackled by law enforcement agencies and the criminal justice system. Although zero tolerance focused on proactive, aggressive street policing, the phrase has come to refer to a more generally diffuse response to crime in which local and national authorities (local and national government, agencies and bodies) and the wider COMMUNITY also have a role to play.

Much of the intellectual and theoretical justification for zero-tolerance approaches can be traced back to James Q. Wilson and George L. Kelling's 1982 article in *Atlantic Monthly*, 'Broken windows: the police and neighborhood safety', which argued that failure to tackle small acts of disorder, anti-social behaviour, VANDALISM and other minor offences can lead to a downward spiral of public fear, disaffection and, consequently, more crime and disorder. The cumulative effects of anti-social behaviour and minor offences can be experienced and seen by the 'law-abiding' community as actually very serious issues, impacting very negatively on senses of individual safety and security and on collective and individual senses of community well-being and quality of life, heightening the fear of crime. Popular and media interest in the idea was heightened with William Bratton, commissioner of the New York Police Department. Bratton, who had experimented with zero-tolerance approaches in Boston, extolled the virtues of this approach and promised to make New York a crime-free city. The approach enjoyed considerable international popularity as its proponents argued that the strategy had led to significant reductions in crime, including murder, in New York. However, high-profile cases of police violence and more considered analysis of crime trends have seen a more cautious approach to zero-tolerance policing adopted in the United States and other countries.

Zero tolerance – partly due to its intellectual and political heritage and partly due to its objectives – is often associated with the centre-right of the political spectrum. However, its appeal is wide and it has proved popular with centre-left and 'Third Way' political groupings such as Britain's New Labour, informing much of this party's anti-crime strategy such as the 1998 Crime and Disorder Act.

Further reading

Bowling, B. (1999) 'The rise and fall of New York murder: zero tolerance or crack's decline?', *British Journal of Criminology* 39(4): 531–54.

Bratton, W.J. and Dennis, N. (eds) (1997) *Zero Tolerance: Policing a Free Society*, London: IEA Publications.

Bratton, W.J. and Knoblach, P. (1998) *Turnaround: How America's Top Cop Reversed the Crime Epidemic*, New York: Random House.

RICHARD HUGGINS

ZONING

Zoning is the most common regulatory device local governments use to help carry out URBAN plans. Zoning ordinances divide the jurisdiction into geographically defined *zones* and specify what can and cannot be built in each zone. Zoning ordinances commonly specify permitted uses, height limits, permitted bulk, and setbacks from lot lines.

Zoning is intended to assure that land uses in GENERAL PLANS are achieved and incompatible uses are separated from each other. Zoning ordinances can help relate land use and transportation and can help rationalize INFRASTRUCTURE development. Zoning laws now exist for almost all US cities and most other large cities throughout the world.

Zoning arose because of conflicting land uses. Before zoning, nothing prevented someone from building a tallow-rendering factory near a residential area. If the owner of a house was offended by smoke or odours from the tallow-rendering factory, he could sue the factory owner for creating a nuisance and perhaps recover money damages or close the factory down. But as cities grew and incompatible land uses proliferated, it became increasingly clear that dealing with incompatible land uses after the fact and on a case-by-case basis was far less effective than thinking ahead about what uses belonged where and imposing systematic proactive regulations to achieve desired ends.

In 1916, New York City adopted a comprehensive zoning law, drafted by Edward Bassett, specifying the use, height and bulk of buildings in different parts of New York City. Other cities followed suit, including Euclid, Ohio. In 1926, in *Euclid v. Ambler Realty Corporation*, the US Supreme Court held zoning was constitutional, though individual unfair applications of zoning law might not be constitutional.

During the 1920s, the US Department of Commerce under the direction of Secretary of Commerce and later President Herbert Hoover drafted a non-binding, but enormously influential, Standard State Zoning Enabling Act (SZEA). After the US Supreme Court upheld the constitutionality of zoning, many cities in the United States adopted zoning laws based on the SZEA.

Under the model proposed, the state government could pass state *enabling legislation* authorizing cities and counties to zone. Cities and counties may then adopt zoning ordinances. In the 1920s and 1930s, some cities began zoning before they had a city PLANNING COMMISSION or a general plan. Today, the historic illogic of regulating first and planning later has been turned around so that cities now plan first and then adopt zoning as one device to achieve their plans.

In most communities, a city (or county) planning commission oversees zoning. Most have a local *board of zoning appeals* to handle issues involving conditional use permits, and challenges to the zoning laws for individual properties. Decisions of the planning commission and the board of zoning appeals may usually be appealed to the governing body of the jurisdiction and then to the state courts.

Today, most zoning laws specify residential, commercial, industrial, agricultural and other zones. Each of these zones may be further subdivided. For example, an R-1 zone might permit only single-family detached residences, an R-2 zone single-family homes and duplexes, an R-3 zone those uses and also apartment buildings with eight or fewer units, etc.

Zoning ordinances have two parts: a map and text. The map shows where zones are located. The text describes uses permitted in the zones. The heart of the zoning ordinance text consists of zoning district regulations. The text of a zoning ordinance also usually contains definitions, general provisions, special development standards, and provisions related to administration and enforcement. As conditions or the values of local elected officials change, zoning ordinances may be modified by changing the map, the text, or both.

In addition to permitted uses, a zone may specify uses that *may* be permitted if the owner obtains a *conditional use permit* (CUP) from the planning commission. The planning commission has discretion to grant or deny CUPs depending upon the attributes of a specific parcel. They often attach parking, landscaping, hour-of-use, design or other conditions before granting a CUP. CUPs are for a limited duration and then must be reviewed and renewed.

Most zoning ordinances permit *non-conforming uses* that existed before the ordinance was adopted or changed for a certain time in order to minimize hardship.

Over time, zoning has grown increasingly sophisticated. Modern ordinances allow for *variances* from the zoning law where a strict application of the law would create hardship, as well as *conditional uses*. Contract zoning, floating zones and planned unit developments provide additional flexibility. Today, zoning may be used to achieve discriminatory ends or require dull development patterns, but inclusionary zoning can promote housing for a variety of income groups and household types. Zoning may be used to promote

mixed-use areas, protect endangered species, foster TRANSIT-ORIENTED DEVELOPMENT and otherwise promote modern planning practice.

Despite the near universality of zoning, the philosophical debate about whether it is desirable continues. Conservative planners, economists and jurists argue that the private market would achieve everything that zoning does and do it better than zoning. Liberals attack much zoning for economically segregating cities and creating dull designs. Political debates about zoning also continue. Among the most important recent zoning controversies is the issue of EXCLUSIONARY ZONING. Some communities make it impossible for low- and moderate-income households to live in the COMMUNITY by zoning all or almost all their land for large-lot, single-family homes. Others try to ensure that they will get only the most desirable development by overzoning for commercial and industrial uses and underzoning for residences, providing for seniors-only housing (to avoid the costs of schools), or otherwise externalizing costs onto their neighbours.

Further reading

Babcock, R. and Siemon, C.L. (1985) *The Zoning Game Revisited*, Boston, MA: Oelgeschlager, Gunn & Hain.

Frug, G. (1999) *City Making*, Princeton, NJ: Princeton University Press.

Porter, D., Phillips, P.L. and Lassar, T.J. (1988) *Flexible Zoning: How It Works*, Washington, DC: Urban Land Institute.

Wright, R. and Gitelman, M. (2000) *Land Use in a Nutshell*, 4th edn, Minneapolis, MN: Wadsworth West.

SEE ALSO: exclusionary zoning; fiscal zoning; floor area ratio; general plan

RICHARD LEGATES

Index

Note: Page references in **bold** indicate the main entry for a topic.